CHEMISTRY

AN INTEGRATED APPROACH

CATHERINE E. HOUSECROFT
EDWIN C. CONSTABLE

LONGMAN

Addison Wesley Longman Limited
Edinburgh Gate
Harlow
Essex CM20 2JE
England
and Associated Companies throughout
the World.

ISBN 0 582 25342 X

British Library Cataloguing-in-Publication Data
A catalogue record for this book is
available from the British Library

Set by 30 in Times NRMT
Produced by Longman Asia Limited, Hong Kong

Also available
Self–Study Workbook ISBN 0582 27430 3

Elements

Element	Symbol	Atomic number	Relative atomic mass / g mol^{-1}	Element	Symbol	Atomic number	Relatve atomic mass / g mol^{-1}	Element	Symbol	Atomic number	Relative atomic mass / g mol^{-1}
Actinium	Ac	89	227.03	Hafnium	Hf	72	178.49	Promethium	Pm	61	146.92
Aluminium	Al	13	26.98	Helium	He	2	4.00	Protactinium	Pa	91	231.04
Americium	Am	95	241.06	Holmium	Ho	67	164.93	Radium	Ra	88	226.03
Antimony	Sb	51	121.75	Hydrogen	H	1	1.008	Radon	Rn	86	222
Argon	Ar	18	39.95	Indium	In	49	114.82	Rhenium	Re	75	186.2
Arsenic	As	33	74.92	Iodine	I	53	126.90	Rhodium	Rh	45	102.91
Astatine	At	85	210	Iridium	Ir	77	192.2	Rubidium	Rb	37	85.47
Barium	Ba	56	137.34	Iron	Fe	26	55.85	Ruthenium	Ru	44	101.07
Berkelium	Bk	97	249.08	Krypton	Kr	36	83.80	Samarium	Sm	62	150.35
Beryllium	Be	4	9.01	Lanthanum	La	57	138.91	Scandium	Sc	21	44.96
Bismuth	Bi	83	208.98	Lawrencium	Lr	103	262	Selenium	Se	34	78.96
Boron	B	5	10.81	Lead	Pb	82	207.19	Silicon	Si	14	28.09
Bromine	Br	35	79.91	Lithium	Li	3	6.94	Silver	Ag	47	107.87
Cadmium	Cd	48	112.40	Lutetium	Lu	71	174.97	Sodium	Na	11	22.99
Caesium	Cs	55	132.91	Magnesium	Mg	12	24.31	Strontium	Sr	38	87.62
Calcium	Ca	20	40.08	Manganese	Mn	25	54.94	Sulfur	S	16	32.06
Californium	Cf	98	252.08	Mendelevium	Md	101	258.10	Tantalum	Ta	73	180.95
Carbon	C	6	12.01	Mercury	Hg	80	200.59	Technetium	Tc	43	98.91
Cerium	Ce	58	140.12	Molybdenum	Mo	42	95.94	Tellurium	Te	52	127.60
Chlorine	Cl	17	35.45	Neodymium	Nd	60	144.24	Terbium	Tb	65	158.92
Chromium	Cr	24	52.01	Neon	Ne	10	20.18	Thallium	Tl	81	204.37
Cobalt	Co	27	58.93	Neptunium	Np	93	237.05	Thorium	Th	90	232.04
Copper	Cu	29	63.54	Nickel	Ni	28	58.69	Thulium	Tm	69	168.93
Curium	Cm	96	244.07	Niobium	Nb	41	92.91	Tin	Sn	50	118.71
Dysprosium	Dy	66	162.50	Nitrogen	N	7	14.01	Titanium	Ti	22	47.90
Einsteinium	Es	99	252.09	Nobelium	No	102	259	Tungsten	W	74	183.85
Erbium	Er	68	167.26	Osmium	Os	76	190.2	Uranium	U	92	238.03
Europium	Eu	63	151.96	Oxygen	O	8	16.00	Vanadium	V	23	50.94
Fermium	Fm	100	257.10	Palladium	Pd	46	106.4	Xenon	Xe	54	131.30
Fluorine	F	9	19.00	Phosphorus	P	15	30.97	Ytterbium	Yb	70	173.04
Francium	Fr	87	223	Platinum	Pt	78	195.08	Yttrium	Y	39	88.91
Gadolinium	Gd	64	157.25	Plutonium	Pu	94	239.05	Zinc	Zn	30	65.37
Gallium	Ga	31	69.72	Polonium	Po	84	210	Zirconium	Zr	40	91.22
Germanium	Ge	32	72.59	Potassium	K	19	39.10				
Gold	Au	79	196.97	Praseodymium	Pr	59	140.91				

To

Philby and Isis
who never allowed there to be a dull
moment during the writing of this book

Summary of contents

Contents

4 HETERONUCLEAR DIATOMIC MOLECULES

5 POLYATOMIC MOLECULES: SHAPES AND BONDING

6 IONS

12 THERMODYNAMICS AND ELECTROCHEMISTRY

13 *p*-BLOCK AND HIGH OXIDATION STATE *d*-BLOCK ELEMENTS

14 POLAR ORGANIC COMPOUNDS

15 RINGS

16 COORDINATION COMPLEXES OF THE *d*-BLOCK METALS

17 CARBONYL COMPOUNDS

APPENDICES

Preface

This book is entitled *Chemistry: An Integrated Approach*. What are our aims and what do we mean by this title? And why should you use this book in preference to any of the many similarly titled volumes on the shelf?

During our lecturing experience, we have often been struck that the same material has been taught at various points in a chemistry course, very often using differing conventions and notation. In at least one course, thermodynamics was taught at least four times in the first year. Another common problem is that of dealing with material 'rigorously' too early in a course. Is it necessary to be able to deal with the mathematical solutions to the Schrödinger equation in order to appreciate the character and utility of an atomic and molecular orbital-based scheme for an understanding of chemical bonding?

In *Chemistry: An Integrated Approach*, concepts are introduced at the level at which they are required at appropriate points in what is, primarily, a first year university textbook. However, all concepts are treated in such a way that there is no need to 'unlearn' anything that you encounter in this book in subsequent studies. We have not made over-simplistic (i.e. incorrect) approximations in order to make the subject more palatable.

We feel strongly that chemistry is a single subject and this is reflected in the content of many of the chapters. We have attempted not to have chapters exclusively identifiable as 'organic', 'inorganic' or 'physical', although in many chapters there is a predominance of one topic. We feel in particular that the distinction between 'organic' and 'inorganic' structural and synthetic chemistry is artificially divisive and this text is designed to encourage students not to partition parts of their chemistry course into 'closed boxes', only to be opened when an examination question demands it. We have included numerous examples of 'real-life' situations including environmental, biological and industrial applications that help to illustrate that inorganic, organic and physical chemistries are not mutually exclusive.

Chemistry: An Integrated Approach has been extensively cross-referenced and this should be especially helpful to those using the book to support teaching of first year chemistry within the remit of the three traditional branches of chemistry or courses designed on a modular system. For example, Chapters 8, 14, 15 and 17 provide four units of organic chemistry that could be combined into a single organic course or be separated into modular units. Cross-referencing to details of IR, UV–VIS and NMR spectroscopy in Chapter 9 will then enable the applications of these physical techniques to be easily appreciated.

And so to the vexed topic of mathematics. After a great deal of heart-searching we have taken a very low-key approach to mathematical derivation. We have not assumed any detailed knowledge of calculus and have introduced simple concepts related to this at various points – this is particularly apparent in Chapter 10 which deals with kinetics. We feel very strongly that an understanding of the consequences of a law or formula is far more important at this stage than the ability to reproduce rigorous mathematical derivations. This is most obvious in Chapter 12 in which we have approached thermodynamics by discussing experimental data rather than beginning with a string of mathematical expressions. We recognize that this approach will meet with a mixed response, but it was driven by the very varied attainments in mathematics that we note in beginning chemistry students.

We have taken as our starting point for this text an assumption of the *core* English A-level syllabus (as it stood in 1994) but Chapter 1 provides a refresher course built around the fundamental concepts and laws that are needed later in the book. One topic that is missing here is Brønsted acids and bases and this resumé (with worked examples to revise relevant calculations) is given in Chapter 11 where we discuss the chemistry of hydrogen.

Accompanying the main text is a *Self-Study Workbook* and here you will find further worked examples, problems and additional explanatory material, as well as some new topics including mass spectrometry, kinetic theory of gases, indicators, buffers and multistep organic syntheses which extend discussions from the text. The workbook is organized by chapter to parallel the main text so that relevant problem sets can easily be found to support general chemistry, modular or traditional course structures.

Unless otherwise stated, physical data are cited mainly from the *Handbook of Chemistry and Physics* (D.R. Lide, Editor in chief), 74th ed., CRC Press, Boca Raton, FL, and A.J. Gordon and R.A. Ford, *The Chemist's Companion: A Handbook of Practical Data, Techniques and References*, Wiley, New York (1972). Three-dimensional structural figures have been drawn using coordinates taken from the Cambridge Crystallographic Data Base, implemented through the ETH, Zürich. Nomenclature follows IUPAC recommendations so far as is practical, but we recognize that many trivial names remain in everyday use in the laboratory. These are introduced along with the systematic names and Appendix 13 provides a reference list of some of the most commonly encountered chemicals which are usually known by names other than their systemic ones.

There are many people to thank. For taking the time to read the drafts of this text we thank in particular Malcolm Gerloch, Martin Mays, John Maier, Diane Smith and the panel of reviewers set up by Addison Wesley Longman whose comments and suggestions have been invaluable. We have spent much effort keeping our nomenclature in line with IUPAC recommendations and we thank Zoë Lewin for answering our queries. For help in obtaining several of the more obscure literature articles, we must thank Cheryl Cook and Monika Burkhard. Of course, writing has periods in which the pen (or rather the word processor) seems to go dead, and we have been bullied along by the ever cheerful faxes and e-mail messages from those at Addison Wesley Longman – Kathy Hick, Paula Turner, Pauline Gillett, Kevin Ancient, Martin Klopstock, Nila Patel, Ros Amery and, in particular, Chris Leeding, Alex Seabrook and Jane Glendening.

All of the writing of this book has been in the company of the two feline members of our family – Philby and Isis – who have spent the hours sitting and watching the screen-saver, sleeping, playing, jumping on the keyboard, or generally making it plain that their meal-times were more important than yet another hour of writing. Their relaxed attitude to life has done a lot to keep us sane when writing deadlines seemed impossible.

We have enjoyed writing *Chemistry: An Integrated Approach* and hope that you will enjoy reading it and find the approach useful. We ask student and lecturer alike to let us know what you think: HOUSECROFT@UBACLU.UNIBAS.CH or CONSTABLE@UBACLU.UNIBAS.CH

Catherine E. Housecroft
Edwin C. Constable
Basel 1996

About the authors

Courtesy of Victor Snieckus

Chemistry: An Integrated Approach draws on the authors' experience over 18 years of teaching chemistry in the United Kingdom, mainland Europe and North America to school, college and university students. After her PhD in Durham, Catherine E. Housecroft began her teaching career at Oxford Girls' High School and went on to teach general chemistry to nursing students at St Mary's College, Notre Dame, Indiana, and to first year undergraduate students at the University of New Hampshire in the USA. She returned to the UK as a teaching Fellow of Newnham College and Lector in Trinity College, Cambridge, later becoming a Royal Society Research Fellow and lecturer in inorganic chemistry at the University of Cambridge where she has chaired the teaching committee in the chemistry department. After his DPhil in Oxford, Edwin C. Constable moved to the University of Cambridge where he became a university lecturer in inorganic chemistry; he has lectured in, and tutored, both inorganic and organic chemistry, being a teaching Fellow of Robinson College.

Currently both authors hold posts at the University of Basel in Switzerland where Catherine Housecroft is a lecturer in the Institute for Inorganic Chemistry and her husband, Edwin Constable, is Professor of Inorganic Chemistry and Director of the Institute.

Dr Housecroft has both a theoretical and experimental inorganic research background and regularly combines the two in studies of the solid state and electronic structures of main-group and metal cluster molecules, an area in which she has published over 100 research papers. Professor Constable's research interests span both organic and inorganic chemistry in the area of supramolecular chemistry and he has published over 200 research papers on this and related topics. Both are established authors, and are Associate Editors of *Coordination Chemistry Reviews* and Professor Constable is Associate Editor of *Chemical Communications*.

Acknowledgements

We are grateful to the following for permission to reproduce copyright material:

Fig 6.2a: Reprinted with permission from Wahl, A. C. (1966) *Science*, **151**, p. 961. Copyright 1966 American Association for the Advancement of Science. Fig 6.2a,b: *Journal of Chemical Physics*, American Institute of Physics, and Professor Richard F. W. Bader. Fig 6.22: *Zeitschrift fuer Naturforschung, Section A: Physical Sciences*, Verlag der Zeitschrift fuer Naturforschung. Fig 7.14: *Chemistry Reviews*, Gordon and Breach Publishers.

Whilst every effort has been made to trace the owners of copyright material, we take this opportunity to offer our apologies to any copyright holders whose rights we may have unwittingly infringed.

1 Some basic concepts

Topics

1.1 WHAT IS CHEMISTRY AND WHY IS IT IMPORTANT?

Matter, be it animal, vegetable or mineral, is composed of chemical elements or combinations thereof. Over a hundred elements are known, although not all are abundant by any means. The vast majority of these elements occur naturally, but some, such as technetium and curium, are man-made.

Chemistry is involved with the understanding of the properties of the elements, how they interact with one another, and how the combination of these elements gives compounds which may undergo chemical changes to generate new compounds.

Life has evolved systems that depend on carbon as a fundamental element; carbon, hydrogen, nitrogen and oxygen are particularly important in biological systems. A true understanding of biology and molecular biology must be based upon a full knowledge of the structures, properties and reactivities of the molecular components of life. This basic knowledge comes from the study of chemistry.

The line between physical and chemical sciences is also a narrow one. Take, for example, the rapidly expanding field of superconducting materials – compounds which possess negligible resistance to the flow of electrons. Typically, this property persists only at very low temperatures but if the superconducting materials are to find general application, they must operate at ambient temperatures. Although a physicist may make the conductivity measurements, it is the preparation and study of the *chemical* composition of the materials that drives the basic research area.

Superconductors: see Box 7.4

Chemistry plays a pivotal role in the natural sciences. It provides the essential basic knowledge for applied sciences, such as astronomy, materials science, chemical engineering, agriculture, medical sciences and pharmacology. After considering these points, the answer to the question '*why study chemistry as a first year university science?*' should not cause the reader much problem. Whatever your final career destination within the scientific world, an understanding of chemical concepts is essential.

It is traditional to split chemistry into the three branches of inorganic, organic and physical; theoretical chemistry may be regarded as a division of the physical discipline. However, the overlap between these branches of the subject is significant and real. Let us consider a case study – the Wacker process, in which ethene is oxidized to ethanal (equation 1.1).

Oxidation: see Section 1.16

$$\underset{\text{ethene}}{C_2H_4} + [PdCl_4]^{2-} + H_2O \rightarrow \underset{\substack{\text{ethanal}\\\text{(acetaldehyde)}}}{CH_3CHO} + 2HCl + 2Cl^- + Pd \qquad (1.1)$$

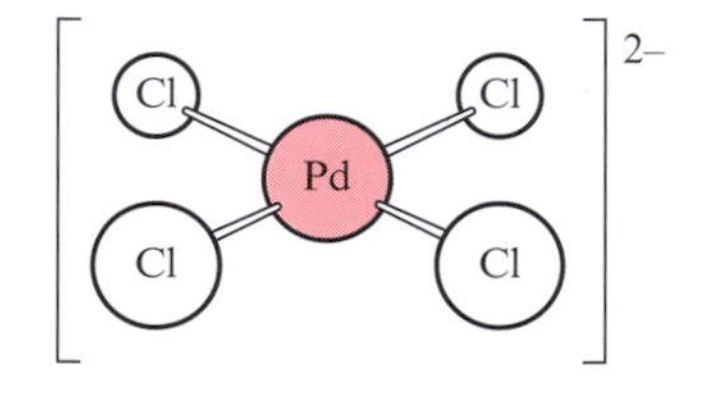

(1.1)

Until recently, this was a significant industrial method for the preparation of ethanal. The reaction occurs in water in the presence of the tetrachloropalladate(II) ion, $[PdCl_4]^{2-}$ **1.1**. The palladium(0) formed in reaction 1.1 is converted back to $[PdCl_4]^{2-}$ by reaction with copper(II) and chloride ions (equation 1.2); the copper(I) so formed (as $[CuCl_2]^-$) is reoxidized by O_2 (equation 1.3). In this way, the $[PdCl_4]^{2-}$ and Cu^{2+} function as *catalysts* since they undergo no *overall* change through the cycle of reactions.

Catalyst: see Section 10.7

$$2Cu^{2+} + Pd + 8Cl^- \rightarrow [PdCl_4]^{2-} + 2[CuCl_2]^- \quad (1.2)$$

$$4Cu^+ + 4H^+ + O_2 \rightarrow 4Cu^{2+} + 2H_2O \quad (1.3)$$

The basic reaction (ethene to ethanal) is an organic one; the catalyst is inorganic; physical chemistry is needed to interpret the kinetics and understand the mechanism of the reaction and to work out the most effective conditions under which it should take place in an industrial setting. All three branches of chemistry are used to understand the Wacker process. This is just one of a myriad of cases in which a discussion of a chemical process or concept requires the interplay of two or more areas of chemistry.

In *Chemistry: An Integrated Approach*, topics have been arranged to provide a natural progression through the chemistry that a first year science undergraduate student needs to assimilate. There are no artificial barriers placed between areas of chemistry that might otherwise be deemed inorganic, organic or physical. In this way, we hope that you will learn to apply the concepts to new situations more easily.

The aim of the rest of this chapter is to revise some essential topics and to provide you with a source of reference for basic definitions.

1.2 WHAT IS IUPAC?

As chemistry continues to expand as a subject and as the number of known chemical compounds continues to grow at a dramatic rate, it becomes increasingly vital that a set of ground rules be accepted for the naming of compounds. Not only accepted, but, probably more importantly, *used* by chemists.

The International Union of Pure and Applied Chemistry (IUPAC) is, among other things, responsible for making recommendations for the naming of both inorganic and organic compounds, as well as for the numbering system and the collective names that should be in common use for groups of elements in the periodic table.

Throughout this text, we will be using recommended IUPAC nomenclature wherever possible, but old habits die hard and a host of trivial names persists in the chemist's vocabulary. Where use of only the IUPAC nomenclature may cause confusion or may not be generally recognized, we have introduced both the recommended and trivial names. Some basic rules and revision for organic and inorganic nomenclature are dealt with in Section 1.24. As new classes of compounds are introduced in this textbook, we detail the systematic methods of naming them.

Trivial names in common use: see Appendix 13

1.3 SI UNITS

A system of internationally standardized and recognized units is equally important as a system of compound names. The Système International d'Unités (SI units) provides us with the accepted system of measurement.

Base SI quantities

There are seven base SI quantities (Table 1.1). In addition there are two supplementary units, the radian and the steradian. The *radian* is the SI unit of angle and its symbol is *rad*. For solid geometry, the unit of solid angle is the *steradian* with the unit symbol *sr*.

An important feature of SI units is that they are completely self-consistent. From the seven base units, we can derive all other units as shown in worked examples 1.1, 1.2 and 1.5. Some of the most commonly used derived units have their own names, and selected units of this type, relevant to chemistry, are listed in Table 1.2.

Derived SI units

Whenever you are using an equation to calculate a physical quantity, the units of the new quantity can be directly determined by substituting into the same equation the units of each component. Consider the volume of a rectangular box of sides a, b and c (equation 1.4). The SI unit of length is the metre (Table 1.1) and so the SI unit of volume can be determined as in equation 1.5. This is a simple example but makes the point that, *so long as you stick to the base SI units*, you can derive the units of any quantity that you need.

$$\text{Volume of box} = \text{length} \times \text{width} \times \text{height} = a \times b \times c = abc \quad (1.4)$$

$$\begin{aligned}\text{SI unit of volume} &= (\text{SI unit of length}) \times (\text{SI unit of width}) \\ &\quad \times (\text{SI unit of height}) \\ &= \text{m} \times \text{m} \times \text{m} = \text{m}^3 \quad (1.5)\end{aligned}$$

Table 1.1 The base quantities of the SI system.

Physical quantity	*Symbol for quantity*	*Base unit*	*Unit symbol*
Mass	m	kilogram	kg
Length	l	metre	m
Time	t	second	s
Electrical current	I	ampere	A
Temperature (thermodynamic)	T	kelvin	K
Amount of substance	n	mole	mol
Luminous intensity	I_v	candela	cd

Degree Celsius, °C, (instead of kelvin) is an older unit for temperature which is still in use; note the correct use of the notation °C and K, not °K.

Table 1.2 Some derived units of the SI system with particular names.

Unit	*Name of unit*	*Symbol*	*Relation to base units*
Energy	joule	J	$kg\,m^2\,s^{-2}$
Frequency	hertz	Hz	s^{-1}
Force	newton	N	$kg\,m\,s^{-2}$
Pressure	pascal	Pa	$kg\,m^{-1}\,s^{-2}$
Electric charge	coulomb	C	A s
Capacitance	farad	F	$A^2\,s^4\,kg^{-1}\,m^{-2}$
Electromotive force (emf)	volt	V	$kg\,m^2\,s^{-3}\,A^{-1}$
Resistance	ohm	Ω	$kg\,m^2\,s^{-3}\,A^{-2}$

Some older units still persist, notably atmospheres, atm (instead of pascals), for pressure, and calories, cal (instead of joules), for energy, but on the whole, SI units are well established and are used internationally.

Worked example 1.1 ***Defining units***
Determine the SI unit of density.

Density is mass per unit volume:

$$\text{Density} = \frac{\text{Mass}}{\text{Volume}}$$

$$\therefore \text{ the SI unit of density} = \frac{\text{SI unit of mass}}{\text{SI unit of volume}} = \frac{\text{kg}}{\text{m}^3} = \text{kg m}^{-3}$$

Large and small numbers

Scientific numbers have a habit of being either extremely large or extremely small. Take a typical carbon–carbon (C–C) single bond length. In the standard SI unit of length, this is 0.000 000 000 154 m. Writing this number of zeros is clumsy and there is always the chance of error when the number is being copied. The distance can be usefully rewritten as 1.54×10^{-10} m. This is a much neater and more easily read way of detailing the same information about the bond length. It would be even better if we could eliminate the need to write down the '$\times 10^{-10}$' part of the statement. Since bond lengths are usually of the same order of magnitude, it would be useful simply to be able to write '1.54' for the carbon–carbon distance. This leads us to the unit of the ångström (Å) – the C–C bond distance is 1.54 Å. Unfortunately, the ångström is *not* an SI unit although it *is* in common use. The SI asks us to choose from one of the accepted multipliers listed in Table 1.3. The distance

Table 1.3 Multiplying prefixes for use with SI units; the ones that are most commonly used are given in red.

Factor	*Name*	*Symbol*	*Factor*	*Name*	*Symbol*
10^{12}	tera	T	10^{-2}	centi	c
10^{9}	giga	G	10^{-3}	milli	m
10^{6}	mega	M	10^{-6}	micro	μ
10^{3}	kilo	k	10^{-9}	nano	n
10^{2}	hecto	h	10^{-12}	pico	p
10	deca	da	10^{-15}	femto	f
10^{-1}	deci	d	10^{-18}	atto	a

of 1.54×10^{-10} m is equal to either 154×10^{-12} m or 0.154×10^{-9} m, and from Table 1.3 we see that these are equal to 154 pm or 0.154 nm respectively. Both are acceptable within the SI. Throughout this book we have chosen to use picometres (pm) as our unit of bond distance.

Consistency of units

A **word of warning** about the use of the multipliers in Table 1.3. In a calculation, you must work in a consistent manner. For example, if a density is given in g cm^{-3} and a volume in m^3, you *cannot* mix these values, e.g. when determining a mass (equation 1.6):

$$\text{Mass} = \text{Volume} \times \text{Density} \qquad (1.6)$$

You must first scale one of the values to be consistent with the other as is shown in worked example 1.2. The most foolproof way of overcoming this frequent source of error is to convert all units into the scale defined by the base units (Table 1.1) *before* you substitute the values into an equation. We shall see many examples of this type of correction in worked examples throughout the book.

Worked example 1.2 ***Units in calculations***

Calculate the volume occupied by 10.0 g of mercury if the density of mercury is 1.36×10^4 kg m^{-3} at 298 K.

The equation which relates mass (m), volume (V) and density (ρ) is:

$$\rho = \frac{m}{V} \qquad \text{or} \qquad V = \frac{m}{\rho}$$

Before substituting in the numbers provided, we must obtain consistency amongst the units.

Density is given in kg m^{-3} but the mass is given in g, and should be converted to kg:

$$10.0 \text{ g} = 10.0 \times 10^{-3} \text{ kg} = 1 \times 10^{-2} \text{ kg}$$

Now we are a position to calculate the volume occupied by the mercury:

$$V = \frac{m}{\rho} = \frac{1 \times 10^{-2}}{1.36 \times 10^{4}} = 7.35 \times 10^{-7}\ \mathrm{m^3}$$

Check the units by the following substitution:

$$V = \frac{m}{\rho} = \frac{\mathrm{kg}}{\mathrm{kg\ m^{-3}}} = \mathrm{m^3}$$

1.4 THE PROTON, ELECTRON AND NEUTRON

The basic particles of which atoms are constituted are the *proton*, the *electron* and the *neutron*.§ Some key properties of the proton, electron and neutron are given in Table 1.4. A neutron and a proton have approximately the same mass and, relative to these, the electron can be assumed to have negligible mass. The charge on a proton is of equal magnitude, but opposite sign, to that on an electron and so the combination of *equal numbers* of protons and electrons results in an assembly that is neutral overall. A neutron, as its name suggests, is neutral – it has no charge. The arrangements and energies of electrons in atoms and ions are discussed in Chapter 2.

1.5 THE ELEMENTS

An element is matter, all of whose atoms are alike in having the same positive charge on the nucleus.

The recommended IUPAC definition of an *element* states that 'an element is matter, all of whose atoms are alike in having the same positive charge on the nucleus'. Each element is given a symbol and these are predominantly internationally accepted even though the names of the elements themselves are subject to linguistic variation.

Table 1.4 Properties of the proton, electron and neutron

	Proton	*Electron*	*Neutron*
Charge / C	$+1.602 \times 10^{-19}$	-1.602×10^{-19}	0
Charge number (relative charge)	1	-1	0
Rest mass / kg	1.673×10^{-27}	9.109×10^{-31}	1.675×10^{-27}
Relative mass	1837	1	1839

§ These are the particles considered fundamental by chemists, although particle physicists have demonstrated that there are yet smaller building blocks. This continual subdivision recalls the lines of Jonathan Swift:

'So, naturalists observe, a flea
Hath smaller fleas that on him prey;
And these have smaller fleas to bite 'em,
And so proceed *ad infinitum*.'

Box 1.1 The elements: names with meanings

The origin of the names of the chemical elements and their variation between languages is fascinating. A wealth of chemical history and fact is hidden within the names, although changing from one language to another may obscure this information. Three examples illustrate this.

Mercury (Hg)

The silver-white, liquid metal mercury was known to the ancient Greeks as *liquid silver*. This name evolved into the Latin *hydrargyrum* from which our present-day symbol *Hg* arises. Mercury used to be known as *quicksilver* in English (*quick* means alive) and the element is called *Quecksilber* in German. The word *mercury* refers to the messenger of the Roman gods – anyone who has tried to 'catch' a globule of mercury will appreciate the analogy.

Cobalt (Co)

The *d*-block element cobalt (Co) is known as *Kobalt* in German. Nickel and cobalt occur in varying amounts in iron ores. When these ores were processed by early miners, ores containing significant amounts of cobalt gave inferior metallic products. These ores became known as *false ores* and were named after the malicious imps or *Kobold* who supposedly lived in the mines.

Tungsten (W)

Tungsten has the unexpected symbol W. In German, tungsten is called *Wolfram* and this refers to its ore *Wolframite*. The other common ore is *Scheelite*, which was named after Scheele, the discover of tungsten. Scheele was Swedish, and before the ore was called Scheelite, it was known as *tungsten* which translates as 'heavy stone'.

Metals, non-metals and semi-metals

Elements can be classified as metals, non-metals or semi-metals. The names of *most* metals end in '-ium', e.g. lithium, sodium, magnesium, calcium, aluminium, scandium, chromium, titanium, vanadium, hafnium, ruthenium, rhodium, iridium, osmium and palladium. Common exceptions include tin, lead, iron, cobalt, copper, zinc, tungsten, platinum, silver and gold. Under IUPAC recommendations, the term 'semi-metal' is preferred over 'metalloid'. We look further at what constitutes a 'semi-metal' in Chapter 7.

Allotropes

Some elements exist in more than one structural form and this property is called *allotropy*. Consider carbon – two commonly quoted *allotropes* of carbon are graphite and diamond (Figure 1.1), both of which possess infinite lattice structures. Both allotropes consist only of atoms of carbon and both burn in an excess of O_2 to give only carbon dioxide, CO_2. The physical appearances of these two allotropes are, however, dramatically different.

Plate 1 A scientist wearing a 'space-suit' as part of the strictly controlled 'clean-room' conditions that are required during the production of precious-metal-based pharmaceuticals such as cisplatin and carboplatin: see Box 16. 6 on anti-cancer drugs. [Malcolm Fielding, Johnson Matthey plc/Science Photo Library]

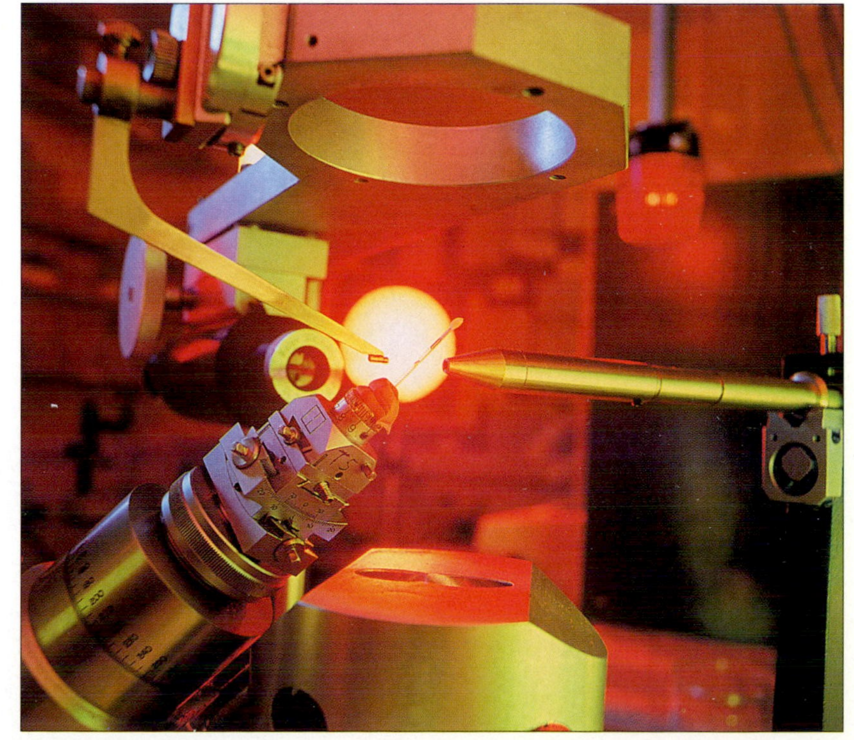

Plate 2 In Section 3.2 we discuss diffraction methods, which are used to determine the structures of compounds and provide bond distance and angle data. This photograph shows the central part of an X-ray diffractometer. The crystal is encapsulated in a thin glass tube (visible at the centre of the picture) and is positioned so as to be within the beam of X-rays. After striking the crystal, the X-rays are diffracted. [James King-Holmes/OCMS/Science Photo Library]

Plate 3 (above) Beryl , $Be_3Al_2Si_6O_{18}$, is a silicate mineral (see Section 13.6) which occurs in some granites and pegmatites. Crystals of beryl possess a hexagonal prismatic habit and are usually pale green. Gem varieties include emerald and aquamarine. [Geoscience Features Picture Library]

Plate 4 (above) The fullerene C_{60} can be prepared from carbon rods using solar energy. The glowing rods at the centre of the picture have been heated by sunlight concentrated into the synthesis cell by an array of mirrors. The black deposit on the walls of the cell is composed of fullerenes. We discuss some aspects of C_{60} in Section 7.9. [Philippe Plailly/Eurelios/Science Photo Library]

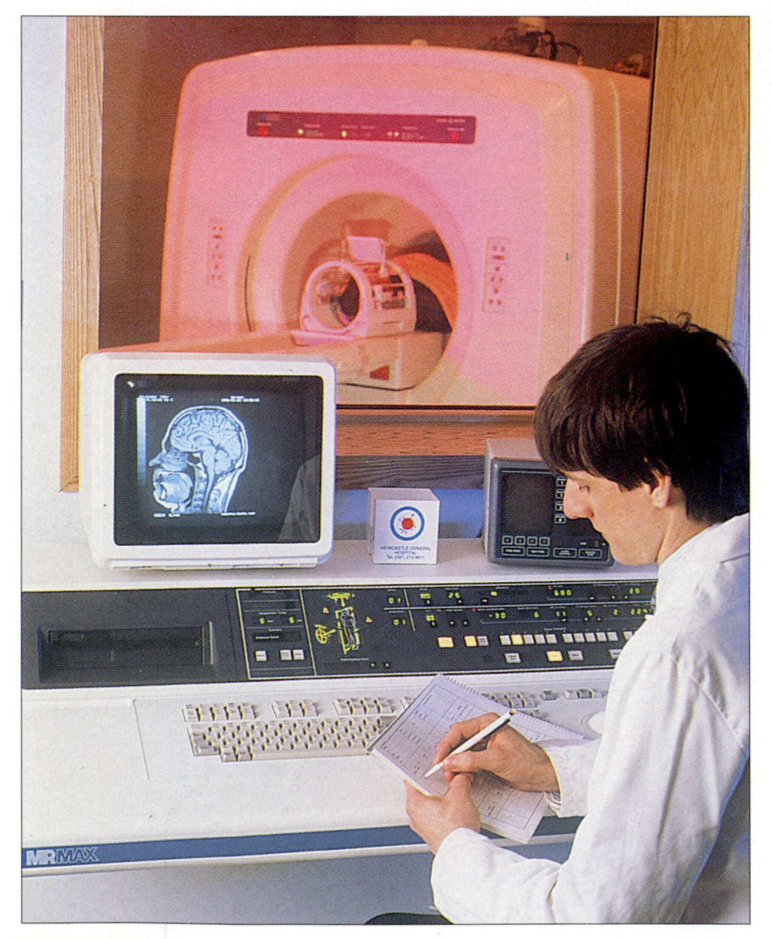

Plate 5 (left) In Chapter 9, we look at the experimental method of nuclear magnetic resonance (NMR) spectroscopy. Magnetic resonance imaging (MRI) uses NMR spectroscopy to measure the concentration of water in living organisms. The technique can be used to give pictures of internal organs, allowing doctors to study patients by non-intrusive methods. The photograph shows a man undergoing an MRI brain scan; the radiographer is viewing an image of the patient's brain. [Simon Fraser/Dept. of Neuroradiology Newcastle General Hospital/Science Photo Library]

Plate 6 Blast-off for the space shuttle *Columbia*. In Box 11.3 we look at the roles of liquid dihydrogen, liquid dioxygen, aluminium, ammonium perchlorate, and carbon–carbon composites in the space shuttle. [NASA/© Corel Corporation 1993]

Plate 7 Acid rain is an environmental problem originating from emissions of sulfur dioxide into the atmosphere. The photograph shows the effects of acid rain – devastation in a conifer forest on the border between Germany and the Czech Republic. Acid rain is discussed further in Box 13.14. [Geoscience Features Picture Library]

Plate 8 A LIDAR (light detection and ranging) air pollution unit uses a laser beam to probe the atmosphere (here, at Albuquerque, USA) and produces a two- or three-dimensional map showing the concentration of pollutants. LIDAR may also be used to study the emissions from volcanoes. The uses of spectroscopic methods to study air pollution are described in Box 9.4. [Los Alamos National Laboratory/ Science Photo Library]

Plate 9 A sulfuric acid plant in Billingham, UK; the acid is manufactured using the Contact Process (see Section 13.10). Sulfuric acid is a vitally important chemical in industry (see Appendix 14) and applications include fertilizers, and dye and synthetic fibre manufacturing (see Figure 13.31). [Martin Bond/ Science Photo Library]

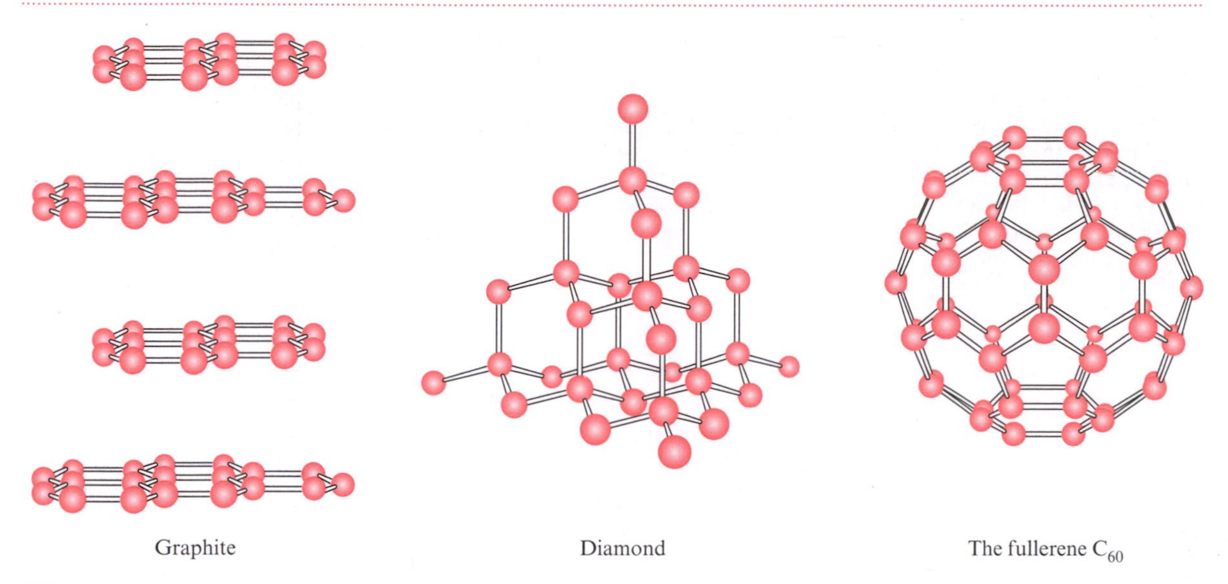

1.1 The structures of three allotropes of carbon; C_{60} is representative of the group of fullerenes.

C_{60}: see Section 7.9

Diamond is thermodynamically unstable at room temperature and pressure with respect to graphite but, fortunately, the interconversion is extremely slow and diamond is termed *metastable*. In the 1980s, further allotropes of carbon called the fullerenes were discovered. These are present in soot to the extent of a few percent by weight, and the commonest component consists of discrete C_{60} molecules (Figure 1.1). Other common elements that exhibit allotropy are tin, phosphorus, arsenic, oxygen, sulfur and selenium.

Allotropes of an element are different structural modifications of that element.

1.6 STATES OF MATTER

Solids, liquids and gases

At a given temperature, an element is in one of three *states of matter* – solid, liquid or vapour (Figure 1.2). A vapour is called a gas when it is above its *critical temperature*.

In a *solid*, the atoms are often arranged in a regular fashion; the solid possesses a fixed volume (at a stated temperature and pressure) and shape. A *liquid* also has a fixed volume at a given temperature and pressure, but has no definite shape. It will flow into a container and will adopt the shape of this vessel. The particles (atoms or molecules) in a *gas* move at random and occupy a large volume. A gas has no fixed shape.

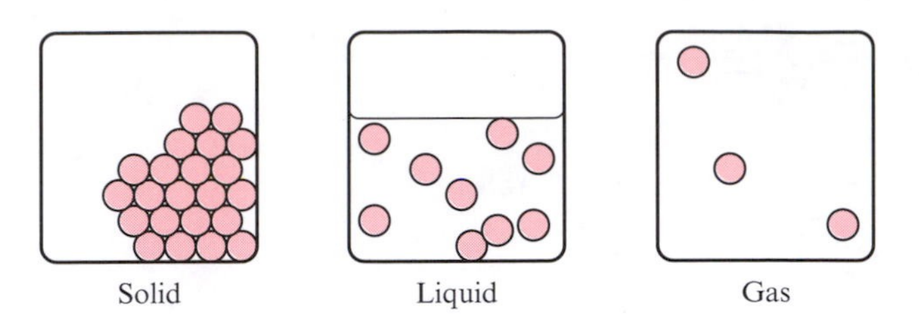

1.2 The arrangement of particles (atoms or molecules) in a solid, a liquid and a gas. For the solid, the surface is determined by the solid itself. The shape of the liquid is controlled by the shape of the container; the liquid has a definite surface. The atoms or molecules of the gas are free to move throughout the container.

In Figure 1.2, the surface boundary of the solid is determined by the solid itself and is not dictated by the container. On the other hand, unless restricted by a container, the boundary of a gas is continually changing; this permits two (or more) gases to mix in a process called *diffusion* (Figure 1.3). In a liquid, the shape of the liquid mimics the shape of the container, but at the upper surface of the liquid the phenomenon of *surface tension* operates and this defines the surface boundary. When liquids mix they are said to be *miscible*; hexane and octane are miscible, as are water and ethanol. If liquids do not mix (e.g. water and oil, water and hexane), they are *immiscible*.

If two liquids mix, they are miscible. Immiscible liquids form distinct layers on top of each other.

In chemical equations, the states of substance are often included and the standard abbreviations are:

solid	(s)
liquid	(l)
gas	(g)
aqueous solution	(aq)

The use of (aq) refers to an aqueous solution; this is *not* a state of matter.

Phases

The three states of a substance are *phases*, but within one state there may be more than one phase. For example, each allotrope of carbon is a different phase but each is in the solid state. Each phase exists under specific conditions of temperature and pressure and this information is represented in a *phase diagram*. We explore the use of a simple phase diagram in Box 1.2.

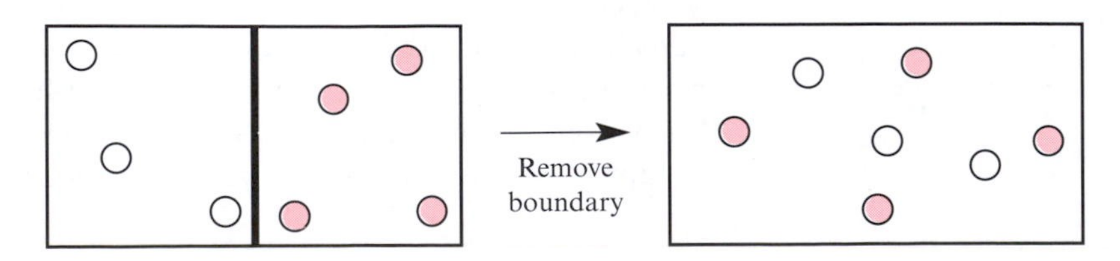

1.3 A gas is trapped within the boundaries of the vessel in which it is held. Two gases can mix together (*diffuse*) if the boundary between them is removed.

Box 1.2 A simple phase diagram and the distinction between a gas and a vapour

The terms *gas* and *vapour* are often used interchangeably. However, there is a real distinction between them. A vapour, but not a gas, can be *liquefied* by increasing the pressure at a constant temperature. This may at first seem contradictory and to understand it we need to consider a simple one-component phase diagram. One type of *phase diagram* describes the variation in phase of a system as a function of temperature and pressure.

The diagram below shows a simple phase diagram for a compound X. The solid lines represent the boundaries between solid X (phase A), liquid X (phase B), and the vapour or gas state (phase C). The broken black line represents phase changes that will occur at a pressure of 10^5 Pa (1 bar) as the temperature is raised. Solid X first melts ($A \rightarrow B$) and then vaporizes ($B \rightarrow C$). Now, follow the broken red line from the bottom to the top of the diagram. This represents what happens as you increase the pressure at a constant temperature (298 K). We start in phase C and crossing the boundary $C \rightarrow B$ has the physical effect of liquefying the vapour. Further increase in pressure leads to solidification ($B \rightarrow A$).

The temperature T_c is the *critical temperature*. Above this temperature, it is no longer possible to liquefy the 'vapour', now called a *gas*; i.e. above temperature T_c it is no longer possible to cross the $C \rightarrow B$ boundary by increasing the pressure. The *critical point* is shown by the red dot on the diagram and is defined by both a temperature and pressure.

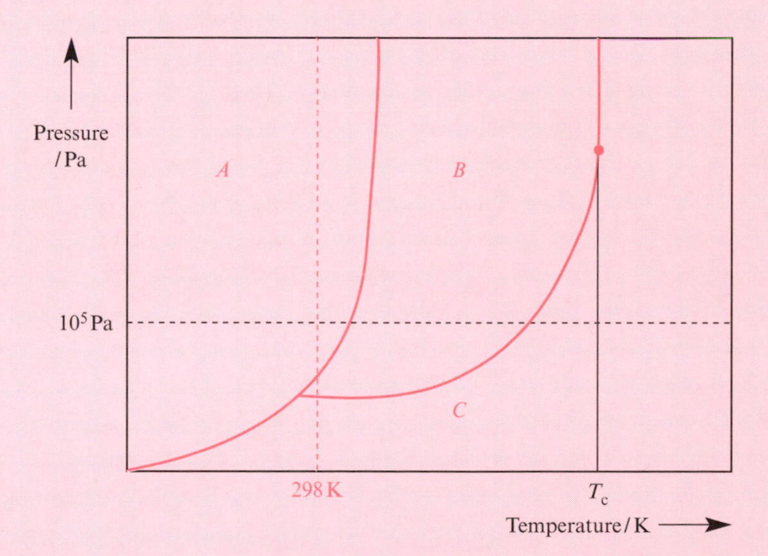

Questions

Choose any temperature on the temperature axis of the diagram and place a straight edge vertically so you can follow the changes in phase of X as a function of pressure at this particular temperature. At the temperature you have chosen, is phase C correctly described as a gas or a vapour? Repeat the exercise at a different temperature. Now choose any pressure on the pressure axis. Follow a horizontal line across the diagram and determine the phase changes that occur as you raise the temperature at your selected (fixed) pressure.

Changes of phase

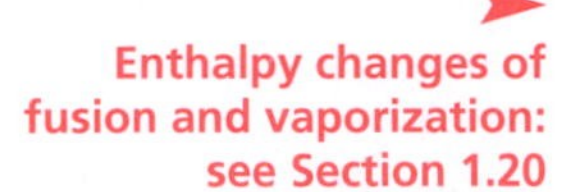

At a given pressure (or temperature), the transformation from one phase to another takes place at a particular temperature (or pressure). Usually, we are concerned with transformations at atmospheric pressure. An element (or a single allotrope if we are dealing with an element that exhibits allotropy) changes from a solid to a liquid at its *melting point* (mp) and from a liquid to a vapour at its *boiling point* (bp). We should refer to the third state as a vapour, and not a gas, until the critical temperature is reached. The melting and boiling points for selected elements are given in Table 1.5. Carbon and sulfur have not been included in this table because neither undergoes a simple change of state. When heated at a particular pressure, some elements transform directly from the solid to the vapour phase and this process is called *sublimation*. Iodine readily sublimes when it is heated.

Table 1.5 Melting and boiling points (at atmospheric pressure) and appearance of selected elements.

Element (allotrope)	*Symbol*	*Melting point / K*	*Boiling point / K*	*Physical appearance at 298 K*
Aluminium	Al	933	2730	White metal
Bromine	Br	266	332.5	Brown-orange liquid
Calcium	Ca	1112	1767	Silver-white metal
Chlorine	Cl	172	239	Green-yellow gas
Chromium	Cr	2173	2963	Silver metal
Cobalt	Co	1768	3373	Silver-blue metal
Copper	Cu	1356	2843	Reddish metal
Fluorine	F	54	85	Very pale yellow gas
Gold	Au	1337	3081	Yellow metal
Helium	He	–	4.2	Colourless gas
Hydrogen	H	14	20	Colourless gas
Iodine	I	387	458	Black solid
Iron	Fe	1808	3023	Silver-grey metal
Lead	Pb	600	2024	Blue-grey metal
Lithium	Li	453.5	1620	Silver-white metal
Magnesium	Mg	922	1378	Silver-white metal
Manganese	Mn	1517	2333	Silver metal
Mercury	Hg	234	630	Silver liquid
Nickel	Ni	1728	3193	Silver metal
Nitrogen	N	63	77	Colourless gas
Oxygen	O	54	90	Colourless gas
Phosphorus (white)	P	317	553.5	White solid
Potassium	K	336	1038	Silver-white metal
Silicon	Si	1693	~3553	Shiny, blue-grey solid
Silver	Ag	1234	2428	Silver-white metal
Sodium	Na	371	1154	Silver-white metal
Tin (white)	Sn	505	2896	Silver metal
Zinc	Zn	692.5	1180	Silver metal

1.7 ATOMS AND ISOTOPES

Atoms and atomic number

An *atom* is the smallest unit quantity of an element that is capable of existence, either alone or in chemical combination with other atoms of the same or another element. It consists of a positively charged *nucleus* and negatively charged electrons. The simplest atom is hydrogen which is made up of one proton and one electron. The proton of a hydrogen atom is its nucleus, but the nucleus of any other atom consists of protons *and* neutrons.

An *atom* is the smallest unit quantity of an element that is capable of existence, either alone or in chemical combination with other atoms of the same or another element.

All atoms are neutral, with the positive charge of the nucleus exactly balanced by the negative charge of a number of electrons equal to the number of protons. The electrons are situated outside the nucleus. Each atom is characterized by its *atomic number*, Z. A shorthand method of showing the atomic number and mass number, A, of an atom along with its symbol, E, is used:

$$\begin{array}{lcc} \text{mass number} \rightarrow & & \\ \text{element symbol} \rightarrow & {}^{A}_{Z}\mathrm{E} & {}^{59}_{27}\mathrm{Co} \\ \text{atomic number} \rightarrow & & \end{array}$$

Atomic number = Z = number of protons in the nucleus = number of electrons

Mass number = A = number of protons + number of neutrons

Relative atomic mass

The mass of an atom is concentrated in the nucleus where the protons and neutrons are found. If we were to add together the actual masses of protons and neutrons present, we would always be dealing with very small, non-integral numbers, and for convenience, a system of *relative atomic masses* is used. The *atomic mass unit* (u) has a value of $\approx 1.660 \times 10^{-27}$ kg, and this corresponds closely to the mass of a proton or a neutron (Table 1.4). Effectively, the mass of each proton and neutron is taken to be one atomic mass unit. The scale of relative atomic masses (A_r) is based upon measurements taken for carbon with all atomic masses stated *relative to* $^{12}C = 12.0000$.

Isotopes

For a given element, there may be more than one type of atom, and these are *isotopes* of the element. But, *do not confuse isotopes with allotropes* Allotropes are different structural forms of a *bulk* element arising from different spatial arrangements of the atoms (Figure 1.1). Isotopes are atoms of

Box 1.3 Artifically produced isotopes and β-particle emission

Some isotopes, in particular those of the heaviest elements in the periodic table, are produced by the bombardment of one nucleus by particles which induce nuclear fusion. Typical particles used for the bombardment are neutrons.

An example of an artifically produced isotope is the formation of $^{239}_{94}Pu$, an isotope of plutonium, in a series of nuclear reactions beginning with the isotope of uranium $^{238}_{92}U$. The bombardment of $^{238}_{92}U$ with neutrons results in nuclear fusion:

$$^{238}_{92}U + ^{1}_{0}n \rightarrow ^{239}_{92}U$$

The relative atomic mass of the product is one atomic mass unit greater than that of the initial isotope because we have added a neutron to it. The isotope $^{239}_{92}U$ undergoes spontaneous loss of a β-particle (defined below) to form an isotope of neptunium ($^{239}_{93}Np$) which again undergoes β-decay to give $^{239}_{94}Pu$:

$$^{239}_{92}U \rightarrow ^{239}_{93}Np + \beta^-$$

$$^{239}_{93}Np \rightarrow ^{239}_{94}Pu + \beta^-$$

β-*Decay*

β-Particle emission (β-decay) occurs when an *electron* is lost *from the nucleus* by a complex process which effectively 'turns a neutron into a proton'.[§] The mass of the nucleus undergoing β-decay does not change because the β-particle has negligible mass; the atomic number of the nucleus undergoing the emission increases by one since, effectively, the nucleus has gained a proton:

$$^{1}_{0}n \rightarrow ^{1}_{1}p + \beta^-$$

§ This notation is not strictly correct. The decay of a nucleus by the loss of a β-particle is a complex process, the study of which belongs within the remit of the particle physicist. Of interest to us here is the fact that a nucleus which undergoes β-decay retains the same atomic mass but increases its atomic number by one.

the same element with *different number of neutrons*. Some isotopes which do not occur naturally may be produced artificially.

An element is characterized by the number of protons and, to keep the atom neutral, this must equal the number of electrons. However, the number of neutrons can vary. For example, hydrogen possesses three isotopes. The most abundant by far (99.984%) is protium, ^{1}H, which has one proton and one electron but no neutrons. The second isotope is deuterium ^{2}H (also given the symbol D) which has one proton, one electron and one neutron. Deuterium is present naturally at an abundance of 0.0156% and is sometimes called 'heavy hydrogen'. Tritium (^{3}H or T) occurs as less than 1 in 10^{17} atoms in a sample of natural hydrogen and is radioactive. The atomic mass of naturally occurring hydrogen reflects the presence of all three isotopes and is a weighted mean of the masses of the isotopes present. The relative atomic mass of hydrogen is 1.0080, and the value is close to 1 because the isotope ^{1}H with mass number 1 makes up 99.984% of the natural mixture of isotopes.

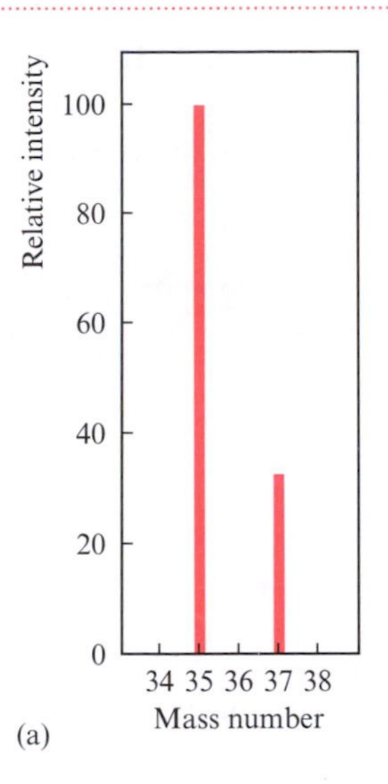

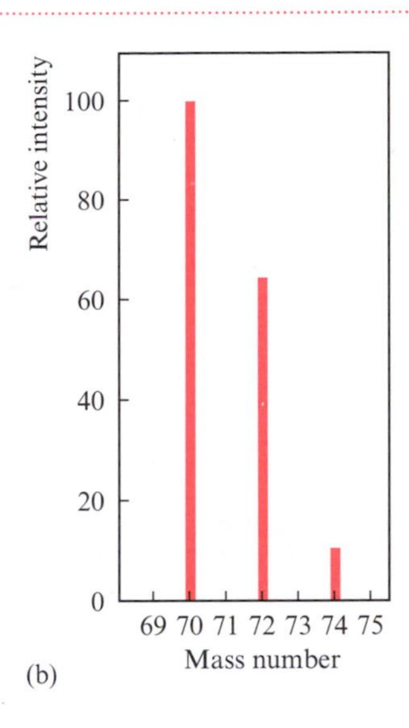

1.4 Mass spectrometric traces for (a) atomic chlorine and (b) molecular Cl_2.

Other examples of elements which exist as mixtures of isotopes are lithium ($^{6}_{3}Li$ and $^{7}_{3}Li$), chlorine ($^{35}_{17}Cl$ and $^{37}_{17}Cl$) and copper ($^{63}_{29}Cu$ and $^{65}_{29}Cu$). Elements which occur naturally as only one type of atom are *monotopic* and include phosphorus ($^{31}_{15}P$) and cobalt ($^{59}_{27}Co$). Isotopes can be separated by *mass spectrometry* and Figure 1.4a shows the isotopic distribution in atomic chlorine. In the mass spectrum of Cl_2 (Figure 1.4b), three peaks are observed and are assigned to the possible combinations of the two isotopes of chlorine. [Exercise: Using the information in Figure 1.4a, account for the mass numbers and the relative intensities of the peaks in Figure 1.4b.]

Isotopes of an element have the same atomic number, *Z*, but different atomic masses.

Worked example 1.3 ***Relative atomic mass***

Calculate the relative atomic mass of naturally occurring magnesium if the distribution of isotopes is 78.7% $^{24}_{12}Mg$, 10.1% $^{25}_{12}Mg$ and 11.2% $^{26}_{12}Mg$.

The relative atomic mass of magnesium is the weighted mean of the relative atomic masses of the three isotopes:

$$\text{Relative atomic mass} = \left(\frac{78.7}{100}\times 24\right) + \left(\frac{10.1}{100}\times 25\right) + \left(\frac{11.2}{100}\times 26\right)$$

$$= 24.325$$

1.8 THE MOLE AND THE AVOGADRO CONSTANT

In Table 1.1 we saw that the SI unit of 'amount of substance' is the *mole*. This unit can apply to any substance and hence it is usual to find the statement 'a mole of *x*' where *x* might be electrons, atoms or molecules. In one mole of substance there are $\approx 6.022 \times 10^{23}$ particles, and this number is called the Avogadro constant or number, L. It is defined as being the number of atoms of carbon in exactly 12 g of a sample of isotopically pure $^{12}_{6}C$. Since L is the number of particles in a mole of substance, its units are mol^{-1}.

In one mole of substance there are the Avogadro number, L, of particles:

$L \approx 6.022 \times 10^{23}$ mol^{-1}

1.9 GAS LAWS AND IDEAL GASES

In this section we summarize some of the important laws which apply to *ideal gases*; we look more carefully at the *kinetic theory of gases* in the accompanying workbook.

Pressure and Boyle's Law

Pressure is defined as force per unit area (equation 1.7) and the derived SI units of pressure are pascals (Pa) (Table 1.2).

$$\text{Pressure} = \frac{\text{Force}}{\text{Area}} \quad (1.7)$$

Although atmospheric pressure is the usual working condition in the laboratory (except when we are specifically working under conditions of reduced or high pressures), a pressure of 1 bar (10^5 Pa) has been defined by the IUPAC as the *standard pressure*.§ Of course, this may not be the exact pressure of working, and in that case appropriate corrections must be made as we show below.

When a fixed mass of a gas is compressed (the pressure on the gas is increased) at a constant temperature, the volume of the gas decreases. The pressure and volume are related by Boyle's Law (equation 1.8), and from this inverse relationship we see that doubling the pressure halves the volume, and halving the pressure doubles the volume of a gas.

$$\text{Pressure} \propto \frac{1}{\text{Volume}}$$

$$P \propto \frac{1}{V} \quad \text{at constant temperature} \quad (1.8)$$

§ Until 1982, the standard pressure was 1 atmosphere (1 atm = 101 300 Pa) and this pressure remains in use in many textbooks and tables of physical data. The bar is a 'non-standard' unit in the same way that the ångström is defined as 'non-standard'.

Charles's Law

Boyle's Law is obeyed only under conditions of constant temperature because the volume of a gas is dependent both on pressure and temperature. The volume and temperature of a fixed mass of gas are related by Charles's Law (equation 1.9) and the direct relationship shows that (at constant pressure), the volume doubles if the temperature is doubled.

$$\text{Volume} \propto \text{Temperature}$$
$$V \propto T \quad \text{at constant pressure} \tag{1.9}$$

A combination of Boyle's and Charles's Laws gives us a relationship between the pressure, volume and temperature of a fixed mass of gas (equation 1.10), and this can be rewritten in the form of equation 1.11.

$$P \propto \frac{T}{V} \tag{1.10}$$

$$\frac{PV}{T} = \text{constant} \tag{1.11}$$

Corrections to a volume of gas from one set of pressure and temperature conditions to another can be done using equation 1.12; each side of the equation is equal to the *same* constant.

$$\frac{P_1V_1}{T_1} = \frac{P_2V_2}{T_2} \tag{1.12}$$

Worked example 1.4 ***Dependence of gas volume on temperature and pressure***

If the volume of a sample of helium is 0.0227 m^3 at 273 K and 10^5 Pa, what is its volume at 293 K and 1.04 × 10^5 Pa?

The relevant equation is:

$$\frac{P_1V_1}{T_1} = \frac{P_2V_2}{T_2}$$

First check that the units are consistent: V in m^3, T in K, P in Pa. (In fact, *in this case* inconsistencies in units of P and V would cancel out – why?)

$P_1 = 10^5$ Pa $\quad V_1 = 0.0227$ m^3 $\quad T_1 = 273$ K

$P_2 = 1.04 \times 10^5$ Pa $\quad V_2 = ?$ $\quad T_2 = 293$ K

$$\frac{P_1V_1}{T_1} = \frac{P_2V_2}{T_2}$$

$$\frac{10^5 \times 0.0227}{273} = \frac{1.04 \times 10^5 \times V_2}{293}$$

$$V_2 = \frac{10^5 \times 0.0227 \times 293}{273 \times 1.04 \times 10^5} = 0.0234 \text{ m}^3$$

Ideal gases

In real situations, we work with 'real' gases, but it is convenient to assume that most gases behave as though they were *ideal*. *If* a gas is ideal, it obeys the *ideal gas law* (equation 1.13) in which the constant in equation 1.11 is the *molar gas constant*, R, and the quantity of gas to which the equation applies is one mole.

$$\frac{PV}{T} = R = 8.314 \text{ J mol}^{-1} \text{ K}^{-1} \qquad (1.13)$$

For n moles of gas, we can rewrite the ideal gas law as equation 1.14.

$$\frac{PV}{T} = nR \quad \text{or} \quad PV = nRT \qquad (1.14)$$

The molar gas constant has the same value for *all gases*, whether they are atomic (Ne, He), molecular and elemental (O_2, N_2) or compounds (CO_2, H_2S, NO).

We have already seen that IUPAC has defined *standard pressure* as 10^5 Pa (1 bar) and a value of *standard temperature* has been defined as 273.15 K. For our purposes, we often use 273 K as the standard temperature.§ The volume of one mole of an ideal gas under conditions of standard pressure and temperature is 0.0227 m^3 or 22.7 dm^3 (equation 1.15). We consider the consistency of the units in this equation in worked example 1.5.

$$\begin{aligned} \text{Volume of 1 mole of ideal gas} &= \frac{nRT}{P} \\ &= \frac{1 \times 8.314 \times 273}{10^5} = 0.0227 \text{ m}^3 \end{aligned} \qquad (1.15)$$

The molar volume of an ideal gas under conditions of standard pressure (10^5 Pa) and temperature (273 K) is 22.7 dm^3.

You should notice that this volume differs from the 22.4 dm^3 with which you may be familiar! A molar volume of 22.4 dm^3 refers to a standard pressure of 1 atm (101 300 Pa), and 22.7 dm^3 refers to a standard pressure of 1 bar (100 000 Pa).

Worked example 1.5 ***Finding a derived SI unit***

Determine the SI unit of the molar gas constant, *R*.

The ideal gas law is:

$$PV = nRT$$

where P = pressure, V = volume, n = number of moles of gas, T = temperature.

$$R = \frac{PV}{nT}$$

§ The standard temperature of 273 K is *different* from the *standard state temperature* of 298 K used in thermodynamics, see Section 1.17 and Chapter 12.

$$\text{The SI unit of } R = \frac{(\text{SI unit of pressure}) \times (\text{SI unit of volume})}{(\text{SI unit of quantity}) \times (\text{SI unit of temperature})}$$

The SI unit of pressure is Pa but this is a *derived* unit. In base units, Pa = kg m^{-1} s^{-2} (Tables 1.1 and 1.2):

$$= \frac{(\text{kg m}^{-1}\,\text{s}^{-2}) \times (\text{m}^3)}{(\text{mol}) \times (\text{K})}$$

$$= \text{kg m}^2\,\text{s}^{-2}\,\text{mol}^{-1}\,\text{K}^{-1}$$

Whilst this unit of R is correct in terms of the base SI units, it can be simplified because the joule, J, is defined as kg m^2 s^{-2} (Table 1.2).

The SI unit of R = J mol^{-1} K^{-1} or J K^{-1} mol^{-1}

Exercise: Now look back at equation 1.14 and, by working in SI *base units*, show that a volume in m^3 is consistent with using values of pressure in Pa, temperature in K and R in J K^{-1} mol^{-1}.

Worked example 1.6 ***Determining the volume of a gas***

Calculate the volume occupied by 1 mole of carbon dioxide at 300 K and 1 bar pressure. (R = 8.314 J K^{-1} mol^{-1})

Assuming that the gas is ideal, the equation to use is:

$$PV = nRT$$

First ensure that the units are consistent – the pressure needs to be in Pa:

$$1 \text{ bar} = 10^5 \text{ Pa}$$

To find the volume, we first rearrange the ideal gas equation:

$$V = \frac{nRT}{P}$$

$$V = \frac{nRT}{P} = \frac{1 \times 8.314 \times 300}{10^5}$$

$$= 0.0249 \text{ m}^3$$

$$= 24.9 \text{ dm}^3$$

NB: You should develop a habit of stopping and thinking whether a numerical answer to a calculation is sensible. Is its magnitude reasonable? In this example, we know that the molar volume of any ideal gas at 10^5 Pa and 273 K is 22.7 dm^3, and we have calculated that at 10^5Pa and 300 K, the volume of 1 mole of gas is 24.9 dm^3. This increased volume appears to be consistent with a relatively small rise in temperature.

Dalton's Law of Partial Pressures

In a mixture of gases, the total pressure exerted by the gas on its surroundings is the sum of the *partial pressures* of the component gases. This is Dalton's Law of Partial Pressures and can be expressed as in equation 1.16.

$$P = P_A + P_B + P_C + \dots \qquad (1.16)$$

where P = total pressure, P_A = partial pressure of gas A, etc.

The partial pressure of each gas is directly related to the number of moles of gas present. Equation 1.17 gives the relationship between the partial pressure of a component gas (P_X) and the total pressure P of a gas mixture.

$$\text{Partial pressure of component X} = \frac{\text{Moles of X}}{\text{Total moles of gas}} \times \text{Total pressure} \qquad (1.17)$$

Worked example 1.7 ***Partial pressures***

At 290 K and 10^5 Pa, a 25 dm^3 sample of gas contains 0.35 moles of argon and 0.61 moles of neon. (a) Are these the only components of the gas mixture? (b) What are the partial pressures of the two gases? ($R = 8.314\ J\ K^{-1}\ mol^{-1}$)

Part (a): The ideal gas law can be used to find the *total* number of moles of gas present:

$$PV = nRT$$

The volume must be converted to m^3: 25 dm^3 = 25×10^{-3} m^3.

$$n = \frac{PV}{RT} = \frac{10^5 \times 25 \times 10^{-3}}{8.314 \times 290} = 1.04$$

Since there are only 0.35 moles of argon and 0.61 moles of neon, there must be 0.08 moles of one (or more) other gaseous components.

Part (b): Now that we know the total moles of gas, we can determine the partial pressures of argon (P_{Ar}) and neon (P_{Ne}):

$$\text{Partial pressure of component X} = \frac{\text{moles of X}}{\text{total moles of gas}} \times \text{total pressure}$$

$$\text{For argon: } P_{Ar} = \frac{0.35}{1.04} \times 10^5 = 33654 \text{ Pa}$$

$$\text{For neon: } P_{Ne} = \frac{0.61}{1.04} \times 10^5 = 58654 \text{ Pa}$$

One use of partial pressures is in the determination of equilibrium constants, K_p, for gaseous systems and we look further at this in Section 1.22 and Chapter 12.

1.10 THE PERIODIC TABLE

The arrangement of elements in the periodic table

The elements are arranged in the periodic table (Figure 1.5) in numerical order according to the number of protons possessed by each element. The division into *groups* places elements with the same number of valence electrons into vertical columns within the table. Under IUPAC recommendations, the groups are labelled from 1 to 18 (Arabic numbers) and blocks of elements are named as in Figure 1.5.

Valence electrons: see Section 2.19

The *d*-block elements are also referred to as the *transition elements*, although the elements zinc, cadmium and mercury (group 12) are not, by IUPAC ruling, included as transition metals since neither the atoms nor the ions possess partially filled *d*-orbitals. The vertical groups of three *d*-block elements are called *triads*. The names lanthanoid and actinoid are preferred over the names lanthanide and actinide because the ending '-ide' usually implies a negatively charged ion. However, the terms lanthanide and actinide are still in common use. The lanthanoids and actinoids are collectively termed the *f*-block elements. IUPAC-recommended names for some of the groups are given in Table 1.6.

Inorganic nomenclature: see Section 1.24

Valence electrons

It is important to know the positions of elements in the periodic table. A fundamental property of an element is the *ground state valence electron configuration* (discussed further in Chapter 2) and this can usually be determined for any element by considering its position in the periodic table.

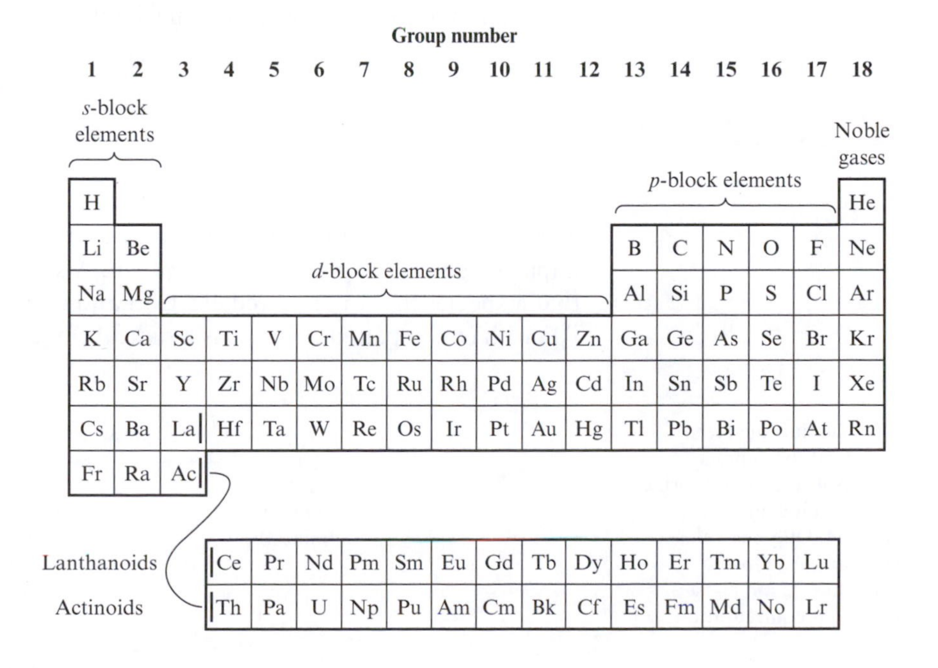

1.5 The periodic table showing *s*-, *p*- and *d*-block elements and the lanthanoids and actinoids. A complete periodic table with more detailed information is given in the inside front cover of the book.

Table 1.6 IUPAC-recommended names for sections and groups in the periodic table.

Group number	*Recommended name*
1 (except H)	Alkali metals
2	Alkaline earth metals
15	Pnicogens[a]
16	Chalcogens
17	Halogens
18	Noble gases
1 (except H), 2, 13, 14, 15, 16, 17, 18	Main group elements

[a]The name pnicogen is proposed but is not IUPAC-approved: G. J. Leigh (ed.), *IUPAC Nomenclature of Inorganic Chemistry: Recommendations 1990*, Blackwell Scientific Publications, Oxford (1990).

For example, elements in groups 1 or 2 have one or two electrons in the outer (valence) shell, respectively. After the *d*-block (which contains ten groups), the number of valence electrons can be calculated by subtracting ten from the group number. Nitrogen (N) is in group 15 and has five valence electrons; tellurium (Te) is in group 16 and has six valence electrons. Trends in the behaviour of the elements follow periodic patterns (*periodicity*).

Learning the periodic table by heart is not usually necessary, but it is extremely useful to develop a feeling for the groupings of elements. To know that selenium is in the same group as oxygen will immediately tell you that there will be some similarity between the chemistries of these two elements. Care is needed though – there is a significant change in properties in descending a group. Consider the elements in group 14: carbon is a non-metal which forms compounds in which carbon normally exhibits a valency of 4, whereas lead (Pb) is a metal with two common oxidation states, +4 and +2. Such trends will be considered in Chapters 7 and 13.

1.11 RADICALS AND IONS

Radicals, anions and cations

The presence of one or more unpaired electrons in an atom or molecule imparts upon it the property of a *radical*. A superscript $^{\bullet}$ is used to signify that a species has an unpaired electron and is a radical. The neutral atom $^{19}_{9}F$, with one unpaired electron, is a radical (Figure 1.6).

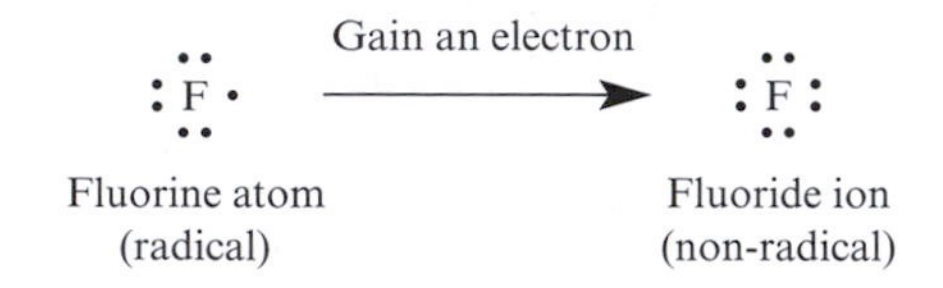

1.6 The fluorine atom (a radical) becomes a negatively charged fluoride ion when it gains one electron. Only the valence electrons are shown. The fluoride ion has a noble gas configuration.

Octet rule: see Section 2.20

The fluorine atom readily accepts one electron (Figure 1.6 and equation 1.18) to give an ion with a *noble gas configuration*.

$$F^{\bullet}(g) + e^- \rightarrow F^-(g) \quad (1.18)$$

A radical possesses at least one unpaired electron.

There is now a charge imbalance between the positive charge of the fluorine nucleus which has nine protons and the negative charge of the ten electrons. On gaining an electron, the neutral fluorine radical becomes a negatively charged *fluoride ion*. The change in name from fluor*ine* to fluor*ide* is significant – the ending '*-ide*' signifies the presence of a negatively charged species. A negatively charged ion is called an *anion*.

The loss of one electron from a neutral atom generates a positively charged ion – a *cation*. A sodium atom may lose an electron to give a sodium cation (equation 1.19) and the positive charge arises from the imbalance between the number of protons in the nucleus and the electrons.

$$Na^{\bullet}(g) \rightarrow Na^+(g) + e^- \quad (1.19)$$

An anion is a negatively charged ion, and a cation is a positively charged ion.

Although in equations 1.18 and 1.19 we have been careful to indicate that the neutral atom is a radical, it is usually the case that *atoms* of an element are written without specific reference to the radical property. For example, equation 1.20 means exactly the same as equation 1.19.

$$Na(g) \rightarrow Na^+(g) + e^- \quad (1.20)$$

The terms *dication*, *dianion*, *trication* and *trianion*, etc. are used to indicate that an ion carries a specific charge. A dication carries a 2+ charge (e.g. Mg^{2+}, Co^{2+}), a dianion has a 2– charge (e.g. O^{2-}, Se^{2-}), a trication bears a 3+ charge (e.g. Al^{3+}, Fe^{3+}), a trianion has a 3– charge (e.g. N^{3-}, PO_4^{3-}).

Worked example 1.8 ***Ion formation***

With reference to the periodic table (Figure 1.5), predict the most likely ions to be formed by the following elements: Na, Ca, Br.

Firstly, find the position of each element in the periodic table. Remember that elements with fewer than four valence electrons will tend to *lose* electrons and elements with more than four valence electrons will tend to *gain* electrons so as to form a noble gas configuration.

Na is in group 1.
Na has one valence electron.
Na will easily lose one electron to form an [Ne] configuration.
Na will form an Na^+ ion.

Ca is in group 2.
Ca has two valence electrons.
Ca will lose two electrons to form an [Ar] configuration.
Ca will form a Ca^{2+} ion.

Br is in group 17.
Br has seven valence electrons.
Br will tend to gain one electron to give a [Kr] configuration.
Br will form a Br^- ion.

1.12 MOLECULES AND COMPOUNDS: BOND FORMATION

Covalent bond formation

A *molecule* is a discrete neutral species resulting from *covalent* bond formation between two or more atoms.

A molecule is a discrete neutral species containing a covalent bond or bonds between two or more atoms.

When two radicals combine, *pairing of the two electrons* may result in the formation of a covalent bond, and the species produced is molecular. Equation 1.21 shows the formation of molecular dihydrogen from two hydrogen atoms (i.e. radicals).

$$2H^{\bullet} \rightarrow H_2 \qquad (1.21)$$

Homonuclear and heteronuclear molecules

The molecule H_2 is a *homonuclear diatomic* molecule. *Diatomic* refers to the fact that H_2 consists of two atoms, and *homonuclear* indicates that the molecule consists of identical atoms. Molecular hydrogen, H_2, should be referred to as *dihydrogen* to distinguish it from atomic hydrogen, H. Other homonuclear molecules include dioxygen (O_2), dinitrogen (N_2), difluorine (F_2), trioxygen (O_3) and octasulfur (S_8). Some allotropes have trivial names – for example, O_3 is called ozone.

Homonuclear diatomics: see Chapter 3

A *heteronuclear* molecule consists of more than one type of element. Carbon monoxide, CO, is a *heteronuclear diatomic* molecule. When a molecule contains three or more atoms, it is called a *polyatomic* molecule; carbon dioxide, CO_2, methane, CH_4, and ethanol, C_2H_5OH, are polyatomic, although CO_2 may also be called a *triatomic* molecule.

Covalent versus ionic bonds

The important difference between a covalent and an ionic bond is the average distribution of the bonding electrons between the nuclei. The electrons which form a covalent bond are 'shared' *between* nuclei. This is represented in Figure 1.7a for the Cl_2 molecule which contains a Cl–Cl single bond. Since the two atoms are identical, the two bonding electrons are located symmetrically between the two chlorine nuclei.

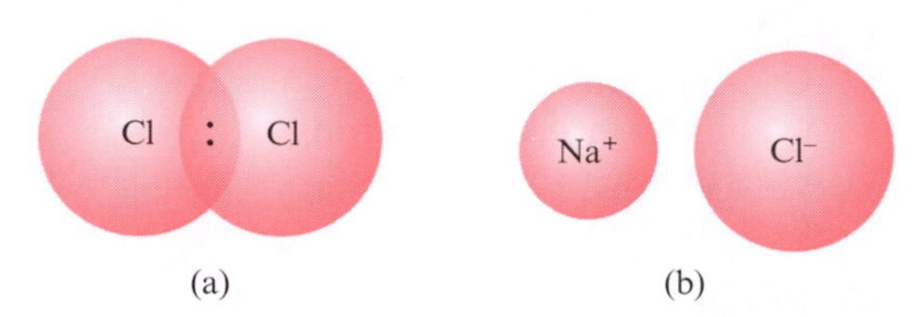

1.7 Representations of (a) the covalent bond in the Cl_2 molecule in which each chlorine atom provides one electron, and (b) the ionic bond formed between a positively charged sodium ion and a negatively charged chloride ion. In practice, whereas the Cl_2 molecule is discrete, the pair of ions in NaCl will be part of an ionic lattice.

The complete transfer of the bonding pair of electrons to one of the nuclei results in the formation of an *ionic bond*. Figure 1.7b schematically shows the situation for a single pair of ions in a sodium chloride lattice; there is a region between the ions in which the electron density approaches zero. Although Figure 1.7b shows an isolated pair of Na^+ and Cl^- ions, this is not a real situation for solid sodium chloride. Positive and negative ions attract one another and the net result is the formation of a three-dimensional *ionic lattice*.

Ionic lattices: see Chapter 6

Molecules and compounds

In a covalent bond, electrons are shared between atoms. In an ionic bond, transfer of one of more electrons from one atom to another occurs.

We must make a careful distinction between the use of the terms *molecule* and *compound*. Compounds are *neutral* and include both covalent and ionic species – NaF (composed of Na^+ and F^- ions), CO (a covalent diatomic molecule) and SF_6 (a covalent polyatomic molecule) are all compounds. Since a molecule is a *discrete neutral species*, of these three compounds, only CO and SF_6 are molecules.

1.13 MOLECULES AND COMPOUNDS: RELATIVE MOLECULAR MASS AND MOLES

Relative molecular mass

M_r is also referred to as the molecular weight of a compound

The relative molecular mass (M_r) of a compound is found by summing the relative atomic masses of the constituent atoms. For example, a mole of carbon monoxide CO has a relative molecular mass of 28 (since the relative atomic masses of C and O are 12 and 16), and M_r for carbon dioxide CO_2 is 44.

Moles of compound

The relative molecular mass gives the mass (in g) of *one mole of a compound* and equation 1.22 gives the relationship between the mass (in g) and the number of moles.

$$\text{Number of moles} = \frac{\text{Mass in grams}}{\text{Relative molecular mass}} \tag{1.22}$$

Worked example 1.9 ***Relative molecular mass and moles***

How many moles of molecules are present in 3.48 g of acetone? (A_r C = 12, O = 16, H = 1)

First, write down the formula of acetone:

$CH_3C(O)CH_3$

$M_r = (3 \times 12) + (1 \times 16) + (6 \times 1) = 58$

$$\text{Number of moles} = \frac{\text{Mass in grams}}{M_r} = \frac{3.48}{58} = 0.06$$

1.14 CONCENTRATIONS OF SOLUTIONS

Molarity

Reactions in the laboratory are often carried out in solution, and the *concentration* of the solution tells us how much of a compound or ion is present in a given volume of the solution. The SI unit for concentration is mol m^{-3} but it is accepted (and usual) practice to quote concentration in mol dm^{-3}. A solution of volume 1 dm^3 containing one mole of dissolved substance (the *solute*) is called a one molar (1 M) solution.

Concentrations and amount of solute

In practice, we do not always work with one molar solutions, and the moles of solute in a given solution can be found using equation 1.23 or, if the volume is in cm^3 as is often the case in the laboratory, equation 1.24.

$$\text{Number of moles} = \text{Volume (in dm}^3) \times \text{Concentration (in mol dm}^{-3}) \tag{1.23}$$

$$\text{Number of moles} = \frac{\text{Volume (in cm}^3) \times \text{Concentration (in mol dm}^{-3})}{1000} \tag{1.24}$$

Worked example 1.10 *Solution concentrations*

If 1.17 g of sodium chloride is dissolved in 100 cm³ of water, what is the concentration of the solution? (A_r Na = 23, Cl = 35.5)

First, we need to find the number of moles of NaCl in 1.17g:

$$M_r = 23 + 35.5 = 58.5$$

$$\text{Moles of NaCl} = \frac{1.17}{58.5} = 0.02$$

Next, we must ensure that the units for the next part of the calculation are consistent. The volume is in cm^3 and so we can use a rearranged form of equation 1.24 to find concentration:

$$\text{Concentration} = \frac{\text{Number of moles} \times 1000}{\text{Volume (in cm}^3)}$$

$$\text{Concentration} = \frac{0.02 \times 1000}{100} = 0.2 \text{ mol dm}^{-3}$$

Worked example 1.11 ***Solution concentrations***

What mass of potassium iodide (KI) must be dissolved in 50 cm³ of water to give a solution of concentration 0.05 M? (A_r K = 39, I = 127)

First we must find the number of moles of KI in 50 cm³ 0.05 mol dm⁻³ solution:

$$\text{Number of moles} = \frac{\text{Volume (in cm}^3) \times \text{Concentration (in mol dm}^{-3})}{1000}$$

$$\text{Number of moles KI} = \frac{50 \times 0.05}{1000} = 2.5 \times 10^{-3}$$

Now convert the moles to mass:

$$\text{Number of moles} = \frac{\text{Mass in grams}}{M_r} \quad \text{or} \quad \text{Mass} = \text{Moles} \times M_r$$

For KI, $M_r = 39 + 127 = 166$:

$$\text{Mass of KI} = (2.5 \times 10^{-3}) \times 166 = 0.415 \text{ g.}$$

1.15 REACTION STOICHIOMETRY

When we write a balanced or *stoichiometric* equation, we state the molecular ratio in which the reactants combine and the corresponding ratios of products. In reaction 1.25, one molecule of pentane reacts with eight molecules of dioxygen to give five molecules of carbon dioxide and six molecules of water. This also corresponds to the ratio of *moles* of reactants and products.

$$\underset{\text{pentane}}{C_5H_{12}(g)} + 8O_2(g) \rightarrow 5CO_2(g) + 6H_2O(l) \quad (1.25)$$

Worked example 1.12 ***Amounts of reactants for complete reaction***

What mass of zinc is required to react completely with 30 cm³ 1 M hydrochloric acid? (A_r Zn = 65)

This is the reaction of a metal with acid to give a salt and H_2. First write a stoichiometric equation:

$$Zn(s) + 2HCl(aq) \rightarrow ZnCl_2(aq) + H_2(g)$$

Now find the number of moles of HCl in the solution:

$$\text{Number of moles} = \frac{\text{Volume (in cm}^3) \times \text{Concentration (in mol dm}^{-3})}{1000}$$

$$\text{Number of moles of HCl} = \frac{30 \times 1}{1000} = 3 \times 10^{-2}$$

Now look at the balanced equation – one mole of Zn reacts with two moles of HCl and therefore 3×10^{-2} moles of HCl will react with 1.5×10^{-2} moles of Zn.

We can now determine the mass of zinc needed:

$$\text{Mass (in grams)} = \text{Moles} \times M_r = 1.5 \times 10^{-2} \times 65 = 0.975 \text{ g}$$

Worked example 1.13 ***Amount of products formed***

How much CO_2 can be formed when 0.3 g of carbon is burnt under conditions of standard temperature and pressure? (A_r C = 12; volume of one mole of gas at 10^5 Pa and 273 K = 22.7 dm^3)

First, write a balanced equation:

$$C(s) + O_2(g) \rightarrow CO_2(g)$$

One mole of carbon atoms gives one mole of CO_2 (assuming complete combustion).

Determine the number of moles of carbon:

$$\text{Number of moles} = \frac{\text{Mass in grams}}{A_r} = \frac{0.3}{12} = 0.025$$

(We use A_r in the equation because we are dealing with atoms of an element.)

At standard temperature (273 K) and pressure (10^5 Pa), one mole of CO_2 occupies 22.7 dm^3.

Volume occupied by 0.025 moles of $CO_2 = 0.025 \times 22.7 = 0.5675 \text{ dm}^3$

[*Exercise*: If the product were carbon *monoxide*, what volume of gas would be produced?]

1.16 OXIDATION AND REDUCTION, AND OXIDATION STATES

Oxidation and reduction

When an element or compound burns in dioxygen to give an oxide, it is *oxidized* (equation 1.26).

$$2Mg(s) + O_2(g) \rightarrow 2MgO(s) \qquad (1.26)$$

Conversely, if a metal oxide reacts with dihydrogen and is converted to the metal, then the oxide has been reduced (equation 1.27).

$$CuO(s) + H_2(g) \xrightarrow{\text{heat}} Cu(s) + H_2O(g) \qquad (1.27)$$

In reaction 1.26, O_2 is the *oxidizing agent* and in reaction 1.27, H_2 is the *reducing agent*.

Although we often think of oxidation in terms of *gaining oxygen* and reduction in terms of *losing oxygen*, we should consider other definitions of these processes. A *reduction* may involve *gaining hydrogen*; in equation 1.28, chlorine is reduced and hydrogen is oxidized.

$$Cl_2(g) + H_2(g) \rightarrow 2HCl(g) \qquad (1.28)$$

Loss of hydrogen may correspond to *oxidation* – for example, chlorine is oxidized when HCl is converted to Cl_2.

Oxidation and reduction may also be defined in terms of electron transfer – *electrons are gained in reduction* processes and are *lost in oxidation* reactions (equations 1.29 and 1.30).

$$S + 2e^- \rightarrow S^{2-} \qquad \textit{Reduction} \qquad (1.29)$$

$$Zn \rightarrow Zn^{2+} + 2e^- \qquad \textit{Oxidation} \qquad (1.30)$$

It is, however, sometimes difficult to apply these simple definitions to a reaction. For example, in the reaction of sodium with water (equation 1.31), what is happening to the water?

$$2Na(s) + 2H_2O(l) \rightarrow 2NaOH(aq) + H_2(g) \qquad (1.31)$$

Oxidation states

The concept of *oxidation states* and the changes in oxidation states that occur during a reaction provide a way of recognizing oxidation and reduction processes. Oxidation states are assigned to each atom of an element in a compound and *are a formalism*, although for ions such as Na^+, we can associate the charge of 1+ with an oxidation state of +1.

The oxidation state of an *element* is taken to be zero. This applies to both atomic (e.g. He) and molecular elements (e.g. H_2, P_4, S_8).

In order to assign oxidation states to atoms in a *compound*, we must follow a set of rules, but be careful! The basic rules are as follows:

- The sum of the oxidation states of the atoms in a neutral compound is zero.
- The sum of the oxidation states of the atoms in an ion is equal to the charge on the ion (e.g. in the sulfate ion $[SO_4]^{2-}$, the sum of the oxidation states of S and O must be –2).
- The oxidation state of hydrogen is +1 when it combines with a non-metal and –1 if it combines with a metal.
- The oxidation state of fluorine in a compound is always –1.
- The oxidation state of chlorine, bromine and iodine is *usually* –1 (exceptions are interhalogen compounds and oxides – see Chapter 13).
- The oxidation state of oxygen in a compound is usually –2.
- The oxidation state of a group 1 metal in a compound is +1.
- The oxidation state of a group 2 metal in a compound is +2.
- Metals from the *d*-block will usually have positive oxidation states (exceptions are some low-oxidation state compounds – see Section 16.14).

Added to these rules are the notions that most elements in groups 13, 14, 15 and 16 can have variable oxidation states. Some of the problems encountered are exemplified for carbon in Box 8.8. In reality, it is essential to have a full picture of the structure of a compound before oxidation states can be assigned.

Worked example 1.14 ***Working out oxidation states***

What are the oxidation states of each element in the following: KI, $FeCl_3$, Na_2SO_4, $[NO_3]^-$?

KI: The group 1 metal will be in oxidation state +1. This is consistent with the iodine being in oxidation state –1, and the sum of the oxidation states is 0.

$FeCl_3$: Chlorine is usually in oxidation state –1, and since there are three Cl atoms, the oxidation state of the iron must be +3 to give a neutral compound.

Na_2SO_4: Of the three elements, S can have a variable oxidation state and so we should deal with this element *last*. Na is in group 1 and will have an oxidation state of +1. Oxygen is usually in an oxidation state of –2. The oxidation state of the sulfur atom is determined by ensuring that the sum of the oxidation states is zero:

$$(2 \times \text{Oxidation state of Na}) + (\text{Oxidation state of S}) + (4 \times \text{Oxidation state of O}) = 0$$

$$(+2) + (\text{Oxidation state of S}) + (-8) = 0$$

$$\text{Oxidation state of S} = 0 + 8 - 2 = +6$$

$[NO_3]^-$: Oxygen is usually in an oxidation state of –2. The overall charge is –1 and therefore:

$$(\text{Oxidation state of N}) + (3 \times \text{Oxidation state of O}) = -1$$

$$\text{Oxidation state of N} = -1 + 6 = +5$$

Changes in oxidation states

Oxidation is defined as an *increase* (more positive) and reduction is a *decrease* (more negative) in oxidation state. In equation 1.29, the change in oxidation state of the sulfur from 0 to –2 is reduction, and in reaction 1.30, the change in oxidation state of the zinc from 0 to +2 is oxidation.

In a reduction–oxidation (*redox*) reaction, the change in oxidation number for the oxidation and reduction steps *must balance*. In reaction 1.32, iron is oxidized to iron(II) and chlorine is reduced to chloride ion. The net increase in oxidation state of the iron *for the stoichiometric reaction* must balance the net decrease in oxidation state for the chlorine.

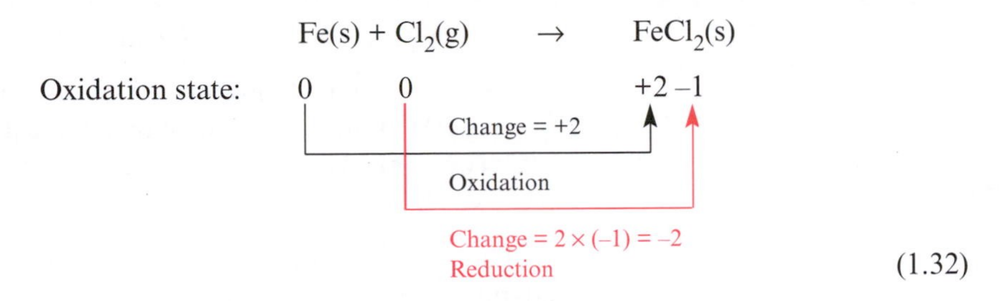

(1.32)

Worked example 1.15 ***Changes in oxidation states in a reaction***

Which elements undergo oxidation and reduction during the reaction:

$2K(s) + 2H_2O(l) \rightarrow 2KOH(aq) + H_2(g)$?

First think about the structures of H_2O and KOH:

H—O—H K^+ [$^-$O—H]

and then assign oxidation states:

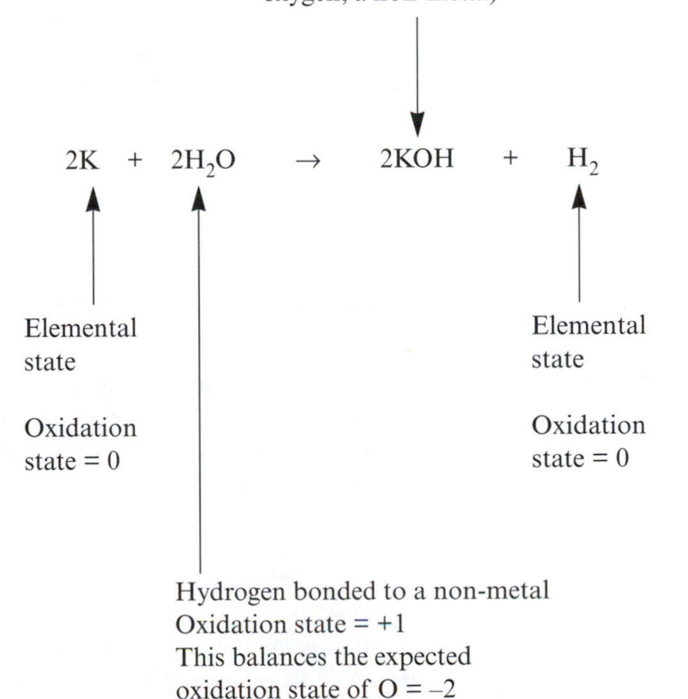

Now look at the *changes in oxidation states*.

K Both K atoms change from 0 to +1; oxidation.
H Two H atoms remain in an oxidation state of +1
Two of the H atoms are reduced from +1 to 0.
O No change (–2 to –2)

NB: The net increase in oxidation numbers balances the net decrease – oxidation balances reduction.

1.17 THERMOCHEMISTRY

Much of chemistry is concerned with chemical reactions. The factors that govern whether a reaction will or will not be observed fall into two categories: *thermodynamic* and *kinetic*. Thermodynamic concepts relate to the energetics of a system, while kinetics deal with the speed at which a reaction occurs. Observations of the reaction kinetics are related to the mechanism of the reaction, and this describes the way in which we believe that the atoms and molecules behave during the reaction.

Change in enthalpy of a reaction

A concept already known to you is the *change in enthalpy*, ΔH. *Thermochemistry* deals with the measurement of enthalpy changes – heats of reaction. The *enthalpy change* that accompanies a reaction is the amount of heat energy liberated or taken in as a reaction proceeds at a temperature T, and the *standard enthalpy change* of a reaction refers to the enthalpy change when all of the reactants and products are in their *standard states*. The notation for this *thermochemical quantity* is $\Delta_r H^o(T)$ where:

Δ means 'change in',
subscript $_r$ stands for 'reaction',
H is the symbol for enthalpy,
superscript ^{o} means 'standard state conditions',
(T) means 'at temperature T'.

This type of notation is found for a wide range of thermochemical (and other thermodynamic) functions that we see later on.

The *standard state of a substance* is its most stable state under a pressure of 1 bar (10^5 Pa) and at some specified temperature T. Most commonly, T is taken to be 298.15 K, and the notation for the standard enthalpy change of a reaction at 298.15 K becomes $\Delta_r H^o$(298.15 K). It is usually sufficient to indicate $\Delta_r H^o$(298 K). *Do not confuse standard thermodynamic temperature with that used for the standard temperature conditions of a gas (Section 1.9).*

Enthalpy changes can be measured experimentally by *calorimetric methods* or can be calculated by using tabulated values of the standard enthalpies of formation.

By definition, a negative value of ΔH corresponds to heat energy given out during a reaction, and a positive value indicates that heat energy is taken in from the surroundings. When heat energy is lost to the surroundings, a reaction is said to be *exothermic* (negative ΔH); in an *endothermic* reaction, heat energy is taken in (positive ΔH).

Heat energy is given out in an exothermic reaction (ΔH is negative).
Heat energy is taken in an endothermic process (ΔH is positive).

Standard enthalpy of formation

The *standard enthalpy of formation* of a compound, $\Delta_f H^o(298\ K)$, relates to the enthalpy change that accompanies the formation of a compound in its standard state from its constituent elements, each being in its standard state.

$\Delta_f H^o(298\ K)$ for an element in its standard state is zero.

The standard state of an element is the thermodynamically most stable form of that element under a pressure of 1 bar (10^5 Pa) and at a temperature of 298 K. The one exception is phosphorus for which the standard state is taken to be white phosphorus rather than the thermodynamically more stable red and black forms.§ *By definition, the standard enthalpy of formation of an element in its standard state is zero.*

The formation of methane from carbon and hydrogen is shown in equation 1.33. The standard state of hydrogen is gaseous dihydrogen and for carbon, the standard state is graphite. The enthalpy change that accompanies this reaction at 298 K corresponds to the standard enthalpy of formation of methane. The negative sign corresponds to an exothermic reaction.

$$2H_2(g) + C(\text{graphite}) \rightarrow CH_4(g)$$
$$\Delta_f H^o(298\ K)\ CH_4(g) = -75\ \text{kJ per mole of methane} \qquad (1.33)$$

Calculating standard enthalpies of reaction

For a general reaction in which reactants A and B combine to give products C and D (equation 1.34), the standard enthalpy change accompanying the reaction is $\Delta_r H^o(298\ K)$ and is the difference between the sum of the standard enthalpies of formation of the products and the sum of the standard enthalpies of formation of the reactants (equation 1.35).

$$A + B \rightarrow C + D \quad \text{Standard enthalpy change} = \Delta_r H^o(298\ K) \qquad (1.34)$$

Σ means summation

$$\Delta_r H^o(298\ K) = \Sigma\Delta_f H^o(298\ K)_{\text{products}} - \Sigma\Delta_f H^o(298\ K)_{\text{reactants}} \qquad (1.35)$$

The summation of the standard enthalpy values gives a measure of the enthalpy of the reactants and products. If the heat content of the products is greater than that of the reactants, the reaction is endothermic (positive $\Delta_r H^o$)

§ The definition of standard state and the exceptional case of phosphorus have been laid down by the National Bureau of Standards.

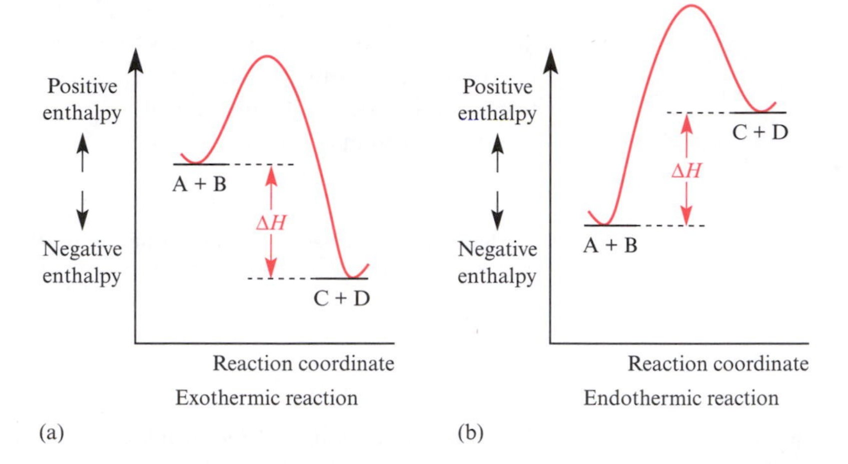

1.8 Reaction profiles for (a) an exothermic reaction and (b) an endothermic reaction.

and if the heat content of the reactants is greater than that of the products, the reaction is exothermic ($-\Delta_r H^o$). This is shown schematically in Figure 1.8; such a diagram shows the *reaction profile*.

An everyday example of an exothermic reaction is the burning of a fuel such as butane, C_4H_{10} (equation 1.36).

$$C_4H_{10}(g) + {}^{13}/_2O_2(g) \rightarrow 4CO_2(g) + 5H_2O(l) \tag{1.36}$$

The value for $\Delta_r H^o(298\ K)$ for reaction 1.36 is found from the standard enthalpies of formation of the products and reactants. Values of $\Delta_f H^o(298\ K)$ $C_4H_{10}(g)$, $O_2(g)$, $CO_2(g)$ and $H_2O(l)$ are -126, 0, -393.5 and $-286\ kJ\ mol^{-1}$ respectively and equation 1.37 shows how $\Delta_r H^o(298\ K)$ for reaction 1.36 is determined.

$$\begin{aligned}\Delta_r H^o(298\ K) &= \Sigma\Delta_f H^o(298\ K)_{products} - \Sigma\Delta_f H^o(298\ K)_{reactants}\\ &= [4 \times (-393.5)] + [5 \times (-286)] - (-126)\\ &= -2878\ kJ\ mol^{-1}\end{aligned} \tag{1.37}$$

Burning a substance in O_2 refers to the process of *combustion* and the enthalpy change is the *enthalpy of combustion*, $\Delta_c H$.

Worked example 1.16 ***Standard enthalpy of reaction***

What is the enthalpy change for the reaction of gaseous Cl_2 with ethene to give 1,2-dichloroethane:

$$C_2H_4(g) + Cl_2(g) \rightarrow C_2H_4Cl_2(l)$$

if the values of $\Delta_f H^o(298\ K)$ of $C_2H_4(g)$ and 1,2-$C_2H_4Cl_2(l)$ are +52.5 and −167 kJ mol^{-1} respectively?

At 298 K, $Cl_2(g)$ is the standard state of the element and $\Delta_f H^o(298\ K) = 0$.

The stoichiometric equation for the reaction shows that one mole of ethene gives one mole of 1,2-dichloroethane.

At 298 K: $\Delta_r H^o = [\Delta_f H^o\ C_2H_4Cl_2(l)] - [\Delta_f H^o\ C_2H_4(g)]$
$= -167 - 52.5$
$= -219.5$ kJ *per mole of reaction*

The phrase *per mole of reaction* means that we are quoting the enthalpy change with respect to the reaction as it is written. If we had two moles of ethene in the reaction:

$$2C_2H_4(g) + 2Cl_2(g) \rightarrow 2C_2H_4Cl_2(l)$$

the enthalpy change would be twice as great (–439 kJ).

Hess's Law of Constant Heat Summation

The calculations above make use of *Hess's Law of Constant Heat Summation* which states that the enthalpy change on going from reactants to products is independent of the reaction path taken.

What we were effectively doing in equation 1.37 was considering the formation of butane and dioxygen from their constituent elements and also the formation of carbon dioxide and water from their constituent elements. These two enthalpy changes form part of a thermochemical cycle, which is completed by considering reaction 1.36. The full cycle is shown in equation 1.38.

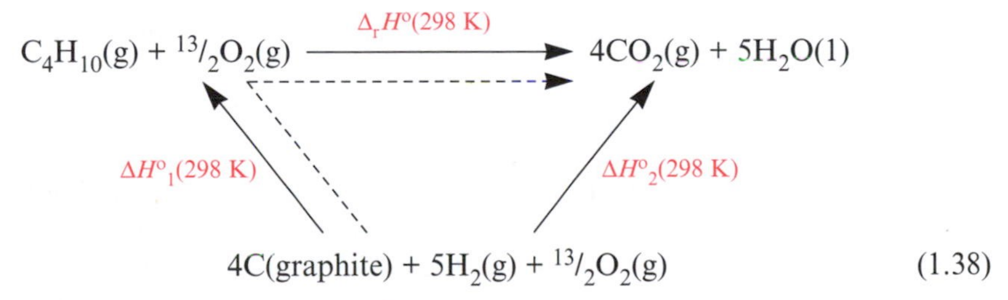

(1.38)

Both the right-hand arrow and the dotted arrow take us from the elements in their standard states to carbon dioxide and water. Application of Hess's Law links the three enthalpy changes that make up the thermochemical cycle (equation 1.39).

$$\Delta_r H^o + \Delta H^o_1 = \Delta H^o_2 \quad \text{(at 298 K)} \qquad (1.39)$$

By rearranging this equation, we can find $\Delta_r H^o$ in terms of the difference between the sums of the standard enthalpies of formation of products (ΔH^o_2) and reactants (ΔH^o_1) exactly as we did in equation 1.37. Hess's Law is particularly useful where we have more complex situations to consider such as the determination of lattice energies (Section 6.16) or enthalpy changes associated with the dissolution of salts (Section 12.11).

Hess's Law of Constant Heat Summation states that the enthalpy change on going from reactants to products is independent of the reaction path taken.

1.18 BOLTZMANN DISTRIBUTION OF MOLECULAR ENERGIES

Activation energy

Figure 1.9 illustrates that in starting from reactants A and B there is an enthalpy barrier to overcome before the reaction can proceed in a 'downhill' direction towards products C and D. This initial barrier is called the *activation enthalpy* and measures the *kinetic* barrier to the reaction (Figure 1.9). The higher the barrier, the less likely it is that the reaction will occur. It is more convenient to talk in terms of the activation *energy*, $E_{activation}$, and we elaborate on this discussion in Section 10.6.

Reaction kinetics: see Chapter 10

The *activation energy*, $E_{activation}$, of a reaction is the energy in excess of that possessed by the ground state that is required for the reaction to proceed.

Distribution of molecular energies

Consider a sample of gaseous H_2 in a closed container at a fixed temperature. The molecules are in continual motion, but some molecules are moving faster than others. The kinetic energy of a molecule is related to the molecular velocity by equation 1.40, and distributions of both molecular speeds and energies are described by a *Boltzmann distribution curve* (Figure 1.10).

$$E = \frac{mv^2}{2} \quad m = \text{mass}, v = \text{velocity} \tag{1.40}$$

The curve is asymmetrical and shows that there are more molecules with lower energies than there are with higher energies. When a chemical reaction takes place, the reactant molecules must possess a certain minimum energy – the activation energy. This is shown as an arbitrary value in Figure 1.10. Only molecules possessing energies equal to or greater than $E_{activation}$ are able to react.

Temperature dependence of $E_{activation}$: see Section 10.6

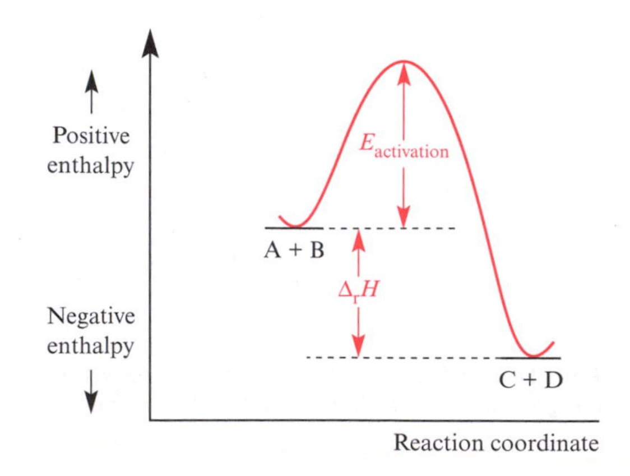

1.9 Reaction profile for an exothermic reaction showing the activation enthalpy, $E_{activation}$. (In Chapter 10, we see that the activation *enthalpy* is approximately equal to the activation *energy*.)

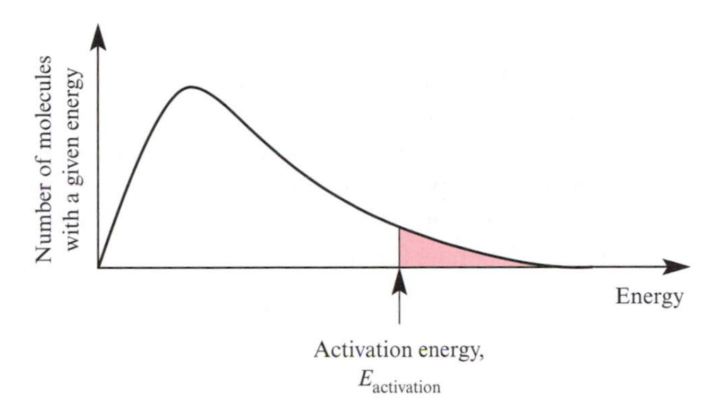

1.10 The Boltzmann distribution of molecular energies: for a given temperature there are fewer molecules with higher than lower energies. The activation energy (which is reaction- and reaction-path dependent) is the minimum energy required before a reaction can proceed.

1.19 THERMODYNAMIC AND KINETIC STABILITY

A term that is commonly (and often inconsistently) used is 'stable'. It is meaningless to say that something is stable or unstable unless you specify '*with respect to what*'. Consider hydrogen peroxide, H_2O_2. This compound is a liquid at room temperature and a solution can be purchased in a bottle as a hair bleach. Because of this you may think that H_2O_2 is a 'stable compound'. However, the conditions under which H_2O_2 is stored are critical. It readily decomposes (equation 1.41) and the standard enthalpy change for this reaction is -98.2 kJ mol^{-1}. The process is slow, but in the presence of some surfaces or alkali, decomposition is rapid, even explosive. We say that H_2O_2 is *unstable with respect to the formation of* H_2O *and* O_2.

$$2H_2O_2(l) \rightarrow 2H_2O(l) + O_2(g) \tag{1.41}$$

We have already mentioned that diamond is thermodynamically unstable with respect to graphite, but that the interconversion takes place very, very slowly.

1.20 ENTHALPIES OF FUSION AND VAPORIZATION

Enthalpy of fusion

When a solid melts, heat energy is *needed* for the phase change. For example, when a metal melts, the solid state lattice (which is a rigid framework) is broken, although the atoms are not completely separated from each other (Figure 1.2). The enthalpy change for melting is called the enthalpy of fusion and its symbol is $\Delta_{fus}H$. When the liquid solidifies, heat energy is released as bonds are formed, and the enthalpy change is $-\Delta_{fus}H$. Whereas melting is an endothermic process, solidification is an exothermic process. Equation 1.42 shows the phase transitions for copper.

$$Cu(s) \underset{-13 \text{ kJ mol}^{-1}}{\overset{+13 \text{ kJ mol}^{-1}}{\rightleftharpoons}} Cu(l) \tag{1.42}$$

Enthalpy of vaporization

The enthalpy change that accompanies the transition from a liquid to a vapour is called the enthalpy of vaporization, $\Delta_{vap}H$. When a liquid metal is vaporized, heat energy is needed to separate the atoms – vaporization is an endothermic process. Conversely, when the vapour of a metal condenses to a liquid, heat energy is released as some bonds are formed. Equation 1.43 shows the phases changes for lead.

$$\mathrm{Pb(l)} \underset{-180\ \mathrm{kJ\ mol^{-1}}}{\overset{+180\ \mathrm{kJ\ mol^{-1}}}{\rightleftharpoons}} \mathrm{Pb(g)} \tag{1.43}$$

1.21 INTERMOLECULAR INTERACTIONS

Intermolecular interactions versus covalent bonds

In the previous section, we considered phases changes for a metal in terms of the separation of the metal *atoms*. In a *molecular compound* such as methane **1.2**, the atoms in the molecule are held together by covalent bonds.

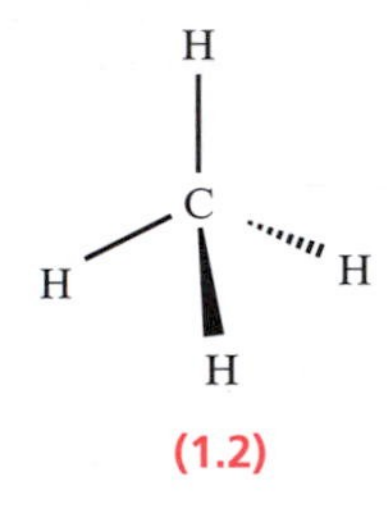

(1.2)

In the vapour state, these molecules are well separated and can be regarded as having little effect on one another. This is one of the criteria of an ideal gas, but in reality gas behaviour deviates from ideality because the molecules interact with each other. When methane is liquefied, the molecules come closer together, and when the liquid is solidified, an ordered structure is formed in which there are *intermolecular interactions* between the CH_4 molecules. The vapour→liquid and liquid→solid phase changes are endothermic, but the intermolecular interactions are only weak: $\Delta_{vap}H$ = 8 kJ mol^{-1} and $\Delta_{fus}H$ = 1 kJ mol^{-1}. During the phases changes, *the covalent bonds within the CH_4 molecule remain intact.*

Table 1.7 Types of intermolecular forces

Interaction	*Acts between*	*Typical energy / kJ mol^{-1}*	*Further discussion*
London dispersion forces	Most molecules	≤2	Section 2.21 (induced dipoles; interatomic forces)
Dipole–dipole interactions	Polar molecules	2	
Ion–dipole interactions	Ions and polar molecules	15	Sections 11.8 and 12.11
Hydrogen bonds	H atom and electronegative atom (N, O, F)	25	Section 11.8

Types of intermolecular interactions

The strengths of intermolecular (*van der Waals*) interactions vary depending upon their precise nature (Table 1.7). The weakest interactions occur between the electron clouds of adjacent molecules and are called *London dispersion forces*. It is these interactions that operate between molecules of methane in the solid state.

We discuss intermolecular interactions in several sections in this book but for now, the important point to remember is that the enthalpies of fusion and vaporization (and melting and boiling points) of molecular species reflect the extent of intermolecular interaction. When we move to an ionic solid in which ions interact with one another through electrostatic forces, the amount of energy needed to separate the ions is often greater than that needed to separate covalent molecules. Enthalpies of fusion of ionic solids are far higher than those of covalent solids.

1.22 EQUILIBRIUM CONSTANTS AND LE CHATELIER'S PRINCIPLE

Le Chatelier's principle

Consider the reaction of orange dichromate ion with hydroxide ion (equation 1.44).

$$\underset{\text{orange}}{[Cr_2O_7]^{2-}(aq)} + 2[OH]^-(aq) \rightleftharpoons \underset{\text{yellow}}{2[CrO_4]^{2-}(aq)} + H_2O(l) \qquad (1.44)$$

Upon reaction with $[OH]^-$ ion, the dichromate ion is converted into the yellow chromate ion, but in the presence of acid, the chromate ion is converted back into orange dichromate. By altering the concentration of acid or alkali, the *equilibrium* can be pushed towards the left- or right-hand side.

Le Chatelier's principle states that when an external change is made to a system in equilibrium, the system will change in order to compensate for the change.

This is an example of Le Chatelier's principle which states that when an external change is made to a system in equilibrium, the system will change in order to compensate for the change. In reaction 1.44, if hydroxide ion is added to the equilibrium, the reaction will move in the direction in which $[OH]^-$ is consumed – that is, to the right-hand side. Conversely, if acid is added, the equilibrium will move to consume the acid – to the left-hand side. [Question: How is the acid used up?]

Position of equilibrium

If we know the *position of equilibrium*, we know the extent to which reactants (left-hand side) predominate over the products (right-hand side), or the products predominate over the reactants. This information is provided in the *equilibrium constant* and we must consider the cases of solution and gaseous equilibria separately.

Solution equilibrium constant, K_c

If an equilibrium consists of solution species, then the equilibrium constant can be expressed in terms of the concentrations of each component of the equilibrium mixture. For reaction 1.45, the equilibrium constant K_c is given by equation 1.46 where the square brackets stand for the concentration of the species.

$$aA + bB \rightleftharpoons cC + dD \qquad (1.45)$$

where the components are liquids or in aqueous solution.

$$K_c = \frac{[C]^c[D]^d}{[A]^a[B]^b} \qquad (1.46)$$

Worked example 1.17 ***Equilibrium constant determination***

Acetic acid and ethanol react as follows:

$$\underset{\text{acetic acid}}{CH_3CO_2H} + \underset{\text{ethanol}}{CH_3CH_2OH} \rightleftharpoons CH_3CO_2C_2H_5 + H_2O$$

When one mole of acetic acid reacts with one mole of ethanol, and the reaction mixture is allowed to reach equilibrium at 298 K, 0.67 moles of $CH_3CO_2C_2H_5$ are present. What is the equilibrium constant?

We must first find the number of moles of each component in the equilibrium mixture and then find their concentrations. Let the total volume be V dm^3.

	CH_3CO_2H	+ CH_3CH_2OH	$\rightleftharpoons$ $CH_3CO_2C_2H_5$	+ H_2O
Initial moles:	1	1	0	0
Moles at equilibrium:	0.33	0.33	0.67	0.67
Concentration at equilibrium:	$\frac{0.33}{V}$	$\frac{0.33}{V}$	$\frac{0.67}{V}$	$\frac{0.67}{V}$

The equilibrium constant is given by:

$$K_c = \frac{\left(\frac{0.67}{V}\right)\left(\frac{0.67}{V}\right)}{\left(\frac{0.33}{V}\right)\left(\frac{0.33}{V}\right)} = \frac{(0.67)^2}{(0.33)^2} = 4.12$$

In this case the total volume V cancels but this will not always be the case.

Gas phase reactions: the equilibrium constant, K_p

Partial pressures: see Section 1.9

For a gas phase reaction, the equilibrium constant is expressed in terms of the partial pressures of the components. For reaction 1.47, the equilibrium constant is given by equation 1.48.

$$aA(g) + bB(g) \rightleftharpoons cC(g) + dD(g) \qquad (1.47)$$

$$K_p = \frac{(P_C)^c (P_D)^d}{(P_A)^a (P_B)^b} \qquad (1.48)$$

Worked example 1.18 ***Gas phase equilibrium constant determination***

Dihydrogen and diiodine react as follows:

$$H_2(g) + I_2(g) \rightleftharpoons 2HI(g)$$

At 400 K, $K_p = 6.3$. If two moles of H_2 and two moles of I_2 are mixed and the total pressure is 10^5 Pa, what is the equilibrium composition?

We must first find the number of moles of each component in the equilibrium mixture and then find their partial pressures.

$$\text{Partial pressure of component X} = \frac{\text{Moles of X}}{\text{Total moles of gas}} \times \text{Total pressure}$$

Let the moles of HI at equilibrium be $2x$ – bear in mind the stoichiometry of the reaction when you assign this unknown.

x moles of H_2, and x moles of I_2 produce $2x$ moles of HI.

	$H_2(g)$	+	$I_2(g)$	$\rightleftharpoons$	$2HI(g)$
Initial moles:	2		2		0
Moles at equilibrium:	$(2-x)$		$(2-x)$		$2x$

Total moles of gas at equilibrium $= (2-x) + (2-x) + 2x = 4$

Partial pressures at equilibrium: $P_{H_2} = P_{I_2} = \dfrac{(2-x)}{4} \times 10^5$ Pa

$$P_{HI} = \frac{2x}{4} \times 10^5 \text{ Pa}$$

The equilibrium constant is given by:

$$K_p = \frac{(P_{HI})^2}{(P_{H_2})(P_{I_2})} = 6.3$$

$$\frac{\left(\frac{2x}{4} \times 10^5\right)^2}{\left(\frac{2-x}{4} \times 10^5\right)\left(\frac{2-x}{4} \times 10^5\right)} = \frac{\left(\frac{2x}{4} \times 10^5\right)^2}{\left(\frac{2-x}{4} \times 10^5\right)^2} = 6.3$$

$$\frac{(2x)^2}{(2-x)^2} = 6.3$$

$$4x^2 = 6.3 \times (4 - 4x + x^2) = 25.2 - 25.2x + 6.3x^2$$

$$0 = 2.3x^2 - 25.2x + 25.2$$

This is a quadratic of the form:

$$ax^2 + bx + c = 0$$

and can be solved using the equation

$$x = \frac{-b \pm \sqrt{b^2 - 4ac}}{2a}$$

$$x = \frac{25.2 \pm \sqrt{25.2^2 - (4 \times 2.3 \times 25.2)}}{2 \times 2.3}$$

There are two solutions, but only one is meaningful – why?

$$x = 1.1$$

At equilibrium there are 2.2 moles HI, 0.9 moles H_2 and 0.9 moles I_2.

1.23 EMPIRICAL, MOLECULAR AND STRUCTURAL FORMULAE

Empirical and molecular formulae

The empirical formula of a compound gives the ratio of atoms of elements that combine to make the compound. However, this is not necessarily the same as the molecular formula which tells you the number of atoms of the constituent elements in line with the relative molar mass of the compound. The relationship between the empirical and molecular formula of a compound is illustrated using ethane, in which the ratio of carbon : hydrogen atoms is 1:3. This means that the *empirical formula* of ethane is CH_3. The relative molecular mass of ethane is 30, corresponding to two CH_3 units per molecule – the molecular formula is C_2H_6. In methane, on the other hand, the empirical formula is CH_4 and this corresponds directly to the molecular formula.

Worked example 1.19 ***Empirical and molecular formulae***

An alkane of general formula C_nH_{2n+2} contains 83.7% carbon by weight. Suggest the identity of the alkane. (A_r C = 12, H = 1).

Let the formula of the compound be C_xH_y. The percentage composition of the alkane is 83.7% C and 16.3% H.

$$\text{\% of carbon by weight} = \frac{\text{Mass of carbon}}{\text{Total mass}} \times 100$$

$$\text{\% of hydrogen by weight} = \frac{\text{Mass of hydrogen}}{\text{Total mass}} \times 100$$

For a mole of the alkane, the total mass = relative molecular mass = M_r.

$$\% \text{C} = 83.7 = \frac{(12 \times x)}{M_r} \times 100$$

$$\% \text{H} = 16.3 = \frac{(1 \times y)}{M_r} \times 100$$

We do not know M_r, but we can write down the *ratio* of moles of C:H atoms in the compound – this is the empirical formula. From above:

$$M_r = \frac{12x}{83.7} \times 100 = \frac{y}{16.3} \times 100$$

$$\frac{y}{x} = \frac{100 \times 16.3 \times 12}{100 \times 83.7} = 2.33$$

The *empirical formula* of the alkane is $CH_{2.33}$ or C_3H_7.

The compound must fit into the family of alkanes of formulae C_nH_{2n+2}, and this suggests that the *molecular formula* is C_6H_{14}.

The working above sets the problem out in full; in practice, the empirical formula can be found as follows:

$\%\text{C} = 83.7 \qquad A_r\ \text{C} = 12$

$\%\text{H} = 16.3 \qquad A_r\ \text{H} = 1$

$$\text{Ratio C:H} = \frac{83.7}{12} : \frac{16.3}{1} \approx 7 : 16.3 = 1 : 2.33$$

Structural formulae

Neither the empirical nor the molecular formula provides information about the way in which the atoms of a molecule are connected. The molecular formula H_2S does not indicate the arrangement of the three atoms in a molecule of hydrogen sulfide, but the structural formula **1.3** *is* informative. We could also have arrived at this structure by considering the number of valence electrons available for bonding.

Bonding and structure: see Chapters 3–5

For some molecular formulae, it is possible to connect the atoms in more than one reasonable way and the molecule is said to possess *isomers*. An example is C_4H_{10} for which two structural formulae **1.4** and **1.5** can be drawn.

Isomers: see Sections 5.11, 8.4, 8.6 and 16.5

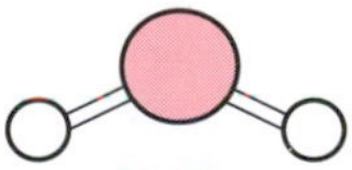

(1.3) 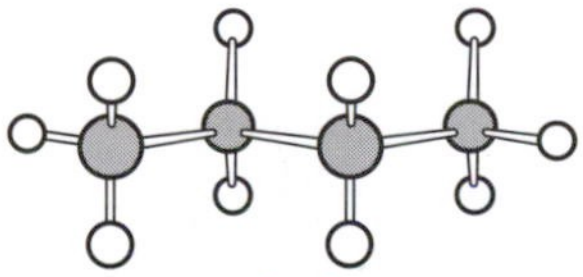 (1.4) 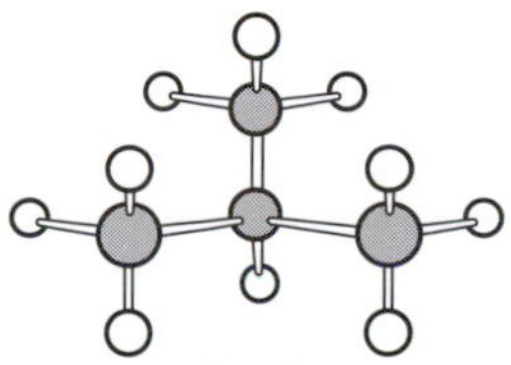 (1.5)

1.24 BASIC NOMENCLATURE

In this section we outline some fundamental nomenclature and set out some important IUPAC ground rules for organic and inorganic compounds. We summarize some widely used 'trivial' names of compounds and you should become familiar with these as well as with their IUPAC counterparts. Detailed nomenclature rules and the specifics of organic chain numbering are found in Chapters 8, 14, 15 and 17.

Basic organic classifications: straight chain alkanes

The general formula for an alkane with a straight chain backbone is C_nH_{2n+2}; the molecule is *saturated* and contains only C–C and C–H bonds. The simplest member of the family is *methane* CH_4, **1.6**. The name methane carries with it two pieces of information:

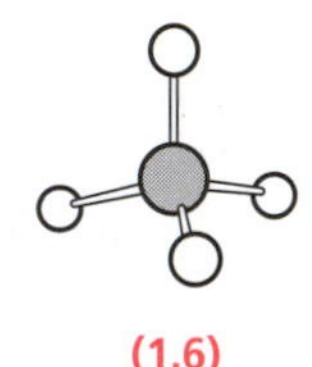

(1.6)

- *meth-* is the prefix that tells us there is one C atom in the carbon chain.
- the ending *-ane* denotes that the compound is an alkane.

The names of organic compounds are composite with the *stem* telling us the number of carbon atoms in the main chain of the molecule. These are listed in the middle column of Table 1.8. For a straight chain *alkane*, the name is completed by adding *-ane* to the prefix in the table. Compound **1.7** is propane and **1.8** is heptane.

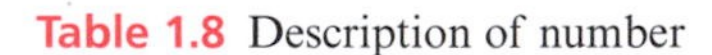

Table 1.8 Description of number

Number	*Stem used to give the number of C atoms in an organic carbon chain*	*Prefix used to describe the number of groups or substituents*
1	meth-	mono-
2	eth-	di-
3	prop-	tri-
4	but-	tetra-
5	pent-	penta-
6	hex-	hexa-
7	hept-	hepta-
8	oct-	octa-
9	non-	nona-
10	dec-	deca-
11	undec-	undeca-
12	dodec-	dodeca-
13	tridec-	trideca-
14	tetradec-	tetradeca-
15	pentadec-	pentadeca-
16	hexadec-	hexadeca-
17	heptadec-	heptadeca-
18	octadec-	octadeca-
19	nonadec-	nonadeca-
20	icos-	icosa-

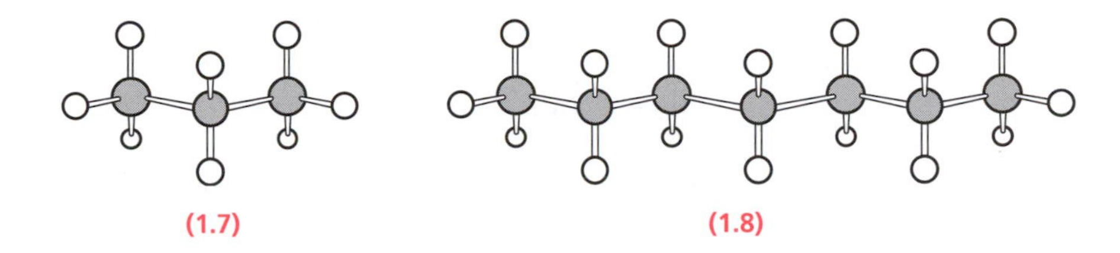

(1.7) (1.8)

Basic organic classifications: functional groups

A *functional group* in a molecule imparts a characteristic reactivity to the compound. The functional group in an alkene is the C=C double bond and, in an alcohol, the functional group is the –OH unit. The organic functional groups that we will describe in this book are listed in Table 1.9. The presence of most of these groups is recognized by using an instrumental technique such as infrared, electronic or nuclear magnetic resonance spectroscopy – see Chapter 9.

Basic inorganic nomenclature

The aim of the IUPAC nomenclature is to provide for a compound or an ion a name that is unambiguous. One problem that we face when dealing with some inorganic elements is the possibility of a variable oxidation state. A simple example is that of distinguishing between the two common oxides of carbon, i.e. carbon monoxide and carbon dioxide. The use of 'mono-' and 'di-' indicates that the compounds are CO and CO_2 respectively. The accepted numerical prefixes are listed in Table 1.8. Note that 'di-' should be used in preference to 'bi-'.

Binary compounds 1

A binary compound is composed of only *two* types of elements. We deal first with cases where there is no ambiguity over oxidation states of the elements present. Examples of such binary compounds include NaCl, CaO, HCl, Na_2S and $MgBr_2$. The formula should be written with the more electropositive element (often a metal) placed first. The names follow directly:

NaCl	sodium chloride
CaO	calcium oxide
HCl	hydrogen chloride
Na_2S	sodium sulfide
$MgBr_2$	magnesium bromide

There is no need to write 'disodium sulfide' or 'magnesium dibromide' because there is usually no ambiguity about the oxidation state of the *s*-block metal.

A binary compound is composed of two types of elements.

Table 1.9 Selected functional groups for organic molecules.

Name of functional group	*Functional group*	*Example; where it is in common use, a trivial name is given in red*
Alcohol	—OH	Ethanol (CH_3CH_2OH)
Aldehyde	—C(=O)H	Ethanal (CH_3CHO) Acetaldehyde
Ketone	—C(=O)R R ≠ H	Propanone ($CH_3C(O)CH_3$) Acetone
Carboxylic acid	—C(=O)O—H	Ethanoic acid (CH_3CO_2H) Acetic acid
Ester	—C(=O)O—R e.g. R = alkyl	Ethyl ethanoate ($CH_3CO_2C_2H_5$) Ethyl acetate
Ether	R'—O—R R = R' or R ≠ R'	Diethyl ether ($C_2H_5OC_2H_5$)
Amine	$—NH_2$	Ethylamine ($CH_3CH_2NH_2$)
Amide	$—C(=O)NH_2$	Ethanamide (CH_3CONH_2) Acetamide
Halogenoalkane	—X X = F, Cl, Br, I	Bromoethane (CH_3CH_2Br)
Acid chloride	—C(=O)Cl	Ethanoyl chloride (CH_3COCl) Acetyl chloride
Nitrile	—C≡N	Ethanenitrile (CH_3CN) Acetonitrile
Nitro compound	$—NO_2$	Nitromethane (CH_3NO_2)

Anions

The endings *'-ide'*, *'-ite'* and *'-ate'* generally signify an anionic species. Some examples are listed in Table 1.10. The endings '-ate' and '-ite' *tend* to indicate the presence of oxygen in the anion (i.e. an oxoanion) and are used for anions which are derived from oxoacids; e.g. the oxoanion derived from sulfuric acid is a sulfate.

There is more than one accepted method of distinguishing between the different oxoanions of elements such as sulfur, nitrogen and phosphorus (Table 1.10). Older names such as sulfate, sulfite, nitrate and nitrite are still accepted within the IUPAC guidelines. It is more informative, however, to incorporate the oxidation state of the element that is combining with oxygen, and an alternative name for sulfate is tetraoxosulfate(VI). This shows not only the oxidation state of the sulfur atom, but the number of oxygen atoms as well. A third accepted alternative is to use the name tetraoxosulfate(2–). In *Chemistry: An Integrated Approach*, we have made every effort to stay within the IUPAC recommendations while retaining the most common alternatives, e.g. sulfate.

Oxidation states

The oxidation state is very often indicated by using the Stock system of Roman numerals. The numeral is always an integer and is placed after the name of the element to which it refers; Table 1.10 shows its application to

Table 1.10 Names of some common anions. In some cases, more than one name is accepted by IUPAC.

Formula of anion	*Name of anion*
H^-	Hydride
$[OH]^-$	Hydroxide
F^-	Fluoride
Cl^-	Chloride
Br^-	Bromide
I^-	Iodide
O^{2-}	Oxide
S^{2-}	Sulfide
Se^{2-}	Selenide
N^{3-}	Nitride
N_3^-	Azide
P^{3-}	Phosphide
$[CN]^-$	Cyanide
$[NH_2]^-$	Amide
$[OCN]^-$	Cyanate
$[SCN]^-$	Thiocyanate
$[SO_4]^{2-}$	Sulfate *or* tetraoxosulfate(VI)
$[SO_3]^{2-}$	Sulfite *or* trioxosulfate(IV)
$[NO_3]^-$	Nitrate *or* trioxonitrate(V)
$[NO_2]^-$	Nitrite *or* dioxonitrate(III)
$[PO_4]^{3-}$	Phosphate *or* tetraoxophosphate(V)
$[PO_3]^{3-}$	Phosphite *or* trioxophosphate(III)
$[ClO_4]^-$	Perchlorate *or* tetraoxochlorate(VII)
$[CO_3]^{2-}$	Carbonate *or* trioxocarbonate(IV)

some oxoanions. The oxidation number can be zero, positive or negative.[§] An oxidation state is assumed to be positive unless otherwise indicated by the use of a negative sign, Thus, (III) is taken to read '(+III)'; but for the negative state, write (–III).

In a formula, the oxidation state is written as a superscript (e.g. $[Mn^{VII}O_4]^-$) but in a name, it is written on the line (e.g. iron(II) bromide). Its use is important when the name could be ambiguous (see below).

Binary compounds 2

We look now at binary compounds where there could be an ambiguity over the oxidation state of the more electropositive element (often a metal). Examples of such compounds include $FeCl_3$, SO_2, SO_3, ClF, ClF_3, $NiSO_4$ and $SnCl_2$. Simply writing 'iron chloride' does not distinguish between the chlorides of iron(II) and iron(III), and for $FeCl_3$ it is necessary to write iron(III) chloride. Another accepted name is iron trichloride.

The oxidation state of sulfur in SO_2 can be seen immediately in the name sulfur(IV) oxide, but also acceptable is the name sulfur dioxide. Similarly, SO_3 can be named sulfur(VI) oxide or sulfur trioxide.

Accepted names for ClF, ClF_3, $NiSO_4$ and $SnCl_2$ are:

ClF	chlorine(I) fluoride or chlorine monofluoride
ClF_3	chlorine(III) fluoride or chlorine trifluoride
$NiSO_4$	nickel(II) sulfate
$SnCl_2$	tin(II) chloride or tin dichloride

Cations

Cations of metals where the oxidation state does not usually vary, notably the *s*-block elements, may be named by using the name of the metal itself (e.g. sodium ion, barium ion), although the charge may be indicated (e.g. sodium(I) ion or sodium(1+) ion, barium(II) ion or barium(2+) ion).

Where there may be an ambiguity, the charge must be shown (e.g. iron(II) or iron(2+) ion, copper(II) or copper(2+) ion, thallium(I) or thallium(1+) ion).[†]

Table 1.11 The names of some common, non-metallic cations.

Formula of cation	*Name of cation*
H^+	Hydrogen ion
$[H_3O]^+$	Oxonium ion
$[NH_4]^+$	Ammonium ion
$[NO]^+$	Nitrosyl ion
$[NO_2]^+$	Nitryl ion
$[N_2H_5]^+$	Hydrazinium ion

§ A zero oxidation state is signified by 0, although this is not a Roman numeral.

† An older form of nomenclature which is commonly encountered still uses the suffix '-ous' to describe the lower oxidation state and '-ic' for the higher one. Thus, copper(I) is cuprous and copper(II) is cupric. This system is only unambiguous when the metal only exhibits two oxidation states.

The names of polyatomic cations are introduced as they appear in the textbook but Table 1.11 lists some of the most common inorganic, non-metallic cations with which you may already be familiar. Look for the ending '-ium'; this often signifies the presence of a cation, although remember that '-ium' is a common ending in the name of elemental metals (see Section 1.5). Many metal ions, in particular those in the d-block, occur as complex ions; these are described in Chapter 16.

1.25 FINAL COMMENTS

The aim of this first chapter is to provide a point of reference for basic chemical definitions, ones which we assume you have encountered before you begin a first year university chemistry course. If you find later in the book that a concept appears to be 'assumed', you should find some revision material to help you in Chapter 1 and in the accompanying study guide. Section 1.24 gives some basic guidelines for naming organic and inorganic compounds, and more detailed nomenclature appears as the book progresses.

We have deliberately *not* called Chapter 1: 'Introduction'. There is often a tendency to pass through chapters so-labelled without paying attention to them. In this text, Chapter 1 is designed to help you and to remind you of basic issues.

PROBLEMS

1.1 What is 0.0006 m in (a) mm, (b) pm, (c) cm, (d) nm?

1.2 A typical C=O double bond distance in an organic aldehyde is 122 pm. What is this in nm?

1.3 The relative molecular mass of NaCl is 58.4 g mol^{-1} and its density is 2.16 g cm^{-3}. What is the volume of 1 mole of NaCl in m^3?

1.4 The equation $E = h\nu$ relates the Planck constant (h) to energy and frequency. Determine the SI units of the Planck constant.

1.5 Kinetic energy is given by the equation: $E = \frac{1}{2}mv^2$. By going back to the base SI units, show that the units on the left- and right-hand sides of this equation are compatible.

1.6 Calculate the relative atomic mass of a sample of naturally occurring boron which contains 19.9% $^{10}_{5}B$ and 80.1% $^{11}_{5}B$.

1.7 The mass spectrum for molecular bromine shows three lines for the parent ion, Br_2^+. The isotopes for bromine are $^{79}_{35}Br$ (50%) and $^{81}_{35}Br$ (50%). Explain why there are three lines and predict their mass values and relative intensities. Predict what the mass spectrum of HBr would look like; isotopes of hydrogen are given in Section 1.7.

1.8 Convert the volume of each of the following to conditions of standard temperature (273 K) and pressure (1 bar = 10^5 Pa) and give your answer in m^3 in each case:
(a) 30.0 cm^3 of CO_2 at 290 K and 101 325 Pa (1 atm)
(b) 5.30 dm^3 of H_2 at 298 K and 100 kPa (1 bar)
(c) 0.300 m^3 of N_2 at 263 K and 102 kPa
(d) 222 m^3 of CH_4 at 298 K and 200 000 Pa (2 bar)

1.9 The partial pressure of helium in a 50 dm^3 gas mixture at 285 K and 10^5 Pa is 4×10^4 Pa. How many moles of helium are present? (Volume of one mole of ideal gas at 273 K, 10^5 Pa = 22.7 dm^3).

1.10 A 20 dm^3 sample of gas at 273 K and 2 bar pressure contains 0.5 moles N_2 and 0.7 moles Ar. What is the partial pressure of each gas, and are there any other gases in the sample? (Volume of one mole of ideal gas at 273 K, 10^5 Pa (1 bar) = 22.7 dm^3.)

1.11 Determine the number of moles present in each of the following: (a) 0.44 g PF_3, (b) 1 dm^3 gaseous PF_3 at 293 K and 2×10^5 Pa, (c) 3.48 g MnO_2, (d) 0.042 g $MgCO_3$. (A_r P = 31, F = 19, Mn = 55, Mg = 24, O = 16, C = 12; volume of one mole of ideal gas at 273 K, 10^5 Pa = 22.7 dm^3.)

1.12 What mass of solid must be dissolved in 100 cm^3 of solution to give the following concentrations of solutions: (a) 0.01 mol dm^{-3} KI; (b) 0.2 mol dm^{-3} NaCl; (c) 0.05 mol dm^{-3} Na_2SO_4? (A_r K = 39, I = 127, Na = 23, Cl = 35.5, S = 32, O = 16.)

1.13 With reference to the periodic table, write down the likely formulae of compounds formed between: (a) sodium and iodine, (b) magnesium and chlorine, (c) magnesium and oxygen, (d) calcium and fluorine, (e) lithium and nitrogen, (f) calcium and phosphorus, (g) sodium and sulfur and (h) hydrogen and sulfur.

1.14 *Use the information in the periodic table* to predict the likely formulae of the oxide, chloride, fluoride and hydride formed by aluminium.

1.15 Give balanced equations for the formation of each of the compounds in problems 1.13 and 1.14 from their constituent elements.

1.16 What do you understand by each of the following terms: proton, electron, neutron, nucleus, atom, radical, ion, cation, anion, molecule, covalent bond, compound, isotope, allotrope?

1.17 Suggest whether you think each of the following species will exhibit covalent or ionic bonding. Which of the species are compounds and which are molecular: (a) NaCl; (b) N_2; (c) SO_3; (d) KI; (e) NO_2; (f) Na_2SO_4; (g) $[MnO_4]^-$; (h) CH_3OH; (i) CO_2; (j) C_2H_6; (k) HCl; (l) $[SO_4]^{2-}$?

1.18 Determine the oxidation state of nitrogen in each of the following oxides: (a) N_2O; (b) NO; (c) NO_2, (d) N_2O_3; (e) N_2O_4; (f) N_2O_5. What assumption have you made? Look at Figure 13.18 and structure **13.30** (Chapter 13) – does the knowledge of structure influence your assignment of oxidation states?

1.19 In each reaction below, assign the oxidation and and reduction steps, and, for (b)–(g), show that the changes in oxidation states for the oxidation and reduction processes balance:

(a) $Cu^{2+}(aq) + 2e^- \rightarrow Cu(s)$
(b) $Mg(s) + H_2SO_4(aq) \rightarrow MgSO_4(aq) + H_2(g)$
(c) $2Ca(s) + O_2(g) \rightarrow 2CaO(s)$
(d) $2Fe(s) + 3Cl_2(g) \rightarrow 2FeCl_3(s)$
(e) $Cu(s) + 2AgNO_3(aq) \rightarrow Cu(NO_3)_2(aq) + 2Ag(s)$
(f) $CuO(s) + H_2(g) \rightarrow Cu(s) + H_2O(g)$
(g) $[MnO_4]^-(aq) + 5Fe^{2+}(aq) + 8H^+(aq) \rightarrow Mn^{2+}(aq) + 5Fe^{3+}(aq) + 4H_2O(l)$

1.20 (a) In a compound, oxygen is usually assigned an oxidation state of –2. What is the oxidation state in the allotropes O_2 and O_3, in the compound H_2O_2, and in the ions $[O_2]^{2-}$ and $[O_2]^+$? (b) What are the oxidation and reduction steps during the decomposition of hydrogen peroxide (equation 1.41)?

1.21 What is meant by each of the terms enthalpy change, standard enthalpy of formation and enthalpy of fusion? What is meant by *standard state*? Explain exactly what is meant by the notation $\Delta_f H^o(298\ K)$, $\Delta_{fus}H$, $\Delta_{vap}H$ and $\Delta_c H$.

1.22 Using the data in Table 1.12, calculate the standard enthalpies of reaction at 298 K for (a) the combustion of ethene; (b) the dehydration (loss of H_2O) of ethanol to give ethene; and (c) the reaction of Br_2 with ethene to give 1,2-dibromoethane. Assume that all the compounds are in their standard states at 298 K.

Table 1.12 Data required for problem 1.22.

Compound	*Formula and state*	$\Delta_f H^o(298\ K)$ / kJ mol^{-1}
Carbon dioxide	$CO_2(g)$	–393.5
Water	$H_2O(l)$	–286
Ethene	$C_2H_4(g)$	+52
Ethanol	$C_2H_5OH(l)$	–278
1,2-Dibromoethane	$C_2H_4Br_2(l)$	–81

1.23 When one mole of N_2 and three moles H_2 combine and the reaction:

$$N_2(g) + 3H_2(g) \rightleftharpoons 2NH_3(g)$$

has reached equilibrium at 500 K and a pressure of 1 bar, the equilibrium mixture contains 0.48 moles of NH_3. What is the value of K_p?

1.24 For the equilibrium:

$$H_2(g) + Cl_2(g) \rightleftharpoons 2HCl(g)$$

the value of K_p at 1500 K is 4.1×10^3. If the total pressure is 10^5 Pa, and initially there are 0.5 moles of each of H_2 and Cl_2, how many moles of HCl will be present when the reaction has reached equilibrium?

1.25 Give a systematic name for each of the following compounds: (a) Na_2CO_3; (b) $FeBr_3$; (c) $CoSO_4$; (d) $BaCl_2$; (e) Fe_2O_3; (f) $Fe(OH)_2$; (g) LiI; (h) KCN; (i) KSCN; (j) Ca_3P_2.

1.26 Write down the formula of each of the following compounds: (a) nickel(II) iodide; (b) ammonium nitrate; (c) barium hydroxide; (d) iron(III) sulfate; (e) iron(II) sulfite; (f) aluminium hydride; (g) lead(IV) oxide; tin(II) sulfide.

1.27 How many atoms make up the carbon chain in (a) octane, (b) hexane, (c) propane, (d) decane, (e) butane?

2 Atoms and atomic structure

Topics

2.1 THE IMPORTANCE OF ELECTRONS

We saw in Chapter 1 that chemistry is concerned with atoms and the species which are formed by their combination. In this chapter, we are concerned with some intimate details of atomic structure. You will learn about a description of atomic structure that is pervasive throughout modern chemistry, and which we will extend to give an understanding of the structure of multi-atomic ions and molecules.

Nucleus, proton, electron: see Section 1.4

Atoms consist of nuclei and electrons. The electrons surround the central nucleus which is in turn composed of protons and neutrons; only in the case of protium (^{1}H) is the nucleus devoid of neutrons.

Most chemistry is concerned with the *behaviour of electrons*; the behaviour of nuclei is more properly the realm of the nuclear chemist or physicist. Ions are formed by the gain or loss of electrons; covalent bonds are formed by the sharing of electrons. The number of electrons in an atom of an element controls the *chemical* properties of the element.

Empirical observations led to the grouping together of sets of elements with similar chemical characteristics and to the construction of the periodic table. A periodic table is presented on the inside front cover of this book.

We now come to the critical question. If the chemical properties of elements are governed by the electrons of the atoms of the elements, why do elements in the same group of the periodic table *but with different total numbers of electrons* behave in a similar manner? This question leads to the concept of *electronic structure* and a need to understand the organization of electrons in atoms.

2.2 THE CLASSICAL APPROACH TO ATOMIC STRUCTURE

The road to the present atomic theory has developed from classical mechanics to quantum mechanics. The transition was associated with a crisis in relating theory to experimental observations and with fundamental changes in the understanding of science in the late 19th and early 20th centuries.

The simplest models of atomic structure incorporate no detailed description of the organization of the positively charged nuclei and the negatively charged electrons. The structure merely involves electrostatic attractions between oppositely charged particles. A series of experiments established a model of the atom in which a central positively charged nucleus was surrounded by negatively charged electrons. Consider the consequences of having a positively charged nucleus with static electrons held some distance away. The opposite charges of the electrons and the protons mean that the electrons will be attracted to the nucleus and will be pulled towards it. The only forces opposing this will be the electrostatic *repulsions* between the similarly charged electrons. This model is therefore not consistent with the idea that electrons are found in a region of space *distant* from the nucleus.

This led to attempts to describe the atom in terms of electrons *in motion* about a nucleus. In 1911, Ernest Rutherford proposed an atom consisting of a positively charged nucleus around which electrons move in circular orbits.

Box 2.1 Newton's Laws of Motion

The First Law

A body continues in its state of rest or of uniform motion in a straight line unless acted upon by an external force.

The Second Law

The rate of change of momentum of a body is proportional to the applied force and takes place in the direction in which the force acts.

The Third Law

For every action, there is an equal and opposite reaction.

In a classical model of such an atom, the electrons obey Newton's Laws of Motion, and Rutherford's proposals were flawed because the electron would be attracted towards the nucleus and would plunge towards it. The orbits could not be maintained. Additionally, this classical picture can only describe completely the relative positions and motions of the nucleus and the electron in the hydrogen atom. In atoms with more than one electron it is impossible to solve algebraically the equations describing their motion. Even for hydrogen however, the Rutherford description was not able to account for some experimentally observed spectroscopic properties.

Atomic spectrum of hydrogen: see Section 2.16

Box 2.2 Atomic theory as a sequence in history

1801 Young demonstrates the **wave properties** of light.

1888 Hertz discovers that radio waves are produced by accelerated electrical charges; this indicates that light is **electromagnetic radiation**.

1900 Rayleigh and Jeans attempt to calculate the energy distribution for **black-body radiation**, but their equation leads to the **'ultraviolet catastrophe'** (see Box 2.3).

1900 Planck states that electromagnetic radiation is **quantized**, i.e. the radiation is emitted or absorbed only in discrete amounts ($E = h\nu$).

1905 Einstein considers that light waves also exhibit **particle-like behaviour**; the particles are called **photons** and have energies $h\nu$. (The name *photon* did not appear until 1926.)

1909 Rutherford, Geiger and Marsden show that when α-particles strike a piece of gold foil, some are deflected (an α-particle is an He^{2+} ion). Rutherford suggests that an atom contains a positively charged nucleus surrounded by negatively charged electrons.

1911 Rutherford proposes an atom consisting of a positively charged nucleus around which the electron moves in a circular orbit; the model is flawed because the orbit would collapse as the electron was attracted towards the nucleus (see Section 2.2).

1913 Bohr proposes a model for the hydrogen atom in which an electron moves around the nucleus in an **orbit with a discrete energy**. Other orbits are possible, also with discrete energies (see Section 2.3).

1924 De Broglie suggests that all particles, including electrons, exhibit *both* particle and wave properties; this is **wave–particle duality** (see Section 2.5).

1926 The **Schrödinger wave equation** is developed (see Section 2.7).

1927 Davisson and Germer experimentally confirm de Broglie's theory.

1927 Heisenberg states that, because of wave–particle duality, it is impossible to determine simultaneously *both* the position and momentum of a microscopic particle; this includes an electron. The statement is the **uncertainty principle** (see Section 2.6).

2.3 THE BOHR ATOM – STILL A CLASSICAL PICTURE

In 1913, Niels Bohr developed a *quantized model* for the atom. In the Bohr atom, electrons move in circular, planetary-like orbits about the nucleus. The basic assumption which made the Bohr atom different from previous models was that *the energy of an electron in a particular orbit remains constant and that only certain energies are allowed.* By classical theory, an electron cannot move in a *circular* orbit unless there is a force holding it on this path – if there is no attractive force acting on the electron, it will escape from the orbit (think what happens if you swing a ball tied on the end of a string in circles and then let go of the string). Figure 2.1 shows an electron moving with a velocity v in a circular path about the fixed nucleus. Velocity is a *vector* quantity and for any given point in the path of the electron, the velocity will have a direction tangential to the circle.

A *scalar* quantity possesses magnitude only, e.g. mass.

A *vector* quantity possesses both magnitude and direction, e.g. velocity.

In the Bohr atom, the outward force exerted on the electron as it tries to escape from the circular orbit is *exactly* balanced by the inward force of attraction between the negatively charged electron and the positively charged nucleus. Thus, the electron can remain away from the nucleus. Bohr built into his atomic model the idea that different orbits were possible and that the orbits could *only* be of certain energies. This was a very important advance in the development of atomic theory. By allowing the electrons to move *only* in particular orbits, Bohr had *quantized* his model. The model was successful in terms of being consistent with some atomic spectroscopic observations but the Bohr atom fails for other reasons.

***Quantization* is the packaging of a quantity into units of a discrete size.**

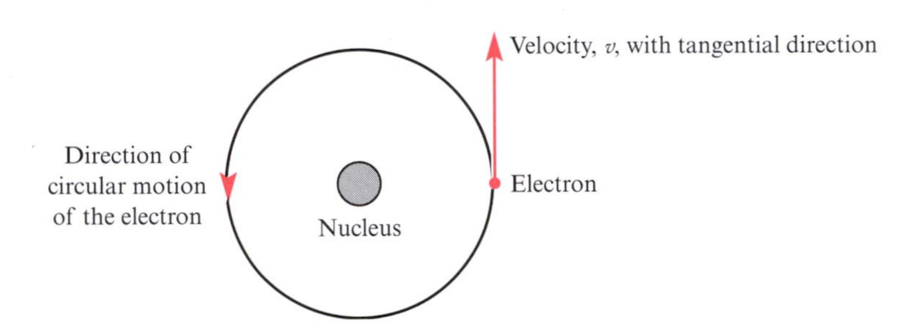

2.1 The Bohr atom. The negatively charged electron is involved in circular motion about a fixed positively charged nucleus. There is an attractive force, but also an outward force due to the electron trying to escape from the orbit.

2.4 QUANTA

In the preceding paragraphs we have considered the organization of the electrons in an atom in terms of classical mechanics and of 'classical' quantum mechanics. In the late 19th century, an important (historical) failure of the classical approach was recognized – the so-called 'ultraviolet catastrophe'. This failure led Max Planck to develop the idea of *discrete energies* for electromagnetic radiation. That is to say, the energy is *quantized.*

Equation 2.1 gives the relationship between energy E and frequency ν. The proportionality constant is h, the Planck constant ($h = 6.626 \times 10^{-34}$ J s).

$$E = h\nu \qquad (2.1)$$

$E = h\nu$

The units of energy E are joules (J), of frequency ν are s^{-1} or Hz, and h is the Planck constant, 6.626×10^{-34} J s.

Box 2.3 Black-body radiation and the 'ultraviolet catastrophe'

An ideal *black-body* absorbs all radiation of all wavelengths which falls upon it. When heated, a black-body emits radiation.

If one applies the laws of classical physics to this situation (as was done by Rayleigh and Jeans), the results are rather surprising. One finds that oscillators of very short wavelength, λ, should radiate strongly at room temperature. The intensity of radiation is predicted to increase continuously, never passing through a maximum. This result disagrees with experimental observations which show a maximum with decreasing intensity to higher or lower wavelengths. Very short wavelengths correspond to ultraviolet light, X-rays and γ-rays. Hence the term the 'ultraviolet catastrophe'.

The situation can only be rectified if one moves away from classical ideas. This was the approach taken by Max Planck. He proposed that the electromagnetic radiation had associated with it only *discrete energies*. The all-important equation arising from this is:

$$E = h\nu$$

where E is the energy, ν is the frequency and h is the Planck constant. This equation means that the energy of the emitted or absorbed electromagnetic radiation must have an energy which is a multiple of $h\nu$.

Remember that frequency, ν, and wavelength, λ, are related by the equation:

$$c = \lambda\nu$$

where c is the speed of light.

2.5 WAVE–PARTICLE DUALITY

Planck's model assumed that light was an electromagnetic wave. In 1905, Albert Einstein proposed that electromagnetic radiation could exhibit particle-like behaviour. However, it became apparent that while electromagnetic radiation possessed some properties that were fully explicable in terms of classical particles, other properties could only be explained in terms of waves. These 'particle-like' entities became known as *photons* and from equation 2.1, we see that each photon of light of a given frequency ν has an energy $h\nu$.

Now let us return to the electrons in an atom. In 1924, Louis de Broglie argued that if radiation could exhibit the properties of both particles *and* waves (something that defeats the laws of classical mechanics), then so too could electrons, and indeed so too could *every* moving particle. This phenomenon is known as *wave–particle duality*.

The de Broglie relationship (equation 2.2) shows that a particle (and this includes an electron) with momentum mv (m = mass and v = velocity of the particle) has an associated wave of wavelength λ. Thus, equation 2.2 combines the concepts of classical momentum with the idea of wave-like properties.

$$\lambda = \frac{h}{mv} \qquad (2.2)$$

where h = the Planck constant.

And so, we have begun to move towards *quantum (or wave) mechanics* and away from classical mechanics. Let us see what this means for the behaviour of electrons in atoms.

2.6 THE UNCERTAINTY PRINCIPLE

For a particle with a relatively large mass, wave properties are unimportant and the position and motion of the particle can be defined and measured almost exactly. However, for an electron with a tiny mass, this is not the case. Treated in a classical sense, an electron can move along a defined path in the same way that a ball moves when it is thrown. However, once we give the electron wave-like properties, it becomes impossible to know exactly both the position and the momentum of the electron *at the same instant in time*. This is Heisenberg's *uncertainty principle*.

We must now think of the electron in a new and different way and consider the *probability* of finding the electron in a given volume of space, rather than trying to define its exact position and momentum. The probability of finding an electron at a given point in space is determined from the function ψ^2 where ψ is the *wavefunction*. A wavefunction is a mathematical function which tells us in detail about the behaviour of an electron-wave.

The probability of finding an electron at a given point in space is determined from the function ψ^2 where ψ is the *wavefunction*.

Now we must look for ways of saying something about values of ψ and ψ^2, and this leads us on to the *Schrödinger wave equation*.

2.7 THE SCHRÖDINGER WAVE EQUATION

The equation

One form of the Schrödinger wave equation is shown in equation 2.3.

$$\mathcal{H}\psi = E\psi \qquad (2.3)$$

When first looking at equation 2.3, you might ask: 'Can one divide through by ψ and say that $\mathcal{H} = E$?' This seems a reasonable question, but in order to answer it (and the answer is 'no'), we need to understand what the various components of equation 2.3 mean.

Look first at the right-hand side of equation 2.3. We have already stated that ψ is a wavefunction. E is the total energy associated with the wavefunction ψ. The left-hand side of equation 2.3 represents a mathematical operation upon the function ψ. $\mathcal{H}$ is called the *Hamiltonian operator*.

Eigenvalues and eigenvectors

Equation 2.3 can be expressed more generally as in equation 2.4. The equation is set up so that, having 'operated' on the function (in our case, ψ) the answer comes out in the form of a scalar quantity (in our case, E) multiplied by the same function (in our case, ψ). For such a relationship, the function is

Box 2.4 Operators in mathematics

Just as the name suggests, a mathematical operator performs a mathematical operation on the function in question.

Differentiation is one type of mathematical operation. For example:

$$\frac{d}{dx}$$

instructs you to differentiate a variable with respect to x.

Let the variable to be operated upon be y where $y = 8x$. Find $\frac{dy}{dx}$.

$$\frac{dy}{dx} = \frac{d(8x)}{dx} = 8$$

The Hamiltonian operator, $\mathcal{H}$, is a complex differential operator and is given by:

$$-\frac{h^2}{8\pi^2 m}\left(\frac{\partial^2}{\partial x^2} + \frac{\partial^2}{\partial y^2} + \frac{\partial^2}{\partial z^2}\right) + V(x, y, z)$$

in which $\frac{\partial^2}{\partial x^2}$, $\frac{\partial^2}{\partial y^2}$ and $\frac{\partial^2}{\partial z^2}$ are *partial differentials* (see Box 12.1) and $V(x,y,z)$ is a potential energy term. The complicated nature of $\mathcal{H}$ arises from the need to describe the position of the electron in three-dimensional Cartesian space.

In equation 2.3, the Hamiltonian operator is acting on the wavefunction ψ.

One form of the Schrödinger wave equation is:

$$\mathcal{H}\psi = E\psi$$

$\mathcal{H}$ is the Hamiltonian operator, the wavefunction ψ is an eigenfunction and the energy E is an eigenvalue.

called an *eigenfunction* and the scalar quantity is called the *eigenvalue*. Applied to equation 2.3, this means that ψ is an eigenfunction and E is an eigenvalue, whilst $\mathcal{H}$ is the operator.

$$\text{Operator working on a function} = (\text{scalar quantity}) \times (\text{the original function}) \qquad (2.4)$$

What information is available from the Schrödinger equation?

It is difficult to grasp the physical meaning of the Schrödinger equation, but the aim of this discussion is not to study it in detail but merely to illustrate that the equation can be set up for a given system and can be solved (either exactly or approximately) to give values of ψ and hence ψ^2. From the Schrödinger equation, we can find energy values that are associated with particular wavefunctions. The quantization of the energies is built into the mathematics of the Schrödinger equation.

A very important point to remember is that the Schrödinger equation can *only* be solved exactly for a two-body problem, i.e. for a species which contains a nucleus and only one electron (e.g. H, He^+ or Li^{2+}) – these are all *hydrogen-like* systems.

The wavefunction ψ is a solution of the Schrödinger equation and describes the behaviour of an electron in the region in space of the atom called the *atomic orbital*. This result can be extended to give information about a *molecular orbital* in a molecular species.

Quantum numbers: see Section 2.10

An atomic orbital is usually described in terms of three integral *quantum numbers*. The principal quantum number, n, is a positive integer with values lying between limits 1 and ∞. Two further quantum numbers, l and m_l, may be derived and a combination of these three quantum numbers defines a unique *orbital*.

A wavefunction ψ is a mathematical function that contains detailed information about the behaviour of an electron (electron-wave).

The region of space defined by such a wavefunction is termed an *atomic orbital*.

Each atomic orbital may be uniquely labelled by a set of three quantum numbers: n, l and m_l.

Radial and angular parts of the wavefunction

So far, we have described the position of the electron by its Cartesian coordinates (x, y, z). It is very often convenient to use a different coordinate system which separates a radial distance coordinate (r) from the angular components (θ and ϕ) (Figure 2.2). These are called *spherical polar coordinates*. We will not be concerned with the mathematical manipulations used to transform a set of Cartesian coordinates into spherical polar coordinates. However, it is useful to be able to define separate wavefunctions for the radial (r) and angular (θ,ϕ) components of ψ. This is represented in equation 2.5 where $R(r)$ and $A(\theta,\phi)$ are new radial and angular wavefunctions respectively.

$$\psi_{\text{Cartesian}}(x,y,z) \equiv \psi_{\text{radial}}(r)\psi_{\text{angular}}(\theta,\phi) = R(r)A(\theta,\phi) \quad (2.5)$$

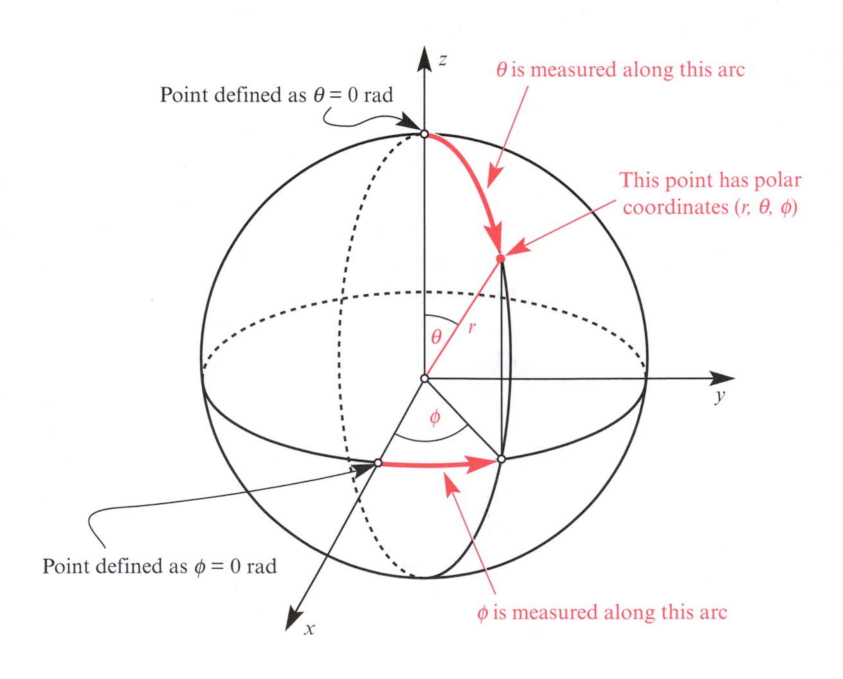

2.2 Definition of the polar coordinates (r, θ, ϕ) for a point shown here in red; r is the radial coordinate and θ and ϕ are angular coordinates. Cartesian axes (x, y and z) are also shown. The point at the centre of the diagram is the origin.

An atomic wavefunction ψ consists of a radial, $R(r)$, and an angular component, $A(\theta,\phi)$.

The radial component in equation 2.5 is dependent upon the quantum numbers n and l, whereas the angular component depends on l and m_l. Hence the components should really be written as $R_{n,l}(r)$ and $A_{l,m_l}(\theta,\phi)$. In this book we shall simplify the expressions. The notation $R(r)$ refers to the radial part of the wavefunction, and $A(\theta,\phi)$ refers to the angular part of the wavefunction.

2.8 PROBABILITY DENSITY

The functions ψ^2, $R(r)^2$ and $A(\theta,\phi)^2$

The function ψ^2§ is proportional to the *probability density* of the electron at a point in space. By considering values of ψ^2 at points in the volume of space about the nucleus, it is possible to define a surface boundary which encloses that region of space in which the electron will spend, say, 95% of its time. By doing this, we are effectively drawing out the boundaries of an atomic orbital. Remember that ψ^2 may be described in terms of the radial and angular components $R(r)^2$ and $A(\theta,\phi)^2$.

The 1*s* atomic orbital

Atomic orbitals: see Section 2.11

The lowest energy solution to the Schrödinger equation for the hydrogen atom leads to the 1*s* orbital. The 1*s* orbital is spherical in shape; it is *spherically symmetric* about the nucleus. If we look only at the probability of finding the electron as a function of distance from the nucleus, we can consider only the radial part of the wavefunction $R(r)$. Values of $R(r)^2$ are largest near to the nucleus and then become very small as we move along a radius centred on the nucleus. Since the chance of the electron being far out from the nucleus is *minutely* small, we draw the boundary surface for the orbital so as to enclose 95% of the probability density of the total wavefunction ψ^2. Figure 2.3 gives a plot of $R(r)^2$ against the distance, r, from the nucleus. The curve for $R(r)^2$ only reaches zero as the radius approaches ∞.

Normalization

Wavefunctions are usually *normalized* to unity. This means that the probability of finding the electron somewhere in space is taken to be unity, i.e. 1. In other words, the electron has to be somewhere!

Mathematically, the normalization is represented by equation 2.6. This effectively states that we are integrating ($\int$) over all space ($\mathrm{d}\tau$) and that the total integral of ψ^2 (which is a measure of the probability density) must be unity.†

$$\int \psi^2 \mathrm{d}\tau = 1 \qquad (2.6)$$

§ Although here we use ψ^2, we ought really to write $\psi\psi^*$ where ψ^* is the complex conjugate of ψ. In one dimension (say x), the probability of finding the electron between the limits of x and $(x + \mathrm{d}x)$, (where $\mathrm{d}x$ is an extremely small change in x) is proportional to the function $\psi(x)\psi^*(x)\mathrm{d}x$. In three dimensions this leads to the use of $\psi\psi^*\mathrm{d}\tau$ in which we are considering the probability of finding the electron in a volume element $\mathrm{d}\tau$. Using only the radial part of the wavefunction, the function becomes $R(r)R^*(r)$.

† Strictly we should write equation 2.6 as $\int \psi\psi^* \mathrm{d}\tau = 1$ where ψ^* is the complex conjugate of ψ.

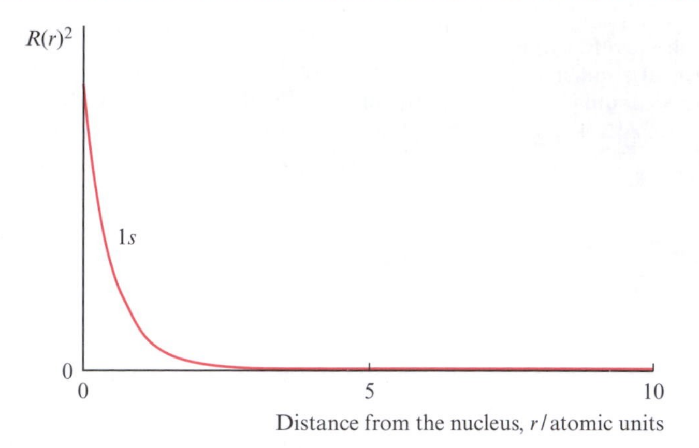

2.3 A plot of $R(r)^2$, as a function of distance, r, from the atomic nucleus. This plot refers to the case of a spherically symmetrical 1*s* orbital.

2.9 THE RADIAL DISTRIBUTION FUNCTION, $4\pi r^2 R(r)^2$

Another way of representing information about the probability density is to plot a *radial distribution function*. The relationship between $R(r)^2$ and the radial distribution function is given in equation 2.7.

$$\text{Radial distribution function} = 4\pi r^2 R(r)^2 \quad (2.7)$$

The advantage of using this new function is that it represents the probability of finding the electron in a *spherical shell* of radius r and thickness (r + dr), where (r + dr) may be defined as (r + a small increment of radial distance) (Figure 2.4). Thus, instead of considering the probability of finding the electron as a function of distance from the nucleus as illustrated in Figure 2.3, we are now able to consider the complete region in space in which the electron resides.

Figure 2.5 shows the radial distribution function for the 1s orbital. Notice that it is zero at the nucleus; this contrasts with the situation for $R(r)^2$ (Figure 2.3). The difference arises from the dependence of the radial distribution function upon r; at the nucleus, r is zero, and so $4\pi r^2 R(r)^2$ must equal zero.

We return to radial distribution functions in Section 2.13

The radial distribution function is given by the expression $4\pi r^2 R(r)^2$. It gives the probability of finding an electron in a spherical shell of radius r and thickness (r + dr). The radius r is measured from the nucleus.

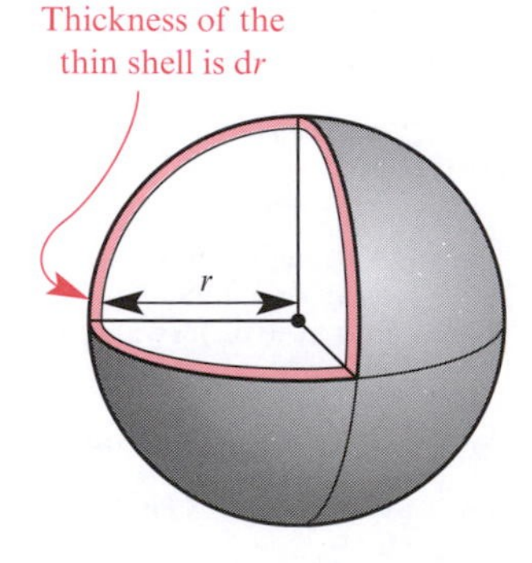

2.4 A view into a spherical shell of inner radius r (centred on the nucleus) and thickness dr, where dr is a very small increment of the radial distance r.

2.5 The radial distribution function $4\pi r^2 R(r)^2$ for the 1s atomic orbital. This function describes the probability of finding the electron in a three-dimensional sense.

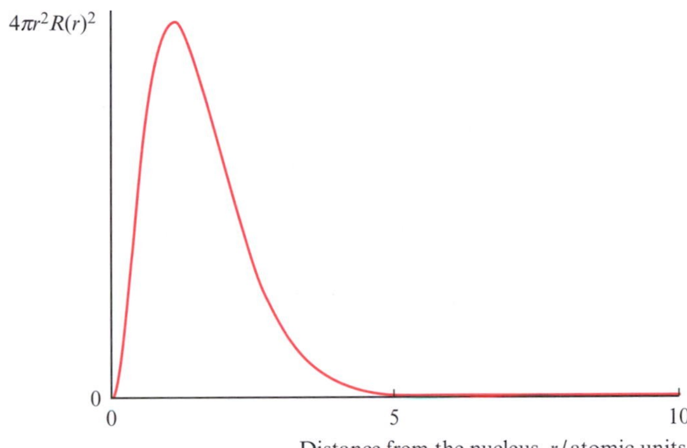

2.10 QUANTUM NUMBERS

An atomic orbital is defined by a unique set of three quantum numbers (n, l and m_l).

An electron in an atom is defined by a unique set of four quantum numbers (n, l, m_l and m_s).

In several parts of the preceding discussion, we have mentioned *quantum numbers*. Now we look more closely at these important numbers.

As we have seen, the effects of quantization are that only certain electronic energies, and hence orbital energies, are permitted. Each orbital is described by a set of quantum numbers n, l and m_l. A fourth quantum number, m_s, gives information about the spin of an electron in an orbital. Every orbital in an atom has a unique set of *three* quantum numbers and each electron in an atom has a unique set of *four* quantum numbers.

The principal quantum number n

Principal quantum number,

n = 1, 2, 3, 4, 5 ... ∞.

The principal quantum number n may have any positive integral value between 1 and ∞. The number n corresponds to the orbital energy level or 'shell'. For a particular energy level there may be sub-shells, the number of which is defined by the quantum number l.

The orbital quantum number l

The quantum number l is also known as the azimuthal quantum number

Orbital quantum number, l = 0, 1, 2, 3, 4 ... (n – 1).

$n = 1$	$l = 0$
$n = 2$	$l = 0, 1$
$n = 3$	$l = 0, 1, 2$
$n = 4$	$l = 0, 1, 2, 3$

The quantum number l is called the orbital quantum number. For a given value of the principal quantum number, the allowed values of l are positive integers lying between 0 and (n – 1). Thus, if n = 3, the permitted values of l are 0, 1 and 2. Each value of l corresponds to a particular type of atomic orbital.

Knowing the value of l provides detailed information about the region of space in which an electron may move; specifically, it describes the shape of the orbital. The value of l also determines the angular momentum of the electron within the orbital, i.e. its *orbital angular momentum*.

Box 2.5 Angular momentum

Angular momentum = Mass × Angular velocity

An electron in an atomic orbital possesses both orbital and spin angular momenta. Both types of angular momenta are quantized.

Angular velocity is the angle, α, turned through per second by a particle travelling on a circular path. The SI units of angular velocity are radians per second.

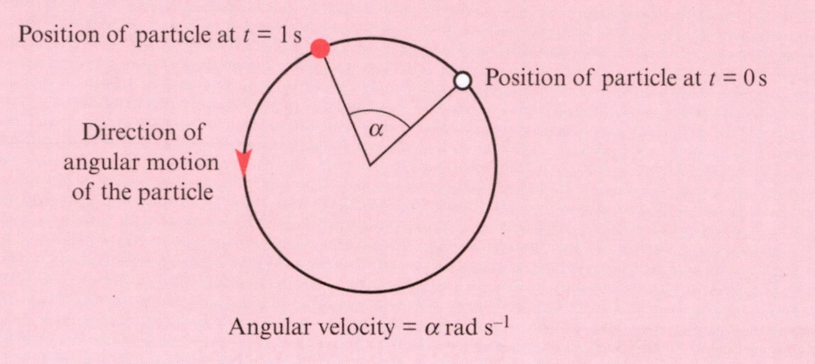

Magnetic quantum number $m_l = -l, (-l + 1), \ldots 0, \ldots (l - 1), l$

The magnetic quantum number m_l

The magnetic quantum number m_l relates to the directionality of an orbital and has values which are integers between $+l$ and $-l$. If $l = 1$, the allowed values of m_l are -1, 0 and $+1$.

Worked example 2.1 ***Deriving quantum numbers and what they mean***

Derive possible sets of quantum numbers for $n = 2$, and explain the significance of these sets of numbers.

Let $n = 2$.
Remember that a value of n defines an energy (or principal) level.
Possible values of l lie in the range 0 to $(n - 1)$.
Therefore, for $n = 2$, $l = 0$ and 1.
This means that the $n = 2$ level gives rise to two sub-levels, one with $l = 0$ and one with $l = 1$.
Now determine the possible values of the quantum number m_l: values of m_l lie in the range $-l \ldots 0 \ldots +l$.
The sub-level $l = 0$ has associated with it one value of m_l:

$m_l = 0$ for $l = 0$

The sub-level $l = 1$ has associated with it three values of m_l:

$m_l = -1, 0$ or $+1$ for $l = 1$

The possible sets of quantum numbers for $n = 2$ are:

n	l	m_l
2	0	0
2	1	-1
2	1	0
2	1	$+1$

The physical meaning of these sets is that for the level $n = 2$, there will be two *types* of orbital (because there are two values of l). For $l = 1$, there will be three orbitals *all of a similar type* but with *different directionalities*. (Details of the orbitals are given in Section 2.11.)

The spin quantum number s and the magnetic spin quantum number m_s

In a classical model, an electron may be considered to spin about an axis passing through it and so possesses spin angular momentum in addition to the orbital angular momentum discussed above. The spin quantum number s determines the *magnitude of the spin angular momentum* of an electron and can *only* have a value of $\frac{1}{2}$. The magnetic spin quantum number m_s determines the *direction of the spin angular momentum* of an electron and has values of $+\frac{1}{2}$ or $-\frac{1}{2}$.

Magnetic spin quantum number $m_s = \pm\frac{1}{2}$.

An atomic (or a molecular) orbital can contain a maximum of two electrons. The two values of m_s correspond to labels for the two electrons that can be accommodated in any one orbital. When two electrons occupy the same orbital, one possesses a value of $m_s = +\frac{1}{2}$ and the other $m_s = -\frac{1}{2}$. We say that this corresponds to the two electrons having *opposite* spins.

We return to quantum numbers in Section 2.18

Worked example 2.2 ***Deriving a set of quantum numbers which uniquely define a particular electron***

Derive a set of quantum numbers that describes an electron in an atomic orbital with $n = 1$.

First, determine how many orbitals are possible for $n = 1$.
For $n = 1$, possible values for l lie between 0 and $(n - 1)$.
Therefore, for $n = 1$, only $l = 0$ is possible.

Now determine possible values of m_l.
Possible values of m_l lie in the range $-l \ldots 0 \ldots +l$.
Therefore for $l = 0$, the only possible value of m_l is 0.

The orbital that has been defined has the quantum numbers:

$$n = 1, l = 0, m_l = 0.$$

This orbital can contain up to two electrons.
Therefore, each electron is uniquely defined by one of the two following sets of quantum numbers:

$$n = 1, l = 0, m_l = 0, m_s = +\tfrac{1}{2}$$

or

$$n = 1, l = 0, m_l = 0, m_s = -\tfrac{1}{2}.$$

2.11 ATOMIC ORBITALS

Types of atomic orbitals

For an *s* orbital, $l = 0$
For a *p* orbital, $l = 1$
For a *d* orbital, $l = 2$
For an *f* orbital, $l = 3$

In the previous section we saw that the quantum number l defines a particular type of atomic orbital. The four types of atomic orbital most commonly encountered are the *s*, *p*, *d* and *f* orbitals. The letters *s*, *p*, *d* and *f* are simply used here as labels. A value of $l = 0$ corresponds to an *s* orbital, $l = 1$ refers to a *p* orbital, $l = 2$ refers to a *d* orbital and $l = 3$ corresponds to an *f* orbital.

The distinction between these four types of atomic orbital comes from their shapes and symmetries, and the shape of the orbital is governed by the quantum number l.

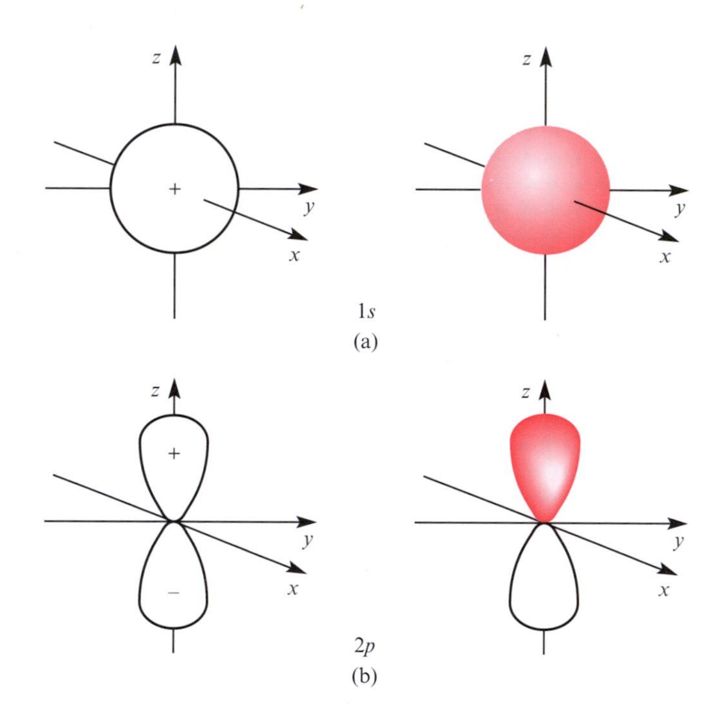

2.6 Two ways of representing the phase of a wavefunction. The examples shown are (a) a 1*s* orbital for which there is no phase change across the orbital, and (b) a 2*p* orbital (here the $2p_z$) for which there is one phase change.

Box 2.6 Orbital labels

The labels *s*, *p*, *d* and *f* have their origins in the words 'sharp', 'principal', 'diffuse' and 'fundamental'. These originally referred to the characterisics of the lines observed in the atomic spectrum of hydrogen (see Section 2.16). Nowadays, it is best to regard them merely as labels.

Further types of atomic orbital are labelled: *g*, *h*, etc.

Value of l	0	1	2	3	4	5
Label of atomic orbital	*s*	*p*	*d*	*f*	*g*	*h*

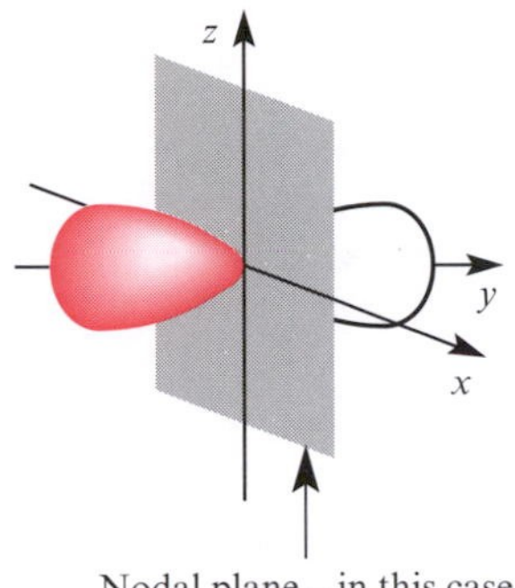

2.7 The phase change in an atomic p orbital. In this case, the orbital is the p_y orbital and the nodal plane lies in the xz plane.

An s orbital is spherically symmetric about the nucleus and the boundary surface of the orbital has a constant *phase*. That is, the amplitude of the wavefunction associated with the boundary surface of the s orbital is always either positive or negative (Figure 2.6a).

For a p orbital, there is *one* phase change with respect to the surface boundary of the orbital, and this occurs at the so-called *nodal plane* (Figure 2.7). Each part of the orbital is called a *lobe*. The phase of a wavefunction is designated as having either a positive or negative amplitude, or by shading (or not) an appropriate lobe as indicated in Figure 2.6b. A nodal plane in an orbital corresponds to a node in a transverse wave (Figure 2.8).

The amplitude of the wavefunction associated with an *s* orbital is *either* positive *or* negative. The surface boundary of an *s* orbital has no change of phase, and the orbital has *no* nodal plane.

The surface boundary of a *p* orbital has *one* change of phase, and possesses *one* nodal plane.

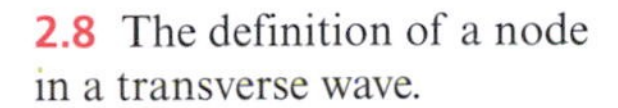

2.8 The definition of a node in a transverse wave.

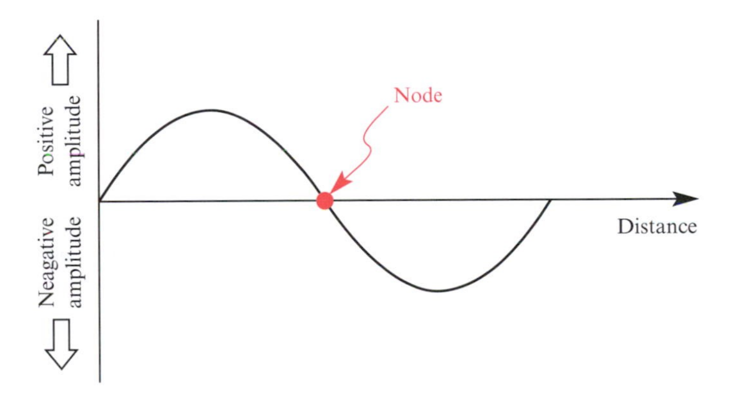

The numbers and shapes of orbitals of a given type

For an s orbital, the orbital quantum number, l, is 0. This corresponds to only one value of m_l: $l = 0$, $m_l = 0$. For a given principal quantum number, n, there is only one s orbital, and the s orbital is said to be *singly degenerate*. The s orbital is spherically symmetric (Figure 2.9).

For a p orbital, the orbital quantum number l has a value of 1. This leads to three possible values of m_l, since, for $l = 1$, $m_l = +1, 0, -1$. The physical meaning of this statement is that the Schrödinger equation gives three solutions for p orbitals for a given value of n when $n \geq 2$. (Why does n have to be greater than 1?) The three solutions are conventionally drawn as shown in Figure 2.9. The surface boundaries of the three np orbitals are identical but the orbitals are directed at right angles to one another. A set of p orbitals is said to be *triply* or *three-fold degenerate*. By normal convention, one p orbital

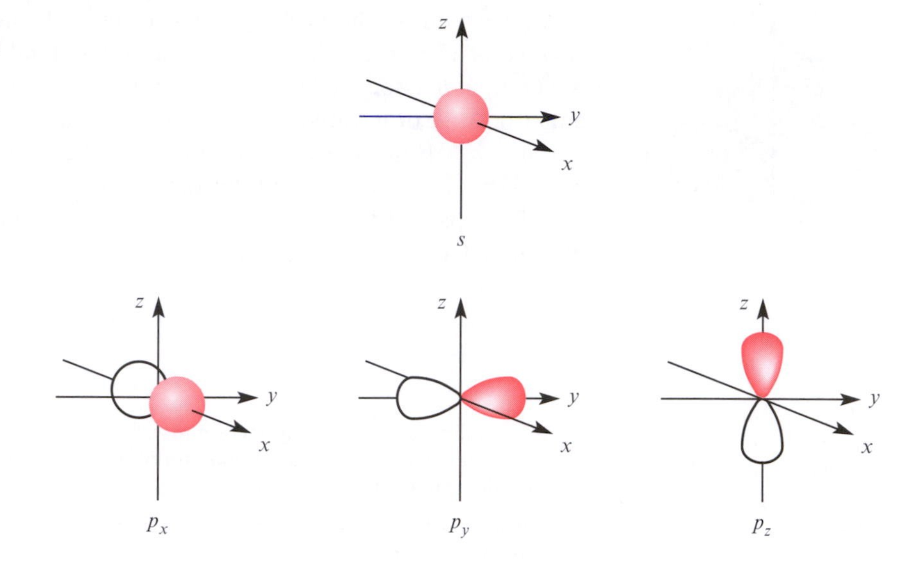

2.9 For a given value of the principal quantum number n there is one s orbital. For a given value of n when $n \geq 2$, there are three p orbitals.

points along the x axis and is called the p_x orbital, while the p_y and p_z orbitals point along the y and z axes respectively; this, in effect, defines the x, y and z axes!

We discuss the shapes of *d* orbitals in Section 16.1

For a d orbital, the orbital quantum number l is 2. This leads to five possible values of m_l (for $l = 2$, $m_l = +2, +1, 0, -1, -2$). The Schrödinger equation gives five real solutions for d orbitals for a given value of n when $n > 2$. (Why must n be greater than 2?) A set of d orbitals is said to be *five-fold degenerate*.

Degenerate orbitals possess the same energy

Sizes of orbitals

For a given atom, a series of orbitals of the same symmetry (*identical values of l and m_l*) but with different values of the principal quantum number (e.g. 1s, 2s, 3s, 4s...) differ in their relative sizes (volumes or spatial extent). The larger the value of n, the larger the orbital. This is shown in Figure 2.10 for a series of ns atomic orbitals, although note that the relationship is not linear. The increase in size also corresponds to the orbital being more *diffuse*.

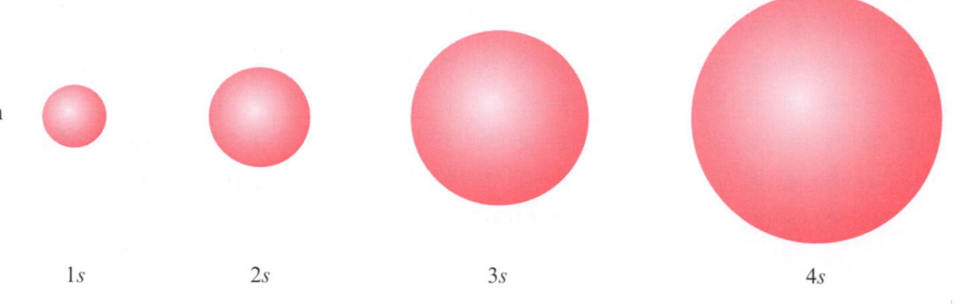

2.10 The increase in size of an ns atomic orbital exemplified for $n = 1, 2, 3$ and 4. The 4s orbital is *more diffuse* than the 3s, and this in turn is more diffuse than the 2s or the 1s orbitals.

2.12 RELATING ORBITAL TYPES TO THE PRINCIPAL QUANTUM NUMBER

For a given value of the principal quantum number n, an allowed series of values of l and m_l can be determined as described in Section 2.10. It follows that the number of orbitals and types allowed for a given value of n can therefore be determined. We distinguish between orbitals of the same type but with different principal quantum numbers by writing ns, np, nd ... orbitals.

Consider the case for $n = 1$

If $n = 1$, then $l = 0$ and $m_l = 0$.
The value of $l = 0$ corresponds to an s orbital which is singly degenerate. There is only one allowed value of m_l ($m_l = 0$).
Thus, for $n = 1$, the allowed atomic orbital is the 1s.

Consider the case for $n = 2$

If $n = 2$, then $l = 1$ or 0.
Take each value of l separately.
For $l = 0$, there is a single s orbital, designated as the 2s orbital.
For $l = 1$, the allowed values of m_l are +1, 0 and –1, corresponding to a set of three p orbitals.
Thus, for $n = 2$, the allowed atomic orbitals are the 2s, $2p_x$, $2p_y$ and $2p_z$.

Consider the case for $n = 3$

If $n = 3$, then $l = 2$, 1 or 0.
Take each value of l separately.
For $l = 0$, there is a single s orbital, designated as the 3s orbital.
For $l = 1$, the allowed values of m_l are +1, 0 and –1, corresponding to a set of three p orbitals.
For $l = 2$, the allowed values of m_l are +2, +1, 0, –1 and –2, corresponding to a set of five d orbitals.
Thus, for $n = 3$, the allowed atomic orbitals are the 3s, $3p_x$, $3p_y$, $3p_z$ and five 3d orbitals.

The labels for the five 3d atomic orbitals are given in Section 16.1

n	Atomic orbitals allowed	Total number of orbitals	Total number of electrons
1	one s	1	2
2	one s, three p	4	8
3	one s, three p, five d	9	18

2.13 MORE ABOUT RADIAL DISTRIBUTION FUNCTIONS

So far, we have focused on the differences between different types of atomic orbitals in terms of their different quantum numbers, and have stated that the shape of an atomic orbital is governed by the quantum number l. For the specific case of the 1s orbital, we have also considered the radial part of the wavefunction (Figures 2.3 and 2.5). Now we look at how the *radial* distribution function, $4\pi r^2 R(r)^2$, varies as a function of r for atomic orbitals other than the 1s orbital.

First, we consider the 2s orbital. Figure 2.10 showed that the surface boundaries of the 1s and 2s orbitals are similar. However, a comparison of the radial distribution functions for the two orbitals (Figure 2.11) shows these to be *dissimilar*. Whereas the radial distribution function, $4\pi r^2 R(r)^2$, for the 1s orbital has one maximum, that for the 2s orbital has two maxima and the function falls to zero between them. The point at which $4\pi r^2 R(r)^2 = 0$ is called a *radial node*. Notice that the value of zero at the radial node is due to the function $R(r)$ equalling zero, and not $r = 0$ as was true at the nucleus.

Note that as we are dealing with a function containing $R(r)^2$, the graph can never have negative values. The wavefunction $R(r)$ can have both positive and negative amplitudes.

At a radial node, the radial distribution function $4\pi r^2 R(r)^2 = 0$.

Figure 2.11 also shows the radial distribution function for the 3s orbital. The number of radial nodes increases in going from the 2s to the 3s orbital. The trend continues, and thus the 4s and 5s orbitals have three and four radial nodes respectively. The presence of the radial nodes in s orbitals with $n > 1$ means that these orbitals are made up of 'layers', rather like an onion is made up of concentric skins.

Radial distribution functions for the 3s, 3p and 3d orbitals are shown in Figure 2.12. Patterns in numbers of radial nodes as a function of n and l are given in Table 2.1.

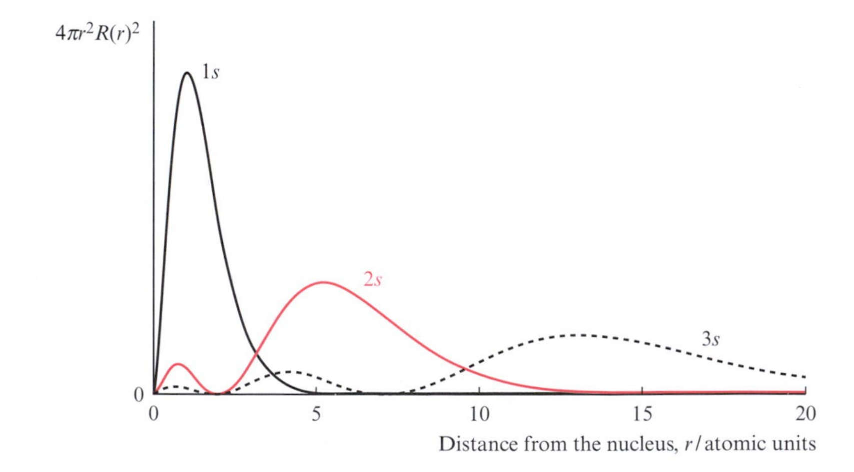

2.11 Radial distribution functions, $4\pi r^2 R(r)^2$, for the 1s, 2s and 3s atomic orbitals of hydrogen. Compare the number of radial nodes in these diagrams with the data in Table 2.1.

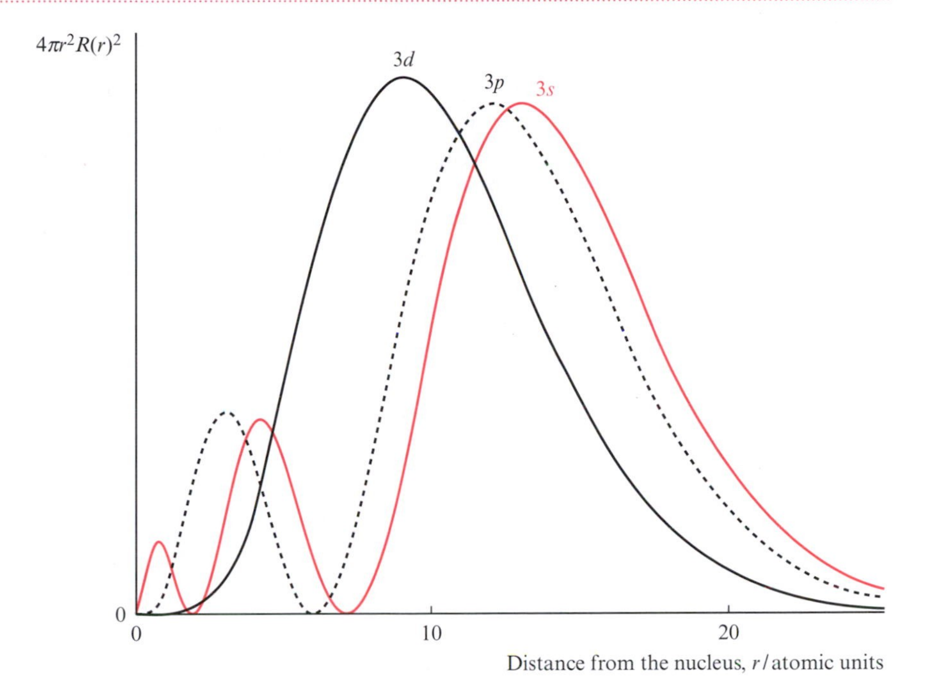

2.12 Radial distribution functions, $4\pi r^2 R(r)^2$, for the 3*s*, 3*p* and 3*d* atomic orbitals of hydrogen. Compare these diagrams with the data in Table 2.1.

Table 2.1 Number of radial nodes as a function of orbital type and principal quantum number, *n*. The pattern continues for $n > 4$.

n	$s\ (l = 0)$	$p\ (l = 1)$	$d\ (l = 2)$	$f\ (l = 3)$
1	0			
2	1	0		
3	2	1	0	
4	3	2	1	0

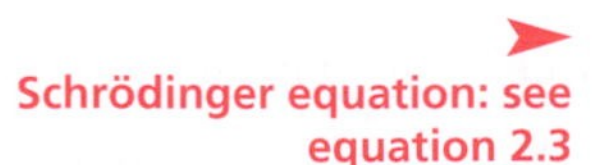

2.14 APPLYING THE SCHRÖDINGER EQUATION TO THE HYDROGEN ATOM

Schrödinger equation: see equation 2.3

In this section we shall look in more detail at the solutions of the Schrödinger equation relating to the hydrogen atom. This is one of the few systems for which the Schrödinger equation can be solved almost exactly. You may think that this is restrictive, given that there are over one hundred elements in addition to vast numbers of molecules and ions. However, approximations can be made in order to apply quantum mechanical equations to systems larger than hydrogen-like systems.

We will not go through the mathematics of solving the Schrödinger equation but will merely consider the solutions. After all, it is the results of this equation that a chemist uses.

The wavefunctions

Solving the Schrödinger equation for ψ gives information that allows the orbitals for the hydrogen atom to be constructed. Each solution of ψ is in the form of a complicated mathematical expression which contains components that describe the radial part, $R(r)$, and the angular part, $A(\theta,\phi)$. Each solution corresponds to a particular set of n, l and m_l quantum numbers.

Some of the solutions for values of ψ are given in Table 2.2. These will give you some idea of the mathematical forms taken by solutions of the Schrödinger equation.

Table 2.2 Solutions of the Schrödinger equation for the hydrogen atom which define the $1s$, $2s$ and $2p$ atomic orbitals

Atomic orbital	n	l	m_l	Radial part of the wavefunction, $R(r)$	Angular part of wavefunction, $A(\theta,\phi)$
$1s$	1	0	0	$2e^{-r}$	$\frac{1}{2\sqrt{\pi}}$
$2s$	2	0	0	$\frac{1}{2\sqrt{2}}(r-2)e^{-\left(\frac{r}{2}\right)}$	$\frac{1}{2\sqrt{\pi}}$
$2p_x$	2	1	+1	$\frac{1}{2\sqrt{6}}re^{-\left(\frac{r}{2}\right)}$	$\frac{\sqrt{3}(\sin\theta\cos\phi)}{2\sqrt{\pi}}$
$2p_z$	2	1	0	$\frac{1}{2\sqrt{6}}re^{-\left(\frac{r}{2}\right)}$	$\frac{\sqrt{3}(\cos\theta)}{2\sqrt{\pi}}$
$2p_y$	2	1	−1	$\frac{1}{2\sqrt{6}}re^{-\left(\frac{r}{2}\right)}$	$\frac{\sqrt{3}(\sin\theta\sin\phi)}{2\sqrt{\pi}}$

Energies

The energies for the hydrogen atom that come from the Schrödinger equation represent energies of orbitals (energy levels) and can be expressed according to equation 2.8.

$$E = -\frac{k}{n^2} \tag{2.8}$$

where E = energy, n = principal quantum number and k = a constant.

The constant k has a value of $1.312 \times 10^3\,\text{kJ}\,\text{mol}^{-1}$. Thus, we can determine the energy of the $1s$ orbital from equation 2.8 by substituting in a value of $n = 1$. Similarly for $n = 2, 3, 4....$

Note that, in this case, only one energy solution will be forthcoming from equation 2.8 for each value of n. This means that *for the hydrogen atom*, an electron possesses the same energy no matter which of a set of atomic orbitals with the same principal quantum number it occupies. Such orbitals are *degenerate*. For example, the energy is the same when the single electron of a hydrogen atom is in a $2s$ or $2p$ orbital. Similarly, equal energy states are

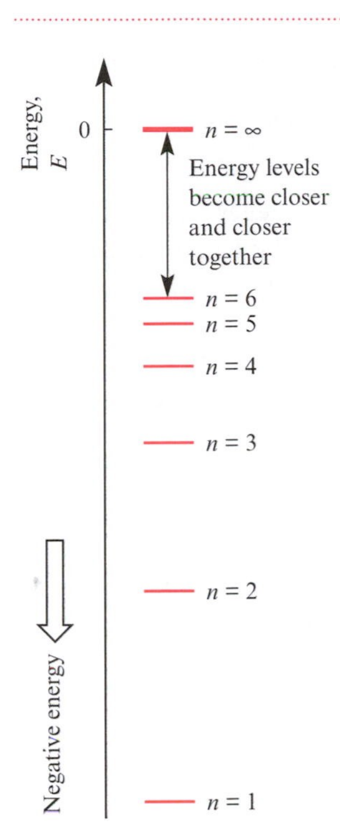

2.13 A schematic (not to scale) representation of the energy solutions of the Schrödinger equation for the hydrogen atom. The energy levels between $n = 6$ and infinity (the continuum) are not shown.

observed when the electron is in any of the 3*s*, 3*p* and 3*d* atomic orbitals of the hydrogen atom. (Note that the electron in the hydrogen atom can only occupy an orbital other than the 1s orbital when the atom is in an *excited state*; see below.)

Figure 2.13 shows a representation of the energy solutions of the Schrödinger equation for the hydrogen atom. *These solutions are peculiar to the hydrogen atom*. The energy levels get closer together as the value of *n* increases, and this result is a general one for all other atoms.

Where does the electron reside in the hydrogen atom?

Once the orbitals and their energies have been determined for the hydrogen atom, the electron can be accommodated in the *orbital in which it has the lowest energy*. This corresponds to the *ground state* for the hydrogen atom and is the most stable state.

Hence, the single electron in the hydrogen atom resides in the 1*s* orbital. This is represented in the notational form $1s^1$ – we say that the *ground state electronic configuration* of the hydrogen atom is $1s^1$.

Other 'hydrogen-like' systems include He^+ and Li^{2+}. Their electronic configurations are equivalent to that of the H atom, although the nuclear charges of these 'hydrogen-like' systems differ from each other.

The type and occupancy of an atomic orbital is represented in the form:

ns^x or np^x or nd^x etc.

where n = principal quantum number and x = number of electrons in the orbital or orbital set.

$1s^1$ means that there is one electron in a 1*s* atomic orbital.

Occupying atomic orbitals with electrons: see Section 2.18

The electron in hydrogen can be promoted (raised in energy) to an atomic orbital of higher energy than the 1*s* orbital. For example, it might occupy the 2*p* orbital. Such a state is called an *excited state*.

2.15 PENETRATION AND SHIELDING

An atom or ion with one electron is a called a 'hydrogen-like' species.

An atom or ion with more than one electron is called a 'many-electron' species.

What effects do electrons in an atom have on one another?

So far, we have only considered interactions between a single electron and the nucleus. As we go from the hydrogen atom to atoms with more than one electron, it is necessary to consider the effects that the electrons have on each other.

In Section 2.13 we looked at the radial properties of different atomic orbitals. Now let us consider what happens when we place electrons into these orbitals. There are two types of electrostatic interaction at work:

- electrostatic attraction between the nucleus and an electron;
- electrostatic repulsion between electrons.

In the hydrogen-like atom, the $2s$ and $2p$ orbitals are degenerate, as are all orbitals with the same principal quantum number. The electron can only ever be in one orbital at once. An orbital which is empty is a *virtual orbital* and has no physical significance.

Now take the case of an atom containing two electrons with one electron in each of the $2s$ and $2p$ orbitals. In order to get an idea about the regions of space which these two electrons occupy, we need to consider the radial distribution functions for the two orbitals. These are drawn in Figure 2.14. The presence of the radial node in the $2s$ orbital means that there is a region of space relatively close to the nucleus in which the $2s$ electron is likely to be found. In effect, an electron in the $2s$ orbital will spend more time nearer to the nucleus than an electron in a $2p$ orbital. This is described as *penetration*. The $2s$ orbital *penetrates* more than the $2p$ orbital.

In the hydrogen atom, with only one electron, the energy is identical whether the electron is in the $2s$ or the $2p$ orbital. The slightly greater average distance of the electron in the $2s$ orbital from the nucleus than the $2p$ orbital is compensated by the radial node in the $2s$ orbital.

What happens when we have electrons in *both* of the $2s$ and $2p$ orbitals? The presence of the radial node in the $2s$ orbital means that there is a region of negative charge (due to the electron in the $2s$ orbital) between the nucleus and the average position of the electron in the $2p$ orbital. The potential energy of the electron is given by equation 2.9 – it is related to the electrostatic attraction between the positively charged nucleus and the negatively charged electron.

$$E \propto \frac{1}{r} \qquad (2.9)$$

where E is the Coulombic (potential) energy and r is the distance between two point charges.

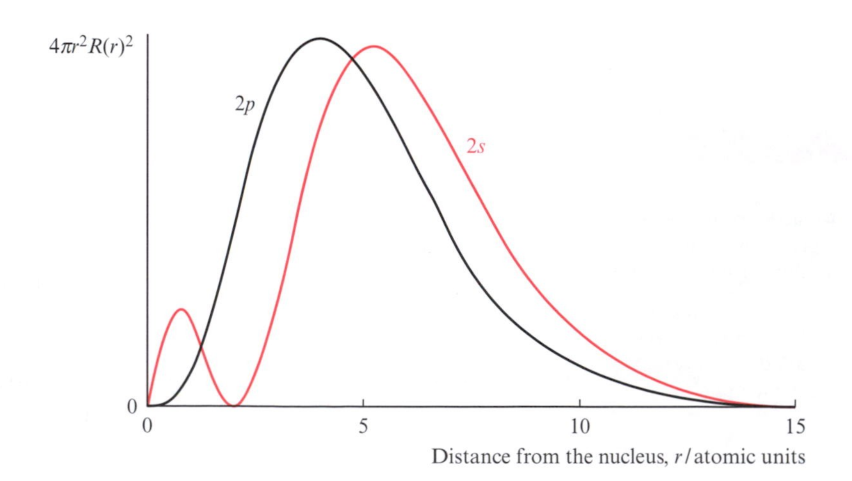

2.14 Radial distribution functions, $4\pi r^2 R(r)^2$, for the $2s$ and $2p$ atomic orbitals.

However, the positive charge (of the nucleus) experienced by the electron in the 2*p* orbital is 'diluted' by the presence of the 2*s* electron density close to the nucleus. The 2*s* electron *screens* or *shields* the 2*p* electron. The nuclear charge experienced by the electron in the 2*p* orbital is less than that experienced by the electron in the 2*s* orbital. Taking into account these effects, it becomes necessary in many-electron atoms to replace the nuclear charge Z by the *effective nuclear charge* Z_{eff}. The value of Z_{eff} for a given atom varies for different orbitals and depends on how many electrons there are and in which orbitals they reside. Values of Z_{eff} can be estimated on an empirical basis using Slater's rules.§

The effective nuclear charge, Z_{eff}, is a measure of the positive charge experienced by an electron taking into account the shielding of the other electrons.

The 2*s* orbitals are more *penetrating* and they *shield* the 2*p* orbitals. Electrons in the 2*p* orbitals experience a lower electrostatic attraction and are therefore higher in energy (less stabilized) than electrons in 2*s* orbitals. Similar arguments place the 3*d* orbitals higher in energy than the 3*p* which in turn are higher than the 3*s* (Figure 2.15).

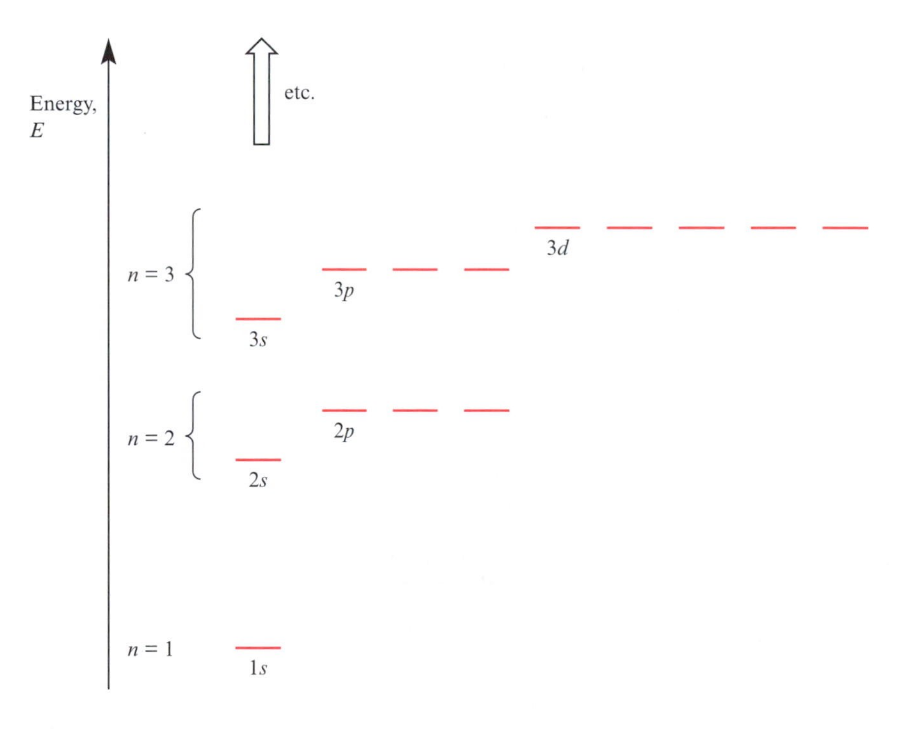

2.15 A schematic (not to scale) representation of the energy solutions (for n = 1, 2 and 3) of the Schrödinger equation for a many-electron atom.

§ We shall not elaborate further on Slater's rules; for further details see: W. L. Jolly, *Modern Inorganic Chemistry*, MacGraw-Hill, New York, p. 17 (1984).

Changes in the energy of an atomic orbital with atomic number

In this section, we briefly consider how the energy of a particular atomic orbital changes as the atomic number increases – for example, the 2*s* atomic orbital is of a different energy in lithium from that in other elements.

As atomic number increases, the nuclear charge must increase, and for an electron in a given *ns* or *np* orbital, the *effective nuclear charge* also increases. This results in a decrease in atomic orbital energy – the trend is relatively smooth, although the relationship between *ns* or *np* atomic orbital energy and atomic number is *non-linear*. Although the energies of *nd* and *nf* atomic orbitals generally decrease with increasing atomic number, the situation is rather complex, and we shall not discuss it further here.§

2.16 THE ATOMIC SPECTRUM OF HYDROGEN AND SELECTION RULES

When the single electron in the 1*s* orbital of the hydrogen atom is *excited* (that is, it is given energy), it may be *promoted* to a higher energy state. The new state is transient and the electron will fall back to a lower energy state, emitting energy as it does so. Evidence for the discrete nature of orbital energy levels comes from an observation of *spectral lines* in the emission spectrum of hydrogen (Figure 2.16). The emphasis here is on the fact that *single lines are observed*; the emission is not continuous over a range of frequencies. Similar single frequencies are also observed in the absorption spectrum.

Some of the electronic transitions which give rise to frequencies observed in the emission spectrum of atomic hydrogen are shown in Figure 2.17. Notice that some transitions are *not* allowed – the *selection rules* given in equations 2.10 and 2.11 must be obeyed.

Selection rules tell us which spectroscopic transitions are allowed: see also Section 9.6

In equation 2.10, Δn is the change in the value of the principal quantum number n, and this selection rule means that there is no restriction on transitions between different principal quantum levels. In equation 2.11, Δl is a change in the value of the orbital quantum number l and this selection rule

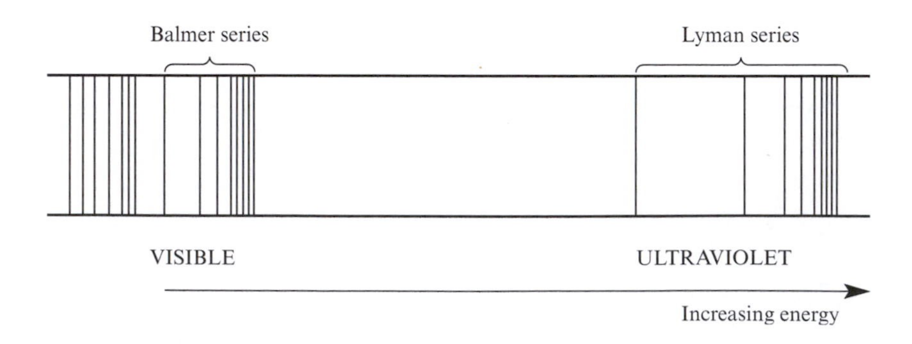

2.16 Part of the emission spectrum of atomic hydrogen. Groups of lines have particular names, for example the Balmer and Lyman series.

§ For a detailed discussion see: K. M. Mackay and R. A. Mackay, *An Introduction to Modern Inorganic Chemistry*, 4th edn, Blackie, Glasgow and London (1989).

Box 2.7 Absorption and emission spectra

The ground state of an atom (or other species) is one in which the electrons are in the lowest energy arrangement.

When an atom (or other species) absorbs electromagnetic radiation, discrete electronic transitions occur and an *absorption spectrum* is observed.

When energy is provided (e.g. as heat or light) one or more electrons in an atom or other species may be promoted from its ground state level to a higher energy state. This *excited state* is transient and the electron falls back to the ground state. An *emission spectrum* is thereby produced.

Absorption and emission transitions can be distinguished by using the following notation:

Emission: (high energy level) $\rightarrow$ (low energy level)
Absorption: (high energy level) $\leftarrow$ (low energy level)

Spectral lines in both absorption and emission spectra may be designated in terms of frequency ν:

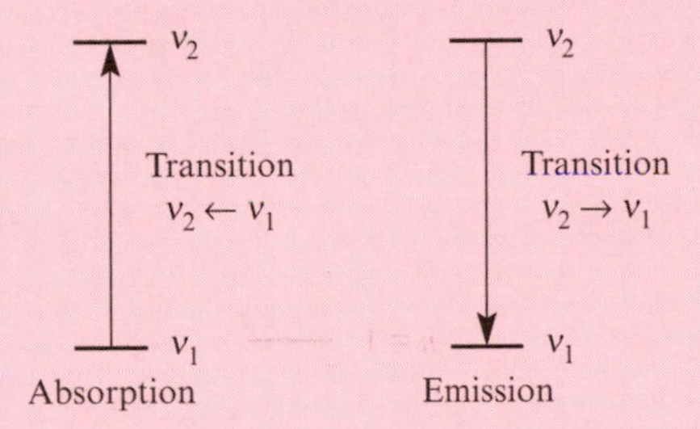

Two useful relationships are:

$$E = h\nu$$

$$\nu = \frac{c}{\lambda}$$

where E = energy, ν = frequency, λ = wavelength, c = speed of light and h = Planck constant.

places a restriction on this transition: l can only change by unity. Equation 2.10 is known as the *Laporte selection rule*, and corresponds to a change of angular momentum by one unit.

$$\Delta n = 0, 1, 2, 3, 4 \ldots \qquad (2.10)$$

$$\Delta l = +1 \text{ or } -1 \qquad \text{Laporte selection rule} \qquad (2.11)$$

Basic selection rules for electronic spectra:

Δn = 0, 1, 2, 3, 4 ...

Δl = +1 or −1 Laporte selection rule

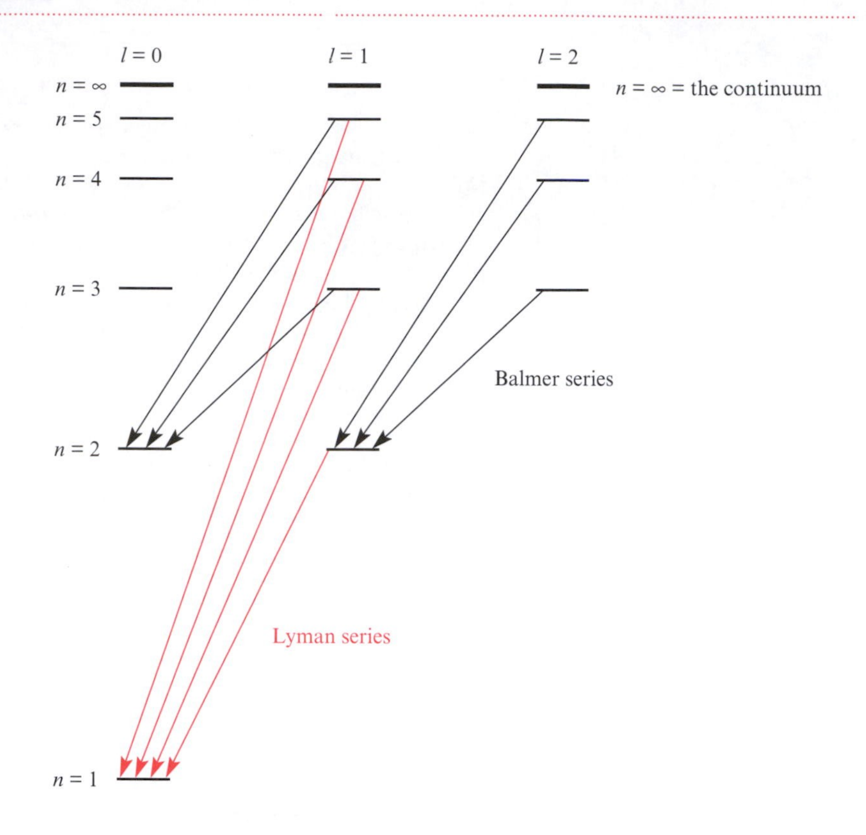

2.17 Some of the allowed transitions that make up the Lyman (shown in red) and Balmer (shown in black) series in the emission spectrum of atomic hydrogen. Note that the selection rules in equations 2.10 and 2.11 must be obeyed.

For example, a transition from a level with $n = 3$ to one with $n = 2$ is only allowed if, at the same time, there is a change in l of ± 1. In terms of atomic orbitals this corresponds to a transition from the $3p \rightarrow 2s$ or $3d \rightarrow 2p$ orbitals but *not* $3s \rightarrow 2s$ or $3d \rightarrow 2s$. The latter are disallowed because they do not obey equation 2.11. (What is Δl for each of these last two transitions?) Of course, in the hydrogen atom, atomic levels with the same value of n are degenerate and so a transition in which $\Delta n = 0$ (say, $2p \rightarrow 2s$) does not give rise to a change in energy, and cannot give rise to an observed spectral line.

Degenerate energy levels in the H atom: see Section 2.14

The lines in the emission spectrum of atomic hydrogen fall into several discrete series, two of which are illustrated (in part) in Figure 2.17. The series are defined in Table 2.3. The relative energies of the transitions mean that

Table 2.3 Series of lines observed in the emission spectrum of the hydrogen atom.

Name of series	$n' \rightarrow n$	*Region in which transitions are observed*
Lyman	$2 \rightarrow 1$, $3 \rightarrow 1$, $4 \rightarrow 1$, etc.	Ultraviolet
Balmer	$3 \rightarrow 2$, $4 \rightarrow 2$, $5 \rightarrow 2$, etc.	Visible
Paschen	$4 \rightarrow 3$, $5 \rightarrow 3$, $6 \rightarrow 3$, etc.	Infrared
Brackett	$5 \rightarrow 4$, $6 \rightarrow 4$, $7 \rightarrow 4$, etc.	Far infrared
Pfund	$6 \rightarrow 5$, $7 \rightarrow 5$, $8 \rightarrow 5$, etc.	Far infrared

each set of spectral lines appears in a different part of the electromagnetic spectrum (see Appendix 4). The Lyman series with lines of type $n' \rightarrow 1$ ($2 \rightarrow 1$, $3 \rightarrow 1$, $4 \rightarrow 1$ ...) contains energy transitions which correspond to frequencies in the ultraviolet region. The Balmer series with lines of type $n' \rightarrow 2$ ($3 \rightarrow 2$, $4 \rightarrow 2$, $5 \rightarrow 2$ ...) contains transitions which correspond to frequencies in the visible region. Of course, in the hydrogen atom, the *nd*, *np* and *ns* orbitals possess the same energy and the transitions $np \rightarrow (n-1)s$ and $nd \rightarrow (n-1)p$ occur at the same frequency for the same value of n.

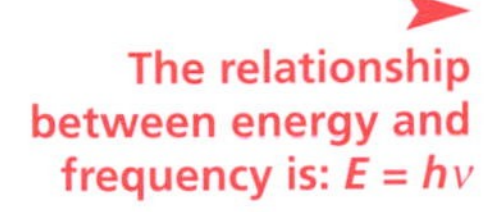

The Lyman series of spectral lines is of particular significance because it may be used to determine the ionization energy of hydrogen. That is, the energy required to remove the 1*s* electron completely from the atom. The frequency of lines in the Lyman series gives values of ν corresponding to the spectral transitions $n' \rightarrow n = 2 \rightarrow 1, 3 \rightarrow 1, 4 \rightarrow 1$. The series can be extrapolated to find a value of ν corresponding to the transition $\infty \rightarrow 1$. The method is shown in worked example 2.3.

The first ionization energy of an atom is the energy required to completely remove the most easily separated electron from the atom in the gaseous state. For an atom X, it is defined for the process:

$$X(g) \rightarrow X^+(g) + e^-$$

Worked example 2.3 ***Determination of the ionization energy of hydrogen.***

The frequencies of some of the spectral lines in the Lyman series of atomic hydrogen are 2.466, 2.923, 3.083, 3.157, 3.197, 3.221 and 3.237 $\times$ 10^{15} Hz. Using these data, calculate the ionization energy of atomic hydrogen.

The data correspond to spectral transitions, ν.

The aim of the question is to find the ionization energy of atomic hydrogen and this corresponds to a value of ν for the transition $n' \rightarrow n = \infty \rightarrow 1$.

The level $n' = \infty$ (the continuum) corresponds to the electron being completely removed from the atom. As we approach the level $n = \infty$, differences between *successive* energy levels become progressively smaller until, at the continuum, the difference between levels is zero. At this point, the energy of the electron is independent of the nucleus and it is no longer quantized. Spectral transitions to the level $n = 1$ from levels approaching $n = \infty$ will have virtually identical energies.

In order to determine the ionization energy of the hydrogen atom, we need first to find the point at which the *difference* in energies between successive spectral transitions approaches zero. This corresponds to the convergence limit of the frequencies.

First, calculate the *differences*, $\Delta\nu$, between successive values of ν for the data given. These are:

$$(2.923 - 2.466) \times 10^{15} = 0.457 \times 10^{15}\ \text{Hz}$$
$$(3.083 - 2.923) \times 10^{15} = 0.160 \times 10^{15}\ \text{Hz}$$
$$(3.157 - 3.083) \times 10^{15} = 0.074 \times 10^{15}\ \text{Hz}$$
$$(3.197 - 3.157) \times 10^{15} = 0.040 \times 10^{15}\ \text{Hz}$$
$$(3.221 - 3.197) \times 10^{15} = 0.024 \times 10^{15}\ \text{Hz}$$
$$(3.237 - 3.221) \times 10^{15} = 0.016 \times 10^{15}\ \text{Hz}$$

Now, plot these differences against *either* the higher *or* the lower value of $\Delta\nu$. (Here we use the lower values).

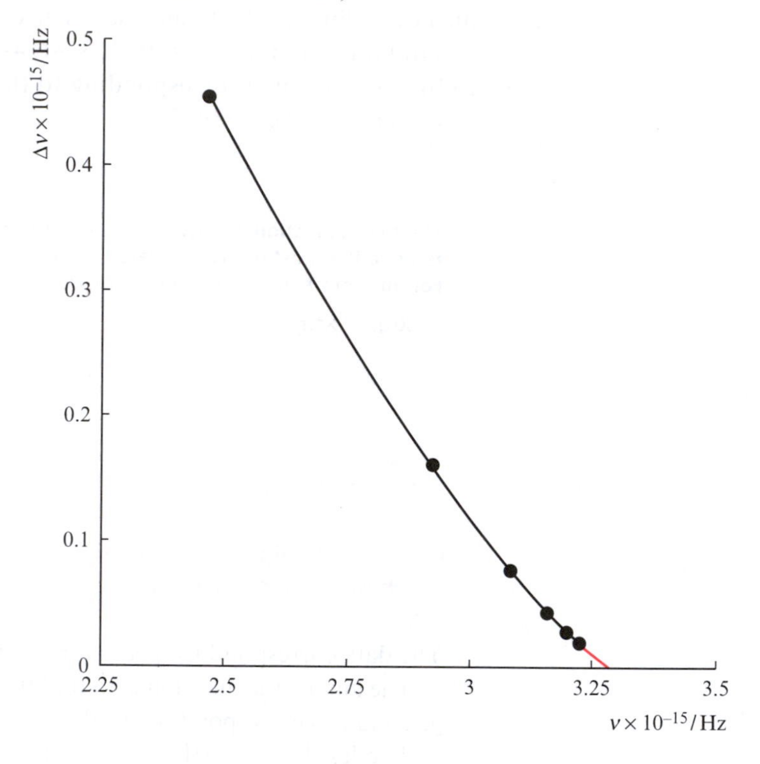

The curve converges at a value of $\nu = 3.275 \times 10^{15}$ Hz. Therefore, the value of E which corresponds to this value of ν is:

$$E = h\nu = (6.626 \times 10^{-34}\ \text{J s}) \times (3.275 \times 10^{15}\ \text{s}^{-1})$$

But this value is per atom.

The ionization energy should be given in units of energy per mole, which means we need to multiply by the Avogadro number (the number of atoms per mole).

Ionization energy
of hydrogen $= (6.626 \times 10^{-34}) \times (3.275 \times 10^{15}) \times (6.022 \times 10^{23})\ \text{J mol}^{-1}$
$= 1.310 \times 10^{6}\ \text{J mol}^{-1}$
$= 1310\ \text{kJ mol}^{-1}$

2.17 MANY-ELECTRON ATOMS

We have already seen that a neutral atom with more than one electron is called a *many-electron atom* and, clearly, all but the hydrogen atom fall into this category. As we have seen, the electrostatic repulsion between electrons in a many-electron atom has to be considered, and solving the Schrödinger equation exactly for such atoms is impossible. Various approximations have to be made in attempting to generate solutions of E and ψ for a many-electron atom, and for atoms, at least, numerical solutions of high accuracy are possible.

We shall not delve into this problem further, except to stress the important result from Section 2.15. The effects of *penetration* and *shielding* mean that in all atoms with more than one electron, orbitals with the same value of n but different values of l possess different energies (Figure 2.15).

2.18 THE *AUFBAU* PRINCIPLE

The word '*aufbau*' is German and means 'building up'. The *aufbau* principle provides a set of rules for determining the ground state electronic structure of atoms. We are concerned for the moment only with the occupancies of atomic (rather than molecular) orbitals.

The *aufbau* principle is used in conjunction with two other principles – Hund's rules and the Pauli exclusion principle. It can be summarized as follows.

- Orbitals are filled in order of energy with the lowest energy orbitals being filled first.

Degenerate = same energy

- If there is a set of degenerate orbitals, pairing of electrons in an orbital cannot begin until *each* orbital in the set contains one electron. Electrons singly occupying orbitals in a degenerate set have the same (*parallel*) spins. This is Hund's rule.

- No two electrons in an atom can have exactly the same set of n, l, m_l and m_s quantum numbers. This is the Pauli exclusion principle. This means that each orbital can accommodate a maximum of two electrons with different m_s values (different spins).

The first of the above statements needs no further clarification. The second statement can be exemplified by considering two electrons occupying a degenerate set of p orbitals. Electrons may be represented by arrows, the orientation of which indicates the direction of spin ($m_s = +\frac{1}{2}$ or $-\frac{1}{2}$). Figure 2.18a shows the *correct* way to arrange the two electrons according to Hund's rule. The other possibilities shown in Figures 2.18b and 2.18c are disallowed by Hund's rule (although they do represent possible excited states of the atom).

The Pauli exclusion principle prevents two electrons with the same spins from entering a single orbital. If they were to do so, they would have the same set of four quantum numbers. In Figure 2.19, the allowed and disallowed arrangement of electrons in a 1s orbital are depicted.

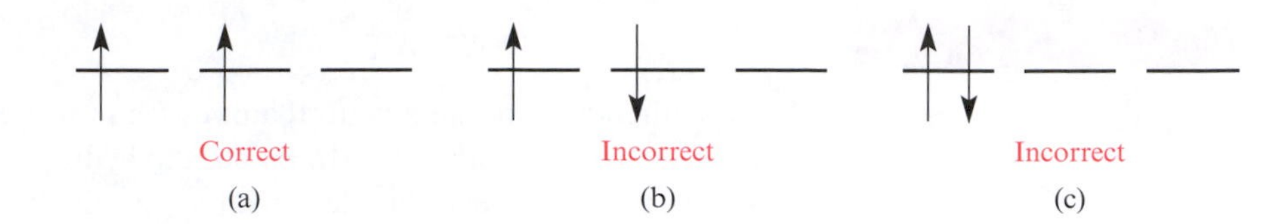

2.18 Hund's rule: two electrons in a degenerate set of p orbitals must be in separate orbitals and have parallel (the same) spins in the ground state.

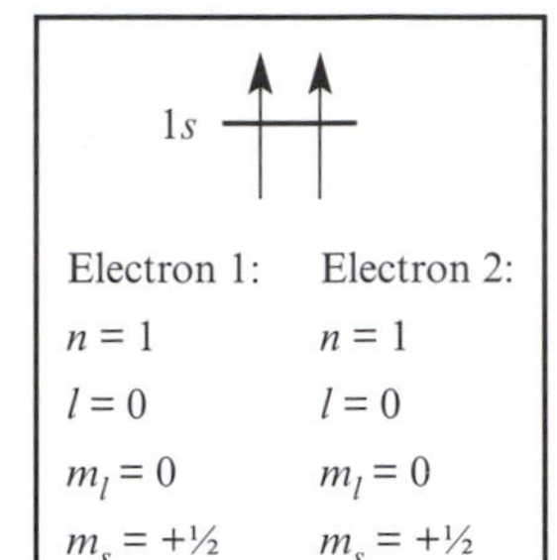

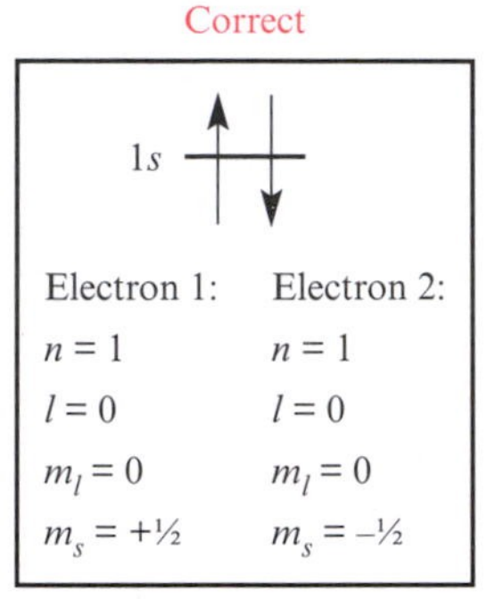

2.19 The Pauli exclusion principle: two electrons occupying a 1s atomic orbital can only have non-identical sets of the four quantum numbers n, l, m_l and m_s if their spins are different.

The Pauli exclusion principle states that no two electrons in an atom can have the same set of n, l, m_l and m_s quantum numbers.

This means that every electron in an atom is uniquely defined by its set of four quantum numbers.

Hund's rule states that when filling a degenerate set of orbitals, pairing of electrons cannot begin until *each* orbital in the set contains one electron. Electrons singly occupying orbitals in a degenerate set have parallel spins.

The *aufbau* principle is a set of rules which must be followed when placing electrons in atomic or molecular orbitals. The *aufbau* principle combines Hund's rules and the Pauli exclusion principle with the following additional facts:

- **Orbitals are filled in order of increasing energy.**
- **An orbital is fully occupied when it contains two electrons.**

Worked example 2.4 ***Using the aufbau principle***

Determine the arrangement of the electrons in an atom of nitrogen in its ground state.

The atomic number of nitrogen is 7, and there are seven electrons to be considered.

The lowest energy orbital is the 1*s* ($n = 1$; $l = 0$; $m_l = 0$).

The maximum occupancy of the 1*s* orbital is two electrons.

The next lowest energy orbital is the 2*s* ($n = 2$; $l = 0$; $m_l = 0$).

The maximum occupancy of the 2*s* orbital is two electrons.

The next lowest energy orbitals are the three making up the degenerate set of 2*p* orbitals ($n = 2$; $l = 1$, $m_l = +1, 0, -1$).

The maximum occupancy of the 2*p* orbitals is six electrons, but only three remain to be accommodated; (seven electrons in total for the nitrogen atom; four in the 1*s* and 2*s* orbitals).

The three electrons will occupy the 2*p* orbitals so as to obey Hund's rule. Each electron enters a separate orbital and the three electrons have parallel spins.

The arrangement of electrons in a nitrogen atom in its ground state is represented by:

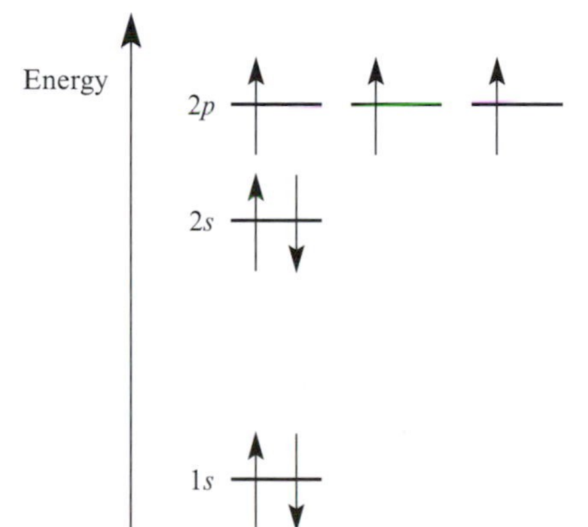

2.19 ELECTRONIC CONFIGURATIONS

Ground state electronic configurations

The *aufbau* principle provides us with a method of predicting the order of filling atomic orbitals with electrons and gives us the ground state arrangement of the electrons – the *ground state electronic configuration* – of an atom. The result may be represented diagrammatically as is shown in Figure 2.20 for the arrangement of electrons in the ground state of a carbon atom ($Z = 6$).

Z = atomic number

The ground state electronic configuration of an atom is its lowest energy state.

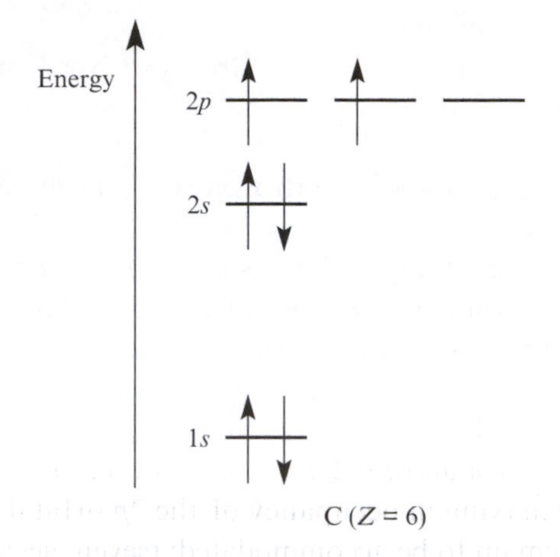

2.20 The arrangement of electrons in the atomic orbitals of carbon in the ground state.

Another way to present the electronic configuration is given below. For the hydrogen atom, we saw in Section 2.14 that the occupancy of the 1*s* atomic orbital by a single electron could be indicated by the notation $1s^1$. This can be extended to give the electronic configuration of any atom. Thus, for carbon, the configuration shown in Figure 2.20 can be written as $1s^2 2s^2 2p^2$. Similarly, for helium and lithium (Figure 2.21) the configurations are $1s^2$ and $1s^2 2s^1$ respectively.

The order of occupying atomic orbitals in the ground state of an atom *usually* follows the sequence (lowest energy first):

$$1s < 2s < 2p < 3s < 3p < 4s < 3d < 4p < 5s < 4d < 5p < 6s < 5d \approx 4f < 6p < 7s < 6d \approx 5f$$

We emphasize that this series may vary for some atoms because the energy of electrons in orbitals is affected by the nuclear charge, the presence of other electrons in the same orbital or in the same sub-level, and the overall charge. We shall mention this point again when we look at aspects of the *d*-block metal chemistry in Chapter 16.

The *usual* ordering (lowest energy first) of atomic orbitals is:

$1s < 2s < 2p < 3s < 3p < 4s < 3d < 4p < 5s < 4d < 5p < 6s < 5d \approx 4f < 6p < 7s < 6d \approx 5f$

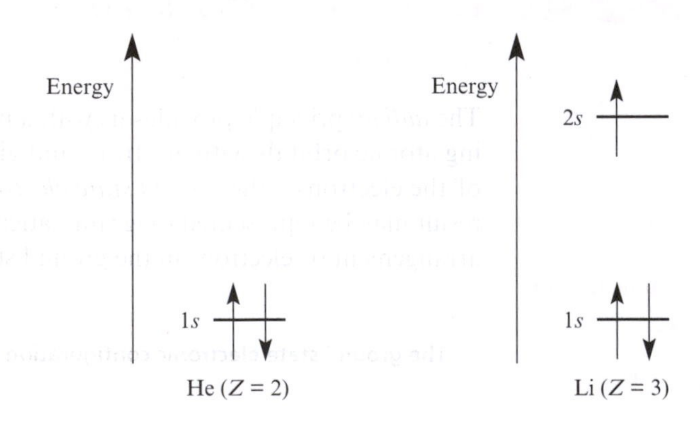

2.21 The arrangement of electrons in the atomic orbitals of helium and lithium in the ground state.

Table 2.4 Ground state electronic configurations for the first 20 elements

Atomic number	*Element symbol*	*Electronic configuration*	*Shortened form of the notation for the electronic configuration*
1	H	$1s^1$	$1s^1$
2	He	$1s^2$	$1s^2$ = [He]
3	Li	$1s^2 2s^1$	[He]$2s^1$
4	Be	$1s^2 2s^2$	[He]$2s^2$
5	B	$1s^2 2s^2 2p^1$	[He]$2s^2 2p^1$
6	C	$1s^2 2s^2 2p^2$	[He]$2s^2 2p^2$
7	N	$1s^2 2s^2 2p^3$	[He]$2s^2 2p^3$
8	O	$1s^2 2s^2 2p^4$	[He]$2s^2 2p^4$
9	F	$1s^2 2s^2 2p^5$	[He]$2s^2 2p^5$
10	Ne	$1s^2 2s^2 2p^6$	[He]$2s^2 2p^6$ = [Ne]
11	Na	$1s^2 2s^2 2p^6 3s^1$	[Ne]$3s^1$
12	Mg	$1s^2 2s^2 2p^6 3s^2$	[Ne]$3s^2$
13	Al	$1s^2 2s^2 2p^6 3s^2 3p^1$	[Ne]$3s^2 3p^1$
14	Si	$1s^2 2s^2 2p^6 3s^2 3p^2$	[Ne]$3s^2 3p^2$
15	P	$1s^2 2s^2 2p^6 3s^2 3p^3$	[Ne]$3s^2 3p^3$
16	S	$1s^2 2s^2 2p^6 3s^2 3p^4$	[Ne]$3s^2 3p^4$
17	Cl	$1s^2 2s^2 2p^6 3s^2 3p^5$	[Ne]$3s^2 3p^5$
18	Ne	$1s^2 2s^2 2p^6 3s^2 3p^6$	[Ne]$3s^2 3p^6$ = [Ar]
19	K	$1s^2 2s^2 2p^6 3s^2 3p^6 4s^1$	[Ar]$4s^1$
20	Ca	$1s^2 2s^2 2p^6 3s^2 3p^6 4s^2$	[Ar]$4s^2$

By combining this sequence of orbitals with the rules stated in the *aufbau* principle, we can write down the ground state electronic configurations of most atoms. Those for elements with atomic numbers 1 to 20 are given in Table 2.4. As we progress down the table, there is clearly a repetition corresponding to the lowest energy electrons which occupy filled quantum levels. Thus, an abbreviated form of the electronic configuration can be used. Here, only electrons entering new quantum levels (new values of n) are emphasized, and the inner levels are indicated by the previous group 18 element (He, Ne or Ar).

For the time being we shall leave ground state electronic configurations for elements with atomic number > 20. This is the point at which the sequence of orbitals given above begins to become less reliable.

Valence and core electrons

We now introduce the terms *valence electronic configuration* and *valence electron* which are commonly used and have significant implication. The valence electrons of an atom are those in the outer (highest energy) quantum levels and it is these electrons which are primarily responsible for determining the chemistry of an element. Consider sodium ($Z = 11$) – the ground electronic configuration is $1s^2 2s^2 2p^6 3s^1$. The principal quantum shells with $n = 1$ and 2 are fully occupied and electrons in these shells are referred to as *core electrons*. Sodium has one valence electron (the $3s$ electron) and the chemistry of sodium reflects this.

Valence electrons and periodicity: see Section 2.22

Worked example 2.5 ***Determining a ground state electronic configuration***

Determine the ground state electronic configuration for sodium ($Z = 11$).

There are 11 electrons to be accommodated.
The sequence of atomic orbitals to be filled is $1s < 2s < 2p < 3s$.
The maximum occupancy of an *s* level is two electrons.
The maximum occupancy of the *p* level is six electrons.
Therefore the ground state electronic configuration for sodium is:

$1s^2 2s^2 2p^6 3s^1$ or $[Ne]3s^1$

2.20 THE OCTET RULE

Inspection of the ground state electronic configurations listed in Table 2.4 shows a clear pattern which is emphasized in the right-hand column of the table. This notation makes use of filled principal quantum shells as 'building blocks' within the ground state configurations of later elements. This trend is one of *periodicity* – repetition of the configurations of the outer electrons but for a different value of the principal quantum number.

We see a repetition of the sequences $ns^{(1 \text{ to } 2)}$ and $ns^2np^{(1 \text{ to } 6)}$. The total number of electrons that can be accommodated in orbitals with $n = 1$ is only two, but for $n = 2$ this number increases to eight (i.e. $2s^2 2p^6$). The number eight is also a feature of the total number of electrons needed to completely fill the 3*s* and 3*p* orbitals. This is the basis of the *octet rule*.

As the name§ suggests, the octet rule has its origins in the observation that atoms of the *s*- and *p*-blocks have a tendency to lose, gain or share electrons in order to end up with eight electrons in their outer shell.

The ground state electronic configuration of fluorine is $[He]2s^2 2p^5$. This is one electron short of the completed configuration $[He]2s^2 2p^6$. Fluorine readily gains an electron to form the fluoride ion F^-, the ground state configuration of which is $[He]2s^2 2p^6$. An atom of nitrogen ($[He]2s^2 2p^3$) requires three electrons to give the $[He]2s^2 2p^6$ configuration and N^{3-} may be formed. On the other hand, a carbon atom ($[He]2s^2 2p^2$) would have to form the C^{4-} ion in order to achieve the $[He]2s^2 2p^6$ configuration and this is energetically unfavourable. The answer here is for the carbon atom to *share* four electrons with other atoms, thereby completing its octet without the need for ion formation.

Formation of ions: see Chapter 6

An atom is obeying the octet rule when it gains, loses or shares electrons to give an *outer* shell containing eight electrons with the configuration ns^2np^6.

§ The word 'octet' is derived from the Latin *'octo'* meaning eight.

Despite its successes, we can see that the concept of the octet rule is rather limited. We can apply it satisfactorily to the principal quantum shell with $n = 2$. Here the quantum shell is fully occupied when it contains eight electrons. But there are exceptions even here: the elements lithium and beryllium which have outer $2s$ electrons (Table 2.4) tend to *lose* electrons so as to possess helium-like ($1s^2$) ground state electronic configurations, and molecules such as BF_3 with six electrons in the valence shell are stable.

What about the principal quantum shell with $n = 3$? Here, sodium and magnesium (Table 2.4) tend to *lose* electrons to possess a neon-like configuration and in doing so they obey the octet rule. Aluminium with a ground state configuration of $[Ne]3s^23p^1$ may lose three electrons to become Al^{3+} (i.e. neon-like) or it may participate in bond formation so as to achieve an octet through sharing electrons. Care is needed though! Aluminium also forms compounds in which it does not formally obey the octet rule, for example $AlCl_3$. Atoms of elements with ground state configurations between $[Ne]3s^23p^2$ and $[Ne]3s^23p^6$ may also share or gain electrons to achieve octets, but there is a complication. For $n = 3$, there is another sub-shell – the $3d$ orbitals – and this leads to the possibility of expanding the octet.

Aluminium halides: see Section 13.4

Expansion of the octet: see Section 5.16

Thus, the octet rule is indeed a useful tool for predicting ion and covalent bond formation, but it is restricted to a relatively small number of elements. However, the *principle* of the octet rule is a very important one and the idea can be extended to an 18 electron rule which takes into account the filling of *ns*, *np* and *nd* sub-levels.

Worked example 2.6 ***The octet rule***

Suggest why fluorine forms only the fluoride F_2 (i.e. molecular difluorine), but oxygen can form a difluoride OF_2. (For F, $Z = 9$; for O, $Z = 8$.)

First, write down the ground state electronic configurations of F and Cl:

F $1s^22s^22p^5$
O $1s^22s^22p^4$

A fluorine atom can achieve an octet of electrons by sharing one electron with another fluorine atom. In this way, the principal quantum level for $n = 2$ is completed.

Oxygen can complete an octet of outer electrons by sharing two electrons with two fluorine atoms in the molecule OF_2. The octet of each F atom is also completed by each sharing one electron with the oxygen atom.

2.21 MONATOMIC GASES

The noble gases

In looking at ground state electronic configurations, we saw that configurations with filled principal quantum shells appeared as 'building blocks' within the ground state configurations of heavier elements. The elements which possess these filled principal quantum shells all belong to one group in the periodic table – group 18. These are the so-called *noble gases*.[§]

The ground state electronic configurations of the noble gases are given in Table 2.5. Note that each configuration (except that for helium) contains an ns^2np^6 configuration for the highest value of n – for example, the configuration for argon ends $3s^23p^6$. The ns^2np^6 configuration is often referred to as an 'inert (noble) gas configuration'. Do not worry that the heaviest elements possess filled d and f shells in addition to filled s and p orbitals.

The fact that each noble gas has a filled outer np shell means that the group 18 elements exist in the elemental state as monatomic species – there is no driving force for bond formation between atoms.

Forces between atoms

Interatomic interactions: see Section 1.21

$\Delta_{vap}H$: see Section 1.21

The noble gases (with the exception of helium) solidify only at extremely low temperatures. The melting points of neon and argon are 24.5 K and 84 K respectively. These low values indicate that the *interatomic forces* in the solid state are very weak indeed and are easily overcome when the solid changes into a liquid. The very narrow range of temperatures over which the group 18 elements are in the liquid state is significant. The net interatomic forces in the liquid state are extremely weak and the ease with which vaporization occurs is apparent from the very low values of the enthalpy changes for this process (Table 2.6).

What is the nature and form of the forces between atoms? Let us consider two atoms of argon. Each is spherical with a central positively charged nucleus surrounded by negatively charged electrons. At large separations, there will only be very small interactions between the atoms. However, as two argon atoms come closer together, the electron clouds begin to repel each other. At very short internuclear distances, there will be strong repulsive forces as a result of electron–electron repulsions. A good mathematical

Table 2.5 Ground state electronic configurations of the noble gases

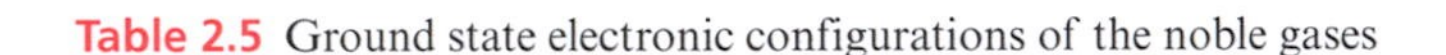

Noble gas	*Symbol*	*Atomic number*	*Ground state electronic configuration*
Helium	He	2	$1s^2$
Neon	Ne	10	$1s^22s^22p^6$
Argon	Ar	18	$1s^22s^22p^63s^23p^6$
Krypton	Kr	36	$1s^22s^22p^63s^23p^64s^23d^{10}4p^6$
Xenon	Xe	54	$1s^22s^22p^63s^23p^64s^23d^{10}4p^65s^24d^{10}5p^6$
Radon	Rn	86	$1s^22s^22p^63s^23p^64s^23d^{10}4p^65s^24d^{10}5p^66s^24f^{14}5d^{10}6p^6$

§ Although 'inert gas' is commonly used, 'noble gas' is the IUPAC recommendation; see Section 1.10

Table 2.6 Some physical properties of the noble gases

Element	*Melting point / K*	*Boiling point / K*	$\Delta_{fus}H$ / $kJ\ mol^{-1}$	$\Delta_{vap}H$ / $kJ\ mol^{-1}$	*Van der Waals radius* (r_v) / *pm*	*First ionization energy / kJ mol*$^{-1}$
Helium	– [a]	4.2	–	0.1	99	2372
Neon	24.5	27	0.3	2	160	2080
Argon	84	87	1.1	6.5	191	1520
Krypton	116	120	1.4	9	197	1351
Xenon	161	165	1.8	13	214	1170
Radon	202	211	–	18	–	1037

[a] Helium cannot be solidified under any conditions of temperature and pressure.

approximation of the repulsive force shows that it depends on the internuclear distance d as stated in equation 2.12. The energy potential is related to the force and varies according to equation 2.13. This repulsive potential (defined with a positive energy) between two argon atoms is shown by the black curve in Figure 2.22.

$$\text{Repulsive force between atoms} \propto \frac{1}{d^{13}} \tag{2.12}$$

$$\text{Energy potential due to repulsion between atoms} \propto \frac{1}{d^{12}} \tag{2.13}$$

If this repulsive force were the only one operative in the system, then elements such as the noble gases would never form a solid lattice. We need to recognize that there is an attractive force between the atoms which opposes the electronic repulsion.

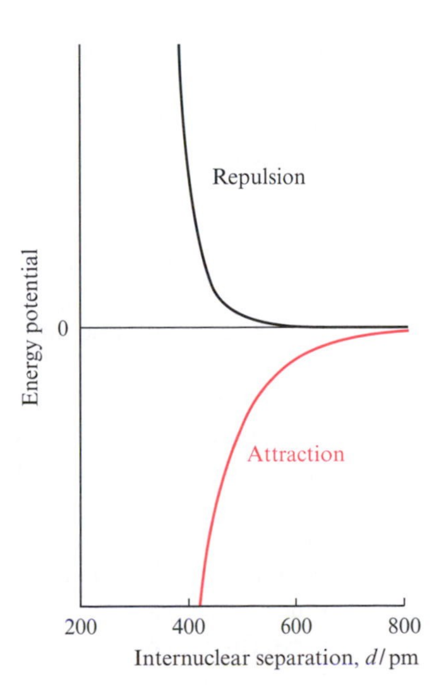

2.22 The repulsive (black line) and attractive (red line) energy potentials between two atoms of argon as a function of the internuclear distance, *d*.

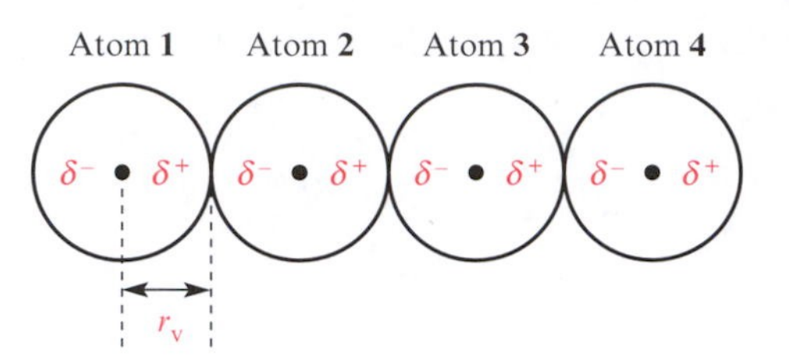

2.23 A dipole set up in atom **1** induces a dipole in atom **2** which induces one in **3**, and so on. The distance r_v is the van der Waals radius = half the distance of closest approach of two non-bonded atoms.

We may envisage this attractive force as arising in the following way. Within the argon atom, the positions of the nucleus and the electrons are not fixed. At any given time, the centre of negative charge density may not coincide with the centre of the positive charge, i.e. with the nucleus. This results in a net, instantaneous dipole. This dipole is transient, but it is sufficient to *induce* a dipole in a neighbouring atom (Figure 2.23) and electrostatic dipole–dipole interactions between atoms result. Since the dipoles are transient, each separate interaction will also be transient, but the net result is an attractive force between the argon atoms.

Dipole moments: see Section 4.11

In an atom or molecule, an asymmetrical distribution of charge leads to the formation of a *dipole*. One part of the atom or molecule has a greater share of the negative (or positive) charge than another.

Dipoles may be transient (as in an atom of argon) or may be permanent (as in a molecule of hydrogen chloride).

The attractive force due to the dipole–dipole interactions may be represented as shown in equation 2.14. This leads to an energy potential which varies with internuclear distance d (equation 2.15) and this attractive potential (defined with a negative energy) between atoms of argon is shown by the red curve in Figure 2.22.

$$\text{Attractive force between atoms} \propto \frac{1}{d^7} \tag{2.14}$$

$$\text{Energy potential due to attraction between atoms} \propto \frac{1}{d^6} \tag{2.15}$$

Finally, in Figure 2.24, we show the overall potential (repulsion plus attraction) between argon atoms as a function of the internuclear separation d. At large separations, the attractive force is dominant. As the two argon atoms come together, the attractive potential increases to a given point of maximum stabilization (lowest energy). However, as the atoms come yet closer together, the electronic repulsion between them increases. At very short distances, the electron–electron repulsion is dominant and highly unstable arrangements with *positive* potentials result.

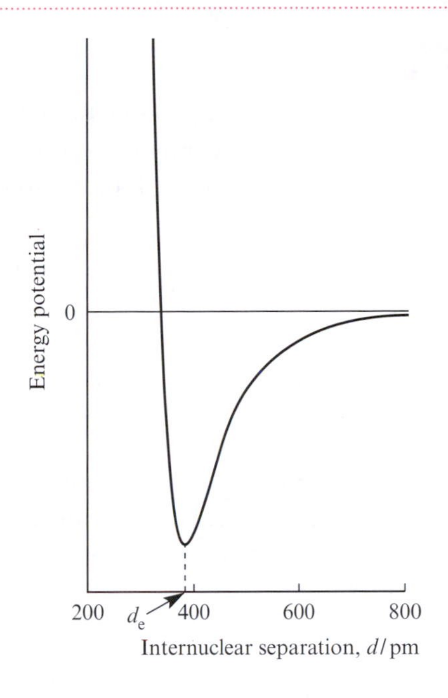

2.24 The overall energy potential of two argon atoms as a function of the internuclear distance, d. The value of d at the energy minimum (d_e) corresponds to *twice* the van der Waals radius of an argon atom.

Van der Waals radius

The value of d at the point in Figure 2.24 where the potential is at a minimum corresponds to the optimum distance, d_e, between two argon atoms. This is a *non-bonded separation*. Note that d_e is ***not*** the point at which the attractive and repulsive forces are equal. This point is reached when the potential is zero.

It is convenient to consider d_e in terms of an atomic property – the value $^{d_e}/_2$ is defined as the *van der Waals radius*, r_v, of an atom of argon. The value of r_v is *half* of the distance of closest internuclear separation of adjacent argon atoms. Values of the van der Waals radii for the noble gases are given in Table 2.6. In the particular case of a monatomic noble gas, these radii can be used to give a good idea of atomic size. The atomic volume may be estimated using the van der Waals radius; the volume of the spherical atom is then $^4/_3\pi r_v^{\ 3}$. The increase in atomic volume on descending group 18 is represented in Figure 2.25.

The van der Waals radius, r_v, of an atom X is measured as half of the distance of closest approach of two non-bonded atoms of X.

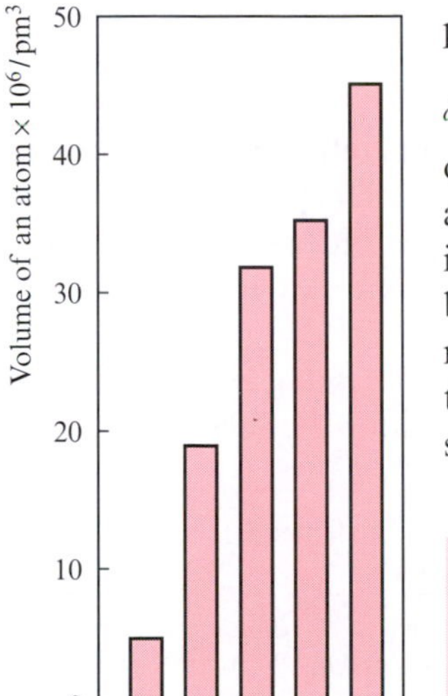

2.25 Atomic volumes for the noble gases determined using the van der Waals radius of each atom. The volume increases as group 18 is descended.

2.22 PERIODICITY

We end this chapter by briefly considering the consequences of the patterns in electronic configurations on the chemistry of the elements, and single out elements in groups 1 and 18 to illustrate some important points.

The noble gases

We have already noted similarities amongst the noble gases, and have emphasized that they are all members of the same *group* of the periodic table. All these elements possess fully occupied atomic orbitals in their outer quantum level. The fact that each noble gas has an outer ns^2np^6 configuration indicates that they are obeying the octet rule.

The physical and chemical properties of the noble gases reflect the fact that they are stable entities. However, for xenon, and to a lesser extent for krypton, the presence of the completed octet does *not* result in a total lack of chemical reactivity. Some reactions involving xenon are given in equations 2.16–2.18 and the reasons for the ability of xenon and krypton to undergo such reactions are discussed in Section 13.13.

$$Xe + F_2 \rightarrow XeF_2 \quad (2.16)$$

$$3XeF_4 + 6H_2O \rightarrow XeO_3 + 2Xe + \tfrac{3}{2}O_2 + 12HF \quad (2.17)$$

$$XeF_6 + H_2O \rightarrow XeOF_4 + 2HF \quad (2.18)$$

Group 1: the alkali metals

s-Block elements: see Chapter 11
p-Block elements: see Chapter 13
d-Block elements: see Chapters 13 and 16

Patterns in the chemical behaviour of the elements reflect the number and arrangement in orbitals of their valence electrons. The periodic table contains 18 groups of elements arranged in three blocks (excluding the lanthanoid and actinoid elements). The three blocks are called the *s*-, *p*- and *d*-blocks of elements.

In the *s*-block, elements are characterized by possessing either a valence electronic configuration of ns^1 (group 1) or ns^2 (group 2). Selected physical

Table 2.7 Physical and chemical data for group 1 elements (alkali metals)

Element	*Symbol*	*Melting point / K*	*Boiling point / K*	*Electronic configuration*	*First ionization energy / kJ mol⁻¹*	*Ion formed*	*Formula of chloride*
Lithium	Li	453.5	1615	$[He]2s^1$	520	Li^+	LiCl
Sodium	Na	371	1154	$[Ne]3s^1$	496	Na^+	NaCl
Potassium	K	336	1032	$[Ar]4s^1$	419	K^+	KCl
Rubidium	Rb	312	961	$[Kr]5s^1$	403	Rb^+	RbCl
Caesium	Cs	301.5	978	$[Xe]6s^1$	376	Cs^+	CsCl

and chemical data for the elements in group 1 (the alkali metals) are listed in Table 2.7. All these elements are solid at 298 K, but the melting points are relatively low, indicating that the interatomic forces in the solid state are overcome quite easily – in some parts of the world, caesium is a liquid at ambient temperatures. The ranges of temperature over which the group 1 elements remain in the liquid state are much greater than those of the noble gases (Table 2.6).

The group 1 elements behave in a similar manner to one another, just as the group 18 elements resemble each other. The values of the first ionization energies of the group 1 elements (Table 2.7) are all relatively§ low and indicate that ionization (equation 2.19) occurs easily. This is consistent with the ground state valence configurations of the group 1 elements being of the form ns^1. Figure 2.26 compares the first ionization energies of the group 1 and group 18 elements. For an atom of a noble gas to undergo reaction 2.19, a filled quantum shell must be disrupted – ionization to a singly charged cation is an unfavourable process.

Ionization energy: see Section 6.3

Trends in ionization energies: see problems 2.11 and 2.12

$$M(g) \rightarrow M^+(g) + e^- \quad (2.19)$$

The formation of M^+ ions by the group 1 elements is seen throughout their chemistry, and equations 2.20–2.23 illustrate just a few reactions of these elements.

$$2Na + Cl_2 \rightarrow 2NaCl \quad (2.20)$$

$$2K + 2H_2O \rightarrow 2KOH + H_2 \quad (2.21)$$

$$4Na + O_2 \rightarrow 2Na_2O \quad (2.22)$$

$$Li_2CO_3 \xrightarrow{\text{heat}} Li_2O + CO_2 \quad (2.23)$$

More detailed discussions of periodicity follow in Chapters 11 and 13.

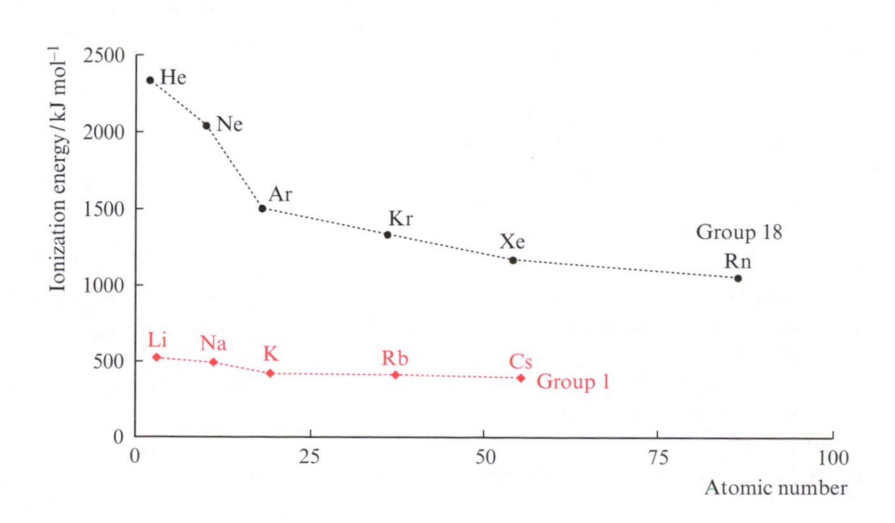

2.26 Trends in the first ionization energies for group 1 and group 18 elements.

§ We are considering the group 1 elements in relation to elements in other groups of the periodic table.

SUMMARY

This chapter has dealt with some difficult concepts, but they are essential to an understanding of the chemistry of the elements. As we introduce other topics in this book, there will be plenty of opportunity to review these concepts and this should provide you with a feeling as to why we have introduced them so early.

Do you know what the following terms mean?

- vector
- scalar
- quantized
- photon
- wavefunction
- eigenfunction
- eigenvalue
- principal quantum numbers n, l, m_l and m_s
- radial distribution function
- atomic orbital
- nodal plane
- lobe (of an orbital)
- phase (of an orbital)
- surface boundary of an orbital
- degenerate orbitals
- non-degenerate orbitals
- diffuse (with reference to an atomic orbital)
- penetration
- shielding
- effective nuclear charge
- hydrogen-like atom or ion
- many-electron atom
- first ionization energy
- *aufbau* principle
- Hund's rules
- Pauli exclusion principle
- ground state electronic configuration
- noble (inert) gas configuration
- octet rule
- core electrons
- valence electrons
- van der Waals radius
- induced dipole moment

You should now be able:

- to describe briefly what is meant by wave–particle duality
- to describe briefly what is meant by the uncertainty principle
- to discuss briefly what is meant by a solution of the Schrödinger wave equation
- to describe the significance of the radial and angular parts of the wavefunction
- to state what is meant by the principal quantum number
- to state what is meant by the orbital and spin quantum numbers and relate their allowed values to a value of the principal quantum number
- to describe an atomic orbital and the electron(s) in it in terms of values of quantum numbers
- to discuss briefly why we tend to use *normalized* wavefunctions
- to sketch graphs for the radial distribution functions of the 1*s*, 2*s*, 3*s*, 4*s*, 2*p*, 3*p* and 3*d* atomic orbitals; what do these graphs mean?
- to outline the shielding effects that electrons have on one another
- to draw the shapes, indicating phases, of *ns* and *np* atomic orbitals
- to distinguish between np_x, np_y and np_z atomic orbitals
- to explain why in a hydrogen-like atom, atomic orbitals with the same principal quantum number are degenerate
- to explain why in a many-electron atom, occupied orbitals with the same value of n but different l are non-degenerate
- to estimate the ionization energy of a hydrogen atom given a series of frequency values from the Lyman series of spectral lines
- to describe how to use the *aufbau* principle to determine the ground state electronic configuration of an atom
- to write down the usual ordering of the atomic orbitals from 1*s* to 4*p*. Which has the lowest energy? What factors affect this order?
- given the atomic number of an element (up to $Z = 20$), to write down in notational form the ground state electronic configuration
- given the atomic number of an element (up to $Z = 20$), to draw an energy level diagram to show the ground state electronic configuration
- to explain how an energy potential curve such as that in Figure 2.24 arises; what is the significance of the energy minimum?
- to discuss the meaning of periodicity with reference to the elements in groups 1 and 18.

PROBLEMS

2.1 Distinguish between a vector and a scalar quantity, and give examples of each.

2.2 Calculate the wavelengths of electromagnetic radiation with frequencies of (a) 2×10^{13} Hz, (b) 4.5×10^{15} Hz, and (c) 2.1×10^{17} Hz. By referring to the spectrum in Appendix 4, assign each wavelength to a particular type of radiation (e.g. X-rays).

2.3 Determine the possible values of the quantum numbers l and m_l corresponding to $n = 4$. From your answer, deduce the number of possible $4f$ orbitals.

2.4 Give the sets of four quantum numbers that uniquely define each electron in an atom of (a) ${}^{4}_{2}He$ and (b) ${}^{10}_{5}B$, each in its ground state.

2.5 Determine the energy of the levels of the hydrogen atom for $n = 1$, 2 and 3. (Hint: see equation 2.8.)

2.6 Using the data from worked example 2.3, plot the values of Δv against the higher frequency values and determine the ionization energy of hydrogen.

2.7 Explain what you understand by (a) wavefunction, (b) radial and angular parts of the wavefunction, (c) atomic orbital, (d) radial node, (e) penetration and shielding, and (f) radial probability function.

2.8 Determine the ground state electronic configurations of Be ($Z = 4$), F ($Z = 9$), P ($Z = 15$) and K ($Z = 19$) giving your answers both in (a) a notational form and (b) the form of an energy level diagram.

2.9 What do you understand by the 'octet rule'?

2.10 Phosphorus forms the chlorides PCl_3 and PCl_5. Are these compounds predicted from the octet rule?

2.11 Use the data in Tables 2.6 and 2.7 and in Figure 2.26 to predict the first ionization energy of francium, Fr.

2.12 Refer to Figure 2.26. Suggest reasons for the trends in decreasing first ionization energies for (a) the group 1 elements and (b) the noble gases.

3 Homonuclear covalent bonds

Topics

3.1 INTRODUCTION

A molecule is a discrete neutral species resulting from the formation of a covalent bond or bonds between two or more atoms.

In Section 1.12 we specified what we mean by a molecule and we stated that a covalent bond is formed when electrons are *shared* between nuclei. In this chapter we develop the ideas of covalent bonding.

We shall be concerned for the moment only with *homonuclear bonds* – bonds formed between atoms of the *same type*, for example the C–C bond in ethane (C_2H_6), the N–N bond in hydrazine (N_2H_4) or the O–O bond in hydrogen peroxide (H_2O_2) (Figure 3.1). Figure 3.2 shows examples of

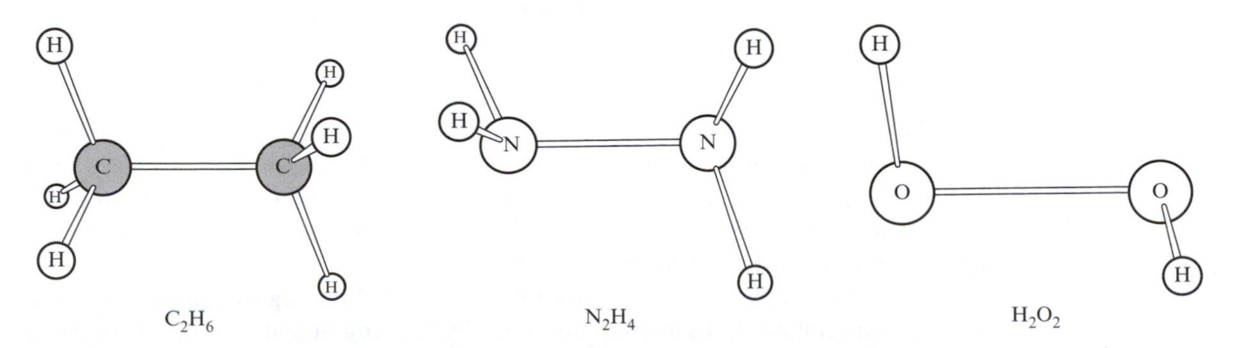

3.1 Molecules that contain one *homonuclear bond*: ethane (C_2H_6), hydrazine (N_2H_4) and hydrogen peroxide (H_2O_2).

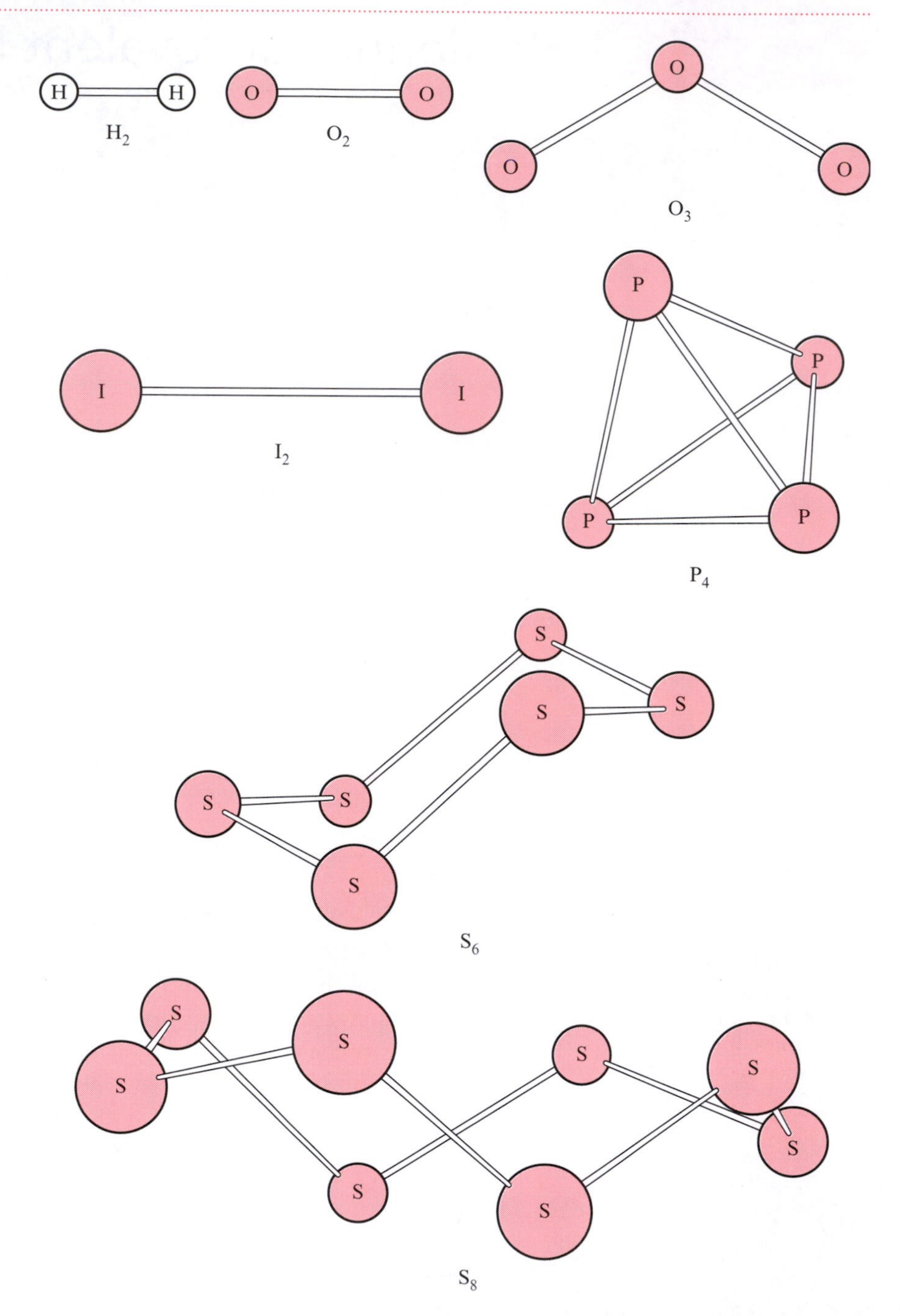

3.2 Examples of covalent homonuclear molecules.

homonuclear molecules – molecules in which all the *atoms are the same*. Each of these molecules is a molecular form of a particular element. In cases where more than one molecular form is known, these are allotropes (e.g. O_2 and O_3) of that element.

Allotropes: see Section 1.5

We shall approach the question of covalent bonding by considering some experimental observables and then by looking at some of the theoretical models which have been developed to describe the bonds.

The concept of bonding arises from observations that, within molecules, the distances between adjacent atoms are usually significantly shorter than those expected from the van der Waals radii of the atoms involved. Although we discussed the van der Waals radii of only monatomic elements in Chapter 2, a similar parameter may be defined for most other elements. In practice, this parameter comes from measurements of the closest *inter*molecular contacts in a series of molecules.

This suggests that distance is a very important parameter in describing atomic arrangements. As far as a bond is concerned, the bond length (or to be strictly accurate, the time-averaged distance between the nuclei) is of fundamental importance.

In the next section, we briefly introduce some of the methods which are used for the experimental determination of intramolecular and intermolecular distances. If you do not wish to interrupt the discussion of bonding, move to Section 3.3 and return to Section 3.2 later.

3.2 MEASURING INTERNUCLEAR DISTANCES

There are a number of methods available for determining important structural parameters. These techniques fall broadly into the groups of *spectroscopic* and *diffraction methods*. It is beyond the scope of this book to consider the details of the methods and we merely introduce them and look at some of the structural data that can be obtained.

Vibrational spectroscopy of diatomics: see Section 9.6

An overview of diffraction methods

For solid state compounds, the two methods most commonly used to determine structure are *X-ray* and *neutron diffraction*. *Electron diffraction* is best used to study the structures of gaseous molecules, although the technique is not strictly confined to use in the gas phase. Whereas electrons are negatively charged, X-rays and neutrons are neutral. Accordingly, X-rays and neutrons get 'deeper' into a molecule than do electrons because they do not experience an electrostatic repulsion from the core and valence electrons of the molecules. X-rays and neutrons are said to be more *penetrating* than electrons. Selected comparative details of the three diffraction methods are listed in Table 3.1.

In devising a method based on diffraction to measure internuclear distances, it is important to note that the distances between bonded atoms are typically in the range of 100 to 300 pm. The wavelength of the radiation that is chosen for the diffraction experiment should be similar in magnitude to the distances of interest; further details are given in Box 3.1.

At this point we stress some relevant facts which affect the type of information obtained from the different diffraction techniques:

- Electrons are negatively charged particles and are predominantly diffracted because they interact with the overall electrostatic field resulting from the negatively charged electrons and the positively charged nuclei of a molecule.

Table 3.1 A comparison of electron, X-ray and neutron diffraction techniques.

	Electron diffraction	*X-ray diffraction*	*Neutron diffraction*
Approximate wavelength of beam used / m	10^{-11}	10^{-10}	10^{-10}
Type of sample studied	Usually gases; also liquid or solid surfaces.	Solid (single crystal or powder); rarely solution.	Usually single crystal or powder.
What does the method locate?	Regions of electron density.	Regions of electron density.	Nuclei.
Relative advantages	Allows structural data to be collected for gaseous molecules; neither X-ray nor neutron diffraction methods are suitable for such samples.	Technique is routinely available and relatively straightforward to apply; ionic lattices and molecular solids can be studied; accurate bond parameters can be obtained for non-hydrogen atoms.	Can locate hydrogen atoms; very accurate structural parameters can be determined; both ionic lattices and molecular solids can be studied.
Relative disadvantages	Refining data for large molecules is complicated; hydrogen atom location is not accurate.	Hydrogen atom location is not always accurate.	Expensive; refining data for large molecules is complicated.

Box 3.1 Diffraction methods

Diffraction methods can be used to measure bond distances but the wavelength of the radiation used must be similar to the internuclear distances.

Internuclear bond distances are usually between 100 and 300 pm.

Since the wavelength of X-rays is about 100 pm, the diffraction of X-rays by a solid (usually crystalline) substance can be used to determine bond lengths.

The wavelength of a beam of electrons can be altered by applying an accelerating voltage, and a wavelength of the order of 10 pm is accessible. This is used for electron diffraction experiments.

Fast neutrons generated in a nuclear reactor can be moderated to give a beam of neutrons with a wavelength of the order of 100 pm. In addition, several neutron sources designed specifically for the purpose of diffraction experiments are available.

For further details of diffraction methods, a suitable book is E. A. V. Ebsworth, D. W. H. Rankin and S. Cradock, *Structural Methods in Inorganic Chemistry*, 2nd edn., Blackwell Scientific Publications, Oxford (1991).

- X-rays are electromagnetic radiation. They are scattered *mainly* by the *electrons* in a molecular or ionic array. It is possible to locate the atomic nuclei approximately because the X-rays and electrons are mostly scattered by the *core electrons* which are concentrated around each nucleus. It follows that covalently bound *hydrogen atoms, which do not possess core electrons*, pose a particular problem in X-ray and electron diffraction experiments.

Box 3.2 The difficulty of locating of a hydrogen nucleus by X-ray or electron diffraction

Consider a C–H bond. The ground state electronic configuration of a carbon atom is $1s^22s^22p^2$ and that of a hydrogen atom is $1s^1$. The carbon atom has four valence electrons and two core electrons. The hydrogen atom has one valence electron and no core electrons.

When the hydrogen atom bonds with the carbon atom, the single electron from the hydrogen atom is used in bond formation. This means that there is little electron density concentrated around the hydrogen nucleus:

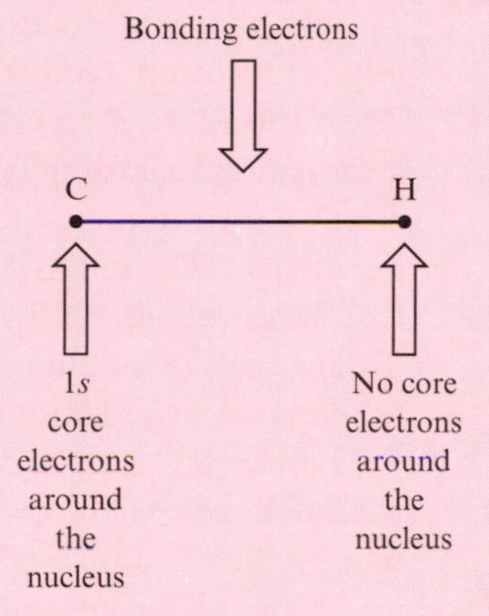

If the length of the C–H bond is being determined by X-ray or electron diffraction, there is a problem. The methods both locate regions of electron density. The positions of the nuclei are inferred from the positions of the electrons.

The carbon nucleus can be located because it has associated with it the core 1*s* electrons, but the hydrogen nucleus cannot normally be directly located.

However, the region of electron density associated with the C–H bond can be found and hence a C–H distance can be estimated but it is shorter than the true value:

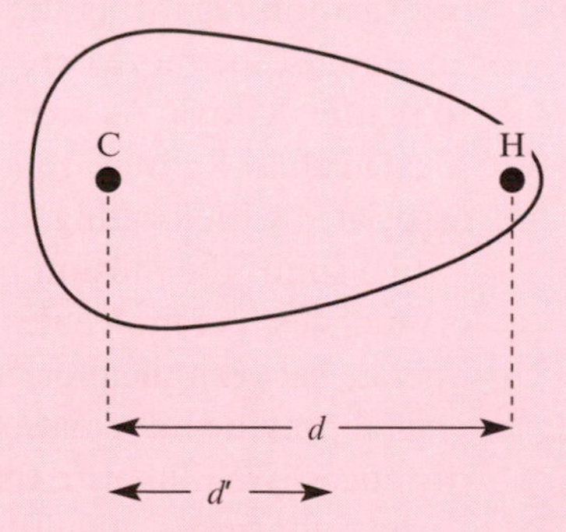

The experimental distance is d' whilst the true internuclear distance is d.

- Neutrons are diffracted by the atomic nuclei, and therefore the positions of nuclei can be found accurately from the results of a neutron diffraction study.

Keep in mind that our goal is to measure internuclear distances. Neutron diffraction would appear to be the method of choice for all solid state studies. However, neutron diffraction is a very expensive experimental method and it is not as readily available as an X-ray diffraction facility. Hence, as a routine method for solid samples, X-ray diffraction usually wins out over neutron diffraction.

Electron diffraction

The gas phase differs from the solid state in that the molecules are continually in motion and there are usually no persistent intermolecular interactions. As a result, diffraction experiments on gases mainly provide information regarding intramolecular distances, and the data collected in an electron diffraction experiment can be analysed in terms of *intra*molecular parameters. The initial diffraction data relate the scattering angle of the electron beam to intensity, and from these results a plot of intensity against distance can be obtained. Peaks in this radial distribution curve correspond to internuclear separations.

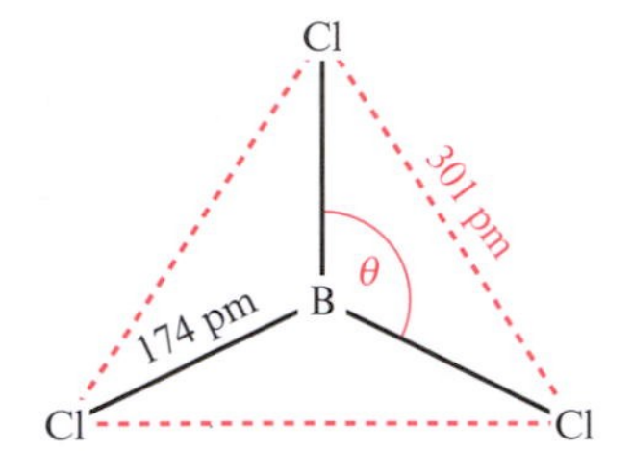

3.3 Intramolecular bonded and non-bonded distances determined from an electron diffraction study of BCl_3; the angle θ is calculated from the distance data.

Consider the electron diffraction results from the determination of the structure of boron trichloride in the gas phase. They provide the bonded and non-bonded distances shown in Figure 3.3; all the B–Cl bonds are of equal length and all the intramolecular non-bonded Cl----Cl distances are equal. The angle θ is determined from the B–Cl and Cl–Cl distances by use of trigonometry: all the Cl–B–Cl angles are equal and $\theta = 120°$. The molecule is planar.

X-ray diffraction

The diffraction of X-rays by the *core* electrons provides the most useful information regarding the positions of the nuclei in a molecule. The heavier an atom is, the more core electrons it possesses, and the greater is its ability to scatter X-rays. As a result, heavier atoms are more easily and accurately located than lighter atoms. We mentioned above the particular difficulty associated with locating hydrogen atoms.

In a solid, the motion of the molecules is restricted. When a compound crystallizes, the molecules form an ordered array and *intermolecular forces* operate between the molecules. The results of an X-ray diffraction study of a crystal provide information about intermolecular as well as intramolecular distances. We illustrate this idea by looking at the solid state structures of various allotropes of sulfur.

Allotropes of sulfur: see Section 7.8

The results of an X-ray diffraction study on crystalline orthorhombic sulfur (α-sulfur, the standard state of sulfur) show that the crystal lattice contains S_8 rings (Figure 3.2).

The S–S bond distances in each ring are equal (206 pm). Table 3.2 lists some allotropes of sulfur and the internuclear S–S distances between pairs of bonded atoms. The reason that all these S–S distances are close in value is that each represents an S–S single covalent bond length, and for a particular

Table 3.2 Internuclear S–S distances between pairs of adjacent atoms in molecules present in some allotropes of sulfur.

Allotrope	*Type of molecule*	*S–S distance / pm*
S_6	Ring	206
S_8 (α-form)	Ring	206
S_8 (β-form)	Ring	205
S_{10}	Ring	206
S_{11}	Ring	206
S_{12}	Ring	205
S_{18}	Ring	206
S_{20}	Ring	205
S_∞	Helical chain	207

homonuclear single bond, the distance is relatively constant. The shortest distances between sulfur atoms in *adjacent* rings in the solid state (intermolecular distances) are larger than the intramolecular S–S bond lengths; 337 pm for S_{11} and 323 pm for S_{10}. These indicate non-bonded interactions.

3.3 THE COVALENT RADIUS OF AN ATOM

Consider a gaseous homonuclear diatomic molecule such as Cl_2. There is a single bond between the two chlorine nuclei and the bond length is 199 pm. It would be convenient to have an *atomic* parameter which we could use to describe the size of an atom when it participates in covalent bonding. We define the *single bond covalent radius,* r_{cov}, of an atom X as half of the internuclear distance in a typical homonuclear X–X single bond. In the case of chlorine, the single bond covalent radius is 99 pm.

The covalent radius, r_{cov}, for an atom X is taken as half of the internuclear distance in a homonuclear X–X bond.

The radius is defined for a particular type of bond. Values appropriate for single, double and triple bonds differ.

As X–X single bonds in molecules are not all exactly the same length, we often use data from a range of compounds to obtain an average r_{cov} value. An estimate for the covalent radius of sulfur can be obtained by taking half of the average S–S bond length obtained for various allotropes of sulfur; the average S–S distance from the data in Table 3.2 is 206 pm which gives a single covalent bond radius for a sulfur atom of 103 pm.

In Table 3.3, values of single bond covalent radii of hydrogen and elements in the *p*-block are compared with those of the van der Waals radii. Since the van der Waals radii are measured from non-bonded distances and

p-Block elements: see Chapter 13

Table 3.3 Van der Waals and covalent radii for hydrogen and atoms in the *p*-block. These values represent the 'best' values from a wide range of compounds containing the elements.

	Element	*Van der Waals radius / pm*	*Covalent radius / pm*
	H	120	37[a]
Group 13	B	208	88
	Al	–	130
	Ga	–	122
	In	–	150
	Tl	–	155
Group 14	C	185	77
	Si	210	118
	Ge	–	122
	Sn	–	140
	Pb	–	154
Group 15	N	154	75
	P	190	110
	As	200	122
	Sb	220	143
	Bi	240	152
Group 16	O	140	73
	S	185	103
	Se	200	117
	Te	220	135
Group 17	F	135	71
	Cl	180	99
	Br	195	114
	I	215	133

[a] Sometimes it is more appropriate to use a value of 30 pm in organic compounds.

the covalent radii are determined from bonded interactions, the former are larger than the latter for a given element. Down any group of the periodic table, the values for both sets of radii generally increase (Figure 3.4).

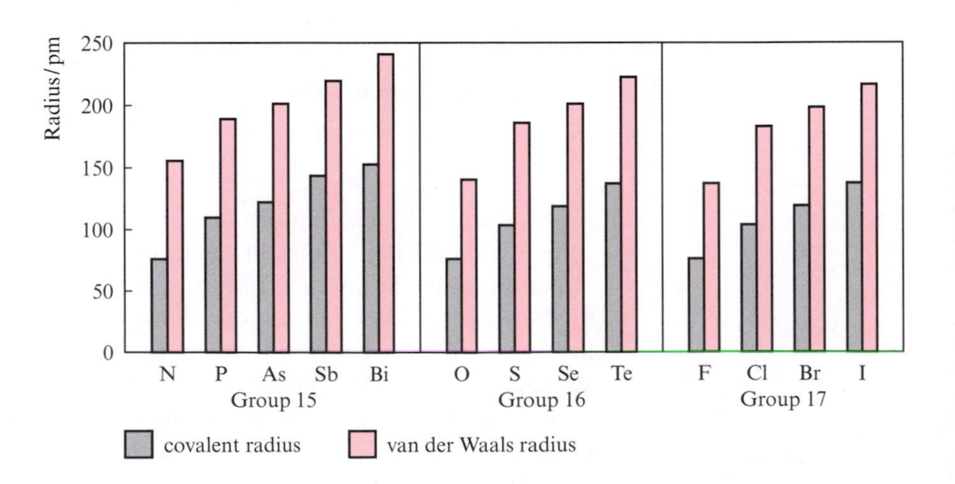

3.4 Trends in covalent and van der Waals radii upon descending groups 15, 16 and 17.

3.4 AN INTRODUCTION TO BOND ENERGY: THE FORMATION OF THE DIATOMIC MOLECULE H_2

Let us move now from the lengths of bonds to their energies. Just as we compared covalent (bonded) distances with van der Waals (non-bonded) ones, we begin our discussion of energy by comparing the energy potential associated with bringing together two hydrogen atoms with that of bringing two atoms of argon together. For argon, we considered two opposing factors – the attractive potential due to the induced dipole–dipole interactions and a repulsive term due to interatomic electron–electron interactions. Can we use a similar approach to describe the interaction between two hydrogen atoms?

Energy potential for two argon atoms: see Section 2.21

There is a fundamental difference between the approach of two argon atoms and two hydrogen atoms. Whatever the distance between the argon atoms, the strongest interatomic *attractive* forces are of the induced dipole–dipole type. There is no net covalent bonding resulting from the sharing of electrons between the two argon nuclei. In contrast, as two hydrogen atoms approach each other, electron sharing becomes possible and a covalent H–H bond can be formed.

We can rationalize this by considering the number of electrons in the valence shell of each atom. Each argon atom possesses the noble gas configuration $[Ne]3s^2 3p^6$. There is no driving force for a change of electron configuration.

In contrast, each hydrogen atom possesses a $1s^1$ configuration. Just as there is a particular stability associated with a noble gas ns^2np^6 configuration (an octet), the $1s^2$ configuration of helium is also stable. There are two ways in which a hydrogen atom could attain the [He] configuration – it could gain an electron and become the H^- anion with a $1s^2$ ground state configuration, or it could *share* an electron with another atom. The simplest atom with which it could share electrons is another hydrogen atom and, in this way, the molecule H_2 is formed (Figure 3.5).

Octet rule: see Section 2.20

Box 3.3 The ground state configuration of helium

The [He] configuration fits into one of the general patterns we observe in the periodic table. Each group 1 metal readily loses an electron to give a cation possessing an inert gas configuration:

$$\underset{[Ar]4s^1}{K} \rightarrow \underset{[Ar]}{K^+} + e^-$$

$$\underset{[Ne]3s^1}{Na} \rightarrow \underset{[Ne]}{Na^+} + e^-$$

$$\underset{[He]2s^1}{Li} \rightarrow \underset{[He]}{Li^+} + e^-$$

Helium does not obey the octet rule *per se*, but it does possess a fully occupied principal quantum level. The $n = 1$ principal shell is unique in being full when it contains only two electrons.

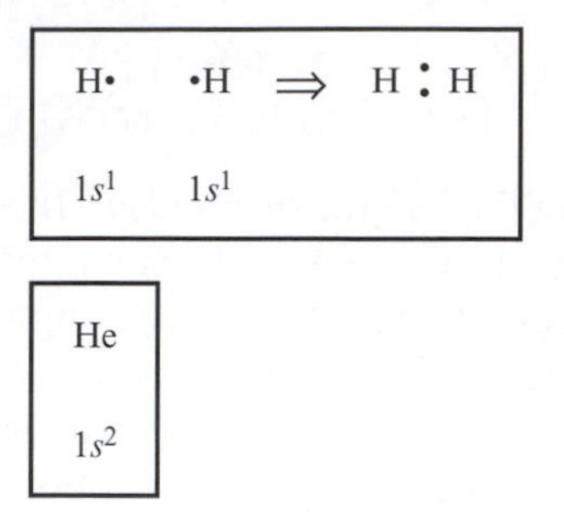

3.5 A hydrogen atom has a valence electronic configuration of $1s^1$ and shares an electron with another hydrogen atom to form H_2. In doing so, each H atom becomes like an atom of the noble gas helium which has a valence configuration of $1s^2$.

As the two hydrogen atoms approach one another, electron sharing becomes more favourable, and we start to form a covalent bond. Figure 3.6 shows an energy potential as a function of internuclear distance. Although this curve is superficially similar to the one in Figure 2.24, the energy terms of which it is composed are rather different and we can characterize four main components:

1. repulsion between electrons
2. attraction between an electron and a proton in the *same* atom
3. attraction between an electron and a proton in the *other* atom; and
4. repulsion between the protons.

These interactions are represented diagrammatically in Figure 3.7.

Strictly speaking, terms (2) and (3) are only meaningful at large internuclear distances because, as the atoms come close together and covalent bond formation occurs, it is not reasonable to associate a particular electron with a particular proton.

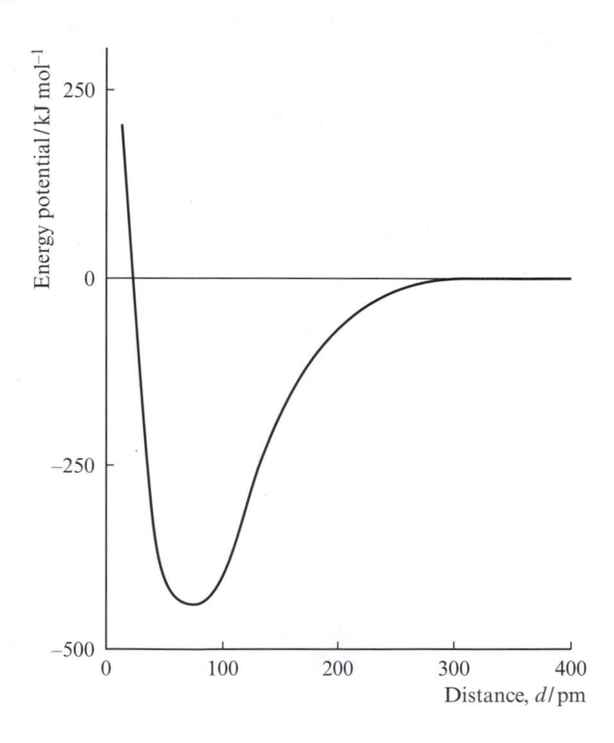

3.6 The potential energy curve that describes the approach of two hydrogen atoms to their equilibrium separation.

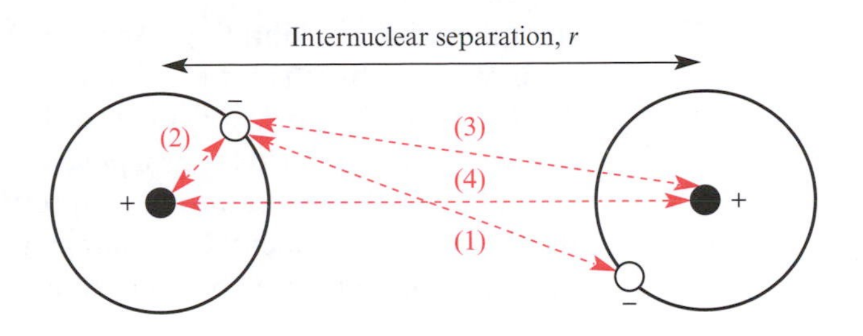

3.7 The approach of two hydrogen atoms. Each contains one proton and one electron. Interactions will be (1) electron–electron (repulsive), (2) electron–proton within the same atom (attractive), (3) electron–proton between different atoms (attractive) and (4) proton-proton (repulsive).

What does the curve in Figure 3.6 tell us? Experimentally we find that the standard state of the element hydrogen is H_2 molecules (dihydrogen) rather than H atoms. The energy minimum should correspond to the equilibrium distance of the two hydrogen nuclei in H_2. The equilibrium distance is the position of lowest (most negative) energy for the system, and this internuclear distance corresponds to the H–H covalent bond distance and is found to be 74 pm.[§]

3.5 BOND DISSOCIATION ENERGIES AND ENTHALPIES

In the preceding section we introduced the H–H covalent bond in the H_2 molecule by considering the equilibrium distance and energy of two approaching H atoms. We could draw a curve similar to Figure 3.6 for the formation of any other homonuclear X–X (or heteronuclear X–Y) covalent bond. So far, we have concentrated upon the distance axis – how far apart are the atoms? But we also need to know something about the energy axis – how much energy is involved in the formation of the H–H covalent bond or how much energy does it take to break (*dissociate*) the H–H bond?

The problem of internal energy and enthalpy

The value of the energy corresponding to the minimum in Figure 3.6 is 458 kJ mol^{-1}. Unfortunately, there is a problem – the value of 458 kJ mol^{-1} does *not* correspond exactly to the amount of energy released when the two hydrogen atoms form an H_2 molecule or, conversely, to the amount of energy required to break the H–H bond in H_2.

The curve in Figure 3.6 corresponds to the *internal energy* of the system (ΔU) measured at a temperature of 0 K. However, it is often more convenient to consider enthalpies (ΔH) at 298 K. In practice, the conversion of ΔU to ΔH involves only ≈3 kJ mol^{-1}. (The terms ΔU and ΔH will be discussed in greater detail in Chapter 12 – at the moment, all that you need to note is that *the difference between ΔU and ΔH for a given system is very small.*) The real problem arises from the fact that even at 0 K, the two hydrogen nuclei are not

§ Actually, because of the asymmetry of the curve, the experimentally measured bond length in H_2 does not coincide exactly with the distance corresponding to the curve minimum.

stationary, but are vibrating. We discuss vibrational states in Chapter 9, but for now, note that the lowest vibrational state of the H_2 molecule lies about 26 kJ mol^{-1} higher in energy than the bottom of the curve in Figure 3.6. This residual energy is called the *zero point energy*. Any experimental measurement will relate to the real molecule and to the lowest energy state that this can reach. Thus, experimental measurements do not 'reach' the very bottom of the potential energy well but only access the lowest energy vibrational state.

ΔU and ΔH: see Section 12.2

ΔU and ΔH are related by the equation:

$\Delta U = \Delta H - P\Delta V$ at constant pressure

where P = pressure and V = volume.

Homonuclear bond dissociation in a diatomic molecule

The most convenient parameter to describe the energy associated with the H–H bond in H_2 is the energy needed to convert an H_2 molecule in its lowest vibrational energy level to two hydrogen atoms (equation 3.1). Table 3.4 lists possible internal energy (ΔU) or enthalpy (ΔH) terms that may be used to describe this process.

$$H_2(g) \rightarrow 2H(g) \tag{3.1}$$

Since the values of ΔH and ΔU are so close, here and for other bond dissociation processes, we will not consistently distinguish between them. It is more convenient to deal with enthalpy values and, thus, for H_2, we associate a value of 436 kJ mol^{-1} with the cleavage of the H–H bond.

Table 3.4 Internal energy changes (ΔU) and the enthalpy change (ΔH) used to describe the dissociation of gaseous H_2. The relationship between ΔU and ΔH is $\Delta U = \Delta H - P\Delta V$ where P = pressure and V = volume.

Quantity	*Process described*	*Value / kJ mol^{-1}*
ΔU (0 K)	$H_2(g) \rightarrow 2H(g)$ at 0 K	432
ΔU (298 K)	$H_2(g) \rightarrow 2H(g)$ at 298 K	433
ΔH (298 K)	$H_2(g) \rightarrow 2H(g)$ at 298 K	436

The bond dissociation energy (ΔU) and the bond dissociation enthalpy (ΔH) for an X–X diatomic molecule refer to the process:

$$X_2(g) \rightarrow 2\,X(g)$$

at a given temperature (often 0 K or 298 K). The value is defined for a gas phase process.

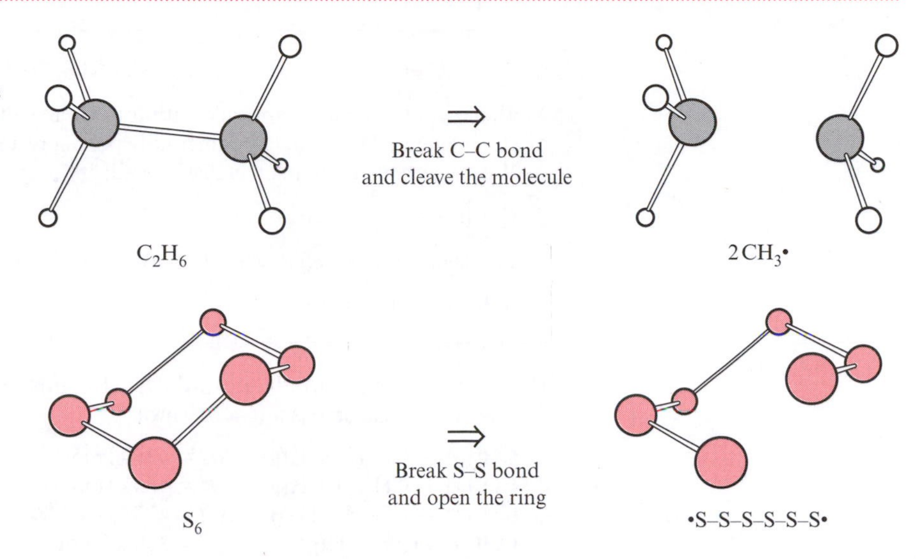

3.8 Bond cleavage in a molecule such as ethane will produce two separate radicals but in a cyclic molecule such as S_6 will cause ring opening.

Homonuclear bond dissociation in a polyatomic molecule

In polyatomic molecules, the dissociation bond energy describes the energy needed to break one particular bond. Equations 3.2–3.4 illustrate the processes for the cleavage of the homonuclear bond in each of the molecules shown in Figure 3.1. Figure 3.8 shows this more explicitly for C–C bond cleavage in C_2H_6.

$$C_2H_6(g) \rightarrow 2\,CH_3^{\bullet}(g) \quad (3.2)$$

$$N_2H_4(g) \rightarrow 2\,NH_2^{\bullet}(g) \quad (3.3)$$

$$H_2O_2(g) \rightarrow 2\,HO^{\bullet}(g) \quad (3.4)$$

Figure 3.8 also shows the result of homonuclear bond cleavage in a cyclic molecule. When one S–S bond is broken in an S_6 ring, the ring opens to form a chain and each terminal sulfur atom has an unpaired electron. Further S–S bond cleavage can occur to give S_x fragments. The enthalpy associated with each S–S bond breakage will be characteristic of that particular process. This introduces the idea that bond dissociation enthalpies for chemically equivalent bonds may vary and this is discussed further in Box 3.4. It is not always convenient to use such individual enthalpy values but rather average bond enthalpies.

$D = \Delta_{diss}H$

We distinguish between a bond dissociation enthalpy which refers to a particular process and a value which is an average of several such enthalpies by using the symbols D and $\bar{D}$ respectively. Thus, for ethane, the bond dissociation enthalpy, D, for the cleavage of the C–C single bond (equation 3.2) is 376 kJ mol^{-1}. However, after considering a wide range of values of D for the cleavage of single C–C bonds in different molecules or in different environments in the same molecule, we arrive at an average value, $\bar{D}$, of about 346 kJ mol^{-1}. Values of $\bar{D}$ for a given bond vary slightly between different tables of

Box 3.4 Average and individual bond dissociation enthalpies

Methane is a tetrahedral molecule with four chemically equivalent C–H bonds of equal length. This is an experimentally proven result.

When methane is completely dissociated:

$$CH_4(g) \rightarrow C(g) + 4H(g)$$

we can determine an *average* value for $\bar{D}$(C–H) from the enthalpy of atomization:

$$\Delta_a H^o = 1664\,\text{kJ mol}^{-1}$$
$$\bar{D}(\text{C–H}) = \tfrac{1}{4} \times 1664 = 416\,\text{kJ mol}^{-1}$$

However, if we *sequentially* break the C–H bonds, then the enthalpy change for each individual step is not $416\,\text{kJ mol}^{-1}$:

$$CH_4(g) \rightarrow CH_3(g) + H(g) \quad \Delta H = 435\,\text{kJ mol}^{-1}$$
$$CH_3(g) \rightarrow CH_2(g) + H(g) \quad \Delta H = 460\,\text{kJ mol}^{-1}$$
$$CH_2(g) \rightarrow CH(g) + H(g) \quad \Delta H = 427\,\text{kJ mol}^{-1}$$
$$CH(g) \rightarrow C(g) + H(g) \quad \Delta H = 338\,\text{kJ mol}^{-1}$$

The conclusion is that, although the bonds in CH_4 are chemically equivalent and have equal strengths, their strength will not be the same as the strength of the three C–H bonds in CH_3, the two C–H bonds in CH_2, or the C–H bond in CH.

data, depending upon the number of data averaged and the exact compounds used to obtain them, and separate values must be determined for single, double or triple (C≡C) bonds. For carbon–carbon bonds $\bar{D}$(C–C) = $346\,\text{kJ mol}^{-1}$, $\bar{D}$(C=C) = $598\,\text{kJ mol}^{-1}$ and $\bar{D}$(C≡C) = $813\,\text{kJ mol}^{-1}$.

A bond dissociation enthalpy, $\bar{D}$, for a bond X–X represents an average value for the enthalpy of dissociation.

Separate values must be determined for single (X–X), double (X=X), or triple (X≡X) bonds.

3.6 THE STANDARD ENTHALPY OF ATOMIZATION OF AN ELEMENT

The standard enthalpy of atomization

Consider the bond dissociation process given in equation 3.1. The dissociation of the H–H bond corresponds to the formation of H atoms and this energy is also known as the enthalpy of atomization. The *standard* enthalpy of atomization of an element ($\Delta_a H^o$) is the enthalpy change when one mole of gaseous atoms is formed from the element in its standard state. For H_2 this is defined for the process given in equation 3.5 whilst for mercury and nickel it is defined according to equations 3.6 and 3.7 respectively.

Standard state: see Section 1.17

$$\tfrac{1}{2}H_2(g) \rightarrow H(g) \quad \Delta_a H^o = 218\,kJ\,mol^{-1} \tag{3.5}$$

$$Hg(l) \rightarrow Hg(g) \quad \Delta_a H^o = 61\,kJ\,mol^{-1} \tag{3.6}$$

$$Ni(s) \rightarrow Ni(g) \quad \Delta_a H^o = 430\,kJ\,mol^{-1} \tag{3.7}$$

Whilst some elements are solid under standard state conditions, others are liquid or gas. If the standard state of the element is a solid, then the value of the standard enthalpy of atomization (defined at 298 K) includes contributions from:

The standard enthalpy of atomization, $\Delta_a H^o$, of an element is the enthalpy change (at 298 K) when one mole of gaseous atoms is formed from the element in its standard state. The SI units are kJ per mole of gaseous atoms formed.

- the enthalpy needed to raise the temperature of the element from 298 K to the melting point;
- the enthalpy of fusion of the element (i.e. the transition from solid to liquid);
- the enthalpy needed to raise the temperature of the element from the melting point to the boiling point (T K);
- the enthalpy of vaporization of the element (i.e. the transition from liquid to vapour);
- any enthalpy change associated with the bond breaking of polyatomic gas phase species; and
- the enthalpy change on going from the gas phase monatomic element at T K to 298 K.

Table 3.5 Selected enthalpies of atomization of the elements ($\Delta_a H^o$ (298 K)).

Element	*Standard state of element*	$\Delta_a H^o / kJ\,mol^{-1}$
As	As(s)	302
Br	Br_2(l)	112
C	C(graphite)	717
Cl	Cl_2(g)	121
Cu	Cu(s)	338
F	F_2(g)	79
H	H_2(g)	218
Hg	Hg(l)	61
I	I_2(s)	107
K	K(s)	90
N	N_2(g)	473
Na	Na(s)	108
O	O_2(g)	249
P	P_4(white)	315
S	S_8(rhombic)	277
Sn	Sn(white)	302

Appendix 10 gives a full list of values of $\Delta_a H^o$

Selected values of standard enthalpies of atomization of the elements are listed in Table 3.5.

Worked example 3.1 ***Contributions to the heat of atomization***

Determine the heat of atomization of mercury by using Hess's Law of constant heat summation. Information about mercury that is required:

bp = 655 K
heat capacity (C_p) for liquid mercury = 27.4 J mol^{-1} K^{-1}
heat capacity (C_p) for gaseous mercury = 20.8 J mol^{-1} K^{-1}
enthalpy of vaporization = 59.1 kJ mol^{-1}.

The standard enthalpy of atomization is defined for the formation of one mole of gaseous atoms from mercury in its standard state – a liquid.

The molar heat capacity (C_P) tells you how much heat energy is needed to raise the temperature of one mole of a substance by one kelvin at constant pressure (see Section 12.3).

Set up a cycle which accounts for all the steps in taking mercury from a liquid at 298 K (standard conditions) to gaseous atoms at 298 K:

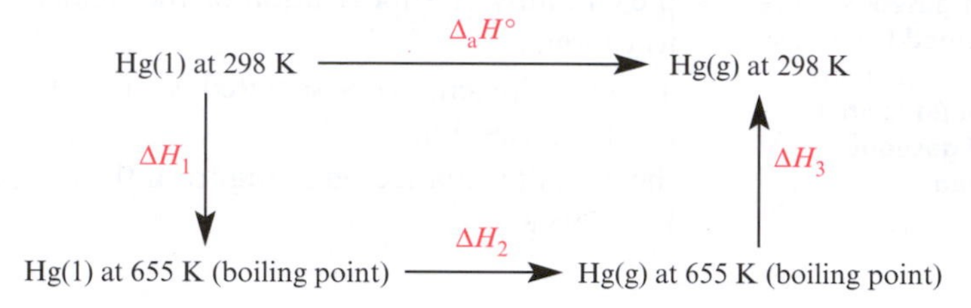

The enthalpy contributions are:

ΔH_1 = enthalpy required to raise the temperature of 1 mole Hg(l) from 298 to 655 K
= (Heat capacity for liquid) × (change in temperature)
= 27.4 × (655 – 298)
= 9782 J mol^{-1}
= 9.8 kJ mol^{-1}

ΔH_2 = enthalpy of vaporization
= 59.1 kJ mol^{-1}

ΔH_3 = enthalpy released when the temperature of 1 mole Hg(g) is lowered from 655 to 298 K
= – [(Heat capacity for gas) × (change in temperature)]
= – [20.8 × (655 – 298)]
= – 7426 J mol^{-1}
= – 7.4 kJ mol^{-1}

By Hess's Law:

$$\Delta_a H^\circ = \Delta H_1 + \Delta H_2 + \Delta H_3$$
$$= 9.8 + 59.1 - 7.4$$
$$= 61.5 \text{ kJ mol}^{-1}$$

Enthalpy of atomization and bond dissociation enthalpy for a gaseous diatomic molecule

The relationship between the enthalpy of atomization and the bond dissociation enthalpy for a diatomic molecule in *the gas phase* is important.

The definition of the bond dissociation enthalpy is in terms of breaking a bond. Thus D is defined per mole of bonds broken (equation 3.8). For dihydrogen, $D = 436\,\text{kJ mol}^{-1}$. On the other hand, the enthalpy of atomization is defined per mole of gaseous atoms formed (equation 3.9). This difference in definition is a common cause of error in calculations.

$$H_2(g) \rightarrow 2H(g) \quad D = 436\,\text{kJ mol}^{-1} \text{ (per mole of bonds broken)} \tag{3.8}$$

$$\tfrac{1}{2}H_2(g) \rightarrow H(g) \quad \Delta_a H^o = 218\,\text{kJ mol}^{-1} \text{ (per mole of gaseous atoms formed)} \tag{3.9}$$

For a gaseous diatomic molecule:

Bond dissociation enthalpy = 2 × (standard enthalpy of atomization)

$$D = 2 \times \Delta_a H^o$$

3.7 DETERMINING BOND ENTHALPIES FROM STANDARD HEATS OF FORMATION

Vibrational spectroscopy for diatomics: see Section 9.6

Hess's Law: see Section 1.17

The bond dissociation enthalpy of the bond in a diatomic molecule can be measured thermochemically or by using spectroscopic methods. However, for larger molecules, direct measurements of bond dissociation enthalpies are not always possible.

A method that provides values for the dissociation enthalpies of bonds of a given type uses Hess's Law of constant heat summation. Thermochemical data that *can* be measured experimentally are standard enthalpies of combustion of elements and molecules and from these, standard enthalpies of formation ($\Delta_f H^o$) can be obtained. These data may be used to determine bond dissociation enthalpies. It is important to keep in mind that values of D obtained this way are *derived values* – they depend upon other bond dissociation enthalpies as shown below.

Suppose that we wish to determine the bond enthalpy term of an N–N single bond. This bond cannot be studied in isolation since N_2 possesses an N≡N triple bond. So, we turn to a simple compound containing an N–N bond, N_2H_4 (Figure 3.1). This molecule is composed of one N–N and four N–H bonds. If we were to dissociate the molecule completely into gaseous atoms, then all of these bonds would be broken and the enthalpy change would be the sum of the bond enthalpy terms as shown in equation 3.10. Note here the use of D(N–N), but $\bar{D}$(N–H). This emphasizes the fact that we are dealing with one N–N bond but take an average value for the N–H bond enthalpy.

$$N_2H_4(g) \rightarrow 2N(g) + 4H(g) \quad \Delta_a H^o = D(N–N) + 4\bar{D}(N–H) \qquad (3.10)$$

However, in trying to determine D(N–N), we have introduced another unknown quantity, $\bar{D}$(N–H). The ammonia molecule is only composed of N–H bonds and the complete dissociation of NH_3 provides an *average* value of the N–H bond enthalpy (equation 3.11). The value of $\Delta_a H^o$ for the process shown in equation 3.11 can be determined from the standard enthalpy of formation of ammonia (equation 3.12) by using Hess's Law (equation 3.13). The standard of enthalpies of atomization for nitrogen and hydrogen are 473 and 218 kJ mol^{-1} respectively.

$$NH_3(g) \rightarrow N(g) + 3H(g) \qquad \Delta_a H^o = 3\bar{D}(N–H) \qquad (3.11)$$

$$\tfrac{1}{2}N_2(g) + \tfrac{3}{2}H_2(g) \rightarrow NH_3(g) \quad \Delta_f H^o = -46\,kJ\,mol^{-1} \qquad (3.12)$$

$$\begin{array}{ccc} NH_3(g) & \xrightarrow{3\,\bar{D}(N–H)} & N(g) + 3H(g) \\ \Delta_f H^o \nwarrow & & \nearrow \Delta_a H^o(N) + 3\Delta_a H^o(H) \\ & \tfrac{1}{2}N_2(g) + \tfrac{3}{2}H_2(g) & \end{array} \qquad (3.13)$$

From equation 3.13:

$$\begin{aligned} 3\,\bar{D}(N–H) &= \Delta_a H^o(N) + 3\Delta_a H^o(H) - \Delta_f H^o(NH_3, g) \\ 3\,\bar{D}(N–H) &= 473 + (3 \times 218) - (-46) \\ &= 1173 \\ \bar{D}(N–H) &= 391\,kJ\,mol^{-1} \end{aligned}$$

This result can now be used to find D(N–N) by setting up a Hess cycle based on equation 3.10:

$$\begin{array}{ccc} N_2H_4(g) & \xrightarrow{D(N–N) + 4\,\bar{D}(N–H)} & 2N(g) + 4H(g) \\ \Delta_f H^o \nwarrow & & \nearrow 2\Delta_a H^o(N) + 4\Delta_a H^o(H) \\ & N_2(g) + 2H_2(g) & \end{array} \qquad (3.14)$$

The standard enthalpy of formation of gaseous hydrazine is 95 kJ mol^{-1}.
From equation 3.14:

$$\begin{aligned} D(N–N) + 4\,\bar{D}(N–H) &= 2\Delta_a H^o(N) + 4\Delta_a H^o(H) - \Delta_f H^o(N_2H_4, g) \\ D(N–N) + 4\,\bar{D}(N–H) &= (2 \times 473) + (4 \times 218) - 95 \\ &= 1723 \\ D(N–N) &= 1723 - 4\,\bar{D}(N–H) \end{aligned}$$

and substituting in the value we calculated above for $\bar{D}$(N–H) gives:

$$\bar{D}(N–H) = 1723 - (4 \times 391) = 159\,kJ\,mol^{-1}$$

Table 3.6 Some typical values of bond dissociation enthalpies for single covalent homonuclear bonds. Some values (e.g. for the H–H bond) are measured directly from the dissociation of a diatomic molecule; some values (e.g. for the C–C bond) are average values for a range of C–C single bonds in molecules and are thus $\bar{D}$ values; some values (e.g. for the N–N bond) depend on the transferability of other bond enthalpy contributions.

Bond X–X	*D(X–X) / kJ mol^{-1}*	*Bond X–X*	*D(X–X) / kJ mol^{-1}*
H–H	436	O–O	146
C–C	346	S–S	266
Si–Si	196	Se–Se	193
Ge–Ge	163	F–F	159
Sn–Sn	152	Cl–Cl	242
N–N	159	Br–Br	193
P–P	200	I–I	151
As–As	177		

We have made one major assumption in this series of calculations – that the average N–H bond enthalpy term can be *transferred* from ammonia to hydrazine. Ammonia and hydrazine are closely related molecules and the assumption is reasonable. However, the transfer of bond enthalpy contributions between molecules of different types should be treated with caution. Naturally, $\bar{D}$ values for single bonds cannot be used for multiple bonds. Such bonds have very different bond dissociation enthalpies.

Some typical values for the bond dissociation enthalpies of homonuclear single covalent bonds are listed in Table 3.6. Some, such as *D*(H–H), *D*(F–F) and D(Cl–Cl), are directly measured from the dissociation of the appropriate gaseous diatomic molecule. Others can be measured from the dissociation of a small molecule, e.g. $\bar{D}$(S–S) from the atomization of gaseous S_8 (Figure 3.2). Values such as that for the N–N bond rely upon the method of bond enthalpy transferability described above.

3.8 THE NATURE OF THE COVALENT BOND IN H_2

In the previous sections, we have considered two experimental observables by which a bond such as that in H_2 can be described: the internuclear distance (which gives a measure of the covalent radius of a hydrogen atom) and the bond dissociation enthalpy. The experimental evidence is that two hydrogen atoms come together to form a bond in which the internuclear separation is 74 pm and the bond dissociation enthalpy is 436 kJ mol^{-1}.

We have also seen that the driving force for two H atoms combining to give a molecule of H_2 is the tendency of each atom to become 'helium-like'. Each hydrogen atom shares one electron with another hydrogen atom.

The covalent bond in dihydrogen can be represented in a Lewis structure, but is more fully described using other methods: the valence bond (VB) and molecular orbital (MO) approaches are the best known. The following sections deal with these methods.

3.9 LEWIS STRUCTURE OF H_2

Structures **3.1** and **3.2** show two representations of the H_2 molecule and are referred to as *Lewis structures*. In structure **3.1**, the electrons are represented by dots§ but the diagram can be simplified by drawing a line to represent the bond between the two atoms as in **3.2**.

H : H (3.1)   H — H (3.2)

Although these Lewis structures show the connection of one hydrogen atom to the other, they do not provide information about the exact character of the bonding pair of electrons, nor about the region of space that they occupy.

Box 3.5 Lewis structures

In 1916, G. N. Lewis presented a simple, but informative, method of describing the arrangement of valence electrons in molecules. The method uses dots (or dots and crosses) to represent the number of valence electrons associated with the nuclei. The nuclei are designated by the symbol of the element.

If possible, all electrons in a molecule should appear in pairs. Single electrons signify a radical species.

Example: Cl_2

The ground state electronic configuration of Cl is $[Ne]3s^23p^5$.

Thus, Cl has seven valence electrons.

In forming the Cl_2 molecule, a Cl atom shares one electron with another Cl atom to form an argon-like octet.

The Lewis structure for Cl_2 is:

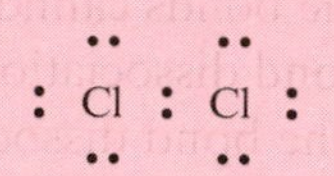

:C̤̈l : C̤̈l :

and this can be simplified to show that Cl–Cl covalent bond and the *lone pairs* of electrons:

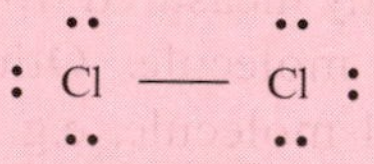

:C̤̈l — C̤̈l :

Lewis structures are extremely useful for giving connectivity patterns – that is, the bonding arrangement between atoms in a molecule.

Some other examples of Lewis structures are:

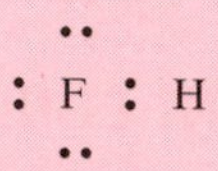

HF, 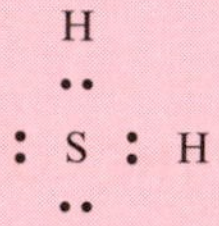 H_2S, 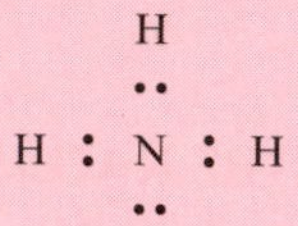 NH_3, 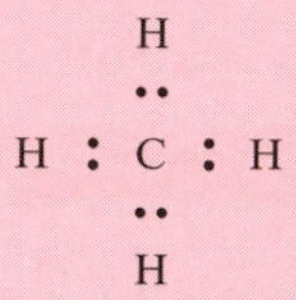 CH_4, 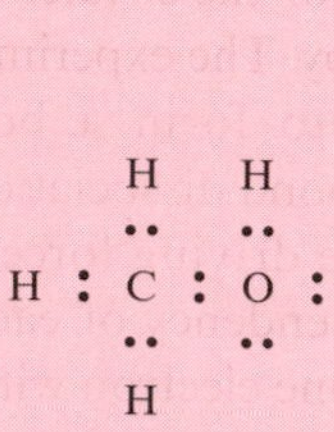 CH_3OH, 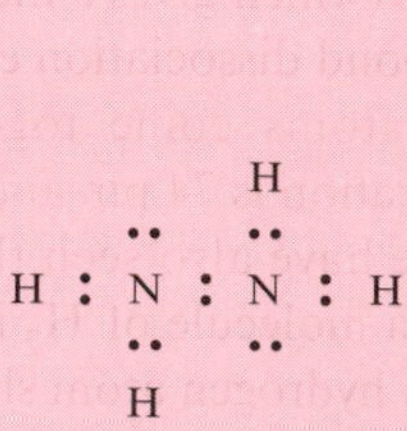 N_2H_4, 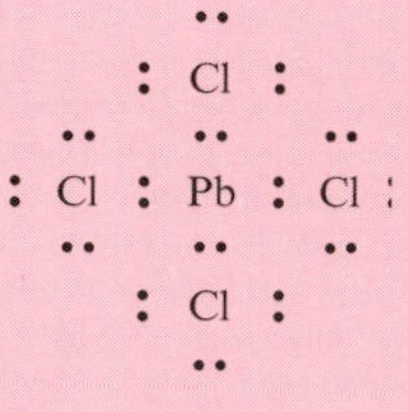 $PbCl_4$

§ Dots and crosses are sometimes used to distinguish between electrons from the two adjacent atoms. Remember, however, that *in practice* electrons are *indistinguishable* from one another.

3.10 THE PROBLEM OF DESCRIBING ELECTRONS IN MOLECULES

The problem

In Section 2.7, we discussed the way in which the Schrödinger equation described the dynamic behaviour of electrons in atoms, and we pointed out that it was only possible to obtain exact mathematical solutions to the Schrödinger equation in the case of hydrogen-like species. If we are to use the Schrödinger equation to describe the behaviour of *electrons in molecules*, we have the same problem – we cannot obtain exact solutions to the wave-equations when we have more than one nucleus and more than one electron. Certainly, we can simplify the problem by assuming that the movement of the nuclei is minimal, but this does not overcome the basic problem.

Our aim is to obtain molecular wavefunctions ($\psi_{molecule}$) that describe the behaviour of electrons in molecules. What can we do?

Two methods of approaching the problem

Two main methods are used to make approximations to molecular wavefunctions – *valence bond* and *molecular orbital* theories. These have different starting assumptions but, if successful, they should give similar results. To introduce these methods we need to get a little ahead of ourselves and consider some of the problems that exist with molecules containing more than two atoms. Water makes a good example.

(3.3) (3.4)

Structures **3.3** and **3.4** give Lewis representations of a water molecule in terms of two O–H bonds and two lone pairs of electrons. The basis of a Lewis structure is the identification of a bond as a pair of electrons shared between two atoms. This is the starting point for the bonding model called *valence bond theory*.

Valence bond theory: see Section 3.11

The valence bond (VB) model starts from the chemically familiar viewpoint that a molecule is composed of a number of discrete bonds; to all intents and purposes, the various bonds exist in isolation from one another and each may be described in terms of interactions between two electrons and two nuclei. This is the so-called *two-centre two-electron* bonding model and the bonding interactions are said to be *localized*. The VB approach attempts to write a separate wavefunction for each of the discrete two-electron interactions, and then derives a total molecular wavefunction by combining them.

Molecular orbital theory: see Section 3.12

The molecular orbital (MO) approach does *not* start with the assumption that electrons are localized into two-centre two-electron bonds. In MO theory, molecular wavefunctions are constructed which encompass any or all of the atomic nuclei in the molecule. An interaction which spreads over more than two nuclei is described as being *delocalized* or *multi-centred*.

A bonding interaction between two nuclei is *localized*. If it spreads over more than two nuclei, the interaction is *delocalized*.

The MO method is mathematically complex and is not as intuitive as VB theory. In the water molecule, it is far more convenient to think in terms of localized O–H bonds rather than multi-centre H---O---H interactions. However, there are many compounds for which the MO approach provides a satisfying description of the bonding. In many respects, this is similar to the situation with bond enthalpies. As chemists, we know that the four C–H bonds in methane are equivalent, but as we showed in Box 3.4, if the bonds are broken sequentially, we obtain a different bond enthalpy value for each step.

Both the VB and MO methods are approximations for obtaining the molecular wavefunction, $\psi_{molecule}$. Both methods must approach the same 'true' ψ_{molecule} and should, ultimately, give equivalent wavefunctions.

Valence bond (VB) theory assigns electrons to two-centre bonds or to atomic based orbitals.

Molecular orbital (MO) theory allows electrons to be delocalized over the entire molecule.

3.11 VALENCE BOND (VB) THEORY

General overview of valence bond theory

In valence bond theory, a description of the bonding within a diatomic molecule is determined by considering the perturbation that the two atoms have on one another as they penetrate one another's regions of space. In practice this means that we attempt to write an approximation to the 'real' wavefunction for the molecule in terms of a combination of wavefunctions describing individual two-electron interactions. The details of the mathematics are not relevant to our discussion.§

The consideration of *electrons* from the start is an important characteristic of VB models.

The bonding in H_2 – an initial approach

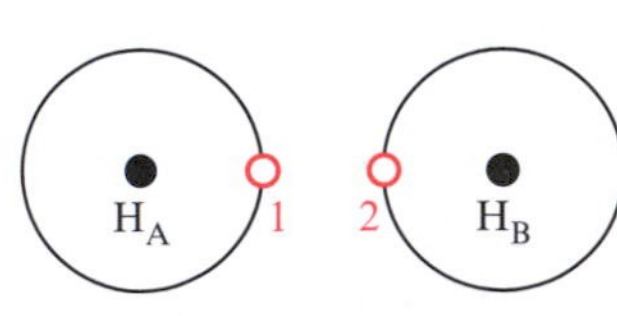

3.9 Labelling scheme used for the nuclei and electrons in a valence bond treatment of the bonding in H_2.

Consider the formation of an H_2 molecule. Each hydrogen atom consists of a proton and one electron. In Figure 3.9 the two nuclei are labelled H_A and H_B and the two electrons are labelled 1 and 2. When the two H atoms are well separated from one another, electron 1 will be wholly associated with H_A, and electron 2 will be fully associated with H_B. Let this situation be described by the wavefunction ψ_1.

Although we have given each electron a different label, they are actually indistinguishable from one another. Therefore, when the atoms are close

§ For more in-depth discussion, see: R. McWeeny, *Coulson's Valence*, 3rd edn, Oxford University Press, Oxford (1979).

together, we cannot tell which electron is associated with which nucleus – electron 2 could be with H_A and electron 1 with H_B. Let this situation be described by the wavefunction ψ_2.

An overall description of the system containing the two hydrogen atoms ($\psi_{covalent}$) can be written in the form of equation 3.15.

$$\psi_{covalent} = \psi_1 + \psi_2 \qquad (3.15)$$

The wavefunction $\psi_{covalent}$ defines an energy. When this is calculated as a function of the internuclear separation H_A–H_B, it provides an energy minimum of 303 kJ mol^{-1} at an internuclear distance of 87 pm for H_2. Using equation 3.15, therefore, does *not* give an answer in very good agreement with either the experimental internuclear H–H distance (74 pm) or the bond dissociation energy (436 kJ mol^{-1}) and some refinement of the method is clearly needed.

The bonding in H_2 – a refinement of the initial picture

Although we have so far allowed each of the nuclei H_A and H_B to be associated with *either* of the electrons 1 and 2, we should also allow for a situation in which one or other of the nuclei might be associated with *both* electrons.

There are four situations that can arise as H_A and H_B come close together:

- (nucleus H_A with electron 1) and (nucleus H_B with electron 2),
- (nucleus H_A with electron 2) and (nucleus H_B with electron 1),
- (nucleus H_A with both electrons 1 and 2) and (nucleus H_B with no electrons)
- (nucleus H_A with no electrons) and (nucleus H_B with both electrons 1 and 2).

Whilst the first two situations retain neutrality at each atomic centre, the last two describe the *transfer* of an electron from one nucleus to the other. This has the effect of producing two ions, H^+ and H^-. If H_A has no electrons, H_A becomes $H_A{}^+$, and H_B becomes $H_B{}^-$. If H_A has both electrons, H_A becomes $H_A{}^-$, and H_B becomes $H_B{}^+$. There is an equal chance of forming $[H_A{}^-H_B{}^+]$ or $[H_A{}^+H_B{}^-]$ since the two hydrogen atoms are identical.

The effect of allowing for contributions from $[H_A{}^-H_B{}^+]$ or $[H_A{}^+H_B{}^-]$ is to build into the valence bond model the possibility that the wavefunction that describes the bonding region between the two hydrogen atoms may have an *ionic contribution* in addition to the *covalent contribution* that we have already described. This is represented in equation 3.16.

$$\psi_{molecule} = \psi_{covalent} + [c \times \psi_{ionic}] \qquad (3.16)$$

The coefficient c indicates the relative contribution made by the wavefunction ψ_{ionic} to the overall wavefunction $\psi_{molecule}$.

With regard to the H_2 molecule, equation 3.16 means that we should write two possible structures to represent the covalent and ionic contributions to the bonding description. Structures **3.5a** and **3.5b** are called *resonance structures* and a double headed arrow is used to represent the resonance between them.

$$\mathrm{H{-}H} \longleftrightarrow \mathrm{H^+ \quad H^-}$$

(3.5a) (3.5b)

More about resonance structures: see Section 5.17

It is very important to realize that the purely covalent and the purely ionic forms of the H_2 molecule *do not exist separately*. We are merely trying to describe the H_2 molecule by the wavefunction $\psi_{molecule}$ in terms of contributions from the two extreme bonding models represented by $\psi_{covalent}$ and ψ_{ionic}.

By trying different values of c in equation 3.16, an estimate of the internuclear distance which corresponds to the energy minimum of the system can be obtained which comes close to the experimental bond distance. If the wavefunctions are normalized, a value of $c \approx 0.25$ gives an H–H internuclear distance of 75 pm, in close agreement with the experimentally determined value. The predicted bond dissociation energy is 398 kJ mol^{-1}. Further mathematical refinement can greatly improve upon this value at the expense of the simple physical picture.

Normalization: see Section 2.8

3.12 MOLECULAR ORBITAL (MO) THEORY

General overview of molecular orbital theory

Molecular orbital theory differs from the valence bond method in that it does not start from an assumption of localized two-centre bonding. In the MO approach, we attempt to obtain wavefunctions for the entire molecule. The aim is to calculate regions in space that an electron might occupy which encompass the molecule as a whole – these are *molecular orbitals*. In extreme cases, this procedure permits the electrons to be delocalized over the molecule, while in others, the results approximate to localized two-centre two-electron orbitals.

The usual starting point for the MO approach to the bonding in a molecule is a consideration of the atomic orbitals that are available. An approximation which is commonly used within MO theory is known as the *linear combination of atomic orbitals* (LCAO). In this method, wavefunctions (ψ_{MO}) approximately describing the molecular orbitals are constructed from the atomic wavefunctions (ψ_{AO}) of the constituent atoms. Remember that the atomic wavefunctions are themselves usually approximations obtained from the hydrogen atom.

Interactions between any two atomic orbitals will be:

- allowed if the symmetries of the atomic orbitals are compatible with one another,
- efficient if the region of overlap of the two atomic orbitals is significant, and
- efficient if the atomic orbitals are relatively close in energy.

The full meaning of the first point will become clearer later in this chapter. With reference to the second point, we should spend a moment considering the meaning of the term 'overlap'. This word is used extensively in MO theory, and the overlap between two orbitals is often expressed in terms of the *overlap integral*, S. Figure 3.10 illustrates two $1s$ atomic orbitals. The orbitals in Figure 3.10a are close but there is no common region of space –

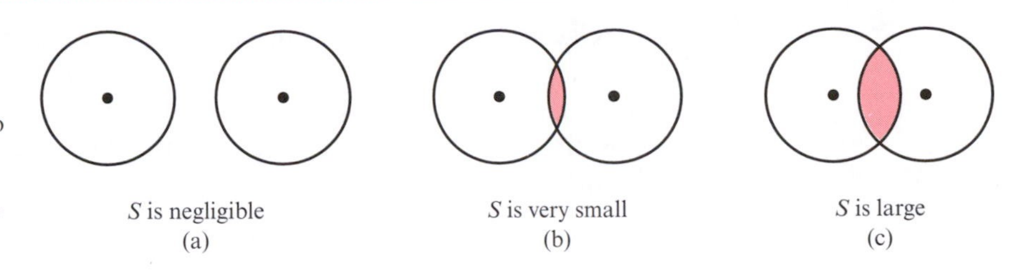

3.10 Schematic drawing to illustrate the meaning of orbital overlap and the overlap integral S: (a) the two $1s$ orbitals effectively do not overlap; (b) there is only a very little overlap; (c) the two $1s$ atomic orbitals overlap efficiently. The two atomic nuclei are represented by the central dots.

the probability of finding an electron in one orbital very rarely coincides with the probability of finding an electron in the other orbital. The overlap integral is effectively zero between these two orbitals.§ In Figure 3.10b, there is a very small common region of space between the two orbitals. The value of S is small, and this situation is not satisfactory for the formation of an effective bonding molecular orbital. In Figure 3.10c, the overlap region is significant and the overlap integral will have a value $0 < S < 1$; it cannot equal unity. (Keep in mind that the nuclei cannot come too close together because they repel one another.) The situation sketched in Figure 3.10c represents good orbital overlap.

In order to understand the significance of the third point above, look back at Figure 2.10 and think about what would happen if we combined a $1s$ with a $4s$ atomic orbital (AO), as opposed to a $1s$ with another $1s$ AO. The $1s$ and $4s$ orbitals have different energies and different spatial extents. Remember that we need to maximize orbital overlap in order to obtain significant interaction between two atomic orbitals.

The interactions between atomic orbitals may be described by linear combinations of the atomic wavefunctions. Figure 3.11 summarizes what happens when two transverse waves combine either in- or out-of-phase. An in-phase combination leads to constructive interference and, if the initial waves are identical, the amplitude of the wave doubles. An out-of-phase combination causes destructive interference and, if the initial waves are identical, the wave is destroyed.

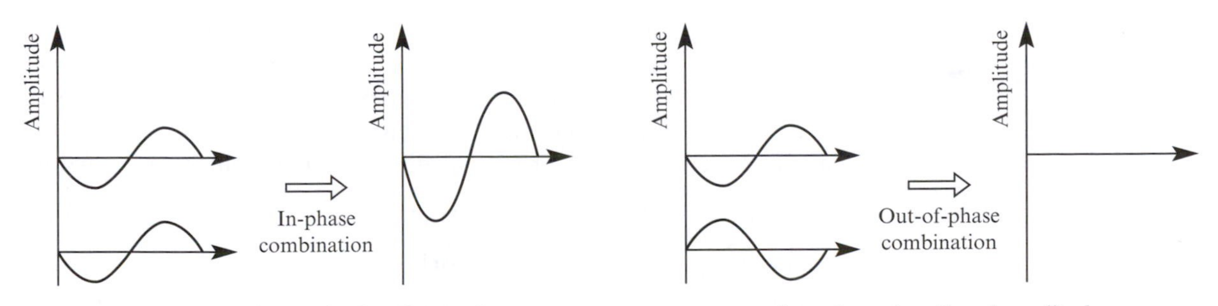

3.11 Constructive and destructive interference between two transverse waves of equal wavelength and amplitude.

§ The overlap integral S cannot be precisely zero; the surface boundaries of the orbitals do not represent 100% probability of finding the electron as discussed in Chapter 2.

Now let us move from transverse waves to atomic orbitals – in a similar manner, 'electron waves' (orbitals) may be in- or out-of-phase. Equations 3.17 and 3.18 represent the in- and out-of-phase linear combinations of two wavefunctions, ψ_1 and ψ_2, which describe two atomic orbitals. Two molecular orbitals result.

$$\psi_{\text{MO(in-phase)}} = N \times [\psi_1 + \psi_2] \quad (3.17)$$

$$\psi_{\text{MO(out-of-phase)}} = N \times [\psi_1 - \psi_2] \quad (3.18)$$

The coefficient N is the normalization factor; this adjusts the equation so that the probability of finding the electron somewhere in space is unity.

A general result that should be remembered is that *the number of molecular orbitals generated will always be equal to the initial number of atomic orbitals* – this becomes important when we construct molecular orbital diagrams in the following sections.

In the construction of a molecular orbital diagram:

$$\left(\begin{array}{l}\textbf{The number of molecular}\\ \textbf{orbitals (MOs) generated}\end{array}\right) = \left(\begin{array}{l}\textbf{The number of atomic}\\ \textbf{orbitals (AOs) used}\end{array}\right)$$

Constructing the molecular orbitals for H_2

Let us now look at the bonding in the H_2 molecule using the MO method.

Each hydrogen atom possesses one $1s$ atomic orbital. The MO description of the bonding in the H_2 molecule is based on allowing the two $1s$ orbitals to overlap when the two hydrogen nuclei are within bonding distance. The two $1s$ atomic orbitals are allowed to overlap because they have the same symmetries. We shall return to the question of symmetry when we consider the bonding in other diatomic molecules.

Let the wavefunction associated with one H atom be $\psi(1s)_A$ and the wavefunction associated with the second H atom be $\psi(1s)_B$. Using equations 3.17 and 3.18, we can write the new wavefunctions that result from the linear combinations of $\psi(1s)_A$ and $\psi(1s)_B$ and these are given in equations 3.19 and 3.20. Since the H_2 molecule is a homonuclear diatomic, each of the atomic wavefunctions will contribute *equally* to each molecular wavefunction.

$$\psi_{\text{MO(in-phase)}} = N \times [\psi(1s)_A + \psi(1s)_B] \quad (3.19)$$

$$\psi_{\text{MO(out-of-phase)}} = N \times [\psi(1s)_A - \psi(1s)_B] \quad (3.20)$$

Each of the new molecular wavefunctions, $\psi_{\text{MO(in-phase)}}$ and $\psi_{\text{MO(out-of-phase)}}$, has an energy associated with it which is dependent on the distance apart of the two hydrogen nuclei. As Figure 3.12 shows, the energy of $\psi_{\text{MO(in-phase)}}$ is always lower than that of $\psi_{\text{MO(out-of-phase)}}$. Further, if we look at the varia-

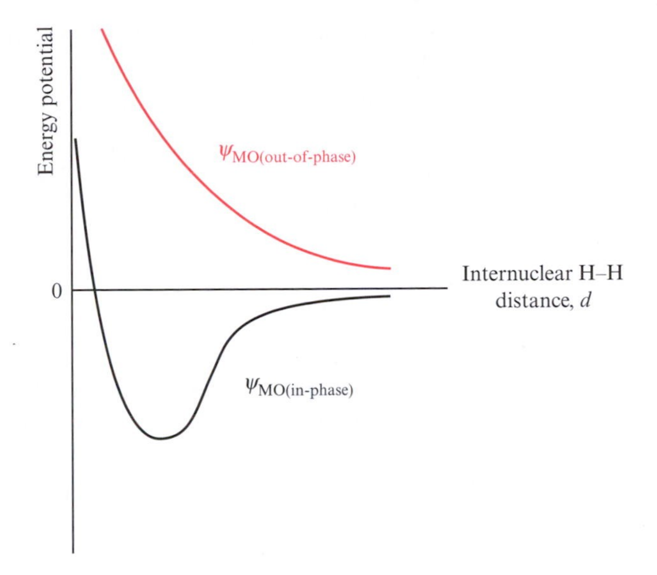

3.12 The variation in the energy of the molecular wavefunctions $\psi_{MO(in-phase)}$ and $\psi_{MO(out-of-phase)}$ for the combination of two hydrogen $1s$ atomic orbitals as a function of H...H separation.

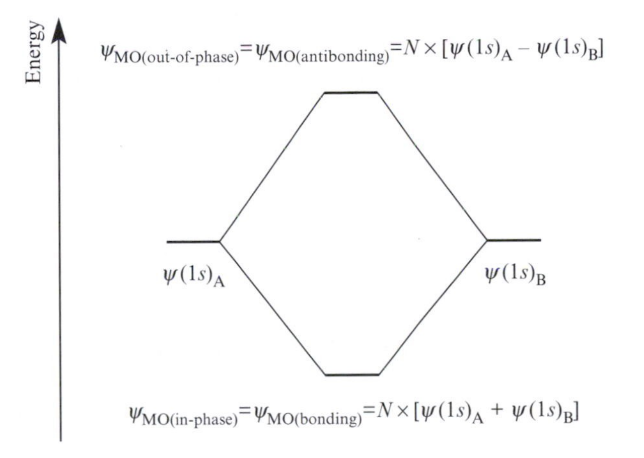

3.13 In-phase and out-of-phase combinations of two $1s$ atomic orbitals lead to two molecular orbitals at low (stabilized) and high (destabilized) energies respectively.

tion of the energy of $\psi_{MO(in-phase)}$ as a function of internuclear distance, we see that the energy curve has a minimum value corresponding to maximum stability. In contrast, the curve describing the variation in the energy of $\psi_{MO(out-of-phase)}$ never reaches a minimum, and represents an increasingly less stable situation as the two nuclei come closer together.

The wavefunction $\psi_{MO(in-phase)}$ corresponds to a *bonding molecular orbital* and $\psi_{MO(out-of-phase)}$ describes an *antibonding molecular orbital*. The combination of $\psi(1s)_A$ and $\psi(1s)_B$ to give $\psi_{MO(bonding)}$ and $\psi_{MO(antibonding)}$ is represented schematically in Figure 3.13 in which $\psi_{MO(bonding)}$ is stabilized with respect to the two atomic orbitals and $\psi_{MO(antibonding)}$ is destabilized.§

§ The amount of energy by which the antibonding MO is destabilized with respect to the atomic orbitals is slightly greater than the amount of energy by which the bonding MO is stabilized. This phenomenon is important but its discussion is beyond the scope of this book.

For the combination of two 1*s* atomic orbitals, the normalization factor is $^{1}/_{\sqrt{2}}$ and so we can now rewrite equations 3.19 and 3.20 in their final forms of equations 3.21 and 3.22.

$$\psi_{\text{MO(bonding)}} = \frac{1}{\sqrt{2}} \times [\psi(1s)_A + \psi(1s)_B] \quad (3.21)$$

$$\psi_{\text{MO(antibonding)}} = \frac{1}{\sqrt{2}} \times [\psi(1s)_A - \psi(1s)_B] \quad (3.22)$$

Putting the electrons in the H_2 molecule

So far in this MO approach, we have not mentioned the electrons! In simple MO theory, the molecular orbitals are constructed and once their energies are known, the *aufbau* principle is used to place the electrons in them.

Aufbau principle: see Section 2.18

The H_2 molecule contains two electrons, and by the *aufbau* principle, these occupy the lowest energy molecular orbital and have opposite spins. The MO energy level diagram for H_2 is shown in Figure 3.14. Significantly, the two electrons occupy the *bonding MO*, whilst the antibonding MO remains empty.

What do the molecular orbitals for H_2 look like?

The bonding and antibonding molecular orbitals of H_2 are drawn schematically in Figure 3.15. The combinations of the two 1*s* orbitals are most simply represented by two overlapping circles, either in- or out-of-phase. This is shown on the left-hand side of Figures 3.15a and 3.15b and is a representation that we shall use elsewhere in this book.

On the right-hand side of Figures 3.15a and 3.15b we show the molecular orbitals in more detail. The diagram representing the bonding orbital for the H_2 molecule illustrates that the overlap between the two atomic orbitals gives a region of space in which the two bonding electrons may be found. This

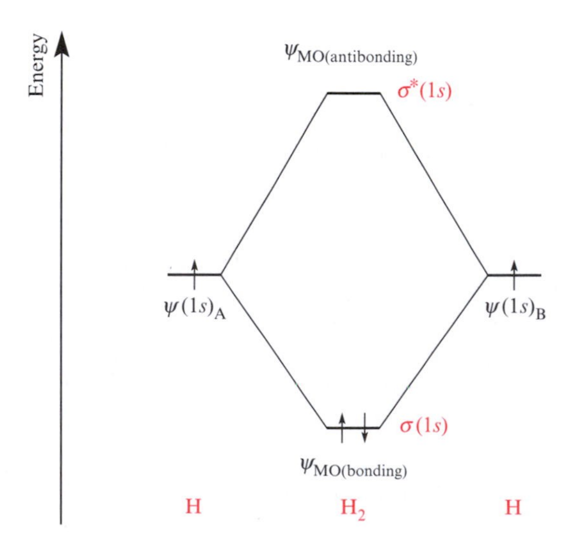

3.14 The bonding and antibonding molecular orbitals in H_2 formed by the linear combination of two 1*s* atomic orbitals. By the *aufbau* principle, the two electrons in H_2 occupy the bonding MO. The labels '$\sigma(1s)$' and '$\sigma^*(1s)$' are explained at the end of sections 3.12 and 3.15.

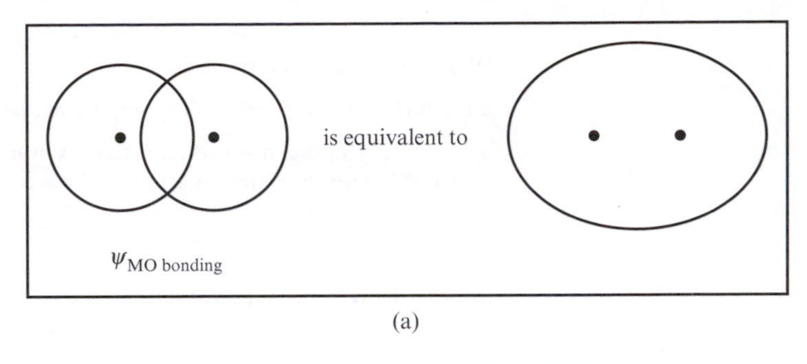

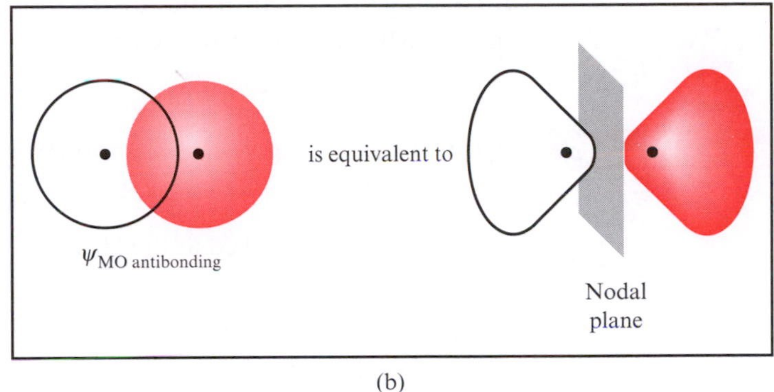

3.15 Schematic representations of (a) the bonding and (b) the antibonding molecular orbitals in the H_2 molecule. The H nuclei are represented by black dots.

corresponds to the H–H bond seen in the VB model. The negatively charged electrons will be found predominantly between the two nuclei, thus reducing internuclear repulsion.

The antibonding orbital for the H_2 molecule is shown in Figure 3.15b. The most important feature is the nodal plane passing *between the two H nuclei*. This means that, were this orbital to be occupied, the probability of finding the electrons at any point in this nodal plane would be zero. The outcome is that the probability of finding the electrons between the nuclei in the antibonding MO is less than between two non-interacting hydrogen atoms, and much less than in the bonding orbital. Accordingly, there is an increase in internuclear repulsion.

Labelling the molecular orbitals in H_2

Some new notation has appeared in Figure 3.14 – the bonding MO is labelled $\sigma(1s)$ and the antibonding MO is designated $\sigma^*(1s)$. These labels are used to provide information about the symmetry of the orbital.

A molecular orbital has σ-symmetry if it is symmetrical with respect to a line joining the two nuclei – there is no phase change when the orbital is rotated about this internuclear axis.

A σ^*-molecular orbital must meet *two* requirements:

More about σ and σ^* labels: see Section 3.15

- In order to keep the σ label, the MO must satisfy the requirement that if it is rotated about the internuclear axis, there is no phase change.
- In order to take the * label, there must be a nodal plane *between* the nuclei.

With respect to a pair of nuclei:

a σ orbital *('sigma-orbital')* is symmetrical about the internuclear axis,

a σ^* orbital *('sigma-star orbital')* is also symmetrical about the internuclear axis, but in addition there is a nodal plane between the nuclei.

Using this new notation, we can write the ground state electronic configuration of the H_2 molecule as $\sigma(1s)^2$. This shows that we have formed a σ-bonding molecular orbital by the overlap of two 1s atomic orbitals, and that the MO contains two electrons.

3.13 WHAT DO THE VB AND MO THEORIES TELL US ABOUT THE MOLECULAR PROPERTIES OF H_2?

Now that we have looked at the H_2 molecule in several ways, we should consider whether the different bonding models paint the same picture.

- The *Lewis structure* shows that the H_2 molecule contains a single covalent H–H bond and the valence electrons are paired.
- *Valence bond theory* shows that the bond in the H_2 molecule can be described by a wavefunction which has both covalent and ionic contributions. The dominant contribution comes from the covalent resonance form. The two electrons in H_2 are paired.
- *Molecular orbital theory* models H_2 on a predominantly covalent basis. There is a localized H–H bonding MO and the two electrons are paired.

Thus, we find that all the models give us similar insights into the bonding in H_2.

3.14 HOMONUCLEAR DIATOMIC MOLECULES OF THE FIRST ROW ELEMENTS – THE *s*-BLOCK

In this and later sections in this chapter we construct diagrams like Figure 3.14 for diatomic molecules X_2 where X is an element in the first row of the periodic table. For each molecule, we use the molecular orbital approach to predict some of the properties of X_2 and compare these with what is known experimentally. We begin by considering the bonding in the diatomic molecules Li_2 and Be_2. The former exists as a gas phase molecule, whilst the latter is not known. Can we rationalize the instability of Be_2?

Li_2

A lithium atom has a ground state electronic configuration of [He]$2s^1$. Figure 3.16 shows the combination of the $2s$ atomic orbitals to give the MOs in Li_2. There is a clear similarity between this and the formation of H_2 (Figure 3.14), although the principal quantum number has changed from 1 to 2.

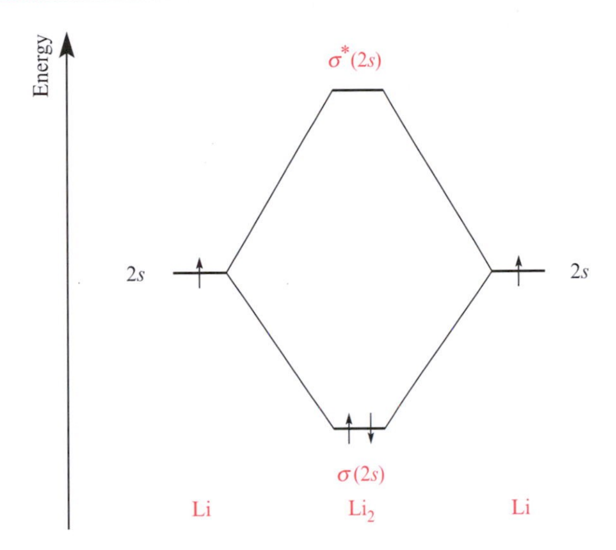

3.16 The bonding and antibonding molecular orbitals in Li_2 formed by the linear combination of two 2s atomic orbitals. The 1s orbitals (with core electrons) have been omitted.

From Figure 3.16, the electronic configuration of Li_2 is $\sigma(2s)^2$. Fully, the electronic configuration is:

$$\sigma(1s)^2\sigma^*(1s)^2\sigma(2s)^2$$

However, of the six electrons in the Li_2 molecule, the four in the $\sigma(1s)$ and $\sigma^*(1s)$ orbitals are core electrons and the two electrons in the $\sigma(1s)$ MO 'cancel' out the two electrons in the $\sigma^*(1s)$ MO. We need only consider the fate of the valence electrons – those occupying the $\sigma(2s)$ orbital.

Thus, MO theory predicts that the Li_2 molecule possesses two electrons which occupy a bonding molecular orbital – the Li_2 molecule has a single covalent bond. This is the same answer as the one we obtained by drawing the Lewis structures **3.6** or **3.7**. In VB theory, this corresponds to a localized Li–Li σ-bond.

Li : Li (3.6)   Li — Li (3.7)

Be_2

The ground state configuration of Be is [He]$2s^2$. Figure 3.17 shows the combination of the 2s atomic orbitals to give the MOs in Be_2. The orbitals themselves are similar to those in Li_2, and we can again ignore the contribution of the core 1s electrons. There are now four electrons to be accommodated in MOs derived from the 2s AOs. From Figure 3.17, the electron configuration of Be_2 is:

$$\sigma(2s)^2\sigma^*(2s)^2$$

Each of the bonding and antibonding MOs contains two electrons. The result is that there would be no net bonding in a Be_2 molecule – this is consistent with its instability.

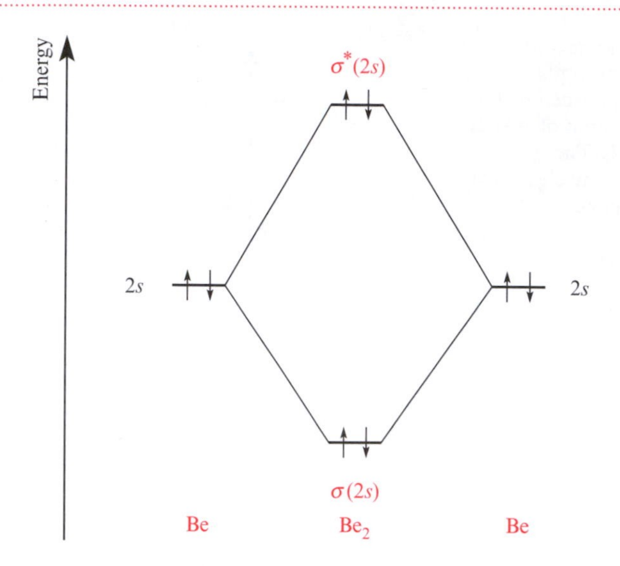

3.17 A molecular orbital diagram for the formation of Be_2 from two Be atoms. The 1*s* orbitals (with core electrons) have been omitted.

Before moving on to the other elements in the first row of the periodic table, we need to introduce two further concepts – the orbital overlap of *p* atomic orbitals, and bond order.

3.15 ORBITAL OVERLAP OF *p* ATOMIC ORBITALS

We saw in section 3.12 that the overlap of two *s* atomic orbitals led to two molecular orbitals of σ- and σ*-symmetries. Each of the *s* atomic orbitals is spherically symmetrical and so the overlap of two such AOs is *independent* of their relative orientations. To convince yourself of this, look at Figure 3.13 – rotate one of the 1*s* orbitals about an axis passing through the nucleus. Does it affect the overlap with the other 1*s* AO? Repeat the exercise but rotate the orbital in another direction.

The situation with *p* atomic orbitals is different. Consider two nuclei placed on the *z* axis§ at their equilibrium (bonding) separation. The $2p_z$ atomic orbitals point directly at one another. They can overlap either in- or out-of-phase as shown in Figures 3.18a and 3.18b respectively. The resultant bonding molecular orbital is represented schematically at the right-hand side of Figure 3.18a; this is a σ orbital. The corresponding antibonding MO has a nodal plane passing between the nuclei (Figure 3.18b) – this plane is *in addition* to the nodal plane present in each 2*p* atomic orbital at the nucleus. The antibonding MO is a σ* orbital.

Recall that a σ orbital is symmetrical with respect to rotation about the internuclear axis

Now consider the overlap of the $2p_x$ AOs (Figures 3.18c and 3.18d). Since we have defined the approach of the nuclei to be along the *z* axis, the two $2p_x$ AOs can overlap only in a sideways manner. This overlap is less efficient than

§ The choice of the *z* axis is an arbitrary one but it is usual to have it running between the nuclei.

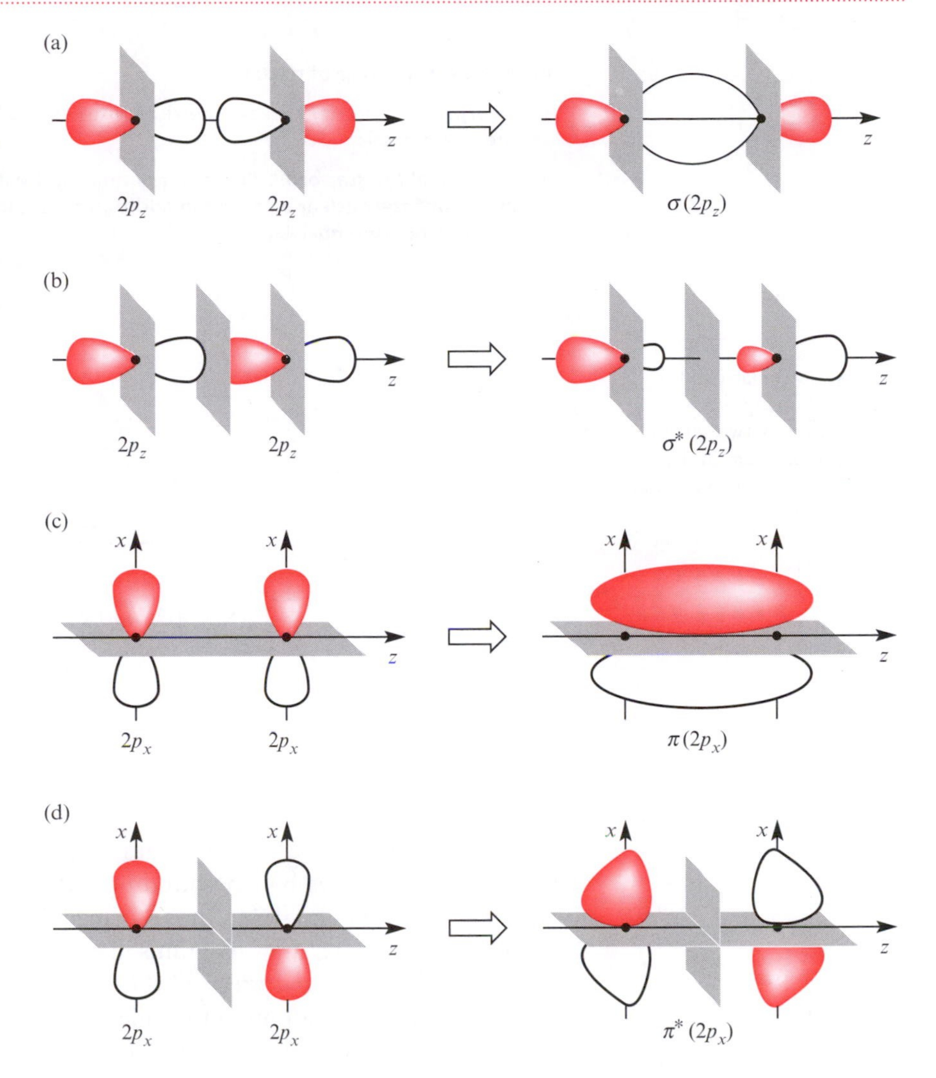

3.18 The overlap of two $2p$ atomic orbital: (a) direct overlap along the z-axis to give a σ-MO (bonding); (b) the formation of the σ^*-MO (antibonding); (c) sideways overlap of two $2p_x$ atomic orbitals to give a π-MO (bonding); (d) the formation of π^*-MO (antibonding). Atomic nuclei are marked in black, and nodal planes in grey.

the direct overlap of the two $2p_z$ AOs (i.e. the overlap integral S is smaller). The result of an in-phase combination of the $2p_x$ AOs is the formation of a bonding MO which retains the nodal plane of each individual AO (Figure 3.18c). This MO is called a π orbital. The π molecular orbital is asymmetrical with respect to rotation about the internuclear axis – if you rotate the orbital about this axis, there is a change of phase.

The result of an out-of-phase combination of two $2p_x$ AOs is the formation of an antibonding MO which retains the nodal plane of each individual AO and has a nodal plane passing between the nuclei (Figure 3.18d). This MO is labelled as a π^*-orbital.

A molecular orbital has π^* symmetry if it meets the following *two* criteria:

- In order to take the π label, there must be a change of phase when the orbital is rotated about the internuclear axis.
- In order to take the * label, there must be a nodal plane *between* the nuclei.

With respect to a pair of nuclei:

- **a π orbital (*'pi orbital'*) is asymmetrical with respect to rotation about the internuclear axis,**
- **a π^* orbital (*'pi star-orbital'*) is also asymmetrical with respect to rotation about the internuclear axis, and in addition possesses a nodal plane between the two nuclei.**

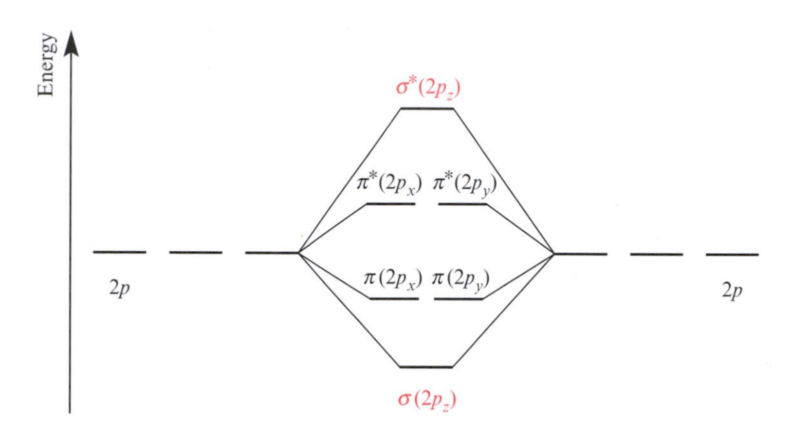

3.19 A molecular orbital diagram showing the combination of two sets of degenerate $2p_x$, $2p_y$ and $2p_z$ atomic orbitals to give σ and π bonding and antibonding MOs. The atomic nuclei lie on the z-axis.

The combination of the two $2p_y$ atomic orbitals is equivalent to the overlap of the two $2p_x$ AOs but the bonding and antibonding MOs that are formed will lie at right angles to those formed from the combination of the $2p_x$ AOs. As an exercise, draw diagrams like those in Figure 3.18 to describe the formation of $\pi(2p_y)$ and $\pi^*(2p_y)$ MOs.

In a homonuclear diatomic molecule, the combination of two sets of degenerate $2p$ AOs (one set per atom) leads to one σ- and two π-bonding MOs and one σ^*- and two π^*-antibonding MOs. (Remember that six AOs must give rise to six MOs.) The only difference between the two π-MOs is their directionality, and consequently, they are energetically degenerate. The same applies to the antibonding MOs. Figure 3.19 shows an energy level diagram that describes the combination of the two sets of degenerate $2p$ AOs; as we shall see, the relative energies of the MOs may vary.

3.16 BOND ORDER

The bond order of a covalent bond gives a measure of the interaction between the nuclei. The three most common categories are the single (bond order = 1), the double (bond order = 2) and the triple (bond order = 3) bond, although bond orders are not restricted to integral values.

Fractional bond orders: see Section 3.22

It is easy to understand what is meant by the single bond in H_2, but we need to be a little more specific when we talk about the bond order of other molecules. The bond order of a covalent bond can be found by using equation 3.23.[§]

$$\text{Bond order} = \tfrac{1}{2} \times \left[\begin{Bmatrix}\text{Number of} \\ \text{bonding electrons}\end{Bmatrix} - \begin{Bmatrix}\text{Number of anti-} \\ \text{bonding electrons}\end{Bmatrix}\right] \quad (3.23)$$

For H_2, Figure 3.14 shows that there are two bonding electrons and no antibonding electrons, giving a bond order in H_2 of 1. Similarly, from Figure 3.16 the bond order of the bond in Li_2 is 1 (equation 3.24). Notice that we only need to consider the valence electrons; you can use equation 3.23 to confirm why in Section 3.14 we said that the core electrons 'cancel out'.

$$\text{Bond order in } Li_2 = \tfrac{1}{2} \times (2 - 0) = 1 \quad (3.24)$$

For Be_2, Figure 3.17 illustrates that the number of bonding electrons is two but there are also two electrons in an antibonding orbital. Therefore, from equation 3.23, the bond order is zero. This corresponds to no net bond between the two beryllium atoms, a result that is consistent with the fact that Be_2 has not been observed experimentally.

The bond order of a covalent bond may be found by considering an MO diagram:

$$\text{Bond order} = \frac{1}{2} \times \left[\begin{Bmatrix}\text{Number of} \\ \text{bonding electrons}\end{Bmatrix} - \begin{Bmatrix}\text{Number of anti-} \\ \text{bonding electrons}\end{Bmatrix}\right]$$

3.17 RELATIONSHIPS BETWEEN BOND ORDER, BOND LENGTH AND BOND ENTHALPY

In Section 3.5, the discussion of bond enthalpies touched on the fact that the bond dissociation enthalpy associated with a carbon–carbon triple bond (C≡C) is greater than that of a carbon–carbon double (C=C) bond, and this is, in turn, greater than that of a carbon-carbon single (C–C) bond. This point is made in Figure 3.20 where we plot bond dissociation enthalpy against bond order, and also correlate the bond order and bond dissociation enthalpy with bond distance. The following trends emerge:

- bond dissociation enthalpy increases as bond order increases;
- bond distance decreases as bond order increases;
- bond dissociation enthalpy decreases as bond distance increases.

The graphs also illustrate that these trends apply to nitrogen–nitrogen and oxygen–oxygen bonds, and in fact the trends are general among the *p*-block elements.

§ Any electrons in non-bonding orbitals are ignored; we return to non-bonding MOs in Section 4.5.

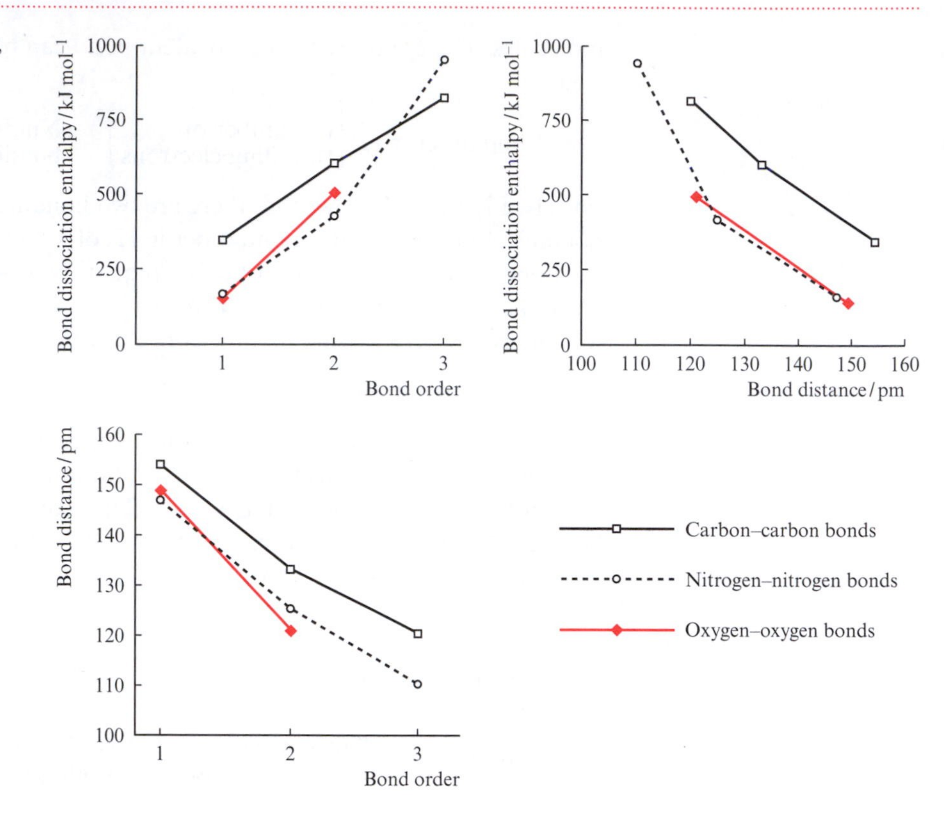

3.20 Trends in bond order, bond length and bond dissociation enthalpies for carbon–carbon, nitrogen–nitrogen and oxygen–oxygen bonds.

3.18 HOMONUCLEAR DIATOMIC MOLECULES OF THE FIRST ROW *p*-BLOCK ELEMENTS: F_2 AND O_2

In this section and Section 3.20 we look at the bonding in the diatomic molecules B_2, C_2, N_2, O_2 and F_2 using the molecular orbital and valence bond approaches, and compare these results with those obtained by drawing a simple Lewis structure for each molecule. The formation of Li_2 and Be_2 involved the overlap of 2*s* atomic orbitals, but that of B_2, C_2, N_2, O_2 and F_2 involves the overlap, not only of 2*s* AOs, but also of 2*p* orbitals.

Throughout the following discussion we shall ignore the core (1*s*) electrons of each atom for reasons already discussed.

F_2

Experimental facts: The standard state of fluorine is the diamagnetic gas F_2.

The ground state electronic configuration of a fluorine atom is [He]$2s^2 2p^5$. Structures **3.8** and **3.9** show two Lewis representations of F_2; each fluorine atom has an octet of valence electrons. A single F–F covalent bond is predicted by this approach. The valence bond method also describes the F_2 molecule in terms of a single F–F bond.

$$:\ddot{\underset{..}{F}}:\ddot{\underset{..}{F}}: \qquad :\ddot{\underset{..}{F}}—\ddot{\underset{..}{F}}:$$

(3.8) (3.9)

We can construct an MO diagram for the formation of F_2 by considering the linear combination of the atomic orbitals of the two fluorine atoms (Figure 3.21). There are 14 valence electrons and, by the *aufbau* principle, these occupy the MOs as shown in Figure 3.21. All the electrons are paired and this is consistent with the observed diamagnetism of F_2.

From Figure 3.21, the electronic configuration of F_2 may be written as:

$$\sigma(2s)^2\sigma^*(2s)^2\sigma(2p_z)^2\pi(2p_x)^2\pi(2p_y)^2\pi^*(2p_x)^2\pi^*(2p_y)^2$$

How is this configuration altered if core electrons are included?

The bond order in F_2 can be determined by using equation 3.23 and the single bond (equation 3.25) corresponds to the value obtained from the Lewis structure.

$$\text{Bond order in } F_2 = \tfrac{1}{2} \times (8 - 6) = 1 \qquad (3.25)$$

Thus, the same conclusions about the bonding in the F_2 molecule are reached by drawing a Lewis structure or by approaching the bonding using VB or MO theories.

A species is *diamagnetic* if all of its electrons are paired.

A species is *paramagnetic* if it contains one or more unpaired electrons.

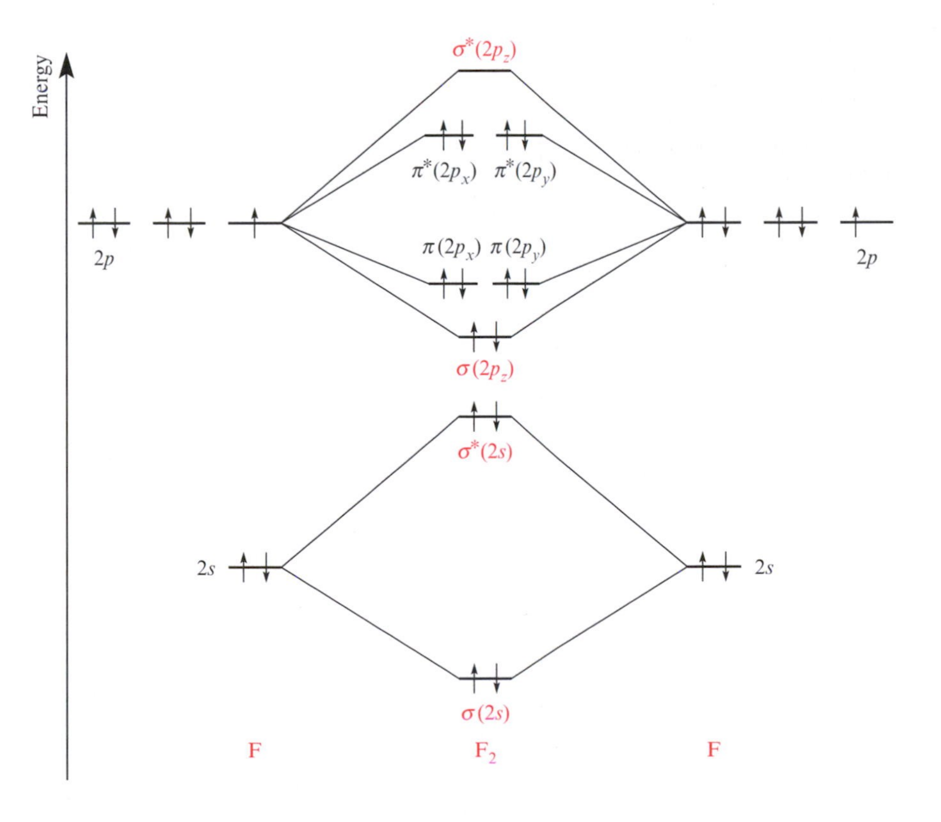

3.21 A molecular orbital diagram to show the formation of F_2. The 1s AOs (with core electrons) have been omitted. The F nuclei lie on the z-axis.

O_2

Experimental facts: Dioxygen is a gas at 298 K and condenses to form a blue liquid at 90 K. The O_2 molecule is paramagnetic – it is a diradical.

The ground state electronic configuration of an oxygen atom is [He]$2s^2 2p^4$. An MO diagram for the formation of O_2 from two oxygen atoms is shown in Figure 3.22. There are 12 valence electrons to be accommodated and, by Hund's rule, the last two electrons must singly occupy each of the degenerate $\pi^*(2p_x)$ and $\pi^*(2p_y)$ molecular orbitals. This picture is consistent with the experimental observation that the O_2 molecule is a diradical – it possesses two unpaired electrons.

From Figure 3.22, the electronic configuration of O_2 may be written as:

$$\sigma(2s)^2\sigma^*(2s)^2\sigma(2p_z)^2\pi(2p_x)^2\pi(2p_y)^2\pi^*(2p_x)^1\pi^*(2p_y)^1$$

$$\text{Bond order in } O_2 = \tfrac{1}{2} \times (8 - 4) = 2 \qquad (3.26)$$

Now let us consider a Lewis structure for O_2. The ground state configuration of each oxygen atom is [He]$2s^2 2p^4$ and there are two unpaired electrons per atom as shown in diagram **3.10**. The Lewis picture of the molecule pairs these electrons up to give an O=O double bond in agreement with the bond order calculated in equation 3.26. Each oxygen atom in Lewis structures **3.11** and **3.12** obeys the octet rule.

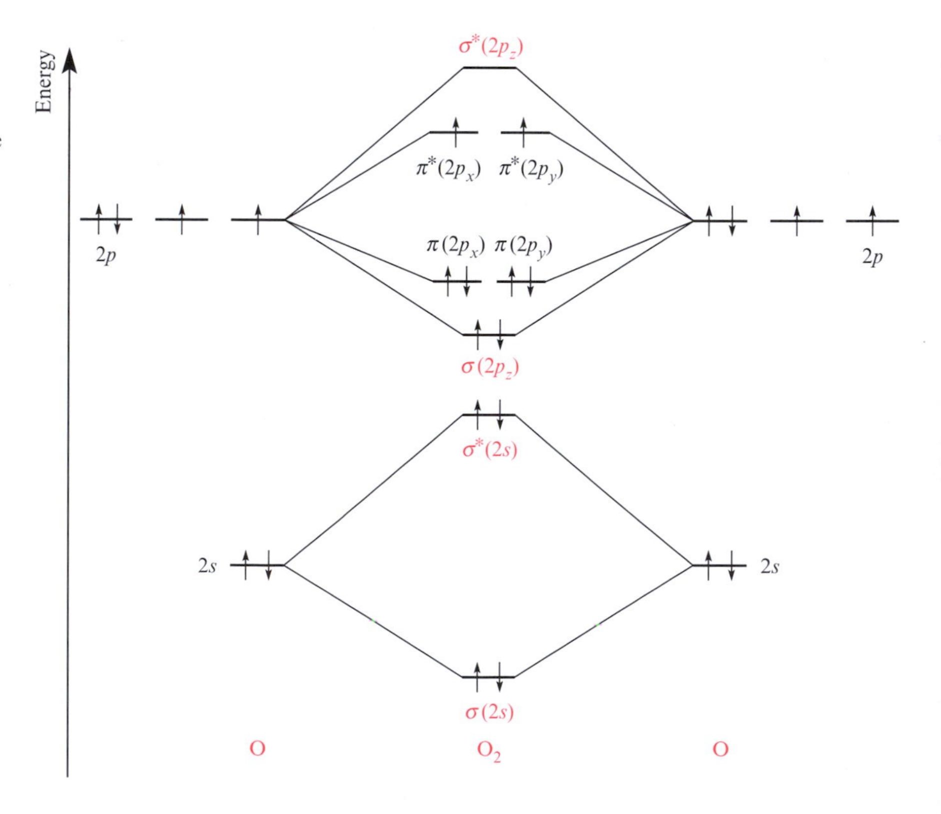

3.22 A molecular orbital diagram to show the formation of O_2. The 1*s* AOs (with core electrons) have been omitted. The O nuclei lie on the *z*-axis.

(3.10) (3.11) (3.12)

However, there is a problem – the Lewis structure does *not* predict the presence of two unpaired electrons in the O_2 molecule. Simple valence bond theory also gives this result. In contrast, the MO description of O_2 gives results about the bond order and the pairing of electrons which are consistent with the experimental data. The explanation of the diradical character of O_2 represents one of the classic successes of MO theory. The two unpaired electrons follow naturally from the orbital diagram.

3.19 ORBITAL MIXING AND σ–π CROSSOVER

Before we can progress to the bonding in B_2, C_2 and N_2, we must deal with two new concepts: orbital mixing and σ–π crossover. However, understanding the reasoning for these phenomena is not critical to a qualitative discussion of the MO diagrams for these diatomics, and you may wish to proceed to Section 3.20, returning to this section later.

When we construct an MO diagram using a linear combination of atomic orbitals, our initial approach allows overlap only between like atomic orbitals. Figure 3.23 shows the overlap between the 2*s* AOs of two identical atoms to give $\sigma(2s)$ and $\sigma^*(2s)$ MOs, and the overlap between the 2*p* atomic orbitals to give $\sigma(2p)$, $\pi(2p)$, $\pi^*(2p)$ and $\sigma^*(2p)$ MOs. The ordering of the molecular orbitals is approximate. Note that the $\sigma(2p)$ lies at lower energy than the $\pi(2p)$ levels.

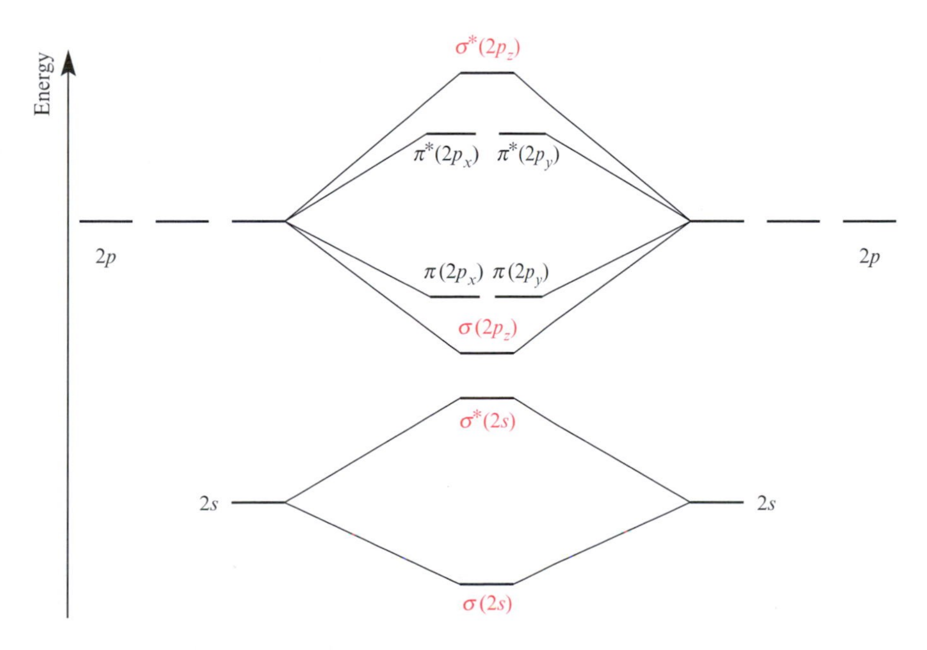

3.23 A molecular orbital diagram showing the approximate ordering of molecular orbitals generated by the combination of 2*s* and 2*p* atomic orbitals. This diagram is applicable to the formation of homonuclear diatomic molecules involving late (O and F) first row *p*-block elements, with the nuclei lying on the *z*-axis.

We can now introduce a new concept. If AOs of similar symmetry and energy can mix to give MOs, can MOs of similar symmetry and energy also mix? The answer is yes.

In Figure 3.23, we have two σ orbitals [$\sigma(2s)$ and $\sigma(2p)$] and two σ^* orbitals [$\sigma^*(2s)$ and $\sigma^*(2p)$]. At the beginning of the first row of the *p*-block elements, the 2*s* and 2*p* AOs are relatively close together in energy ($\approx$ 550 kJ mol^{-1} for boron), whereas by the end they are very much further apart ($\approx$ 2100 kJ mol^{-1} for fluorine). It turns out that the *molecular* orbitals derived from the 2*s* and 2*p* atomic orbitals are closer together for the earlier *p*-block elements. If the orbitals have similar energies and similar symmetries, they can mix. The results of mixing $\sigma(2s)$ with $\sigma(2p)$, and $\sigma^*(2s)$ with $\sigma^*(2p)$ are shown in Figure 3.24.

s–p separation: see Section 2.15

Generally, if the 2*s* and 2*p* AOs are close enough in energy, mixing of the $\sigma(2s)$ and $\sigma(2p)$ molecular orbitals occurs and results in the $\sigma(2p)$ orbital lying higher in energy than the $\pi(2p)$ levels. This is actually the case in the earlier *p*-block elements (B, C and N). Naturally, the mixing of orbitals will also occur with the later elements, but the energy mismatch makes the perturbation less, and the $\sigma(2p)$ still lies below the $\pi(2p)$.

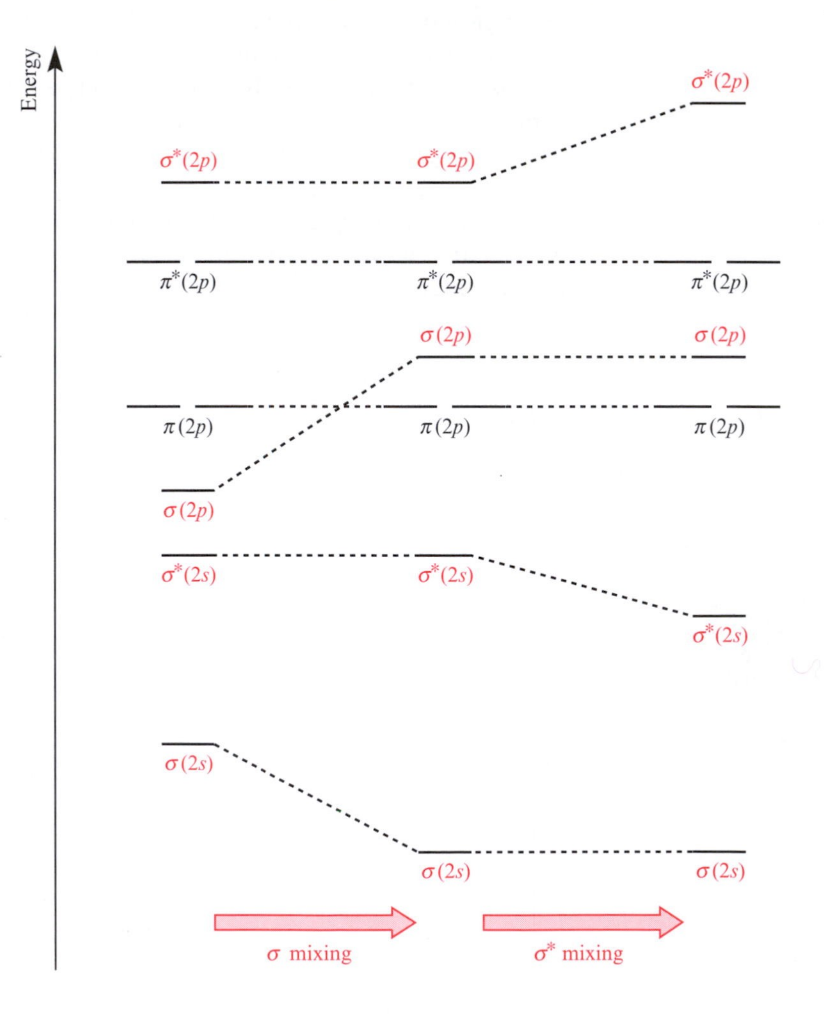

3.24 The effects of orbital mixing on the ordering of the MOs in a first row diatomic molecule. We consider the effects of the σ and σ^* mixing sequentially.

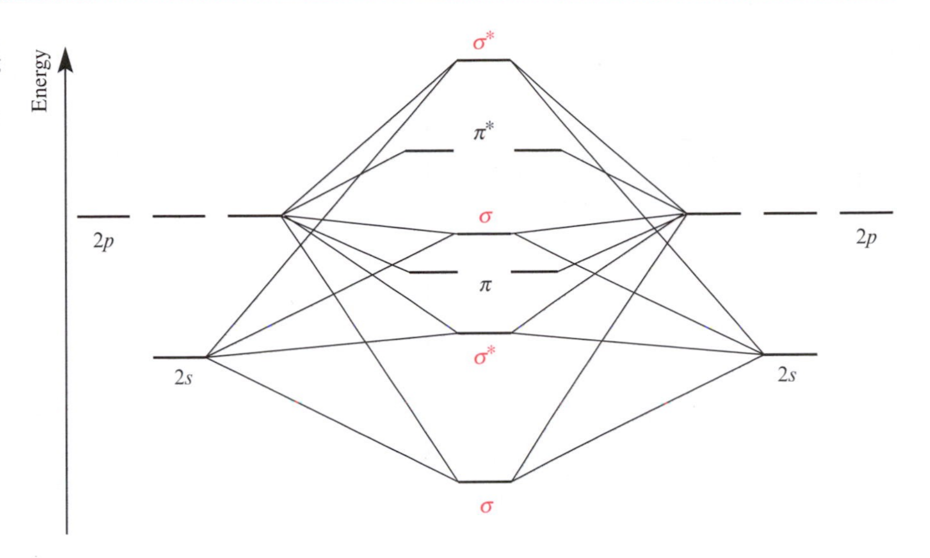

3.25 A molecular orbital diagram showing the ordering of molecular orbitals generated by the combination of $2s$ and $2p$ atomic orbitals *and* allowing for the effects of orbital mixing. This diagram is applicable to the formation of homonuclear diatomic molecules involving early (B, C and N) first row p-block elements.

Finally, in Figure 3.25 we show a full MO diagram for the combination of $2s$ and $2p$ atomic orbitals in which we include the effects of orbital mixing. Notice that the character of the atomic orbitals involved in the mixing is spread into a number of molecular orbitals. We return to this theme in Section 4.14 when we discuss the bonding in carbon monoxide. It is also important to see in Figure 3.25 that although atomic orbitals are now involved in several molecular orbitals, the total number of MOs still equals the total number of AOs.

Now we are in a position to move on to a discussion of the bonding in B_2, C_2 and N_2 and in it, we use the orbital energy levels obtained from Figure 3.25.

3.20 HOMONUCLEAR DIATOMIC MOLECULES OF THE FIRST ROW p-BLOCK ELEMENTS: B_2, C_2 AND N_2

B_2

Experimental fact: The vapour phase of elemental boron contains paramagnetic B_2 molecules.

The ground state electronic configuration of a boron atom is $[He]2s^22p^1$. An MO diagram for the formation of a B_2 molecule is shown in Figure 3.26. Each boron atom provides three valence electrons. Four electrons occupy the $\sigma(2s)$ and $\sigma^*(2s)$ MOs. This leaves two electrons. For the B_2 molecule to be paramagnetic, these electrons must singly occupy each of the degenerate $\pi(2p_x)$ and $\pi(2p_y)$ orbitals. We have experimental evidence from magnetic measurements for the σ–π crossover in B_2, as we discuss more fully for C_2.

From Figure 3.26, the electronic configuration of B_2 may be written as:

$$\sigma(2s)^2\sigma^*(2s)^2\pi(2p_x)^1\pi(2p_y)^1$$

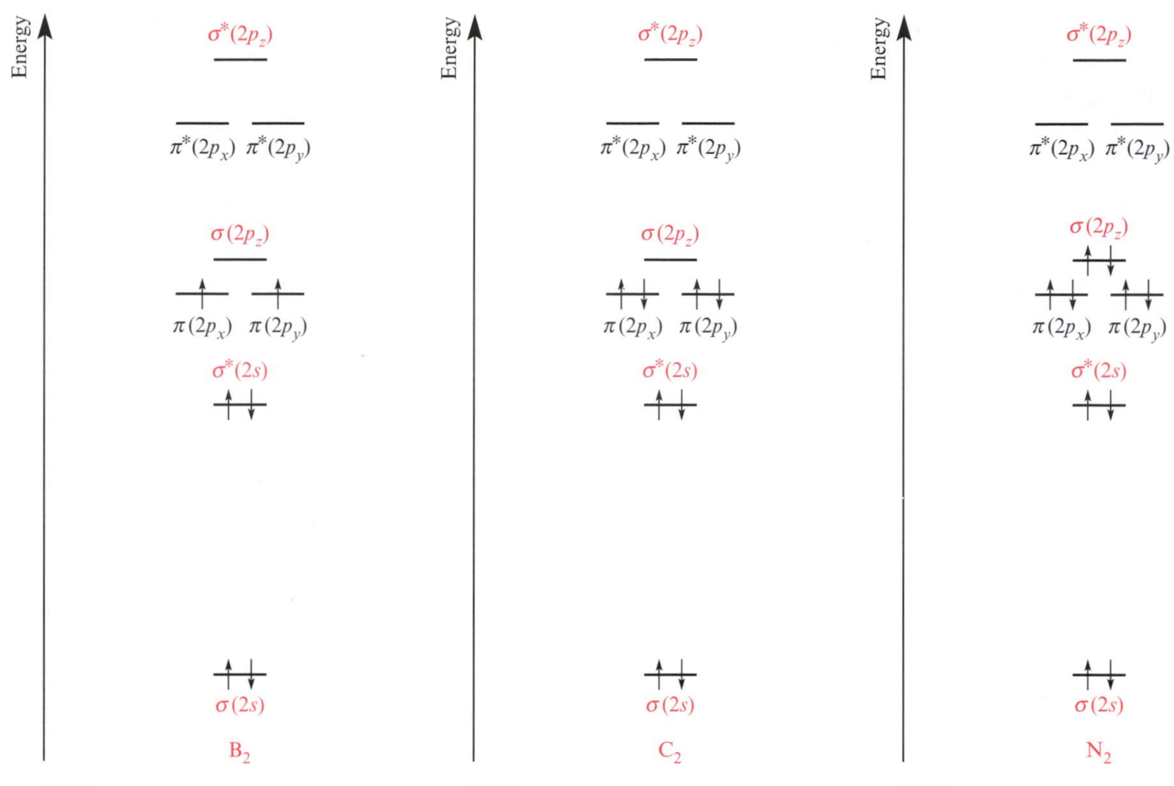

3.26 A molecular orbital diagram for the formation of B_2. The $\sigma(1s)$ and $\sigma^*(1s)$ MOs (with core electrons) have been omitted.

3.27 A molecular orbital diagram for the formation of C_2. The $\sigma(1s)$ and $\sigma^*(1s)$ MOs (with core electrons) have been omitted.

3.28 A molecular orbital diagram for the formation of N_2. The $\sigma(1s)$ and $\sigma^*(1s)$ MOs (with core electrons) have been omitted.

We predict a bond order for the boron–boron bond in B_2 of 1 (equation 3.27), i.e. a single bond is predicted.

$$\text{Bond order in } B_2 = \tfrac{1}{2} \times (4 - 2) = 1 \qquad (3.27)$$

Now let us draw a Lewis structure for B_2. The problem here is that the three electrons of each boron atom will be expected to pair up to give a triple bond as in structures **3.13** and **3.14** – an octet cannot be achieved. This bond order is not in accord either with the molecular orbital result or with the experimental data. An alternative Lewis structure can be drawn in either of the forms **3.15** or **3.16**. Here, a single bond is drawn, leaving a lone pair of electrons per boron atom. Although this result is more acceptable in terms of the B–B bond itself, it still does not predict that the B_2 molecule will be paramagnetic – remember that the Lewis approach will always give *paired* electrons whenever possible.

B ⁝ B **(3.13)**   B≡B **(3.14)**   : B : B : **(3.15)**   : B—B : **(3.16)**

For B_2, we see that drawing a Lewis structure does not yield a satisfactory description of the electrons in the molecule. On the other hand, MO theory can be used successfully to rationalize the bond order and the paramagnetism of this diatomic molecule.

C_2

Experimental facts: The C_2 molecule is a gas phase species and is diamagnetic.

The ground state electronic configuration of a carbon atom is $[He]2s^2 2p^2$. Lewis structures for C_2 are given in **3.17** and **3.18**, and show a C=C double bond and one lone pair of electrons per carbon atom. All electrons are paired.

: C ⁞ C :   : C=C :

(3.17)   **(3.18)**

An MO diagram for the formation of the C_2 molecule is shown in Figure 3.27. There are eight valence electrons to be accommodated in the MOs which arise from the combination of the carbon $2s$ and $2p$ atomic orbitals. Four electrons occupy the $\sigma(2s)$ and $\sigma^*(2s)$ MOs. If the $\pi(2p_x)$ and $\pi(2p_y)$ levels are lower in energy than the $\sigma(2p_z)$ level, then the last four electrons will occupy the degenerate set of orbitals as two pairs. This gives rise to a diamagnetic molecule and is in agreement with the experimental observations. Thus, the magnetic data for C_2 provide evidence for the σ–π crossover. If the energy of the $\sigma(2p_z)$ level were lower than that of the $\pi(2p_x)$ and $\pi(2p_y)$ levels, the electronic configuration for C_2 would be $\sigma(2s)^2\sigma^*(2s)^2\sigma(2p_z)^2\pi(2p_x)^1\pi(2p_y)^1$, predicting (incorrectly) a paramagnetic species.

From Figure 3.27, the electronic configuration of C_2 may be written as:

$$\sigma(2s)^2\sigma^*(2s)^2\pi(2p_x)^2\pi(2p_y)^2$$

The bond order of the carbon-carbon bond in C_2 may be determined from Figure 3.27 and equation 3.28.

$$\text{Bond order in } C_2 = \tfrac{1}{2} \times (6 - 2) = 2 \qquad (3.28)$$

In conclusion, we see that a C=C double bond and complete pairing of electrons in C_2 can be rationalized in terms of a Lewis structure, VB and MO theories. We use the term 'rationalized' rather than 'predicted' in regard to the MO approach because of the uncertainties associated with the σ-π crossover; at a simple level of calculation, whether or not this crossover will occur cannot be predicted. However, we *can* rationalize the fact that C_2 is diamagnetic by allowing for such a change in energy levels.

N_2

Experimental facts: The N_2 molecule is relatively unreactive; dinitrogen is often used to provide an inert atmosphere for experiments in which reagents and/or products react with components in the air, usually O_2 or water

Table 3.7 Experimental data for homonuclear diatomic molecules containing elements of the first row of the periodic table.

Molecule	*Bond distance / pm*	*Bond dissociation enthalpy / kJ mol^{-1}*	*Magnetic data*
Li_2	267	110	Diamagnetic
Be_2 (not observed)	–	–	–
B_2	159	297	Paramagnetic
C_2	124	607	Diamagnetic
N_2	110	945	Diamagnetic
O_2	121	498	Paramagnetic
F_2	141	159	Diamagnetic

vapour. The bond dissociation enthalpy of N_2 is particularly high and the bond length is short (Table 3.7). The N_2 molecule is diamagnetic.

Let us first construct a Lewis structure for N_2. The ground state electronic configuration of a nitrogen atom is [He]$2s^2 2p^3$. The three $2p$ electrons are unpaired (as represented in **3.19**) and when two nitrogen atoms come together, a triple bond is required so that each nitrogen atom completes its octet of valence electrons. This is represented in the Lewis structures **3.20** and **3.21**. Similarly, the valence bond model for the N_2 molecule shows a localized N≡N triple bond and one lone pair per nitrogen atom.

(3.19) (3.20) (3.21)

An MO diagram for the formation of N_2 is shown in Figure 3.28. There are ten valence electrons to be accommodated and filling the MOs by the *aufbau* principle gives rise to complete pairing of the electrons. This is consistent with the experimental observation that the N_2 molecule is diamagnetic.

For N_2, the presence (or not) of the σ–π crossover makes no difference to predictions about the pairing of electrons or to the bond order, but its existence is suggested by the results of *photoelectron spectroscopic* studies of N_2.

Photoelectron spectroscopy: see Box 3.6

From Figure 3.28, the electronic configuration of N_2 may be written as:

$$\sigma(2s)^2\sigma^*(2s)^2\pi(2p_x)^2\pi(2p_y)^2\sigma(2p_z)^2$$

and from equation 3.29, the bond order in the N_2 molecule is found to be 3. This multiple bond order is in keeping with the high bond dissociation enthalpy of N_2 and the short internuclear distance (Table 3.7).

$$\text{Bond order in } N_2 = \tfrac{1}{2} \times (8 - 2) = 3 \qquad (3.29)$$

In conclusion, the triple bond in N_2 and its diamagnetic character can be correctly predicted by use of a Lewis structure or by the VB or MO approaches.

Box 3.6 Photoelectron spectroscopy (PES)

Photoelectron spectroscopy is a technique used to study the energies of *occupied* atomic and molecular orbitals.

An atom or molecule is irradiated with electromagnetic radiation of energy, E, and this causes electrons to be ejected from the sample. Each electron has a characteristic *binding energy* and, to be ejected from the atom or molecule, the electron must absorb an amount of energy that is equal to its binding energy.

If we measure the energy of the electron as it is ejected, then this excess energy will be given by:

$$\text{Excess energy of electron} = E - (\text{binding energy of electron})$$

providing, of course, that E is greater than the binding energy of the electron. Hence, we can measure the binding energy.

Koopmans' theorem relates the binding energy of an electron to the orbital in which it resides. Effectively, it allows the binding energy to be a measure of the orbital energy.

Photoelectron spectroscopy is an important experimental method which permits us to measure the binding energies of both valence and core electrons. From these energies, we can gain insight into the ordering of molecular orbitals for a given atomic or molecular system. We mention photoelectron spectroscopy again in Section 5.22.

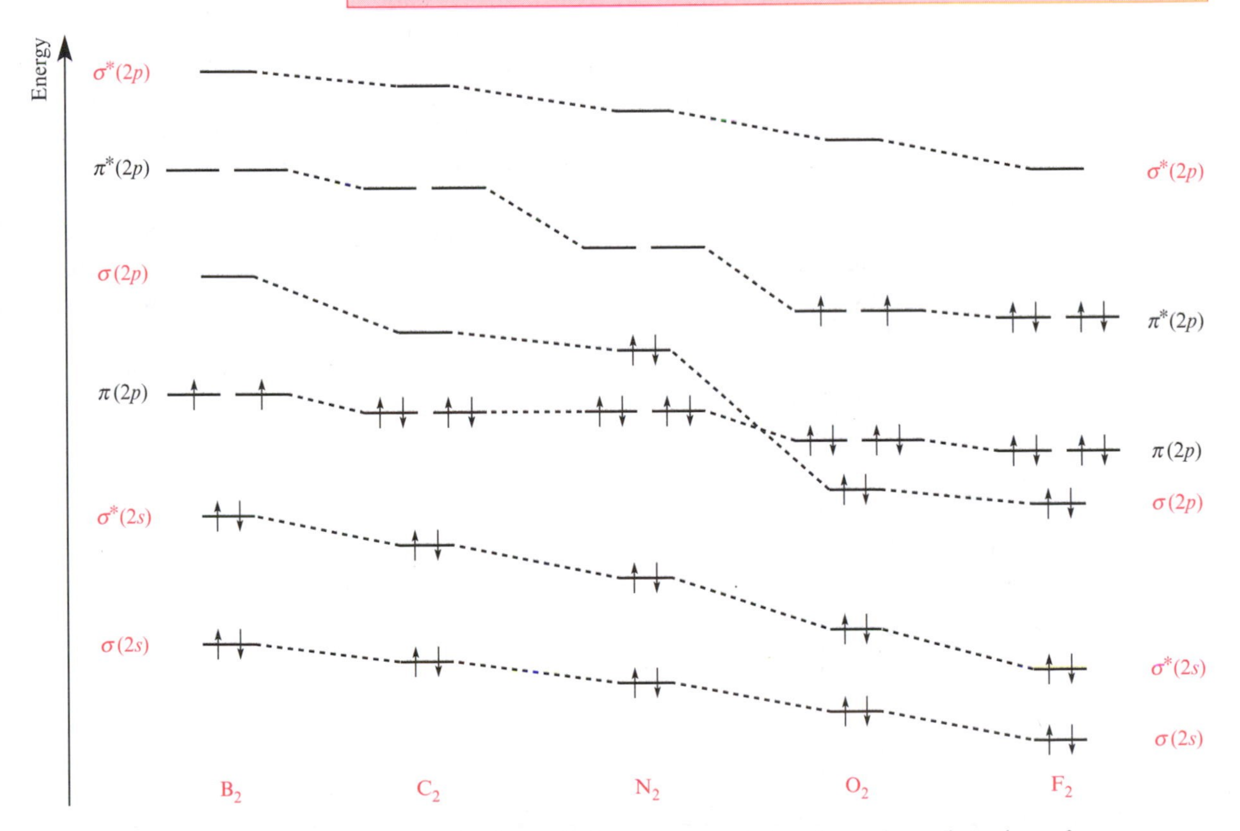

3.29 Changes in the energy levels of the molecular orbitals and the ground state electronic configurations of homonuclear diatomic molecules involving first row p-block elements.

The first row *p*-block elements: a summary

Let us briefly summarize what happens to the bonding in the X_2 diatomic molecules for X = B, C, N, O and F. Figure 3.29 illustrates the changes in the energy levels of the molecular $\sigma(2s)$, $\sigma^*(2s)$, $\sigma(2p)$, $\sigma^*(2p)$, $\pi(2p)$ and $\pi^*(2p)$ orbitals in going from B_2 to F_2.

The main points to note are:

- the effects of molecular orbital mixing which lead to the σ–π crossover;
- the use of the *aufbau* principle to give ground state electronic configurations;
- the use of the MO diagram to determine bond order;
- the use of the MO diagram to explain diamagnetic or paramagnetic behaviour.

3.21 PERIODIC TRENDS IN THE HOMONUCLEAR DIATOMIC MOLECULES OF THE FIRST ROW ELEMENTS

In this section we summarize the information gained about the homonuclear diatomic molecules of the first row elements – Li_2 to F_2, excluding 'Be_2'.

Figure 3.30 shows the trends in X–X bond distances and bond dissociation enthalpies in the X_2 molecules as we cross the periodic table from left to right. From B_2 to N_2, there is a decrease in bond length which corresponds to an increase in the bond dissociation enthalpy of the X–X bond. From N_2 to F_2, the X–X bond lengthens and weakens. These trends follow changes in the bond order in the sequence B_2 (single) < C_2 (double) < N_2 (triple) > O_2 (double) > F_2 (single).

Note that we *cannot exactly* correlate particular values of the bond dissociation enthalpy and distance with a given bond order. Factors such as internuclear and interelectron repulsion play an important part in determining the observed values of these parameters. For example, the Li_2 molecule has a single bond and is thus similar to B_2 and F_2. Yet the bond distance of 267 pm in Li_2 is far greater than that of either B_2 or F_2. Correspondingly, the bond

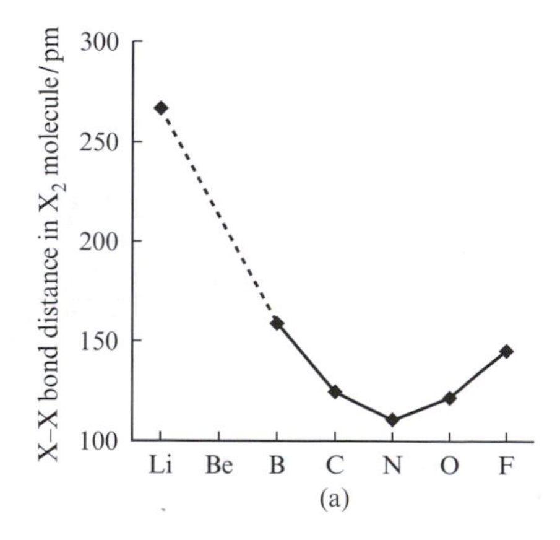

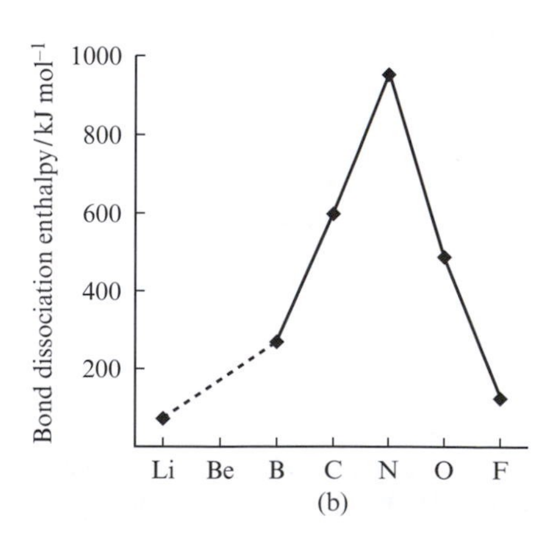

3.30 The trends in (a) the X–X bond distances and (b) the X–X bond dissociation enthalpy for X_2 molecules containing the first row elements (X = Li to F).

dissociation enthalpy of Li_2 is very low. The difference lies in the radial extent of the orbitals involved in bond formation. The 2*s* and 2*p* orbitals of the boron atom experience a greater effective nuclear charge than the 2*s* orbitals of the lithium atom, and are thus more contracted. In order to get an efficient overlap, the atoms in B_2 must be closer together than the atoms in Li_2.

3.22 THE DIATOMIC SPECIES O_2, $[O_2]^+$, $[O_2]^-$ AND $[O_2]^{2-}$

Once an MO diagram has been constructed for a homonuclear diatomic molecule, it can be used to predict the properties of (or rationalize experimental data for) species derived from that molecule. We exemplify this by looking at O_2, $[O_2]^+$, $[O_2]^-$ and $[O_2]^{2-}$, some physical properties of which are listed in Table 3.8.

Figure 3.22 showed an MO diagram for the formation of the O_2 molecule. The configuration $\sigma(2s)^2\sigma^*(2s)^2\sigma(2p_z)^2\pi(2p_x)^2\pi(2p_y)^2\pi^*(2p_x)^1\pi^*(2p_y)^1$ supports the presence of an O=O double bond and fact that the O_2 molecule is paramagnetic.

What happens if we oxidize or reduce O_2? The addition of one or more electrons corresponds to *reduction* and the removal of one or more electrons is *oxidation*. We can use the MO diagram for O_2 to monitor changes (bond order, bond distance, bond dissociation enthalpy and electron pairing) that occur to the diatomic species by adding or removing electrons. The addition of electrons will follow the *aufbau* principle; the removal of electrons follows the same rules but in reverse.

The one-electron oxidation of the O_2 molecule (equation 3.30) corresponds to the removal of one electron from a π^* orbital in Figure 3.22. Inspection of the MO diagram shows that the $[O_2]^+$ cation will be paramagnetic with one unpaired electron. It is also a radical species and is known as a *radical cation*.

Oxidation and reduction may be defined in terms of the loss and gain of an electron or electrons.

In an oxidation process, one or more electrons are lost.

In a reduction process, one or more electrons are gained.

Table 3.8 Experimental data for dioxygen and derived diatomic species.

Diatomic species	*Bond distance / pm*	*Bond dissociation enthalpy / kJ mol^{-1}*	*Magnetic data*
O_2	121	498	Paramagnetic
$[O_2]^+$	112	644	Paramagnetic
$[O_2]^-$	132	360	Paramagnetic
$[O_2]^{2-}$	149	149	Diamagnetic

$$O_2 \rightarrow [O_2]^+ + e^- \tag{3.30}$$

$$\text{Bond order in } [O_2]^+ = \tfrac{1}{2} \times (8 - 3) = 2.5 \tag{3.31}$$

In going from O_2 to $[O_2]^+$, the bond order increases from 2.0 to 2.5 (equations 3.26 and 3.31). This is reflected in an increase in the bond dissociation enthalpy and a decrease in the oxygen–oxygen bond distance (Table 3.8).

A radical cation is a cation that possesses an unpaired electron.

A radical anion is an anion that possesses an unpaired electron.

The one- and two-electron reductions of O_2 are shown in equations 3.32 and 3.33. By considering the molecular orbitals in Figure 3.22, we can see that the superoxide ion, $[O_2]^-$, is formed by the addition of one electron to one of the π^* orbitals. The $[O_2]^-$ ion is paramagnetic (it is a *radical anion*) and has a bond order of 1.5. The data in Table 3.8 support the weakening and lengthening of the oxygen–oxygen bond on going from O_2 to $[O_2]^-$.

$$O_2 + e^- \rightarrow [O_2]^- \tag{3.32}$$

$$O_2 + 2e^- \rightarrow [O_2]^{2-} \tag{3.33}$$

Oxides, peroxides and superoxides: see Sections 11.12 and 13.10

The addition of two electrons to O_2 to give $[O_2]^{2-}$ (the peroxide ion) results in the formation of a diamagnetic species, and the bond order becomes 1. Again, the experimental data concur with this. The dianion $[O_2]^{2-}$ has a longer and weaker bond than either of O_2 or $[O_2]^-$.

3.23 GROUP TRENDS AMONGST HOMONUCLEAR DIATOMIC MOLECULES

Now we turn from trends amongst diatomic molecules in the first row of the periodic table to trends within *groups of elements*, focusing first on the alkali metals and the halogens. We then consider group 15 in which the changes in X–X bond enthalpies in going down the group cause dramatic effects in elemental structure.

Group 1: alkali metals

Hydrogen is excluded from this discussion

Each group 1 element has a ground state valence electronic configuration of ns^1. Homonuclear diatomic molecules of the alkali metals occur in the gas phase.

The formation of the Li_2 molecule from two lithium atoms was illustrated in Figure 3.16. Similar diagrams can be constructed for the formation of Na_2, K_2, Rb_2, Cs_2 and Fr_2, although in practice, the experimental data (Table 3.9) on these species become sparse as the group is descended.

Molecular orbital theory predicts that each group 1 diatomic molecule has a bond order of one. This is also the conclusion when Lewis structures are drawn or when the bonding is considered using VB theory – each alkali metal atom has one valence electron and will share one electron with another group 1 atom.

The bond dissociation enthalpies and bond distances for Li_2, Na_2 and K_2 (Table 3.9) indicate that as the group is descended, the M–M bond gets

Table 3.9 Experimental data for homonuclear diatomic molecules of the group 1 elements.

Diatomic species, X_2	*Ground state electronic configuration of atom X*	*Bond distance / pm*	*Bond dissociation enthalpy / kJ mol^{-1}*
Li_2	$[He]2s^1$	267	110
Na_2	$[Ne]3s^1$	308	74
K_2	$[Ar]4s^1$	390	55
Rb_2	$[Kr]5s^1$	–	49
Cs_2	$[Xe]6s^1$	–	44

longer and weaker. The trend in enthalpies continues for Rb_2 and Cs_2. This reflects the change in the principal quantum number of the *ns* atomic orbital and the spatial properties of the orbitals. We reiterate how the radial distribution function varies amongst the *ns* atomic orbitals; remember that this gives the probability of finding the electron in a spherical shell of radius *r* and thickness (r + dr) from the nucleus. Figure 2.11 showed radial distribution functions for 1*s*, 2*s* and 3*s* atomic orbitals; the valence orbitals of Li and Na are the 2*s* and 3*s* atomic orbitals).

Spatial properties of orbitals: see Figure 2.10

Group 17: halogens

Each member of group 17 has an ns^2np^5 valence electronic configuration. Lewis structures for Cl_2, Br_2 and I_2 resemble that drawn for F_2 in structures **3.8** and **3.9**; each molecule has a single bond. An MO diagram for each of the Cl_2, Br_2 and I_2 molecules is similar to Figure 3.21, except that the principal quantum numbers are different and this affects the relative energies of the orbitals. In each case, a bond order of 1 is calculated for the X_2 molecule, and we find that each diatomic species is diamagnetic.

Table 3.10 gives some data for F_2, Cl_2, Br_2 and I_2. The relative values of the distances and enthalpies are in agreement with the notion of single X–X bonds. However, instead of a *smooth* trend of decreasing bond dissociation enthalpies and increasing bond distances, we see that the F–F bond is actually much weaker than would be expected. This is emphasized in the graph in Figure 3.31. The additional weakening is usually attributed to repulsion between the lone pairs of electrons on the two fluorine atoms. Although

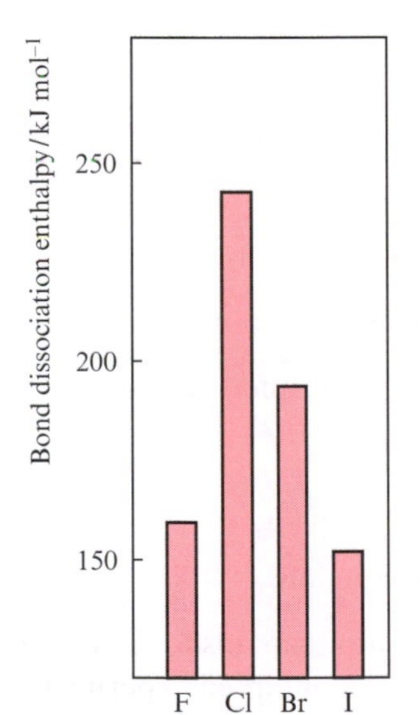

3.31 The trend in the X–X bond dissociation enthalpy for X_2 molecules where X is a group 17 element.

Table 3.10 Experimental data for homonuclear diatomic molecules of the group 17 elements.

Diatomic species, X_2	*Ground state electronic configuration of atom X*	*Bond distance / pm*	*Bond dissociation enthalpy / kJ mol^{-1}*
F_2	$[He]2s^22p^5$	141	159
Cl_2	$[Ne]3s^23p^5$	199	242
Br_2	$[Ar]4s^24p^5$	228	193
I_2	$[Kr]5s^25p^5$	267	151

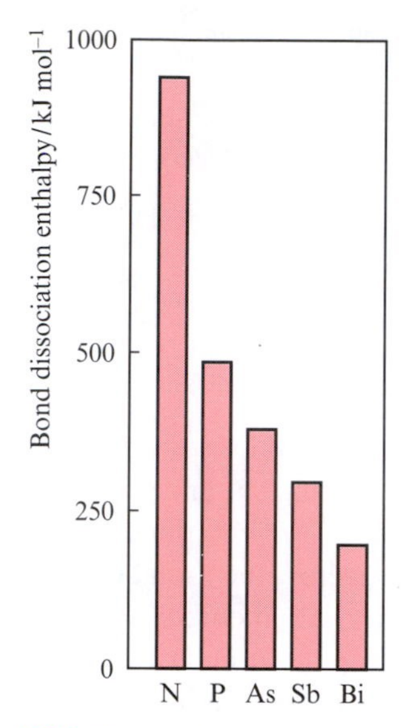

3.32 The trend in the X–X bond dissociation enthalpy for X_2 molecules where X is a group 15 element.

each X atom in each X_2 molecule has three lone pairs of electrons, those in F_2 will experience the greatest repulsion because of the shorter distance between the nuclei.

Group 15

The trend in the bond dissociation enthalpies of the molecules N_2, P_2, As_2, Sb_2 and Bi_2 is shown in Figure 3.32. There is a sharp decrease in the value of the dissociation enthalpy in going from N_2 to P_2 and then less dramatic decreases in going from P_2 through to Bi_2.

Let us step back for a moment and consider the choices open to atoms of group 15 elements when they combine with one another. The situation is illustrated in Figure 3.33. The presence of three unpaired electrons suggests that *either* a diatomic molecule with a triple bond *or* a tetraatomic molecule with six single bonds can form.

Bond enthalpy is an important factor in determining what degree of molecular aggregation will occur. If we have four nitrogen atoms in the gas phase, they may come together to form two molecules of N_2 (equation 3.34) or one molecule of N_4 (equation 3.35).

$$4N(g) \rightarrow 2N_2(g) \qquad (3.34)$$

$$4N(g) \rightarrow N_4(g) \qquad (3.35)$$

When two moles of N_2 form, two triple bonds are made and the associated enthalpy change is:

$$\begin{aligned}\Delta H^o[4N(g) \rightarrow 2N_2(g)] &= -\{2 \times D(N{\equiv}N)\} \\ &= -(2 \times 945) \\ &= -1890\,\text{kJ per mole of reaction}\end{aligned}$$

When one mole of N_4 forms, six single bonds are made and the associated enthalpy change is:

The value of *D*(N–N) comes from Table 3.6

$$\begin{aligned}\Delta H^o[4N(g) \rightarrow N_4(g)] &= -\{6 \times D(N{-}N)\} \\ &= -(6 \times 159) \\ &= -954\,\text{kJ per mole of reaction}\end{aligned}$$

Thus, it is more favourable to form two moles of $N_2(g)$ than it is to form one mole of $N_4(g)$. In fact, the N_4 molecule has not been observed experimentally and the particularly strong N≡N triple bond is responsible for the N_2 molecule being the standard state of this element.

X = N

X = P, As, Sb or Bi

3.33 Atoms of a group 15 may come together to form a diatomic molecule with a triple bond, or a tetraatomic molecule with six single bonds. The P_4, As_4, Sb_4 and Bi_4 molecules are formed under drastically different conditions.

Now let us repeat the exercise for phosphorus.

$$4P(g) \rightarrow 2P_2(g) \tag{3.36}$$

$$4P(g) \rightarrow P_4(g) \tag{3.37}$$

For the formation of two moles of P_2, the enthalpy change is:

$$\begin{aligned}\Delta H^o[4P(g) \rightarrow 2P_2(g)] &= -\{2 \times D(P{\equiv}P)\}\\ &= -(2 \times 490)\\ &= -980\,\text{kJ per mole of reaction}\end{aligned}$$

For the formation of one mole of P_4, the enthalpy change is:

The value of D(P–P) comes from Table 3.6

$$\begin{aligned}\Delta H^o[4P(g) \rightarrow P_4(g)] &= -\{6 \times D(P{-}P)\}\\ &= -(6 \times 200)\\ &= -1200\,\text{kJ per mole of reaction}\end{aligned}$$

This time it is more favourable for the gaseous P_4 molecule to form, although the enthalpy difference between the two forms of phosphorus is not as great as between the two forms of nitrogen. The standard state of phosphorus (white phosphorus) consists of P_4 molecules packed in a solid state array.

Above 827 K, P_4 dissociates into P_2 molecules (equation 3.38) and, from the data calculated above, we would estimate that ΔH^o for this process is 220 $kJ\,mol^{-1}$ – experimentally, it is found to be 217 $kJ\,mol^{-1}$.

$$P_4(g) \rightarrow 2P_2(g) \tag{3.38}$$

In the gas phase, arsenic exists as As_4 molecules. At relatively low temperatures, antimony vapour contains mainly Sb_4 molecules; Sb_4, Sb_2, Bi_4 and Bi_2 molecules have been observed in noble gas matrices. However, all these species are inaccessible at 298 K; the standard states of these elements possess covalent lattice structures similar to that of black phosphorus. We will look further at these structures in Chapter 7.

Worked example 3.2 *Allotropes of oxygen and sulfur*

Suggest reasons why oxygen forms an O_2 molecule rather than a cyclic structure such as O_6, whereas sulfur forms S_6 rather than an S_2 molecule at 298 K and atmospheric pressure.

Bond dissociation enthalpy data provided:
D(O–O) = 146 $kJ\,mol^{-1}$
D(S–S) = 266 $kJ\,mol^{-1}$
D(O=O) = 498 $kJ\,mol^{-1}$
D(S=S) = 425 $kJ\,mol^{-1}$

Consider the formation of O_2 and O_6 molecules from oxygen atoms. In order to make the comparison meaningful, we must consider equal numbers of oxygen atoms:

$$6O(g) \rightarrow 3O_2(g)$$
$$6O(g) \rightarrow O_6(g)$$

For the formation of three moles of $O_2(g)$:

$$\begin{aligned}\Delta H^o &= -[3 \times D(O{=}O)] \\ &= -(3 \times 498) \\ &= -1494 \text{ kJ per mole of reaction}\end{aligned}$$

For the formation of 1 mole of $O_6(g)$:

$$\begin{aligned}\Delta H^o &= -[6 \times D(O{-}O)] \\ &= -(6 \times 146) \\ &= -876 \text{ kJ per mole of reaction}\end{aligned}$$

Hence, on enthalpy grounds O_2 molecule formation is favoured over O_6 ring formation. (Remember that ΔH is not the only factor involved – see Chapter 12.)

For sulfur, we compare the enthalpies of reaction for the following two processes:

$$6S(g) \rightarrow 3S_2(g)$$
$$6S(g) \rightarrow S_6(g)$$

For the formation of three moles of $S_2(g)$:

$$\begin{aligned}\Delta H^o &= -[3 \times D(S{=}S)] \\ &= -(3 \times 425) \\ &= -1275 \text{ kJ per mole of reaction}\end{aligned}$$

For the formation of 1 mole of $S_6(g)$:

$$\begin{aligned}\Delta H^o &= -[6 \times D(S{-}S)] \\ &= -(6 \times 266) \\ &= -1596 \text{ kJ per mole of reaction}\end{aligned}$$

Hence, on enthalpy grounds S_6 ring formation is favoured over S_2 molecule formation. (Other cyclic molecules of type S_n are also viable as illustrated in Table 3.2.)

SUMMARY

In this chapter the theme has been bonding in homonuclear molecules.

Do you know what the following terms mean?

- covalent bond
- homonuclear
- covalent radius
- bond dissociation enthalpy
- enthalpy of atomization
- Lewis structure
- valence bond theory
- molecular orbital theory
- linear combination of atomic orbitals
- localized bond
- delocalized bond
- σ and π bonds
- bonding and antibonding molecular orbitals
- bond order
- diamagnetic
- paramagnetic
- radical cation
- radical anion

You should now be able:

- to discuss briefly methods of determining bond distances
- to relate bond dissociation enthalpy values to enthalpies of atomization
- to distinguish between an individual and an average bond dissociation enthalpy value
- to use Hess's Law of constant heat summation to estimate bond dissociation enthalpies from enthalpies of formation and atomization
- to draw Lewis structures for simple covalent molecules
- to discuss the principles of MO and VB theories
- to discuss the bonding in H_2 in terms of MO and VB theories and compare the two models
- to construct MO diagrams for homonuclear diatomic molecules of the *s* and *p* block elements, and make use of experimental data to assess whether your diagram is realistic
- to relate bond order *qualitatively* to bond enthalpy and bond distance for homonuclear bonds
- to discuss trends in homonuclear bond enthalpies and bond distances across the first row of elements in the periodic table and down groups of the *s* and *p* block elements.

PROBLEMS

3.1 State what you understand by each of the following: (a) covalent radius, (b) bond dissociation enthalpy and (c) the standard enthalpy of atomization of an element.

3.2 (a) In an X-ray diffraction experiment, what causes the diffraction of the electromagnetic radiation? (b) Explain why X-ray diffraction results give a value of 103 pm for the length of a localized two-centre two-electron B–H bond in the $[B_2H_7]^-$ ion whilst the same bond is found to be 118 pm in length from the results of a neutron diffraction experiment.

3.3 State, with reasons, whether you can reasonably transfer a value of the stated bond dissociation enthalpy between the following pairs of molecules: (a) the carbon–carbon bond enthalpy between ethane (C_2H_6) and ethene (C_2H_4), (b) the nitrogen–nitrogen bond enthalpy between hydrazine (N_2H_4) and N_2F_4, and (c) the oxygen–hydrogen bond enthalpy between water and hydrogen peroxide (H_2O_2).

3.4 Using the bond dissociation enthalpy data in Table 3.6, estimate values for the enthalpy of atomization for each of the following gas phase molecules: (a) P_4; (b) S_8; (c) As_4; (d) I_2. (The structures of these molecules occur in the chapter.)

3.5 Using the answer obtained for P_4(g) in question 3.4 and the standard enthalpy of atomization listed in Table 3.5, estimate the standard enthalpy of formation (at 298 K) of *gaseous* P_4.

3.6 The standard enthalpies of formation (at 298 K) for methane (CH_4) and ethane (C_2H_6) are –75 and –85 kJ mol^{-1} respectively. Determine values for the single C–C and C–H bond dissociation enthalpies at 298 K given that the standard enthalpies of atomization of carbon and hydrogen are 717 and 218 kJ mol^{-1} respectively. Compare your answers with the values listed in Table 3.6 and comment on the origins of any differences.

3.7 By considering only valence electrons in each case, draw Lewis structures for: (a) Cl_2, (b) Na_2, (c) S_2, (d) NF_3, (e) HCl, (f) BH_3 and (g) SF_2.

3.8 By considering only valence electrons in each case, draw Lewis structures for the following molecules: (a) methane (CH_4); (b) bromomethane (CH_3Br); (c) ethane (C_2H_6); (d) ethanol (C_2H_5OH); (e) ethene (C_2H_4).

3.9 How many Lewis structures can you draw for C_2? In the text, we only considered one structure; suggest why the other structure(s) that you have drawn is (are) unreasonable. [Hint: Think about the distribution of the bonding electrons in space.]

3.10 By using an MO approach to the bonding, show that the formation of the molecule He_2 would not be favourable.

3.11 Construct a *complete* MO diagram for the formation of Li_2 showing the involvement of both the core and valence electrons of the two lithium atoms. Determine the bond order in Li_2. Does the inclusion of the core electrons affect the value of the bond order?

3.12 When Li_2 undergoes a one-electron oxidation, what species is formed? Use the MO diagram you have drawn in problem 3.11 and the data about Li_2 in Table 3.9 to say what you can about the bond order, the bond dissociation enthalpy, the bond length and the magnetic properties of the oxidized product.

3.13 Using an MO approach, rationalize the trends in the data given in Table 3.11. Will the $[N_2]^-$ and $[N_2]^+$ ions be diamagnetic or paramagnetic?

Table 3.11 Data for problem 3.13.

Diatomic species	*Bond distance / pm*	*Bond dissociation enthalpy / kJ mol^{-1}*
N_2	110	945
$[N_2]^-$	119	765
$[N_2]^+$	112	841

3.14 Table 3.12 lists the bond dissociation enthalpies of diatomic molecules of the group 16 elements. Suggest reasons for the observed trend. Are the values of the bond distances in O_2 and S_2 consistent with the bond dissociation enthalpies for these molecules?

Table 3.12 Data for problem 3.14.

Diatomic species	*Bond distance / pm*	*Bond dissociation enthalpy / kJ mol^{-1}*
O_2	121	498
S_2	189	425
Se_2	217	333
Te_2	256	260

4 Heteronuclear diatomic molecules

Topics

4.1 INTRODUCTION

In Chapter 3 we discussed *homonuclear* bonds such as those in homonuclear diatomic molecules (H_2 or N_2) or homonuclear polyatomic molecules (P_4 or S_8), or homonuclear bonds in *heteronuclear* molecules (the C–C bond in C_2H_6 or the O–O bond in H_2O_2). In this chapter we are concerned with heteronuclear molecules with an emphasis on diatomic species.

In a homonuclear diatomic molecule X_2, the two X atoms contribute equally to the bond. The molecular orbital diagrams drawn in Chapter 3 possess a symmetrical appearance, and each molecular orbital contains equal contributions from the constituent atomic orbitals of each atom.

A *heteronuclear bond* is one formed between *different atoms;* examples are the C–H bonds in methane and ethane, the B–F bonds in boron trifluoride and the N-H bonds in ammonia (Figure 4.1). Since the bulk of 'chemistry' deals with molecules and ions containing more than one type of atom, it is clear that an understanding of the interactions and bonding between atoms of different types is crucial to an understanding of such species. Later in this book we shall encounter many heteronuclear *polyatomic* molecules, but we begin our discussion by considering the bonding in the diatomic molecules HF, LiF and LiH.

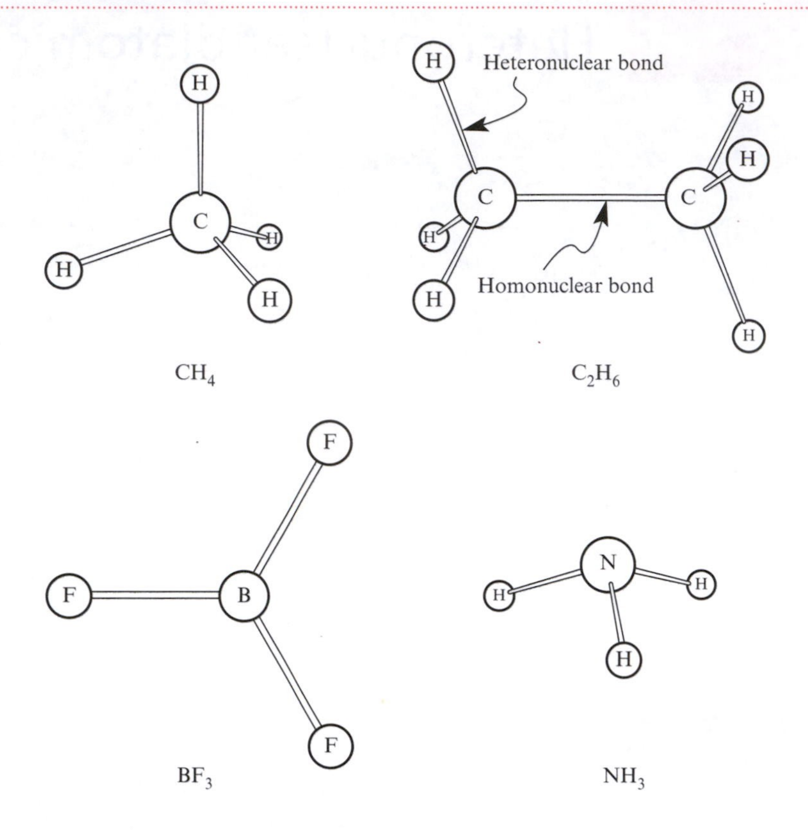

4.1 Molecules with heteronuclear bonds: methane (CH_4), ethane (C_2H_6), boron trifluoride (BF_3) and ammonia (NH_3).

4.2 LEWIS STRUCTURES FOR HF, LiF AND LiH

Lewis structures for hydrogen, lithium and fluorine atoms are shown in Figure 4.2a along with their ground state electronic configurations.

The elements hydrogen and lithium have one valence electron, while fluorine has seven. Just as we formed a covalent bond by sharing electrons between two hydrogen atoms to give the H_2 molecule, and between two fluorine atoms to give F_2, so we can pair together a hydrogen atom and a fluorine atom to give the HF molecule. Similarly, we can predict the formation of the molecules LiF and LiH and draw Lewis structures for these heterodiatomic molecules as shown in Figure 4.2b; this predicts that each molecule possesses a single bond.

In the gas phase, HF, LiF and LiH can each exist as molecular species. However, as we shall see later, at room temperature and pressure, the bonding is not always as simple as the Lewis approach predicts.

(a)

H·	Li·	·F: (seven dots)
$1s^1$	$[He]2s^1$	$[He]2s^22p^5$

(b)

H : F :	Li : F :	Li : H
H—F :	Li—F :	Li—H

4.2 (a) Lewis diagrams and ground state electronic configurations for hydrogen, lithium and fluorine; (b) Lewis structures for hydrogen fluoride, lithium fluoride and lithium hydride.

4.3 THE VALENCE BOND APPROACH TO THE BONDING IN HF, LiF AND LiH

What differences are involved in considering the homonuclear diatomic H_2 and a heteronuclear molecule XY?

In Chapter 3 we showed how valence bond theory can be applied to the H_2 molecule and how the bonding can be described by the resonance structures **4.1, 4.2** and **4.3**. The two resonance forms **4.2** and **4.3** are equivalent in energy.

VB approach to H_2: see Section 3.11

$$\mathrm{H{-}H} \longleftrightarrow \mathrm{H_A^+\,H_B^-} \longleftrightarrow \mathrm{H_A^-\,H_B^+}$$

(4.1) (4.2) (4.3)

A similar approach can be used to describe the bonding in heteronuclear diatomic molecules. However, there is an important distinction to be made between homo- and heteronuclear diatomics. In a diatomic molecule XY, there will be a resonance structure **4.4** that represents the covalent contribution to the overall bonding picture and *two different* resonance structures, **4.5** and **4.6**, that represent the ionic contributions.

$$\mathrm{X{-}Y} \longleftrightarrow \mathrm{X^+\,Y^-} \longleftrightarrow \mathrm{X^-\,Y^+}$$

(4.4) (4.5) (4.6)

The relative importance of the three resonance structures depends upon their relative energies, and each of structures **4.4**, **4.5** and **4.6** will make a different contribution to the overall bonding in the molecule XY. Let us apply this idea to HF, LiF and LiH.

Hydrogen fluoride

The valence bond approach suggests three resonance forms for the hydrogen fluoride molecule, as shown in structures **4.7, 4.8** and **4.9**.

$$\mathrm{H{-}F} \longleftrightarrow \mathrm{H^+\,F^-} \longleftrightarrow \mathrm{H^-\,F^+}$$

(4.7) (4.8) (4.9)

The covalent form **4.7** implies that the F and H atoms *share* two electrons so that each atom has a noble gas configuration; forming a covalent single bond between the atoms in the HF molecule allows the hydrogen atom to gain a helium-like outer shell while the fluorine atom becomes neon-like.

In order to assess the relative energies of the ionic forms, consider the ground state configuration of the fluorine atom. Resonance structure **4.8** indicates that an electron has been transferred from the H atom to the F atom (equations 4.1 and 4.2).

$$\mathrm{H} \rightarrow \mathrm{H^+} + \mathrm{e^-} \qquad (4.1)$$

$$\mathrm{F} + \mathrm{e^-} \rightarrow \mathrm{F^-} \qquad (4.2)$$

The ground state electron configuration of fluorine is $[He]2s^22p^5$ and gaining one electron provides a neon-like configuration – $[He]2s^22p^6 \equiv [Ne]$. The F^- ion obeys the octet rule. The ionic form **4.8** appears to be a reasonable alternative to the covalent structure **4.7**.

Resonance structure **4.9** indicates that an electron has been transferred from fluorine to hydrogen (equations 4.3 and 4.4).

$$F \rightarrow F^+ + e^- \quad (4.3)$$

$$H + e^- \rightarrow H^- \quad (4.4)$$

Although the formation of the H^- ion gives a helium-like configuration ($1s^2 \equiv [He]$), the energetic *advantage* of this step is not sufficient to offset the energetic *disadvantage* of forming F^+. The loss of an electron from a fluorine atom to give the F^+ cation means that a species with six valence electrons (a sextet) is generated. This is expected to be *less* stable than the F^- anion with an octet. The preference for gaining an electron is far, far greater than that for losing one – reaction 4.2 is favoured over reaction 4.3. (At the end of Chapter 6, problem 6.6 asks you to consider the energetics of reactions 4.1 to 4.4.)

Accordingly, it would not be reasonable to expect a resonance structure involving F^+ to feature strongly (if at all) in a description of the bonding in hydrogen fluoride, and the contribution made by the structure $H^- F^+$ is negligible. Thus, the bonding in hydrogen fluoride can be effectively described in terms of a covalent structure and an ionic form in which H^+ and F^- ions are involved.

Lithium fluoride and lithium hydride

Valence bond theory can be used in a similar manner to that outlined for HF to describe the bonding in the LiF and LiH molecules. Resonance structures for these molecules are shown in Figure 4.3.

In LiF, we can once again disregard the resonance structure involving the F^+ ion, and the bonding in lithium fluoride can be described in terms of a covalent resonance structure and an ionic resonance form.

In LiH, three resonance structures can be drawn, all of which contribute to the overall bonding in the compound. The lithium and hydrogen atoms (Figure 4.2) possess ns^1 ground state configurations and can readily *lose* an electron to form the Li^+ or H^+ ions. Each of these ions possesses an ns^0 configuration. But hydrogen could also *gain* an electron to give H^- which is a helium-like species.

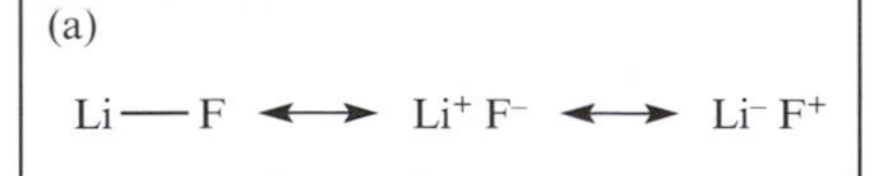

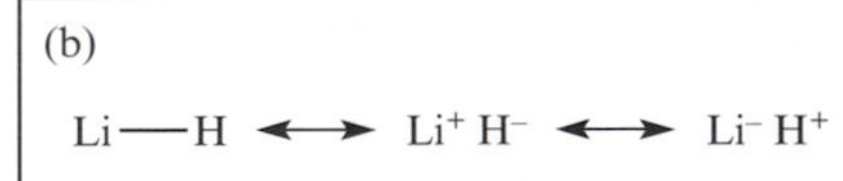

4.3 Resonance structures for the molecules (a) LiF and (b) LiH. For LiF, the right-hand structure provides a negligible contribution to the overall bonding picture.

VB theory: conclusions

Valence bond theory applied to the molecules HF, LiF and LiH gives bonding descriptions in terms of covalent and ionic resonance structures. However, in a given set of resonance structures, some forms are more important than others. In general, a qualitative treatment in terms of the stability of possible ions proves to be successful, and the method provides a good intuitive description of the bonding in simple molecules. Of course, it *is* possible to approach the problem more rigorously and estimate the contributions that each resonance structure makes to the overall bonding. For us, however, it is sufficient merely to consider their relative merits in the way discussed above.

4.4 THE MOLECULAR ORBITAL APPROACH TO THE BONDING IN A HETERONUCLEAR DIATOMIC MOLECULE

In Chapter 3 we used molecular orbital theory to describe the bonding in homonuclear diatomic molecules. The MO diagrams we constructed had a symmetrical appearance; atomic orbitals of the same type but belonging to different atoms had the same energy as seen, for example, for F_2 in Figure 3.21. Each MO of a molecule X_2 contains equal atomic orbital contributions from each atom X.

Effective nuclear charge: see Section 2.15

Now consider a heteronuclear molecule XY. In this chapter we take the case where the atoms X and Y are different, but with atomic number less than 10. A fundamental difference between X and Y is that the effective nuclear charge, Z_{eff}, experienced by the electrons in the valence shells of the two atoms is not the same. If Y is in a *later* group of the periodic table than X, then $Z_{eff}(Y) > Z_{eff}(X)$ for an electron in a given atomic orbital. As Z_{eff} increases across a row of the periodic table, the energy of a particular AO is lowered (more negative energy). This is illustrated schematically in Figure 4.4

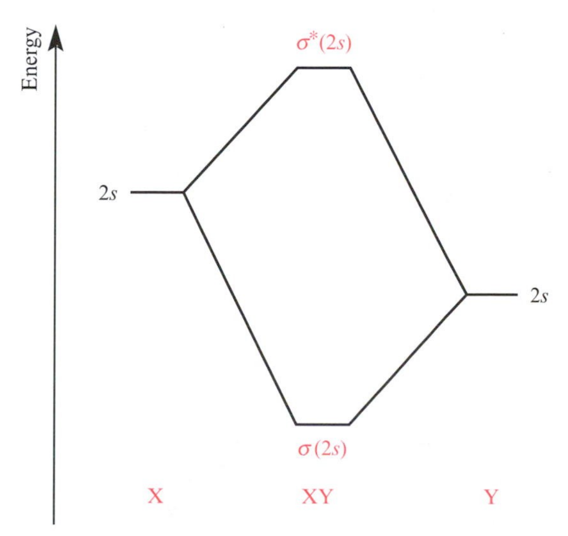

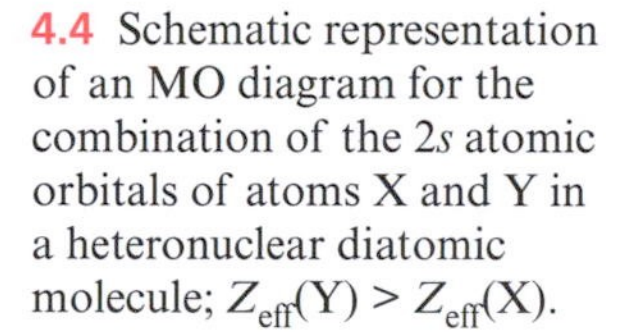
4.4 Schematic representation of an MO diagram for the combination of the 2*s* atomic orbitals of atoms X and Y in a heteronuclear diatomic molecule; $Z_{eff}(Y) > Z_{eff}(X)$.

for the combination of the 2*s* AOs of atoms X and Y to form the $\sigma(2s)$ and $\sigma^*(2s)$ MOs of the molecule XY.

An important feature of Figure 4.4 is that the energy of the bonding MO is closer to that of the 2*s* AO of atom Y than to that of the 2*s* AO of atom X. Conversely, the energy of the antibonding MO is closer to that of the 2*s* AO of atom X than to that of the 2*s* AO of atom Y. This has an effect on the nature of the molecular orbitals as illustrated in Figure 4.5. The 2*s* AO of atom Y makes a *larger* contribution to the bonding $\sigma(2s)$ orbital than does the 2*s* AO of atom X. For the antibonding MO, the situation is reversed and the 2*s* AO of atom Y makes a *smaller* contribution to the $\sigma^*(2s)$ MO than does the AO of atom X. We say that the $\sigma(2s)$ MO possesses more 2*s character* from Y and the $\sigma^*(2s)$ MO has more 2*s character* from X. In Figures 4.5a and 4.5b, the larger or smaller 2*s* contributions are represented pictorially; notice how this affects the final shape of the molecular orbitals and renders them asymmetrical.

The wavefunction for an MO in a heteronuclear diatomic molecule must show the contribution that each constituent AO makes. This is done by using *coefficients*, c_n. In equation 4.5, the coefficients c_1 and c_2 indicate the composition of the MO in terms of the atomic orbitals ψ_X and ψ_Y. The relative values of c_1 and c_2 are dictated by the effective nuclear charges of the atoms X and Y.

$$\psi_{MO} = \{c_1 \times \psi_X\} + \{c_2 \times \psi_Y\} \qquad (4.5)$$

Equation 4.6 gives an expression for the wavefunction that describes the bonding orbital in Figures 4.4 and 4.5.

$$\psi_{MO} = \{c_1 \times \psi_{X(2s)}\} + \{c_2 \times \psi_{Y(2s)}\} \qquad \text{where } c_2 > c_1 \qquad (4.6)$$

The fact that coefficient c_2 is greater than c_1 corresponds to there being more 2*s* character from Y than X with the result that the MO (Figure 4.5a) is distorted in favour of atom Y.

One final point remains. In Chapter 3 we noted the importance of *energy matching* between atomic orbitals as a criterion for an efficient orbital inter-

(a)

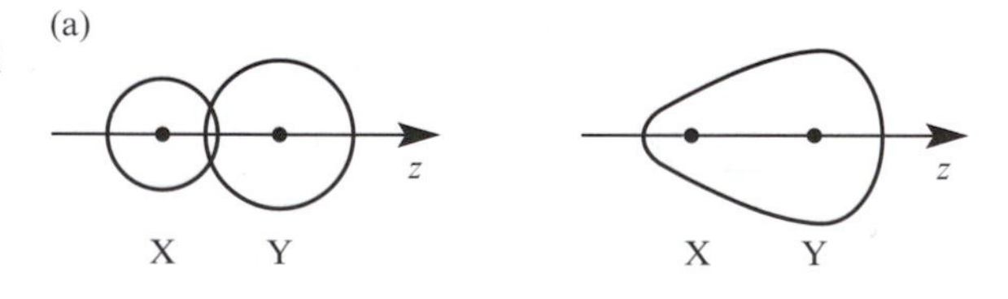

(b)

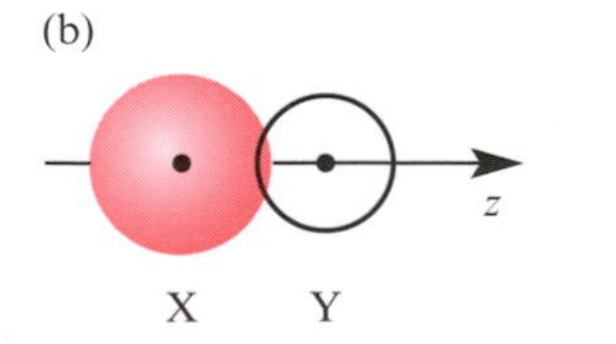

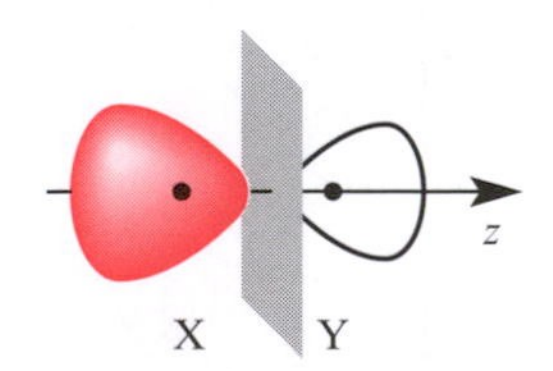

4.5 Schematic representations of (a) the $\sigma(2s)$ MO and (b) the $\sigma^*(2s)$ MO of a heteronuclear diatomic molecule XY. The nuclei are shown by black dots.

action. In the case of the homonuclear diatomic molecules, this did not give any problems when the discussion was restricted to overlap between like orbitals – in H_2, the energies of the two hydrogen 1*s* AOs were the same. In a heteronuclear diatomic there will always be differences between the orbital energies associated with the two different atoms, and it follows that there may be cases where the energy difference may be too great for effective interaction to be achieved. We illustrate this point later in the discussion of the bonding in LiF and HF.

4.5 THE MOLECULAR ORBITAL APPROACH TO THE BONDING IN LiH, LiF AND HF

Two important preliminary comments

- In using molecular orbital theory to describe the bonding in molecules at a relatively simple level, we only consider the overlap of the *valence atomic orbitals*.
- Molecular orbital theory describes the bonding in a *covalent* manner. It does not explicitly build into the picture ionic contributions as is done in valence bond theory.

Lithium hydride

The ground state configurations of hydrogen and lithium are $1s^1$ and $[He]2s^1$ respectively. The hydrogen 1*s* orbital is lower in energy than the lithium 2*s* AO. The symmetries of the 1*s* AO of hydrogen and the 2*s* AO of lithium are compatible and overlap can occur. The combination of these AOs to give two molecular orbitals is shown in Figure 4.6.

There are two valence electrons and, by the *aufbau* principle, these will have opposite spins and occupy the σ-bonding MO. The bond order in the LiH molecule is 1 (equation 4.7).

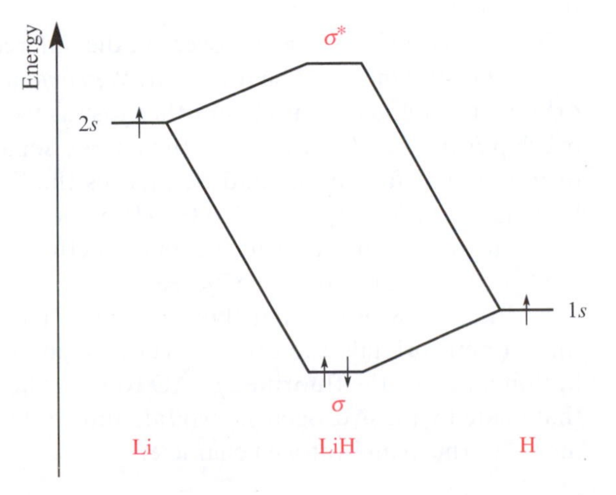

4.6 An MO diagram for the formation of the LiH molecule. Only the valence atomic orbitals and electrons are shown.

Bond order: see Section 3.16

$$\text{Bond order in LiH} = \tfrac{1}{2} \times (2 - 0) = 1 \tag{4.7}$$

Figure 4.6 shows that the hydrogen 1*s* AO contributes more to the σ-bonding MO in LiH than does the lithium 2*s* AO. This means that the σ-bonding MO exhibits more hydrogen than lithium character; the orbital has approximately 74% hydrogen character and 26% lithium character, and this is represented pictorially in Figure 4.7. As a result, the electrons spend a greater amount of their time 'associated' with the hydrogen than the lithium nucleus.

The wavefunction given in equation 4.8 describes the σ-bonding MO in LiH.

$$\psi(\text{LiH bonding}) = \{c_1 \times \psi_{\text{Li}(2s)}\} + \{c_2 \times \psi_{\text{H}(1s)}\} \tag{4.8}$$

where $c_1 < c_2$

It follows from Figure 4.6 that the σ* MO has more lithium than hydrogen character. [Draw an appropriate representation of the σ* MO in LiH.]

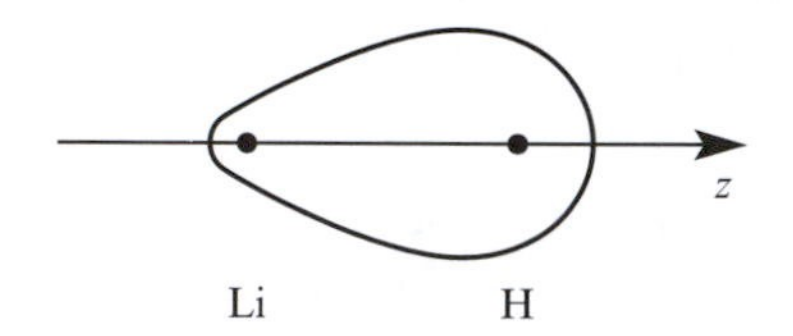

4.7 A representation of the σ-bonding MO of LiH.

Hydrogen fluoride

The ground state configurations of hydrogen and fluorine are $1s^1$ and $[\text{He}]2s^22p^5$ respectively. Before we consider the overlap of the atomic orbitals, we must think about their relative energies. In going from hydrogen to fluorine, the effective nuclear charge increases considerably, and this significantly lowers the energies of the fluorine atomic orbitals with respect to hydrogen 1*s* AO. Thus, despite the difference in principal quantum number, the fluorine 2*s* and 2*p* orbitals lie at lower energies than the hydrogen 1*s* AO. This is shown in Figure 4.8; the energy axis in the figure is broken in order to signify the fact that the fluorine 2*s* orbital is actually even lower in energy than shown.

Remember that for efficient overlap, the difference in energy between two atomic orbitals must *not* be too great. We *could* allow the H 1*s* and the F 2*s* orbitals to interact with one another just as we did in LiH, but although overlap *is allowed by symmetry*, the energy separation is too large. Thus, overlap is not favourable and this leaves the fluorine 2*s* AO unused – it becomes a *non-bonding orbital* in the HF molecule.

A *non-bonding molecular orbital* is one which has neither bonding nor antibonding character overall.

The hydrogen and fluorine nuclei are defined to lie on the *z* axis, and so the hydrogen 1*s* and fluorine $2p_z$ AOs can overlap as illustrated in Figure 4.9a. Figure 4.8 shows that there is a satisfactory energy match between these atomic orbitals and overlap occurs to give σ and σ* MOs. The contribution made by the fluorine $2p_z$ AO to the σ-bonding MO is greater than that made by the hydrogen 1*s* orbital, and so the bonding orbital contains more fluorine than hydrogen character.

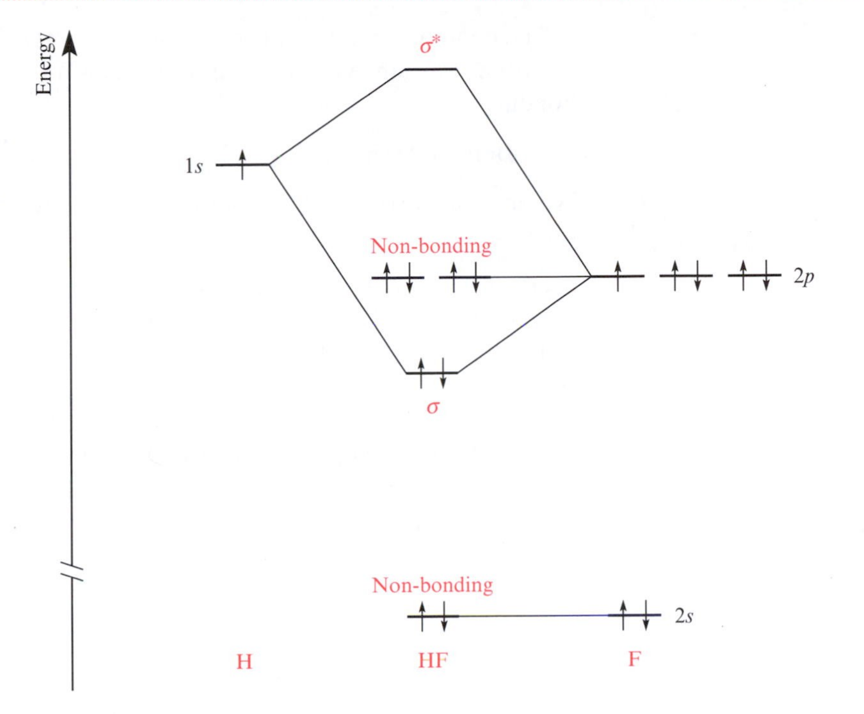

4.8 An MO diagram for the formation of HF. Only the valence atomic orbitals and electrons are shown. The break in the vertical (energy) axis signifies that the energy of the fluorine $2s$ AO is much lower than is actually shown.

The two remaining fluorine $2p$ AOs lie at right angles to the H–F internuclear axis and have no net bonding overlap with the hydrogen $1s$ orbital; Figure 4.9b illustrates this for the $2p_x$ AO. Thus, the fluorine $2p_x$ and $2p_y$ AOs become non-bonding orbitals in the HF molecule.

There are eight valence electrons in the HF molecule and, by the *aufbau* principle, they occupy the MOs as shown in Figure 4.8. Only two electrons occupy a molecular orbital which has H–F bonding character; the other six are in non-bonding MOs and are localized on the fluorine atom. These three non-bonding pairs of electrons are analogous to the lone pairs drawn in the Lewis structure for the HF molecule in Figure 4.2.

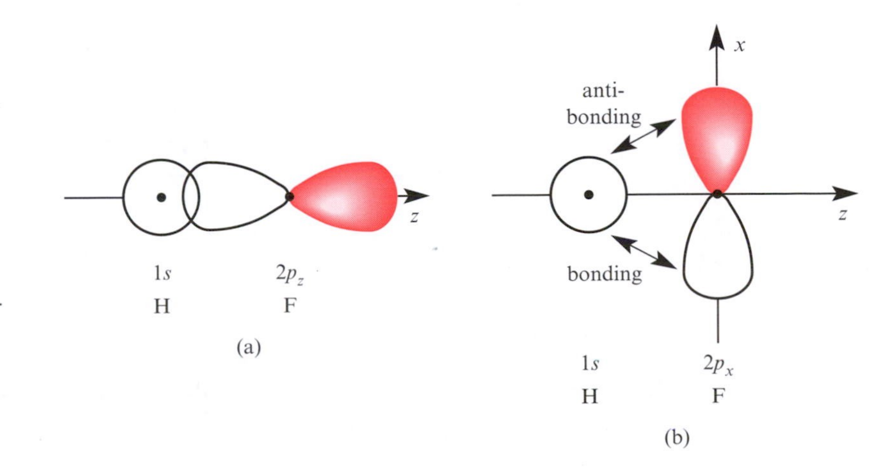

4.9 (a) The symmetry-allowed overlap of the hydrogen $1s$ AO and the fluorine $2p_z$ AO; note that the H and F nuclei lie on the z axis. (b) The combination of the hydrogen $1s$ AO and the fluorine $2p_x$ AO results in equal in-phase and out-of-phase interactions, and the overall interaction is non-bonding. A similar situation arises for the $2p_y$ orbital.

The bond order in hydrogen fluoride (equation 4.9) can be determined in the usual way since electrons in non-bonding MOs have no effect on the bonding:

$$\text{Bond order in HF} = \tfrac{1}{2} \times (2 - 0) = 1 \qquad (4.9)$$

Electrons in non-bonding MOs do not influence the bond order.

We can summarize the MO approach to the bonding in the HF molecule as follows:

- Overlap occurs between the hydrogen $1s$ and the fluorine $2p_z$ AOs to give one σ-bonding MO and one σ^*-antibonding MO.
- The fluorine $2p_x$ and $2p_y$ orbitals have no net overlap with the hydrogen $1s$ AO and are non-bonding orbitals.
- The fluorine $2s$ orbital can overlap with the hydrogen $1s$ AO on symmetry grounds but the mismatch in atomic orbital energies means that the overlap is poor – the fluorine $2s$ orbital is effectively non-bonding.
- Of the eight valence electrons in the HF molecule, two occupy an H–F bonding MO and six occupy non-bonding orbitals localized on the fluorine atom.
- The character of the σ-bonding MO in HF is biased towards the fluorine atom.

Lithium fluoride

The MO approach to the bonding in LiF closely follows that described for HF; the major difference is that we are dealing with an [He]$2s^1$ ground state configuration for lithium in place of the $1s^1$ configuration for hydrogen.

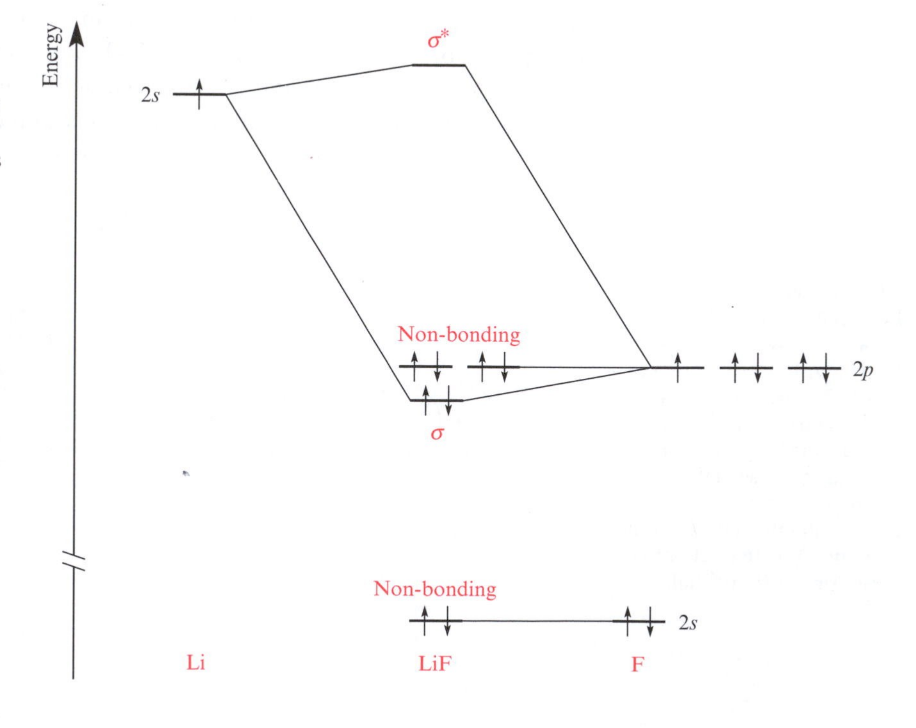

4.10 An MO diagram for the formation of the LiF molecule. Only the valence atomic orbitals and electrons are shown. The break in the vertical (energy) axis signifies that the energy of the fluorine $2s$ AO is lower than is actually shown.

Figure 4.10 gives an MO diagram for the formation of lithium fluoride; the lithium and fluorine nuclei lie on the z axis, as shown in structure **4.10**. As in HF, the fluorine $2s$ AO lies at low energy and generates a non-bonding orbital in the LiF molecule. The fluorine $2p_x$ and $2p_y$ AOs are also non-bonding as they were in HF.

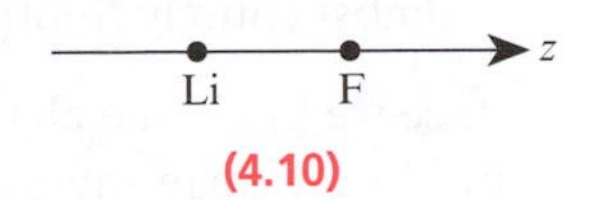

(4.10)

Overlap can occur between the lithium $2s$ AO and the fluorine $2p_z$ AO. However, the energy separation between these orbitals is significant and the interaction is not particularly effective. This is apparent in Figure 4.10 – the σ-bonding MO is stabilized *only slightly* with respect to the fluorine $2p_z$ AO. A comparison of Figures 4.8 and 4.10 shows that the stabilization of the σ-bonding orbital is smaller in LiF than in HF. What does this mean in terms of the LiF molecule?

The wavefunction that describes the bonding MO in Figure 4.10 is given in equation 4.10.

$$\psi(\text{LiF bonding}) = \{c_1 \times \psi_{\text{Li}(2s)}\} + \{c_2 \times \psi_{\text{F}(2p_z)}\} \qquad (4.10)$$

where $c_1 << c_2$

If the coefficient c_1 is very small, then the wavefunction describes an orbital which is weighted very much in favour of the fluorine atom – in equation 4.10, the term $\{c_2 \times \psi_{\text{F}(2p_z)}\}$ is *much* greater than $\{c_1 \times \psi_{\text{Li}(2s)}\}$. This means that in the bonding MO, there is a much greater probability of finding the electrons in the vicinity of fluorine, rather than lithium.

In conclusion, molecular orbital theory gives a picture for the bonding in LiF that shows one MO with bonding character; this MO is occupied by two electrons and the bond order in the molecule is one. There are three filled, non-bonding orbitals which are localized on the F atom; these are analogous to the lone pairs of electrons drawn in the Lewis structure of LiF (Figure 4.2). The bonding MO in the LiF diatomic molecule has a predominantly fluorine character (≈ 97%). Thus, although the MO approach is built upon a covalent model, the 'answer' is telling us that the bonding in LiF may be described as *tending towards being ionic*. We return to LiF in Section 6.2.

MO theory: conclusions

- The MO approach to the bonding in the LiH, HF and LiF molecules shows each to possess a bond order of one. In each molecule, the bonding MO is a σ-orbital.
- The bonding MO in LiH is formed by the overlap of the Li $2s$ and the H $1s$ atomic orbitals. In HF, the bonding MO is formed by the overlap of the H $1s$ and one of the F $2p$ orbitals. In LiF, the character of the bonding MO comes from the Li $2s$ and one of the F $2p$ orbitals.

- The use of different atomic orbitals on different nuclei can lead to dramatic differences in the percentage contributions that the atomic orbitals make to the bonding molecular orbitals. In LiH the bonding MO contains more hydrogen than lithium character. In HF, the bonding MO contains more fluorine than hydrogen character. In LiF, the bonding MO possesses almost entirely fluorine character.

Once we know the character of the single bonding MO in each of LiH, HF and LiF, we can say something about the electron distribution in the bonds in these molecules. In contrast to the symmetrical distribution of electron density between the nuclei in a homonuclear diatomic molecule, in LiH, HF and LiF it is shifted in favour of one of the atoms. In LiH, the maximum electron density is closer to the H than the Li nucleus whereas in HF it is closer to the F nucleus. Attempts to rationalize these results lead us eventually to the concept of electronegativity and to polar bonds, but first we look at the bond dissociation enthalpies of heteronuclear bonds.

4.6 BOND DISSOCIATION ENTHALPIES OF HETERONUCLEAR BONDS

In Section 3.7, we discussed the way in which Hess's Law of constant heat summation could be used to estimate values for bond dissociation enthalpies. We were primarily concerned with estimating the bond dissociation enthalpies of *homonuclear* single bonds, but to do so, it was necessary to involve the bond dissociation enthalpies of heteronuclear bonds – an estimate for D(N–N) in N_2H_4 depended upon transferring a value of $\bar{D}$(N–H) from NH_3 to N_2H_4.

$\bar{D}$ = average bond dissociation enthalpy

Thus, we are already familiar with the idea that average bond dissociation enthalpy values for heteronuclear bonds can be obtained from standard enthalpies of atomization. This is shown for CCl_4 (**4.11**), and 1,2-$C_2H_4Br_2$ (**4.12**) in equations 4.11 and 4.12.

(4.11)

(4.12)

$$\Delta_a H^o \; CCl_4(g) = 4 \times \bar{D}(C–Cl) \tag{4.11}$$

$$\Delta_a H^o \; C_2H_4Br_2(g) = [4 \times \bar{D}(C–H)] + D(C–C) + [2 \times \bar{D}(C–Br)] \tag{4.12}$$

In Section 3.6, we emphasized the relationship between the bond dissociation enthalpy, D, and the standard enthalpy of atomization for a gaseous homonuclear diatomic molecule. This is illustrated for fluorine in equation 4.13.

$$\Delta_a H^o \text{ F} = \tfrac{1}{2} \times D(\text{F–F}) \qquad (4.13)$$

Is it feasible to use a *similar* strategy to estimate the bond dissociation enthalpies of heteronuclear bonds? For an X–Y single bond, can we estimate values that are 'characteristic' of the contributions made to D(X–Y) by atoms X and Y? A logical approach is to take half of D(X–X) and allow this to be a measure of the contribution made to the bond enthalpy D(X–Y) by atom X. Similarly for Y, let half of D(Y–Y) be a measure of the contribution made to D(X–Y) by atom Y. Estimates for values of D(X–Y) can then be made using equation 4.14.

$$D(\text{X–Y}) = [\tfrac{1}{2} \times D(\text{X–X})] + [\tfrac{1}{2} \times D(\text{Y–Y})]$$

or (4.14)

$$D(\text{X–Y}) = \tfrac{1}{2} \times [D(\text{X–X}) + D(\text{Y–Y})]$$

But, are we justified in applying an additivity rule of this type? This is best answered by comparing some estimates made using equation 4.14 with experimentally determined bond dissociation enthalpies (Table 4.1). For some of the molecules (BrI, BrCl and HI) the model we have used gives good estimates of the bond dissociation enthalpies, but for others (HF and HCl) the discrepancies are large. Why does the model fail in some cases?

The breakdown of this simple model (the non-additivity of bond dissociation enthalpy contributions for heteronuclear bonds) led Linus Pauling to outline the concept of *electronegativity*.

Table 4.1 Bond dissociation enthalpies, D, of homonuclear and heteronuclear diatomic molecules. Experimental values for D(X–X) and D(Y–Y) are used (equation 4.14) to estimate values of D(X–Y). Comparison is then made with experimental data. The value ΔD in the final column is the difference between the experimental D(X–Y) and the estimated value, i.e. the difference between the value in column 6 in the table and the value in column 5 (see equation 4.15).

X	*Y*	*D for X–X bond in X_2 / kJ mol^{-1}*	*D for Y–Y bond in Y_2 / kJ mol^{-1}*	*$\frac{1}{2} \times [D(X–X) + D(Y–Y)]$ / kJ mol^{-1}*	*Experimental D(X–Y) in XY / kJ mol^{-1}*	*ΔD (from equation 4.15)*
H	F	436	159	298	570	272
H	Cl	436	242	339	432	93
H	Br	436	193	315	366	51
H	I	436	151	294	298	4
F	Cl	159	242	201	256	55
Cl	Br	242	193	218	218	0
Cl	I	242	151	197	211	14
Br	I	193	151	172	179	7

Box 4.1 Linus Pauling (1901–1994)

Roger Ressmeyer, Starlight / Science Photo Library

You probably associate the name Pauling with electronegativity values but his contributions to chemistry and to science more generally *far* exceed χ^P. His imagination and genius have left a permanent mark on the underlying theories of modern chemistry.

Linus Carl Pauling was born in Oregon in 1901 and received a degree in chemical engineering in 1922. From there, his interests focused on X-ray diffraction methods and the solving of crystal structures. The year 1926 saw Pauling working with Schrödinger and Bohr, trying to get to grips with quantum theory and the structure of the atom. In the 1920s, the structures of silicate minerals were beginning to unfold, but their complexity brought with it challenges to structural perception. In 1929, Pauling set out rules based on ionic radii and charges which helped to rationalize the silicate family. His preoccupation at this time with the chemical bond brought sense to many aspects of bonding that had previously not been understood, and he combined quantum theory with the more qualitative theories of, for example, G. N. Lewis. Pauling was also the advocate of resonance structures, realizing that chemical properties could be best understood in terms of a combination of various structural forms. The establishment of Pauling's valence bond theory provided chemists with a new foundation on which to build their chemical models and, in the late 1930s, Pauling's book *The Nature of the Chemical Bond* opened the eyes of many to the world of chemical bonding and structure. This book is still one of the masterpieces of the chemical literature.

Biochemistry has also benefited from Pauling's gift of imagination and it was he who first considered the α-helix as a motif in protein structure, and his ideas were confirmed through X-ray diffraction experiments in Cambridge, UK, in the late 1940s. In 1954, for his contributions to chemistry, Pauling received the first of two Nobel prizes – the second was awarded in 1963 for his involvement in world peace efforts. He was an adamant anti-nuclear campaigner and was, for a time, at loggerheads with the US government. From 1966 onwards, Pauling turned his attention to vitamin C and for many of his later years he dedicated himself to it. He was convinced that vitamin C was the cure for all ills including cancer and he took huge doses daily.

4.7 ELECTRONEGATIVITY – PAULING VALUES (χ^P)

Deriving Pauling electronegativity values

When we compare the estimated and experimental values of D(X–Y) listed in Table 4.1, we see that the estimated value is often too small. In 1932, Pauling suggested that these differences (equation 4.15) could be rationalized in terms of his valence bond model. Specifically, he was concerned with the discrepancy, ΔD (Table 4.1), between the experimentally determined value for the X–Y bond and the average of the values for X_2 and Y_2.

$$\Delta D = [D(\text{X–Y})_{\text{experimental}}] - \{\tfrac{1}{2} \times [D(\text{X–X}) + D(\text{Y–Y})]\} \qquad (4.15)$$

Pauling proposed that ΔD was a measure of the 'ionic' character of the X–Y bond. In valence bond terms, this corresponds to a measure of the importance of resonance structures X^+Y^- and X^-Y^+ in the description of the bonding in molecule XY – how much more important are X^+Y^- and X^-Y^+ in the bonding picture of XY than are X^+X^- and Y^-Y^+ in X_2 and Y_2 respectively? Pauling then suggested that the resonance forms X^+Y^- and X^-Y^+ were not equally important. (We have justified this suggestion for HF in Section 4.3.)

Let us assume that the form X^+Y^- predominates. Remember, this does *not* imply that the molecule XY actually exists as the pair of ions X^+ and Y^-. All we are saying is that the electrons in the X–Y bond are drawn more towards atom Y than they are towards atom X. That is, in the XY molecule, atom Y has a greater *electron-withdrawing power* than atom X.

Pauling then defined a term that he called *electronegativity* (χ^P)§ which he described as 'the power of an atom *in a molecule* to attract electrons to itself'.This approach treats Pauling electronegativity as an *atomic* property rather than a property of a bond. It is important to remember that electronegativity is a property of an *atom in a molecule*, and is not a fundamental property of an isolated atom.

Electronegativity, χ^P, was defined by Pauling as 'the power of an atom in a molecule to attract electrons to itself '.

Although we approached the idea of electronegativity by considering the energy involved in the formation of the diatomic XY, it is clear that this concept is equally valid when considering X–Y bonds in larger molecules.

Pauling values of electronegativities (Table 4.2) are calculated as follows. The difference ΔD (equation 4.15) is usually measured in units of kJ mol^{-1}. Values of χ^P relate to thermochemical data given in the non-SI units of electron volts (eV). The value ΔD was used to estimate the difference in electronegativities between atoms X and Y (equation 4.16) where ΔD is found from equation 4.15 and is given in eV.

1 eV = 96.5 kJ mol^{-1}

$$\chi^P(\text{Y}) - \chi^P(\text{X}) = \sqrt{\Delta D} \qquad (4.16)$$

By solving equation 4.16 for a range of different combinations of atoms X and Y, Pauling estimated a self-consistent set of electronegativity values.

§ The superscript in the symbol χ^P distinguishes Pauling (P) electronegativity values from those of other scales.

Table 4.2 Pauling electronegativity values (χ^P) for selected elements of the periodic table. For some elements which exhibit a variable oxidation state, χ^P is given for a specific state. It is conventional not to assign units to values of electronegativity.

Group 1	Group 2		Group 13	Group 14	Group 15	Group 16	Group 17
H 2.2							
Li 1.0	Be 1.6		B 2.0	C 2.6	N 3.0	O 3.4	F 4.0
Na 0.9	Mg 1.3		Al(III) 1.6	Si 1.9	P 2.2	S 2.6	Cl 3.2
K 0.8	Ca 1.0		Ga(III) 1.8	Ge(IV) 2.0	As(III) 2.2	Se 2.6	Br 3.0
Rb 0.8	Sr 0.9	(*d*-block elements)	In(III) 1.8	Sn(II) 1.8 Sn(IV) 2.0	Sb 2.1	Te 2.1	I 2.7
Cs 0.8	Ba 0.9		Tl(I) 1.6 Tl(III) 2.0	Pb(II) 1.9 Pb(IV) 2.3	Bi 2.0	Po 2.0	At 2.2

Pauling worked with a set of numbers that gave $\chi^P(H) \approx 2$ and $\chi^P(F) = 4$. The larger the value of χ^P, the greater the electron-attracting (withdrawing) power of the atom – fluorine is the most electronegative atom in Table 4.2.

Over the years since Pauling introduced this concept, more accurate thermochemical data have become available and values of χ^P have been updated. Table 4.2 lists values of χ^P in current use. In the same way that thermochemical data for X–X and Y–Y bonds in the molecules X_2 and Y_2 may be used, so may values for X–X and Y–Y *bonds in molecules* (e.g. the C–C bond in C_2H_6 or the O–O bond in H_2O_2) be used to obtain values of χ^P. This last extension allows us to define electronegativity values for elements in various oxidation states.

Worked example 4.1 ***Estimating the bond dissociation enthalpy of a heteronuclear covalent bond***

Estimate the dissociation enthalpy of the bond in iodine monochloride, ICl, given that:

$D(Cl\text{–}Cl) = 242$ kJ mol^{-1}
$D(I\text{–}I) = 151$ kJ mol^{-1}
$\chi^P(Cl) = 3.2$
$\chi^P(I) = 2.7$.

The equations needed are 4.15 and 4.16:

$$\Delta D = D(\text{X–Y}) - \{\tfrac{1}{2} \times [D(\text{X} - \text{X}) + D(\text{Y} - \text{Y})]\}$$
$$\chi^P(\text{Y}) - \chi^P(\text{X}) = \sqrt{\Delta D}$$

These equations use values of D in eV.
First, find ΔD.

$$\begin{aligned}\sqrt{\Delta D} &= \chi^P(\text{Cl}) - \chi^P(\text{I}) \\ &= 3.2 - 2.7 \\ &= 0.5 \\ \Delta D &= (0.5)^2 \\ &= 0.25 \text{ eV}\end{aligned}$$

Converting to kJ mol^{-1}:

$$\begin{aligned}\Delta D &= 0.25 \times 96.5 \\ &= 24.1 \text{ kJ mol}^{-1}\end{aligned}$$

Now, find D(I–Cl):

$$\begin{aligned}\Delta D &= D(\text{X–Y}) - \{\tfrac{1}{2} \times [D(\text{X–X}) + D(\text{Y–Y})]\} \\ 24.1 &= D(\text{I–Cl}) - \{\tfrac{1}{2} \times [D(\text{I–I}) + D(\text{Cl–Cl})]\} \\ D(\text{I–Cl}) &= 24.1 + \{\tfrac{1}{2} \times (151 + 242)\} \\ &= 220.6 \text{ kJ mol}^{-1}\end{aligned}$$

4.8 THE DEPENDENCE OF ELECTRONEGATIVITY ON OXIDATION STATE AND BOND ORDER

Oxidation state

Some elements exhibit more than one oxidation state, and electronegativity values are dependent on the oxidation state – in Table 4.2, some of the values of χ^P are given, not just for the element, but for the element in a defined oxidation state, e.g. χ^P Sn(II) = 1.8 and Sn(IV) = 2.0.

Oxidation states: see Section 1.16

Consider the group 14 element lead. This is a metal and may use all of its valence electrons to form lead(IV) compounds such as $PbCl_4$. On the other hand, it often uses only two valence electrons to form lead(II) compounds such as $PbCl_2$ and PbI_2. The electron-withdrawing power of a lead(IV) centre is greater than that of a lead(II) centre and this is apparent in the electronegativity values.

In general, *electronegativity values for a given element increase as the oxidation state becomes more positive.*

Bond order

The electronegativity of a carbon atom involved in covalent single bonds is not the same as that of a carbon atom involved in multiple bonding.

(4.13) 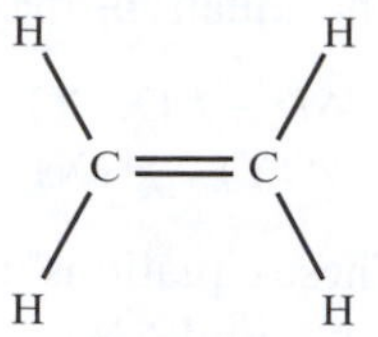 (4.14) 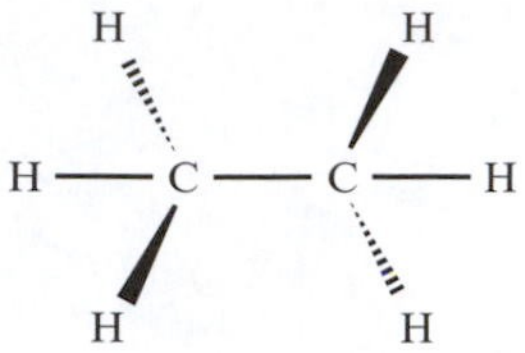 (4.15)

In the hydrocarbons C_2H_2 (**4.13**), C_2H_4 (**4.14**) and C_2H_6 (**4.15**), the carbon–carbon bond strength varies and so one expects different χ^P values. On the other hand, Table 4.2 only lists one value for $\chi^P(C)$! The differences between such electronegativity values are often quite small, but the discrepancy can be sufficient to give some spurious results if the values are used to estimate how electron density is distributed between two atoms. We return to this problem in Section 4.13.

4.9 AN OVERVIEW OF THE BONDING IN HF

In this section, we review the ideas developed so far in this chapter by considering the bonding in hydrogen fluoride, combining the ideas of the Lewis approach, VB and MO theories, bond dissociation enthalpies and electronegativities.

- The resonance structure H^+F^- contributes to the bonding significantly more than does the form H^-F^+.
- The MO approach to the bonding in HF indicates that the character of the σ-bonding MO is weighted in favour of the fluorine atom.
- In both VB and MO theories, the greater electron-withdrawing power of the fluorine atom means that the distribution of electron density along the H–F bond-vector is unequal and is significantly displaced towards the fluorine atom.
- An estimate of the bond dissociation enthalpy of the H–F bond in HF by a method of simple additivity proves to be unsatisfactory (Table 4.1).
- Pauling's concept of electronegativity allows the estimated value of D(H–F) to be corrected for the added strength resulting from a degree of ionic character in the bond. The electronegativity difference between F and H is 1.8 (Table 4.2) and thus, by combining equations 4.15 and 4.16, a corrected value for D(H–F) of 610 kJ mol^{-1} is obtained. This compares with an experimental value of 570 kJ mol^{-1}. Typically, bond dissociation enthalpies calculated in this way are within ±10% of the experimental values.

4.10 OTHER ELECTRONEGATIVITY SCALES

Although Pauling's electronegativity scale works well, it has a number of problems.

Values of χ^P are arrived at in an empirical manner. Why should we set the difference in electronegativities between two atoms to the *square root* of the difference in bond dissociation enthalpies? If χ^P is a property of atoms in molecules, what controls it? How do Pauling electronegativity values relate to electron distributions in molecules?

In an attempt to answer these questions, various other electronegativity scales have been devised based on a variety of atomic properties. We consider two such scales.

Mulliken electronegativity values (χ^M)

Consider a diatomic molecule XY. The average position of the electron density in the bond will depend upon the relative electron-withdrawing powers of the atoms X and Y. There will be a *tendency* for each atom to possess *either* a positive *or* negative charge. This idea leads to Mulliken's method of estimating electronegativity values. Mulliken suggested that an estimate of the electronegativity of an atom could be obtained from atomic properties relating to the formation of cations and anions. The appropriate properties are the first ionization energy and the first electron affinity of an element. We deal with these quantities in Chapter 6, but for convenience definitions are also given here.

The first ionization energy (IE_1) of an atom is the energy change which is associated with the removal of the first valence electron (equation 4.17); the internal energy change is measured for the gas phase process.

$$X(g) \rightarrow X^+(g) + e^- \qquad (4.17)$$

The first electron affinity (EA_1) of an element is *minus* the internal energy change ($EA = -\Delta U(0\ K)$ which accompanies the gain of one electron (equation 4.18); the energy change is measured for the gas phase process.

$$Y(g) + e^- \rightarrow Y^-(g) \qquad (4.18)$$

Mulliken's proposal followed from Pauling's definition of electronegativity – *the power of an atom in a molecule to attract electrons to itself* – and is consistent with the valence bond concept that the ionic resonance forms X^+Y^- or X^-Y^+ may contribute significantly to the bonding in the heteronuclear diatomic molecule XY.

Mulliken electronegativity values are determined using equation 4.19, but numbers calculated from this relationship are *not directly comparable* with Pauling electronegativities, although they may be adjusted to a 'Pauling-compatible' scale.

$$\chi^M = \frac{IE_1 + EA_1}{2}$$

where IE_1 and EA_1 are in eV (1 eV = 96.5 kJ mol^{-1}) (4.19)

Allred-Rochow electronegativity values (χ^{AR})

Allred and Rochow devised another electronegativity scale based upon the effective nuclear charge, Z_{eff}, of an atom. Effective nuclear charges may be calculated by using Slater's rules.

Slater's rules: see Section 2.15

The Allred–Rochow electronegativities, χ^{AR}, are determined using equation 4.20 which scales values of χ^{AR} so that they are *directly comparable* with those of χ^{P}.

$$\chi^{AR} = \left(3590 \times \frac{Z_{eff}}{r_{cov}^{2}}\right) + 0.744$$

where r_{cov} is in pm. (4.20)

4.11 POLAR DIATOMIC MOLECULES

A homonuclear diatomic molecule such as H_2 or N_2 possesses a symmetrical distribution of electronic charge, and the two ends of the molecule are indistinguishable. Such molecules are said to be *non-polar*.

A heteronuclear diatomic molecule is composed of two different atoms. If these atoms exhibit different electron-withdrawing powers, the electron density will be shifted towards the more electronegative atom. Such molecules are said to be *polar* and possess an *electric dipole moment* (μ). The inclusion of the term electric is needed to distinguish the property from a magnetic dipole but is not always included.

Magnetic moment: see Section 16.13

A *polar* diatomic molecule is one in which the average charge distribution is shifted towards one end of the molecule. Such molecules possess an *electric dipole moment*, μ. A dipole moment is a vector quantity.

The SI units of μ are coulomb metres (C m); however, for convenience, debye units are often used (1 D = 3.336×10^{-30} C m).

In a polar diatomic molecule, one end of the molecule is negatively charged *with respect to* the other end. This is illustrated in Figure 4.11 for HBr. The notation δ^{+} and δ^{-} shows that there is a partial charge separation in the molecule but it is not a quantitative measure – it does not indicate the extent to which the separation occurs. A barred arrow (Figure 4.11c) may be used to indicate the direction of the dipole moment. Values of μ for some heteronuclear diatomic molecules containing single bonds are listed in Table 4.3.

Electronegativity values can be used to predict whether or not a molecule will be polar and to estimate the magnitude of the dipole moment. Consider hydrogen bromide. The electronegativities, χ^{P}, of H and Br are 2.2 and 3.0 respectively. This tells us that the Br atom withdraws electrons more strongly than does the H atom, and we would predict a dipole moment in accordance with that shown in Figure 4.11.

Box 4.2 Dipole moments

Dipole moments arise as a consequence of electronic charges.

In a diatomic molecule XY in which the atoms have different electron withdrawing powers, let the partial charges on the atoms be $+q$ and $-q$:

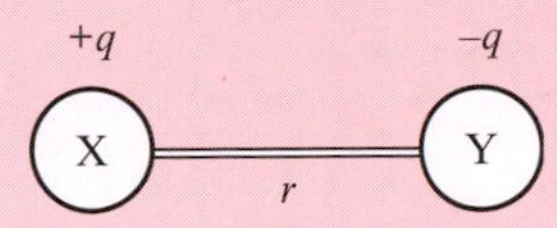

The molecule is neutral overall and so the *magnitudes* of the charges are the same, but the signs are different.

Let the separation between the two centres of charge be r.

The dipole moment, μ, of the molecule is given by:

$$\mu = (\text{electronic charge}) \times (\text{distance between the charges})$$

The magnitude of the charge shown in the above diagram is q, but the electronic charge is $(q \times e)$ – the factor e, the charge on the electron, scales the charge to an atomic level.

$$\mu = q \times e \times r$$

Charge is measured in coulombs and the distance, r, is in metres.

The SI units of μ are coulomb metres (C m) but for convenience, μ is often given in units of debyes (D) where $1\ \text{D} = 3.336 \times 10^{-30}$ C m.

Problem

The dipole moment of HF is 1.83 D. The H–F bond length is 92 pm. Estimate the atomic charge distribution in the HF molecule.

(Charge on the electron = 1.602×10^{-19} C)

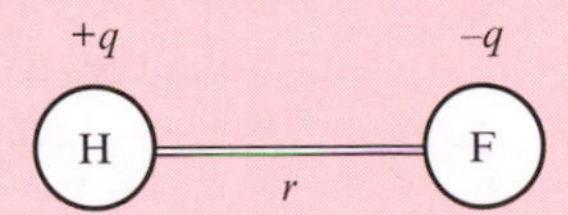

The dipole moment is given by:

$$\mu = q \times e \times r$$

where q is the atomic charge

If we assume that the centres of charge coincide with the H and F nuclei, then r is equal to the bond distance.

The H–F bond distance = 92 pm = 9.2×10^{-11} m

$$q = \frac{\mu}{e \times r} \quad \text{where } \mu \text{ is in C m.}$$

$$= \frac{1.83 \times 3.336 \times 10^{-30}}{1.602 \times 10^{-19} \times 9.2 \times 10^{-11}}$$

$$= 0.4$$

This value indicates that the charge in the HF molecule is distributed such that the fluorine atom has effectively gained 0.4 of an electron and the hydrogen atom has lost 0.4 of an electron with respect to the neutral atoms.

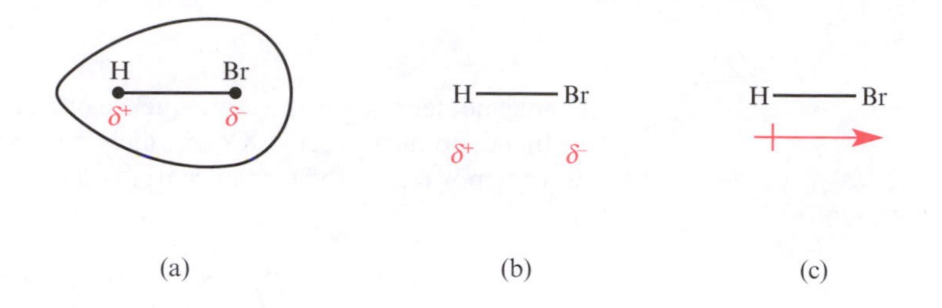

4.11 (a) The polarity of the HBr molecule arises from an asymmetrical charge distribution; (b) this creates a polar bond; (c) the dipole can be denoted by a barred arrow which points towards the more negatively charged end of the molecule.

Table 4.3 Dipole moments (μ) of some diatomic molecules in the gas phase. Values are given in debye units (D); these are not SI units but are used for convenience. (1 D = 3.336×10^{-30} C m)

Molecule XY	μ / D	*Polarity* (δ^+ *and* δ^-)
HF	1.83	H (δ^+) F (δ^-)
HCl	1.11	H (δ^+) Cl (δ^-)
HBr	0.83	H (δ^+) Br (δ^-)
HI	0.45	H (δ^+) I (δ^-)
ClF	0.89	Cl (δ^+) F (δ^-)
BrCl	0.52	Br (δ^+) Cl (δ^-)
BrF	1.42	Br (δ^+) F (δ^-)
ICl	1.24	I (δ^+) Cl (δ^-)
IBr	0.73	I (δ^+) Br (δ^-)
LiH	5.88	Li (δ^+) H (δ^-)
LiF	6.33	Li (δ^+) F (δ^-)

$$\overset{\delta^+}{\text{H}}-\overset{\delta^-}{\text{X}}$$

(4.16)

The difference in electronegativities between the atoms in a heteronuclear diatomic molecule gives some indication of the magnitude of the dipole moment. In the series of hydrogen halides HF, HCl, HBr and HI, χ^P decreases in the order $\chi^P(\text{F}) > \chi^P(\text{Cl}) > \chi^P(\text{Br}) > \chi^P(\text{I})$. In each case, χ^P(halogen) > χ^P(H). Accordingly, we predict that the most polar molecule in the series is HF and the least polar is HI, but in each hydrogen halide, HX, the dipole moment is in the direction shown in structure **4.16**.

Electronegativity values must be used with caution in molecules containing multiple bonds – care must be taken to choose values appropriate to the oxidation state and atoms involved.

4.12 ISOELECTRONIC SPECIES

In this section, we take an aside to introduce a valuable concept – that of *isoelectronic* species.

The number of valence electrons in the N_2 molecule is ten – five from each nitrogen atom. The CO molecule also has ten valence electrons – four from carbon and six from oxygen. In fact, N_2 and CO have the same *total* (valence + core) number of electrons and are said to be *isoelectronic*. The term isoelectronic means 'the same number of electrons'.

Two species are *isoelectronic* if they possess the same *total* number of electrons, that is, {core + valence} electrons.

The concept of isoelectronic species is extremely useful and is one to which we shall return at various points in this book.

Box 4.3 Isoelectronic species

Using the periodic table is a valuable (and quick) way of seeing if two species are isoelectronic.

Part of the *p*-block is shown below:

Group 13 ns^2np^1	Group 14 ns^2np^2	Group 15 ns^2np^3	Group 16 ns^2np^4	Group 17 ns^2np^5
B $Z = 5$	C $Z = 6$	N $Z = 7$	O $Z = 8$	F $Z = 9$
Al $Z = 13$	Si $Z = 14$	P $Z = 15$	S $Z = 16$	Cl $Z = 17$
Ga $Z = 31$	Ge $Z = 32$	As $Z = 33$	Se $Z = 34$	Br $Z = 35$

Moving one place to the right in a given row in the table adds one valence electron: fluorine has one more valence electron than oxygen.

Moving one place to the left in a given row in the table means that there is one valence electron fewer: phosphorus has one less valence electron than sulfur.

For elements in a given group, the number of *valence electrons* remains the same but the number of *core electrons* changes: nitrogen has a ground state configuration of $1s^22s^22p^3$ and phosphorus has a configuration of $1s^22s^22p^63s^23p^3$. This means that elements in the group are *not* strictly isoelectronic with each other. They differ in the number of core electrons but possess the same number of valence electrons. However, because of their close relationship, we can loosely say that they are isoelectronic, adding the proviso that they are 'isoelectronic with respect to the valence electrons'.

Worked example 4.2 ***Isoelectronic atoms and ions***

Confirm that the F atom and the O^- ion are isoelectronic (F $Z = 9$; O $Z = 8$).

The fluorine atom has a ground state electronic configuration [He]$2s^22p^5$.
The oxygen atom has a ground state configuration [He]$2s^22p^4$.
The O^- ion has a ground state configuration [He]$2s^22p^5$.
Therefore, the O^- ion is isoelectronic with the fluorine atom.

Worked example 4.3 ***Isoelectronic molecules and ions***

Show that the species CO, $[CN]^-$ and N_2 are isoelectronic. (C $Z = 6$; N $Z = 7$; O $Z = 8$).

An oxygen atom and an N^- ion both possess 8 electrons.
The molecule CO possesses $(6 + 8) = 14$ electrons.
The anion $[CN]^-$ possesses $(6 + 8) = 14$ electrons.
The molecule N_2 possesses $(7 + 7) = 14$ electrons.
The species CO, $[CN]^-$ and N_2 are all isoelectronic.

4.13 THE BONDING IN CO BY THE LEWIS AND VB APPROACHES

Bonding in C_2 and O_2: see Sections 3.18 and 3.20

Structures **4.17** and **4.18** show Lewis structures for the homonuclear diatomics C_2 and O_2. The fact that a double bond is common to both these molecules suggests that we could form a similar molecule between a carbon and an oxygen atom – structure **4.19** shows CO, carbon monoxide.

:C=C: (4.17)   :Ö=Ö: (4.18)   :C=Ö: (4.19)

The Lewis structure **4.19** is not the most satisfactory of pictures for the CO molecule. The formation of a C=O double bond gives the oxygen atom an octet, but the carbon atom only has six electrons in its outer shell. The observation that CO is isoelectronic with N_2 is an important key in being able to understand how the CO molecule overcomes this problem.

A nitrogen atom possesses five valence electrons. In the N_2 molecule **4.20**, two nitrogen atoms combine to form a triple bond and this gives each atom an octet of valence electrons.

:N≡N: (4.20)

A nitrogen atom is isoelectronic with both C^- and O^+, and so, by comparing CO with N_2, it is possible to write down a Lewis structure for CO by combining C^- and O^+. This is done in Figure 4.12; compare this with structure **4.20**.

The two resonance structures that describe carbon monoxide are shown in structures **4.21** and **4.22**. Note that structure **4.21** involves a C=O double bond whilst structure **4.22** involves a triple bond. In structure **4.22**, each atom has an octet of electrons.

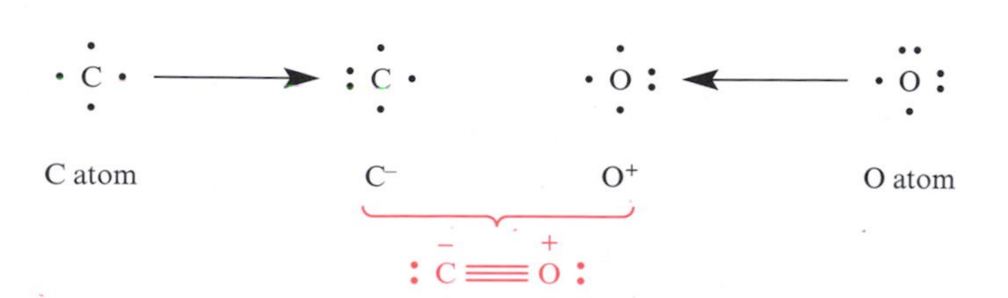

4.12 Using the isoelectronic relationship between N_2 and CO, the bonding in CO could be viewed in terms of the combination of C^- and O^+.

$$:\mathrm{C}{=}\ddot{\mathrm{O}}: \longleftrightarrow :\overset{-}{\mathrm{C}}{\equiv}\overset{+}{\mathrm{O}}:$$

(4.21) (4.22)

If structure **4.22** contributes significantly to the overall bonding in CO, it suggests that the molecule should be polar *but with the carbon atom being* δ^-. (This contrasts sharply with what you might have expected simply by looking at the values of the electronegativities of carbon and oxygen in Table 4.2 and emphasizes the dependence of χ on bond order.) The observed dipole moment of CO is 0.11 D. This is a comparatively low value, and indicates that CO is almost non-polar – the carbon-end of the molecule is only slightly negative with respect to the oxygen-end. This has important consequences for the chemistry of CO.

Chemistry of CO: see Sections 13.6 and 16.14

If structure **4.21** is important, then the bond dissociation enthalpy of CO and the carbon–oxygen bond distance should be consistent with a double bond. Structure **4.22** suggests a stronger bond with a bond order of 3. Table 4.4 lists some relevant data for CO and also for N_2 – remember that N_2 has a triple bond. The bond distance in CO is similar to that in N_2, and the bond is even stronger. These data support the presence of a triple bond.

Box 4.4 Carbon monoxide – a toxic gas

Carbon monoxide is a very poisonous gas and its toxicity arises because it binds more strongly to the iron centres in haemoglobin than does O_2 and thereby prevents the uptake and transport of dioxygen in the bloodstream. The structure of one of the chains of the mammalian dioxygen-binding protein haemoglobin is shown in diagram (a) below. Hydrogen atoms have been omitted from this structure for simplicity. The iron is bound by a cyclic nitrogen-donor molecule in a unit called haem (or heme), and this portion of the molecule is highlighted in the shaded circle. An enlargement of the environment about the iron centre is shown in diagram (b) where the iron centre can be seen to be attached to five nitrogen atoms. When O_2 or CO binds to the iron, it occupies a site so that the iron centre is near-octahedrally sited.

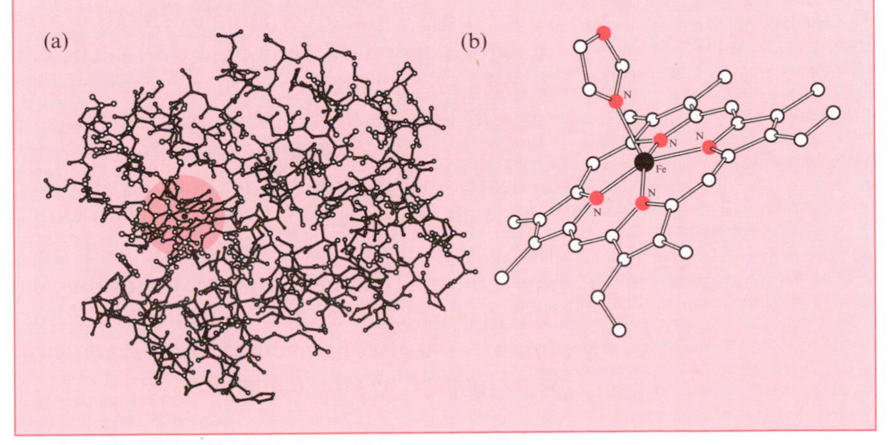

Table 4.4 Some properties of CO compared with those of N_2.

Property	*CO*	*N_2*
Bond distance / pm	113	110
Bond dissociation enthalpy / kJ mol^{-1}	1076	945
Dipole moment / D	0.11	0
Magnetic properties	Diamagnetic	Diamagnetic

Diamagnetism: see Section 3.18

In conclusion, valence bond theory describes the bonding in carbon monoxide in terms of resonance structures which show either a double or triple bond. All of the electrons are paired, in keeping with the observed diamagnetism of the compound. The observed properties of CO (the strong, short bond, and the slight polar nature, $C^{\delta-}\,O^{\delta+}$, of the molecule) can be rationalized in terms of contributions from *both* of the two resonance structures **4.21** and **4.22**. In addition, note that any dipole moment predicted from Pauling electronegativity values opposes the charge separation indicated by resonance structure **4.22**.

4.14 THE BONDING IN CARBON MONOXIDE BY MO THEORY

An initial approach

We now consider the bonding in carbon monoxide using molecular orbital theory and build upon details that we discussed in Chapter 3.

Initial points to note are:

- The effective nuclear charge of oxygen atom is greater than that of a carbon atom.
- The $2s$ AO of oxygen lies at lower energy than the $2s$ AO of carbon.
- The $2p$ AOs of oxygen lie at lower energy than the $2p$ AOs of carbon.
- The $2s$–$2p$ energy separation in oxygen is greater than that in carbon.

A linear combination of atomic orbitals of carbon and oxygen gives rise to the MO diagram shown in Figure 4.13. The bonding MOs [$\sigma(2s)$, $\sigma(2p)$ and $\pi(2p)$] all have more oxygen than carbon character. In particular, Figure 4.13 suggests that the $\sigma(2s)$ MO is predominantly oxygen $2s$ in character.

To a first approximation, this MO diagram is satisfactory – it correctly indicates that the bond order in CO is three. It also predicts that the molecule is diamagnetic and this is consistent with experimental data. However, this simple picture gives an incorrect ordering of the MOs in the molecule and indicates that the σ-MOs are definitively either bonding or antibonding. You may wish to leave the discussion of the bonding of CO at this point, but for the sake of completeness, the remainder of this section deals with the consequences of orbital mixing.

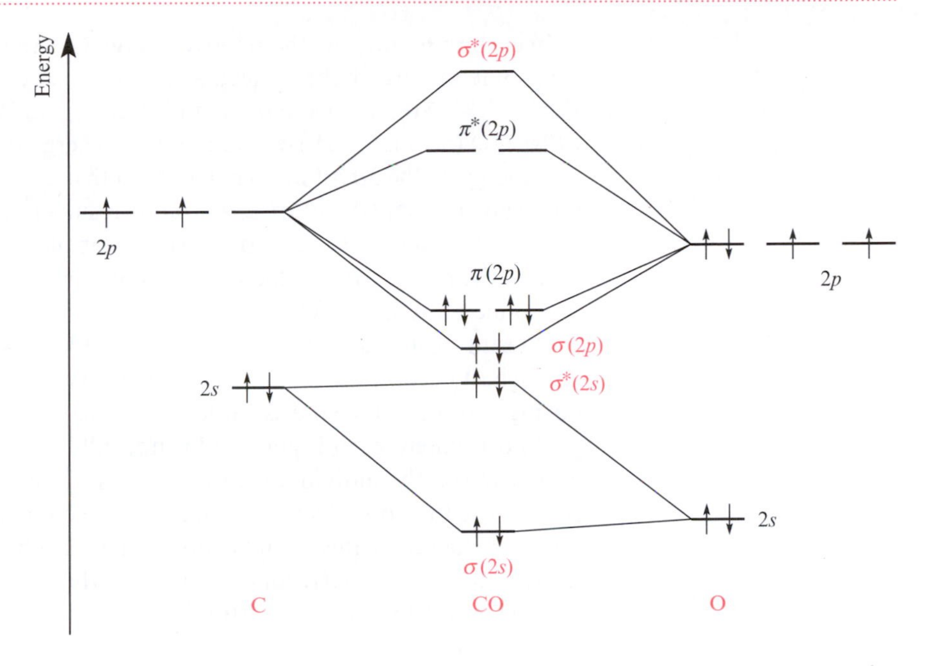

4.13 An MO diagram for the formation of CO which allows only overlap between 2*s*(C) and 2*s*(O) and between 2*p*(C) and 2*p*(O) atomic orbitals.

The effects of orbital mixing

σ–π crossover: see Section 3.19

In section 3.19, we discussed the consequences of orbital mixing. If the 2*s* and 2*p* atomic orbitals of each atom in a homonuclear diatomic molecule are close enough in energy, then mixing of the σ(2*s*) and σ(2*p*) *molecular* orbitals can occur. Similarly, mixing of the σ*(2*s*) and σ*(2*p*) molecular orbitals can occur. This results in a re-ordering of the MOs.

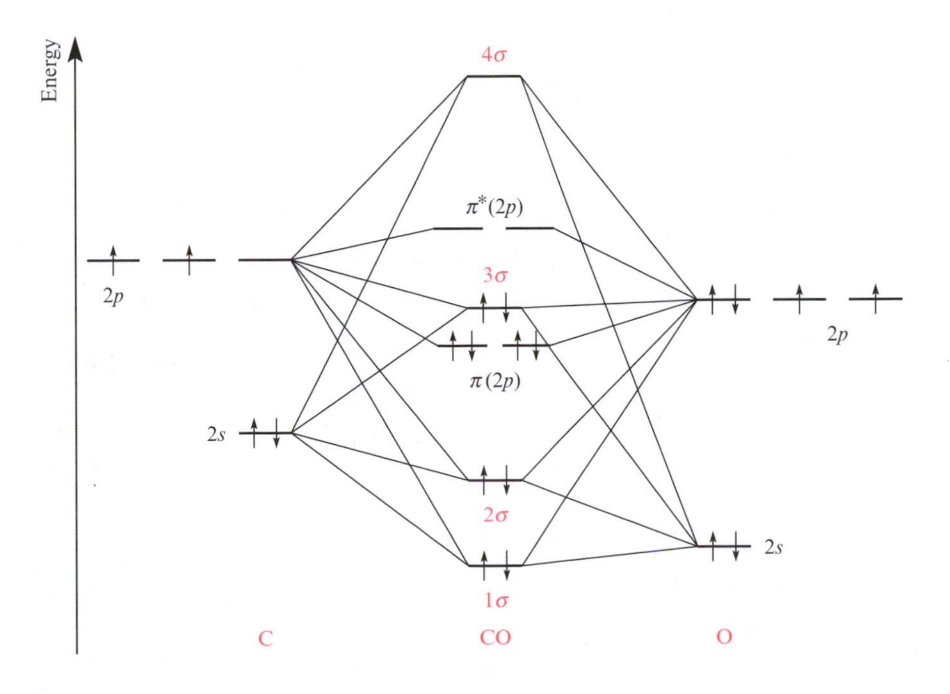

4.14 An MO diagram for the formation of CO which allows for orbital mixing between the σ(2*s*) and σ(2*p*) and between the σ*(2*s*) and σ*(2*p*). Compared with Figure 4.13, one major change is the σ–π crossover which results in the highest occupied MO being of σ-symmetry.

In carbon monoxide, the relative energies of the atomic orbitals are such that some mixing of the molecular orbitals *can* occur. This is illustrated in Figure 4.14. Mixing of the $\sigma(2s)$ and $\sigma(2p)$ orbitals lowers the energy of the $\sigma(2s)$ MO (now labelled 1σ) and raises the energy of $\sigma(2p)$ (now labelled 3σ). The energy of the 3σ orbital is higher than that of the $\pi(2p)$ orbitals. Similarly, mixing of the $\sigma^*(2s)$ and $\sigma^*(2p)$ orbitals results in a lowering of $\sigma^*(2s)$ (now labelled 2σ) and a raising of $\sigma^*(2p)$ (now labelled 4σ). As a consequence, all four of the σ-type MOs in the CO molecule contain character from the $2s$ and $2p$ atomic orbitals of each of the carbon and oxygen atoms. The most significant result (seen when comparing Figures 4.13 and 4.14) is the change-over in the highest occupied MO from one of π- to one of σ-symmetry. The degree of mixing will control the precise ordering of the 1σ and 2σ orbitals.

Unfortunately, even Figure 4.14 is not fully instructive because it does not clearly reveal the individual contributions made by the carbon and oxygen AOs to the four σ-molecular orbitals. These contributions are unequal, and the consequence of this complication is that the character of the σ-symmetry MOs is not readily partitioned as being either bonding (σ) or antibonding (σ^*). This is the reason these MOs have been labelled as 1σ, 2σ, 3σ and 4σ in Figure 4.14.

Two important features emerge when the MO bonding picture for CO is more fully explored:

- The dominant oxygen $2s$ character of the 1σ MO means that this orbital is essentially non-bonding with respect to the carbon–oxygen interaction.
- The 3σ MO contains only a small percentage of oxygen $2s$ and $2p$ character – this means that the 3σ MO is effectively a carbon-centred orbital and can also be considered to be almost non-bonding in character.

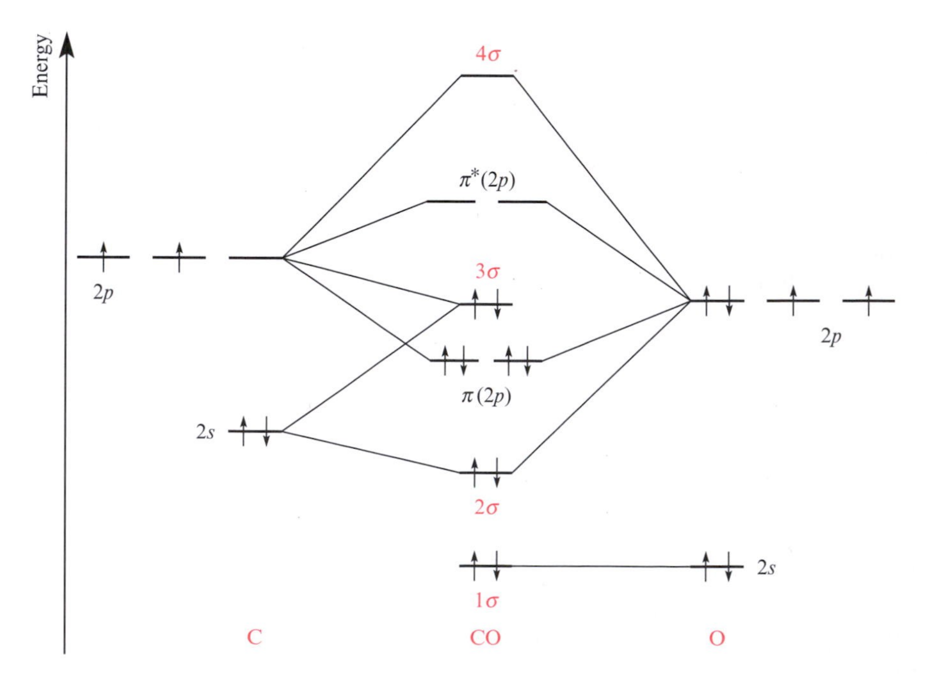

4.15 An MO diagram for CO which allows for the effects of orbital mixing but also recognizes the fact that not all the AOs will make major contributions to all the σ- MOs. A major change in going from Figure 4.14 to 4.15 is that the 1σ and 3σ MOs become non-bonding. This is actually an over-simplification.

These results lead us to draw a revised MO picture for carbon monoxide – Figure 4.15. Note that this diagram is *over-simplified* but gives a chemically useful view of the bonding in the molecule.

In the second point above, we observed that the 3σ MO is essentially a non-bonding orbital. This is apparent in Figure 4.15 – there are no lines connecting the 3σ MO to the oxygen atomic orbitals. Notice, though, that we have slightly refined our depiction of a non-bonding orbital. In earlier diagrams we have shown a non-bonding orbital to be at the same energy as an atomic orbital. Look, for example, at the levels of the non-bonding MOs compared to the fluorine $2p$ AOs in Figure 4.8. However, the non-bonding 3σ orbital in the CO molecule is *not* at the same energy as any of the carbon atomic orbitals. It lies at the *weighted average* of the energies of the carbon $2s$ and $2p$ AOs.

Summary of the molecular orbital view of the bonding in CO

If we use Figure 4.15 as our premise, then molecular orbital theory suggests the following about the bonding in the CO molecule:

- There are six electrons occupying the 2σ and $\pi(2p)$ bonding MOs. The other four electrons occupy non-bonding orbitals (1σ and 3σ). The 4σ and $\pi^*(2p)$ MOs are strongly antibonding.
- The bond order in CO is three (equation 4.21).

$$\text{Bond order in CO} = \tfrac{1}{2} \times (6 - 0) = 3 \qquad (4.21)$$

- CO is diamagnetic – this agrees with experimental observations.
- The highest occupied MO in CO has predominantly carbon character and contains contributions both from the carbon $2s$ and $2p$ AOs, and this results in an outward pointing lobe as shown in Figure 4.16. In Section 16.14 we will discuss some chemical properties of CO that rely, in part, on the nature of this MO and on the fact that the lowest lying *empty* MOs are of π-symmetry.
- It is *not* easy to say anything about the polarity of the CO molecule from MO theory at this rather simple and qualitative level. The charge distribution in a molecule follows from the summed character of the *occupied* molecular orbitals; the empty orbitals cannot affect this molecular property. In the occupied 1σ MO, the electron density is oxygen centred ($O^{\delta-}$) whilst in the 2σ and 3σ MOs there is a greater probability of finding the electrons nearer to the carbon nucleus ($C^{\delta-}$). Both the $\pi(2p)$ MOs will have more oxygen than carbon character ($O^{\delta-}$).

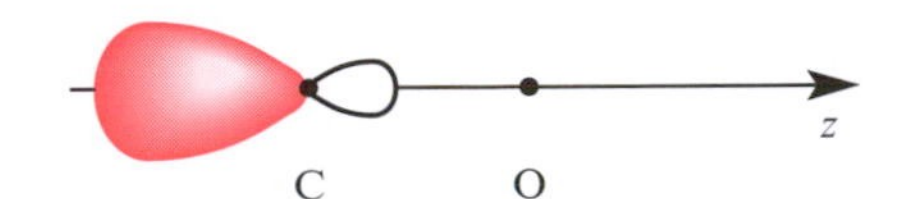

4.16 A schematic representation (an over-simplification, but a chemically useful one) of the highest occupied MO of carbon monoxide. The black dots show the positions of the C and O nuclei.

4.15 $[CN]^-$ AND $[NO]^+$: TWO IONS THAT ARE ISOELECTRONIC WITH CO

The cyanide ion $[CN]^-$, and the nitrosyl cation $[NO]^+$ are well-known chemical species.

Carbon possesses four valence electrons and the nitrogen atom has five. In the cyanide ion, there are ten valence electrons [4 + 5 + 1 (from the negative charge) = 10].

Nitrogen possesses five valence electrons and the oxygen atom has six. In the nitrosyl cation, there are thus ten valence electrons [5 + 6 – 1 (due to the positive charge) = 10].

Carbon, nitrogen and oxygen all have the same number of core electrons. The ions $[CN]^-$ and $[NO]^+$ are thus isoelectronic. They are also isoelectronic with the molecules CO and N_2.

We have already seen that there are some similarities and some differences between the bonding descriptions of the homonuclear diatomic molecule N_2 and the heteronuclear CO. How do the ions $[CN]^-$ and $[NO]^+$ fit into this series? Is the bond order the same in each species? Are the magnetic properties the same? What difference does the presence of an external charge make to the bonding in these molecules? Is it possible to predict on which atom the charge in the $[CN]^-$ and $[NO]^+$ ions will be localized?

Bear in mind throughout this section that C and N^+ are isoelectronic with each other, N and O^+ are isoelectronic, and N^- and O are isoelectronic.

Valence bond theory

Resonance structures for $[NO]^+$ and $[CN]^-$ can readily be drawn by using those of CO as a basis. The C atom can be replaced by the isoelectronic N^+, and the O atom by N^- as shown in Figure 4.17. Structure **II** for each species can be compared to the Lewis structure for N_2; remember that an N atom is isoelectronic with both C^- and O^+.

We have already seen that the observation of a short, strong bond in CO suggests that resonance structure **II** contributes significantly to the bonding in this molecule and emphasizes the similarity between the bonding in CO and N_2. The bond distance in the $[CN]^-$ anion is 114 pm and in the $[NO]^+$

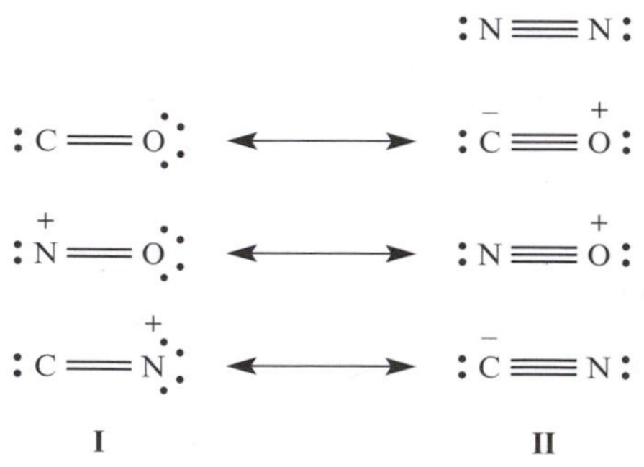

4.17 Resonance structures for $[CN]^-$ and $[NO]^+$ can readily be derived from those of N_2 and CO by considering the isoelectronic relationships between C and N^+, N and O^+, and N^- and O.

cation is 106 pm. These values compare with distances of 110 pm and 113 pm in N_2 and CO respectively, and are consistent with a significant contribution to the bonding in each ion from resonance structure **II**.

The small dipole moment in CO can be rationalized by allowing contributions from both resonance structures **I** and **II** to partly offset one another, making allowances too for the differences in electronegativity between the two atoms. It is, however, difficult to use the valence bond method to *predict* the polarity of the carbon–oxygen bond. Each of the two resonance forms for $[NO]^+$ suggests a different polarity for the ion (Figure 4.17) – **I** implies that the nitrogen atom carries the positive charge whilst **II** suggests that it is localized on the oxygen atom. The same problem arises if we try to predict whether carbon or nitrogen carries the negative charge in the $[CN]^-$ ion.

Molecular orbital theory

Despite the differences between the MO diagrams for N_2 (Figure 3.28) and CO (Figure 4.15), the results of MO theory are consistent with the observed diamagnetism, and a bond order of 3 in each case. The charge distribution in each molecule follows from the character of the occupied molecular orbitals. In N_2, the MOs possess equal contributions from both atoms and the bond is non-polar (μ = 0 D). In carbon monoxide, differences in the relative energies of the carbon and oxygen atomic orbitals cause distortions in the MOs and as a result the carbon–oxygen bond is polar (μ = 0.11 D).

The heteronuclear nature of $[NO]^+$ and $[CN]^-$ suggests that it might be better to use CO as a starting point rather than N_2 when discussing the bonding in these ions using MO theory. In many regards this viewpoint is correct, but it is still difficult to come up with the 'right' answer when we are only using a qualitative MO picture. As a first approximation it is acceptable to use the rather simple MO picture drawn in Figure 4.13 to depict the bonding in any heteronuclear diatomic species that is isoelectronic with CO. The answers may not be 'correct' in the sense that some predictions may not agree exactly with experimental results. However, with experience, it is possible to see how changes to the atomic orbital energies can influence the energies and the composition of molecular orbitals and, therefore, molecular properties.

Our initial approach is to use the molecular orbital diagram in Figure 4.13, keeping in mind the fact that the electronegativity ordering of O > N > C (Table 4.2) means that the oxygen AOs will be lower in energy than those of nitrogen, which will, in turn, be lower than those of carbon. We can now make the predictions:

- the bond order in CO, $[CN]^-$ and $[NO]^+$ will be three;
- the species CO, $[CN]^-$ and $[NO]^+$ will be diamagnetic.

The same results are obtained by using the more complicated MO diagram in Figure 4.15. However, more sophisticated calculations reveal some subtle changes. Molecular orbital 1σ which is tending towards being non-bonding in CO, has a greater bonding character in $[NO]^+$ and $[CN]^-$. Molecular orbital 2σ which is bonding in CO, becomes antibonding in $[NO]^+$ and $[CN]^-$. The orbital 3σ, which is predominantly a carbon-centred lone pair in CO, has more bonding character in $[NO]^+$ and $[CN]^-$ but still possesses an outward-pointing lobe (on carbon in $[CN]^-$ and on nitrogen in $[NO]^+$).

4.16 $[NO]^+$, NO AND $[NO]^-$

We end this chapter by considering the closely related species $[NO]^+$, NO and $[NO]^-$. Note that NO is isoelectronic with $[O_2]^+$, whilst $[NO]^-$ is isoelectronic with O_2; you may find it useful to compare some of the new results in this section with those detailed in Section 3.22 for the dioxygen species. Some properties of NO and $[O_2]^+$ are listed for comparison in Table 4.5.

As in Section 4.15 where we considered the ion $[NO]^+$, it is possible to use Figure 4.13 as the basis for a discussion of the bonding in NO and $[NO]^-$. Nitrogen monoxide (NO) has one more valence electron than $[NO]^+$, and by the *aufbau* principle, this will occupy one of the $\pi^*(2p)$ orbitals shown in Figure 4.13. The NO molecule therefore has an unpaired electron and is predicted to be paramagnetic. This agrees with experimental data – NO is a radical.

The bond order in NO may be determined using equation 4.22. The increase in bond length in going from $[NO]^+$ (106 pm) to NO (115 pm) is consistent with the decrease in bond order from 3 to 2.5.

$$\text{Bond order in NO} = \tfrac{1}{2} \times (6 - 1) = 2.5 \qquad (4.22)$$

The odd electron occupies an antibonding orbital; this MO will have more nitrogen than oxygen character and therefore our results suggest that the odd electron is more closely associated with the N than the O atom. Of course, the extent to which this statement is true depends on the relative energies of the N and O atomic orbitals.

Box 4.5 Nitrogen monoxide (nitric oxide) in biology

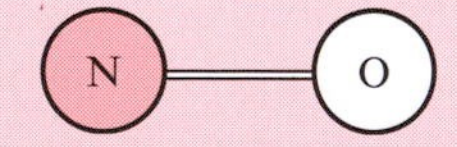

Nitrogen monoxide NO is now known to play an important role in biology and this area of research is one of current interest. Nitrogen monoxide is a small molecule and can easily diffuse through cell walls. It behaves as a messenger-molecule in biological systems, and appears to play an active role in mammalian functions such as the regulation of blood pressure, muscle relaxation, and neuro-transmission.

One of its remarkable properties is that NO appears to be cytotoxic – that is, it is able to destroy particular cells – and has an effect on the ability of the body's immune system to kill tumour cells.

An article that expands this discussion is 'The biological roles of nitric oxide' by A. R. Butler, *Chemistry & Industry*, p. 828 (1995).

Table 4.5 Some properties of nitrogen monoxide, NO, compared with those of the iso-electronic cation $[O_2]^+$.

Property	*NO*	$[O_2]^+$
Bond distance / pm	115	112
Bond dissociation enthalpy / kJ mol^{-1}	631	644
Magnetic properties	Paramagnetic	Paramagnetic
Dipole moment / D	0.16	0

The anion $[NO]^-$ has two more valence electrons than $[NO]^+$. By the *aufbau* principle, these will singly occupy each of the $\pi^*(2p)$ orbitals shown in Figure 4.13. The $[NO]^-$ ion will be paramagnetic, and will resemble the O_2 molecule. The bond order is found to be two (equation 4.23).

$$\text{Bond order in } [NO]^- = \tfrac{1}{2} \times (6 - 2) = 2 \qquad (4.23)$$

SUMMARY

In this chapter the main topic has been bonding in heteronuclear diatomic molecules.

Do you know what the following terms mean?

- heteronuclear
- resonance structure
- orbital character
- coefficient (in respect of a wavefunction)
- orbital energy matching
- non-bonding molecular orbital
- electronegativity
- polar molecule
- (electric) dipole moment
- isoelectronic

You should now be able:

- to summarize the salient differences between homonuclear and heteronuclear diatomic molecules
- to discuss the bonding in HF, LiH, LiF, CO, $[NO]^+$, $[CN]^-$, NO and $[NO]^-$ in terms of VB theory
- to discuss the bonding in HF, LiH, LiF, CO, $[NO]^+$, $[CN]^-$, NO and $[NO]^-$ in terms of MO theory
- to draw schematic representations of the molecular orbitals for heteronuclear diatomic molecules and to interpret these diagrams
- to use bond dissociation enthalpies to estimate enthalpies of atomization
- to discuss briefly the concept of electronegativity
- to define the terms (electric) dipole moment and polar molecule
- to use Pauling electronegativity values to determine if a diatomic molecule will be polar
- to write down sets of isoelectronic diatomic molecules and ions
- to use your knowledge about the bonding in one molecule or ion to comment on the bonding in a molecule or ion which is isoelectronic with the first

PROBLEMS

4.1 The wavefunction for a bonding MO in a heteronuclear diatomic molecule AB is given by the equation:

$$\psi(\text{MO}) = (0.9 \times \psi_A) + (0.3 \times \psi_B).$$

What can you deduce about the relative values of Z_{eff} for atoms A and B?

4.2 Give resonance structure(s) for the hydroxyl ion $[OH]^-$. What description of the bonding in this ion does valence bond theory give?

4.3 Bearing in mind that $[OH]^-$ is isoelectronic with HF, construct an approximate MO diagram to describe the bonding in $[OH]^-$. What is the bond order in $[OH]^-$? Does the MO picture of the bonding differ from that obtained in the previous question using VB theory?

4.4 Calculate the bond dissociation enthalpy of chlorine monofluoride, ClF, if the bond dissociation enthalpies of F_2 and Cl_2 are 159 and 242 kJ mol^{-1} respectively, $\chi^P(\text{F}) = 4.0$, and $\chi^P(\text{Cl}) = 3.2$ (1 eV = 96.5 kJ mol^{-1}). Compare your answer with the experimental value.

4.5 Equations 4.15 and 4.16 can also be applied to polyatomic molecules. Using data in Tables 3.6 and 4.2, calculate the bond dissociation enthalpy of the (a) O–H, (b) C–Cl and (c) S–F bonds.

4.6 Using the bond dissociation enthalpies you have found in problem 4.5, estimate the enthalpies of atomization of (a) H_2O, (b) CCl_4 and (c) SF_4.

4.7 Use values of the first ionization energies and first electron affinities listed in Appendices 8 and 9 to calculate values for the Mulliken electronegativity, χ^M, of N, O, F, Cl, Br and I. Present these values graphically to show the trends in values of χ^M across the first row of the *p*-block and down group 17. How do these *trends* compare with similar trends in values of χ^P (Table 4.2)?

4.8 Write down five ions (cations or anions) that are isoelectronic with neon.

4.9 Explain whether or not the two species in each of the following pairs are isoelectronic with one another: (a) He and Li^+, (b) $[Se_2]^{2-}$ and Br_2; (c) NO and $[O_2]^+$; (d) F_2 and ClF; (e) N_2 and P_2; (f) P^{3-} and Cl^-.

4.10 *This question requires you to have read the material in Box 4.2*: Estimate the charge distribution in CO if the dipole moment of the molecule is 0.11 D, and the C≡O bond distance is 113 pm. The carbon atom carries the δ^- charge.

5 Polyatomic molecules: shapes and bonding

Topics

5.1 INTRODUCTION

Molecular geometry

Our discussion of molecules and molecular ions in Chapters 3 and 4 focused on diatomic species, but descriptions of the shapes of these molecules were omitted. This omission was for a good reason – diatomic molecules *must be linear*! However, only a few of the total number of molecules and molecular ions that are known are diatomic and in this chapter we look at *polyatomic* species, i.e. those containing *three or more atoms*. Although a few polyatomic molecules or ions are linear (e.g. CO_2, HC≡N, $[I_3]^-$, HC≡CC≡CH), most are not.

A polyatomic molecule or ion contains three or more atoms.

In order to gain a picture of the three-dimensional structure of a polyatomic molecule, it is convenient to use the geometry around the central atom as the basic structural descriptor. Table 5.1 lists some common geometries for molecules of formula XY_n. For most, several geometries are possible. We shall describe these geometries in more detail later, but Table 5.1 gives a useful checklist.

Table 5.1 Molecular geometries for polyatomic molecules of formula XY_n.

Formula XY_n	*Coordination number of atom X*	*Geometrical descriptor*	*Spatial representation*	*Ideal angles* ($\angle$Y–X–Y) */ degrees*
XY_2	2	Linear	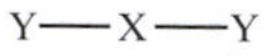	180
XY_2	2	Bent or V-shaped	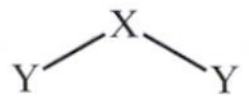	Variable ($\neq$ 180)
XY_3	3	Trigonal planar	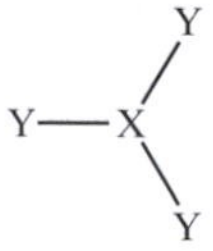	120
XY_3	3	Trigonal pyramidal	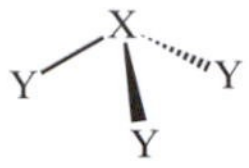	variable (< 120)
XY_3	3	T-shaped	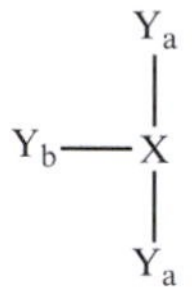	$\angle Y_a–X–Y_b = 90$ $\angle Y_a–X–Y_a = 180$ (the atoms lie in a plane)
XY_4	4	Tetrahedral	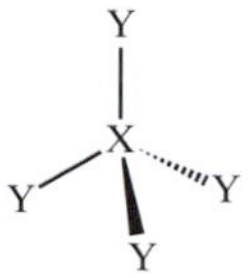	109.5
XY_4	4	Square planar	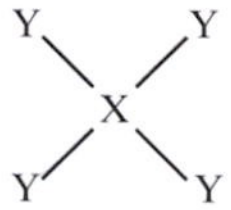	90
XY_4	4	Disphenoidal	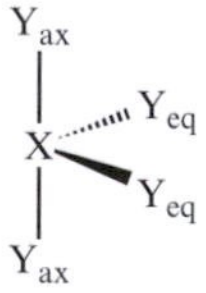	$\angle Y_{ax}–X–Y_{eq} = 90$ $\angle Y_{eq}–X–Y_{eq} = 120$
XY_5	5	Trigonal bipyramidal	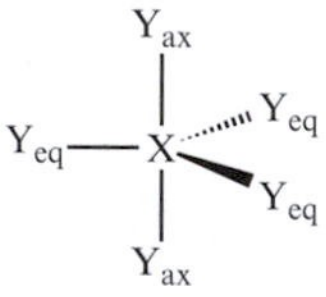	$\angle Y_{ax}–X–Y_{eq} = 90$ $\angle Y_{eq}–X–Y_{eq} = 120$
XY_5	5	Square-based pyramidal	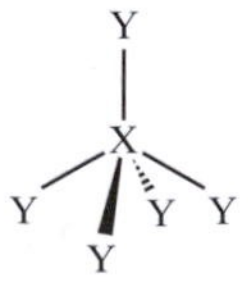	Variable
XY_6	6	Octahedral	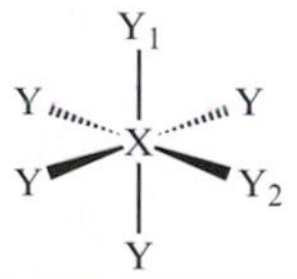	$\angle Y_1–X–Y_2 = 90$[a]

[a] Although Y_1 and Y_2 are distinguished to define the angle, all Y atoms are equivalent.

Bond angles

In describing a molecular structure, the bond distances and bond angles are of particular significance. The angular relationships between atoms in a molecule conveniently describe the molecular geometry, and a bond angle is defined between three bonded atoms.§ For example, in a trigonal planar XY_3 molecule, each of the angles ∠Y–X–Y is 120° (Figure 5.1a).

Molecular shapes are often described in terms of *idealized geometries* and the corresponding *ideal bond angles* are given in the right-hand column of Table 5.1 for common structural types. Note that it is *not* possible to define ideal angles for all the geometries.

In practice, many molecules are described in terms of one of the shapes listed in Table 5.1 *without* actually possessing the ideal bond angles. A common reason for deviation from an ideal geometry is that the substituents in a molecule may not be identical. The molecules BF_3 and BCl_3 both possess *ideal* trigonal planar structures, but BCl_2F and $BClF_2$ are *approximately* trigonal planar. The chlorine atoms are larger than the fluorine atoms and this difference contributes towards deviations from the ideal 120° bond angles (Figure 5.1b). The bond distances, *d*, also differ: *d*(B–Cl) > *d*(B–F).

Molecules with more than one 'centre'

Table 5.1 focuses on XY_n molecules, each with an atom X at its centre. Many polyatomic molecules possess more than one 'centre'; ethane, **5.1**, contains

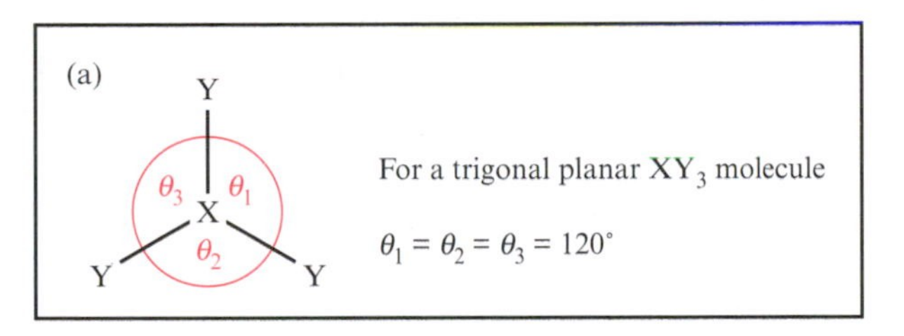

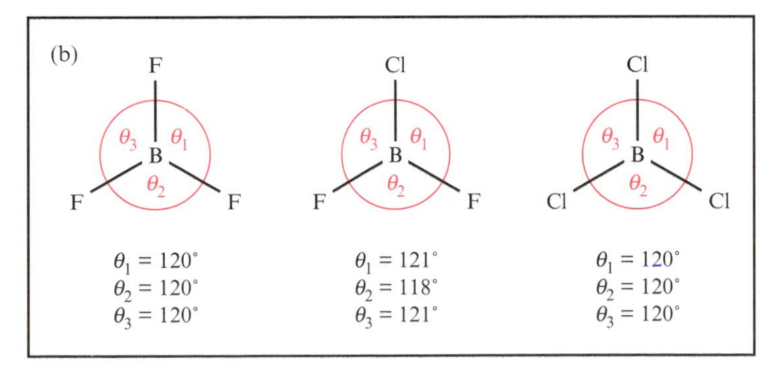

5.1 (a) A bond angle is defined as the angle between three bonded atoms. In a trigonal planar molecule XY_3, the ideal angle ∠Y–X–Y is 120°. (b) The bond angle is 120° if all the substituents are the same, but if the substituents are different, the bond angles may not be exactly 120°; the geometry is then *approximately trigonal planar*.

§ We are not concerned here with *torsion angles* which define, for four atoms, the degree of twist around the central bond and the deviation from planarity for the four atoms that this generates.

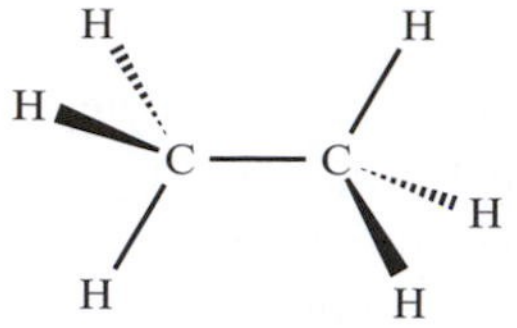

(5.1)

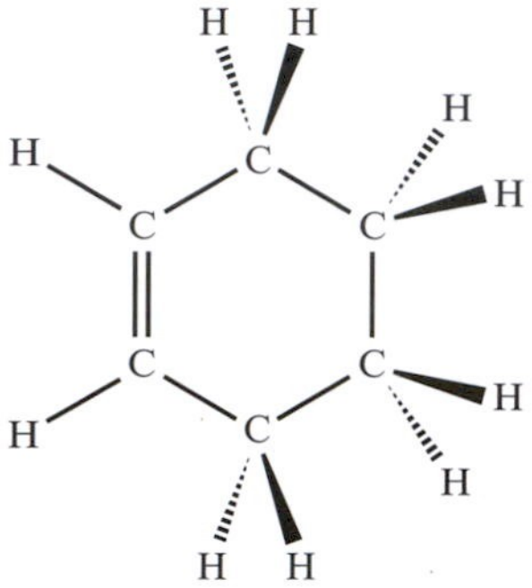

(5.2)

two carbon atoms, each of which has a tetrahedral arrangement of atoms about it, and cyclohexene, **5.2**, contains six carbon centres, four of which are tetrahedral and two of which are trigonal planar. The shape of large molecules may more conveniently be described in terms of the geometries about a variety of atomic centres as we discuss in Section 5.3.

Coordination number

The *coordination number* of an atom defines the number of atoms or groups of atoms which are attached to it. It does *not* rely upon a knowledge of the bond types (single, double, or triple) or whether the bonds are localized or delocalized.

The *coordination number* of an atom defines the number of atoms (or groups of atoms) which are attached to it.

The geometries listed in Table 5.1 give examples of coordination numbers ranging from two to six for the central atom X in the molecules XY_n. In BF_3 **(5.3)** and $[CO_3]^{2-}$ **(5.4)**, the coordination number of the central atom is three even though the bond types are different. Similarly, in CH_4 **(5.5)** and SO_2Cl_2 **(5.6)**, the coordination number of the central atom is four.

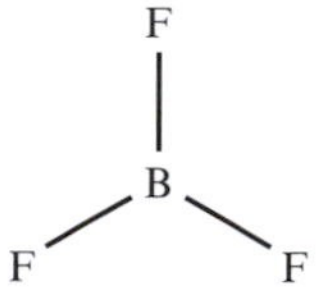

(5.3) 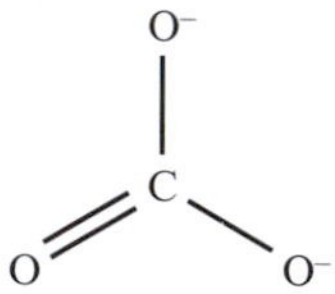 (5.4) 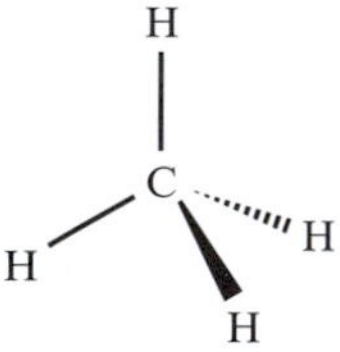 (5.5) 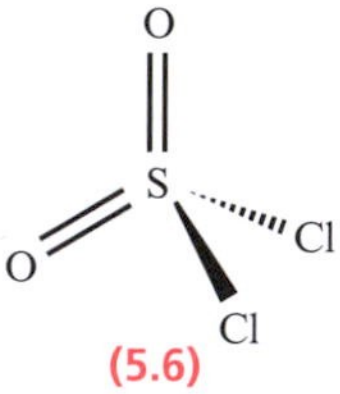 (5.6)

Some aims of this chapter

In this chapter, we describe the shapes of molecules and ions XY_n with a single central atom X, and where Y is an atom or a molecular group. How does the increased molecular complexity involved in going from a diatomic to a polyatomic molecule affect such properties as the molecular dipole moment? We also consider ways of approaching the bonding in polyatomic species. We begin by taking a brief 'tour' around the *p*-block, and summarize the diversity of molecular shapes that are observed, starting with triatomic species.

5.2 THE GEOMETRIES OF TRIATOMIC MOLECULES

O=C=O

(5.7)

H—C≡N

(5.8)

H–O–H (bent)

(5.9)

O=S=O (bent)

(5.10)

A triatomic species may possess a *linear* or *bent* geometry.§ Examples of linear molecules are carbon dioxide, **5.7**, and hydrogen cyanide, **5.8**, and bent molecules include water, **5.9**, and sulfur dioxide, **5.10**.

Linear triatomic molecules and ions

Figure 5.2 shows the structure of the carbon dioxide molecule, the nitryl cation† $[NO_2]^+$, the azide anion $[N_3]^-$ and the cyanate ion $[NCO]^-$. These species are isoelectronic and possess linear structures – they are *isostructural*. Replacing the oxygen atom in the $[NCO]^-$ anion by sulfur gives the thiocyanate anion $[NCS]^-$. *With respect to their valence electrons*, $[NCO]^-$ and $[NCS]^-$ are isoelectronic and possess analogous structures (Figure 5.2). Further examples of species which are isoelectronic with respect to their valence electrons and are isostructural are the linear polyhalide anions $[ICl_2]^-$ (**5.11**) and $[I_3]^-$ (**5.12**).

Isostructural means 'having the same structure'.

Isoelectronic: see Section 4.12.

$[Cl—I—Cl]^-$

(5.11)

$[I—I—I]^-$

(5.12)

The important message here is that we expect isoelectronic species to possess similar structures; this is true for polyatomic species in general and is not restricted to triatomics. Since the covalent radii of atoms differ, the bond distances in isoelectronic molecules or ions will not necessarily be equal, but the geometry at the central atom, which is defined by bond angles, will be the same or very similar.

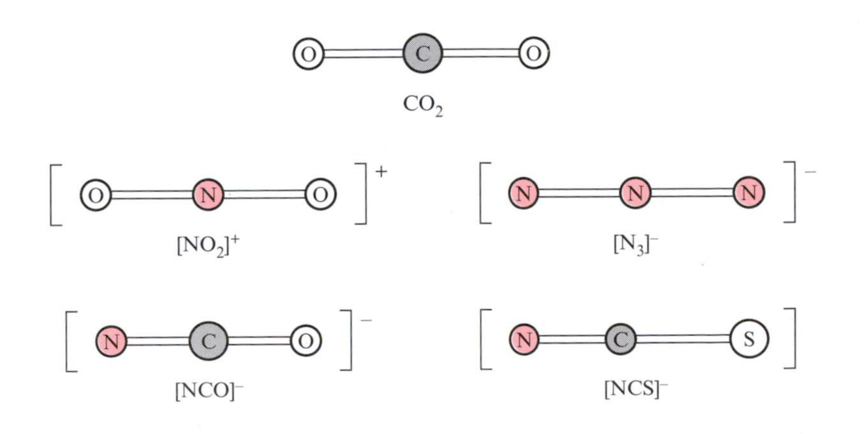

5.2 The isoelectronic species CO_2, $[NO_2]^+$, $[N_3]^-$ and $[NCO]^-$ all possess linear structures. The cyanate $[NCO]^-$ and thiocyanate $[NCS]^-$ anions have the same number of valence electrons and may be considered to be isoelectronic *(with respect to their valence electrons)*.

§ Other terms are often used for the geometry of a bent triatomic molecule; these include non-linear and V-shaped.

† Older nomenclature calls $[NO_2]^+$ the nitronium ion; this is still in common use.

However, when comparing molecules and ions that are isoelectronic *only with respect to their valence electrons*, the expectation of isostructural behaviour may *not* hold. A classic example is seen when we compare CO_2 with SiO_2: at room temperature, carbon dioxide is a linear molecule but SiO_2 has a giant lattice structure containing silicon atoms in tetrahedral environments. On the other hand, some species do conform as we have seen above.

CO_2 and SiO_2: see Section 13.6

Molecules or ions which are isoelectronic are usually isostructural.

However, species which have the same structure are *not* necessarily isoelectronic.

Bent molecules and ions

Figure 5.3 shows some bent molecules. Notice that the bond angle (measured as angle α in structure **5.13**) has no characteristic value. The *exact* geometry of a bent molecule is only known if the bond angle α has been measured experimentally since the term 'bent' can be used for any triatomic XY_2 with $0 < \alpha < 180°$.

(5.13)

Three of the molecules in Figure 5.3 are related to one another: OF_2 and SF_2 are isoelectronic with respect to their valence electrons, whilst H_2O and OF_2 are related because each possesses a central oxygen atom which forms two single bonds. In addition, SF_2 and SO_2 appear to be related in that the sulfur atom has a coordination number of two in each case. However, the S–F bonds in SF_2 are single bonds while in the sulfur dioxide molecule, S=O double bonds are present.

We saw above that the polyhalide anions $[ICl_2]^-$ and $[I_3]^-$ are linear. The polyhalide cations $[ICl_2]^+$ and $[I_3]^+$ are also known but possess bent structures. A change of structure accompanying a change in charge is not unique.

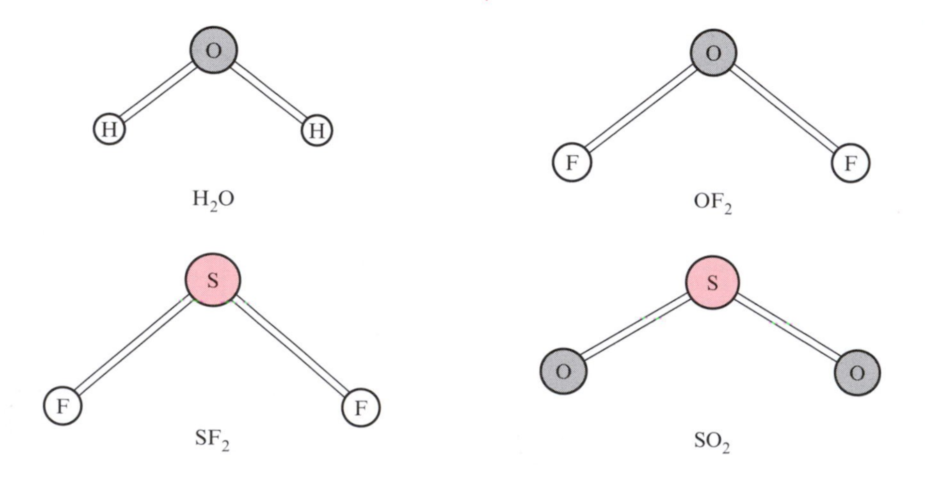

5.3 Some bent molecules illustrating a range of bond angles: H_2O (∠H–O–H = 104.5°), F_2O (∠F–O–F = 103°), SF_2 (∠F–S–F = 98°) and SO_2 (∠O–S–O = 119°).

For example, the $[NO_2]^+$ cation (Figure 5.2) is linear and yet *both* NO_2 and $[NO_2]^-$ are bent, although their bond angles are quite different: ∠O–N–O = 134° in NO_2 (Figure 5.4) and 115° in the $[NO_2]^-$ ion.

Some asymmetrical bent triatomic species are shown in Figure 5.5. The structure of hypochlorous acid (HOCl) is similar to that of H_2O, with one chlorine atom replacing a hydrogen atom; another member of this group is Cl_2O (related to OF_2 in Figure 5.3). Replacing the chlorine atom in ClNO by another group 17 atom gives the structurally similar molecules FNO and BrNO, but with different bond angles (Figure 5.6). Trends in angles can often be correlated with the size and/or the electronegativity of the atoms.

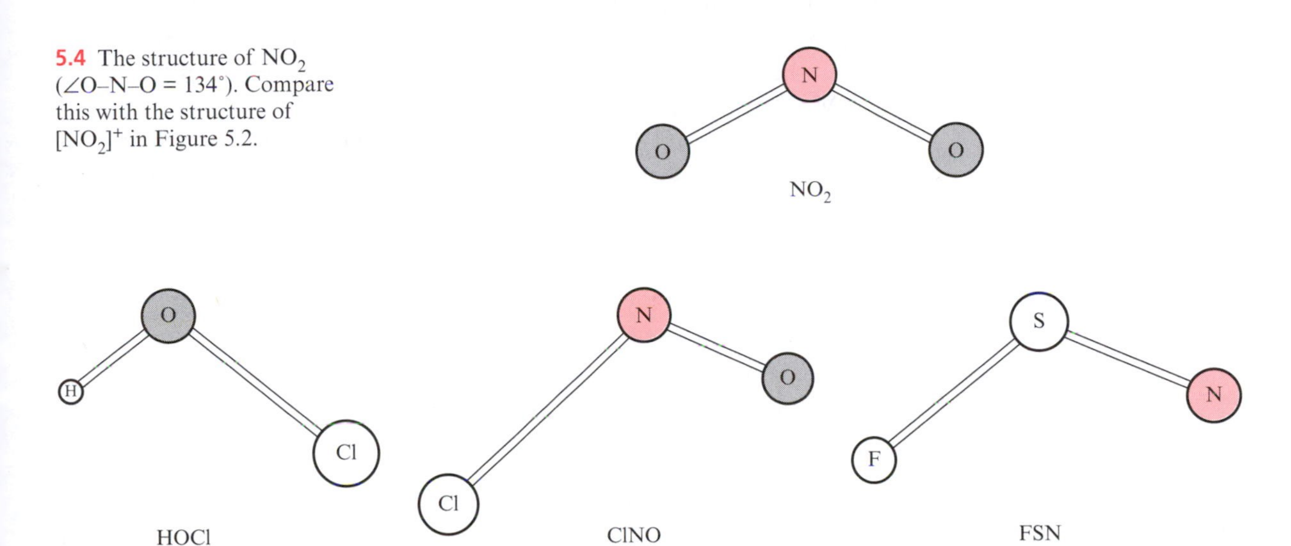

5.4 The structure of NO_2 (∠O–N–O = 134°). Compare this with the structure of $[NO_2]^+$ in Figure 5.2.

5.5 Some asymmetrical bent triatomic molecules: HOCl (∠H–O–Cl = 102.5°), ClNO (∠O–N–Cl = 113°) and NSF (∠N–S–F = 117°).

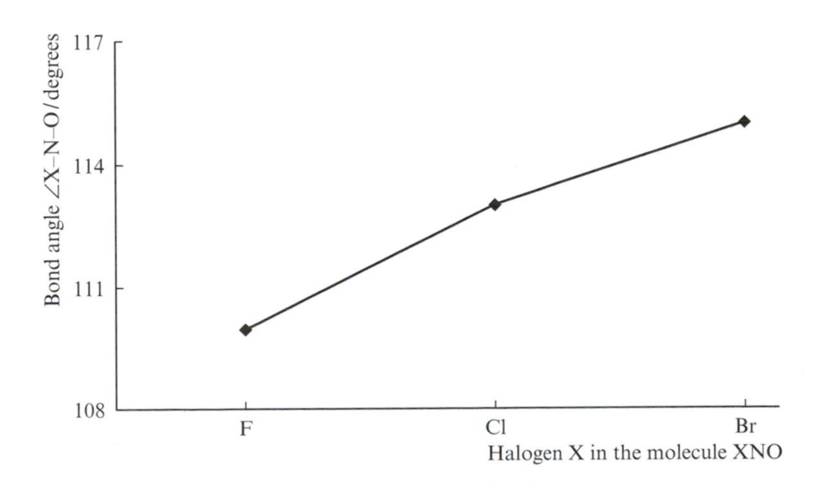

5.6 In the molecules FNO, ClNO and BrNO, the angle ∠X–N–O increases as the halogen atom X becomes larger.

5.3 MOLECULES LARGER THAN TRIATOMICS BUT WHICH MAY BE DESCRIBED AS HAVING LINEAR OR BENT GEOMETRIES

Although linear and bent geometries are usually associated with triatomic molecules or ions, the terms can also be usefully applied to larger molecules.

A molecule or ion which contains a backbone of three atoms is readily related to a triatomic species. Figure 5.7a shows that hydrogen cyanide is a linear triatomic molecule, and that acetonitrile (MeCN) is related to hydrogen cyanide – the hydrogen atom in HCN has been replaced by a methyl group, and the methyl group may be regarded as a 'pseudo-atom'. It is useful to consider MeCN as having a CCN backbone as this allows the molecular geometry to be described easily – acetonitrile contains a *linear* CCN backbone.

Me = CH_3 = methyl group

Similarly, we can relate the structure of dimethyl sulfide (Me_2S) to that of H_2S as shown in Figure 5.7b. Me_2S retains a bent geometry at the sulfur atom, but the bond angle changes from 92° in H_2S to 99° in Me_2S. The compound MeSH has a similar structure, but with ∠C–S–H = 96.5°. A related set of compounds is H_2O, MeOH and Me_2O, each of which is bent with respect to the oxygen centre. Note the way in which this description has been phrased; the emphasis is on the geometry *with respect to each oxygen centre*. The description says nothing about the geometry at each carbon centre.

Alkynes of the general formula RC≡CR (R is an organic group) possess linear tetraatomic backbones; the first carbon atom in each R group is counted into the backbone. Figure 5.8 shows the structures of HC≡CH and MeC≡CMe. The change from hydrogen to methyl substituents does not alter the fundamental linear geometry of the central part of the molecule.

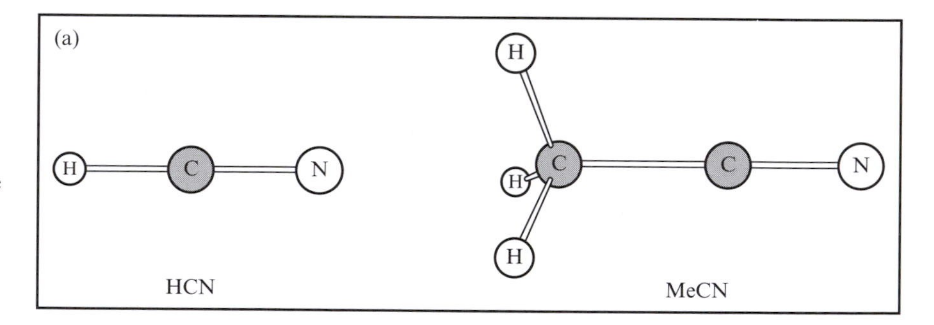

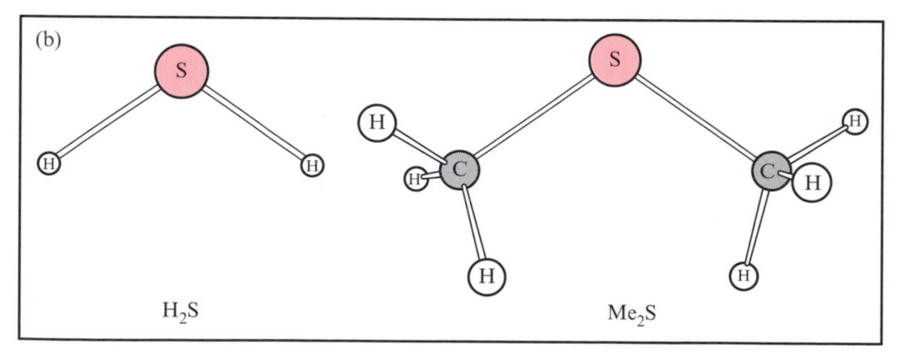

5.7 The structural relationships between (a) hydrogen cyanide (HCN) and acetonitrile (MeCN) and (b) H_2S and Me_2S. If the methyl group is regarded as a 'pseudo-atom', MeCN can be considered to possess a linear CCN backbone, and Me_2S can be described as a bent molecule.

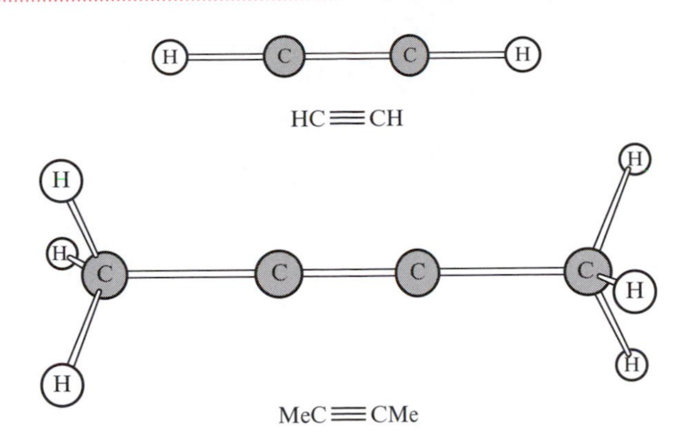

5.8 The structural relationship between the alkynes HC≡CH and MeC≡CMe; HC≡CH is truly linear, but MeC≡CMe has a linear CCCC backbone.

Remember, though, that describing MeC≡CMe as 'linear' only specifies the geometry with respect to the four central carbon atoms. On the other hand, the HC≡CH molecule *is* truly linear.

The small number of examples in this section make an important and useful point – the geometries of large molecules can often be adequately described by considering the *local geometry* at individual atomic centres.

5.4 GEOMETRIES OF MOLECULES WITHIN THE *p*-BLOCK: THE FIRST ROW

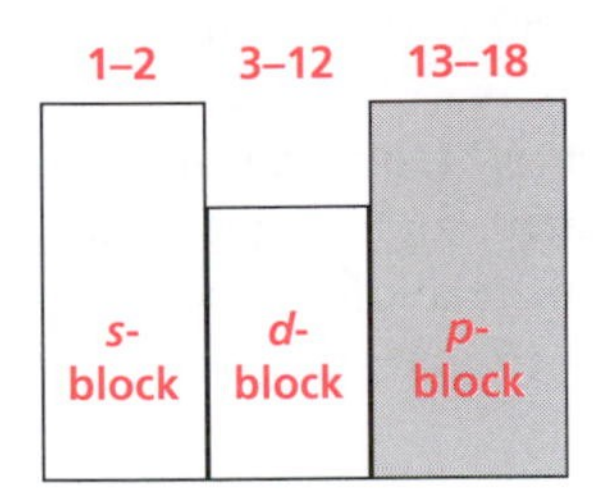

In this section we consider the shapes of molecules which contain a *p*-block element as the central atom, and confine our survey to compounds of the type XY_n with one central atom and one type of substituent. Figure 5.9 shows the five geometries that are common for such molecules in which X is one of the first row elements boron, carbon, nitrogen or oxygen; Table 5.2 lists some species which adopt these structures. The first row element fluorine is rarely found at the centre of polyatomic molecules and is usually in a *terminal* site as in BF_3, CF_4, NF_3 and OF_2.

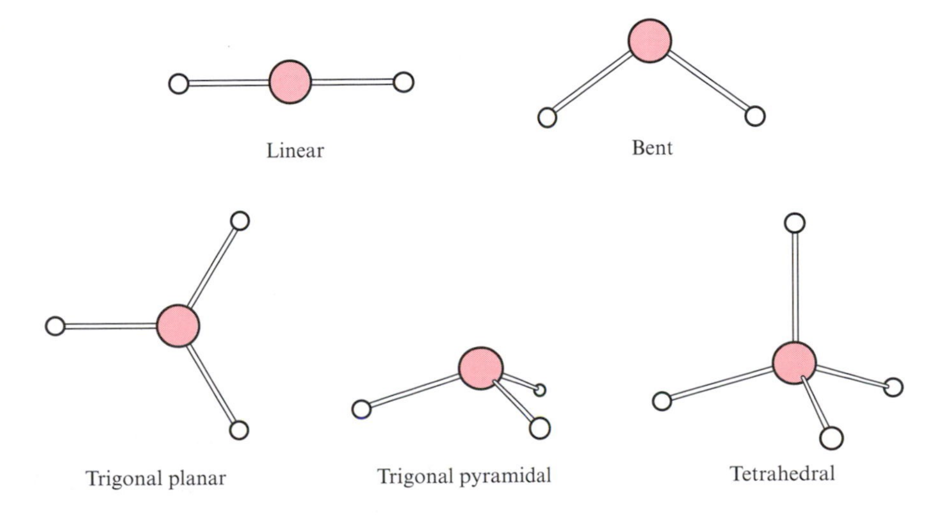

5.9 The common five shapes for molecules of type XY_n where X is an element from the first row of the *p*-block.

Table 5.2 Examples of molecules and ions of general formula XY_n, $[XY_n]^{m+}$ or $[XY_n]^{m-}$ in which X is a first row *p*-block element (boron to fluorine). The atom X is shown in red.

Shape (refer to Table 5.1 and Figure 5.9)	*Examples (Me = methyl = CH_3)*
Linear	CO_2, $[NO_2]^+$, $[NCO]^-$, $[N_3]^- = [NNN]^-$
Bent	H_2O, OF_2, HOF, NO_2, $[NO_2]^-$, $[NH_2]^-$,
Trigonal planar	BF_3, BCl_3, BBr_3, BMe_3, $[BO_3]^{3-}$, $[CO_3]^{2-}$, $[NO_3]^-$
Pyramidal	NH_3, NF_3, NMe_3, $[H_3O]^+$
Tetrahedral	$[BH_4]^-$, $[BF_4]^-$, CH_4, CMe_4, CF_4, $[NH_4]^+$, $[NMe_4]^+$

Boron (group 13)

Three-coordinate compounds containing one boron centre are trigonal planar, and examples include BF_3, BCl_3, BMe_3 and $[BO_3]^{3-}$. Tetrahedral boron centres are seen in a range of anions such as $[BH_4]^-$ and $[BF_4]^-$.

Carbon (group 14)

See also Section 5.14

For molecules containing carbon, one of three geometries is generally observed – linear, trigonal planar or tetrahedral. Linear carbon centres are observed in the triatomic species CO_2, $[NCO]^-$ and HCN. A trigonal planar carbon centre is present in $H_2C{=}O$, $Cl_2C{=}O$, HCO_2H (**5.14**) and $[CO_3]^{2-}$.

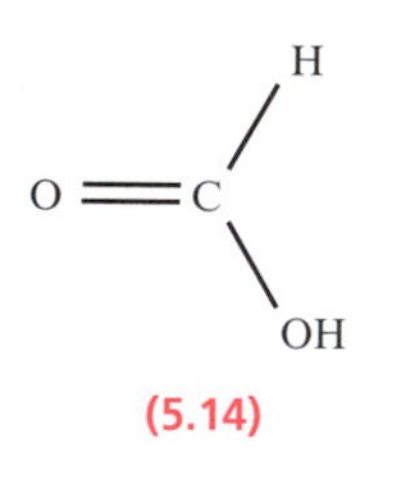

(5.14)

Tetrahedrally coordinated carbon is widely represented, for example in saturated alkanes such as methane, CH_4. The isoelectronic relationship between CH_4 and $[BH_4]^-$ suggests that they will be isostructural (Figure 5.10).

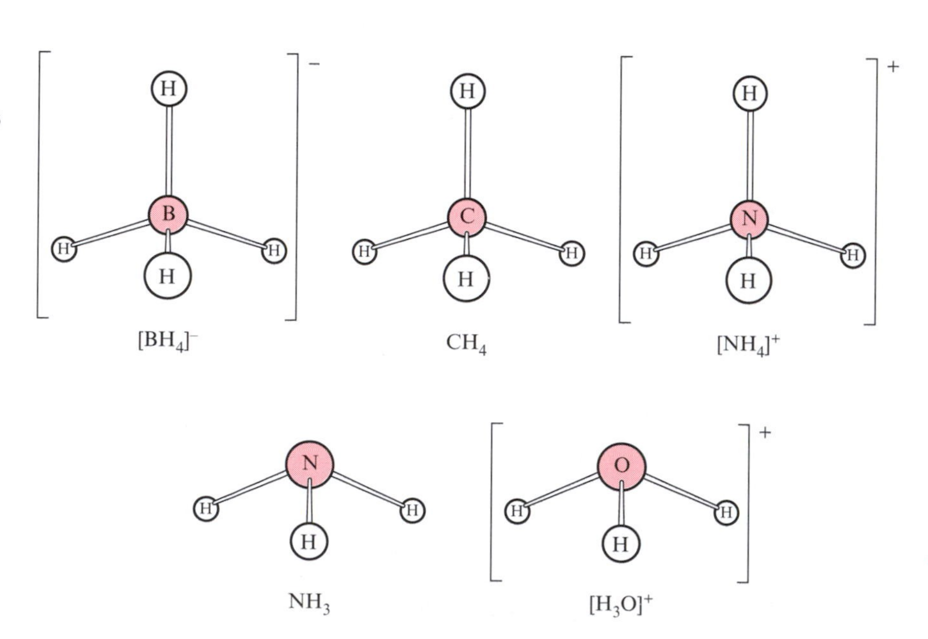

5.10 The species $[BH_4]^-$, CH_4 and $[NH_4]^+$ are isoelectronic and isostructural; NH_3 and $[H_3O]^+$ are also isoelectronic and isostructural.

Nitrogen (group 15)

Common geometries for nitrogen atoms are linear (Figure 5.2), bent (Figure 5.4), trigonal planar, trigonal pyramidal and tetrahedral.

In neutral three-coordinate compounds, the trigonal pyramidal geometry is usual for nitrogen – for example NH_3 (Figure 5.10) and NF_3. The family can be extended to include organic amines with the general formulae RNH_2, R_2NH and R_3N (**5.15**) where R is an organic group.

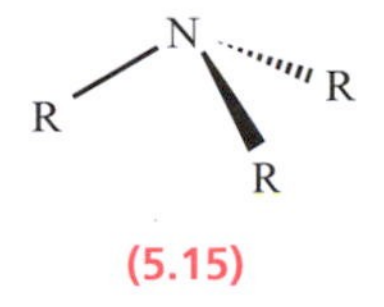

(5.15)

The ammonium ion $[NH_4]^+$ is isoelectronic with CH_4 and $[BH_4]^-$ (Figure 5.10) and is tetrahedral. Trigonal planar nitrogen is seen in the nitrate ion $[NO_3]^-$, and the isoelectronic principle establishes a structural relationship between $[NO_3]^-$, $[CO_3]^{2-}$ and $[BO_3]^{3-}$.

Oxygen (group 16)

Simple molecules with oxygen as the central atom are bent (Figure 5.3). In three-coordinate cations, oxygen is trigonal pyramidal, and the simplest example is the oxonium ion $[H_3O]^+$ which is isoelectronic with NH_3 (Figure 5.10).

Overview of the first row of the *p*-block

- Compounds of the first row *p*-block elements tend to possess one of the five shapes shown in Figure 5.9.
- A greater range of geometries is observed for boron, carbon and nitrogen than for oxygen and fluorine.
- A change in the overall charge often results in a change in geometry, e.g. NO_2 to $[NO_2]^-$.

5.5 HEAVIER *p*-BLOCK ELEMENTS

p-Block chemistry: see Chapter 13

The elements that are discussed in this section are shown in Figure 5.11, and the relevant molecular shapes are shown in Figures 5.9 and 5.12; Table 5.3 lists examples of each type of structure.

5.11 The heavier *p*-block elements (excluding the noble gases – group 18).

	Group 13	Group 14	Group 15	Group 16	Group 17
2nd row	**Al** aluminium	**Si** silicon	**P** phosphorus	**S** sulfur	**Cl** chlorine
3rd row	**Ga** gallium	**Ge** germanium	**As** arsenic	**Se** selenium	**Br** bromine
4th row	**In** indium	**Sn** tin	**Sb** antimony	**Te** tellurium	**I** iodine
5th row	**Tl** thallium	**Pb** lead	**Bi** bismuth	**Po** polonium	**At** astatine

5.12 Molecular shapes (in addition to those shown in Figure 5.9) observed for molecules of type XY_n where X is a heavier *p*-block element.

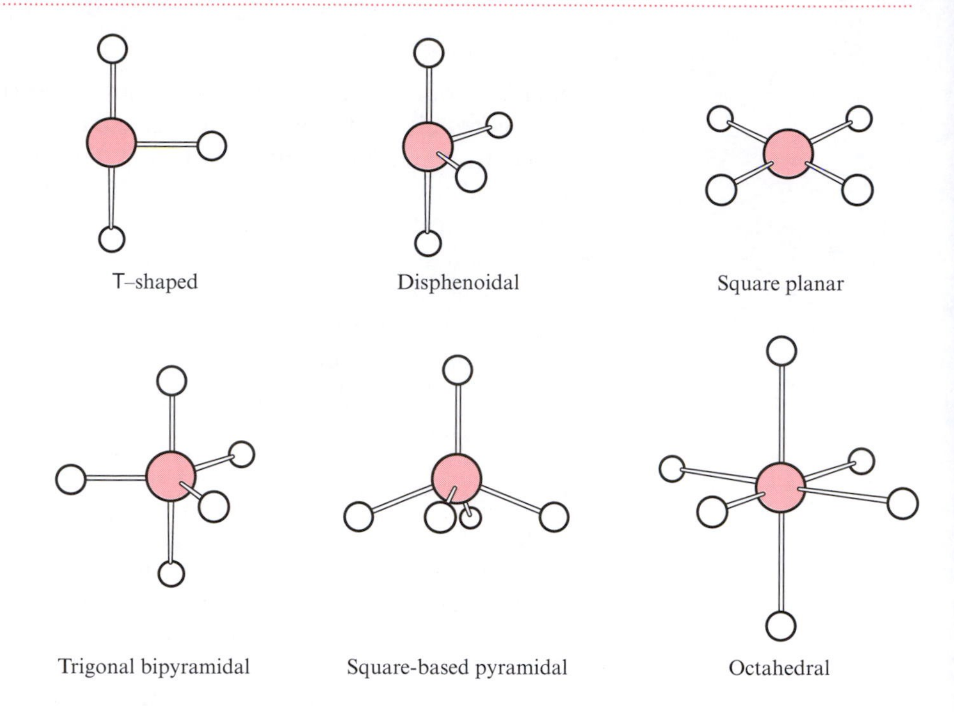

Table 5.3 Examples of molecules of general formula XY_n in which X is a heavy *p*-block element (see Figures 5.9 and 5.12). The atom X is shown in red.

Shape	*Examples*
Linear	$[ICl_2]^-$, $[I_3]^- = [I–I–I]^-$
Bent	$SnCl_2$, H_2S, SF_2, SO_2, H_2Se, $SeCl_2$, $TeCl_2$
Trigonal planar	$AlCl_3$, $AlBr_3$, SO_3
Pyramidal	PF_3, PCl_3, AsF_3, $SbCl_3$, PMe_3, $AsMe_3$, $[PO_3]^{3-}$, $[SO_3]^{2-}$, $[SeO_3]^{2-}$, $[ClO_3]^-$
T-shaped	ClF_3, BrF_3
Tetrahedral	SiH_4, GeH_4, SiF_4, $SiCl_4$, GeF_4, $SnCl_4$, $PbCl_4$, $[AlH_4]^-$, $[GaBr_4]^-$, $[InCl_4]^-$, $[TlI_4]^-$, $[PBr_4]^+$, $[PO_4]^{3-}$, $[SO_4]^{2-}$, $[ClO_4]^-$
Disphenoidal	SF_4, $[PBr_4]^-$, $[ClF_4]^+$, $[IF_4]^+$
Square planar	$[ClF_4]^-$, $[BrF_4]^-$, $[ICl_4]^-$
Trigonal bipyramidal	PCl_5, AsF_5, SbF_5, SOF_4, $[SnCl_5]^-$
Square-based pyramidal	ClF_5, BrF_5, IF_5, $[SF_5]^-$, $[TeF_5]^-$, $[SbF_5]^{2-}$
Octahedral	SF_6, SeF_6, $Te(OH)_6$, $[IF_6]^+$, $[PF_6]^-$, $[AsCl_6]^-$, $[SbF_6]^-$, $[SnF_6]^{2-}$, $[AlF_6]^{3-}$

Group 13

Temperature dependent structure of $AlCl_3$: see Section 13.4

At high temperatures, the aluminium trihalides $AlCl_3$, $AlBr_3$ and AlI_3 exist as molecular species with trigonal planar structures. However, a tetrahedral environment is more common for aluminium, gallium and indium atoms and is observed in a range of anions including $[AlH_4]^-$, $[AlCl_4]^-$, $[GaBr_4]^-$ and $[InCl_4]^-$. These two shapes resemble those seen in compounds of boron, but simple species containing heavier group 13 elements also exhibit geometries with higher coordination numbers – for example, the ions $[AlF_6]^{3-}$ (**5.16**) and $[Ga(H_2O)_6]^{3+}$ are octahedral.

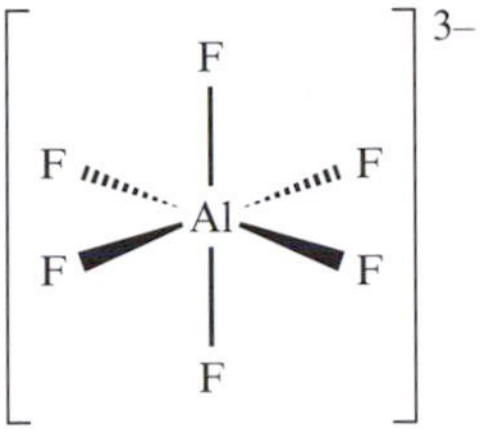

(5.16)

Group 14

In the gas phase, $SnCl_2$ is bent, but other tin(II) compounds generally have more complex structures. Many compounds of group 14 elements have formulae reminiscent of carbon compounds but caution is needed – CO_2 and SiO_2 may appear to be similar but their structures are very different.

CO_2 and SiO_2: see Section 13.6

Silane, SiH_4, is tetrahedral like CH_4, and the same isoelectronic and isostructural relationship exists between SiH_4 and $[AlH_4]^-$ as between CH_4 and $[BH_4]^-$. Similarly, GeH_4 is isoelectronic and isostructural with $[GaH_4]^-$. Other tetrahedral molecules include SiF_4, GeF_4, $SnCl_4$ and $PbCl_4$.

Within group 14, coordination numbers greater than four are seen in the $[SnCl_5]^-$ (**5.17**) and $[Me_2SnCl_3]^-$ anions (which have trigonal bipyramidal structures) and in the octahedral $[SnF_6]^{2-}$ dianion.

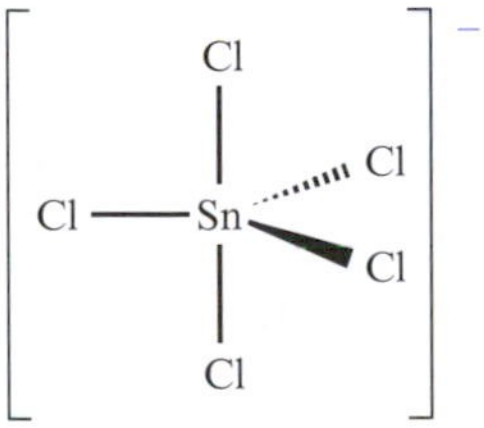

(5.17)

Group 15

Structural diversity is further extended in the compounds formed by the heavier group 15 elements.

The trigonal pyramid is a common shape for species containing phosphorus, arsenic and antimony and examples include PF_3, AsF_3, $SbCl_3$, PMe_3 (**5.18**) and $[PO_3]^{3-}$. Although our discussion is concentrating on molecules and ions with a single group 15 atomic centre, it is worth remembering that each phosphorus atom in the P_4 molecule is in a trigonal pyramidal site (Figure 3.2).

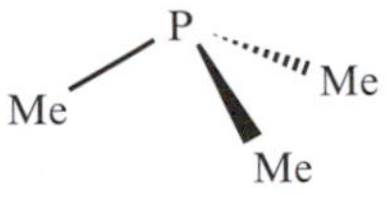

(5.18)

A coordination number of four for a group 15 element is usually associated with a tetrahedral geometry but the disphenoidal structure (Figure 5.12) is also observed. The ions $[PCl_4]^+$, $[PBr_4]^+$ and $[PO_4]^{3-}$ are tetrahedral, but the change of charge in going from $[PBr_4]^+$ to $[PBr_4]^-$ is associated with a change in shape and the anion is disphenoidal, as shown in structure **5.19**.

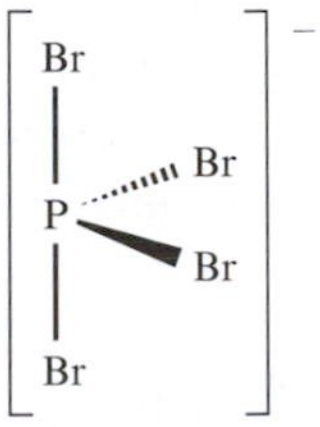

The number of five-coordinate species is large. In the gas phase, the pentahalides PF_5, AsF_5 and $SbCl_5$ are trigonal bipyramidal, but a change in charge leads to a change in shape – in the solid state, the $[SbF_5]^{2-}$ ion has a square-based pyramidal structure (Figure 5.13).

(5.19)

5.13 In $[SbF_5]^{2-}$ the fluorine atoms are in a square-based pyramidal arrangement around the central antimony atom.

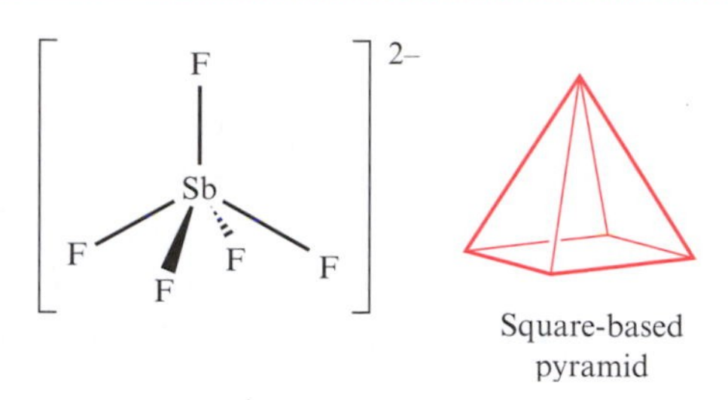

Square-based pyramid

Octahedral structures are common for anionic species in group 15 and examples include $[PF_6]^-$, $[AsF_6]^-$ and $[SbF_6]^-$.

Group 16

For oxygen, only bent and trigonal pyramidal shapes are favoured, but for the heavier elements in group 16, the variety of observed geometries is far greater.

Bent molecules include H_2S, SF_2, SO_2, $SeCl_2$ and $TeCl_2$. Trigonal planar molecules are exemplified by SO_3, (**5.20**), whilst trigonal pyramidal structures are observed for $SOCl_2$, $SeOCl_2$, $[SO_3]^{2-}$ and $[TeO_3]^{2-}$. Once again, a change of shape accompanies a change in charge as is seen in going from SO_3 to $[SO_3]^{2-}$.

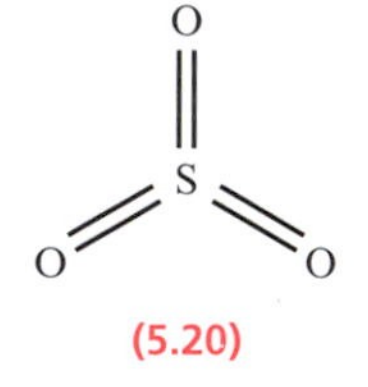

(5.20)

Four-coordinate sulfur, selenium and tellurium atoms are found in both tetrahedral and disphenoidal environments. The $[SO_4]^{2-}$ ion is tetrahedral, but SF_4, SeF_4 and Me_2TeI_2 possess disphenoidal geometries. For five-coordinate atoms, both the trigonal bipyramidal (e.g. SOF_4, **5.21**) and square-based pyramidal (e.g. $[SF_5]^-$) geometries are observed. Octahedral molecules include SF_6, SeF_6 (**5.22**) and TeF_6.

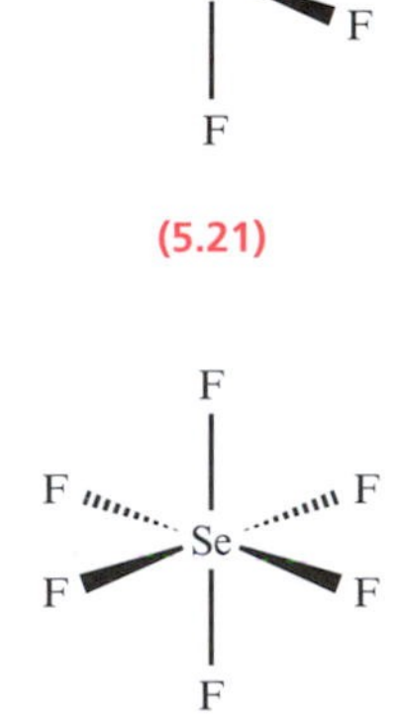

(5.22)

Group 17

Unlike fluorine, the heavier halogens are found in a wide range of geometrical environments. Iodine is the central atom in the linear anion $[I_3]^-$ and the bent cation $[I_3]^+$. For three-coordinate atoms, both T-shaped and trigonal pyramidal species (Figure 5.12) are known; ClF_3 is an example of a T-shaped molecule although the two F–Cl–F bond angles, which should ideally be 90°, are actually 87°. The anion $[ClO_3]^-$ is trigonal pyramidal.

Elements from group 17 may adopt one of *three* different four-coordinate structures. Tetrahedral anions include $[ClO_4]^-$ and $[IO_4]^-$, while the cations $[ClF_4]^+$ and $[IF_4]^+$ are disphenoidal in shape; $[ClF_4]^+$ is isoelectronic with SF_4 (see above). On going from $[ClF_4]^+$ to $[ClF_4]^-$, there is a change in structure from a disphenoidal to a square planar geometry.

Five-coordinate halogen atoms provide examples of both trigonal bipyramidal (e.g. ClO_2F_3) and square-based pyramidal (e.g. ClF_5 and IF_5) shapes, whilst the cation $[IF_6]^+$ is an example of a six-coordinate, octahedral species.

Overview of the heavier p-block elements

- Compounds of the heavier elements in groups 13 to 17 of the *p*-block show a greater range of shapes than do the first row elements of each respective group.
- The variety of molecular shapes *increases* on moving across the *p*-block from group 13 to 17.
- A change in the overall charge of a species usually causes a change in geometry.

5.6 VALENCE-SHELL ELECTRON-PAIR REPULSION (VSEPR)

In the previous sections we have discussed the shapes of molecules and molecular ions purely as observable facts, and we have noted some trends in the geometrical environments found for certain elements within both groups and periods in the *p*-block. In this section, we consider a method for predicting or rationalizing the shape of a molecule or ion.

The Valence-Shell Electron-Pair Repulsion (VSEPR) model was initially proposed by Sidgwick and Powell in 1940, and further developed by Nyholm and Gillespie. The basis of the VSEPR model is the consideration of the *repulsive forces between pairs of valence electrons*, and it is assumed that *only the electrons in the valence shell influence the molecular geometry*.

Minimizing inter-electron repulsion

The VSEPR model assumes that for maximum stability pairs of valence electrons will be as far apart from one another as possible. In this way, inter-electron repulsions are minimized. Assume that in a molecule XY_n, the atoms Y are all bonded to a central atom X and that all the valence electrons are involved in bonding. Let each X–Y bond be a single bond; i.e. one pair

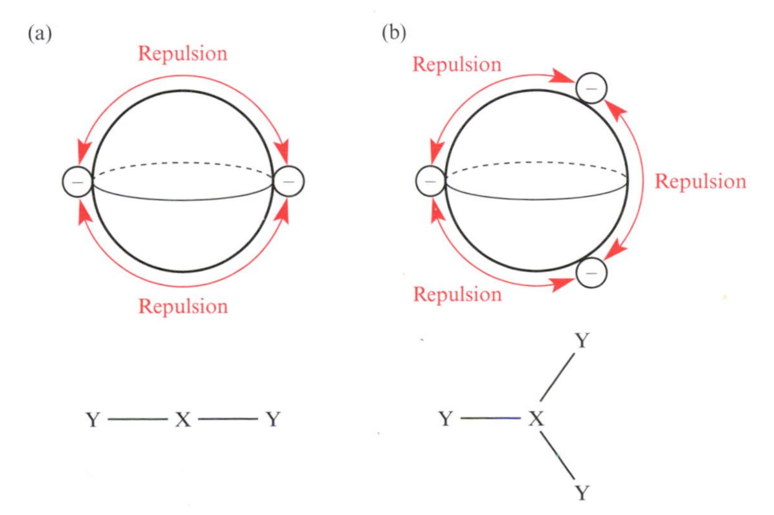

5.14 Pairs of valence electrons can be considered as point charges. They will repel one another and move as far away from one another as possible to minimize the energy of the system. (a) Two points of negative charge take up positions at opposite sides of the sphere and this results in a linear arrangement of atoms in the molecule XY_2 if all the valence electrons in the molecule are used for bonding. (b) The mutual repulsion between three points of negative charge results in a triangular arrangement for minimum energy; this gives a trigonal planar arrangement of Y atoms in the molecule XY_3 if all the valence electrons in the molecule are used for bonding. (Compare Figure 5.16.)

of electrons is associated with each bond. For a minimum energy system, the relative positions of the n pairs of electrons can be determined by considering each pair of electrons as a point of negative charge, and placing these points on the surface of a sphere.

For $n = 2$, the repulsive interactions between the two point charges are at a minimum if the charges are at opposite sides of the sphere (Figure 5.14a). This corresponds to a *linear* arrangement for the atoms in the molecule XY_2.

For $n = 3$, the total inter-electron repulsion between the three point charges is minimized when the charges are placed at the corners of a triangle (Figure 5.14b). This corresponds to a *trigonal planar* arrangement for the atoms in the molecule XY_3.

Similarly, for $n = 4$, 5 and 6, the geometries which correspond to minimum inter-electron repulsions are the tetrahedron, the trigonal bipyramid and the octahedron. Figure 5.15 illustrates how these polyhedra relate to the *tetrahedral, trigonal bipyramidal* and *octahedral* shapes of the XY_4, XY_5 and XY_6 molecules.

Energy differences between structures: see Section 5.12

Notice that, although in Table 5.1 we listed more than one geometrical possibility for $n = 2$, 3, 4 and 5, only one geometry corresponds to the lowest energy structure in each case.

Lone pairs of electrons

Not all valence electrons are necessarily involved in bonding. The Lewis structure for H_2O (**5.23**) reveals that there are two lone pairs in the valence shell of the oxygen atom. On the other hand, in CH_4 (**5.24**) all of the valence electrons are used for bonding.

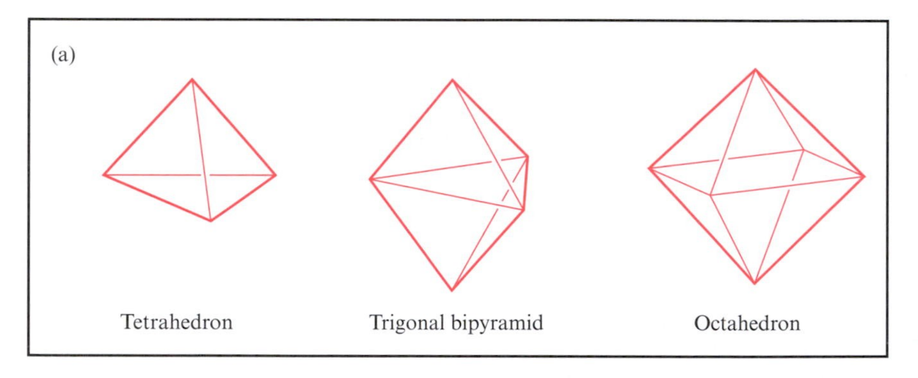

(b)
Tetrahedral
Trigonal bipyramidal
Octahedral

5.15 For four, five or six points of negative charges, repulsions are minimized when they lie at the corners of a tetrahedron, a trigonal bipyramid or an octahedron, respectively. For molecules XY_n in which all the valence electrons are used for bonding, this corresponds to tetrahedral, trigonal bipyramidal or octahedral shapes for the molecules XY_4, XY_5 and XY_6, respectively.

H : Ö :
H

(5.23)

H
H : C : H
H

(5.24)

Inter-electron repulsions in the valence shell of the central atom in a molecule involve *both lone and bonding pairs* of electrons, and so the presence of one or more lone pairs influences the molecular shape. Lewis structures for BH_3 and NH_3 are shown on the left-hand side of Figures 5.16a and 5.16b. The boron atom uses all of its valence electrons to form three B–H single bonds, but the nitrogen atom has five valence electrons, and one lone pair remains after the three N–H bonds have been formed. There are *three* pairs of electrons in the valence shell of the boron atom in BH_3, and by the VSEPR model, the molecule will be trigonal planar (Figure 5.16a). In NH_3, there are *four* pairs of electrons in the valence shell of the nitrogen atom, and the VSEPR model predicts that the molecular shape will be based upon a tetrahedron with the lone pair of electrons occupying one of the vertices (Figure 5.16b). The NH_3 molecule may be described as being *derived from* a tetrahedron, although the nitrogen and three hydrogen atoms actually define a trigonal pyramid (Figure 5.9).

Relative magnitudes of inter-electron repulsions

In XY_n compounds which contain *only* bonding pairs of electrons in the valence shell of the central atom, we can assume that the electron–electron repulsions are all equal provided that the bonds are all identical.

- Multiple bonds: multiple bonds contain greater densities of electrons (more negative charge) than single bonds, and therefore, inter-electron repulsions involving multiple bonds are greater than those involving single bonds (Figure 5.17).
- Lone pairs: if one or more lone pairs of electrons are present, the relative magnitudes of the inter-electron repulsions are assumed to follow the sequence:

  lone pair–lone pair > lone pair–bonding pair > bonding pair–bonding pair

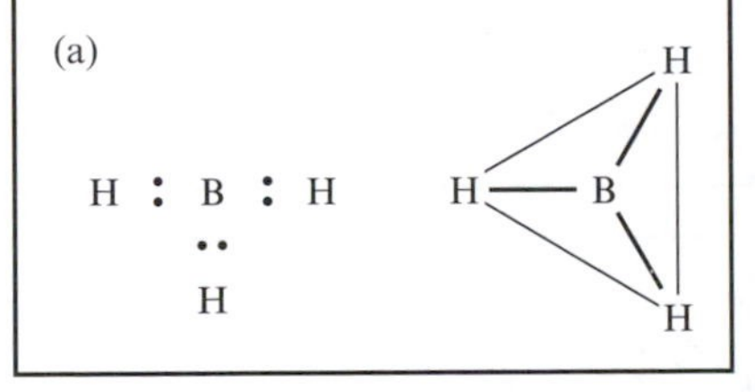

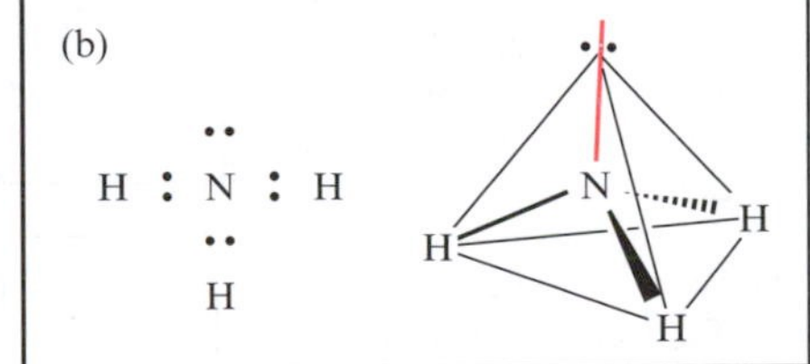

5.16 The application of the VSEPR model to predict the shapes of the BH_3 and NH_3 molecules. Note the role of the lone pair of electrons in NH_3.

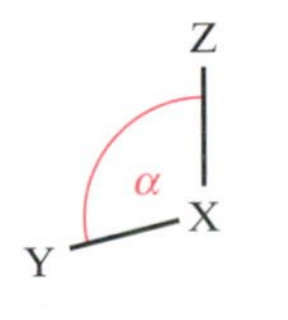

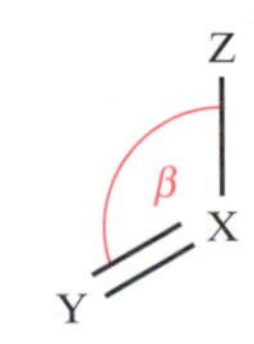

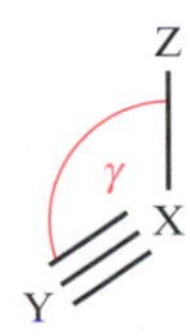

$\angle\gamma > \angle\beta > \angle\alpha$

5.17 Multiple bonds contain greater electron density than single bonds, and so inter-electron repulsions involving multiple bonds are greater than those involving single bonds.

These differences in repulsions affect the detailed geometry of a molecule as illustrated in the series CH_4, NH_3 and H_2O. The total number of inter-electron repulsive interactions in each molecule is six (Figure 5.18a). In the CH_4 molecule, these are all between bonding pairs of electrons, but in the NH_3 and H_2O molecules, lone pair–bonding pair or lone pair–lone pair interactions are involved. The result of increased inter-electron repulsion is a decrease in the bond angle as shown in Figure 5.18b.

The *Valence-Shell Electron-Pair Repulsion* (VSEPR) model is used to predict or rationalize molecular shapes. The model considers only the repulsion between the electrons in the valence shell of the central atom in the molecule or molecular ion.

- Pairs of valence electrons are arranged so as to *minimize* inter-electron repulsions.
- The relative magnitudes of repulsive forces between pairs of electrons follow the order:

  lone pair–lone pair > lone pair–bonding pair > bonding pair–bonding pair

- For n = 2 to 6 pairs of valence electrons, the ideal geometries are:

  n = 2 linear
  n = 3 trigonal planar
  n = 4 tetrahedral
  n = 5 trigonal bipyramidal
  n = 6 octahedral

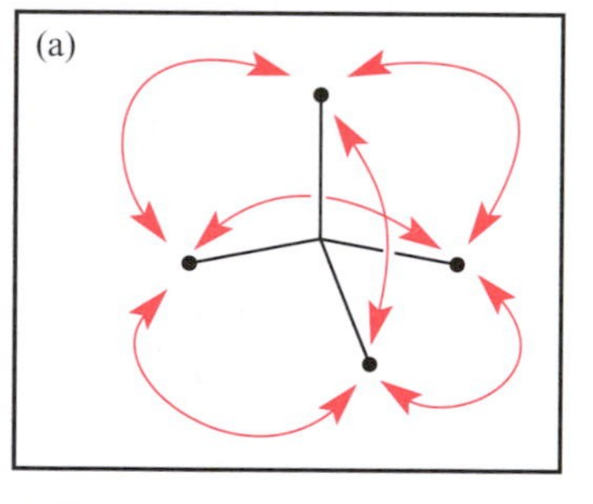

(b)

$\theta = 109.5°$ $\theta = 107.5°$ $\theta = 104.5°$

5.18 (a) In a tetrahedral arrangement of four electron pairs, there are six repulsive interactions. (b) In the CH_4 molecule, there are six bonding pair–bonding pair repulsions and all the H–C–H bond angles are equal. In NH_3, there are three lone pair–bonding pair repulsions and three bonding pair–bonding pair repulsions. The H–N–H angle in NH_3 is therefore smaller than the H–C–H angle in CH_4. In the H_2O molecule, there is one lone pair–lone pair repulsion, one bonding pair–bonding pair repulsion, and four lone pair–bonding pair repulsions. The H–O–H angle is smaller than the H–N–H angle in NH_3 or the H–C–H angle in CH_4.

Application of the VSEPR model to species involving *p*-block elements

A molecular geometry can be predicted by using the valence-shell electron-pair model by following the procedure set out below.

1. Draw a Lewis structure for the molecule or ion, and determine the number of bonding and lone pairs of electrons in the valence shell of the central atom.
2. The 'parent' geometry (Table 5.4) is determined by the number of points of negative charge – a 'point of charge' is taken to be the electrons involved in a single bond, a multiple bond or a lone pair.
3. If all bonds are single bonds, the amount of distortion away from the ideal bond angles (see Table 5.1) for the arrangement determined in step (2) can be estimated by using the sequence of relative inter-electron repulsions:
   lone pair–lone pair > lone pair–bonding pair > bonding pair–bonding pair
4. If the electrons are involved in bonds other than single bonds, account must be taken of the fact that:
   repulsion due to the electrons in a triple bond > repulsion due to the electrons in a double bond > repulsion due to the electrons in a single bond
5. In a trigonal bipyramid, lone pairs of electrons usually occupy the equatorial rather than axial sites. A multiple bond also tends to occupy an equatorial site.
6. In an octahedron, two lone pairs are opposite (rather than next) to one another.

Example 1: H_2S

Sulfur is in group 16 and has six valence electrons.
The hydrogen atom has one valence electron.
The Lewis structure for H_2S is:

$$\mathrm{H} : \ddot{\underset{\cdot\cdot}{\mathrm{S}}} : \quad \text{(with the second H below S)}$$

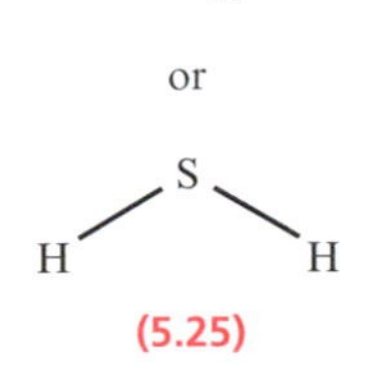

(5.25)

There are four points of negative charge (two bonding pairs and two lone pairs of electrons).

The parent shape is a tetrahedron. The H_2S molecule is therefore derived from a tetrahedron and the three atoms define a bent structure **5.25**.

The ∠H–S–H should be less than 109.5°. (The experimentally determined angle is 92.1°.)

Table 5.4 Basic arrangements of points of charge (≡ bonds or lone pairs of electrons) that are used to predict molecular geometries using the VSEPR model. Ideal angles are listed in Table 5.1.

Number of points of negative charge	*Arrangement*
2	Linear
3	Trigonal planar
4	Tetrahedral
5	Trigonal bipyramidal
6	Octahedral

Example 2: $[NH_4]^+$

Nitrogen is in group 15 and has five valence electrons.

The positive charge may be assigned to the nitrogen atom; N^+ has four valence electrons.

The hydrogen atom has one valence electron.

The Lewis structure for the $[NH_4]^+$ ion is:

$$\left[\begin{array}{ccc} & H & \\ & \cdot\cdot & \\ H & :N: & H \\ & \cdot\cdot & \\ & H & \end{array}\right]^+$$

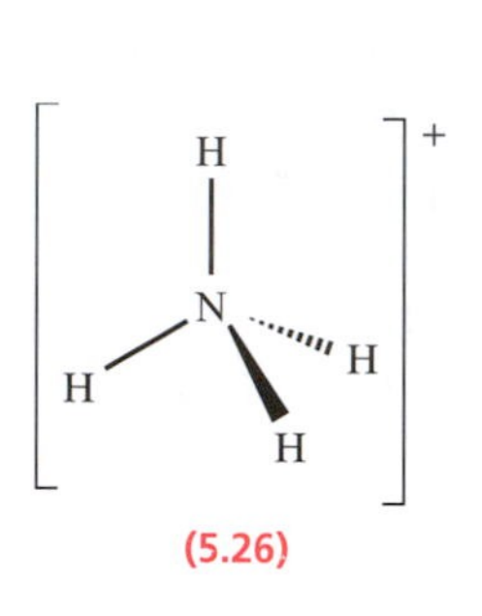

(5.26)

There are four points of negative charge (four bonding pairs of electrons).

The parent shape is a tetrahedron.

The $[NH_4]^+$ ion therefore has the tetrahedral structure **5.26** and the ∠H–N–H will be 109.5°. Compare the predicted result with the structure drawn in Figure 5.10.

Example 3: CO_2

Carbon is in group 14 and has four valence electrons. Oxygen is in group 16 and has six valence electrons.

The Lewis structure for the CO_2 molecule is:

O :: C :: O

The carbon atom is involved in two double bonds; there are two points of negative charge.

O═C═O

(5.27)

The CO_2 molecule (**5.27**) is linear. Compare the predicted result with the structure in Figure 5.2. [Why do the lone pairs of electrons on the oxygen atoms not affect the molecular shape?]

Example 4: SO_2

Sulfur and oxygen are both in group 16 and each has six valence electrons.

The Lewis structure for the SO_2 molecule is:

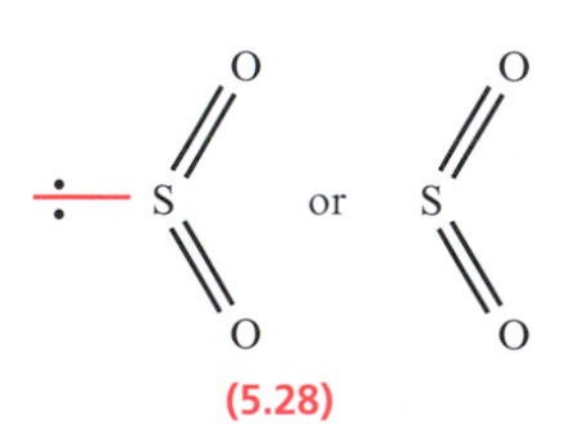

(5.28)

There are three points of negative charge around the sulfur atom consisting of one lone pair of electrons, and two points of charge which are associated with the S=O double bonds.

The geometry of the SO_2 molecule **5.28** is derived from a trigonal planar arrangement, the ideal angles for which are 120°. In structure **5.28** it is not easy to assess the relative magnitudes of the lone pair–double bond and double bond–double bond inter-electron repulsions. The experimental value for the angle ∠O–S–O is 119°.

Example 5: SF_4

Sulfur is in group 16 and has six valence electrons. Fluorine is in group 17 and has seven valence electrons.

The Lewis structure for the SF_4 molecule is:

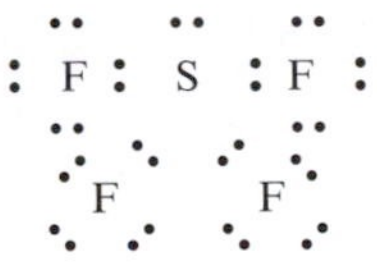

There are five points of negative charge around the sulfur atom consisting of four bonding pairs and one lone pair of electrons.

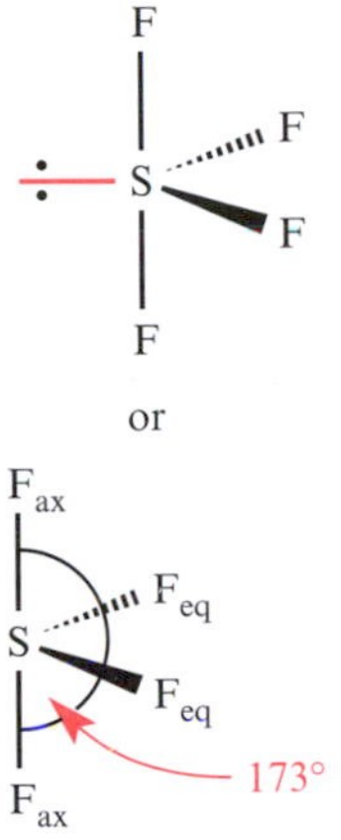

The parent geometry is trigonal bipyramidal and the lone pair occupies an equatorial site in order to minimize repulsive interactions. The atoms in the SF_4 molecule define the disphenoidal geometry **5.29**. There are two different fluorine environments – the axial (F_{ax}) and equatorial (F_{eq}) positions. [Do the fluorine lone pairs have an effect on the molecular shape?]

The F_{eq}–S–F_{eq} bond angle defined by the fluorine atoms in the equatorial plane will be larger than the angles $\angle F_{ax}$–S–F_{eq}. Taking into account the relative lone pair–lone pair, lone pair–bonding pair and bonding pair–bonding pair inter-electron repulsions, we can predict bond angles as follow:

$$\angle F_{ax}\text{–S–}F_{ax} \approx 180^\circ \text{ or } < 180^\circ$$
$$90^\circ < \angle F_{eq}\text{–S–}F_{eq} < 120^\circ$$

The experimentally determined bond angles in SF_4 are $\angle F_{ax}$–S–F_{ax} = 173° and $\angle F_{eq}$–S–F_{eq} = 102°.

(5.29)

Example 6: $[ICl_4]^-$

Iodine and chlorine are both in group 17 and each has seven valence electrons. The negative charge may be assigned to the central iodine atom; I^- has eight valence electrons.

The Lewis structure for the $[ICl_4]^-$ ion is:

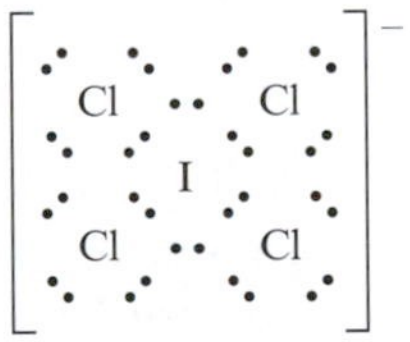

There are six points of negative charge around the iodine atom (four bonding pairs of electrons and two lone pairs of electrons).

The geometry of the $[ICl_4]^-$ ion **5.30** is derived from an octahedron. The lone pairs will be opposite one another in order to minimize their mutual repulsions. The five atoms define a square planar structure.

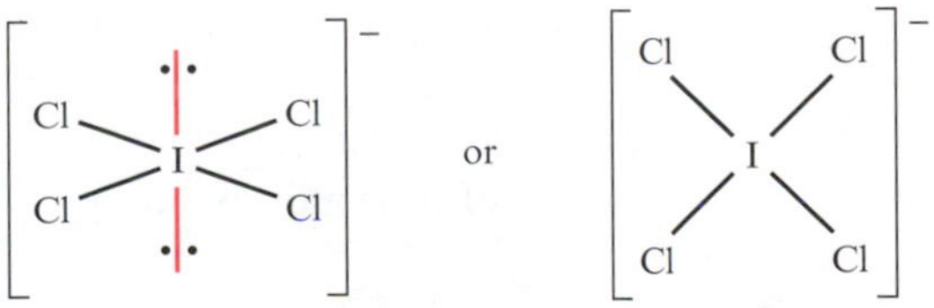

(5.30)

5.7 THE VSEPR MODEL: SOME AMBIGUITIES

The advantage of the VSEPR model is that it is simple to apply. It is a successful model and its use is widespread. However, there are some species in which ambiguities can arise, and in this section we look at several such cases.

Chlorine trifluoride

The structure of the interhalogen compound chlorine trifluoride can be predicted as follows.

The Lewis structure of ClF_3 is:

```
      ••
    : F :
 ••   ••   ••
: F : Cl : F :
 ••  •• ••  ••
```

There are five centres of negative charge around the chlorine atom, and the geometry of ClF_3 is based on a trigonal bipyramid.

The problem now is 'which sites do the lone pairs prefer?' The alternatives are given below and the problem comes in assessing the relative merits of each.

(5.31)

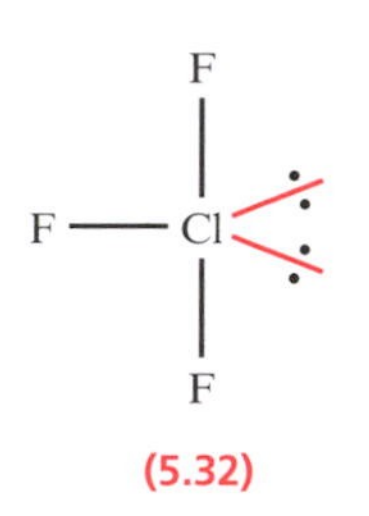

(5.32)

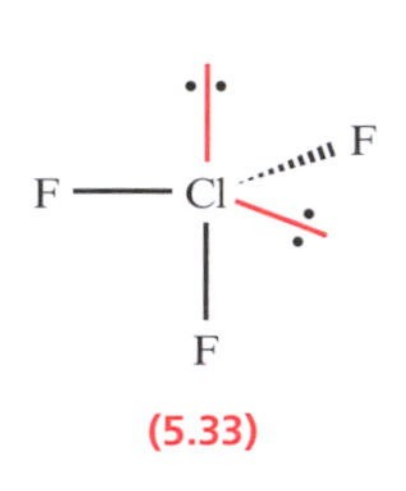

(5.33)

1. Place the two lone pairs as far apart as possible as shown in structure **5.31**. This minimizes the lone pair–lone pair interaction, but maximizes the number of 90° lone pair–bonding pair interactions.

2. Place the two lone pairs in the equatorial plane to give structure **5.32**. This places the two lone pairs of electrons at 120° to one another. In addition, there are four lone pair–bonding pair repulsive interactions acting at 90° and two acting at 120°, plus two 90° bonding pair–bonding pair interactions.

3. Place the two lone pairs at 90° to each other, **5.33**. This is the worst possible arrangement in terms of repulsive interactions because the lone pairs of electrons are at 90° to one another. Hence, structure **5.33** can be discounted.

Structures **5.31** and **5.32** are possible but it is extremely difficult to judge between them qualitatively in terms of minimizing the repulsive energy of the system. The experimentally determined structure of ClF_3 shows the molecule to be T-shaped, with F_{ax}–Cl–F_{eq} bond angles of 87°. This result is consistent with structure **5.32** – the lone pairs are placed in the equatorial plane of the trigonal bipyramid. The reduction in F_{ax}–Cl–F_{eq} bond angle from an ideal value of 90° to an experimental one of 87° may be interpreted in terms of lone pair–bonding pair repulsions.

The anions $[SeCl_6]^{2-}$ and $[TeCl_6]^{2-}$

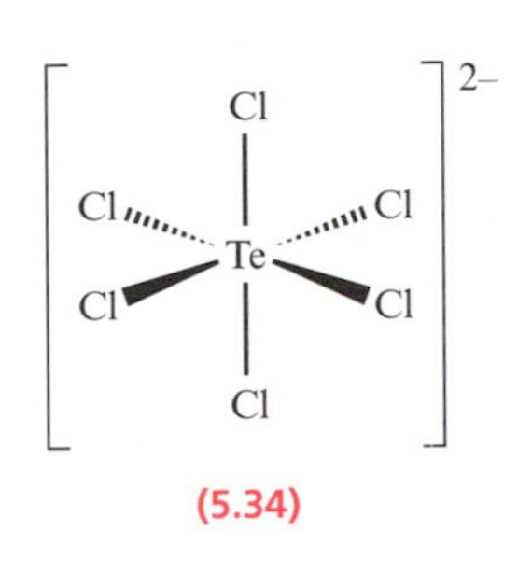

(5.34)

Selenium and tellurium are in group 16 and each possesses six valence electrons. In the ions $[SeCl_6]^{2-}$ and $[TeCl_6]^{2-}$, the valence shell of the central atom contains 14 electrons (six from the group 6 element, one from each chlorine atom, and two from the dinegative charge). The VSEPR model predicts that each ion possesses a structure based upon a seven-coordinate geometry, with the lone pair occupying one site. However, the observed structure of each of the $[SeCl_6]^{2-}$ and $[TeCl_6]^{2-}$ (**5.34**), anions is that of a *regular octahedron*. This experimental result suggests that the lone pair of electrons does not influence the shape in the way that the VSEPR model predicts it should. Such a lone pair is said to be *stereochemically inactive*.

We cannot readily predict the structures of $[SeCl_6]^{2-}$ and $[TeCl_6]^{2-}$, but we can rationalize them in terms of the presence of a stereochemically inactive pair of electrons. Stereochemically inactive lone pairs of electrons are usually observed for the heaviest members of a periodic group, and the tendency for valence shell *s* electrons to adopt a non-bonding role in a molecule is known as the *inert pair effect*.

When the presence of a lone pair of electrons influences the shape of a molecule or ion, the lone pair is said to be *stereochemically active*.

If the geometry of a molecule or ion is *not* affected by the presence of a lone pair of electrons, then the lone pair is *stereochemically inactive*.

The tendency for the pair of valence *s* electrons to adopt a non-bonding role in a molecule or ion is know as the *inert pair effect*.

5.8 THE KEPERT MODEL

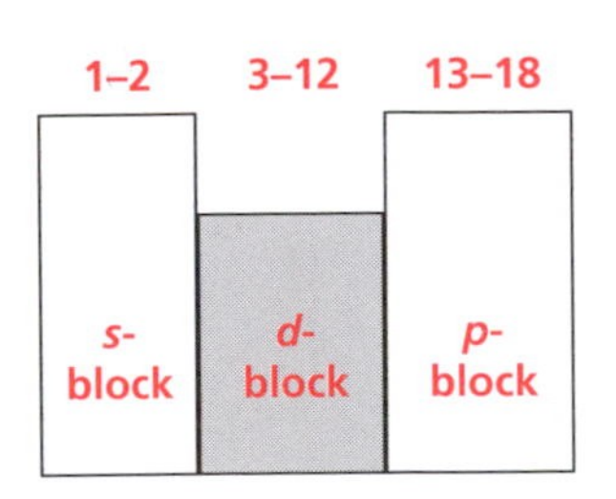

Ligand: see Section 16.3

The VSEPR model is most commonly used for the prediction and rationalization of the shapes of molecules and ions containing central *p*-block atoms. Kepert developed the VSEPR model so that it is applicable to molecules containing a *d*-block metal as the central atom. Consider a molecule of formula ML_n where M is a *d*-block metal and L is a *ligand* – a group attached to the metal centre.

In Kepert's model, the metal lies at the centre of a sphere and the ligands are free to move over the surface of the sphere. The ligands are considered to repel one another in much the same way that the point charges repel one another in the VSEPR model. Kepert's approach predicts the relative positions of the ligands in a molecule ML_n in an analogous manner to the way in which the VSEPR model predicts the relative positions of the groups Y in a molecule XY_n. Ions of the type $[ML_n]^{m+}$ or $[ML_n]^{m-}$ can be treated similarly. The polyhedra listed in Table 5.4 apply both to molecules containing *p*-block and *d*-block atomic centres.

The prediction of the shapes of the *d*-block molecules is considered by Kepert to be *independent* of the ground state electronic configuration of the metal centre. The model differs from the VSEPR model in that it ignores the presence of any lone pairs of electrons in the valence shell of the metal centre.

The Kepert model predicts the shapes of *d*-block metal compounds ML_n by considering the repulsions between the groups L. The model ignores lone pairs of electrons.

5.9 APPLICATION OF THE KEPERT MODEL

Linear geometry

$[NC — Au — CN]^-$

(5.35)

In the anion $[Au(CN)_2]^-$, two cyanide groups are bonded to the gold centre. The repulsion between them is minimized if they are 180° apart. The anion is predicted to have the linear NC–Au–CN framework **5.35**, and this agrees with the observed structure in which the bond angle ∠C–Au–C is 180°.

The $[HgCl_2]^-$ and $[AuCl_2]^-$ anions have similar structures to that of $[Au(CN)_2]^-$.

Trigonal planar geometry

When three groups are present around the metal centre as in $[Cu(CN)_3]^{2-}$, the structure is predicted to be trigonal planar so that repulsions between the cyanide groups are minimized. This agrees with the observed structure, shown in Figure 5.19.

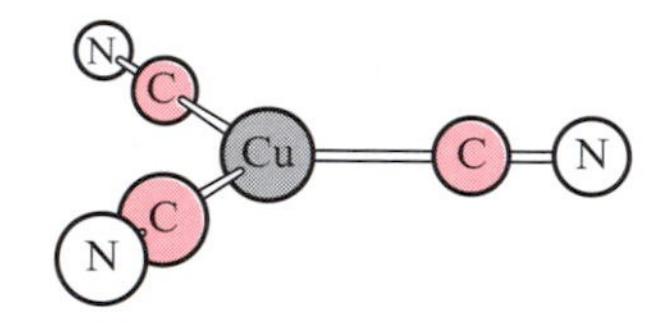

5.19 The trigonal planar structure of the copper(I) ion $[Cu(CN)_3]^{2-}$.

Tetrahedral geometry

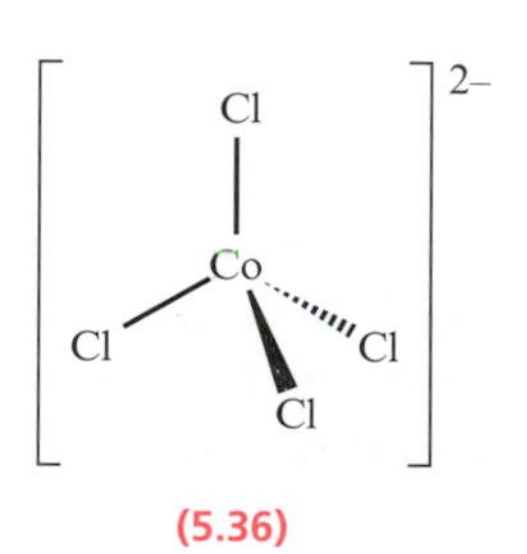

(5.36)

Tetrahedrally coordinated *d*-block metals are reasonably common and an example is the $[CoCl_4]^{2-}$ ion in which four chloride groups surround one cobalt(II) centre. In the tetrahedral geometry **5.36**, repulsions between the chlorides are minimized.

Similarly, tetrahedral structures are predicted (and found) for the ions $[MnO_4]^-$, $[CrO_4]^{2-}$, $[NiCl_4]^{2-}$ and $[VCl_4]^-$ and for the neutral compounds $ZrCl_4$ and OsO_4.

Five-coordinate geometries

(5.37)

For the cadmium(II) species $[Cd(CN)_5]^{3-}$, the trigonal bipyramidal structure **5.37** is predicted so that repulsion between the five cyanide groups is minimized. Similarly, structure **5.38** is predicted for the iron(0) compound $[Fe(CO)_5]$.

In the VSEPR model, lone pairs of electrons preferentially occupy the equatorial sites of a trigonal bipyramidal structure because repulsions involving lone pairs of electrons are relatively large. Similarly, in the Kepert model, a sterically demanding (bulky) group is expected to occupy such a site to minimize inter-ligand repulsions.

Me = CH_3 = methyl
Ph = C_6H_5 = phenyl
PMe_2Ph = dimethylphenylphosphine

Structure **5.39** is predicted for the nickel(II) compound $[NiCl_2(PMe_2Ph)_3]$ – the PMe_2Ph groups are bulky organic phosphines and lie in the three equatorial sites.

The predicted geometry of the compound $[NiBr_3(PMe_2Ph)_2]$ is shown in structure **5.40** and Figure 5.20 illustrates the structure as determined by X-ray diffraction. Repulsions between the five groups are minimized if the two phosphines are remote from one another. In addition, because the bromides are also sterically demanding, bromide–bromide repulsions are minimized if the Br–Ni–Br bond angles are 120° rather than 90°; (see problem 5.7 at the end of this chapter).

(5.38)

(5.39)

(5.40)

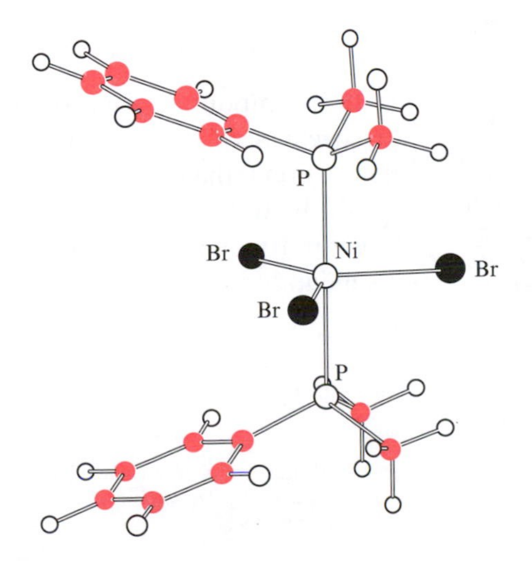

5.20 The observed structure of the nickel(III) compound $[NiBr_3(PMe_2Ph)_2]$ determined by X-ray diffraction.

(5.41)

Octahedral geometry

Among molecular species involving the *d*-block elements, the octahedral geometry is extremely common and is the geometry expected from the Kepert model, for example, for the cobalt(II) ion $[Co(H_2O)_6]^{2+}$ (**5.41**). Similar structures are predicted (and found) for $[Ni(H_2O)_6]^{2+}$, $[Fe(CN)_6]^{3-}$, $[Fe(CN)_6]^{4-}$ and $[NiF_6]^{2-}$.

5.10 AN EXCEPTION TO THE KEPERT MODEL: THE SQUARE PLANAR GEOMETRY

Although the Kepert model succeeds in predicting the correct geometry for many molecular species involving the *d*-block elements, the simple form of the theory described here can *never* predict the formation of a square planar structure for a four-coordinate species.

In the VSEPR model, a square planar structure such as $[ICl_4]^-$ **5.30** is derived from an octahedral arrangement which contains four bonding pairs and two lone pairs of electrons in the valence shell of the central atom. However, Kepert's model focuses on the repulsions between ligands, and ignores lone pairs of electrons. For a four-coordinate species, repulsive interactions (i.e. *steric effects*) will *always* favour a tetrahedral geometry.

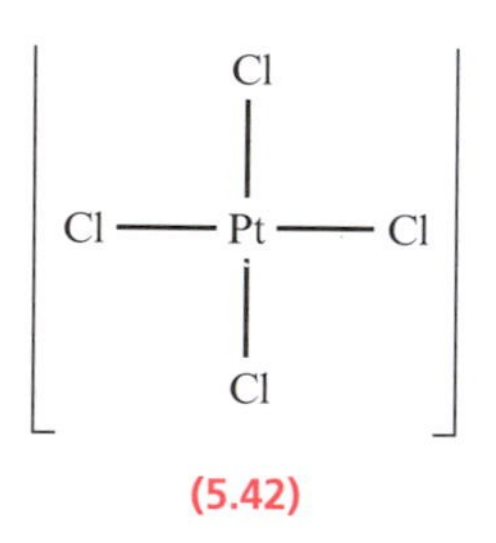

(5.42)

For $[CoCl_4]^{2-}$, the tetrahedral shape **5.36** gives Cl–Co–Cl bond angles of 109.5°. This structure is both predicted by Kepert and is observed in practice. However, the anion $[PtCl_4]^{2-}$ possesses the square planar structure **5.42** with Cl–Pt–Cl bond angles of 90°. Other square planar anions are $[Ni(CN)_4]^{2-}$, $[PdCl_4]^{2-}$ and $[AuCl_4]^-$.

Many square planar molecules and ions, each with a *d*-block metal centre, are know and their structures are clearly *not* controlled by steric effects.§

5.11 GEOMETRICAL ISOMERISM

If two compounds have the same formulae and the same structural framework, but differ in the spatial arrangement of different atoms or groups about a central atom or a double bond, then the compounds are *geometrical isomers.*

If two compounds have the same molecular formula but differ in the arrangement of different atoms or groups of atoms about a central atom or double bond, then the compounds are *geometrical isomers*.

We illustrate geometrical isomerism by looking at the arrangements of atoms in trigonal bipyramidal, square planar and octahedral species and in compounds containing a double bond.

§ For a discussion of electronic effects relating to square planar structures, see: M. Gerloch and E.C. Constable, *Transition Metal Chemistry: The Valence Shell in d-Block Chemistry*, VCH, Weinheim (1994).

Trigonal bipyramidal structures

In the trigonal bipyramid, there are two different sites that atoms or groups can occupy, the *axial and equatorial sites*, as shown for $[Fe(CO)_5]$ in structure **5.43**.

(5.43)

When a carbonyl group in $[Fe(CO)_5]$ is replaced by another group such as PMe_3, two possible structures can be drawn. Structures **5.44** and **5.45** for $[Fe(CO)_4(PMe_3)]$ are *geometrical isomers*.

(5.44)

(5.45)

The trigonal bipyramidal molecule PCl_2F_3 has three geometrical isomers depending upon whether the two chlorine atoms are arranged in the two axial sites **5.46**, two equatorial positions **5.47**, or one axial and one equatorial **5.48**. (The three isomers can also be described in terms of the positions of the fluorine atoms.) As we saw for $[NiCl_2(PMe_2Ph)_3]$ **5.39**, one isomer may be favoured over another.

(5.46) (5.47) (5.48)

Square planar structures

In the anion $[PtCl_4]^{2-}$ **5.42**, all four chlorides are equivalent. If one chloride is replaced by a new group, only one product can be formed. Thus, $[PtCl_3(NH_3)]^-$ **5.49** has no geometrical isomers.

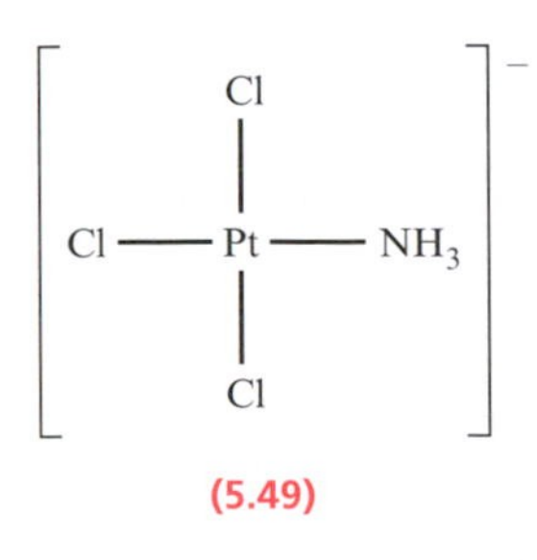

(5.49)

If two groups are introduced, they could be positioned next to or opposite one another. This gives two geometrical isomers for $[PtCl_2(NH_3)_2]$ as shown in structures **5.50** and **5.51**. Isomer **5.50** is called the *cis*-isomer and **5.51** is the *trans*-isomer of $[PtCl_2(NH_3)_2]$.§ These structural differences are of more than academic interest. The *cis*-isomer of $[PtCl_2(NH_3)_2]$ is the drug cisplatin which is effective for the treatment of certain forms of cancer; in contrast, *trans*-$[PtCl_2(NH_3)_2]$ is orders of magnitude less active.

§ The prefixes *cis* and *trans* are in common use, but IUPAC has introduced a system of configuration indexes.

NH_3
Cl — Pt — NH_3
Cl

(5.50)

Cl
H_3N — Pt — NH_3
Cl

(5.51)

The structures of *cis*-$[PtCl_2(PMePh_2)_2]$ and *trans*-$[NiBr_2(PMe_2Ph)_2]$ have been determined by X-ray diffraction and are shown in Figure 5.21.

A square planar species of general formula XY_2Z_2 has two geometrical isomers.

In *cis*-XY_2Z_2, the two Y groups (and thus the two Z groups) are next to each other, and in *trans*-XY_2Z_2, the two Y groups (and thus the two Z groups) are in opposite sites.

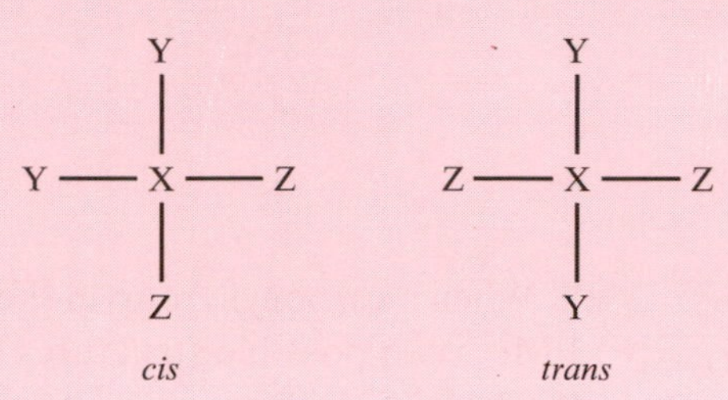

$[MoCl_5(H_2O)]^{2-}$ (Cl, Cl, Cl, Mo, H_2O, Cl, Cl; 2–)

(5.52)

Octahedral structures

In each of the octahedral species SF_6, $[Co(H_2O)_6]^{2+}$ and $[MoCl_6]^{3-}$ the six groups around the central atom are the same. If one group is replaced by another, a single product can be formed, and $[MoCl_5(H_2O)]^{2-}$ **5.52** does not possess any geometrical isomers.

If we introduce *two* groups, two geometrical isomers may be formed. The cation $[CoCl_2(NH_3)_4]^+$ can have the chlorides in a *cis* (**5.53**) or *trans* (**5.54**) arrangement.

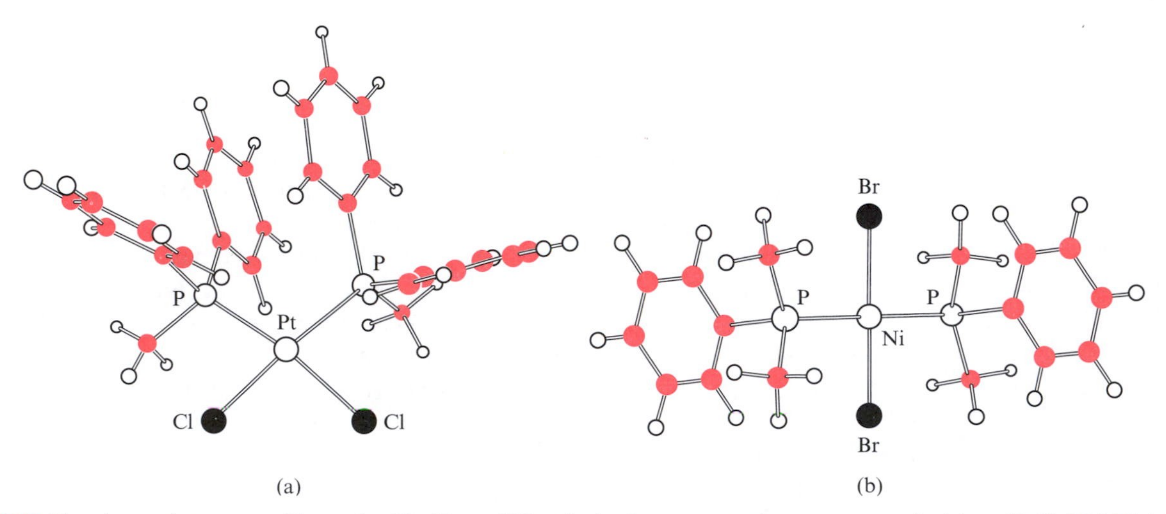

5.21 The observed structures (determined by X-ray diffraction) of two square planar compounds: (a) *cis*-$[PtCl_2(PMePh_2)_2]$ and (b) *trans*-$[NiBr_2(PMe_2Ph)_2]$.

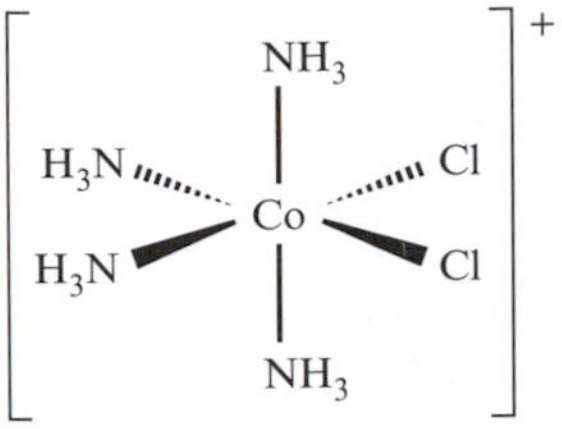

NH_3
H_3N Co Cl
H_3N Cl
NH_3 $]^+$

(5.53)

Cl
H_3N Co NH_3
H_3N NH_3
Cl $]^+$

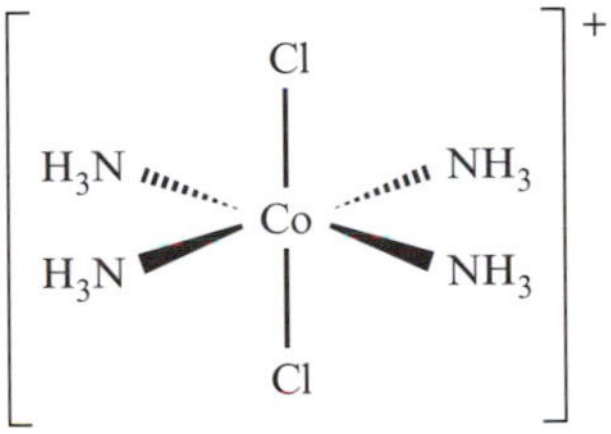

(5.54)

An octahedral species of general formula XY_2Z_4 has two geometrical isomers.

In *cis*-XY_2Z_4, the two Y groups are in adjacent sites, whereas in *trans*-XY_2Z_4, the two Y groups are in opposite sites.

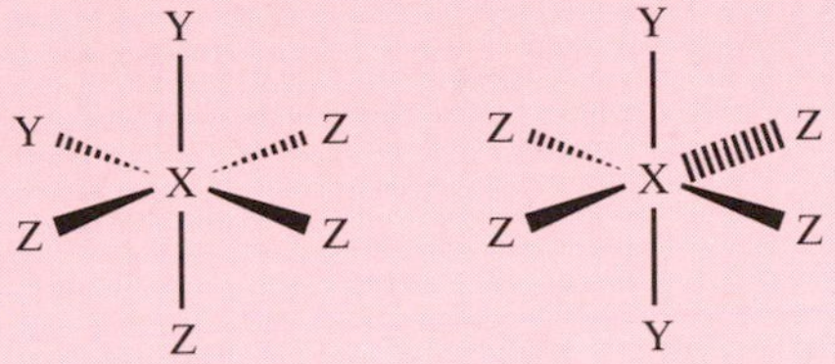

Y; Y Z; X; Z Z; Z (*cis*)

Y; Z Z; X; Z Z; Y (*trans*)

cis *trans*

If an octahedral compound has the general formula XY_3Z_3, two arrangements of the Y and Z atoms or groups are possible as Figure 5.22 shows. The prefix *fac* (*or facial*) means that the three Y (and three Z) groups define one face of an octahedron. The prefix *mer* (or *meridional*) indicates that the three Y (or Z) groups are coplanar with the metal centre.

An octahedral species of general formula XY_3Z_3 has two geometrical isomers.

In *fac*-XY_3Z_3, the three Y groups (and the three Z groups) define a face of the octahedron, whereas in *mer*-XY_3Z_3, the three Y groups (and the three Z groups) are coplanar with the metal centre.

Y; Y Z; X; Y Z; Z (*fac*)

Z; Y Y; X; Y Z; Z (*mer*)

fac *mer*

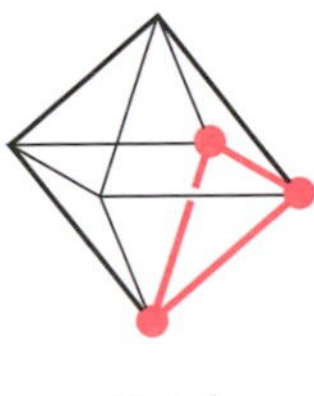

Facial arrangement

fac-isomer

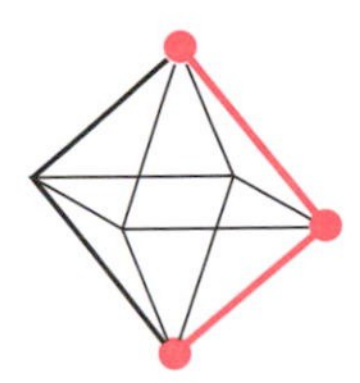

Meridional arrangement

mer-isomer

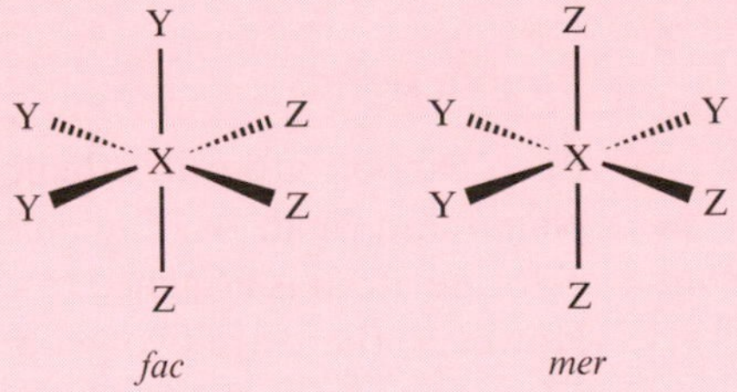

5.22 In an octahedral compound of general formula XY_3Z_3, two arrangements of the Y and Z atoms or groups are possible: the *fac*- and *mer*-isomers.

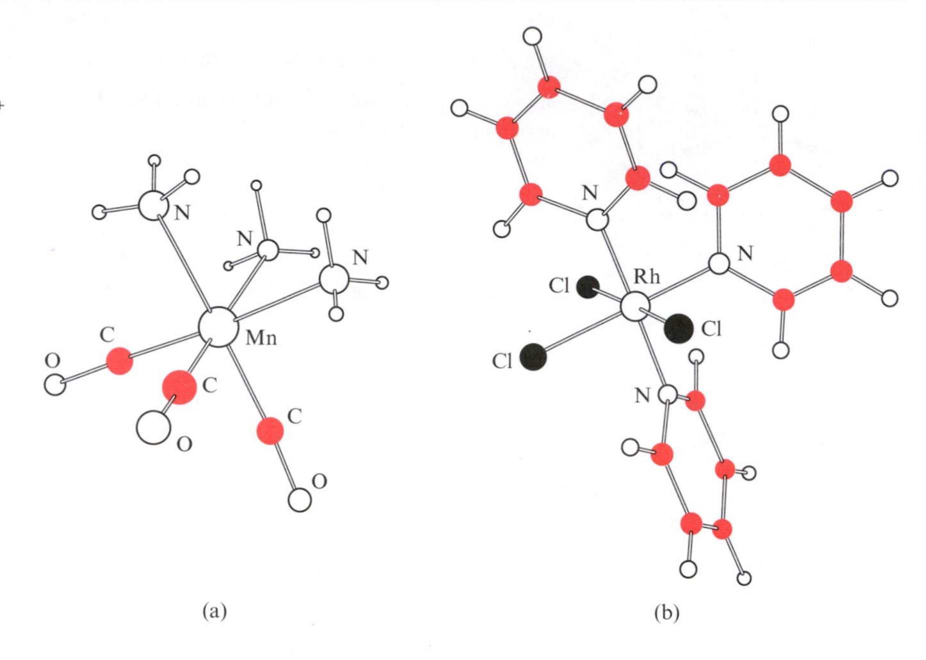

5.23 The octahedral structures (determined by X-ray diffraction) of (a) the cation *fac*-$[Mn(CO)_3(NH_3)_3]^+$ and (b) the compound *mer*-$[RhCl_3(py)_3]$ (where py = pyridine, C_5H_5N).

The structures of *fac*-$[Mn(CO)_3(NH_3)_3]^+$ and *mer*-$[RhCl_3(py)_3]$ (where py is the abbreviation for pyridine) have been determined by X-ray diffraction methods and are shown in Figure 5.23. Notice that a *fac*-arrangement of the CO groups in the $[Mn(CO)_3(NH_3)_3]^+$ cation necessarily means that the three NH_3 groups are also in a *fac* arrangement. Similarly, if the chloride groups in $[RhCl_3(py)_3]$ are in a *mer*-arrangement, the pyridine groups must also be in a *mer*-arrangement.

Pyridine: see Section 15.12

Double bonds

Nomenclature of alkenes: see Section 8.2

Each carbon atom in ethene **5.55** is trigonal planar and the molecule is planar and rigid. In 1,2-dichloroethene, the nomenclature indicates that one chlorine atom is attached to each of the two carbon atoms in the molecule. However, there are two geometrical isomers of 1,2-dichloroethene. In the (*Z*)-isomer **5.56**, the two chlorine atoms are on the *same* side of the double bond, and in (*E*)-1,2-dichloroethene **5.57**, the two chlorines are attached on *opposite* sides of the double bond.§

(5.55)
Atom numbers shown in red

(5.56)

(5.57)

§ In an older nomenclature, (*Z*) was termed '*cis*' and (*E*) was termed '*trans*'. These names are still commonly encountered in organic chemistry.

This type of isomerism is not restricted to organic alkenes. An inorganic compound which shows a similar isomerism is N_2F_2. The bent geometry at each nitrogen atom is caused by the presence of a lone pair of electrons and gives rise to the formation of the geometrical isomers **5.58** and **5.59**.

(5.58)

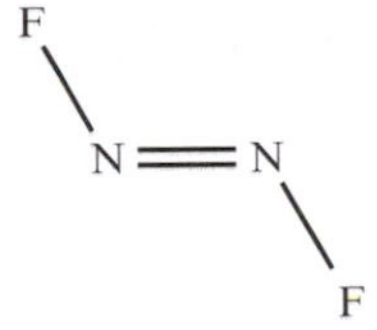

(5.59)

5.12 TWO STRUCTURES THAT ARE CLOSE IN ENERGY: THE TRIGONAL BIPYRAMID AND SQUARE-BASED PYRAMID

In the solid state

In the solid state, the structure of the $[Ni(CN)_5]^{3-}$ anion depends upon the cation present. In some salts, the $[Ni(CN)_5]^{3-}$ ion is closer to being a square-based pyramid **5.60** than a trigonal bipyramid **5.61**. This illustrates that these two five-coordinate structures are often close in energy and this is further emphasized by the observation that in the hydrated salt $[Cr(en)_3][Ni(CN)_5]\cdot 1.5H_2O$, in which the cation is $[Cr(en)_3]^{3+}$, two forms of the $[Ni(CN)_5]^{3-}$ ion are present. Four of the anions, in the arrangement in which they occur in a single crystal of $[Cr(en)_3][Ni(CN)_5]\cdot 1.5H_2O$, are shown in Figure 5.24. Careful inspection of the bond angles reveals just how similar the structures labelled 'trigonal bipyramid' and 'square-based pyramid' actually are!

en is the abbreviation for $H_2NCH_2CH_2NH_2$: see Chapter 16, Table 16.2

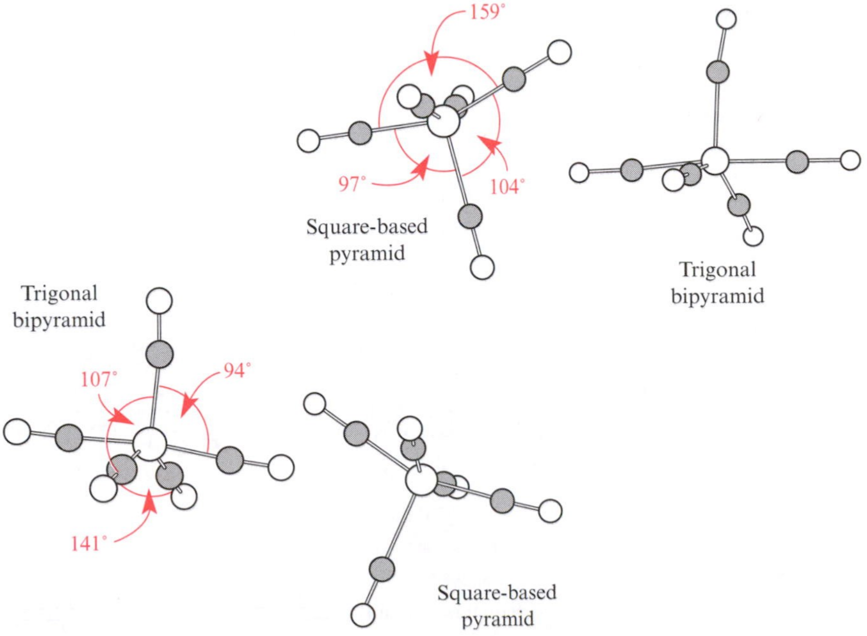

5.24 Four of the $[Ni(CN)_5]^{3-}$ anions, shown in the arrangement in which they occur in the crystal lattice of $[Cr(en)_3][Ni(CN)_5]\cdot 1.5H_2O$ (en is the abbreviation for $H_2NCH_2CH_2NH_2$). The $[Cr(en)_3]^{3+}$ cations and the water molecules of crystallization have been omitted from the diagram for clarity. Evaluation of the bond angles illustrates how similar the structures labelled 'trigonal bipyramid' and 'square-based pyramid' actually are.

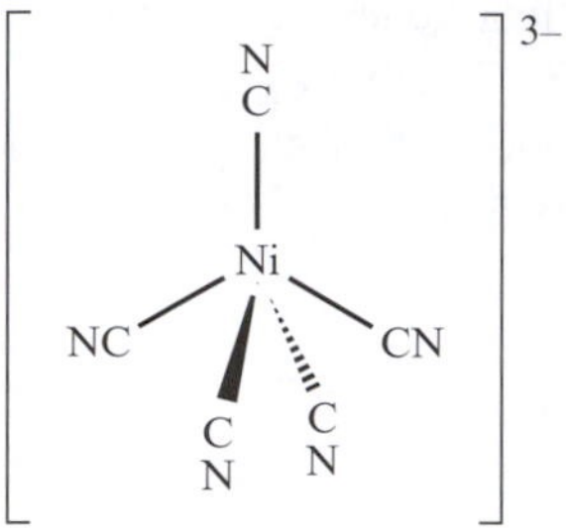

(5.60)

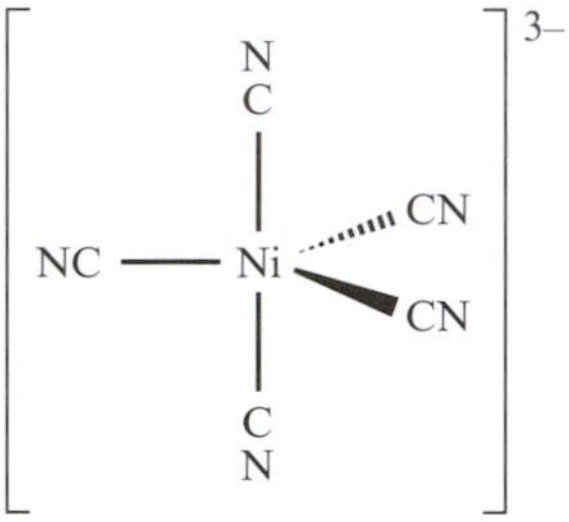

(5.61)

Although ligand–ligand repulsions are minimized in a trigonal bipyramidal arrangement, only a minor change to the structure is needed to turn it into a square-based pyramid. In a trigonal bipyramidal molecule, the two axial groups are related by a bond angle (subtended at the central atom) of 180°. The three equatorial groups are related by bond angles (subtended at the central atom) of 120°. Consider what happens if the angle between the two axial ligands is made a little smaller (< 180°), and the angle between two of the equatorial ligands is enlarged a little (>120°). This is shown in Figure 5.25 – the result is to convert the trigonal bipyramid into a square-based pyramid.

The small energy difference between the trigonal bipyramidal and square-based pyramidal structures means that, in the solid state, the nature of a five-coordinate molecule or ion can be influenced by the forces operating between the species in the crystal. There are many five-coordinate species with structures that are best described as lying somewhere between the trigonal bipyramid and square-based pyramid.

In solution: Berry pseudo-rotation

In solution, the small energy difference between the trigonal bipyramidal and square-based pyramidal structures for a given five-coordinate species can have a dramatic effect. In solutions of PF_5 or $[Fe(CO)_5]$ there are fewer restrictions on the molecular motion than in the solid state, and it becomes possible for the process shown in Figure 5.25 to become a *dynamic one* – the process happens in real time. Moreover, further angular distortion can occur to regenerate a trigonal bipyramidal structure. This low-energy process is called *Berry pseudo-rotation*.

Berry pseudo-rotation is the name given to a low-energy process that interconverts trigonal bipyramidal and square-based pyramidal structures with the effect that the axial and equatorial groups of the trigonal bipyramidal structure are exchanged.

An important consequence of Berry pseudo-rotation is the interconversion of axial and equatorial groups in the trigonal bipyramid (Figure 5.26). Repeated structural interconversions of this type allow *all* five of the groups attached to the central atom to exchange places, and the molecule is said to be *fluxional* or *stereochemically non-rigid*.

How do we *observe* fluxional behaviour? Consider the five-coordinate molecule XY_5. If the *rate of exchange* of the axial and equatorial atoms Y is *faster than the timescale* of the experimental technique used, the experiment will never be able to distinguish between the two types of Y atom. Hence, the

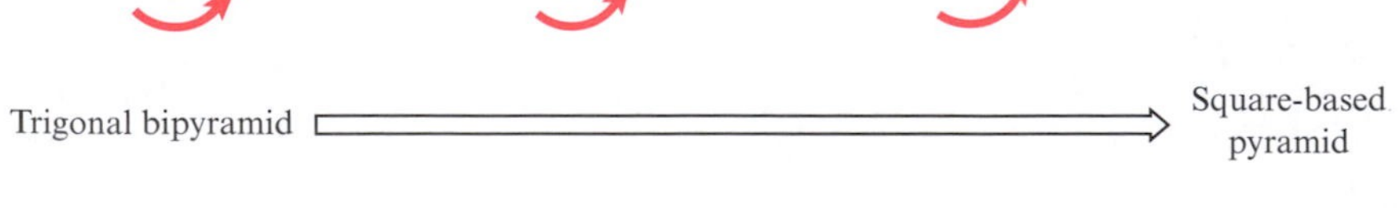

Trigonal bipyramid → Square-based pyramid

5.25 The conversion of a trigonal bipyramidal molecule into a square-based pyramidal one is achieved by an angular distortion. This process usually requires only a small amount of energy. The view of the square-based pyramid is not the conventional one; compare with Figure 5.12.

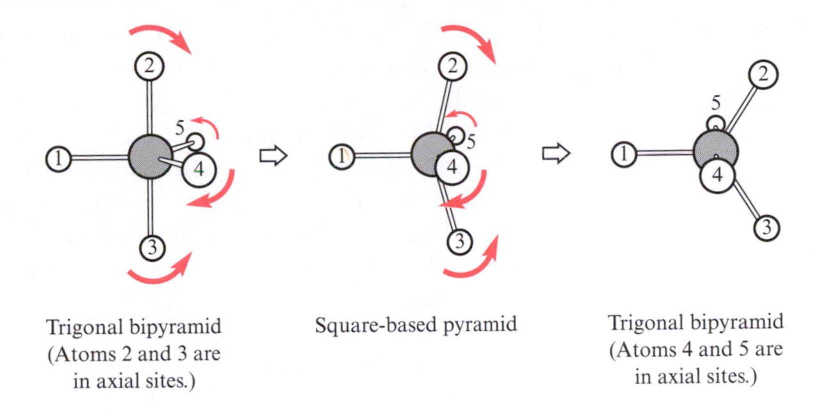

5.26 The Berry pseudo-rotation mechanism. This interconverts one trigonal bipyramidal structure into another via a square-based pyramidal intermediate. The numbering scheme emphasizes that axial and equatorial sites in the trigonal bipyramid are interchanged. The first step in the process is shown in more detail in Figure 5.25.

Timescale: see Section 9.3

axial and equatorial atoms will appear to be equivalent. If the rate of exchange of the axial and equatorial atoms Y is *slower* than the timescale of the experimental technique used, the experiment *should* be able to distinguish between the two types of Y atom, and the axial and equatorial atoms will appear to be different.

A molecule or molecular ion is *fluxional* or *stereochemically non-rigid* if atoms or groups in the system are involved in a dynamic process such that they exchange places.

'NMR active' means that the nucleus can be observed using NMR spectroscopy

A technique that is often used to study stereochemically non-rigid systems is nuclear magnetic resonance (NMR) spectroscopy. In this section, we deal only with the result of such experiments, and more details of NMR spectroscopy are given in Chapter 9.

Consider the molecule $[Fe(CO)_5]$ **5.38**. The carbon nucleus ^{13}C is *NMR active* and *each environment* (axial ^{13}C in the two axial CO groups or equato-

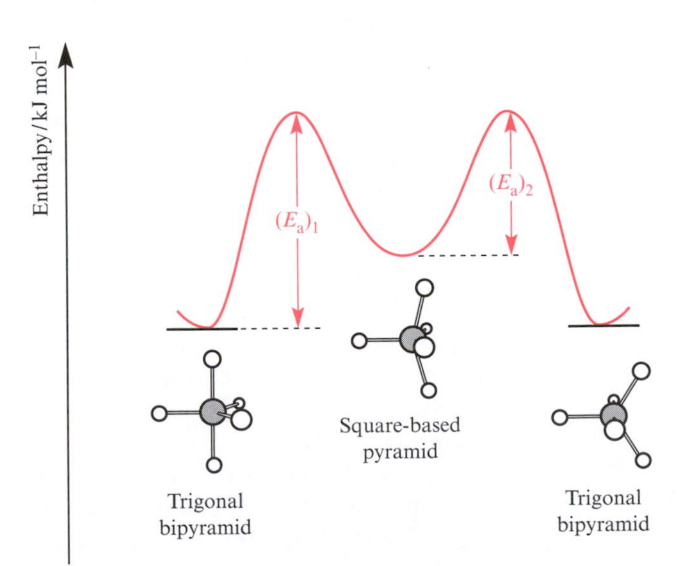

5.27 A schematic representation of the enthalpies of activation (E_a) for the interconversion of a trigonal bipyramidal into a square-based pyramidal structure and, then, on to another trigonal bipyramidal structure. The energies of the two trigonal bipyramidal structures are the same. The square-based pyramidal structure is represented at a higher energy level.

rial ^{13}C in the three equatorial CO groups) should give rise to a different spectroscopic signal. At room temperature, sufficient thermal energy is available and interconversion between structures is facile – the ^{13}C nuclei all appear to be identical, and only one NMR spectroscopic signal is seen.

Each structural interconversion possesses a characteristic activation enthalpy (Figure 5.27). At lower temperatures, when less energy is available, fewer molecules will possess enough energy to overcome the activation barrier. As the temperature of the system is lowered, it may be possible to *freeze out* the fluxional process, although this is *not* the case for $[Fe(CO)_5]$ – even at 103 K, the axial and equatorial CO groups are still exchanging their positions on the NMR timescale.

5.13 SHAPE AND MOLECULAR DIPOLE MOMENTS

In Chapter 4 we discussed polar molecules and (electric) dipole moments (μ). In a polar diatomic, one end of the molecule is negatively charged *with respect to* the other end. The situation with polyatomic molecules is more complex, and requires careful consideration.

Linear molecules

A vector has magnitude *and* direction.

Consider a molecule of CO_2. From the Pauling electronegativity values $\chi^P(C) = 2.6$ and $\chi^P(O) = 3.4$, we predict that *each carbon–oxygen bond* in the molecule is polar ($C^{\delta+}O^{\delta-}$). However, the *molecular* dipole moment of CO_2 is zero. This is understood if we consider the shape of the molecule *and* recall that a dipole moment is a vector quantity. Figure 5.28a shows that, although each bond is polar, the two dipole moments act equally in opposite (at 180°) directions. The net result is that the molecule is non-polar ($\mu = 0$).

The CO_2 (Figure 5.28a) and OCS (Figure 5.29a) molecules are both linear, but the former is symmetrical about the carbon atom while the latter is not. The Pauling electronegativity values (Figure 5.29a) suggest that OCS will have a net *molecular dipole moment* acting in the direction shown (the oxygen end of the molecule is δ^-). The observed value of μ for the OCS molecule is 0.72 D (Table 5.5).

δ^- δ^+ δ^-
O=C=O

(a)

δ^+
S
δ^- O δ^- O
molecular dipole moment

(b)

5.28 (a) In a CO_2 molecule, each carbon–oxygen bond is polar but the two dipole moments (which are vectors) cancel out. (b) In an SO_2 molecule, each sulfur–oxygen bond is polar and the bent structure means that the two bond dipoles reinforce one another. The direction of the *molecular* dipole moment is shown in red.

Bent molecules

Dipole moments for several bent molecules are included in Table 5.5. First, we compare the dipole properties of CO_2 and SO_2. Although sulfur and carbon have the same Pauling electronegativity values ($\chi^P = 2.6$), the dipole moments of CO_2 and SO_2 are zero and 1.63 respectively. This dramatic difference is due to a difference in molecular shape. Figure 5.28 shows that the *bond dipoles* in CO_2 oppose one another and cancel each other out, but in SO_2 they reinforce one another, leading to a significant *molecular dipole*

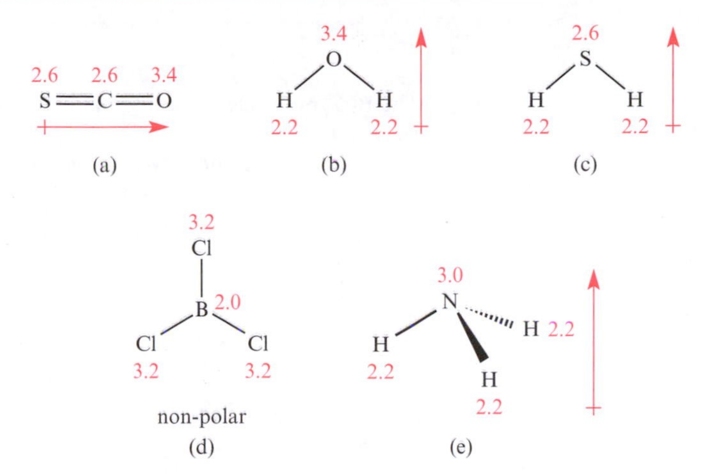

5.29 Polar and non-polar polyatomic molecules. The red numbers correspond to values of the electronegativity (χ^P) for each atom. The red barred-arrows represent the direction of the molecular dipole moment.

moment. This example emphasizes how important it is to differentiate between *bond* and *molecular* polarities.

The ability of water to function as a solvent is partly due to the presence of a molecular dipole. The values listed in Table 5.5 show that the dipole moment for H_2O is relatively large, but in going from H_2O to H_2S the molecular dipole moment decreases from 1.85 D to 0.97 D. Although the shape of the molecule is unchanged (Figure 5.29), the bond polarities decrease and hence the resultant molecular dipole decreases.

Trigonal planar and trigonal pyramidal molecules

A trigonal planar molecule such as BCl_3 (Figure 5.29d) has no net molecular dipole moment. Although each B–Cl bond is polar, the three bond dipoles cancel each other out.

The molecules NCl_3, NF_3, NH_3, PCl_3, PF_3 and PH_3 all possess trigonal pyramidal structures, and are all polar (Table 5.5). However, the magnitude and direction of each molecular dipole moment depends on the relative electronegativities of the atoms in each molecule and on the presence of lone

Table 5.5 Dipole moments for selected polyatomic molecules.

Molecule	*Molecular shape*	*Dipole moment / D*
BCl_3	Trigonal planar	0
NCl_3	Trigonal pyramidal	0.39
NF_3	Trigonal pyramidal	0.24
NH_3	Trigonal pyramidal	1.47
PCl_3	Trigonal pyramidal	0.56
PF_3	Trigonal pyramidal	1.03
PH_3	Trigonal pyramidal	0.57
H_2O	Bent	1.85
H_2S	Bent	0.97

Molecule	*Molecular shape*	*Dipole moment / D*
OF_2	Bent	0.30
SF_2	Bent	1.05
SF_4	Disphenoidal	0.63
CF_4	Tetrahedral	0
CO_2	Linear	0
OCS	Linear	0.72
SO_2	Bent	1.63
NO_2	Bent	0.32
N_2O	Linear	0.16

Box 5.1 Confirmation that the molecule BCl_3 is non-polar

The BCl_3 molecule has a trigonal planar structure, and each Cl–B–Cl bond angle is 120°.

The Pauling electronegativity values are:

$\chi^P(B) = 2.0$
$\chi^P(Cl) = 3.2$

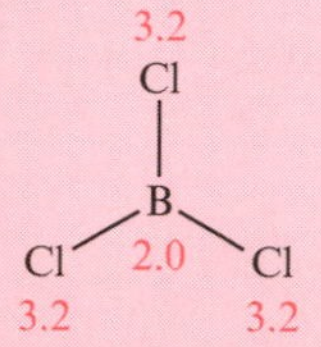

Each B–Cl bond is polar, and each bond dipole is a vector quantity.
The three bond dipoles have the same magnitude but act in different directions.
Let each vector be $\boldsymbol{V}$. The three vectors act as follows:

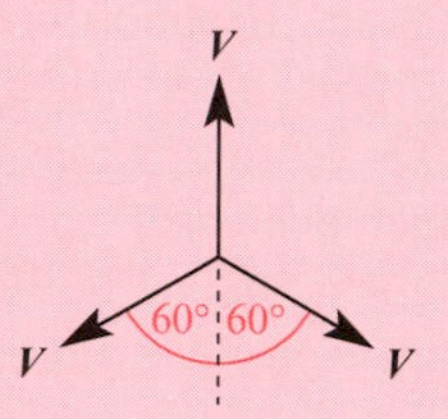

Resolving the vectors into two directions gives:

In an upward direction: total vector = $\boldsymbol{V}$

In a downward direction: total vector = $2 \times (\boldsymbol{V} \times \cos 60) = 2 \times \left(\frac{\boldsymbol{V}}{2}\right) = \boldsymbol{V}$

Therefore, two equal vectors act in opposite directions and cancel each other out.
The net molecular dipole is zero.

Conclusion

The BCl_3 molecule does not possess a dipole moment.

pairs. Consider NH_3 (μ = 1.47 D) (Figure 5.29e). Each N–H bond is polar in the sense $N^{\delta-}$–$H^{\delta+}$ and the nitrogen atom also carries a lone pair of electrons. The resultant molecular dipole is such that the nitrogen atom has a partial negative charge ($N^{\delta-}$). Although NF_3 is structurally related to NH_3, its molecular dipole is smaller (μ = 0.24). Each N–F bond is polar in the sense $N^{\delta+}$–$F^{\delta-}$ but the net dipole due to the three bond dipole moments is almost cancelled by the effect of the lone pair of electrons on the nitrogen atom.

Four-coordinate structures

A tetrahedral molecule XY_4 with four identical Y atoms or groups is non-polar. In CF_4, for example, the four bond dipoles oppose one another, and there is no net molecular dipole moment. Similarly, a square-planar molecule XY_4 in which the Y atoms or groups are identical has no overall dipole moment.

A four-coordinate molecule with a disphenoidal structure *does* have a dipole moment – the dipole moment for SF_4 (structure **5.29**) is 0.63 D. Each S–F bond is polar and the two bond dipoles that act along the axial directions cancel each other. However, in the equatorial plane there is a net dipole moment.

Five- and six-coordinate species

An octahedral molecule of formula XY_6 (in which the Y groups are the same) is non-polar, because, although each X–Y *bond* may be polar, pairs of bond dipoles oppose one another: $\mu = 0$ for SF_6.

A trigonal bipyramidal molecule XY_5 (in which the Y atoms or groups are identical) does not possess a molecular dipole. In PF_5, the two axial P–F bond dipole moments cancel out, and the three equatorial P–F bond dipoles oppose each other with a net cancellation of the dipole. It is more difficult to predict whether a square-based pyramidal molecule XY_5 will have a dipole moment, since much depends upon the exact molecular geometry. The molecular dipole moment of IF_5 is 2.18 D. Here, resolution of the individual bond dipole moments (vectors) into two directions does not lead to equal and opposite vectors.

In a polyatomic molecule, the presence of a molecular dipole moment depends *both* on the presence of bond dipole moments *and* the shape of the molecule.

A summary of molecular dipole moments

- Molecular geometry has an important influence on the property of molecular dipole moments.
- The presence of polar bonds in a molecule does not necessarily mean that the molecule will possess a dipole moment.
- A casual glance at the electronegativity values of the atoms in a molecule may not be enough to establish whether or not a molecule is polar, or in what direction the dipole acts.

5.14 CARBON: AN ELEMENT WITH ONLY THREE COMMON GEOMETRIES

In this chapter, we have considered the geometries that are available for two-, three-, four-, five- and six-coordinate species. For the heavier group 15, 16 and 17 elements and for *d*-block metals, a larger range of coordination numbers and molecular shapes is possible. However, we have also seen that elements in the first row of the *p*-block have fewer structural choices available. We now briefly consider carbon in order to appreciate the restrictions that its geometrical preferences impose on it. This has an effect on the chemistry of this element as we shall see in later chapters.

R—C≡C—R

(5.62)

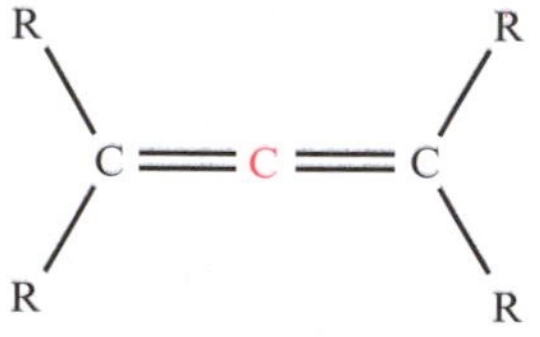

$R_2C=C=CR_2$

(5.63)

Linear environments

Apart from diatomic and triatomic systems such as CO, $[CN]^-$, CO_2, CS_2 and HCN, two-coordinate carbon atoms are most commonly encountered in alkynes **5.62** and allenes§ **5.63** (R is any organic substituent). A linear structure is forced on the atom by the presence of a triple bond (in the alkyne) or two adjacent double bonds (in the allene), and the structural restriction imposes *rigidity* in the molecular backbone.

Trigonal planar environments

Two of the carbon atoms in structure **5.63** are three-coordinate and trigonal planar. This is a characteristic of an alkene as shown in structure **5.64**.

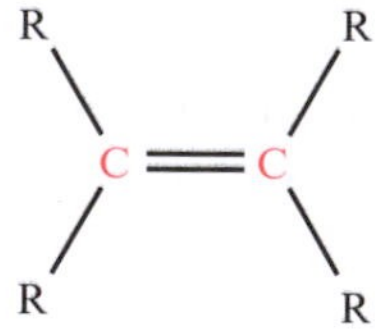

$R_2C=CR_2$

(5.64)

The presence of *one* double bond is a feature common to most carbon atoms that possess trigonal planar shapes – this may be a C=C double bond as in the alkenes, or a C=O double bond as in the carbonyl compounds that we describe in Chapter 17. One of the fundamental building blocks of aromatic chemistry is the benzene ring **5.65** which contains six carbon atoms, each of which is trigonal planar. The effect of the geometrical restriction is that the ring itself is planar. Planar rings are observed in a multitude of molecular systems that are related to benzene, e.g. pyridine **5.66** and naphthalene **5.67**.

Aromatic compounds: see Chapter 15

Two allotropes of carbon also have bulk structures containing trigonal planar carbon atoms – graphite consists of layers of fused six-membered rings, while in fullerenes, there is some deviation from strict planarity

§ An allene is a type of *diene* in which the two double bonds are in adjacent positions in the carbon chain.

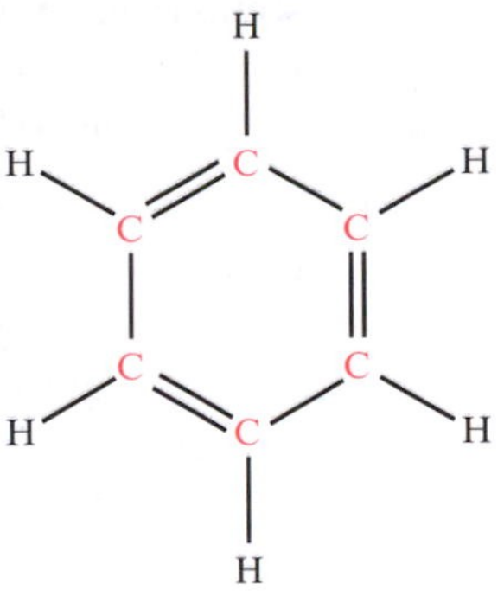

(5.65)

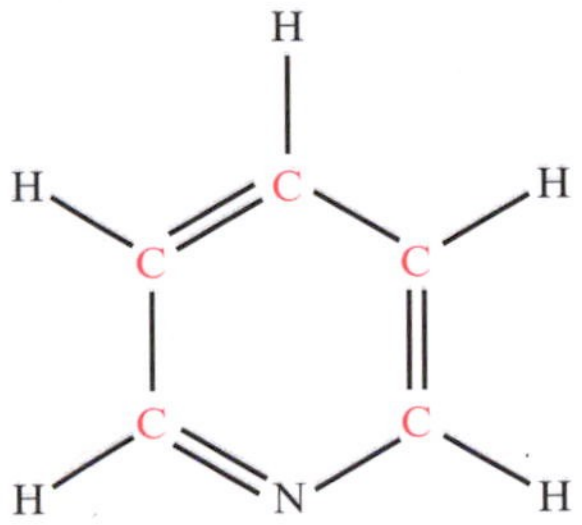

(5.66)

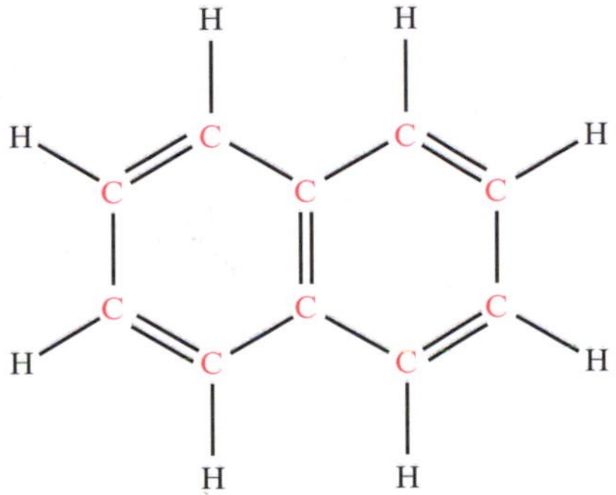

(5.67)

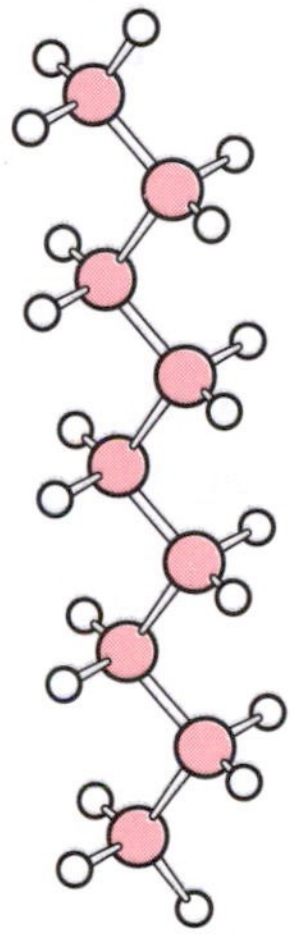

(5.68)

because the six-membered rings are fused with five-membered rings to generate approximately spherical C_n (e.g. $n = 60$) structures.

C_{60} and graphite: see Sections 7.9 and 7.10

Tetrahedral environments

Saturated hydrocarbons (alkanes) and their derivatives contain tetrahedral carbon atoms, and in a so-called 'straight chain' alkane, their presence leads to a zigzag backbone as is illustrated by the structure of nonane, C_9H_{20} **5.68**. Such an organic chain is far more flexible than one containing three- or two-coordinate carbon atoms. This has important consequences in terms of the number of possible arrangements of the chain.

Conformation of alkanes: see Section 8.5

5.15 MOLECULAR SHAPE AND THE OCTET RULE

In Sections 5.4 and 5.5 we saw that:

- in the first row of the *p*-block, boron, carbon and nitrogen show a greater variation in geometrical environment than do oxygen and fluorine;
- heavier elements in groups 13 to 17 of the *p*-block show a greater range of geometries in their compounds than do the first row elements of each group;
- for the heavier *p*-block elements, the range of structures *increases* in going from group 13 to 17.

In this and the following sections we show how the first of these observations can be rationalized in terms of the octet rule, and how the octet rule must be reconsidered in order that we can understand the second two observations.

Fluorine

The ground state electronic configuration of fluorine is [He]$2s^2 2p^5$. The addition of only one electron is required to complete an octet in the valence shell. A fluorine atom forms one single bond and the structural restrictions that we have seen follow from this.

Oxygen

The ground state electronic configuration of oxygen is [He]$2s^2 2p^4$. An octet is completed if oxygen adds two electrons to its valence shell. This can be done either by forming:

- two single bonds (e.g. H_2O, OF_2); or
- one double bond (e.g. O=O, O=C=O)

We can rationalize the limited geometries in which we find oxygen atoms in terms of the octet rule. The formation of two single bonds leads to a bent molecular geometry (e.g. H_2O). When an oxygen atom is involved in the formation of a double bond, it will be *terminally* attached to another atom (e.g. CO_2).

At first glance it may appear that the oxygen atom cannot form three single bonds but we saw in Figure 5.10 the structure of the trigonal pyramidal $[H_3O]^+$ ion. Three-coordinate species in which oxygen is the central atom have one thing in common – *they are cationic*.

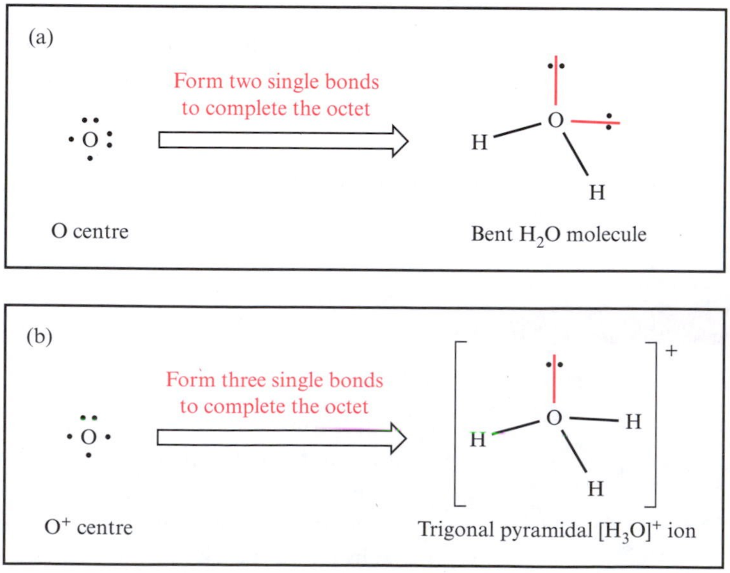

(a)

Form two single bonds to complete the octet

O centre

Bent H_2O molecule

(b)

Form three single bonds to complete the octet

O^+ centre

Trigonal pyramidal $[H_3O]^+$ ion

5.30 (a) A *neutral* oxygen atom obeys the octet rule when it forms two single bonds. (b) A *cationic* oxygen centre obeys the octet rule when it forms three single bonds. The localization of the positive charge on the oxygen centre is a formalism used to aid the interpretation of the bonding in the $[H_3O]^+$ ion (see Box 5.2).

If we formally localize the positive charge in $[H_3O]^+$ on the oxygen centre, then the geometry of the ion can be explained in terms of the O^+ ion (and not the O atom) obeying the octet rule. Figure 5.30 explains this point. In Figure 5.30a, the *neutral oxygen centre* gains an octet if it forms *two* single bonds, and the VSEPR model predicts that the H_2O molecule has a bent structure. Figure 5.30b shows that the *cationic oxygen centre* achieves an octet if it forms *three* single bonds, and the VSEPR model predicts that the $[H_3O]^+$ cation is trigonal pyramidal in shape.

To achieve an octet of valence electrons, a neutral oxygen atom forms two bonds and an O^+ centre forms three bonds.

Box 5.2 Formal localization of charge on a central atom

In sections 5.14 and 5.15, we discuss factors that determine the number of bonds that an atom of a *p*-block element can form. In some cases, the number appears at first glance to be larger than we might initially think possible. In these cases we can adopt a formalism which involves localizing an overall charge on the central atom.

In $[H_3O]^+$, the oxygen centre forms three O–H bonds. This is readily understood in terms of an O^+ centre (and not a neutral O atom) obeying the octet rule.

In $[BF_4]^-$, there are four B–F single bonds. Their formation can be rationalized in terms of a B^- centre attaining an octet of electrons.

In $[SiF_6]^{2-}$, the six Si–F bonds that are formed can be rationalized if we allow the silicon centre to carry a formal 2– charge; the Si^{2-} centre has six valence electrons and can form six single bonds. Note that the silicon atom is not obeying the octet rule in $[SiF_6]^{2-}$. (How many electrons are there in the valence shell of the silicon atom?)

Whilst this method of working is simple and extremely useful, it is important to realise that *it is only a formalism*. We are not implying that the central atom actually carries all of the overall charge in the ion concerned. In the $[H_3O]^+$, the charge distribution is *not* such that the oxygen centre necessarily bears a charge of +1.

It is also important to realize that the formal charge assignment does *not* imply anything about the oxidation state of the central atom.

In $[H_3O]^+$, the oxidation state of the oxygen centre is –2; each hydrogen atom has an oxidation state of +1.

In $[BF_4]^-$, the oxidation state of each fluorine centre is –1 and that of the boron centre is +3.

In $[SiF_6]^{2-}$, the oxidation states of the silicon and fluorine centres are +4 and –1 respectively.

Nitrogen

Like oxygen and fluorine, nitrogen obeys the octet rule, and the structures of nitrogen-containing species *must* be rationalized within this rule.

A nitrogen atom has five valence electrons and can achieve an octet by forming:

- three single bonds (e.g. NH_3, NF_3, NMe_3);
- one double bond and one single bond (e.g. Cl–N=O, F–N=O);
- one triple bond (e.g. H–C≡N, $[C{\equiv}N]^-$).

The geometries then follow as a matter of course. Remembering that there is a lone pair of electrons present on the nitrogen atom, three single bonds lead to a trigonal pyramidal geometry (Figure 5.31a), one double bond and one single bond gives a bent molecule, and a triple bond provides a linear environment for the nitrogen atom.

The number of species in which nitrogen is four-coordinate is large. We need only consider the wide range of salts containing ammonium $[NH_4]^+$ or tetramethylammonium $[NMe_4]^+$ ions to realize that four-coordinate nitrogen is important in the chemistry of this element. Bearing in mind that *nitrogen obeys the octet rule*, we rationalize the formation of four single bonds by considering an N^+ centre rather than a neutral N atom. This is shown in Figure 5.31b and the VSEPR model predicts that the $[NH_4]^+$ cation is tetrahedral.

To achieve an octet of valence electrons, a neutral N atom forms three bonds, and an N^+ centre forms four bonds.

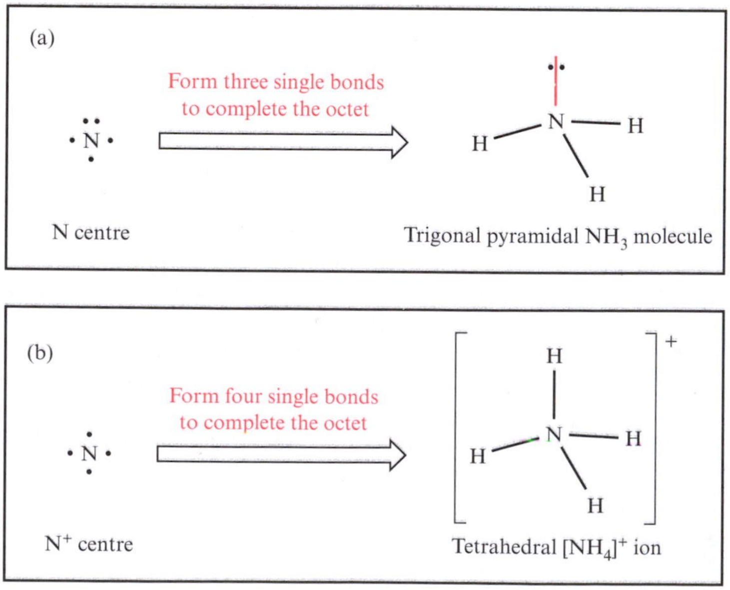

(a)
Form three single bonds to complete the octet
N centre
Trigonal pyramidal NH_3 molecule

(b)
Form four single bonds to complete the octet
N^+ centre
Tetrahedral $[NH_4]^+$ ion

5.31 (a) A *neutral* nitrogen atom obeys the octet rule when it forms three single bonds. (b) A *cationic* nitrogen centre obeys the octet rule when it forms four single bonds.

Carbon

Application of the octet rule shows that carbon ([He]$2s^22p^2$) requires four electrons to complete a noble gas configuration, i.e. [He]$2s^22p^6$ ≡ [Ne]. The four electrons can be gained by forming:

- four single bonds (e.g. CH_4); or
- two single bonds and one double bond (e.g. $O{=}CCl_2$); or
- one single and one triple bond (e.g. H–C≡N); or
- two double bonds (e.g. O=C=O).§

These bonding options correspond to the three structural preferences we have seen for carbon – tetrahedral, trigonal pyramidal and linear – and the fact that *carbon obeys the octet rule* accounts for the limited geometries that we observe for this element.

Boron

Neutral molecules which contain a single boron centre are often trigonal planar in shape and this follows directly from the ground state electronic configuration of boron ([He]$2s^22p^1$). The three valence electrons can be involved in the formation of three single bonds, but in so doing boron does *not* achieve an octet.

Species such as $[BH_4]^-$ and $[BF_4]^-$ which contain a four-coordinate boron centre have a feature in common – *they are anionic*. The B^- centre is isoelectronic with a carbon atom and has four valence electrons. It can form four single bonds and so achieve an octet. The tetrahedral shapes of $[BH_4]^-$ and $[BF_4]^-$ follow directly from the VSEPR model.

5.16 THE EXPANSION OF THE OCTET

The problem

Achieving an octet of valence electrons is synonymous with the formation of a maximum of four single bonds. However, as we have seen earlier in this chapter, the heavier *p*-block elements are able to form compounds in which:

- the *p*-block element has more than eight electrons in its valence shell (e.g. ClF_3, SF_4), and
- the coordination number is greater than four (e.g. SF_6, PCl_5).

Why are the heavier *p*-block elements able to achieve what the first row elements cannot?

§ The formation of a quadruple bond is, in theory, an option but is not observed.

The valence bond solution to an expanded octet

The first row of the *p*-block contains elements with the ground state electronic configuration of $2s^2 2p^m$ where $m = 1–6$, and the octet rule arises because there is a certain stability associated with having this principal quantum shell fully occupied.

The second row of the *p*-block contains elements with the ground state electronic configuration of $3s^2 3p^m$ where $m = 1–6$. For a species in which the octet rule is obeyed, the 3*p* level is completely filled. However, there is another level which has the same principal quantum number – the 3*d* level. This level is fully occupied when it contains ten electrons. This means that a new rule could be defined for the second row *p*-block elements since a limiting configuration of $3s^2 3p^6 3d^{10}$ (18 electrons) may be obtained. Thus, it is possible for *each* of the heavier *p*-block elements shown in Figure 5.11 to *expand its valency* beyond four and the heavier elements are chemically distinct from their first row counterparts.

Although we have suggested that the 'target' after bond formation is either a configuration of $ns^2 np^6$ or $ns^2 np^6 nd^{10}$, in practice, the number of electrons in the valence shell does not *have to be 8 or 18*. The *maximum* number of electrons which can be accommodated in the valence shell is 18. In practice, this is rarely achieved, but electron counts *between* 8 and 18 *are* commonly observed.

The number of single bonds which can be formed reflects the number of electrons in the valence shell. Iodine forms the neutral fluorides IF (although it is unstable), IF_3, IF_5 and IF_7 as well as the ions $[IF_2]^-$, $[IF_2]^+$, $[IF_4]^+$ and $[IF_6]^+$. In Chapter 13, we consider the chemistry of the *p*-block elements and give examples of compounds in which the valence shell of the *p*-block element contains more than eight electrons.

Elements in the first row of the *p*-block obey the octet rule in their compounds. An exception is boron which often only has a sextet of valence electrons.

The heavier elements of the *p*-block *may* obey the octet rule in their compounds, but may also expand their valence shell to a *maximum* of 18 electrons.

5.17 VALENCE BOND THEORY: RESONANCE STRUCTURES

In Chapters 3 and 4, we introduced valence bond and molecular orbital models for diatomic molecules. In this and the following sections we extend these models to polyatomic species. We deal firstly with VB theory and apply it to H_2O, BF_3 and $[SO_4]^{2-}$.

Equivalence of bonds

In the water molecule, the O–H bond distances are equal, and any bonding description for the H_2O molecule must be in accord with this experimental observation. Similarly, experimental data show that the three B–F bonds in the BF_3 molecule are equivalent, as are the four sulfur–oxygen bonds in $[SO_4]^{2-}$.

Resonance structures

Resonance structures: see Sections 3.11 and 4.3

Resonance structures for the H_2O molecule are drawn in Figure 5.32a. The resonance forms labelled 'covalent' and 'ionic' treat the molecule in a symmetrical manner and therefore the equivalence of the oxygen–hydrogen interactions is maintained. '*Partial ionic*' resonance structures for the water molecule can also be drawn but two such structures are needed to produce equivalent oxygen–hydrogen interactions in the molecule.

Even for a triatomic molecule (which is actually a very 'small' molecule), drawing out a set of resonance structures is tedious. Are all of the structures necessary? Do they all contribute to the same extent? In order to assess this, we have to use chemical intuition. Three points are crucial:

- charge separation should reflect the relative electronegativities of the atoms,
- we should avoid building up large positive or negative charges on any one atom,
- adjacent charges should be of opposite sign (two adjacent positive or two adjacent negative charges will have a destabilizing effect).

Consider again the resonance structures in Figure 5.32a bearing in mind the electronegativities $\chi^P(O) = 3.4$ and $\chi^P(H) = 2.2$. The ionic resonance structure will contribute negligibly because the charge separation in the molecule is unreasonably large. Thus, the three resonance structures shown in Figure 5.32b provide an approximate, but adequate, description of the bonding in H_2O. (Question: In Figure 5.32a, we could have included resonance forms containing O^+ and H^- ions – why did we ignore them?)

The bonding in the BF_3 and CH_4 molecules can be similarly represented by a series of resonance structures. Figure 5.33a illustrates just how tedious

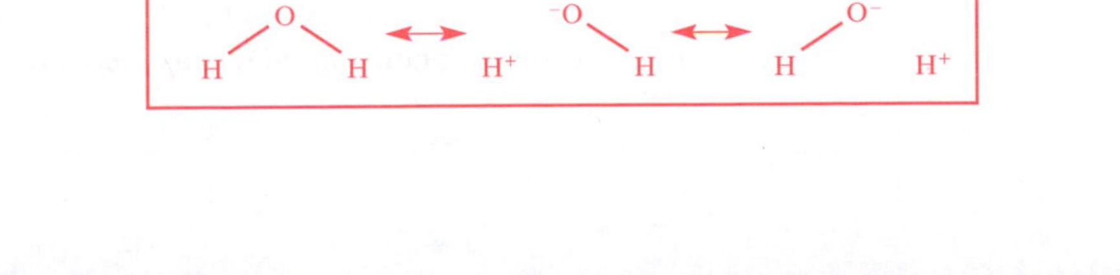

(a) 'covalent' ⟷ 'partial ionic' ⟷ 'partial ionic' ⟷ 'ionic'

(b)

5.32 (a) Resonance structures for H_2O. (b) The bonding can be adequately described in terms of only three resonance structures.

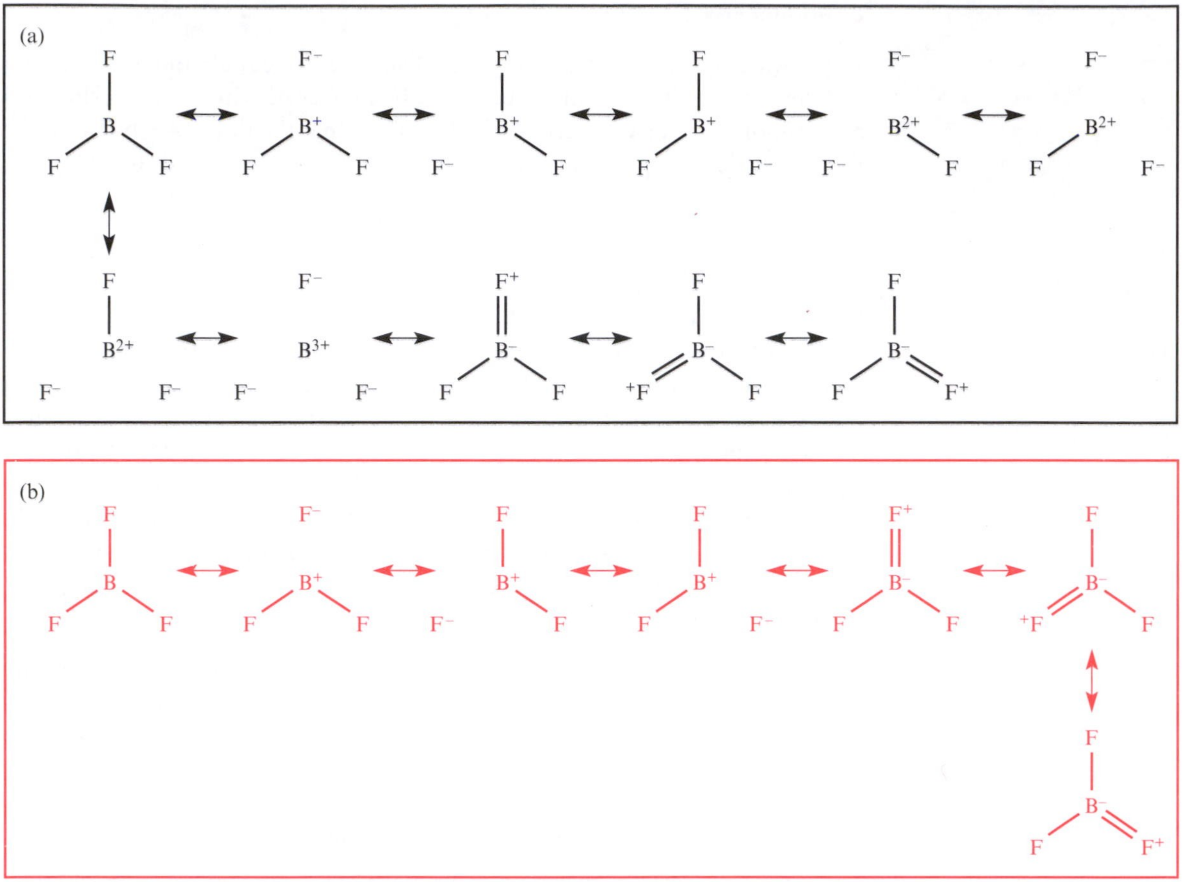

5.33 (a) Resonance structures for the BF_3 molecule. (b) Only some of the resonance structures contribute significantly and the bonding picture can be approximated to include only these forms.

the process is becoming! Boron trifluoride is still a 'small' molecule and yet there are 11 possible resonance structures. We can group the resonance structures for BF_3 according to the charge on the central atom and eliminate some of them on the grounds that they will contribute only negligibly to the overall bonding description. On this basis, structures with a multiple charge on the boron atom may be neglected. Figure 5.33b shows the resonance structures that will contribute the most to the bonding in the BF_3 molecule, and this approximate picture is quite satisfactory. Note that we require *all three members of a set* of related partial ionic structures if we are to maintain B–F bond equivalence.

BF_3: see also Section 5.20

When drawing resonance structures for an ionic species, we have the additional problem of where to localize the overall charge. Figure 5.34 shows a set of resonance structures for the sulfate ion which gives a reasonable charge distribution and indicates that the four sulfur–oxygen interactions are equivalent. Other structures can be drawn, but none would contribute as significantly to the bonding in $[SO_4]^{2-}$ as the six forms shown here.

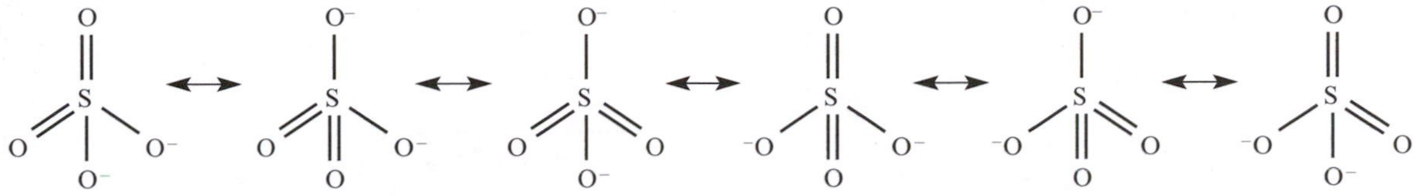

5.34 Resonance structures for the sulfate ion.

Worked example 5.1 ***Resonance structures for the nitrate ion***

Draw resonance structures for the nitrate ion and predict its shape.

The nitrate ion is $[NO_3]^-$.

Nitrogen is in group 15 and has five valence electrons.

Oxygen is in group 16 and has six valence electrons.

Both are first row elements and will obey the octet rule.

The (major contributing) resonance structures for the nitrate ion are:

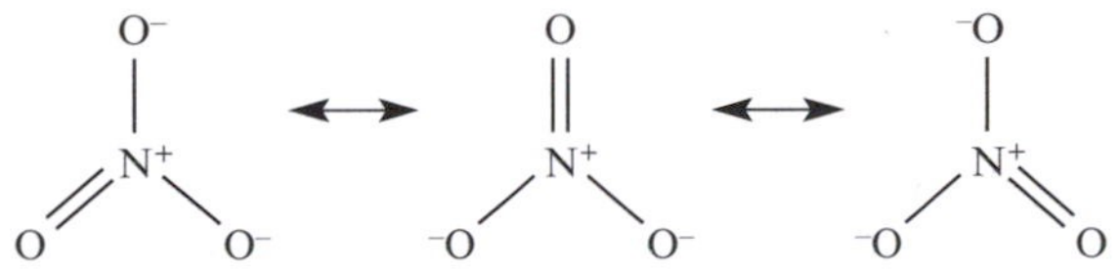

The ion will have a trigonal planar geometry.

It is tempting to write down a structure involving two N=O bonds and one N–O⁻ bond but this is ***incorrect****. Why?*

5.18 VALENCE BOND THEORY AND HYBRIDIZATION

In earlier parts of this chapter we used the ground state electronic configuration to determine the number of valence electrons available to an atom, and hence to say how many bonds that atom can form. We also noted that the tendency for a first row element in the *p*-block to achieve an octet arises from a desire to occupy the 2*s* and 2*p* AOs completely.

In this section, we consider how VB theory views the participation of the atomic orbitals in bond formation in polyatomic molecules and ions.

The directional properties of atomic orbitals

Shapes of atomic orbitals: see Section 2.11

With the exception of the spherically symmetric *s* orbital, an atomic orbital has directional properties. The question is: when atoms are bonded together, are their relative positions restricted by the directional properties of the atomic orbitals involved in the bonding?

Consider the formation of a trigonal planar molecule such as BF_3. The valence shell of the boron atom consists of the 2*s* and three 2*p* atomic orbitals.

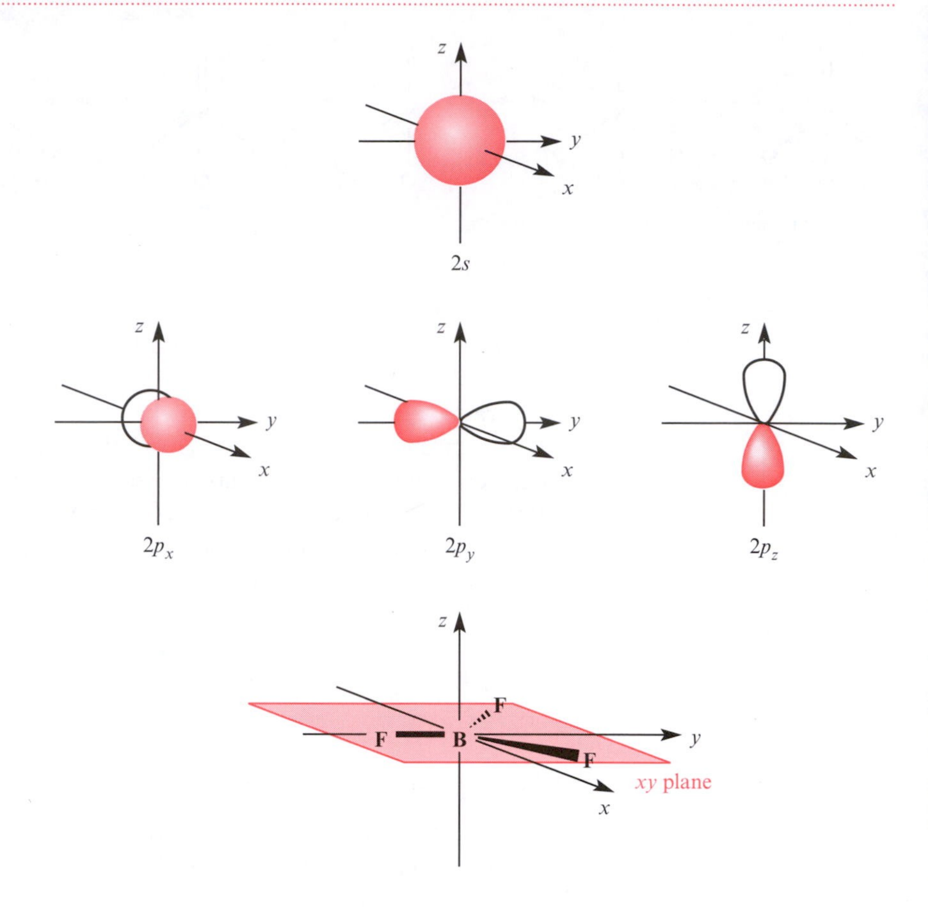

5.35 In order to approach the bonding in a trigonal planar molecule such as BF_3, we need to consider the valence atomic orbitals of a boron atom, and define a convenient relationship between the relative positions of the atoms in BF_3 and the Cartesian axis system. It is not possible for all three of the B–F bond vectors to coincide simultaneously with the directions of the three 2*p* AOs.

These are shown in Figure 5.35; whilst the 2*s* AO is spherically symmetrical, each 2*p* orbital lies on one of the three Cartesian axes. Figure 5.35 also shows a convenient relationship between the relative positions of the atoms in a BF_3 molecule and the Cartesian axes. The boron atom has been placed at the origin and one fluorine atom lies on the $-y$ axis. The molecule lies in the xy plane. Although one B–F bond vector coincides with the $2p_y$ atomic orbital, the other two B–F bond vectors do *not* coincide with either of the remaining two 2*p* AOs. Similar difficulties arise with other molecular shapes.

We can now define the problem. How do we describe the bonding in a molecule in terms of localized orbitals when the directional properties of the valence atomic orbitals do not match the bond vectors defined by the atoms in the molecule?

Hybridization of atomic orbitals

In the remainder of this section, we describe a method that allows us to generate *spatially directed orbitals* which may be used to produce *localized bonds* in a valence bond scheme.These are known as *hybrid orbitals*, and may be used to construct bonding orbitals in exactly the same way as atomic orbitals. Although hybridization (orbital mixing) is a useful way to describe bonding within the valence bond framework, there is *no* implication that it is an *actual* process.

Hydrization means 'mixing'

Hybridization of atomic orbitals can be considered *if* the component atomic orbitals are close in energy. The character of the hybrid orbitals depends on the AOs involved and the percentage contribution that each makes. The number of hybrid orbitals generated is equal to the number of AOs used; this reiterates the rule in MO theory that equates the number of MOs formed to the initial number of AOs.

The simplest case is where we have one *p* atomic orbital and one *s* atomic orbital, both from the same principal quantum level (2*s* and 2*p*, or 3*s* and 3*p*, etc.). Figure 5.36 shows that when the *s* and *p* AOs combine in-phase, the orbital will be reinforced (constructive interference). Where the *s* and *p* atomic orbitals combine out-of-phase, the orbital will be diminished in size (destructive interference). The net result is that two hybrid orbitals are formed which retain some of the directional properties of the atomic *p* orbital. These hybrid orbitals are denoted *sp* and each *sp* hybrid orbital contains 50% *s* and 50% *p* character.

We can repeat the exercise with a combination of one 2*s* and two 2*p* atomic orbitals. Since we begin with three atomic orbitals, we form three

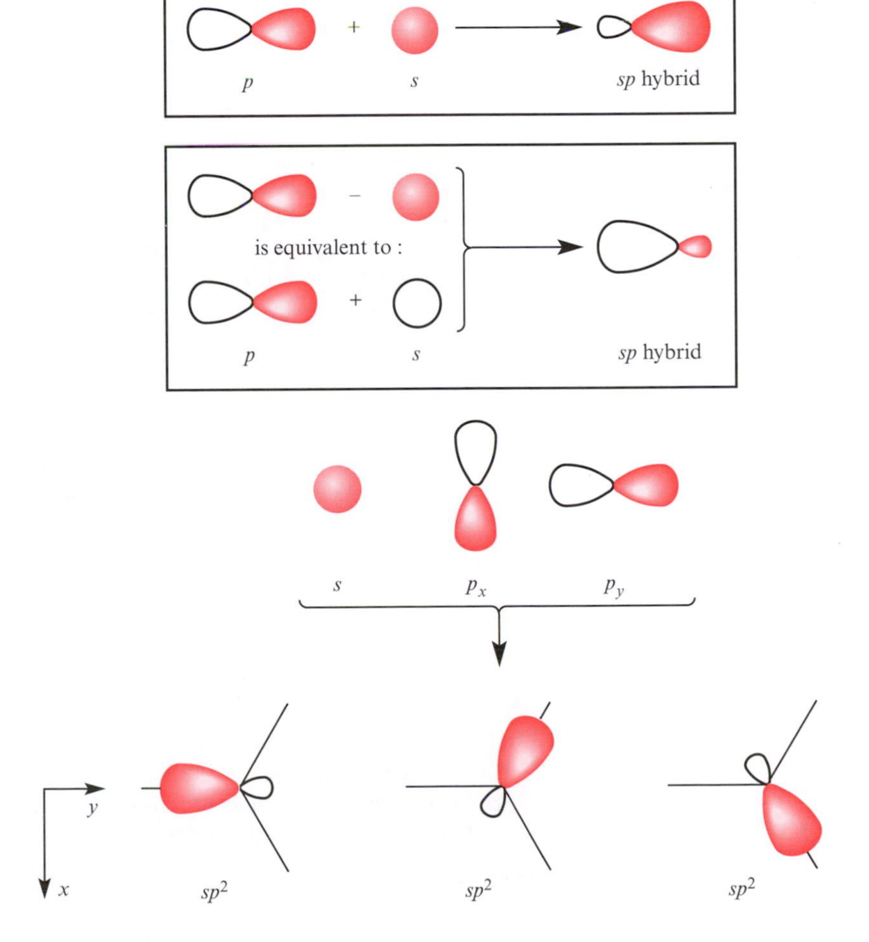

5.36 The formation of two *sp* hybrid orbitals from in- or out-of-phase combinations of an *s* and a *p* atomic orbital. The effect of altering the phase of the orbital (the *s* AO) is shown pictorially in the lower part of the figure.

5.37 The combination of one 2*s* and two 2*p* atomic orbitals to generate three equivalent sp^2 hybrid orbitals.

equivalent hybrid orbitals as shown in Figure 5.37. The resultant orbitals are called sp^2 hybrids and they lie in the same plane (here, the xy plane because we chose to hybridize the s, p_x and p_y AOs) and define a trigonal planar arrangement.

Box 5.3 Directionality of sp^2 hybrid orbitals

Figure 5.37 shows that a combination of s, p_x and p_y atomic orbitals gives three sp^2 hybrid orbitals which are identical except for the directions in which they point. How do the directionalities arise?

First, remember that each p orbital has a directional property – the p_x orbital lies on the x axis and the p_y orbital lies on the y axis.

The three hybrid orbitals lie in the xy plane. There is no contribution from the p_z AO and therefore the hybrid orbitals do not include any z-component.

The process of combining the s, p_x and p_y atomic orbitals can be broken down as follows:

1. The character of the s AO must be divided equally between the three hybrid orbitals.
2. Each hybrid orbital must end up with the same amount of p character.
3. Let us fix the direction of one of the hybrid orbitals to coincide with one of the initial p AOs, say the p_y AO (this is an arbitrary choice). The first sp^2 hybrid orbital therefore contains one third s and two-thirds p_y character:

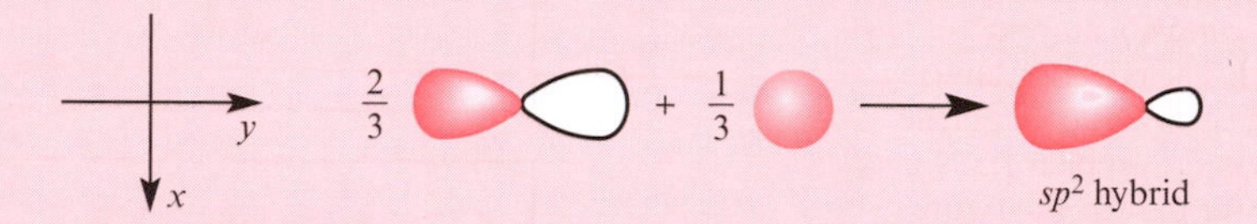

4. The remaining components are one third of the p_y AO, all of the p_x AO, and two thirds of the s AO. These must be combined to give two equivalent hybrid orbitals. The directionality of each hybrid is determined by components from both the p_x and p_y AOs. When s character is added in, two sp^2 hybrid orbitals result:

(Remember that changing the sign of a p AO alters its direction.)

Note that the fact that the sp^2 hybrid orbitals lie in the xy plane is wholly arbitrary. They could equally well contain p_y and p_z components (and lie in the yz plane) or p_x and p_z components (and lie in the xz plane).

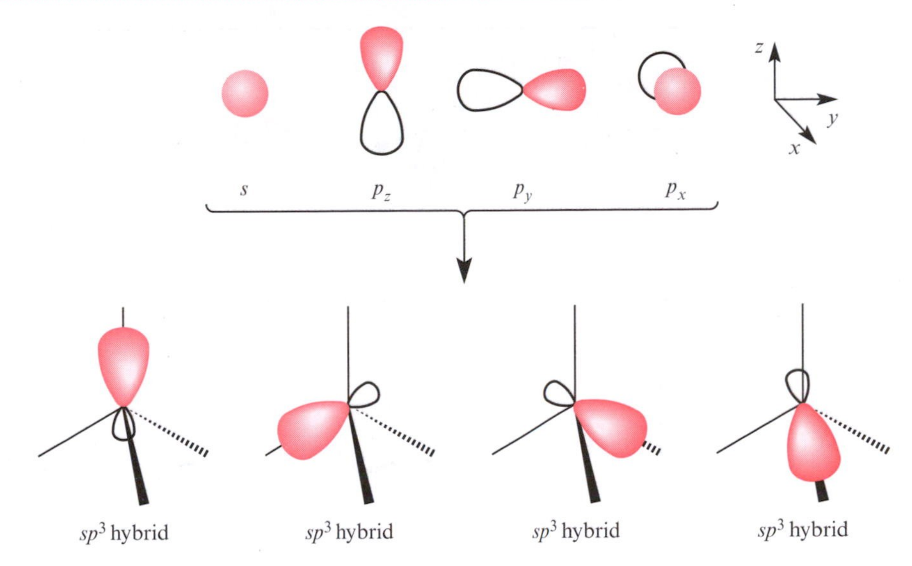

5.38 The combination of one $2s$ and three $2p$ atomic orbitals to generate four equivalent sp^3 hybrid orbitals.

A set of *four* equivalent sp^3 hybrid orbitals is formed by combining the s, p_x, p_y and p_z atomic orbitals as shown in Figure 5.38. Together, they define a tetrahedral arrangement, and each sp^3 hybrid orbital possesses 25% s character and 75% p character.

For the principal quantum level with $n = 2$, only sp, sp^2 and sp^3 hybrid orbitals can be formed because only s and p atomic orbitals are available. But, for higher principal quantum levels, d atomic orbitals are also present.

The mixing of s, p_x, p_y, p_z and d_{z^2} atomic orbitals gives a set of five sp^3d hybrid orbitals. The spatial disposition of the sp^3d hybrid orbitals corresponds to a trigonal bipyramidal arrangement (Figure 5.39a). This set of hybrid orbitals is unusual because, unlike the other sets described here, the sp^3d hybrid orbitals are *not* equivalent and divide into two groups: axial and equatorial.

A combination of s, p_x, p_y, p_z, d_{z^2} and $d_{x^2-y^2}$ atomic orbitals generates six sp^3d^2 hybrid orbitals, and the spatial arrangement of these hybrids is octahedral (Figure 5.39b).

The characters of atomic orbitals can be mixed to generate *hybrid* orbitals. Each set of hybrid orbitals is associated with a particular shape:

sp	linear
sp^2	trigonal planar
sp^3	tetrahedral
sp^3d	trigonal bipyramidal
sp^3d^2	octahedral

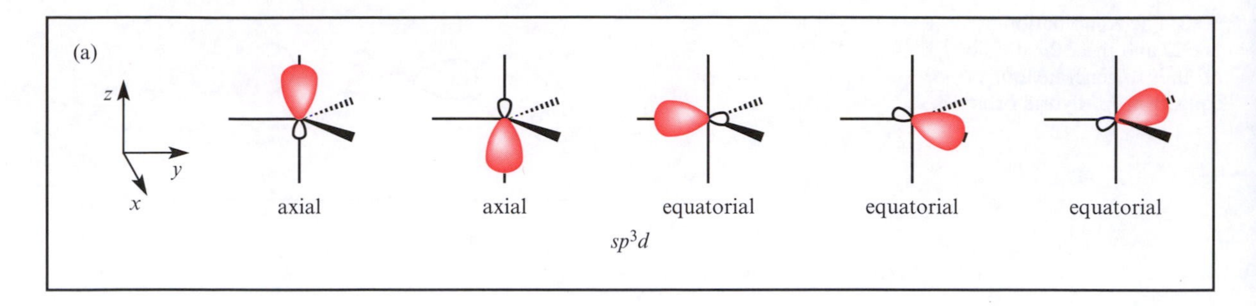

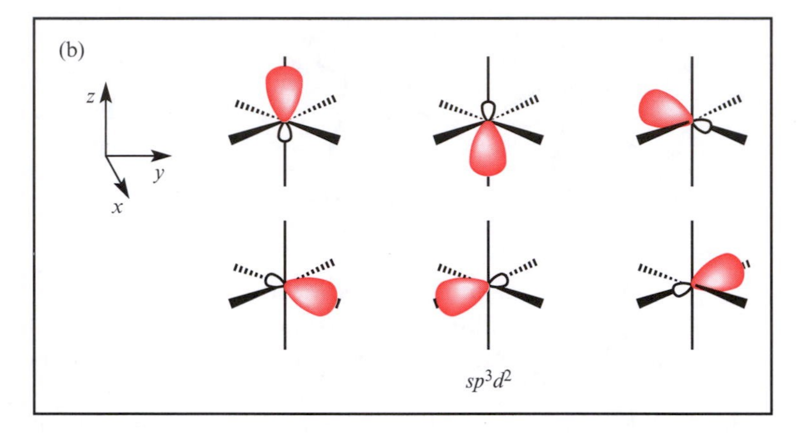

5.39 (a) A set of five sp^3d hybrid orbitals and (b) a set of six sp^3d^2 hybrid orbitals, shown in relationship to the Cartesian axis set.

Box 5.4 Hybridization means mixing

The term hybridization is used not only for orbitals but also for biological mixing. Hybrid plants are often developed to combine the natural properties of different species. For example, oilseed rape (the crop responsible for the bright yellow fields in many parts of the UK) is grown for its edible oil, harvested in high yields. Wild honesty is a plant from which small amounts of long-chain organic acids such as erucic (docosenoic) acid and nervonic (tetracosenoic) acid can be extracted – these are *fatty acids* (see Section 17.7). Their importance lies in the cosmetics industry, other industrial applications and in potential uses as drugs. By producing a hybrid plant from oilseed rape and honesty, it is hoped to produce a plant that will give higher yields of the long-chain fatty acids.

5.19 HYBRIDIZATION AND MOLECULAR SHAPE IN THE *p*-BLOCK

In the previous section, we saw that each particular set of hybridized orbitals had an associated spatial arrangement. Now we consider the use of hybridization schemes to describe the bonding in some molecules containing *p*-block atoms. By matching the geometries of the hybrid orbital sets with molecular shapes we see that in a linear triatomic molecule (e.g. CO_2) the central atom is *sp* hybridized, in a trigonal planar molecule (e.g. BF_3) the central atom is sp^2 hybridized, and in a tetrahedral molecule (e.g. CH_4) the central atom is sp^3 hybridized.

A most important point is that *a molecule does not adopt a particular shape because the central atom possesses a particular hybridized set of orbitals.* Hybridization is a convenient model within VB theory which successfully accounts for an observed molecular geometry.

A hybrid orbital can be regarded as a domain in which a pair of electrons resides. In this way, we can see a clear link between hybridization and the VSEPR model. For linear ($[I_3]^-$), trigonal planar (BBr_3), tetrahedral ($SiCl_4$), trigonal bipyramidal (PCl_5) or octahedral (SF_6) molecular species, the hybridization of the central atom may be considered to be *sp*, sp^2, sp^3, sp^3d or sp^3d^2, respectively. However, we have described molecules and ions that possess other shapes – bent, T-shaped, disphenoidal, square planar and

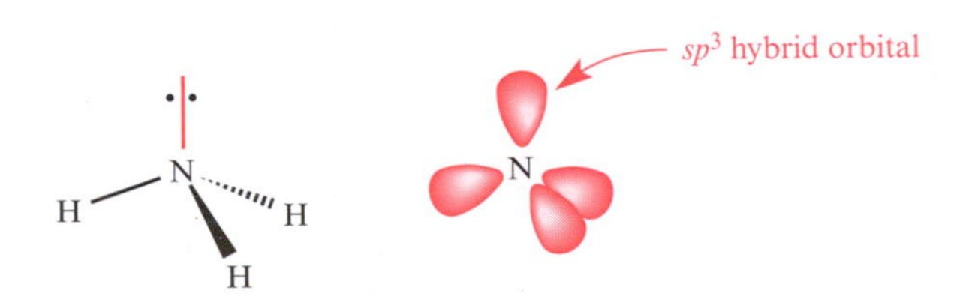

5.40 The VSEPR model predicts that the geometry of the NH_3 molecule is based on a tetrahedral arrangement of electron pairs, with the four atoms defining a trigonal pyramidal shape. The nitrogen atom is sp^3 hybridized with the lone pair of electrons occupying one of the sp^3 orbitals.

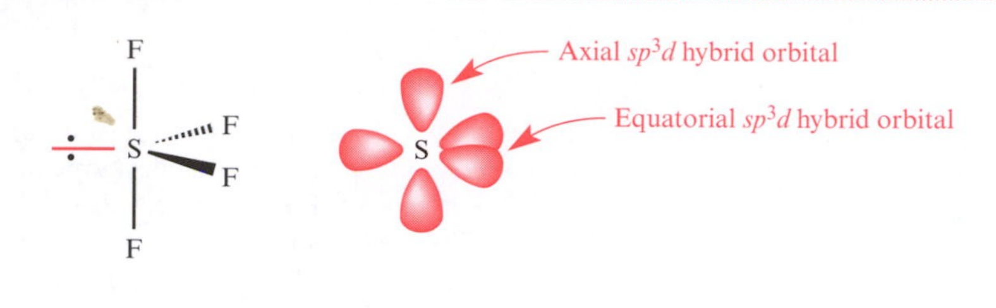

5.41 The VSEPR model predicts that the geometry of the SF_4 molecule is based on a trigonal bipyramidal arrangement of electron pairs with the atoms defining a disphenoidal shape. The sulfur atom is sp^3d hybridized with the lone pair of electrons occupying one of the equatorially directed sp^3d orbitals.

square-based pyramidal. Within the VSEPR model, these geometries arise because of lone pairs of electrons within the valence shell of the central atom. In the hybrid orbital approach, a lone pair of electrons occupies an orbital domain in the same way as bonding pair of electrons.

Consider a molecule of NH_3. The VSEPR model predicts that the molecular shape is derived from a tetrahedral arrangement of electron pairs (Figure 5.16b). It follows that the nitrogen atom may be sp^3 hybridized with the lone pair of electrons occupying one of the sp^3 orbitals (Figure 5.40).

Similarly, VSEPR theory predicts that the structure of the SF_4 molecule is derived from a trigonal bipyramidal array of one lone and four bonding pairs of electrons. This corresponds to an sp^3d hybridized sulfur atom (Figure 5.41) in which one equatorial orbital is occupied by the lone pair of electrons.

5.20 HYBRIDIZATION: THE ROLE OF UNHYBRIDIZED ATOMIC ORBITALS

'Left-over' atomic orbitals

A set of one *s* and three *p* (p_x, p_y and p_z) atomic orbitals can be involved in one of three types of hybridization. If an atom is sp^3 hybridized, all four of the AOs are used in the formation of the four hybrid orbitals. However, in the case of an sp^2 hybridized atom, the atom retains one pure *p* AO, whilst an *sp* hybridized atom has two 'left-over' *p* AOs. Similarly, when an atom is sp^3d or sp^3d^2 hybridized, either four or three (respectively) atomic *d* orbitals on the central atom remain unhybridized.

In this section we consider the roles that these 'left-over' *p* and *d* atomic orbitals play in bonding.

The σ-bonding framework

Methane contains four *single* C–H bonds. The CH_4 molecule is tetrahedral and the carbon atom is considered to be sp^3 hybridized, and the mutual orientations of the four sp^3 hybrid orbitals coincide with the orientations of the four C–H single bonds (Figure 5.42). Each hybrid orbital overlaps with a hydrogen 1*s* AO, and two valence electrons (one from carbon and one from

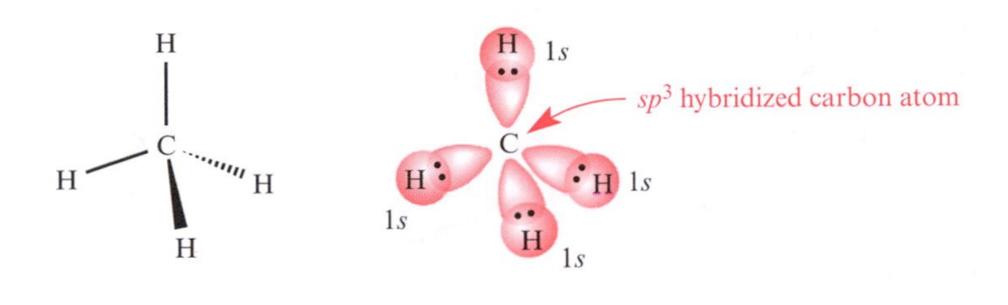

5.42 Valence bond theory in CH_4. The mutual orientations of the four sp^3 hybrid orbitals centred on the carbon atom coincide with the orientations of the four C–H single bonds. Each hybrid orbital overlaps with a hydrogen 1*s* atomic orbital, to give four localized C–H σ-bonds.

hydrogen) occupy each resultant localized bonding orbital. Each C–H single bond is therefore a *localized σ-bond*, and the set of four such orbitals describes a *σ-bonding framework*. This completely describes the bonding situation for CH_4 within the VB model. Similarly, VB theory gives a description of the bonding in the NH_3 molecule in terms of three N–H σ-bonds and one lone pair of electrons.

A trigonal planar molecule: BH_3

The boron atom in the BH_3 molecule is sp^2 hybridized. Valence bond theory describes the bonding in the BH_3 molecule in terms of the overlap between the boron sp^2 hybrid orbitals and the hydrogen 1*s* atomic orbitals (Figure 5.43). Each resultant bonding orbital contains two electrons (one from boron and one from hydrogen) and defines a localized B–H σ-bond. In BH_3, the boron atom retains an unhybridized, empty 2*p* atomic orbital.

A trigonal planar molecule: BF_3

In BF_3, the σ-bonding framework is similar to that in BH_3. Each localized B–F σ-bond is formed by the overlap of a boron sp^2 hybrid orbital and an orbital from the fluorine atom; each resultant σ-bonding orbital is occupied by two electrons.

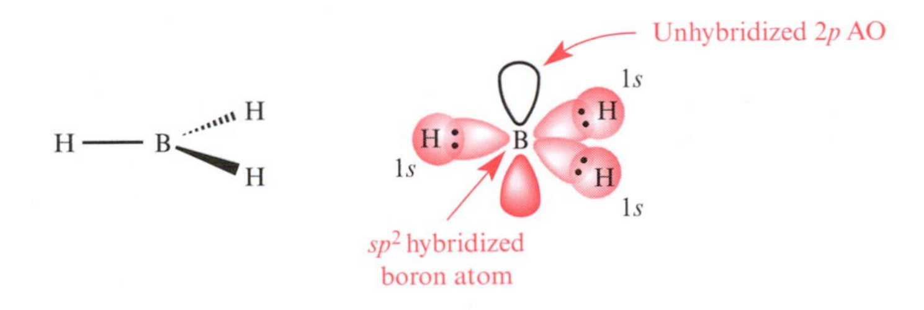

5.43 Valence bond theory in BH_3. The mutual orientations of the three sp^2 hybrid orbitals centred on the boron atom coincide with the orientations of the B–H single bonds. Each hybrid orbital overlaps with a hydrogen 1*s* atomic orbital giving three localized B–H σ-bonds. If the molecule lies in the *xy* plane, then the unused AO is a $2p_z$ orbital.

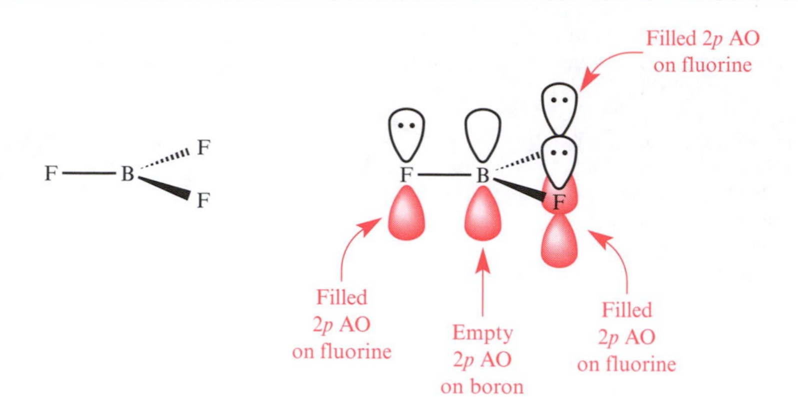

5.44 Valence bond theory in BF_3. After the formation of the σ-bonding framework using an sp^2 hybridization scheme, the remaining 2*p* AO on the boron atom is of the correct symmetry to overlap with an occupied 2*p* AO of a fluorine atom. A π-bond is therefore formed.

As in BH_3, once the σ-bonding framework has been formed in the BF_3 molecule, an unhybridized, empty 2*p* atomic orbital remains on the boron atom. However, *unlike* the situation in BH_3, in BF_3 each fluorine atom possesses three lone pairs, and one may be considered to occupy a 2*p* AO with the same orientation as the unused 2*p* AO on the boron atom (for example, they may be $2p_z$ AOs). Figure 5.44 illustrates that it is possible for the boron atom to form a π-bond with any one of the fluorine atoms by overlap between the boron and fluorine 2*p* atomic orbitals. Look back at the resonance structures for BF_3 in Figure 5.33. One group of three resonance structures includes B=F double bond character.

The valence bonding picture for the BF_3 molecule illustrates a general point: *unhybridized p orbitals can be involved in π-bonding*. Related examples are BCl_3, $[BO_3]^{3-}$, $[CO_3]^{2-}$ and $[NO_3]^-$. These species all possess trigonal planar structures in which the central atom may be sp^2 hybridized. In each case, the central atom possesses an unhybridized 2*p* AO which can overlap with a *p* AO on an adjacent atom to form a π-bond. (Question: Which AOs are involved in π-bond formation in BCl_3 and $[CO_3]^{2-}$?)

A linear molecule: CO_2

The carbon atom in the linear CO_2 molecule is *sp* hybridized and retains two 2*p* atomic orbitals (Figure 5.45). The σ-bonding framework in the molecule

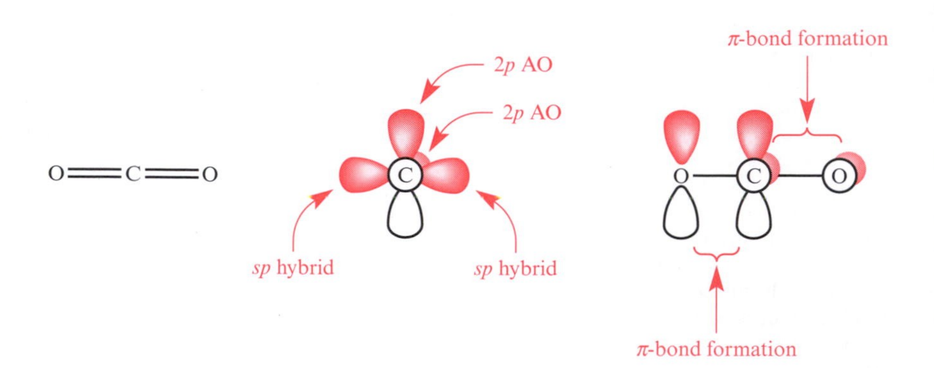

5.45 Valence bond theory in CO_2. After forming the σ-bonding framework using an *sp* hybridization scheme, the two remaining 2*p* AOs on the carbon atom can be used to form two carbon–oxygen π-bonds. Each oxygen atom uses a 2*p* AO for π-bonding.

involves two *sp* hybrids on carbon which overlap with orbitals on the two oxygen atoms. Each localized σ-bonding orbital contains two electrons (one from carbon and one from oxygen).

The two remaining 2*p* atomic orbitals on the carbon atom lie at right angles to one another and each overlaps with a 2*p* AO on one of the oxygen atoms. This produces two carbon–oxygen π-bonds (Figure 5.45) and each π-bonding orbital is occupied by two electrons (formally, one from carbon and one from oxygen).

The overall result is the formation of two carbon–oxygen double bonds, each consisting of a σ- and a π-component.

π-Bonding involving *d* atomic orbitals

In five- and six-coordinate species containing the heavier *p*-block elements, the central atom can be sp^3d or sp^3d^2 hybridized, respectively. In each case, some *d* orbitals are 'left-over' and are available for π-bonding.

For example, the σ-bonding framework of the trigonal bipyramidal SOF_4 molecule (Figure 5.46) can be described in terms of the overlap between the five sp^3d hybrid orbitals of the sulfur atom and orbitals on either the fluorine or oxygen atoms. Each localized σ-bonding orbital is occupied by two electrons. Let the molecule be oriented such that the sulfur–oxygen σ-bond lies on the *y* axis and the axial S–F bonds lie on the *z* axis. Of the four pure *d* atomic orbitals available on the sulfur atom, the d_{yz} orbital is correctly oriented to overlap with the oxygen $2p_z$ AO, and this gives rise to the formation of a sulfur–oxygen π-bonding orbital. There are 12 bonding electrons SOF_4 and these are involved in five σ- bonds and one π-bond.

Hybridization: a summary

For the example in each category below, draw out a bonding scheme using the hybrid orbital approach. Remember that each σ-bond is a single bond.

First row *p*-block elements

- sp^3 hybridization is consistent with four σ-bonds and a tetrahedral shape (e.g. CF_4).
- sp^2 hybridization is consistent with three σ-bonds and a trigonal planar geometry, and there is the possibility of π-bonding involving the remaining 2*p* AO on the central atom (e.g. $[CO_3]^{2-}$).

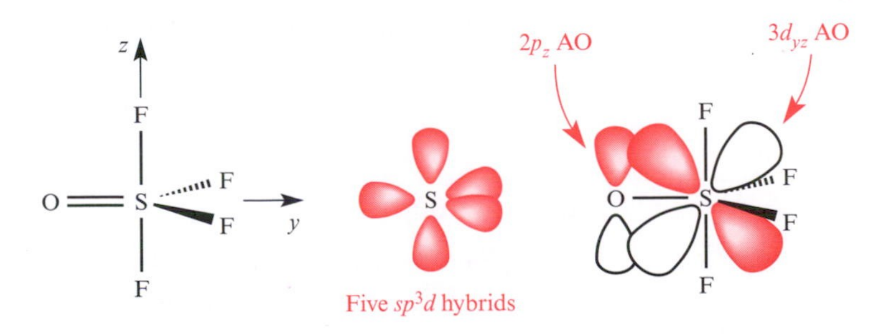

5.46 The σ-bonding framework in SOF_4 can be described in terms of an sp^3d hybridized sulfur atom. One residual 3*d* AO on the sulfur atom can overlap with an oxygen 2*p* AO to give a localized sulfur–oxygen π-bond.

- *sp* hybridization is consistent with two σ-bonds and a linear geometry, and there is the possibility of π-bonding involving the remaining two 2*p* AOs on the central atom (e.g. $[NO_2]^+$).

Heavier *p*-block elements

- sp^3 hybridization is consistent with four σ-bonds and a tetrahedral shape, and π-bonding is possible if *d* atomic orbitals are used (e.g. $[ClO_4]^-$).
- sp^2 hybridization is consistent with three σ-bonds (single bonds) and a trigonal planar geometry, and there is the possibility of π-bonding involving the remaining 2*p* AO on the central atom (e.g. SO_3); *d* atomic orbitals are also available.
- *sp* hybridization is consistent with two σ-bonds and a linear geometry (e.g. $[ICl_2]^-$); π-bonding is possible involving the remaining *p* and *d* AOs on the central atom.
- sp^3d hybridization is consistent with five σ-bonds and a trigonal bipyramidal geometry (e.g. PF_5); π-bonding is possible if *d* atomic orbitals are used (e.g. SOF_4).
- sp^3d^2 hybridization is consistent with six σ-bonds and an octahedral geometry (e.g. SF_6); π-bonding is possible if *d* atomic orbitals are used (e.g. IOF_5).

5.21 MOLECULAR ORBITAL THEORY AND POLYATOMIC MOLECULES

Polyatomic molecules create a problem

In this book, we only look briefly at the way in which MO theory is used to approach the bonding in polyatomic molecules. A key question is how to represent the problem. If we look back at MO diagrams such as Figure 3.14 (the formation of H_2) and Figure 4.8 (the formation of HF), we notice that each side of each diagram represents one of the two atoms of the diatomic molecule. What happens if we wish to represent an MO diagram for a polyatomic molecule? Any diagram which represents the composition of MOs in terms of contributions from various AOs will become very complicated, and, probably, unreadable!

An approach that is commonly used is to resolve the MO description of a polyatomic molecule into a *two-component* problem. Instead of looking at the bonding in CH_4 in terms of the interactions of the 2*s* and 2*p* AOs of the carbon atom with the individual 1*s* AOs of the hydrogen atoms (this would be a five-component problem), we consider the way in which the atomic orbitals of the carbon atom interact with the *set* of four hydrogen atoms. This is a two-component problem, and is called a *ligand group orbital* approach.

Ligand group orbitals

Methane is tetrahedral and to make an MO bonding analysis for CH_4 easier, it is useful to recognize that the tetrahedron is related to a cube as shown in

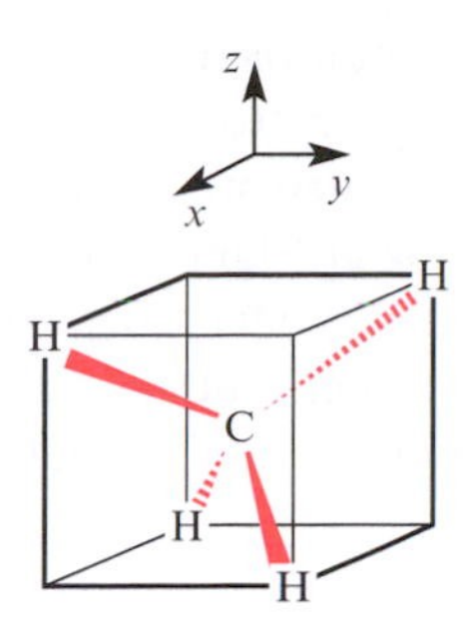

5.47 The relationship between the tetrahedral shape of CH_4 and a cubic framework. Each edge of the cube runs parallel to one of the three Cartesian axes.

Figure 5.47. This conveniently relates the positions of the hydrogen atoms to the Cartesian axes, with the carbon atom at the centre of the cube.

The valence orbitals that the carbon atom possesses are the 2*s*, $2p_x$, $2p_y$ and $2p_z$ AOs (Figure 5.48a). Remember that the 2*s* AO is spherically symmetric. The orientations of the 2*p* AOs are related to a cubic-framework; the same axis set is used in Figures 5.47 and 5.48.

Consider now the four 1*s* AOs that the four hydrogen atoms contribute. Each 1*s* AO has two possible phases and, when the *four orbitals are taken as a group*, various combinations of phases are possible. With four AOs, we can construct four *ligand group orbitals* (LGOs) as shown Figure 5.48b. The in-phase combination of the four 1*s* AOs is labelled as LGO(1). We now make use of the relationship between the arrangement of the hydrogen atoms and the cubic framework. Each of the *xy*, *xz* and *yz* planes bisects the cube in a dif-

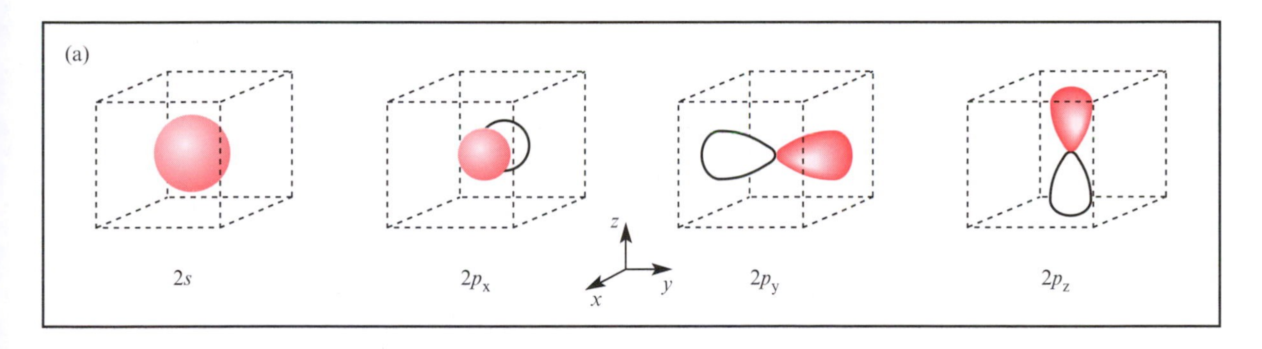

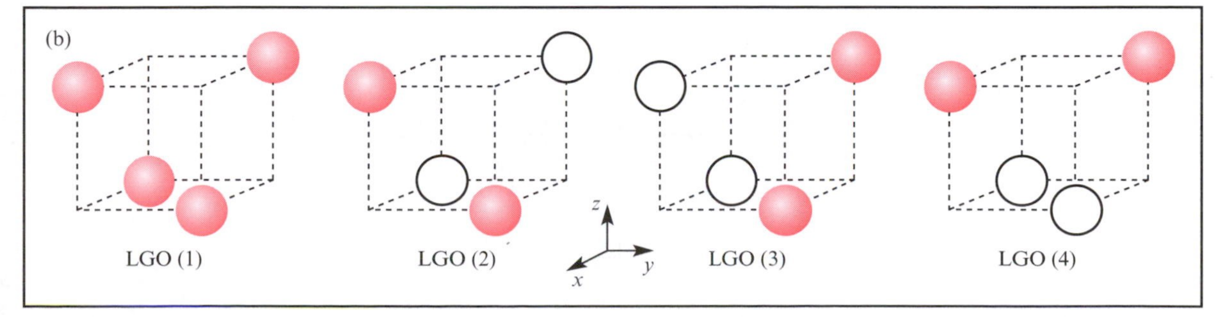

5.48 (a) The 2*s*, $2p_x$, $2p_y$ and $2p_z$ AOs are the valence orbitals of carbon. (b) The four hydrogen 1*s* AOs combine to generate four ligand group orbitals (LGOs).

ferent direction. In Figure 5.47, mentally sketch the *yz* plane through the cube – two hydrogen atoms lie above this plane, and two lie below. Similarly, the four hydrogen atoms are related in pairs across each of the *xy* and *xz* planes. Now look at Figure 5.48b. In LGO(2), the four 1*s* AOs are drawn as two pairs, one pair on each side of the *yz* plane. The AOs in one pair are in-phase with one another, but are out-of-phase with the orbitals in the second pair. Ligand group orbitals LGO(3) and LGO(4) can be constructed similarly.

The number of ligand group orbitals formed = the number of atomic orbitals used

Combining the atomic orbitals of the central atom with the ligand group orbitals

The next step in the ligand group orbital approach is to 'match up' the valence orbitals of the carbon atom with LGOs of the four hydrogen atoms. The criterion for matching is symmetry. Each carbon AO has to 'find a matching partner' from the set of LGOs.

The carbon 2*s* AO has the same symmetry as LGO(1). If the 2*s* AO is placed at the centre of LGO(1), overlap will occur between the 2*s* orbital and *each* of the hydrogen 1s orbitals. This interaction leads to one bonding MO in the CH_4 molecule: $\sigma(2s)$. This is the lowest-lying MO in Figure 5.49. It possesses C–H bonding character which is *delocalized over all four of the C–H interactions*.

The carbon $2p_x$ AO has the same symmetry as LGO(2), and if it is placed at the centre of LGO(2), overlap will occur between the $2p_x$ orbital and *each* of the hydrogen 1*s* orbitals. This interaction leads to a second bonding MO

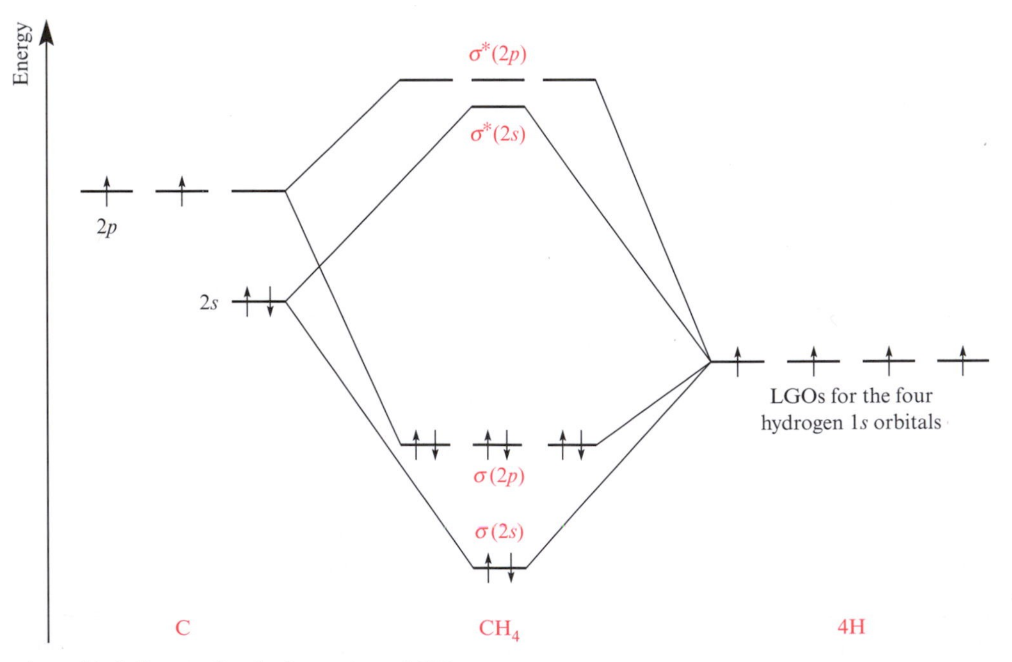

5.49 Molecular orbital diagram for the formation of CH_4.

in the methane molecule: $\sigma(2p_x)$. This MO contains a nodal plane coincident with the *yz* plane. Similarly, we can match-up the carbon $2p_y$ AO with LGO(3), and the $2p_z$ AO with LGO(4). These combinations generate two more bonding molecular orbitals. The three MOs that are formed from the combinations of the carbon 2*p* AOs and LGOs(2) to (4) are equivalent to one another except for their orientations. These MOs are therefore degenerate and are labelled $\sigma(2p)$ in Figure 5.49. Each MO contains C–H bonding character which is *delocalized over all four of the C–H interactions*.

There are therefore four bonding MOs (and hence four antibonding MOs) which describe the bonding in CH_4. In each orbital, the C–H bonding character is *delocalized*.

Putting in the electrons

As usual in MO theory, the last part of the operation is to count up the available valence electrons and place them in the molecular orbitals according to the *aufbau* principle.

In CH_4 there are eight valence electrons and these occupy the MOs as shown in Figure 5.49. All the valence electrons in the molecule are paired and CH_4 is predicted to be diamagnetic, in keeping with experimental data.

5.22 HOW DO THE VB AND MO PICTURES OF THE BONDING IN METHANE COMPARE?

The valence bond and molecular orbital descriptions of methane

Valence bond theory describes the bonding in the CH_4 molecule in terms of an sp^3 hybridized carbon atom, and four equivalent σ-bonding orbitals. Each orbital is *localized*, meaning that each C–H bond is described by one σ-bonding orbital.

Molecular orbital theory also produces a description of CH_4 which involves four molecular orbitals. However, unlike the VB model, the MO approach gives one unique MO and a set of three degenerate MOs (Figure 5.49). In each MO, the C–H bonding character is *delocalized over all four of the C–H vectors*.

Are these two pictures really different? If so, is one of them wrong? Indeed, is either of them realistic?

Photoelectron spectroscopy

One of the major differences between the results of the VB and MO treatments of the bonding in the CH_4 molecule is that the former suggests that there are four equivalent MOs (equal energy) while the latter is consistent with there being two molecular orbital energy levels, one associated with a unique MO and one with a set of three degenerate orbitals.

In Box 3.6, we briefly mentioned the experimental technique of photoelectron spectroscopy in the context of probing the energies of molecular orbitals. This method can be used to investigate the question of the MO energy levels in CH_4 and the photoelectron spectroscopic data support the results of MO theory. But, does this mean that the VB model gives the wrong 'answer'?

VB versus MO theory

Molecular orbital theory describes the bonding in the CH_4 molecule in terms of four MOs. The unique $\sigma(2s)$ MO is spherically symmetric and provides equal bonding character in all of the four C–H interactions. The $\sigma(2p_x)$, $\sigma(2p_y)$ and $\sigma(2p_z)$ MOs are related to one another by rotations through 90°. Because they are degenerate, we must consider them *as a set and not as individual orbitals*. Taken together, they describe the four C–H bonds equally. Thus, the MO picture of CH_4 is of a molecule with four equivalent C–H bonds. This result arises *despite* the fact that the four MOs are not all identical.

The sp^3 hybrid model (VB theory) of CH_4 describes four equivalent C–H bonds in terms of four *localized* σ-bonds. The associated bonding orbitals are of equivalent energy.

Both the VB and MO approaches are bonding *models*. Both models achieve a goal of showing that the CH_4 molecule contains four equivalent bonds. While, the MO model appears to give a more realistic representation of the energy levels associated with the bonding electrons, the VB method is simpler to apply. While MO theory can be applied to small and large molecules alike, it very quickly goes beyond the 'back-of-an-envelope' level of calculation.

SUMMARY

The theme of this chapter has been the shapes of and bonding in polyatomic molecules.

Do you know what the following terms mean?

- polyatomic
- coordination number
- isostructural
- trigonal planar
- trigonal pyramidal
- tetrahedral
- trigonal bipyramidal
- square-based pyramidal
- octahedral
- square planar
- disphenoidal
- VSEPR model
- Kepert model
- stereochemically inactive lone pair
- inert pair effect
- geometrical isomerism
- stereochemically non-rigid
- Berry pseudo-rotation
- molecular dipole moment
- expansion of the octet
- hybrid orbital

You should now be able:

- to provide examples of molecules and ions which possess two-, three-, four-, five- or six-coordinate structures
- to discuss why boron, carbon and nitrogen exhibit a greater structural diversity than fluorine and oxygen
- to discuss why the heavier *p*-block elements in a given group show a greater structural diversity than the first row member of that group
- for the *p*-block elements, to discuss when the octet rule is obeyed, and when and how the valence shell can be expanded
- to predict the shapes of species containing a central *p*-block or *d*-block element
- to draw out geometrical isomers for square planar, trigonal bipyramidal and octahedral species and for molecules containing a double bond.
- to discuss the limited range of geometrical environments in which carbon atoms are found
- to outline the mechanism that exchanges axial and equatorial atoms or groups in a five-coordinate molecule
- to determine whether a polyatomic molecule is likely to possess a dipole moment
- to draw a set of resonance structures for a simple polyatomic molecule or molecular ion
- to discuss what is meant by orbital hybridization
- to relate hybridization schemes to molecular shapes
- to use hybridization schemes to descibe the bonding in simple polyatomic molecules containing both single and double bonds

PROBLEMS

5.1 Write down Lewis structures for CO_2, $[NO_2]^+$, $[N_3]^-$ and N_2O.

5.2 Using the VSEPR model, predict the shape of the anion $[I_3]^-$. Also predict, with reasoning, the shapes of the anions $[ICl_2]^-$, $[IBr_2]^-$, $[ClF_2]^-$ and $[BrF_2]^-$. (Hint: The central atom in each is the one *lowest* in group 17.)

5.3 Taking the linear structure of CO_2 and the bent structure of SO_2 as your starting points, use the isoelectronic principle to predict the structures of the following species:

(a) $[NO_2]^+$; (b) $[NO_2]^-$; (c) O_3;

(d) CS_2; (e) N_2O; (f) $[NCS]^-$.

5.4 Use the VSEPR model to predict the shapes of the following molecules:

(a) BBr_3, (b) NMe_3; (c) SCl_2; (d) $[I_3]^+$;

(e) $BrCl_3$; (f) PF_5; (g) $PClF_4$; (h) $[NMe_4]^+$;

(i) XeF_4; (j) XeF_2 (Me=CH_3).

5.5 Use the Kepert model to predict the geometries of the following molecular species containing *d*-block metal centres:

(a) $[AuBr_2]^-$; (b) $[FeF_6]^{3-}$; (c) $[Ag(CN)_2]^-$;

(d) $[Ni(NH_3)_6]^{2+}$; (e) $[TiCl_4]$; (f) $[MoS_4]^{2-}$;

(g) $[Cr(CO)_6]$; (h) $[Ni(CO)_4]$; (i) $[VCl_3(NMe_3)_2]$.

5.6 Draw out the possible geometrical isomers for

(a) $CH_3CH{=}CHCH_3$; (b) $[PdCl_2(PPh_3)_2]$;

(c) $[FeH(CO)_4]^-$; (d) $[W(CO)_4(NCMe)_2]$;

(e) $[CrCl_3(PEt_3)_3]$; (f) $[IrBr_2(Et)(PMe_3)_3]$.

(Et = ethyl = C_2H_5; MeCN, see Figure 5.7, bonds to the metal through the N atom.)

5.7 What structure does the Kepert model predict for a five-coordinate compound of general formula ML_5? Based on this geometry, how many isomers are possible for the compound $[NiBr_3(PMe_2Ph)_2]$? Which of these isomers is likely to exhibit the minimum inter-ligand repulsions? Give reasons for your choice.

5.8 Predict the structures of the following molecules. Which of the molecules do you expect to possess a dipole moment? (Pauling electronegativities are listed in Table 4.2.)

(a) BF_3; (b) PF_3; (c) PF_5; (d) $BBrF_2$; (e) IF_5;

(f) ClF_3; (g) H_2Se; (h) SO_3.

5.9 Table 5.6 gives values of the dipole moments of a series of related organic molecules. Rationalize the trends in these values, given that values of χ^P for C, H, Cl and F are 2.6, 2.2, 3.2 and 4.0, respectively.

Table 5.6 Data for problem 5.9.

Molecule	*Dipole moment / D*
CH_4	0
CF_4	0
CCl_4	0
CH_3F	1.86
CH_2F_2	1.98
CHF_3	1.65
CH_3Cl	1.89
CH_2Cl_2	1.60
$CHCl_3$	1.04

5.10 Outline how the combination of carbon 2*s* and 2*p* atomic orbitals leads to hybrid orbitals that may be used to describe the bonding in the $[CO_3]^{2-}$ ion.

5.11 Write down the hybridization of the central atom in each of the following molecules:

(a) SiH_4; (b) NF_3; (c) PF_5; (d) BF_3;

(e) $[CoF_6]^{3-}$; (f) IF_3; (g) $TiCl_4$.

5.12 Consider the molecule CO_2.

(a) Use the VSEPR model to predict its shape.

(b) Draw resonance structures for this molecule.

(c) Describe the bonding in the molecule in terms of a hybridiziation scheme. (Hint: Remember that all bonding schemes must give the same final answer.)

6 Ions

Topics

6.1 INTRODUCTION

In the previous three chapters, we have discussed covalent bonding. The electron density in a homonuclear bond is concentrated mid-way between the two atoms in the bond, but in a heteronuclear covalent bond, the electron density may be displaced towards one of the atoms according to the relative electronegativities of the atoms.

A review of the HF molecule

Within the valence bond model, resonance structures represent different bonding descriptions of a molecule. In a molecule such as HF, the difference between the electronegativities of hydrogen ($\chi^P = 2.2$) and fluorine ($\chi^P = 4.0$) means that the resonance structure describing an ionic form contributes significantly to the overall bonding model. When we used molecular orbital theory to describe the bonding in the HF molecule, the differences in energy between the hydrogen and fluorine atomic orbitals meant that the σ-bonding MO possessed considerably more fluorine than hydrogen character. The electron density in the H–F bond is displaced towards the fluorine atom and the bond is polar. In both of these bonding models, the fluorine atom in HF carries a partial negative charge (δ^-) and the hydrogen atom bears a partial positive charge (δ^+).

Resonance structures for HF: see Section 4.3

MO approach to HF: see Section 4.5

Approaching the bonding in sodium fluoride by VB and MO theories

In the gas phase, sodium fluoride contains NaF *molecules* (or more correctly, NaF formula units), although in aqueous solution, the liquid or solid state, Na^+ and F^- *ions* are present.

The difference between the electronegativities of sodium ($\chi^P = 0.9$) and fluorine ($\chi^P = 4.0$) is larger than the difference between hydrogen and fluorine. When we write out resonance structures for sodium fluoride, the *ionic form* (Na^+F^-) predominates.

An MO diagram for the formation of an NaF molecule is shown in Figure 6.1. The 3*s* valence AO of the sodium atom lies at higher energy than the valence 2*s* and 2*p* AOs of fluorine. The energy matching of these orbitals is poor, and the bonding MO contains far more fluorine than sodium character. We are *tending towards* a situation in which this MO has so little sodium character that it is almost non-bonding in character.§ There are eight valence electrons (one from the sodium atom and seven from the fluorine atom) and these occupy the orbitals in NaF according to the *aufbau* principle. The important consequence of this is that three of the filled MOs in NaF possess

Orbital energy matching: see Sections 3.12 and 4.4.

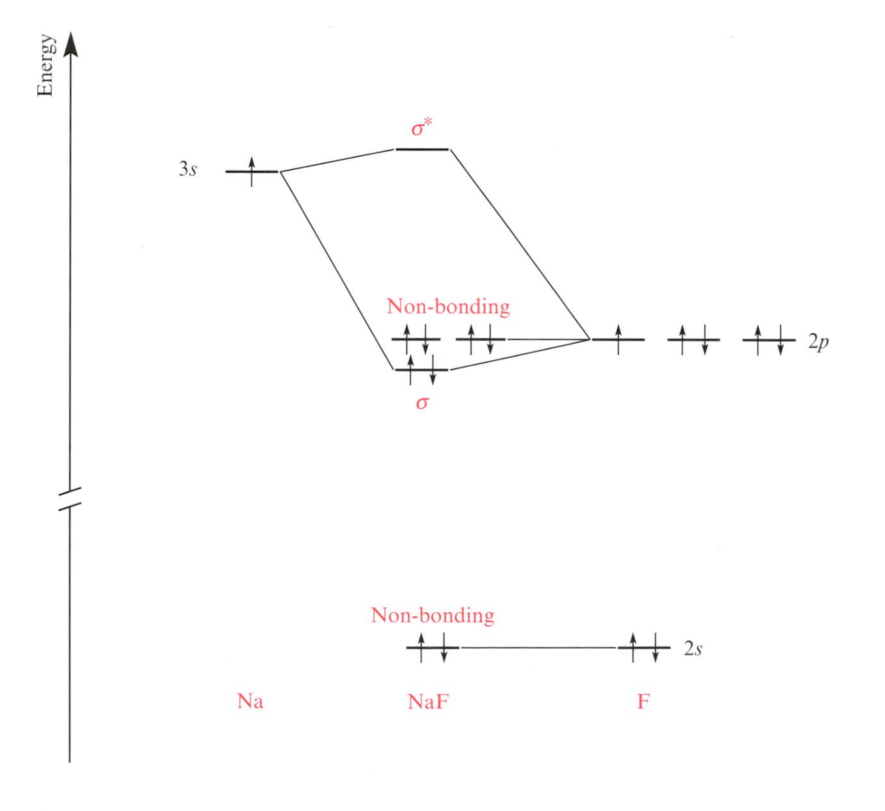

6.1 An MO diagram for the formation of an NaF molecule. Only the valence atomic orbitals and electrons are shown. The break in the vertical (energy) axis signifies that the energy of the fluorine 2*s* AO is much lower than is actually shown. The σ-bonding MO contains mainly fluorine character.

§ We are using 'non-bonding' in the sense usually adopted in discussions of MO diagrams; specifically, there is little sharing of electrons between the nuclei. The charge separation results in 'ionic bonding'.

only fluorine character and one filled MO possesses *virtually all fluorine* character. It follows that the electron density between the sodium and fluorine nuclei is displaced so far in one direction that we are tending towards the point at which one electron has been transferred from the sodium atom to the fluorine atom. If the transfer is complete, the result is the formation of a fluoride ion (F^-) and a sodium ion (Na^+).

Ions

In Section 1.11, we reviewed ion formation. A fluorine atom (with a ground state electronic configuration $[He]2s^22p^5$) may accept one electron to form a fluoride anion (F^-) with the noble gas configuration $[He]2s^22p^6$ (or [Ne]). The loss of one electron from a neutral sodium atom (ground state electronic configuration $[Ne]3s^1$) generates a positively charged cation (Na^+) with the noble gas configuration of [Ne].

An anion is a negatively charged ion and a cation is a positively charged ion.

In the same way that van der Waals forces or dipole–dipole interactions arise between atoms or molecules which possess dipole moments (transient or permanent), so there are strong electrostatic interactions between spherically symmetrical ions of opposite charge. In sodium fluoride, there are attractive interactions between the Na^+ cations and *all* the nearby F^- ions. Similarly, there are electrostatic (attractive) interactions between the F^- anions and *all* nearby Na^+ cations. An electrostatic interaction depends upon distance and not upon direction. Ultimately this leads to the assembly of an *ionic lattice*.

This chapter is concerned with *ions* and the enthalpy changes that accompany the processes of cation and anion formation. What forces operate between oppositely charged ions? How much energy is associated with these interactions? How are ions arranged in the solid state?

Ionic bonding is a *limiting model*, just as a covalent model is. However, just as valence bond theory can be improved by including ionic resonance structures, we see later in this chapter that the ionic model can be adjusted to take account of some covalent character. In fact, very few systems can be considered to be purely ionic or purely covalent. The bonding in many species is more appropriately described as lying somewhere between these two limits.

6.2 ELECTRON DENSITY MAPS

The distinction between a fully covalent bond X–Y and an ionic interaction X^+Y^- is made by considering the regions of electron density around the nuclei X and Y. We are already familar with the idea that a covalent bond is synonymous with the presence of 'shared' electron density. The distribution of the electron density between the atomic centres in a molecule can be investigated by means of appropriate calculations using forms of the Schrödinger equation. Such results can be represented in the form of a contour map. An *electron density map* is composed of contour lines which connect points of equal electron density. Contours drawn close to the nucleus relate mainly to the presence of core electron density.

Schrödinger equation: see Section 2.7

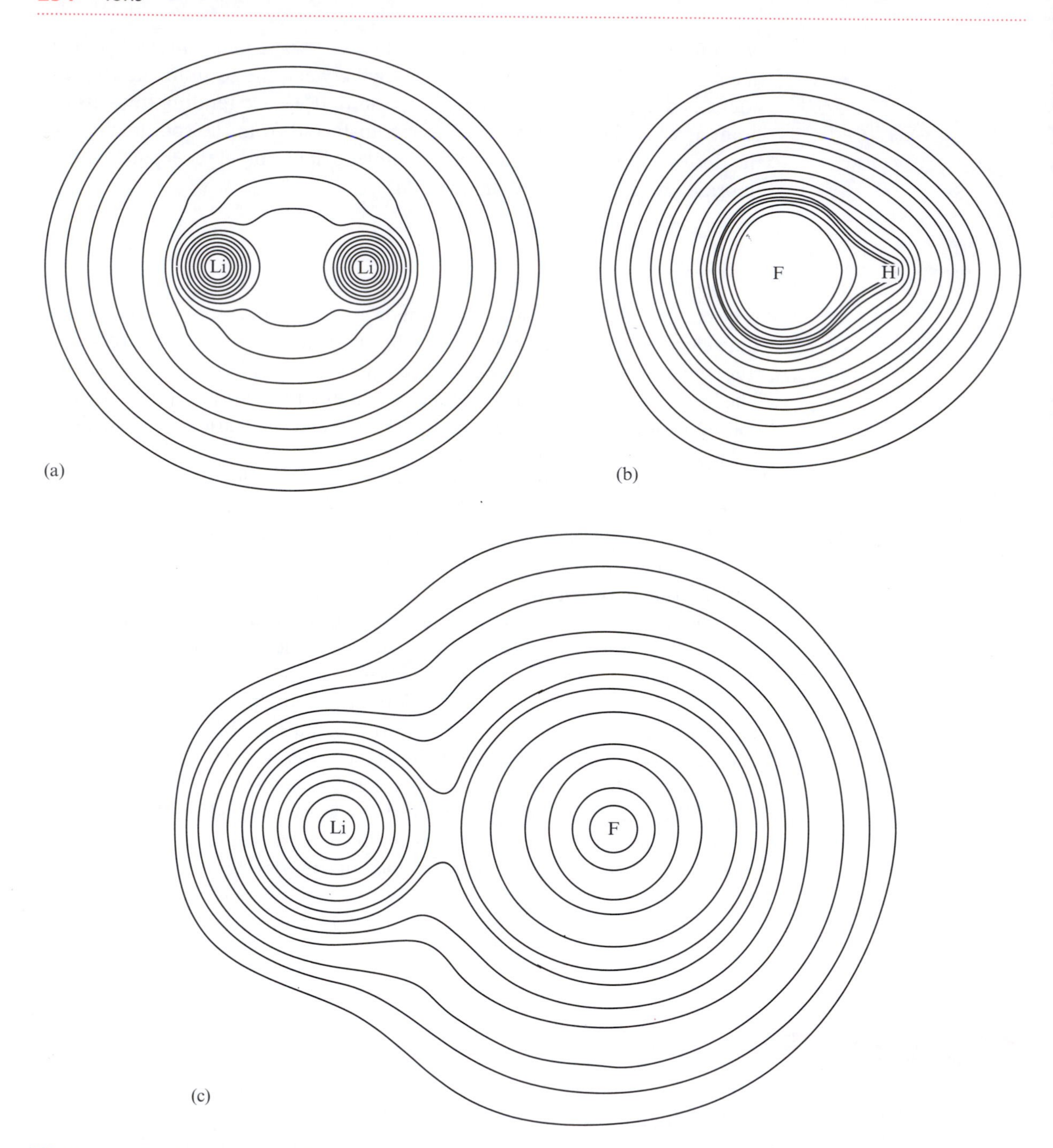

6.2 Electron density maps for (a) Li_2, (b) HF and (c) LiF. In Li_2, the distribution of electron density is symmetrical and a region of electron density is situated between the two lithium nuclei. This corresponds to a region of Li–Li bonding electron density. Hydrogen fluoride is a polar molecule, and the electron density is displaced towards the fluorine centre, shown on the left-hand side of the figure. A region of bonding electron density between the H and F nuclei is apparent. In LiF, the circular electron density contours correspond to sections through two spherical regions of electron density. The tendency is towards the formation of spherical Li^+ and F^- ions.
[From: A.C. Wahl, *Science*, **151**, 961 (1966) (Figure 6.2a); R.F.W. Bader, *et al.*, *J. Chem. Phys.*, **47**, 3381 (1967) (Figure 6.2b); R.F.W. Bader, *et al.*, *J. Chem. Phys.*, **49**, 1653 (1968) (Figure 6.2c).]

Figure 6.2 shows three electron density maps. Each is drawn as a section through a molecule and indicates the electron density distribution in that plane. Figure 6.2a shows an electron density map for the homonuclear diatomic molecule Li_2. The map has a symmetrical appearance, with the electron density at one Li centre mirrored at the other. There is a region of electron density located symmetrically between the two atomic nuclei and this represents the *bonding electron density*; the Li–Li bond is mainly covalent in character.

An electron density map for HF is shown in Figure 6.2b. The electron density contours in this diagram are typical of those in a polar covalent bond. The map is asymmetrical and the region of electron density is ovate (i.e. 'egg-shaped'). More electron density is associated with the fluorine centre than the hydrogen centre. There is, however, still a region of electron density between the nuclei, and this corresponds to the electron density in the covalent bond.

Figure 6.2c illustrates the electron density distribution in an LiF molecule. Now, we can best describe the electron density as being concentrated in two circular regions (that is, spherical in three dimensions). Each region is centred around one of the two nuclei. The peripheral contour lines join points of only low electron density. The bonding in LiF may be described as *tending towards being ionic*.

The electron density map in Figure 6.2c illustrates the same situation that we discussed for sodium fluoride in Section 6.1. As the covalent interaction between two atoms diminishes, the tendency is towards the formation of two *spherical* ions. In the extreme case, the electron density map will appear as shown in Figure 6.3. To a first approximation, we can assume that an ion is spherical, but later in this chapter we discuss to what extent this description is valid.

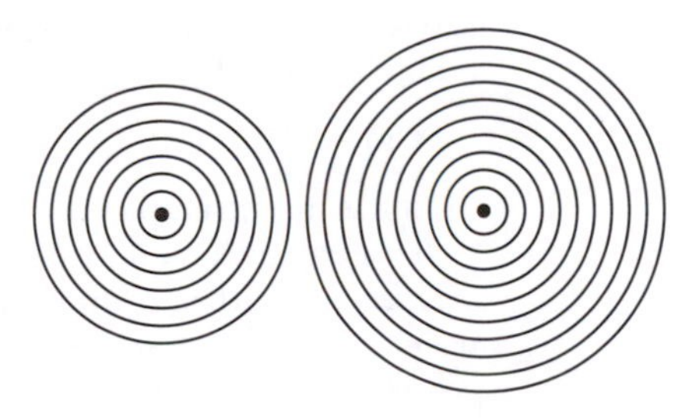

6.3 An idealized electron density map showing sections through a pair of spherical ions.

6.3 IONIZATION ENERGY

In Section 4.10 we briefly mentioned ionization energies in the context of the Mulliken electronegativity scale, and we defined the first ionization energy (IE_1) of an atom as the internal energy change at 0 K (ΔU(0 K)) associated with the *removal* of the first valence electron (equation 6.1). The energy change is defined for a gas phase process. The units are kJ mol^{-1} or electron volts (eV).

1 eV = 96.5 kJ mol^{-1}

$$X(g) \rightarrow X^+(g) + e^- \qquad (6.1)$$

The first ionization energy (IE_1) of a gaseous atom is the internal energy change at 0 K (ΔU(0 K)) associated with the removal of the first valence electron:

$$X(g) \rightarrow X^+(g) + e^-$$

For thermochemical cycles, an associated change in *enthalpy* is used:

$$\Delta H(298\ K) \approx \Delta U(0\ K)$$

Hess cycle: see Section 1.17

We often need to incorporate the values of ionization energies into Hess cycles and it is then convenient to use a change in enthalpy (ΔH(298 K)) rather than an internal energy change (ΔU(0 K)). We have already come across this problem in Section 3.5, and discuss the differences between ΔU and ΔH more fully in Chapter 12. The difference between an ionization energy and the associated enthalpy change is very small and values of IE can be used in enthalpy cycles provided that a very high degree of accuracy is not required.

The second ionization energy (IE_2) of an atom refers to the process defined in equation 6.2 – it is also the first ionization step of cation X^+. Successive ionizations can also occur; the third ionization energy (IE_3) of atom X refers to the loss of the third electron (equation 6.3), and so on. We are generally concerned only with the removal of electrons from the valence shell of an atom. Removing core electrons requires considerably more energy. For example, with an atom in group 2 of the periodic table, the first

Box 6.1 Ionization energies and associated enthalpy values

The first ionization energy (IE_1) of an element X is the *internal energy change* at 0 K (ΔU(0 K)) which is associated with the removal of the first valence electron; it is defined for a gas phase process:

$$X(g) \rightarrow X^+(g) + e^-$$

Most thermochemical cycles involve enthalpy changes at 298 K (ΔH(298 K)).

If we assume that the gaseous atoms of element X and the gaseous ions X^+ are *ideal* gases, then ΔU(0 K) and ΔH(298 K) are related as shown below; R is the molar gas constant (8.314×10^{-3} kJ K^{-1} mol^{-1}) and ΔT is the difference between the two temperatures (0 K and 298 K):

$$\Delta H(298\ K) = \Delta H(0\ K) + (\tfrac{5}{2} \times R \times \Delta T)$$
$$\Delta H(298\ K) = \Delta H(0\ K) + (\tfrac{5}{2} \times 8.314 \times 10^{-3} \times 298)$$
$$\Delta H(298\ K) = \Delta H(0\ K) + 6.2\ kJ\ mol^{-1}$$

Typically, ionization energies referring to the removal of valence electrons are of the order of 10^3 kJ mol^{-1} and the addition of ≈ 6 kJ mol^{-1} makes little difference to the value. It is therefore acceptable for most purposes to use tabulated values of ionization energies (ΔU(0 K)) when an enthalpy of ionization (ΔH(298 K)) is required:

$$IE = \Delta U(0\ K) \approx \Delta H(298\ K)$$

and second ionization energies refer to the removal of the two valence electrons, whilst higher values of *IE* refer to the removal of core electrons.

$$X^{+}(g) \rightarrow X^{2+}(g) + e^{-} \quad (6.2)$$

$$X^{2+}(g) \rightarrow X^{3+}(g) + e^{-} \quad (6.3)$$

Values of some ionization energies for the elements are listed in Appendix 8.

6.4 TRENDS IN IONIZATION ENERGIES

First ionization energies across the first period

The first period runs from lithium (group 1) to neon (group 18). The ground state electronic configurations of the two *s* block elements in the period (lithium and beryllium) are $[He]2s^1$ and $[He]2s^2$ respectively, and of the six *p*-block elements (boron to neon) are $[He]2s^22p^1$, $[He]2s^22p^2$, $[He]2s^22p^3$, $[He]2s^22p^4$, $[He]2s^22p^5$ and $[He]2s^22p^6$.

Figure 6.4 shows the trend in the first ionization energies (IE_1) as the period is crossed; the value of IE_1 for helium (the noble gas which precedes lithium in the periodic table) is also included in the figure. Three features in the graph should be recognized:

- there is a sharp decrease in the value of IE_1 in going from He to Li;
- there is a *general* increase in the values of IE_1 in crossing the period;
- two apparent discontinuities in the general increase from Li to Ne occur.

These observations can be rationalized in terms of the arrangement of the valence electrons.

Helium to lithium

The first ionization energy for helium refers to the process given in equation 6.4. An electron is removed from a filled 1*s* shell, but as we have already seen there is some special stability associated with the $1s^2$ configuration.

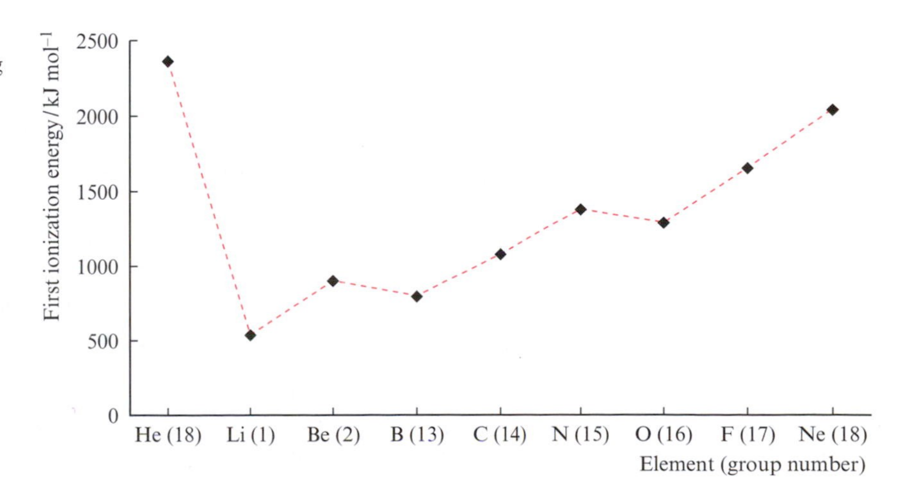

6.4 The trend in first ionization potentials on going from helium ($Z = 2$) to neon ($Z = 10$). The elements from lithium to neon make up the first period.

Stability of $1s^2$ configuration: see Section 3.4

Removing an electron destroys the noble gas configuration and requires a relatively large amount of energy.

$$\underset{1s^2}{\mathrm{He(g)}} \rightarrow \underset{1s^1}{\mathrm{He^+(g)}} + \mathrm{e^-} \qquad (6.4)$$

The ground state electronic configuration of lithium is [He]$2s^1$ and the first ionization energy is the energy required to remove the $2s$ electron. This is the only electron in the valence shell and removing it is a relatively easy process.

Lithium to neon

In going from lithium to neon, two significant changes occur: firstly, the valence shell is filled with electrons in a stepwise manner and secondly the effective nuclear charge increases and the atomic orbitals contract. It follows that it becomes increasingly difficult (more energy is needed) to remove one electron from the valence shell on going from an atom with the [He]$2s^1$ ground state configuration (lithium) to one with the [He]$2s^22p^6$ configuration (neon).

Orbital contraction: see Section 3.21

Beryllium to boron

In Section 2.15, we stated that the nuclear charge experienced by an electron in the $2p$ orbital is less than that experienced by an electron in the $2s$ orbital *in the same atom* because electrons in the $2s$ AO shield electrons in the $2p$ orbital. In going from beryllium to boron, the ground state electronic configuration changes from [He]$2s^2$ to [He]$2s^22p^1$. But we must be careful. The number of protons also increases by one. The *real* increase in the nuclear charge does *not*, however, outweigh the effect of shielding, and it is easier to remove the $2p$ electron from the boron atom than one of the $2s$ electrons from the berylium atom.

Nitrogen to oxygen

The trend of increasing ionization energies across a row in the p-block is interrupted between nitrogen and oxygen. The slight fall in values between these two elements is associated with the fact that nitrogen possesses the ground state electronic configuration [He]$2s^22p^3$ (Figure 6.5). The $2p$ level is *half-full* and this imparts a degree of stability to the nitrogen atom.[§] Thus, it is slightly harder to remove an electron from the valence shell of the nitrogen atom than might otherwise be expected.

First ionization energies across the second period

Figure 6.6 shows the trend in first ionization energies on going from neon to argon. The dramatic fall from neon to sodium corresponds to the difference between removing an electron from the full octet in neon, and removing an electron from a singly occupied $3s$ atomic orbital in sodium.

From sodium to argon, the second period is crossed. The same pattern of values in ionization energies is observed in Figure 6.6 as in Figure 6.4 and,

§ The origin of this stability is complex. We merely note that there is a particular stability associated with half-filled configurations such as p^3 and d^5.

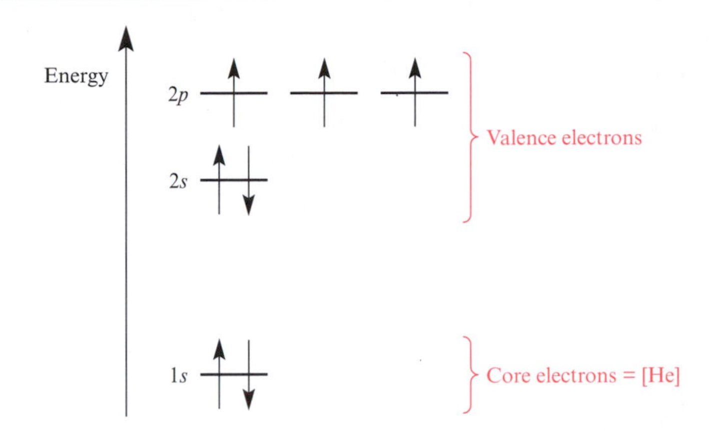

6.5 The ground state electronic configuration of nitrogen. Note that the $2p$ level is half-full.

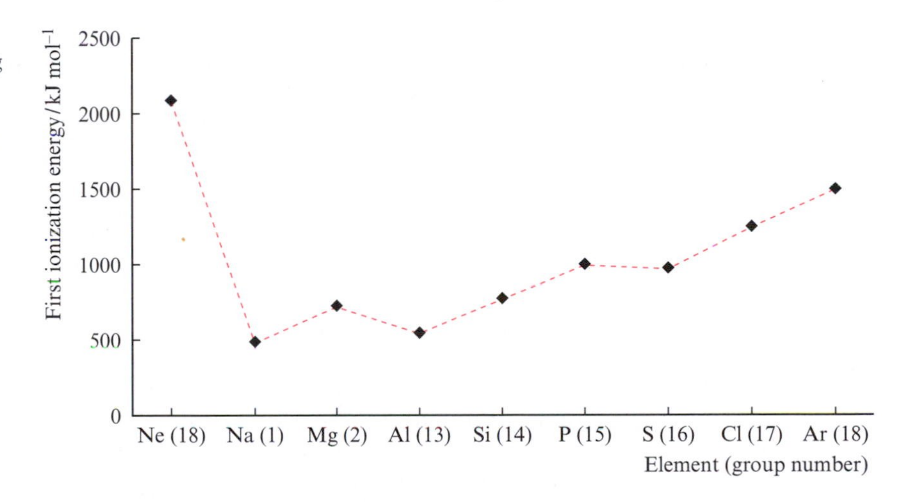

6.6 The trend in first ionization potentials on going from neon ($Z = 10$) to argon ($Z = 18$). The elements from sodium to argon make up the second period.

not surprsingly, the same arguments can be applied to explain the trend for the second period as the first, except that now we are dealing with the 3*s* and 3*p* atomic orbitals.

First ionization energies across the third period

The third period includes elements from the *d*-block as well as from the *s*- and *p*-blocks. The first ionization energies for the elements from argon (the last element in the second period) to krypton (the last element in the third period) are plotted in Figure 6.7. Compare this figure with Figures 6.4 and 6.6: the same pattern in values of IE_1 is observed on crossing the first, second and third rows of the *s*- and *p*-blocks.

The elements from scandium (Sc) to zinc (Zn) possess the ground state electronic configurations $[Ar]4s^2 3d^1$ to $[Ar]4s^2 3d^{10}$ respectively. In these elements, the marked similarity in the first ionization energies is associated with the closeness in energy of the 4*s* and 3*d* AOs. The rise on going from copper to zinc can be attributed to the particular stability of the zinc atom, the ground

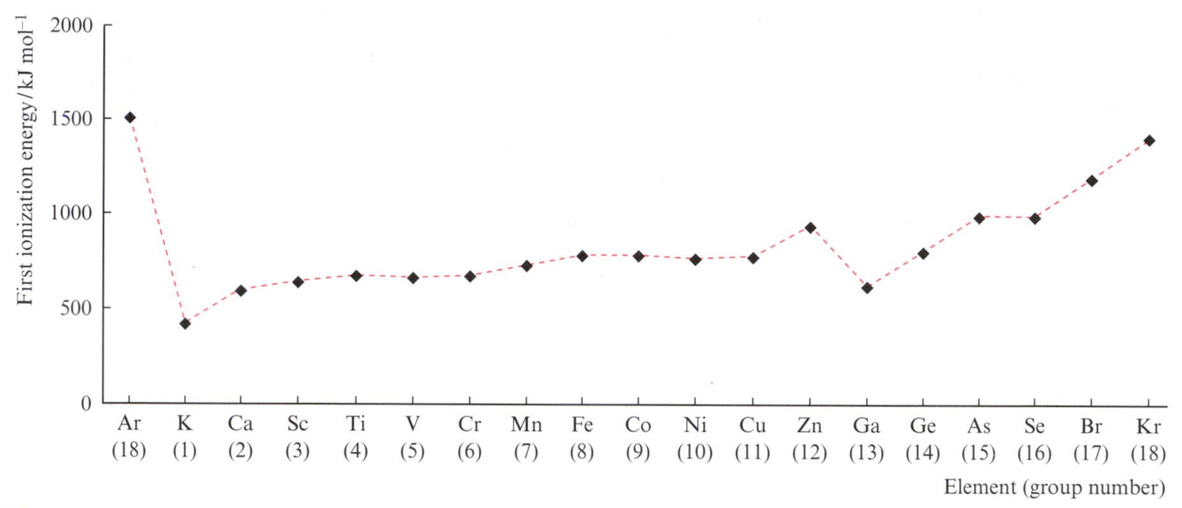

6.7 The trend in first ionization potentials on going from argon ($Z = 18$) to krypton ($Z = 36$). The elements from potassium to krypton make up the third period. The elements in groups 1 and 2 are in the *s* block, elements from groups 3 to 12 inclusive are in the *d*-block, and elements in groups 13 to 18 inclusive are in the *p*-block.

state electronic configuration of which is [Ar]$4s^2 3d^{10}$ with the 4*s* and 3*d* levels fully occupied. We return to *d*-block metal ions in Chapter 16.

First ionization energies down a group

If we look more closely at the graphs in Figures 6.4, 6.6 and 6.7, we observe that, although the *trends* for the first ionization energies of the *s*- and *p*-block elements are similar, the plots are displaced to lower energies on going from Figure 6.4 to 6.6, and from Figure 6.6 to 6.7. This new trend corresponds to a *general decrease in the values of the first ionization energy as a group is descended.*

Consider the elements in group 1. Values of IE_1 for lithium, sodium, potassium, rubidium and caesium are plotted in the left-hand graph in Figure 6.8.

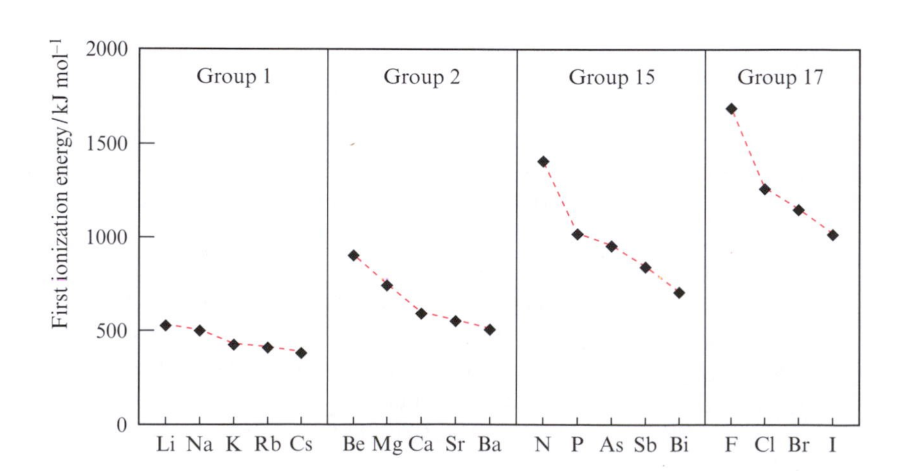

6.8 First ionization potentials decrease as a group in the *s*- or *p*-block is descended.

A gradual decrease in values is observed. The first ionization of lithium involves the removal of one electron from the 2*s* AO; for sodium, a 3*s* electron is lost, while for potassium, rubidium or caesium, a 4*s*, 5*s* or 6*s* electron is removed, respectively. The decrease in *IE*s reflects an increase in the distance between the nucleus and the valence electron with an increase in the principal quantum number.

We can apply the same arguments to explain the trends observed in the values of IE_1 for the elements in groups 2, 15 and 17 (Figure 6.8).

Successive ionization energies for an atom

So far, we have been concerned with the formation of singly charged cations (equation 6.1). For an alkali metal with a valence ground state configuration of ns^1, the loss of *more* than one electron requires a considerable amount of energy. The first and second ionizations of sodium involve the processes shown in equations 6.5 and 6.6.

$$\text{First ionization:}\quad \text{Na(g)} \rightarrow \text{Na}^+\text{(g)} + \text{e}^- \qquad (6.5)$$

$$\text{Second ionization:}\quad \text{Na}^+\text{(g)} \rightarrow \text{Na}^{2+}\text{(g)} + \text{e}^- \qquad (6.6)$$

The value of IE_1 for sodium is 496 kJ mol^{-1} but IE_2 is 4563 kJ mol^{-1}. This almost ten-fold increase is due to the fact that the second electron must be removed:

1. from a positively charged ion, and
2. from the fully occupied 2*p* level.

Discussions above about neon cover point (2) – Ne and Na^+ are isoelectronic. Point (1) addresses the point that the outer electrons in the Na^+ ion experience a greater effective nuclear charge than the outer electron in the sodium atom.

The first five ionization energies for sodium are plotted in Figure 6.9. Sharp increases are seen as successive core electrons are removed.

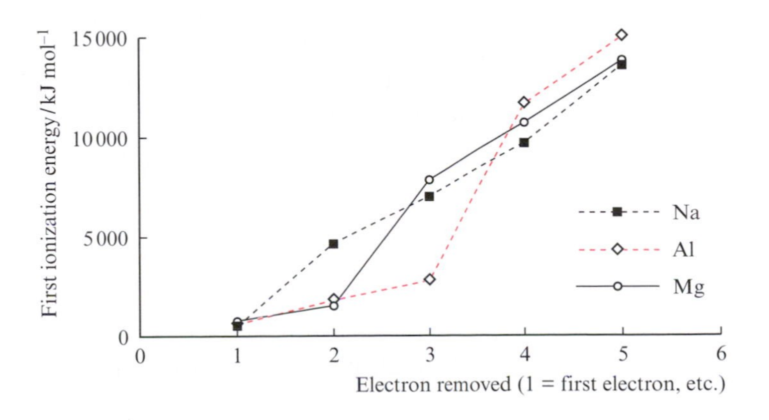

6.9 A plot of the first five ionization potentials for sodium (ground state configuration [Ne]3s^1), magnesium (ground state configuration [Ne]3s^2) and aluminium (ground state configuration [Ne]3$s^2$3p^1).

Consider now the successive ionizations of a magnesium atom, the ground state electronic configuration of which is [Ne]$3s^2$. The first five ionization energies are plotted in Figure 6.9; the first three values refer to the processes given in equations 6.7–6.9.

$$\text{First ionization:} \quad Mg(g) \rightarrow Mg^{+}(g) + e^{-} \tag{6.7}$$

$$\text{Second ionization:} \quad Mg^{+}(g) \rightarrow Mg^{2+}(g) + e^{-} \tag{6.8}$$

$$\text{Third ionization:} \quad Mg^{2+}(g) \rightarrow Mg^{3+}(g) + e^{-} \tag{6.9}$$

The plot of successive ionization energies for magnesium in Figure 6.9 shows an initial increase which is followed by a much sharper rise. The discontinuity occurs *after* the second electron has been removed (equation 6.8). Whereas the first two electrons are removed from the valence $3s$ shell, successive ionizations involve the removal of core electrons.

Similarly, a plot (Figure 6.9) of the first five ionization energies for aluminium (ground state electronic configuration [Ne]$3s^2 3p^1$) shows a gradual increase in energy corresponding to the loss of the three valence electrons, followed by a much sharper rise as we begin to remove core electrons.

In both magnesium and aluminium, it is harder to remove the second valence electron than the first because the second electron is being removed from a positively charged ion. Similarly, the energy required to remove the third valence electron from aluminium involves a contribution that reflects the work done in overcoming the attraction between the effective nuclear charge in the Al^{2+} ion and the electron to be removed (equation 6.10).

$$Al^{2+}(g) \rightarrow Al^{3+}(g) + e^{-} \tag{6.10}$$

Although it is tempting to rationalize the observed oxidation states of sodium, magnesium and aluminium (and of other elements) in terms of only the *successive* ionization energies, this is actually dangerous. As an exercise, you should calculate (using *IE* values from Appendix 8) the *sums* of the first two *IE*s for sodium (the *total* energy needed to form the Na^{2+} ion from a gaseous Na atom) and the first three *IE*s for aluminium (the *total* energy needed to form the Al^{3+} ion from a gaseous Al atom). Is it still obvious why, in compounds of these elements, the Al^{3+} ion may be present but the Na^{2+} ion is not? And why, for example, are Al^{2+} compounds very rare when the expenditure of energy to form this ion is less than that involved in forming Al^{3+}? The answer is that ionization energies are not the only factor governing the stability of a particular oxidation state. This reasoning will become clearer later in the chapter.

6.5 ELECTRON AFFINITY

Definition and sign convention

It is unfortunate (and certainly confusing) that the sign convention used for electron affinity values is the opposite of the normal convention used in thermodynamics. The first electron affinity (EA_1) is *minus* the energy internal change ($EA = -\Delta U(0\ K)$) which accompanies the gain of one electron by a *gaseous* atom (equation 6.11). The second electron affinity of atom Y refers to reaction 6.12.

$$Y(g) + e^- \rightarrow Y^-(g) \tag{6.11}$$

$$Y^-(g) + e^- \rightarrow Y^{2-}(g) \tag{6.12}$$

In a Hess cycle, it is the enthalpy change ($\Delta H(298\ K)$) associated with a process such as reaction 6.11 or 6.12 that is appropriate, rather than the change in internal energy ($\Delta U(0\ K)$). A correction can be made, but the difference between corresponding values of $\Delta H(298\ K)$ and $\Delta U(0\ K)$) is small. Thus, values of electron affinities are usually used directly in thermochemical cycles, remembering of course to ensure that the sign is appropriate: a negative enthalpy (ΔH), but a positive electron affinity (EA), corresponds to an exothermic process.§

Values of the enthalpy changes ($-EA$) associated with the attachment of an electron to an atom or negative ion are listed in Appendix 9. The units are kJ mol^{-1} or electron volts (eV).

1 eV = 96.5 kJ mol^{-1}

The first electron affinity (EA_1) of an atom is *minus* the internal energy change at 0 K ($-\Delta U(0\ K)$) which is associated with the gain of one electron by a gaseous atom:

$$Y(g) + e^- \rightarrow Y^-(g)$$

For thermochemical cycles, an associated *enthalpy* is used:

$$\Delta H(298\ K) \approx \Delta U(0\ K) = -EA$$

Worked example 6.1 ***Enthalpy of formation of an anion***

The bond dissociation enthalpy (at 298 K) of chlorine is 242 kJ mol^{-1} and the first electron affinity (EA) of chlorine is 3.61 eV. Determine the standard enthalpy of formation (at 298 K) of the chloride ion. (1 eV = 96.5 kJ mol^{-1}.)

First, ensure that the standard sign convention is used, and that the units are consistent. The data given refer to the following processes.

Bond dissociation:

$Cl_2(g) \rightarrow 2Cl(g)$
$D = 242$ kJ mol^{-1}

This is an *enthalpy* value, with a conventional use of sign. Breaking the Cl–Cl bond is an endothermic process.

§ Some tables of data list enthalpy values and others, electron affinities. Particular attention must be paid to such data before the numbers are used in calculations.

Gaining an electron:

$Cl(g) + e^- \rightarrow Cl^-(g)$
$EA = 3.61\ \text{eV} = (3.61 \times 96.5)\ \text{kJ mol}^{-1} = 348\ \text{kJ mol}^{-1}$

Firstly, the sign convention for *EA* is the reverse of the convention used in thermodynamics. When a chlorine atom gains one electron, the process is exothermic.

Secondly, *EA* is not an enthalpy term, but is an internal energy, specifically:

$$-EA = \Delta U(0\ \text{K}) \approx \Delta H(298\ \text{K})$$

$$\Delta H(298\ \text{K}) \approx -EA \approx -348\ \text{kJ mol}^{-1}$$

The aim of the question is to find the *standard enthalpy of formation* ($\Delta_f H^o$ at 298 K) of the chloride ion. This is the enthalpy change for the process:

$\frac{1}{2}Cl_2(g) + e^- \rightarrow Cl^-(g)$
standard
state

Set up a Hess cycle:

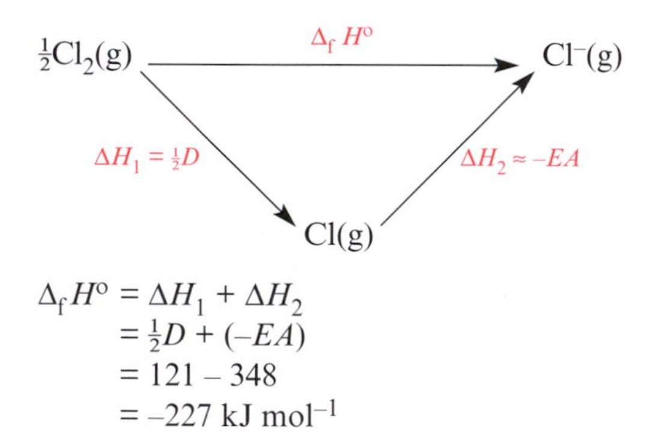

$$\begin{aligned}\Delta_f H^o &= \Delta H_1 + \Delta H_2 \\ &= \tfrac{1}{2}D + (-EA) \\ &= 121 - 348 \\ &= -227\ \text{kJ mol}^{-1}\end{aligned}$$

The enthalpy changes accompanying the attachment of the first and second electrons

Values of the enthalpy changes which accompany the attachment of one electron to a range of *s*- and *p*-block *atoms* are included in Table 6.1. With one exception, the values are negative indicating that the reactions are exothermic; for nitrogen, the enthalpy change is approximately zero. Enthalpy changes for the attachment of an electron to the gaseous O^- and S^- anions are also given: both are *endothermic* reactions.

When an electron is added to an *atom*, there is repulsion between the incoming electron and the valence shell electrons, but there will also be an attraction between the nucleus and the extra electron. In general, the overall enthalpy change for the process shown in equation 6.11 is negative.

Table 6.1 Approximate enthalpy changes (ΔH(298 K)) associated with the gain of one electron to a gaseous atom or anion. The values are actually internal energy changes (ΔH(0 K)) as explained in the text.

Process	$\approx \Delta H / kJ\ mol^{-1}$
$H(g) + e^- \rightarrow H^-(g)$	−73
$Li(g) + e^- \rightarrow Li^-(g)$	−60
$Na(g) + e^- \rightarrow Na^-(g)$	−53
$K(g) + e^- \rightarrow K^-(g)$	−48
$N(g) + e^- \rightarrow N^-(g)$	≈ 0
$P(g) + e^- \rightarrow P^-(g)$	−72
$O(g) + e^- \rightarrow O^-(g)$	−141
$O^-(g) + e^- \rightarrow O^{2-}(g)$	+798
$S(g) + e^- \rightarrow S^-(g)$	−201
$S^-(g) + e^- \rightarrow S^{2-}(g)$	+640
$F(g) + e^- \rightarrow F^-(g)$	−328
$Cl(g) + e^- \rightarrow Cl^-(g)$	−349
$Br(g) + e^- \rightarrow Br^-(g)$	−325
$I(g) + e^- \rightarrow I^-(g)$	−295

When an electron is added to an *anion*, the repulsive forces are significant and energy must be provided to overcome this repulsion. As a result, the enthalpy change is positive.

Now consider the formation of a gaseous oxide ion O^{2-} from an atom of oxygen (equation 6.13). The enthalpy change for the overall process is the sum of the values of ΔH for the two individual electron attachments (equation 6.14).

$$O(g) + 2e^- \rightarrow O^{2-}(g) \quad (6.13)$$

$$\Delta H[O(g) \rightarrow O^{2-}(g)] = \Delta H[O(g) \rightarrow O^-(g)] + \Delta H[O^-(g) \rightarrow O^{2-}(g)] \quad (6.14)$$

$$= -141 + 798$$
$$= +657 \text{ kJ mol}^{-1}$$

The formation of the oxide ion from an oxygen atom in the gas phase is therefore a highly endothermic process and may appear unfavourable. In Section 6.16 (by examining the formation of KCl) we can appreciate that this endothermic change must be offset by an exothermic one if solid state oxides are to be formed.

6.6 ELECTROSTATIC INTERACTIONS BETWEEN IONS

Consider the formation of a gas phase *ion-pair*, X^+Y^-. It is important to remember that in any isolable chemical compound, a *cation cannot occur without an anion*.§ In a simple process, an electron lost in the formation of a

§ In one or two remarkable cases, the anion is not a *chemical* species, but an electron. An example is a solution of sodium in liquid ammonia – see Section 11.12.

The term *ion-pair* refers to a single pair of oppositely charged ions; in sodium chloride, Na^+Cl^- is an ion-pair

singly charged cation, X^+, is transferred to another atom to form an anion, Y^-. The overall reaction is given in equation 6.15.

$$\left.\begin{array}{l} X(g) \rightarrow X^+(g) + e^- \\ Y(g) + e^- \rightarrow Y^-(g) \end{array}\right\} \Rightarrow XY(g) \qquad (6.15)$$

Electrostatic (Coulombic) attraction

A spherical ion can be treated as a *point charge*. Electrostatic (or *Coulombic*) interactions operate between these point charges; oppositely charged ions attract one another, while similarly charged ions repel each other. Although isolated pairs of ions are not usually encountered, we initially consider the attraction between a pair of Na^+ and Cl^- ions for the sake of simplicity. Equation 6.16 shows the formation of an Na^+ ion and a Cl^- ion, and the subsequent formation of an isolated ion-pair in the gas phase.

$$\left.\begin{array}{l} Na(g) \rightarrow Na^+(g) + e^- \\ Cl(g) + e^- \rightarrow Cl^-(g) \end{array}\right\} \Rightarrow NaCl(g) \qquad (6.16)$$

A spherical ion can be treated as a *point charge*.

Electrostatic (Coulombic) interactions operate between ions. Oppositely charged ions attract one another and ions of like charge repel each other.

The change in internal energy, $\Delta U(0\ K)$, which is associated with the electrostatic attraction between the two isolated ions Na^+ and Cl^- in NaCl is given by equation 6.17, and the units of $\Delta U(0\ K)$ from this equation are joules.

$$\text{For an isolated ion-pair: } \Delta U(0\ K) = -\left(\frac{|z_+| \times |z_-| \times e^2}{4 \times \pi \times \varepsilon_0 \times r}\right) \qquad (6.17)$$

where:

$|z_+|^\S$ = modulus of the positive charge (for Na^+, $|z_+| = 1$; for Ca^{2+}, $|z_+| = 2$)

$|z_-|^\S$ = modulus of the negative charge (for Cl^-, $|z_-| = 1$; for O^{2-}, $|z_-| = 2$)

e = charge on the electron = 1.602×10^{-19} C

ε_0 = permittivity of a vacuum = 8.854×10^{-12} F m^{-1}

r = internuclear distance between the ions (units = m)

If we have a *mole of sodium chloride* in which each Na^+ ion interacts with *only one* Cl^- ion, and each Cl^- ion interacts with *only one* Na^+ ion, equation

§ The modulus of a real number is its positive value; although the charges z_+ and z_- have positive and negative values respectively, $|z_+|$ and $|z_-|$ are both positive.

Avogadro constant: see Section 1.8

6.17 is corrected by multiplying by the Avogadro constant (L = 6.022 × 10^{23} mol^{-1}) to give equation 6.18, in which the units of ΔU(0 K) are J mol^{-1}.

$$\text{For a mole of ion-pairs: } \Delta U(0\text{ K}) = -\left(\frac{L \times |z_+| \times |z_-| \times e^2}{4 \times \pi \times \varepsilon_0 \times r}\right) \qquad (6.18)$$

Worked example 6.2 ***Coulombic attractions between ions***

Calculate the internal energy change, ΔU(0 K), which is associated with the attractive forces operating between the cations and anions in a mole of ion-pairs of sodium chloride in the gas phase, if the internuclear distance between an Na^+ and Cl^- ion in each ion-pair is 236 pm.

Data required: $L = 6.022 \times 10^{23}$ mol^{-1}, $e = 1.602 \times 10^{-19}$ C and $\varepsilon_0 = 8.854 \times 10^{-12}$ F m^{-1}.

$$\text{For a mole of ion-pairs: } \Delta U(0\text{ K}) = -\left(\frac{L \times |z_+| \times |z_-| \times e^2}{4 \times \pi \times \varepsilon_0 \times r}\right) \text{ J mol}^{-1}$$

but, remember that in this equation, the distance r is in m:

r = 236 pm = 2.36×10^{-10} m

$$\Delta U(0\text{ K}) = \left(\frac{6.022 \times 10^{23} \times 1 \times 1 \times (1.602 \times 10^{-19})^2}{4 \times 3.142 \times 8.854 \times 10^{-12} \times 2.36 \times 10^{-10}}\right)$$

$$= -588\ 502 \text{ J mol}^{-1}$$

$$= -588.5 \text{ kJ mol}^{-1}$$

Coulombic repulsion

If ion-pairs are close to each other, then one cation in one ion-pair will repel cations in other ion-pairs, and anions will repel other anions.

The internal energy associated with Coulombic repulsions can be calculated using a form of equation 6.17 (or equation 6.18 on a molar scale). However, while the internal energy associated with the *attractive* forces has a *negative* value (stabilizing), that associated with the *repulsive* forces has a *positive* value (destabilizing). If we are dealing only with repulsive interactions, equation 6.19 is appropriate.

$$\text{For two similarly charged ions: } \Delta U(0\text{ K}) = +\left(\frac{|z_1| \times |z_2| \times e^2}{4 \times \pi \times \varepsilon_0 \times r}\right) \qquad (6.19)$$

where $|z_1|$ and $|z_2|$ are the moduli of the charges on the two ions.

The repulsive forces between ions in different ion-pairs become important *only* when the ion-pairs are relatively close together, and so for the gas phase ion-pairs that we have been discussing, we can effectively ignore Coulombic repulsions. However, in the solid state their contribution *is* important.

6.7 IONIC LATTICES

Ions in the solid state

In gaseous sodium chloride, each ion-pair consists of one Na^+ cation and one Cl^- anion. In a 'thought-experiment', we can condense the gas phase ion-pairs to give solid sodium chloride. The cations and anions will now be close to each other, and if we consider a purely electrostatic model with ions as point charges, then each Na^+ cation experiences Coulombic attractive forces from all nearby Cl^- anions. Conversely, each Cl^- anion is attracted to all the nearby Na^+ cations. At the same time, any chloride ions which are close will repel one another, and any Na^+ ions which approach one another will be repelled.

Ionic lattices can be described in terms of the close-packing of spheres: see Chapter 7

The net result of the electrostatic attractions and repulsions is the formation of an ordered array of ions called an *ionic lattice* in which the internal energy associated with the electrostatic forces is minimized. This is achieved by the systematic arrangement of ions such that, wherever possible, ions of *opposite* charge are adjacent.

For many compounds which contain simple ions, the ions are arranged so as to give one of several general *ionic lattice-types*. Each lattice-type is described by a generic name – that of a compound (e.g. sodium chloride or caesium chloride) or a mineral that possesses that particular structure (e.g. fluorite, rutile, zinc blende or wurtzite). In this chapter we have chosen some of the more common lattice-types to illustrate the different ways in which ions can be arranged in the solid state.[§] For example, a description of the sodium chloride lattice not only indicates the way in which Na^+ and Cl^- ions are arranged in crystalline NaCl, but it also serves to describe the way in which Ca^{2+} and O^{2-} ions are arranged in solid CaO.

In the solid state, a compound which is composed of ions forms an ordered structure called an *ionic lattice*. There are a number of common lattice-types which are adopted by compounds containing simple ions.

The unit cell

The smallest repeating unit in an ionic lattice is a *unit cell*.

An ionic lattice may be extended indefinitely in three dimensions, and is composed of an infinite number of repeating units. Such a building-block is called a *unit cell* and must be large enough to carry *all* the information necessary to construct *unambiguously* an infinite lattice (see problem 6.10). We illustrate how the concept of a unit cell works by considering several common lattice types in the following sections.

§ For detailed information about a far wider range of ionic lattices, see: A. F. Wells, *Structural Inorganic Chemistry*, 5th edn, Oxford University Press, Oxford (1984).

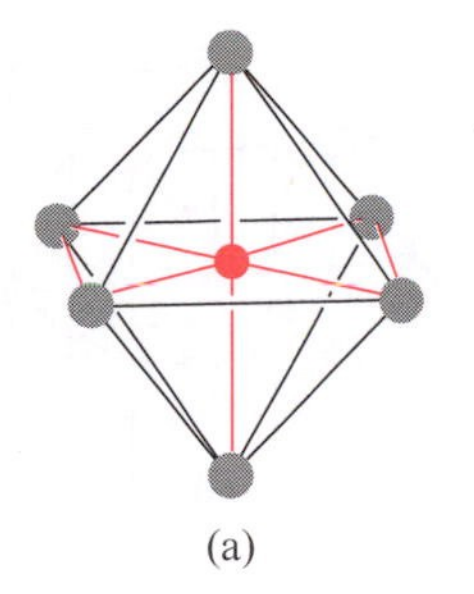

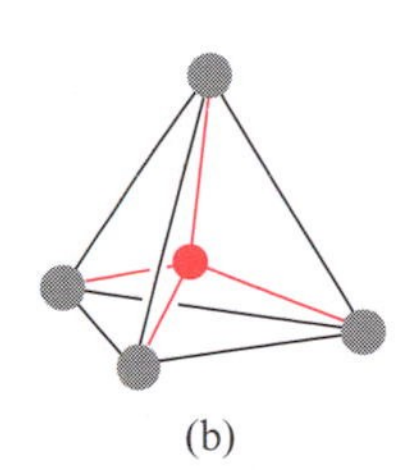

6.10 (a) A six-coordinate ion in an octahedral environment; (b) a four-coordinate ion in a tetrahedral environment.

Coordination number

The coordination number of an ion in a lattice is the number of closest ions of opposite charge.

In the following sections, we shall look at the structures of NaCl, CsCl, CaF_2, TiO_2 and ZnS. Within the lattices of these compounds, ions are found with coordination numbers of three, four, six and eight. Figure 6.10a shows the octahedral environment around a six-coordinate ion (shown in red). In Figure 6.10b, the red ion has a coordination number of four — four ions of opposite charge (the grey ions) are arranged tetrahedrally around the central ion.

The *coordination number* of an ion in a lattice is the number of closest ions of opposite charge.

6.8 THE SODIUM CHLORIDE (ROCK SALT) LATTICE

Figure 6.11a shows one unit cell of the sodium chloride structure. The lattice is *cubic*, with alternating Na^+ (shown in red) and Cl^- ions (shown in grey).

The unit cell has a chloride ion at its centre.§ There are six Na^+ ions immediately next to this Cl^- ion, making the chloride ion six-coordinate and octahedrally sited (Figure 6.11b).

Now consider an Na^+ ion at the centre of a square face in Figure 6.11a. If we extend the lattice by adding another unit cell to that face (Figure 6.11c), the coordination sphere around that sodium ion is completed. It too is octahedral (Figure 6.11d), and the coordination number of the Na^+ ion is six.

Stoichiometry: see Section 1.15

Selected compounds that possess the same lattice-type in the solid state as sodium chloride are listed in Table 6.2. Note that each has a 1 : 1 stoichiometry – the ratio of cations to anions is one to one.

§ An alternative unit cell could be drawn with a sodium ion at the centre, since all of the sites in the lattice are equivalent.

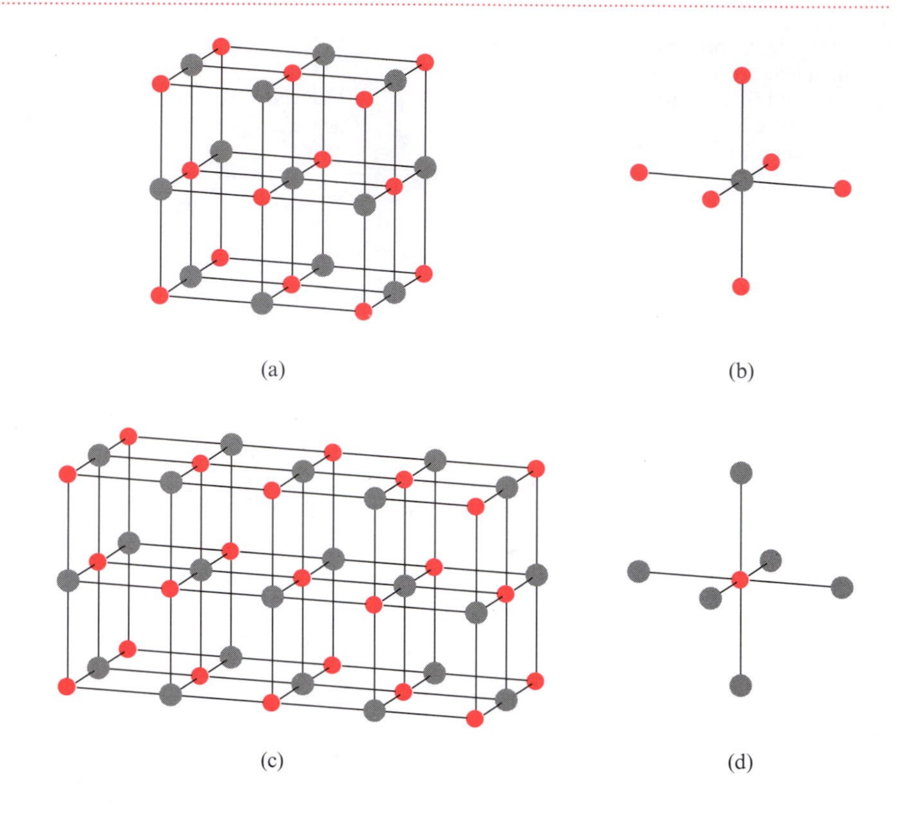

6.11 (a) A unit cell of the sodium chloride lattice; Na^+ ions are shown in red, Cl^- ions are shown in grey. This unit cell is drawn placing a chloride ion at the centre; equally, we could construct a unit cell with a sodium ion at the centre. (b) The coordination number of each chloride ion is six. (c) Two units cells of sodium chloride with one shared face. (d) The coordination number of each sodium ion is six.

Table 6.2 Selected compounds which possess a sodium chloride-type structure in the solid state (see Figure 6.11) under room temperature conditions.

Compound	*Formula*	*Cation*	*Anion*
Sodium chloride	NaCl	Na^+	Cl^-
Sodium fluoride	NaF	Na^+	F^-
Sodium hydride	NaH	Na^+	H^-
Lithium chloride	LiCl	Li^+	Cl^-
Potassium bromide	KBr	K^+	Br^-
Potassium iodide	KI	K^+	I^-
Silver fluoride	AgF	Ag^+	F^-
Silver chloride	AgCl	Ag^+	Cl^-
Magnesium oxide	MgO	Mg^{2+}	O^{2-}
Calcium oxide	CaO	Ca^{2+}	O^{2-}
Barium oxide	BaO	Ba^{2+}	O^{2-}
Iron(II) oxide	FeO	Fe^{2+}	O^{2-}
Magnesium sulfide	MgS	Mg^{2+}	S^{2-}
Lead(II) sulfide	PbS	Pb^{2+}	S^{2-}

6.9 DETERMINING THE STOICHIOMETRY OF A COMPOUND FROM THE LATTICE: NaCl

We stated in section 6.7 that we must be able to generate an infinite lattice from the information provided in the unit cell. It follows that the stoichiometry of an ionic compound can be determined from its unit cell.

In Figure 6.11a, the number of Na^+ ions does not equal the number of Cl^- ions. However, since the diagram is of a unit cell, it *must* indicate that the ratio of $Na^+ : Cl^-$ ions in sodium chloride is 1 : 1. Let us look more closely at the structure. The important feature is that in the NaCl lattice, the unit cell shown in Figure 6.11a is immediately surrounded by six other unit cells. One of these adjacent cells is shown in Figure 6.11c.

When unit cells are fused together, ions which are at points of fusion are *shared* between more than ion cell. We can identify three different categories of sites of fusion and these are illustrated for a cubic unit cell in Figure 6.12.

The left-hand side of Figure 6.12a shows two unit cells, each of which has an ion sited in the *centre of a face*. When these units cells are joined together, the ion is *shared between two cells*. Figure 6.12b shows that for an ion sited in

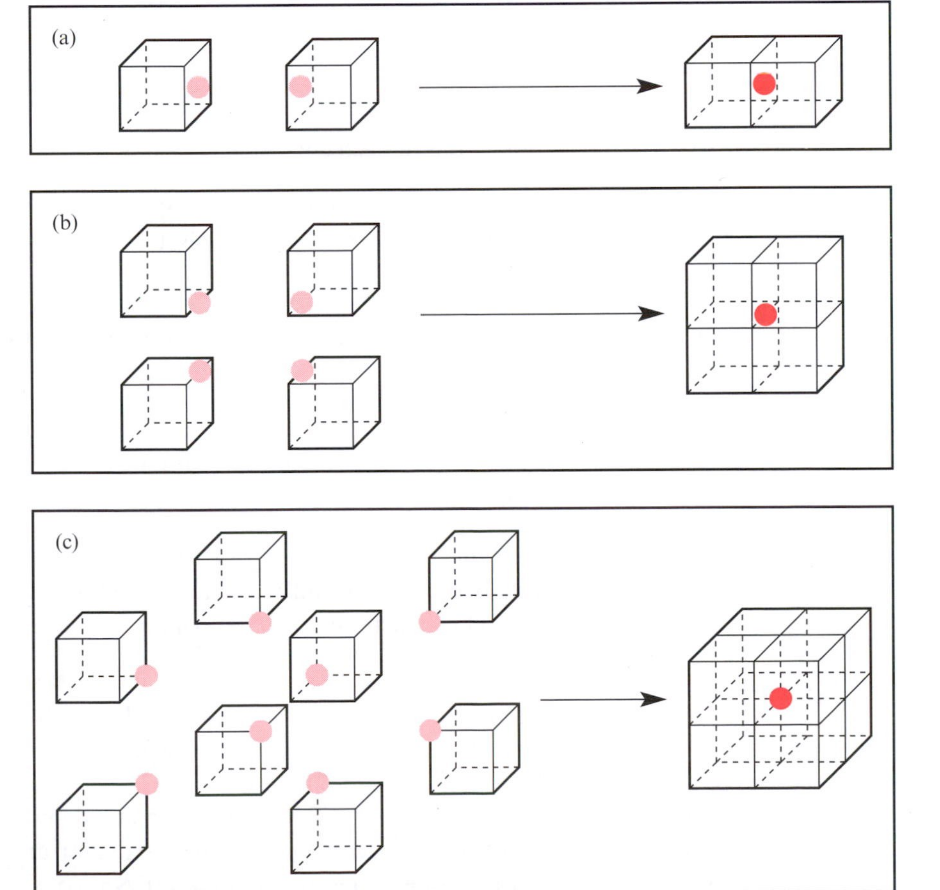

6.12 The sharing of ions between cubic unit cells. (a) An ion at the centre of a face is shared between two unit cells. (b) An ion in the middle of an edge is shared between four unit cells. (c) An ion in a corner site is shared between eight unit cells.

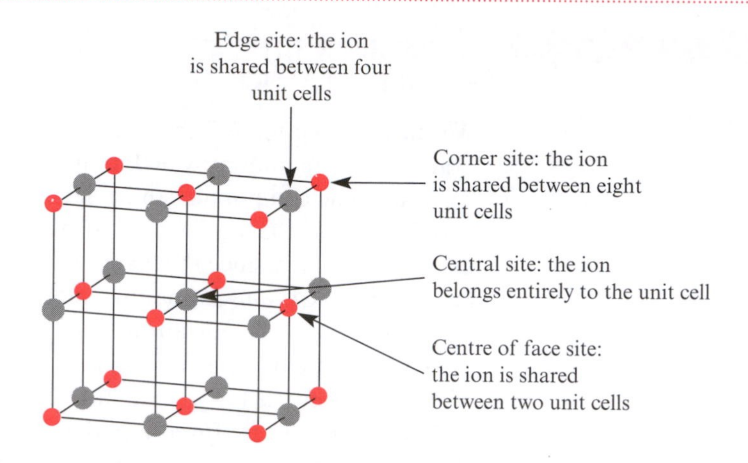

6.13 In a unit cell of sodium chloride, there are four different sites in which an ion can reside. There is one unique site at the centre of the unit cell; this ion (a Cl^- ion shown in grey) belongs entirely to the unit cell. There are 6 sites at the centres of faces, 12 sites situated in the middle of edges and 8 corner sites.

the middle of an *edge*, the combination of unit cells means that the ion is *shared between four cells*. In Figure 6.12c, we consider what happens to an ion at the *corner* of a unit cell, and when we construct a larger part of the lattice around the unit cell, this ion is *shared between eight unit cells*.

Let us now apply this to the unit cell of NaCl shown in Figure 6.13. Firstly, we must identify the different types of site. These are:

- the unique central site;
- 6 sites, one at the centre of each face;
- 12 sites, one in the middle of each edge; and
- 8 sites, one at each corner.

Next, we must determine what fraction of each ion actually 'belongs' to this one unit cell — how does the fusion of adjacent unit cells affect the ions in the different sites?

The unique central site This ion is not shared with any other unit cell. In Figure 6.13, the central Cl^- ion belongs wholly to the unit cell shown.

The sites at the centres of the faces An ion in this site is shared between two unit cells. In Figure 6.13, there are six such ions (all Na^+ cations) and half of each ion belongs to the unit cell shown.

The edge sites When unit cells are fused together to form an infinite lattice, each edge site is shared between four unit cells. In Figure 6.13, there are 12 such ions, all Cl^- anions. One quarter of each ion belongs to the unit cell shown.

The corner sites When unit cells are placed together in a lattice, each corner site is shared between eight unit cells. In Figure 6.13, there are eight such Na^+ ions, and one-eighth of each ion belongs to the unit cell shown.

The stoichiometry of sodium chloride is determined by summing the fractions of ions which belong to one particular unit cell and Table 6.3 shows how this is done. The total number of Na^+ ions belonging to the unit cell is

Table 6.3 Determination of the stoichiometry of sodium chloride from a consideration of the structure of the unit cell.

Site (see Figure 6.13)	*Number of Na^+ ions*	*Number of Cl^-*
Central	0	1
Centre of face	$(6 \times \frac{1}{2}) = 3$	0
Edge	0	$(12 \times \frac{1}{4}) = 3$
Corner	$(8 \times \frac{1}{8}) = 1$	0
Total	**4**	**4**

four and this is also the number of Cl^- ions in the unit cell. The ratio of Na^+: Cl^- ions is 4 : 4, or 1 : 1, and it follows that the empirical formula of sodium chloride is NaCl.

6.10 THE CAESIUM CHLORIDE LATTICE

Body-centred cubic structure: see also Section 7.3

A unit cell of caesium chloride, CsCl, is shown in Figure 6.14a. This is a *body-centred cubic* unit. In the figure, the Cs^+ ion (shown in red) is the central ion and is eight-coordinate.

In Figure 6.14b, the lattice is extended into parts of adjacent unit cells to illustrate the environment of the chloride ion (shown in grey). It too is eight-coordinate, and is situated at the centre of a cubic array of caesium ions. This confirms that we could have drawn a unit cell for the CsCl lattice with either a Cs^+ or a Cl^- ion at the centre of the cube.

We can confirm the stoichiometry of caesium chloride from the arrangement of ions in the unit cell as follows. In Figure 6.14a, the Cs^+ ion is in the centre of the unit cell and it belongs in its entirety to this cell. There are eight Cl^- ions, each in a corner site and so the number of Cl^- ions belonging to a unit cell is $(8 \times \frac{1}{8}) = 1$. The ratio of Cs^+ : Cl^- ions is 1 : 1.

In the solid state, thallium(I) chloride also adopts a CsCl-type lattice with the thallium(I) centres replacing the Cs^+ ions.

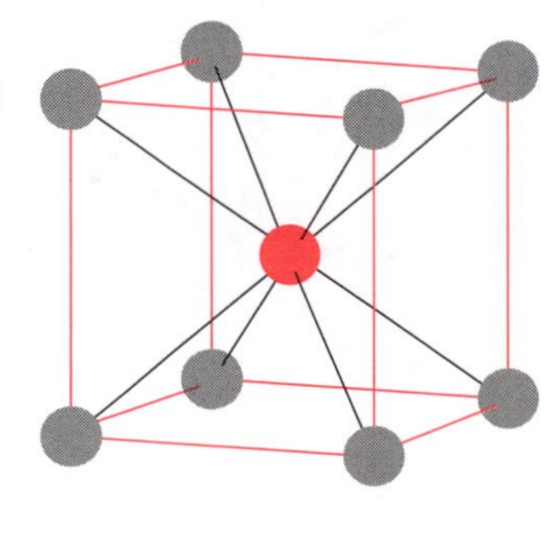

(a)

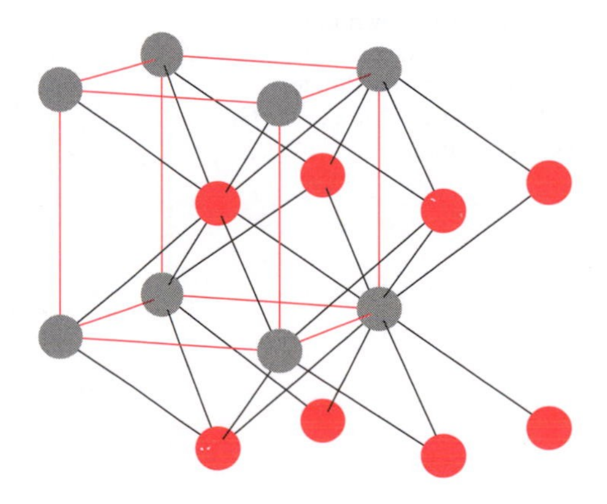

(b)

6.14 (a) A unit cell of the caesium chloride lattice; Cs^+ ions are shown in red and Cl^- ions in grey. The central Cs^+ ion is eight-coordinate. (b) An extension of the unit cell shows that each Cl^- ion is also eight-coordinate.

6.11 THE FLUORITE (CALCIUM FLUORIDE) LATTICE

The mineral fluorite is calcium fluoride, CaF_2, a unit cell of which is shown in Figure 6.15a. Note that the arrangement of the Ca^{2+} ions (shown in red) is the same as the arrangement of the Na^+ ions in NaCl (Figure 6.11a). However, the relationship between the positions of the F^- (shown in grey) and Ca^{2+} ions in fluorite is different from that of the Cl^- and Na^+ ions in NaCl.

Each fluoride ion in fluorite is in a four-coordinate, tetrahedral site (Figure 6.15b). The coordination number of the Ca^{2+} ions can only be seen if we extend the structure into the next unit cell, and this is best done by considering one of the Ca^{2+} ions at the centre of a face. It is at the centre of a cubic arrangement of F^- ions and is eight-coordinate (Figure 6.15c).

The number of Ca^{2+} and F^- ions per unit cell in fluorite are calculated in equations 6.20 and 6.21 (see problem 6.11). The ratio of $Ca^{2+} : F^-$ ions is therefore 4 : 8, or 1 : 2, and this confirms a formula for calcium fluoride of CaF_2.

$$\text{Number of } Ca^{2+} \text{ ions per unit cell of fluorite} = (6 \times \tfrac{1}{2}) + (8 \times \tfrac{1}{8}) = 4 \quad (6.20)$$

$$\text{Number of } F^- \text{ ions per unit cell of fluorite} = (8 \times 1) = 8 \quad (6.21)$$

Some compounds which adopt a fluorite-type of lattice in the solid state are listed in Table 6.4.

Table 6.4 Selected compounds with a fluorite-type structure (a CaF_2 lattice) in the solid state (see Figure 6.15) under room temperature conditions.

Compound	*Formula*	*Compound*	*Formula*
Calcium fluoride	CaF_2	Lead(II) fluoride	β-PbF_2
Barium fluoride	BaF_2	Zirconium(IV) oxide	ZrO_2
Barium chloride	$BaCl_2$	Hafnium(IV) oxide	HfO_2
Mercury(II) fluoride	HgF_2	Uranium(IV) oxide	UO_2

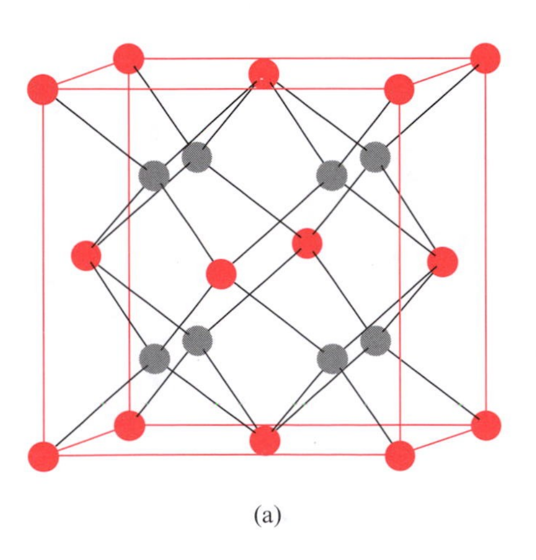

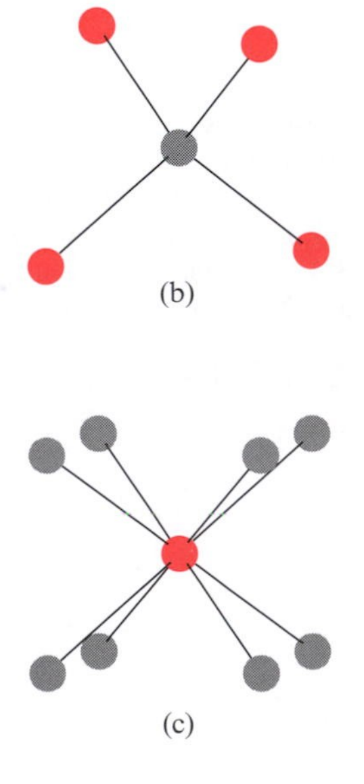

6.15 (a) A unit cell of the fluorite (CaF_2) lattice; Ca^{2+} ions are shown in red, Cl^- ions in grey. (b) Each Cl^- ion is four-coordinate, and (c) each Ca^{2+} ion is eight-coordinate.

6.12 THE RUTILE (TITANIUM(IV) OXIDE) LATTICE

Cuboid: see Appendix 15

The three lattices discussed so far possess cubic unit cells. In contrast, the solid state structure of the mineral rutile (titanium(IV) oxide, TiO_2) has a unit cell which is a *cuboid* (Figure 6.16a).

Eight titanium(IV) centres are sited at the corners of the unit cell and one resides in the centre. Each of these is six-coordinate, in an octahedral environment (Figure 6.16b). The structure must be extended in several directions in order to appreciate that the titanium centres in the corner sites of the unit cell are octahedrally coordinated in the infinite lattice. Each oxide ion is three-coordinate and is essentially trigonal planar (Figure 6.16c).

The stoichiometry of rutile can be confirmed from the solid state structure. The method of calculating the number of titanium(IV) and oxide ions in the unit cell is shown in equations 6.22 and 6.23. The ratio of titanium : oxygen centres in rutile is therefore 2 : 4, or 1 : 2. This confirms a formula for titanium(IV) oxide of TiO_2.

$$\text{Number of Ti(IV) centres per unit cell of rutile} = (1 \times 1) + (8 \times \tfrac{1}{8}) = 2 \quad (6.22)$$

$$\text{Number of O}^{2-}\text{ ions per unit cell of rutile} = (4 \times \tfrac{1}{2}) + (2 \times 1) = 4 \quad (6.23)$$

Table 6.5 lists some compounds which adopt a rutile-type of lattice in the solid state.

Table 6.5 Selected compounds which possess a rutile-type structure (a TiO_2 lattice) in the solid state (see Figure 6.16) under room temperature conditions.

Compound	*Formula*	*Compound*	*Formula*
Titanium(IV) oxide	TiO_2	Lead(IV) oxide	PbO_2
Manganese(IV) oxide	β-MnO_2	Magnesium fluoride	MgF_2
Tin(IV) oxide	SnO_2	Iron(II) fluoride	FeF_2

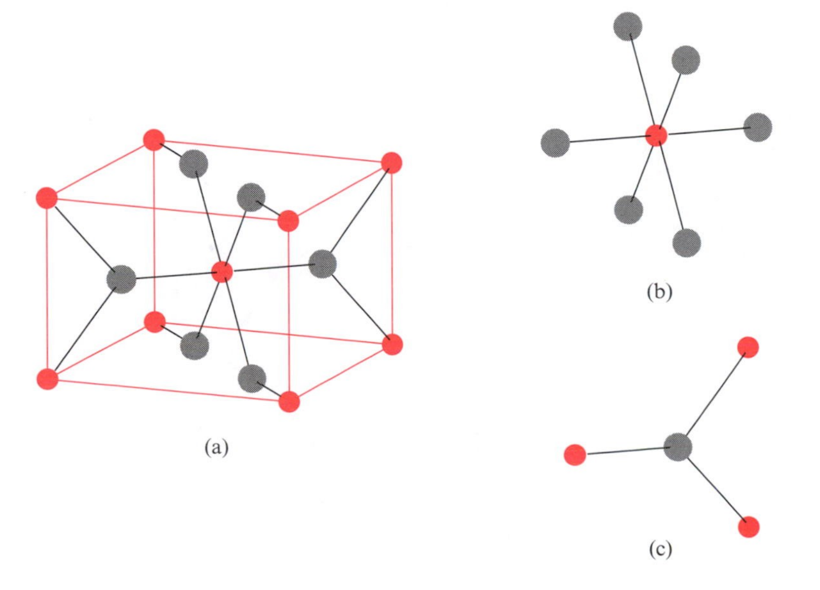

6.16 (a) A unit cell of the rutile (TiO_2) structure with the titanium centres shown in red and oxide ions in grey. (b) The coordination number of each titanium(IV) centre is six, and (c) each oxide ion is three-coordinate.

Box 6.2 Titanium(IV) oxide: Brighter than white

Titanium(IV) oxide is a white crystalline, non-toxic compound which is thermally stable. Its applications as a white pigment are widespread – TiO_2 is responsible for the brightness of 'brilliant white' paintand white paper. It has now replaced toxic lead-, antimony- or zinc-containing compounds which have previously been used as pigments. The essential optical properties that make TiO_2 so valuable are that it does not absorb visible light (see Section 9.5) and that it has an extremely high refractive index.

6.13 THE STRUCTURES OF ZINC(II) SULFIDE

There are two structural forms of zinc(II) sulfide (ZnS) in the solid state – the lattice-type adopted by one is called *zinc blende*, and that of the other is *wurtzite*.

The zinc blende lattice

A unit cell of zinc blende is drawn in Figure 6.17a. If we compare this with Figures 6.11a and 6.15a, we see that the Zn^{2+} ions (shown in red) in zinc blende are arranged in the same way as the Na^+ ions in NaCl and the Ca^{2+} ions in CaF_2. However, the arrangement of the S^{2-} ions (shown in grey) in zinc blende differs from those of the anions in either NaCl or CaF_2. Each sulfide ion in zinc blende is four-coordinate (tetrahedral) (Figure 6.17b).

None of the zinc(II) ions lies in a central site in the unit cell, and so the coordination environment can only be seen if the lattice is extended. Each Zn^{2+} ion is four-coordinate (tetrahedral) as shown in Figure 6.17c.

The stoichiometry of zinc blende can be confirmed by considering the structure of the unit cell. From equations 6.24 and 6.25, the ratio of Zn^{2+} : S^{2-} ions is 4 : 4, or 1 : 1, giving an empirical formula of ZnS.

$$\text{Number of } Zn^{2+} \text{ ions per unit cell of zinc blende} = (6 \times \tfrac{1}{2}) + (8 \times \tfrac{1}{8}) = 4 \quad (6.24)$$

$$\text{Number of } S^{2-} \text{ ions per unit cell of zinc blende} = (4 \times 1) = 4 \quad (6.25)$$

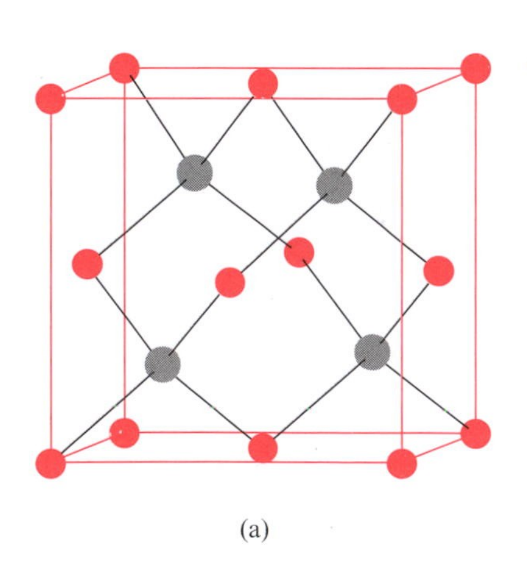

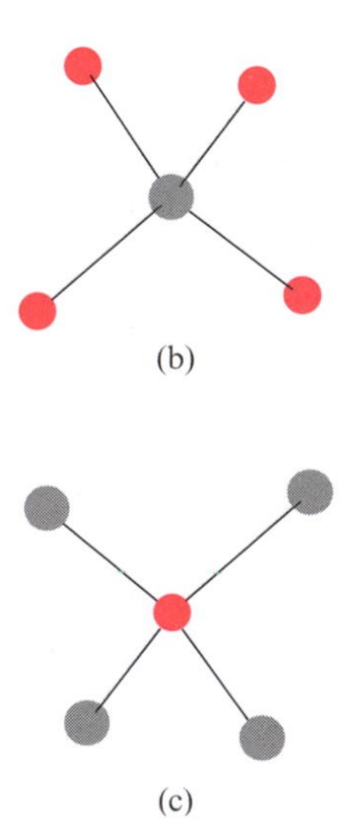

6.17 (a) A unit cell of the zinc blende lattice with the Zn^{2+} ions shown in red and the S^{2-} ions in grey. (b) Each sulfide ion is four-coordinate and (c) each Zn^{2+} ion is four-coordinate. Both sites are tetrahedral.

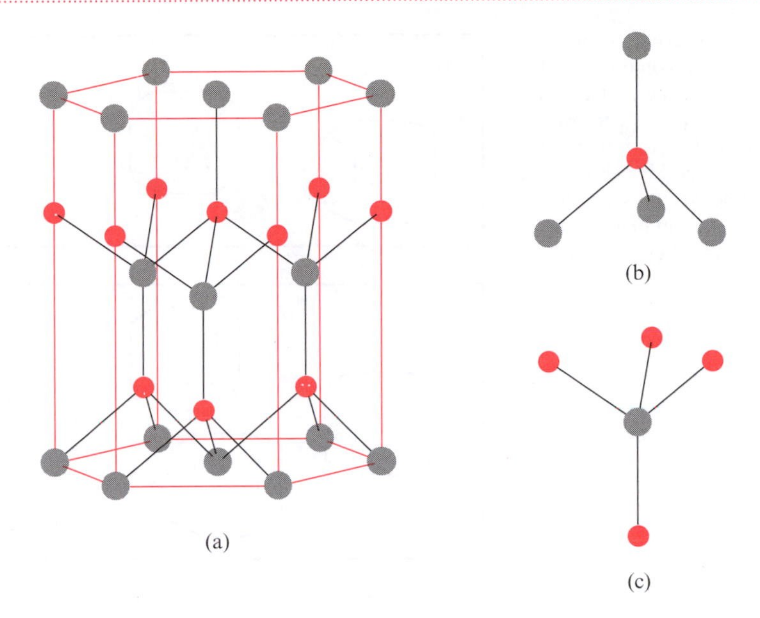

6.18 (a) A unit cell of the wurtzite lattice; the zinc ions are shown in red and the sulfide ions in grey. (b) Each sulfur centre is four-coordinate and (c) each Zn^{2+} ion is four-coordinate. Both sites are tetrahedral.

The wurtzite lattice

Hexagonal close-packed structure: see Section 7.2

Figure 6.18a shows a unit cell of the mineral wurtzite with the Zn^{2+} ions in red and the S^{2-} ions in grey. This compound provides an example of a structure with a *hexagonal unit cell*. Each zinc(II) cation and sulfide anion in wurtzite is four-coordinate (tetrahedral) (Figures 6.18b and 6.18c).

The stoichiometry of wurtzite may be confirmed from the solid state structure by using equations 6.26 and 6.27. Note that in an infinite lattice containing hexagonal unit cells, ions in the centre of faces are shared between two unit cells (Figure 6.19a), ions on vertical-edges are shared between three unit cells (Figure 6.19b), and ions in corner sites are shared between six unit cells (Figure 6.19c).

$$\text{Number of } Zn^{2+} \text{ ions per unit cell of wurtzite} = (6 \times \tfrac{1}{3}) + (4 \times 1) = 6 \qquad (6.26)$$

$$\text{Number of } S^{2-} \text{ ions per unit cell of wurtzite} = (2 \times \tfrac{1}{2}) + (12 \times \tfrac{1}{6}) + (3 \times 1) = 6 \qquad (6.27)$$

The ratio of Zn^{2+} : S^{2-} ions in wurtzite is 6 : 6, or 1 : 1, giving an empirical formula of ZnS.

6.14 SIZES OF IONS

Determining the internuclear distances in an ionic lattice

X-Ray diffraction: see Section 3.2

X-Ray diffraction may be used to determine the solid state structure of an ionic compound and gives the distances *between the nuclei*. Thus, the results of an X-ray diffraction experiment give an Na–Cl distance in solid NaCl of 281 pm, and an Na–F distance of 231 pm in crystalline NaF. Each distance corresponds to the equilibrium separation of the closest, oppositely charged ions in the respective crystal lattices.

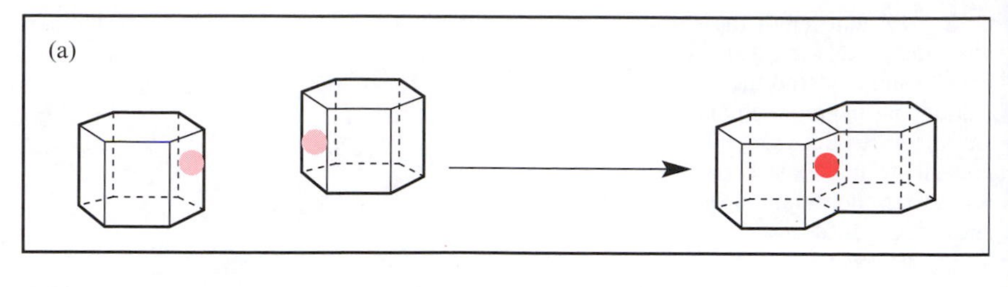

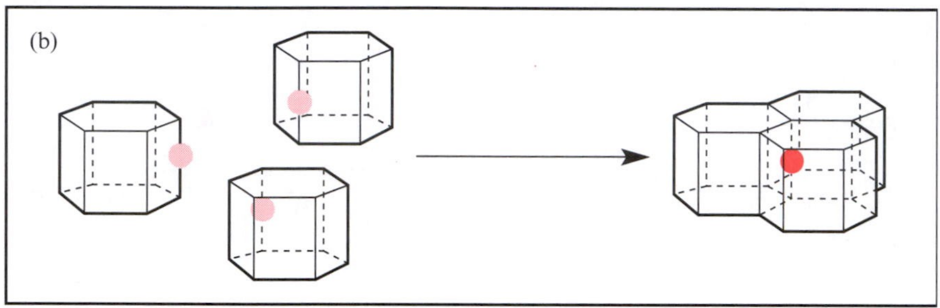

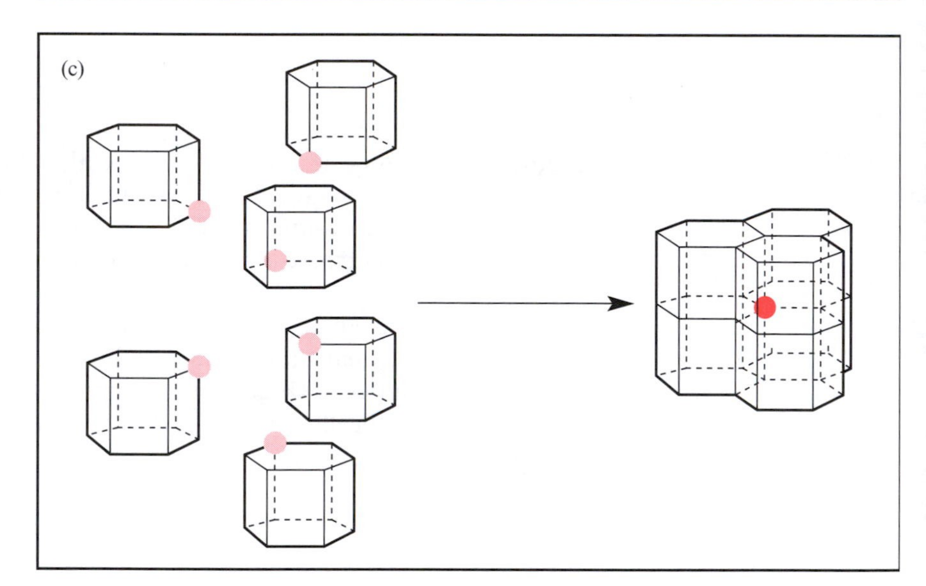

6.19 The sharing of ions between hexagonal unit cells. (a) An ion at the centre of a face is shared between two units cells. (b) An ion in the middle of the edge shown is shared between three unit cells. (c) An ion in a corner site is shared between six unit cells.

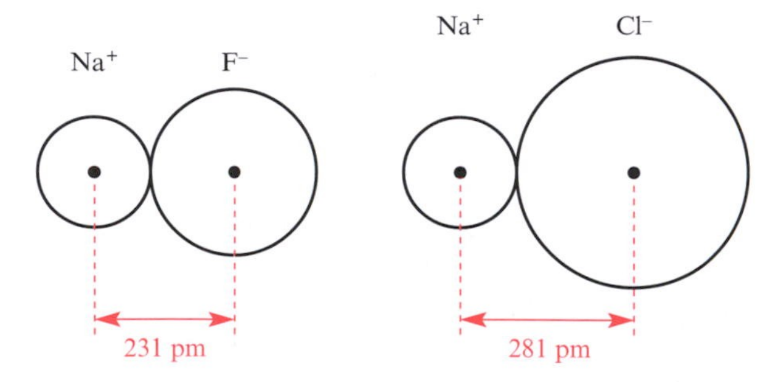

6.20 The results of X-ray diffraction experiments give internuclear distances. In sodium fluoride, the internuclear distance is 231 pm. In sodium chloride, it is 281 pm. If we assume that the ions are spherical and touch each other, then the sum of the ionic radii equals the internuclear distance in each case.

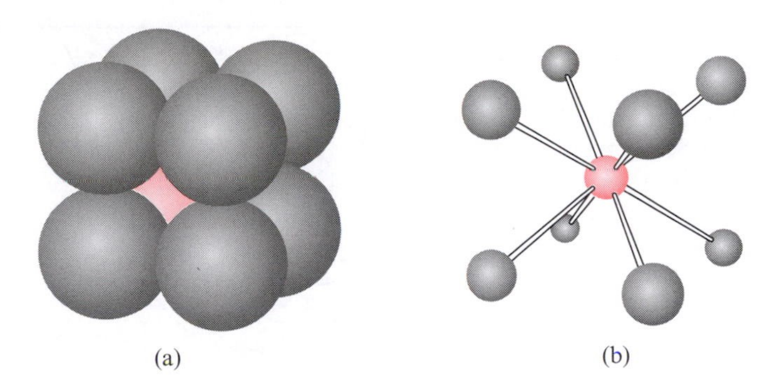

6.21 A space-filling diagram of a unit cell of CsCl is shown in (a). Compare this picture with the representation of the same unit cell in (b). The space-filling diagram is more realistic, but the lattice structure is easier to see if we use the type of diagram shown (b).

In the electrostatic model, we assume that the ions are spherical with a finite size. We can also assume that in the solid state lattice, ions of opposite charge touch one another. This is shown in Figure 6.20 for the Na^+F^- and Na^+Cl^- ion-pairs. Note that the diagrams of the NaCl, CsCl, CaF_2, TiO_2 and ZnS lattices drawn earlier show the ions with significant gaps between them; this representation is used simply for the purposes of clarity. The ions effectively occupy the space in the lattice as is illustrated in Figure 6.21 for a unit cell of CsCl.

Ionic radius

We have previously defined the van der Waals and covalent radii of atoms. For an ion, a new parameter, the *ionic radius* (r_{ion}) may be derived from X-ray diffraction data. As Figure 6.20 shows however, there is a problem. The experimental data only give the internuclear distance which is the *sum of the ionic radii of the cation and anion* (equation 6.28).

$$\text{Internuclear distance between a cation and the closest anion in a lattice} = r_{cation} + r_{anion} \qquad (6.28)$$

In some cases, electron density contour maps have been obtained. The contour map in Figure 6.22 shows the presence of two different types of ions, Na^+ and Cl^- ions. The positions of the nuclei coincide with the centres of the spherical§ regions of electron density. The distance between the nuclei of two adjacent Na^+ and Cl^- ions is the internuclear separation (281 pm). Between the two nuclei, the electron density falls to a minimum value as can be seen in Figure 6.22. In a model compound, the minimum electron density would be zero. This point corresponds to the 'boundary' between the Na^+ and Cl^- ions, and we can partition the internuclear distance into components due to the radius of the cation (r_{cation}) and the radius of the anion (r_{anion}). For NaCl, we can then write equation 6.29. If we can accurately measure the electron density between the nuclei, we can use the 'boundary' to find the radius of the two ions.

§ The contour map is a section through the lattice and the circles in such maps correspond to sections through spherical regions of electron density.

6.22 An electron density contour map of part of a sodium chloride lattice showing two Na^+ and two Cl^- ions. [From G. Schoknecht, *Z. Naturforsch., Teil A*, **12**, 983 (1957).]

$$281\,\text{pm} = r_{Na^+} + r_{Cl^-} \tag{6.29}$$

In general, estimated values of ionic radii are not obtained by the method described above. Instead, given a sufficiently large data set of internuclear distances, it is possible to calculate a self-consistent set of ionic radii by assuming the relationship given in equation 6.28 and setting up series of simultaneous equations. We assume that if an ion is in a similar environment in two ionic compounds (e.g. octahedral coordination), then its radius is transferable from one compound to another. We must also assume that, in at least one compound in the set of compounds, the anion–anion distance can be measured experimentally. The need for this assumption is explained in Box 6.3.

Values of ionic radii for selected ions are listed in Table 6.6. The ionic radius of a cation or anion may vary with coordination number; for example, r_{ion} (Ca^{2+}) is 100 pm for a six-coordinate ion and 112 pm for an eight-coordinate ion. The values in Table 6.6 all refer to six-coordinate ions.

For an element that exhibits a variable oxidation state, more than one type of ion is observed. This is commonly seen for the *d*-block metals, for example, both Fe^{2+} and Fe^{3+} are known. A consideration of the effective nuclear charges in these species shows why the Fe^{3+} ion (r_{ion} = 55 pm) is smaller than the Fe^{2+} ion (r_{ion} = 61 pm). Several other pairs of ions are listed in Table 6.6.

Box 6.3 Obtaining a self-consistent set of ionic radii

The following compounds all possess a sodium chloride-type of lattice in the solid state. For each, the internuclear distance between a cation and the closest anion has been measured by using X-ray diffraction methods.

Compound	*Formula*	*Internuclear distance / pm*
Lithium chloride	LiCl	257
Sodium chloride	NaCl	281
Lithium fluoride	LiF	201
Sodium fluoride	NaF	231

Using equation 6.28, we can write down a set of four simultaneous equations:

$$r_{Li^+} + r_{Cl^-} = 257$$
$$r_{Na^+} + r_{Cl^-} = 281$$
$$r_{Li^+} + r_{F^-} = 201$$
$$r_{Na^+} + r_{F^-} = 231$$

but there are not enough data to solve these equations and find the four ionic radii. Introducing another compound necessarily introduces another variable and does not improve the situation.

In order to solve the equations, we need to estimate the value of one of the radii by another means. If we choose a lattice which contains a *large anion* and a *small cation* we can make the assumption that, not only do the closest anions and cations touch one another, but so do the closest anions. This is shown below, for LiCl, where the assumption is reasonably valid.

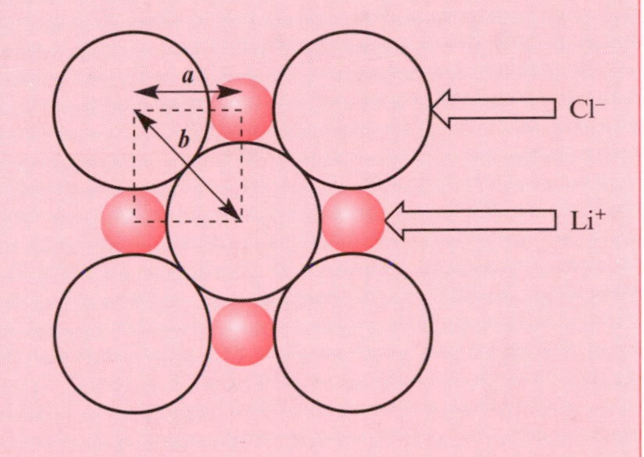

The internuclear distances ***a*** and ***b*** can both be found experimentally:

$$\boldsymbol{a} = r_{Li^+} + r_{Cl^-} = 257 \text{ pm} \qquad \text{(i)}$$

$$\boldsymbol{b} = 2 \times r_{Cl^-} = 363 \text{ pm} \qquad \text{(ii)}$$

From equation (ii):

$$r_{Cl^-} = 181.5 \text{ pm}$$

and, substituting this value into equation (i) gives:

$$r_{Li^+} = 257 - 181.5 = 75.5 \text{ pm}$$

As an exercise, use these results in the set of four simultaneous equations given above to see what happens when you solve the equations to find the radii of the Na^+ and F^- ions. Use all the data to double check the answers. You will find that the answers are not fully consistent with one another and that a larger data set is required to give reliable average values.

Table 6.6 Ionic radii for selected ions. The values are given for six-coordinate ions. The ionic radius varies with the coordination number of the ion (see Section 6.14). A more complete list is provided in Appendix 6.

Anions		Cations	
Anion	*Ionic radius* (r_{ion}) / *pm*	*Cation*	*Ionic radius* (r_{ion}) / *pm*
F^-	133	Li^+	76
Cl^-	181	Na^+	102
Br^-	196	K^+	138
I^-	220	Rb^+	149
O^{2-}	140	Cs^+	170
S^{2-}	184	Mg^{2+}	72
Se^{2-}	198	Ca^{2+}	100
N^{3-}	171	Al^{3+}	54
		Ti^{3+}	67
		Cr^{2+}	73
		Cr^{3+}	62
		Mn^{2+}	67
		Fe^{2+}	61
		Fe^{3+}	55
		Co^{2+}	65
		Co^{3+}	55
		Ni^{2+}	69
		Cu^{2+}	73
		Zn^{2+}	74

The ionic radius (r_{ion}) of an ion provides a measure of the size of a spherical ion.

Internuclear distance between a cation and the closest anion in a lattice $= r_{cation} + r_{anion}$

For a given ion, r_{ion} may vary with the coordination number.

Group trends in ionic radii

In Section 2.21 we noted that the van der Waals radii of the noble gas atoms increased on going down group 18 and Figure 3.4 illustrated an increase in values of the covalent radii of atoms on descending groups 15, 16 and 17. Similar trends are observed for series of ionic radii (Table 6.6), provided that we are comparing the radii of ions with a constant charge, for example M^{2+}

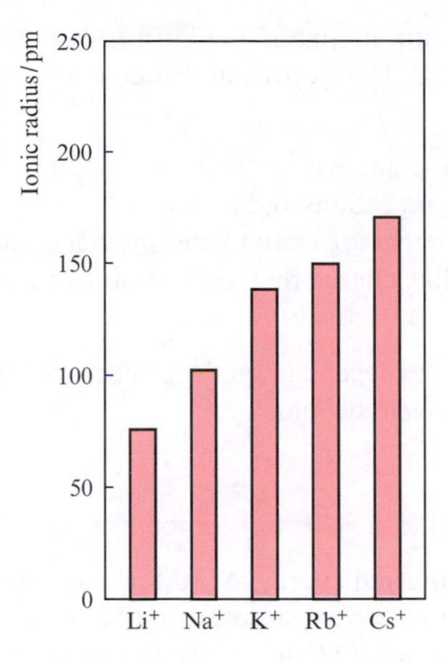

6.23 The increase in ionic radii of the alkali metal ions on descending group 1.

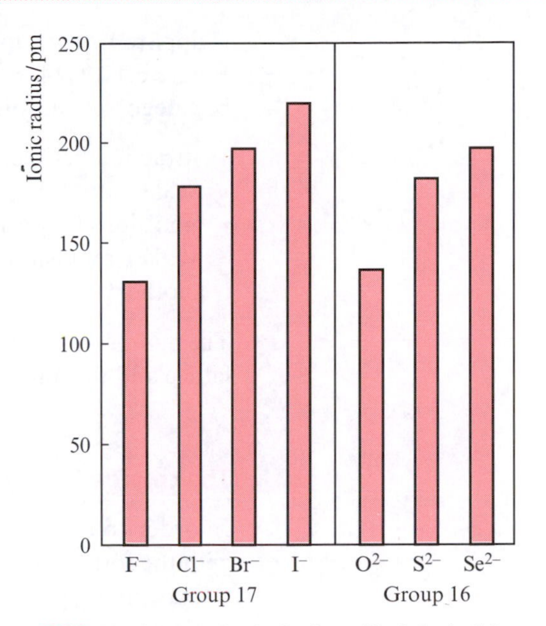

6.24 The increase in the ionic radii of the halide ions on descending group 17, and of the dianions of the group 16 elements.

ions. Figure 6.23 shows that, upon descending group 1, the ionic radius of the M^+ ions increases. A similar trend is seen for the halide anions (X^-) in group 17, and for the dianions of the group 16 elements (Figure 6.24).

6.15 LATTICE ENERGY – A PURELY IONIC MODEL

The *lattice energy*, $\Delta U(0\ \text{K})$, of an ionic compound is the change in internal energy that accompanies the formation of one mole of the solid from its constituent gas phase ions at a temperature of 0 K. It can be estimated by assuming that the solid state lattice of an ionic compound contains *spherical* ions. We shall examine the extent to which this approximation is true in Sections 6.17 and 6.18. In order to calculate the lattice energy, *all* attractive and repulsive interactions between ions must be considered.

Coulombic forces in an ionic lattice

In Section 6.6, we discussed the electrostatic interactions between two oppositely charged ions. For a mole of ion-pairs, the change in internal energy associated with the Coulombic attraction between the ions *within* an ion-pair was given by equation 6.18. We assumed that each ion-pair was isolated and that there were no *inter*-ion-pair forces.

Once the ions are arranged in a solid state lattice, each ion will experience both attractive and repulsive Coulombic forces from surrounding ions. This is apparent if we look at a particular ion in any of the lattices shown in Figure 6.11, 6.14, 6.15, 6.16, 6.17 or 6.18. The electrostatic interactions can be categorized as follows:

- attraction between adjacent cations and anions;
- weaker attractions between more distant cations and anions;
- repulsions between ions of like charge that are close to one another; and
- weaker repulsions between ions of like charge that are distant from one another.

The number and magnitude of these forces depend upon the geometry of the lattice and the internuclear distance between the ions.

The change in internal energy associated with Coulombic forces

In order to determine the change in internal energy, $\Delta U(0\ \text{K})$, associated with the formation of an ionic lattice, it is convenient to consider the formation as taking place according to equation 6.30 – that is, the aggregation of *gaseous* ions to form a solid state lattice.

$$X^+(g) + Y^-(g) \rightarrow XY(s) \qquad (6.30)$$

In forming the ionic lattice, there are contributions to $\Delta U(0\ \text{K})$ from Coulombic attractive (equation 6.18) and repulsive (equation 6.19) interactions. Equations 6.18 and 6.19 are of the same form and can be readily combined. The number and relative magnitudes of the Coulombic interactions, and whether these are attractive or repulsive, are taken into account by using a factor known as the *Madelung constant*, A. Table 6.7 lists values of A for the lattice-types we have previously described.

We can now write equation 6.31, where $\Delta U(0\ \text{K})$(Coulombic) refers to the lowering of internal energy associated with the Coulombic forces which are present when a mole of the ionic solid XY forms from a mole of isolated gaseous X^+ ions and a mole of isolated gaseous Y^- ions. As energy is released upon the formation of the lattice, the value of $\Delta U(0\ \text{K})$(Coulombic)

Table 6.7 Madelung constants, A, for some common lattice-types. The values are calculated by considering the geometrical relationships between the ions in the lattice as illustrated for sodium chloride in Box 6.4. Values of A are simply numerical and have no units.

Lattice-type	*Lattice shown in Figure*	*A*
Sodium chloride (NaCl)	6.11	1.7476
Caesium chloride (CsCl)	6.14	1.7627
Fluorite (CaF_2)	6.15	2.5194
Rutile (TiO_2)	6.16	2.408
Zinc blende (β-ZnS)	6.17	1.6381
Wurtzite (α-ZnS)	6.18	1.6413

Box 6.4 Determining the value of a Madelung constant

The Madelung constant takes into account the different Coulombic forces, both attractive and repulsive, that act on a particular ion in a lattice.

Consider the NaCl lattice shown here. We stated in Section 6.8 that each Na^+ ion and each Cl^- ion is six-coordinate. The coordination number gives the number of closest ions of opposite charge. In the diagram opposite, the central Cl^- ion is surrounded by six Na^+ ions, each at a distance r, and this is the internuclear distance measured by X-ray diffraction (281 pm). This applies to *all* chloride ions in the lattice since each is in the same environment. Similarly, each Na^+ ion is at a distance of 281 pm (r) from six Cl^- ions.

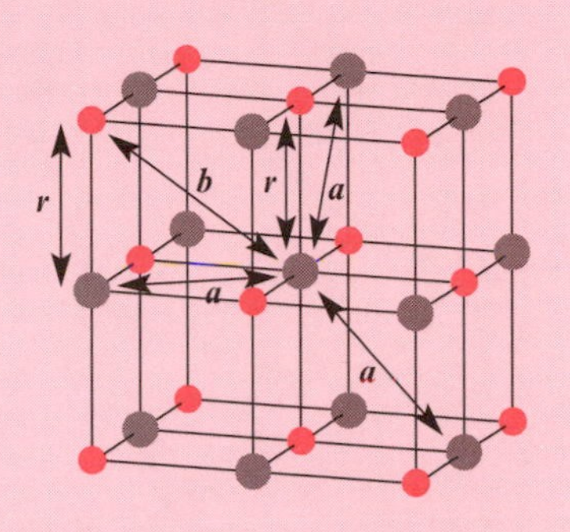

Now consider which other ions are close to the central Cl^- ion. Firstly, there are 12 Cl^- ions, each at a distance a from the central ion, and the Cl^- ions repel one another. All 12 Cl^- ions are shown in the diagram, but only three distances a are indicated. Distance a is related to r by the equation:

$$a^2 = r^2 + r^2 = 2r^2$$
$$a = r\sqrt{2}$$

Next, there are eight Na^+ ions, each at a distance b from the central Cl^- ion giving rise to attractive forces. Distance b is related to r by the equation:

$$b^2 = r^2 + a^2$$
$$b^2 = r^2 + (r\sqrt{2})^2$$
$$b^2 = 3r^2$$
$$b = r\sqrt{3}$$

Further attractive and repulsive interactions occur, but as the distances involved increase, the Coulombic interactions decrease. Remember that the energy associated with a Coulombic force is inversely proportional to the distance (see equation 6.17).

The Madelung constant, A, contains terms for all the attractive and repulsive interactions experienced by a given ion, and so for the NaCl lattice, the Madelung constant is given by:

$$A = 6 - (12 \times \frac{1}{\sqrt{2}}) + (8 \times \frac{1}{\sqrt{3}}) - \dots$$

where the series will continue with additional terms for interactions at greater distances. Note that r is not included in this equation. The value of A calculated in the equation is for *all* sodium chloride-*type* lattices, and r varies depending upon the ions present (see Table 6.2). However, the *ratios* of the distances r:a:b will always be $1:\sqrt{2}:\sqrt{3}$. The inverse relationships arise, as stated previously, from the fact that the energies associated with the Coulombic forces depend on the inverse of the distance between the point charges.

When all terms are accounted for, we obtain a value of $A = 1.7476$ for the NaCl-type lattice.

Question: Only the first three terms in the series to calculate A for NaCl are given above. What are the next two terms in the series?

is negative. In equation 6.31, the units of ΔU(0 K)(Coulombic) are J mol^{-1}. For an ionic lattice:

$$\Delta U(0\text{ K})\text{ (Coulombic)} = -\left(\frac{L \times A \times |z_+| \times |z_-| \times e^2}{4 \times \pi \times \varepsilon_0 \times r}\right) \qquad (6.31)$$

where $|z_+|$ = modulus of the positive charge
$|z_-|$ = modulus of the negative charge
e = charge on the electron = 1.602×10^{-19} C
ε_0 = permittivity of a vacuum = 8.854×10^{-12} F m^{-1}
r = internuclear distance between the ions (units = m)
A = Madelung constant (see Table 6.7)

Born forces

In a real ionic lattice, the ions are *not* the point charges assumed in the electrostatic model but have finite size. In addition to the electrostatic forces discussed above, there also Born (repulsive) forces between adjacent ions. These arise from the electron–electron and nucleus–nucleus repulsions.

The internal energy that is associated with the Born forces is related to the internuclear distance between adjacent ions according to equation 6.32.

$$\Delta U(0\text{ K})\text{ (Born)} \propto \frac{1}{r^n} \qquad (6.32)$$

where r = internuclear distance and n = Born exponent.

The value of n depends upon the electronic configuration of the ions involved, and values are listed in Table 6.8. Thus, for potassium chloride which consists of argon-like ions, the Born exponent is 9. For sodium chloride which contains neon- and argon-like ions (Na^+ is isoelectronic with Ne; Cl^- is isoelectronic with Ar), n is $(\frac{7}{2} + \frac{9}{2}) = 8$.

➤ K^+, Cl^- and Ar are isoelectronic

➤ Na^+ and Ne are isoelectronic; Cl^- and Ar are isoelectronic

Table 6.8 Values of the Born exponent, n, given for an ionic compound XY in terms of the electronic configuration of the ions [X^+][Y^-]. Values of n are numerical and have no units. The value of n for an ionic compound in which the ions have *different* electronic configurations is taken by averaging the component values. For example, for NaF, n = 7, but for LiF, $n = \frac{5}{2} + \frac{7}{2} = 6$.

Electronic configuration of the ions in an ionic compound XY	*Examples of ions*	*n*
[He][He]	H^-, Li^+	5
[Ne][Ne]	F^-, O^{2-}, Na^+, Mg^{2+}	7
[Ar][Ar], or [3d^{10}][Ar]	Cl^-, S^{2-}, K^+, Ca^{2+}, Cu^+	9
[Kr][Kr] or [4d^{10}][Kr]	Br^-, Rb^+, Ba^{2+}, Ag^+	10
[Xe][Xe] or [5d^{10}][Xe]	I^-, Cs^+, Au^+	12

The Born–Landé equation

Equation 6.32 can be modified by including a contribution to $\Delta U(0\ \text{K})$ due to the Born repulsions, and the new expression is given in equation 6.33.

$$\text{Lattice energy} = \Delta U(0\ \text{K}) = -\left(\frac{L \times A \times |z_+| \times |z_-| \times e^2}{4 \times \pi \times \varepsilon_0 \times r}\right) \times \left(1 - \frac{1}{n}\right) \quad (6.33)$$

Given a chemical formula for an ionic compound, equation 6.33 can be readily applied provided we know the internuclear distance, r, and the lattice-type. Both these data can be obtained from the results of X-ray diffraction experiments. Since the Madelung constant is defined for a particular lattice-type, this piece of information must be available in order to use equation 6.33 (Table 6.7). A value of r can be estimated by summing the radii of the cation and anion (equation 6.28), and question 6.13 at the end of the chapter concerns this problem.

The *lattice energy* of an ionic compound is the change in internal energy that accompanies the formation of one mole of a solid from its constituent gas-phase ions at a temperature of 0 K.

***Assuming a model for an ionic lattice which consists of spherical ions,* values of lattice energy can be estimated by using the Born–Landé equation:**

$$\text{Lattice energy} = \Delta U(0\ \text{K}) = -\left(\frac{L \times A \times |z_+| \times |z_-| \times e^2}{4 \times \pi \times \varepsilon_0 \times r}\right) \times \left(1 - \frac{1}{n}\right)$$

Worked example 6.3 *Lattice energy*

Estimate the change in internal energy which accompanies the formation of one mole of CsCl from its constituent gaseous ions, assuming an electrostatic model for the solid state lattice.

Data required: $L = 6.022 \times 10^{23}\ \text{mol}^{-1}$; $A = 1.7627$; $e = 1.602 \times 10^{-19}$ C; $\varepsilon_0 = 8.854 \times 10^{-12}\ \text{F m}^{-1}$; Born exponent for CsCl = 10.5; internuclear Cs–Cl distance = 356 pm.

The change in internal energy (the lattice energy) is given by the Born–Landé equation:

$$\Delta U(0\ \text{K}) = -\left(\frac{L \times A \times |z_+| \times |z_-| \times e^2}{4 \times \pi \times \varepsilon_0 \times r}\right) \times \left(1 - \frac{1}{n}\right)$$

r must be in m: 356 pm = 3.56×10^{-10} m.

$$\Delta U(0\ \text{K}) = -\left(\frac{6.022 \times 10^{23} \times 1.7627 \times 1 \times 1 \times (1.602 \times 10^{-19})^2}{4 \times 3.142 \times 8.854 \times 10^{-12} \times 3.56 \times 10^{-10}}\right) \times \left(1 - \frac{1}{10.5}\right)$$

$$= -\left(\frac{2.724 \times 10^{-14}}{3.961 \times 10^{-20}}\right) \times 0.905$$

$$= -622\,373\ \text{J mol}^{-1}$$

$$= -622\ \text{kJ mol}^{-1}$$

6.16 LATTICE ENERGY – EXPERIMENTAL DATA

We have seen how a purely ionic model can be used to *estimate* values of lattice energies and in this section we show how values may be experimentally determined.

Using Hess's Law: The Born–Haber cycle

Hess's Law: see Section 1.17

Lattice energies are not measured directly, and you can appreciate why if you consider the definition given in Section 6.15. Instead, use is made of Hess's Law. For example, although the lattice energy for potassium chloride is the change in internal energy at 0 K for reaction 6.34, we could also define a *lattice enthalpy change* ($\Delta_{lattice}H^o$) as the enthalpy change that accompanies this reaction under standard conditions. Using Hess's Law, $\Delta_{lattice}H^o$ can be determined from equation 6.35.

$$K^+(g) + Cl^-(g) \rightarrow KCl(s) \qquad (6.34)$$

$$\Delta_{lattice}H^o(KCl,s) = \Delta_f H^o(KCl,s) - \Delta_f H^o(K^+,g) - \Delta_f H^o(Cl^-,g) \qquad (6.35)$$

Values of $\Delta_f H^o$: see Appendix 11

On the right-hand side of equation 6.35, the standard enthalpy of formation of solid KCl (as for a wide range of inorganic compounds) is readily available from tables of thermochemical data. The standard enthalpies of formation of gaseous ions can be determined by setting up appropriate Hess cycles (equations 6.36 and 6.37).

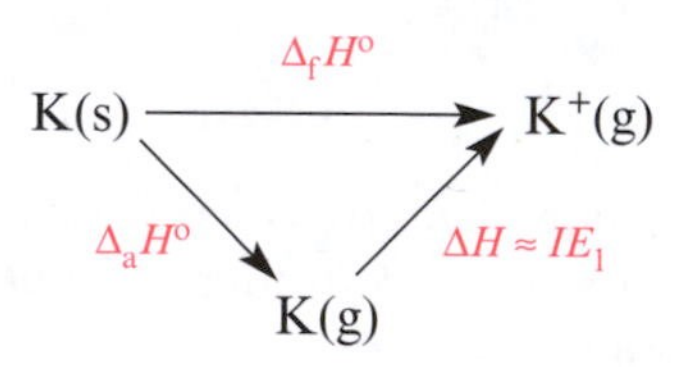

$$\Delta_f H^o(K^+,g) = \Delta_a H^o(K,g) + IE_1(K,g) \qquad (6.36)$$

$\Delta_f H^o$

$\frac{1}{2}Cl_2(g)$ → $Cl^-(g)$

$\Delta H = \frac{1}{2}D$ $\quad$ $\Delta_{EA}H \approx -EA_1$

Cl(g)

$$\Delta_f H^o(Cl^-,g) = \tfrac{1}{2}D(Cl_2,g) + \Delta_{EA}H(Cl,g) \qquad (6.37)$$

where $\Delta_{EA}H$ is the *enthalpy change* associated with the attachment of an electron.

$\Delta_a H^o$ and bond dissociation enthalpies: see Section 3.6

Values of standard enthalpies of atomization, $\Delta_a H^o$, bond dissociation energies and ionization energies are readily available for the elements, but only a relatively few electron affinities have been measured accurately.

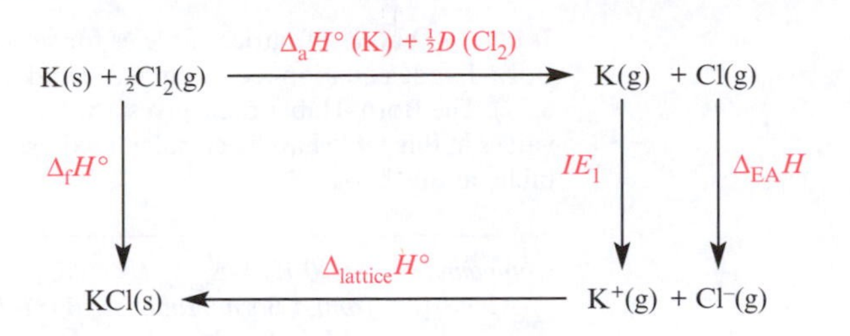

6.25 A Born–Haber cycle for potassium chloride.

We can now combine equations 6.35, 6.36 and 6.37 to give a single Hess cycle called a *Born–Haber cycle* (Figure 6.25) and from it we can find $\Delta_{lattice}H^o$ (equations 6.38 and 6.39). The value determined for $\Delta_{lattice}H^o$ in this way represents an *experimental value* since it is derived from experimentally determined data.

$$\Delta_f H^o(KCl,s) = \Delta_a H^o(K,g) + \tfrac{1}{2}D(Cl_2,g) + IE_1(K,g) + \Delta_{EA}H(Cl,g) + \Delta_{lattice}H^o KCl(s) \quad (6.38)$$

Equation 6.38 can be rearranged to give equation 6.39 for the lattice enthalpy.

$$\Delta_{lattice}H^o(KCl,s) = \Delta_f H^o(KCl,s) - \Delta_a H^o(K,g) - \tfrac{1}{2}D(Cl_2,g) - IE_1(K,g) - \Delta_{EA}H(Cl,g) \quad (6.39)$$

We have seen previously that both ionization energies and electron affinities are changes in internal energy, but, since the necessary corrections are relatively small, they can be approximated to changes in enthalpy. Similarly, $\Delta_{lattice}H^o \approx \Delta U(0\ K)$.

Thus for KCl, the lattice energy can be determined (equation 6.40) from equation 6.39 using experimental values for $\Delta_f H^o(KCl,s) = -437\ kJ\ mol^{-1}$, $\Delta_a H^o(K,g) = 90\ kJ\ mol^{-1}$, $D(Cl_2,g) = 242\ kJ\ mol^{-1}$, $IE_1(K,g) = 419\ kJ\ mol^{-1}$ and $\Delta_{EA}H\ (Cl,g) = -349\ kJ\ mol^{-1}$.

$$\Delta U(0\ K) \approx \Delta_{lattice}H^o(KCl,s) = -437 - 90 - 121 - 419 + 349 = -718\ kJ\ mol^{-1} \quad (6.40)$$

6.17 A COMPARISON OF LATTICE ENERGIES DETERMINED BY THE BORN–LANDÉ EQUATION AND THE BORN–HABER CYCLE

Table 6.9 gives values for the lattice energies of some ionic compounds, determined both from the Born–Landé equation, $\Delta U(0\ K)$(Born–Landé), and by using a Born–Haber cycle, $\approx \Delta U(0\ K)$(Born–Haber). Compounds have been grouped in the table according to periodic relationships and cover the halides of two group 1 metals (Na and K) and a group 11 metal (Ag). With the exception of silver(I) iodide, each compound in the table has an NaCl-type of lattice; AgI has a wurtzite-type structure at room temperature and pressure.

The difference between $\Delta U(0\ K)$(Born–Haber) and $\Delta U(0\ K)$(Born–Landé) for each compound is expressed as a percentage of the experimental value (the last column of Table 6.9). The agreement between the values of the lat-

Table 6.9 Values of lattice energies for selected compounds determined using the Born–Landé equation (equation 6.33) and a Born–Haber cycle (Figure 6.25 and equation 6.39). The Born–Haber cycle gives a value of $\Delta_{lattice}H^o$ which approximates to $\Delta U(0\ K)$; values in this table have been calculated using thermochemical data from appropriate tables in this book.

Compound	$\Delta U(0\ K) \approx \Delta_{lattice}H^o$ *from a Born–Haber cycle* / kJ mol^{-1}	$\Delta U(0\ K)$ *from the Born–Landé equation* / kJ mol^{-1}	$[\Delta U(0\ K)$ *(Born–Haber)* $- \Delta U(0\ K)$ *(Born–Landé)*$]$ / kJ mol^{-1}	*Percentage difference in values of* $\Delta U(0\ K)^a$
NaF	–931	–901	–30	3.2%
NaCl	–786	–756	–30	3.8%
NaBr	–736	–719	–17	2.3%
NaI	–671	–672	+1	≈ 0
KF	–827	–798	–29	3.5%
KCl	–718	–687	–31	4.3%
KBr	–675	–660	–15	2.2%
KI	–617	–622	+5	0.8%
AgF	–972	–871	–101	10.4%
AgCl	–915	–784	–131	14.3%
AgBr	–888	–758	–130	14.6%
AgI	–858	–737	–121	14.1%

[a] The percentage difference is calculated using the equation $\left(\frac{\Delta U(0\ K)\ (\text{Born–Haber}) - \Delta U(0\ K)\ (\text{Born–Landé})}{\Delta U(0\ K)(\text{Born–Haber})}\right)$

tice energies are fairly good for the sodium and potassium halides, and this suggests that the discrete-ion model assumed for the Born–Landé equation is reasonably appropriate for these compounds.[§]

For each of the silver(I) halides, the Born–Landé equation underestimates the lattice energy by more than 10%. This suggests that the discrete-ion model is *not* appropriate for the silver(I) halides.

We mentioned in Section 6.1, that just as the covalent bonding model can be improved by allowing for ionic contributions to a bond, an ionic model can similarly be improved by allowing for some covalent character. This is exactly the problem that we have encountered with the silver(I) halides.

6.18 POLARIZATION OF IONS

We end this chapter by reconsidering the assumpion that ions are spherical. Is this always true?

Consider a small ion such as Li^+, and assume that it is spherical. Since the surface area of the Li^+ ion is small, the charge density is relatively large (equation 6.41).

§ Using the Born–Landé equation is not the only method of calculating lattice energies. More sophisticated equations are available which will improve the answer. However, the pattern of variation noted in Table 6.9 remains the same. For more detailed discussion see: W. E. Dasent, *Inorganic Energetics*, 2nd edn, Cambridge University Press, Cambridge (1984).

$$\text{Charge density of an ion} = \frac{\text{Surface area of the ion}}{\text{Charge on the ion}} \tag{6.41}$$

Now consider a large anion such as I^- where the charge density is low. If the iodide ion is close to a cation with a high charge density, the charge distribution might be distorted and may no longer be spherical. The distortion means that the centre of electron density in the iodide anion no longer coincides with the position of the nucleus, and the result is a dipole moment, induced by the adjacent, small cation. The electron density is distorted towards the cation. The iodide anion is said to be readily *polarizable*. The cation that causes the polarization to occur is said to be *polarizing*.

Induced dipole: see Section 2.21

Charge density of an ion = $\dfrac{\text{Surface area of the ion}}{\text{Charge on the ion}}$

Small cations such as Li^+, Mg^{2+}, Al^{3+}, Fe^{3+} and Ti^{4+} possess high charge densities, and each has a high *polarizing power*.

Anions such as I^-, H^-, N^{3-}, Se^{2-} and P^{3-} are easily polarizable. As we descend a group, for example from fluoride to iodide where the charge remains constant, the polarizability increases as the size of the ion increases.

When an ion (usually the cation) induces a dipole in an adjacent ion (usually the anion), an ion–dipole interaction results.

SUMMARY

In this chapter we have been concerned with the formation of cations and anions from gaseous atoms and the interactions between ions in the solid state.

Do you know what the following terms mean?

- electron density map
- ionization energy
- electron affinity
- ion-pair
- Coulombic interaction
- ionic lattice
- unit cell
- coordination number (in an ionic solid)
- ionic radius
- lattice energy
- Madelung constant
- Born exponent
- Born–Landé equation
- Born–Haber cycle
- charge density of an ion

You should now be able:

- to discuss why there is a relationship between the pattern in the values of successive ionization energies for a given atom and its electronic structure
- to describe the structures of sodium chloride, caesium chloride, fluorite, rutile and zinc(II) sulfide
- to deduce the stoichiometry of a compound given a diagram of a unit cell of the solid state lattice
- to write down the Born–Landé equation and know how to use it to estimate lattice energies
- to appreciate the assumptions that are inherent in the Born–Landé equation
- to set up a Born–Haber cycle and determine the lattice energy of a compound, or, given a lattice energy, to use a Born–Haber cycle to find another quantity such as the electron affinity of an atom
- to appreciate why the values of lattice energies calculated by the Born–Landé equation may or may not agree with experimental values

PROBLEMS

6.1 With reference to the oxygen atom, explain what you understand by the (a) first ionization energy, (b) second ionization energy, (c) first electron affinity, (d) second electron affinity and (e) enthalpy change associated with the attachment of an electron.

6.2 The first five ionization energies of a gaseous atom X are 589, 1148, 4911, 6494 and 8153 kJ mol^{-1}. To what group is this element likely to belong?

6.3 The first ionization energies of oxygen ($Z = 8$), fluorine ($Z = 9$), neon ($Z = 10$) and sodium ($Z = 11$) are 1312, 1681, 2080 and 495 kJ mol^{-1} respectively. Rationalize the trend in these values in terms of the electronic structures of these elements.

6.4 Calculate the enthalpy change that accompanies the process:

$$\frac{1}{2}Li_2(g) \rightarrow Li^+(g) + e^-$$

given that the bond dissociation enthalpy for dilithium is 110 kJ mol^{-1} and IE_1 for lithium is 520 kJ mol^{-1}.

6.5 Calculate the enthalpy change that accompanies the reaction:

$$\frac{1}{2}Li_2(g) + e^- \rightarrow Li^-(g)$$

given that the bond dissociation enthalpy for Li_2 is 110 kJ mol^{-1} and EA_1 for lithium is 60 kJ mol^{-1}.

6.6 Calculate the enthalpy changes, ΔH, for the reactions:

$$H^{\bullet}(g) + F^{\bullet}(g) \rightarrow H^+(g) + F^-(g) \qquad \text{(i)}$$
$$H^{\bullet}(g) + F^{\bullet}(g) \rightarrow H^-(g) + F^+(g) \qquad \text{(ii)}$$

and comment on the results in the light of the discussion about resonance structures in Section 4.3. (Data required: See Table 6.1 and Appendices 8 and 9).

6.7 Using the data in Table 6.1, determine the enthalpy change that accompanies the formation of a gaseous S^{2-} ion from a gaseous sulfur atom. Comment on the sign and magnitude of your answer in terms of the electronic configurations of the S atom and the S^- and S^{2-} ions.

6.8 Distinguish between the terms *van der Waals radius, covalent radius* and *ionic radius*. How would you estimate the ionic radius of a sodium ion? Point out what other information you would need and what assumptions you would make.

6.9 Show that equation 6.19 is dimensionally correct. (Hint: refer to Table 1.2.)

6.10 Figure 6.11 shows a unit cell of the sodium chloride lattice. Why could the unit cell not be smaller, for example with only four ion-pairs defining a cube?

6.11 Derive equations 6.20 and 6.21 in terms of the positions of the ions in the unit cell of calcium fluoride.

6.12 Suggest reasons why the distance between adjacent Na and Cl nuclei is shorter (236 pm) in NaCl in the gas phase than in the solid state (281 pm).

6.13 (a) Draw the structures of the sulfate(VI) $[SO_4]^{2-}$, and tetrafluoroborate $[BF_4]^-$ ions and rationalize why they can be considered to be approximately spherical. (b) Given the lattice energies of sodium sulfate and sodium tetrafluoroborate, suggest a method to estimate the radii of the $[SO_4]^{2-}$ and $[BF_4]^-$ ions. State clearly what assumptions (if any) you would have to make, and also what other information you would require.

6.14 Using values in Table 6.8, determine the Born exponents for the following ionic compounds: (a) NaCl; (b) LiCl; (c) AgF; (d) MgO.

6.15 In worked example 6.3, the lattice energy for CsCl was calculated using the experimental internuclear distance of 356 pm. Calculate the lattice energy for the same structure using values of $r_{ion}Cs^+$ =170 pm and $r_{ion}Cl^-$ = 181 pm. Rationalize the difference between the two values of ΔU(0 K).

6.16 Calculate the lattice energy for magnesium oxide using both the Born–Landé equation and a Born–Haber cycle; MgO has an NaCl-type lattice. All the data needed for the Born–Landé method are to be found in Chapter 6; thermochemical data are listed in Appendices 8–11. Use the data obtained to comment on whether a spherical-ion model is appropriate for MgO.

7 Elements

Topics

- close-packing of spheres
- solid structures of group 18 elements
- solids with molecular units
- solids with infinite covalent lattices
- metals
- metallic bonding

7.1 INTRODUCTION

In Section 2.21, we discussed the van der Waals forces that operate between atoms of a noble gas and noted that the weakness of these forces is reflected in some of the physical properties of the noble gases (Table 2.6). We have already observed that these elements only form solids at very low temperatures, and that helium can only be solidified under pressures greater than atmospheric pressure. When a group 18 element solidifies, the atoms form an ordered structure in which they are *close-packed*. Before considering the specifics of the solid state structures of the noble gases, we discuss what is meant by the *close-packing of spheres*. This concept may be familiar to some, but not all, readers and a brief discussion of the topic is given in Section 7.2. In Section 7.3, we consider simple and body-centred cubic packing of spheres.

In this chapter we consider not only elemental solids that consist of close-packed atoms, but also the structures of those elements which contain molecular units, small or large, or infinite covalent lattices in the solid state.

7.2 CLOSE-PACKING OF SPHERES

Hexagonal and cubic close-packing

Suppose we have a rectangular box and place in it some spheres of equal size. If we impose a restriction that there must be a *regular arrangement* of the spheres, then the most efficient way in which to cover the floor of the box is to pack the spheres as shown in Figure 7.1. This is part of a *close-packed arrangement*, and spheres not on the edge of the assembly are in contact with six other spheres within the single layer. Figure 7.1 emphasizes that a motif of hexagons is visible.

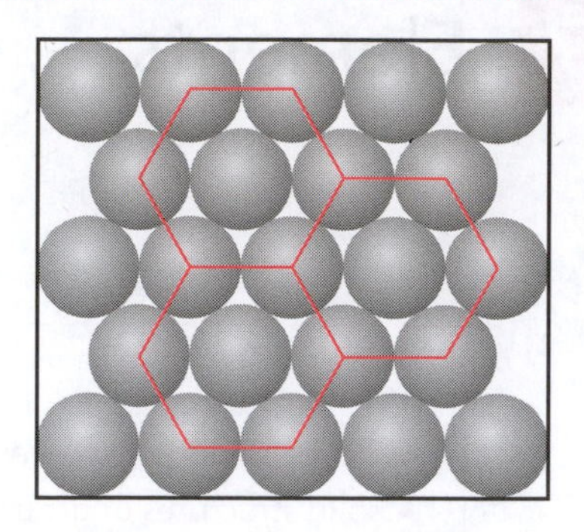

7.1 The aim is to cover the floor of a rectangular box with spheres which are of equal size, and the criterion is that the arrangement must be an ordered one. The most efficient arrangement is a *close-packed* one. The repeating hexagonal motif is shown in red.

Close-packing of spheres represents the most efficient use of space.

If we now add a second layer of close-packed spheres to the first, allowing the spheres in the second layer to rest in the hollows between those in the first layer, there is only enough room for every *other hollow* to be occupied (Figure 7.2). When a third layer of spheres is added, there is again only room to place them in every other hollow, but now there are *two distinct sets of hollows*. In the first set, the hollows lie directly *over the spheres* of the first layer. These are the hollows marked A in the top diagram in Figure 7.3. The hollows of the second set lie *over the unoccupied hollows* of the first layer. These are the hollows marked C at the top of Figure 7.3. Depending upon which set of hollows is occupied as the third layer of spheres is put in place, one of two close-packed structures results.

In Figure 7.3a, the third layer of spheres lies directly over the top of the first layer and this is emphasized in a side-view of the same arrangement in Figure 7.3c. The layers are labelled A and B, and the repetition of layers gives rise to an ABABAB... arrangement as we continue packing the spheres in this manner.

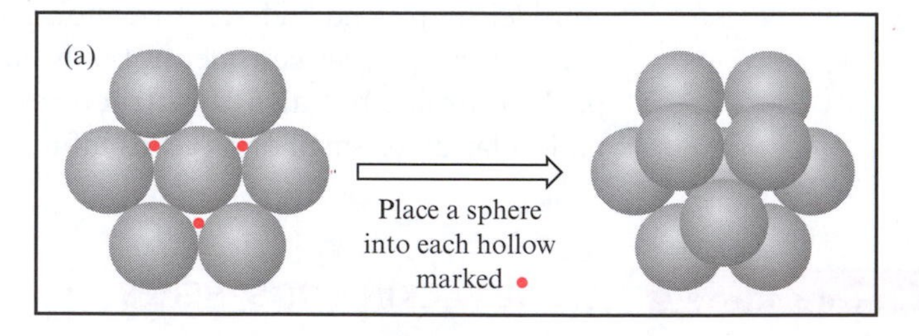

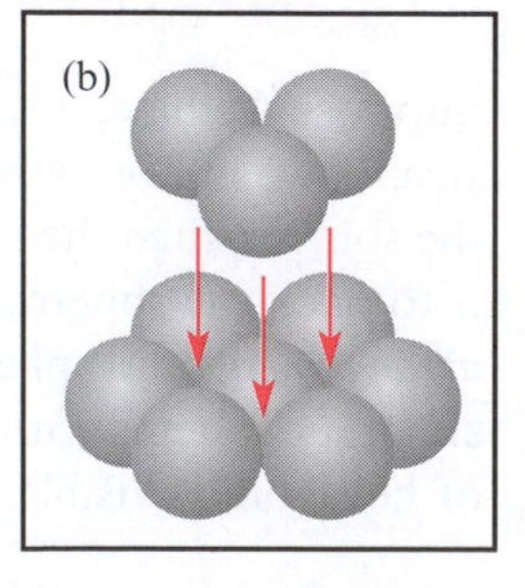

7.2 When three spheres are arranged in a triangle and touch one another, there is a hollow at the centre of the triangle. A single layer of close-packed spheres possesses hollows which form a regular pattern. (a) There are six hollows in between the seven close-packed spheres. When a second layer of spheres is placed on top of the first so that the new spheres lie in the hollows in the first layer, there is only room for every other hollow to be occupied. This is emphasized in (b) which gives a side view of the arrangement as the second layer of spheres is added to the first.

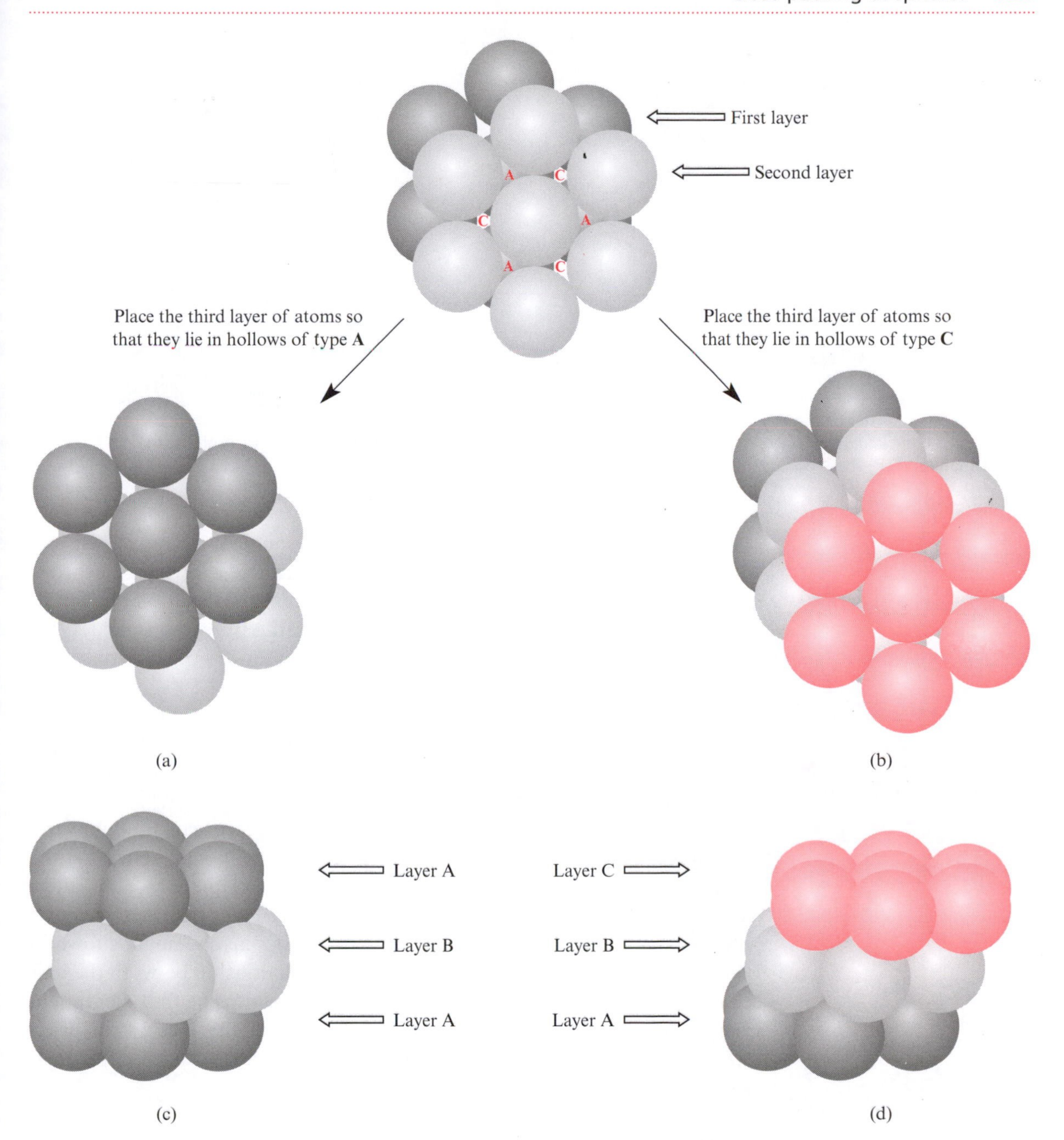

7.3 The second layer in a close-packed array of spheres possesses two types of hollow. Those labelled **A** lie directly over spheres in the first layer, while a hollow of type **C** lies over a hollow in the first layer. This leads to two possible arrangements of the spheres in the third layer. The sequences (a) ABA and (b) ABC viewed from above, and (c) ABA and (d) ABC viewed from the side. Notice in (d) that the ABC sequence of layers generates a plane consisting of repeating *square-units*.

In Figures 7.3b and 7.3d, the third layer of spheres does not lie directly over the first layer. The layers are labelled A, B and C, and the repetition of layers produces an ABCABC... arrangement.

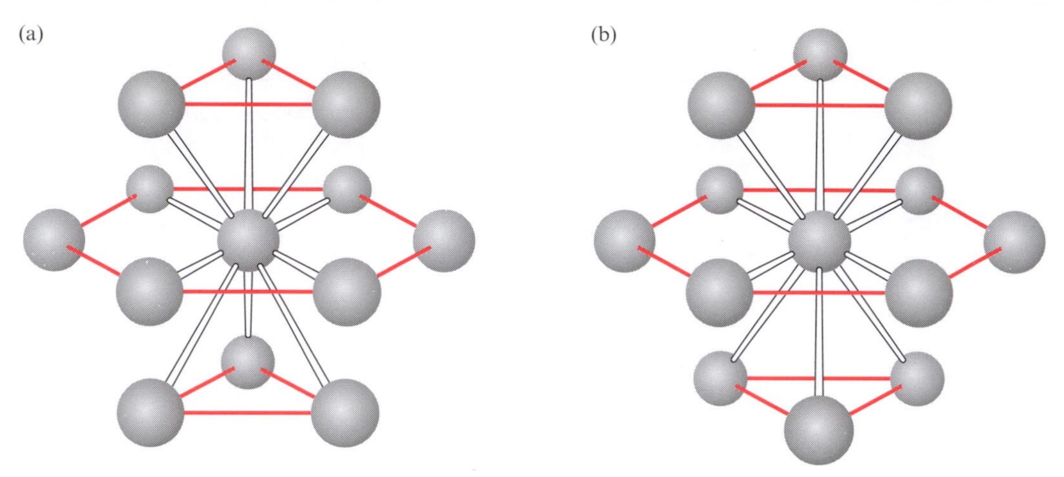

7.4 In a close-packed assembly of spheres, each sphere has 12 nearest-neighbours. This is true for both the (a) ABA arrangement and (b) the ABC arrangement.

In both of these close-packed structures, each sphere is in contact with six spheres within its layer plus three below and three above, giving 12 nearest-neighbours, i.e. each sphere is 12-coordinate. Figure 7.4a shows this for an ABA arrangement and Figure 7.4b depicts it for an ABC arrangement. Note that the two figures differ only in the mutual orientations of the two sets of three atoms at the top and bottom of the diagrams.

The ABABAB... arrangement is called *hexagonal close-packing* (hcp) of spheres, while the ABCABC... arrangement is known as *cubic close-packing* (ccp). The unit cells that characterize these arrangements of spheres are shown in Figure 7.5.

➤ Unit cell: see Section 6.7

The unit cell shown in Figure 7.5a illustrates the presence of the cubic unit that lends its name to cubic close-packing. It consists of eight spheres at the corners of the cube, with a sphere at the centre of each square-face. An alternative name for cubic close-packing is a *face-centred cubic* (fcc) arrangement.

It is easy to relate Figure 7.5b to the ABABAB... arrangement shown in Figure 7.3 but it is not so easy to recognize the relationship between the unit cell in Figure 7.5a and ABCABC... layer arrangement. The first diagram in Figure 7.6a shows three adjacent face-centred cubic units. If the structure is rotated through 45°, the close-packed layers discussed above become apparent.

Spheres of equal size may be close-packed in (at least) two ways.

In hexagonal close-packing (hcp), layers of close-packed spheres pack in an ABABAB... pattern.

In cubic close-packing (ccp) or face-centred cubic (fcc) arrangement, layers of close-packed spheres are in an ABCABC... pattern.

In both an hcp or ccp array, each sphere has 12 nearest-neighbours.

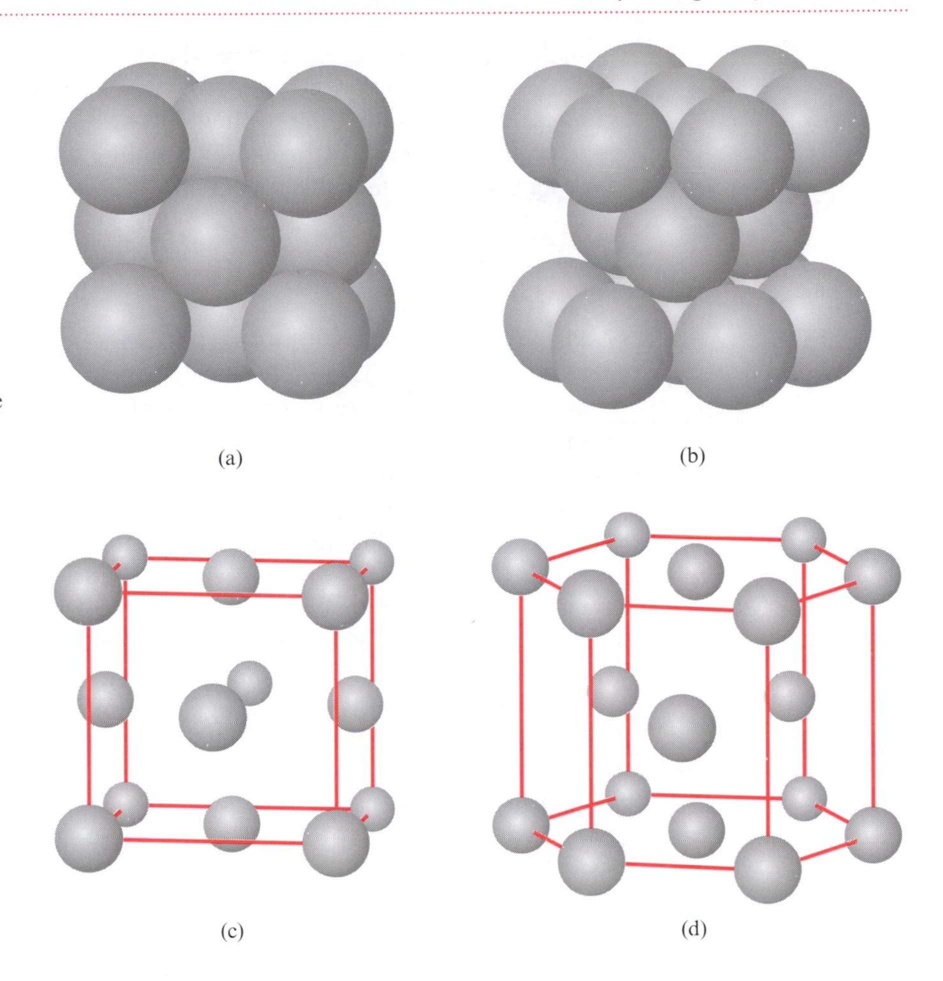

7.5 The ABCABC... and ABABAB... close-packed arrangements of spheres are termed (a) *cubic* and (b) *hexagonal* close-packing respectively. If the spheres are 'pulled apart', (c) the *cubic* and (d) the *hexagonal* units become clear. Cubic close-packing is also called a *face-centred cubic* arrangement. In diagrams (a) and (c), each face of the cubic unit possesses a sphere at its centre.

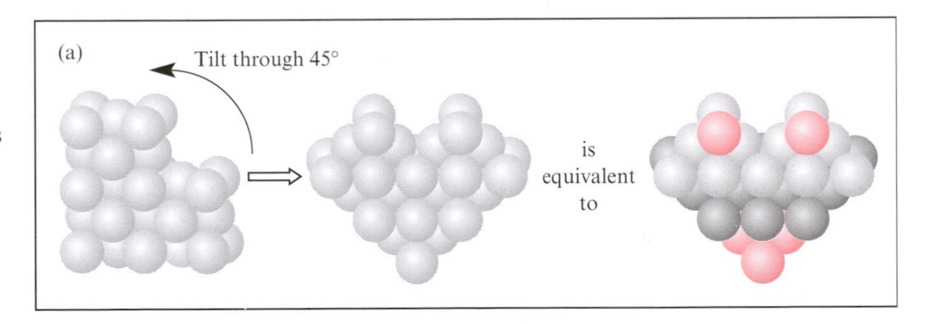

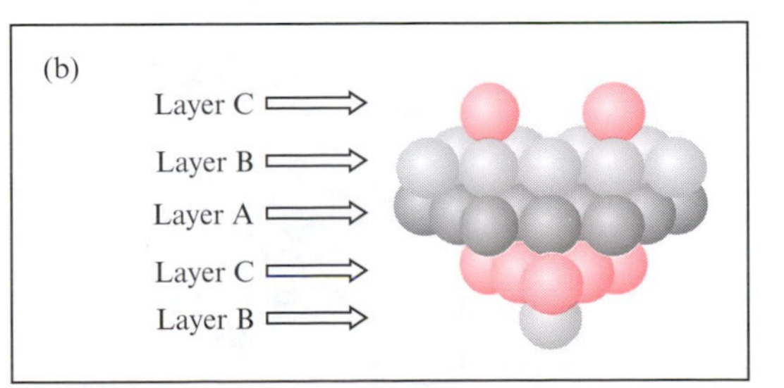

7.6 (a) The relationship between face-centred cubic units (three are shown) and the ABCABC... sequence of layers of close-packed spheres can be seen by tilting the cubic-units through 45°. (b) The last structure from figure (a) is redrawn to show the ABCABC... sequence of layers more clearly.

In the last diagram in Figure 7.6a, the spheres are colour-coded according to the layer in which they belong. This same diagram is redrawn in a different orientation in Figure 7.6b to show the A, B and C layers more clearly. The relationship is best confirmed by building a three-dimensional model or using a computer modelling program.

Interstitial holes

Placing one sphere on top of three others in a close-packed array gives a tetrahedral unit (Figure 7.7a) inside which is a cavity called an *interstitial hole* – specifically in this case, a *tetrahedral hole*. In an hcp arrangement, all the interstitial holes are tetrahedral.

A ccp array also contains tetrahedral holes but in addition there are *octahedral holes* (Figure 7.7b). Both types of interstitial hole are present in the unit cell shown in Figure 7.5a, and this is further explored in Box 7.1.

An hcp arrangement of sphere contains only tetrahedral holes, but in a ccp (fcc) array, both octahedral and tetrahedral holes are present.

It is not necessarily obvious why octahedral holes should arise in a ccp, but not an hcp, arrangement, but if we look at Figures 7.3 and 7.6, we see that the sequence of three repeating layers in the ccp structure leads to planes of spheres in square-motifs. Such a plane is visible on the left-hand side of Figure 7.3d, and is viewed 'face-on' in the second two diagrams in Figure 7.6a. Stacking a sphere above and below each square array leads to the formation of an octahedral hole (Figure 7.7b). There is no substitute for model-building to clarify these arguments!

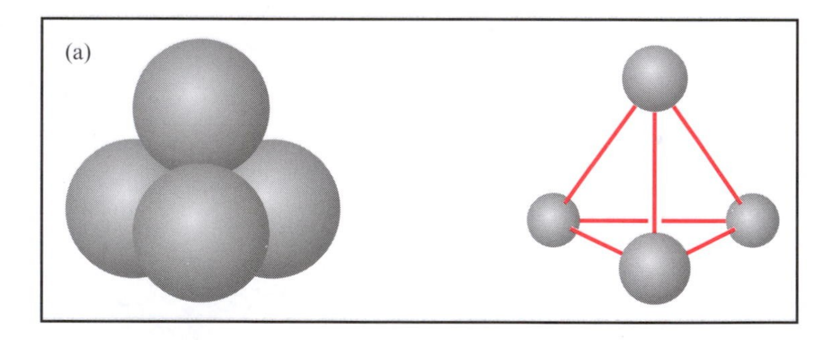

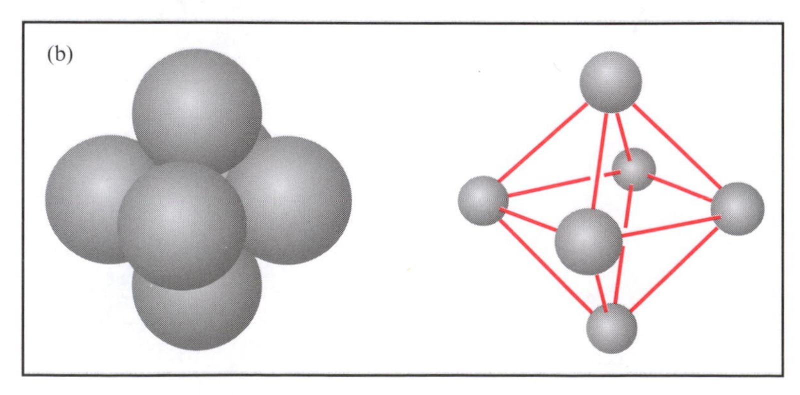

7.7 (a) A tetrahedral hole (found in both ccp and hcp arrangements) is formed when one sphere rests in the hollow between three others which are close-packed. (b) An octahedral hole is formed when two spheres stack above and below a square-unit of spheres. Such an arrangement is found in a ccp assembly of spheres.

Box 7.1 An alternative description of ionic lattices

A face-centred cubic (fcc) arrangement of spheres (cubic close-packing) is shown below. The left-hand diagram emphasizes the octahedral hole which is present at the centre of the fcc unit. The right-hand diagram shows one tetrahedral hole; there are eight within the fcc unit.

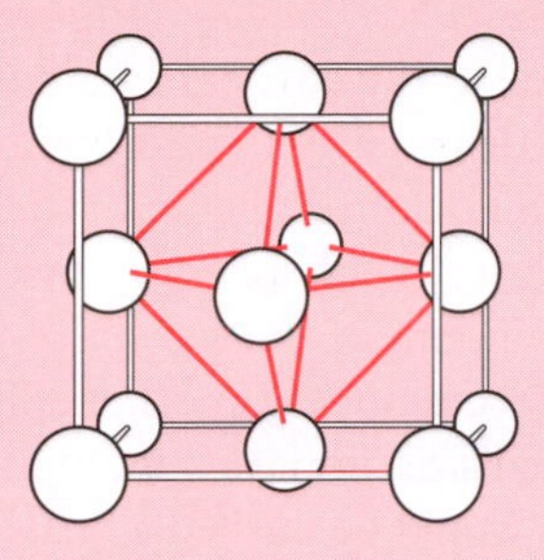

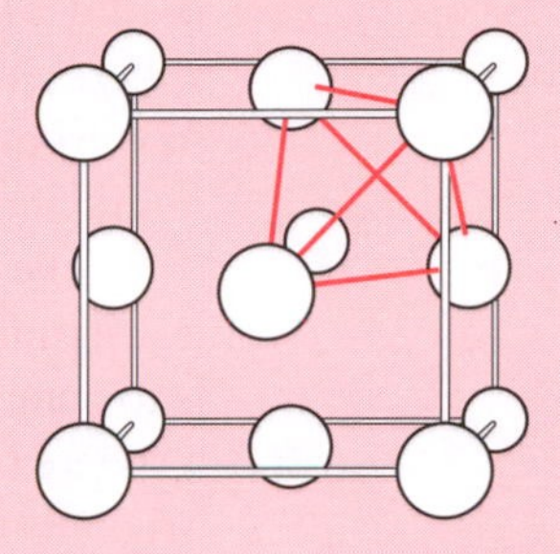

For an array which is made up of spheres of equal sizes, the interstitial holes remain empty.

An ionic lattice contains spheres (i.e. ions) which are *not* all the same size. In many cases (e.g. NaCl) the anions are larger than the cations, and it is convenient to consider a close-packed array of anions with the smaller cations in the interstitial holes.

Consider a cubic close-packed array of sulfide *anions*. They repel one another when they are in close proximity, but if zinc *cations* are placed in some of the interstitial holes, we can generate an ionic lattice which is stabilized by cation–anion attractive forces. A unit cell of the lattice of zinc blende (ZnS) is shown below:

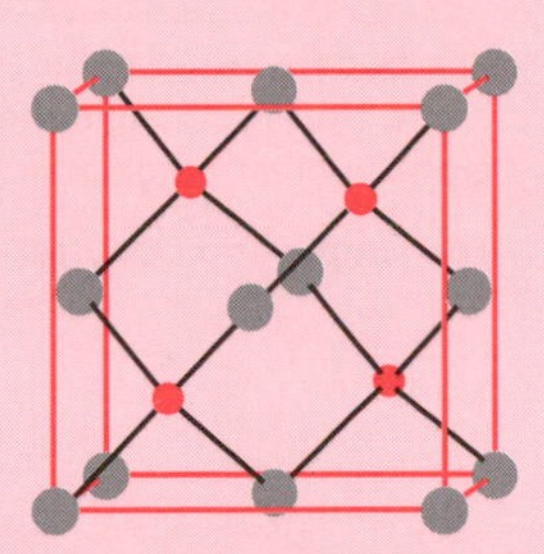

The S^{2-} ions are in an fcc arrangement. *Half of the tetrahedral holes* are occupied by Zn^{2+} ions. The organization of the cations is such that every other tetrahedral hole is filled with a cation.

This description of the structure of zinc blende relies on the idea that *small cations* can fit into the interstitial holes which lie between *large anions which are close packed*. Note that the anions thus define the corners of the unit cell; compare this with the unit cell of zinc blende drawn in Figure 6.17 – here the Zn^{2+} cations are shown at the corners of the unit cell. Do the two representations of the unit cell of zinc blende provide the same information about the overall lattice structure?

Similarly, it is possible to describe the structures of NaCl and CaF_2 in terms of cations occupying holes in a close-packed array of anions and we consider this approach in the accompanying workbook.

7.3 SIMPLE CUBIC AND BODY-CENTRED CUBIC PACKING OF SPHERES

In cubic or hexagonal close-packed assemblies, each sphere has 12 nearest-neighbours, and spheres are packed efficiently so that there is the minimum amount of 'wasted' space. It is, however, possible to pack spheres of equal size in an ordered assembly which is *not* close-packed and one such method is that of *simple cubic packing* (Figure 7.8). The unit cell contains eight spheres arranged at the corners of a cube, and when these units are placed next to one another in an extended array, each sphere has six nearest-neighbours and is in an octahedral environment.

Simple cubic packing of spheres provides an ordered arrangement but there is a significant amount of unused space. Each interstitial hole (a 'cubic' hole) is too small to accommodate another sphere of equal size. If we were to force a sphere (of equal size) into each interstitial hole in the simple cubic array, the original spheres would be pushed apart slightly. The result is the formation of a *body-centred cubic* (bcc) arrangement. The difference between simple and body-centred cubic structures can be seen by comparing Figures 7.8a and 7.9a. Although bcc packing makes better use of the space available than simple cubic packing, it is still less efficient than that in ccp or hcp assemblies.

The repeating unit in a bcc assembly is shown in Figure 7.9b and the number of nearest-neighbours in the bcc arrangement is eight (Figure 7.9c).

The repetition of cubic units of spheres of equal size gives simple cubic packing, and each sphere has six nearest-neighbours.

In a body-centred cubic (bcc) arrangement, the unit cell consists of a cube of eight, non-touching spheres with one sphere in the centre. Each sphere has eight nearest-neighbours.

The simple and body-centred cubic arrangements are *not* close-packed.

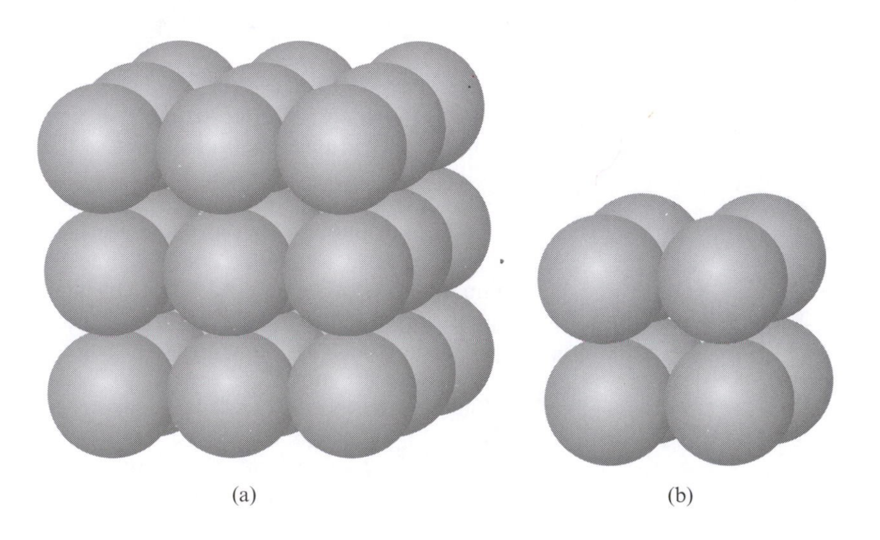

7.8 (a) The arrangement of spheres of equal size in a simple cubic packing. (b) The repeating unit in this arrangement consists of one sphere at each corner of a cube.

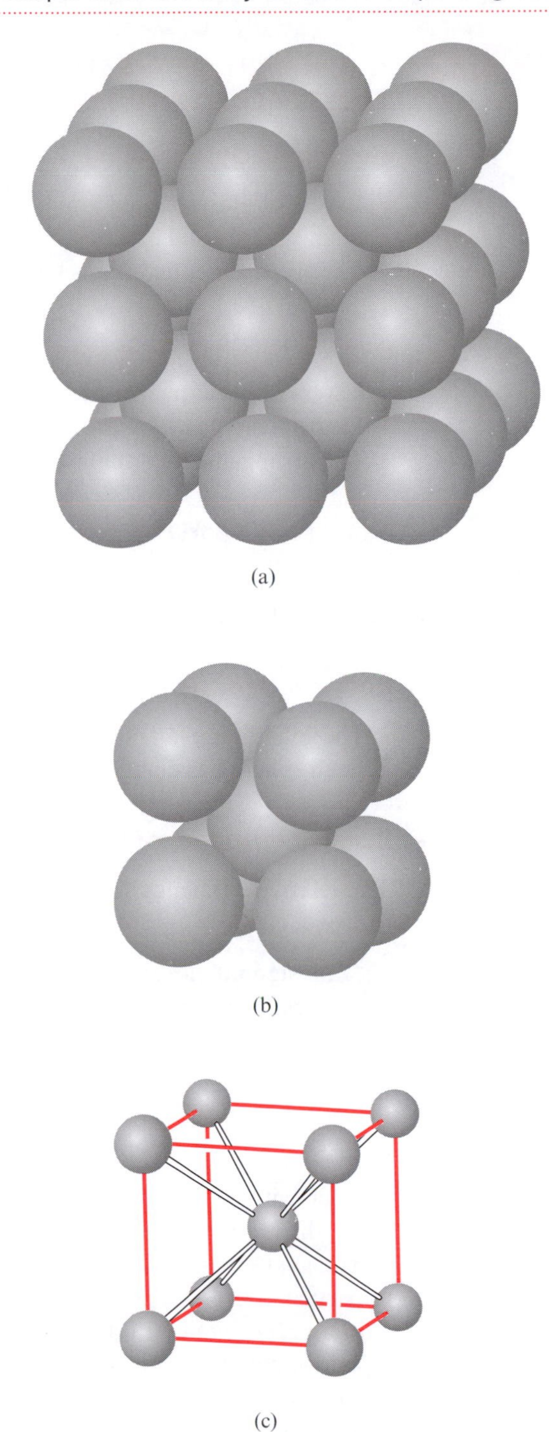

7.9 (a) The packing of spheres of equal size in a body-centred cubic (bcc) arrangement. (b) The unit cell in a bcc arrangement of spheres is a cube with one sphere at the centre. The central sphere touches each corner sphere, but the corner spheres do *not* touch one another. (c) Each sphere in a bcc arrangement has eight nearest neighbours.

7.4 A SUMMARY OF THE SIMILARITIES AND DIFFERENCES BETWEEN CLOSE-PACKED AND NON-CLOSE-PACKED ARRANGEMENTS

Hexagonal close-packing (Figures 7.5a and 7.5c)

- Layers of close-packed atoms are arranged in an ABABAB... manner.
- Each sphere has 12 nearest-neighbours.
- The arrangement contains tetrahedral interstitial sites.

Cubic close-packing or face-centred cubic (Figures 7.5b and 7.5d)

- Layers of close-packed atoms are arranged in an ABCABC... manner.
- Each sphere has 12 nearest-neighbours.
- The arrangement contains tetrahedral *and* octahedral interstitial sites.

Simple cubic packing (Figure 7.8)

- The spheres are *not* close-packed.
- Each sphere has six nearest-neighbours.

Body-centred cubic packing (Figure 7.9)

- The spheres are *not* close-packed.
- Each sphere has eight nearest-neighbours.

Relative efficiency of packing

The relative efficiency with which spheres of equal size are packed follows the sequence:

> hexagonal close-packing = cubic close-packing > body-centred cubic packing > simple cubic packing

7.5 CRYSTALLINE AND AMORPHOUS SOLIDS

X-ray diffraction: see Section 3.2

In a *crystalline* solid, atoms, molecules or ions are packed in an ordered manner, with a unit cell that is repeated throughout the crystal lattice. For an X-ray diffraction experiment, a single crystal is usually required. If a single crystal shatters, it may cleave along well-defined *cleavage planes*. This leads to particular crystals possessing characteristic shapes.

In an *amorphous* solid, the particles are not arranged in an ordered or repetitive manner. Crushing an amorphous solid leads to the formation of a *powder*, whereas crushing crystals leads to *microcrystalline* materials; however, microcrystals may look like powders to the naked eye!

In a crystalline solid, atoms, molecules or ions are packed in an ordered lattice with a characteristic unit cell.

7.6 SOLID STATE STRUCTURES OF THE GROUP 18 ELEMENTS

The elements in group 18 are referred to as noble gases, and it is not usual to think of them in other states. The group 18 elements solidify only at low temperatures (Table 2.6) and the enthalpy change that accompanies the fusion (melting) of one mole of each element is very small, indicating that the van der Waals forces between the atoms in the solid state are very weak. In the crystalline solid, the atoms of each group 18 element are close-packed. Cubic close-packing is observed for the atoms of each of solid neon, argon, krypton and xenon, whilst an hcp structure is observed for atoms of solid helium.

Box 7.2 Liquid helium: an important coolant

Although the group 18 elements are usually encountered in the gas phase, they have a number of important applications. One such is the use of liquid helium as a coolant.

Liquid nitrogen (bp 77 K) is frequently used as a coolant in the laboratory or in industry. However, it is not possible to reach extremely low temperatures by using liquid N_2 alone; liquefaction of gaseous N_2 under pressure provides a liquid at a temperature just below the boiling point. In order to reach lower temperatures it is necessary to use a liquid with a boiling point which is much lower that of N_2, and liquid helium is widely used for this purpose. As normally found, helium boils at 4.2 K, and the use of liquid helium is the most important method for reaching temperatures which approach absolute zero. Below 2.2 K, isotopically pure ^4He undergoes a transformation into the so-called He II. This is a remarkable liquid which possesses a viscosity close to zero, and a thermal conductivity which far exceeds that of copper.

Until recently it was necessary to resort to cooling potential superconducting materials in liquid helium in order to observe the superconductivity, although higher-temperature superconductors are now known (see Box 7.4).

7.7 ELEMENTAL SOLIDS CONTAINING DIATOMIC MOLECULES

Figure 3.2 showed a selection of covalent homonuclear molecules (H_2, O_2, O_3, I_2, P_4, S_6 and S_8). Each is a molecular form of an element. In the gas phase, these molecules are separate from one another, but in the solid state, they pack together with van der Waals forces operating between them. In this and the next two sections we consider the solid state structures of H_2, elements from groups 17 and 16, an allotrope of phosphorus, and one group of allotropes of carbon – all are molecular solids and *non-metals*.

Dihydrogen and difluorine

When gaseous H_2 is cooled to 20.4 K, it liquefies.§ Further cooling to 14.0 K results in the formation of solid dihydrogen. Even approaching absolute zero

§ Here, and throughout the chapter, we consider phase changes at atmospheric pressure, unless otherwise stated.

(0 K), molecules of H_2 possess sufficient energy to rotate about a point in the solid state lattice. Figure 7.10 shows that the result is that each H_2 molecule is described by a single sphere, the centre of which coincides with the midpoint of the H–H bond. Solid H_2 possesses an hcp arrangement of such spheres, each of which represents one H_2 molecule. It is only possible to apply the model of close-packed spheres *because the H_2 molecules are rotating at the temperature at which the solid state structure has been determined.*

Molecular F_2 solidifies at 53 K. Below 45 K, the molecules of F_2 can freely rotate, and the structure is described as distorted close-packed, with each F_2 molecule being represented by a sphere.§ The enthalpy of fusion of F_2 is 0.5 kJ mol^{-1} and this low value suggests that only van der Waals forces must be overcome in order to melt the solid.

The situation described for crystalline H_2 and F_2 is unusual. In the solid state, molecules of most elements or compounds are *not* freely rotating. Thermal motion such as the vibration of bonds does occur, but the positions of the atoms in a molecule can often be defined to a reasonable degree of accuracy. This means that the packing of spheres is not an appropriate model for most solid state structures because the component species are not spherical. It is applicable to elements of group 18 because they are monatomic, to H_2 and F_2 because the molecules are freely-rotating, and to metals.

Metal lattices: see Section 7.11

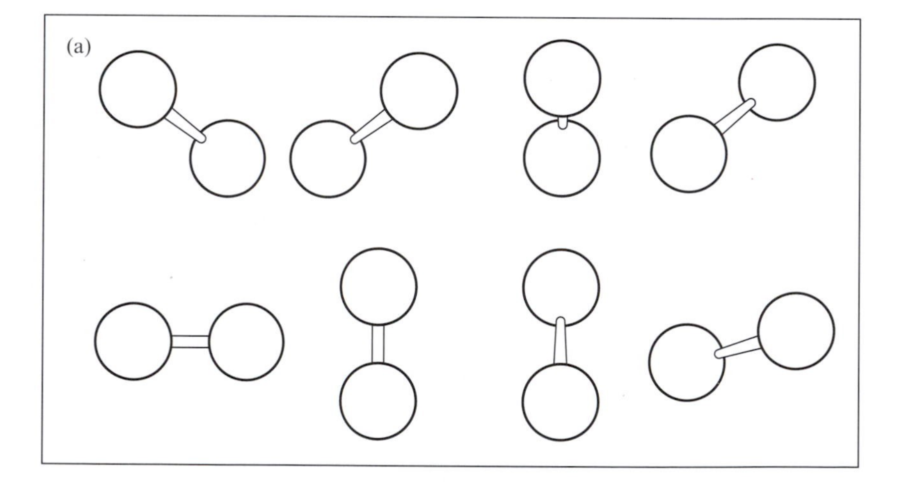

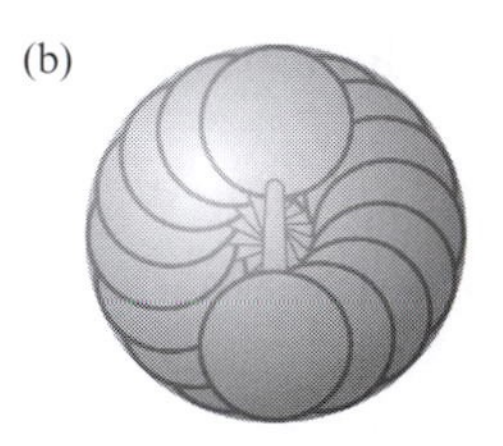

7.10 Molecules of H_2 are freely-rotating in the solid state. (a) Some possible orientations for H_2 molecules with respect to the mid-point of the H–H bond. (b) Different orientations are superimposed to illustrate the fact that each rotating molecule can be represented as a sphere; the centre of the sphere coincides with the mid-point of the H–H bond.

§ Above 45 K, a second phase with a more complicated structure exists.

Dichlorine, dibromine and diiodine (group 17)

Some physical and structural properties of Cl_2, Br_2 and I_2 are given in Table 7.1. At 298 K (1 atm pressure), I_2 is a solid, but Br_2 is a liquid and Cl_2 a gas. The solid elements share common structures that differ from those of F_2.

In the crystalline state, molecules of Cl_2 (or Br_2 or I_2) are arranged in a zigzag pattern within a layer (Figure 7.11) and these layers of molecules are stacked together. There are three characteristic distances in the structure which are particularly informative. Consider the structure of solid Cl_2. Within a layer (part of which is shown in Figure 7.11), the *intra*molecular Cl–Cl distance is 198 pm (***a*** in Figure 7.11). The measured Cl–Cl bond distance is twice the covalent radius (Table 7.1). Also, within a plane, we can measure *inter*molecular Cl···Cl distances, and the shortest such distance (***b*** in Figure 7.11) is 332 pm. This is shorter than twice the van der Waals radius of chlorine and suggests that there is some degree of interaction between the Cl_2 molecules in a layer. The shortest *inter*molecular Cl···Cl distance *between* layers of molecules is 374 pm. The phenomenon becomes more pronounced in going from Cl_2 to Br_2, and from Br_2 to I_2 as the distances in Table 7.1 indicate. Note also that the I–I bond length in solid I_2 is longer than in a gaseous molecule (Table 7.1) although there is little change in either the

Table 7.1 Some physical and structural properties of dichlorine, dibromine and diiodine. For details of the solid state structure of these elements, refer to the text and Figure 7.11.

Element	*Melting point / K*	*$\Delta_{fus}H$ / kJ mol^{-1}*	*Covalent radius (r_{cov}) / pm*	*Van der Waals radius (r_v) / pm*	*Intramolecular distance, **a** in Figure 7.11 / pm*	*Intermolecular distance within a layer, **b** in Figure 7.11 / pm*	*Intermolecular distance between layers / pm*	*Intramolecular distance for molecule in the gaseous state / pm*
Chlorine	171.5	6.4	99	180	198	332	374	199
Bromine	265.8	10.6	114	195	227	331	399	228
Iodine	386.7	15.5	133	215	272	350	427	267

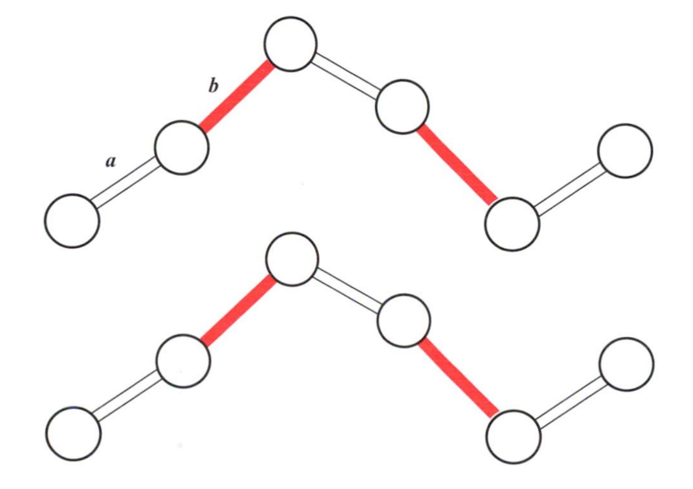

7.11 The solid state structure of Cl_2, Br_2 and I_2 consists of X_2 (X = Cl, Br or I) molecules arranged in zigzag chains within layers. Part of one layer is shown. Values of the intramolecular X–X distance, ***a***, and the intermolecular distance, ***b***, are listed in Table 7.1.

Cl–Cl or Br–Br bond length in going from gaseous to solid Cl_2 or Br_2. In solid I_2, the bonding interaction between molecules is at the expense of some bonding character within each I_2 molecule.

7.8 ELEMENTAL MOLECULAR SOLIDS IN GROUPS 15 AND 16

Sulfur (group 16)

Sulfur forms S–S bonds in a variety of cyclic and chain structures and Table 3.2 listed a range of allotropes of sulfur. One allotrope is S_6 which has a cyclic structure with a *chair conformation* (Figure 7.12). When S_6 crystallizes, the rings pack together efficiently to give a solid which is the highest density form of elemental sulfur (2.2 g cm^{-3}). Only van der Waals forces operate between the rings.

7.12 Two views of an S_6 molecule – one view emphasizes the chair conformation of the ring.

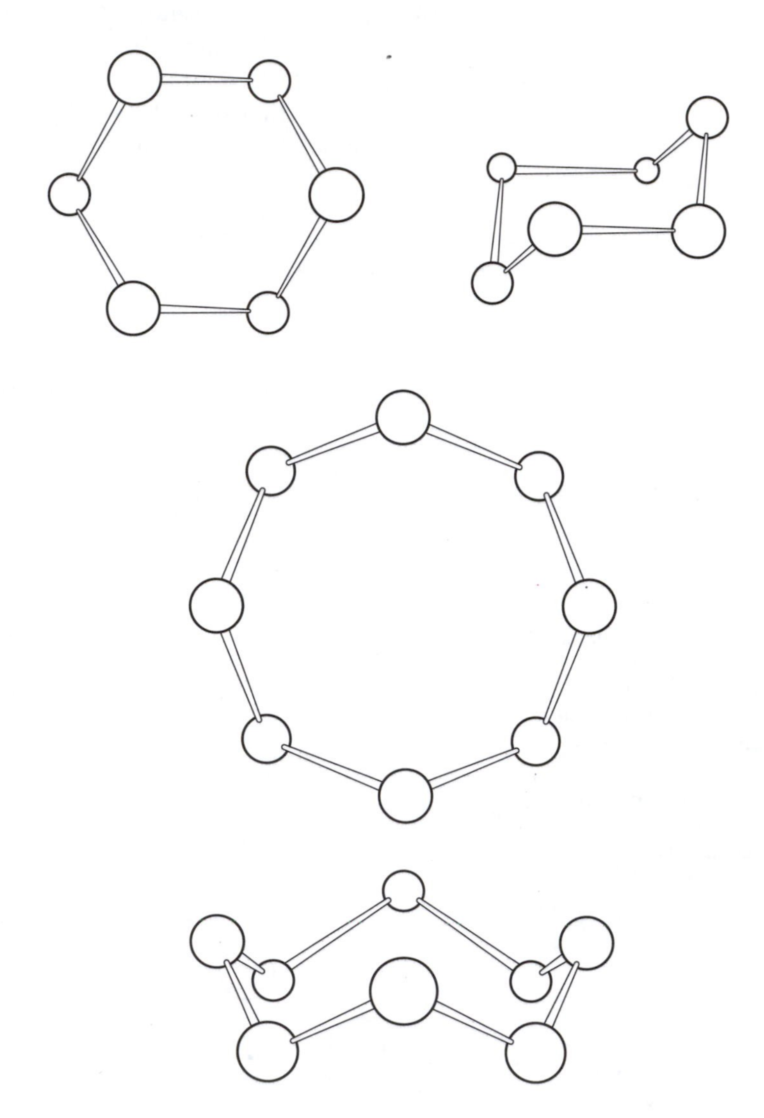

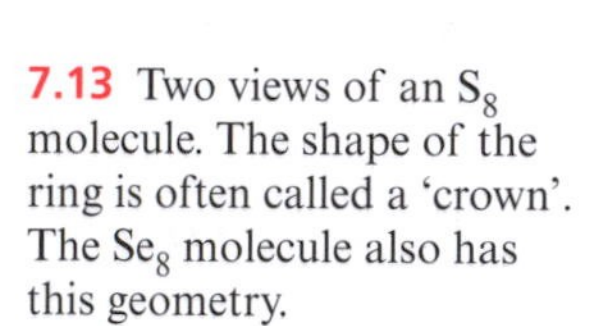

7.13 Two views of an S_8 molecule. The shape of the ring is often called a 'crown'. The Se_8 molecule also has this geometry.

Chair and boat conformers: see Section 15.2

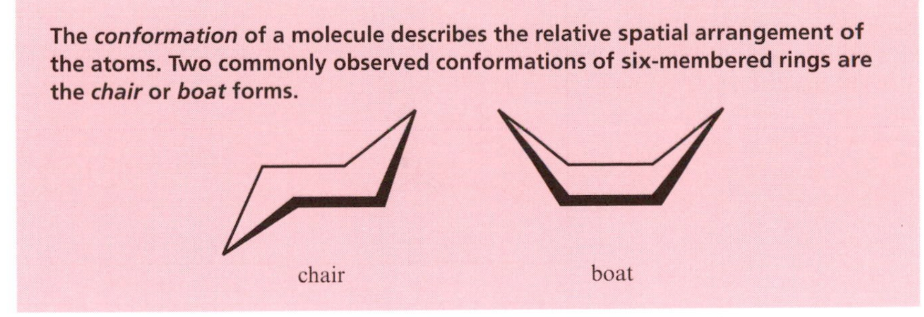

$r_{cov}(S) = 103$ pm.

Crystalline orthorhombic sulfur (the α-form, and the standard state of the element) consists of S_8 rings (Figure 7.13) which are packed together with van der Waals interactions between the rings. The average S–S bond length within each ring is 206 pm, consistent with the presence of single bonds. The organization of the S_8 rings in the crystalline state is shown in Figure 7.14. The rings do *not* simply stack immediately on top of each other.

Monoclinic sulfur (the β-form) also contains S_8 rings but these are less efficiently packed in the solid state (density = 1.94 g cm^{-3}) than are those in orthorhombic sulfur (density = 2.07 g cm^{-3}). When orthorhombic sulfur is heated to 368 K, a reorganization of the S_8 rings in the lattice occurs and the solid transforms into the monoclinic form. However, *single* crystals of orthorhombic sulfur can be rapidly heated to 385 K when they melt instead of undergoing the orthorhombic to monoclinic transformation. If crystallization takes place at 373 K, the S_8 rings adopt the structure of monoclinic sulfur, but the crystals must be cooled rapidly to 298 K. On standing at room temperature, monoclinic sulfur crystals change into the orthorhombic allotrope within a few weeks.

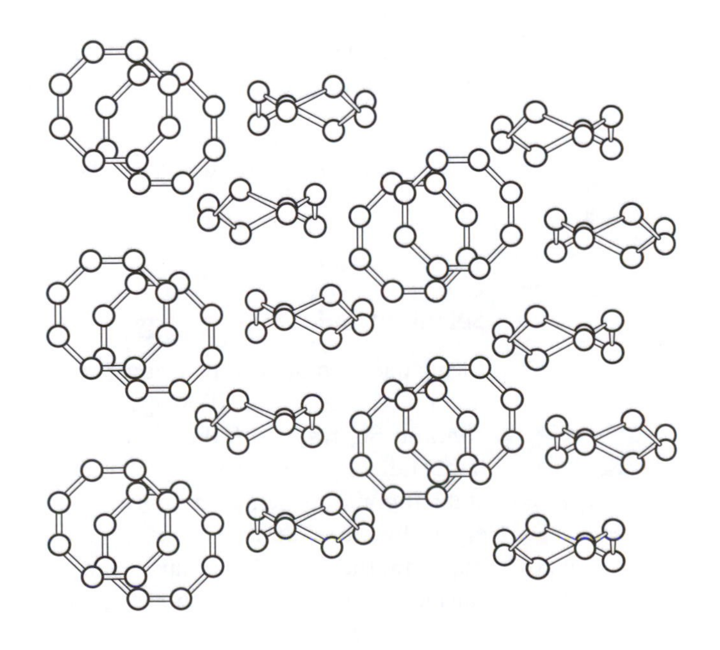

7.14 The arrangement of S_8 rings in the solid state of orthorhombic sulfur (the standard state of the element). From: B. Meyer, *Chem. Rev.*, **76**, 367 (1976).

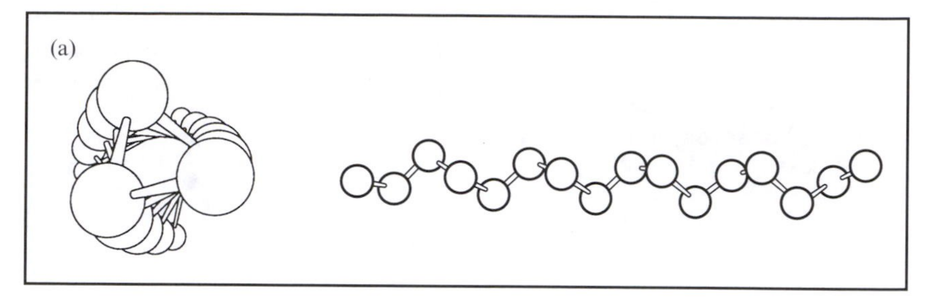

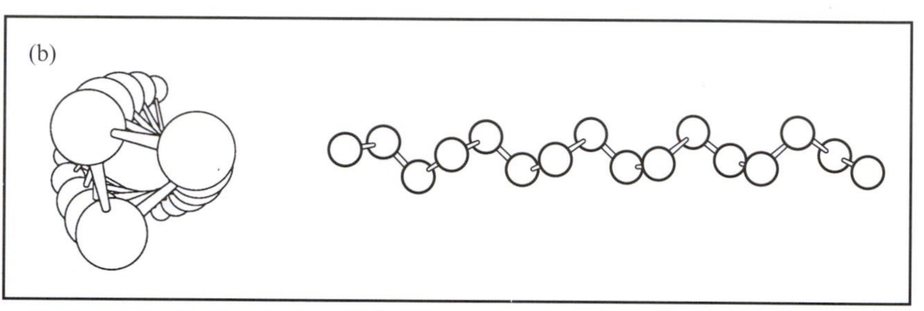

7.15 A strand of helical sulfur (S_∞) has a handedness: (a) a right-handed helix and (b) a left-handed helix. The two chains are non-superimposable.

Most allotropes of sulfur contain cyclic units (Table 3.2) but in some, S_x chains of various lengths are present. Each chain contains S–S single bonds and forms a helical structure (Figure 7.15). An important property of a helix is its *handedness*. It can turn in either a right-handed or a left-handed manner, each form is distinct from the other and they cannot be superimposed.

There are different forms of *poly*catena*sulfur* which contain mixtures of rings and chains, and these include rubbery and plastic sulfur. Filaments of these can be drawn from molten sulfur, their compositions alter with time, and at 298 K, transformation into orthorhombic sulfur eventually occurs. Two examples of well-characterized allotropes which contains helical chains are fibrous and laminar sulfur. In fibrous sulfur, the chains lie parallel to one another and equal numbers of left- and right-handed helices are present. In laminar sulfur, there is some criss-crossing of the helical chains.

The prefix *catena* is used within the IUPAC nomenclature to mean a chain structure.

Selenium and tellurium (group 16)

Elemental selenium and tellurium form both rings and helical chains, and selenium possesses several allotropes. Crystalline monoclinic selenium is red and contains Se_8 rings with the same crown shape as S_8 (Figure 7.13). The standard state of the element is grey (or metallic) selenium, and in the crystalline state, it contains helical chains of selenium atoms (Se_∞). Tellurium has one crystalline form and this contains helical Te_∞ chains. In both this and grey selenium, the axes of the chains lie parallel to each other, and a view through each lattice down the axes shows the presence of a hexagonal network.

Se_∞ or Te_∞ = chain of infinite length

Phosphorus (group 15)

Standard state: see Section 1.17

The standard state of phosphorus is 'white phosphorus'. This allotrope is, uniquely, *not* the thermodynamically most stable state of the element. The most stable crystalline form of the element is black phosphorus, and this, and red phosphorus, are described in Section 7.10.

Crystalline white phosphorus contains tetrahedral P_4 molecules (Figure 7.16). The intramolecular P–P distance is 221 pm, consistent with the presence of P–P single bonds (r_{cov} = 110 pm).

7.16 The tetrahedral P_4 molecular unit present in white phosphorus. All the P–P distances are equal.

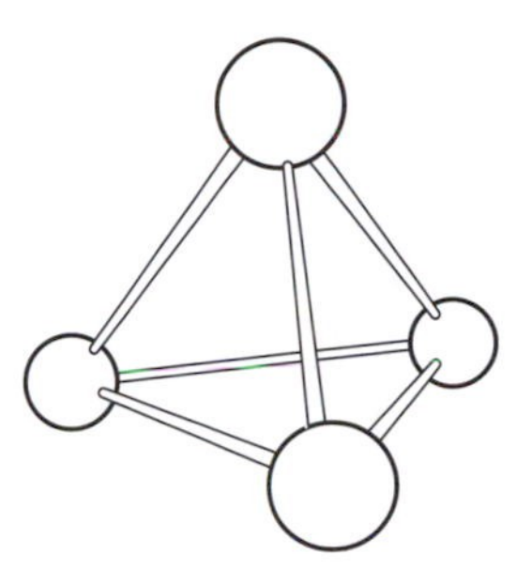

7.9 A MOLECULAR ALLOTROPE OF CARBON: C_{60}

Diamond and graphite: see Section 7.10

The allotropes of carbon that have, in the past, been most commonly cited are diamond and graphite. Since the mid-1980s, new allotropes of carbon – the *fullerenes* – have been recognized.

The fullerenes are discrete molecules, and the most widely studied is C_{60} (Figure 7.17a). The spherical shell of 60 atoms is made up of five- and six-membered rings and the carbon atoms are equivalent. Each five-membered ring (a pentagon) is connected to five six-membered rings (hexagons). No five-membered rings are adjacent to each other.

Restricted geometry of carbon: see Section 5.14

The geometry about a carbon atom is usually either linear, trigonal planar or tetrahedral, and although apparently complex, the structure of C_{60} complies with this restriction. Each carbon atom in C_{60} is covalently bonded to

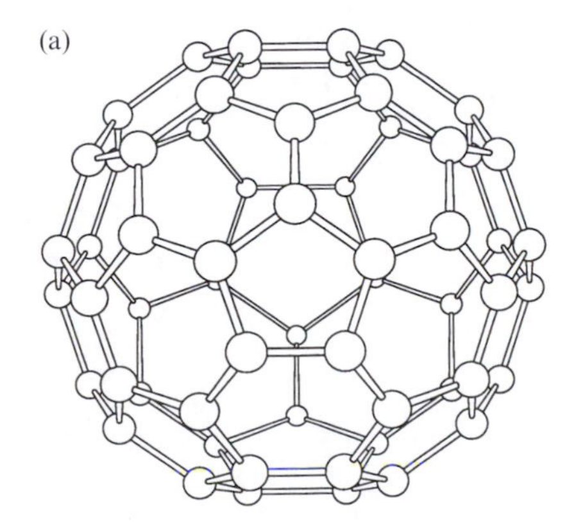

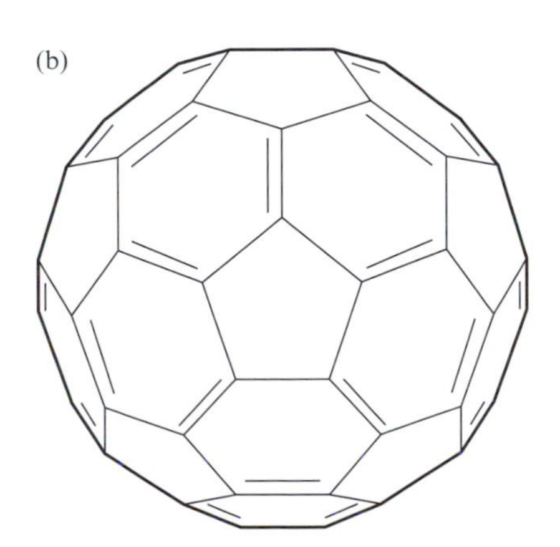

7.17 One of the fullerenes – C_{60}. (a) The C_{60} molecule is made up of fused 5- and 6-membered rings of carbon atoms which form an approximately spherical molecule.
(b) A representation of C_{60} showing only the upper surface (in the same orientation as in (a)) illustrating the localized single and double carbon–carbon bonds.

Box 7.3 Why the name fullerene?

The architect Robert Buckminster Fuller has designed geodesic domes such as the one built at EXPO '67 (an international exhibition) in Montreal. The geodesic dome was constructed of hexagonal motifs but on its own, the network can only lead to a planar sheet. The placement of pentagonal panels at intervals in the structure leads to a curvature of the surface, sufficient to construct a dome. The structure of C_{60} is also represented in a football (soccer ball), which has pentagonal (often black) and hexagonal (often white) panels. C_{60} – buckminsterfullerene – has also been christened 'bucky-ball'.

The name 'fullerene' has been given to the class of near spherical C_n allotropes which include C_{60}, C_{70} and C_{84}.

A very readable article that conveys the excitement of the discovery of C_{60} has been written by Harold W. Kroto: 'C_{60}: Buckminsterfullerene, The Celestial Sphere that Fell to Earth.' *Angewandte Chemie, International Edition in English*, (1992) vol. 31, p.111.

In October 1996, Sir Harry Kroto of Sussex University, UK, and Professors Richard Smalley and Robert Curl of Rice University, USA, were awarded the Nobel Prize for Chemistry for their pioneering work on C_{60}.

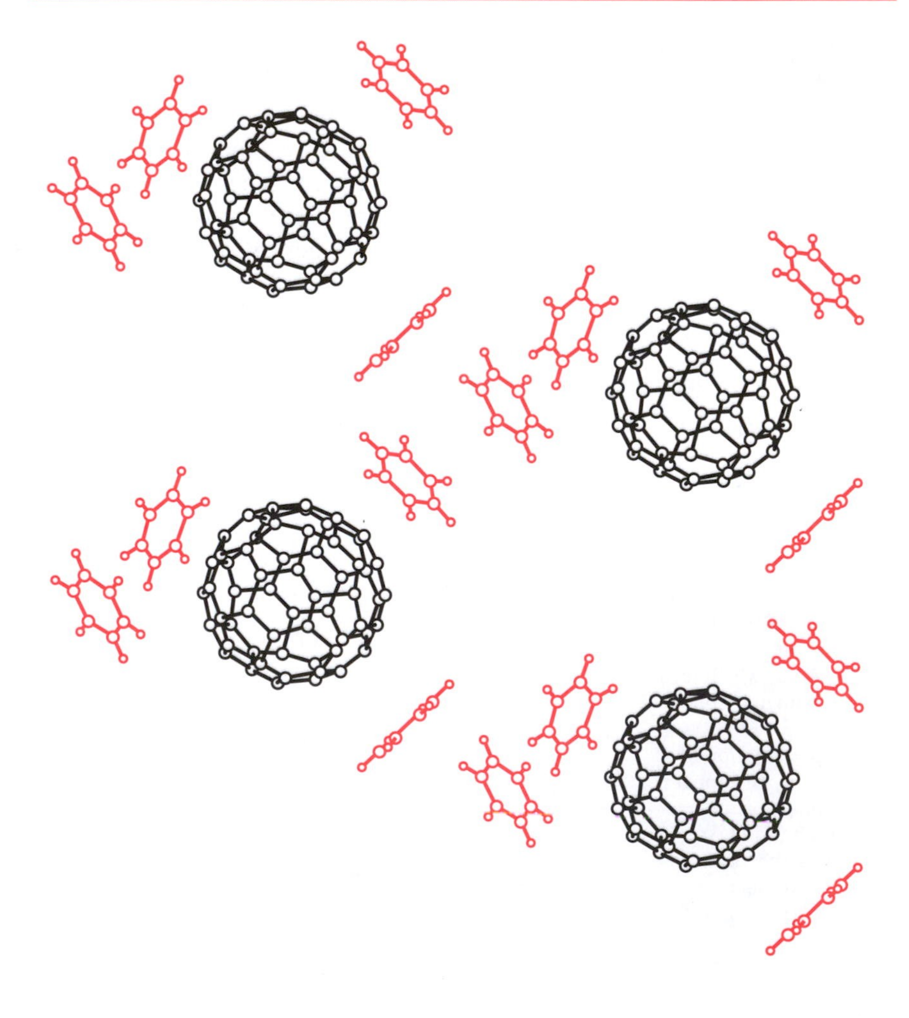

7.18 Part of the solid state structure of $C_{60}{\cdot}4C_6H_6$. The C_{60} molecules form an ordered array with the benzene molecules in between them. The formula $C_{60}{\cdot}4C_6H_6$ indicates that one mole of C_{60} crystallizes with four moles of benzene; this ratio is apparent in the diagram.

Box 7.4 Superconductivity: Alkali metal fullerides M_3C_{60}

Alkali metals, M, reduce C_{60} to give fulleride salts of type $[M^+]_3[C_{60}]^{3-}$ and at low temperatures, some of these compounds become *superconducting*. A superconductor is able to conduct electricity without resistance and, until 1986, no compounds were known that were superconductors above 20 K. The temperature at which a material becomes superconducting is called its critical temperature (T_c), and in 1987, this barrier was broken – *high-temperature superconductors* were born. Many high-temperature superconductors are metal oxides, for example $YBa_2Cu_4O_8$ (T_c = 80 K), $YBa_2Cu_3O_7$ (T_c = 95 K), $Ba_2CaCu_2Tl_2O_8$ (T_c = 110 K) and $Ba_2Ca_2Cu_3Tl_2O_{10}$ (T_c = 128 K).

The M_3C_{60} fulleride superconductors are structurally simpler than the metal oxide systems, and can be described in terms of the alkali metal cations occupying the interstitial holes in a lattice composed of close-packed C_{60} cages. Each $[C_{60}]^{3-}$ anion is approximately spherical and a close-packing of spheres approach is valid. In K_3C_{60} and Rb_3C_{60}, the $[C_{60}]^{3-}$ cages are arranged in a face-centred cubic (fcc) arrangement:

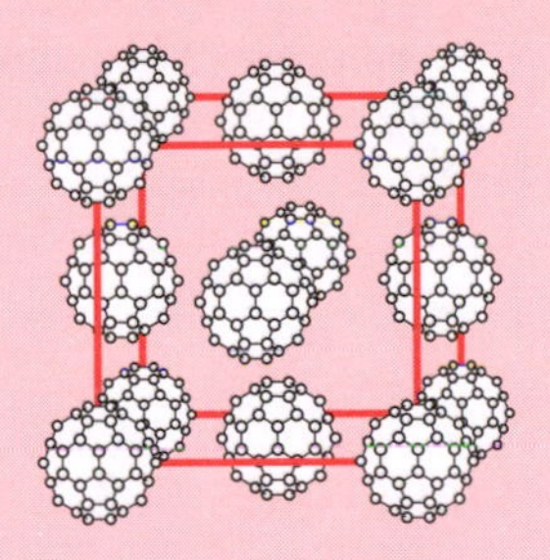

If you look back at Box 7.1, you will see that the fcc unit cell contains an octahedral hole and eight tetrahedral holes. There are also 12 octahedral holes shared between adjacent unit cells. The alkali metal cations in K_3C_{60} and Rb_3C_{60} completely occupy the octahedral (black) and tetrahedral (red) holes:

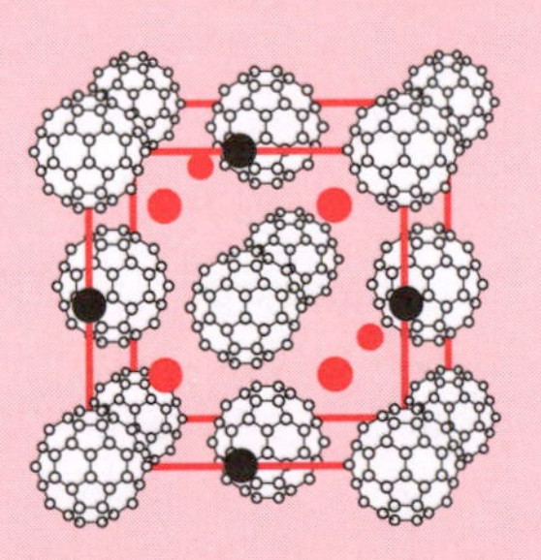

The values of T_c for K_3C_{60} and Rb_3C_{60} are 18 K and 28 K respectively, but for Cs_3C_{60} (in which the C_{60} cages adopt a body-centred cubic lattice), T_c = 40 K. Cs_3C_{60} is (at present) the highest temperature superconductor of this family of alkali metal fullerides. [What kind of interstitial holes can the Cs^+ ions occupy in the bcc lattice?] Na_3C_{60} is structurally related to its potassium and rubidium analogues, but it is not superconducting. This area of fullerene chemistry is actively being pursued with hopes of further raising the T_c barrier.

A series of well-illustrated articles describing various aspects of superconductivity can be found in *Chemistry in Britain*, (1994) vol. 30, pp.722–748.

three others in an approximately trigonal planar arrangement. Since the surface of the C_{60} molecule is relatively large, the deviation from planarity at each carbon centre is small. The C–C bonds in C_{60} fall into two groups – the bonds at the junctions of two hexagonal rings (139 pm) and those at the junctions of a hexagonal and a pentagonal ring (145 pm). Figure 7.17b shows the usual representation of C_{60}, with carbon–carbon double and single bonds.

In the solid state at 298 K, the spherical C_{60} molecules are arranged in a close-packed structure. However, most single crystal X-ray diffraction studies of C_{60} have involved solvated samples rather than the pure solid element. For example, C_{60} is soluble in benzene (C_6H_6), and single crystals grown by evaporating solvent from a solution of C_{60} in benzene have the composition $C_{60}{\cdot}4C_6H_6$. Figure 7.18 shows part of the crystal lattice of $C_{60}{\cdot}4C_6H_6$. The C_{60} molecules are arranged in an ordered manner with the benzene molecules occupying the spaces between them.

When crystals of a substance are grown from a solution, they may contain *solvent of crystallization*, the presence of which is indicated in the molecular formula.

7.10 STRUCTURES OF SOLIDS CONTAINING INFINITE COVALENT LATTICES

Some non-metallic elements in the *p*-block crystallize with *infinite lattice* structures. (These are also called giant, or extended, lattices.) Diamond and graphite are well-known examples and are described below along with allotropes of boron, silicon, phosphorus, arsenic and antimony. When these elements melt, *covalent bonds* are broken.

Boron (group 13)

The standard state of boron is the β-rhombohedral form. The structure of this allotrope is complex and we begin the discussion instead with α-rhombohedral boron. The basic building-block of both α- and β-rhombohedral boron is an icosahedral B_{12}-unit (Figure 7.19). Each boron atom is covalently bonded to another five boron atoms within the icosahedron, despite the fact that a boron atom only has three valence electrons. The bonding within each B_{12}-unit is *delocalized* and it is important to remember that the B–B connections in Figure 7.19 are *not* two-centre two-electron bonds.

The structure of α-rhombohedral boron consists of B_{12}-units arranged in an approximately cubic close-packed manner. The boron atoms of the icosahedral unit lie on a spherical surface, and so the close-packing of spheres is an appropriate way in which to describe the solid state structure. However, unlike the close-packed arrays described earlier, the 'spheres' in α-rhombohedral boron are *covalently linked* to each other. Part of the structure (one

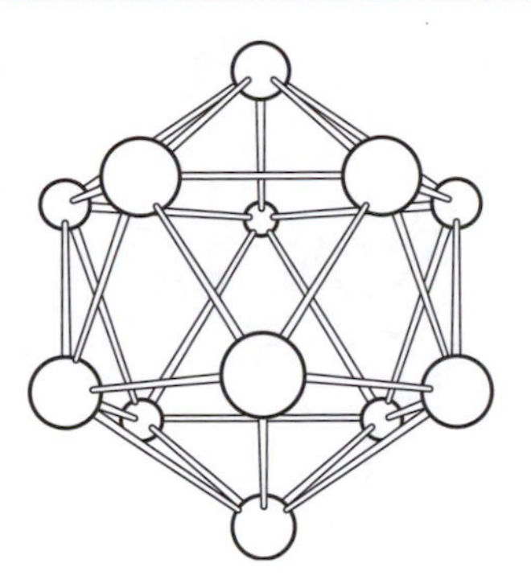

7.19 The B_{12}-icosahedral unit that is the fundamental building block in both α- and β-rhombohedral boron. These allotropes possess infinite covalent lattices in the solid state.

layer of the infinite lattice) is shown in Figure 7.20, and such layers are arranged in an ABCABC... fashion (Figures 7.3 to 7.5). The presence of B–B covalent bonding interactions between the B_{12}-units distinguishes this as a infinite covalent lattice rather than a true close-packed assembly.

The structure of β-rhombohedral boron consists of B_{84}-units, linked together by B_{10}-units. Each B_{84}-unit is conveniently described in terms of three subunits – $(B_{12})(B_{12})(B_{60})$. At the centre of the B_{84}-unit is a B_{12}-icosahedron (Figure 7.21a) and radially attached to each boron atom is another boron atom (Figure 7.21b). The term 'radial' is used to signify that the bonds which connect the second set of 12 boron atoms to the central B_{12}-unit point outwards from the centre of the unit. The $(B_{12})(B_{12})$ sub-unit so-formed lies inside a B_{60}-cage which has the same geometry as a C_{60} molecule (compare Figures 7.21c and 7.17a). The whole B_{84}-unit is shown in Figure 7.21d.

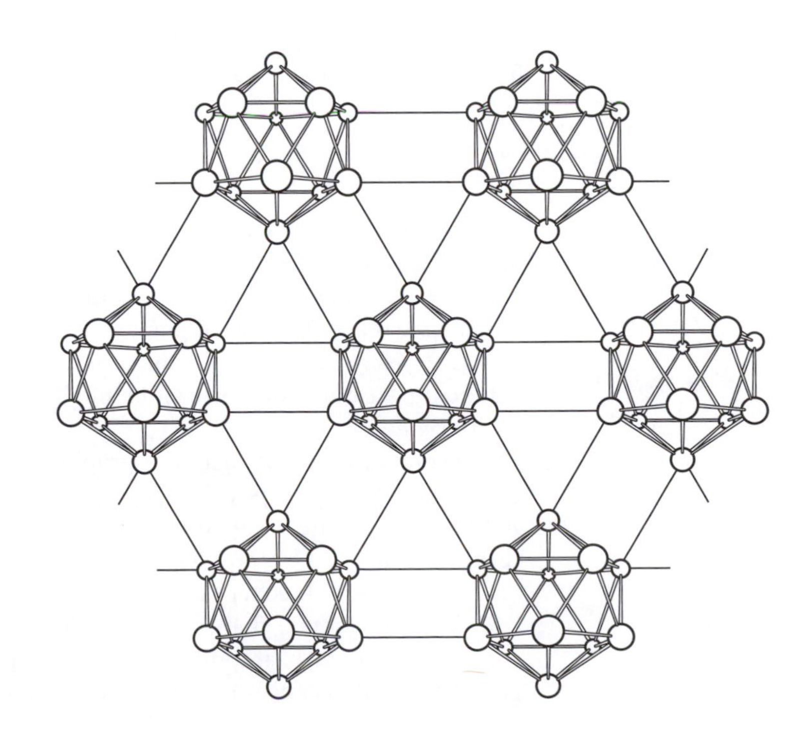

7.20 Part of one layer of the infinite lattice of α-rhombohedral boron. The building-blocks are B_{12}-icosahedra. The overall stucture may be considered to consist of spheres in a cubic close-packed arrangement. Delocalized, covalent bonding interactions between the B_{12}-units support the framework of the infinite lattice making it rigid.

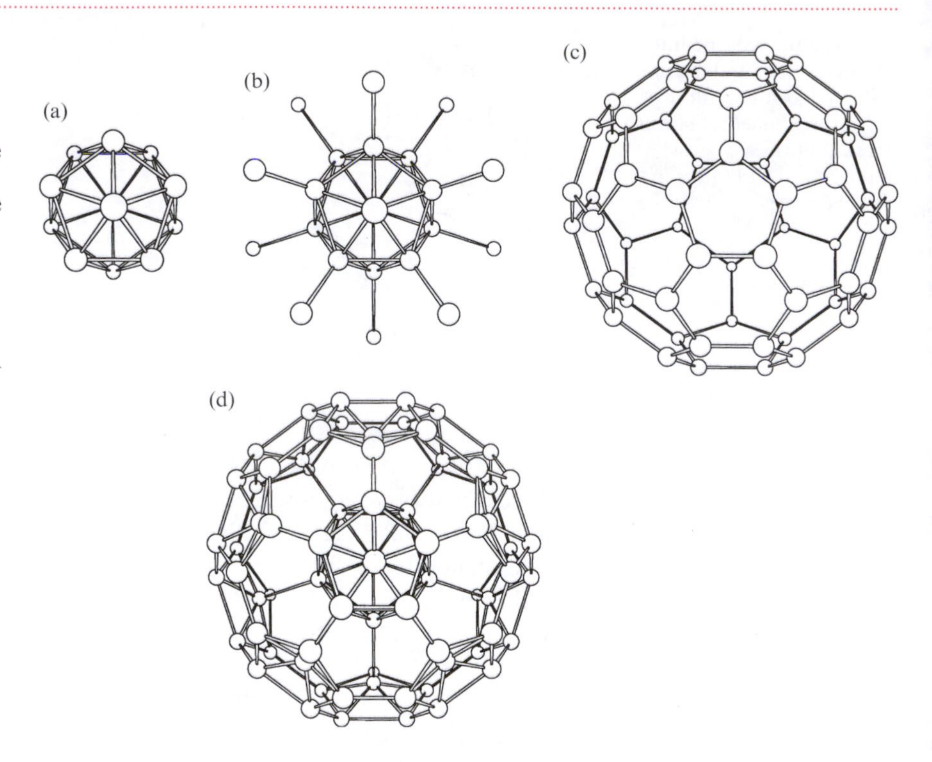

7.21 The construction of the B_{84}-unit that is the main building block in the infinite lattice of β-rhombohedral boron. (a) In the centre of the unit is a B_{12}-icosahedron. (b) To each boron atom in the central icosahedron, another boron atom is attached. (c) A B_{60}-cage is the outer 'skin' of the B_{84}-unit. (d) The final B_{84}-unit contains three, covalently bonded sub-units – $(B_{12})(B_{12})(B_{60})$.

Notice that each of the boron atoms of the second B_{12}-unit is connected to a pentagonal ring of the B_{60}-sub-unit. In the infinite lattice structure of β-rhombohedral boron, the B_{84}-units shown in Figure 7.21d are interconnected by B_{10}-units. The whole network is extremely rigid, as is that of the α-rhombohedral allotrope, and crystalline boron is very hard, and has a particularly high melting point (2453 K for β-rhombohedral boron).

Carbon (group 14)

Diamond and graphite are allotropes of carbon and possess infinite covalent lattices in the solid state. They differ remarkably in their physical appearance and properties. Crystals of diamond are transparent, extremely hard and are highly prized for jewellery. Crystals of graphite are black and have a slippery feel.

Figure 7.22 shows part of the crystalline structure of diamond. Each tetrahedral carbon atom forms four single C–C bonds and the overall lattice is very rigid.

The standard state of carbon is graphite. Strictly, this is α-graphite, since there is another allotrope called β-graphite. The structures of both α- and β-graphite are infinite covalent lattices, composed of parallel planes of fused hexagonal rings. In α-graphite, the assembly contains two repeating layers, whereas there are three repeat units in the β-form. Heating β-graphite above 1298 K brings about a change to α-graphite.

The structure of α-graphite ('normal' graphite) is shown in Figure 7.23. The C–C bond distances *within* a layer are equal (142 pm) and the distance

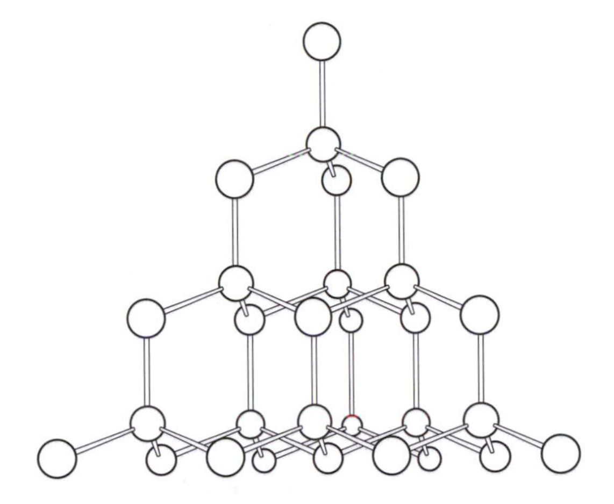

7.22 Part of the infinite covalent lattice of diamond, an allotrope of carbon.

Box 7.5 The relationship between the structure of diamond and zinc blende

The structure of zinc blende (ZnS) was discussed in Section 6.13, and the unit cell was shown in Figure 6.17. If all the zinc(II) and sulfide centres in this structure are replaced by carbon atoms, the unit cell shown below results:

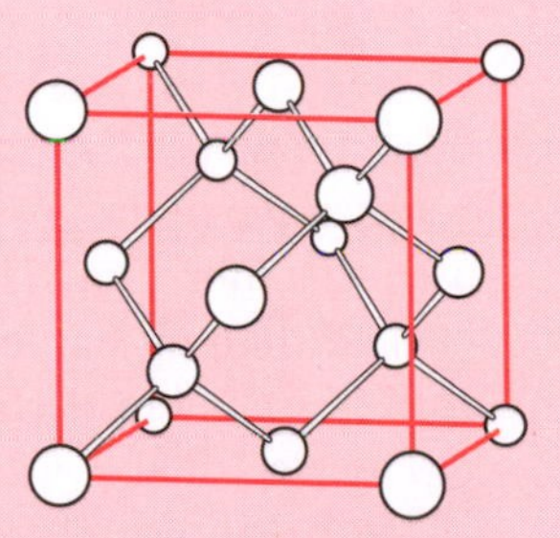

If we view this unit cell from a different angle, the same representation of the diamond lattice that we showed in Figure 7.22 becomes apparent:

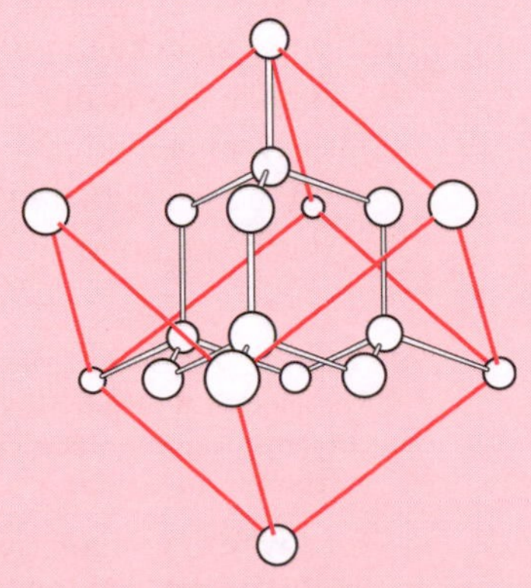

We can therefore see that the diamond and zinc blende structures are related.

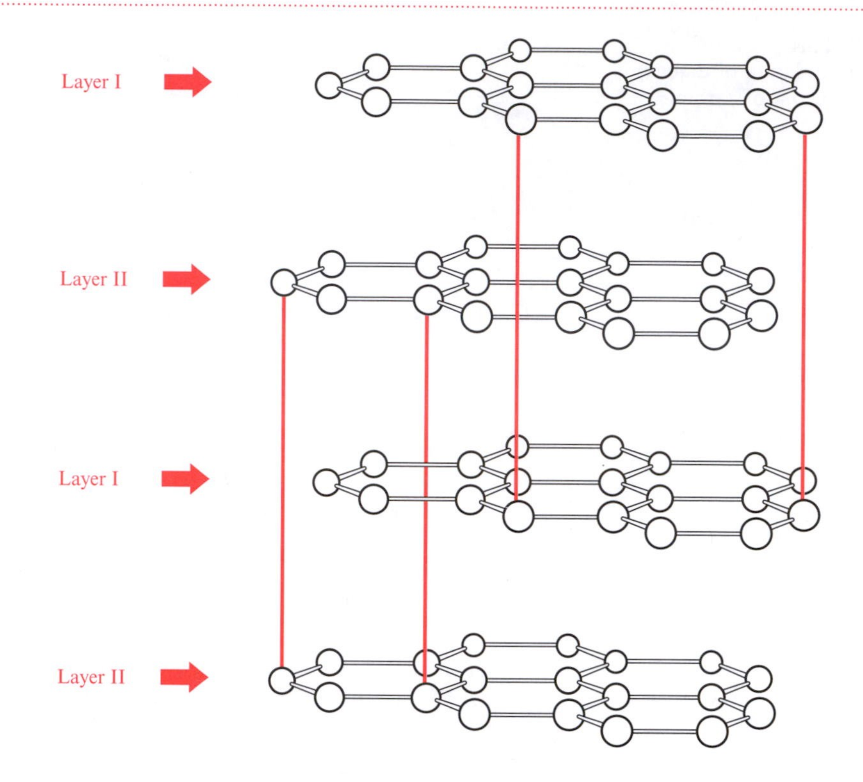

7.23 Part of the infinite covalent lattice of normal (α) graphite. This allotrope is the standard state of carbon. There are two repeating layers consisting of fused hexagonal rings. The red lines between the layers indicate which carbon atoms lie directly over which other atoms.

$r_{cov}(C) = 77$ pm
$r_v(C) = 185$ pm

between two adjacent layers is 335 pm, indicating that *inter*-layer interactions are weak. The physical nature of graphite reflects this; it cleaves readily *between* the planes of hexagonal rings. This property allows graphite to be used as a lubricant. The carbon atoms in normal graphite are less efficiently packed than in diamond; the densities of α-graphite and diamond are 2.3 and 3.5 g cm^{-3} respectively.

The electrical conductivity of graphite is an important property of this allotrope, and can be explained in terms of the structure and bonding. The valence electronic configuration of carbon is $[He]2s^22p^2$, and as each carbon atom forms a covalent bond to each of three other atoms in the same layer, one valence electron remains unused. The odd electrons may be conducted through the planes of hexagonal rings. The electrical resistivity of α-graphite is 1.3×10^{-5} Ω m at 293 K in a direction through the planes of hexagonal rings, but is about 1 Ω m in a direction perpendicular to the planes. The electrical resistivity of diamond is 1×10^{11} Ω m, and diamond is an excellent *insulator*. All valence electrons in diamond are involved in localized C–C single bond formation.

The electrical resistivity of a material measures its resistance (to an electrical current). A good electrical conductor has a very low resistivity, and the reverse is true for an insulator.

For a wire of uniform cross section, the resistivity (ρ) is given in units of ohm metre (Ω m) where the resistance is in ohms, and the length of the wire is measured in metres.

Box 7.6 Carbon nanotubes

We usually think of graphite sheets as being flat (see Figure 7.23) but research over the last few years has shown that it is possible to make 'tube-like' structures about 10 nm in diameter consisting of rolled graphite sheets – these are known as *nanotubes*. Carbon nanotubes can now be made on a macroscopic scale – an electric arc between a graphite anode and graphite cathode in a helium atmosphere leads to the formation of the tubes on the cathode surface. Recent collaborative work between research groups in Switzerland and Brazil has resulted in the production of a single layer of carbon nanotubes aligned vertically on a polymer surface. The monolayer can be made to emit electrons and these can be detected at an anode. The potential application of this device is in the construction of flat-screen visual displays such as those in portable lap-top computers. It is hoped that carbon nanotubes will be one answer to the enhancement of the quality of such flat-screens.

An article that presents the three-dimensional beauty of nanotubes and other C_n-species is: 'Beyond C_{60}: Graphite Structures for the Future', *Chemical Society Reviews*, (1995) vol. 24, p.341.

Silicon, germanium and tin (group 14)

In the solid state at 298 K silicon crystallizes with a diamond-type lattice (Figure 7.22). Germanium and the grey allotrope of tin also adopt this infinite structure but the character of these elements puts them into the category of *semi-metals* rather than non-metals. We discuss these allotropes further in Section 7.11, but it is instructive here to compare some of their physical properties (Table 7.2), particularly their electrical resistivities, to understand why a distinction is made between the group 14 elements.

Table 7.2 Some physical and structural properties of diamond, silicon, germanium and grey tin. All share a common infinite lattice type (Figure 7.22).

Element	*Appearance of crystalline solid*	*Melting point / K*[a]	*Enthalpy of fusion / kJ mol^{-1}*	*Density / g cm^{-3}*	*Interatomic distance in the crystal lattice / pm*	*Electrical resistivity / Ω m (temperature)*
Carbon (diamond)	Transparent	3820	105	3.5	154	1×10^{11} (293 K)
Silicon	Blue-grey, lustrous	1683	40	2.3	235	1×10^{-3} (273 K)
Germanium	Grey-white, lustrous	1211	35	5.3	244	0.46 (295 K)
Tin (grey allotrope)	Dull grey	–	–	5.75	280	11×10^{-8} (273 K)

[a]The grey allotrope of tin is the low-temperature form of the element. Above 286 K the white form is stable; this melts at 505 K.

Phosphorus, arsenic, antimony and bismuth (group 15)

White phosphorus: see Section 7.8

Although it is defined as the standard state of the element, white phosphorus is a metastable state. Other allotropes of phosphorus are either amorphous, or possess infinite covalent lattices, but on melting, all allotropes give a liquid state containing P_4 molecules.

Amorphous red phosphorus is more dense and less reactive than white phosphorus. Crystallization of red phosphorus in molten lead produces a monoclinic allotrope called Hittorf's (violet) phosphorus. The solid state structure is a complicated infinite lattice and two views of the repeat unit are shown in Figure 7.24. Units of this type are connected end-to-end to form chain-like arrays of three-coordinate phosphorus atoms. These chains lie parallel to each other forming layers, but within a layer the chains are not bonded together. An infinite lattice is created by placing the layers one on top of another, such that an atom of the type labelled P' in Figure 7.24 is covalently bonded to another similar atom (labelled P") in an adjacent sheet. The chains in one layer lie at right angles to the chains in the next bonded layer, giving a criss-cross network overall. The P–P bond distances in the lattice are all similar ($\approx$ 222 pm) and are consistent with single covalent bonds.

Black phosphorus is the most thermodynamically stable form of the element. The solid state structure of the rhombohedral form of black phosphorus consists of a hexagonal net of P_6-rings (Figure 7.25) and hexagonal P_6-units are also present in the orthorhombic form of the element.

In the solid state, the rhombohedral allotropes of arsenic, antimony and bismuth are isostructural with the rhombohedral form of black phosphorus. Down group 15, there is a tendency for the coordination number of each atom to change from three (atoms within a layer) to six (three atoms within a layer and three in the next layer). These allotropes of arsenic, antimony and bismuth are known as the 'metallic' forms, and the metallic character of the element increases as group 15 is descended.

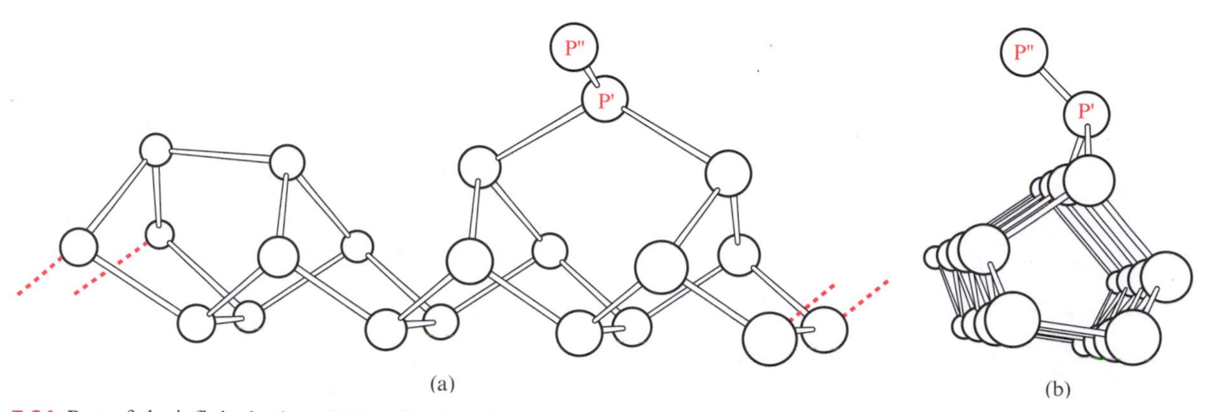

7.24 Part of the infinite lattice of Hittorf's phosphorus. (a) Part of the chain-like arrays of atoms; the repeat unit contains 21 atoms, and atoms P′ and P″ are equivalent atoms in adjacent chains; the chains are linked through the P′–P″ bond. (b) The same unit viewed from the end, emphasizing the channels that run through the structure.

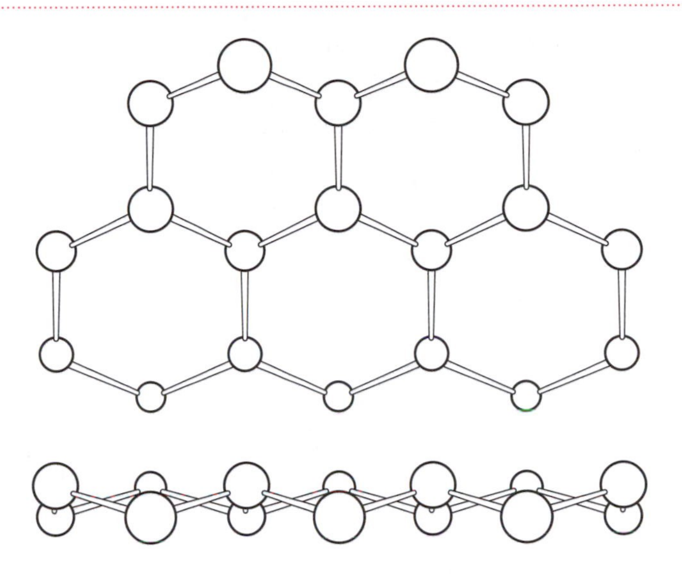

7.25 Layers of puckered six-membered rings are present in the structures of black phosphorus and the rhombohedral allotropes of arsenic, antimony and bismuth, all group 15 elements.

7.11 THE STRUCTURES OF METALLIC ELEMENTS AT 298 K

Elements in groups 1 and 2 (the *s*-block) and the *d*-block are metallic in character. In the *p*-block, a diagonal line *approximately* separates non-metallic from metallic elements (Figure 7.26) although the distinction is *not* clear-cut. The solid state structures of metals are readily described in terms of the packing of their atoms. Table 7.3 lists lattice types for metals of the *s*- and *d*-blocks, and also gives the melting points of these elements.

s-Block metals

In the solid state at 298 K, the atoms of each alkali metal (group 1) are packed in a body-centred cubic arrangement (Figure 7.9). All these metals are soft and have relatively low melting points (Table 7.3). The enthalpies of fusion are correspondingly low, decreasing down group 1 from 3.0 kJ mol^{-1} for lithium to 2.1 kJ mol^{-1} for caesium.

With the exception of barium which has a bcc lattice, the alkaline earth metals (group 2) possess hexagonal close-packed structures, and their melting points are higher than those of the group 1 metals.

Group 13	Group 14	Group 15	Group 16	Group 17	Group 18
B	C	N	O	F	Ne
Al	Si	P	S	Cl	Ar
Ga	Ge	As	Se	Br	Kr
In	Sn	Sb	Te	I	Xe
Tl	Pb	Bi	Po	At	Rn

7.26 A 'diagonal' line is often drawn through the *p*-block to indicate *approximately* the positions of the non-metals (shown in grey) and the metals (shown in red). The distinction is not clear-cut and elements lying along the line may show the characteristics of both, being classed as semi-metals.

Table 7.3 Structures (at 298 K) and melting points (K) of the metallic elements:
◆ = hexagonal close-packed
□ = cubic close-packed (face-centred cubic)
● = body-centred cubic.

1	2	3	4	5	6	7	8	9	10	11	12
Li ● 454	Be ◆ 1560										
Na ● 371	Mg ◆ 923										
K ● 337	Ca ◆ 1115	Sc ◆ 1814	Ti ◆ 1941	V ● 2183	Cr ● 2180	Mn see text 1519	Fe ● 1811	Co ◆ 1768	Ni □ 1728	Cu □ 1358	Zn ◆ 693
Rb ● 312	Sr ◆ 1050	Y ◆ 1799	Zr ◆ 2128	Nb ● 2750	Mo ● 2896	Tc ◆ 2430	Ru ◆ 2607	Rh □ 2237	Pd □ 1828	Ag □ 1235	Cd ◆ 594
Cs ● 301	Ba ● 1000	La ◆ 1193	Hf ◆ 2506	Ta ● 3290	W ● 3695	Re ◆ 3459	Os ◆ 3306	Ir □ 2719	Pt □ 2041	Au □ 1337	Hg see text 234

d-Block metals

At 298 K, the structures of the metals in the *d*-block are either hcp, ccp or bcc, with the exceptions of mercury and manganese (see below). Table 7.3 shows that the lattice type adopted depends, in general, on the position of the metal in the periodic table. For most of the *d*-block metals, a close-packed structure is observed at 298 K, and a bcc structure is present as a high temperature form of the element. Iron is unusual in that it adopts a bcc structure at 298 K, transforms to an fcc lattice at 1179 K, and reverts to a bcc structure at 1674 K.

The melting points of the metals in groups 3 to 11 are far higher than those of the *s*-block metals, and enthalpies of fusion are correspondingly higher. Two *d*-block elements are worthy of special note. The first is mercury. All three metals in group 12 stand out amongst the *d*-block elements in possessing relatively low melting points, but mercury is well known for the fact that it is a liquid at 298 K. Its enthalpy of fusion is 2.3 kJ mol^{-1}, a value that is atypical of the *d*-block elements but is, rather, similar to those of the alkali metals. In the crystalline state, mercury atoms are arranged in a distorted simple cubic lattice (Figure 7.8). The second metal of interest is manganese. Atoms in the solid state are arranged in a complex cubic lattice in such a way that there are four atom types with coordination numbers of 12, 13 or 16. The reasons for this deviation from one of the more common structural types are not simple to understand.

Metals and semi-metals in the *p*-block

The physical and chemical properties of the heavier elements in groups 13, 14 and 15 indicate that these elements are metallic; elements which are intermediate between metals and non-metals are termed semi-metals, for example germanium. The structures described below are those observed at 298 K.

Elemental boron: see Section 7.10

Aluminium possesses a cubic close-packed lattice, typical of a metallic element. The melting point (933 K) is only slightly higher than that of magnesium, the element preceding it in the periodic table, and is dramatically lower than that of boron (2453 K). Atoms of thallium form an hcp structure typical of a metal, and indium possesses a distorted close-packed arrangement of atoms. The solid state structure of gallium is not so easily described; there is one nearest-neighbour (247 pm) and six other close atoms at distances between 270 and 279 pm. Gallium has a low melting point (303 K) which means that it is a liquid metal in some places in the world but a solid in others! The crystalline state of gallium is in between that of a metal and a molecular solid containing Ga_2 units.

The 'diagonal' line in Figure 7.26 passes through group 14 between silicon and germanium suggesting that we might consider silicon to be a non-metal, and germanium a metal. But the distinction is not clear-cut. In the solid state, both elements have the same infinite covalent lattice as diamond, but their electrical resistivities are significantly lower than that of diamond (Table 7.2), indicating metallic behaviour. The heaviest element, lead, possesses a ccp lattice. The intermediate element is tin. White (β) tin is the stable allotrope at 298 K, but at temperatures below 286 K, this transforms into the grey α-form which has a diamond-type lattice (Table 7.2). The structure of white tin is related to that of the grey allotrope by a distortion of the lattice such that each tin atom goes from having four to six nearest-neighbours. The density of white tin (7.31 g cm^{-3}) is greater than that of the grey allotrope (5.75 g cm^{-3}) and this is an unusual observation in going from a low to higher temperature elemental form. The transition from white to grey tin is quite slow, but it can be dramatic. In the 19th century, military uniforms used tin buttons which crumbled in exceptionally cold winters. Similarly, in 1851, the citizens of Zeitz were alarmed to discover that the tin organ pipes in their church had crumbled to powder!

The structure of bismuth was described in Section 7.10.

7.12 METALLIC RADIUS

Van der Waals radius: see Section 2.21
Covalent radius: see Section 3.3
Ionic radius: see Section 6.14

The *metallic radius* is half of the distance between the *nearest-neighbour* atoms in a solid state metallic lattice, and Table 7.4 lists metallic radii for the *s*- and *d*-block elements. Atom size increases down each of groups 1 and 2. In each of the triads of the *d*-block elements, there is generally an increase in radius in going from the first to second row element, but very little change in size in going from the second to the third row metal. This latter observation is due to the presence of a filled 4*f* level, and the so-called *lanthanoid con-*

Table 7.4 Metallic radii (pm) of the *s*- and *d*-block metals

1	2	3	4	5	6	7	8	9	10	11	12
Li 157	Be 112										
Na 191	Mg 160										
K 235	Ca 197	Sc 164	Ti 147	V 135	Cr 129	Mn 137	Fe 126	Co 125	Ni 125	Cu 128	Zn 137
Rb 250	Sr 215	Y 182	Zr 160	Nb 147	Mo 140	Tc 135	Ru 134	Rh 134	Pd 137	Ag 144	Cd 152
Cs 272	Ba 224	La 187	Hf 159	Ta 147	W 141	Re 137	Os 135	Ir 136	Pt 139	Au 144	Hg —

traction – the first row of lanthanoid elements lies between lanthanum (La) and hafnium (Hf). The poorly shielded 4*f* electrons are relatively close to the nucleus and have little effect on the observed radius.

The metallic radius is half of the distance between the *nearest-neighbour* atoms in a solid state metal lattice.

7.13 METALLIC BONDING

Metals are electrical conductors

One physical property that characterizes a metal is its low electrical resistivity – that is, a metal conducts electricity very well. With the exception of mercury, all elements which are metallic at 298 K are solid. Although the packing of atoms is a convenient means of describing the solid state structure of metals, it gives no feeling for the communication that there must be between the atoms. Communication is implicit in the property of electrical conductivity, since electrons must be able to flow through an assembly of atoms in a metal. In order to understand why metals are such good electrical conductors, we must first consider the bonding between atoms in a metal.

An *electrical conductor* offers a low resistance (measured in ohms, Ω) to the flow of an electrical current (measured in amperes, A). An *insulator* offers a high resistance.

A 'sea of electrons'

An early approach to metallic bonding (the Drude–Lorentz theory) was to consider a model in which the valence electrons of each metal atom were free to move in the crystal lattice. Thus, instead of simply being composed of neutral atoms, the metal lattice is considered to be an assembly of positive ions (the nuclei surrounded by their core electrons) and electrons (the valence electrons). When a potential difference[§] is applied across a piece of metal, the valence electrons move from high to low potential and a current flows.

In this model, a metallic element is considered to consist of positive ions (arranged, for example, in a close-packed manner) and a 'sea of electrons'. The theory gives a satisfactory general explanation for the conduction of electricity but cannot account for the detailed variation of electrical conductivities amongst the metallic elements. Several other theories have been described, of which *band theory* is the most general.

Band theory

Band theory follows from a consideration of the energies of the molecular orbitals of an assembly of metal atoms. In constructing the ground state electronic configuration of a molecule, we have previously applied the *aufbau* principle and have arranged the electrons so as to give the lowest energy state. In the case of a degeneracy, this may give singly occupied highest lying molecular orbitals (for example in the B_2 and O_2 molecules).

A molecular orbital diagram that describes the bonding in a metallic solid is characterized by having a large number of orbitals which are very close in energy. In this case, they form a continuous set of energy states called a *band*. These arise as follows. In an LCAO approach, we consider the interaction between similar AOs, for example 2*s* with 2*s*, and 2*p* with 2*p*. Figure 7.27 shows the result of the interaction of 2*s* AOs for different numbers of lithium atoms. *The energies of these 2s AOs are the same*. If two Li atoms combine, the overlap of the two 2s AOs leads to the formation of two MOs. If there are three lithium atoms, three MOs are formed, and if there are four metal atoms, four MOs result. For an assembly containing *n* lithium atoms, there must be *n* molecular orbitals, but because the 2*s* AOs were all of the same energy, the energies of the resultant MOs are very close together and can be described as a *band* of orbitals. The occupation of the band depends upon the number of valence electrons available. Each Li atom provides one valence electron and the band shown in Figure 7.27 is half-occupied. This leads to a delocalized picture of the bonding in the metal, and the metal–metal bonding is *non-directional*.

LCAO: see Section 3.12

When metal atoms have more than one type of AO in the valence shell, correspondingly more bands are formed. If two bands are close together

§ Use of the equation $V = IR$ (potential difference = current × resistance) is exemplified in the accompanying workbook.

7.27 The interaction of two $2s$ AOs in Li_2 leads to the formation of two MOs. If there are three lithium atoms, three MOs are formed, and so on. For Li_n, there are n molecular orbitals, but because the $2s$ AOs were all of the same energy, the energies of these MOs are very close together and are described as a *band* of orbitals.

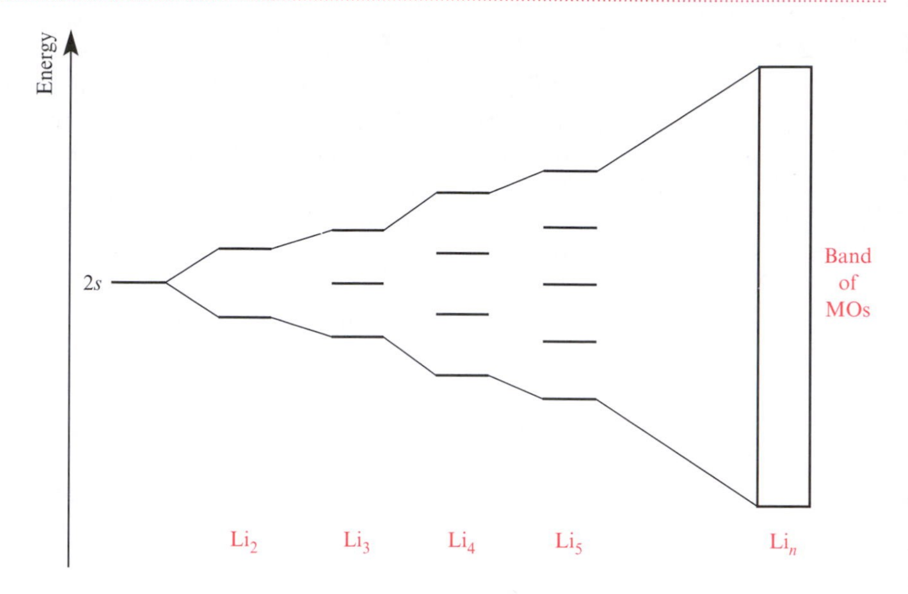

they will overlap, giving a single band in which there is mixed orbital character (for example s and p character). Some bands will be separated from other bands by defined energy gaps, as shown in Figure 7.28. The lowest band is fully occupied with electrons, while the highest band is empty. The central band is partially occupied with electrons, and because the energy states that make up the band are so close, the electrons can move between energy states *in the same band*. In the bulk metal, electrons are therefore mobile. If a potential difference is applied across the metal, the electrons move in a direction from high to low potential and a current flows. The energy gaps *between* bands are relatively large, and it is the presence of a *partially occupied* band that characterizes a metal.

7.28 The molecular orbitals that describe the bonding in a bulk metal are very close in energy and are represented by bands. Bands may be fully occupied with electrons (shown in black), unoccupied (shown in white), or partially occupied. The figure shows a schematic representation of these bands for a metal.

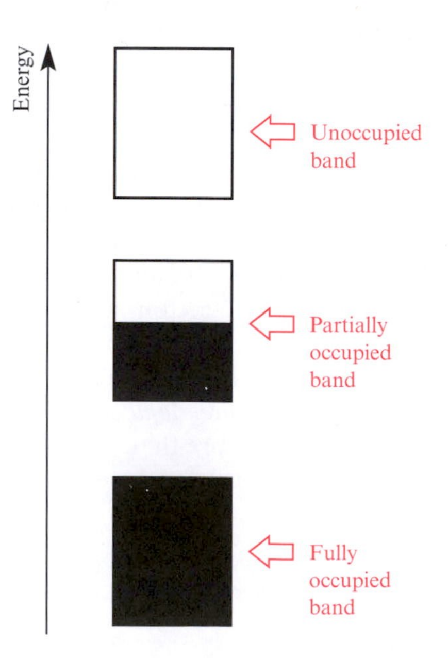

A *band* is a group of MOs which are extremely close in energy. The energy differences are so small that the system behaves as if a continuous, non-quantized variation of energy within the band is possible.

A *band gap* occurs when there is a significant energy difference between two bands.

Semiconductors

In going down each of groups 13, 14, 15 and 16, there is a transition from a non-metal to a metal, passing through some intermediate stage characterized by the *semi-metals*. As we have already seen for the group 14 elements, the metal/non-metal boundary is not well defined.

Figure 7.29 gives a representation of the bonding situation for these allotropes. The MO diagram for bulk diamond can be represented in terms of a fully occupied and an unoccupied band. There is a large band gap (520 kJ mol^{-1}) and diamond is an insulator. The situation for silicon and germanium can be similarly represented, but now the band gaps are much smaller (106 and 64 kJ mol^{-1} respectively). In grey tin, only 8 kJ mol^{-1} separates the filled and empty bands, and here the situation is approaching that of a single band which is partly occupied. The conduction of electricity in silicon, germanium and grey tin depends upon *thermal population* of the upper band (the *conduction band*) and these allotropes are classed as *semiconductors*. As the temperature is increased, some electrons will have sufficient energy to make the transition from the lower to higher energy band. The smaller the band gap, the greater the number of electrons that will possess sufficient energy to make the transition, and the greater the electrical conductivity.

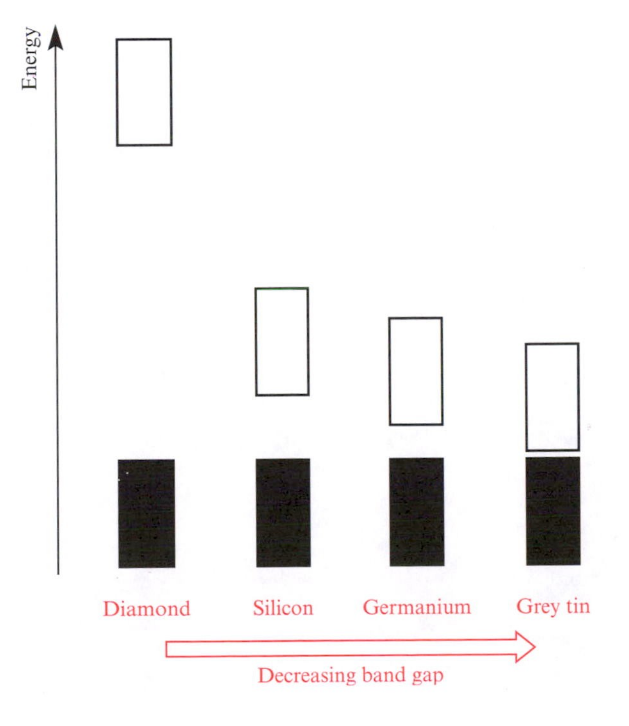

7.29 The energy difference (the band gap) between occupied (shown in black) and unoccupied (shown in white) bands of MOs decreases in going down group 14. This allows a change from non-metallic towards metallic character in going down the group.

SUMMARY

In this chapter, we have discussed the structures of elements in the solid state.

Do you know what the following terms mean?

- close-packing of spheres
- cubic close-packing
- hexagonal close-packing
- simple cubic lattice
- body-centred cubic lattice
- interstitial hole
- crystalline solid
- amorphous solid
- chair and boat conformers of a six-membered ring
- handedness of a helical chain
- catena-
- electrical resistivity
- insulator
- metallic radius
- metallic bonding
- band gap
- semiconductor

You should be able:

- to discuss how the close-packing of spheres can give rise to at least two assemblies
- to discuss the relationship between simple and body-centred cubic packing of spheres
- to state how many nearest-neighbours an atom has in ccp, hcp, simple cubic and bcc arrangements
- to compare the efficiencies of packing in ccp, hcp, simple cubic and bcc arrangements
- to appreciate *why* the packing of spheres is an appropriate model for some but not all solid state lattices
- to distinguish between intra- and inter-molecular bonds in solid state structures (are both types of bonding always present?)
- to describe structural variation amongst the allotropes of boron, carbon, phosphorus and sulfur
- to describe similarities and differences in the solid state structures of the elements of group 17
- to give examples of metals with different types of lattice structures
- to describe using simple band theory how a metal can conduct electricity

PROBLEMS

7.1 When spheres of an equal size are close-packed, what are the features that characterize whether the arrangement is hexagonal or cubic close-packed? Draw a representation of the repeat unit for each arrangement.

7.2 What is an *interstitial hole*? What types of holes are present in hcp and ccp arrangements?

7.3 How are spheres organized in a body-centred cubic arrangement? How does the body-centred cubic arrangement differ from a simple cubic one?

7.4 What is meant by a *nearest-neighbour* in an assembly of spheres? How many nearest-neighbours does each sphere possess in a (a) cubic close-packed, (b) hexagonal close-packed, (c) simple cubic and (d) body-centred cubic arrangement?

7.5 Write down the ground state electronic configuration of chlorine, sulfur and phosphorus, and use these data to describe the bonding in each of the molecular units present in (a) solid dichlorine, (b) orthorhombic sulfur, (c) fibrous sulfur and (d) white phosphorus.

7.6 What is meant by electrical resistivity? Use the data in Table 7.5 to discuss the statement: '*All metals are good electrical conductors.*'

Table 7.5 Table of data for problem 7.6. All the resistivities are measured for pure samples at 273 K unless otherwise stated.

Element	*Electrical resistivity / Ω*
Copper	1.5×10^{-8}
Silver	1.5×10^{-8}
Aluminium	2.4×10^{-8}
Iron	8.6×10^{-8}
Gallium	1.4×10^{-7}
Tin	3.9×10^{-7}
Mercury	9.4×10^{-7}
Bismuth	1.1×10^{-6}
Manganese	1.4×10^{-6}
Silicon	1.0×10^{-3}
Boron	1.8×10^{4}
Phosphorus	1.0×10^{9} (293 K)

7.7 Briefly discuss allotropy with respect to carbon. What is the standard state of this element? Explain why α-graphite and diamond show such widely differing electrical resistivities, and suggest why the electrical resistivity in a graphite rod is direction-dependent.

7.8 What is meant by *solvent of crystallization*? The crystallization of C_{60} from benzene and diiodomethane yields solvated crystals of formula $C_{60}{\cdot}xC_6H_6{\cdot}yCH_2I_2$. If the loss of solvent leads to a 32.4% reduction in the molar mass, what is a possible stoichiometry of the solvated compound?

7.9 What lattice structure is typical of an alkali metal at 298 K? Table 7.6 lists values of the enthalpies of fusion and vaporization for the alkali metals. Describe what is happening in each process in terms of interatomic interactions. Use the data in Table 7.6 to plot graphs which show the trends in melting point, and enthalpies of fusion and atomization down group 1. How do you account for the trends observed, and any relationships that there may be between them?

Table 7.6 Table of data for problem 7.9.

Alkali metal	*Melting point / K*	*Enthalpy of fusion / kJ mol⁻¹*	*Enthalpy of atomization / kJ mol⁻¹*
Lithium	454	3.0	162
Sodium	371	2.6	108
Potassium	337	2.3	90
Rubidium	312	2.2	82
Caesium	301	2.1	78

7.10 Comment on the *relative* values of the metallic (197 pm) and ionic radii (100 pm) of calcium.

7.11 A localized covalent σ-bond is directional, whilst metallic bonding is non-directional. Discuss the features of covalent and metallic bonding that lead to this difference.

7.12 Using simple band theory, describe the differences between electrical conduction in a metal such as lithium and in a semiconductor such as germanium.

8 Alkanes, alkenes and alkynes

Topics

8.1 INTRODUCTION

Organic chemistry is the study of compounds of carbon. Conventionally, elemental carbon and compounds such as carbon monoxide, carbon dioxide and carbonates lie within the realm of the inorganic chemist. In this text we try to show that common themes run through the chemistry of 'organic' and 'inorganic' compounds, and that such divisions are artificial and often unhelpful. Today, interdisciplinary research is common and the divisions between organic chemistry, inorganic chemistry, biochemistry and biology are not clear-cut and should certainly not be regarded as being so. Real progress in science is made when researchers from different areas communicate their ideas and knowledge to one another.

What is a hydrocarbon?

➤ Aromatic and cyclic compounds: see Chapter 15

Hydrocarbon compounds contain only carbon and hydrogen, and are conveniently grouped according to whether they are *aliphatic* or *aromatic*, and whether the aliphatic compound is *saturated* or *unsaturated* (Figure 8.1). The word 'aliphatic' was originally used to describe fats and related compounds,

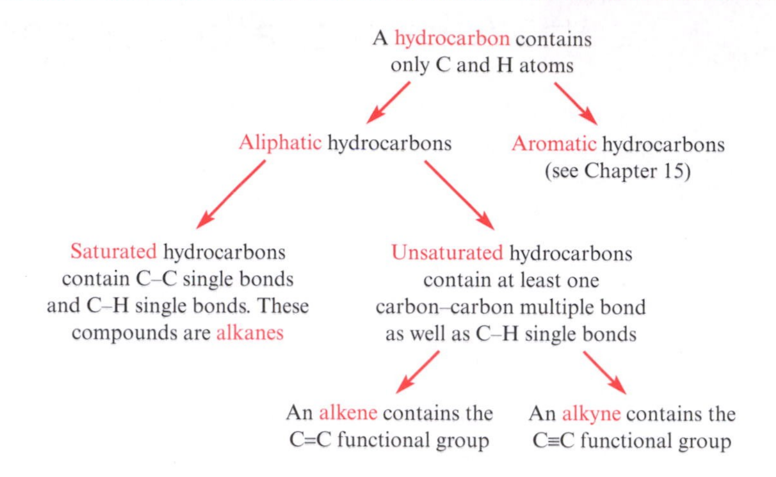

8.1 Classification of hydrocarbon compounds.

but is now used to refer to hydrocarbons containing open carbon chains (acyclic compounds) or compounds with cyclic structures but with properties similar to the open-chain hydrocarbons. In this chapter we are concerned only with hydrocarbons which are aliphatic and acyclic.

In an acyclic hydrocarbon, the carbon atoms are connected in an open chain. If they are in a ring, the compound is cyclic.

Carbon compounds are usually categorized by the *functional group(s)* they contain, and these groups undergo characteristic reactions. Alkenes and alkynes contain C=C and C≡C functional groups respectively. *Saturated* aliphatic hydrocarbons do not possess any functional groups, and the old name for a hydrocarbon was *paraffin*, derived from Latin and meaning 'little affinity'. The general formula for an acyclic alkane is C_nH_{2n+2} – every carbon atom forms four single bonds.

A functional group is one which undergoes characteristic reactions.

Geometry and hybridization

Hydrocarbons may contain tetrahedral (sp^3), trigonal planar (sp^2) or linear (sp) carbon atoms depending upon the presence or not of carbon–carbon multiple bonds and some examples are listed in Table 8.1. Problem 8.8 at the end of this chapter reviews hybridization in hydrocarbons.

Restricted geometries for carbon: see Section 5.14

Table 8.1 Geometries and hybridizations at carbon centres in hydrocarbon molecules (see also Section 5.14).

	Alkane	*Alkene*	*Allene*	*Alkyne*
Geometry at carbon centre	Tetrahedral	Trigonal planar	Linear	Linear
Hydridization of carbon centre	sp^3	sp^2	sp	sp
Example:				
Name	Ethane	Ethene	Propadiene	Ethyne
Molecular formula	C_2H_6	C_2H_4	C_3H_4	C_2H_2
Structural formulae	CH_3CH_3	$H_2C{=}CH_2$	$H_2C{=}C{=}CH_2$	$HC{\equiv}CH$

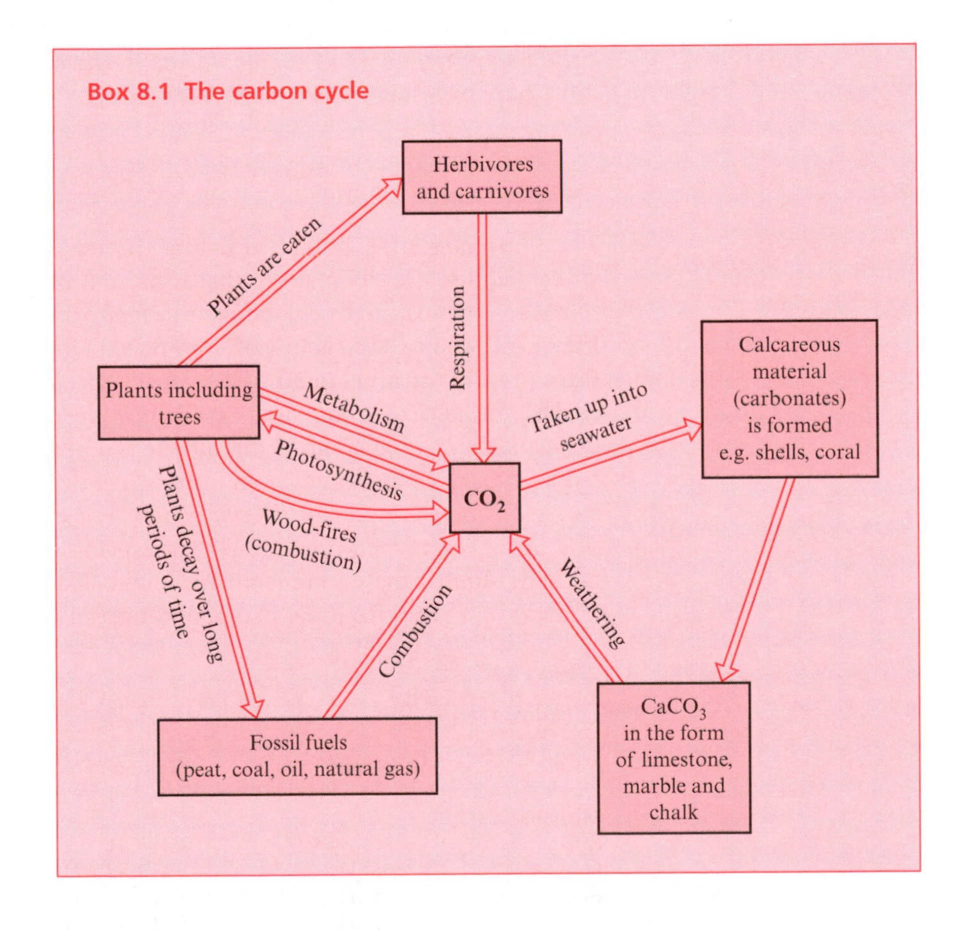

8.2 NOMENCLATURE

In Sections 1.23 and 1.24 we outlined the differences between empirical, molecular and structural formulae, and some basic nomenclature of organic compounds. Although we use IUPAC nomenclature in this book, many trivial names are still used and some are listed in Appendix 13.

The three name endings that are relevant in this chapter are:

- -ane (saturated hydrocarbon);
- -ene (unsaturated hydrocarbon with a C=C double bond);
- -yne (unsaturated hydrocarbon with a C≡C triple bond).

Straight chain alkanes

The name for a *straight chain* alkane follows directly from the number of carbon atoms in the chain. Table 1.8 lists the prefixes that are used to indicate the number of carbon atoms in a chain. Table 8.2 lists several straight chain alkanes, giving both molecular and structural formulae. These hydrocarbons are members of a *homologous series* in which adjacent members differ only by a single CH_2 (*methylene*) group.

Branched chain alkanes

The *root-name* of a branched chain alkane is derived from the longest *straight* chain of carbon atoms that is present in the compound. Remember that a so-called straight chain may not *look* straight. This is emphasized in Figure 8.2 where the structure of a branched alkane is considered – there are three possible straight chains. In Figures 8.2a and 8.2c, the straight (red) chains contain six carbon atoms, while in Figure 8.2b, seven atoms make up the main chain – the name of this alkane is derived from *heptane*.

The longest chain is numbered from one end, this being chosen to give the *lowest position-numbers for the side chains*. For the alkane in Figure 8.2b, the side chain is attached to atom C(4) irrespective of the end from which we begin to number the seven-membered chain. The name for a side chain (*substituent*) is derived from the parent alkane and the general name is an *alkyl*

Table 8.2 Nomenclature for selected straight chain alkanes.

Name	*Molecular formula*	*Structural formula*
Methane	CH_4	CH_4
Ethane	C_2H_6	CH_3CH_3
Propane	C_3H_8	$CH_3CH_2CH_3$
Butane	C_4H_{10}	$CH_3CH_2CH_2CH_3$
Pentane	C_5H_{12}	$CH_3CH_2CH_2CH_2CH_3$
Octane	C_8H_{18}	$CH_3CH_2CH_2CH_2CH_2CH_2CH_2CH_3$
Decane	$C_{10}H_{22}$	$CH_3CH_2CH_2CH_2CH_2CH_2CH_2CH_2CH_2CH_3$

Box 8.2 'Straight' chains in carbon chemistry

The tendency of carbon to form carbon–carbon bonds means that carbon chemistry frequently involves chains and rings of carbon atoms.

The words 'straight' and 'linear' are commonly used to describe chains of atoms. A straight chain is often represented in a manner that suggests that the carbon atoms are joined together in a strictly linear sense. For example, although the structure of hexane might be shown as:

each carbon atom is sp^3 hybridized (tetrahedral) and better representations of the 'straight' chain are:

Additionally, rotation about the C–C single bonds permits the chain to 'curl-up':

The term 'straight' means only that the carbon atoms are bonded in a linear sequence, and tells us nothing about the conformation of the chain. When the sequence of the atoms in the carbon backbone is not continuous, the chain is said to be 'branched'.

group – methyl is CH_3 (Me), ethyl is C_2H_5 (Et), but see also Figure 8.7 in Section 8.4. The correct name for the hydrocarbon in Figure 8.2b is 4-ethylheptane. Note two features of this name:

- There is a hyphen between the position number and the name of the substituent.
- There is *no* space between the substituent name and the root-name.

If more than one substituent is present, their names are given in alphabetical order, each with its respective position number. The prefixes di-, tri- and tetra- are used to show the presence of more than one substituent of the

8.2 The IUPAC name for an organic compound is based upon the longest straight chain in the molecule. The diagram shows a branched alkane containing nine carbon atoms. There are three ways to trace out a carbon chain, and the chain indicated in red contains (a) six C atoms, (b) seven C atoms, and (c) six C atoms.

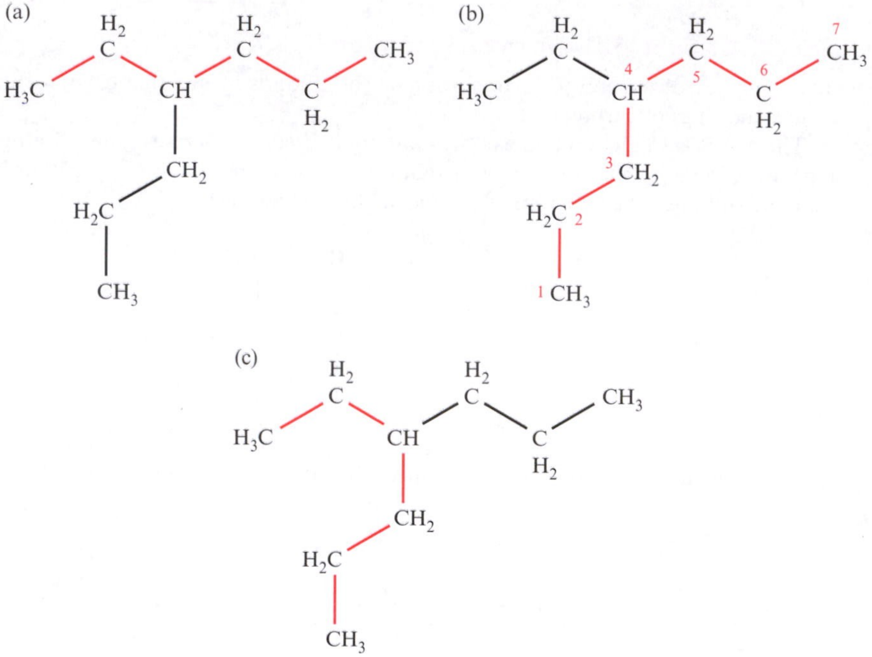

same type – dimethyl means 'two methyl groups' and tetraethyl means 'four ethyl groups'. The alphabetical ordering of the alkyl groups takes priority over the multipliers, e.g. 4-ethyl-3,3-dimethyloctane, *not* 3,3-dimethyl-4-ethyloctane. Table 8.3 lists examples, and a short-hand way of drawing diagrammatic structural formulae is introduced. Only the carbon framework is indicated, and it is assumed that the coordination number of each carbon atom is made up to four by hydrogen atoms. A *single line* drawn as a branch means that the substituent is a methyl group (not a hydrogen atom).

The basic rules for naming a branched chain alkane are summarized as follows:

1. **Find the longest straight chain in the molecule; this gives the root-name.**
2. **The longest chain is numbered from one end, chosen to give the lowest position-numbers for the substituents.**
3. **Prefix the root-name with the substituent descriptors, and if more than one is present, give them in alphabetical order.**

Straight chain alkenes

The first task in naming an alkene is to locate the longest straight chain containing the double bond to give the root-name. When one double bond is introduced into a hydrocarbon chain, the name of the compound must indicate:

- the number of carbon atoms in the chain, *and*
- the position of the alkene functionality.

Table 8.3 Nomenclature and structural formulae for branched chain alkanes.

Diagrammatic structural formula	*Abbreviated diagrammatic structural formula*	*Other structural formulae*	*Compound name*
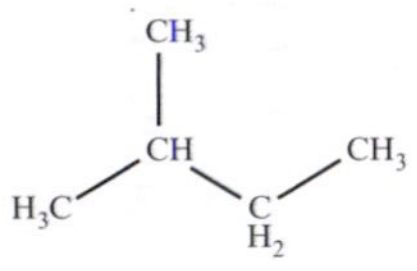	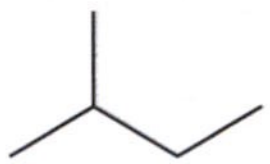	$CH_3CH(CH_3)CH_2CH_3$ $CH_3CHMeCH_2CH_3$	2-methylbutane
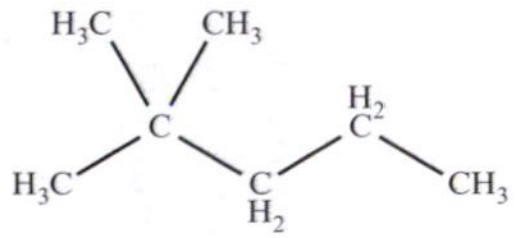	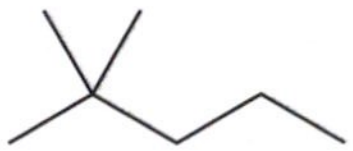	$CH_3C(CH_3)_2CH_2CH_2CH_3$ $CH_3CMe_2CH_2CH_2CH_3$	2,2-dimethylpentane
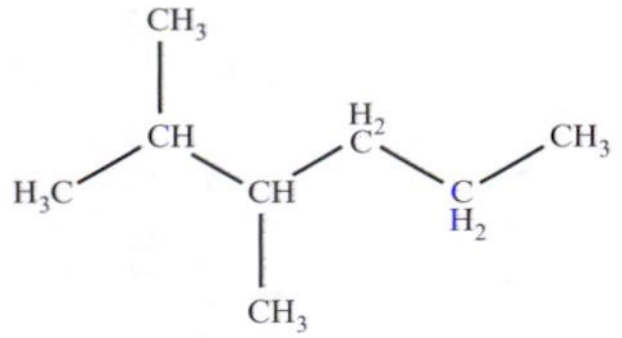	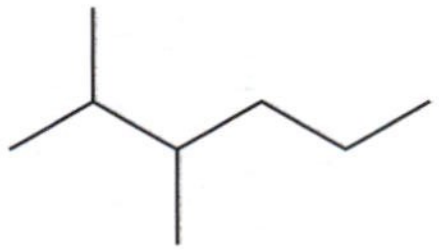	$CH_3CH(CH_3)CH(CH_3)CH_2CH_2CH_3$ $CH_3CHMeCHMeCH_2CH_2CH_3$	2,3-dimethylhexane
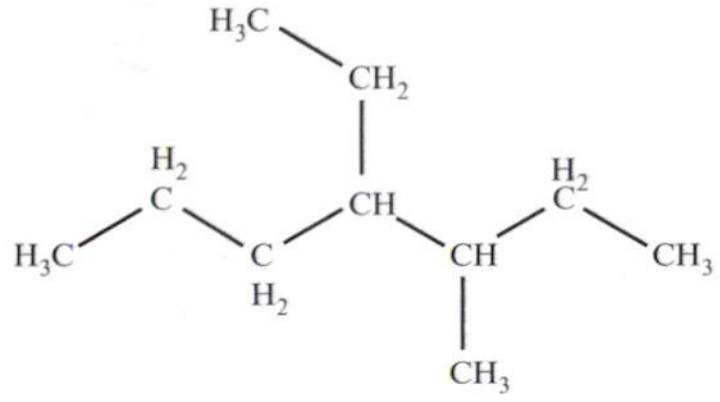	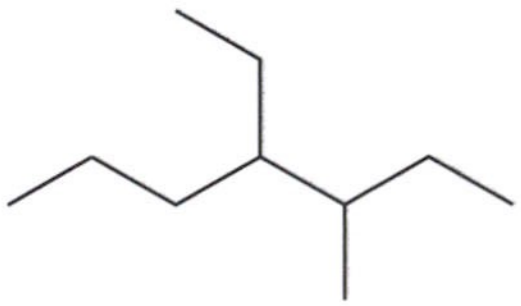	$CH_3CH_2CH_2CH(C_2H_5)CH(CH_3)CH_2CH_3$ $CH_3CH_2CH_2CHEtCHMeCH_2CH_3$	4-ethyl-3-methylheptane (*not* 3-methyl-4-ethylheptane)

8.3 Naming geometrical isomers of alkenes.

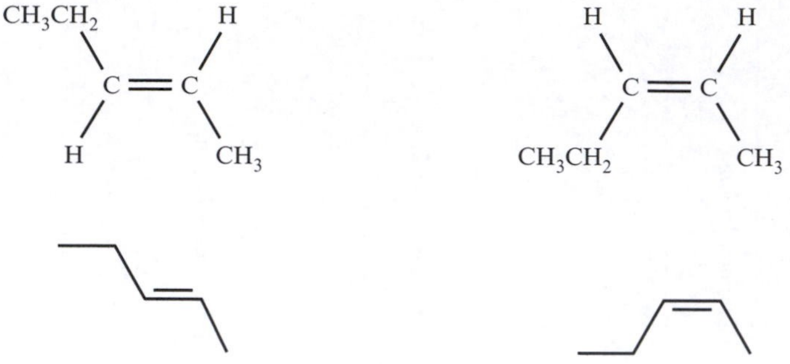

CH_3CH_2 H C=C H CH_3

H H C=C CH_3CH_2 CH_3

(*E*)-pent-2-ene

(*Z*)-pent-2-ene

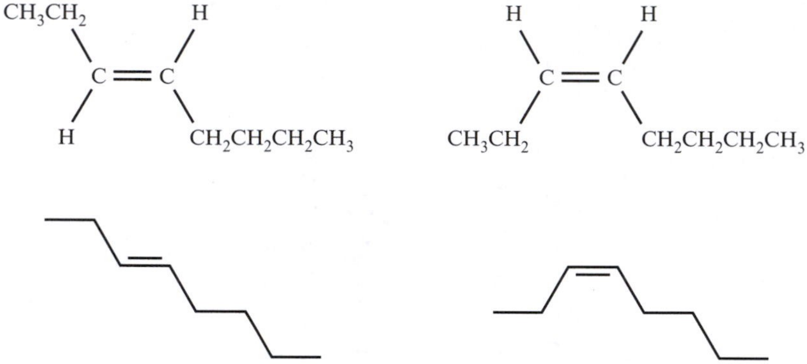

CH_3CH_2 H C=C H $CH_2CH_2CH_2CH_3$

H H C=C CH_3CH_2 $CH_2CH_2CH_2CH_3$

(*E*)-oct-3-ene

(*Z*)-oct-3-ene

The carbon chain is numbered so that the position of the double bond is described by the lowest possible site-number. In structure **8.1**, the chain has five carbon atoms and the root name is *pent-*. The chain is numbered from the right-hand end because the C=C double bond is nearer to this end. The compound is pent-2-ene *not* pent-3-ene. In **8.2**, the eight carbon atoms give a root-name of *oct-* and atom numbering begins at the left-hand end of the chain. (Why?) The compound is oct-3-ene.

$$\overset{5}{C}H_3\overset{4}{C}H_2\overset{3}{C}H{=}\overset{2}{C}H\overset{1}{C}H_3$$

(8.1)

$$\overset{1}{C}H_3\overset{2}{C}H_2\overset{3}{C}H{=}\overset{4}{C}H\overset{5}{C}H_2\overset{6}{C}H_2\overset{7}{C}H_2\overset{8}{C}H_3$$

(8.2)

(*Z*)- and (*E*)-isomers: see Section 5.11

The alkenes **8.1** and **8.2** also possess geometrical isomers and the formulae shown do not specify a particular isomer. In Figure 8.3, the prefixes (*E*)- and (*Z*)- are used to indicate the exact structures of the alkenes.

Dienes and trienes

When two double bonds are present in a straight chain, the hydrocarbon is called a diene – this is a particular type of alkene. The rules for naming a diene follow those outlined above, but include the use of two site-numbers and the prefix 'di'. Similarly, a triene contains three alkene functionalities. Examples are shown in Figure 8.4, along with the use of the (*Z*) and (*E*) prefixes to distinguish between the geometrical isomers of hepta-2,4-diene.

A diene contains two C=C double bonds.

An allene is a particular type of diene – one in which the two double bonds are attached to the same carbon atom. The simplest example is propadiene (Table 8.1) and this compound has the trivial name of 'allene'.

An allene contains two C=C double bonds which share a common carbon atom (–C=C=C–).

Alkynes

The rules for naming alkynes are similar to those for alkenes, although no geometrical isomers are possible. Thus, $CH_3C{\equiv}CCH_3$ is but-2-yne, and $CH_3CH_2C{\equiv}CH$ is but-1-yne (*not* but-3-yne).

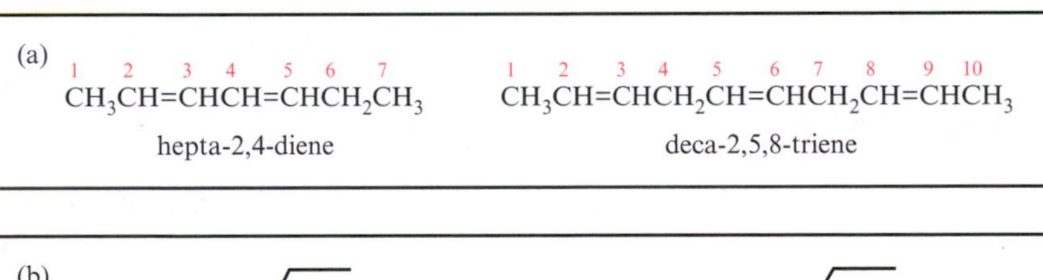

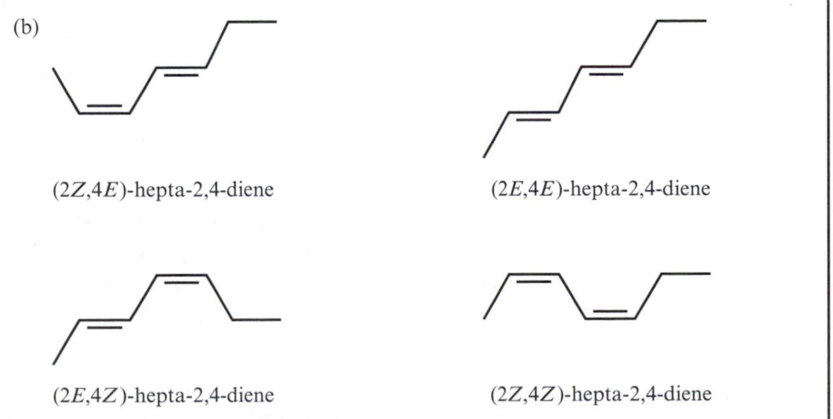

8.4 (a) Nomenclature for a diene and a triene.
(b) Geometrical isomers of hepta-2,4-diene; the prefixes contain both numerical *and* (*Z*) or (*E*) descriptors to ensure there is no ambiguity.

8.3 PRIMARY, SECONDARY, TERTIARY AND QUATERNARY CARBON ATOMS

A carbon atom is often designated as primary, secondary, tertiary or quaternary depending upon the number of other carbon atoms to which it is connected:

- A *primary* carbon atom is connected to one other carbon atom.
- A *secondary* carbon atom is connected to two other carbon atoms.
- A *tertiary* carbon atom is connected to three other carbon atoms.
- A *quaternary* carbon atom is connected to four other carbon atoms.

For example, ethane contains two primary carbon atoms. Propane **8.3** contains one secondary and two primary carbon atoms; in 2-methylpropane **8.4**, there is one tertiary and three primary carbon atoms, and in 2,2-dimethylpropane **8.5**, there is one quaternary and four primary carbon atoms.

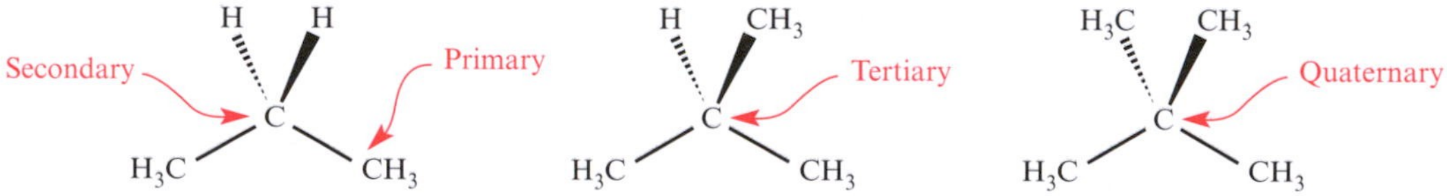

(8.3) (8.4) (8.5)

The distinction between the different types of centres is particularly useful when it comes to reactivity patterns of aliphatic compounds, and these labels are used frequently.

8.4 STRUCTURAL ISOMERISM

We have already seen that alkenes (but not alkanes and alkynes) may possess *geometrical isomers*. However, all three types of compounds may show *structural isomerism* arising because the carbon atoms may be connected in different ways.

Alkanes

The alkanes CH_4, C_2H_6 and C_3H_8 have *no* structural isomers. [Draw their structural formulae – is there any ambiguity about the way in which the carbon atoms are connected?] For an alkane of formula C_nH_{2n+2} with $n > 3$, more than one structural formula can be drawn, and the number of possibilities increases dramatically with increasing n (Figure 8.5). Figure 8.6 illustrates the structural isomers of C_6H_{14}; a mixture of these used as a solvent is often referred to *hexanes*, but only one carries the IUPAC name of hexane.

8.5 The number of isomers (structural + optical) for an alkane of formula C_nH_{2n+2} increases dramatically with the number of carbon atoms, n. The presence of one structure for each of methane, ethane and propane is counted to be one isomer to be consistent with the other data. Optical isomers are discussed in Section 8.6.

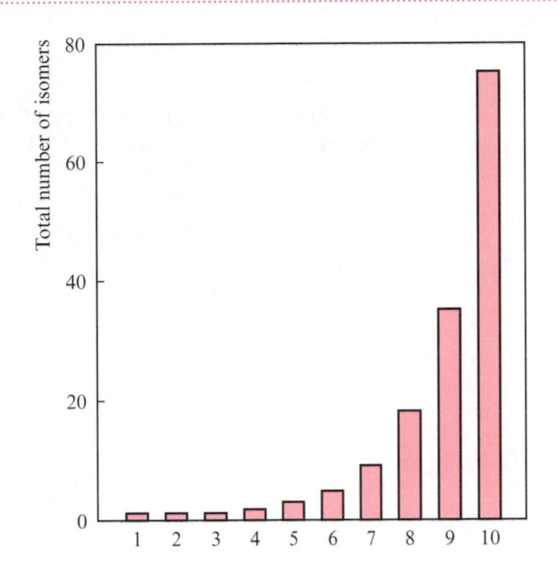

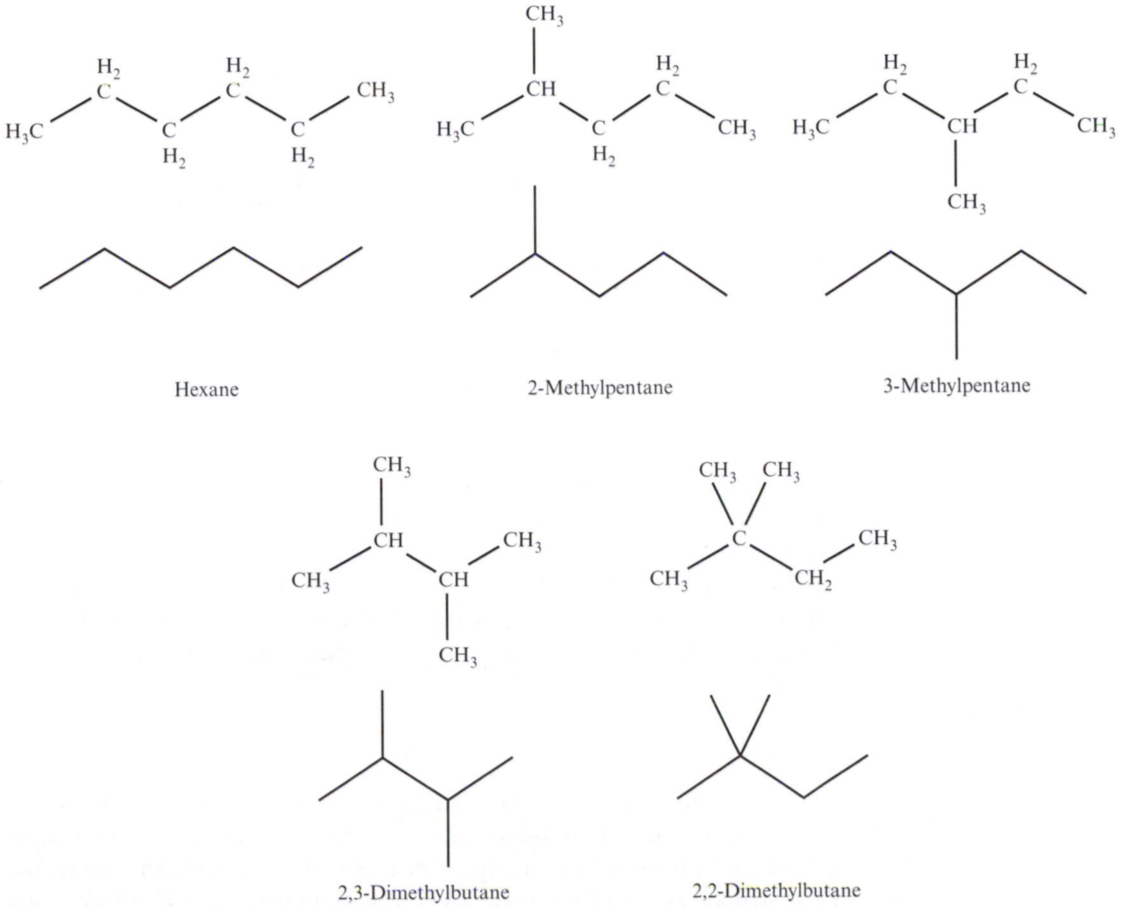

H_3C H_2C CH_2 H_2C CH_2 CH_3

Hexane

CH_3 H_3C CH CH_2 H_2C CH_3

2-Methylpentane

H_3C H_2C CH H_2C CH_3 CH_3

3-Methylpentane

CH_3 CH_3 CH CH CH_3 CH_3

2,3-Dimethylbutane

CH_3 CH_3 CH_3 C CH_2 CH_3

2,2-Dimethylbutane

8.6 Structural isomers of C_6H_{14}; only isomers with non-cyclic structures are included.

Box 8.3 Gas chromatography: 40–60 petroleum ether

'40–60 Petroleum ether' is not an ether at all but a fraction of petroleum with a boiling point range of 40–60°C (313–333 K). The components of 40–60 petroleum ether are alkanes and the mixture can be analysed by using *gas chromatography* (GC). The trace below shows the result of passing a sample of 40–60 petroleum ether through a gas chromatograph. Each peak represents a particular hydrocarbon and the relative areas of the peaks indicates the relative amounts of each compound in the mixture. The retention time of a fraction depends on the molecular mass and whether the chain is straight or branched. Typical components of 40–60 petroleum ether are isomers of pentane and hexane.

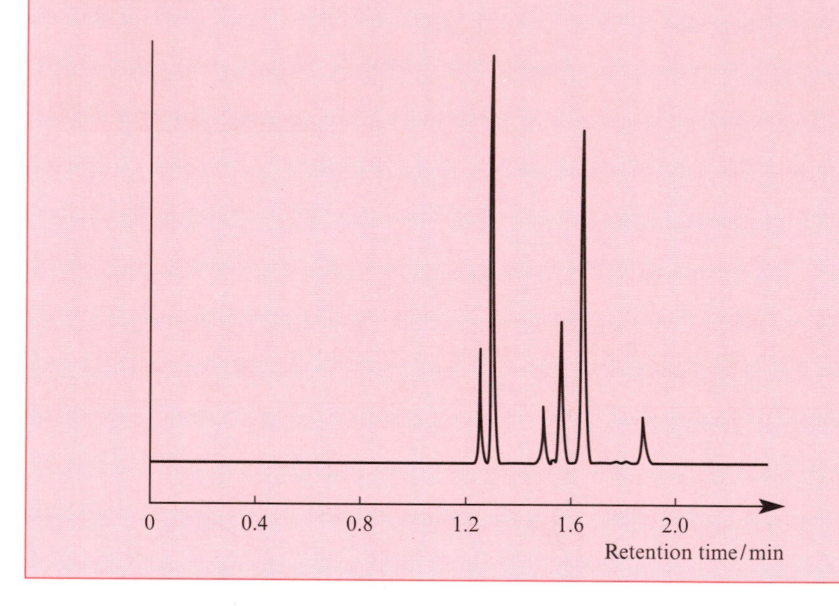

Alkyl substituents

Methyl and ethyl substituents are derived from methane and ethane respectively (Figure 8.7a) and there are no isomers of these alkyl groups. In forming a propyl group from propane, there are two different possibilities (Figure 8.7b) and in going from butane (which has two structural isomers) to a butyl group, there are four different isomers that can be formed (Figure 8.7c). For longer chain alkanes, the situation becomes increasingly complicated.

Alkenes and alkynes

For an unsaturated hydrocarbon, structural isomerism may arise because the C=C or C≡C bond may be in one of several positions. Ethene, propene, ethyne and propyne have no structural isomers (confirm this by drawing their structures), but for molecules containing more carbon atoms, isomers are possible.

8.7 Alkyl substituents. (a) There are no structural isomers for methyl and ethyl substituents, but there are (b) two isomers of the propyl group and (c) four isomers of the butyl group. How do the names *sec*-butyl and *tert*-butyl arise? (Look at Section 8.3.)

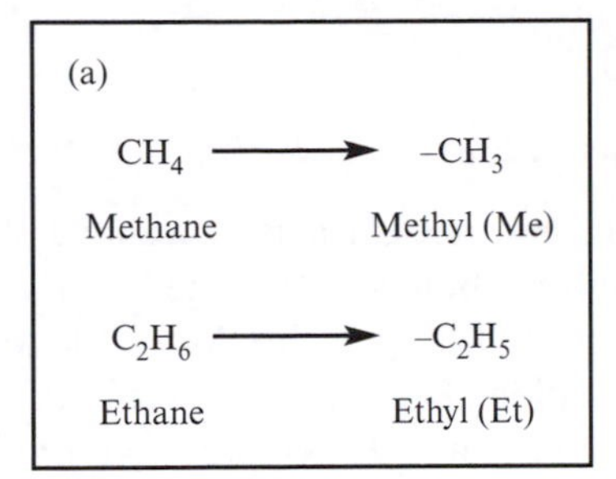

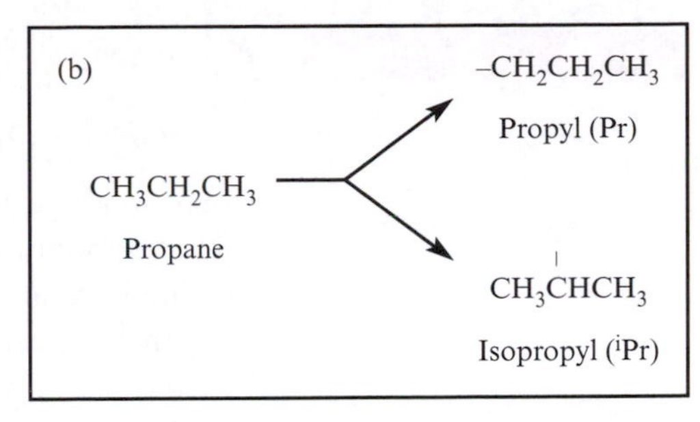

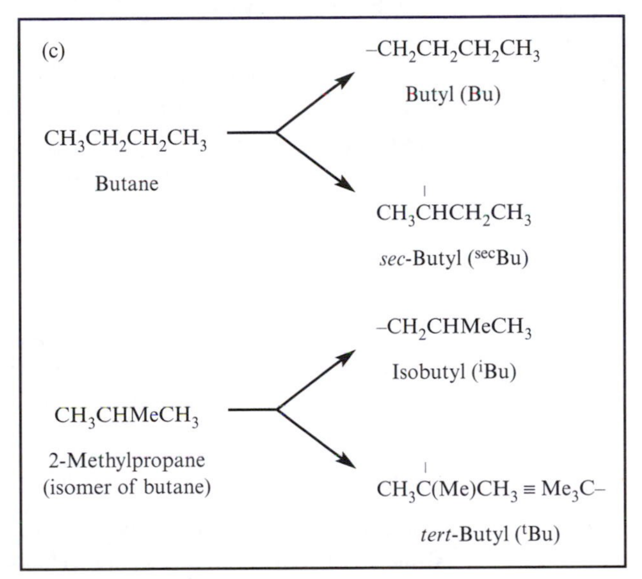

Structures **8.6, 8.7** and **8.8** are structural isomers of C_6H_{12} which differ only in the position of the C=C bond; **8.7** and **8.8** also have (*E*)- and (*Z*)-isomers – what are their structures? Further isomers may be drawn with branched chains.

Isomers with cyclic structures can also be drawn for C_6H_{12}: see Chapter 15

$CH_2{=}CHCH_2CH_2CH_2CH_3$
hex-1-ene
(8.6)

$CH_3CH{=}CHCH_2CH_2CH_3$
hex-2-ene
(8.7)

$CH_3CH_2CH{=}CHCH_2CH_3$
hex-3-ene
(8.8)

8.5 CONFORMATION

Staggered, eclipsed and skew conformations

Rotation about the internuclear axis of a single bond is usually possible, but rotation about multiple bonds is restricted because of the π-component. By looking back at Figure 3.18, you should be able to appreciate why this is.

Rotation can usually occur about a single bond but not about a multiple bond.

In an alkane, rotation about *all* the single bonds in the molecule can occur. In C_2H_6, rotation about a C–H bond does not affect the shape of the molecule, but rotation about the C–C bond alters the *relative* orientations of the two methyl groups. These different arrangements are called *conformations; conformers of a molecule can be interconverted by rotation about single bonds.*

The two conformers of ethane that represent the two extreme positions of the CH_3 groups with respect to one another are shown in Figure 8.8a. Between the *staggered* and *eclipsed* conformations lie an infinite number of *skew conformers*.

If two different arrangements of atoms in a molecule can be interconverted by rotations about single bonds, the arrangements are *conformers*.

Newman projections, sawhorse drawings, and the use of wedges

There are several schematic ways of representing conformations and these are shown in Figures 8.8b–d with C_2H_6 as the example. In the Newman projection, the *stereochemistry* is shown by looking along the C–C bond; atoms attached to the carbon atom closest to you are drawn in front of a circle, and atoms attached to the more distant carbon atom lie behind the circle. In a 'sawhorse' diagram of C_2H_6, the C–C bond is drawn at an angle, and, by convention, the carbon atom at the lower left-hand corner is nearer to you. A third method is to use solid wedges, with either hashed wedges or dashed lines. The C–C bond is drawn to lie in the plane of the paper. Solid wedges represent bonds that point up from the paper, and hashed wedges represent bonds pointing away from you. Use of these schemes allows the stereochemistry of complicated molecules to be shown in a relatively unambiguous manner.

The stereochemistry of a molecule describes the spatial arrangement of the atoms in that molecule.

Steric energy changes associated with bond rotation

Steric interactions: see Section 5.9

When an ethane molecule is in the staggered conformation, the steric interactions between hydrogen atoms on adjacent carbon atoms are at a minimum – this represents the minimum *steric energy*. As rotation about the C–C bond occurs, steric interactions increase and the steric energy varies as shown in Figure 8.9. The eclipsed conformation is the energy maximum. In a full 360° rotation about the C–C bond, the molecule passes through three identical eclipsed and three identical staggered conformations.

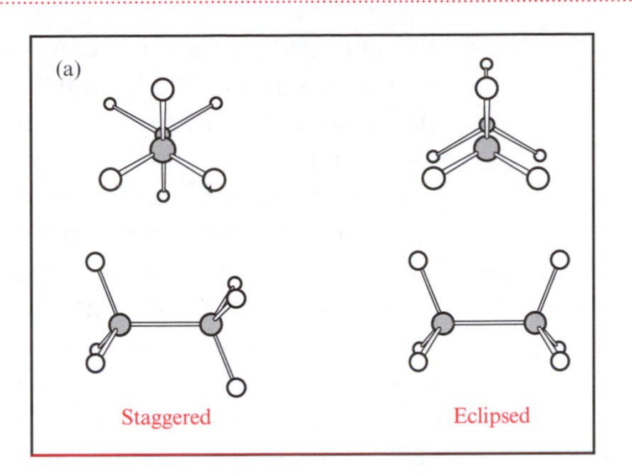

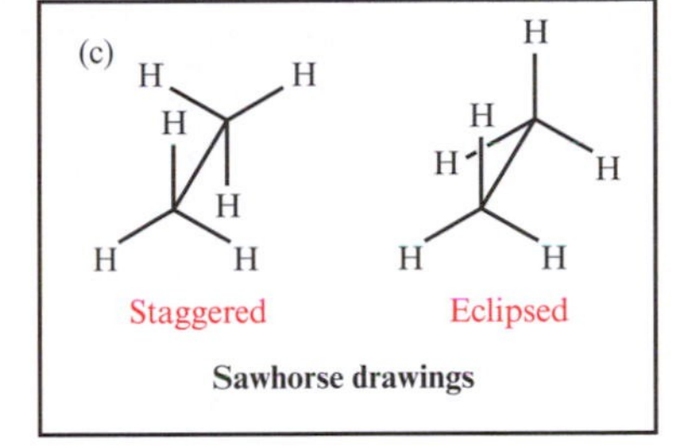

(b) Newman projections: Staggered, Eclipsed

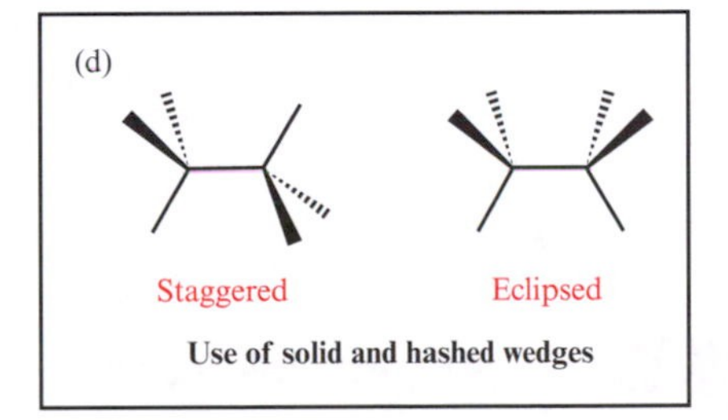

8.8 Conformers of ethane. (a) 'Ball-and-stick' diagrams of a molecule of C_2H_6 showing two views of each of the staggered and eclipsed conformers. Three methods are in common use for representing conformations. (b) In a Newman projection, atoms in front of the circle point towards you. (c) In a sawhorse diagram, the lower left-hand carbon atom is, by convention, closer to you. (d) Solid wedges and hashed wedges (or dashed lines) indicate, respectively, atoms or groups that point towards you or away from you; the remaining lines are in (or close to being in) the plane of the paper.

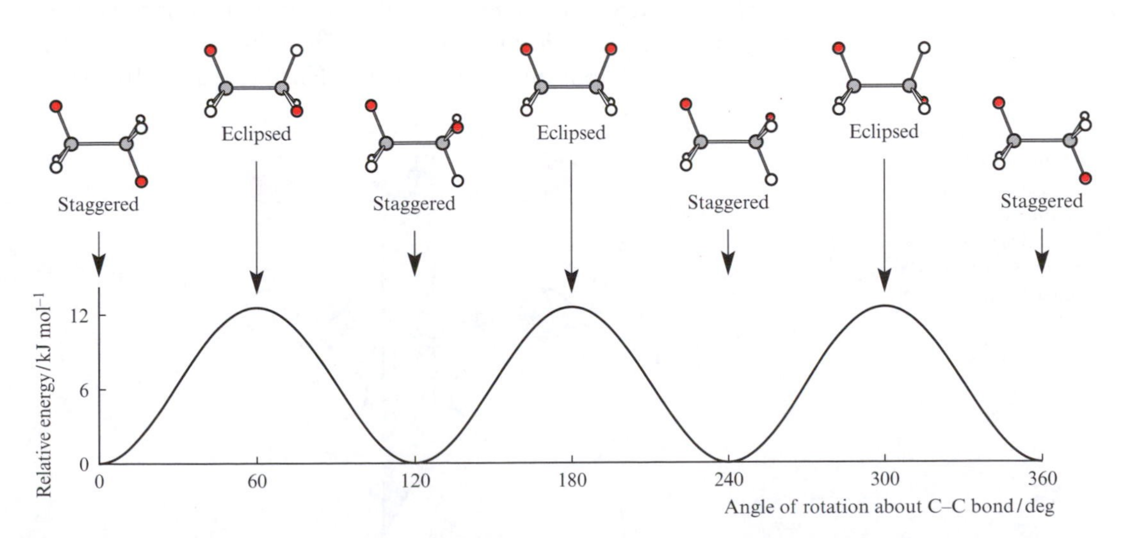

8.9 The relative steric energy of ethane changes as one methyl group rotates with respect to the other about the C–C bond. Two of the six equivalent hydrogen atoms are marked in red to highlight their relationship throughout the rotation.

The energy difference between the staggered and eclipsed conformations in ethane is about 12.5 kJ mol^{-1}. This is relatively small and means that, at 298 K, we can consider the ethane molecule to be freely rotating about the C–C bond.

If we replace the hydrogen atoms in Figure 8.9 by alkyl groups, a staggered conformation remains the most favourable, but the barrier to rotation increases. It is important to recognize that the ability of an alkane molecule to undergo C–C bond rotation means that the so-called straight chains (*extended conformation*) can in fact fold up into 'balls'. The flexibility of an alkane chain is best illustrated by using molecular models or computer molecular modelling programs, but structure **8.9** illustrates a conformer of heptane in which the carbon backbone is *not* extended.

See problem 8.9

See also Box 8.2

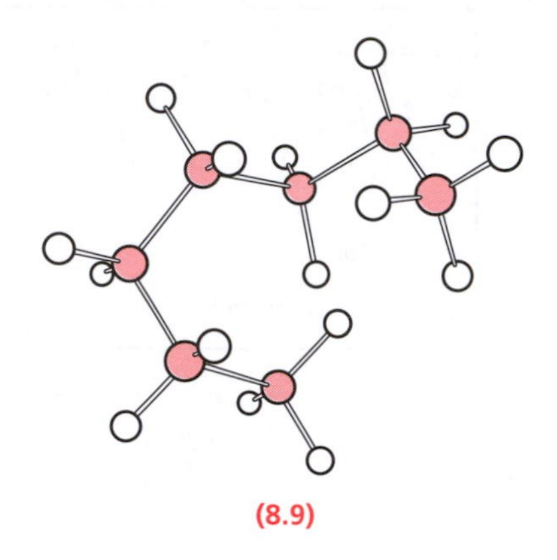

(8.9)

8.6 CHIRAL COMPOUNDS

A compound which is *not superimposable* upon its mirror image is described as being *chiral*. Many such compounds contain an *asymmetric carbon atom* – a carbon atom with four different groups attached (Figure 8.10). The object

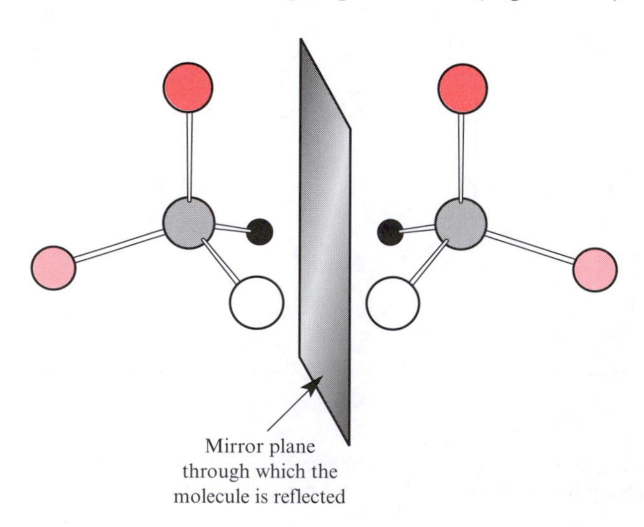

8.10 An asymmetric carbon atom has four different groups attached to it. The molecules on the left- and right-hand sides of the diagram are non-superimposable mirror images; they are optical isomers.

and its mirror image are called *optical isomers* or *enantiomers* – these are a particular pair of *stereoisomers.* Optical isomers differ only in their interactions with other chiral objects. Hands and gloves are chiral; your right hand will fit into a right-handed (but not a left-handed) glove. Helical chains are also chiral.

Helical S_∞: see Section 7.8

Optical isomers or enantiomers are a pair of stereoisomers, the structures of which are not superimposable on their mirror images.

The real test for chirality is non-superimposability of the mirror image on the original object. There are a number of quick tests which allow you to check whether a molecule is likely to be chiral, but these are *not infallible*:

- The presence of an asymmetric carbon atom has been widely used as a test in the past. However, many chiral compounds are known which do *not* contain an asymmetric carbon atom, but more importantly, many compounds containing two or more asymmetric carbon atoms are *not* chiral.
- Look for a plane of symmetry. Compounds *with* a plane of symmetry are *not* chiral. However, there is an exception to this test – some special types of molecule do not possess a plane of symmetry but are, nonetheless, achiral.

Achiral = non-chiral

Box 8.4 Plane of symmetry

Molecular symmetry is usually described in terms of *symmetry operators* such as axes of rotation and planes of symmetry.

If a molecule is symmetrical with respect to a plane that is drawn through it, then it is said to contain a plane of symmetry. A molecule may contain more than one plane of symmetry:

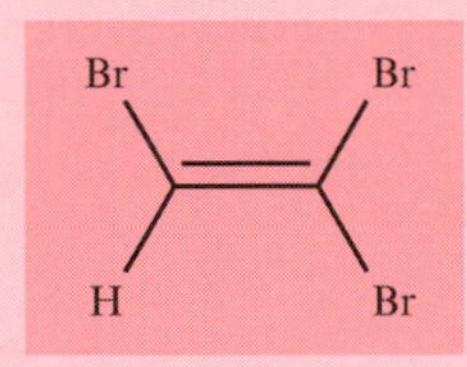

The molecule and the plane of symmetry both lie in the plane of the paper.

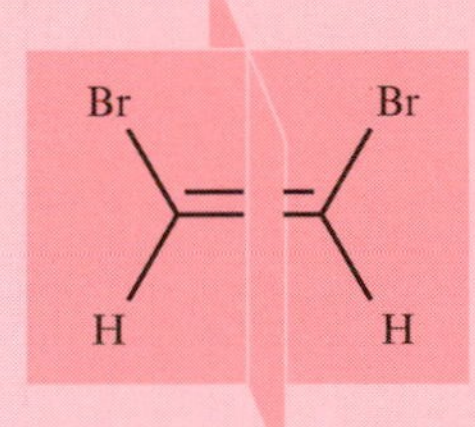

The molecule lies in the plane of the paper. One plane of symmetry is in the plane of the paper. The second plane passes through the C=C bond as shown.

None of the following molecules contains a plane of symmetry, and all three compounds are chiral. Note, however, that in only one example is there an asymmetric carbon atom. Compare these results with the 'tests for chirality' given in the text.

H
Me C Br
F

Me S O
Et

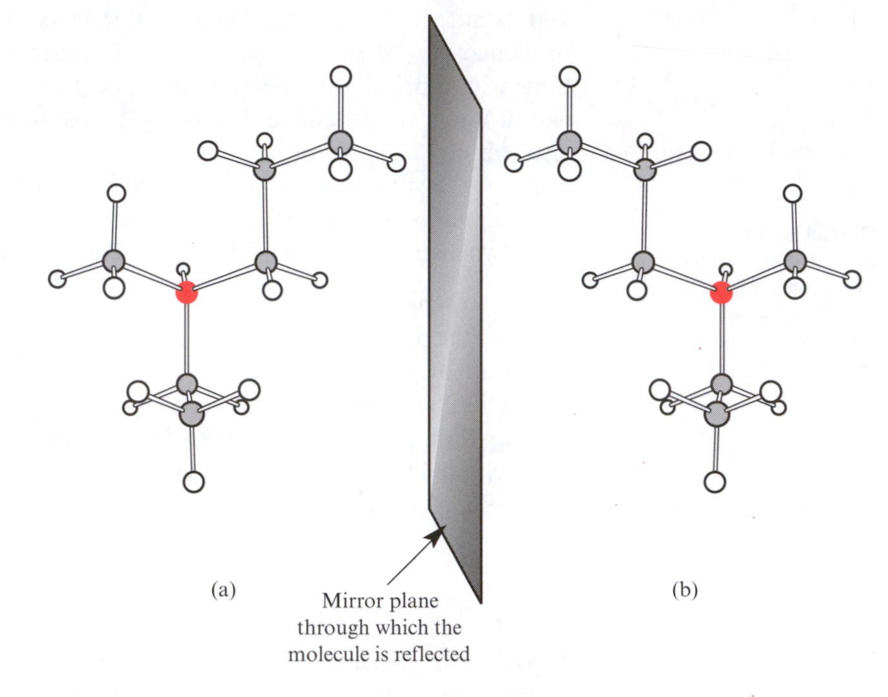

8.11 (a) and (b) are optical isomers of 3-methylhexane. The carbon atom shown in red is a chiral centre. The optical isomers are related to each other by reflection through a mirror plane, and are non-superimposable. This is made clearer in diagram (c), where we begin with the isomer shown in (b), and rotate it about the vertical C–C bond indicated. The resultant conformer is *not* equivalent to isomer (a). There is no substitute for model building or using computer modelling programs to clarify this point.

(c)

Rotation of molecule

Box 8.5 Chirality and life

As you look around you in your everyday life, you become aware that we live in a chiral world. On the macroscopic level, some clothes have a handedness, the tendrils of vines grow in a particular direction, and water goes down the plughole in a chiral, circular motion. At a molecular level, the body is optimized to the use of only *one* enantiomer or diastereomer (see Box 14.3) of biological building blocks – sugars and amino acids. You might think that using the 'wrong' diastereomer of a sugar (one that the body could not metabolize) might provide an easy way to diet. Usually, however, the sweet taste is also specific to the biologically occurring form.

In 3-methylhexane, one of the carbon atoms (shown in red in Figure 8.11) has a hydrogen atom, a methyl group, an ethyl group and a propyl group attached to it. The two optical isomers of 3-methylhexane are shown in Figures 8.11a and 8.11b. Figure 8.11c shows that isomer (b) cannot be converted back to (a) by simple rotation, and so the enantiomers are not superimposable – look carefully at the perspective in Figure 8.11c.

If the position and nature of a substituent destroys a plane of symmetry in a molecule, then the act of substitution usually causes the molecule to become chiral. For alkanes that already possess a large number of structural isomers, this significantly increases the total number of isomers. In Figure 8.5, *all* possible enantiomers were included in the total number of isomers.

Worked example 8.1 *Chiral and achiral alkanes*

Explain why heptane and 4-methylheptane are not chiral, whereas 3-methylheptane is.

The heptane molecule contains a plane of symmetry passing through the atoms of the central methylene group – atom C(4). When a methyl group is substituted at this position, its presence does not destroy the plane of symmetry (remember that there is effectively free rotation about the C(4)–C(Me) bond). Neither heptane nor 4-methylheptane is chiral. However, by placing the methyl group at the 3-position, the plane of symmetry in the molecule is destroyed. In the chiral compound, one carbon atom has four different groups attached to it.

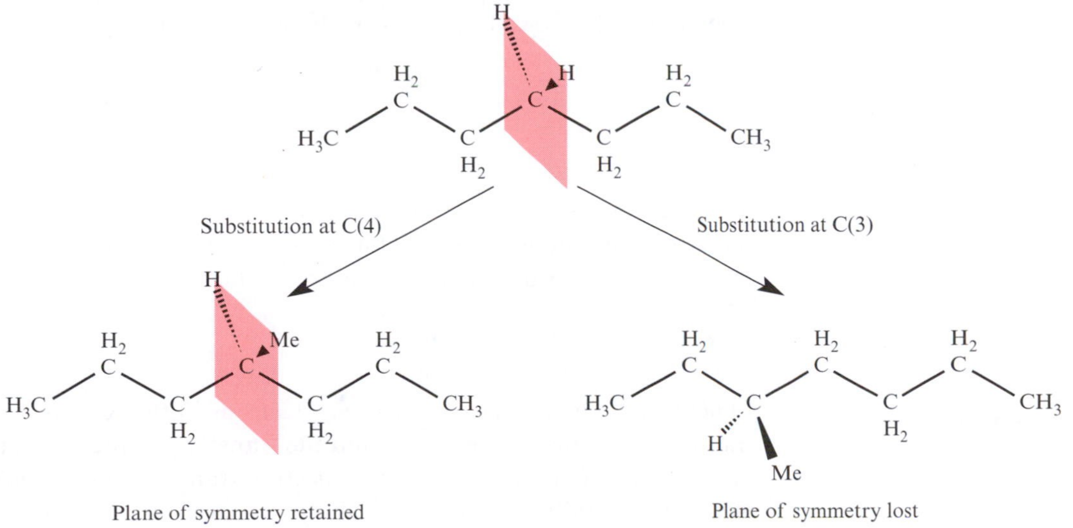

Substitution at C(4)

Substitution at C(3)

Plane of symmetry retained

Plane of symmetry lost

8.7 SOME GENERAL FEATURES OF A REACTION MECHANISM

Before we discuss some of the reactions of hydrocarbons, we must consider what we mean by *reaction mechanism*. This term is used extensively in chemistry, and describes the way in which the course of a reaction proceeds from reactants to products at a molecular level.

In this section, we define some of the terms used in descriptions of reaction mechanisms. Many others will appear later but the following include some of the most fundamental features. Chapter 10 considers reaction kinetics in more detail.

We cannot stress enough that a mechanism is merely a convenient way of thinking about a chemical change. A mechanism can never be proven; it can only be shown to be consistent with the experimental facts.

Activation energy

Activation energy: see Sections 1.18 and 10.6

The activation energy is the energy barrier which must be overcome if a reaction is to proceed. Consider reaction 8.1.

$$A + B \rightarrow C + D \qquad (8.1)$$

In order for compounds A and B to react with one another to give C and D, bonds may be broken or formed. The precise sequence of events by which these bonds are broken or made is described as a *reaction pathway*.

There are many possible pathways by which A and B can react to give C and D, and each of these has a characteristic activation energy. There is a direct relationship between $E_{\text{activation}}$ and the speed or rate of a reaction. The reaction usually takes the pathway with the lowest enthalpy barrier.§

Bond cleavage

Bond cleavage can occur either *homolytically* or *heterolytically*. In the homolytic cleavage of an X–Y bond, one electron is transferred to each of the two atoms and is represented by a 'half-headed' arrow. In equation 8.2, two radicals are formed as a result of homolytic bond cleavage.

$$X{-}Y \longrightarrow X^{\bullet} + Y^{\bullet} \qquad (8.2)$$

In heterolytic bond cleavage, the two electrons in the X–Y bond are both transferred to one of the atoms, and the transfer is represented by a 'full-headed' arrow (equation 8.3). A two-electron transfer usually *generates* ions (equation 8.4), although this is not always the case – in equation 8.5, there is no change in the total number of ions.

§ This statement is not strictly correct. In practice, molecules are crossing all possible barriers. However, the majority of products arise from the crossing of the lowest energy barrier.

$$X{-}Y \longrightarrow X^{+} + Y^{-} \tag{8.3}$$

$$Me_3C{-}Cl \longrightarrow Me_3C^{+} + Cl^{-} \tag{8.4}$$

$$H_2O^{+}{-}H \longrightarrow H{-}O{-}H + H^{+} \tag{8.5}$$

Homolytic bond cleavage **of a single bond X–Y means a one-electron transfer to each atom and the formation of radicals X• and Y•.**

Heterolytic bond cleavage **of a single bond X–Y means a two-electron transfer to either X or Y, and (usually) the formation of ions.**

When drawing arrows to represent electron transfers, their origin and destination must be shown clearly. Arrows should never be vaguely positioned.

A one-electron transfer is represented by a 'half-headed' arrow

A two-electron transfer is represented by a 'full-headed' arrow

Electrophiles and nucleophiles

Heterolytic bond cleavage is the net consequence of a reaction with an electrophile or nucleophile.

An *electrophile* is 'electron-loving' and seeks electrons. It is attracted to an electron-rich centre with a negative or δ^- charge. A *nucleophile* is a species which can donate electrons and is attracted to a centre with a positive or a δ^+ charge.

An *electrophile* is electron seeking; a *nucleophile* is electron-donating.

Examples of some common electrophiles and nucleophiles are given in Table 8.4.

Table 8.4 Some common electrophiles and nucleophiles.

Name of electrophile	*Formula*
Hydrogen cation	H^+
Chlorine cation	Cl^+
Bromine cation	Br^+
Iodine cation	I^+
Methyl cation	Me^+
Nitryl cation[a]	$[NO_2]^+$

Name of nucleophile	*Formula*
Hydride anion	H^-
Chloride anion	Cl^-
Bromide anion	Br^-
Methyl anion	Me^-
Hydroxide anion	$[OH]^-$
Water	H_2O

[a] The older name (still in common use) is nitronium ion.

Transition states and intermediates

In a reaction with a single step (for example, that shown in the profile in Figure 1.9), there is one associated activation energy. However, a reaction pathway may involve several steps, each with a characteristic value of $E_{\text{activation}}$. The reaction profile in Figure 8.12 describes the reaction between A and B, and the reaction pathway has two steps. The activation energy for the first step is $E_{\text{activation}}(1)$. After this first stage, an *intermediate* species is formed, which can then undergo further reaction (with an activation energy of $E_{\text{activation}}(2)$) to form the final products, C and D. Each of the energy maxima represents the energy of a *transition state*. In Figure 8.12, $\{TS(1)\}^{\ddagger}$ is the transition state during the first step of the reaction, and $\{TS(2)\}^{\ddagger}$ is the transition state associated with the second step. The braces ('curly brackets') with the superscripted 'double-dagger', ‡, are used to indicate that the species is a transition state.

The distinction between a *transition state* and an *intermediate* is important. A transition state occurs at an energy maximum along the reaction profile, whereas an intermediate occurs at a local energy minimum. Transition states *cannot* be isolated, whereas an intermediate species can be detected, and may be isolable.

A *transition state* occurs at an energy maximum, and cannot be isolated. An *intermediate* occurs at a local energy minimum, and can be detected and, perhaps, isolated.

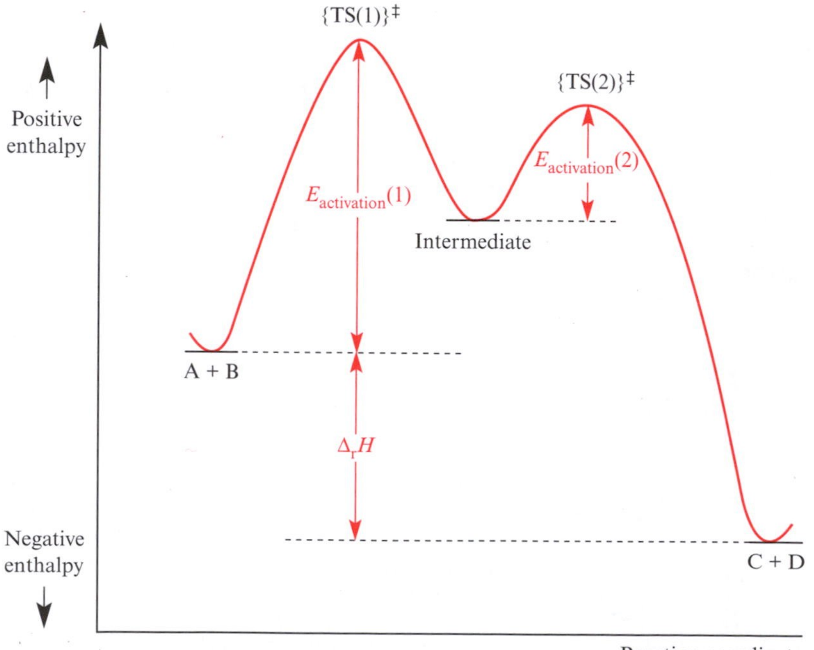

8.12 Profile for the two-step reaction of A with B, to give C and D. The first and second transition states are labelled $\{TS(1)\}^{\ddagger}$ and $\{TS(2)\}^{\ddagger}$ respectively. The overall enthalpy change during the reaction is $\Delta_r H$.

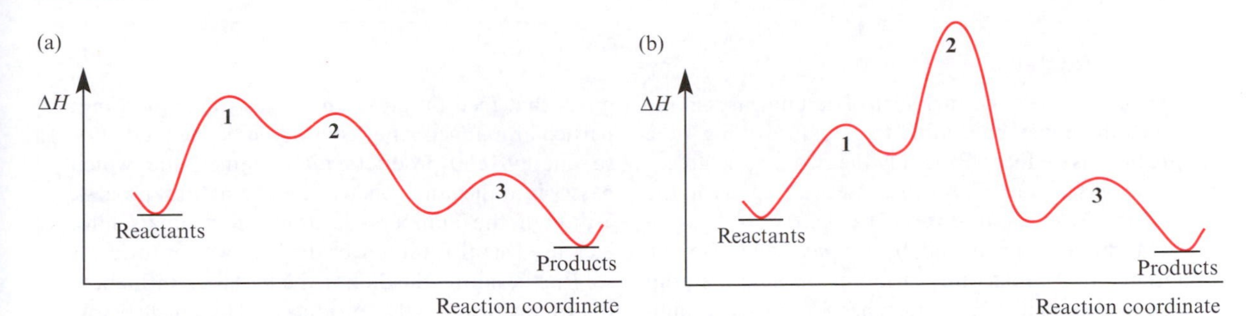

8.13 Schematic representations of reaction profiles for two three-step reactions. In (a), step **1** is the rate-determining step (slow) because it has the highest $E_{activation}$; steps **2** and **3** are relatively fast. In (b) step **2** is the slow step.

Rate-determining step

The *rate-determining step* (RDS) is the slowest step in a reaction.

Rate-determining step: see Section 10.9

In a reaction pathway with a series of steps, it is not necessarily the first step that has the highest activation energy. The larger the value of $E_{activation}$, the more difficult it is for the reaction to proceed and, in proceeding along a reaction profile, any step that has a high activation energy slows the reaction down. The step with the highest $E_{activation}$ is called the *rate-determining step* (RDS), or *slow step*.

Figure 8.13 shows two reaction profiles for three-step reactions. In Figure 8.13a, step **1** has the highest activation energy and is the rate-determining step. Once over this barrier, steps **2** and **3** proceed faster than step **1**. In Figure 8.13b, the second step in the pathway is the rate-determining step, while steps **1** and **3** are faster.

8.8 PHYSICAL PROPERTIES OF ALKANES

We could consider the physical properties of alkanes from a range of viewpoints. For example, how do the properties vary through a series of structural isomers? If we are looking for *trends* however, then it is most useful to consider a homologous series.

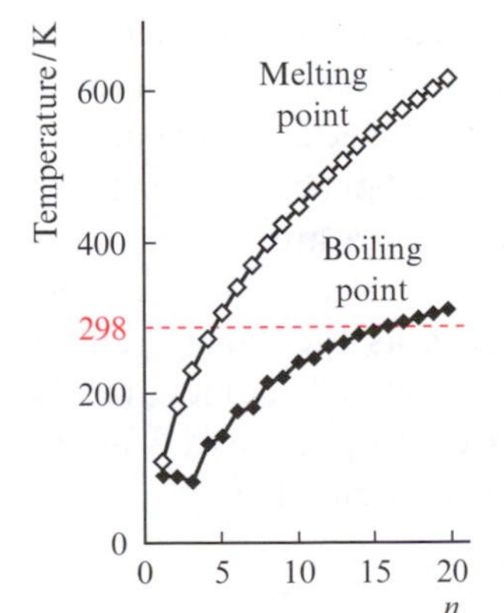

8.14 Trends in melting and boiling points of the alkanes C_nH_{2n+2} as n increases.

A homologous series of alkanes

The Pauling electronegativities of carbon and hydrogen are 2.6 and 2.2 respectively and the C–H bond dipole moment is small, and we regard the bond as relatively non-polar.

Alkanes are essentially non-polar molecules, and in both the solid and liquid states, the intermolecular interactions are weak van der Waals forces. Figure 8.14 shows the trends in melting and boiling points along the series of alkanes C_nH_{2n+2} from $n = 1$ to 20. Note that at 298 K and atmospheric pressure, methane, ethane, propane and butane are gases, while straight-chain alkanes from pentane to heptadecane are liquids, and higher alkanes are solids. Figure 8.15 shows *ball-and-stick* and *space-filling* diagrams of a pentane molecule in

Box 8.6 Methane: a 'greenhouse gas'

Increases in the concentrations of certain gases in the atmosphere may be leading to global warming – the 'greenhouse effect'. Probably the two major 'greenhouse gases' are CO_2 (discussed along with the 'greenhouse effect' in Box 13.2) and CH_4.

Methane is produced by the *anaerobic* (in the absence of O_2) decomposition of organic material and is released into the atmosphere; its escape as bubbles from wetlands gave rise to the name 'marsh gas'. Flooded areas (e.g. rice paddy fields) are a source of large quantities of CH_4. Methane also originates from ruminants – animals such as cows, sheep and goats that feed on grass and digest their food in a particular manner. After conversion of the foodstuffs to energy, end-products include methane which passes into the atmosphere. This is a natural process, although the numbers of domestic animals have increased significantly over the last two centuries or so. Furthermore, the diet of the livestock influences the quantities of CH_4 produced. The chart below summarizes the emission of methane into the atmosphere from animals and humans in 1984 – domestic cattle are responsible for 71% of the total emission. (Data from *Chemistry & Industry,* (1992) p. 334.)

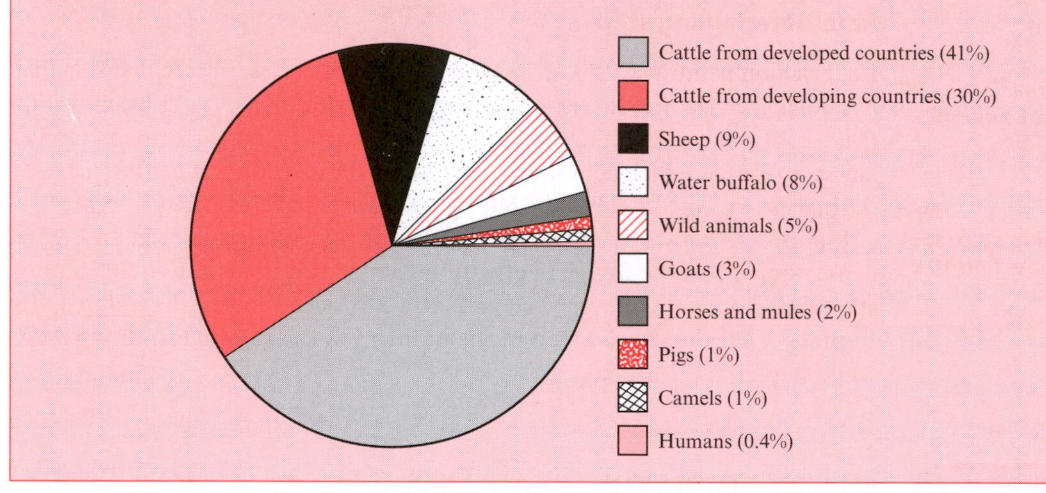

the extended conformation, and the 'space-filler' emphasizes its overall rod-like shape. In the solid and liquid phases, van der Waals forces operate between the atoms on the periphery of the molecule and those of its neighbours. Figure 8.16 illustrates the way in which molecules of icosane are packed in the crystal lattice. The general rise in melting points with chain length reflects the fact that the intermolecular interactions will be greater as the surface area of the molecule increases.

Density of $H_2O = 1.00\ g\ cm^{-3}$

The densities of straight-chain alkanes increase as the carbon chain lengthens, but the values level off at about $0.8\ g\ cm^{-3}$, and all alkanes are less dense than water. In keeping with the fact they are non-polar and water is polar, alkanes are *immiscible* with water, and when a liquid alkane is mixed with pure water, the alkane settles as the upper layer.

If two liquids are *miscible*, they form a solution. If they are *immiscible*, they form two layers with the more dense liquid as the lower layer.

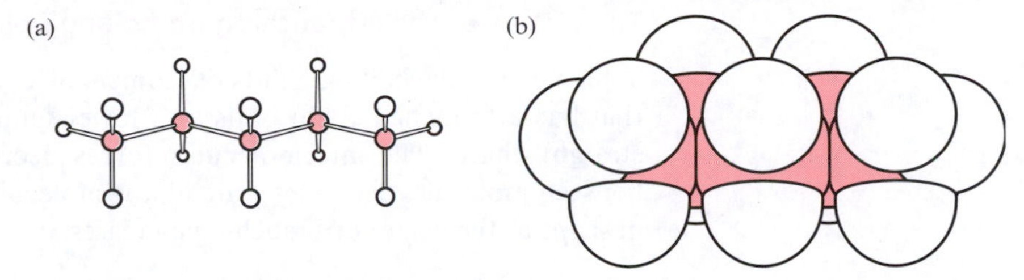

8.15 Two representations of a pentane molecule shown in an extended conformation. (a) A 'ball-and -stick' diagram and (b) a 'space-filling' diagram. In the space-filling model, the size of the atoms represents the region of space in which the electron will spend, say, 95–99% of its time. Various representations based upon covalent or van der Waals radii may be used. The ball-and-stick picture is a convenient way of representing the three-dimensional structure of the molecule.

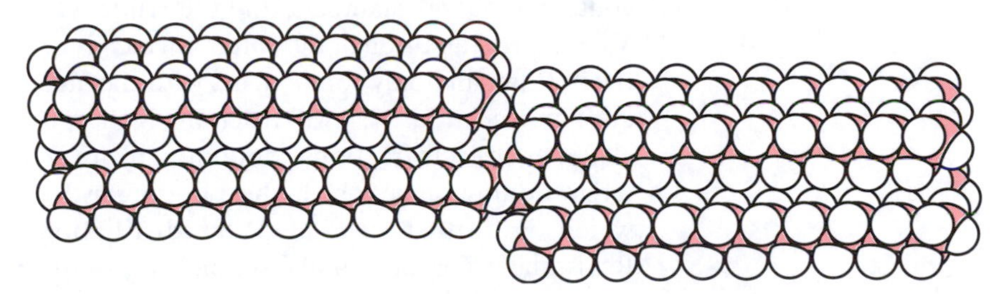

8.16 In the crystal lattice, molecules of icosane ($C_{20}H_{42}$) are in the extended conformation and lie parallel to one another.

Table 8.5 Melting and boiling points (measured at atmospheric pressure) of hexane and its structural isomers.

Isomer	*Carbon backbone*	*Boiling point / K*
Hexane		342
2-Methylpentane		333
3-Methylpentane		336
2,3-Dimethylbutane		331
2,2-Dimethylbutane		323

The effects of chain-branching on boiling points

Table 8.5 lists the boiling points of isomers of C_6H_{14}. In general, an isomer that has a branched chain boils at a lower temperature than one with a straight chain. The intermolecular forces decrease as the contact area between molecules decreases – an alkane molecule becomes more spherical in shape as the degree of branching increases.

8.9 INDUSTRIAL INTERCONVERSIONS OF HYDROCARBONS

Refining

After extraction from natural sources, crude oil is *refined*. This separates the oil into the main alkane components (gases, motor fuels, kerosene, diesel oil, wax, lubricants and bitumen) and involves fractional distillation and gas separation processes. The fractionating columns used for distillation each contain about 40 chambers, and the crude oil is introduced at about 600 K. Vapours rise up through the column which is cooler in higher regions than at the bottom, and only components with the highest boiling points (lowest relative molecular masses) will reach the highest chambers in the column. Each chamber collects fractions containing hydrocarbon components with a different boiling range. The highest boiling hydrocarbons remain as a residue which is further separated by distillation under reduced pressure. Redistillation of any initial fraction leads to further separation into fractions with narrower boiling point ranges.

Cracking

Polymers: see Boxes 8.11 to 8.13

The fractions obtained from refining do not all contain a commercially viable distribution of hydrocarbons – too many higher molecular weight residues remain. Such fractions must be subject to 'cracking', a process that breaks down the long hydrocarbon chains into shorter ones and produces alkenes (not present in crude oil) which are required as starting materials in the polymer industry. Cracking can be carried out by heating at high temperatures and pressures (thermal cracking) or in the presence of a catalyst (catalytic cracking). Thermal cracking is used to treat both liquid fractions or solid residue, but catalytic methods are suitable only for the liquid components, and a typical catalyst is aluminium oxide.

Reforming

Fractions to be used as fuels should burn as smoothly as possible, and this requirement leads to the process of reforming in which unbranched alkanes are converted in branched compounds. Reforming is simply isomerization. When straight-chain alkanes are combusted in an engine, premature ignition gives rise to 'knocking'. The octane number of a motor fuel indicates whether

Box 8.7 Wine as a monitor of lead pollution

A group of researchers in France and Belgium has used the vintage wine Châteauneuf-du-Pape to monitor lead pollution from motor engine emissions. Grapevines grown at the side of French roads have been the source of grapes for this wine for many years. Wines made between 1962 and 1991 have been analysed for their lead content and the concentration of tetraethyl lead Et_4Pb (the anti-knock additive) has varied over the period as shown in the bar chart.

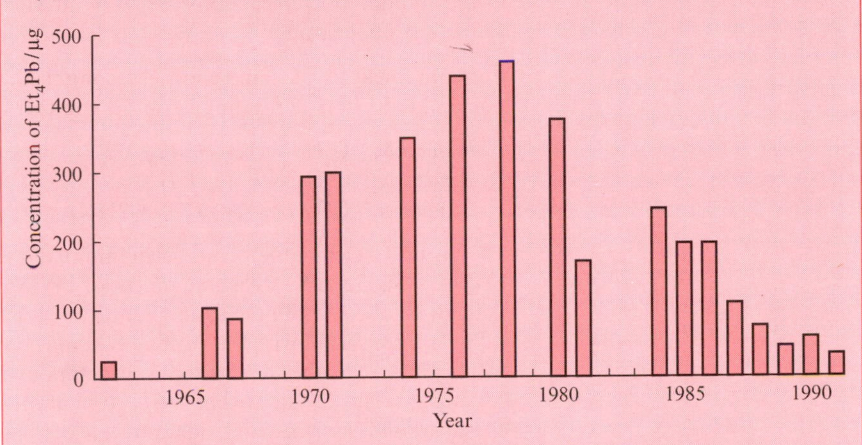

The initial rise follows an increase in the amount of traffic passing the vineyard. The fall coincides with the introduction of unleaded fuels. At their peak, the lead levels were a possible health risk. (Data from: R. Lobinski *et al.*, *Nature*, (1994) vol. 370, p. 24)

or not the fuel is resistant to knocking; the octane scale runs from 0 to 100 with 100 assigned to a fuel which is completely resistant. Anti-knocking agents such as tetraethyl lead (Et_4Pb) can be added but lead compounds are environmental pollutants and modern motor engines are designed to run on unleaded fuels.

8.10 REACTIONS OF ALKANES

Alkanes commonly undergo the following general types of reactions:

- **combustion;**
- **cracking (see Section 8.9);**
- **radical substitutions.**

Combustion

See Section 1.17

Alkanes burn in O_2 to give carbon dioxide and water (equations 8.6 and 8.7) and such combustions are exothermic.

Box 8.8 Oxidation states in carbon compounds

It is not usual to assign formal oxidation states to carbon but since we classify some organic reactions as oxidations or reductions, it is useful to mention briefly the range of formal oxidation states that are available to the carbon atom.

Oxidation is:

- the loss of hydrogen atoms, or
- the addition of oxygen atoms, or
- the loss of electrons

Reduction is:

- the addition of hydrogen atoms, or
- the loss of oxygen atoms, or
- the addition of electrons.

It is convenient to 'pick and choose' among these various definitions of oxidation and reduction, depending upon the context:

Reduction: $C_2H_4 + H_2 \rightarrow C_2H_6$
Oxidation: $C_3H_8 \rightarrow C_3H_6 + H_2$
Oxidation: $CH_4 + 2O_2 \rightarrow CO_2 + 2H_2O$

The group 14 elements silicon, germanium and tin form *hydrides* of formula EH_4 (E = Si, Ge or Sn). An assignment of the formal oxidation state of +4 for element E follows from the assignment of –1 for the hydride group, and these values are consistent with the Pauling electronegativities of the elements (Table 4.2). CH_4 is a member of this series of compounds, and the values of $\chi^P(C) = 2.6$ and $\chi^P(H) = 2.2$ show that the carbon atom carries a slight δ^- charge. In CH_4, carbon is assigned an oxidation state of –4 and hydrogen, +1.

If the hydrogen atoms in an alkane are substituted by halogen atoms, the situation is reversed ($\chi^P(C) = 2.6$, $\chi^P(F) = 4.0$, $\chi^P(Cl) = 3.2$, $\chi^P(Br) = 3.0$, $\chi^P(I) = 2.7$) and carbon is assigned an oxidation state of +4 (CF_4, CF_2Cl_2, CCl_4). In CH_3Cl, the oxidation state of the carbon is –2, as it is in methanol, CH_3OH. Carbon–carbon bonds may be neglected in the formal counting process (if you are worried by this, think what the oxidation state of fluorine is in F_2). The oxidation state of carbon in C_2H_6 is –3. In CO, the oxidation states of O and C are –2 and +2 respectively.

Question: What is the oxidation state of carbon in each of the following: CH_3F, CH_2Cl_2, CH_3CH_2OH, $Cl_2C{=}O$, CO_2, $[CO_3]^{2-}$?

$$CH_4 + 2O_2 \rightarrow CO_2 + 2H_2O \tag{8.6}$$

$$C_{11}H_{24} + 17O_2 \rightarrow 11CO_2 + 12H_2O \tag{8.7}$$

In practice, combustion is incomplete and varying amounts of elemental carbon and other carbon compounds are formed in addition to CO_2. The presence of elemental carbon and large aromatic molecules results in the well-known 'sooty' flame observed when higher hydrocarbons burn. When the amount of O_2 is limited (e.g. burning fuels in a closed room), combustion can lead to carbon monoxide – formation of CO from the partial combustion of hydrocarbons may prove fatal if it occurs in faulty domestic heating appliances.

Worked example 8.2 ***Determining the formula of a hydrocarbon***

When it is fully combusted, an acyclic alkane X gives 211.2 g CO_2 and 97.2 g H_2O. Suggest a possible identity for X.
[Relative atomic masses: C = 12, O = 16, H = 1]

The molecular formula of a hydrocarbon compound gives the ratio of moles of C : H atoms; this can be found from the moles of CO_2 and H_2O formed upon combustion.

$$\text{Moles} = \frac{\text{Mass}}{\text{Relative molecular mass}}$$

$$\text{Moles of } CO_2 = \frac{211.2}{44} = 4.8$$

$$\text{Moles of } H_2O = \frac{97.2}{18} = 5.4$$

Number of moles of C in the quantity of compound X combusted $= \text{moles } CO_2 = 4.8$

Number of moles of H in the quantity of compound X combusted $= 2 \times \text{moles } H_2O = 10.8$

$$\text{Ratio moles C : H} = 4.8 : 10.8 = 1 : 2.25$$

The formula of X is thus $C_nH_{2.25n}$, but we know that X is an acyclic alkane, and has the formula C_nH_{2n+2}.

$$2n + 2 = 2.25n$$

$$0.25n = 2$$

$$n = \frac{2}{0.25} = 8$$

Thus, X is C_8H_{18} (octane or an isomer of octane).

Halogenation

If a mixture of CH_4 and Cl_2 is left in sunlight, a reaction occurs to yield hydrogen chloride and members of the family of chloroalkanes $CH_{4-n}Cl_n$ where n = 1, 2, 3 or 4. Equations 8.8–8.11 summarize the formation of chloromethane followed by sequential reactions with Cl_2; $h\nu$ stands for 'irradiation' and its use indicates that reagents are irradiated in order to induce a reaction. This is a *photolysis reaction*, and the radiation used usually has a wavelength in the range 200–800 nm.

Electromagnetic spectrum: see Appendix 4

A *photolysis* reaction takes place upon irradiation (symbolized by $h\nu$)

$$CH_4 + Cl_2 \xrightarrow{h\nu} CH_3Cl + HCl \tag{8.8}$$

$$CH_3Cl + Cl_2 \xrightarrow{h\nu} CH_2Cl_2 + HCl \tag{8.9}$$

$$CH_2Cl_2 + Cl_2 \xrightarrow{h\nu} CHCl_3 + HCl \tag{8.10}$$

$$CHCl_3 + Cl_2 \xrightarrow{h\nu} CCl_4 + HCl \tag{8.11}$$

These are *substitution* reactions. The chlorine atoms have been substituted for the hydrogen atoms in CH_4 – this particular substitution is a *chlorination* reaction. Bromination of methane occurs when Br_2 and CH_4 are photolysed, but no such reaction occurs with I_2. The reaction with F_2 is explosive and proceeds even in the absence of light. The relative order of reactivities of the halogens with methane is:

$$F_2 >> Cl_2 > Br_2 >> I_2$$

Although there is no reaction between CH_4 and I_2, we include I_2 in the series because this order of reactivities is encountered in many halogenation reactions.

In a substitution reaction, one atom or group is exchanged for another.

The fluorination, chlorination and bromination of higher alkanes lead to complex mixtures of substituted products. As the number of carbon atoms in the chain increases, the number of possible sites of substitution increases. As an exercise, write down the possible products of the monochlorination of octane.

8.11 THE CHLORINATION OF METHANE: A RADICAL CHAIN REACTION

In this section we discuss the mechanism by which the chlorination of CH_4 occurs. Why is it necessary to irradiate the mixture of gases? The mechanisms described are general for the chlorination and bromination of other alkanes, although reactions with F_2 may proceed by a different mechanism. The reaction between CH_4 and Cl_2 is an example of a *radical chain reaction*.

We distinguish between linear and branched chains in Section 10.11

General features of a radical chain reaction

A radical chain reaction involves a series of steps, classified as being one of:

- initiation, or
- propagation, or
- termination.

In the *initiation step*, free radicals are *produced* from one or more of the reactants. The generation of these radicals may result from heating or irradiating the reaction mixture, or by adding a chemical *initiator*. Alternatively, a 'stable' free radical may be added as the initiator. Equation 8.12 shows a common initiation step in which radicals $X^{\bullet}$ are formed from X_2 by homolytic bond cleavage on irradiating X_2 with light of an appropriate wavelength:

$$X_2 \xrightarrow{h\nu} 2\,X^{\bullet}$$ *Initiation step* (8.12)

Organic radicals are often (but not necessarily) short-lived. As they collide with other species in the system, reactions can occur, and these may result in either propagation or termination. If a radical collides with a non-radical species, then a *propagation step* may occur in which a new reactive radical is generated (equation 8.13). The newly generated radical can now react with a second species to generate a third radical (equation 8.14). Although the non-radical reacting species (RY) in equations 8.13 and 8.14 are shown as being the same, this does not have to be the case (see below).

$$X^{\bullet} + RY \rightarrow RX + Y^{\bullet}$$ *Propagation step* (8.13)

$$Y^{\bullet} + RY \rightarrow Y_2 + R^{\bullet}$$ *Propagation step* (8.14)

In a *termination step*, radicals are *removed* from the reaction without the generation of other reactive radicals. An example of a termination step is shown in equation 8.15 and in this case, it is the reverse of the initiation step (equation 8.12). It is equally possible for $X^{\bullet}$ to combine with any other radicals which might be present in the reaction system.

$$2X^{\bullet} \rightarrow X_2$$ *Termination step* (8.15)

The overall sequence of events combine to give a *chain reaction*, with the propagation steps keeping the chain 'alive'. The concentration of radicals in the reaction system at any given time is low, and the interaction of a radical with a non-radical species (*propagation*) is far more likely than the reaction between two radicals (*termination*). After the initiation process a large number of propagation steps can occur before radicals are removed in a termination step. Very often, the generation of a single radical allows the formation of many thousands of product molecules before the chain is terminated. The combustion of alkanes and other organic compounds is a radical chain reaction.

Radical chain reaction
1. initiation (radicals formed)
2. propagation (radicals react and more are formed)
3. termination (radicals are removed from the reaction)

The mechanism of the chlorination of methane

The absorption of light by Cl_2 causes homolytic cleavage of the Cl–Cl bond (equation 8.16).

$$Cl_2 \xrightarrow{h\nu} 2Cl^{\bullet}$$ (8.16)

Collisions can occur between chlorine radicals and methane molecules and a sequence of propagation steps begins which includes those shown in equations 8.17–8.20. In each of these steps, a radical is consumed and another radical is formed. In equations 8.17 and 8.19, a $Cl^{\bullet}$ radical abstracts a hydrogen atom ($H^{\bullet}$) from the reactant to give HCl.

$$Cl^{\bullet} + CH_4 \rightarrow HCl + CH_3^{\bullet} \tag{8.17}$$

$$CH_3^{\bullet} + Cl_2 \rightarrow CH_3Cl + Cl^{\bullet} \tag{8.18}$$

$$Cl^{\bullet} + CH_3Cl \rightarrow HCl + CH_2Cl^{\bullet} \tag{8.19}$$

$$CH_2Cl^{\bullet} + Cl_2 \rightarrow CH_2Cl_2 + Cl^{\bullet} \tag{8.20}$$

Termination of the chain occurs when two radical species combine. Equations 8.21–8.23 show three possible steps, but others involving chloroalkane radicals (e.g. $CH_2Cl^{\bullet}$) can also occur.

$$2Cl^{\bullet} \rightarrow Cl_2 \tag{8.21}$$

$$Cl^{\bullet} + CH_3^{\bullet} \rightarrow CH_3Cl \tag{8.22}$$

$$2CH_3^{\bullet} \rightarrow C_2H_6 \tag{8.23}$$

Equation 8.21 is the reverse of the initiation step (equation 8.16), while equation 8.22 illustrates the formation of a desired product. Reactions of the type shown in equation 8.23 complicate the process by extending the carbon chain. The ethane molecule may participate in a reaction similar to that shown in equation 8.17, and this may ultimately lead to further extension of the carbon backbone and the production of higher alkanes. A radical chain reaction is, therefore, *not* specific.

8.12 THE CHLORINATION OF PROPANE AND 2-METHYLPROPANE

For alkane chains which possess more than two carbon atoms, an additional complication arises – the chlorine radical can attack at one of several different sites. We exemplify the problem by considering the reactions of Cl_2 with propane and 2-methylpropane.

Propane and dichlorine

When Cl_2 and C_3H_8 are irradiated, the initiation is reaction 8.16. Propagation steps then follow one of two sequences, depending upon whether the hydrogen at a primary or secondary carbon centre is involved (equations 8.24–8.26).

$$CH_3CH_2CH_3 \xrightarrow{Cl^{\bullet}} \begin{cases} HCl + CH_3CH_2CH_2^{\bullet} \\ HCl + CH_3\dot{C}HCH_3 \end{cases} \tag{8.24}$$

Secondary (pointing to the CH_2 carbon); Primary, Primary (pointing to the two CH_3 carbons)

Box 8.9 The depletion of the ozone layer

The 'ozone layer' is a layer in the atmosphere that is 15–30 km above the Earth's surface. Ozone absorbs strongly in the ultraviolet (UV) region of the spectrum and the ozone layer shields the earth from UV radiation from the Sun. One possible effect of UV radiation on humans is skin cancer.

One group of pollutants from the Earth are chlorofluorocarbons (CFCs). These are used as propellants in aerosols, as refrigerants, in air conditioners, as solvents and in foams used in furnishings. After use, the CFCs escape into the atmosphere, and in the stratosphere they can undergo photochemical reactions, such as:

$$CCl_2F_2 \xrightarrow{\text{UV light}} Cl^{\bullet} + CClF_2^{\bullet}$$

A radical chain reaction ensues, and reactions with ozone include:

$$O_3 + Cl^{\bullet} \rightarrow O_2 + ClO^{\bullet}$$

The devastating result is that the ozone layer is slowly being depleted. This fact was first recognized in the 1970s, and action was taken swiftly in some countries regarding legislating against the use of CFCs. Although so-called 'ozone-friendly' aerosols and other products are now increasingly available, the problem of the depletion of the ozone layer remains. Emissions of CFCs from Earth *must* be drastically reduced if we are to remain protected from the intense UV radiation of the Sun.

In 1987, the 'Montreal Protocol for the Protection of the Ozone Layer' was established and called for the phasing out of CFCs. Taking the 1986 European consumption of CFCs as a standard (100%), the graph below shows how the usage was reduced between 1986 and 1993, and a large contribution to this was a drastic limitation in the uses of CFCs in aerosols. (Data from *Chemistry & Industry*, (1994) p. 323.

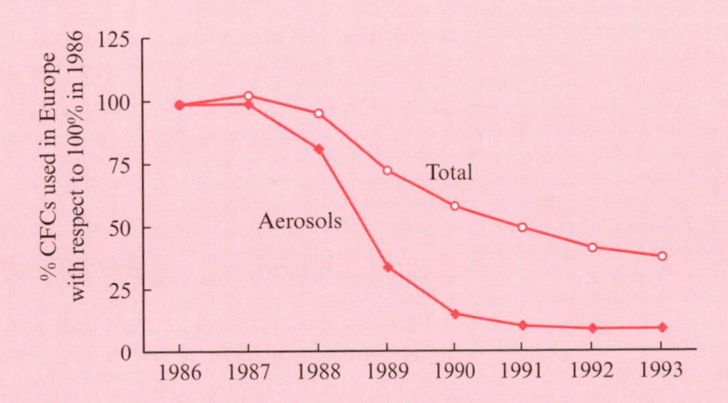

$$CH_3CH_2CH_2^{\bullet} + Cl_2 \rightarrow \underset{\text{1-chloropropane}}{CH_3CH_2CH_2Cl} + Cl^{\bullet} \quad (8.25)$$

$$CH_3\dot{C}HCH_3 + Cl_2 \rightarrow \underset{\text{2-chloropropane}}{CH_3CHClCH_3} + Cl^{\bullet} \quad (8.26)$$

In the final product mixture, 2-chloropropane predominates over 1-chloropropane, despite the fact that the *chance* of a collision between a chlorine radical and a hydrogen atom attached to a primary carbon atom is higher than that between a chlorine radical and a hydrogen atom attached to a secondary carbon atom. The abstraction of a hydrogen radical to give an alkyl radical with a secondary carbon atom is preferred to abstraction of a hydrogen radical to give an alkyl radical with a primary carbon atom.

2-Methylpropane and dichlorine

After the initiation step (reaction 8.16), chlorine radicals can react with 2-methylpropane (equation 8.27) and further propagation steps give two isomeric products (equations 8.28 and 8.29).

$$(CH_3)_2CHCH_3 \xrightarrow{Cl^{\bullet}} \begin{cases} HCl + CH_3CH(CH_3)CH_2^{\bullet} \\ HCl + CH_3\dot{C}(CH_3)CH_3 \end{cases} \quad (8.27)$$

$$CH_3CH(CH_3)CH_2^{\bullet} + Cl_2 \rightarrow \underset{\text{1-chloro-2-methylpropane}}{CH_3CH(CH_3)CH_2Cl} + Cl^{\bullet} \quad (8.28)$$

$$CH_3\dot{C}(CH_3)CH_3 + Cl_2 \rightarrow \underset{\text{2-chloro-2-methylpropane}}{CH_3CCl(CH_3)CH_3} + Cl^{\bullet} \quad (8.29)$$

Experimental data show that the formation of 2-chloro-2-methylpropane predominates over 1-chloro-2-methylpropane, and illustrates that abstraction of a hydrogen radical to give an alkyl radical with a tertiary carbon atom is preferred to abstraction of a hydrogen radical to give an alkyl radical with a primary carbon atom.

Radicals with primary, secondary and tertiary carbon atoms

The reactions described above lead us to a general, important result. In a radical reaction, a *radical which contains a tertiary carbon atom is formed in preference to one with a secondary carbon atom, and one with a secondary carbon atom is formed in preference to one with a primary carbon atom*. This has an effect on the distribution of products in a chain reaction, but also affects the rate at which the reaction proceeds.

The experimental data suggest that the stability of the radicals is in the order:

tertiary ($R_3C^\bullet$) > secondary ($R_2HC^\bullet$) > primary ($RH_2C^\bullet$) > methyl ($H_3C^\bullet$)

R represents a general alkyl substituent

In the radical, the carbon atom has only seven electrons in its valence shell and is electron-deficient. The observed pattern of stabilities suggests that alkyl groups are better able to stabilize an electron-deficient centre than hydrogen atoms. In effect, we say that the alkyl group is *electron-releasing* and in a C–R bond, the electrons are polarized towards the carbon in the sense $C^{\delta-}–R^{\delta+}$.

Alkyl groups, R, are *electron-releasing*, and the stability of radicals is in the order:

$R_3C^\bullet > R_2HC^\bullet > RH_2C^\bullet > H_3C^\bullet$

8.13 REACTIONS OF ALKENES 1: SOME OXIDATION AND ADDITION REACTIONS

Like alkanes, alkenes burn in dioxygen to give water and carbon dioxide (equation 8.30).

$$CH_3CH_2CH{=}CH_2 + 6O_2 \rightarrow 4CO_2 + 4H_2O \qquad (8.30)$$

Whereas the reactions of alkanes are usually *substitutions*, the C=C functionality in an alkene undergoes *addition* reactions of the type shown in equation 8.31.

$$\text{>C=C<} + AB \longrightarrow \text{>C(A)–C(B)<} \qquad (8.31)$$

Typical reactions of alkenes are:

- **electrophilic addition;**
- **radical addition;**
- **polymerization.**

This section gives examples of some oxidation and addition reactions of the alkene functionality, and the reaction mechanisms are discussed in Section 8.14. Further reactions are described in the following sections.

Hydrogenation

The *hydrogenation* of an alkene involves the addition of dihydrogen, and converts an unsaturated hydrocarbon into a saturated one; it is an important method of synthesizing alkanes (equation 8.32).

Catalyst: see Sections 10.7

$$CH_3CH{=}CH_2 + H_2 \xrightarrow{\text{catalyst}} CH_3CH_2CH_3 \qquad (8.32)$$

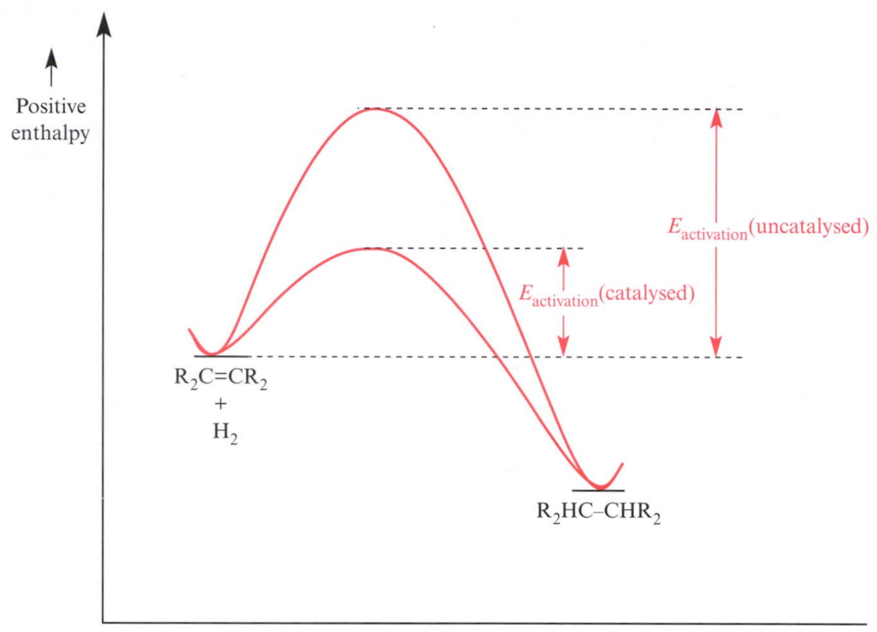

8.17 The addition of H_2 to alkenes has a high activation barrier. This schematic representation of the reaction profile shows that in the presence of a suitable catalyst, $E_{activation}$ is lowered.

The presence of a catalyst (often a nickel, palladium or platinum metal surface) is necessary since the activation energy for the reaction between H_2 and an alkene is high. Figure 8.17 schematically illustrates the role of the catalyst; the activation energy for the pathway from the alkene ($R_2C{=}CR_2$) to the hydrogenated product (R_2CHCHR_2) is lowered, making the reaction faster. The use of a metal surface as a catalyst for a reaction between gases, or between a gas and a liquid, is an example of a *heterogeneous catalyst*.

A hydrogenation reaction is an addition of dihydrogen.

Addition of halogens

Alkenes react with Cl_2 or Br_2 in addition reactions to give dichloro- or dibromo-derivatives respectively (equations 8.33 and 8.34). In the products, the halogen atoms are attached to *adjacent* carbon atoms — such compounds are also known as *vicinal dihalides*.

$$CH_3CH_2CH{=}CHCH_3 + Cl_2 \rightarrow \underset{\text{2,3-dichloropentane}}{CH_3CH_2CHClCHClCH_3} \qquad (8.33)$$

$$CH_2{=}CHCH_2CH_3 + Br_2 \rightarrow \underset{\text{1,2-dibromobutane}}{CH_2BrCHBrCH_2CH_3} \qquad (8.34)$$

Reactions with F_2 are extremely violent, while those with I_2 are extremely slow or fail altogether. Chlorination and bromination take place at or below 298 K without the need for irradiation. This suggests that the reaction does not involve radicals, but if a source of radicals *is* available, addition still occurs but may be complicated by competing radical substitution reactions.

The decolorization of an aqueous solution of dibromine (bromine water) is commonly used as a qualitative test for the presence of a C=C double bond, although it is by no means definitive. Mixed products are obtained (see the discussion of the reactions of alkenes with Br_2 and H_2O later in this section).

Addition of hydrogen halides

Equation 8.35 shows the reaction between an alkene and hydrogen chloride, hydrogen bromide or hydrogen iodide. The substrate is a symmetrical alkene – when HBr is added, the product can only be 2-bromobutane.

$$CH_3CH{=}CHCH_3 + HX \rightarrow CH_3CH_2CHXCH_3 \qquad (8.35)$$

X = Cl, Br or I

With asymmetrical alkenes, two products are possible as shown for the addition of HBr to propene (equation 8.36), although in practice, 2-bromopropane is the major product. We return to the mechanism of this reaction in the next section, but for now, note that the reaction favours attachment of the halide to the *secondary* carbon atom in the product.

$$CH_3CH{=}CH_2 + HBr \rightarrow \begin{cases} CH_3CH_2CH_2Br \text{ (1-bromopropane)} & \textit{Minor} \\ CH_3CHBrCH_3 \text{ (2-bromopropane)} & \textit{Major} \end{cases} \qquad (8.36)$$

Similarly, when HCl reacts with 2-methylpropene, the major product is 2-chloro-2-methylpropane – the preference is for the chloro-group to be attached to the *tertiary* (rather than the primary) carbon atom.

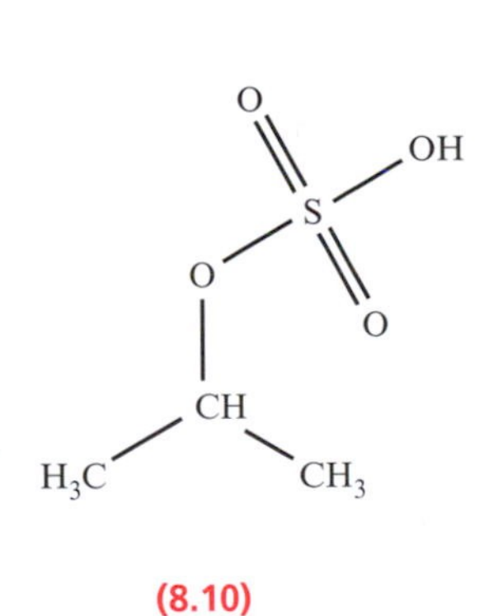

(8.10)

(8.11)

Addition of sulfuric acid and water

Alkenes react with cold, concentrated sulfuric acid (equation 8.37). The product is an alkyl hydrogensulfate, and has the structure shown in **8.10**.

$$CH_3CH{=}CH_2 + H_2SO_4 \rightarrow CH_3CH(OSO_3H)CH_3 \qquad (8.37)$$

The reaction involves the addition of an O–H bond across the C=C double bond, and is thus similar in type to the addition of the hydrogen halides described above. The product shows the same preference for the attachment of the oxygen atom to the secondary carbon atom as did the bromine atom above.

Alkyl hydrogensulfates are not usually isolated, but if water is added and the mixture is heated, *hydrolysis* occurs (equation 8.38) and an *alcohol* is formed. Compound **8.10** is hydrolysed to **8.11** in a *nucleophilic substitution* reaction – the nucleophile is water (*not* hydroxide ion).

Nucleophile: see Section 8.7; details of nucleophilic substitutions: see Chapter 14

When an X group is replaced by an OH group as a result of reaction with H_2O, the process is termed *hydrolysis*.

$$CH_3CH(OSO_3H)CH_3 + H_2O \longrightarrow \underset{\text{propan-2-ol}}{CH_3CH(OH)CH_3} + H_2SO_4 \quad (8.38)$$

Together, reactions 8.37 and 8.38 give a route from an alkene to an alcohol in which the sulfuric acid functions as a catalyst; H_2SO_4 is a reagent in the first reaction but is regenerated in the second. If the desired product is the alcohol, the two steps can be carried out simultaneously by using water in the presence of acid (equation 8.39).

$$CH_3CH{=}CH_2 + H_2O \xrightarrow{H^+ \text{ catalyst}} CH_3CH(OH)CH_3 \quad (8.39)$$

Dehydration of alcohols: see Section 14.10

Reaction 8.39 may be driven in either direction – heating an alcohol with strong acids generates alkenes amongst other products.

Reactions with halogens and water

If dichlorine or dibromine *and* water react together with an alkene, the reactions proceed as shown in equations 8.40 and 8.41.

$$CH_3CH{=}CH_2 \xrightarrow{Cl_2,\ H_2O} \underset{\text{1-chloropropan-2-ol}}{CH_3CH(OH)CH_2Cl} \quad (8.40)$$

$$CH_3CMe{=}CH_2 \xrightarrow{Br_2,\ H_2O} \underset{\text{1-bromo-2-methylpropan-2-ol}}{CH_3CMe(OH)CH_2Br} \quad (8.41)$$

The hydroxy (OH) group shows a preference to be attached to the secondary or tertiary carbon atom. Formation of the vicinal dihalide may be competitive with the products in reactions 8.40 and 8.41 as is observed when aqueous Br_2 is used as a test for an alkene.

Formation of diols

The oxidation of an alkene by cold alkaline potassium permanganate§ leads to the formation of a *vicinal diol* (equation 8.42).

$$CH_3CH_2CH{=}CH_2 \xrightarrow{\text{cold, alkaline } K[MnO_4]} \underset{\text{butane-1,2-diol}}{CH_3CH_2CH(OH)CH_2OH} \quad (8.42)$$

§ IUPAC accepts *permanganate* as one name for the anion $[MnO_4]^-$, and this is in common use. The systematic name is tetraoxomanganate(VII).

An intermediate of the general type shown in structure **8.12** is proposed; the alkene has been oxidized whilst the manganese centre has been reduced.

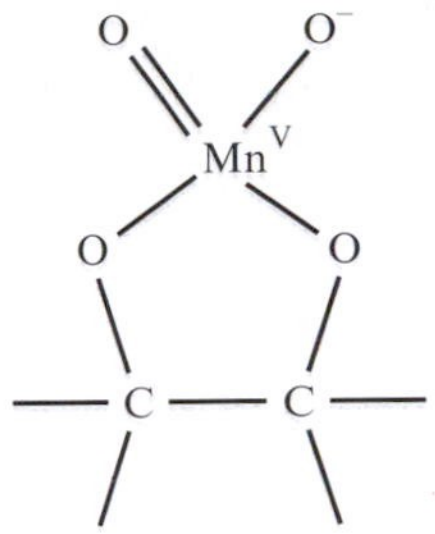

(8.12)

The older name for a diol is a *glycol* and this terminology is still used. The glycol was named by adding 'glycol' to the name of the alkene from which it was derived: ethane-1,2-diol is also called ethylene glycol, ethylene being the older name for ethene. Ethylene glycol is encountered in its role as an 'anti-freeze' additive to automobile radiator water; it possesses a low melting point (261 K), high boiling point (471 K) and is completely miscible with water.

Reaction with ozone

The reaction of an alkene with O_3 leads to a cyclic *ozonide*. A cyclic intermediate which is an isomer of the ozonide shown in equation 8.43 is involved in the reaction.

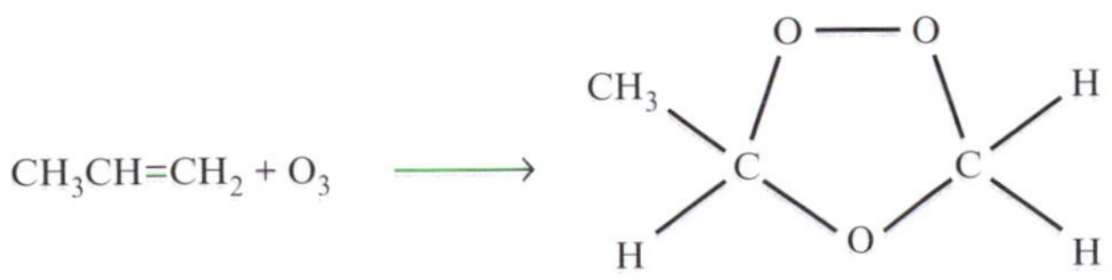

$$CH_3CH{=}CH_2 + O_3 \longrightarrow \qquad (8.43)$$

An important application of this reaction lies not in the isolation of ozonides (which are often explosive), but in the products of their hydrolysis. Reductive hydrolysis of ozonides gives aldehydes or ketones, depending upon the substituents present in the original alkene. Equations 8.44 and 8.45 give two examples.

Aldehydes and ketones: see Table 1.9 and Chapter 17

Box 8.10 Polyethylene glycol and the *Mary Rose*

The sunken Tudor warship *Mary Rose* was brought to the surface in 1982. An immediate problem, as with other salvaged wooden wrecks and wooden artefacts that have been in the sea or wet places, was how to preserve the timbers of the ship as they began to dry out. A method of conservation for wood that has been preserved through history in a wet state is to use a treatment involving polyethylene glycol. This is a waxy compound and when it impregnates the wood, it helps to prevent the deterioration of the drying wood.

$$\text{(ozonide of propene)} \xrightarrow{\text{Zn, H}_2\text{O}} CH_3CHO + HCHO \quad (8.44)$$

ethanal methanal

$$\text{(ozonide of 2-methylbut-1-ene)} \xrightarrow{\text{Zn, H}_2\text{O}} CH_3CH_2C(O)CH_3 + HCHO \quad (8.45)$$

butan-2-one methanal

8.14 THE MECHANISM OF ELECTROPHILIC ADDITION

Additions can occur either by radical or electrophilic mechanisms, and in this section we consider the latter. An electrophilic mechanism is favoured when the reaction is carried out in a *polar solvent*, whilst a radical mechanism requires an initiator.

After considering the details of the mechanism you should be able to understand many of the experimentally observed results described in the previous section.

Addition of HX to a symmetrical double bond

An electrophile is 'electron-seeking'

Electrophiles (for example H^+) are attracted towards the region of electron density in the C=C double bond, and in the first step of electrophilic addition, the two electrons in the π-bond are transferred to the electrophile to form a new σ-bond (equation 8.46).

$$H_2C{=}CH_2 + H^+ \longrightarrow H_2\overset{+}{C}{-}CH_2H \quad (8.46)$$

carbenium ion

A carbenium ion has the general formula $[R_3C]^+$ where R is an H atom or organic group.

The carbon atom that does *not* form the new σ-bond is electron deficient, with only six electrons in its valence shell. It bears a formal positive charge, and is called a *carbenium ion* or a *carbocation*. (The older name, still in use, is *carbonium ion*.) The carbenium ion is an intermediate in the reaction. The carbenium ion is susceptible to attack by a nucleophile, for example, Br^- (equation 8.47).

A nucleophile donates electrons

$$\mathrm{Br^-} + \mathrm{H_2\overset{+}{C}{-}CH_2(H)} \longrightarrow \underset{\text{bromoethane}}{\mathrm{H_2C(Br){-}CH_2(H)}} \qquad (8.47)$$

It is not necessary to add the electrophile and the nucleophile separately. Often the nucleophile is generated as a consequence of the formation of the carbenium ion. For example, the HBr molecule is polar ($H^{\delta+}$—$Br^{\delta-}$) and the hydrogen end of the molecule is attracted to the C=C double bond. This polarizes the HBr molecule still further and heterolytic cleavage of the H–Br bond occurs. Effectively, H^+ is made available and reaction 8.46 follows.

Figure 8.18 shows a representation of the profile for the reaction of ethene with hydrogen bromide. The formation of the carbenium ion intermediate is the rate-determining step of the reaction. Subsequent reaction with bromide ion occurs in a faster step. The first transition state, {TS(1)}‡, may be described in terms of a species which is part-way between the reactants and the intermediate. The second transition state, {TS(2)}‡ in Figure 8.18, may similarly be considered as being a species in which the bromide ion has *begun to form* a bond with the carbenium ion, for example a 'stretched bond' may be envisaged as in structure **8.13**.

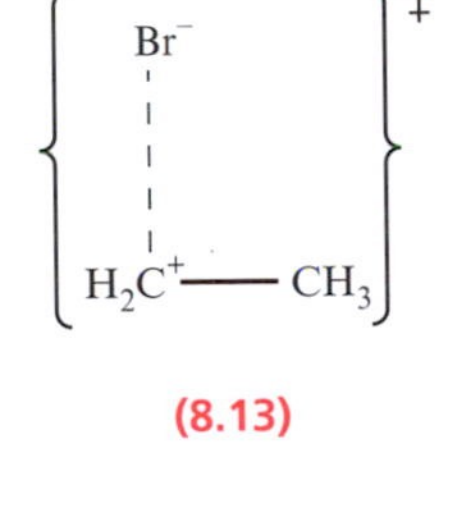

(8.13)

The mechanism described here can be extended to a range of other electrophilic additions in which HBr is replaced by, for example, HCl, HI and $HOSO_3H$ (i.e. H_2SO_4).

Addition of HX to an asymmetrical double bond

When an electrophile attacks an *asymmetrical alkene*, there are two possible pathways (equation 8.48). In this example, one route leads to a primary carbenium ion whilst the other gives a secondary carbenium ion.

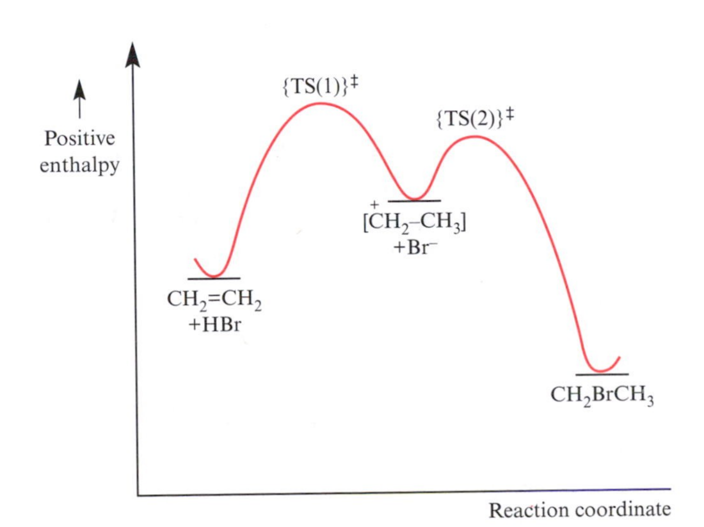

8.18 A schematic representation of the reaction profile for the electrophilic addition of HBr to ethene. The rate-determining step is the formation of the carbenium ion. This intermediate is at a *local energy minimum*.

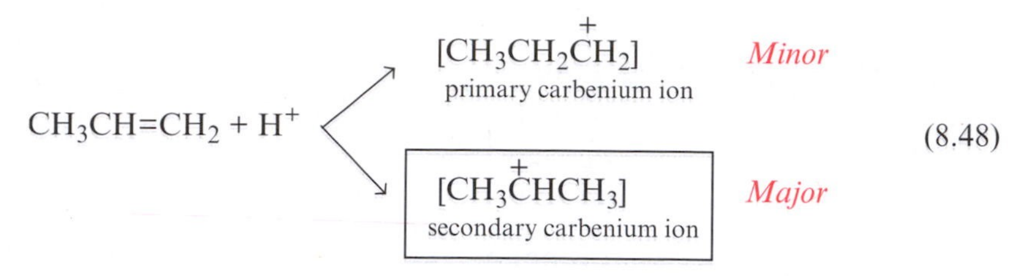

(8.48)

We have seen that radicals are stabilized by electron-releasing alkyl groups, and in the same way, electron deficient carbenium ions are stabilized by electron-releasing alkyl groups attached to the positively charged carbon centre. Thus, *a tertiary carbenium ion is more stable than a secondary carbenium ion, and a secondary carbenium ion is more stable than a primary carbenium ion.* The order of stabilities of carbenium ions is:

$$[R_3C]^+ > [R_2CH]^+ > [RCH_2]^+ > [CH_3]^+$$

In equation 8.48, the $[CH_3\overset{+}{C}HCH_3]$ ion is the more stable intermediate.

The relative stabilities of carbenium ions follow the order:

$[R_3C]^+ > [R_2CH]^+ > [RCH_2]^+ > [CH_3]^+$

where R is an alkyl group.

In the reaction of propene with HBr, the dominant product is 2-bromopropane (rather than 1-bromopropane, equation 8.36) and this follows from the preferred formation of the secondary carbenium ion (equation 8.49). The reaction is said to be *regioselective* because, although the nucleophilic alkene has two sites which can react with the electrophile (H^+), one product is formed preferentially.

$$CH_3CH{=}CH_2 + H^+ \longrightarrow [CH_3\overset{+}{C}HCH_3] \xrightarrow{Br^-} CH_3CHBrCH_3 \qquad (8.49)$$

Most HX molecules are polarized in the sense $H^{\delta+}$—$X^{\delta-}$, and *the preferential attachment of the hydrogen atom to the carbon atom which bears the most hydrogen atoms is referred to as Markovnikov addition.* This behaviour holds for the addition of a variety of HX molecules (HI, HCl, $HOSO_3H$) to asymmetric alkenes.

$HOSO_3H = H_2SO_4$

When a reaction that *could* proceed in more than one manner is observed to proceed only or predominantly in one way, it is said to be regioselective.

The acid-catalysed addition of water

Equation 8.39 showed the acid-catalysed addition of H_2O to an alkene. In principle, the hydrogen atom of a polar O–H bond could provide the electrophile, but in practice it is necessary to use additional acid to generate the intermediate carbenium ion.

Water is polar and a nucleophile, and it reacts with the carbenium ion (equation 8.50). The positive charge is transferred first to the oxygen atom, and then to a hydrogen atom. The transfer of the charge takes place as a pair of electrons is donated from the oxygen atom to the C^+ centre, and then from the O–H bond to the O^+ centre. This results in the loss of H^+ and regeneration of the catalyst.

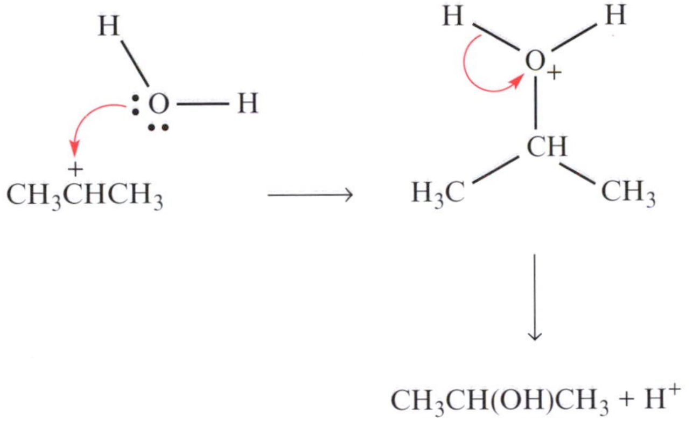

$$CH_3\overset{+}{C}HCH_3 + H_2O \longrightarrow (CH_3)_2CH{-}\overset{+}{O}H_2 \longrightarrow CH_3CH(OH)CH_3 + H^+ \qquad (8.50)$$

The acid-catalysed hydration of an alkene is an example of Markovnikov addition.

The addition of dichlorine and dibromine

Induced dipole: see Section 2.21

Dichlorine and dibromine are both non-polar. However, the electrons in the C=C double bond can induce a dipole in either Cl_2 (Figure 8.19) or Br_2, thereby encouraging heterolytic bond cleavage in the rate-determining step of the reaction between an alkene and the halogen. Equation 8.51 shows that heterolytic cleavage of the Cl–Cl bond is followed by the addition of a Cl^+ electrophile (with associated formation of a carbenium ion) and the addition of a Cl^- nucleophile.

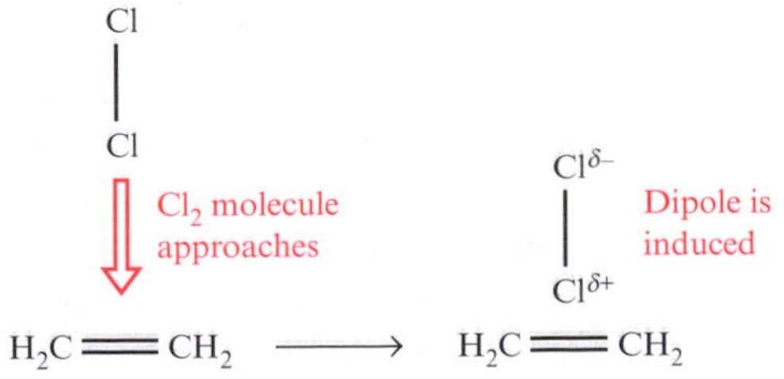

8.19 Dichlorine is non-polar but as Cl_2 approaches the C=C double bond, a dipole is induced by the electrons in the double bond.

$$H_2C{=}CH_2 + Cl{-}Cl \longrightarrow H_2\overset{+}{C}{-}CH_2Cl + Cl^- \longrightarrow CH_2ClCH_2Cl \quad \text{(1,2-dichloroethane)} \qquad (8.51)$$

When an alkene $CR_2{=}CR_2$ reacts with Br_2, there is experimental evidence that the intermediate ion possesses structure **8.14** – this is a *bromonium ion* and is closely related to the intermediate in equation 8.51.

(8.14)

Syn- and *anti-*addition

So far we have not addressed the question of the *stereochemistry* of addition reactions. During the addition of HX, the two carbon atoms in the original C=C double bond change from sp^2 to sp^3 hybridization. Figure 8.20 shows that the H and X atoms of HX can add to an alkene molecule either from the same (*syn*-addition) or different (*anti*-addition) sides. If all the substituents on the alkene are the same (A = B in Figure 8.20), it makes no difference to the stereochemistry of the product whether *syn-* or *anti-*addition occurs.

If A ≠ B in Figure 8.20, *syn*-addition of HX pushes *all* of the A and B substituents on to the same side of the double bond. Of course, free rotation about the C–C *single* bond can occur, and one staggered conformer of the final product is shown in Figure 8.20a. When HX undergoes *anti*-addition, one set of A and B substituents is pushed down and one set is pushed up, giving an arrangement of substituents in the product (Figure 8.20b) which is *different* from that formed as a result of *syn*-addition.

In most cases, the nucleophile adds to the alkene *after* the electrophile. In an intermediate carbenium ion such as the bromonium ion **8.14**, the electrophile blocks one side of the carbon–carbon bond and forces the nucleophile to enter from the opposite side: dibromine undergoes *anti*-addition to alkenes. In an intermediate carbenium ion such as **8.15**, the nucleophile may attack from either side of the planar (sp^2) carbenium centre, giving rise to *either syn-* or *anti-*addition. If the carbenium ion has a rela-

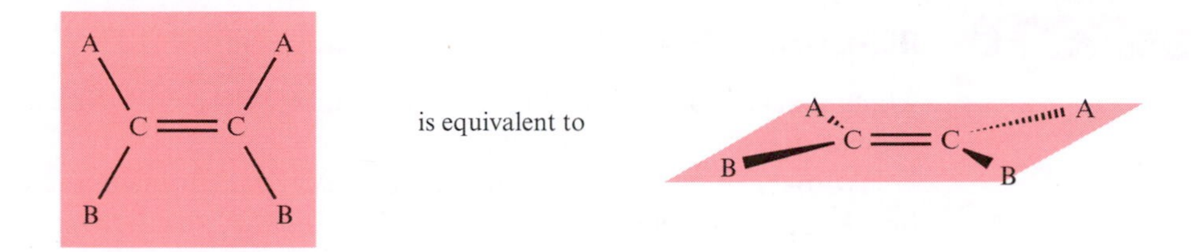

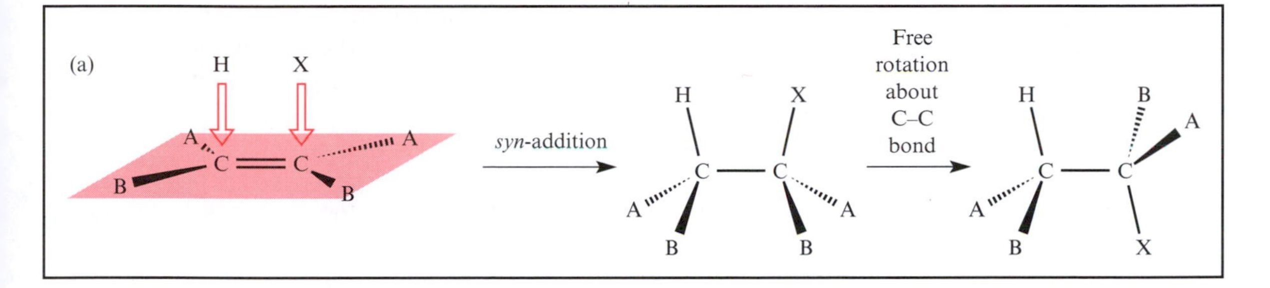

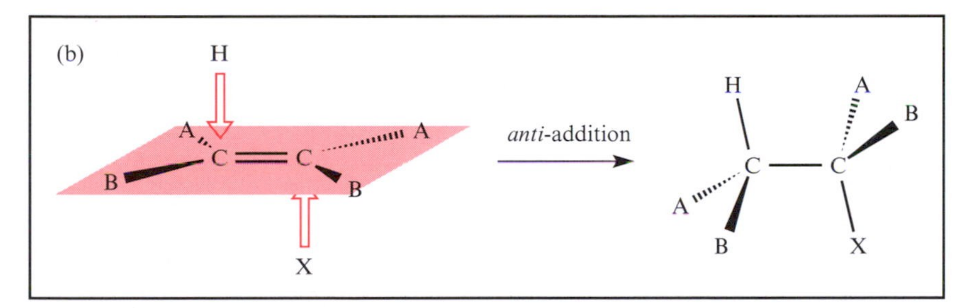

8.20 An alkene molecule ABC=CAB is planar (sp^2 hybridized carbon). When the H and X atoms of HX add to the C=C double bond, they can do so (a) from the same side (*syn*-addition) or (b) from different sides (*anti*-addition). The presence of the two A and two B substituents on the alkene means that the stereochemistries of the products of the *syn*- and *anti*-additions are not the same.

tively long lifetime, rotation about the C–C single bond can occur and the addition will be *non-specific*, although factors such as steric demands might favour *syn*- or *anti*-addition.

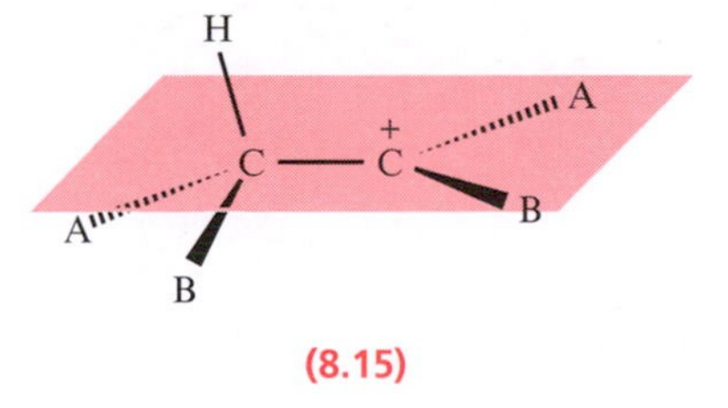

(8.15)

We give further examples of *syn*- and *anti*-addition in the accompanying workbook, and also discuss the consequences of having asymmetric carbon atoms in the products of alkene additions.

8.15 REACTIONS OF ALKENES 2: RADICAL SUBSTITUTION

Many alkenes contain alkyl groups in addition to the C=C functionality, and radical substitutions are possible.

Hydrogen atoms attached directly to the carbon atoms in the C=C double bond are called *vinylic hydrogens* (Figure 8.21) and these are hard to abstract. However, *a hydrogen atom that is attached to the carbon atom adjacent to a C=C double bond (an allylic hydrogen) is more easily abstracted.*

In the radical reaction of propene with Br_2, the propagation steps 8.52 and 8.53 lead to the formation of a bromo-substitution product.

$$CH_3CH{=}CH_2 + Br^{\bullet} \rightarrow HBr + \dot{C}H_2CH{=}CH_2 \qquad (8.52)$$

$$\dot{C}H_2CH{=}CH_2 + Br_2 \rightarrow CH_2BrCH{=}CH_2 + Br^{\bullet} \qquad (8.53)$$

Resonance structures: see Section 5.17

The $\dot{C}H_2CH{=}CH_2$ radical is called the *allyl radical*. Its relative stability can be attributed to the contributions made by the two resonance structures **8.16**. In general, the more resonance structures that can be written, the more stable a molecule, radical or ion is.

(8.16)

Under special circumstances it is also possible to observe radical *addition* to alkenes, and such reactions tend to give *anti-Markovnikov* products. This means that HX adds to the C=C double bond in the opposite sense to a Markovnikov addition.

A vinylic hydrogen atom is attached to a carbon atom of a C=C bond.

An allylic hydrogen atom is attached to an *sp*³ hybridized carbon atom adjacent to a C=C bond.

Vinylic hydrogen atoms

Allylic hydrogen atoms

8.21 Hydrogen atoms attached to carbon atoms in a C=C double bond are called vinylic hydogens, and those attached to the adjacent carbon atoms are *allylic hydrogen atoms*. In this example, R and R′ are alkyl groups.

8.16 REACTIONS OF ALKENES 3: RADICAL ADDITION POLYMERIZATION

A polymer is a macromolecule containing repeating monomer units.

A *polymer* is a macromolecule that consists of repeating units. Alkenes undergo *addition polymerization*, a process in which the alkene units (monomers) add to one another to yield a polymer. Polymers are an everyday part of our lives and Figure 8.22 exemplifies this with the varied and widespread uses of polyvinylchloride (PVC), the monomer for which is chloroethene.

The polymerization of an alkene, such as ethene, may be a radical reaction requiring an initiator such as a peroxide. Equation 8.54 shows the formation of radicals from a general initiator $Y_2^{\bullet}$.

$$Y_2 \xrightarrow{h\nu} 2\,Y^{\bullet} \qquad \textit{Initiation step} \qquad (8.54)$$

When the radicals collide with alkene molecules (the monomers), propagation of the radical chain reaction begins. Equation 8.55 shows how attack of a radical at a carbon atom in ethene leads to homolytic cleavage of the carbon–carbon π-bond. Note that this propagation step does *not* involve the abstraction of a group by the attacking radical – this contrasts with the first propagation step shown in equation 8.18 for the reaction between methane and a chlorine radical which involved the abstraction of $H^{\bullet}$.

$$Y^{\bullet} + H_2C{=}CH_2 \longrightarrow YCH_2CH_2^{\bullet} \qquad \textit{Propagation step} \qquad (8.55)$$

The propagation of the polymerization reaction continues according to equations 8.56 and 8.57, and the polymeric chain grows further in sequential radical reactions of this type.

$$YCH_2CH_2^{\bullet} + CH_2{=}CH_2 \rightarrow YCH_2CH_2CH_2CH_2^{\bullet} \qquad \textit{Propagation step} \qquad (8.56)$$

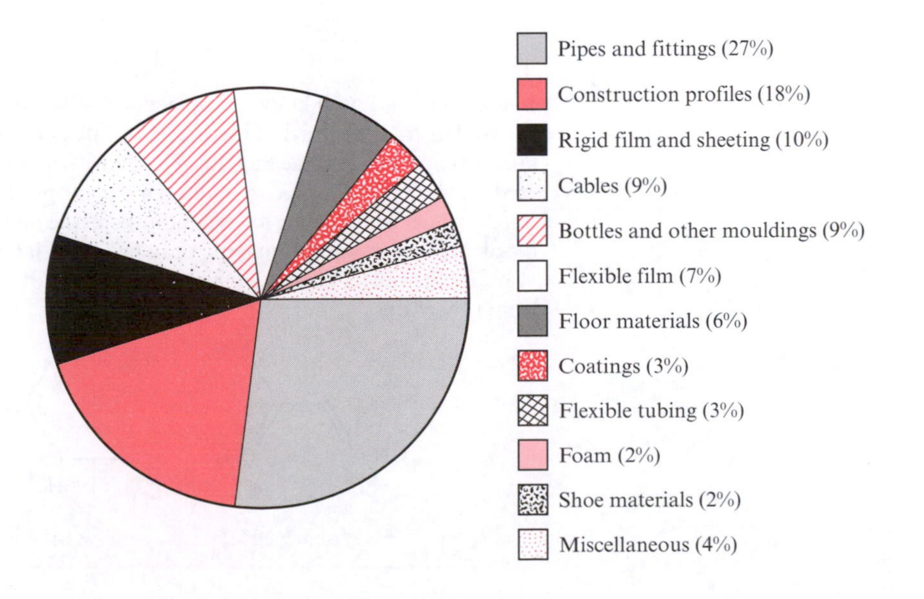

8.22 Uses of PVC in Western Europe in 1995. (Data from *Chemistry & Industry*, (1996) p.119.)

$$YCH_2CH_2CH_2CH_2^{\bullet} + CH_2{=}CH_2 \rightarrow YCH_2CH_2CH_2CH_2CH_2CH_2^{\bullet}$$
Propagation step (8.57)

Alternatively, a non-radical product can be formed in a propagation or termination step (equations 8.58 and 8.59).

$$YCH_2CH_2CH_2CH_2CH_2CH_2^{\bullet} + Y_2 \rightarrow YCH_2CH_2CH_2CH_2CH_2CH_2Y + Y^{\bullet}$$
Propagation step (8.58)

$$2YCH_2CH_2CH_2CH_2^{\bullet} \rightarrow YCH_2CH_2CH_2CH_2CH_2CH_2CH_2CH_2Y$$
Termination step (8.59)

The polymerization process is summarized in equation 8.60, where n is any integer between 1 and ∞. The length of the polymeric chain depends upon the point in the reaction when termination occurs.

$$n\,H_2C{=}CH_2 \longrightarrow Y{-}[CH_2{-}CH_2]_n{-}Y \quad (8.60)$$

Monomer Polymer

An interesting complication occurs if the monomer is a diene. Consider the attack by a radical $Y^{\bullet}$ on one of the C=C double bonds in octa-1,7-diene (equation 8.61).

$$Y^{\bullet} + CH_2{=}CH(CH_2)_4CH{=}CH_2 \longrightarrow YCH_2\dot{C}H(CH_2)_4CH{=}CH_2 \quad (8.61)$$

Box 8.11 Latex rubber: natural sources

Latex rubber is a well known polymer with widespread uses, and its source is the rubber trees of Brazil. However, this latex contains proteins which cause allergic responses in some humans. Now a new source of latex has been discovered – a desert bush called guayule. This plant produces (*Z*)-1,4-polyisoprene (latex) which contains far lower amounts of the allergenic proteins. There are disadvantages, however: guayule only produces latex for three months of the year and the rubber cannot be tapped from the bark in the same way that it can from rubber trees.

$$H_2C{=}CH{-}C(Me){=}CH_2 \qquad [{-}CH_2{-}CH{=}C(Me){-}CH_2{-}]_n$$

isoprene (*Z*)-1, 4-polyisoprene

The radical produced in this propagation step can undergo reaction with another molecule of the monomer or can undergo an *intramolecular* reaction (equation 8.62). The cyclic species so-formed can become involved in the polymerization process, or can react with the initiator in a termination step.

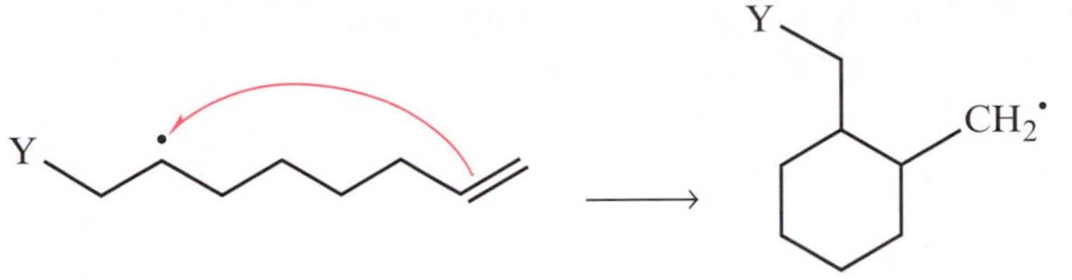

(8.62)

It is not difficult to envisage how many problems there are associated with the control of a free radical polymerization process which has an alkene or a diene as the monomer. Nonetheless, radical addition polymerization of alkenes remains an important method of producing polymers on an industrial scale.

Another method for polymerizing alkenes is *cationic polymerization*, for example, the polymerization of dec-1-ene for application in the lubricant industry.

Box 8.12 Polymers for contact lenses

Polymeric material for soft contact lenses must be optically transparent, chemically stable, be able to withstand being constantly wet, possess suitable mechanical properties, and be permeable to O_2 – the cornea obtains most of its O_2 supply from the atmosphere rather than the blood stream. Such a range of requirements provides challenges to polymer chemists.

Compound **1** is a monomer used in the manufacture of polymers for contact lenses. Its radical polymerization results in the formation of a polymer which is a *hydrogel* – it absorbs water but at the same time is insoluble in water. Water retention in a contact lens is important because there is a relationship between water content and O_2 permeability. Designing lenses for longer wearing periods has been a relatively recent problem for the contact lens industry. These polymers must possess enhanced O_2 permeability characteristics – this can be solved either by manufacturing thinner lenses or by increasing the water content of the hydrogel. The latter option is preferable because reducing the thickness of a lens reduces its strength and renders it prone to tearing. Copolymerization of monomer **1** with highly hydrophilic ('*water-loving*') monomers such as **2** or **3** has proved to be a successful approach to increasing the water content.

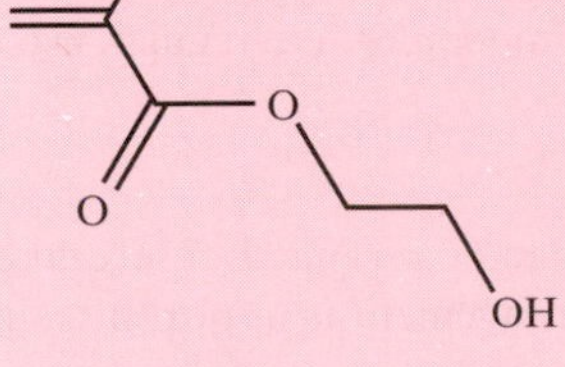

1 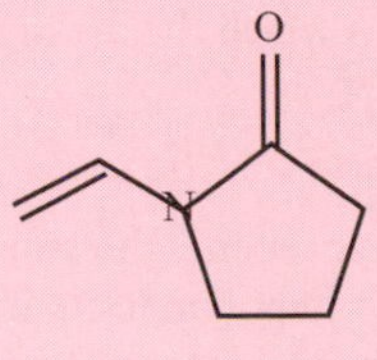 **2** 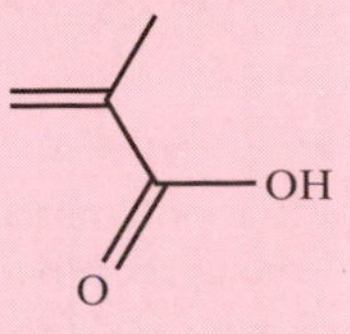 **3**

Box 8.13 Addition polymers

Some commercial polymers are listed below along with the monomer unit from which they are made.

Commercial name of polymer	*Name of monomer*	*Structure of monomer*
Polythene (polyethylene, PET)	Ethene	$H_2C{=}CH_2$
PVC (polyvinylchloride)	Chloroethene	$H_2C{=}CHCl$
Polypropylene (polypropene, PP)	Propene	$H_2C{=}CH(CH_3)$
Teflon (PTFE)	Tetrafluoroethene	$F_2C{=}CF_2$
Polystyrene	Phenylethene (styrene)	$H_2C{=}CH(C_6H_5)$
Perspex	Methyl 2-methylpropenoate (methyl methacrylate)	$H_2C{=}C(CH_3)C({=}O)OCH_3$

Copolymers

All of the above are examples of *homopolymers* – only one monomer is used in preparing the polymer. If more than one monomer is used, a *copolymer* is formed. Copolymerization is one method of 'tuning' the properties of a polymer to meet commercial needs. Three types of copolymer in which the monomer units are represented as A and B are:

- an alternating copolymer in which the monomer sequence is ABABABABAB;
- a block copolymer which may possess a sequence such as AAABBBAAABBB; and
- a random copolymer in which the arrangement of monomers is irregular, e.g. ABBABAABA.

Isotactic, syndiotactic and atactic polymers

In a polymer such as polypropene, the substituents can be arranged in several different ways with respect to the carbon backbone of the polymer chain. This arrangement is important, since it affects the way that the chains pack together in the solid state.

continues ▶

In an *isotactic* arrangement, all the methyl groups in polypropene are on the same side of the chain:

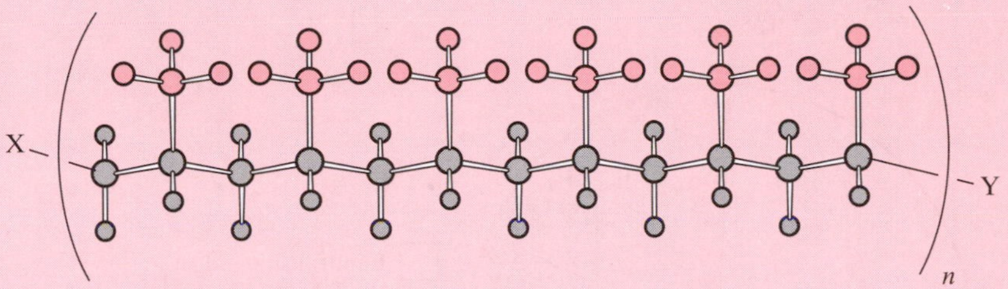

X– (...)$_n$ –Y

In an *syndiotactic* arrangement, the methyl groups are regularly arranged in an alternating sequence along the carbon chain:

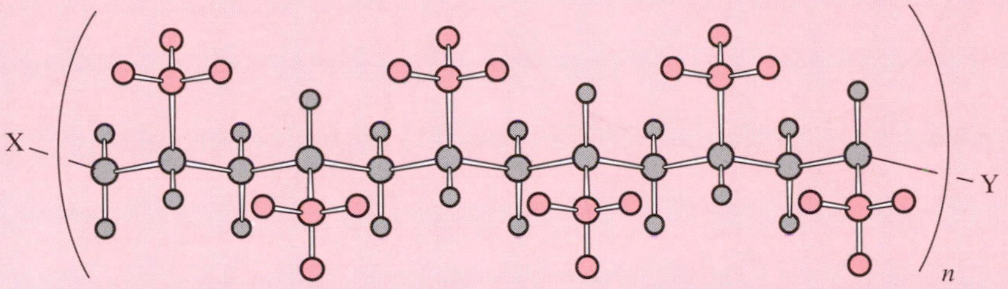

X– (...)$_n$ –Y

In an *atactic* arrangement, the methyl groups are randomly positioned along the polymer chain:

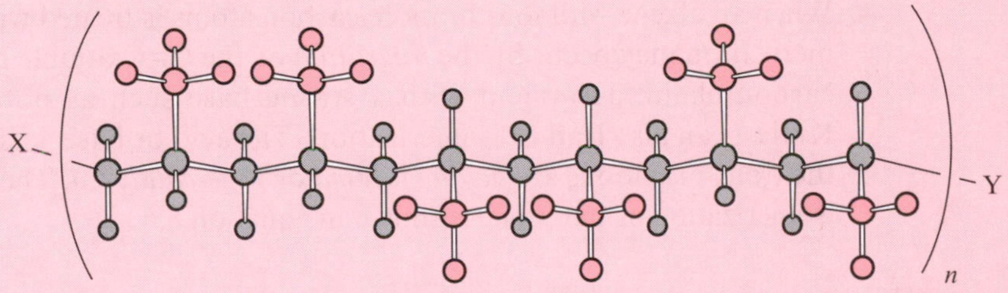

X– (...)$_n$ –Y

In the solid state, molecules of both isotactic and syndiotactic polypropene pack efficiently and the material is crystalline. Atactic polypropene, on the other hand, is soft and elastic. Whereas isotactic and syndiotactic polypropene can be used, for example, as a material for plastic pipes and sheeting, atactic polypropene has no such commercial application.

Question: Why is it not possible to convert the syndiotactic or atactic arrangements into the isotactic arrangement by rotating about C–C single bonds in the carbon backbone of the polymer?

Recycling of plastics

Polymers have an enormous range of applications, as Figure 8.22 illustrated for PVC. In the modern world, life without polymers and plastics seems unthinkable, but at the end of their life, how do we destroy the tonnes of man-made materials that we have created? The recycling of waste plastics is an environmental issue and European countries have varying programmes for recycling. In 1990, Western Europe recycled only 7% of its

continues ▶

waste plastic as the pie chart below indicates – attitudes are changing, but there is still a long way to go. The market for biodegradable, photodegradable or inherently degradable plastics is increasing. (Data from *Chemistry & Industry*, (1992) p. 401.)

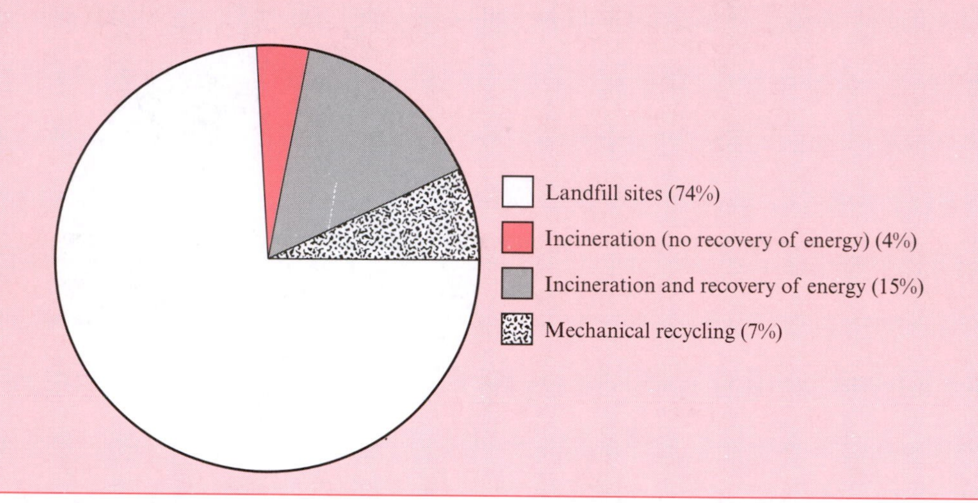

8.17 REACTIONS OF ALKENES 4: DOUBLE BOND MIGRATION AND ALKENE ISOMERIZATION

When an alkene with four or more carbon atoms is treated with an acid, isomerization may occur by the *migration* of the C=C double bond along the carbon chain. Treatment with a strong base such as potassium amide, KNH_2, can also lead to isomerization. The acid or base acts as a catalyst – the transformations are *acid-catalysed* or *base-catalysed*. The acid-catalysed isomerization of but-1-ene is shown in equation 8.63.

$$CH_3CH_2CH{=}CH_2 \xrightarrow{H^+ \text{ catalyst}} CH_3CH{=}CHCH_3 \qquad (8.63)$$

Electronegativities of different carbon centres: see Section 8.20

The tendency is for the double bond to migrate so as to increase the number of alkyl groups attached to it, and therefore decrease the number of directly attached hydrogen atoms. The preferences follow directly from the greater electronegativity of sp^2 over sp^3 carbon atoms; the electron-releasing alkyl groups should be attached to the sp^2 centres.

The longer the carbon chain, the more isomers can be formed by migration of the C=C functionality. In practice, mixtures of isomers are usually obtained with the thermodynamically preferred isomer predominating.

In a migration reaction, an atom or functional group 'moves' from one position in the molecule to another.

The mechanism of acid-catalysed double bond migration

The addition of acid to an alkene produces a carbenium ion as we have already seen; pent-1-ene reacts with H^+ to give a secondary carbenium ion in preference to a primary one (equation 8.64). In the absence of a nucleophile to complete an addition reaction (e.g. reaction 8.47), the carbenium ion can lose a hydrogen ion and form either pent-1-ene (equation 8.65) or pent-2-ene (equation 8.66). The greater the number of alkyl groups attached to the double bond, the more stable is the alkene, and pent-2-ene is the favoured product. The combination of reactions 8.64 and 8.66 results in alkene isomerization by C=C double bond migration.

$$CH_3CH_2CH_2CH{=}CH_2 + H^+ \rightarrow CH_3CH_2CH_2\overset{+}{C}HCH_3 \qquad (8.64)$$

$$CH_3CH_2CH_2\overset{+}{C}H{-}CH_2(H) \longrightarrow \underset{\text{pent-1-ene}}{CH_3CH_2CH_2CH{=}CH_2} + H^+ \qquad (8.65)$$

$$CH_3CH_2CH(H){-}\overset{+}{C}HCH_3 \longrightarrow \underset{\text{pent-2-ene}}{CH_3CH_2CH{=}CHCH_3} + H^+ \qquad (8.66)$$

The mechanism of base-catalysed double bond migration

Not only can the allylic hydrogen atom be abstracted in a radical reaction, it is also susceptible to attack by a base – *an allylic hydrogen atom is relatively acidic*. In reaction 8.67, the base, B^-, removes a hydrogen ion from the carbon atom adjacent to the C=C double bond. The intermediate that is formed is called a *carbanion*.

$$R{-}CH(H)CH{=}CH_2 + B^- \longrightarrow R\overset{-}{C}HCH{=}CH_2 + BH \qquad (8.67)$$

A carbanion has the general formula $[R_3C]^-$ where R is an H atom or organic substituent.

Resonance structures: see Section 5.17

The allyl carbanion so-formed is resonance-stabilized by the contributing structures **8.17** with the result that the π -bonding is *delocalized* over the three carbon centres. Structure **8.18** is an alternative way of representing the resonance pair **8.17**.

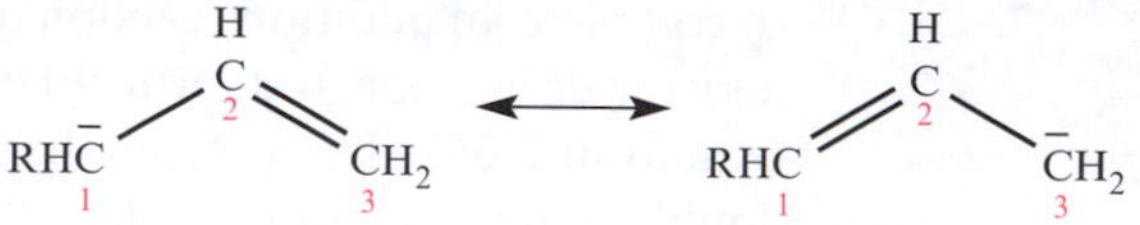

$$R\bar{H}C^{1}-\overset{H}{C}^{2}{=}CH_2^{3} \longleftrightarrow RHC^{1}{=}\overset{H}{C}^{2}-\bar{C}H_2^{3}$$

(8.17)

$$RHC\overset{H}{\cdots C \cdots}CH_2 \quad (-)$$

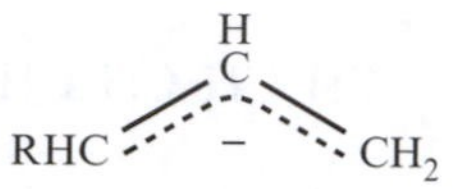

(8.18)

An allylic hydrogen atom is relatively acidic.

The carbanion can be protonated at either atom C(1) or C(3), and the favoured product will be the alkene with the greater number of R groups attached to the C=C double bond. This follows directly from the electronegativity difference between sp^2 and sp^3 hybridized carbon atoms. An sp^2 carbon atom is more electronegative that an sp^3 centre, and the most stable product will be the one with the most electron-releasing alkyl groups attached to the sp^2 carbon atom. The overall result is the migration of the C=C double bond towards the centre of the carbon chain (equation 8.68).

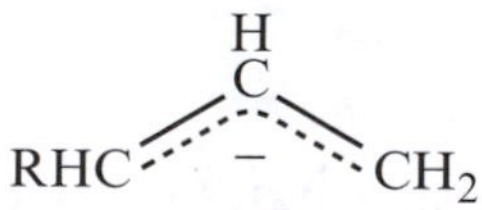

$$[RHC\overset{H}{\cdots C \cdots}CH_2]^- + BH \longrightarrow RCH{=}CHCH_3 + B^- \qquad (8.68)$$

The source of the proton is BH, the protonated base produced in equation 8.67 – the catalyst B^- is regenerated in equation 8.68.

8.18 REACTIONS OF ALKENES 5: HYDROBORATION

Hydroboration with BH_3

When we write 'BH_3' in this section, we are referring either to B_2H_6 or to a Lewis base adduct: see Section 11.5

The reaction of an alkene with BH_3 leads to the *addition* of the B–H bond to the C=C double bond (equation 8.69). This reaction is called *hydroboration*.

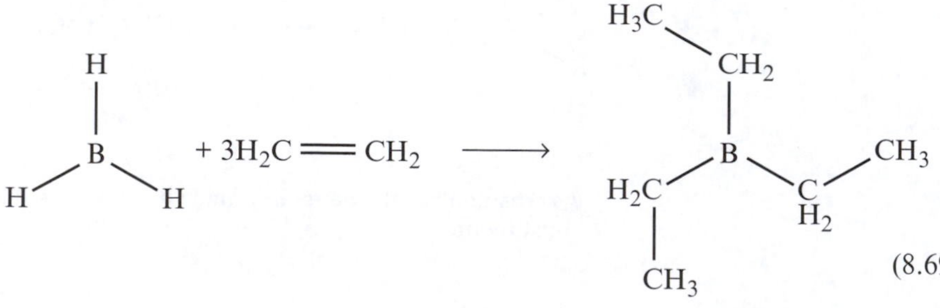

$$BH_3 + 3H_2C{=}CH_2 \longrightarrow B(CH_2CH_3)_3 \qquad (8.69)$$

triethylborane

If the alkene is *sterically hindered* by the presence of organic substituents, only one, or perhaps two, B–H bonds will be involved in the hydroboration reaction; equation 8.70 shows an example.

$$BH_3 + (CH_3)_2C{=}C(CH_3)_2 \longrightarrow H{-}C(CH_3)_2{-}C(CH_3)_2{-}BH_2 \quad (8.70)$$

Using reactions of this type, organoborane compounds of the general type RBH_2 and R_2BH (R = alkyl) can be formed. Further reactions with other alkenes yield organoboranes containing different alkyl substituents. Addition of a borane to an asymmetrical alkene is regioselective, *with the boron atom attaching itself to the carbon atom of the alkene that possesses the most hydrogen atoms*. This is an anti-Markovnikov addition, and happens because hydrogen is more electronegative than boron. The boron–hydrogen bond is polarized in the sense $B^{\delta+}{-}H^{\delta-}$, and the hydrogen is, in this case, *not the electrophile*. The addition is not 'breaking the rules' and is only anti-Markovnikov in terms of the position in which the H atom finds itself after addition.

Organoboranes are oxidized when treated with hydrogen peroxide and sodium hydroxide to give alcohols (equation 8.71), and this reaction represents an important use of the boron-containing compounds.

Hydroboration in alcohol synthesis: see Section 14.9

$$\underset{\text{triethylborane}}{(CH_3CH_2)_3B} + 3H_2O_2 \xrightarrow{[OH]^-} 3CH_3CH_2OH + \underset{\text{boric acid}}{B(OH)_3} \quad (8.71)$$

8.19 REACTIONS OF ALKYNES 1: ADDITIONS

Like alkanes and alkenes, alkynes burn in oxygen to give carbon dioxide and water (equation 8.72).

$$2CH_3C{\equiv}CCH_3 + 11O_2 \rightarrow 8CO_2 + 6H_2O \quad (8.72)$$

Alkynes can undergo addition reactions to form derivatives of alkenes, and then alkanes. The first step of the reaction can lead to an isomeric mixture of products (equation 8.73).

$$H{-}C{\equiv}C{-}H \xrightarrow{X_2} \underset{(Z)\text{-isomer}}{(H)(X)C{=}C(H)(X)} + \underset{(E)\text{-isomer}}{(H)(X)C{=}C(X)(H)} \quad (8.73)$$

Addition of dihydrogen

(*Z*)- and (*E*)-isomers: see Section 5.11

The addition of H_2 to give alkenes is readily achieved in the presence of *d*-block metal catalysts such as nickel, palladium or platinum. *Selective reduction* to (*Z*)- or (*E*)-alkenes may be achieved in various ways. For example, reaction with H_2 in the presence of a $Pd/BaSO_4$ catalyst gives (*Z*)-isomers while use of sodium in liquid ammonia as the reducing agent gives primarily (*E*)-isomers. However, the sodium/liquid NH_3 route is not a successful method of reducing a –C≡CH (as opposed to a –C≡CR) group (see Section 8.20).

There is the added complication that reduction of an alkyne to an alkene may be followed by reduction to an alkane. A useful catalyst for selectively reducing a C≡C triple bond to a C=C double bond is the Lindlar catalyst which consists of palladium metal supported on a $CaCO_3/PbO$ surface.

Addition of halogens and hydrogen halides

The mechanisms of the additions of halogens and hydrogen halides to alkynes are similar to those of the alkenes. With Br_2, propyne reacts according to equation 8.74, and there is a preference for the (*Z*)-isomer of the alkene.

$$\underset{\text{propyne}}{CH_3C{\equiv}CH} \xrightarrow{Br_2} \underset{(Z)\text{-1,2-dibromopropene}}{(Z)\text{-}CH_3CBr{=}CHBr} \xrightarrow{Br_2} \underset{\text{1,1,2,2-tetrabromopropane}}{CH_3CBr_2CHBr_2} \qquad (8.74)$$

Under the experimental conditions for electrophilic addition, hydrogen halides undergo Markovnikov addition to alkynes. A second Markovnikov addition leads specifically to the two halo-substituents being *attached to the same carbon atom* (equations 8.75 and 8.76). Such a compound is also called a *gem*-dihalide.

gem = geminal

$$CH_3C{\equiv}CH \xrightarrow{HCl} CH_3CCl{=}CH_2 \xrightarrow{HCl} \underset{\text{2,2-dichloropropane}}{CH_3CCl_2CH_3} \qquad (8.75)$$

$$CH_3C{\equiv}CH \xrightarrow{HCl} CH_3CCl{=}CH_2 \xrightarrow{HBr} \underset{\text{2-bromo-2-dichloropropane}}{CH_3CClBrCH_3} \qquad (8.76)$$

8.20 REACTIONS OF ALKYNES 2: ALKYNES AS ACIDS

The dependence of the electronegativity of carbon on hybridization

In Table 4.2, we listed the Pauling electronegativity for carbon as 2.6, and that for hydrogen as 2.2. However, we added a note of caution, indicating that the electronegativity of an atom is dependent upon oxidation state and bond order. As the hydridization of a carbon atom changes from sp^3 to sp^2 to sp, the electronegativity changes (Figure 8.23). The increase is significant,

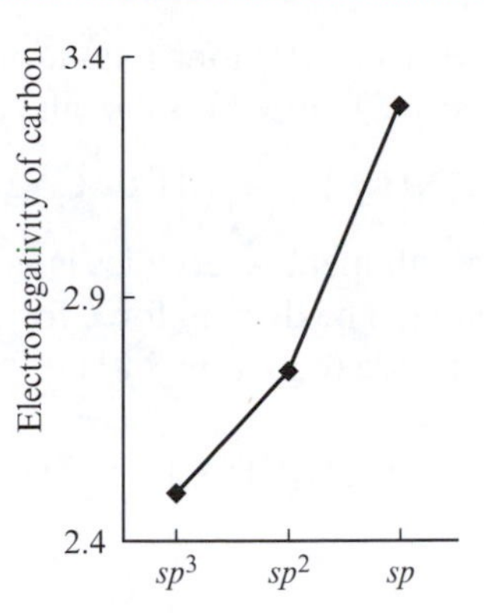

8.23 A plot of the electronegativity of carbon against the hybridization of the carbon atom. These values are calculated from ionization potential and electron affinity data and are on the Mulliken–Jaffé scale of electronegativities; they are scaled so as to be comparable with Pauling values.

Table 8.6 Relative acidities of C–H bonds.

Compound	pK_a	*Hybridization of the carbon atom*
CH_4	48	sp^3
$H_2C{=}CH_2$	44	sp^2
$HC{\equiv}CH$	25	sp

and means that a C–H bond in an alkyne is more polar than in an alkene, and this in turn is more polar than one in an alkane.

These electronegativity values predict that an $[RC{\equiv}C]^-$ anion should be more stable than an $[R_2C{=}CR]^-$ anion, which in turn should be more stable than an $[R_3C]^-$ anion. This is indeed the case and approximate pK_a values are given in Table 8.6. Remember that the acidity of the allylic hydrogen atom on the sp^3 carbon centre in $R_2C{=}CRCH_2R$ is due to the particular stability of the resonance-stabilized allyl carbanion **8.18**.

$pK_a = -\log K_a$

There is evidence for the acidity of the $-C{\equiv}C-H$ group in some of the reactions of alkynes, but we stress that these are *very weak acids*, and ethyne (equation 8.77) is a weaker acid than water. Since ethyne is such a weak acid, the $[HC{\equiv}C]^-$ anion must be a strong base.

Review of acids and bases: see Section 11.9

$$H{-}C{\equiv}C{-}H \rightleftharpoons [H{-}C{\equiv}C]^- + H^+ \quad (8.77)$$

Only terminal alkynes are acidic, because the acidity is a property of the $-C{\equiv}CH$ group.

Acetylide salts

Terminal alkynes react with sodium (equation 8.78) or potassium amide (equation 8.79) to give alkali metal salts of acetylides.

$$2CH_3CH_2C{\equiv}CH + 2Na \rightarrow 2[CH_3CH_2C{\equiv}C]^-Na^+ + H_2 \quad (8.78)$$

$$CH_3C{\equiv}CH + K[NH_2] \rightarrow [CH_3C{\equiv}C]^-K^+ + NH_3 \quad (8.79)$$

Reaction also occurs with silver(I) ions (equation 8.80), although when dry, silver acetylides are explosive. Copper(I) salts can be formed in similar reactions.

$$CH_3C{\equiv}CH + [Ag(NH_3)_2]^+ \rightarrow CH_3C{\equiv}CAg + NH_3 + [NH_4]^+ \quad (8.80)$$

An important use of alkali metal acetylides is in the preparation of alkynes with longer carbon chains. The driving force for the reactions is the elimination of an alkali metal halide (equation 8.81).

$$[CH_3CH_2C{\equiv}C]^-Na^+ + CH_3CH_2Cl \rightarrow CH_3CH_2C{\equiv}CCH_2CH_3 + NaCl \quad (8.81)$$

8.21 REACTIONS OF ALKYNES 3: ALKYNE COUPLING

When terminal alkynes are heated with copper(II) salts in pyridine **8.19**, a coupling reaction can take place to give a *diyne* (equation 8.82). This is a quite general reaction with wide application and equation 8.83 shows how it can be used to form a cyclic hexyne starting from a terminal diyne.

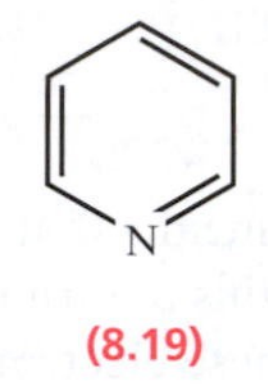

(8.19)

A diyne contains two C≡C functionalities.

$$2CH_3CH_2C{\equiv}CH \xrightarrow{Cu^{2+}\text{ in pyridine}} CH_3CH_2C{\equiv}C{-}C{\equiv}CCH_2CH_3 \quad (8.82)$$

Cu^{2+} in pyridine

(8.83)

SUMMARY

This chapter has been concerned with the chemistry of acyclic alkanes, alkenes and alkynes. In addition to some of the chemistry of these compounds, we have introduced many concepts that are important in organic chemistry generally.

Do you know what the following terms mean?

- miscible and immiscible
- primary, secondary, tertiary and quaternary carbon centres
- conformer
- asymmetric carbon atom
- chiral compound
- optical isomer (enantiomer)
- stereoisomer
- homologous series of alkanes
- 'ball-and-stick' and 'space-filling' diagrams
- homolytic and heterolytic bond cleavage
- intermediate
- transition state
- electrophile
- nucleophile
- photolysis
- hydrolysis
- radical substitution
- radical polymerization
- electrophilic addition
- regioselective

You should be able:

- to discuss structural and geometrical isomerism in aliphatic hydrocarbon compounds
- to describe the differences between staggered, eclipsed and skew conformations, and the steric energy associated with rotation about C–C bonds
- to draw Newman projections and sawhorse diagrams, and to use wedges to illustrate the conformation of a given molecule
- to discuss what is realistically meant by a 'straight chain' hydrocarbon
- to discuss whether a chiral compound *has* to possess an asymmetric carbon centre, and what is meant by the 'handedness' of a helical chain
- to give two 'tests of chirality' (are they infallible?)
- to recognize whether a carbon atom is being oxidized or reduced (or not) in a reaction
- to distinguish between an intermediate complex and a transition state in a reaction, and to discuss their location on a reaction profile
- to discuss typical reactions of alkanes and outline the mechanism by which the radical substitutions occur
- to discuss typical reactions of alkenes and outline the mechanism by which the electrophilic additions occur
- to distinguish between Markovnikov and anti-Markovnikov addition, and give examples of both, explaining why the particular type of addition occurs
- to explain when and why migration of a C=C double bond occurs
- to explain what is meant by *syn*- and *anti*-addition to an alkene, and discuss the circumstances under which the two types of addition occur
- to discuss typical reactions of alkynes, and comment on the weakly acidic nature of terminal alkyne

PROBLEMS

8.1 Write down the names of the straight chain alkanes which have the following formulae: (a) C_3H_8; (b) C_7H_{16}; (c) $C_{12}H_{26}$; (d) $C_{20}H_{42}$.

8.2 Write down the molecular formulae of the following straight chain alkanes: (a) hexane; (b) nonane; (c) decane; (d) tetradecane.

8.3 Draw out structural formulae for the following alkanes: (a) 2-methylpentane; (b) 3-ethyloctane; (c) 2,2-dimethylbutane, (d) 2,2,4,4-tetramethylhexane. How many *types* of carbon atoms are there in each compound?

8.4 Why is 1,1-dimethylpropane not the correct name for $Me_2CHCH_2CH_3$? What is the systematic name for this compound?

8.5 What is the systematic name for $CH_3CH_2CHMeCHMeCH_2CH_2CH_2CHMeCH_3$?

8.6 Give names for the following groups or substituents: (a) CH_3; (b) C_2H_5; (c) C_4H_9; (d) Me_3C; (e) Me_2CH.

8.7 Refer to the isomers of C_6H_{14} in Figure 8.6. (a) Why is there no isomer labelled 4-methylpentane? (b) Why is there no isomer labelled 3,3-dimethylbutane? (c) Why are there no isomers based on a propane chain? (d) Why is cyclohexane not a structural isomer of hexane?

8.8 In each of the following molecules, give the hybridization of each numbered carbon centre:

(a) $\overset{1}{C}H_3\overset{2}{C}H=\overset{3}{C}H\overset{4}{C}H_2\overset{5}{C}H_3$;

(b) $\overset{1}{C}H_3\overset{2}{C}\equiv\overset{3}{C}\overset{4}{C}H_3$;

(c) $\overset{1}{C}H_3\overset{2}{C}H=\overset{3}{C}=\overset{4}{C}H\overset{5}{C}H_2\overset{6}{C}H_2\overset{7}{C}H=\overset{8}{C}H_2$;

(d) $\overset{1}{C}H_3\overset{2}{C}Me=\overset{3}{C}Me\overset{4}{C}H_2\overset{5}{C}H_2\overset{6}{C}Me_2\overset{7}{C}H_3$

8.9 Figure 8.24 shows the variation in steric energy as rotation occurs around the C(2)–C(3) bond in butane. The energies are relative to a value of zero taken for a staggered conformation (angle = 0°). (a) Draw Newman projections of the conformation of butane at points A, B, C and D marked in the figure. [Hint: Consider the projection to be along the C(2)–C(3) bond.] (b) Explain the shape of the graph in Figure 8.24.

8.10 What is being oxidized or reduced in each of the following reactions?

(a) $CH_3CH_2Cl + 2Na \rightarrow C_4H_{10} + 2NaCl$

(b) $C_2H_6 + Cl_2 \rightarrow CH_3CH_2Cl + HCl$

(c) $2NO + O_2 \rightarrow 2NO_2$

(d) $CH_3C\equiv CH + H_2 \rightarrow CH_3CH=CH_2$

(e) $2CH_3CH_2C\equiv CH + 2K \rightarrow 2[CH_3CH_2C\equiv C]K + H_2$

[Hint: refer to Box 8.8]

8.11 What is meant by each of the following terms: (a) intermediate; (b) transition state; (c) electrophile; (d) heterolytic bond cleavage; (e) homolytic bond cleavage; (f) stereochemistry; (g) regioselective.

8.12 A plot of $-\Delta_c H$ against n for alkanes of formula C_nH_{2n+2} is shown in Figure 8.25. Suggest why this graph is linear.

8.13 Why do you think that O_2 has a dramatic effect upon many radical reactions? [Hint: refer to Chapter 3.]

8.14 Nuclear fission processes are often described as chain reactions. In what way do they differ from the chain reaction between CH_4 with Cl_2?

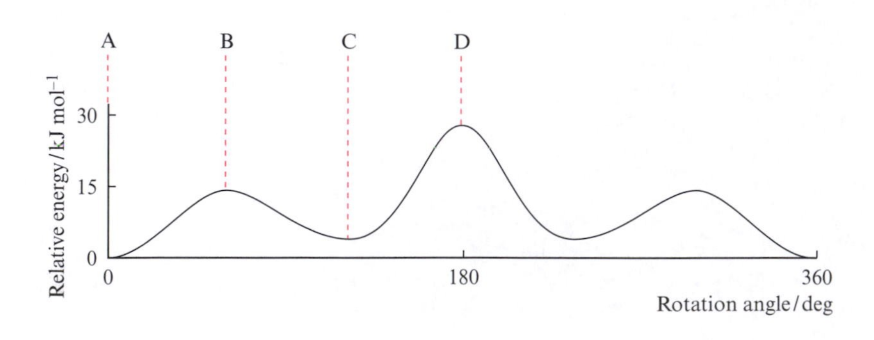

8.24 For problem 8.9.

8.25 For problem 8.12.

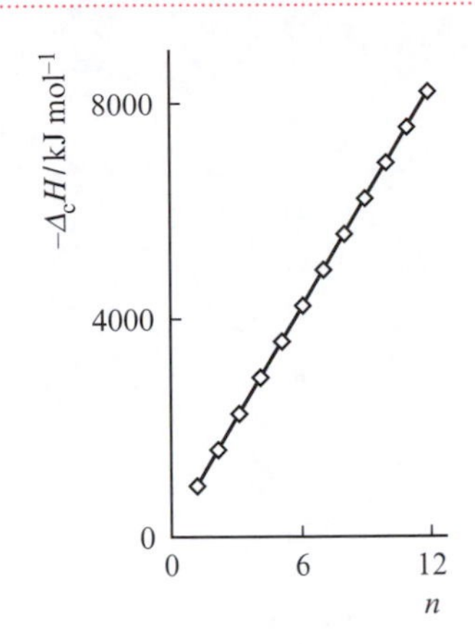

8.15 Equations 8.17–8.20 give four of the propagation steps which occur during the chlorination of methane. These show the formation of only two members of the family of chloroalkanes $CH_{4-n}Cl_n$. Write down additional propagation steps that show how the other members of the family may be formed. What other chloroalkanes might be formed as secondary products of the chlorination of methane?

8.16 Values for the average bond dissociation enthalpies of the C=C, C–C, H–H and C–H bonds are 598, 346, 436 and 412 kJ mol^{-1} respectively. Determine the enthalpy change associated with the hydrogenation of ethene. Compare your answer with that determined using the standard enthalpies of formation of ethene (+52.5 kJ mol^{-1}) and ethane (–83.8 kJ mol^{-1}).

8.17 Suggest a mechanism by which HCl reacts with 2-methylbut-1-ene in the absence of irradiation. Indicate, with reasoning, the relative product distribution that you might expect.

8.18 What products do you expect to form when pent-2-ene (a) reacts with dibromine, (b) is burned in dioxygen, (c) reacts with ozone followed by treatment with water in the presence of zinc, (d) reacts with hydrogen iodide, and (e) is mixed with dichlorine and irradiated.

8.19 Equation 8.40 showed the reaction of propene with Cl_2/H_2O and we noted that the alcohol functionality showed a preference to be attached to the secondary carbon atom. Suggest a mechanism that is consistent with the experimental results.

8.20 9-Borabicyclo[3.3.1]nonane (commonly called 9-BBN) is a selective hydroborating reagent:

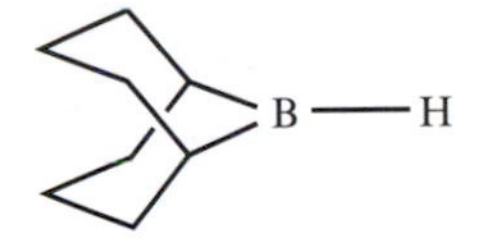

9-BBN

Starting from a suitable *cyclic diene*, suggest a method of preparing 9-BBN. What other isomer(s) might be formed in the reaction?

8.21 Suggest why the use of $Pd/BaSO_4$ as the catalyst in the reaction of H_2 with an alkyne leads to (*Z*)- rather than (*E*)-alkenes. Explain why the use of sodium in liquid ammonia is not usually a successful method of reducing *terminal* alkynes.

9 Spectroscopy

Topics

- Spectroscopic techniques
- Timescales
- Beer–Lambert Law
- Colorimetry
- Vibrational spectroscopy (infrared)
- Electronic spectroscopy (UV–VIS)
- Nuclear magnetic resonance spectroscopy

9.1 WHAT IS SPECTROSCOPY?

Atomic spectrum of hydrogen: see Section 2.16; Photoelectron spectroscopy: see Box 3.6

Spectroscopic methods of analysis are an everyday part of modern chemistry, and we have already mentioned the atomic spectrum of hydrogen and photoelectron spectroscopy. There are many different spectroscopic techniques (Table 9.1) and by using different types of spectroscopy, it is possible to investigate many aspects of atomic and molecular structure.

In this chapter we consider infrared (IR), electronic (or ultraviolet–visible, UV–VIS) and nuclear magnetic resonance (NMR) spectroscopies – these are the techniques that you are most likely to use in the laboratory.

The aim of our discussion is *not* to delve deeply into the theory of spectroscopy, but rather to provide information that will assist in the practical application of these techniques.

Absorption and emission spectra

Absorption and emission: see Box 2.7

The terms *absorption and emission* are fundamental to discussions of spectroscopy; we distinguished between absorption and emission *atomic spectra*. In general, the absorption of electromagnetic radiation by an atom, molecule or ion causes a transition from a lower to a higher energy level. (We shall deal more explicitly with what is meant by a 'level' later in this chapter). Remember from Chapter 2 that only *certain* transitions are allowed. The energy of the radiation absorbed gives the energy difference between the two levels. Figure 9.1 shows a schematic diagram of some components of an *absorption spectrophotometer* used to measure the absorption of electromagnetic radiation. The 'source' provides electromagnetic radiation covering a range of frequencies.

$E = h\nu$

Table 9.1 Some important spectroscopic techniques; note that electron, X-ray and neutron diffraction methods (Chapter 3) and mass spectrometry (discussed in the accompanying workbook) are *not* spectroscopic techniques.

Name of technique	*Comments*
Atomic absorption spectroscopy	Used for elemental analysis; observes the absorption spectra of atoms in the vapour state.
Electron spin resonance (ESR) spectroscopy *or* electron paramagnetic resonance (EPR) spectroscopy	Used in the study of species with one or more unpaired electrons.
Electronic (ultraviolet–visible, UV–VIS) spectroscopy	Absorption spectroscopy [100–200 nm (vacuum-UV), 200–800 nm (near-UV and visible)] used to study transitions between atomic and molecular electronic energy levels (see Sections 9.9–9.11).
Far infrared spectroscopy	Infrared spectroscopy below $\approx 200\ cm^{-1}$.
Fluorescence spectroscopy	Used to study compounds that *fluoresce, phosphoresce* or *luminesce*; fluorescence is the emission of energy which may follow the absorption of UV or visible radiation, and is a property exhibited only by certain species. The light emitted is usually at longer wavelength than that absorbed.
Infrared (IR) spectroscopy	A form of vibrational spectroscopy; absorptions are usually recorded in the range $200–4000\ cm^{-1}$. Extremely useful as a 'fingerprinting' technique; see Sections 9.6–9.8.
Microwave spectroscopy	Absorption spectroscopy used to study the rotational spectra of gas molecules.
Mössbauer spectroscopy	Absorption of γ-radiation by certain nuclei (e.g. ^{57}Fe and ^{197}Au); used to study the chemical environment, including oxidation state, of the nuclei.
Nuclear magnetic resonance (NMR) spectroscopy	Absorption or emission of radiofrequency radiation used to observe nuclear spin states (see Sections 9.12–9.16); a very powerful analytical tool which is used to elucidate molecular structures and study dynamic behaviour in solution and the solid state.
Photoelectron spectroscopy (PES)	Absorption spectroscopy used to study the energies of occupied atomic and molecular orbitals (see Box 3.6).
Raman spectroscopy	A form of vibrational spectroscopy but with different selection rules from IR spectroscopy; some modes which are IR inactive (see Sections 9.6 and 9.7) are Raman active.

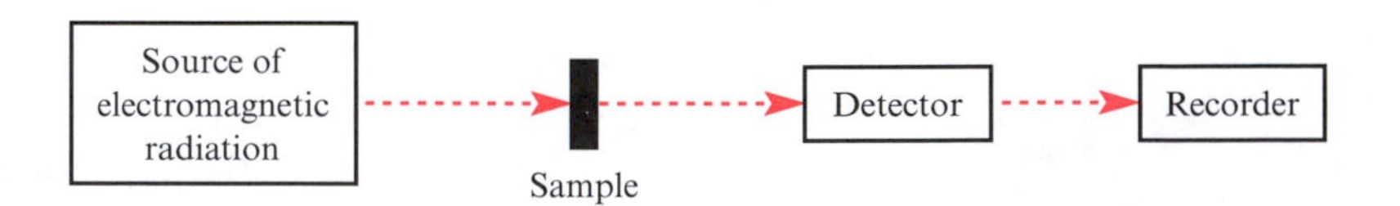

9.1 A schematic representation of parts of an absorption spectrometer; the electromagnetic radiation originates from the source, and is directed at the sample where some radiation is absorbed and some transmitted. The resultant radiation passes on to a detector, and the information is then output in the form of a spectrum.

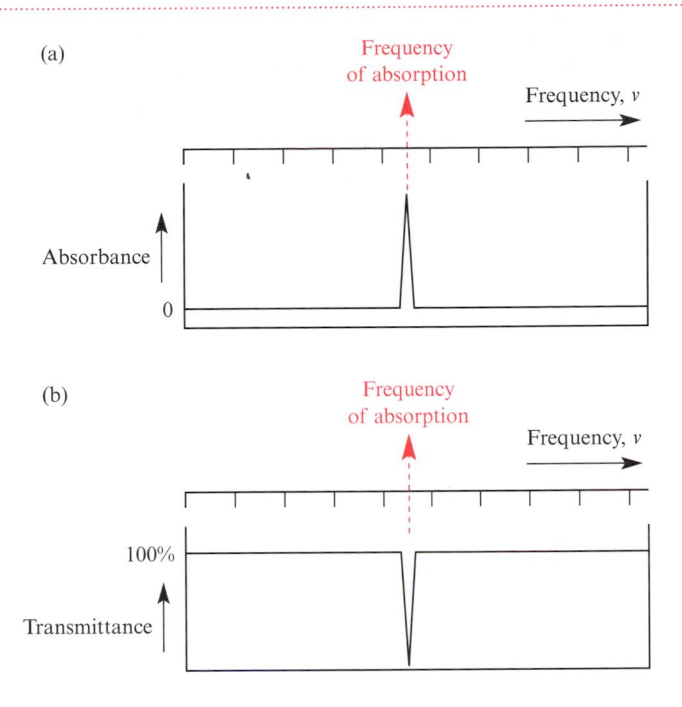

9.2 Schematic representation of an absorption spectrum consisting of a single absorption. The spectrum gives information about the intensity and frequency of the absorption; the intensity can be measured in terms of (a) absorbance or (b) transmittance. The frequency can be converted into a value of energy or wavelength.

When the radiation passes through a sample, some may be absorbed, but some is transmitted, detected and recorded in the form of a spectrum (Figure 9.1). The spectrum is a plot of the *absorbance* or *transmittance* of the radiation against the frequency.

The intensity of an absorption (the reading of absorbance or transmittance in Figure 9.2) depends upon several factors:

- the *probability* of a transition occurring;
- the *populations* of the levels; and
- the amount (concentration) of the sample.

Whereas an absorption spectrum measures the frequency and amount of radiation that has been removed from the initial range of frequencies, an emission spectrum measures the radiation emitted by the *excited state* of a sample. In an *emission spectrometer*, the sample is *excited* (thermally, electrically or by using electromagnetic radiation) to a short-lived higher energy level. As the transition back to a lower energy level occurs, energy is emitted. The spectrum recorded is a plot of the intensity of emission against frequency. The difference in energy between the higher and lower energy levels corresponds to the energy of the emitted radiation. In an emission spectroscopic experiment, it is important to ensure that sufficient time has elapsed between the initial excitation and the time of recording the emission; if the time delay is too long, the transition from higher to lower level will already have occurred and if it is too short, emission may not have taken place.

9.2 THE RELATIONSHIP BETWEEN THE ELECTROMAGNETIC SPECTRUM AND SPECTROSCOPIC TECHNIQUES

The electromagnetic spectrum scale is shown in Appendix 4. Electromagnetic radiation can be described in terms of frequency (ν), wavelength (λ) or energy (E), but another convenient unit is 'wavenumber'. Wavenumbers are the reciprocal of wavelength (equation 9.1) and the units are 'reciprocal centimetres' (cm^{-1}). These units are used as a convenient quantity which is linearly related to energy.

$$\text{Wavenumber } (\bar{\nu}) = \frac{1}{\text{Wavelength}} \tag{9.1}$$

Different spectroscopic methods are associated with different parts of the electromagnetic spectrum, since radiation of a particular energy is associated with certain transitions in an atom or molecule. For example, some of the spectral lines in the Lyman series of atomic hydrogen lie between 2.466×10^{15} and 3.237×10^{15} Hz. This corresponds to the ultraviolet region of the electromagnetic spectrum. In Figure 9.3, the methods of NMR, IR and UV–VIS spectroscopies are related to the electromagnetic spectrum.

Lyman series: see worked example 2.3

Nuclear magnetic resonance spectroscopy is concerned with transitions between different nuclear spin states. Such transitions require little energy (<0.01 kJ mol^{-1}) and can be brought about using radiation from the radiofrequency region of the electromagnetic spectrum.

NMR spectroscopy: see Sections 9.12–9.16

Vibrational spectroscopy is concerned with transitions between vibrational states of a molecule. The energy needed to bring about such transitions is typically in the region of 1 to 100 kJ mol^{-1}, corresponding to the infrared region of the electromagnetic spectrum. This range corresponds to wavenumbers from 100 to 10 000 cm^{-1}. A 'normal' laboratory IR spectrometer operates between 200 and 4000 cm^{-1}.

IR spectroscopy: see Sections 9.6–9.8

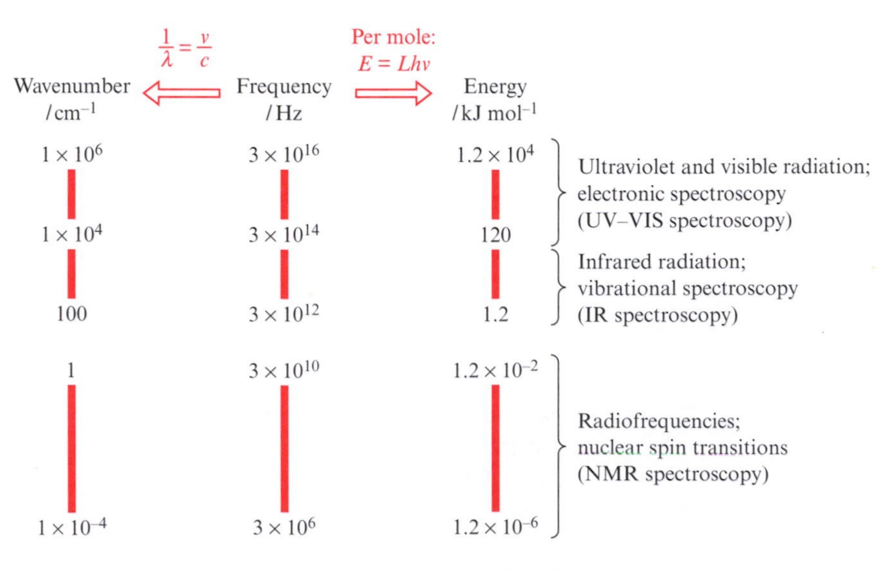

9.3 Part of the electomagnetic spectrum; the radiation can be described in terms of energy, frequency or wavenumber. The relationship to wavelength is shown in Appendix 4. Different regions of the spectrum are associated with different spectroscopic techniques.

Electronic spectroscopy: see Sections 9.9–9.11

Electronic spectroscopy is the study of transitions between electronic energy levels in a molecule. The energy differences correspond to radiation from the UV–VIS region of the electromagnetic spectrum – hence the name UV–VIS *spectroscopy*. The normal range of a laboratory UV–VIS spectrophotometer is 200 to 900 nm; the measurement is often made as a wavelength.

9.3 TIMESCALES

At this point, it is necessary to say something about the spectroscopic *timescales*. This is a complex topic and we are only concerned with one or two aspects. A critical question is: Will a particular spectroscopic technique give us a 'snapshot' of the molecule at a given moment of time or an 'averaged' view?

The conventional 'static' view of a molecule is incorrect – molecules are continually vibrating and rotating, and these motions occur at a rate of 10^{12} to 10^{14} per second (Figure 9.3). If the spectroscopic technique is *faster* than this, we obtain a 'snapshot' of the event, but if it is *slower*, we only see an averaged view of the molecule undergoing its various motions. Electronic spectroscopy gives a 'snapshot' of a molecule in a given vibrational and rotational state whereas NMR spectroscopy often gives an averaged view, since the timescale of the technique is slower than the molecular vibrational and rotational motions.

A second feature refers to the fact that we are not usually looking at a single molecule but at an assembly of molecules. Although electronic spectroscopy gives a 'snapshot' of a particular molecule in a particular vibrational and rotational state, a typical sample may contain 10^{18} molecules, not all of which will be in the same vibrational and rotational states.

Berry pseudo-rotation: see Section 5.12

A final point is that the molecule may be undergoing a dynamic process such as Berry pseudo-rotation, and this creates an additional problem of timescales. Lowering the temperature will slow the dynamic behaviour, and *may* make it slower than the spectroscopic timescale although as we have described for $[Fe(CO)_5]$, even at 103 K, the axial and equatorial CO ligands are exchanging positions and the ^{13}C NMR spectrum shows the presence of only one (an average) ^{13}C environment.

9.4 THE BEER–LAMBERT LAW

For a given compound, the intensity of an absorption depends mainly on the amount of the sample. If a particular frequency of radiation is being absorbed by a molecule, then the more molecules there are, the more radiation of that frequency will be absorbed and less transmitted. The *absorbance* or *optical density* of a sample is related to the *transmittance* (equation 9.2) and this relationship is shown graphically in Figure 9.4.

log = $\log_{10}$

$$\text{Absorbance} = -\log(\text{Transmittance}) \qquad A = -\log T \qquad (9.2)$$

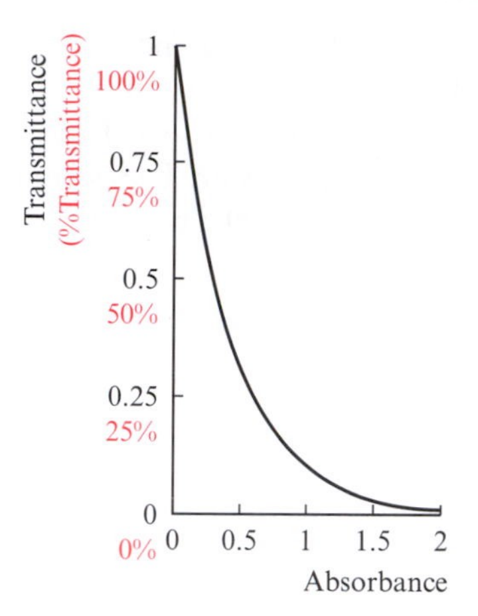

9.4 The relationship between transmittance and absorbance (see equation 9.2). Note that the curve tends to infinity when the transmittance is zero.

Values of transmittance, T, lie between 0 and 1, but experimentally we often express T as a percentage. From equation 9.2, 100% transmittance corresponds to zero absorbance, and for zero transmittance, the curve in Figure 9.4 tails off to infinite value of absorbance; for $T = 0.01$ (1%), the absorbance is 2.

The transmittance is equal to the ratio of the intensity of the transmitted radiation (I) to that of the incident radiation (I_0) and combining this with equation 9.2 gives equation 9.3. Figure 9.5 illustrates the relationship between I and I_0 and demonstrates how the absorbance of a sample in solution is determined.

$$A = -\log \frac{I}{I_0} \qquad (9.3)$$

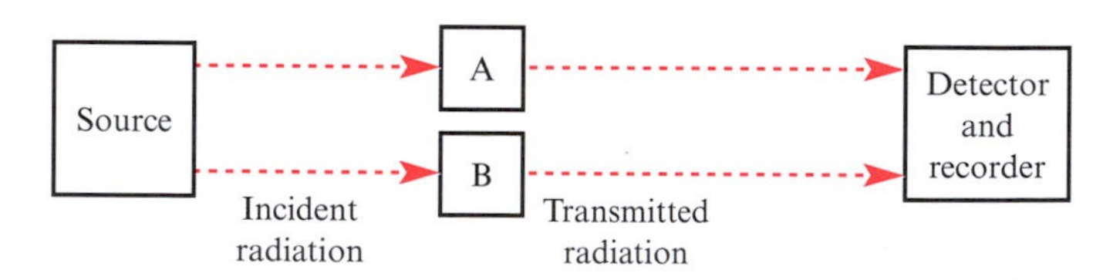

9.5 Solution cell A contains a sample in solution, and cell B contains pure solvent – the *same* solvent as in cell A; the cells have the same path length. If the solvent does *not* absorb any radiation, the intensity of the radiation emerging from cell B is the same as that of the incident radiation (I_0); the sample in cell A absorbs some radiation, and the intensity of the radiation emerging from cell A is that of the transmitted radiation (I). The transmittance of the sample is the ratio of $I{:}I_0$, and absorbance can be determined by using equation 9.3.

In a spectrophotometer, the sample is contained in a *solution cell* of accurately known dimensions. The distance travelled by the radiation through the cell is called the *path length*, ℓ, and, often, the absorbance is related to the concentration (c) and path length by the *Beer–Lambert Law* (equation 9.4) where ε is the *molar extinction* or *absorption coefficient* of the dissolved compound.

$$A = -\log \frac{I}{I_0} = \varepsilon \times c \times \ell \qquad (9.4)$$

The concentration is measured in mol dm^{-3} and the cell path length in cm, giving the units of ε as $dm^3\ mol^{-1}\ cm^{-1}$. The extinction coefficient is a *property of the compound* and is independent of concentration.

The Beer–Lambert Law relates the absorbance of a solution sample to the molar extinction coefficient, the concentration and the cell path length:

$$A = \varepsilon \times c \times \ell$$

Worked example 9.1 ***Use of the Beer–Lambert Law***

Naphthalene is an aromatic compound with the following structure:

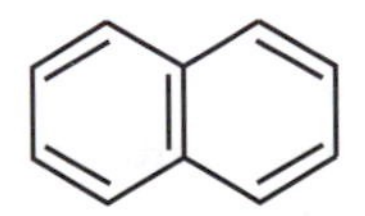

Solutions of naphthalene absorb light of wavelength 312 nm and the extinction coefficient for this transition is 288 $dm^3\ mol^{-1}\ cm^{-1}$. A solution of naphthalene in ethanol in a cell of path length 1 cm gives an absorbance of 1.2. What is the concentration of the solution?

By the Beer–Lambert Law:

$$A = \varepsilon \times c \times \ell$$

Therefore:

$$c = \frac{A}{\varepsilon \times \ell}$$

$$= \frac{1.2}{288 \times 1} = 4.2 \times 10^{-3}\ \text{mol dm}^{-3}$$

9.5 COLORIMETRY

One consequence of the direct relationship between the absorbance and the sample concentration (equation 9.4) is that the Beer–Lambert Law can be applied analytically. For most compounds, the Beer–Lambert Law is obeyed reasonably well in dilute solution. The technique of *colorimetry* is used to determine the concentration of coloured compounds in solution. Not only can one-off measurements be made, but changes in the concentration of a coloured component during a reaction can be followed by monitoring the change in absorbance.

Types of coloured compounds: see Section 9.11 and Chapter 16

Colours

When a coloured compound is dissolved in a solvent, the *intensity* of colour depends on the concentration of the solution, but the *actual* colour depends upon the wavelength of visible light absorbed by the sample.

The colour of a solution depends upon the wavelength of visible light absorbed by the sample, and the intensity of colour depends on the concentration of the solution.

White light consists of a continuous spectrum of electromagnetic radiation from about 400 to 750 nm. When white light is incident on a solution of a compound that absorbs within the visible region, the transmitted light is coloured, and the observed colour depends upon the 'missing' wavelength(s). Copper(II) sulfate solution absorbs orange light and as a result, the solution appears blue – blue is the *complementary colour* of orange. Table 9.2 lists the colours of light, corresponding wavelengths and the complementary colours in the visible spectrum.

Table 9.2 The visible part of the electromagnetic spectrum.

Colour of light absorbed	*Approximate range of wavelengths / nm*	*Colour of light transmitted (the complementary colour of the light absorbed)*
Red	750–620	Green
Orange	620–580	Blue
Yellow	580–560	Violet
Green	560–490	Red
Blue	490–430	Orange
Violet	430–380	Yellow

Worked example 9.2 ***The dependence of absorbance on concentration***

A 0.1 M solution of copper(II) sulfate gives an absorbance of 0.55. What is the absorbance when the concentration is doubled? Assume that the same solution cell is used for the two readings.

The relationship between absorbance and concentration is given by the Beer–Lambert Law:

$$A = \varepsilon \times c \times \ell$$

For a constant path length (constant solution cell) and constant ε, we can write:

$$\frac{A_1}{A_2} = \frac{c_1}{c_2}$$

where A_1 is the absorbance for a concentration c_1, and A_2 is the absorbance for a concentration c_2.

$$A_1 = \frac{c_1 \times A_2}{c_2}$$

$$= \frac{0.2 \times 0.55}{0.1}$$

$$= 1.1$$

The Beer–Lambert Law gives a *linear relationship* between A and c for a given compound *if the path length is constant*.

The colorimeter

Figure 9.5 schematically illustrated how the absorbance of a compound in solution can be measured. In a colorimeter, the 'source' is tuned to provide a particular wavelength – the choice is made based on a knowledge of the wavelength absorbed by the compound or ion under study.

The aim of a colorimetric investigation may be to determine the concentration of a solution or to follow a reaction (see below). A further example of its use is the determination of the stoichiometry of a complex as is illustrated in Section 9.11.

Case Study 1. Analysis of iron(II)

The compound 4,7-diphenyl-1,10-phenanthroline **9.1** is used to analyse for iron(II) ions.

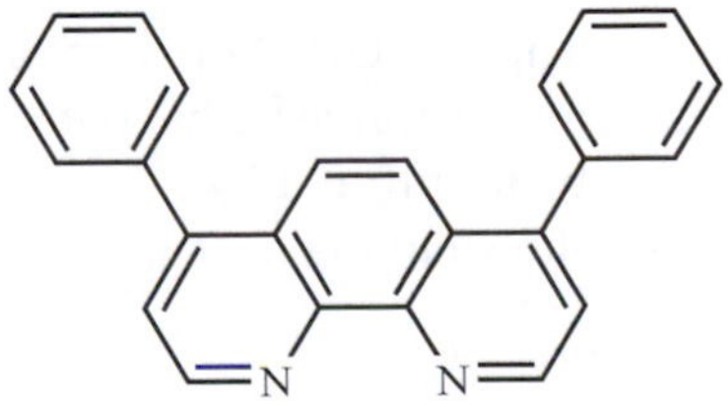

N N

(9.1)

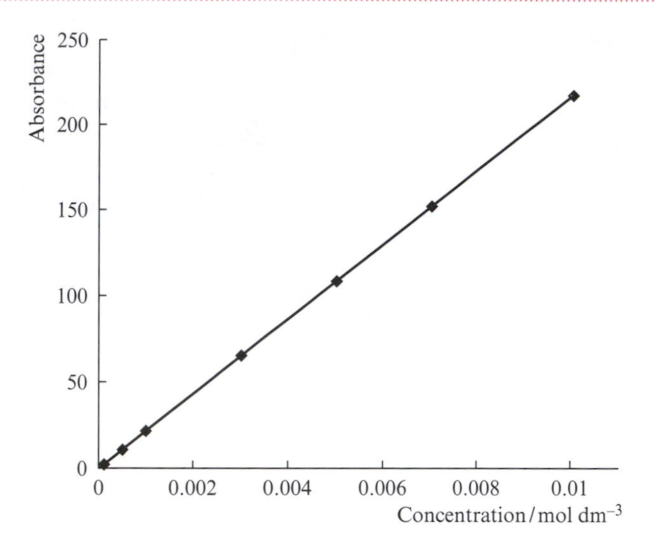

9.6 A plot of absorbance (533 nm) against concentration of a solution of the compound formed between iron(II) ions and 4,7-diphenyl-1,10-phenanthroline. The path length of the solution cell is 1 cm. The plot illustrates the Beer–Lambert Law.

When compound **9.1** is added to a solution containing iron(II) ions, an intensely coloured red compound is formed which absorbs light of wavelength 533 nm and has a molar extinction coefficient of 22 000. If the colorimeter is tuned to 533 nm, the concentration of the red compound can be determined by measuring the absorbance. One mole of this red compound contains one mole of Fe^{2+} ions, and if all the iron(II) present in the solution has reacted with compound **9.1**, the concentration of Fe^{2+} can be determined. Figure 9.6 shows how the absorbance (at 533 nm) varies with the concentration of iron(II) ions. Such a plot can be used to analyse a series of samples containing Fe^{2+} (see problem 9.3); since ε is large, even very low concentrations of Fe^{2+} ions can be measured accurately.

Case Study 2. Following the reaction between copper metal and potassium dichromate(VI)

Reactions involving coloured species may be monitored by using colorimetry, and may involve the appearance or disappearance of a particular coloured species. This can provide valuable information about the rate of a reaction (see Chapter 10).

Copper metal reacts with potassium dichromate(VI) in the presence of acid (equation 9.5) and during the reaction, the red-orange colour of the potassium dichromate is lost and the solution becomes blue-green.

$$3Cu(s) + [Cr_2O_7]^{2-}(aq) + 14H^+ \rightarrow 3Cu^{2+}(aq) + 2Cr^{3+}(aq) + 7H_2O \quad (9.5)$$

The $[Cr_2O_7]^{2-}$ ion absorbs close to 500 nm, while the products absorb at longer wavelengths. With the colorimeter tuned to 500 nm, the reaction can be monitored by taking readings of absorbance as a function of time as shown in Figure 9.7.

9.7 A plot of absorbance (500 nm) against time which follows the disappearance of the $[Cr_2O_7]^{2-}$ ion in its reaction with copper metal in the presence of acid.

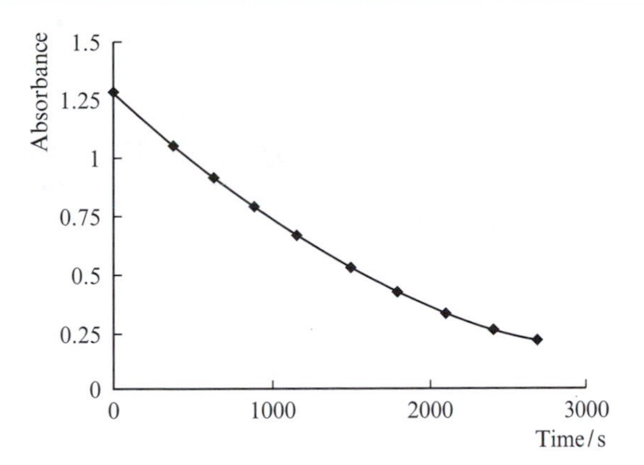

9.6 VIBRATIONAL SPECTROSCOPY 1: DIATOMIC MOLECULES

Vibrational spectroscopy is concerned with the study of molecular vibrations. Infrared (IR) and Raman spectroscopies are two types of vibrational spectroscopy and the former technique is readily available in practical laboratories. An IR spectrum records absorptions of infrared radiation that are associated with the *vibrational modes* of molecules.

The vibration of a diatomic molecule

A diatomic molecule is composed of two bonded atoms and can be likened to a classical model of two objects (corresponding to the atoms) connected by a spring (corresponding to the bonding electrons). Oscillations of the spring correspond to *stretching modes* for the molecule (Figure 9.8). Even at 0 K, the H_2 molecule is vibrating and possesses an internal energy called the zero point energy. This corresponds to the energy of the *lowest vibrational level* or the *vibrational ground state* of the molecule, but by providing the molecule with energy, transitions to higher vibrational levels may be made.

Zero point energy: see Section 3.5

9.8 The stretching mode of a heteronuclear diatomic molecule: (a) the equilibrium position; (b) extension of the bond; and (c) compression of the bond.

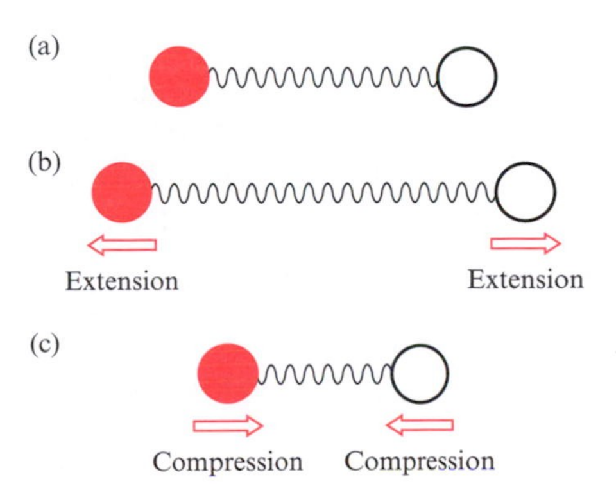

In its vibrational ground state, a molecule still possesses zero point energy.

That is, *the molecule vibrates more vigorously than in the lowest vibrational level*. If sufficient energy is put into the system, the molecule will dissociate.

The analogy between a spring and a molecule is a common one, but we need to think about one important difference that is easily forgotten. When we talk about a *spring*, the rest state is a static one, but a molecule is never stationary – it is always vibrating, even in the vibrational ground state. When we talk about vibrational modes of a molecule we really mean that we are taking the molecule from one vibrational state (usually its vibrational ground state) to the next vibrational (excited) state. Another important difference between the vibrations of a spring and those of a molecule is that the vibrational energy of a molecule is quantized and *only certain transitions are allowed*.

In general, the energy difference between the vibrational ground state and the first excited state is such that at 298 K, most molecules are in their ground state; the number of molecules in a particular state is called the *population* of that state. At 298 K, the vibrational ground state has a large population, and higher states have only small populations. The distribution of molecules between the vibrational states is given by a Boltzmann distribution. In this chapter, we shall assume that the only transition of importance is from the vibrational ground state to the first excited state, and *whenever we refer to a molecular vibration, we are referring to this vibrational transition*.

Boltzmann distribution: see Section 1.18

Let us now return to the diatomic molecule and provide it with energy in the form of IR radiation. The molecule will only undergo a vibrational transition when it absorbs an appropriate frequency of radiation; the frequency of a vibration depends on the strength of the covalent bond in the diatomic molecule. An IR spectrum records the frequency of the vibration, but not every vibrational mode gives rise to an observable absorption band in the IR spectrum. A selection rule must be obeyed – *for a vibrational mode to be infrared active, it must give rise to a change in the molecular dipole moment.*

For a vibrational mode to be infrared active, the vibration must result in a change in molecular dipole moment.

Consider the H_2 molecule – it is non-polar, and when the H–H bond is stretched or compressed, the molecule does *not* gain a dipole moment. This is true for any homonuclear diatomic molecule. The vibrational mode of the H_2 molecule is therefore *IR inactive*. A CO molecule has a dipole moment (μ = 0.11 D), and this changes if the distance between the carbon and oxygen atoms alters (equation 9.6).

Refer to Box 4.3

$$\mu = (\text{electronic charge}) \times (\text{distance between the charges}) \qquad (9.6)$$

As the CO molecule vibrates, a change in dipole moment occurs and this leads to the mode of vibration being *IR active*.

The force constant of a bond

The *force constant*, k, of a bond is related to the bond strength; the stronger the bond, the larger the value of k. In Figures 9.9 and 9.10 we plot the dissociation enthalpies and force constants of the bonds in HF, HCl, HBr and HI. The similarity of the trends in these graphs is clear.

Box 9.1 The force constant of a spring

When a spring vibrates, a force must be applied to counter the vibration and bring the spring back to its rest state. This is the *restoring force* and its magnitude depends on the magnitude of the oscillation and on a property of the spring called the *force constant, k*.

Restoring force = – (k × Displacement of particle from equilibrium position)

The units of k are N m^{-1} (force per unit distance).

The reduced mass of a diatomic molecule

Let us return to the analogy between a diatomic molecule and a spring. If the two objects connected by the spring have very different masses, then during the oscillation of the spring, the smaller mass will move more freely than the larger one – if you connected a spring between a car and a tennis ball and set the spring in motion, the ball would move *far* more easily than the car, and any oscillation of the spring would effectively *only* describe the movement of the ball. However, if the masses of the two particles are similar, then movement of *both* the particles will be significant as the spring oscillates. The more similar the masses, the more equally the two particles contribute to the overall motion.

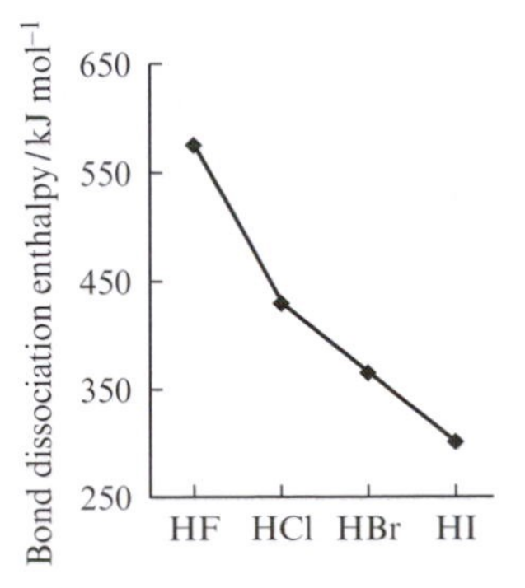

9.9 The trend in bond dissociation enthalpies along the series of the hydrogen halides, HX.

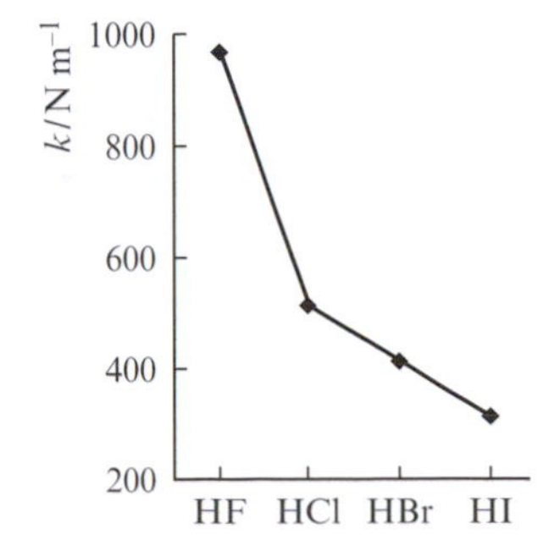

9.10 The trend in force constants for the bonds in the hydrogen halides, HX.

A diatomic molecule may contain two nuclei which are of similar mass (e.g. CO), or very different masses (e.g. HI), and we need to define a quantity that describes the mass of the diatomic molecule in such a way that it also reflects the *relative* masses of the nuclei. This is the *reduced mass*, μ,[§] and for two nuclei of masses m_1 and m_2, the reduced mass is obtained from equation 9.7.

$$\frac{1}{\mu} = \frac{1}{m_1} + \frac{1}{m_2} \tag{9.7}$$

The frequency of the vibration of a diatomic molecule

We can now consider the frequency of the vibration of the molecule. The analogy with the spring is a good place to start because we can apply the principle of *simple harmonic motion*, but must keep in mind the *relative* motions of the two nuclei in the molecule.

The transition from the vibrational ground state to the first excited state gives rise to a *fundamental absorption* in the IR spectrum of a diatomic molecule – assuming that the mode is IR active. The frequency (in Hz) of this absorption is related to the force constant of the bond (in N m^{-1}) and the reduced mass (in kg) of the diatomic molecule (equation 9.8).

$$\nu = \frac{1}{2\pi}\sqrt{\frac{k}{\mu}} \tag{9.8}$$

Wavenumbers: see Section 9.2

However, the units of ν in equation 9.8 are Hz, and the scale on an IR spectrometer is usually in wavenumbers (cm^{-1}). Equation 9.9 gives the relationship between wavenumber and *stretching frequency* and combining equations 9.8 and 9.9 leads to equation 9.10. Worked example 9.3 illustrates how equation 9.10 may be applied.

$$\bar{\nu} = \frac{\nu}{c} \tag{9.9}$$

$$\bar{\nu} = \frac{1}{2\pi \times c}\sqrt{\frac{k}{\mu}} \tag{9.10}$$

It follows that less energy is required to stretch bonds involving heavy elements than light ones, and Figure 9.11 shows the trend in fundamental stretching frequencies of the hydrogen halides. A comparison of Figures 9.9 to 9.11 emphasizes that there is a relationship between the stretching frequency, dissociation enthalpy and force constant of a bond.

§ The symbol μ is used both for reduced mass and dipole moment (as well as other things), but the context should minimize confusion.

Box 9.2 The simple harmonic oscillator: is it a good model for a vibrating bond?

Equation 9.8 gives a relationship between the frequency of the absorption in the IR spectrum of a diatomic molecule, the force constant of the bond and the reduced mass of the molecule. This equation is appropriate if we assume that the molecule mimics a spring and acts as a simple harmonic oscillator. The potential energy well that describes simple harmonic motion is given in diagram (a). In a real diatomic molecule, a plot of energy potential as a function of internuclear distance is as shown in diagram (b), and it is this curve that really describes a molecular vibration.

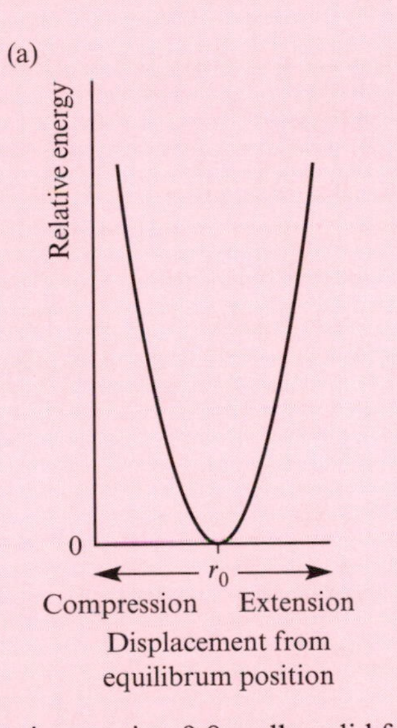

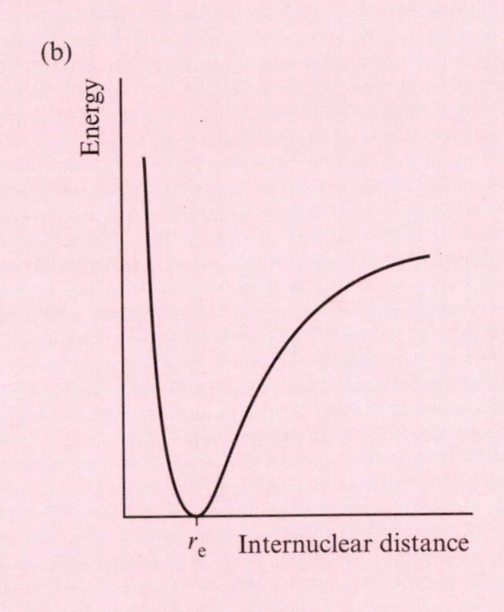

So, is equation 9.8 really valid for a molecular vibration?

Since we are principally concerned with transitions from the vibrational ground state to the first excited state, we are only looking at the lowest section of the potential energy well. In this region, the curves in (a) and (b) are very similar, and we are justified in applying the equation of the simple harmonic oscillator to the molecular vibration.

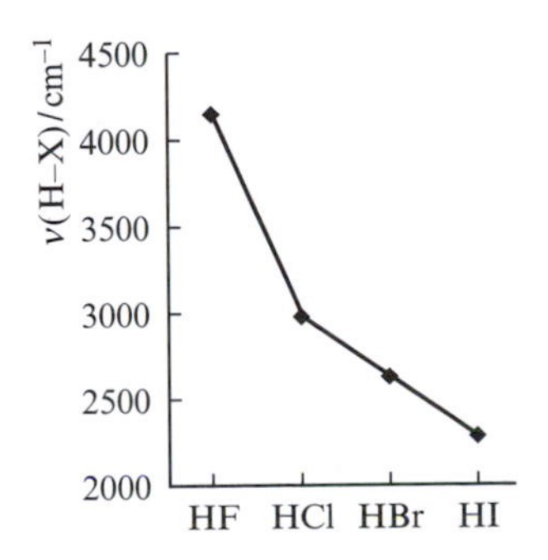

9.11 The trend in values of the fundamental stretching frequencies for the hydrogen halides, HX.

Worked example 9.3 ***Determination of the force constant of the bond in carbon monoxide***

The stretching mode of CO gives rise to an absorption in the IR spectrum at 2170 cm^{-1}. What is the force constant of the bond in the carbon monoxide molecule?

Data: $c = 3 \times 10^8$ m s^{-1}; atomic mass unit = 1.66×10^{-27} kg; relative atomic mass C = 12; O = 16.

The equation needed is

$$\bar{\nu} = \frac{1}{2\pi \times c}\sqrt{\frac{k}{\mu}}$$

and this can be rearranged to give k as the subject of the equation:

$$k = 4\pi^2 \times c^2 \times \bar{\nu}^2 \times \mu$$

We must ensure that the units are consistent.
The wavenumber of the absorption is 2170 cm^{-1}.
For consistency, $c = 3 \times 10^8$ m s^{-1} $= 3 \times 10^{10}$ cm s^{-1}.
The reduced mass, μ is given by the equation:

$$\frac{1}{\mu} = \frac{1}{m_1} + \frac{1}{m_2} = \left(\frac{1}{12 \times 1.66 \times 10^{-27}}\right) + \left(\frac{1}{16 \times 1.66 \times 10^{-27}}\right) = 8.79 \times 10^{25}$$

$$\mu = \frac{1}{8.79 \times 10^{25}}\ \text{kg}$$

Thus, the force constant is given by:

$$k = 4\pi^2 \times c^2 \times \bar{\nu}^2 \times \mu$$

$$= \frac{4 \times 3.142^2 \times (3 \times 10^{10})^2 \times 2170^2}{8.79 \times 10^{25}}$$

$$= 1904\ \text{N m}^{-1}$$

9.7 VIBRATIONAL SPECTROSCOPY 2: POLYATOMIC MOLECULES

Vibrational modes of freedom

Whereas the vibration of a diatomic molecule is restricted to a single stretching mode (Figure 9.8), the vibrational modes of a polyatomic molecule are more complex.

CO_2: a linear triatomic molecule

The carbon dioxide molecule is linear and contains two equivalent C=O bonds. When the molecule vibrates, it can do so in two different ways. Figure 9.12a shows the *symmetric stretch* in which the two oxygen atoms move outwards at the same time and then move in together. Figure 9.12b illustrates

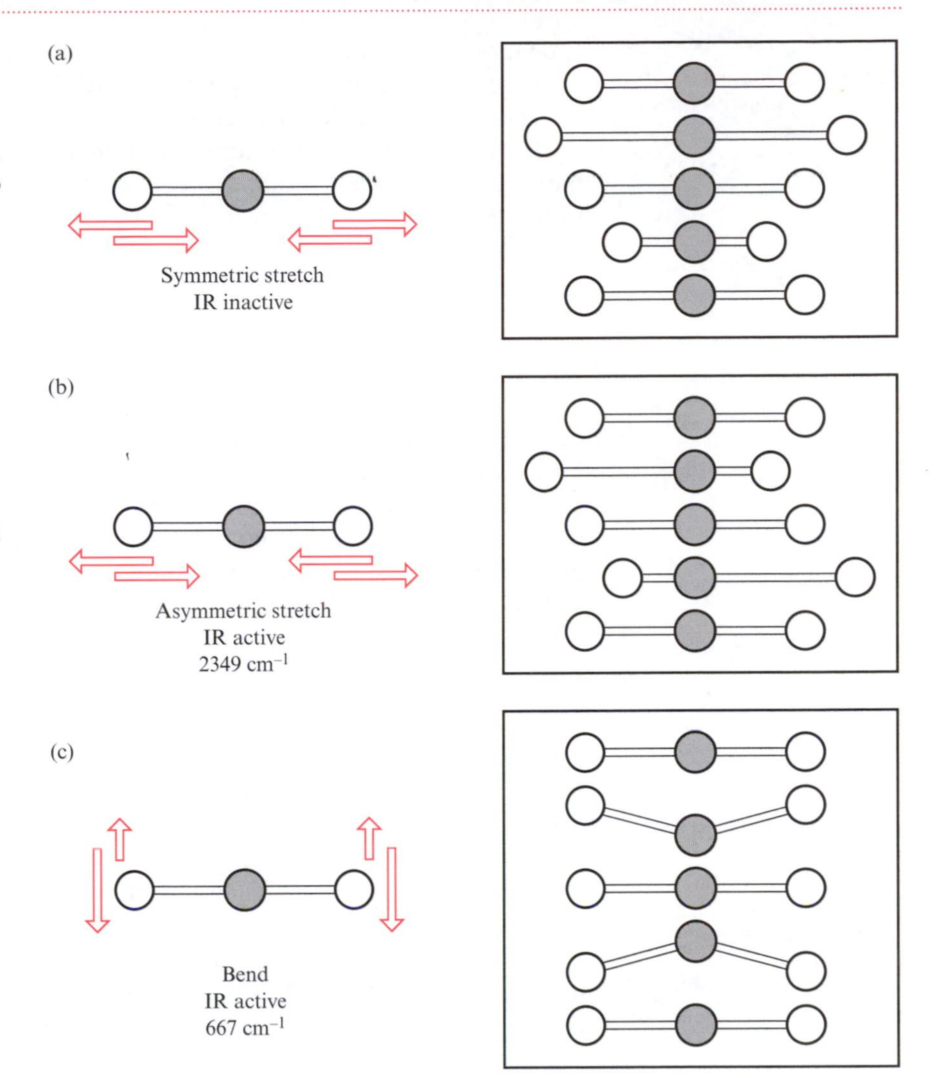

9.12 The vibrational modes of carbon dioxide. (a) The symmetric stretch – the diagram on the left-hand side summarizes the motion; follow the figures from top to bottom in the box to see the overall motion of the CO_2 molecule during the symmetric stretch. (b) The asymmetric stretch is summarized in the diagram on the left-hand side; follow the figures from top to bottom in the box to see the overall motion of the CO_2 molecule during the asymmetric stretch. (c) The bending mode is summarized in the left-hand diagram; in the box, the series of diagrams follows the motion as the molecule bends in the plane of the paper.

the *asymmetric stretch* – here one C=O bond is stretched as the other is compressed. Carbon dioxide does *not* possess a molecular dipole moment, although each C=O bond is polar. During the *symmetric* stretch, the change in dipole moment is zero, and the symmetric stretch is IR inactive. During the *asymmetric* stretch, a change in dipole moment occurs, and you can see why this happens by studying the right-hand side of Figure 9.12b. The asymmetric stretch is therefore IR active. For a linear triatomic that *does* possess a molecular dipole moment (e.g. HCN), both the symmetric and asymmetric stretching modes should be IR active.

$\chi^P(C) = 2.6$, $\chi^P(O) = 3.4$

A linear triatomic molecule also possesses a *bending mode* (Figure 9.12c); this vibration for CO_2 may also be represented in terms of the carbon atom moving up and down with respect to the two oxygen atoms. In Figure 9.12c, the motion is represented in the plane of the paper, but there is an equivalent vibration in a plane which is perpendicular to the plane of the paper. These

Degenerate means 'having the same energy'

two vibrational modes require the same amount of energy and are degenerate. The bending of the CO_2 molecule leads to a change in the dipole moment, and this mode is IR active.

The frequencies of the asymmetric stretching and bending vibrations of the CO_2 molecule are 2349 and 667 cm^{-1} respectively, and this tells us that the molecule requires more energy to undergo an asymmetric stretch than a bend.

H_2O: a bent triatomic molecule

The three vibrational modes for the water molecule are shown in Figure 9.13. Water possesses a molecular dipole moment (μ = 1.85 D) and this changes during the symmetric stretching vibration. This is true for other bent molecules and the symmetric stretch is IR active. The asymmetric stretching mode is also IR active. If the two hydrogen atoms move towards each other and then away from one another, this is called a *scissoring* motion or symmetric bending. A change in molecular dipole moment occurs and an additional absorption is observed in the IR spectrum.

Vibrational degrees of freedom

In the discussion above, we appear to have arbitrarily arrived at the *number* of vibrational modes for the CO_2 and H_2O molecules, and certainly when considering larger molecules, it is difficult to write down all the possible vibrational modes. There are, however, some simple rules from which you can work out the number of allowed vibrations.

A molecule with n atoms has $3n$ degrees of freedom.

If a molecule has n atoms, it possesses $3n$ *degrees of freedom*. The molecule as a whole can move in space and this is described as *translational* motion. Movement is a vector quantity, and translational motion can be described in terms of 3 degrees of freedom relating to 3 axes (for example, x, y and z). The molecule possesses *3 degrees of translational freedom*.

Having allocated 3 degrees of freedom for translational motion, the molecule is left with ($3n$ – 3) degrees of freedom for other types of motion and these are classified either as *rotational* or *vibrational degrees of freedom*. Like

(a)

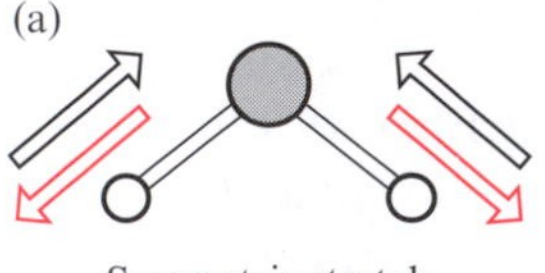

Symmetric stretch
IR active
3657 cm^{-1}

(b)

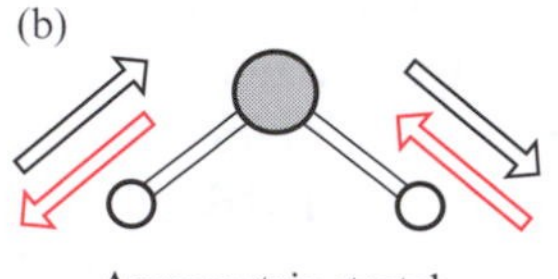

Asymmetric stretch
IR active
3756 cm^{-1}

(c)

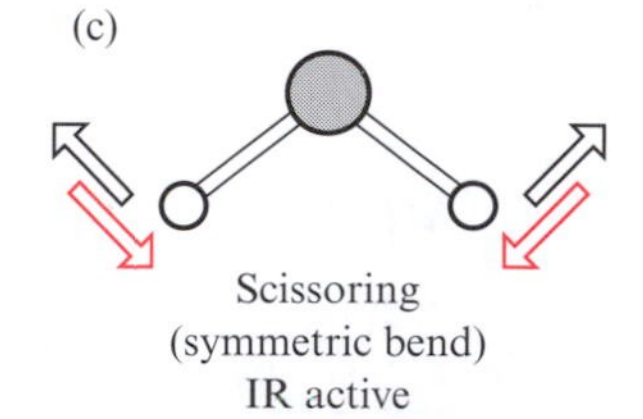

Scissoring
(symmetric bend)
IR active
1595 cm^{-1}

9.13 The vibrational modes of water. (a) In the symmetric stretch, both hydrogen atoms move out at the same time, and then in together. (b) In the asymmetric stretch, one hydrogen atom moves out as the other moves in. (c) During the symmetric bending or scissoring of the molecule, the H–O–H bond angle increases and decreases; this vibration takes place in the plane of the paper.

translational motion, rotations may be described with respect to an axis set. We simply state[§] that for a non-linear molecule there are 3 degrees of rotational freedom, but for a linear molecule, there are only 2 degrees of rotational freedom. It follows that the number of degrees of vibrational freedom is dependent upon the number of atoms in the molecule and on whether the molecule is linear or non-linear. Equations 9.11 and 9.12 summarize these results.

Number of degrees of vibrational freedom for a *non-linear* molecule $= 3n - 6$ (9.11)

Number of degrees of vibrational freedom for a *linear* molecule $= 3n - 5$ (9.12)

VSEPR: see Section 5.6

We can now apply equations 9.11 and 9.12 to several small molecules, the shapes of which are predicted by VSEPR theory.

Example 1: SO_2

SO_2 is a bent molecule. $n = 3$
Number of degrees of vibrational freedom $= 3n - 6 = 3$

Example 2: CO_2

CO_2 is a linear molecule. $n = 3$
Number of degrees of vibrational freedom $= 3n - 5 = 4$
Two of the modes of vibration are degenerate (see earlier in this section).

Example 3: CH_4

CH_4 is a tetrahedral molecule. $n = 5$
Number of degrees of vibrational freedom $= 3n - 6 = 9$

In addition to the stretching and scissoring (bending) modes described for CO_2 and H_2O, larger molecules can undergo *rocking, twisting* and *wagging* vibrations. Detailed discussion of these modes is beyond the scope of this book, but Figure 9.14 summarizes the modes of vibration that a methylene group in an alkane chain can undergo.

9.8 THE USE OF IR SPECTROSCOPY AS AN ANALYTICAL TOOL

We now turn our attention to the everyday use of IR spectroscopy in the laboratory. Compounds may be examined in either the solid state, solution or gas phase. There are two ways of interpreting an IR spectrum. It is always correct to interpret the spectrum as a property of the molecule as a whole, but this is extremely time consuming and only possible for simple molecules. It is common to consider the IR spectrum as a composite of individual absorptions from various components of the molecule. This is related to the functional group model in organic chemistry, which has as a fundamental assumption the fact that a group has similar properties regardless of the molecule in which it is found. Similarly, it is convenient to interpret IR spectra in

§ For further discussion see: P.W. Atkins, *Physical Chemistry*, 5th edn, Oxford University Press, Oxford (1994).

9.14 The vibrational modes of a methylene (CH_2) group in an alkane chain: (a) the symmetric stretching mode; (b) the asymmetric stretching mode; (c) the scissoring mode, (d) the rocking mode (in the plane of the paper); (e) the twisting mode; and (f) the wagging mode (the CH_2 unit moves in front of and behind the plane of the paper).

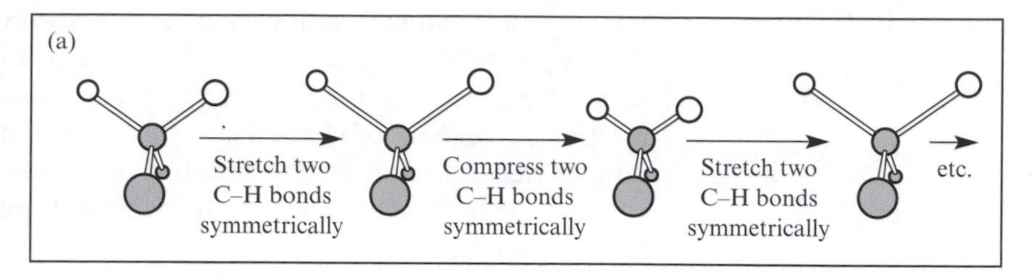

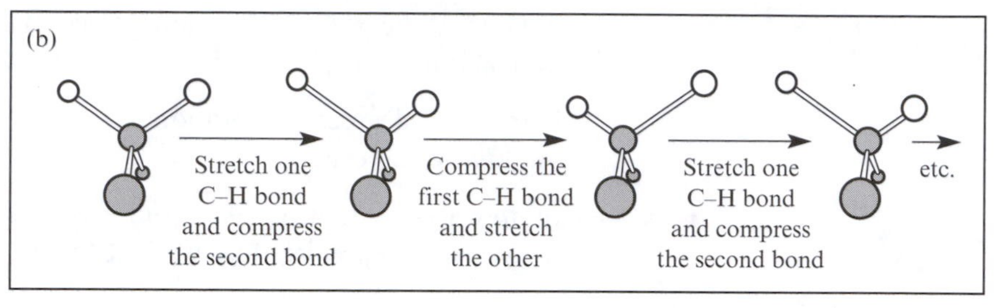

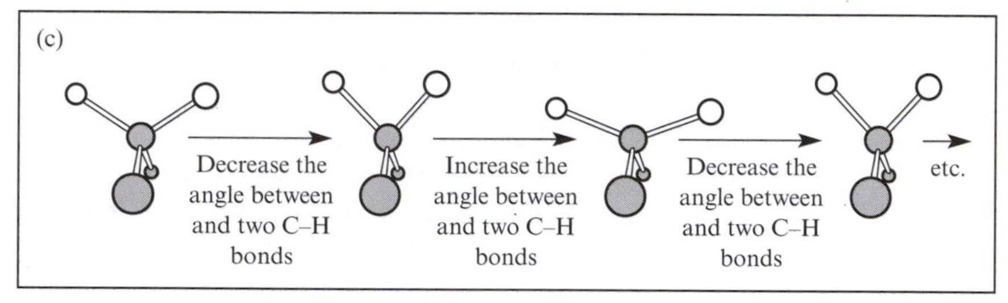

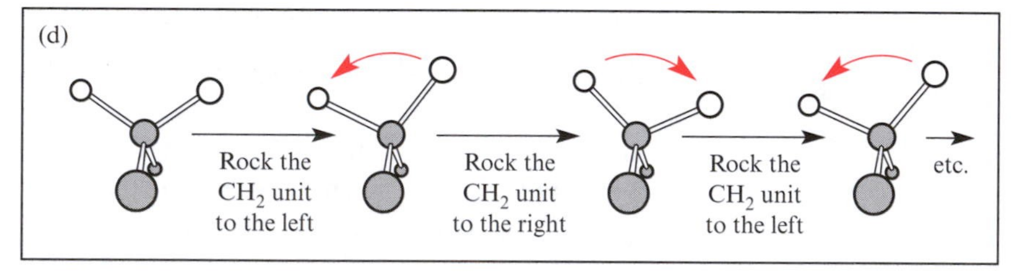

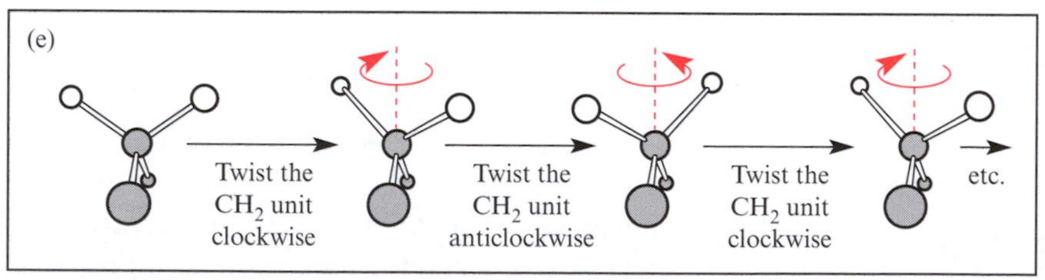

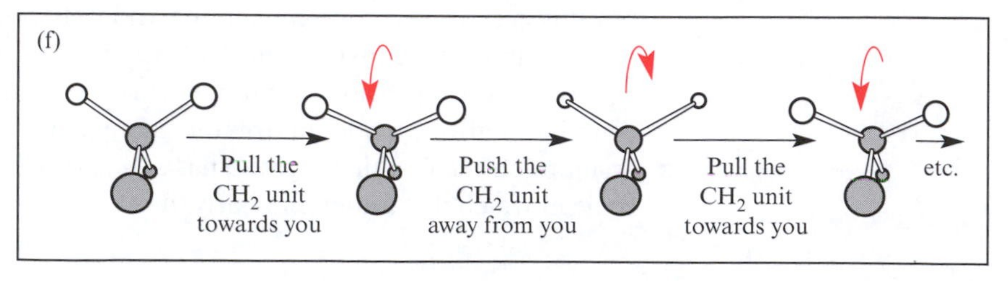

terms of absorptions arising from specific functional groups. *This is an over-simplification*, and is particularly dangerous with 'inorganic' compounds.

The IR spectra of organic compounds possess several characteristic regions:

- those arising from vibrations and other modes of the molecular framework, particularly C–C single bonds, and
- those arising from vibrations of a functional group, in particular those with multiple bonds.

This effectively divides the spectrum into two regions. Above 1500 cm^{-1}, the majority of absorptions are due to stretching modes of multiple bonds to carbon (C=C, C≡C, C=O, C=N, C≡N, etc.). Below 1500 cm^{-1}, in the *fingerprint region*, absorptions due to C–X single bond stretching modes as well as other vibrational modes are found. These distinctions are not absolute, e.g. absorptions due to C–H stretches are found around 3000 cm^{-1}.

The fingerprint region of the IR spectrum

Usually the IR spectrum of a compound contains a series of absorptions below about 1500 cm^{-1} which *together* are diagnostic of that compound. It is not usual for these absorptions to be individually assigned to particular vibrations within the molecule. Consider the IR spectrum of caffeine (the stimulant found in tea and coffee) shown in Figure 9.15. There are numerous absorptions below 1500 cm^{-1}, but it would clearly be difficult to assign them to individual vibrational modes of the molecule. However, this spectrum is an absolute characteristic of the molecule caffeine – no other molecule should exhibit an identical spectrum. This is why this part of the spectrum is known as the *fingerprint region*.

The most effective way to use the fingerprint region of an IR spectrum is to match the frequencies and relative intensities of the absorptions with those of an authentic sample of the compound.

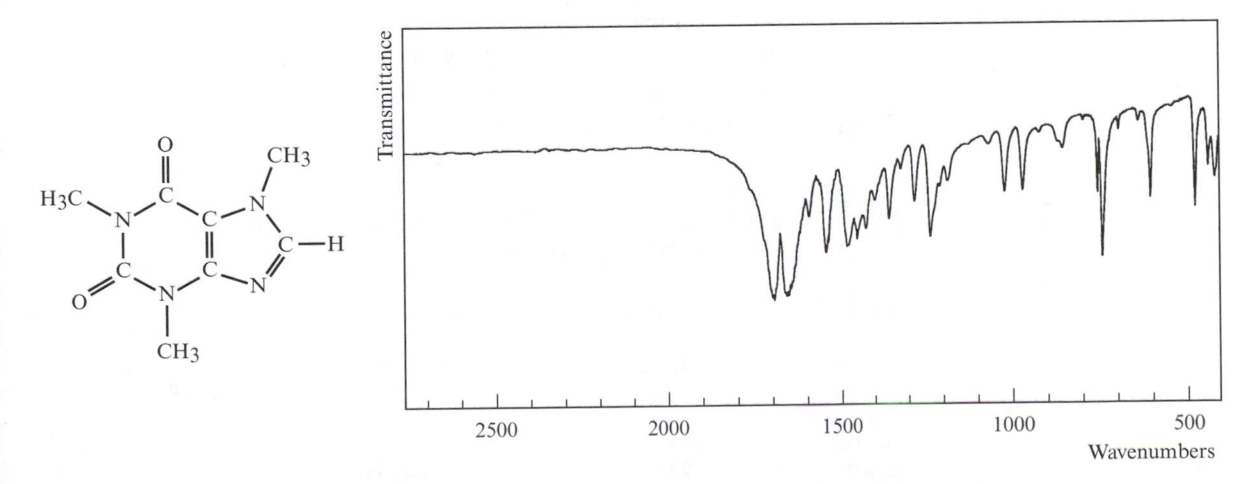

9.15 The IR spectrum of caffeine; the structure consists of two fused rings containing carbon and nitrogen atoms.

Functional groups and the interpretation of IR spectra

Functional groups in a molecule can give rise to absorptions at particular frequencies in the range 400–4000 cm^{-1}. It is easiest to detect (and assign) functional group absorptions which do not fall within the fingerprint region.

Carbonyl compounds: see Chapter 17

We take as an example the carbonyl group C=O which occurs in a range of compounds, including aldehydes **9.2**, ketones **9.3**, carboxylic acids **9.4**, acid chlorides **9.5**, esters **9.6** and carboxylates **9.7**.

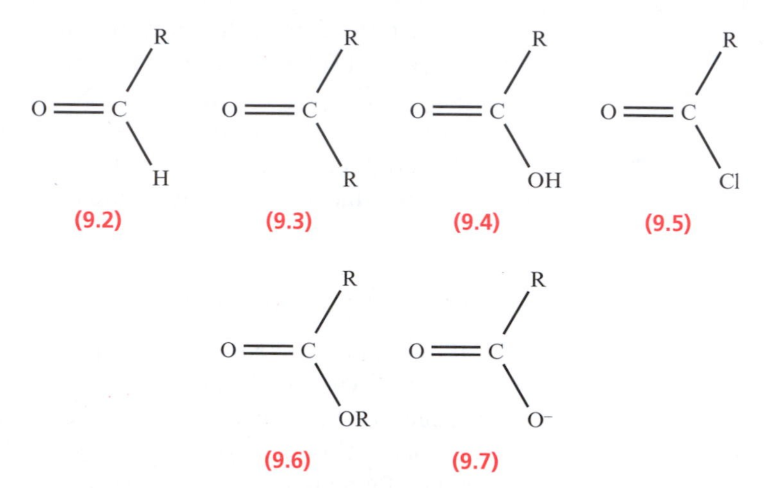

Electron-releasing: see Section 8.12

The stretching of the C=O double bond typically gives rise to a *strong* absorption around 1700 cm^{-1} (see Figure 17.6) but the exact frequency depends upon the strength of the bond and this is in turn affected by the other groups in the molecule or ion. Let us take the aldehyde RCHO as a reference point. On going from this to a ketone, the H is changed for an *electron-releasing alkyl group* and the carbon atom of the C=O group will become less δ^+. The carbonyl bond polarization ($C^{\delta+}$–$O^{\delta-}$) is therefore smaller in the ketone than in the aldehyde, and as a consequence stretching the C=O bond in the ketone is easier than in the aldehyde. The carbonyl absorptions for acetone and ethanal are shown in Figure 9.16.

$\chi^P(Cl) = 3.2$, $\chi^P(H) = 2.2$

Now compare the aldehyde RCHO with the corresponding acid chloride RCOCl. Chlorine is *electron-withdrawing* and the carbon atom of the C=O group carries a greater δ^+ charge in the acid chloride than in the aldehyde.

Table 9.3 Approximate ranges for carbonyl (C=O) stretching frequencies in compounds in structures **9.2** to **9.7** with R = alkyl.

Compound type	*Structure number*	*Absorption range / cm^{-1}*
Aldehyde	**9.2**	1740–1720
Ketone	**9.3**	1725–1705
Carboxylic acid	**9.4**	1725–1700
Acid chloride	**9.5**	1815–1790
Ester	**9.6**	1750–1735
Carboxylate	**9.7**	1610–1550

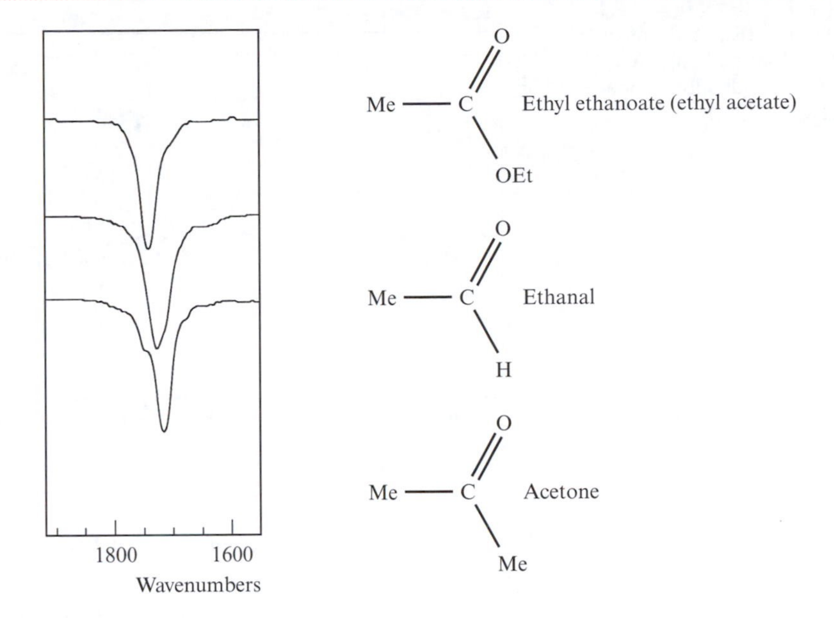

9.16 The carbonyl absorptions for acetone, ethanal and ethanoate (ethyl acetate).

This means that the C=O bond has increased ionic character and is more difficult to stretch – the stretching frequency increases accordingly (Table 9.3).

In an ester, there are two effects. The oxygen atom of the OR group is more electronegative than carbon and this tends to increase the δ^+ charge on carbonyl C atom. However, the lone pair on the OR oxygen atom can be delocalized (equation 9.13) with the effect of weakening the carbonyl carbon–oxygen bond. The overall effect is that the absorption due to the carbonyl stretch in an ester may not be much different from that in the corresponding aldehyde (Figure 9.16).

$$\mathrm{O{=}C(R)(\ddot{O}R)} \longleftrightarrow \mathrm{\bar{O}{-}C(R){=}\overset{+}{O}R} \qquad (9.13)$$

Table 9.4 Approximate ranges for stretching frequencies of some functional groups involving multiple bonds to carbon.

Functional group	*Typical frequency range of absorption band / cm^{-1}*
$R_2C{=}CR_2$	1680–1620
$RC{\equiv}CR$	2260–2100
$R_2C{=}NR$	1690–1640
$RC{\equiv}N$	2260–2200
$R_2C{=}O$	1815–1550 (see Table 9.3)
$R_2C{=}S$	1200–1050

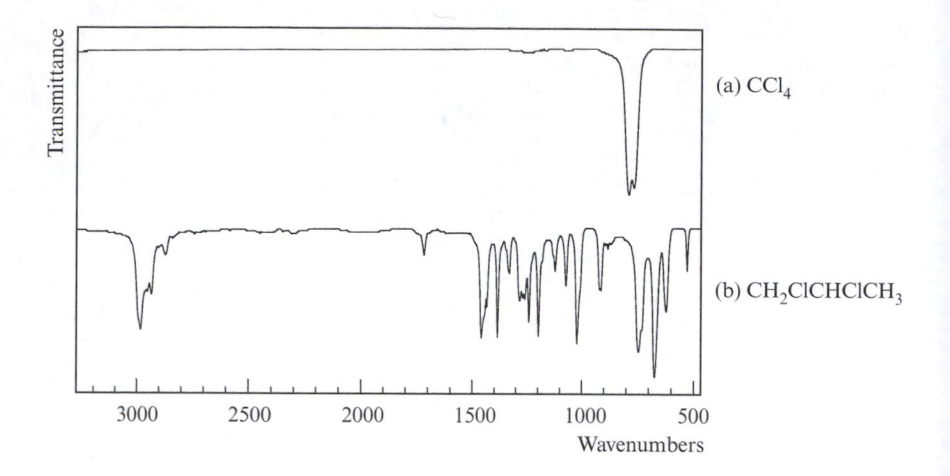

9.17 The IR spectra of (a) tetrachloromethane, CCl_4, and (b) 1,2-dichloropropane, $CH_2ClCHClCH_3$.

Table 9.5 Approximate ranges for stretching frequencies of carbon–halogen bonds.

Group	*Typical frequency range of absorption band / cm^{-1}*
C–F	1400–1000
C–Cl	800–600
C–Br	750–500
C–I	≈ 500

Some other functional groups with multiple bonds to carbon include C≡C, C=C, C≡N, C=N and C=S and typical frequency ranges for the absorptions due to the stretching of these bonds are listed in Table 9.4.

Many functional groups contain single bonds, and absorptions due to the vibrations of these bonds often fall within the fingerprint region and may not easily be assigned. The stretching frequencies of carbon–halogen bonds are listed in Table 9.5. Whereas the absorptions due to C–Cl stretching are readily observed in the IR spectrum of CCl_4 (Figure 9.17a), they disappear into the many bands of the fingerprint region in the spectrum of 1,2-dichloropropane (Figure 9.17b).

Some of the more readily observed absorptions are due to the stretching modes of O–H, N–H, S–H and C–H bonds (Table 9.6). Figure 9.18 shows the IR spectra of propan-1-ol and propan-2-ol. In each, the band near 3340 cm^{-1} is assigned to the O–H stretch. The broadness of the bands arises from the extensive degree of hydrogen bonding that is present between the alcohol molecules and is a common feature of absorptions associated with O–H bonds. Although it would be difficult to assign all the bands in the spectrum of either compound, they exhibit quite different, but characteristic, IR spectra and can be distinguished provided that authentic samples are available for comparison. Absorptions due to N–H bonds occur in a similar part of the IR spectrum to those of O–H bonds.

Alcohol nomenclature: see Section 14.9

In Figures 9.17b and 9.18, the absorptions around 3000 cm^{-1} are due to C–H stretches. The values in Table 9.6 indicate that the hybridization of the

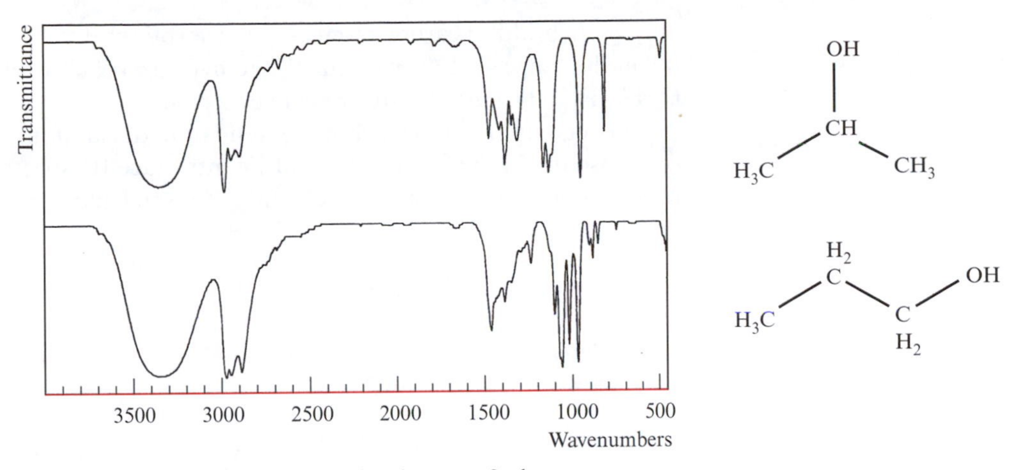

9.18 The IR spectra of propan-1-ol and propan-2-ol.

Table 9.6 Approximate ranges for stretching frequencies of O–H, N–H, S–H and C–H bonds.

Group	*Typical frequency range of absorption band / cm^{-1}*
O–H	3600–3200 (broadened by hydrogen bonding)
N–H	3500–3300
S–H	2600–2350 (weak absorption)
C–H (sp^3 carbon)	2950–2850
C–H (sp^2 carbon)	3100–3010
C–H (sp carbon)	≈ 3300

Electronegativities of carbon atoms: see Section 8.20

carbon atom influences the stretching frequency of the C–H bond. This is because the polarity (and strength) of a C–H bond decreases in the order C(sp)–H > C(sp^2)–H > C(sp^3)–H. Figure 9.19 shows part of the IR spectra of benzene and ethylbenzene. Benzene is a planar molecule and contains

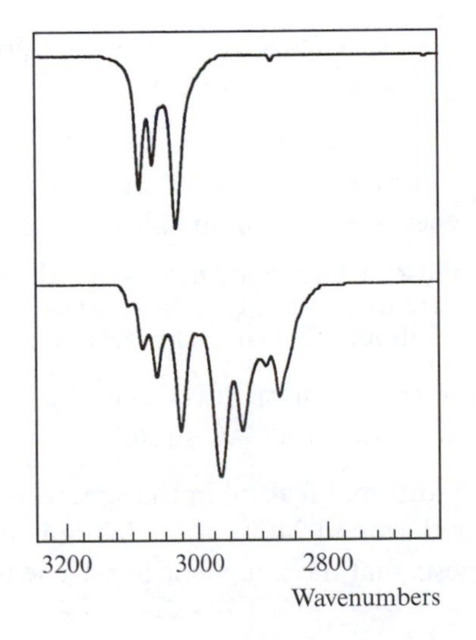

9.19 The IR spectra of benzene and ethylbenzene; only the regions around 3000 cm^{-1} (the C–H stretching region) are shown.

only sp^2 hybridized carbon atoms – the absorptions due to the C–H stretches fall in the range ≈ 3120–3000 cm^{-1}. In ethylbenzene, absorptions due to *both* C(sp^2)–H and C(sp^3)–H stretches are observed.

The accompanying workbook extends the discussion of the use of IR spectroscopy in the laboratory, and we introduce IR spectroscopic absorptions characteristic of other functional groups in Chapters 14, 15 and 17.

Worked example 9.4 ***The IR spectra of hexane and hex-1-ene***

The figure shows the IR spectra of hexane and hex-1-ene. What are the structural features in these molecules that are responsible for the observed absorptions in their spectra?

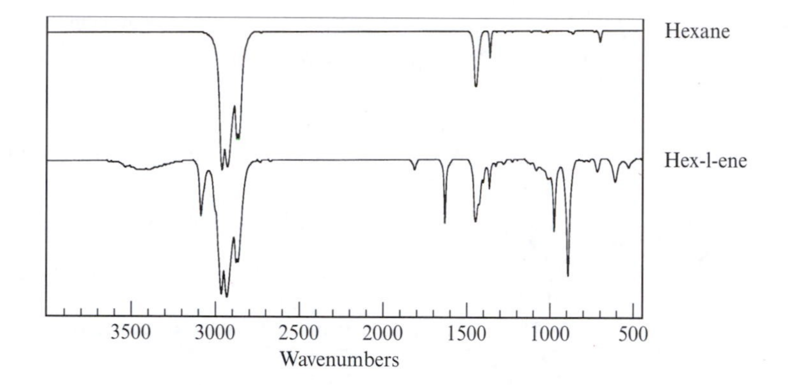

The structures of hexane and hex-1-ene are:

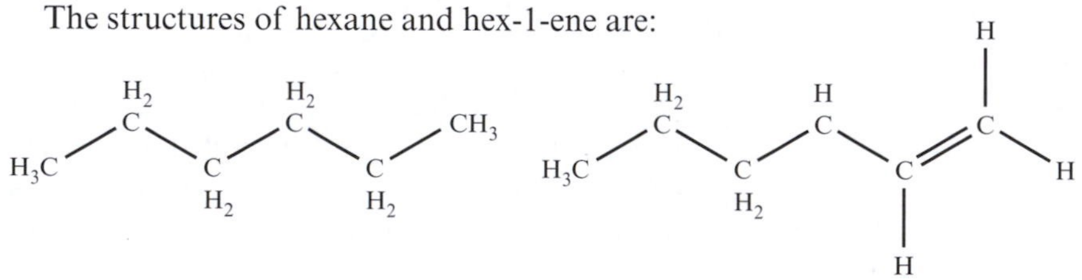

1. The part of each spectrum below 1500 cm^{-1} is the fingerprint region and provides a diagnostic pattern of IR bands for each compound.

2. In the IR spectrum of hexane, the only other feature is the group of absorptions around 3000 cm^{-1} which can be assigned to the C–H stretches; the C atoms are all sp^3 hybridized.

3. In the IR spectrum of hex-1-ene, the strong absorptions around 3000 cm^{-1} are due to the C(sp^3)–H stretches, whilst the less intense absorption at about 3080 cm^{-1} can be assigned to the C(sp^2)–H stretches.

4. In the spectrum of hex-1-ene, the sharp absorption at 1643 cm^{-1} is characteristic of a C=C stretch.

5. An additional feature in the spectrum of hex-1-ene is the broad band centred around 3400 cm^{-1}. This can be assigned to an O–H stretch and suggests that the sample of hex-1-ene is wet!

Worked example 9.5 ***Assignment of some IR spectra.***

The figure below shows the IR spectra of octane, propanenitrile, dodecan-1-ol and heptan-2-one. Assign each spectrum to the correct compound.

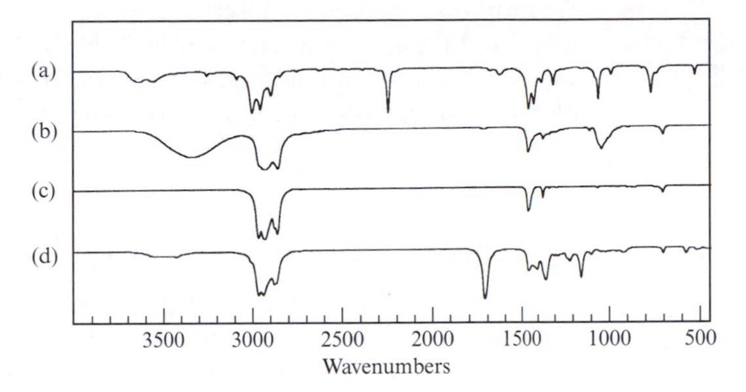

The structures of the compounds are:

Octane	$CH_3CH_2CH_2CH_2CH_2CH_2CH_2CH_3$
Dodecan-1-ol	$CH_3CH_2CH_2CH_2CH_2CH_2CH_2CH_2CH_2CH_2CH_2CH_2OH$
Propanenitrile	$CH_3CH_2C{\equiv}N$
Heptan-2-one	$CH_3CH_2CH_2CH_2CH_2C(=O)CH_3$

Begin the spectral assignments by looking for absorptions which are characteristic of the functional groups present.

Spectrum (a) has a strong absorption at $\approx 2250\ cm^{-1}$. This may be assigned to the C≡N group of propanenitrile. The spectrum also contains absorptions around $3000\ cm^{-1}$ assigned to stretches of the aliphatic C–H bonds.

Spectrum (b) exhibits a broad and strong absorption centred $\approx 3200\ cm^{-1}$ which is characteristic of an alcohol group. The spectrum may be assigned to dodecan-1-ol; C–H stetches around $3000\ cm^{-1}$ are observed and are expected for this compound.

Spectrum (d) shows a strong absorption at $\approx 1700\ cm^{-1}$ indicative of a carbonyl group. This spectrum can be assigned to heptan-2-one. In addition to the fingerprint region, this spectrum shows aliphatic C–H stretches which are expected for this compound.

Above the fingerprint region, spectrum (c) shows only absorptions due to C–H stretches; there appears to be no functional group. This spectrum is assigned to octane.

Conclusions

Spectrum	*Assignment*
(a)	Propanenitrile
(b)	Dodecan-1-ol
(c)	Octane
(d)	Heptan-2-one

9.9

ELECTRONIC SPECTROSCOPY 1: ELECTRONIC TRANSITIONS IN MOLECULES, THE VACUUM-UV AND SOLVENT CHOICES

Atomic spectrum of hydrogen: see Section 2.16

Electronic spectroscopy observes transitions between electronic energy levels, and the atomic spectrum of hydrogen is an example. In this chapter we are concerned with the *electronic spectra of molecules*, and the emphasis of our discussion is on the use of electronic spectroscopy in the laboratory. *Electronic spectra are often reported in terms of wavelength (nm).*

Absorption of ultraviolet and visible light

HOMO = highest occupied molecular orbital;

LUMO = lowest unoccupied molecular orbital.

The molecular orbitals of a molecule may be bonding, non-bonding or antibonding. The bonding and antibonding orbitals may have σ- or π- (or other) symmetry. If a molecule *absorbs* an appropriate amount of energy, an electron from an occupied orbital may be excited to an unoccupied or partially occupied orbital. As the energies of these orbitals are quantized, it follows that each transition is associated with a specific amount of energy. The energy of the electronic transition from the highest occupied molecular orbital (HOMO) to the lowest unoccupied molecular orbital (LUMO) often corresponds to the ultraviolet (UV) or visible (VIS) region of the electromagnetic spectrum – wavelengths between 100 and 800 nm. Normal laboratory UV–VIS spectrophotometers operate between ≈200 and 900 nm (Figure 9.20).

Figure 9.21 shows that an electronic transition should occur when light of energy ΔE is absorbed by a molecule, and the absorption spectrum might be expected to consist of a sharp band (as in Figure 9.2). In practice, however, *molecular* electronic spectra usually consist of *broad absorptions* (Figure 9.22) in contrast to the sharp absorptions of atomic spectra. Why is this?

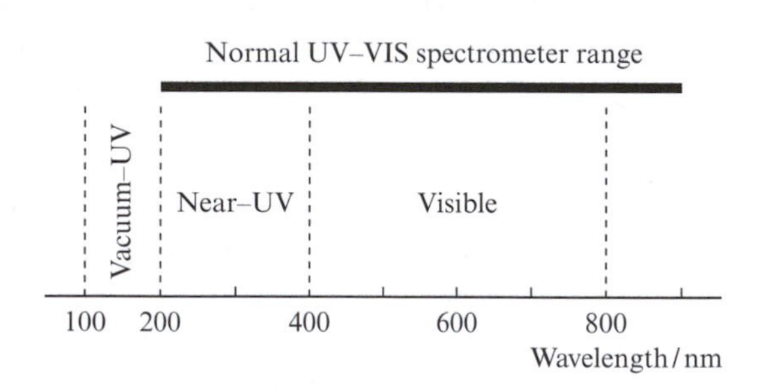

9.20 Regions within the UV–VIS range of the electromagnetic spectrum; the vacuum-UV is also called the far-UV.

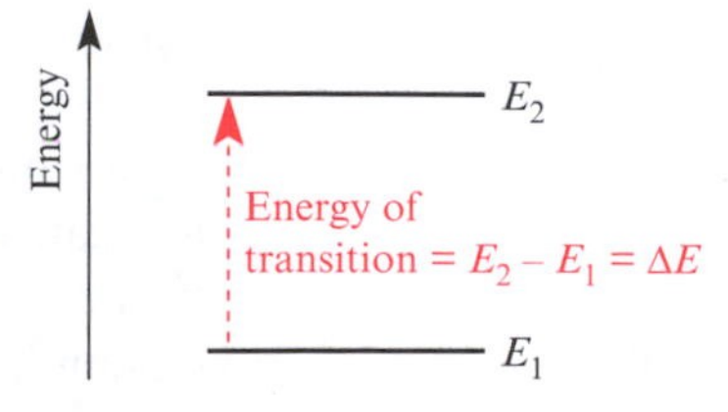

9.21 A transition between two electronic levels may occur when light of energy ΔE is absorbed by a molecule; the energy is quantized.

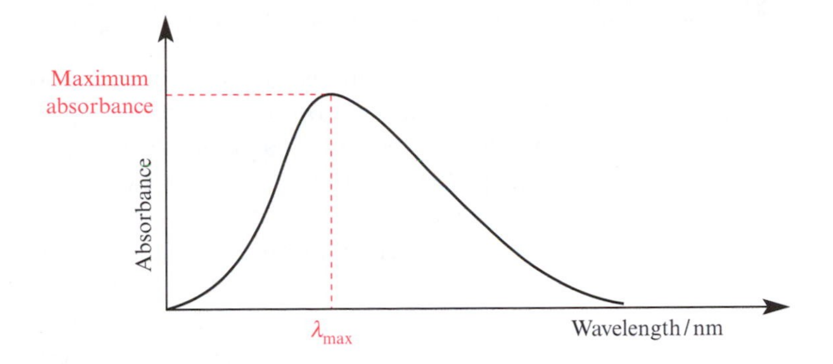

9.22 Absorptions in the electronic spectrum of a molecule or molecular ion are often broad, and cover a range of wavelengths. The absorption is characterized by values of λ_{max} and ε_{max}.

The answer is a matter of differing timescales – the absorption of a photon of light is a fast process ($\approx 10^{-18}$ s) and is considerably faster than any molecular vibrations or rotations. Accordingly, the electronic transition is a 'snapshot' of the molecule in a particular vibrational and rotational state at a particular moment in time. As the energies of the molecular orbitals are dependent on the molecular geometry, it follows that a range of ΔE values (corresponding to the different vibrational and rotational states) will be observed. As the vibrational and rotational energy levels are much more closely spaced than the electronic levels, a broad band is observed in the electronic spectrum. If the rotational and vibrational states are restricted, for example by cooling, sharper spectra are obtained. Similarly, if transitions are essentially localized on a single atomic centre, the spectrum will also be sharp – this is observed in the electronic spectra of lanthanoid compounds.

Although the absorption band covers a range of wavelengths, its position may be described by the wavelength corresponding to the absorption maximum – this is written as λ_{max} and is measured from the spectrum as indicated in Figure 9.22.

Absorptions in electronic spectra exhibit characteristic intensities and the molecular property describing this is the molar extinction coefficient (ε), determined using the Beer–Lambert Law. Equation 9.14 gives the relationship between ε_{max} and the maximum absorbance A_{max}; compare this with equation 9.4.

Beer–Lambert Law: see Section 9.4

$$\varepsilon_{max} = \frac{A_{max}}{c \times \ell} \qquad (9.14)$$

Values of ε_{max} range from close to zero (a very weak absorption) to > 10 000 $dm^3\ mol^{-1}\ cm^{-1}$ (an intense absorption). Absorptions in electronic spectra must be described in terms of both λ_{max} and ε_{max}.

Absorption bands in molecular electronic spectra are often broad, and are described in terms of both λ_{max} (nm) and ε_{max} ($dm^3\ mol^{-1}\ cm^{-1}$).

Electronic transitions in the vacuum-UV

Electronic transitions may originate from occupied bonding (σ or π) or non-bonding molecular orbitals. In general, electronic transitions from σ MOs to σ^* MOs correspond to light in the vacuum-UV region of the spectrum.§ For example, C_2H_6 absorbs light with a wavelength of 135 nm and this transition (written using the notation $\sigma \rightarrow \sigma^*$) is out of the range of the normal UV–VIS spectrophotometer. The energy difference between the π and π^* MOs in ethyne (HC≡CH) is smaller than the σ–σ^* energy gap in ethane, but still corresponds to light in the vacuum-UV region (λ_{max} = 173 nm, ε_{max} = 6000 $dm^3 \, mol^{-1} \, cm^{-1}$). Typically, for an isolated C=C or C≡C bond in a molecule, the $\pi \rightarrow \pi^*$ electronic transitions lie in the vacuum-UV region.

Non-bonding MO: see Section 4.5

A third type of molecular electronic transition is designated $n \rightarrow \sigma^*$, where n stands for non-bonding. Such transitions may occur in a saturated molecule in which an atom carries a lone pair of electrons. Examples include water, alcohols (ROH), amines (RNH_2) and halogenoalkanes (RF, RCl, RBr and RI). In many cases, $n \rightarrow \sigma^*$ electronic transitions occur in the vacuum-UV region, for example, in water (λ_{max} =167 nm, ε_{max} = 500 $dm^3 \, mol^{-1} \, cm^{-1}$) and methanol ($\lambda_{max}$ = 177 nm, ε_{max} = 200 $dm^3 \, mol^{-1} \, cm^{-1}$).

Choosing a solvent for UV–VIS spectroscopy

Solution samples are often used for UV–VIS spectroscopy, and the choice of a solvent is important. Ideally, we need a solvent that is *transparent* in the spectral region of interest. Figure 9.23 shows part of the UV–VIS spectra of four solvents. Acetonitrile is 'transparent' over the range from 200 to 800 nm – on moving to a lower wavelength, it begins to absorb at 200 nm, and this increase corresponds to the onset of a broad absorption (λ_{max} = 167 nm). A wavelength of 200 nm is therefore the lower limit (or cut-off point) for acetonitrile as a solvent in UV–VIS spectroscopy, and it is a good solvent for laboratory

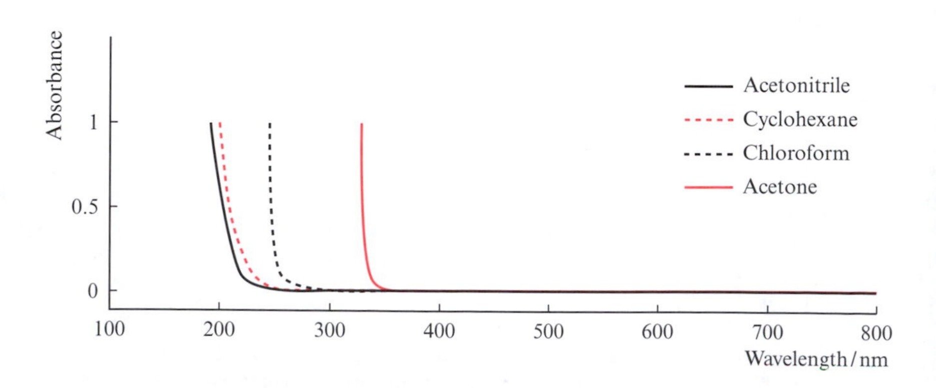

9.23 The electronic spectra (in part) of some common solvents; the path length for each is 1 cm. The increase in absorbance at lower wavelength for each solvent marks the onset of a broad absorption band; e.g. for acetonitrile λ_{max} = 167 nm, and for acetone λ_{max} = 279 nm.

§ The region below ≈200 nm is referred to as the vacuum-UV or far-UV range. It is necessary to evacuate the sample chamber to make measurements because O_2 absorbs in this region.

use. Similarly, water is transparent down to 210 nm (λ_{max} = 167 nm). The lower limit for cyclohexane is also 210 nm, and for chloroform, it is around 240 nm. On the other hand, acetone cuts off at 330 nm and this restricts its use as a solvent. However, this absorption for acetone is relatively weak (ε_{max} = 15 $dm^3 mol^{-1} cm^{-1}$) and it may still be used as a solvent for compounds which possess very much larger extinction coefficients.

9.10 ELECTRONIC SPECTROSCOPY 2: π-CONJUGATION

The dependence of the $\pi \rightarrow \pi^*$ transition on the number and arrangement of C=C bonds

Figure 9.24 illustrates the shift in λ_{max} for the $\pi \rightarrow \pi^*$ transition along the series ethene, buta-1,3-diene and hexa-1,3,5-triene. The corresponding values of ε_{max} are 15 000, 21 000 and 34 600 $dm^3 mol^{-1} cm^{-1}$. The significant relationship between these molecules is that the carbon chain increases, with an incremental addition of *alternating single and double carbon–carbon bonds*. The shift to longer wavelength with increased chain length is a general phenomenon and indicates that the energy difference between the π and π^* levels is decreasing. Why does this happen? To answer the question, we consider how carbon–carbon π-bonding is affected by chain length and the arrangement of single and double bonds, and we approach the problem by sequentially adding carbon atoms to a C_2-chain.

Localized and delocalized π–bonding

In ethene, the π-bond is formed by the in-phase interaction of two 2*p* AOs which lie perpendicular to the plane of the molecule, and the π^* MO results from the out-of-phase interaction of the two 2*p* AOs (Figure 9.25a). The carbon–carbon π-bond is *localized*.

We have already considered the bonding in the allyl anion using valence bond theory (structures **8.17** and **8.18**), but we can also represent it using an MO model. The anion is planar, and after the σ-bonding framework has

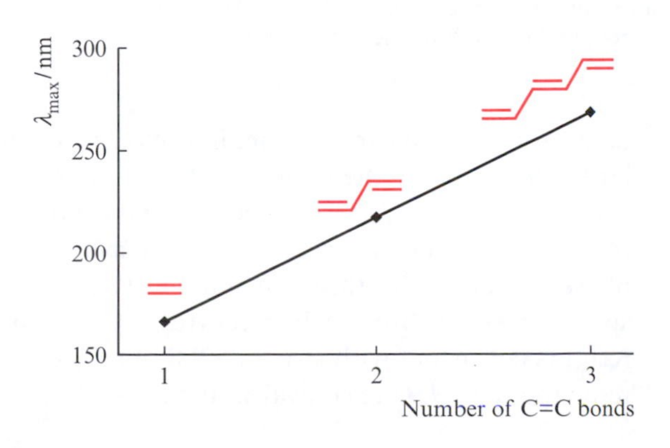

9.24 In the series of compounds ethene, buta-1,3-diene and hexa-1,3,5-triene, λ_{max} for the $\pi \rightarrow \pi^*$ transition shifts to longer wavelength. This means that if the three spectra were overlaid, the absorption would effectively 'move'.

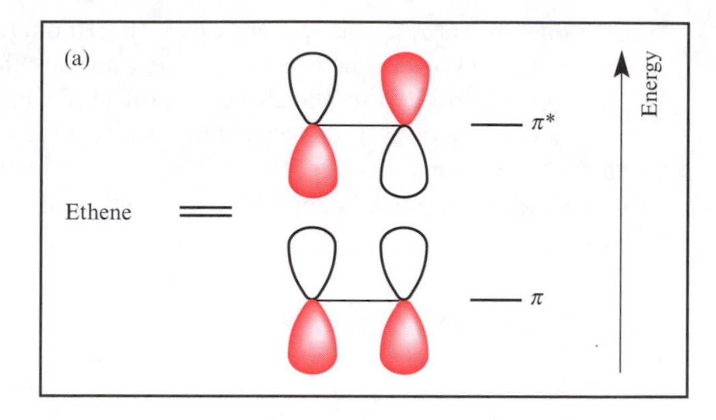

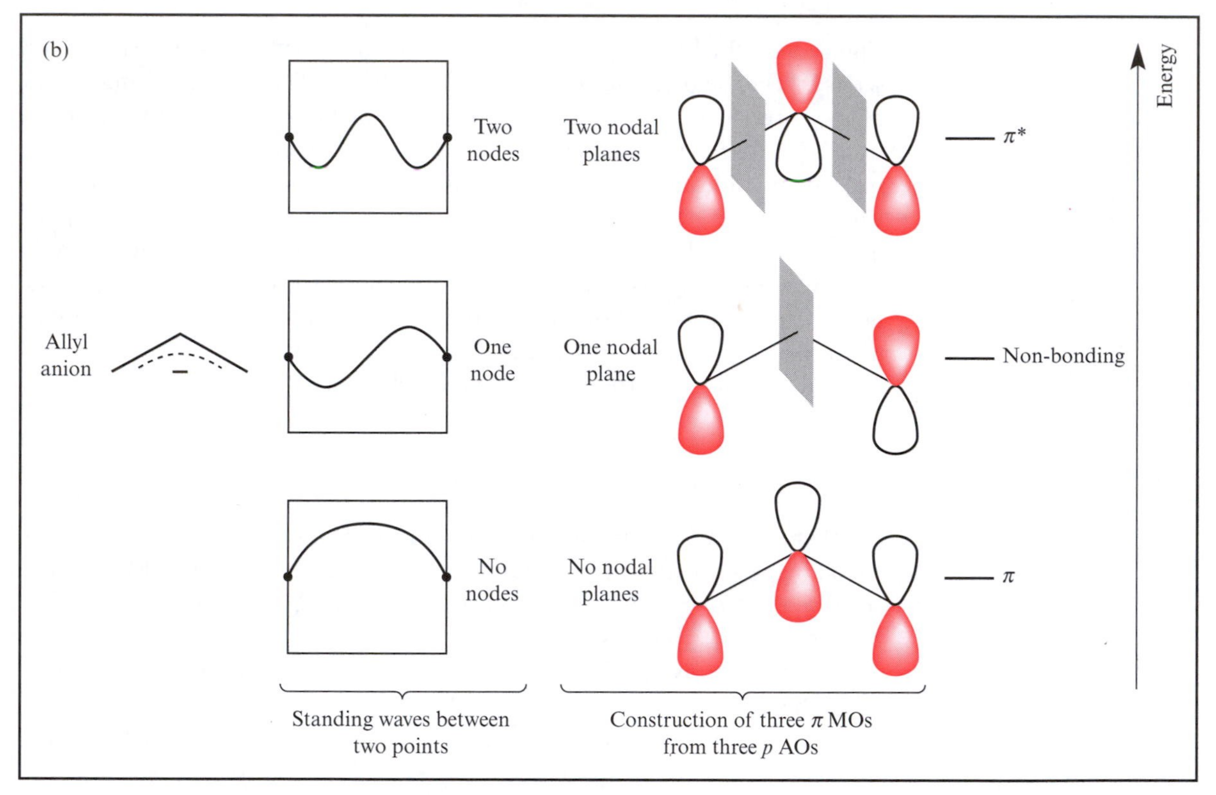

9.25 (a) The π-bonding and antibonding MOs in ethene; (b) the π MOs in the allyl anion can be understood in terms of a standing wave between two fixed points which has no, one or two nodes.

been formed, each carbon atom has one $2p$ AO remaining lying perpendicular to the plane of the molecule. The π-bonding in the allyl anion results from the overlap of these atomic orbitals and *since there are three AOs, we form three π MOs* (Figure 9.25b). The π-bonding MO arises from the in-phase overlap of all three carbon $2p$ AOs, and the π-bonding character is spread over all three carbon centres – the π-bond is *delocalized* and this result is consistent with structure **9.8** obtained using the VB approach. The resonance pair **9.9** are equivalent to **9.8**. The π^* orbital is readily constructed by taking an out-of-phase combination of the three 2*p* AOs and is antibonding between each pair of adjacent carbon atoms.

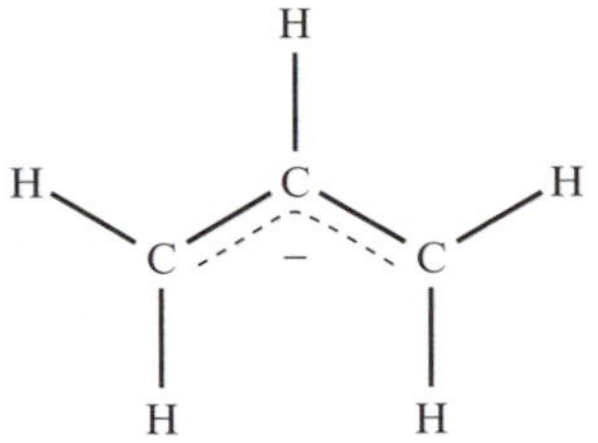

(9.8)

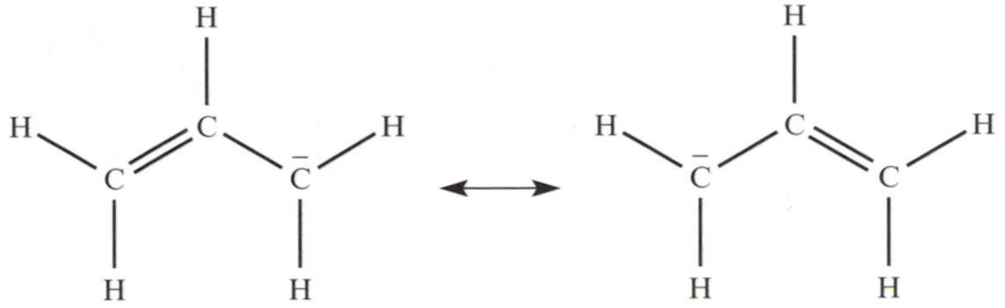

(9.9)

The origins of the third MO are more difficult to understand – Figure 9.25b shows that this MO has no contribution from the central carbon atom and the MO is labelled 'non-bonding'. One way in which we can understand this result is to take the two terminal carbon atoms in the allyl anion as fixed points and consider a standing wave between the two points. There may be no change of sign in the amplitude of the wave (no node) and this wave corresponds to the π-bonding MO in Figure 9.25b. The amplitude of the standing wave could change sign once or twice between the fixed points to give one or two nodes, and these situations correspond to the highest two π MOs shown in Figure 9.25b. The nodal plane in the second MO passes through the central carbon atom and means that there is *no contribution from this* 2*p* AO. This MO is therefore *non-bonding* with respect to the C–C–C framework. In the allyl anion the four π electrons fully occupy the two lowest energy MOs.

π-Conjugation in buta-1,3-diene

We now extend the carbon chain to four atoms and consider the π-bonding in buta-1,3-diene **9.10**.

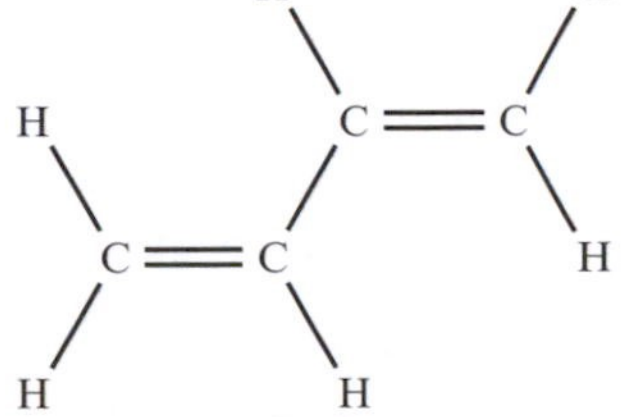

(9.10)

After the formation of the σ-framework in buta-1,3-diene, one 2*p* AO per C atom remains and there must be four π-MOs (Figure 9.26). As in the allyl anion, there is a pattern in the number of nodal planes. There are no nodal planes in the lowest-lying MO and the π-character is delocalized over all four carbon centres. The next MO has one nodal plane and the orbital possesses π-character localized as shown in structure **9.10**. The highest-lying MOs

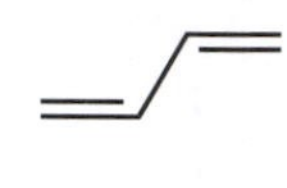

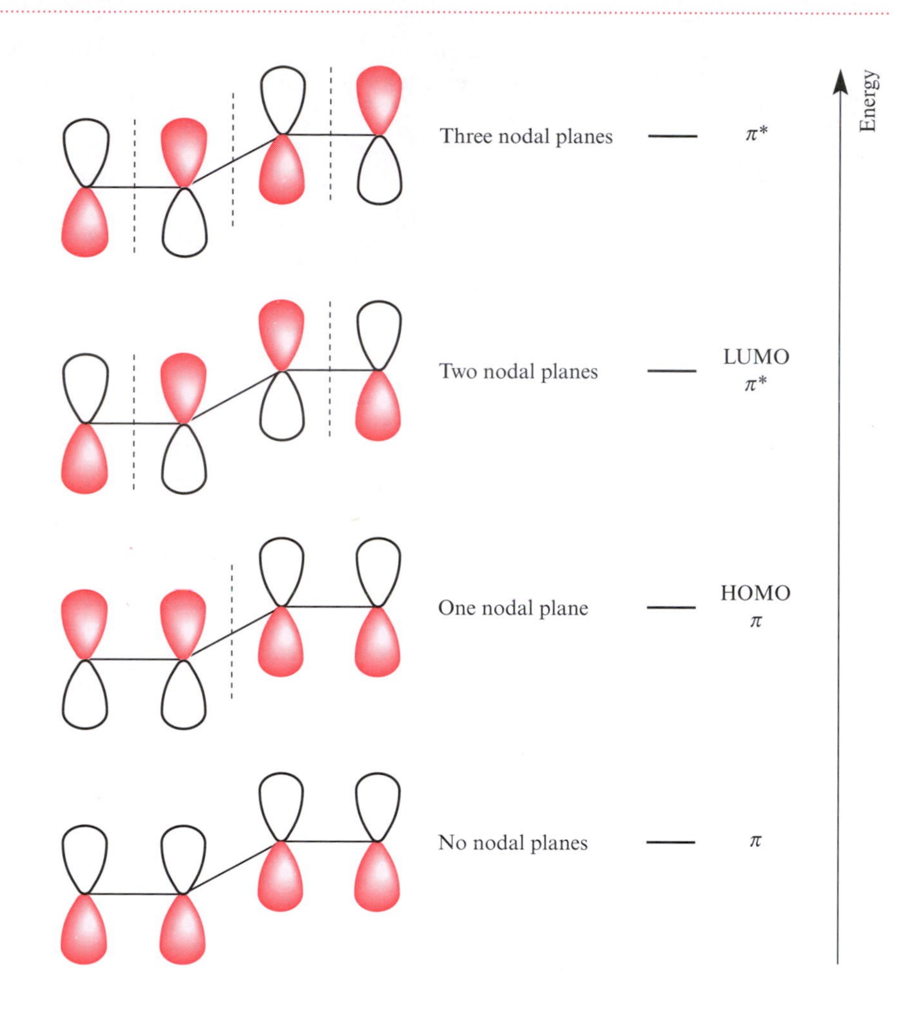

9.26 The π-MOs of buta-1,3-diene.

A polyene contains more than one C=C bond and if the carbon chain contains alternating C–C and C=C bonds, the molecule is conjugated.

have two and three nodal planes respectively. We may label the lower two MOs as π-bonding (these are fully occupied), and the higher two as π^* MOs (which are unoccupied). An important result of this bonding analysis is that, although the name 'buta-1,3-diene' suggests that the π-bonds are localized in particular positions, the π-character is actually spread out over the carbon chain. The molecule is said to be *conjugated*.

The effect of conjugation on the $\pi \rightarrow \pi^*$ transition

A consequence of *conjugation* is that the energy difference between the highest-lying π MO and the lowest-lying π^* MO is lowered, and the associated $\pi \rightarrow \pi^*$ transition shifts to longer wavelength. In buta-1,3-diene, the $\pi \rightarrow \pi^*$ transition corresponds to the HOMO–LUMO separation (Figure 9.26) and the transition is observed in the near-UV part of the spectrum (λ_{max} = 217 nm).

The –C=C–C=C– unit is known as a *chromophore* – it is the group of atoms which is responsible in buta-1,3-diene for the absorption of light in

Table 9.7 Observed values of λ_{max} for some conjugated polyenes. Compare these values with those predicted (see text) using buta-1,3-diene as your starting point.

Compound	λ_{max} / nm	ε_{max} / dm^3 mol^{-1} cm^{-1}
	217	21 000
	227	23 000
	263	30 000
	352	147 000

the UV–VIS spectrum. The addition of another C=C bond increases the π-conjugation and causes a further shift in the $\pi \rightarrow \pi^*$ transition to a longer wavelength (Figure 9.24). This is called a *red shift* (because the shift in absorption is towards the red end of the spectrum, see Table 9.2) or *bathochromic effect*. The approximate shift in wavelength for each additional C=C bond added to the alternating –C=C–C=C– chain is 30 nm. The absorption also becomes more intense (larger ε_{max}).

The addition of an alkyl substituent to a carbon atom in the chromophore also leads to an increase in conjugation, but as this is due to an interaction between the σ-electrons in the alkyl group and the π-electrons in the alkene chain, the effect is only small. The result is a red shift of about 5 nm per alkyl substituent. The data in Table 9.7 illustrate the effects of conjugation in some polyenes.

In a polyene, the π-conjugation and, therefore, the chromophore can be extended by adding another C=C bond or adding an alkyl substituent.

The addition of substituents other than alkyl groups may also affect the position of the absorption maximum in the electronic spectrum. For example, an OR substituent (R = alkyl) causes a red shift of about 6 nm, an SR group, a red shift of around 30 nm, and an NR_2 group, a red shift of ≈ 60 nm. Each of these groups can donate electron density from a lone pair into the carbon π-system, thereby increasing the conjugation in the molecule and extending the chromophore. This is illustrated in resonance structures **9.11**.

$Me_2\ddot{N}$–CH=CH–CH=CH$_2$ ⟷ $Me_2\overset{+}{N}$=CH–CH=CH–$\overset{-}{C}H_2$

(9.11)

Conjugation also occurs in alkynes with alternating C–C and C≡C bonds. As the length of the chromophore increases, the absorption in the electronic spectrum undergoes a red shift and becomes more intense.

Conjugation involving π-electrons is not restricted to carbon–carbon multiple bonds, and other compounds that exhibit absorption spectra in the

Box 9.3 Polyenes as natural colouring agents

Carotenoids are a group of polyenes that are natural colouring agents providing yellow, orange and red hues to a variety of plants and to some animal tissues. Some examples are shown below – notice the structural units that are common to these molecules. Carotenoids fall into two groups: the carotenes and the xanthophylls. Carotenes are hydrocarbons whilst xanthophylls are derivatives of carotenes with hydroxy or other oxygen-containing functionalities.

The red colour in tomatoes is due largely to the presence of lycopene (λ_{max} = 469 nm).

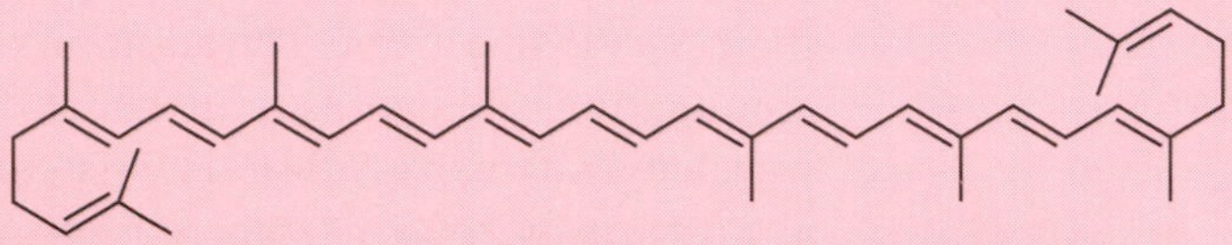

Lycopene

β-Carotene gives rise to the orange colour in carrots and mangoes (λ_{max} = 452 nm); zeaxanthin is also present in mangoes and contributes to the yellow colour of egg yolks.

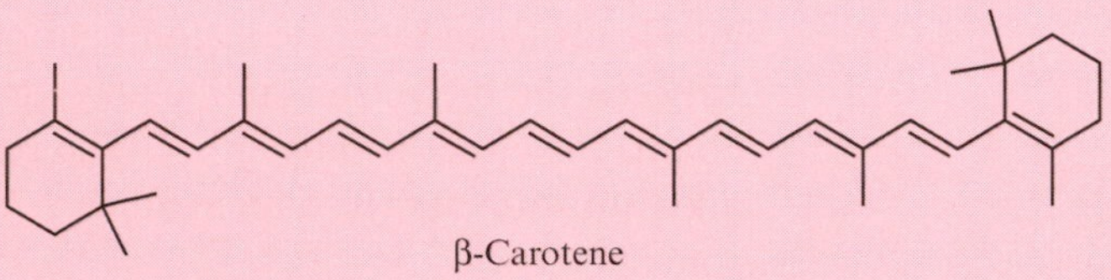

β-Carotene

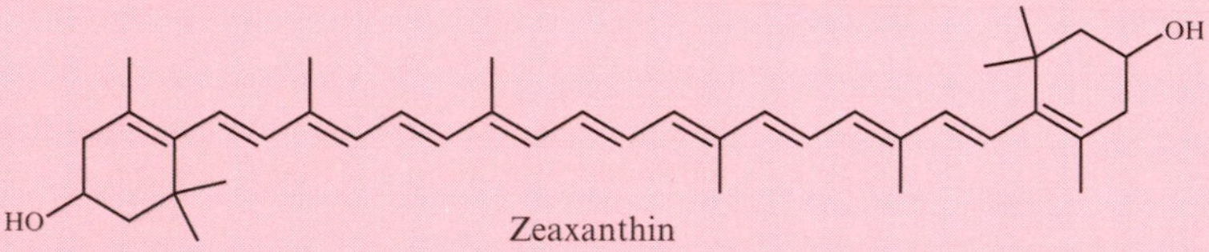

OH

HO

Zeaxanthin

α -Carotene and violaxanthin are present in oranges.

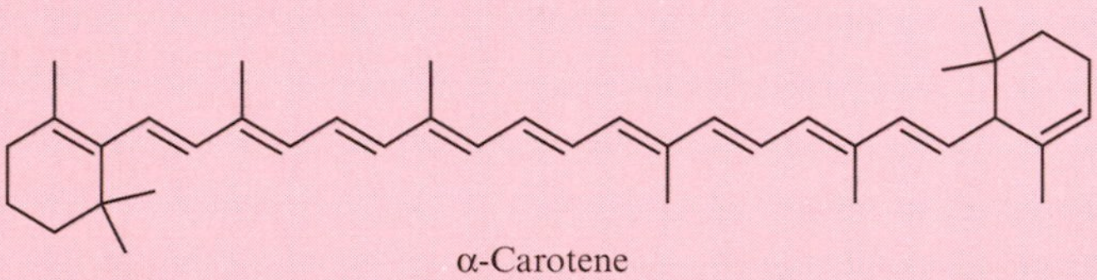

α-Carotene

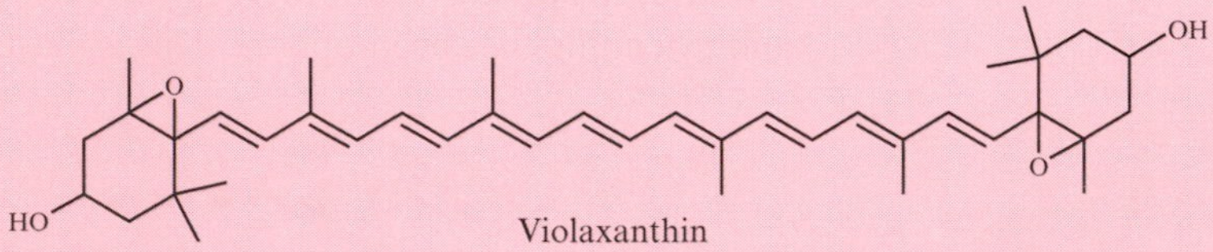

O

OH

HO

O

Violaxanthin

The pink pigments of salmon and lobsters are due to astaxanthin.

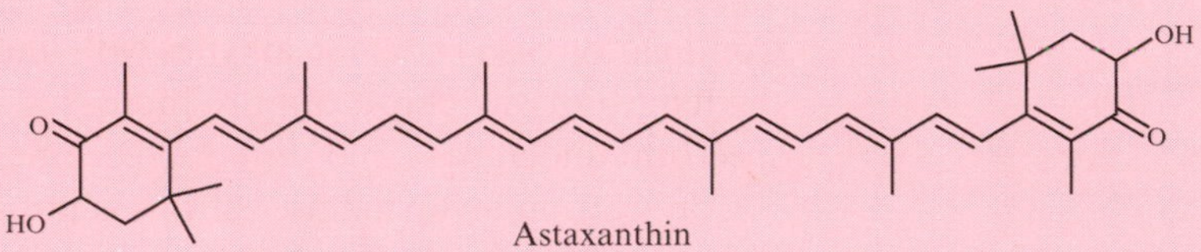

OH

O

O

HO

Astaxanthin

α,β-unsaturated ketones and aldehydes: see Section 17.13

near-UV region include α,β-unsaturated ketones and aldehydes. Structure **9.12** shows the group of atoms that must be present; the carbon atoms of the C=C group are labelled α and β as indicated.

β α O

(9.12)

The π-electrons in an α,β-unsaturated ketone or aldehyde are delocalized in the same way as in an alternating C=C /C–C chain, and this leads to a π→π* transition characterized by an intense absorption around 220 nm. The presence of substituents on the α and β carbon atoms may cause significant shifts in the absorption maximum; e.g. λ_{max} for the π→π* transition in **9.13** is 219 nm (ε_{max} = 3600 $dm^3\ mol^{-1}\ cm^{-1}$) but in **9.14**, it is 235 nm (ε_{max} = 14 000 $dm^3\ mol^{-1}\ cm^{-1}$).

O O

(9.13) **(9.14)**

In addition to the band assigned to the π→π* transition, the electronic spectrum of an α,β-unsaturated ketone or aldehyde contains a less intense absorption which is due to an electronic transition involving an oxygen lone pair of electrons: a π→π* transition. Aldehydes, ketones, acid chlorides, carboxylic acids, esters and azo compounds are amongst those for which π→π* transitions can also be observed, and some typical spectroscopic data are listed in Table 9.8.

A chromophore is the group of atoms in a molecule responsible for the absorption of electromagnetic radiation.

Table 9.8 Selected electronic spectral data for compounds exhibiting $n\rightarrow\pi^*$ transitions.

Compound type (R = alkyl)	λ_{max} / nm	ε_{max} / $dm^3\ mol^{-1}\ cm^{-1}$
$R_2C=O$ (ketone)	270–290	10–20
RHC=O (aldehyde)	290	15
RCOCl (acid chloride)	280	10–15
RC(O)OR' (ester) or RCO_2H (carboxylic acid)	≤200–210	40–100
RN=NR (azo compound)	350–370	10–15

Box 9.4 Spectroscopy and air-quality monitoring

Monitoring air quality is an important method of keeping a watch on atmospheric pollution, and in the UK, a series of automated monitoring stations is now in operation. Spectroscopic methods of analysis are used to monitor levels of CO, O_3, NO_2 and SO_2.

Carbon monoxide has an IR spectroscopic absorption at 2174 cm^{-1}, while ozone can be detected by using UV spectroscopy (λ_{max} = 254 nm). Monitoring the levels of NO_2 is less straightforward: NO_2 is first thermally decomposed to NO and then the *chemiluminescent* reaction between NO and O_3 is followed, light from which can be detected and used to quantify the concentration of NO. Fluorescence spectroscopy is used to measure amounts of SO_2 in the atmosphere.

Other pollutants are monitored by a range of non-spectroscopic methods including gas chromatography (see Box 8.3).

Dust particles also contribute to atmospheric pollution and samples taken in Leeds, UK, during 1982–83 illustrate some of the problem particles. (Data from *Chemistry in Britain*, (1994) vol. 30, p. 987; (1995) vol. 31, p. 131.)

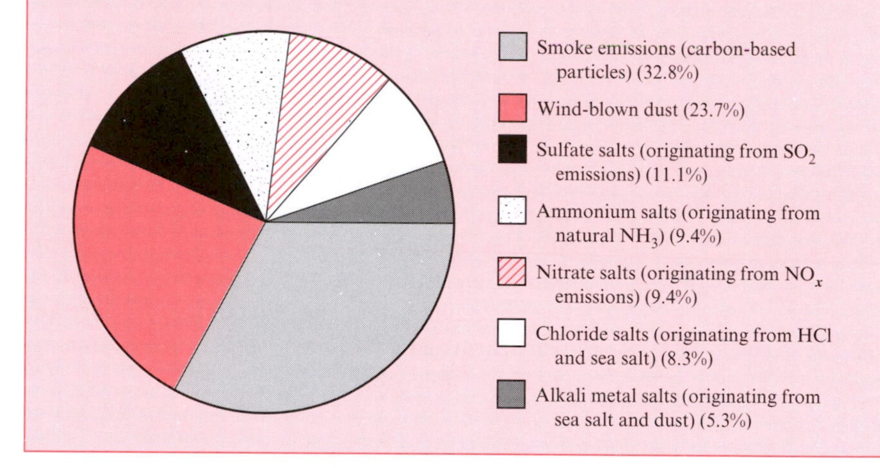

9.11 ELECTRONIC SPECTROSCOPY 3: THE VISIBLE REGION OF THE SPECTRUM

The –N=N– chromophore

The electronic spectra of many organic compounds have absorptions *only* in the near-UV region, and the compounds appear colourless. However, Table 9.8 indicates that azo compounds absorb light of a wavelength that is approaching the ultraviolet–visible boundary. Just as the absorption maximum for polyenes can be shifted by introducing substituents to the carbon chain, so λ_{max} for the –N=N– chromophore can be shifted into the visible region. This has important consequences – most azo compounds are coloured and many are used commercially as dyes. Three examples are shown in Figure 9.27, and the UV–VIS spectrum of methyl orange is shown in Figure 9.28.

The azo chromophore undergoes both $n \rightarrow \pi^*$ and $\pi \rightarrow \pi^*$ transitions but it is the $n \rightarrow \pi^*$ transition that falls close to or in the visible region.

9.27 Examples of azo-dyes, all of which contain the –N=N– chromophore. Para dye and Congo red are used as biological stains, and methyl orange is an acid-base indicator. Congo red is also used to estimate free mineral acids.

Para dye

λ_{max} = 488 nm

Methyl orange

λ_{max} = 464 nm

Congo red

λ_{max} = 497 nm

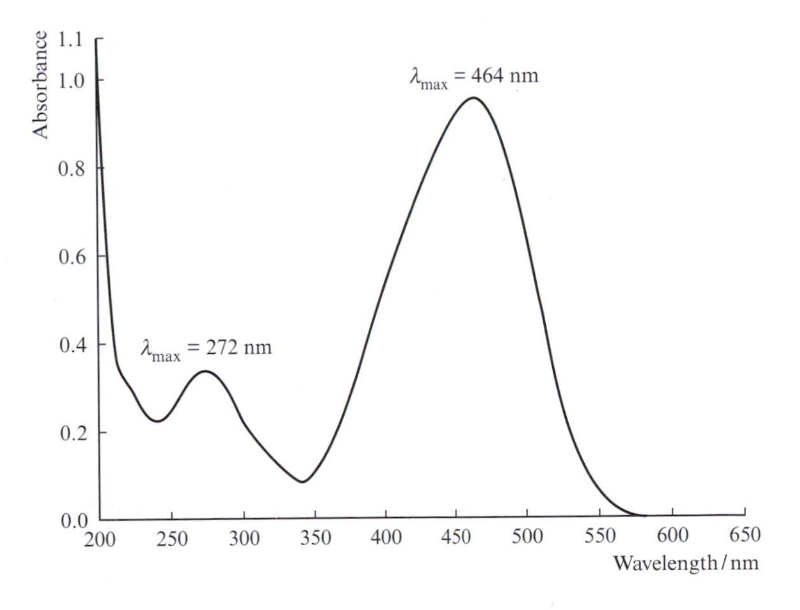

9.28 The UV–VIS spectrum of methyl orange dissolved in water.

Ions of the *d*-block metals

d-Block metal ions: see Chapter 16

Compounds of the *d*-block metals are often coloured – in many cases the colours are pale but characteristic. An aqueous solution of a titanium(III) salt contains the *aqua ion* $[Ti(H_2O)_6]^{3+}$ ion **9.15**. The electronic spectrum of aqueous titanium(III) chloride shows a band with λ_{max} = 510 nm due to

Table 9.9 Colours of some hydrated ions of the *d*-block elements; see also Chapter 16.

Hydrated metal ion	*Observed colour*
$[Ti(H_2O)_6]^{3+}$	Violet
$[V(H_2O)_6]^{3+}$	Green
$[Cr(H_2O)_6]^{3+}$	Purple
$[Cr(H_2O)_6]^{2+}$	Blue
$[Fe(H_2O)_6]^{2+}$	Very pale green
$[Co(H_2O)_6]^{2+}$	Pink
$[Ni(H_2O)_6]^{2+}$	Green
$[Cu(H_2O)_6]^{2+}$	Blue

absorption by the octahedral $[Ti(H_2O)_6]^{3+}$ cation, and the observed colour of aqueous titanium(III) ions is purple. Other metal ions also form octahedral $[M(H_2O)_6]^{n+}$ ions with characteristic colours (Table 9.9).

(9.15)

If the water molecules are replaced by other electron-donating groups, the new species may absorb light of a different wavelength. For example, the addition of ammonia to a solution containing pale blue $[Cu(H_2O)_6]^{2+}$ ions results in the formation of a very dark blue solution.

The electronic transitions that occur are known as '*d–d*' transitions and we discuss them in greater detail in Section 16.10. A key point is that when an ion of a *d-block metal possesses partially filled d orbitals*, electronic transitions between *d* orbitals may occur. A zinc(II) ion has a filled 3*d* shell, and no electronic transitions are possible – solutions of zinc(II) compounds are colourless.

Determination of the stoichiometry of a complex ion by colorimetry: Job's method

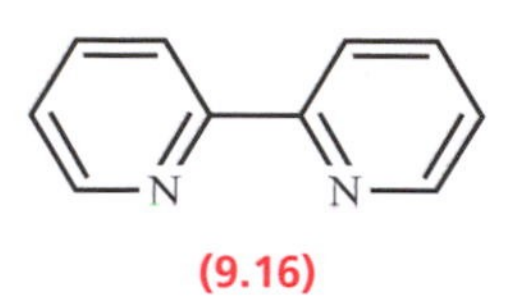

(9.16)

Colorimetry can be used to determine the stoichiometry of a reaction leading to the formation of a coloured species by using *Job's method*. Consider the reaction between iron(II) ions and 2,2'-bipyridine **9.16** which gives a red compound. The aim of the experiment is to determine the ratio of moles of Fe(II) : 2,2'-bipyridine in the coloured product.

Firstly, λ_{max} for the complex is found from its absorption spectrum, and the colorimeter is tuned to this wavelength. A series of solution cells (of constant path length) is prepared with solutions containing different ratios of

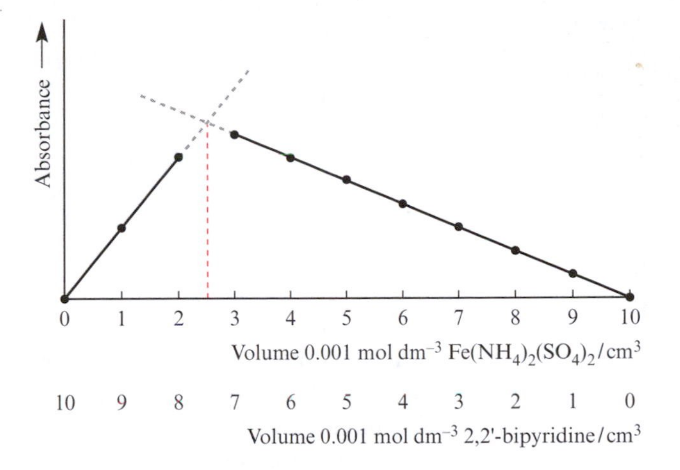

9.29 The use of Job's method to determine the stoichiometry of the reaction between iron(II) ions and 2,2'-bipyridine; the iron(II) ammonium sulfate is the source of Fe^{2+}.

moles of iron(II) to 2,2'-bipyridine, and with a constant total volume as shown on the horizontal axis of the graph in Figure 9.29. A reading of the absorbance for each solution is taken and a graph of absorbance against solution composition is constructed. The maximum absorbance corresponds to the highest concentration of the coloured product in solution. In this case, the maximum absorbance is found by extrapolation and corresponds to an iron(II)-to-2,2'-bipyridine ratio of 1 : 3.

Structure of the 1 : 3 complex: see Figure 16.7

9.12 NUCLEAR MAGNETIC RESONANCE SPECTROSCOPY 1: SOME THEORETICAL ASPECTS

Nuclear spin states

Many nuclei possess a property described as spin. This is a quantum effect which may be interpreted in a classical picture as the nucleus spinning about an axis. The nuclear spin (nuclear angular momentum) is quantized and is conveniently described by the spin quantum number I. The spin quantum number can have values $I = 0, \frac{1}{2}, 1, \frac{3}{2}, 2, \frac{5}{2}$ etc., but in this chapter we are only concerned with nuclei for which $I = \frac{1}{2}$; examples are ^{1}H, ^{13}C, ^{19}F and ^{31}P.

Nuclei possess positive charges, and (if we continue with our classical picture) a spinning charge generates an associated magnetic field, which can interact with an external magnetic field. In the classical picture, a magnetic spin quantum number m describes the *direction* of spin and m has values $+I$, $+(I-1)\ldots-I$; for nuclei with $I = \frac{1}{2}$, $m = +\frac{1}{2}$ or $-\frac{1}{2}$. There is no difference in energy between nuclei spinning in different directions. However, the direction of spin affects the polarity of the associated magnetic field. If an external magnetic field is applied, this could be in the same or opposite direction to that due to the spinning nucleus. The external magnetic field results in an energy difference between the $m = +\frac{1}{2}$ and $-\frac{1}{2}$ spin states (Figure 9.30). If splitting of the spin states into different energies occurs upon the application of a magnetic field, it should be possible to use a spectroscopic method to determine the energy difference.

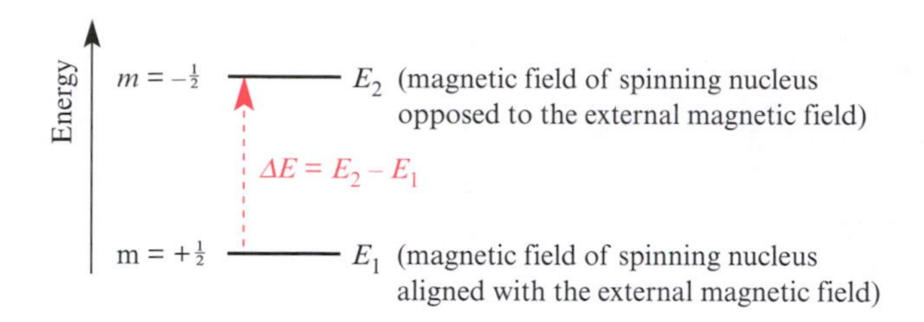

9.30 The splitting of nuclear spin states on the application of an external magnetic field.

The experiment we have just described leads to *nuclear magnetic resonance* (NMR). The energy difference between the $m = +\frac{1}{2}$ and $-\frac{1}{2}$ spin states depends upon the strength of the applied field but is very small at all achievable fields. Typically, the energy difference is less than 0.01 kJ mol^{-1} and electromagnetic radiation is absorbed in the radiofrequency (RF) region. Absorption of an appropriate RF energy results in a nucleus being excited from the lower to higher energy spin state, i.e. from the state $m = +\frac{1}{2}$ to $m = -\frac{1}{2}$. *Relaxation* from the upper to lower spin state then occurs.

RF = radiofrequency

Only certain nuclei are NMR active, and these include ^{1}H, ^{13}C, ^{19}F and ^{31}P.

Recording an NMR spectrum

An NMR spectrum is a plot of absorbance against frequency. The simplest way to record such a spectrum is to place a sample in a magnetic field and scan through the radiofrequencies until an absorption is observed. Although this method is widely used, it has an inherent disadvantage. The energy difference between the $m = +\frac{1}{2}$ and $-\frac{1}{2}$ spin states is very small, and if we consider a Boltzmann distribution we find that the difference in population between the upper and lower energy levels is about one nucleus in every million! This means that NMR spectroscopy is an inherently insensitive technique. Experimentally, we can assess the sensitivity in terms of the *signal-to-noise ratio* of the recorded spectrum.

Boltzmann distribution: see Section 1.18

One way of increasing the signal-to-noise ratio is to record the same spectrum a number of times, and then add these spectra together. For n spectra, the signal increases according to n and the noise by $n^{\frac{1}{2}}$; the improvement in the signal-to-noise ratio is therefore $n^{\frac{1}{2}}$. However, scanning through a range of frequencies over and over again is time-consuming.

We noted above that the energy difference between the lower and upper energy levels was very small. In the same way that the equilibrium populations of these levels is determined by the energy gap, so is the time taken to return to this state after excitation, i.e. *the time for the relaxation process to occur*. The *smaller* the energy gap, the *longer* it takes to return to the equilibrium population. In the case of NMR spectroscopy, the relaxation time may vary from seconds to hours. This places another time constraint upon the addition of spectra, because we must wait for the equilibrium population to be reached between the accumulation of consecutive spectra.

Box 9.5 Signal-to-noise ratios

If a spectrum is 'well resolved', the signals are clearly observable as is any fine structure in them. Another phrase that is applied to spectra is the *signal-to-noise ratio*. This again indicates how well resolved the spectrum is. A poor signal-to-noise ratio means that you may have difficulty deciding what is a signal and what is background noise. Such a spectrum is shown in spectrum (a) below. The reasons for poorly resolved spectra of this type are many, but two common problems are that too little a sample is being used or too few data have been collected. In spectrum (b), the signal-to-noise ratio is high and there are no ambiguities in distinguishing the signal from the background noise.

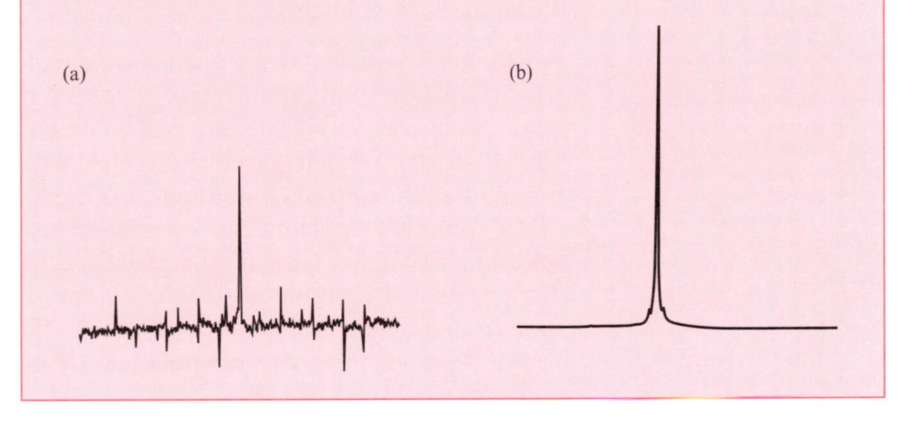

The modern method of recording NMR spectra makes use of some of these effects to provide a partial solution to these problems. We want a plot of intensity of absorption against frequency as an NMR spectrum, but a plot of *emission* against frequency is equivalent, and as the relaxation process is relatively slow, we could just as easily monitor the *emission* of RF after excitation. In itself this would give no advantage — in fact, it would be disadvantageous since the intensity of emission at any given instant will always be lower than any absorption intensity because the absorption is effectively instantaneous and the emission occurs over a period of time (Figure 9.31). However, a plot of *emission intensity against time* contains exactly the same information as a plot of *absorption intensity against frequency* and the mathematical operation of a Fourier transform (FT) interconverts the two sets of data.

FT = Fourier transform

Now comes the important point. The precise frequency of the absorption of a nucleus depends upon its chemical environment and if a compound

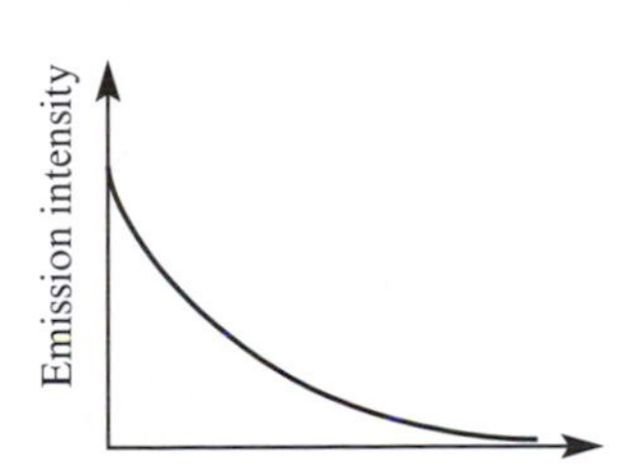

9.31 A plot of energy emission against time from an excited nuclear spin state; the curve is called a *free induction decay* (FID).

contains several different types of environment, the simple experiment described above involves scanning through the RF frequencies and finding all of the absorptions. By using an FT technique, we can use a mixture of frequencies to excite *all* the nuclei in the different environments at once, and then record the *emission* as a function of time. Such a plot is called a free induction decay (FID). A series of FIDs may be recorded and added together to improve the signal-to-noise ratio, and a Fourier transformation applied to give the conventional absorption versus frequency spectrum.

FID = free induction decay

Resonance frequencies

A particular nucleus (^{1}H, ^{13}C etc.) *resonates* at a characteristic frequency – that is, it absorbs radiofrequencies within a certain range. You can draw an analogy between an NMR spectrometer and a radio. Having tuned a radio to a particular frequency you will receive only the station selected because different radio stations broadcast at different frequencies. Similarly, an NMR spectrometer is tuned to a particular resonance frequency to detect a selected NMR active nucleus. The resonance frequencies of some nuclei with $I = \frac{1}{2}$ are listed in Table 9.10. When we record a ^{1}H NMR spectrum (a 'proton spectrum') the *signals* or *resonances* that are observed are due only to the ^{1}H nuclei in the sample. Similarly, a ^{31}P NMR spectrum contains resonances due only to ^{31}P nuclei.

Natural abundance of a nucleus

We have already seen that a nucleus is only NMR active if it possesses a nuclear spin, but another property that is important is the *natural abundance* of the nucleus (Table 9.10). Fluorine possesses one isotope and when a ^{19}F NMR spectrum is recorded, we observe all of the fluorine nuclei present.

Table 9.10 Resonance frequencies (referred to 100 MHz for ^{1}H)[a] for selected nuclei with $I = \frac{1}{2}$.

Nucleus	*Resonance frequency / MHz*	*Natural abundance / %*	*Chemical shift reference*
^{1}H	100	≈ 99.9	Me_4Si
^{19}F	94.0	100	$CFCl_3$
^{31}P	40.48	100	H_3PO_4 (85% aqueous)
^{119}Sn	37.29	8.6	Me_4Sn
^{13}C	25.00	1.1	Me_4Si
^{195}Pt	21.46	33.8	$Na_2[PtCl_6]$
^{29}Si	19.87	4.7	Me_4Si
^{107}Ag	4.05	51.8	Aqueous Ag^+
^{103}Rh	3.19	100	Rh (metal)

[a]NMR spectrometers can operate at different magnetic field strengths and it is convenient to describe them in terms of the RF needed to detect protons. A 100 MHz instrument records ^{1}H NMR spectra at 100 MHz, whilst a 250 MHz spectrometer records the ^{1}H NMR spectra at 250 MHz. A 100 MHz instrument operates with a magnetic field of 2.35 tesla and a 250 MHz at 5.875 tesla.

The situation is similar for hydrogen, because ≈99.9% of natural hydrogen consists of ^{1}H nuclei. When a ^{13}C NMR spectrum is recorded, only 1.1% of the carbon nuclei present can be observed, because the remaining 98.9% are ^{12}C and are not NMR active ($I = 0$).

Although a low natural abundance can be problematical, it does *not* mean that the nucleus is unsuitable for NMR spectroscopy. A greater number of FIDs may be collected in order to obtain an enhanced signal-to-noise ratio, but alternatively a sample can be isotopically enriched.

Chemical shifts and NMR spectroscopic references

The resonance frequency of a ^{1}H nucleus depends on its precise chemical environment, and similar effects are found with other NMR active nuclei. Why is this?

We explained earlier that the energy difference between the $m = +\frac{1}{2}$ and $m = -\frac{1}{2}$ nuclear spin states arose as a result of the interaction between the magnetic fields of the spinning nuclei and an applied external field. However, the *local field*, $\boldsymbol{B}$, experienced at a nucleus is not precisely the same as that applied. This is because all of the bonding electron pairs in the molecule are also charged bodies moving in space, and they generate small local magnetic fields. The field $\boldsymbol{B}$ is the sum of the applied field, $\boldsymbol{B}_0$, and all the smaller fields. The local fields from the electrons vary with the chemical environment of the nucleus. Typically, the differences in field are very small, and variations in absorption frequency of about one part per million are observed, and these differences give rise to different signals in the spectrum. The position of a signal is denoted by a *chemical shift value*, δ, a value that is given relative to the signal observed for a specified reference compound. The standard reference for ^{1}H and ^{13}C NMR spectroscopies is tetramethylsilane (TMS) **9.17**. The 12 hydrogen atoms in TMS are equivalent and there is one proton environment giving rise to one signal in the ^{1}H NMR spectrum of this compound. The chemical shift value of this signal is defined as zero (δ 0). Similarly, there is only one carbon atom environment in TMS and the signal in the ^{13}C NMR spectrum of TMS is also defined as zero.

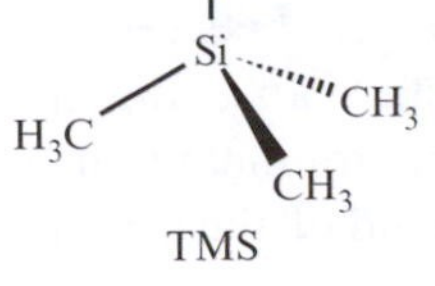

CH_3
Si
H_3C CH_3 CH_3

TMS

(9.17)

The signals in a particular NMR spectrum are always referenced with respect to a standard compound (Table 9.10), but care should be taken because different references are sometimes used. Although 85% aqueous phosphoric acid is a common reference compound in ^{31}P NMR spectroscopy, $P(OMe)_3$

(trimethylphosphite) is sometimes used, and it is important to quote the reference when reporting NMR spectra.

Signals due to particular nuclei in a compound that appear in the NMR spectrum are said to be *shifted with respect to the standard reference signal*. A shift to positive δ is called a 'downfield shift' and a shift to negative δ is an 'upfield shift'.

The chemical shift value δ of a signals in an NMR spectrum is quoted with respect to that of a reference defined at $\delta 0$. A positive chemical shift is *downfield* of $\delta = 0$ and a negative shift is *upfield* of $\delta = 0$.

What exactly is this δ scale? The problem with NMR spectroscopy is that the energy difference between the $m = +\frac{1}{2}$ and $m = -\frac{1}{2}$ spin states depends upon the external magnetic field. Similarly, the difference in frequency of absorptions between nuclei in different environments is also dependent on the applied field. How can we compare data recorded at different magnetic field strengths? The solution is to define a *field-independent parameter* – the chemical shift δ.

We define δ as follows. The frequency difference ($\Delta\nu$), in Hz, between the signal of interest and some defined reference frequency (ν_0) is divided by the absolute frequency of the reference signal. In order to obtain convenient numbers for δ, the ratio is multiplied by 10^6 (equation 9.15). The chemical shift refers to a difference of *parts per million*, and is sometimes reported in ppm rather than a δ value.

$$\delta = \frac{(\nu - \nu_0) \times 10^6}{\nu_0} = \frac{\Delta\nu \times 10^6}{\nu_0} \tag{9.15}$$

Chemical shift ranges

The *chemical shift range* (i.e. the range of chemical shifts over which signals appear in the spectrum) is dependent upon the nucleus. For example, ^1H NMR resonances usually fall in the range δ +15 to –35, whereas ^{13}C NMR spectra are usually recorded over a range of δ +250 to –50, and ^{31}P NMR spectra between the approximate limits δ +300 to –300. However, new chemistry can bring surprises in the form of a nucleus in an unprecedented environment and the spectral 'window' should not be a fixed agenda in the mind of the experimentalist.

Choosing a solvent

Although NMR spectra may be recorded in the solid state, most samples are studied as neat liquids or as solutions. Molecules of most solvents contain carbon and hydrogen atoms and so in the ^1H or ^{13}C NMR spectrum of a solution sample, the solvent naturally gives rise to its own spectrum which is superimposed on that of the sample. As the solvent is usually present in a vast excess with respect to the sample, the signals from the solvent are often the largest features observed in the spectrum.

A ^{1}H NMR spectrum is easily 'swamped' by the effect of the solvent and signals due to the sample may be obscured. The problem is readily solved by using *deuterated solvents* – the ^{1}H nuclei in the solvent molecules are replaced by ^{2}H (D) nuclei. For example, chloroform is a common solvent and can be isotopically labelled with deuterium to give $CDCl_3$. Although the ^{2}H nucleus is NMR active ($I = 1$), its resonance frequency is quite different from that of ^{1}H and therefore the deuterated solvent is effectively 'silent' in a ^{1}H NMR spectrum. Such solvents are commercially available and are usually labelled to an extent of > 99.5%. Residual molecules containing ^{1}H nuclei are present but give rise to relatively low-intensity signals in a ^{1}H NMR spectrum.

9.13 NUCLEAR MAGNETIC RESONANCE SPECTROSCOPY 2: MOLECULES WITH ONE ENVIRONMENT

Nuclear magnetic resonance spectroscopy can be used to 'count' the number of environments of a particular nucleus in a molecule – ^{1}H NMR spectroscopy looks only at proton environments and ^{13}C NMR spectroscopy can be used to count the different types of carbon atoms in a molecule. The number of signals in the spectrum corresponds to the number of different environments.

Molecules with one environment

In this section we illustrate that more information than simply 'one signal means one environment' can be obtained from an NMR spectrum. The chemical shift can give important information about the environment, for example, the possible geometry of the centre, or the functional groups that are attached to it.

Figure 9.32 shows the ^{31}P NMR spectra of PMe_3 and PCl_3. Each spectrum exhibits one signal, consistent with there being one phosphorus environment in each molecule. However, note how the nature of the

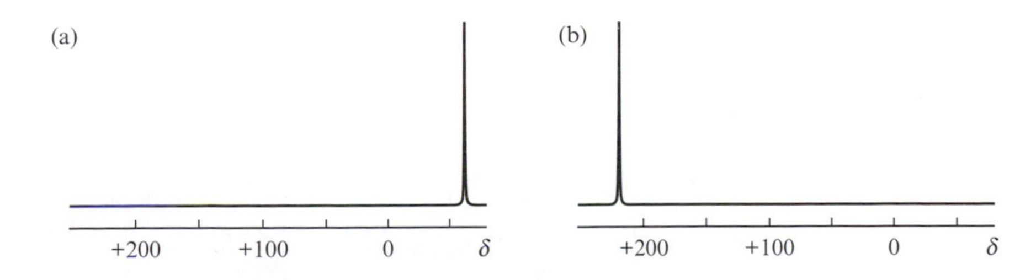

9.32 The ^{31}P NMR spectra of (a) PMe_3 and (b) PCl_3.

substituents affects the chemical shift of the signal – the difference in shift between these two trigonal pyramidal PR_3 molecules is particularly dramatic. For PF_3, PBr_3 and PI_3 (all trigonal pyramidal) the ^{31}P NMR chemical shifts are δ = +96, +228 and +177 respectively. Each of CH_4, CH_3Cl, CH_2Cl_2 and $CHCl_3$ is tetrahedral, but the presence of the chloro-substituents influences the chemical shift of the ^{13}C nucleus: CH_4 (δ +2), CH_3Cl (δ +22), CH_2Cl_2 (δ +54) and $CHCl_3$ (δ +77).

9.14 NUCLEAR MAGNETIC RESONANCE SPECTROSCOPY 3: MOLECULES WITH MORE THAN ONE ENVIRONMENT

Chemical shifts in ^{13}C NMR spectra

We can relate the chemical shift of a ^{13}C NMR signal to the hybridization of the carbon centre – sp^2 carbon atoms are downfield (more positive δ) with respect to *sp* carbon centres, which in turn are downfield of sp^3 carbon atoms. Table 9.11 gives *approximate* ranges for carbon nuclei in some environments in organic molecules. Compounds containing a carbonyl group (C=O) are characterized by a ^{13}C NMR resonance at lowfield. The variety of compounds containing sp^3 hybridized carbon atoms is large and includes cyclic and acyclic alkanes, alcohols (X = OH), halogenoalkanes (X = F, Cl, Br or I), amines (X = NH_2), thiols (X = SH) and ethers (X = OR). The particular differences between ^{13}C NMR chemical shifts for carbon nuclei in alkanes (sp^3 C), alkenes (sp^2 C) and alkynes (*sp* C) are illustrated in Figure 9.33.

^{13}C NMR spectra of carbonyl compounds: see Section 17.4

Signal intensities

The *relative intensities* of NMR signals can assist in their assignments because the ratio of intensities gives information about the *ratio* of the number of different environments in the molecule.

The ^{13}C NMR spectrum of acetone consists of two signals (Figure 9.34) which can readily be assigned from their relative intensities of 1 : 2. *A word*

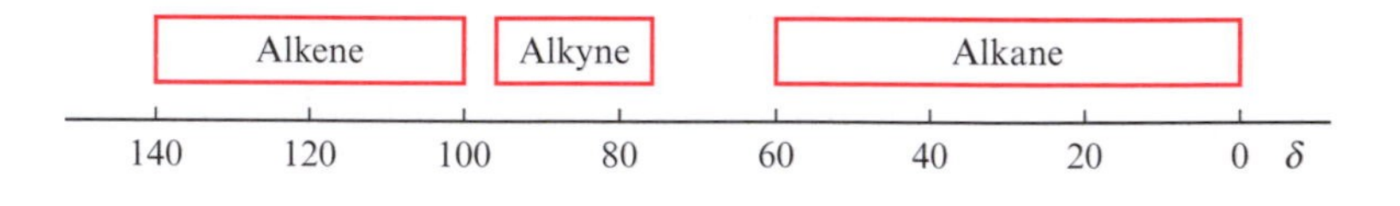

9.33 Typical ^{13}C NMR chemical shift ranges for alkanes, alkenes and alkynes.

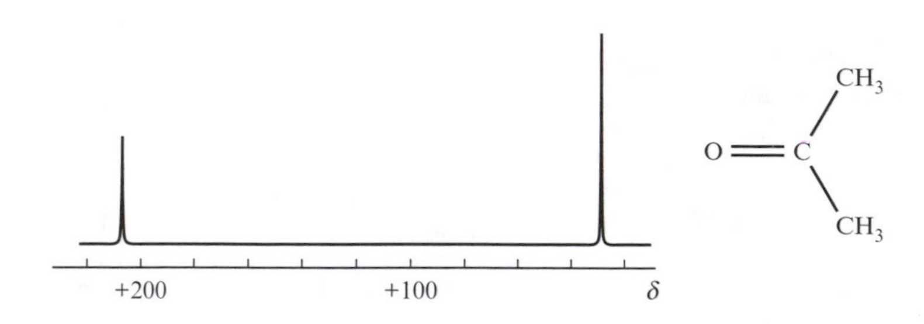

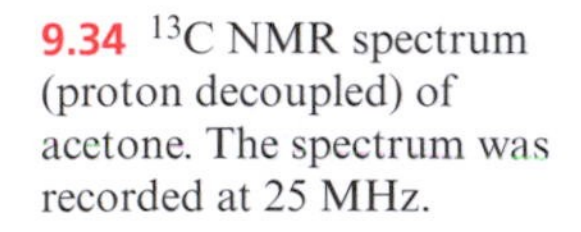

9.34 ^{13}C NMR spectrum (proton decoupled) of acetone. The spectrum was recorded at 25 MHz.

Table 9.11 Approximate shift ranges for carbon environments in ^{13}C NMR spectra (chemical shifts are with respect to TMS = 0; downfield is at positive δ).

Carbon environment		*Hybridization*	*^{13}C chemical shift range*	*Comments*
$R_2C{=}O$	Ketones	sp^2	+230 to +190	
$R(H)C{=}O$	Aldehydes	sp^2	+210 to +185	
$R(HO)C{=}O$	Carboxylic acids	sp^2	+185 to +165	
$R(X)C{=}O$	Amides X = NH_2 Esters X = OR	sp^2	+180 to +155	
$R_2C{=}CR_2$	Alkenes	sp^2	+160 to +100	
(benzene ring)	Aromatic compounds	sp^2	+145 to +110	Note some overlap of the *sp* and sp^2 regions.
$R{-}C{\equiv}N$	Nitriles	*sp*	+125 to +110	
$R{-}C{\equiv}C{-}R$	Alkynes	*sp*	+110 to +75	
$R_3C{-}X$	Aliphatic compounds	sp^3	Some overlap of regions depending upon X. Overall range +90 to 0 Generally: $CH_3X < RCH_2X < R_2CH < R_3CX$ ⟶ more positive δ	
$R_2HC{-}X$	Aliphatic compounds	sp^3		
$RH_2C{-}X$	Aliphatic compounds	sp^3		
$H_3C{-}X$	Aliphatic compounds	sp^3		

of caution however. Signal intensities in ^{13}C NMR spectra can be misleading due to different relaxation effects of the different ^{13}C nuclei. The problem arises from our use of the FT method. Because we collect emission data in the time domain, we should wait between acquiring consecutive data sets until the Boltzmann distribution is achieved for nuclei in *all* environments. If this is not done, we will observe *rapidly* relaxing nuclei more efficiently than those relaxing slowly. In practice, time constraints may mean that a compromise is made in the time that we wait between consecutive excitations of the nuclei and, as a result, the ratios of the signal intensities are not always exactly equal to the ratios of the environments of the nuclei. This is particularly true in ^{13}C NMR spectra.§ Fortunately, chemical shift values can come to our aid in assigning signals.

Relaxation: see Section 9.12

Assigning ^{13}C NMR spectra on the basis of chemical shifts

Refer to Table 9.11

Example 1: Acetonitrile

$\overset{a}{H_3C}-\overset{b}{C}\equiv N$

The ^{13}C NMR spectroscopic chemical shifts are δ +1.3 and +117.7 (Figure 9.35a).
There are two carbon environments:

a aliphatic (sp^3)
b nitrile (sp)

and the nitrile carbon should be downfield of the aliphatic carbon.
Assignment: δ +1.3 is due to carbon *a*
δ +117.7 is due to carbon *b*

Example 2: Ethanal

$\overset{a}{H_3C}-\overset{b}{C}(H)=O$

The observed ^{13}C NMR spectroscopic chemical shifts are δ +30.7 and +199.7 (Figure 9.35b).
There are two carbon environments:

a aliphatic (sp^3)
b aldehyde (sp^2)

and the aldehyde carbon should be downfield of the aliphatic carbon.
Assignment: δ +30.7 is due to carbon *a*
δ +199.7 is due to carbon *b*

§ For more detailed discussion, see: B. K. Hunter and J. K. M. Sanders, *Modern NMR Spectroscopy: A Guide for Chemists*, 2nd edn, Oxford University Press, Oxford (1993).

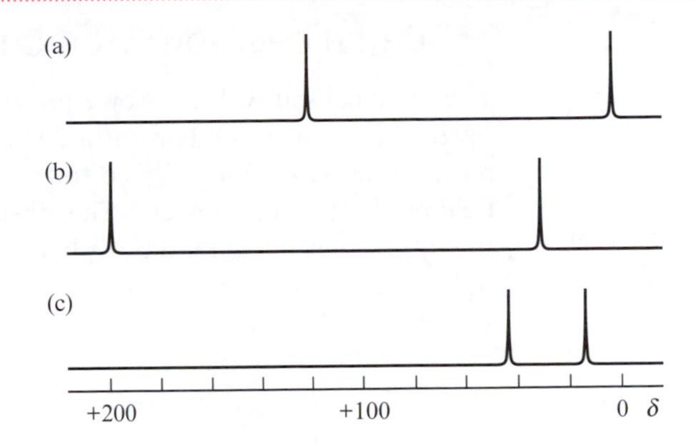

9.35 25 MHz ^{13}C NMR spectra (proton decoupled) of (a) CH_3CN, (b) CH_3CHO and (c) Et_2NH. The same scale applies to all the spectra.

Example 3: Diethylamine

H_3C(*a*)–CH_2(*b*)–N(H)–CH_2(*b*)–CH_3(*a*)

The ^{13}C NMR spectroscopic signals are at δ +15.4 and +44.1 (Figure 9.35c). There are two sp^3 carbon environments:

a terminal CH_3 group
b RCH_2X group

and in general RCH_2X carbon atoms are downfield of terminal CH_3 carbons.
Possible assignments: δ +15.4 is due to carbon *a*
δ +44.1 is due to carbon *b*

9.15 NUCLEAR MAGNETIC RESONANCE SPECTROSCOPY 4: ^{1}H NMR SPECTRA

Chemical environments in ^{1}H NMR spectra

Since the chemical shift range for ^{1}H nuclei is relatively small, there is a significant overlap of the ranges which are characteristic of different proton environments. In Table 9.12 we give some approximate values for ^{1}H NMR spectroscopic signals in selected groups of organic compounds.[§]

Table 9.12 indicates that OH and NH_2 protons may give rise to *broad* signals. This phenomenon arises because O–H and N–H protons undergo *exchange* processes on a timescale that is comparable with the NMR spectroscopic timescale; equation 9.16 illustrates the exchange of the hydrogen atom in ethanol with a proton from water. The rate of exchange can be slowed down by lowering the temperature and this can cause sharpening and shifting of the signals.

Timescales: see Section 9.3

[§] For greater detail, see: D. H. Williams and I. Fleming, *Spectroscopic Methods in Organic Chemistry*, 4th edn, McGraw-Hill, London (1989).

$$CH_3CH_2OH + HOH \rightleftharpoons CH_3CH_2OH + HOH \quad (9.16)$$

Hydrogen bonding: see Section 11.8

The chemical shift value of some protons is dependent on solvent and this is particularly true of O–H protons. The ^1H NMR spectrum of neat ethanol has a signal at δ +5.4 assigned to the OH proton, but this shifts to higher field on dilution in a solvent. This observation is due to hydrogen bonding; the intermolecular interactions between ethanol molecules **9.18** lead to a reduction in the electron density around the hydrogen atom and causes a shift in the ^1H NMR signal to lower field.

Table 9.12 Typical shift ranges for proton environments in ^1H NMR spectra of organic compounds (chemical shifts are with respect to TMS = 0; downfield is at positive δ).

Proton environment		*^1H chemical shift range*	*Comments*
R(HO)C=O	Carboxylic acids	+9 to +13	Often broad and dependent upon solvent
R(H_2N)C=O	Amides	+5 to +12	Broad and dependent upon solvent
R(H)C=O	Aldehydes	+8 to +10	Sharp (usually no confusion with carboxylic acid, amide or amine signals)
CHX_3 (X = Cl or Br)		≈ +7	$CHCl_3$ δ +7.25
C_6H_6	Aromatic compounds	+6 to +10	C_6H_6 δ +7.2
$R_2C=C(H)R$	Alkenes	+4 to +8	
R—C≡C—H	Terminal alkynes	+2.5 to +3	
RNH_2	Amines	+1 to +6	Broad and dependent upon solvent
ROH	Alcohols	+0.5 to +8	Often broad and dependent upon solvent
R_3C—H	Methine	+1.5 to +4.5	
R_2CH_2	Methylene	+1.5 to +4.5	
R—CH_3	Methyl	0 to +4	TMS defined as δ = 0

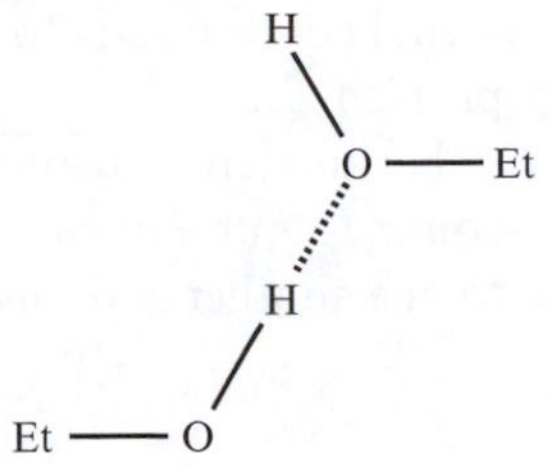

(9.18)

Similarly, hydrogen bonding between carboxylic acid molecules is one of the reasons why these protons are characterized by particularly downfield signals. The ^{1}H NMR spectrum of acetic acid (Figure 9.36) exhibits a broad signal centred at δ +11.4 (due to OH) and a sharp, more intense signal at δ +2.1 (due to CH_3). Acetic acid can form a hydrogen-bonded dimer **9.19** as we discuss further in Section 17.4.

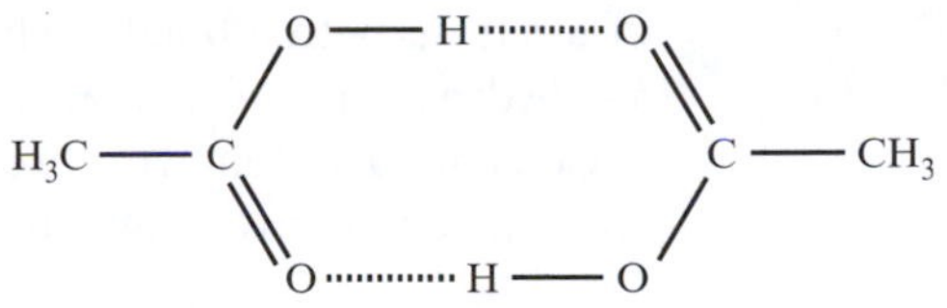

(9.19)

Overlapping of the characteristic regions in ^{1}H NMR spectra may make spectroscopic assignment difficult. In propyne **9.20**, the signals for the C_{alkyne}–H and the methyl group are coincident (δ +1.8).

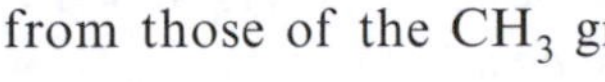

$H-C\equiv C-CH_3$

(9.20)

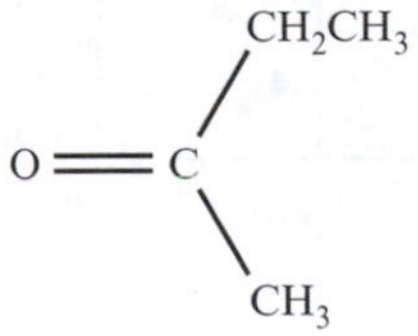

(9.21)

The ^{1}H NMR spectrum of butan-2-one **9.21** has three signals at δ +1.1, +2.1 and +2.5. Although the relative intensities allow us to distinguish the CH_2 resonance from those of the CH_3 groups, how can we determine which

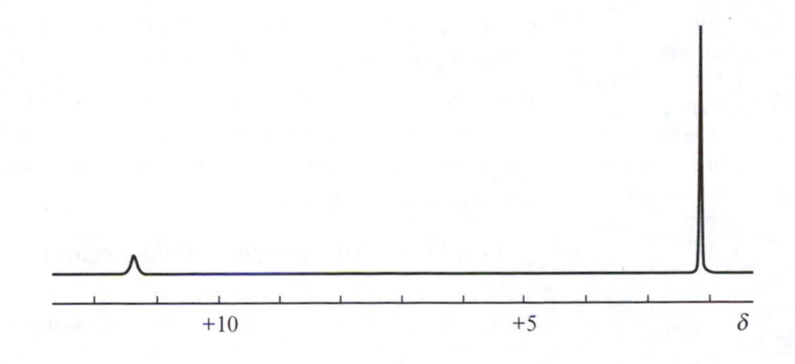

9.36 100 MHz 1H NMR spectrum of acetic acid, CH_3CO_2H.

signal corresponds to which methyl group? (After reading Section 9.16, see problem 9.24.)

Fortunately, a feature of NMR spectroscopy that we have so far ignored now comes to our aid: *nuclear spins which are magnetically inequivalent may couple* to one another and this results in characteristic *splittings* of the NMR signals.

9.16 NUCLEAR MAGNETIC RESONANCE SPECTROSCOPY 5: NUCLEAR SPIN–SPIN COUPLING BETWEEN NUCLEI WITH $I = \frac{1}{2}$

Coupling between two inequivalent ^{1}H nuclei

A hydrogen nucleus may be in one of two spin states ($m = +\frac{1}{2}$, $m = -\frac{1}{2}$) and we stated earlier that the energy difference between these spin states was controlled by the applied magnetic field. Later, we showed that smaller energy differences (giving rise to different chemical shifts) resulted from movement of electrons within bonds. What happens when two *magnetically non-equivalent* hydrogen nuclei in a molecule are close to each other?

Consider a system with two hydrogen atoms H_A and H_B which are in different magnetic environments. Let us introduce a shorthand notation:

H_A with $m = +\frac{1}{2}$ is labelled α(A)
H_A with $m = -\frac{1}{2}$ is labelled β(A)
H_B with $m = +\frac{1}{2}$ is labelled α(B)
H_B with $m = -\frac{1}{2}$ is labelled β(B)

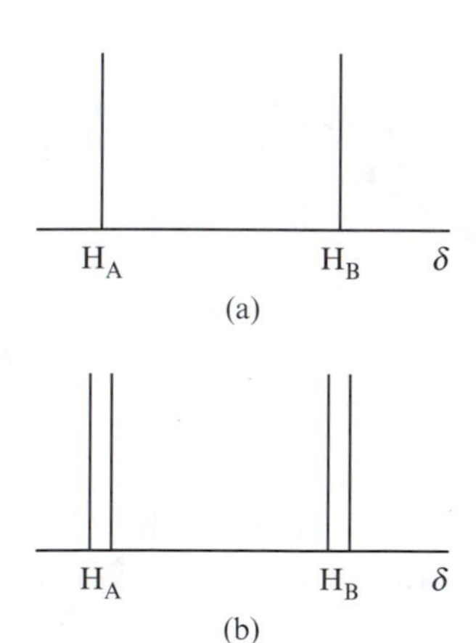

9.37 (a) Two singlet resonances are observed for two protons H_A and H_B if the two nuclei do not couple. (b) If coupling occurs, each resonance is split *by the same amount* to give a doublet.

In the simplest case, the magnetic field experienced by H_A (or H_B) is independent of the spin state of H_B (or H_A). In other words, the local magnetic field generated by the spin of one nucleus is not detected by the other nucleus. The spectrum will consist of two absorptions corresponding to the $\alpha(A) \rightarrow \beta(A)$ and $\alpha(B) \rightarrow \beta(B)$ transitions (Figure 9.37). We say that these two resonances are *singlets* and that there is *no coupling* between H_A and H_B.

Now let nucleus H_A be affected by the magnetic fields associated with H_B. The consequence is that the ^{1}H NMR signal for H_A is *split into two equal lines*, depending on whether it is 'seeing' H_B in the α or β spin state. It follows that the signal for H_B is also split into two lines in an *exactly equal* manner. We say that the H_A and H_B are *coupling* and that the spectrum consists of two *doublets.*

Coupling can only occur *between non-equivalent nuclei*, and the magnitude of the coupling is given by the coupling constant, J, measured in Hz. The coupling constant is independent of the field strength of the NMR spectrometer. It is important that the coupling constant is quoted in Hz; if it were given in δ, then it would vary with the spectrometer field. For example, a coupling constant of 25 Hz corresponds to 0.1δ at 250 MHz and 0.25δ at 100 MHz (equation 9.17).

$$J \text{ (in Hz)} = \Delta\delta \times \text{magnetic field strength of spectrometer (in MHz)} \quad (9.17)$$

where $\Delta\delta = \delta(\text{line 1}) - \delta(\text{line 2})$

Box 9.6 A more rigorous approach to the observation of two singlets for two nuclei with $I = \frac{1}{2}$ that do not couple

There is in fact a selection rule implicit in the discussion in the main text, and this rule if that $\Delta m = \pm 1$.

An alternative way of describing our spin system for nuclei H_A and H_B involves writing the *total* spin states for both nuclei. This gives us a total of four energy levels as shown below. Notice that this refers to the total energy of the system – there is no direct interaction between H_A and H_B.

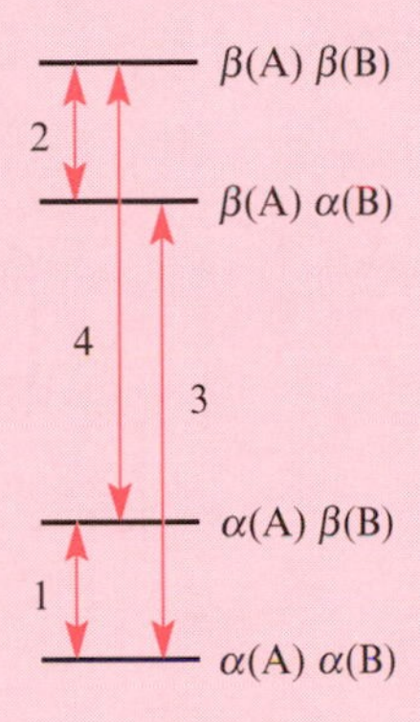

There are four transitions possible when we consider the selection rules $\Delta m(H_A) = \pm 1$, and $\Delta m(H_B) = \pm 1$, and these are:

1 $\alpha(A)\alpha(B) \rightarrow \alpha(A)\beta(B)$ } H_B transitions
2 $\beta(A)\alpha(B) \rightarrow \beta(A)\beta(B)$
3 $\alpha(A)\alpha(B) \rightarrow \beta(A)\alpha(B)$ } H_A transitions
4 $\alpha(A)\beta(B) \rightarrow \beta(A)\beta(B)$

If there is no interaction between the nuclei, the energy difference between $\alpha(A)\alpha(B)$ and $\alpha(A)\beta(B)$ states must be the same as between the $\beta(A)\alpha(B)$ and $\beta(A)\beta(B)$ states, and so the energy of transitions **1** and **2** must be the same. Similarly, transitions **3** and **4** will occur at the same energy, *but* at a different energy from **1** and **2**. In other words we will see two resonances in the NMR spectrum assigned to H_A and H_B, and each is a singlet.

The nuclei which are coupling should be indicated by adding subscripts to the symbol for J; for example if the coupling constant between H_A and H_B is 10 Hz, then we write: $J_{H_AH_B}$ 10 Hz (or J_{HH} 10 Hz).

The coupling constant J is measured in Hz:

J (in Hz) = $\Delta\delta \times$ magnetic field strength of spectrometer (in MHz)

The splitting of the resonance for H_A by its coupling with nucleus H_B is summarized in Figure 9.38a, and measurement of J_{HH} is indicated.

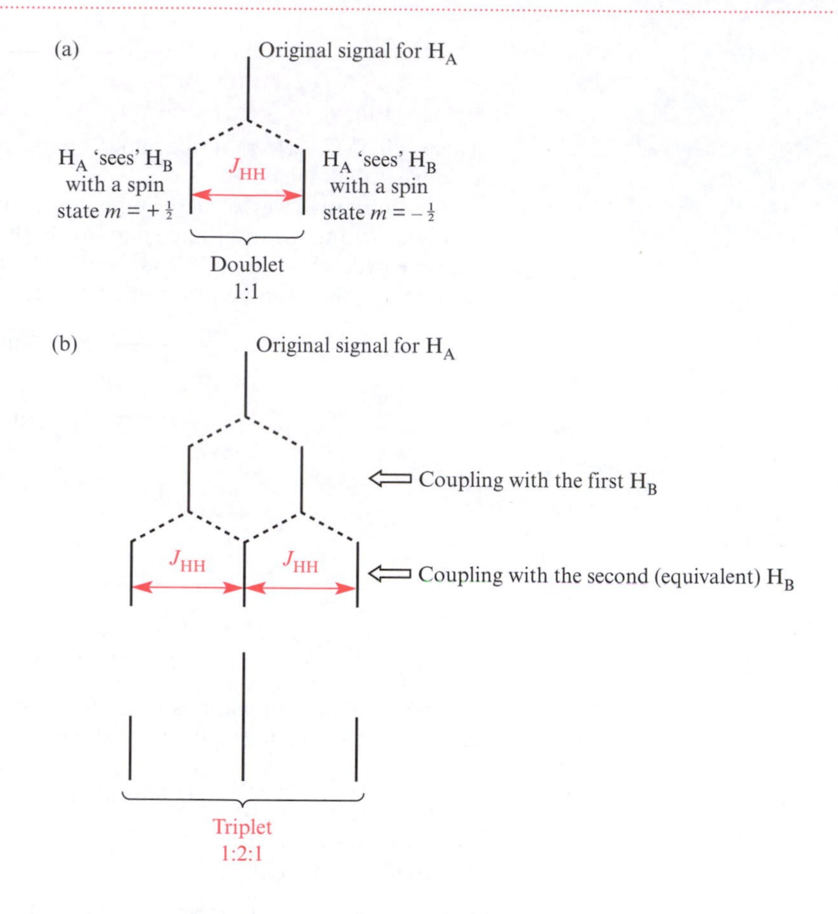

9.38 (a) The formation of a doublet signal for H_A as it couples to one nucleus H_B; the doublet has equal intensity lines, split by an amount equal to the coupling constant, J_{HH}. (b) The formation of a triplet signal for H_A as it couples to two equivalent nuclei $H_{B(1)}$ and $H_{B(2)}$; adjacent lines in the triplet are split by an amount equal to the coupling constant, J_{HH}.

Coupling between more than two inequivalent ^{1}H nuclei

Now consider the case in which H_A can 'see' *two equivalent* nuclei H_B, labelled $H_{B(1)}$ and $H_{B(2)}$. Nucleus H_A can couple with both $H_{B(1)}$ and $H_{B(2)}$ and the process is illustrated in a stepwise form in Figure 9.38b. The result is a three-line pattern (a *triplet*) for the ^{1}H NMR signal for H_A. Each line represents one of the possible combinations of the two spin states of $H_{B(1)}$ and $H_{B(2)}$ as 'seen' by H_A and these are:

$\alpha(B1)\alpha(B2)$
$\alpha(B1)\beta(B2)$ } These are degenerate because the energy difference between $\alpha(B1)$ and $\beta(B1)$ must be the same as that between $\alpha(B2)$ and $\beta(B2)$
$\beta(B1)\alpha(B2)$ }
$\beta(B1)\beta(B2)$

The three lines of the triplet therefore have the relative intensities 1 : 2 : 1 as shown in Figure 9.38b.

In another compound, H_A could couple to three or four equivalent H_B nuclei to give rise to a quartet (1 : 3 : 3 : 1) or a quintet (1 : 4 : 6 : 4 : 1) respectively. For coupling to *equivalent* nuclei with $I = \frac{1}{2}$, the number of components in the signal of a simple system (the *multiplicity of the signal*) is given by equation 9.18, and the intensity ratio of the components can be determined by using a Pascal's triangle (Figure 9.39). The general name for a doublet,

Singlet								1							
Doublet							1		1						
Triplet						1		2		1					
Quartet					1		3		3		1				
Quintet				1		4		6		4		1			
Sextet			1		5		10		10		5		1		
Septet		1		6		15		20		15		6		1	
Octet	1		7		21		35		35		21		7		1

9.39 Pascal's triangle gives the relative intensities of the components of multiplet signals when coupling is between nuclei with $I = \frac{1}{2}$. Note that each row begins and ends with 1. The other entries in a given row in the series may be obtained by summing pairs of numbers in the previous row – the triplet has intensities 1: (1+1) : 1, and the quartet has intensities 1: (1+2): (2+1) : 1. The triangle can thereby be extended (see problem 9.23).

triplet, quartet, etc., is a *multiplet* and this description is often used if the exact nature of the signal cannot be resolved.

$$\text{Multiplicity} = n + 1 \tag{9.18}$$

where n = number of *equivalent* coupled nuclei, each of spin $I = \frac{1}{2}$.

A most important point to remember is that *magnetically equivalent nuclei do not exhibit spin–spin coupling.*

Spin–spin coupling occurs between non-equivalent nuclei, and if *n* is the number of *equivalent coupled* nuclei, each of spin $I = \frac{1}{2}$, the multiplicity of the signal is $n + 1$.

Consider the ^{1}H NMR spectrum of ethanol **9.22**. The spectrum (Figure 9.40) shows three signals: a broad signal at δ +2.6, a triplet at δ +1.2 and a quartet at δ +3.7. The broad signal is assigned to the OH group; note that the expected coupling to the OH group in an alcohol is not usually observed. Protons of type *a* couple to the two equivalent protons *b*, and the signal for protons *a* is a triplet. Protons of type *b* couple to the three equivalent protons *a*, and the signal due to protons *b* is a quartet. The spectrum may be assigned on the basis of coupling, but the relative *integrals* of the two signals

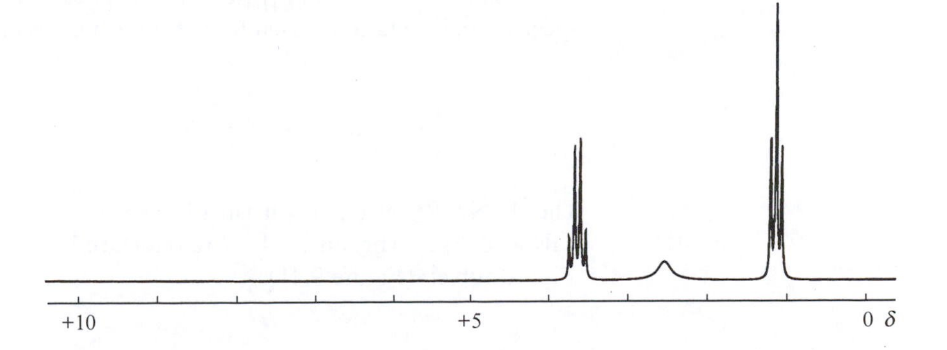

9.40 100 MHz ^{1}H NMR spectrum of ethanol.

($a : b = 3 : 2$) will confirm the answer. Notice that we have used the word *integral* rather than intensity. Because the signal is split by the coupling, it is the sum of the intensities of the component peaks that we must measure – this is best found by measuring the area under the peak and this is the *integral of the signal*.

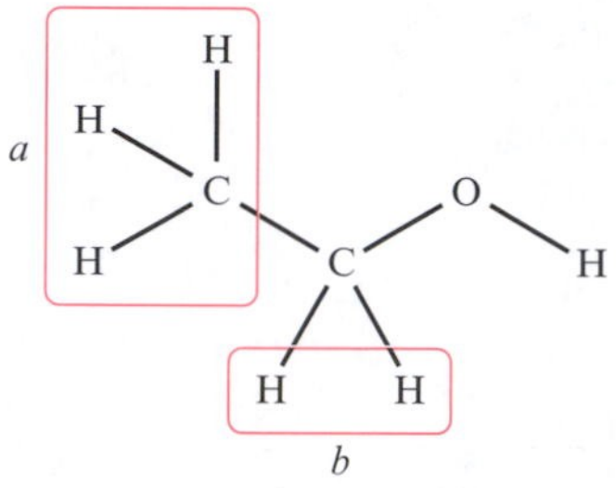

(9.22)

The question arises: 'How far along a carbon chain does coupling occur?' It is difficult to give a definitive answer but a good start is to consider ^1H–^1H coupling through one C–C bond (as in ethanol). In practice, more distant coupling may be observed, and this possibility must be kept in mind when interpreting spectra. Some examples are provided in the accompanying workbook.

If we return to Figure 9.40, both the triplet *and* the quartet have values of J_{HH} 7 Hz. It is no coincidence that the values are the same – both the triplet *and* the quartet arise from coupling between the same sets of nuclei.

The magnitude of coupling constants depends on a variety of factors, and within this book we can only deal with the topic superficially. However, the relationship between geometry and coupling constant in alkenes is diagnostic.

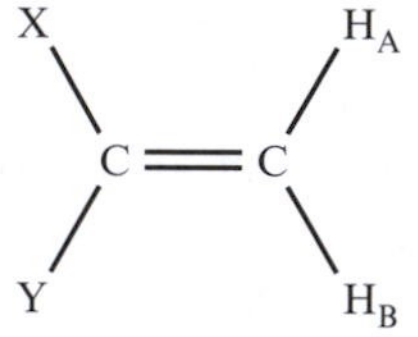

(9.23) 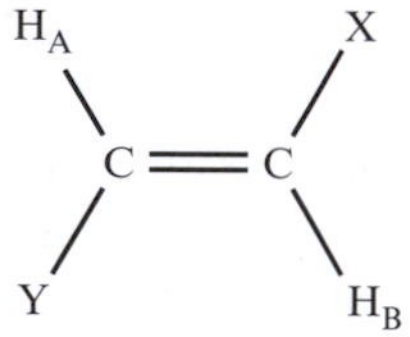 (9.24) 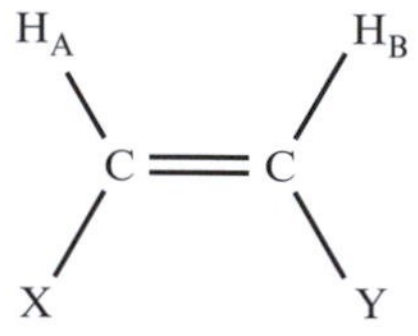 (9.25)

In each of the alkenes **9.23** to **9.25**, substituents X and Y are different from each other, and in each structure, H_A and H_B are magnetically inequivalent and couple to each other. Typically, for **9.23**, J_{HH} 0–2 Hz, for **9.24**, J_{HH} 12–18 Hz, and for **9.25**, J_{HH} 6–12 Hz. These values may allow ^1H NMR spectroscopic data to be used to distinguish between isomers of an alkene.

Interpretation of some spectra involving ^1H–^1H coupling

Example 1: 2-chloropropanoic acid

The ^1H NMR spectrum consists of three signals at δ +11.2 (broad), +4.5 (quartet) and +1.7 (doublet) (Figure 9.41).

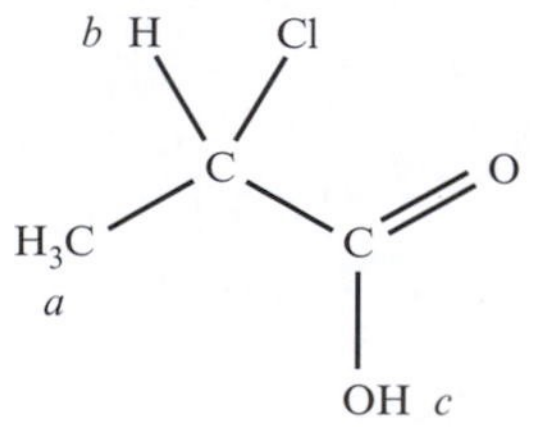

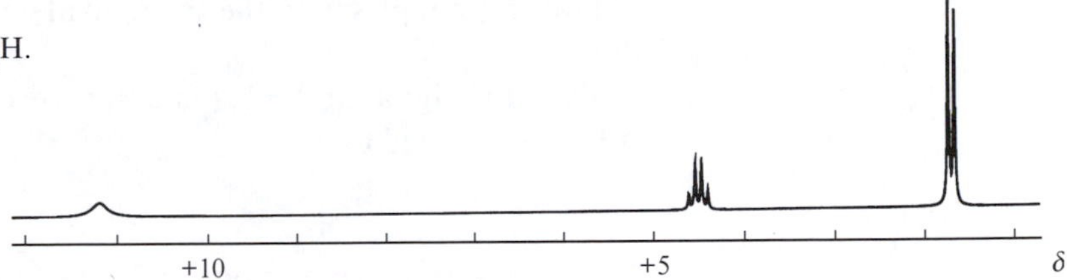

9.41 100 MHz ^{1}H NMR spectrum of 2-chloropropanoic acid, $CH_3CHClCO_2H$.

The downfield signal is assigned to the OH group.

The three methyl protons couple with proton *b*, and the signal for *b* is a quartet.

Proton *b* couples with the methyl protons and the signal for *a* is a doublet. Thus the signals at δ +4.5 and +1.7 can be unambiguously assigned.

Example 2: Propan-1-ol

The ^{1}H NMR spectrum consists of four signals at δ +3.6 (triplet), +2.3 (broad), +1.6 (sextet) and +0.9 (triplet).

The broad signal can be assigned to the OH group.

The signal for the terminal methyl group should be a triplet due to coupling to the two equivalent protons *b*. Since terminal CH_3 groups *usually* come further upfield than methylene (CH_2) groups, we can assign the signal at δ +0.9 to protons *a*.

We are now left with a sextet at δ +1.6 and a triplet at δ +3.6. The best way of tackling the assignment of these two signals is to predict what we expect and try to match the predictions to the observations.

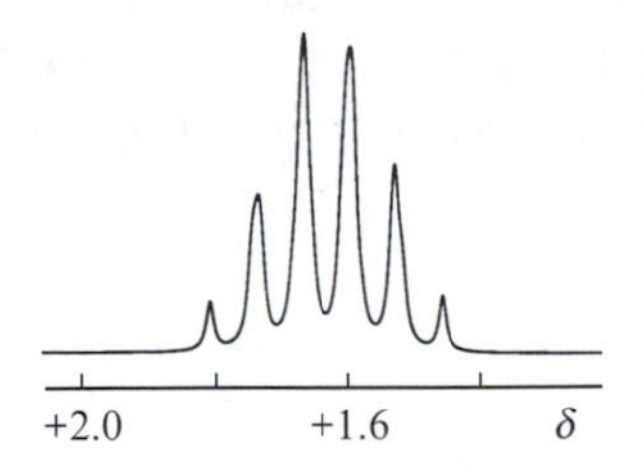

9.42 The signal at δ +1.6 in propan-1-ol is expected to be a sextet because the coupling constants $J_{H_aH_b}$ and $J_{H_bH_c}$ are approximately equal.

The methylene protons *c* will couple with the two equivalent protons *b* to give a triplet. Further coupling to the OH proton is expected but we have already seen in Figure 9.40 that such coupling is not seen in ethanol, and this is actually a general result. We can assign the triplet at δ +3.6 to methylene group *c*.

Protons *b* can couple *both* to three equivalent protons of type *a* and two equivalent protons of type *c*, but the values of $J_{H_aH_b}$ and $J_{H_bH_c}$ will be approximately the same and a sextet (Figure 9.42) due to coupling to five protons is expected.

Question: What would be the effect on the signal shown in Figure 9.42 if $J_{H_aH_b} > J_{H_bH_c}$, or $J_{H_aH_b} < J_{H_bH_c}$?

Example 3: 2-Iodopropane

The ^{1}H NMR spectrum of 2-iodopropane contains a doublet at δ +1.9 and a septet at δ +4.3.

There are two proton environments: *a* (six protons) and *b* (one proton).

Protons of type *a* couple with *b* and the signal for *b* will be a septet.

Proton b couples with the six equivalent nuclei a, and the signal for a is a doublet.

Thus the signal at δ +1.9 is assigned to the methyl groups and that at δ +4.3 is assigned to b.

Chemical and magnetic equivalence and inequivalence

In the above examples, proton sites have been taken to be magnetically inequivalent on the basis of their *chemical* inequivalence. We now make the distinction between nuclei which are *chemically equivalent but magnetically inequivalent*.

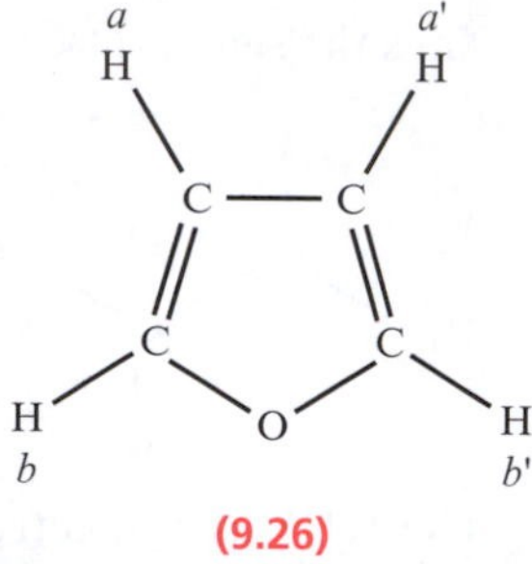

(9.26)

Furan **9.26** contains two proton environments (a and b). The ^{1}H NMR spectrum of furan consists of two signals (as expected for two chemically different ^{1}H environments) *but* the signals (δ +6.6 and +7.4) are both multiplets rather than having simple coupling patterns. This can be explained as follows.

Put yourself in the position of one of the protons of type a – you 'see' one proton b attached to the adjacent carbon centre *and* you see another proton b' that *you think* is different from b. Although the two protons of type b are *chemically equivalent*, they are (in the eyes of proton a) *magnetically inequivalent*. There are two sets of couplings:

- between a and b to give a doublet
- between a and b' to give a doublet

and the signal for a is a *doublet of doublets*.

Similarly, a proton of type b 'thinks' that a is different from a' and so the resonance for b is also a doublet of doublets (Figure 9.43).

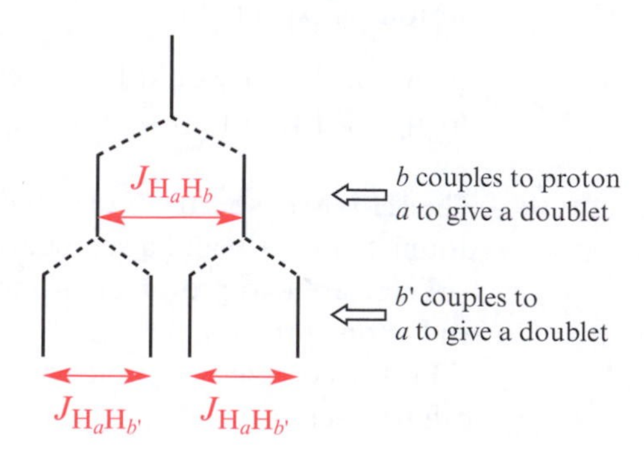

9.43 In furan **9.26**, proton a couples to b and b' to give a doublet of doublets.

Coupling between ^{13}C nuclei

When we discussed ^{13}C NMR spectra, we ignored the possibility of ^{13}C–^{13}C coupling. Why was this? The answer lies in a consideration of the negligible chance of this coupling occurring. Since the natural abundance of ^{13}C is only 1.1%, it follows that the probability of having two ^{13}C nuclei attached to one another in a molecule is very low indeed. We are justified in ignoring J_{CC} *unless* a sample has been isotopically enriched with ^{13}C.

Coupling between ^{1}H and ^{13}C nuclei: heteronuclear coupling

Spin–spin coupling is not restricted to like spins (e.g. ^{1}H–^{1}H). Any non-equivalent nuclei that have $I > 0$ may interact with one another, and to illustrate this we consider coupling between ^{1}H and ^{13}C nuclei. When different nuclei couple, it is called heteronuclear coupling.

The argument as to why we do not observe ^{13}C–^{13}C coupling in a ^{13}C NMR spectrum also applies to ^{1}H–^{13}C coupling *in a ^{1}H NMR spectrum.*

Consider acetone **9.27**. The probability of a ^{1}H NMR nucleus being bonded to a ^{13}C nucleus is very low, and in the ^{1}H NMR spectrum we can ignore the possibility of J_{CH} — the spectrum is a singlet (δ +2.1). On the other hand, since the natural abundance of ^{1}H is ≈99.9%, it is approximately true that *every* ^{13}C nucleus present in the methyl groups of acetone is attached to ^{1}H nuclei. In the ^{13}C spectrum, ^{13}C–^{1}H coupling *is* observed as Figure 9.44 shows.

H_3C–C(=O)–CH_3

(9.27)

Heteronuclear coupling can be suppressed in a spectrum by an instrumental technique called *heteronuclear decoupling* and in the laboratory, you may often encouter *proton-decoupled* ^{13}C NMR spectra. A comparison of Figures 9.34 and 9.44 shows the differences between proton-decoupled and proton-coupled ^{13}C NMR spectra of acetone.

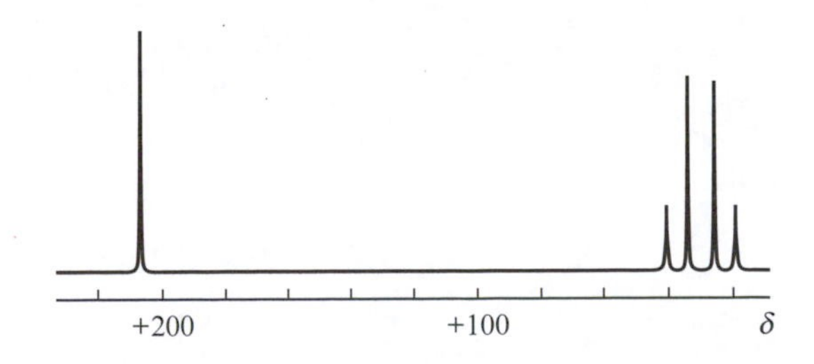

9.44 25 MHz ^{13}C NMR spectrum of acetone; for the quartet, J_{CH} = 172 Hz.

Box 9.7 Satellite peaks in NMR spectra

We have seen that in a ^{1}H NMR spectrum, coupling between ^{13}C and ^{1}H nuclei can be ignored, but that in the ^{13}C NMR spectrum of the same sample, ^{13}C–^{1}H coupling is observed. The difference followed from the different natural abundances of the two nuclei.

Consider now the case of the ^{1}H NMR spectrum of $SnMe_4$ which is drawn below:

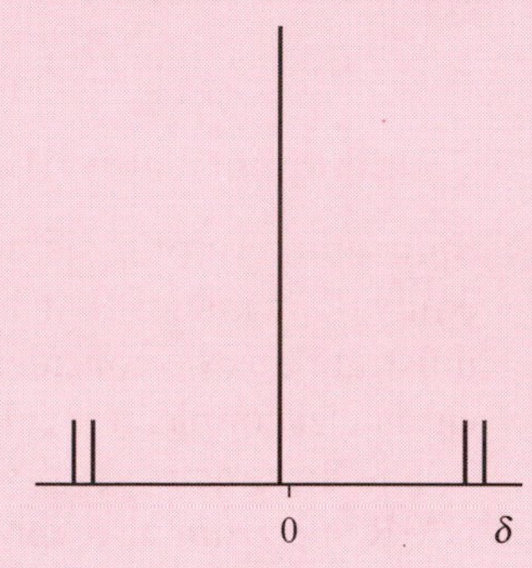

Tin possesses ten isotopes, and two of these are NMR active: ^{117}Sn ($I = \frac{1}{2}$, 7.6%) and ^{119}Sn ($I = \frac{1}{2}$, 8.6%). In a sample of $SnMe_4$ there is a statistical distribution of all the naturally occurring isotopes of tin.

83.8% of the $SnMe_4$ molecules possess Sn nuclei with $I = 0$.
7.6% of the $SnMe_4$ molecules possess ^{117}Sn nuclei.
8.6% of the $SnMe_4$ molecules possess ^{119}Sn nuclei.

This means that in the ^{1}H NMR spectrum of $SnMe_4$, 83.8% of the ^{1}H nuclei give rise to a singlet (no coupling to tin), whilst 7.6% of the protons couple to ^{117}Sn (to give a doublet) and 8.6% couple to ^{119}Sn (also to give a doublet). The $J_{^{119}Sn^{1}H}$ and $J_{^{117}Sn^{1}H}$ coupling constants are 54 Hz and 52 Hz respectively and therefore the two doublets are separated from one another. The doublets are called *satellites*.

Questions: In the spectrum above:

- where is the singlet?
- where are the two doublets?
- where would you measure the values of $J_{^{119}Sn^{1}H}$ and $J_{^{117}Sn^{1}H}$?

SUMMARY

In this chapter, we have described some aspects of spectroscopy, with an emphasis on the use of three techniques in the laboratory.

Do you know what the following terms mean?

- spectroscopic timescale
- absorption spectrum
- emission spectrum
- wavenumber
- Beer–Lambert Law
- absorbance
- transmittance
- molar extinction coefficient
- IR spectroscopy
- vibrational ground state
- zero point energy
- vibrational mode
- reduced mass
- fingerprint region
- UV–VIS spectroscopy
- λ_{max}
- ε_{max}
- conjugated polyene
- chromophore
- ^{1}H NMR spectroscopy
- ^{13}C NMR spectroscopy
- chemical shift value
- signal intensity
- signal integral
- spin–spin coupling
- coupling constant

You should be able:

- to compare the relative energies, frequencies and wavelengths of IR, UV, VIS and RF radiation
- to relate absorbed and transmitted wavelengths, and understand how this relates to the observed colours of solutions
- to use the Beer–Lambert Law to determine concentrations from absorbance or transmittance data
- to use the Beer–Lambert Law to determine the stoichiometry of selected reactions
- to discuss the vibrating spring model in relation to the vibrations of a diatomic molecule.
- to work out for diatomic and triatomic molecules the allowed modes of vibration, and state if (and why) they are IR active
- to interpret the IR spectra of relatively simple compounds in terms of the fingerprint region and functional groups present
- to understand how some electronic spectra arise, e.g. the spectrum of a polyene
- to explain why absorptions in the electronic spectrum of a molecular species are usually broad
- to explain why values of λ_{max} may shift in series of related compounds
- to give examples of compounds that absorb in the near-UV and visible regions
- to describe an application of Job's method
- to list the properties of a nucleus that make it NMR active.
- to discuss briefly why FT-NMR spectroscopy has made an impact on the development of this spectroscopic technique
- to interpret ^{13}C NMR spectra in terms of different chemical environments
- to understand why a *deuterated* solvent is usually used in ^{1}H NMR spectroscopy
- to interpret ^{1}H NMR spectra in terms of different proton environments and simple spin–spin couplings

PROBLEMS

9.1 What are the *relative* energies of transitions observed in vibrational, electronic and nuclear magnetic resonance spectroscopies?

9.2 Aqueous solutions containing nickel(II) ions appear green. What is the approximate wavelength of light absorbed by aqueous nickel(II) ions?

9.3 Refer to Figure 9.6. What does the gradient of the graph tell you? Determine the concentration of iron(II) ions in a solution that gives an absorbance of (a) 20 and (b) 300.

9.4 The fundamental stretching frequency for CO is 2170 cm^{-1} and the force constant for the bond is 1905 N m^{-1}. If the force constant for the bond in NO is 1595 N m^{-1}, determine the frequency of the fundamental vibration of this molecule.

9.5 Determine the reduced mass of the diatomics H_2, HD (D = ^{2}H), $H^{35}Cl$ and $H^{37}Cl$. Determine the ratio of the reduced masses for each pair of isotopomers. Would you expect to see *chemical* differences between (a) H_2 and HD, and (b) $H^{35}Cl$ and $H^{37}Cl$?

9.6 Refer to Figures 9.12b and 9.12c. In what direction does the molecular dipole moment act as the CO_2 molecule (a) stretches asymmetrically and (b) bends? In either case, is a *permanent* molecular dipole moment generated?

9.7 Would each of the symmetric stretching, asymmetric stretching and bending modes of the following molecules be IR active or inactive: (a) HCN; (b) CS_2; (c) OCS; (d) XeF_2?

9.8 The IR spectrum of H_2O has three absorptions that are assigned to the asymmetric stretching mode (3756 cm^{-1}), the symmetric stretching mode (3657 cm^{-1}) and the scissoring mode (1595 cm^{-1}). Upon deuteration to give D_2O, these three absorptions move to 2788, 2671 and 1178 cm^{-1} respectively. How do you account for these observations?

9.9 Rationalize why the values for the carbonyl stretching mode given in Table 9.3 for carboxylic acids and carboxylate ions are generally lower than those of the corresponding aldehydes.

9.10 For each of the following compounds, state the nature of the functional group present and approximately where in the IR spectrum you would expect to observe a characteristic absorption: (a) acetonitrile; (b) butan-2-ol; (c) ethylamine; (d) propanal; (e) pent-1-ene.

9.11 Figure 9.45 shows the IR spectra of high-density polythene, dodecane, hexane and hex-1-ene. (a) Why are the spectra of polythene, dodecane and hexane so similar to each other? (b) What *two* features of the spectrum of hex-1-ene allow you to assign this spectrum to this alkene in preference to the corresponding alkane?

9.12 Figure 9.46 shows four spectra labelled (a) to (d). Assign, with reasoning, each spectrum to one of the compounds cyclohexane, benzene, toluene (methylbenzene) and phenol.

9.45 For problem 9.11.

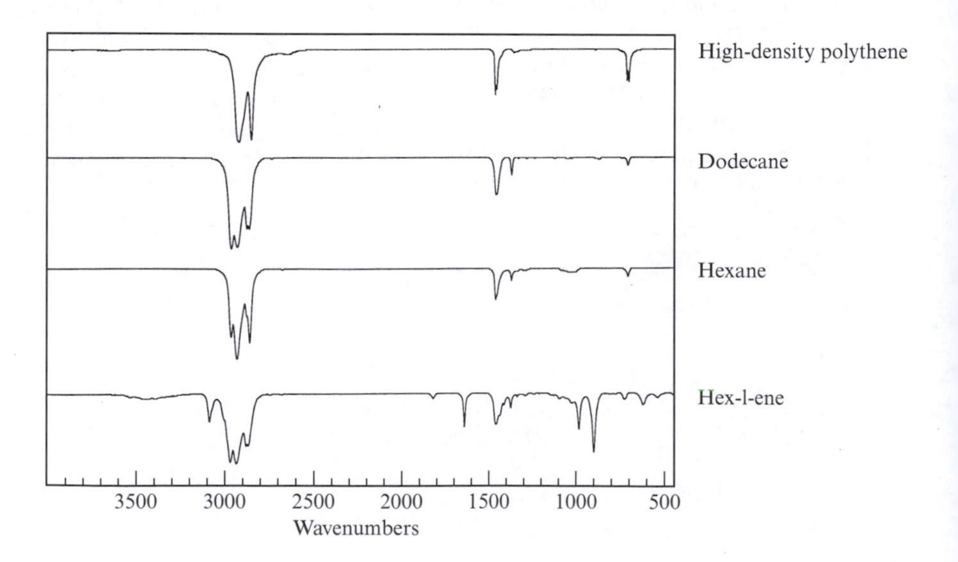

9.46 For problem 9.12.

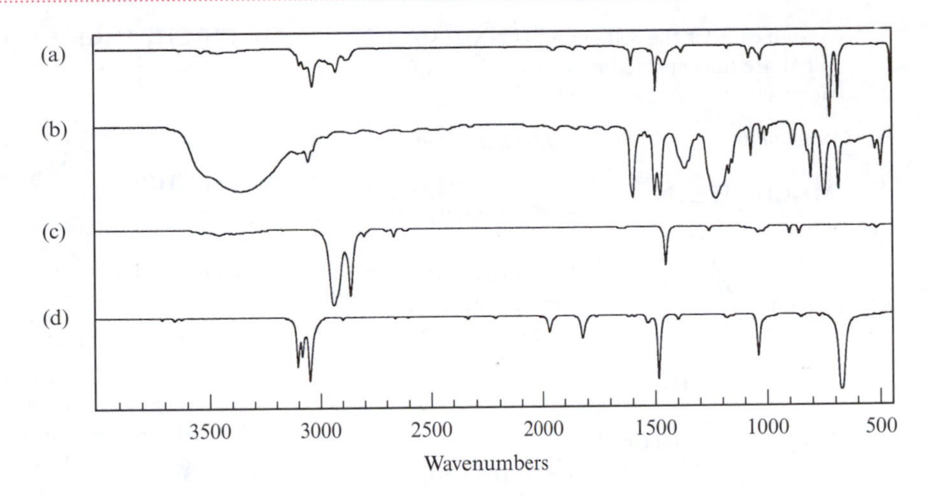

9.13 How could you use IR spectroscopy to distinguish between each of the following pairs of compounds?

(a) $CH_3C(O)CH_3$ and $CH_3CH(OH)CH_3$;
(b) $CH_3CH_2CH{=}CH_2$ and $CH_3CH_2CH_2CH_3$;
(c) $CH_3CH_2CH_2Cl$ and $CH_3CH_2CH_2NH_2$.

For each case, comment on whether you think that ^{13}C NMR spectroscopy would have been as equally diagnostic.

9.14 Using the notation that we introduced in equation 4.5, write three equations that describe the three π-MOs in the allyl anion.

9.15 What is the π-bond order in (a) the allyl cation and (b) the allyl anion?

9.16 A complex X absorbs at 466 nm (ε_{max} = 75 dm^3 mol^{-1} cm^{-1}) and at 338 nm (ε_{max} = 68 dm^3 mol^{-1} cm^{-1}). What are the *relative* absorbances of the two bands in the spectrum of a 0.1 M solution of X? Is your answer dependent upon the path length of the cell in the spectrophotometer?

9.17 What do you understand by the terms (a) absorption, (b) absorbance and (c) transmittance.

9.18 What type of transitions ($\sigma \rightarrow \sigma^*$, $\pi \rightarrow \pi^*$, $n \rightarrow \sigma^*$ or $n \rightarrow \pi^*$) might you expect to give rise to significant features in the electronic spectra of the following compounds: (a) pentane; (b) pent-1-ene; (c) octa-2,4,6-triene; (d) ethanol?

9.19 Using the data in Table 9.10, determine the resonance frequency of ^{31}P nuclei on a 400 MHz spectrometer, and the resonance frequency of ^{13}C nuclei on a 250 MHz instrument.

9.20 The ^{13}C NMR spectrum of acetic acid has signals at δ +20.6 and +178.1. Assign the spectrum.

9.21 The ^{13}C NMR spectrum of 2-methylpropan-2-ol has signals at δ +31.2 and +68.9. Suggest assignments for these signals. What other feature of the spectrum would confirm the assignments?

9.22 The ^{13}C NMR spectrum of the following compound possesses signals of approximately equal intensities at δ +20.5, +52.0, +80.8 and +170.0. Suggest assignments for the spectrum.

$$H_3C{-}C({=}O){-}O{-}CH_2{-}C{\equiv}C{-}CH_2{-}O{-}C({=}O){-}CH_3$$

9.23 Predict the relative intensities and the coupling patterns of the signals in the ^1H NMR spectrum of 2-methylpropane.

9.24 Assign each of the following ^{1}H NMR spectra. What are the expected relative integrals of the signals?

	Compound	*Resonances* / δ
(a)	$Me_2CHC{\equiv}N$	+1.3 (doublet), +2.7 (septet)
(b)	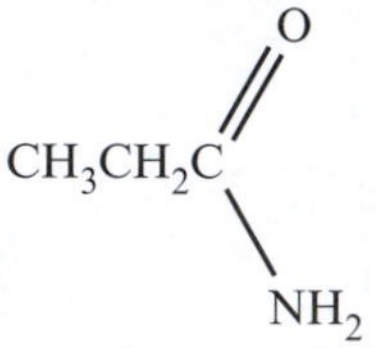 $CH_3CH_2C(=O)NH_2$	+1.1 (triplet), +2.2 (quartet), +6.4 (very broad)
(c)	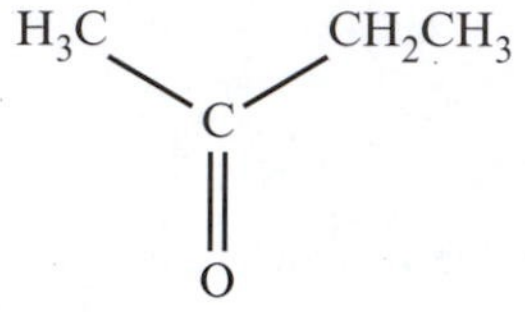 $H_3C-C(=O)-CH_2CH_3$	+1.1 (triplet), +2.1 (singlet), +2.5 (quartet)
(d)	CH_3CH_2–O–CH_2CO_2H	+1.3 (triplet), +3.7 (quartet), +4.1 (singlet), +10.9 (broadened)
(e)	CH_3CHBr_2	+2.5 (doublet), +5.9 (quartet)
(f)	$CH_2ClCH_2CH_2Cl$	+2.2 (quintet), +3.7 (triplet)

9.25 Explain why the ^{1}H NMR spectrum of CH_2ClCF_2Cl shows a triplet (δ +4.0). [Hint: refer to Table 9.10.] What would you predict for the ^{1}H NMR spectrum of CF_3CH_2OH?

10 Reaction kinetics

Topics

10.1 INTRODUCTION

In Section 1.18, we reviewed activation barriers, and in Chapter 8 discussed reaction mechanisms and reaction pathways. Knowledge of the mechanism comes, in part, from a study of the *rate* of the reaction. Such information is concerned with the *reaction kinetics*, and in this chapter we shall discuss this topic in detail. In this introductory section we summarize some facts about reaction rates with which you will probably already be familiar.

Thermodynamics and kinetics

In discussions of the thermodynamics of a reaction, we can talk about various energy terms (see Chapter 12) but for now we restrict our discussion to the enthalpy.

The fact that a reaction is thermodynamically favourable does *not* mean that it will necessarily take place quickly. For example, the reaction of an alkane with dioxygen is exothermic (equation 10.1) but it is not spontaneous – an initiation such as a spark or flame is required. Mixtures of alkanes and dioxygen may be kept unchanged for long periods.

$$C_5H_{12}(l) + 8O_2(g) \rightarrow 5CO_2(g) + 6H_2O(l) \qquad (10.1)$$
$$\Delta_r H = -3509 \text{ kJ per mole of reaction}$$

Many other reactions which do actually proceed (for example the dissolution of ammonium nitrate in water) are endothermic. Clearly the enthalpy change is not the ultimate arbiter of the spontaneity of a chemical reaction and we will see in Chapter 12 that a new energy term, the change in free energy ΔG, is an important factor. However, the fact remains that a knowledge of the thermodynamic changes in a reaction gives no knowledge *per se* about the *rate* of the conversion of starting materials to products.

In Chapter 8, we referred to the *rate-determining step* of a reaction – Figure 8.13 showed reaction coordinates for two three-step reactions, and we saw that even though each of the three steps was exothermic, one of the steps had a large activation enthalpy and this step controlled the rate at which the reaction would occur.

What is the rate of a reaction?

The progress of a reaction can be monitored in terms of the appearance of a product or the disappearance of a reactant as a function of time as shown in Figure 10.1.

The change in concentration of a reactant or product with respect to time is a measure of the rate of a reaction, and this gives units of rate as mol dm^{-3} s^{-1}.

The change in concentration of a reactant or product with respect to time is a measure of the rate of a reaction. The units of rate are mol dm^{-3} s^{-1}.

Consider two different reactions involving a reactant A which are both of the general form given in equation 10.2. The rate at which A disappears depends on the exact reaction concerned and the precise reaction conditions.

$$A \rightarrow \text{products} \qquad (10.2)$$

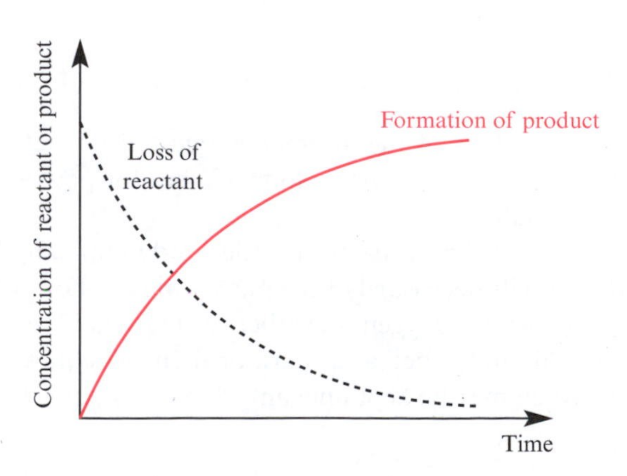

10.1 In a reaction of the type $A \rightarrow B$ (i.e. one mole of A is converted into one mole of B), the rate of disappearance of A should mirror the rate of appearance of B.

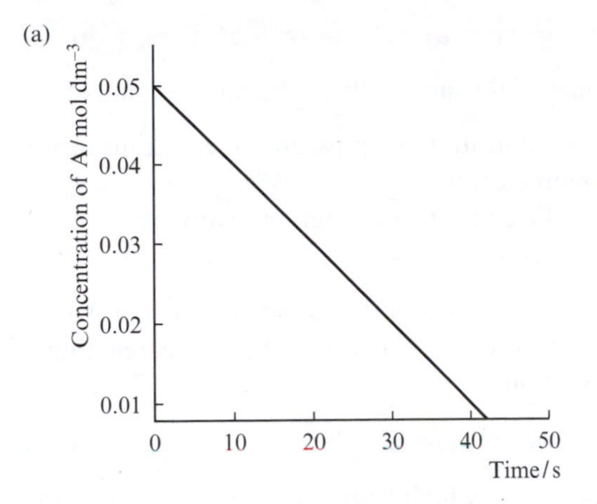

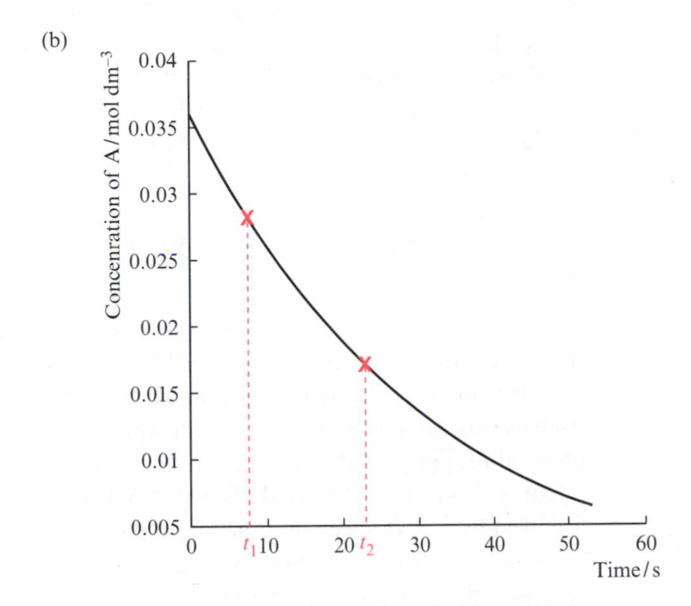

10.2 A plot of the decrease in the concentration of a reactant A against time may be (a) linear, or (b) non-linear. The rate of the disappearance of A is constant in graph (a); the rate is equal to the gradient of the line. In graph (b), the rate varies with time, and is found at various times by drawing tangents to the curve and measuring their gradients; rate at t_1 > rate at t_2.

Figure 10.2a describes a reaction in which the concentration of A decreases linearly during the reaction. The rate at which A disappears is equal to the gradient of the graph, and is constant over the course of the reaction, i.e. it does not depend upon the concentration of A.

Figure 10.2b refers to a reaction in which the concentration of A decreases quickly at the beginning of the reaction but then decreases more slowly as time progresses. Here, the rate of reaction changes with time. At time t_1, the rate is found by drawing a tangent to the curve at the point $X(t_1)$ and finding the gradient of the tangent. At a time t_2, the rate is determined by drawing a second tangent to the curve at a second point, $X(t_2)$.

Problem 10.1 asks you to determine rates of reaction from Figure 10.2

Factors that affect the rate of a reaction

Some of the factors that influence the rate of a reaction are:

- concentration (or pressure for reactions involving gases);
- temperature;
- surface area (for a reaction involving a solid);
- pressure.

Figure 10.3 shows the rate of carbon dioxide evolution from two reaction vessels in which calcium carbonate is reacting with dilute hydrochloric acid (equation 10.3).

$$CaCO_3(s) + 2HCl(aq) \rightarrow CaCl_2(aq) + H_2O(l) + CO_2(g) \qquad (10.3)$$

The rate at which CO_2 is produced is a measure of the rate of the reaction. The two reactions are proceeding at different rates; that represented by plot I proceeds approximately twice as fast as that shown in plot II. Clearly, the conditions in the two reaction vessels are different – the rate may be influenced by the concentration of hydrochloric acid, the surface area of the calcium carbonate (e.g. powder instead of granules) or the temperature.

How do you choose an experimental method for monitoring a reaction?

When choosing a method to follow a reaction, you must look for features that give rise to a *measurable change* between starting materials and products. It is most convenient to use a *non-intrusive* method so that the reaction itself is not perturbed. Conventional spectroscopic methods or some other physical property of the reaction solution such as its refractive index, viscosity or volume may be used. Some reactions proceed too quickly to be followed by *simple* spectroscopic methods.

We are not concerned here with details of particular methods, but in the reactions described in this chapter, we exemplify some possible experimental techniques. Equation 10.3 showed an example of a reaction monitored by the evolution of a gas, and in Section 9.5, a reaction with a change in colour was illustrated.

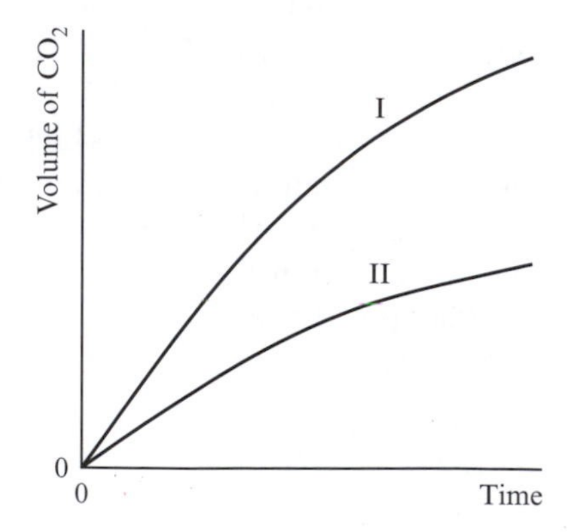

10.3 The rate of the reaction between solid calcium carbonate and hydrochloric acid may be followed by measuring the volume of CO_2 evolved. The rate of reaction I is approximately twice as fast as the rate of reaction II. What differences might there be between the conditions in the two reaction vessels?

10.2 RATE EQUATIONS: THE DEPENDENCE OF RATE ON CONCENTRATION

Rate expressed as a differential equation

Throughout this section we shall be concerned with the general reaction:

$$A \rightarrow \text{products}$$

The rate of this reaction is given by equation 10.4 where the *differential* $-\frac{d[A]}{dt}$ shows the change in the concentration of A with respect to time, t; the concentration of A is denoted as [A].

The concentration of a component A is written as [A].

$$\text{Rate of reaction} = -\frac{d[A]}{dt} \qquad (10.4)$$

Box 10.1 Rate as the derivative $-\frac{d[A]}{dt}$

Consider the reaction:

$$A \rightarrow \text{products}$$

in which the concentration of A, [A], decreases linearly with time as shown here.

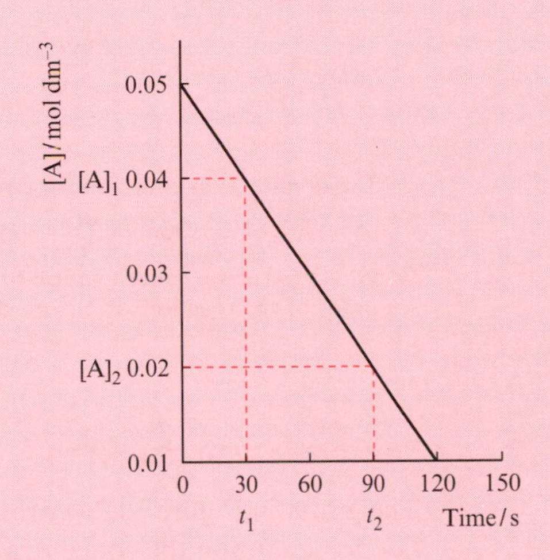

The rate at which A disappears is constant. From the graph we see that in the time interval from t_1 to t_2, the concentration of A has fallen from $[A]_1$ to $[A]_2$.

$$\text{Rate of disappearance of A} = \frac{0.02}{60} = 3.33 \times 10^{-4}\ \text{mol dm}^{-3}\ \text{s}^{-1}$$

The rate of reaction may be expressed in terms of the rate of disappearance of A with time. This is written mathematically as the *negative differential* of [A] with respect to time, t:

$$\text{Rate of reaction} = -\frac{d[A]}{dt}$$

In a reaction:

$$A \rightarrow B + C$$

the rate of disappearance of A is equal to the rate of appearance of B and of C:

$$\text{Rate of reaction} = -\frac{d[A]}{dt} = \frac{d[B]}{dt} = \frac{d[C]}{dt}$$

Note that [A] or $[A]_t$ refers to the *concentration of A at time t*. The initial concentration of A is denoted as $[A]_0$.

The general rate equation

At the beginning of a reaction, the concentration of A is at a maximum – the concentration will decrease during the reaction. Equation 10.5 gives a *general* expression in which *the rate of the reaction depends on the concentration of A*.

$$\text{Rate of reaction} = -\frac{d[A]}{dt} \propto [A]^n \qquad (10.5)$$

The power n in equation 10.5 is called the *order of the reaction with respect to A* and may be zero or have an integral or fractional value. The order shows the exact dependence of the rate on [A], but in this chapter we are primarily concerned with values of n of 0, 1 or 2.

Equation 10.5 is often written in the form shown in equation 10.6 where k is the rate constant for the reaction. The units of the *rate constant* vary with the order of the reaction as discussed below. Whereas the *rate* of a reaction depends upon the concentration of the reactants, the *rate constant* is independent of concentration. It is important that you distinguish between these two terms. For a given concentration, the larger the value of k, the faster the reaction proceeds.

$$\text{Rate of reaction} = -\frac{d[A]}{dt} = k[A]^n \qquad (10.6)$$

The general rate equation for a reaction A $\rightarrow$ products is:

$$\text{Rate of reaction} = -\frac{d[A]}{dt} = k[A]^n$$

where [A] is the concentration of A at time t, k is the rate constant and n is the order of the reaction with respect to A.

Zero order with respect to A

The graph in Figure 10.2a corresponded to a reaction in which the rate did *not* depend upon the concentration of A. That is, the power n in equation 10.6 is zero – remember that a number raised to the power zero is equal to unity, and so $[A]^0 = 1$. Thus, the rate of reaction is equal to a constant k (equation 10.7).

$$\text{Rate of reaction} = -\frac{d[A]}{dt} = k \times [A]^0 = k \qquad (10.7)$$

We say that this reaction is *zero (or zeroth) order with respect to A*. In this case, the units of the rate constant are the same as those of rate: mol dm^{-3} s^{-1}. The rate of the reaction does not vary with time.

When a reaction is zero order with respect to a reactant A, the rate does *not* depend upon the concentration of A and the rate equation is:

$$\text{Rate of reaction} = -\frac{d[A]}{dt} = k$$

The units of k are mol dm^{-3} s^{-1}.

First order with respect to A

If the rate of a reaction is directly proportional to the concentration of reactant A, then the reaction is *first order with respect to A*. The appropriate rate equation is given in equation 10.8.

$$\text{Rate of reaction} = -\frac{d[A]}{dt} = k[A] \qquad (10.8)$$

This means that as [A] decreases during the reaction, the rate of reaction also decreases, and a plot of [A] against time will be non-linear.

Problem 10.3 deals with units of rate constants

The rate constant in equation 10.8 has units of s^{-1}.

When a reaction is first order with respect to a reactant A, the rate depends upon the concentration of A according to the rate equation:

$$\text{Rate of reaction} = -\frac{d[A]}{dt} = k[A]$$

The units of k are s^{-1}.

Second order with respect to A

If the rate of reaction depends upon the square of the concentration of A, then the reaction is *second order with respect to A* – the rate law is given in equation 10.9.

$$\text{Rate of reaction} = -\frac{d[A]}{dt} = k[A]^2 \qquad (10.9)$$

Once again, a plot of [A] against time is non-linear.

The units of the rate constant in equation 10.9 are $dm^3\ mol^{-1}\ s^{-1}$.

When a reaction is second order with respect to a reactant A, the rate equation is:

$$\text{Rate of reaction} = -\frac{d[A]}{dt} = k[A]^2$$

The units of k are $dm^3\ mol^{-1}\ s^{-1}$.

10.3 DOES A REACTION SHOW A ZERO, FIRST OR SECOND ORDER DEPENDENCE ON A?

Once again in this section, we refer to the general reaction 10.10.

$$A \rightarrow \text{products} \qquad (10.10)$$

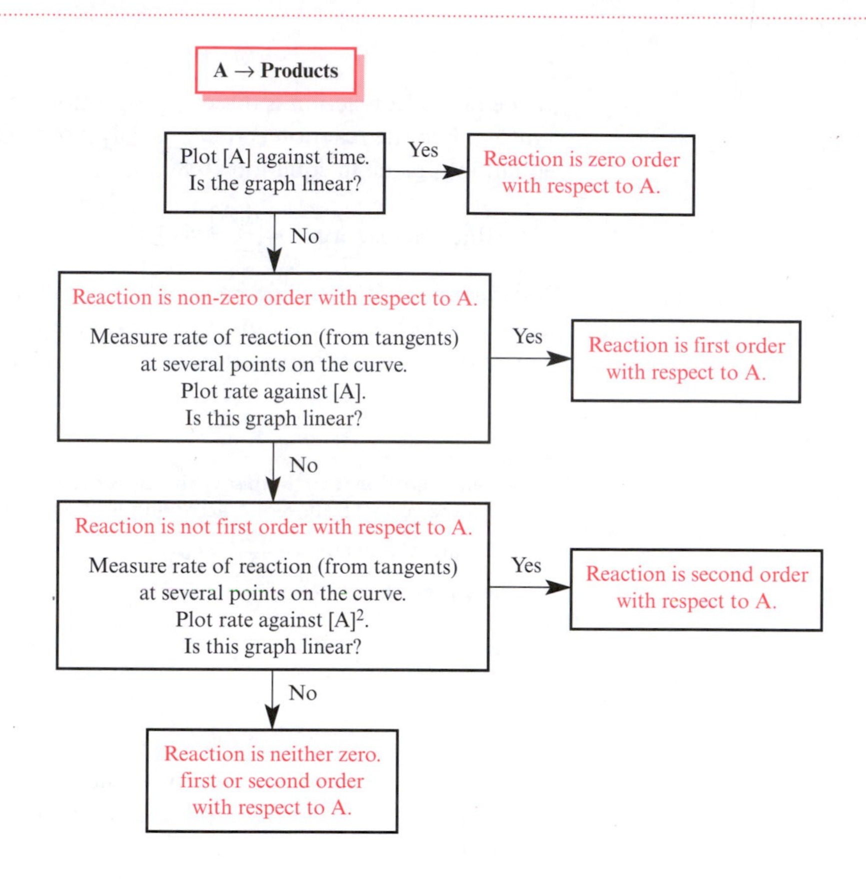

10.4 This flowchart summarizes how to use graphical methods to determine whether a reaction of the type:

A → Products

is zero, first or second order with respect to A.

Graphical methods

If a plot of [A] against time is linear, we may conclude that there is a zero order dependence upon A. On the other hand, if such a plot is non-linear (as in Figure 10.2b), we cannot *immediately* deduce the exact order (other than saying it is non-zero). If we impose a restriction that the order is likely to be first or second, how can we distinguish between them?

There are several ways of processing experimental results, and one method is summarized in the flow diagram in Figure 10.4. The choices of the graphs to be plotted follow from the rate equations 10.8 and 10.9. As we have already seen, the rate of reaction at a given time can be determined from a graph of [A] against time and is equal to the gradient of a tangent drawn to the curve.

Worked example 10.1 ***Thermal decomposition of benzoyl peroxide***

Above 100°C, benzoyl peroxide decomposes. Benzoyl peroxide has a strong absorption in its IR spectrum at 1787 cm^{-1} and the disappearance of this band gives a measure (using the Beer–Lambert Law) of the concentration of the peroxide. Use the data given (which were recorded at 107°C) to determine the rate equation for the decomposition of benzoyl peroxide.

Benzoyl peroxide

Time/min	0	15	30	45	60	75	90	120	150	180
[Benzoyl peroxide] x 10^2 /mol dm^{-3}	2.00	1.40	1.00	0.70	0.50	0.36	0.25	0.13	0.06	0.03

(Data from: M.D. Mosher, S. Vinson, J. Connagahan, R. Forsythe and M.W. Mosher, *Journal of Chemical Education*, (1991) vol. 68, p. 510.)

The initial concentration of the benzoyl peroxide was 2.00×10^{-2} mol dm^{-3}. (Notice how this information is stated in the table.)

First plot the concentration of benzoyl peroxide as a function of time:

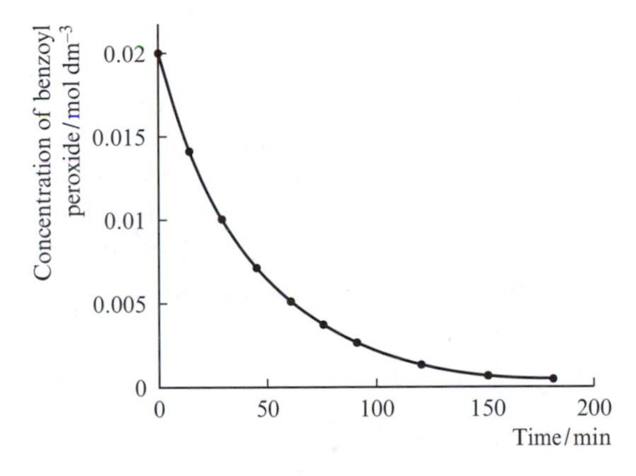

As the graph is non-linear, the decomposition is not zero order, but may be first or second order with respect to benzoyl peroxide. One method of distinguishing between these possibilities is to see if a plot of rate against [A] is linear.

Next, determine the rate of the decomposition at several points – five are sufficient, *but they should be representative of the data as a whole*. The rate of reaction at time t is found by drawing a tangent to the curve at time t. The tangents have negative gradients because we have plotted the variation in [benzoyl peroxide] – since the peroxide is decomposing during the reaction, the negative gradient corresponds to a positive rate of reaction.

Tabulate the new data; you should use the graph above to confirm the values of the rates stated in the table below.

[Benzoyl peroxide] $\times 10^2$/mol dm^{-3}	2.00	1.40	1.00	0.50	0.13
Rate of reaction $\times 10^4$/mol dm^{-3} min^{-1}	4.44	3.27	2.23	1.17	0.25

Now plot the reaction rate against the concentration of the peroxide:

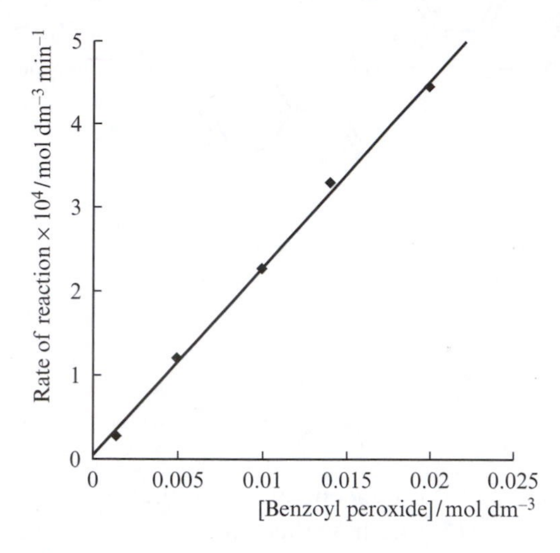

The straight line means that the reaction is first order with respect to benzoyl peroxide and we can write down the rate equation as:

$$\text{Rate of reaction} = -\frac{\text{d[Benzoyl peroxide]}}{\text{d}t} = k\text{[Benzoyl peroxide]}$$

The rate constant k can be found from the second graph – it equals the gradient of the straight line, 2.25×10^{-2} min^{-1}, or 3.75×10^{-4} s^{-1}.

Drawing tangents to a curve is not an accurate method of obtaining your answer, and you will find an alternative method for solving worked example 10.1 later in the chapter.

Half-life method

The time taken for the concentration of reactant A to fall from $[A]_t$ to $\frac{[A]_t}{2}$ (where $[A]_t$ is the concentration of reactant A at a specific time t) is called the *half-life* of the reaction.

For a first order dependence, half-lives measured for the steps:

$$[A]_t \rightarrow \frac{[A]_t}{2}, \quad \frac{[A]_t}{2} \rightarrow \frac{[A]_t}{4}, \quad \frac{[A]_t}{4} \rightarrow \frac{[A]_t}{8}, \quad \frac{[A]_t}{8} \rightarrow \frac{[A]_t}{16}, \text{ etc.}$$

are constant. This is not true for a reaction which is second order with respect to A.

The half-life of a reaction is the time taken for the concentration of reactant A at time t, $[A]_t$, to fall to half of its value, $\frac{[A]_t}{2}$.

Worked example 10.2 ***The decay of a photogenerated species***

The blue mercury(II) complex I shown below is formed by irradiating a related mercury(II) compound, but once formed, this photogenerated species converts back to the original compound. The disappearance of I (λ_{max} = 605 nm, ε = 27 000 $dm^3\ mol^{-1}\ cm^{-1}$) can be followed by UV–VIS spectroscopy, and the concentration of I at time *t* can be found using the Beer–Lambert Law. Use the following table of data to deduce the order of the reaction with respect to I.

Compound **I**

Time / min	0	0.5	1	1.5	2	2.5	3	3.5	4	4.5	5
$[I] \times 10^5$ / mol dm^{-3}	1.447	1.061	0.826	0.612	0.449	0.336	0.249	0.200	0.139	0.106	0.083

(Data from: R. L. Petersen and G. L. Harris, *Journal of Chemical Education*, (1985) vol 62, p. 802.)

The rate equation for the decay of **I** will have the general form:

$$-\frac{d[\mathbf{I}]}{dt} = k[\mathbf{I}]^n$$

First, plot [**I**] against time:

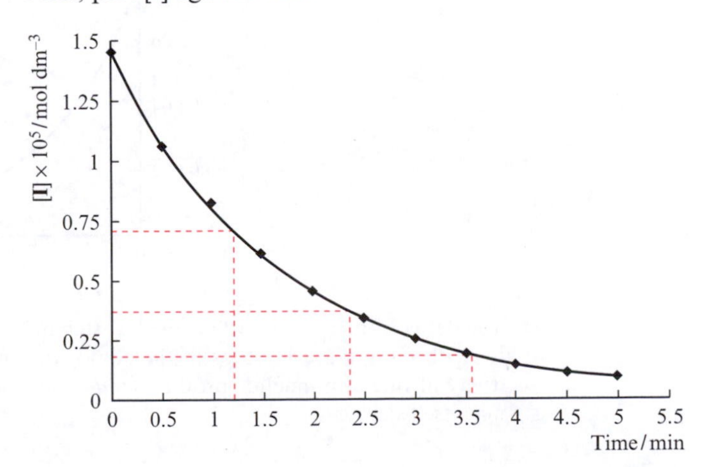

The curve shows that the reaction is not zero order with respect to compound I.

There are various ways to proceed now, and one is to measure several half-lives.

The initial concentration $[\mathbf{I}]_0 = 1.447 \times 10^{-5}$ mol dm^{-3}
From the graph, the first half-life is the time for the concentration to fall from $[\mathbf{I}]_0$ to $\frac{[\mathbf{I}]_0}{2} = 1.2$ min.

The second half-life is the time to go from $\frac{[\mathbf{I}]_0}{2}$ to $\frac{[\mathbf{I}]_0}{4} = 2.35 - 1.2$ min = 1.15 min.
The third half-life is the time to go from $\frac{[\mathbf{I}]_0}{4}$ to $\frac{[\mathbf{I}]_0}{8} = 3.55 - 2.35$ min = 1.2 min.

The three half-lives are effectively equal and the reaction is therefore first order with respect to compound **I**.

Box 10.2 Radioactive decay

The decay of a radionuclide is always a first order process, and a radionuclide therefore possesses a characteristic half-life. This may be a matter of seconds, or it may be years or even thousands of years; the production of artificial nuclides was described in Box 1.3.

Isotope	*Naturally occurring?*	*Half-life*
$^{235}_{92}U$	Yes	7.04×10^8 years
$^{57}_{27}Co$	No	270 days
$^{130}_{55}Cs$	No	30.7 minutes

The figure below illustrates the decay of $^{207}_{81}Tl$ to the stable isotope $^{207}_{82}Pb$. The half-life can be determined by measuring the time taken for the initial concentration of $^{207}_{81}Tl$ to halve. To increase the accuracy of the measurement, several consecutive half-lives should be determined and the average value taken. What is the half-life of $^{207}_{81}Tl$?

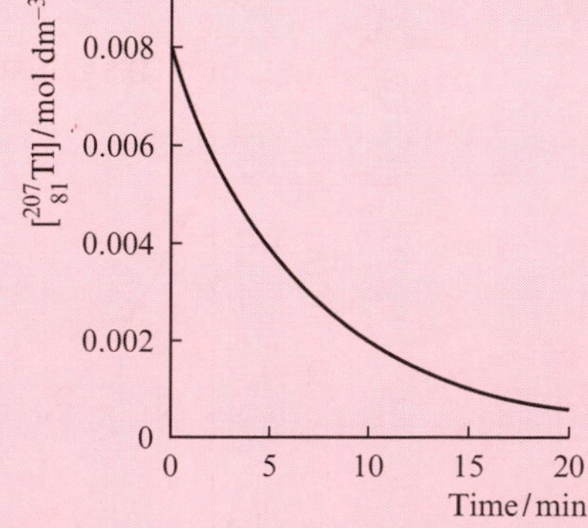

The decay from $^{207}_{81}Tl$ to $^{207}_{82}Pb$ is the last step in a natural decay chain from $^{235}_{92}U$ to $^{207}_{82}Pb$ – there are 11 steps in the chain. In a radioactive decay chain, the decay of one radionuclide produces a *daughter-nuclide* that may be stable or may itself decay.

10.4 RATE EQUATIONS FOR REACTIONS WITH MORE THAN ONE REACTANT

The simple reaction of A that we have used so far is not common in real chemical situations. More often there are two or more reactants and the rate of reaction is dependent upon the concentration of more than one species. We now consider the implications of this for the rate equation. In this section our general reaction is given in equation 10.11.

$$A + B \rightarrow \text{products} \tag{10.11}$$

The general rate equation and orders

The rate of the reaction in equation 10.11, may depend upon:

- [A] only; or
- [B] only; or
- both [A] and [B]; or
- neither [A] nor [B].

The rate of the reaction can be measured by monitoring the disappearance of A or B, or the appearance of the products. Since A and B react in a 1 : 1 ratio, they will be used up at the same rate as each other. A general rate equation is written in equation 10.12.

$$\text{Rate of reaction} = -\frac{d[A]}{dt} = -\frac{d[B]}{dt} = k[A]^n[B]^m \tag{10.12}$$

The reaction is n^{th} order with respect to A, and m^{th} order with respect to B, and the *overall order* of the reaction is $(m + n)$. Compare this with the case of reaction 10.10 where the overall order and the order with respect to A must be identical. The rate constant k refers to the overall rate of the reaction. For example, the rate equation for the hydrolysis of sucrose (reaction 10.13 and Figure 10.5) is given in equation 10.14. The reaction is first order with respect to sucrose and first order with respect to water, and, consequently, is second order overall.

$$\underset{\text{sucrose}}{C_{12}H_{22}O_{11}} + H_2O \rightarrow \underset{\text{glucose}}{C_6H_{12}O_6} + \underset{\text{fructose}}{C_6H_{12}O_6} \tag{10.13}$$

$$\text{Rate of reaction} = k[C_{12}H_{22}O_{11}][H_2O] \tag{10.14}$$

It is important to remember that the stoichiometric equation for a reaction tells you *nothing*, *a priori*, about the rate equation. In reaction 10.13, both of the reactants *do* occur in the rate equation, but this is not always the case. For example, reaction 10.15 does *not* exhibit a rate equation which follows from the stoichiometry (equation 10.16).

See the discussion of molecularity in Section 10.9

$$[MnO_4]^- + 8H^+ + 5Fe^{2+} \rightarrow Mn^{2+} + 4H_2O + 5Fe^{3+} \tag{10.15}$$

$$\text{Rate of reaction} \neq k[MnO_4^-][H^+]^8[Fe^{2+}]^5 \tag{10.16}$$

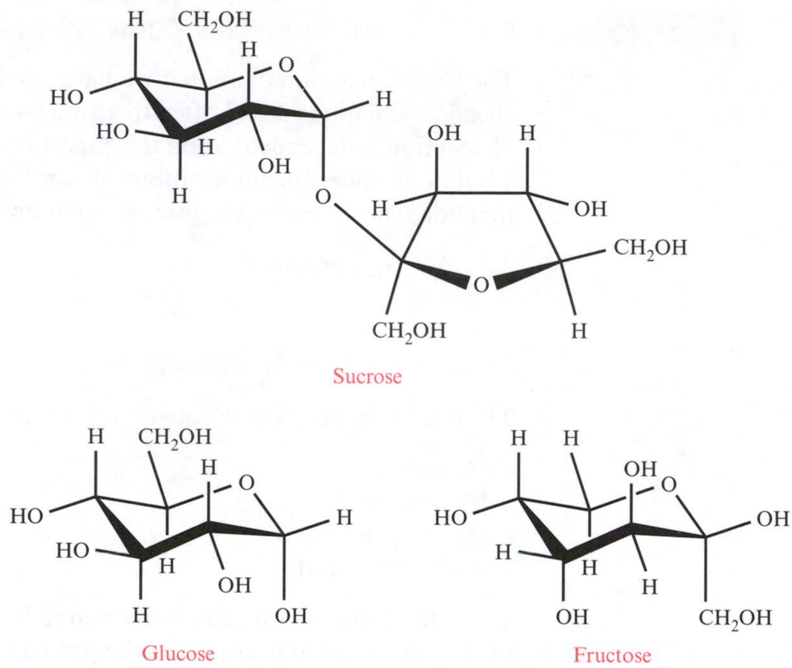

10.5 The structures of sucrose, glucose and fructose.

Sucrose

Glucose

Fructose

The problem of investigating the rate when there is more than one reactant

In a system which has more than one reactant, the concentrations of *all* the reactants decrease with time. In order to deduce the rate equation, we should separately observe the effect that each reactant has on the rate of the reaction. This can be achieved by having all but one of the reactants present in *vast excess* so that as the reaction proceeds, there is no effective change in the concentration of these components.

Consider the reaction of A with B (equation 10.11) and let there be 1×10^{-4} moles of A and 1×10^{-2} moles of B present initially in 10 cm^3 of solution. When half of the initial amount of A has reacted, 9.95×10^{-3} moles of B remain unused. Thus the concentration of B has hardly changed, whereas the concentration of A has fallen significantly. (As an exercise, determine the concentration of A and B at the beginning of the reaction and at the point when half of A has reacted. What is the percentage change in each of the two concentrations?) Any effect on the observed rate will be mainly due to changes in the concentration of A. In order to find out how the rate depends upon [B], the reaction must be repeated either with A in vast excess, or with different concentrations of B (always in a large excess) as will be shown in worked example 10.3.

Consider an example in which the overall rate equation for the reaction between A and B is given in equation 10.17. If B is in vast excess, then [B] is assumed to be a constant (known) value, and the rate equation takes the form shown in equation 10.18.

$$\text{Rate of reaction} = -\frac{d[A]}{dt} = k[A][B] \quad (10.17)$$

$$\text{Rate of reaction} = -\frac{d[A]}{dt} = k'[A] \quad (10.18)$$

See worked examples 10.3–10.5

where $k' = k[B]$. The reaction is said to show *pseudo-first order* kinetics. It is first order with respect to A, and the condition of a constant concentration of B leads to the reaction *appearing* to be first order overall.

10.5 INTEGRATED RATE EQUATIONS

So far we have dealt with rates of reaction in terms of a differential equation, and Figure 10.4 illustrated a graphical strategy for determining simple rate equations. However, determining rates by drawing tangents to curves is not always an accurate method.

The experimental data are often in the form of readings of [A] at various times, t, and therefore it would be useful to know exactly how [A] varies with t rather than having to treat the data as outlined in Figure 10.4. Thus, if the reaction is zero, first or second order with respect to A, what is the *actual* relationship between [A] and t?

These relationships are given by writing the *integrated forms of the rate equations*. Details of the integrations are given in Box 10.3; in the text, we shall simply present the equations and apply them. All expressions refer either to a reaction of the form given in equation 10.2, or to the condition in which all reactants other than A are in a vast excess (see Section 10.4).

Box 10.3 How is an integrated rate equation derived?

The equation for a straight line

The general equation for a straight line is:

$$y = mx + c$$

where x and y are the variables, m is the gradient of the line, and c is the intercept on the y axis.

Zero order with respect to A

For a zero order rate equation, the rate is expressed in the differential form as:

$$\text{Rate of reaction} = -\frac{d[A]}{dt} = k$$

However, experimental results are often in the form of readings of [A] taken at particular times, t. A plot of [A] against t gives a straight line, and the equation

continues ▶

of this line is of the form:

y = m x + c

y → [A]; m → gradient of line; x → t; c → intercept on y axis

But what are the significances of the gradient and the intercept?

The complete equation can be found by *integrating* the differential form of the rate equation. The symbol ∫ means that we are taking the integral of something but we must specify 'integration *with respect to* what?' Thus, $\int dx$ means that we are integrating *with respect to the variable x*.

Both [A] and t are variables, but before we can integrate the differential form of the rate equation, we must separate the variables onto opposite sides of the equation. Thus we have:

$$-d[A] = k\,dt$$

On integration of both sides of the equation we have:

$$-\int d[A] = \int k\,dt$$

$$c - [A] = kt$$

where c is the integration constant.

To find c, we know that at $t = 0$, [A] is the initial concentration $[A]_0$ and substituting these values into the above equation gives:

$$[A]_0 = c$$

and thus:

$$[A]_0 - [A] = kt$$

or

$$[A] = [A]_0 - kt$$

This is therefore the equation for the line obtained in a plot of [A] against t when the reaction is zero order with respect to A.

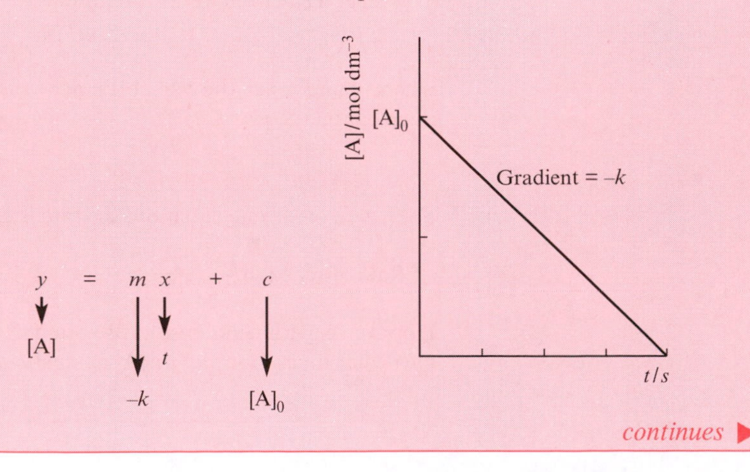

continues ▶

The rate constant can therefore be found directly from the gradient.

Similarly, we can derive the integrated forms of the first and second order rate equations as follows.

First order with respect to A

The derivative form of the rate equation is:

$$-\frac{d[A]}{dt} = k\,[A]$$

After separating the variables, we can write:

$$-\frac{d[A]}{[A]} = k\,dt$$

Integration then gives:

$$-\int \frac{d[A]}{[A]} = \int k\,dt$$

$$c - \ln[A] = kt$$

where c is the integration constant.

At $t = 0$, $[A] = [A]_0$ and substituting these values into the above equation gives:

$$\ln[A]_0 = c$$

and thus:

$$\ln[A]_0 - \ln[A] = kt$$

or

$$\ln[A] = \ln[A]_0 - kt$$

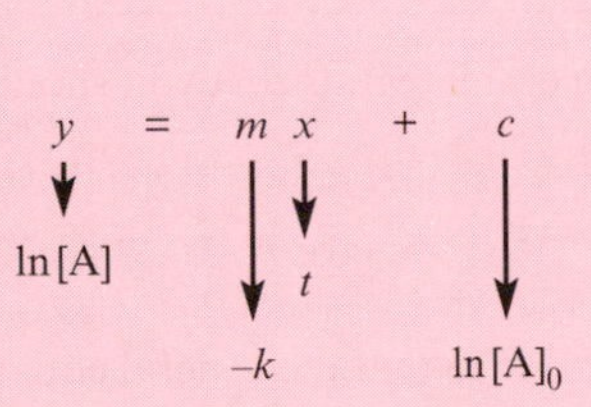

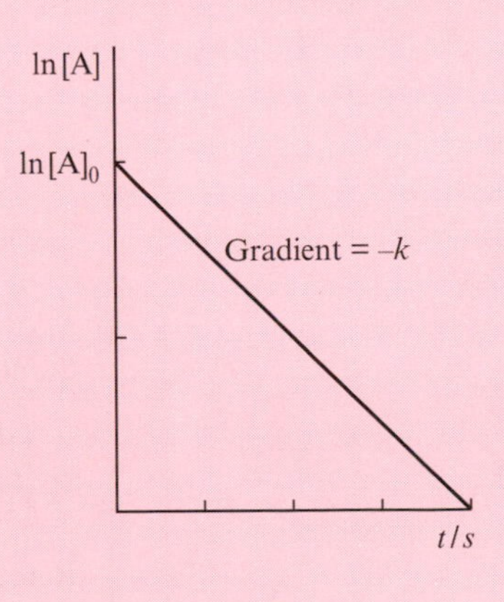

Thus, if a plot of $\ln[A]$ against time is linear, the rate is first order with respect to A, and the rate constant is found from the gradient of the graph.

continues ▶

Second order with respect to A

The derivative form of the rate equation is:

$$-\frac{d[A]}{dt} = k\,[A]^2$$

After separating the variables, this leads to:

$$-\frac{d[A]}{[A]^2} = k\,dt$$

Integration then gives:

$$-\int \frac{d[A]}{[A]^2} = \int k\,dt$$

$$\frac{1}{[A]} = c + kt$$

where c is the integration constant.
At $t = 0$, $[A] = [A]_0$ and substituting these values into the above equation gives:

$$\frac{1}{[A]_0} = c$$

and thus:

$$\frac{1}{[A]} = \frac{1}{[A]_0} + kt$$

Thus, if a plot of $\frac{1}{[A]}$ against time is linear, the rate is second order with respect to A, and the rate constant is found from the gradient of the graph.

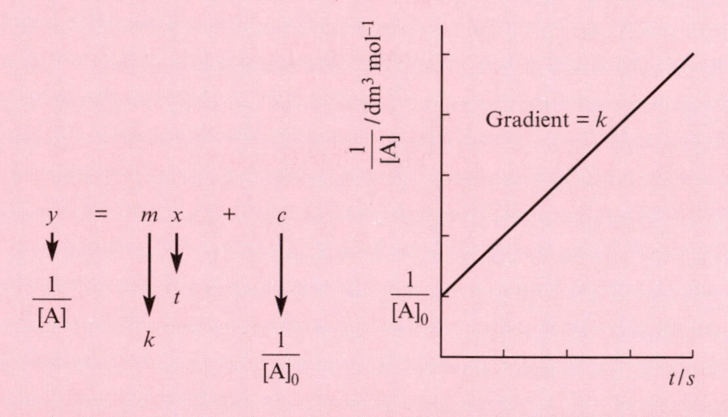

Some general integrals that you should remember:

$\int dx = x$ $\quad\quad$ $\int a\,dx = a\,x$ (a = constant)

$\int \frac{dx}{x} = \ln x$ $\quad\quad$ $\int \frac{dx}{x^2} = -\frac{1}{x}$

Remember also that the general equation for a straight line is:

$$y = mx + c$$

where x and y are variables, m is the gradient, and c is the intercept on the y axis.

Zero order with respect to A

The integrated rate equation for a reaction that is zero order with respect to A has the form given in equation 10.19.

$$[A] = [A]_0 - kt \qquad (10.19)$$

where [A] is the concentration of A at time t, and $[A]_0$ is the initial concentration of A.

A plot of [A] against t is linear and the rate constant, k, can be found from the gradient.

Worked example 10.3 ***The bromination of acetone***

Bromine reacts with acetone in the presence of hydrochloric acid according to the following equation:

$$H_3C{-}C(=O){-}CH_3 + Br_2 \xrightarrow{H^+} H_3C{-}C(=O){-}CH_2Br + HBr$$

The reaction can be monitored by using UV–VIS spectroscopy; aqueous bromine absorbs light with a wavelength of 395 nm.

The rate equation is given by:

$$\text{Rate of reaction} = -\frac{d[Br_2]}{dt} = k[Br_2]^m[CH_3C(O)CH_3]^n[H^+]^p$$

(a) A reaction is carried out in aqueous solution with initial concentrations of acetone, hydrochloric acid and bromine being 1.6, 0.4 and 0.0041 mol dm^{-3} respectively, and the results are tabulated below. Determine the order of the reaction with respect to Br_2.

Time / s	0	10	20	30	40	50	60
$[Br_2] \times 10^3$ / mol dm^{-3}	4.1	3.85	3.6	3.3	3.1	2.85	26

(b) If the acid and acetone are kept in excess but their concentrations are varied, the rate of the reaction alters according to the following results. Use these data along with those in part (a) to derive the overall rate equation.

Acetone / mol dm^{-3}	1.6	1.6	0.8	0.8
$[H^+]$ / mol dm^{-3}	0.4	0.2	0.4	0.2
$[Br_2]_0$ / mol dm^{-3}	0.0041	0.0041	0.0041	0.0041
Relative rate of reaction	1	0.5	0.5	0.25

(Data from: J. P. Birk and D. L. Walters, *Journal Chemical Education*, (1992) vol 69, p. 585.)

Part (a): The conditions of the reaction are such that both the acid and acetone are in vast excess and so the rate equation can therefore be written as:

$$-\frac{d[Br_2]}{dt} = k' [Br_2]^m$$

We are given no clues about the possible dependence of the rate on $[Br_2]$ and so we may begin by plotting $[Br_2]$ against time:

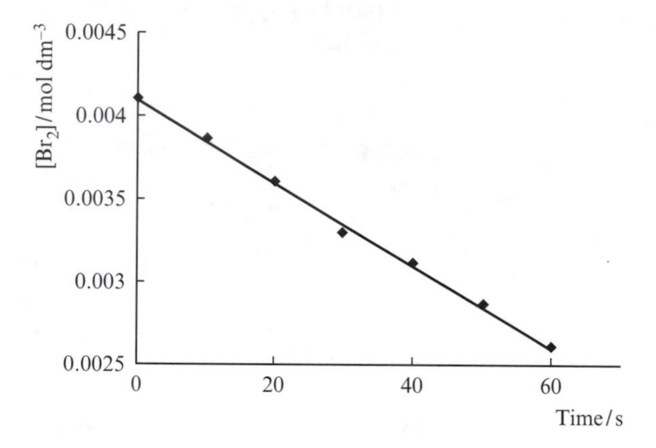

Since this graph is linear, we can deduce that the rate is independent of the concentration of bromine in solution – it is zero order with respect to Br_2.

The rate equation with respect to bromine concentration is therefore:

$$-\frac{d[Br_2]}{dt} = k'[Br_2]^0 = k'$$

Part (b): The overall rate equation is given in the question as:

$$-\frac{d[Br_2]}{dt} = k[Br_2]^m[CH_3C(O)CH_3]^n[H^+]^p$$

but since $m = 0$, we can simplify this to:

$$-\frac{d[Br_2]}{dt} = k[CH_3C(O)CH_3]^n[H^+]^p$$

From the table of data, we see that:

- the rate halves when $[H^+]$ halves;
- the rate halves when $[CH_3C(O)CH_3]$ halves;
- the rate falls by a factor of 4 when both $[H^+]$ and $[CH_3C(O)CH_3]$ are halved.

We can conclude that the rate depends directly on each of $[CH_3C(O)CH_3]$ and $[H^+]$ – the reaction is therefore first order with respect to each component.

The rate equation becomes:

$$-\frac{d[Br_2]}{dt} = k[CH_3C(O)CH_3][H^+]$$

How can we determine the overall rate constant?

For the reaction in part (a), we know both $[CH_3C(O)CH_3]$ and $[H^+]$ and these are assumed to be constant during the reaction.

[*Question*: Work out the percentage change in the concentrations of acetone and H^+ when *all* of the Br_2 has reacted. Is the assumption of constant concentrations valid?]

Thus, for this particular experiment:

$$-\frac{d[Br_2]}{dt} = k[CH_3C(O)CH_3][H^+] = k \times 1.6 \times 0.4$$

The gradient of the graph in part (a) is k':

$$k' = 2.5 \times 10^{-5} \text{ mol dm}^{-3} \text{ s}^{-1}$$

Combining the two expressions for the rate we have:

$$-\frac{d[Br_2]}{dt} = k' = k \times 1.6 \times 0.4$$

Therefore, we can find the overall rate constant, k.

$$k = \frac{k'}{1.6 \times 0.4} = \frac{2.5 \times 10^{-5}}{1.6 \times 0.4}$$

$$= 3.9 \times 10^{-5} \text{ dm}^3 \text{ mol}^{-1} \text{ s}^{-1}$$

[*Question*: Why are the units of k different from those of k' ?]

First order with respect to A

Equation 10.20 gives the integrated form of the rate equation for a reaction that is first order with respect to A.

$$\ln[A] = \ln[A]_0 - kt \qquad (10.20)$$

Plotting $\ln[A]$ against time gives a straight line with an intercept of $\ln[A]_0$. The rate constant may be determined from the gradient of the line.

Worked example 10.4 ***The esterification of trifluoroacetic acid***

Benzyl alcohol reacts with trifluoroacetic acid according to the following equation:

$$\underset{\text{benzyl alcohol}}{C_6H_5CH_2OH} + \underset{\text{trifluoroacetic acid}}{CF_3CO_2H} \longrightarrow \underset{\text{benzyl trifluoroacetate}}{CF_3CO_2CH_2C_6H_5} + \underset{\text{water}}{H_2O}$$

This is an example of the more general reaction:

Alcohol + Carboxylic acid ⟶ Ester + Water

The disappearance of the alcohol can be followed by ^{1}H NMR spectroscopy, and monitoring changes in the integral of the signal for the CH_2 protons. The data below have been recorded at 310 K for a reaction in which the acid is present in a large excess. Confirm that the reaction is first order with respect to the alcohol, and determine the pseudo-first order rate constant.

Time / min	0	6.5	10	15	20	25	30	40	50
[Alcohol] / mol dm^{-3}	1.00	0.67	0.57	0.41	0.30	0.23	0.14	0.07	0.04

(Data from: D. E. Minter and M. C. Villareal, *Journal of Chemical Education*, (1985) vol. 62, p. 911.)

The rate equation for the reaction may be written as:

$$\text{Rate of reaction} = -\frac{\mathrm{d}[C_6H_5CH_2OH]}{\mathrm{d}t} = k[C_6H_5CH_2OH]^n[CF_3CO_2H]^m$$

but with the acid in vast excess the equation is:

$$-\frac{\mathrm{d}[C_6H_5CH_2OH]}{\mathrm{d}t} = k'[C_6H_5CH_2OH]^n$$

We need to confirm that $n = 1$, and also find the pseudo-first order rate constant k'. This is most efficiently achieved by plotting ln [Alcohol] against time.

First, tabulate the necessary data:

Time/min	0	6.5	10	15	20	25	30	40	50
ln [Alcohol]	0	–0.40	–0.562	–0.892	–1.204	–1.470	–1.966	–2.659	–3.219

A graph of ln [Alcohol] against time is linear:

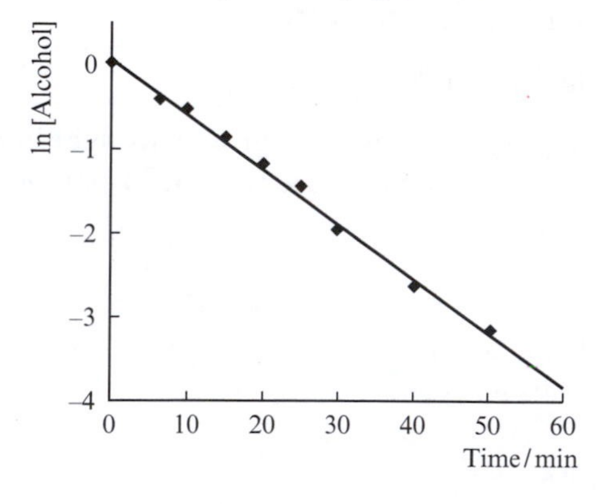

This is consistent with the integrated first order rate equation:

$$\ln[\text{Alcohol}] = \ln[\text{Alcohol}]_0 - k't$$

and the result confirms that the reaction is first order with respect to benzyl alcohol.

The pseudo-first order rate constant k' is found from the slope of the line:

$$k' = 0.066\ \text{min}^{-1} \text{ or } 1.1 \times 10^{-3}\ \text{s}^{-1}$$

Second order with respect to A

The integrated form of the rate equation for a reaction that is second order with respect to A is stated in equation 10.21.

$$\frac{1}{[A]} = \frac{1}{[A]_0} + kt \qquad (10.21)$$

A graph of $\frac{1}{[A]}$ against time is linear, and the intercept is $\frac{1}{[A]_0}$; k is determined from the gradient of the line.

Worked example 10.5 ***The reaction between I_2 and hex-1-ene***

Hex-1-ene reacts with I_2 in acetic acid to give 1,2-diodohexane. The dependence of the rate on $[I_2]$ can be studied if a large excess of hex-1-ene is used. Results recorded at 298 K are tabulated below. Determine the order of the reaction with respect to I_2, and write a rate equation for the reaction. Do your results exclude the involvement of acetic acid in the rate-determining step?

Time / s	0	1000	2000	3000	4000	5000	6000	7000	8000
$[I_2]$ / mol dm^{-3}	0.02	0.0156	0.0128	0.0109	0.0094	0.0083	0.0075	0.0068	0.0062

(Data from: K. W. Field, D. Wilder, A. Utz and K. E. Kolb, *Journal of Chemical Education*, (1987) vol. 64, p. 269.)

The overall rate equation can be written as follows:

$$-\frac{d[I_2]}{dt} = k[I_2]^m[\text{hex-1-ene}]^n$$

but since the hex-1-ene is in excess, we can write:

$$-\frac{d[I_2]}{dt} = k'[I_2]^m$$

Firstly, plot $[I_2]$ as a function of time:

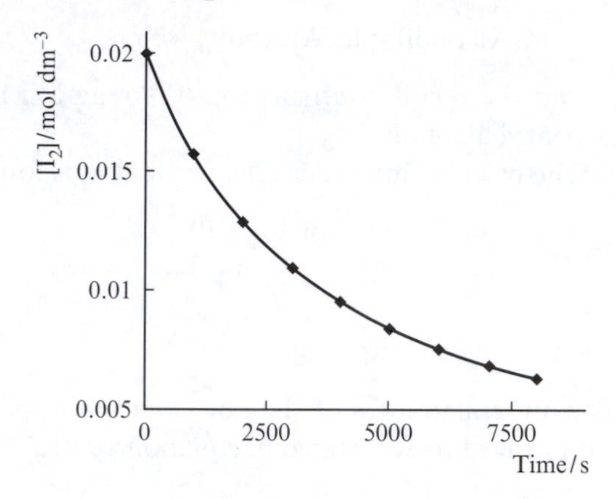

Since this graph is non-linear, we know that $m \neq 0$, but it may be 1 or 2. However, we can rule out the first order dependence by looking at the first two half-lives. The time taken for the initial concentration to halve is ≈3500 s. If the reaction were first order with respect to I_2, it should take another 3500 s for the concentration of I_2 to halve again. This is *not* the case. [*Question*: Give a rough estimate for the second half-life.]

The next step is to test for a second order dependence. Using the integrated form of the rate equation, we have:

$$\frac{1}{[I_2]} = \frac{1}{[I_2]_0} + k't$$

and so we need to plot $\frac{1}{[I_2]}$ against time.

Time/s	0	1000	2000	3000	4000	5000	6000	7000	8000
$\frac{1}{[I_2]}$/mol^{-1} dm^3	50	64	78	92	106	120	134	148	162

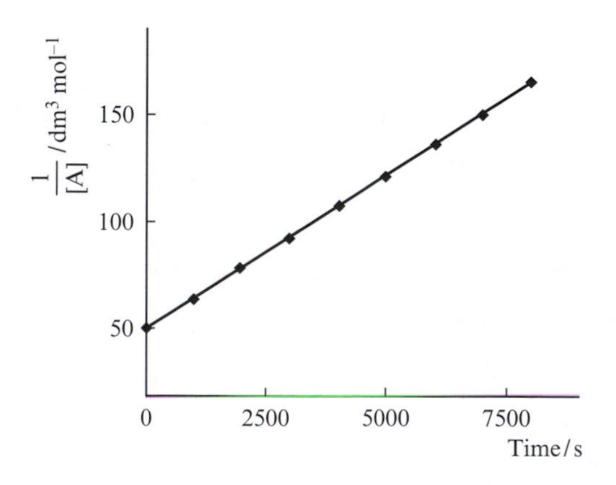

The plot is linear and confirms that the reaction is second order with respect to I_2.

The pseudo-second order rate equation is therefore:

$$-\frac{d[I_2]}{dt} = k'[I_2]^2$$

and we can determine k' from the gradient of the graph.

Gradient = k' = 0.014 $dm^3 mol^{-1} s^{-1}$

The involvement of acetic acid cannot be ruled out even though it is the solvent; it is present in vast excess.

10.6 THE DEPENDENCE OF RATE ON TEMPERATURE: THE ARRHENIUS EQUATION

Boltzmann distributions and activation energy

Boltzmann distribution: see Section 1.18

The Boltzmann distribution of molecular energies is temperature-dependent, and Figure 10.6 shows typical distributions at two temperatures. The number of molecules in a given sample stays constant, and so the area under the two graphs must be the same. However, the maximum of the graph shifts to higher energy at higher temperature – the change in shape of the graph corresponds to there being more molecules with higher energies at the higher temperature.

Energy and enthalpy: see Section 3.5

A reaction can only proceed when molecules possess a certain minimum energy, $E_{activation} \approx \Delta_{activation}H$. For a given reaction, the activation energy (Figure 10.7) essentially has a *fixed* value; $E_{activation}$ is *temperature-independent*. If we mark an arbitrary value on Figure 10.6 as $E_{activation}$, then it is apparent that the number of molecules that possess this, or a higher, energy increases as the temperature is increased. This means that a greater proportion of molecules can react, and there is an *increase in the rate of reaction*. It is often found that a rise in temperature of 10 K leads to an approximate doubling of the rate of reaction; we consider this statement critically later in this section.

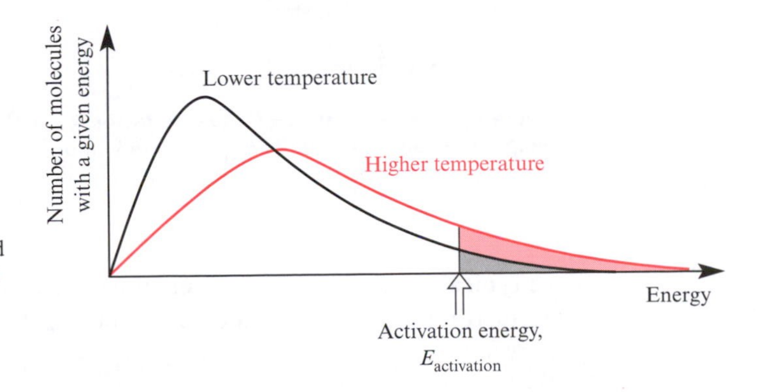

10.6 Boltzmann distributions of molecular energies at two temperatures. For a specified number of molecules, the area under each curve is the same. For a given reaction, the activation energy is *not* dependent on temperature. The pink shaded area overlaps the grey shaded area.

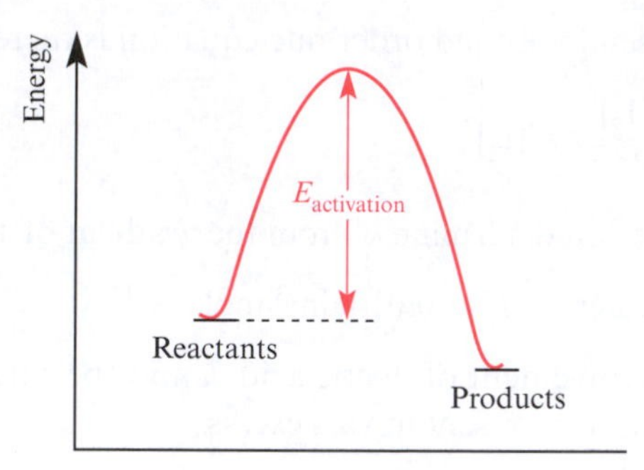

10.7 The reaction profile for a single-step reaction showing $E_{activation}$.

The Arrhenius equation

The discussion above is quantified by the *Arrhenius* relationship (equation 10.22). Notice that the Arrhenius equation involves the *energy* of activation; in Section 3.5 we discussed the difference between an energy and an enthalpy value; here we set $E_{activation} \approx \Delta_{activation}H$.

$$\ln k = \ln A - \frac{E_{activation}}{RT} \qquad (10.22)$$

where: the units of $E_{activation}$ are kJ mol^{-1},
k = rate constant,
A = frequency factor,
R = molar gas constant = 8.314×10^{-3} kJ K^{-1} mol^{-1},
T = temperature (K).

The frequency factor or pre-exponential factor, A, may be taken to be a constant for a given reaction. It has the same units as the rate constant and is related to the rate at which collisions occur between reactant molecules and also to the relative orientation of the reactants.

The Arrhenius equation states:

$$\ln k = \ln A - \frac{E_{activation}}{RT}$$

where k = rate constant, A = frequency factor, R = molar gas constant, and T = temperature and the units of $E_{activation}$ are kJ mol^{-1}.

Some typical values of activation energies are listed in Table 10.1, and the data illustrate the effect that a catalyst has on the value of $E_{activation}$ for a given reaction.

Worked example 10.6 ***Determination of the activation energy for the decomposition of an organic peroxide***

The decomposition of an organic peroxide ROOR is first order with respect to the peroxide. The rate constant varies with temperature as follows:

Temperature / K	**410**	**417**	**426**	**436**
k / s^{-1}	**0.0193**	**0.0398**	**0.0830**	**0.2170**

Determine the energy of activation for the reaction. [$R = 8.314 \times 10^{-3}$ kJ K^{-1} mol^{-1}]

The Arrhenius equation gives a relationshp between k and T:

$$\ln k = \ln A - \frac{E_{\text{activation}}}{RT}$$

and so we need to plot $\ln k$ against $\frac{1}{T}$.

$\frac{1}{T}$ / K^{-1}	0.00244	0.00240	0.00235	0.00229
$\ln k$	–3.948	–3.224	–2.489	–1.528

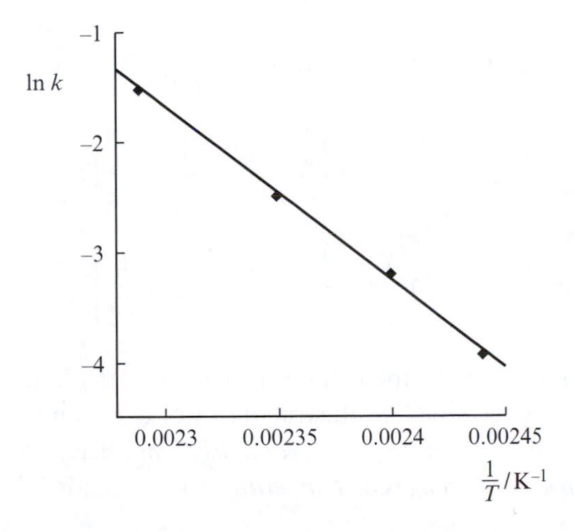

$E_{\text{activation}}$ can be found from the gradient of the line:

$$-\frac{E_{\text{activation}}}{R} = -15\,960$$

$$E_{\text{activation}} = 15\,960 \times 8.314 \times 10^{-3} = 132.7 \text{ kJ mol}^{-1}$$

Table 10.1 Typical values of activation energies. Note the effect that a catalyst has on the value of $E_{activation}$.

Reaction	*Comments*	$E_{activation}$ / kJ mol^{-1}
$2H_2O_2 \rightarrow 2H_2O + O_2$	No catalyst	79
$2H_2O_2 \rightarrow 2H_2O + O_2$	Enzyme catalysed	23
$2NOCl \rightarrow 2NO + Cl_2$		100
$NO + O_3 \rightarrow NO_2 + O_2$		10.5
$2HI \rightarrow H_2 + I_2$	No catalyst	185
$2HI \rightarrow H_2 + I_2$	With a gold catalyst	121
$2HI \rightarrow H_2 + I_2$	With a platinum catalyst	59
$2NH_3 \rightarrow N_2 + 3H_2$	No catalyst	335
$2NH_3 \rightarrow N_2 + 3H_2$	With a tungsten catalyst	162
$C_2H_5Br + [OH]^- \rightarrow C_2H_5OH + Br^-$	Reaction in aqueous solution	90

In addition to allowing us to calculate the activation energy of a reaction, the Arrhenius equation also permits us to estimate how the rate of a reaction changes as a function of temperature. For a given reaction, let us assume that A and $E_{activation}$ are constant and thus (using equation 10.22), we can write equation 10.23. This simplifies to equation 10.24 in which k_n is the rate constant at temperature T_n.

$$\ln k_2 - \ln k_1 = \left[\ln A - \frac{E_{activation}}{RT_2}\right] - \left[\ln A - \frac{E_{activation}}{RT_1}\right]$$

$$= \frac{E_{activation}}{R}\left[\frac{1}{T_1} - \frac{1}{T_2}\right] \qquad (10.23)$$

$$\ln \frac{k_2}{k_1} = \frac{E_{activation}}{R}\left[\frac{1}{T_1} - \frac{1}{T_2}\right] \qquad (10.24)$$

This gives us the *ratio of rate constants* $\frac{k_2}{k_1}$, and a ratio of reaction rates can be determined as illustrated in worked example 10.7.

A word of caution: *reactions do not necessarily obey the Arrhenius equation over wide ranges of temperature* and equation 10.24 must be used with care.

Worked example 10.7 ***The dependence of rate on temperature***

The activation energy for the reaction of hydroxide ion with bromoethane to give ethanol is 90 kJ mol^{-1}. How much faster will the reaction proceed if the temperature is raised from 295 to 305 K? [$R = 8.314 \times 10^{-3}$ kJ K^{-1} mol^{-1}]

The larger the value of k, the faster the reaction rate.

From the Arrhenius equation:

$$\ln \frac{k_2}{k_1} = \frac{E_{activation}}{R}\left[\frac{1}{T_1} - \frac{1}{T_2}\right]$$

$$\ln \frac{k_2}{k_1} = \frac{90}{8.314 \times 10^{-3}} \times \left[\frac{1}{295} - \frac{1}{305} \right]$$

$$= 1.20$$

$$\frac{k_2}{k_1} = 3.32$$

Thus, the rate constant increases by a factor of 3.32 when the temperature is raised from 295 to 305 K.

We have been given no information about the rate equation but for *two reactions with equal initial concentrations of reactants*, the ratio of the rates will be equal to the ratio of the rate constants:

$$\frac{\text{Rate 1}}{\text{Rate 2}} = \frac{k_1[C_2H_5Br]^m[OH^-]^n}{k_2[C_2H_5Br]^m[OH^-]^n}$$

Hence the rate of reaction at 305K is 3.32 times as fast as the rate of reaction at 295 K.

At the beginning of this section, we commented that a rise in temperature of 10 K leads to an approximate doubling of the reaction rate. Notice, however, that equation 10.24 shows that there is also a dependence on $E_{activation}$. Figure 10.8 illustrates how an increase in the temperature from 300 to 310K affects the rate of a reaction for values of $E_{activation}$ ranging from 10 to 150 kJ mol^{-1}. The approximate doubling of rate only appertains to values of $E_{activation} \approx 50$ kJ mol^{-1}; this is a typical value for many reactions in solution.

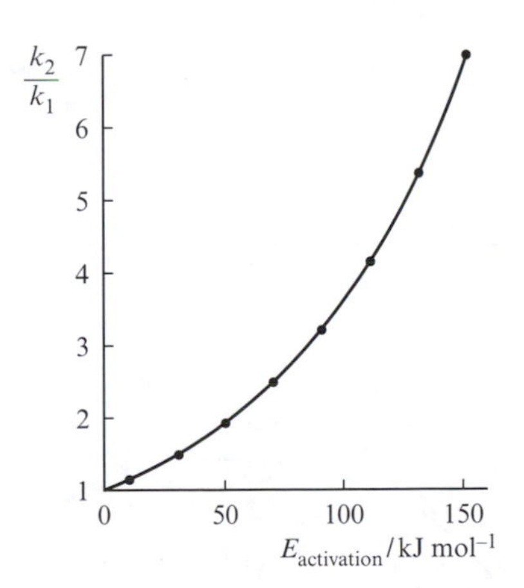

10.8 For a rise in temperature from 300 to 310K, the ratio of the rate constants $\frac{k_2}{k_1}$ for a reaction changes as a function of the energy of activation. The approximate doubling of a reaction rate for a 10 degree temperature rise is only valid for values of $E_{activation}$ of ≈ 50 kJ mol^{-1}. (See problem 10.13).

10.7 CATALYSIS AND AUTOCATALYSIS

A catalyst is a substance that alters the rate of a reaction without appearing in any of the products of that reaction. Catalysts may speed up or slow down a reaction. The latter is known as *negative catalysis* and is critical in stabilizing many commercially important materials. Examples of some catalysed reactions are given in Table 10.1.

A catalyst may speed up or slow down a reaction, and does not appear in the product.

An interesting phenomenon occurs when one of the products of a reaction is able to act as a catalyst for the process. The reaction is said to be *autocatalytic.* The reaction in equation 10.25 begins with only A and B present; one of the products, D, is a catalyst for the reaction and thus once D is present (equation 10.26), the rate of the reaction will increase.

$$\text{Initially:} \qquad A + B \longrightarrow C + D \qquad (10.25)$$

$$\text{Autocatalysed:} \qquad A + B \xrightarrow{D} C + D \qquad (10.26)$$

Figure 10.9 shows how the concentration of $[MnO_4]^-$ varies during reaction with dimethylamine. Initially, the reaction proceeds relatively slowly, but once products are present in solution, the rate at which the manganate(VII) ions disappear increases. The reaction is catalysed by the Mn^{2+} ions generated in the reduction of manganate(VII).

A biological example is the conversion of trypsinogen to trypsin, a process which is catalysed by trypsin. Figure 10.10 shows the appearance of trypsin in the reaction as a function of time. The S-shaped plot is characteristic of an autocatalytic reaction – once the catalyst enters the system, the reaction accelerates.

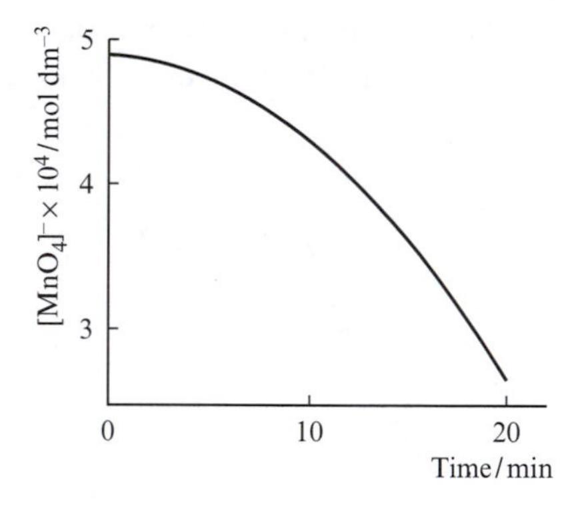

10.9 A plot of the concentration of manganate(VII) against time for the autocatalytic reaction between Me_2NH and $[MnO_4]^-$ in aqueous solution; the catalyst is Mn(II). (Data from: F. Mata-Perez and J. F. Perez-Benito, *Journal of Chemical Education*, (1987) vol. 64, p. 925.)

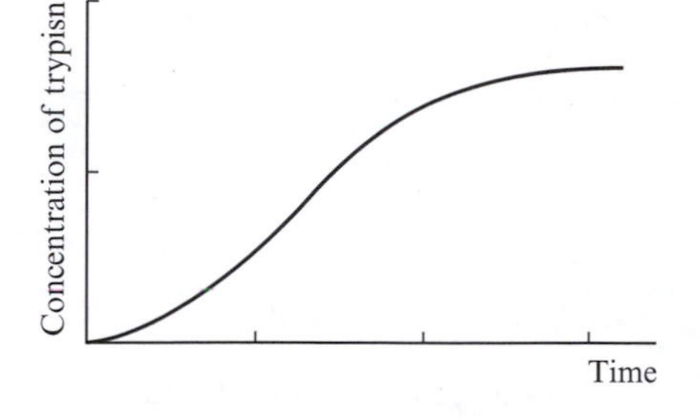

10.10 A plot of [trypsin] against time as trypsinogen is hydrolysed to trypsin. This product catalyses the reaction, and the S-shape of the curve is characteristic of an autocatalytic reaction.

10.8 REVERSIBLE REACTIONS

So far we have only considered reactions which go in one direction, but in many cases, the forward reaction is opposed by the reverse (or back) reaction. In this section we briefly consider the consequences on the rate equation of having two opposing *first order* processes.

Consider the reactions given in equation 10.27, in which the rate constants for the forward and backward reactions are k_1 and k_{-1} respectively.

$$\mathrm{A} \underset{k_{-1}}{\overset{k_1}{\rightleftharpoons}} \mathrm{B} \quad (10.27)$$

The measured rate of disappearance of A does not obey simple first order kinetics because as B is formed, some of it re-forms A in a first order process. The rate of the forward reaction is given by equation 10.28.

$$\text{Rate of reaction} = -\frac{\mathrm{d[A]}}{\mathrm{d}t} = k_1[\mathrm{A}] - k_{-1}[\mathrm{B}] \quad (10.28)$$

Assuming that only A is present initially, then the *initial rate of reaction* is simply $k[\mathrm{A}]_0$ (from equation 10.8, with the initial concentration of A, $[\mathrm{A}]_0$).

The concentration of B at time t is equal to the *difference* in the concentrations of A at times 0 and t (equation 10.29). Note that this relationship is dependent upon the stoichiometry of the reaction; in this case, one mole of A gives one mole of B, and vice versa.

$$[\mathrm{B}] = [\mathrm{A}]_0 - [\mathrm{A}] \quad (10.29)$$

Thus, we can rewrite the rate equation in terms of A as shown in equations 10.30 and 10.31.

$$\text{Rate of reaction} = -\frac{\mathrm{d[A]}}{\mathrm{d}t} = k_1[\mathrm{A}] - k_{-1}\{[\mathrm{A}]_0 - [\mathrm{A}]\} \quad (10.30)$$

or

$$\text{Rate of reaction} = -\frac{\mathrm{d[A]}}{\mathrm{d}t} = (k_1 + k_{-1})[\mathrm{A}] - k_{-1}[\mathrm{A}]_0 \quad (10.31)$$

Remember that $[\mathrm{A}]_0$ is a constant and therefore the second term on the right-hand side of equation 10.31 is also a constant. Thus the reaction shows a first order dependence upon A, and a plot of rate against [A] is linear. Equation 10.31 has the form:

$$y = mx + c$$

and therefore a plot of rate of reaction against [A] is linear with a gradient equal to $(k_1 + k_{-1})$. The value of k_{-1} can be found from the intercept since the initial concentration of A is known. Thus, both rate constants may be determined.

Worked example 10.8 ***A reversible reaction***

Experimental data for the reaction:

$$\mathrm{A} \underset{k_{-1}}{\overset{k_1}{\rightleftharpoons}} \mathrm{B}$$

are tabulated below. The initial concentration of compound A is 0.05 mol dm^{-3}. Use these results to determine the rate constants k_1 and k_{-1}.

[A] / mol dm^{-3}	0.045	0.04	0.03	0.02	0.01
Rate $\times 10^4$ / mol dm^{-3} min^{-1}	8.97	7.94	5.88	3.82	1.76

The reversible reaction stated in the question should obey the rate equation:

$$-\frac{d[A]}{dt} = (k_1 + k_{-1})[A] - k_{-1}[A]_0$$

(from equation 10.31) and has the form: $y = mx + c$.
A plot of the reaction rate against [A] will be linear, with a gradient $m = (k_1 + k_{-1})$ and an intercept on the y axis, $c = -k_{-1}[A]_0$.

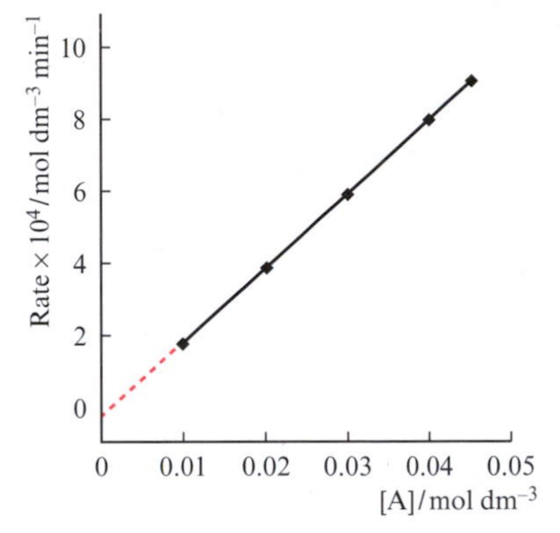

The gradient of the line = $(k_1 + k_{-1}) = 2.06 \times 10^{-2}$ min^{-1}.
After extrapolation, intercept on the y axis $= -k_{-1}[A]_0$

$$= -0.3 \times 10^{-4} \text{ mol dm}^{-3} \text{ min}^{-1}$$

From the intercept:

$$k_{-1} = \frac{0.3 \times 10^{-4}}{[A]_0} = \frac{0.3 \times 10^{-4}}{0.05} = 6 \times 10^{-4} \text{ min}^{-1}$$

Substituting the value for k_{-1} into the equation for the gradient allows us to find k_1:

$$\begin{aligned} k_1 &= (2.06 \times 10^{-2}) - k_{-1} \\ &= (2.06 \times 10^{-2}) - (6 \times 10^{-4}) \\ &= 0.02 \text{ min}^{-1} \end{aligned}$$

(Notice that the rate constant for the back reaction is much smaller than that for the forward reaction. What would be the effect on the reaction if $k_{-1} > k_1$?)

10.9 MOLECULARITY

In this section we move from studying the behaviour of bulk matter to a discussion of reactions at the level of individual molecules. Eventually this will lead us to the microscopic reaction mechanism. We emphasize a crucial point here – the kinetics of a reaction may be experimentally determined, and the *mechanism* of the reaction is inferred from these measurements. *A mechanism cannot be 'proved'* – at best a mechanism is consistent with all of the kinetic data available.

Elementary reactions

Our discussions of rate equations have centred on the two reactions given in equations 10.10 and 10.11. In many reactions, the complete reaction pathway from reactants to products involves a series of steps, each of which is known as an *elementary reaction*. For example, the chain reactions described in Section 8.11 involved several initiation, propagation and termination steps, and from the particular sequence of the elementary reactions, we can formulate a reaction mechanism. The mechanism may involve one or more intermediate species as shown in Figure 10.11 where the overall reaction is $A \rightarrow D$.

Molecularity: unimolecular and bimolecular reactions

The *molecularity* of an elementary reaction step is the *number of molecules or atoms of reactant* taking part. Most elementary reactions involve only one or two species and are known as *unimolecular* or *bimolecular steps* respectively.

The molecularity of an elementary step is the number of molecules or atoms of reactant taking part.

Equations 10.32 and 10.33 show examples of unimolecular reactions. The first is a decomposition (by homolytic bond cleavage) and the second is an isomerization.

$$Br_2 \rightarrow 2Br^{\bullet} \tag{10.32}$$

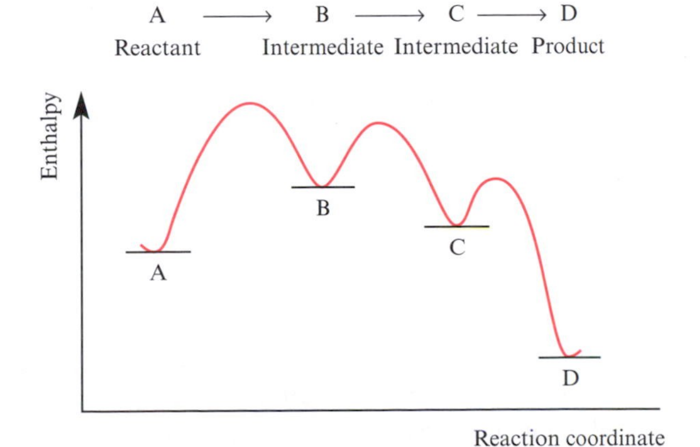

10.11 The reaction profile for a reaction $A \rightarrow D$. The mechanism comprises three elementary steps involving two intermediates, B and C.

$$\underset{\text{H}}{\overset{\text{Cl}}{}}\!\!\text{C}{=}\text{C}\!\!\underset{\text{Cl}}{\overset{\text{H}}{}} \longrightarrow \underset{\text{Cl}}{\overset{\text{H}}{}}\!\!\text{C}{=}\text{C}\!\!\underset{\text{Cl}}{\overset{\text{H}}{}} \qquad (10.33)$$

In a bimolecular step, two species combine to give the product(s). The reacting species may be identical to or different from each other, as equations 10.34 and 10.35 illustrate.

$$2NO_2 \rightarrow N_2O_4 \qquad (10.34)$$

$$CH_3CH_2Br + [OH]^- \rightarrow CH_3CH_2OH + Br^- \qquad (10.35)$$

In Section 10.10, we develop the relationship between elementary steps and overall reaction mechanisms, and in Section 10.11 we consider the reaction between H_2 and Br_2, the mechanism of which involves at least five elementary reactions.

Mechanisms of reactions are also discussed in Chapters 14, 15 and 17

Overall rate equations

For a given reaction, one of the individual elementary steps must be the slowest (rate-determining) stage. The molecularity of this rate-determining step (RDS) will define the observed kinetics for the overall reaction. This is the reason why the terms appearing in the rate equation may differ from the *overall* stoichiometry of the reaction.

RDS stands for rate-determining step.

In many cases, complex rate equations are observed because the individual species involved in the RDS are intermediates rather than starting materials. Normally we report rates of reactions in terms of the rate of loss of starting materials or rate of formation of products and do not include terms involving intermediate species; an example of a complex reaction of this type is discussed in Section 10.11.

10.10 MICROSCOPIC REACTION MECHANISMS AND THE STEADY-STATE APPROXIMATION

We have now seen the role that elementary steps play in defining the mechanism of a reaction. In this section we address the problem of writing rate equations in terms of the *measured* concentrations of starting materials or products.

Writing rate equations for unimolecular and bimolecular steps

Earlier in this chapter we used experimental data to determine *overall rate equations*. We have also seen that overall reaction mechanisms are composed of elementary reactions, and these are usually unimolecular or bimolecular steps. An objective of studying the kinetics of a reaction is to use the experimental data to write an overall reaction mechanism. The problem is tackled by first suggesting a mechanism, i.e. a series of elementary steps, and then writing down the rate equation which is in accord with the mechanism. If this rate equation is consistent with that determined experimentally, then it is possible (but not proven) that the proposed mechanism is correct.

Overall rate equations: see Section 10.4

We now come to a critical point. Although we *cannot* write down an overall rate equation using the stoichiometry of a reaction, we *can* deduce the rate equation for an elementary step from the stoichiometry of that step.

The rate equation for an elementary step can be deduced from the stoichiometry of that step, *but* the overall rate equation *cannot* be written down using the stoichiometry of the overall reaction.

In a unimolecular step such as the decomposition of Cl_2 (equation 10.36), the rate of reaction depends directly on the concentration of the reactant — one molecule of reactant is involved in the step. The decomposition of Cl_2 occurs when the molecule is excited (thermally or photolytically) and the Cl–Cl bond is broken. The rate of reaction for the fission of Cl_2 into two radicals is given in equation 10.37.

$$Cl_2 \rightarrow 2Cl^{\bullet} \tag{10.36}$$

$$\text{Rate of reaction} = k[Cl_2] \tag{10.37}$$

In a bimolecular step, two molecules must collide before a reaction can occur, and therefore the rate of reaction depends on the concentrations of both species. The dimerization of NO_2 is a bimolecular process (equation 10.34) and the rate of dimerization is given by equation 10.38. Notice that the *squared* dependence tells us that *two* molecules are involved.

$$\text{Rate of reaction} = k[NO_2]^2 \tag{10.38}$$

If the bimolecular reaction involves two different molecules, the rate equation will reflect this. Dioxygen is formed when an oxygen radical collides and reacts with an ozone molecule (equation 10.39) and the rate of the reaction depends on both $[O^{\bullet}]$ and $[O_3]$ as equation 10.40 shows.

$$O^{\bullet} + O_3 \rightarrow 2O_2 \tag{10.39}$$

$$\text{Rate of reaction} = k[O^{\bullet}][O_3] \tag{10.40}$$

For a unimolecular step:	**$A \rightarrow B$** **Rate of reaction = $k[A]$**
For a bimolecular step:	**$2A \rightarrow B$** **Rate of reaction = $k[A]^2$**
For a bimolecular step:	**$A + B \rightarrow C$** **Rate of reaction = $k[A][B]$**

Writing the overall rate equation for a two-step reaction

The discussion above assumed a simple elementary step in a reaction. What happens if there is more than one such step? Consider a reaction A → C which proceeds through an intermediate compound B; each step has a characteristic rate constant as shown in equations 10.41 and 10.42.

$$\mathrm{A} \xrightarrow{k_1} \mathrm{B} \qquad (10.41)$$

$$\mathrm{B} \xrightarrow{k_2} \mathrm{C} \qquad (10.42)$$

Now consider the consequences of some possible relationships between k_1 and k_2.

Case 1: $k_1 \ll k_2$

In this case, the rate of formation of B from A is much slower than the conversion of B to C. This means that as soon as any B is formed, it is converted to C. The overall rate of reaction is controlled by the RDS, i.e. A → B, and is given by equation 10.43.

Magnitude of k: see Section 10.2

$$\text{Rate of reaction} = k_1[\mathrm{A}] \qquad (10.43)$$

Case 2: $k_1 \gg k_2$

In this case, the RDS is the formation of C from the intermediate B and so the rate of reaction is given by equation 10.44.

$$\text{Rate of reaction} = k_2[\mathrm{B}] \qquad (10.44)$$

Now we have a problem. Our 'theoretical' rate equation is in terms of intermediate B, while our experimental data are in terms of the concentrations of the starting materials (A) or products (C). It is very unusual to be able to directly follow the concentrations of intermediates with time. In many cases, intermediates are only present in very low concentrations and cannot be measured. We can make use of this in what is known as the steady-state (or stationary-state) approximation. Here, we assume that during most of the reaction period, the concentration of B is at a constant, steady-state value (equation 10.45). Obviously, this approximation is invalid at the very beginning and end of a reaction as Figure 10.12 illustrates.

$$\frac{d[\mathrm{B}]}{dt} = -\frac{d[\mathrm{B}]}{dt} = 0 \qquad (10.45)$$

The steady-state approximation states that the concentration of an intermediate B during most of a reaction is constant:

$$\frac{d[\mathrm{B}]}{dt} = -\frac{d[\mathrm{B}]}{dt} = 0$$

How can we use this approximation? Intermediate B is *generated from* A in a reaction with a rate constant k_1, and it is *converted to* C with a rate constant k_2.

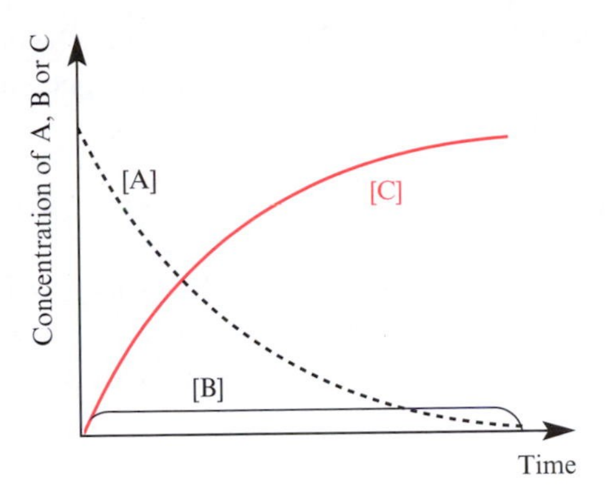

10.12 Consider a reaction $A \rightarrow C$ that proceeds through an intermediate species B. In the steady-state approximation, we assume that the concentration of B is at a constant, steady-state value. This is not valid at the beginning and end of the reaction.

Thus, equation 10.46 gives the rate of formation of B, and rearrangement of this equation leads to an expression for the concentration of intermediate B (equations 10.47 and 10.48).

$$\frac{d[B]}{dt} = k_1[A] - k_2[B] = 0 \tag{10.46}$$

$$k_1[A] = k_2[B] \tag{10.47}$$

$$[B] = \frac{k_1[A]}{k_2} \tag{10.48}$$

But, we know from equation 10.44 that the rate of the reaction is related to the concentration of B, and thus by combining equations 10.44 and 10.48, we have a new expression for the rate of reaction (equation 10.49).

$$\text{Rate of reaction} = k_2[B] = k_2\left(\frac{k_1[A]}{k_2}\right) = k_1[A] \tag{10.49}$$

The important point about this new equation is that it involves the *experimentally measurable* concentration of reactant A.

Writing an overall rate equation when there are two competing steps

Consider now the two-step formation of C from A, the mechanism of which is complicated by some of the intermediate B re-forming A (equations 10.50–10.52). Elementary steps 10.51 and 10.52 are *competitive steps.*

$$A \xrightarrow{k_1} B \tag{10.50}$$

$$B \xrightarrow{k_2} C \tag{10.51}$$

$$B \xrightarrow{k_{-1}} A \tag{10.52}$$

Experimentally, the rate of the reaction may be determined by following the formation of C (equation 10.53) or the disappearance of A; here we consider only the former.

$$\text{Rate of reaction} = \frac{d[C]}{dt} = k_2[B] \quad (10.53)$$

The steady-state approximation allows us to set the rate of change of [B] equal to zero (equation 10.54) and hence we can write the concentration of B in terms of the experimentally measurable concentration of A (equations 10.55 and 10.56).

$$\frac{d[B]}{dt} = k_1[A] - k_2[B] - k_{-1}[B] = 0 \quad (10.54)$$

$$k_1[A] = k_2[B] + k_{-1}[B] = (k_2 + k_{-1})[B] \quad (10.55)$$

$$[B] = \frac{k_1[A]}{(k_2 + k_{-1})} \quad (10.56)$$

Combining equations 10.53 and 10.56 gives expression 10.57 for the overall rate equation which shows that the rate of the reaction depends only upon [A]. But notice that the overall rate constant k' includes terms from all three of the elementary steps.

$$\text{Rate of reaction} = \frac{d[C]}{dt} = \frac{k_2 k_1[A]}{(k_2 + k_{-1})} = k'[A] \quad (10.57)$$

$$\text{where } k' = \frac{k_2 k_1}{(k_2 + k_{-1})}$$

10.11 RADICAL CHAIN REACTIONS

Radical chain reactions: see Section 8.11

The kinetics of radical chain reactions are not simple but the topic is an important one. The chemistries of flames, atmospheric reactions and explosions are some examples of chain reactions. In this section we illustrate the complex nature of reactions involving radicals.

Linear chain processes

In a linear chain, there is no increase in the number of radicals present.

A chain reaction involves initiation and propagation steps in which radicals are generated, and termination steps in which radicals are removed from the system. If there is *no net gain in the number of radicals* formed in the propagation steps, such as in reaction 10.58, the chain is called a *linear chain*.

$$CH_3^{\bullet} + Cl_2 \rightarrow CH_3Cl + Cl^{\bullet} \quad (10.58)$$

Branched chain processes

In some radical reactions, a propagation step may result in a net increase in the number of radicals in the system. Equation 10.59 illustrates the reaction between an $H^\bullet$ radical and a dioxygen molecule in which both a hydroxyl radical and an oxygen atom (a diradical) are formed.

$$H^\bullet + O_2 \rightarrow OH^\bullet + {}^\bullet O^\bullet \quad (10.59)$$

Such a process has the effect of *branching* the chain – the increased number of radicals in the system means that more reaction steps involving radicals can now take place. The rate of the reaction increases dramatically and can result in an explosion. Two examples are gas-phase hydrocarbon oxidations (combustion) and the reaction between H_2 and O_2, initiated by a spark, to give water.

In a branched chain, the number of radicals increases during the reaction.

The dramatic difference between a linear and branched chain is summarized in Figure 10.13.

The reaction between dihydrogen and dibromine

Gaseous H_2 and Br_2 react at 500 K to form hydrogen bromide; the mechanism is a linear chain reaction. The initiation, propagation and termination steps are given in equations 10.60–10.62 and 10.64. Reaction 10.63 is an *inhibition step* – although a new radical is formed, some of the HBr product is used up. Note that reaction 10.63 is the reverse of propagation step 10.61, and the termination step shown in equation 10.64 is the reverse of the initiation reaction 10.60.

$$Br_2 \xrightarrow{k_1} 2Br^\bullet \quad (10.60)$$

$$Br^\bullet + H_2 \xrightarrow{k_2} HBr + H^\bullet \quad (10.61)$$

$$H^\bullet + Br_2 \xrightarrow{k_3} HBr + Br^\bullet \quad (10.62)$$

$$HBr + H^\bullet \xrightarrow{k_{-2}} Br^\bullet + H_2 \quad (10.63)$$

$$2Br^\bullet \xrightarrow{k_{-1}} Br_2 \quad (10.64)$$

This mechanism is consistent with the observed rate law which is given in equation 10.65, and is derived in Box 10.4 on the basis of the elementary steps shown above and by using the steady-state approximation.

$$\text{Rate} = \frac{d[HBr]}{dt} = \frac{2k_2k_3\left(\frac{k_1}{k_{-1}}\right)^{\frac{1}{2}}[Br_2]^{\frac{3}{2}}[H_2]}{k_3[Br_2] + k_{-2}[HBr]} \quad (10.65)$$

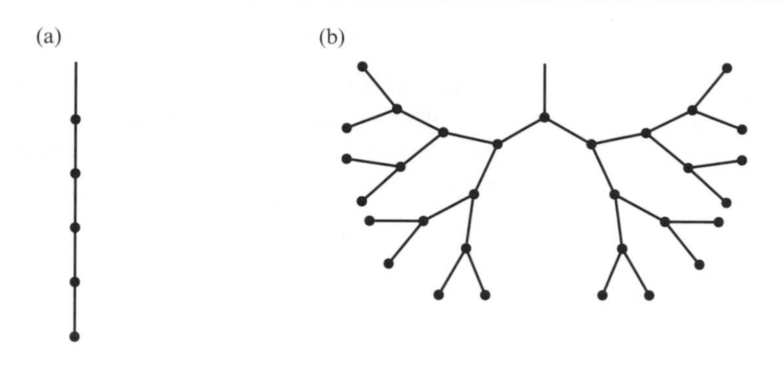

10.13 The difference between a linear and branched radical chain: (a) a linear chain in which each propagation step uses *and* generates one radical, and (b) a branched chain in which each step (after the initial radical is formed) generates two radicals.

Two other elementary steps could have been considered, the initiation step involving the homolytic cleavage of the H–H bond in H_2, and the inhibition step involving the reaction between HBr and a $Br^{\bullet}$ radical. However, the consistency between experimental data and equation 10.65 indicates that neither of these steps is important.

Box 10.4 Use of the steady-state approximation

The elementary reactions shown in equations 10.60–10.64 together comprise the linear chain mechanism for the reaction of H_2 and Br_2 to give HBr. A rate law, which is consistent with experimentally observed kinetic results, can be derived as follows.

HBr is formed in equations 10.61 and 10.62 and consumed in equation 10.63. The rate of formation of HBr is therefore:

$$\frac{d[HBr]}{dt} = k_2[Br^{\bullet}][H_2] + k_3[H^{\bullet}][Br_2] - k_{-2}[H^{\bullet}][HBr] \qquad \textbf{(I)}$$

By the steady-state approximation:

$$\frac{d[Br^{\bullet}]}{dt} = \frac{d[H^{\bullet}]}{dt} = 0$$

$$\frac{d[Br^{\bullet}]}{dt} = 2k_1[Br_2] - k_2[Br^{\bullet}][H_2] + k_3[H^{\bullet}][Br_2] + k_{-2}[H^{\bullet}][HBr] - 2k_{-1}[Br^{\bullet}]^2 = 0 \qquad \textbf{(II)}$$

and

$$\frac{d[H^{\bullet}]}{dt} = k_2[Br^{\bullet}][H_2] - k_3[H^{\bullet}][Br_2] - k_{-2}[H^{\bullet}][HBr] = 0 \qquad \textbf{(III)}$$

The addition of equations **(II)** and **(III)** results in the cancellation of three terms, and we have:

$$2k_1[Br_2] - 2k_{-1}[Br^{\bullet}]^2 = 0$$

Therefore:

$$[Br^{\bullet}]^2 = \frac{2k_1[Br_2]}{2k_{-1}}$$

$$[Br^\bullet] = \left(\frac{k_1}{k_{-1}}\right)^{\frac{1}{2}}[Br_2]^{\frac{1}{2}} \qquad \textbf{(IV)}$$

Equation **(II)** can be rearranged to make $[H^\bullet]$ the subject:

$$[H^\bullet] = \frac{k_2[Br^\bullet][H_2]}{k_3\,[Br_2] + k_{-2}\,[HBr]}$$

and substituting in for $[Br^\bullet]$ with equation **(III)** gives:

$$[H^\bullet] = \frac{k_2\left(\frac{k_1}{k_{-1}}\right)^{\frac{1}{2}}[H_2][Br_2]^{\frac{1}{2}}}{k_3[Br_2] + k_{-2}\,[HBr]} \qquad \textbf{(V)}$$

Now we return to the rate equation **(I)** and make substitutions for $[Br^\bullet]$ and $[H^\bullet]$ using equations **(IV)** and **(V)**. First, note that $[H^\bullet]$ is a factor of the last two terms and thus we can write:

$$\frac{d[HBr]}{dt} = k_2[Br^\bullet][H_2] + [H^\bullet]\{k_3[Br_2] - k_{-2}[HBr]\}$$

$$= k_2\left(\frac{k_1}{k_{-1}}\right)^{\frac{1}{2}}[Br_2]^{\frac{1}{2}}[H_2] + \left(\frac{k_2\left(\frac{k_1}{k_{-1}}\right)^{\frac{1}{2}}[H_2][Br_2]^{\frac{1}{2}}}{k_3[Br_2] + k_{-2}\,[HBr]}\right)\{k_3[Br_2] - k_{-2}[HBr]\}$$

$$= k_2\left(\frac{k_1}{k_{-1}}\right)^{\frac{1}{2}}[Br_2]^{\frac{1}{2}}[H_2]\left(1 + \frac{k_3[Br_2] + k_{-2}\,[HBr]}{k_3[Br_2] + k_{-2}\,[HBr]}\right)$$

$$= k_2\left(\frac{k_1}{k_{-1}}\right)^{\frac{1}{2}}[Br_2]^{\frac{1}{2}}[H_2]\left(\frac{k_3[Br_2] + k_{-2}\,[HBr] + k_3\,[Br_2] - k_{-2}\,[HBr]}{k_3[Br_2] + k_{-2}\,[HBr]}\right)$$

$$= k_2\left(\frac{k_1}{k_{-1}}\right)^{\frac{1}{2}}[Br_2]^{\frac{1}{2}}[H_2]\left(\frac{2k_3[Br_2]}{k_3[Br_2] + k_{-2}\,[HBr]}\right)$$

$$= \frac{2k_2k_3\left(\frac{k_1}{k_{-1}}\right)^{\frac{1}{2}}[Br_2]^{\frac{3}{2}}[H_2]}{k_3[Br_2] + k_{-2}\,[HBr]}$$

This final rate equation can be written in the simpler form:

$$\frac{d[HBr]}{dt} = \frac{k[Br_2]^{\frac{3}{2}}[H_2]}{[Br_2] + k'\,[HBr]}$$

where $k = 2k_2\left(\frac{k_1}{k_{-1}}\right)^{\frac{1}{2}}$ and $k' = \frac{k_{-2}}{k_3}$.

SUMMARY

In this chapter, we have discussed some aspects of reaction kinetics and have introduced some important equations. The use of the worked examples has been a critical part of this chapter – learning to deal with experimental data in studies of reaction rates is one of the keys to understanding this topic. You will have noticed that in many cases there is no one way to begin to solve a problem. For example, if you want to confirm that the reaction:

$A \rightarrow B$

is first order with respect to A, you could:

- use the half-life method; or
- use the integrated rate equation (is a plot ln [A] against time linear?); or
- measure tangents to a curve of [A] against time, and then see if a plot of the rate (the gradient of a tangent) against [A] is linear.

Do you know what the following terms mean?

- rate of a reaction
- rate-determining step
- rate equation
- rate constant
- order of reaction
- half-life
- pseudo-first order
- integrated form of a rate equation
- Boltzmann distribution
- activation energy
- Arrhenius equation
- catalyst
- autocatalysis
- elementary step
- molecularity
- reaction mechanism
- unimolecular step
- bimolecular step
- steady-state approximation
- chain reaction
- linear chain
- branched chain

You should be able:

- to explain what is meant by the rate and the rate equation of a reaction
- to distinguish between the order of a reaction with respect to a particular reactant and the overall order
- for a reaction *A → products*, to write down the differential forms of the rate equations for zero, first or second order dependences on A
- to explain why the units of rate constants are not always the same
- for a reaction *A → products*, to write down (or derive) the integrated forms of the rate equations for zero, first or second order dependences on A
- to treat experimental data to determine whether a reaction is zero, first or second order with respect to a particular reactant
- to determine rate constants from experimental data
- to describe how to design an experiment using pseudo-*n*th order conditions and understand when and why these conditions are necessary
- to distinguish between pseudo-*n*th order and overall rate constants and determine them from experimental data
- to derive the differential form of a rate equation for a simple reversible reaction
- to discuss why the rate of a reaction depends on temperature
- to write down the Arrhenius equation
- to use the variation of rate constant with temperature to determine the activation energy for a reaction
- to briefly explain why a catalyst alters the rate of a reaction
- to describe what is meant by autocatalysis
- to discuss the meanings of the terms elementary reaction and unimolecular and bimolecular steps
- to write down rate equations for unimolecular and bimolecular steps
- to relate a series of (two or three) elementary steps to an overall mechanism and use these details to derive an overall rate equation
- to distinguish between a linear and a branched chain mechanism
- to discuss the validity of the steady-state approximation and state why it is useful
- to relate the rate equation for a *simple* chain reaction to its mechanism (there is scope here to extend this to the derivation of such rate equations).

PROBLEMS

10.1 (a) The data plotted in Figure 10.2a refer to a reaction: $A \rightarrow$ *products*. Determine the rate of this reaction. (b) Figure 10.2b describes a different reaction: $A \rightarrow$ *products*, in which the rate changes as a function of time. Determine the rates of reaction at times t_1 and t_2 marked on the graph.

10.2 For a reaction $A \rightarrow$ *products* which is zero order with respect to A, sketch a graph of (a) rate of reaction against time, and (b) [A] against time. Repeat the exercise for reactions that are first and second order with respect to A.

10.3 For a reaction $A \rightarrow$ *products* which is first order with respect to A, show that the units of the rate constant are s^{-1}. Similarly, confirm that the units of the second order rate constant are $dm^3\ mol^{-1}\ s^{-1}$.

10.4 What is meant by each of the following terms: (a) rate of reaction; (b) differential and integrated forms of a rate equation; (c) order of a reaction (both the overall order and the order with respect to a given reactant); (d) rate constant, pseudo *n*th-order rate constant and overall rate constant; (e) activation energy; (f) catalysis and autocatalysis, (g) unimolecular step; (h) bimolecular step?

10.5 Iron(III) oxidizes iodide according to the following equation:

$$2Fe^{3+} + 2I^- \rightarrow 2Fe^{2+} + I_2$$

Write down a rate equation for this reaction if doubling the iodide concentration increases the rate by a factor of four, and doubling the iron(III) ion concentration doubles the rate.

10.6 2-Chloro-2-methylpropane reacts with water according to the equation:

$$Me_3CCl + H_2O \rightarrow Me_3C(OH) + HCl$$

Changes in concentration of Me_3CCl with time in an experiment carried out at 285 K are as follows:

Time / s	$[Me_3CCl] \times 10^4$ / mol dm^{-3}
20	2.29
25	1.38
30	0.90
34.5	0.61
39.5	0.42
44.5	0.26

(a) Determine the order of this reaction with respect to Me_3CCl.
(b) Calculate a rate constant assuming the rate equation to be of the form:
Rate = $k[Me_3CCl]^n$
(c) Does the rate equation in (b) necessarily mean that only Me_3CCl is involved in the rate-determining step?

(Data from: A. Allen, A.J. Haughey, Y. Hernandez and S. Ireton, *Journal of Chemical Education*, (1991) vol. 68, p. 609.)

10.7 The kinetics of reactions between alkenes and I_2 depend upon the nature of the alkene and the solvent. The data below give the results of the reaction of pent-1-ene with I_2 in two different solvents; the alkene is always in vast excess. Determine the order with respect to iodine and the pseudo-*n*th order rate constant in each reaction.

Reaction I: Solvent = 1,2-dichloroethane

Time / s	$[I_2]$ / mol dm^{-3}
0	0.0200
1000	0.0152
2000	0.0115
3000	0.0087
4000	0.0066
5000	0.0050
6000	0.0038
7000	0.0029
8000	0.0022

Reaction II: Solvent = acetic acid

Time / s	$[I_2]$ / mol dm^{-3}
0	0.0200
1000	0.0163
2000	0.0137
3000	0.0119
4000	0.0105
5000	0.0093
6000	0.0084
7000	0.0077
8000	0.0071

(Data from: K. W. Field, D. Wilder, A. Utz and K. E. Kolb, *Journal of Chemical Education*, (1987) vol. 64, p. 269.)

10.8 Phenyldiazonium chloride is hydrolysed according to the equation:

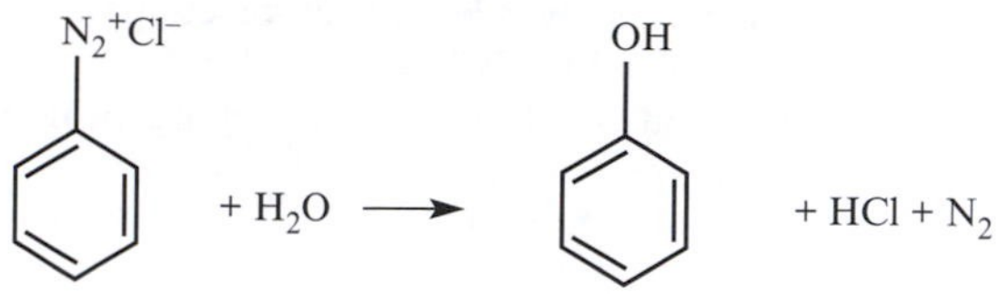

$$C_6H_5N_2^+Cl^- + H_2O \longrightarrow C_6H_5OH + HCl + N_2$$

and the reaction can be followed by the production of dinitrogen. The volume of N_2 produced is a direct measure of the amount of starting material consumed in the reaction. Use the following data to deduce the order of the reaction with respect to phenyldiazonium chloride; the reaction has been carried out in aqueous solution. The final volume of N_2 collected when the reaction has run to completion is 40 cm^3.

Time/min	Volume N_2/cm^3
0	0
5	8.8
10	15.7
20	25.3
30	31.1
40	34.5
50	36.7
60	38.0

10.9 The dimerization of a cyclic dienone **I** is shown below:

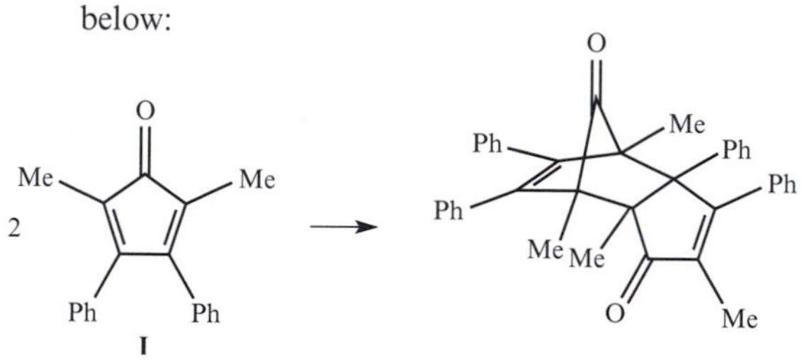

2 **I** ⟶ dimer

Compound **I** absorbs light at 460 nm (ε_{max} = 225 dm^3 mol^{-1} cm^{-1}) and the rate of the reaction has been followed using a colorimeter (path length = 1 cm). Use the tabulated absorbance data to determine (a) the order of the reaction with respect to **I**, and (b) the rate constant.

Time/min	Absorbance
0	1.20
5	1.10
10	1.00
20	0.81
30	0.72
35	0.67
45	0.59
55	0.54
75	0.45
120	0.33

(Data from: H.M. Weiss and K. Touchette, *Journal of Chemical Education*, (1990) vol. 67, p. 707.)

10.10 Phenolphthalein is an acid–base indicator. Below pH 8 it is colourless and can be represented as H_2P, whilst at higher pH values it is pink and has the form $[P]^{2-}$. In strongly alkaline solution, the colour fades as the following reaction occurs:

$$[P]^{2-} + [OH]^- \rightarrow [POH]^{3-}$$

An investigation of the kinetics of this reaction using a colorimeter (with a constant path length = 1 cm) yielded the following data:

Experiment 1: $[OH^-]$ = 0.2 mol dm^{-3}

Time/min	Absorbance
0	0.560
1	0.454
2	0.361
3	0.292
4	0.228

Experiment 2: $[OH^-]$ = 0.1 mol dm^{-3}

Time/min	Absorbance
0	0.560
2	0.449
4	0.361
6	0.301
8	0.247

Experiment 3: $[OH^-] = 0.05$ mol dm^{-3}

Time/min	Absorbance
0	0.560
3	0.468
6	0.415
9	0.368
12	0.313

(a) Why was more than one experiment carried out? Why do you think three rather than two experiments were studied?
(b) What is the order of the reaction with respect to $[P]^{2-}$?
(c) What is the order of the reaction with respect to hydroxide ion?
(d) What other data are needed before the overall rate constant can be determined?

(Data from: L. Nicholson, *Journal of Chemical Education*, (1989) vol. 66, p. 725.)

10.11 For the reaction:

$A \rightarrow B + C$

which of the following would you expect to vary over the temperature range 290 to 320 K: (a) reaction rate; (b) rate constant; (c) $E_{activation}$; (d) $[A]_0 - [A]$?

10.12 Hydrogen peroxide decomposes as follows:

$2H_2O_2 \rightarrow 2H_2O + O_2$

The rate constant for the reaction varies with temperature as follows:

Temperature/K	k/s^{-1}
295	4.93×10^{-4}
298	6.56×10^{-4}
305	1.40×10^{-3}
310	2.36×10^{-3}
320	6.12×10^{-3}

Determine the activation energy for this reaction.

10.13 Figure 10.8 showed how the ratio of the rate constants, $\frac{k_2}{k_1}$ for a reaction changes as a function of the energy of activation and the result was valid for a rise in temperature from 300 to 310K. For values of $E_{activation}$ of 10, 30, 50, 70, 90, 110, 130 and 150 kJ mol^{-1}, determine $\frac{k_2}{k_1}$ for a change in temperature from (a) 320 to 330 K and (b) 420 to 430 K. Plot $\frac{k_2}{k_1}$ as a function of $E_{activation}$. What significant differences are there between your graphs and Figure 10.8? Comment critically on the statement that a rise in temperature of 10 K leads to an approximate doubling of the rate of reaction.

10.14 (a) Explain briefly what you understand by the steady-state approximation, and indicate under what circumstances it is valid.
(b) *This part of the question assumes that the reader has worked through the material in Box 10.4*. The mechanism for the decomposition of ozone can represented as follows:

$$O_3 \xrightarrow{k_1} O_2 + O^\bullet$$

$$O_2 + O^\bullet \xrightarrow{k_{-1}} O_3$$

$$O_3 + O^\bullet \xrightarrow{k_2} 2O_2$$

Assuming the steady-state approximation, show that the overall rate of reaction is given by the expression:

$$\text{Rate of reaction} = -\frac{d[O_3]}{dt} = \frac{2k_1k_2[O_3]^2}{k_{-1}[O_2] + k_2[O_3]}$$

11 Hydrogen and the *s*-block elements

Topics

1	2		13	14	15	16	17	18
H								He
Li	**Be**		B	C	N	O	F	Ne
Na	**Mg**		Al	Si	P	S	Cl	Ar
K	**Ca**		Ga	Ge	As	Se	Br	Kr
Rb	**Sr**	*d*-block	In	Sn	Sb	Te	I	Xe
Cs	**Ba**		Tl	Pb	Bi	Po	At	Rn
Fr								

11.1 INTRODUCTION

In Section 2.22 we introduced periodicity and arranged the elements into families or groups in the periodic table in which properties were related to the ground state atomic electronic configurations. In Chapters 11 and 13 we look at the chemistries of hydrogen and the *s*- and *p*-block elements§ and in addition to describing physical and chemical properties we relate some of the trends in properties.

§ For more detailed coverage see: N. N. Greenwood and A. Earnshaw *Chemistry of the Elements*, Pergamon Press, Oxford (1984).

The title of this chapter 'Hydrogen and the *s*-block elements' emphasizes the fact that, although hydrogen has a ground state electronic configuration of $1s^1$, it is *not* an alkali metal. Some versions of the periodic table place hydrogen at the head of group 1, while others do not associate it with any particular group.

11.2 THE ELEMENT HYDROGEN

Hydrogen is the most abundant element in the universe, and the third most abundant on Earth (after oxygen and silicon). On Earth, it occurs mainly in the form of water or combined with carbon in organic molecules – hydrocarbons, plant and animal material. Dihydrogen is *not* a major constituent of the Earth's atmosphere (Figure 11.1), occurring to an extent of less than one part per million by volume. Light gases such as H_2 and He are readily lost from the atmosphere.

Box 11.1 Inside Jupiter and Saturn

The cores of Saturn and Jupiter are probably composed of metallic hydrogen, although until recently, the metallic state of this element had not been observed on the Earth. Early in 1996, researchers from the Livermore Laboratory in the USA reported that they had subjected a thin layer of liquid H_2 to tremendous pressure and observed changes in conductivity that were consistent with the formation of metallic hydrogen. Now that more is known about the conditions needed for the liquid–solid phase transition, it should be possible to gain a better picture of the composition of the cores of Saturn and Jupiter.

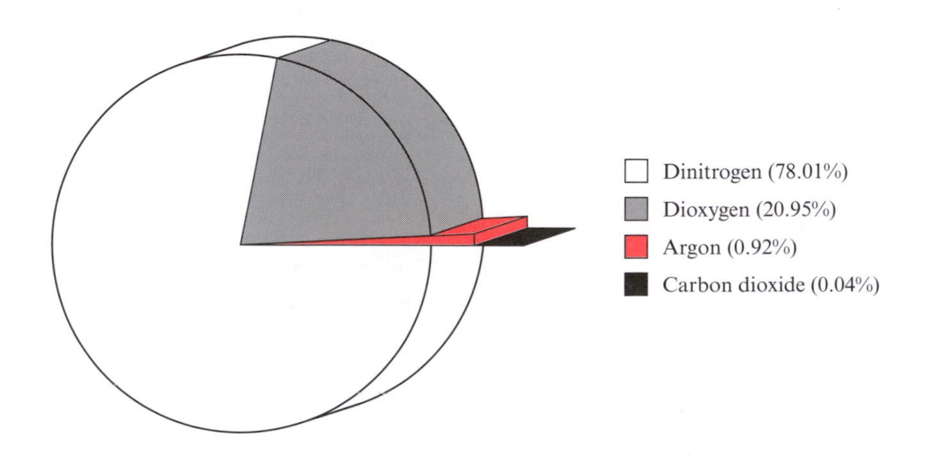

11.1 The principal components (by percentage volume) of Earth's atmosphere. Note that H_2 is not amongst these constituents.

Isotopes of hydrogen

Isotopes: see Section 1.7

Hydrogen possesses three isotopes but ^{1}H is by far the most abundant (99.984%). Although the natural abundance of deuterium, ^{2}H or D, is only 0.0156%, deuterium-labelled compounds are important, e.g. as solvents in NMR spectroscopy. Of particular significance is their use in following what happens to hydrogen atoms in reactions.

We have seen that the stretching frequency of an IR spectroscopic absorption is related to the reduced mass of the system (equation 9.8). Exchanging the hydrogen atom in an X–H bond for a deuterium atom alters the reduced mass and shifts the position of the absorption in the IR spectrum. Consider a C–H bond. The stretching frequency is inversely proportional to the square root of the reduced mass (equation 11.1), and we can write a similar equation for a C–D bond. Combining the two expressions gives equation 11.2.

$$\bar{\nu}_{C-H} \propto \frac{1}{\sqrt{\mu_{C-H}}} \tag{11.1}$$

$$\frac{\bar{\nu}_{C-D}}{\bar{\nu}_{C-H}} = \sqrt{\frac{\mu_{C-H}}{\mu_{C-D}}} \tag{11.2}$$

The relative atomic masses of C, H and D are 12, 1 and 2 respectively; the reduced mass of a C–H bond is calculated in equation 11.3 and similarly we find that μ_{C-D} is 1.714.

$$\frac{1}{\mu_{C-H}} = \frac{1}{m_1} + \frac{1}{m_2} = \frac{1}{12} + 1 = 1.083 \qquad \mu_{C-H} = 0.923 \tag{11.3}$$

An absorption at 3000 cm^{-1} due to a C–H vibration will shift to 2201 cm^{-1} (equation 11.4) upon exchanging the hydrogen for a deuterium atom. (This assumes that the force constants of C–H and C–D bonds are the same.)

Force constant: see Section 9.6

$$\bar{\nu}_{C-D} = \bar{\nu}_{C-H} \times \sqrt{\frac{\mu_{C-H}}{\mu_{C-D}}} = 3000 \times \sqrt{\frac{0.923}{1.714}} = 2201\ cm^{-1} \tag{11.4}$$

This aids the assignment of bands in the IR spectra of organic compounds. For example, N–H, O–H and C–H bonds all absorb around 3000–3500 cm^{-1}. If the organic compound is shaken with D_2O, usually only the OH and NH groups undergo a *deuterium exchange reaction* (equation 11.5).

$$R–XH + D_2O \rightleftharpoons R–XD + HOD \tag{11.5}$$

By seeing which IR spectroscopic bands shift (and by how much) and which remain unchanged, it is possible to separate N–H, O–H and C–H absorptions (see problem 11.1).

Deuterium labelling may also be used to probe the mechanism of a reaction. Consider the C–H bond again. Let us assume that in the rate-determining step of a reaction a particular C–H bond is broken. If this is the case, then labelling the compound with deuterium at that site will result in cleavage of a C–D rather than a C–H bond. The bond dissociation enthalpy of a C–D bond is higher than that of a C–H bond because the zero point

11.2 The zero point energy (the lowest vibrational state) of a C–D bond is lower than that of a C–H bond and this results in the bond dissociation enthalpy of the C–D bond being greater than that of the C–H bond.

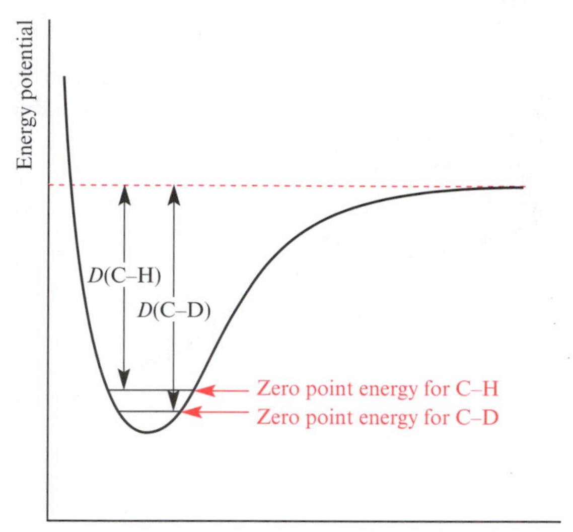

Zero point energy: see Sections 3.5 and 9.6

energy is lowered when the reduced mass of a bond is increased [μ (C–D) > μ (C–H)] as is shown in Figure 11.2. The fact that it requires more energy to break the C–D than the C–H bond should cause the rate-determining step to proceed more slowly – this is called a *kinetic isotope effect* and is quantified by comparing the rate constants (k_H and k_D) for the reactions which involve the non-deuterated and deuterated compounds. A value of the ratio k_H/k_D greater than 1 corresponds to the observation of a kinetic isotope effect.

Physical properties of H_2

Dihydrogen§ is a colourless and odourless gas (mp = 13.7 K, bp = 20.1 K), of particularly low density (0.09 g dm^{-3}) which is almost insoluble in water.

Dihydrogen is quite unreactive at 298 K largely due to the high H–H bond dissociation enthalpy (436 kJ mol^{-1}) and many reactions of H_2 require high temperatures, high pressures or catalysts as we described for the addition of H_2 to ethene in Section 8.13.

Sources of H_2

Industrially, H_2 is prepared by the reaction of carbon or a hydrocarbon (e.g. CH_4) with steam followed by the reaction of the carbon monoxide so-formed with more water vapour (equation 11.6). The mixture of CO and H_2 produced in the first reaction is called *synthesis gas* and the sequence of reactions is called the *water–gas shift reaction*.

§ Each H nucleus in an H_2 molecule could have a nuclear spin of $+\frac{1}{2}$, or $-\frac{1}{2}$,. The forms of H_2 with the spin combinations ($+\frac{1}{2}$, $+\frac{1}{2}$) [or ($-\frac{1}{2}$, $-\frac{1}{2}$)] and ($+\frac{1}{2}$, $-\frac{1}{2}$) are known as *ortho*- and *para*-dihydrogen respectively.

$$\left.\begin{aligned} CH_4(g) + H_2O(g) &\xrightarrow{\text{1200 K, nickel catalyst}} CO(g) + 3H_2(g) \\ CO(g) + H_2O(g) &\xrightarrow{\text{700 K, iron oxide catalyst}} CO_2(g) + H_2(g) \end{aligned}\right\} \quad (11.6)$$

On a laboratory scale, H_2 can be formed by reactions between acids and *electropositive* metals. The success of a particular reaction depends both upon the thermodynamics and kinetics of the process but the electrochemical series can be used *qualitatively* to predict whether the metal should be oxidized by H^+ ions. Table 11.1 lists some reduction half-reactions – at the top of the table oxidation of the metal to the corresponding metal ion is thermodynamically *favourable* with respect to the reduction of H^+ to H_2. Equations 11.7 and 11.8 illustrate two reactions suitable for the preparation of H_2 in the laboratory.

The electrochemical series is quantified in Chapter 12

Table 11.1 The electrochemical series: a qualitative approach to predicting reactions between H^+ and metals, and between H_2 and metal ions. See also Table 12.4.

← Oxidation
$Li^+ + e^- \rightleftharpoons Li$
$K^+ + e^- \rightleftharpoons K$
$Ca^{2+} + 2e^- \rightleftharpoons Ca$
$Na^+ + e^- \rightleftharpoons Na$
$Mg^{2+} + 2e^- \rightleftharpoons Mg$
$Al^{3+} + 3e^- \rightleftharpoons Al$
$Mn^{2+} + 2e^- \rightleftharpoons Mn$
$Cr^{2+} + 2e^- \rightleftharpoons Cr$
$Zn^{2+} + 2e^- \rightleftharpoons Zn$
$Cr^{3+} + 3e^- \rightleftharpoons Cr$
$Fe^{2+} + 2e^- \rightleftharpoons Fe$
$Cr^{3+} + e^- \rightleftharpoons Cr^{2+}$
$Co^{2+} + 2e^- \rightleftharpoons Co$
$Ni^{2+} + 2e^- \rightleftharpoons Ni$
$Sn^{2+} + 2e^- \rightleftharpoons Sn$
$Pb^{2+} + 2e^- \rightleftharpoons Pb$
$2H^+ + 2e^- \rightleftharpoons H_2$
$Cu^{2+} + 2e^- \rightleftharpoons Cu$
$Cu^+ + e^- \rightleftharpoons Cu$
$Fe^{3+} + e^- \rightleftharpoons Fe^{2+}$
$[Hg_2]^{2+} + 2e^- \rightleftharpoons 2Hg$
$Ag^+ + e^- \rightleftharpoons Ag$
$Hg^{2+} + 2e^- \rightleftharpoons Hg$
$Ce^{4+} + e^- \rightleftharpoons Ce^{3+}$
$Co^{3+} + e^- \rightleftharpoons Co^{2+}$
Reduction →

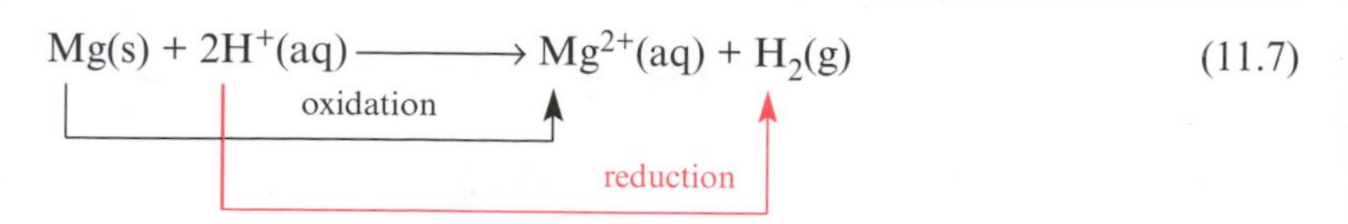

Electropositive metals release H_2 from acids.

$$Zn(s) + 2H^+(aq) \longrightarrow Zn^{2+}(aq) + H_2(g) \quad (11.8)$$

oxidation

reduction

Box 11.2 Chemical equilibria and Le Chatelier's principle: Application to the Haber process

Industrially, dinitrogen may be fixed in the following reversible reaction:

$$3H_2(g) + N_2(g) \rightleftharpoons 2NH_3(g)$$

The forward reaction is exothermic; at 298 K, $\Delta_f H^o NH_3(g) = -45.9$ kJ mol^{-1}. Unfortunately the reaction is extremely slow at room temperature – mixtures of N_2 and H_2 are indefinitely stable. We saw in Chapter 10 that the *rate* of a reaction increases with increasing temperature.

Le Chatelier's principle states that when a change (e.g. pressure or temperature) is made to a system in equilibrium, the equilibrium will tend to change to counteract the change. Thus, in the Haber process, lowering the temperature will cause the equilibrium to shift to the right-hand side – heat is produced and this raises the temperature and opposes the external change. The result can be seen by looking at the temperature dependence of the equilibrium constant, K_p:

$$K_p = \frac{(pNH_3)^2}{(pH_2)^3(pN_2)}$$

where pNH_3, pH_2 and pN_2 are the partial pressures of the three gases in the system.

For this reaction, K_p varies as follows:

$$\left.\begin{array}{ll} 298\text{ K} & K_p = 7.5 \times 10^2 \text{ bar}^{-2} \\ 500\text{ K} & K_p = 0.32 \text{ bar}^{-2} \\ 800\text{ K} & K_p = 3.0 \times 10^{-3} \text{ bar}^{-2} \end{array}\right\} \Longrightarrow NH_3 \text{ production is increased at lower temperature}$$

For the process to be industrially viable, ammonia should be formed both in good yield *and* at a reasonable rate – a low temperature may favour the formation of ammonia but the *rate* at which it is formed would be enhanced by a higher temperature. These two results conflict: at higher temperatures, the reaction is faster but the conversion to ammonia is low. This problem can be resolved in part by considering Le Chatelier's principle again.

In the gas phase reaction, four moles of reactants produce two moles of products. If the pressure of the system is increased, the equilibrium will counter the change by lowering the pressure, i.e. it will tend to move to the right-hand side and more NH_3 will be produced. Remember that the pressure of a gas mixture depends upon the number of molecules present. Thus, working at a higher temperature *and* a higher pressure can produce favourable amounts of ammonia. However, the rate of the reaction is still rather slow, and a catalyst is needed. The final reaction conditions are a temperature of 723 K, a pressure of 202 600 kPa, and Fe_3O_4 mixed with KOH, SiO_2 and Al_2O_3 as the catalyst.

Some uses of H_2

An important use of H_2 is in the industrial *fixation of dinitrogen*, a process by which N_2 is removed from the atmosphere and converted into commercially useful compounds including ammonia. The reversible reaction 11.9 is achieved in the Haber process, and in Box 11.2, we review Le Chatelier's principle by applying it to this system.

$$3H_2(g) + N_2(g) \rightleftharpoons 2NH_3(g) \tag{11.9}$$

A second major use of H_2 is in the manufacture of methanol. Reaction between H_2 and CO takes place at high pressure (25 300 kPa), high temperature ($\approx$ 600K) and in the presence of a catalyst such as Al_2O_3. Methanol is industrially important as an additive in unleaded motor-fuel, a precursor to organic compounds such as methanal (HCHO) and acetic acid, and in the synthesis of plastics and fibres.

$$2H_2(g) + CO(g) \rightarrow CH_3OH(g) \tag{11.10}$$

The very low density of H_2 has, in the past, meant that it was used in balloons, but the high risks of an explosive reaction with O_2 means that helium (which is approximately twice as dense as H_2) is favoured nowadays. A relatively new use is as a rocket fuel, but the storage of H_2 is not simple. It can be stored as a *liquid* but the low boiling point makes this impractical for many purposes. Other methods of 'storing' hydrogen (not necessarily as gaseous or liquid H_2) are being sought and these include some metal hydrides.

Interstitial metal hydrides: see Section 11.4

Box 11.3 The space shuttle

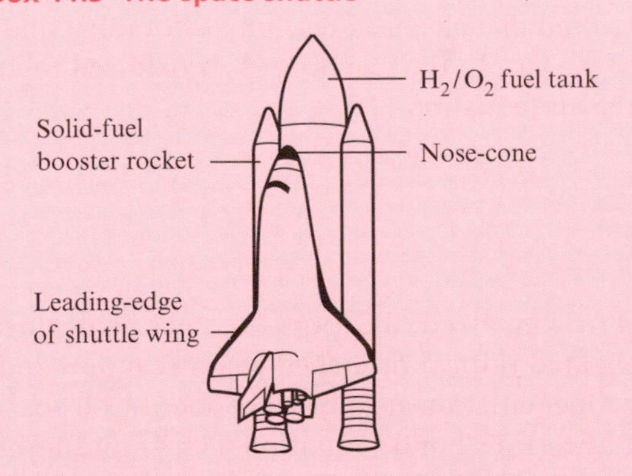

When the space shuttle lifts off, it is attached to the back of a huge fuel tank which contains liquid H_2 (about 1 457 000 litres) and liquid O_2 (about 530 000 litres). The two elements are kept in separate storage compartments inside the tank. The fuel is used up within the initial combustion and the fuel tank is jettisoned, leaving the shuttle to fly on into space. The fuel for the booster rockets of the shuttle is composed of ammonium perchlorate $[NH_4][ClO_4]$, and aluminium — the $[ClO_4]^-$ anions are the source of oxygen for the oxidation of the aluminium. (Perchlorates are discussed in Section 13.12.)

In January 1986, the shuttle *Challenger* began an ill-fated mission. Exhaust gases escaped from a joint in one of the solid-fuel boosters and caused a fire which broke open the H_2/O_2 fuel tank. The resulting explosion and fires destroyed *Challenger* and its crew. The leak is thought to have been caused by a rubber O-ring which failed to keep a seal in the cold January weather.

Lifting off is one thing, but returning to the Earth is another – the space shuttle must be able to withstand the stress and thermal shock of re-entry into the Earth's atmosphere. The nose-cone and leading edges of the wings are constructed of *carbon–carbon composites* which are fibre-reinforced materials with properties that include high strength, rigidity, chemical inertness, thermal stability, high resistance to thermal shock and retention of mechanical properties at high temperatures.

Saturated and unsaturated fats: see Box 17.4

Dihydrogen is also widely used as a reducing agent, and the reduction of C=C double bonds by H_2 is the basis of the formation of saturated oils and fats for human consumption. H_2 can reduce some metal ions to lower oxidation state ions (equation 11.11) or to the elemental state (equation 11.12), and this finds application in the extraction of certain metals from their ores. The electrochemical series (Table 11.1) can be used to predict whether a metal ion may be reduced by H_2. At the top of the table, reduction is thermodynamically *unfavourable* with respect to the oxidation of H_2 to H^+. At the bottom of the table, reduction of the metal ion is thermodynamically *favourable* with respect to the oxidation of H_2 to H^+. Thus, H_2 should reduce silver(I) salts to metallic silver, but will not reduce magnesium(II) ions to magnesium. In fact, H_2 gas reduces hot silver(I) oxide and copper(II) oxide but does not reduce the ions when in solution under ambient conditions. We shall quantify the thermodynamics of these reactions in the next chapter.

$$2Ce^{4+}(aq) + H_2(g) \longrightarrow 2Ce^{3+}(aq) + 2H^+(aq) \qquad (11.11)$$

reduction

oxidation

$$CuO(s) + H_2(g) \xrightarrow{\Delta} Cu(s) + H_2O(g) \qquad (11.12)$$

reduction

oxidation

Oxidation of H_2

The 'pop' that is heard when a flame is placed into the mouth of a test-tube containing H_2 is a familiar sound in a chemical teaching laboratory, and is a common qualitative test for the gas. In the reaction, H_2 is oxidized to water by the reaction with O_2 in the air (equation 11.13).

$$2H_2 + O_2 \rightarrow 2H_2O \qquad (11.13)$$

oxidation

reduction

Radical chain reactions: see Section 10.11

When larger amounts of H_2 are involved, the 'pop' becomes an explosion. The mechanism is a complicated radical branched chain reaction and we give only a simplified form. One initiation step, brought about by a spark, is the homolytic cleavage of the H–H bond (equation 11.14) and another occurs when H_2 and O_2 molecules collide (equation 11.15).

$$H_2 \rightarrow 2H^{\bullet} \qquad \textit{Initiation} \qquad (11.14)$$

$$H_2 + O_2 \rightarrow 2OH^{\bullet} \qquad \textit{Initiation} \qquad (11.15)$$

Branching of the chain then takes place, increasing the number of radicals (equations 11.16 and 11.17). Efficient branching results in a rapid reaction and an explosion.

$$H^{\bullet} + O_2 \rightarrow OH^{\bullet} + {}^{\bullet}O^{\bullet} \qquad \textit{Branching} \qquad (11.16)$$

$${}^{\bullet}O^{\bullet} + H_2 \rightarrow OH^{\bullet} + H^{\bullet} \qquad \textit{Branching} \qquad (11.17)$$

Water is produced (for example) in a propagation step (equation 11.18).

$$OH^{\bullet} + H_2 \rightarrow H_2O + H^{\bullet} \qquad \textit{Propagation} \qquad (11.18)$$

11.3 WHAT DOES *HYDRIDE* IMPLY?

The term 'hydride' conjures up the notion of the hydride ion, H^-, and for compounds of the type MH_x in which M is a *metal,* the hydrogen will either be in the form of H^- or will carry a δ^- charge.

How realistic is the term 'hydride' for a compound formed between hydrogen and a *p*-block element? Table 11.2 lists the Pauling electronegativity values for elements in groups 13 to 17 and highlights whether χ^P for the element is less than, greater than, or equal to $\chi^P(H)$. The red boxes show those elements that are more electronegative than hydrogen. A B–H bond is polar in the sense $B^{\delta+}–H^{\delta-}$ but an N–H bond is polar in the opposite sense $N^{\delta-}–H^{\delta+}$. In PH_3 and AsH_3, the P–H and As–H bonds are effectively non-polar. In groups 13 to 15, with the exception of carbon and nitrogen, the name *hydride* correctly implies that the hydrogen atom is in an oxidation state of –1, although in PH_3 and AsH_3 the situation is rather ambiguous.

As we move to groups 16 and 17, the H atom becomes *less* electronegative than the atom, E, to which it is attached, and the term *hydride* (although correct nomenclature) may appear somewhat misleading for these binary EH_x compounds. For this reason we treat the hydrides of groups 13 to 15 (Sections 11.5–11.7) separately from our discussion of those of the group 16 and 17 elements (Sections 11.10 and 11.11).

Table 11.2 Pauling electronegativity values, χ^P, for elements in groups 13 to 17. A more complete list is given in Table 4.2. For hydrogen $\chi^P = 2.2$. The shading in the table codes the elements according to whether they are more electronegative than H (red), less electronegative than H (white) or have the same value of χ^P (grey).

B 2.0	C 2.6	N 3.0	O 3.4	F 4.0
Al(III) 1.6	Si 1.9	P 2.2	S 2.6	Cl 3.2
Ga(III) 1.8	Ge(IV) 2.0	As(III) 2.2	Se 2.6	Br 3.0
In(III) 1.8	Sn(IV) 2.0	Sb 2.1	Te 2.1	I 2.7
Tl(III) 2.0	Pb(IV) 2.3	Bi 2.0	Po 2.0	At 2.2

11.4 BINARY HYDRIDES OF THE *s*- AND *d*-BLOCK METALS

A binary compound is one composed of only two types of elements.

Many millions of compounds contain hydrogen and it is impossible to discuss them all here. It is convenient to think about hydrogen-containing compounds in terms of the other components of the molecules, and this is adequately illustrated by thinking about organic compounds where C–H bonds are major building blocks. In the first part of this chapter we consider binary hydrides of a range of elements.

Hydrides of the *s*-block metals

The reactions of group 1 or group 2 elements with H_2 at high temperatures lead to the formation of the binary hydrides MH or MH_2 respectively (equations 11.19 and 11.20).

$$\text{Group 1} \qquad 2K + H_2 \xrightarrow{\Delta} 2KH \qquad (11.19)$$

$$\text{Group 2} \qquad Ba + H_2 \xrightarrow{\Delta} BaH_2 \qquad (11.20)$$

NaCl type lattice: see Section 6.8

The electropositive *s*-block metals form ionic compounds, and the group 1 hydrides M^+H^- possess sodium chloride type lattices in the solid state. The radius of the hydride ion varies with the metal cation (Figure 11.3). This is attributed to the relatively weak interaction between the proton and the two 1*s* electrons of the H^- anion and, consequently, the H^- ion is easily deformed. Extreme distortion results in a covalent interaction and the sharing of electrons between H and M.

The hydrides of lithium, beryllium and magnesium show considerable covalent character. Beryllium hydride, BeH_2, has a polymeric structure in which each beryllium centre is four-coordinate (Figure 11.4) and is connected to the next beryllium atom by two two-coordinate or *bridging* hydrogen

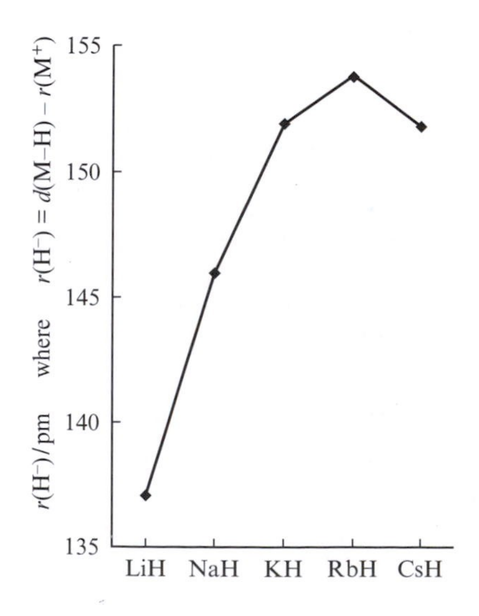

11.3 The apparent ionic radius of the H^- ion is variable and depends on the metal ion present. The graph shows the variation in the radius of the H^- ion in the alkali metal hydrides.

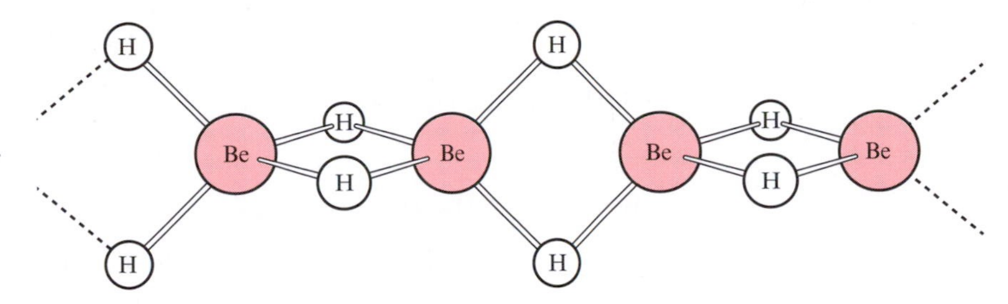

11.4 Part of the chain structure of BeH_2. Each Be–H–Be bridge is a delocalized 3-centre 2-electron (3c–2e) interaction.

atoms. Each H atom possesses only one valence electron and each Be atom has only two valence electrons, and yet the connectivities of H and Be in the polymeric beryllium hydride appear to exceed the bonding capabilities of the two elements. This can be rationalized by describing the bonding in the Be–H–Be bridges as *delocalized*. Each beryllium atom may be considered to be sp^3 hybridized. The 1*s* AO of each hydrogen can overlap with *two* sp^3 hybrid orbitals from different beryllium atoms (Figure 11.5). The distribution of electrons means that each bridge is associated with only two electrons and we have a *three-centre two-electron (3c–2e) interaction*. Beryllium hydride is usually prepared by the thermal decomposition of beryllium alkyls such as $^tBu_2Be$ rather than by the direct reaction of Be with H_2.

^{t}Bu = *tert*-butyl: see Figure 8.7

Reactions of the *s*-block metal hydrides often involve formation of H_2 (equation 11.21) and in many cases the source of H^+ is a weak acid such as H_2O (equation 11.22).

$$H^- + H^+ \rightarrow H_2 \qquad (11.21)$$

$$H^-(aq) + H_2O(l) \rightarrow H_2(g) + [OH]^-(aq) \qquad (11.22)$$

Beryllium hydride is fairly stable in water, but the other *s*-block metal hydrides react rapidly, releasing H_2. NaH reacts violently (equation 11.23) although CaH_2 reacts more slowly and is routinely used as a drying agent for organic solvents (hydrocarbons, ethers, amines and higher alcohols) and also as a means of 'storing' H_2. The controlled release of H_2 can be achieved by contact between CaH_2 and limited amounts of water (equation 11.24).

$$NaH(s) + H_2O(l) \rightarrow NaOH(aq) + H_2(g) \qquad (11.23)$$

$$CaH_2(s) + 2H_2O(l) \rightarrow Ca(OH)_2(aq) + 2H_2(g) \qquad (11.24)$$

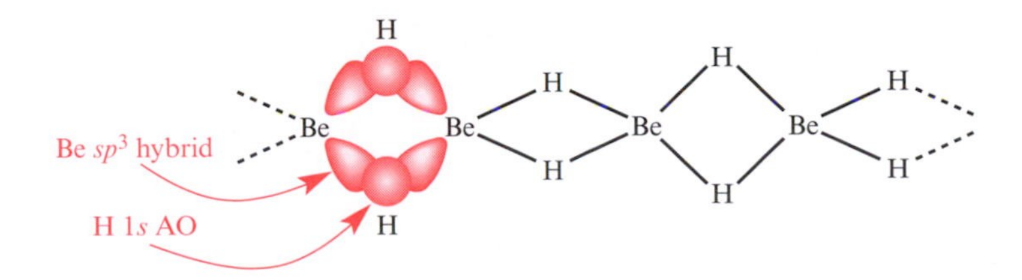

11.5 The formation of two Be–H–Be bridging interactions by the overlap of Be sp^3 and H 1*s* orbitals. Each H atom has one valence electron and each Be atom has two. The electrons are omitted from the diagram, but a pair of electrons is associated with each 3-centre Be–H–Be bridge.

In general, the reactivity towards water increases down each group but also depends on the purity of the metal hydride. In moist air, RbH, CsH and BaH_2 spontaneously ignite.

Hydrides of the *d*-block metals

The binary hydrides of the early *d*-block metals include ScH_2, YH_2, TiH_2 and HfH_2. Hafnium dihydride is a lustrous metallic solid, formed according to equation 11.25.

$$\text{Hf} \xrightarrow{1000\text{ K}} \xrightarrow{\text{cool under H}_2\text{ (300 kPa)}} \text{HfH}_2 \qquad (11.25)$$

Non-stoichiometric compounds are also formed when titanium, zirconium, hafnium and niobium react with H_2. The unusual stoichiometry of compounds such as $TiH_{1.7}$, $Hf_{1.98}$ and $Hf_{2.10}$ arises because the hydrogen atoms are small enough to enter the metal lattice and occupy the interstitial holes. Niobium forms a series of non-stoichiometric hydrides of formula NbH_x ($0 < x \leq 1$) and at low hydrogen content, the body-centred cubic structure of metallic niobium is retained. When these *d*-block metal hydrides are heated, H_2 is released and this property means that they are convenient storage 'vessels' for dihydrogen.

Interstitial holes: see Section 7.2

A non-stoichiometric compound is one which does not have the exact stoichiometric composition expected from the electronic structure, and this is often associated with a *defect* in the crystal lattice.

Anionic *d*-block hydrides include $[ReH_9]^{2-}$, $[TcH_9]^{2-}$ and $[Pt_2H_9]^{5-}$. A neutron diffraction study of $K_2[ReH_9]$ has shown that the $[ReH_9]^{2-}$ dianion has an unusual structure in the solid state (Figure 11.6a) in which there are two hydrogen environments – six prism-corner sites and three capping sites. In solution, the ^{1}H NMR spectrum shows one signal indicating that the dianion is stereochemically non-rigid on the NMR spectroscopic timescale. The salt $Li_5[Pt_2H_9]$ contains the anion $[Pt_2H_9]^{5-}$ (Figure 11.6b) and although a high pressure of H_2 is needed to prepare it, the compound is stable with respect to loss of H_2 at 298 K.

Neutron diffraction: see Section 3.2

Non-rigidity and timescales: see Sections 5.12 and 9.3

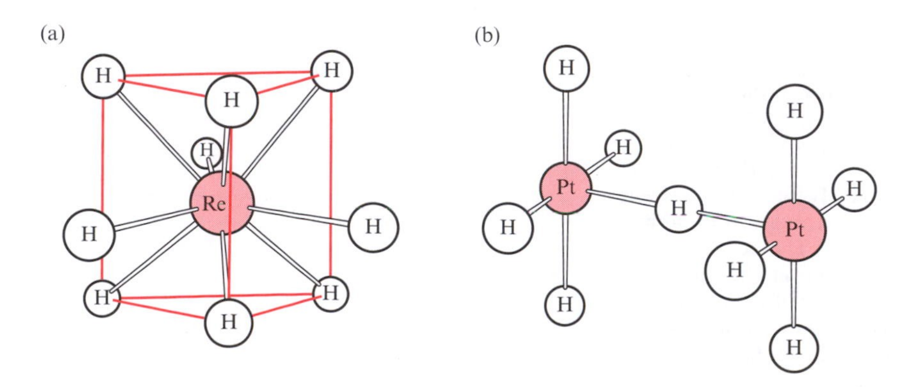

11.6 (a) The tricapped trigonal prismatic structure of the $[ReH_9]^{2-}$ anion in the dipotassium salt; six H atoms define a trigonal prism and each of the three remaining H atoms cap one square face of the prism. (b) The structure of the $[Pt_2H_9]^{5-}$ anion in the pentalithium salt.

11.5 BINARY HYDRIDES OF GROUP 13 ELEMENTS

Boron forms a range of hydrides (*boranes*) including neutral compounds of general formulae B_nH_{n+4} and B_nH_{n+6} and hydroborate dianions of formula $[B_nH_n]^{2-}$. These possess cage-like structures and *delocalized* bonding schemes are usually needed to describe the bonding.

Box 11.4 Boron hydride (borane) clusters

Boron forms a large group of neutral boron hydrides (boranes) which possess three-dimensional *cluster* structures. In each, the bonding is considered to be delocalized, since the connectivity of each boron atom is typically between five and seven, despite the fact that a boron atom only has three valence electrons.

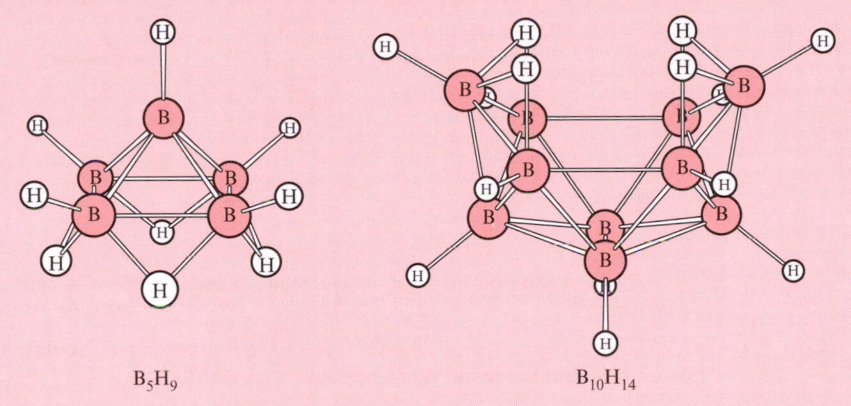

A full discussion of these novel compounds is beyond the scope of this text but further details of these and related species may be found in: N. N. Greenwood and A. Earnshaw, *Chemistry of the Elements*, Pergamon, Oxford, Ch. 6 (1984); C.E. Housecroft, *Boranes and Metallaboranes: Structure, Bonding and Reactivity*, 2nd edn, Ellis Horwood, Hemel Hempstead, (1994); C.E. Housecroft, *Cluster Molecules of the p-Block Elements*, OUP, Oxford, (1994).

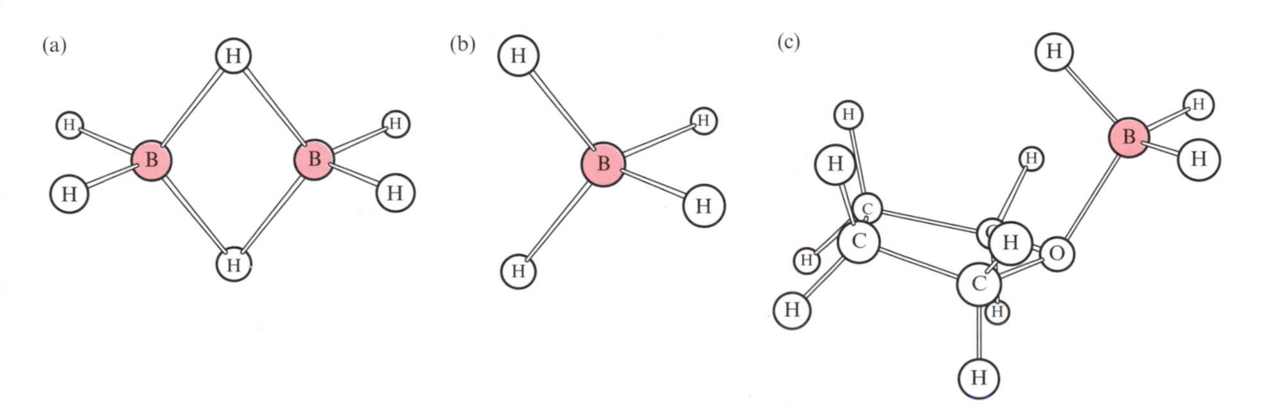

11.7 The structures of (a) B_2H_6, (b) $[BH_4]^-$ and (c) $THF{\bullet}BH_3$. In each, the boron atom is approximately tetrahedral.

$[BH_4]^-$ is also called borohydride

Two small molecular hydrides of boron are diborane[§] B_2H_6 (Figure 11.7a) and the tetrahydroborate(1–) anion $[BH_4]^-$ (Figure 11.7b).

Bonding in BH_3: see Section 5.20

Donor–acceptor compounds: see Chapter 16

Boron has a ground state electronic configuration of $[He]2s^22p^1$ and is expected to form a hydride of formula BH_3. This compound is known in the gas phase but is very reactive. The planar BH_3 molecule possesses an empty $2p$ AO which readily accepts a pair of electrons from a *Lewis base* such as tetrahydrofuran, THF, and in this way boron completes its valence octet of electrons. A Lewis base compound such as THF•BH_3 (Figure 11.7c) is called an *adduct* and can be represented either by the valence bond structure **11.1** or by showing a *dative* or *coordinate bond* from the donor atom of the Lewis base to the boron atom (structure **11.2**). The BH_3 molecule acts as a *Lewis acid* by accepting a pair of electrons.

(11.1) (11.2)

A Lewis base can donate a pair of electrons; a Lewis acid can accept a pair of electrons.

A coordinate bond forms when a Lewis base donates a pair of electrons to a Lewis acid and the resulting compound is an adduct.

3c–2e = 3-centre 2-electron

The $[BH_4]^-$ anion can be considered to be a Lewis base–acid adduct, formed by the donation of a pair of electrons from an H^- ion to BH_3. Diborane is a dimer of BH_3 and the formation of the B–H–B bridges (Figure 11.7a) completes the octet of each boron atom as structure **11.3** shows. Each bridge in B_2H_6 consists of a delocalized 3c–2e bonding interaction and is similar to a Be–H–Be bridge in beryllium hydride (Figure 11.5). Each boron atom in B_2H_6, $[BH_4]^-$ and THF•BH_3 is tetrahedral, consistent with VSEPR theory.

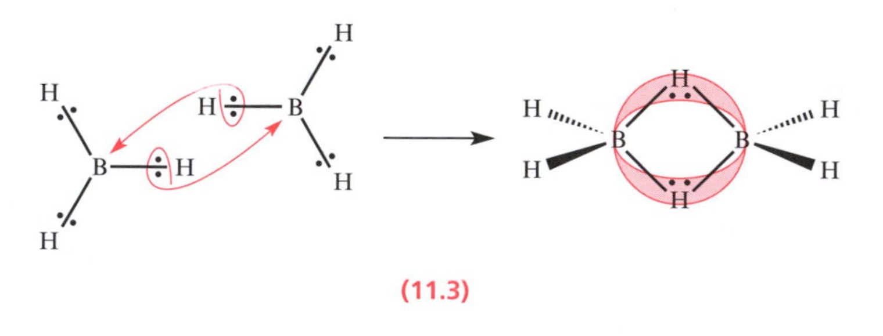

(11.3)

§ Strictly the name should be diborane(6) to indicate the presence of *both* two B atoms and six H atoms.

Diborane and adducts such as THF•BH_3 are used extensively as reducing agents and hydroborating agents. THF•BH_3 is a convenient reagent and is purchased as a solution in THF. Both B_2H_6 and THF•BH_3 are water-sensitive and hydrolyse rapidly (equation 11.26) and the product is trioxoboric acid, commonly called boric acid, $B(OH)_3$ or H_3BO_3 **11.4**.

Hydroboration: see Section 8.18

$$B_2H_6(g) + 6H_2O(l) \rightarrow \underset{\text{boric acid}}{2B(OH)_3(aq)} + 6H_2(g) \qquad (11.26)$$

Tetrahydroborate(1–) salts are also important reducing agents. The sodium salt $Na[BH_4]$ is a white, reasonably air-stable crystalline solid that is available commercially. It can be prepared from sodium hydride (equations 11.27 and 11.28) under water-free conditions. The $[BH_4]^-$ anion is *kinetically* stable towards hydrolysis, but pre-formed, cobalt-doped pellets of $Na[BH_4]$ can be used as a convenient source of H_2 — simply add to water.

See problem 11.5

$$4NaH + BCl_3 \rightarrow Na[BH_4] + 3NaCl \qquad (11.27)$$

$$4NaH + B(OCH_3)_3 \xrightarrow{520\text{ K}} Na[BH_4] + 3NaOCH_3 \qquad (11.28)$$

Although $Na[BH_4]$ is an ionic salt, some metal tetrahydroborates such as $[Al(BH_4)_3]$ are covalent with the hydrogen atoms taking part in 3c–2e bridges between the boron and the metal centres (Figure 11.8).

In comparison with boron, other members of group 13 form fewer hydrides. Aluminium trihydride is a colourless solid; it is unstable if heated above 420 K, and reacts violently with water to produce H_2. In the solid state, AlH_3 units are connected through Al–H–Al bridges. An aluminium analogue of B_2H_6 has not yet been established.

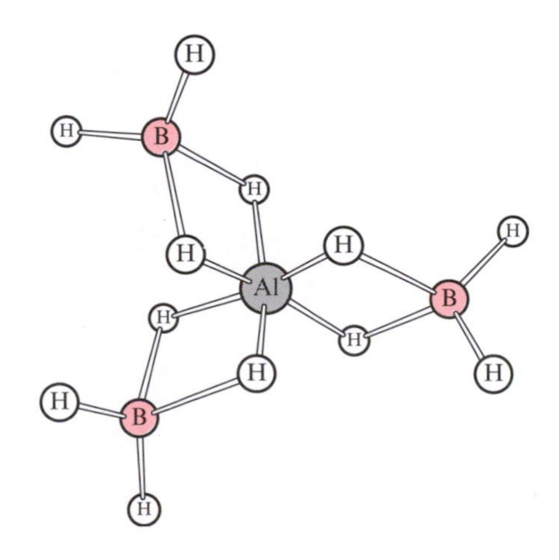

11.8 The structure of $[Al(BH_4)_3]$. The aluminium centre is octahedral and is associated with six 3c–2e Al–H–B bridges.

Like $Na[BH_4]$, $[AlH_4]^-$ salts are widely used as reducing agents; $Li[AlH_4]$ and $Na[AlH_4]$ are available commercially. Lithium tetrahydroaluminate(1–) (also called lithium aluminium hydride or lithal) can be prepared from lithium hydride (equation 11.29) and is a white, crystalline solid. It is very reactive and must be handled in moisture-free conditions; it is soluble in a range of ethers.

$$4LiH + AlCl_3 \rightarrow Li[AlH_4] + 3LiCl \qquad (11.29)$$

Gallium hydride has been an elusive compound but was fully characterized in the early 1990s. It is prepared from gallium(III) chloride in the two-stage

Box 11.5 Lithium aluminium hydride as a reducing agent and a source of hydride

The use of $Li[AlH_4]$ as a *reducing agent* is widespread. Organic reductions include the following:

Starting material	*General reaction*	*Product*
Halogenoalkane	$RX \xrightarrow{Li[AlH_4]} RH$	Alkane
Aldehyde	$R(H)C{=}O \xrightarrow{Li[AlH_4]} RCH_2OH$	Primary alcohol
Ketone	$R_2C{=}O \xrightarrow{Li[AlH_4]} R_2CHOH$	Secondary alcohol
Carboxylic acid	$R{-}C({=}O)OH \xrightarrow{Li[AlH_4]} RCH_2OH$	Primary alcohol
Acid chloride	$R{-}C({=}O)Cl \xrightarrow{Li[AlH_4]} RCH_2OH$	Primary alcohol
Amide	$R{-}C({=}O)NH_2 \xrightarrow{Li[AlH_4]} RCH_2NH_2$	Primary amine
Azide	$RN_3 \xrightarrow{Li[AlH_4]} RNH_3$	Amine
Nitrile	$RC{\equiv}N \xrightarrow{Li[AlH_4]} RCH_2NH_2$	Primary amine
Hydroperoxide	$R{-}O{-}OH \xrightarrow{Li[AlH_4]} ROH$	Alcohol
Peroxide	$R{-}O{-}O{-}R \xrightarrow{Li[AlH_4]} 2ROH$	O–O bond cleavage and formation of alcohols

Lithium aluminium hydride may be used to *convert some halides to hydrides*:

$$\text{Metal or non-metal halide} \xrightarrow{Li[AlH_4]} \text{Metal or non-metal hydride}$$

e.g. $SiCl_4 \xrightarrow{Li[AlH_4]} SiH_4$ *similarly with Ge and Sn halides*

$PCl_3 \xrightarrow{Li[AlH_4]} PH_3$ *similarly with As and Sb halides*

Hydrides may be formed from organometallic compounds:

$$[ZnMe_4]^{2-} \xrightarrow{Li[AlH_4]} [ZnH_4]^{2-}$$

In some cases reduction occurs instead of hydride formation:

$$AgCl \xrightarrow{Li[AlH_4]} Ag$$

Electron diffraction: see Section 3.2

reaction 11.30. Digallane, Ga_2H_6, condenses at low temperature as a white solid which melts at 223 K to give a colourless, viscous liquid; it decomposes above 253 K (equation 11.31). Electron diffraction studies have confirmed that the gas phase structure of Ga_2H_6 resembles that of B_2H_6 (Figure 11.7a).

$$GaCl_3 \xrightarrow{Me_3SiH} H_2Ga(\mu\text{-}Cl)_2GaH_2 \xrightarrow[240\ K]{Li[GaH_4]} Ga_2H_6 \qquad (11.30)$$

$$Ga_2H_6 \rightarrow 2Ga + 3H_2 \qquad (11.31)$$

Both B_2H_6 and Ga_2H_6 react with Lewis bases, and equation 11.32 summarizes the reactions with NH_3 and NMe_3. Whereas two of the small ammonia molecules can attack the *same* boron or gallium centre, reaction with the more sterically demanding NMe_3 tends to follow a different pathway. Competition between these routes (known as asymmetric and symmetric cleavage of the E_2H_6 molecule) may be observed.

$$[EH_2(NH_3)_2]^+[EH_4]^- \xleftarrow[\text{asymmetric cleavage}]{NH_3} H_2E(\mu\text{-}H)_2EH_2 \xrightarrow[\text{symmetric cleavage}]{NMe_3} 2Me_3N{\bullet}EH_3 \qquad (11.32)$$

E = B or Ga

Neutral, binary hydrides of indium and thallium are not known at present. The compounds $Li[EH_4]$ for E = Ga, In and Tl are all thermally unstable – $Li[GaH_4]$ is prepared according to equation 11.33 and decomposes at 320 K, while $Li[InH_4]$ and $Li[TlH_4]$ both decompose around 273 K.

$$4LiH + GaCl_3 \rightarrow Li[GaH_4] + 3LiCl \qquad (11.33)$$

11.6 BINARY HYDRIDES OF GROUP 14 ELEMENTS

Hydrocarbons: see Chapter 8

Semiconductors: see Section 7.13

Hydrocarbons (saturated and unsaturated) might be considered to be 'carbon hydrides' although they are rarely described as such. The contrast between the wealth of hydrocarbons and the small number of silicon hydrides (*silanes*) is dramatic. The silicon analogue of methane SiH_4 is formed by reacting $SiCl_4$ or SiF_4 with $Li[AlH_4]$ and is an important source of pure silicon (equation 11.34) for use in semiconductors. It is a colourless gas (bp 161 K), insoluble in water (although reaction with alkali is violent), and spontaneously inflammable in air – mixtures of SiH_4 and O_2 are explosive (equation 11.35).

$$SiH_4 \xrightarrow{\Delta} Si + 2H_2 \qquad (11.34)$$

$$SiH_4 + 2O_2 \rightarrow SiO_2 + 2H_2O \qquad (11.35)$$

Further members of the family of saturated silanes with the general formula Si_nH_{2n+2} are known up to $n = 8$ both with straight and branched chains, and a mixture of SiH_4 and higher silanes is produced when magnesium silicide (Mg_2Si) reacts with aqueous acid. The structures of the Si_nH_{2n+2} molecules are directly related to those of their carbon (alkane) counterparts. Potassium reacts with SiH_4 (equation 11.36) to form the white, crystalline salt $K[SiH_3]$ which is a useful synthetic reagent.

$$2SiH_4 + 2K \rightarrow 2K[SiH_3] + H_2 \qquad (11.36)$$

The heavier members of group 14 form few binary hydrides. GeH_4 as well as several higher germanes are known; GeH_4 is a colourless gas (bp 184 K) and is insoluble in water and inflammable in air. The reaction of $SnCl_4$ with $Li[AlH_4]$ gives SnH_4 (bp 221 K) but at 298 K it decomposes to its constituent elements. It is unlikely that the lead analogue PbH_4 has been prepared.

11.7 BINARY HYDRIDES OF GROUP 15 ELEMENTS

Ammonia

Ammonia, NH_3, is prepared on an industrial scale by the Haber process (equation 11.9 and Box 11.2) and is used extensively for the manufacture of fertilizers, nitric acid, explosives and synthetic fibres. It is a colourless gas (bp 239 K) with a characteristic smell and is corrosive and an irritant. The trigonal pyramidal structure (Figure 11.9a) is as predicted by VSEPR theory. Combustion in air takes place according to equation 11.37.

$$4NH_3 + 3O_2 \rightarrow 2N_2 + 6H_2O \qquad (11.37)$$

oxidation (N: $NH_3 \rightarrow N_2$); reduction (O: $O_2 \rightarrow H_2O$)

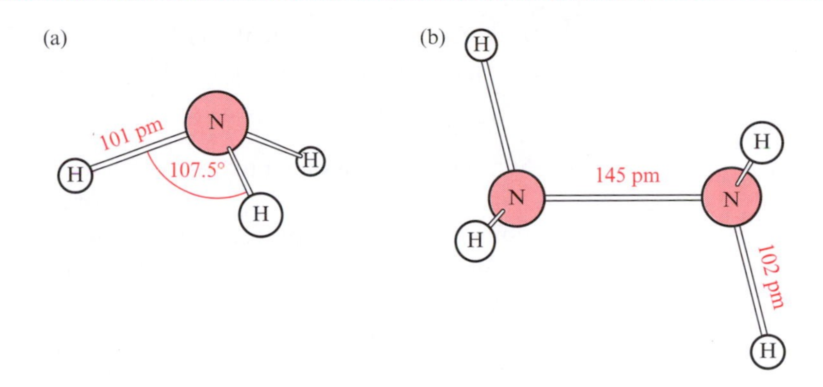

11.9 The structures of (a) ammonia, NH_3, and (b) hydrazine, N_2H_4.

Ammonia is very soluble in water and a concentrated solution (density 0.88 g cm^{-3}) is commercially available. The high solubility is due to the extensive hydrogen bonding that occurs between H_2O and NH_3 molecules. Aqueous solutions are alkaline because of the hydroxide ions that are formed in equilibrium 11.38, and may be neutralized by acids. In the expression for the corresponding equilibrium constant (equation 11.39), the concentration of water (the bulk solvent) is defined as unity.§

Hydrogen bonding: see Section 11.8

$$NH_3(aq) + H_2O(l) \rightleftharpoons [NH_4]^+(aq) + [OH]^-(aq) \quad (11.38)$$

$$K_c = \frac{[NH_4^+][OH^-]}{[NH_3][H_2O]} = \frac{[NH_4^+][OH^-]}{[NH_3]} = 1.8 \times 10^{-5}\ \text{mol dm}^{-3}\ \text{(at 298 K)} \quad (11.39)$$

This value of K is 'small' and means that the concentration of NH_3 in aqueous solution vastly exceeds the concentration of ions. We return to acid–base equilibria in Section 11.9 and to equilibria in the next chapter. Although you may find aqueous solutions of ammonia referred to as 'ammonium hydroxide', no such compound exists in the solid state.

Water-free, liquid ammonia is *self-ionizing* (equation 11.40) and is used extensively as a *non-aqueous solvent*. An interesting property of liquid ammonia is the fact that it dissolves alkali metals to give electrically conducting solutions.

Alkali metals in liquid NH_3: see Section 11.12

$$2NH_3(l) \rightleftharpoons \underset{\text{ammonium cation}}{[NH_4]^+(solv)} + \underset{\text{amide anion}}{[NH_2]^-(solv)} \quad (11.40)$$

If a pure liquid partially dissociates into ions, it is said to be self-ionizing.

Hydrazine

Hydrazine, N_2H_4, is a colourless liquid (mp 275 K, bp 386 K) which is miscible with water and with a range of organic solvents. It is corrosive and toxic, and its vapour forms explosive mixtures with air. Although the formation of

§ $[H_2O] = \dfrac{\text{Molar concentration of } H_2O}{\text{Molar concentration of pure water}} \approx 1.$

N_2H_4 from its constituent elements is an endothermic process (equation 11.41), hydrazine is *kinetically* stable with respect to N_2 and H_2.

$$N_2(g) + 2H_2(g) \rightarrow N_2H_4(l) \quad \Delta_f H^o(298\ K) = +50.6\ kJ\ mol^{-1} \qquad (11.41)$$

Figure 11.9b shows the molecular structure of N_2H_4; it is related to that of C_2H_6 but the presence of two lone pairs (one per N atom) increases the number of possible conformers. The most favourable conformation for the gas phase molecule is the *gauche* form, shown as a Newman projection in structure **11.5**. You might expect a staggered conformer would be preferred but this is not the case.

Conformer: see Section 8.5

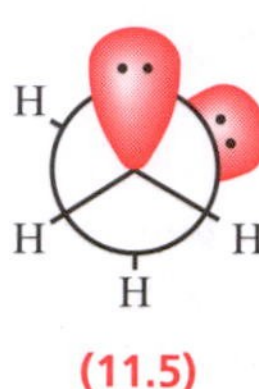

(11.5)

Hydrazine can be prepared from ammonia (equation 11.42) by the *Raschig reaction* upon which its industrial synthesis is based. Hydrazine is used in the agricultural and plastics industries, and also as a rocket fuel. It is a powerful reducing agent, and one application is in the removal of O_2 from industrial water boilers to minimize their corrosion (equation 11.43).

$$2NH_3 + \underset{\text{sodium hypochlorite}}{NaOCl} \rightarrow N_2H_4 + NaCl + H_2O \qquad (11.42)$$

$$N_2H_4 + O_2 \longrightarrow N_2 + 2H_2O \qquad (11.43)$$

oxidation ($N_2H_4 \rightarrow N_2$); reduction ($O_2 \rightarrow H_2O$)

Phosphane

Phosphane is also called phosphine

Phosphane, PH_3, is produced by the action of water on calcium phosphide (equation 11.44) or magnesium phosphide, or by the reaction of PCl_3 with lithium hydride (equation 11.45) or $Li[AlH_4]$. Phosphane is a colourless gas (bp 185 K) with a garlic-like odour, and is extremely poisonous. It has been detected in the atmospheres of Saturn, Jupiter and Uranus.

$$Ca_3P_2 + 6H_2O \rightarrow 2PH_3 + 3Ca(OH)_2 \qquad (11.44)$$

$$PCl_3 + 3LiH \rightarrow PH_3 + 3LiCl \qquad (11.45)$$

Phosphane ignites in dry air at about 420 K (equation 11.46), or spontaneously if P_2H_4 (formed as a by-product in its synthesis) is present. It is a strong reducing reagent.

$$PH_3 + 2O_2 \rightarrow \underset{\text{phosphoric acid}}{H_3PO_4} \qquad (11.46)$$

In a number of respects, PH_3 is rather different from NH_3. It is less soluble than ammonia in water, and dissolves to give neutral solutions. The molecular structure of PH_3 is trigonal pyramidal, as is that of NH_3, but whereas the H–N–H bond angle is 107.5°, the H–P–H angle is only 93°. The former is indicative of sp^3 hybridization and the latter suggests that p orbitals are used in bonding; this difference has an important chemical consequence. The lone pair in the sp^3 hydridized NH_3 molecule is readily available and ammonia is a *reasonably* strong base. In contrast, the lone pair in PH_3 is less available and PH_3 is a very weak base. Although phosphonium, $[PH_4]^+$, salts can be formed, they are decomposed by water (equation 11.47).

$$[PH_4]^+(aq) + H_2O(l) \rightarrow PH_3(g) + [H_3O]^+(aq) \qquad (11.47)$$

Diphosphane

Diphosphane, P_2H_4, is a colourless liquid at room temperature (mp 174 K, bp 340 K), is toxic and spontaneously inflammable; when heated, it forms higher phosphanes. The gas phase structure of the P_2H_4 molecule has a *gauche* conformation like N_2H_4 (Figure 11.9b) with bond distances of P–P = 222 pm and P–H = 145 pm.

Hydrides of arsenic, antimony and bismuth

Arsane, stibane and bismuthane (with the more common names of arsine, stibine and bismuthine) are all unstable gases at 298 K and are toxic. They can be prepared by the reaction of the respective trichlorides with $Na[BH_4]$. Arsane, AsH_3, has a particularly repulsive smell; it has been detected in the atmospheres of Jupiter and Saturn. AsH_3 and SbH_3 are readily oxidized to As_2O_3 and Sb_2O_3 respectively.

One of the old tests (the Marsh test) for the presence of arsenic or antimony involved the generation of the hydrides. The material of interest was treated with zinc dust and acid, and the gas evolved, containing H_2 and either AsH_3 or SbH_3, was passed through a tube that was heated at one point. The thermal instabilities of SbH_3 and AsH_3 meant that they decomposed – SbH_3 is *less* stable than AsH_3 and decomposes in the warm region *before* the flame, i.e. region **A** in Figure 11.10. AsH_3 decomposes in region **B**. The result is that Sb-containing materials give a black deposit (which appears as a mirror) on the wall of the glass tube in region **A** (equation

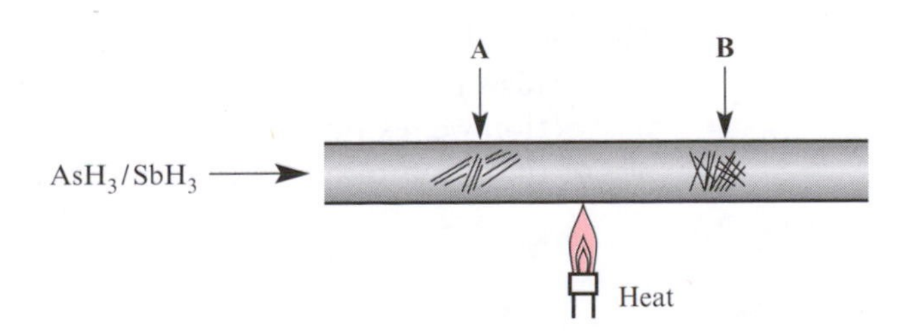

11.10 The differing thermal stabilities of arsine and stibine can be used to distinguish between them. SbH_3 is *less* stable than AsH_3 and decomposes in region **A** whilst AsH_3 decomposes in region **B**.

11.48) whilst As-containing compounds produce a deposit of arsenic in region **B** (equation 11.49).

$$2SbH_3(g) \xrightarrow{\Delta} 2Sb(s) + 3H_2(g) \qquad (11.48)$$

$$2AsH_3(g) \xrightarrow{\Delta} 2As(s) + 3H_2(g) \qquad (11.49)$$

11.8 HYDROGEN BONDING

Before we consider the compounds formed between hydrogen and the group 16 and 17 elements, we must mention two important consequences of having a hydrogen atom bonded to an electronegative atom. In Section 11.9, we review the theory of Brønsted acids and bases, but first we consider hydrogen bonding.

The hydrogen bond

A *hydrogen bond* is an interaction between a hydrogen atom attached to an electronegative atom, and an electronegative atom which possesses a lone pair of electrons. The strongest hydrogen bonds involve the first row elements F, O or N, and structure **11.6** shows the formation of a hydrogen bond between two water molecules.

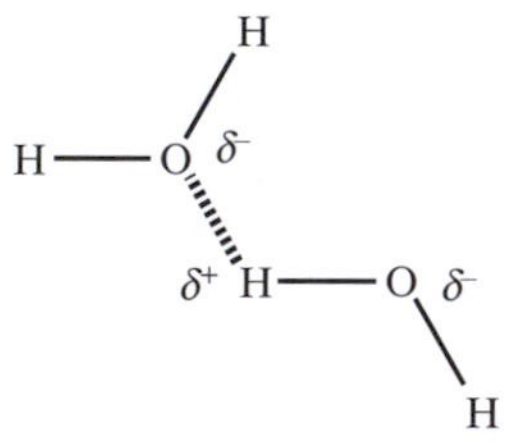

H
H—O δ−
δ+ H—O δ−
H

(11.6)

A hydrogen bond is an interaction between a hydrogen atom attached to an electronegative atom, and an electronegative atom which possesses a lone pair of electrons.

The physical properties of compounds may indicate the presence of hydrogen bonds since such interactions result in the association of molecules. In this section we restrict our discussion to binary hydrides but further examples of the effects of hydrogen bonding are given in later chapters.

Boiling points and enthalpies of vaporization

Figure 11.11a shows the trend in the boiling points of the compounds EH_3 where E is a group 15 element. The trend in enthalpies of vaporization (Figure 11.11b) is the same and this is *not* a coincidence! Trouton's rule (equation 11.50) gives an approximate relationship between $\Delta_{vap}H$ (measured at the boiling point of a liquid) and the boiling point (in K) of the liquid.

Trouton's rule is more fully explained in Section 12.9

$$\frac{\Delta_{vap}H}{bp} \approx 0.088 \text{ kJ K}^{-1} \text{ mol}^{-1} \qquad \textit{Trouton's rule} \qquad (11.50)$$

Figure 11.11c illustrates that similar plots are obtained for the boiling points of hydrides of the group 16 and 17 elements. Although the general trend is an increase down a group, NH_3, H_2O and HF stand out as having anomalously high boiling points.

The transition from a liquid to a vapour involves the cleavage of some (but not necessarily all) of the *intermolecular* interactions. The strength of a hydrogen bond is around 25 kJ mol^{-1} and although this is considerably weaker than a covalent bond, it is strong enough to affect the enthalpy of vaporization and the boiling point of a compound significantly. The unusually high values for NH_3, H_2O and HF are due to the additional energy needed to break the N····H, O····H or F····H hydrogen bonds which are present in the liquid state.

Association in the solid state

The association of molecules through hydrogen bonding in the solid state often leads to well-organized assemblies of molecules. In solid hydrogen

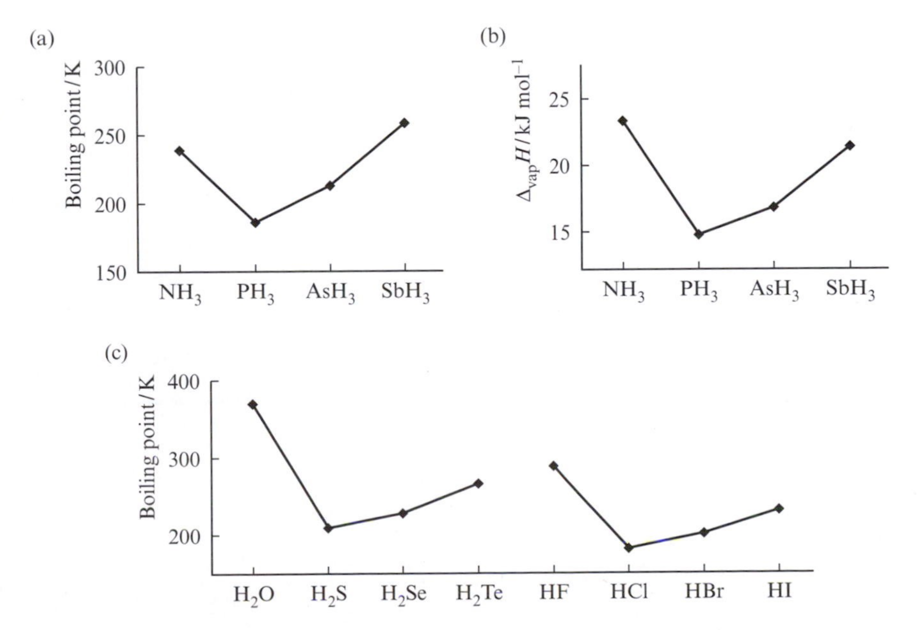

11.11 The trends in (a) boiling points, and (b) $\Delta_{vap}H$ for the group 15 hydrides, and (c) the trends in boiling points for the group 16 and 17 hydrides.

Box 11.6 Hydrogen bonding and DNA

DNA (deoxyribonucleic acid) is the key to life – when a cell divides, the genetic information from the original cell is passed to the new cell by the DNA molecule. The second major role of DNA is in the synthesis of proteins.

DNA is a nucleic acid polymer and its structure consists of two helical chains which interact with one another through hydrogen bonding to form a *double helix*. The backbone of each chain consists of sugar units (each containing a five-membered C_4O-ring) linked by phosphate groups. Attached to each sugar is an organic base. In DNA, there are four different bases: adenine, guanine, cytosine and thymine. The left-hand structure below represents part of the polymeric backbone of DNA and shows the positions of attachment of the bases (shown on the right).

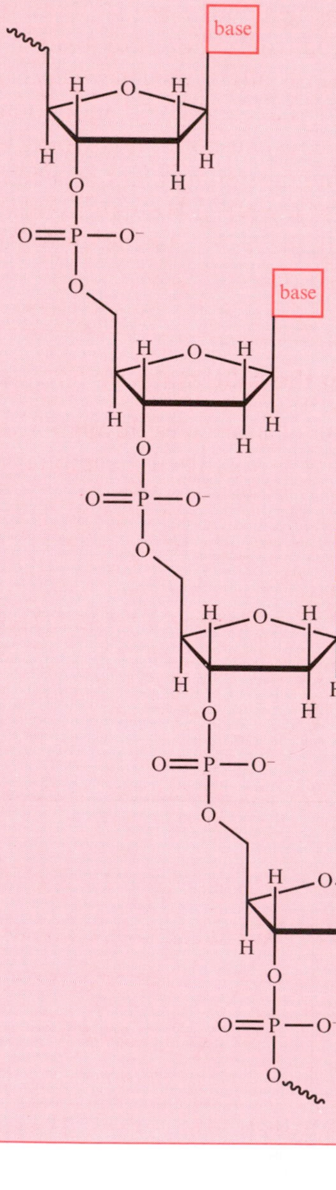

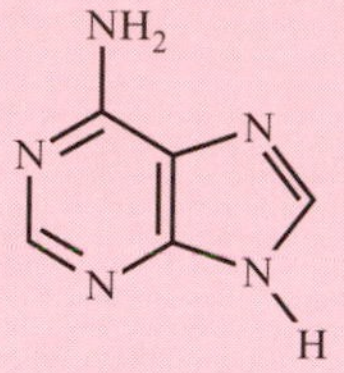

Adenine

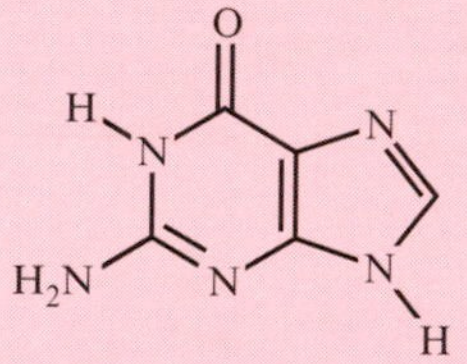

Guanine

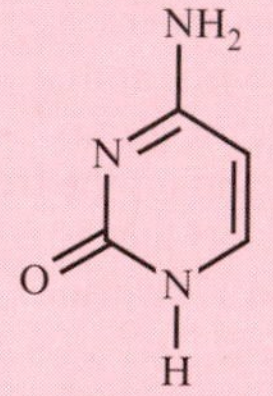

Cytosine

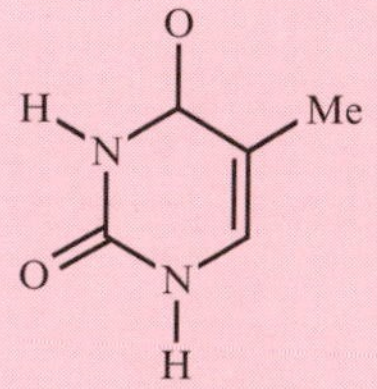

Thymine

The structures of adenine and thymine are *exactly* matched to permit hydrogen bonding interactions between these bases – adenine and thymine are *complementary bases*. Guanine and cytosine are similarly matched.

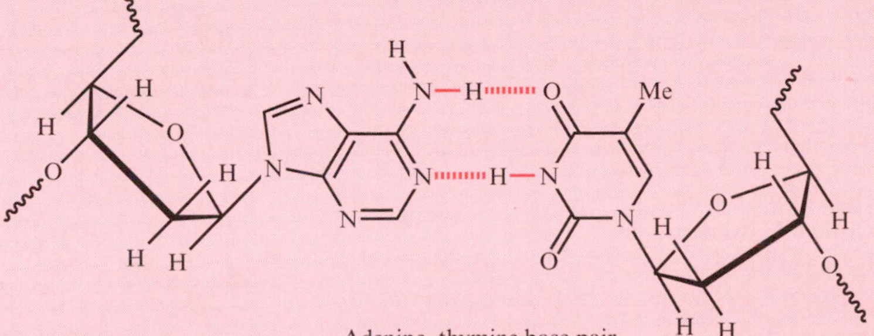

Adenine–thymine base pair

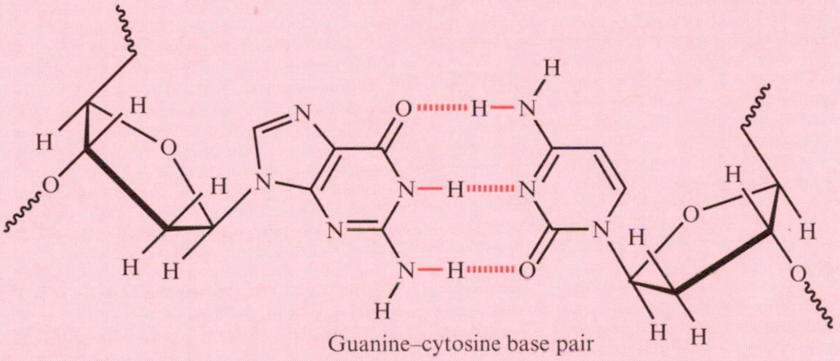

Guanine–cytosine base pair

The result of this base-pairing is that the sequence of the bases in one chain of DNA is *complemented* by a sequence in a second chain, and the two hydrogen-bonded chains combine to form a *double helix*.

Proteins consist of a sequence of amino acids, and the sequence is specific to a particular protein. During protein synthesis, DNA is able to *code* the sequence correctly by transmitting information in the form of its own base sequence. Each amino acid in the protein chain has an associated code of three adjacent DNA bases (a *codon*).

For related information in this book see: ATP (structure **13.35**), heterocyclic organic molecules (Section 15.12), anti-cancer drugs and DNA (Box 16.6).

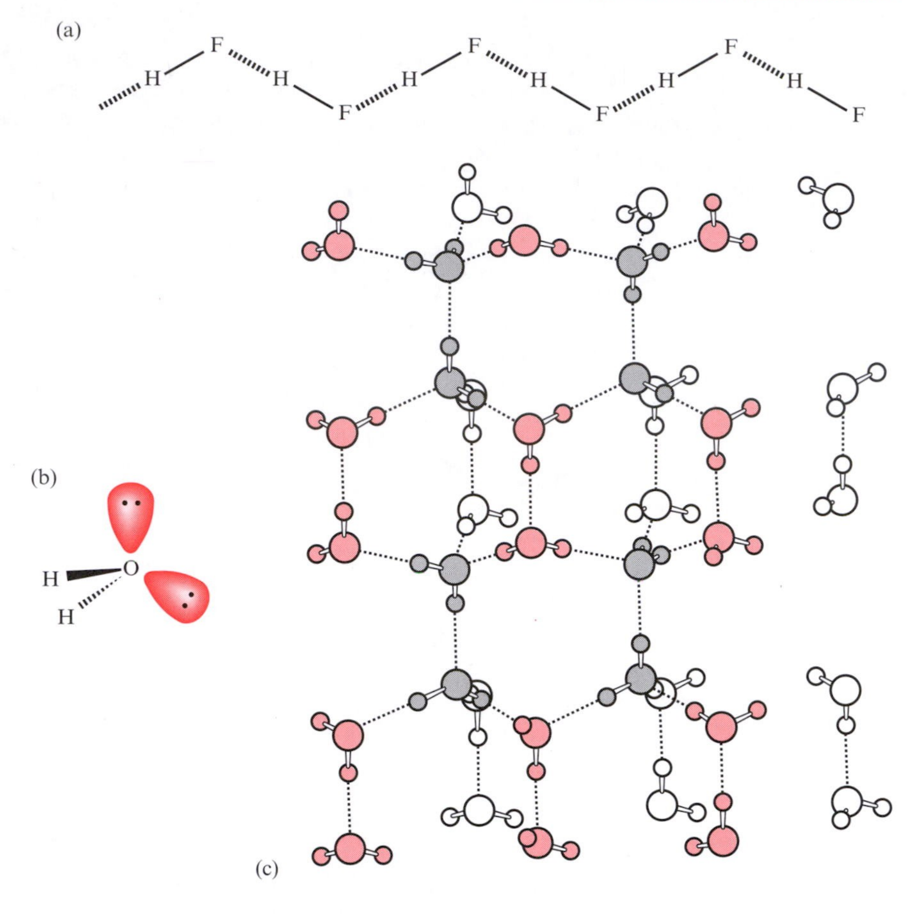

11.12 (a) In the solid state of HF, the molecules are hydrogen bonded together to form zigzag chains; (b) the relative orientations of the two H atoms and the two lone pairs in a molecule of water are perfectly set up to generate a three-dimensional network of hydrogen-bonded molecules in ice as shown in diagram (c). From: L. Pauling, *The Nature of the Chemical Bond*, Cornell University Press, Ithaca (1960).]

fluoride, the HF molecules are arranged in zigzag chains (Figure 11.12a) in which each H atom is in a linear environment.

Intermolecular hydrogen bonding occurs in both water and ammonia in the solid state. In water, the relative orientations of the two lone pairs and the two hydrogen atoms (Figure 11.12b) are set up so that when *linear* O–H····O interactions occur, a three-dimensional structure results. In ice, each oxygen atom is in an approximately tetrahedral environment (Figure 11.12c) with two O–H distances of 101 pm and two of 175 pm. The network structure is a very open one and when ice melts at 273 K, the density *increases* as some (but not all) of the hydrogen bonds are destroyed. The structure is similar to that of diamond (Figure 7.22).

Hydrogen bonding and solvation

Just as H_2O or NH_3 molecules can associate with one another in liquid water or ammonia, so too can H_2O *and* NH_3 molecules when ammonia dissolves in water. Some of the ammonia ionizes in the water (equation 11.38) but most is present as *solvated* NH_3 molecules. Solvation is the intermolecular interaction between a solute and a solvent that leads to the solute dissolving,

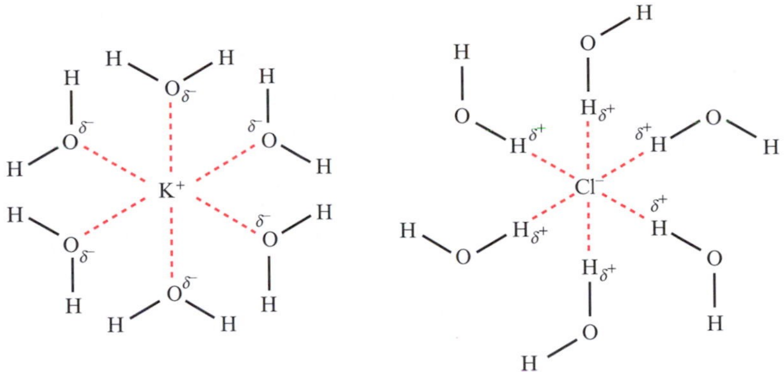

11.13 When solid KCl dissolves in water, the ions are freed from the crystal lattice and are hydrated. Ion–dipole interactions are responsible for the formation of hydration shells around each ion.

Solvation is the intermolecular interaction between a solute and a solvent that leads to the solute dissolving. Hydration is the specific case in which the solvent is water.

and *hydration* is the specific case in which the solvent is water. Many other liquids can be used as solvents, and non-aqueous solvents such as NH_3, HF, N_2O_4 and SO_2 have been widely investigated as they dissolve ionic solids.

The use of water as a solvent is widespread – think how many reactions are carried out in *aqueous solution*! When water behaves as a solvent, some hydrogen bonds between the H_2O molecules must be broken but new ion–dipole or dipole–dipole (electrostatic) interactions are generated as the solute is solvated. To illustrate this, we consider the solvation of potassium chloride in water, and Figure 11.13 schematically illustrates the hydrated K^+ and Cl^- ions in which ion–dipole interactions are indicated by the dotted red lines. The water molecules shown make up the first hydration shell and hydrogen bonds are present between these H_2O molecules and others in the bulk solvent.

Solubility of KCl: see Section 12.11

11.9 ACID–BASE EQUILIBRIA AND pH

Any discussion of the hydrides of the group 16 and, in particular, the group 17 elements necessarily involves acid–base equilibria. Similarly, when we look at the *s*-block elements, we must discuss basic solutions. This section reviews the fundamentals of acid–base equilibria and pH – topics which are probably familiar to most readers.

Brønsted acids and bases, and the acid dissociation constant, K_a

A Brønsted acid is a proton donor, and a Brønsted base is a proton acceptor.

A Brønsted acid is a proton donor (e.g. hydrochloric acid) and a Brønsted base is a proton acceptor (e.g. water). Equation 11.51 shows the reaction between HA and H_2O in which a proton is transferred from the acid to the base.

$$\underset{\text{Brønsted acid}}{HA(aq)} + \underset{\text{Brønsted base}}{H_2O(l)} \rightleftharpoons [H_3O]^+(aq) + A^-(aq) \qquad (11.51)$$

Water can also act as a Brønsted acid and equation 11.52 shows a general reaction with a Brønsted base B.

$$\underset{\text{Brønsted base}}{B(aq)} + \underset{\text{Brønsted acid}}{H_2O(l)} \rightleftharpoons [BH]^+(aq) + [OH]^-(aq) \quad (11.52)$$

If the two reactions above go essentially to completion, then HA is termed a 'strong acid' and B is a 'strong base.' An example of a strong acid is aqueous HCl, and aqueous NaOH is a strong base. In many other cases, the system is in equilibrium as shown for aqueous acetic acid in equation 11.53.

Carboxylic acids: see Chapter 17

$$\underset{\text{acetic acid}}{CH_3CO_2H(aq)} + H_2O(l) \rightleftharpoons \underset{\text{acetate ion}}{[CH_3CO_2]^-(aq)} + [H_3O]^+(aq) \quad (11.53)$$

Aqueous acetic acid CH_3CO_2H is a 'weak' Brønsted acid because the equilibrium lies to the left-hand side. The equilibrium constant is called the *acid dissociation constant*, K_a, and for acetic acid is given by equation 11.54; remember that $[H_2O]$ is taken to be unity.§

$$K_a = \frac{[CH_3CO_2^-][H_3O^+]}{[CH_3CO_2H][H_2O]} = \frac{[CH_3CO_2^-][H_3O^+]}{[CH_3CO_2H]} = 1.7 \times 10^{-5} \text{ mol dm}^{-3} \quad (11.54)$$

Worked example 11.1 ***Concentration of $[H_3O]^+$ ions in a solution of CH_3CO_2H***

Determine the concentration of $[H_3O]^+$ ions in a solution of 0.01 M acetic acid (ethanoic acid). [$K_a = 1.7 \times 10^{-5}$ mol dm^{-3}]

The equilibrium in aqueous solution is:

$$CH_3CO_2H(aq) + H_2O(l) \rightleftharpoons [CH_3CO_2]^-(aq) + [H_3O]^+(aq)$$

and the acid dissociation constant is given by:

$$K_a = \frac{[CH_3CO_2^-][H_3O^+]}{[CH_3CO_2H][H_2O]} = \frac{[CH_3CO_2^-][H_3O^+]}{[CH_3CO_2H]}$$

where the concentrations in the expression are the *equilibrium concentrations*.

Since the concentration of $[CH_3CO_2]^-$ is equal to the concentration of $[H_3O]^+$, we can rewrite the expression for K_a:

$$K_a = \frac{[H_3O^+]^2}{[CH_3CO_2H]}$$

§ NB: Square brackets are used here in two different senses. In standard nomenclature, the formula of a coordination compound (see Chapter 16) is enclosed in square brackets and the charge is placed *outside*, e.g. $[H_3O]^+$. Square brackets are used to mean 'the concentration of something', e.g. $[CH_3CO_2H]$ means the concentration of acetic acid. It is cumbersome to write two sets of brackets when we need to refer to the concentration of an ion and it is usual simply to write the formula of the ion, *including the charge* inside the square brackets, e.g. $[H_3O^+]$ means 'the concentration of $[H_3O]^+$ ions' and actually stands for $[[H_3O]^+]$.

and this can be rearranged so that we can find $[H_3O]^+$:

$$[H_3O^+]^2 = K_a \times [CH_3CO_2H]$$

$$[H_3O^+] = \sqrt{K_a \times [CH_3CO_2H]}$$

The *initial* concentration of CH_3CO_2H is 0.01 mol dm^{-3} but what is the concentration *at equilibrium*? Since CH_3CO_2H is a weak acid, we may make the approximation that the concentration of CH_3CO_2H *at equilibrium* is roughly the same as the original concentration:

$$[CH_3CO_2H] \approx 0.01 \text{ mol dm}^{-3}$$

Since we are given $K_a = 1.7 \times 10^{-5}$ mol dm^{-3}, we can find $[H_3O]^+$:

$$[H_3O^+] = \sqrt{1.7 \times 10^{-5} \times 0.01}$$

$$[H_3O]^+ = 4.1 \times 10^{-4} \text{ mol dm}^{-3}$$

pK_a

Although we often use values of K_a, a very useful and common way of recording this number is as the pK_a value. This is defined as the negative logarithm (to the base 10) of the equilibrium constant (equation 11.55):

log = log$_{10}$

$$pK_a = -\log K_a \qquad (11.55)$$

If pK_a is known, equation 11.55 can be rearranged to find K_a (equation 11.56).

$$K_a = 10^{-pK_a} \qquad (11.56)$$

Note that as the acid strength decreases, K_a decreases, but pK_a increases. For example, CH_3CO_2H is a weaker acid than HCO_2H and the respective values of K_a are 1.7×10^{-5} and 1.6×10^{-4} mol dm^{-3} and of pK_a are 4.77 and 3.80.

Worked example 11.2 ***The relationship between K_a and pK_a***

If the pK_a of hydrocyanic acid (HCN) is 9.31, determine the concentration of $[H_3O]^+$ ions in a 0.02 M solution of the acid.

The appropriate equilibrium is:

$$HCN(aq) + H_2O(l) \rightleftharpoons [H_3O]^+(aq) + [CN]^-(aq)$$

and the equilibrium constant is given by:

$$K_a = \frac{[H_3O^+][CN^-]}{[HCN]}$$

Since $[CN]^- = [H_3O]^+$, we can write:

$$K_a = \frac{[H_3O^+]^2}{[HCN]}$$

and making $[H_3O]^+$ the subject of the equation:

$$[H_3O^+] = \sqrt{K_a \times [HCN]}$$

K_a is related to pK_a by the equation:

$$pK_a = -\log K_a$$

and therefore:

$$K_a = 10^{-pK_a}$$
$$= 10^{-9.31} = 4.9 \times 10^{-10} \text{ mol dm}^{-3}$$

Using the approximation that $[HCN]_{equilibrium} \approx [HCN]_{initial}$ (see worked example 11.1), we can now find $[H_3O]^+$:

$$[H_3O^+] = \sqrt{4.9 \times 10^{-10} \times 0.02}$$
$$= 3.1 \times 10^{-6} \text{ mol dm}^{-3}$$

Neutralization reactions

Equations 11.57 and 11.58 summarize general neutralization reactions, and reactions 11.59 to 11.61 give examples.

$$[H_3O]^+ + [OH]^- \rightarrow 2H_2O \quad (11.57)$$

$$\text{Acid} + \text{Base} \rightarrow \text{Salt} + \text{Water} \quad (11.58)$$

$$HCl(aq) + NaOH(aq) \rightarrow NaCl(aq) + H_2O(l) \quad (11.59)$$

$$H_2SO_4(aq) + 2KOH(aq) \rightarrow K_2SO_4(aq) + 2H_2O(l) \quad (11.60)$$

$$CH_3CO_2H(aq) + NaOH(aq) \rightarrow CH_3CO_2Na(aq) + H_2O(l) \quad (11.61)$$

Conjugate acids and bases

When a Brønsted acid donates a proton, it forms a species which can, in theory, accept the proton back again. Equation 11.62 shows the dissociation of nitrous acid in water.

$$HNO_2(aq) + H_2O(l) \rightleftharpoons [NO_2]^-(aq) + [H_3O]^+(aq) \quad (11.62)$$

In the forward direction, HNO_2 acts as a Brønsted acid and H_2O is a base, while in the reverse direction the acid is $[H_3O]^+$ and the base is $[NO_2]^-$. The $[NO_2]^-$ anion is known as the *conjugate base* of HNO_2, and conversely, HNO_2 is the *conjugate acid* of $[NO_2]^-$. Similarly, $[H_3O]^+$ is the conjugate acid of H_2O, and H_2O is the conjugate base of $[H_3O]^+$. The conjugate acid–base pairs are shown in equation 11.63.

$$HNO_2(aq) + H_2O(l) \rightleftharpoons [NO_2]^-(aq) + [H_3O]^+(aq) \quad (11.63)$$

conjugate acid 1 — conjugate base 2 — conjugate base 1 — conjugate acid 2

conjugate acid–base pair (base 2 / acid 2)

conjugate acid–base pair (acid 1 / base 1)

Remembering that equation 11.63 is an *equilibrium*, it follows that a weak acid must have a strong conjugate base, and vice versa. The acid dissociation constant for HNO_2 is given in equation 11.64.

$$K_a = \frac{[H_3O^+][NO_2^-]}{[HNO_2]} = 4.7 \times 10^{-4}\ \text{mol dm}^{-3} \quad (11.64)$$

This means that the equilibrium lies towards the left-hand side and so HNO_2 is a weak acid and the tendency for its conjugate base $[NO_2]^-$ to accept a proton is high. Table 11.3 lists some conjugate acid–base pairs; note the inverse relationship between the relative strengths of the conjugate acids and the bases.

Table 11.3 Conjugate acid and bases, and values of K_a for the dilute aqueous solutions of acids; strong acids are fully dissociated in aqueous solution.

Acid	*Formula*	*K_a/mol dm^{-3}*	*Conjugate base*	*Formula*
Perchloric acid	$HClO_4$	–	Perchlorate ion	$[ClO_4]^-$
Sulfuric acid	H_2SO_4	–	Hydrogensulfate(1–) ion	$[HSO_4]^-$
Hydrochloric acid	HCl	–	Chloride ion	Cl^-
Nitric acid	HNO_3	–	Nitrate ion	$[NO_3]^-$
Sulfurous acid	H_2SO_3	1.5×10^{-2}	Hydrogensulfite(1–) ion	$[HSO_3]^-$
Hydrogensulfate(1–) ion	$[HSO_4]^-$	1.2×10^{-2}	Sulfate ion	$[SO_4]^{2-}$
Phosphoric acid	H_3PO_4	7.5×10^{-3}	Dihydrogenphosphate(1–) ion	$[H_2PO_4]^-$
Nitrous acid	HNO_2	4.7×10^{-4}	Nitrite ion	$[NO_2]^-$
Acetic acid	CH_3CO_2H	1.7×10^{-5}	Acetate ion	$[CH_3CO_2]^-$
Carbonic acid	H_2CO_3	4.3×10^{-7}	Hydrogencarbonate(1–) ion	$[HCO_3]^-$
Hydrogensulfite(1–) ion	$[HSO_3]^-$	1.2×10^{-7}	Sulfite(2–) ion	$[SO_3]^{2-}$
Dihydrogenphosphate(1–) ion	$[H_2PO_4]^-$	6.2×10^{-8}	Hydrogenphosphate(2–) ion	$[HPO_4]^{2-}$
Hydrocyanic acid	HCN	4.0×10^{-10}	Cyanide ion	$[CN]^-$
Hydrogencarbonate(1–) ion	$[HCO_3]^-$	5.6×10^{-11}	Carbonate ion	$[CO_3]^{2-}$
Hydrogenphosphate(2–) ion	$[HPO_4]^{2-}$	2.2×10^{-13}	Phosphate(3–) ion	$[PO_4]^{3-}$

Increasing strength of acid (upwards) — Increasing strength of conjugate base (downwards)

The self-ionization of water

Water itself is ionized to a very small extent (equation 11.65).

$$2H_2O(l) \rightleftharpoons [H_3O]^+(aq) + [OH]^-(aq) \qquad (11.65)$$

The self-ionization constant for water, K_w, is given by equation 11.66.

$$K_w = [H_3O^+][OH^-] = 1 \times 10^{-14}\ \text{mol}^2\ \text{dm}^{-6}\ (\text{at } 298\ \text{K}) \qquad (11.66)$$

It is convenient to define the term pK_w which, as equation 11.67 shows, bears the same relationship to K_w that pK_a does to K_a.

$$pK_w = -\log K_w = 14.00 \qquad (11.67)$$

Polybasic acids

Acids such as HCl, CH_3CO_2H and HCN are *monobasic acids* because they can lose only one proton per molecule of acid. Some Brønsted acids may be able to lose two, three or more protons. Sulfuric acid is an example of a *dibasic acid*. Equations 11.68 and 11.69 show that the first dissociation goes to completion, but that the second dissociation is an equilibrium.

$$H_2SO_4(aq) + H_2O(l) \rightarrow [H_3O]^+(aq) + [HSO_4]^-(aq) \qquad (11.68)$$

$$[HSO_4]^-(aq) + H_2O(l) \rightleftharpoons [H_3O]^+(aq) + [SO_4]^{2-}(aq) \quad pK_a = 1.92 \qquad (11.69)$$

A reaction with a base may lead to the formation of a sulfate(2–) or to a hydrogensulfate(1–) salt. It is a general trend that the first dissociation constant of a dibasic acid is larger than the second, and this is seen in Table 11.3 for H_2SO_3 and H_2CO_3. This is because it is harder to remove a proton from a negatively charged species than from a neutral one.

Phosphoric acid is tribasic, and the three dissociation steps are shown in equations 11.70 to 11.72. The acid strength of $H_3PO_4 > [H_2PO_4]^- > [HPO_4]^{2-}$.

$$H_3PO_4(aq) + H_2O(l) \rightleftharpoons [H_3O]^+(aq) + [H_2PO_4]^-(aq) \quad pK_a = 2.12 \qquad (11.70)$$

$$[H_2PO_4]^-(aq) + H_2O(l) \rightleftharpoons [H_3O]^+(aq) + [HPO_4]^{2-}(aq) \quad pK_a = 7.21 \qquad (11.71)$$

$$[HPO_4]^{2-}(aq) + H_2O(l) \rightleftharpoons [H_3O]^+(aq) + [PO_4]^{3-}(aq) \quad pK_a = 12.66 \qquad (11.72)$$

The dissociation constant K_b

In an aqueous solution of a *weak base* B, proton transfer is not complete and equation 11.73 shows the general equilibrium and the associated expression for K_b. The weaker the base, the smaller the value of K_b.

$$B(aq) + H_2O(l) \rightleftharpoons [BH]^+(aq) + OH^-(aq)$$

$$K_b = \frac{[BH^+][OH^-]}{[B][H_2O]} = \frac{[BH^+][OH^-]}{[B]} \qquad (11.73)$$

Worked example 11.3 ***Determining the concentration of $[OH]^-$ ions in an aqueous solution of ammonia***

Calculate the concentration of hydroxide ions in a 0.05 M aqueous solution of NH_3. [$K_b = 1.8 \times 10^{-5}$ mol dm^{-3}]

The appropriate equilibrium is:

$$NH_3(aq) + H_2O(l) \rightleftharpoons [NH_4]^+(aq) + [OH]^-(aq)$$

and the equilibrium constant is given by:

$$K_b = \frac{[NH_4^+][OH^-]}{[NH_3][H_2O]}$$

but since $[H_2O] = 1$, this can be simplified to:

$$K_b = \frac{[NH_4^+][OH^-]}{[NH_3]}$$

The concentration of ammonium ions equals the concentration of hydroxide ions, and therefore we can write:

$$K_b = \frac{[OH^-]^2}{[NH_3]}$$

We now make the assumption that $[NH_3]_{equilibrium} \approx [NH_3]_{initial} = 0.05$ mol dm^{-3}. Rearranging so that we can find $[OH^-]$ gives:

$$\begin{aligned}[OH^-] &= \sqrt{K_b \times [NH_3]} \\ &= \sqrt{1.8 \times 10^{-5} \times 0.05} \\ &= 9.5 \times 10^{-4} \text{ mol dm}^{-3}\end{aligned}$$

The relationship between K_a and K_b for a conjugate acid–base pair

We mentioned above that the conjugate base of a weak acid will be relatively strong and we can quantify this by considering the relationship between K_a and K_b for a conjugate acid–base pair. Consider both the dissociation of acetic acid in aqueous solution (equation 11.74) and that of the acetate ion in water (equation 11.75); these are the processes for which K_a and K_b are defined.

$$CH_3CO_2H(aq) + H_2O(l) \rightleftharpoons [H_3O]^+(aq) + [CH_3CO_2]^-(aq)$$

$$K_a = \frac{[H_3O^+][CH_3CO_2^-]}{[CH_3CO_2H]} \qquad (11.74)$$

$$[CH_3CO_2]^-(aq) + H_2O(l) \rightleftharpoons CH_3CO_2H(aq) + [OH]^-(aq)$$

$$K_b = \frac{[CH_3CO_2H][OH^-]}{[CH_3CO_2^-]} \qquad (11.75)$$

If we compare these two expressions with that given in equation 11.66 for K_w, we see that the three equilibrium constants are related by equation 11.76. Similarly, pK_a and pK_b are related by equation 11.77.

$$K_a \times K_b = \left\{\frac{[H_3O^+][CH_3CO_2^-]}{[CH_3CO_2H]}\right\} \times \left\{\frac{[CH_3CO_2H][OH^-]}{[CH_3CO_2^-]}\right\} \quad (11.76)$$

$$= [H_3O^+][OH^-]$$

$$= K_w$$

See problem 11.10

$$pK_a + pK_b = pK_w = 14.00 \quad (11.77)$$

Worked example 11.4 ***The relationship between a conjugate acid–base pair***

The value of K_a for HNO_2 is 4.7×10^{-4} mol dm^{-3}. Determine pK_b for its conjugate base.

There are two ways to tackle this problem.

Method 1: The conjugate base of HNO_2 is $[NO_2]^-$, and the relationship between pK_a for HNO_2 and pK_b for $[NO_2]^-$ is:

$$pK_a + pK_b = 14.00$$

If $K_a = 4.7 \times 10^{-4}$ mol dm^{-3}, $-\log K_a = pK_a = 3.33$. Therefore:

$$pK_b = 14.00 - 3.33 = 10.67$$

Method 2: The conjugate base of HNO_2 is $[NO_2]^-$, and the relationship between K_a for HNO_2 and K_b for $[NO_2]^-$ is:

$$K_a \times K_b = K_w = 1 \times 10^{-14} \text{ mol}^2 \text{ dm}^{-6}$$

$$K_a = 4.7 \times 10^{-4} \text{ mol dm}^{-3}$$

and so:

$$K_b = \frac{1 \times 10^{-14}}{4.7 \times 10^{-4}} = 2.13 \times 10^{-11} \text{ mol dm}^{-3}$$

$$pK_b = -\log K_b = 10.67$$

pH

The concentration of $[H_3O]^+$ ions in a solution is usually given by a pH value as defined in equation 11.78.[§] This relationship is identical to those between K_a and pK_a, K_b and pK_b, and K_w and pK_w.

$$pH = -\log [H_3O^+] \quad (11.78)$$

§ You may also find equation 11.78 written in the form $pH = -\log [H^+]$; in water, protons combine with water molecules and the predominant species are $[H_3O]^+$ ions.

Worked example 11.5 ***pH of a solution of hydrochloric acid***
What is the pH of a 0.05 M solution of hydrochloric acid?

First we must decide whether or not the concentration of the acid that is given can be used directly to find the concentration of $[H_3O]^+$ ions. Hydrochloric acid is fully dissociated in water:

$$HCl(aq) + H_2O(l) \rightarrow [H_3O]^+(aq) + Cl^-(aq)$$

and therefore: $[H_3O^+] = [HCl] = 0.05 \text{ mol dm}^{-3}$.

$$\begin{aligned} pH &= -\log[H_3O^+] \\ &= -\log(0.05) \\ &= 1.30 \end{aligned}$$

Worked example 11.6 ***pH of a solution of methanoic acid***
What is the pH of a 0.05 M solution of methanoic acid? [$K_a = 1.8 \times 10^{-4}$ mol dm^{-3}]

Methanoic acid is a weak acid and is not fully dissociated in aqueous solution:

$$HCO_2H(aq) + H_2O(l) \rightleftharpoons [H_3O]^+(aq) + [HCO_2]^-(aq)$$

and therefore $[H_3O^+] \neq [HCO_2H]$.
To find $[H_3O^+]$ we must consider the equilibrium:

$$K_a = \frac{[H_3O^+][HCO_2^-]}{[HCO_2H]} = 1.8 \times 10^{-4}$$

but since the number of $[H_3O]^+$ ions in solution equals the number of $[HCO_2]^-$ ions, we can write:

$$K_a = \frac{[H_3O^+]^2}{[HCO_2H]}$$

We can make the assumption that:

$$[HCO_2H]_{equilibrium} \approx [HCO_2H]_{initial} = 0.05 \text{ mol dm}^{-3}$$

To find $[H_3O^+]$:

$$\begin{aligned} [H_3O^+] &= \sqrt{K_a \times [HCO_2H]} \\ &= \sqrt{1.8 \times 10^{-4} \times 0.05} \\ &= 3.0 \times 10^{-3} \text{ mol dm}^{-3} \end{aligned}$$

$$\begin{aligned} pH &= -\log[H_3O^+] \\ &= -\log(3.0 \times 10^{-3}) \\ &= 2.52 \end{aligned}$$

Figure 11.14 illustrates the pH scale with examples of strong and weak acids and bases of different concentrations. The pH of an aqueous solution of a

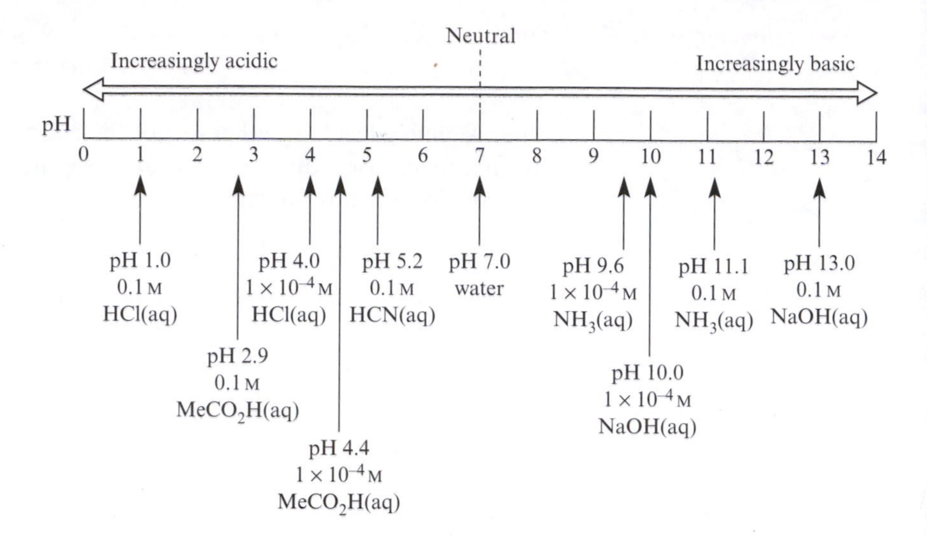

11.14 The pH scale runs below 0 to 14. The chart gives some examples of aqueous acids and bases covering a range of pH values.

base can be determined by first finding the concentration of hydroxide ions in solution and then using the relationship in equation 11.66 to determine the concentration of $[H_3O]^+$ ions. Worked examples 11.7 and 11.8 consider the cases of a strong (fully dissociated) and weak (partially dissociated) base respectively.

Worked example 11.7 *pH of a solution of sodium hydroxide*

What is the pH of a 0.05 M solution of sodium hydroxide?

Sodium hydroxide is fully dissociated in water:

$$NaOH(aq) \rightarrow Na^+(aq) + OH^-(aq)$$

The concentration of hydroxide ions is equal to the initial concentration of sodium hydroxide:

$$[OH^-] = 0.05 \text{ mol dm}^{-3}$$

The pH of the solution is given by the equation:

$$pH = -\log[H_3O^+]$$

and so we need to relate $[H_3O^+]$ to the known value of $[OH^-]$:

$$K_w = [H_3O^+][OH^-] = 1 \times 10^{-14} \text{ mol}^2 \text{ dm}^{-6} \text{ (at 298 K)}$$

$$[H_3O^+] = \frac{1 \times 10^{-4}}{0.05} = 2.0 \times 10^{-13} \text{ mol dm}^{-3}$$

$$\begin{aligned} pH &= -\log[H_3O^+] \\ &= -\log(2.0 \times 10^{-13}) \\ &= 12.7 \end{aligned}$$

Worked example 11.8 ***pH of a solution of methylamine***

What is the pH of a 0.1 M solution of methylamine? [$K_b = 4.6 \times 10^{-4}$ mol dm^{-3}]

Methylamine is a weak base and is partially dissociated in aqueous solution:

$$CH_3NH_2(aq) + H_2O(l) \rightleftharpoons [CH_3NH_3]^+(aq) + [OH]^-(aq)$$

and therefore $[OH^-] \neq [CH_3NH_2]$.

To find $[OH^-]$, consider the equilibrium constant:

$$K_b = \frac{[CH_3NH_3^+][OH^-]}{[CH_3NH_2]}$$

The number of $[CH_3NH_3]^+$ ions in solution equals the number of $[OH]^-$ ions, and so we can write:

$$K_b = \frac{[OH^-]^2}{[CH_3NH_2]}$$

We may make the approximation that $[CH_3NH_2]_{equilibrium} \approx [CH_3NH_2]_{initial}$. Thus $[OH^-]$ is given by:

$$\begin{aligned}[OH^-] &= \sqrt{K_b \times [CH_3NH_2]} \\ &= \sqrt{4.6 \times 10^{-4} \times 0.1} \\ &= 6.8 \times 10^{-3} \text{ mol dm}^{-3}\end{aligned}$$

We now have to find $[H_3O^+]$ and this is related to $[OH^-]$ as follows:

$$K_w = 1 \times 10^{-14} = [H_3O^+][OH^-]$$

$$[H_3O^+] = \frac{1 \times 10^{-14}}{6.8 \times 10^{-3}} = 1.47 \times 10^{-12} \text{ mol dm}^{-3}$$

$$\begin{aligned}pH &= -\log[H_3O^+] \\ &= -\log(1.47 \times 10^{-12}) \\ &= 11.8\end{aligned}$$

Essential equations relating to acid–base equilibria:

For a general weak acid HA in aqueous solution:

$$K_a = \frac{[H_3O^+][A^-]}{[HA]}$$

For a general weak base B in aqueous solution:

$$K_b = \frac{[BH^+][OH^-]}{[B]}$$

$pK_a = -\log K_a$ $\quad$ $K_a = 10^{-pK_a}$

$pK_b = -\log K_b$ $\quad$ $K_b = 10^{-pK_b}$

$K_w = [H_3O^+][OH^-] = 1 \times 10^{-14}$ mol^2 dm^{-6} $\quad$ $pK_w = -\log K_w = 14.00$

$K_w = K_a \times K_b$ $\quad$ $pH = -\log[H_3O^+]$

11.10 HYDRIDES OF GROUP 16 ELEMENTS

Water

Water is the most important hydride of oxygen and its properties colour our description of all other chemical properties. When we talk about substances being soluble or insoluble without qualification, we are usually referring to solubility in water. Water is the commonest liquid on Earth and possesses unique and remarkable properties. The strong hydrogen bonding network within liquid water ensures the large liquid range, and ultimately is responsible for the existence and maintenance of life on Earth.

We have already commented on the self-ionization of water (equation 11.65) and merely mention here that the concept of pH is defined in terms of the products of this self-ionization.

Hydrogen peroxide

Hydrogen peroxide, H_2O_2, is a blue, viscous liquid (mp 272 K, bp 425 K) but is usually encountered as an aqueous solution where the concentration is given in terms of a 'volume' or percentage. A 10 vol. aqueous solution of H_2O_2 means that 1 cm^3 of the solution will liberate 10 cm^3 of O_2 upon decomposition (see below); this corresponds to a 3% solution. Extensive hydrogen bonding occurs both between H_2O_2 molecules in the pure liquid, and between H_2O_2 and H_2O molecules in aqueous solutions. A molecule of H_2O_2 (Figure 11.15) possesses a skew conformation, with a *torsion angle* (the angle between the planes as illustrated in Figure 11.15) of 90° in the solid state; the barrier to rotation about the O–O bond is low and in the gas phase the torsion angle increases to 111°.

Hydrogen peroxide is produced commercially by using compound **11.7** (a derivative of anthraquinone) in a two-step process involving reaction with H_2 (equation 11.79) followed by oxidation with O_2 (equation 11.80). Common uses of H_2O_2 are as an oxidizing agent, and include its roles as a bleach, antiseptic and in pollution control. The use of H_2O_2 as a bleach for hair has led to the expression 'peroxide blonde'.

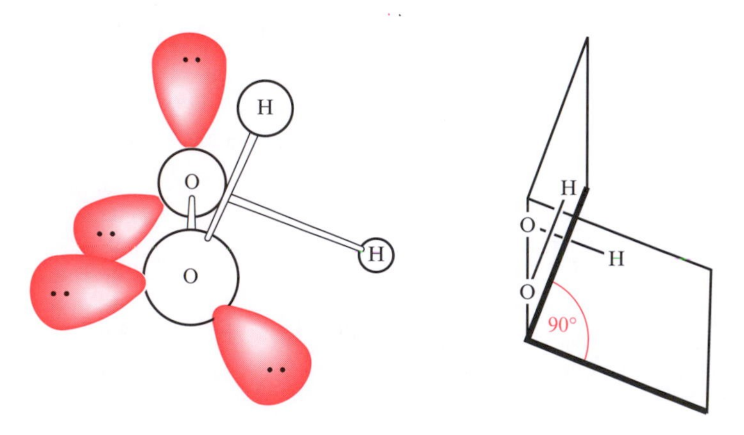

11.15 The structure of H_2O_2. In the gas phase, the O–O bond distance is 147 pm, and the angles O–O–H are 95°; the angle shown as 90° is the *torsion angle*.

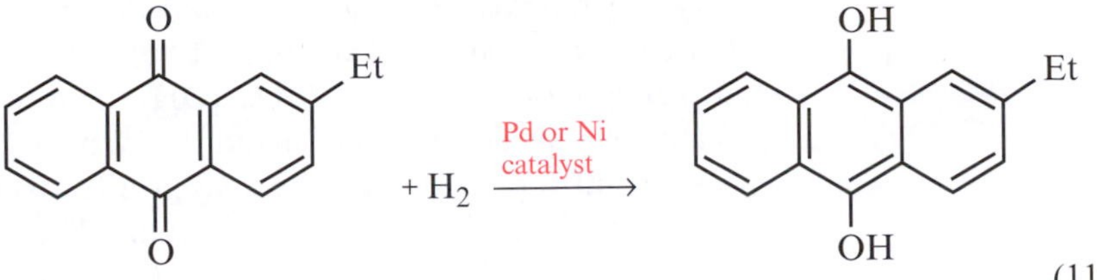

(11.7) $+ H_2 \xrightarrow{\text{Pd or Ni catalyst}}$ (11.79)

$+ O_2 \rightarrow H_2O_2 +$ **(11.7)** (11.80)

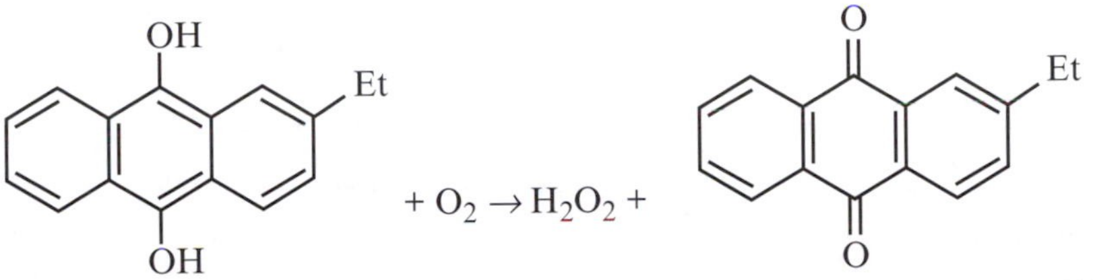

Box 11.7 Catalytic cycles

Equations 11.79 and 11.80 together show the conversion of H_2 to H_2O_2 in the presence of compound **11.7** which acts as a catalyst. The two-step reaction can be represented in the *catalytic cycle* shown below.

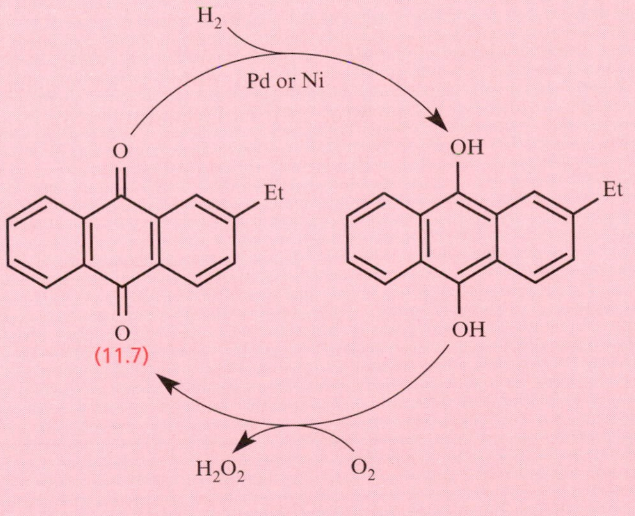

H_2

Pd or Ni

O Et O **(11.7)**

OH Et OH

H_2O_2 O_2

Catalytic cycles of this type are a convenient way of illustrating the key steps of reactions involving a sequence of steps. They are commonly used to describe both industrial and biological processes.

Hydrogen peroxide is *thermodynamically* unstable with respect to decomposition to water and O_2 (equation 11.81) but is *kinetically* stable. The reaction is catalysed by traces of MnO_2, $[OH]^-$ or some metal surfaces. Blood also catalyses this decomposition and this may be the basis of H_2O_2 being used as an antiseptic as the O_2 released effectively kills anaerobic bacteria. Reaction 11.81 is a *disproportionation*. The oxidation state of oxygen in H_2O_2 is –1, and in water, it is –2 and in O_2 it is zero. In reaction 11.81 the oxygen is simultaneously oxidized and reduced.

$$2H_2O_2(l) \rightarrow 2H_2O(l) + O_2(g) \qquad (11.81)$$

reduction
oxidation

A species disproportionates if it undergoes simultaneous oxidation and reduction.

Hydrogen peroxide can act as an oxidizing agent both in acidic or alkaline solutions (reactions 11.82 and 11.83). As soon as any MnO_2 is formed in reaction 11.83, it catalyses the decomposition of H_2O_2 and so the reaction shown is not very efficient.

$$2Fe^{2+} + H_2O_2 + 2H^+ \longrightarrow 2Fe^{3+} + 2H_2O$$

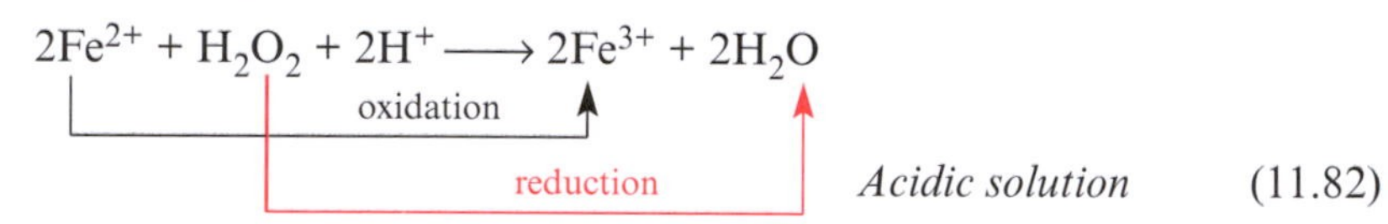

Acidic solution (11.82)

$$Mn(OH)_2 + H_2O_2 \longrightarrow MnO_2 + 2H_2O$$

oxidation
reduction

Alkaline solution (11.83)

Hydrogen peroxide can also act as a reducing agent in both acidic or alkaline solutions, but will only reduce species which are themselves strong oxidizing agents (equations 11.84 and 11.85).

$$2[MnO_4]^- + 5H_2O_2 + 6H^+ \longrightarrow 2Mn^{2+} + 8H_2O + 5O_2$$

reduction
oxidation

Acidic solution (11.84)

$$Cl_2 + H_2O_2 + 2OH^- \longrightarrow 2Cl^- + 2H_2O + O_2$$

reduction
oxidation

Alkaline solution (11.85)

The dissociation of aqueous H_2O_2 (equation 11.86) occurs to a greater extent than does the self-ionization of water, and H_2O_2 is a slightly stronger acid than H_2O.

$$H_2O_2(aq) + H_2O(l) \rightleftharpoons [H_3O]^+(aq) + \underset{\text{hydroperoxide ion}}{[HO_2]^-(aq)} \qquad (11.86)$$

$$K_a = \frac{[H_3O^+][HO_2^-]}{[H_2O_2]} = 1.78 \times 10^{-12}\ \text{mol dm}^{-3} \qquad pK_a = 11.75 \qquad (11.87)$$

Hydrogen sulfide

The hydrides of sulfur are called *sulfanes* but for H_2S, the name hydrogen sulfide is usually used. Hydrogen sulfide is a colourless gas (bp 214 K) with a characteristic smell of bad eggs. It is extremely toxic, and the fact that it anaesthetizes the sense of smell means that the intensity of smell is not a reliable measure of the amount of H_2S present.

Hydrogen sulfide is a natural product of decaying sulfur-containing matter, and it occurs in coal pits, gas wells and sulfur springs. Historically, laboratory preparations used a Kipp's apparatus (which was designed so as to minimize the escape of gaseous H_2S) in which dilute hydrochloric acid reacted with iron(II) sulfide (equation 11.88). The hydrolysis of calcium sulfide (equation 11.89) produces purer H_2S.

$$FeS(s) + 2HCl(aq) \rightarrow H_2S(g) + FeCl_2(aq) \qquad (11.88)$$

$$CaS + 2H_2O \rightarrow H_2S + Ca(OH)_2 \qquad (11.89)$$

The presence of H_2S can be confirmed by its reaction with lead acetate. Filter paper dipped in aqueous $Pb(O_2CCH_3)_2$, and then dried, turns black when it comes into contact with hydrogen sulfide (equation 11.90).

Qualitative test for gaseous H_2S:

$$H_2S(g) + \underset{\text{colourless}}{Pb(O_2CCH_3)_2} \rightarrow \underset{\text{black}}{PbS(s)} + 2CH_3CO_2H \qquad (11.90)$$

The bent structure of H_2S is consistent with VSEPR theory, *but* the H–S–H bond angle of 92° is much smaller than would be predicted. In the liquid state, hydrogen bonding is not very important as we saw from the trends in boiling points in Figure 11.11.

Box 11.8 H_2S: Toxic or life-supporting?

Hydrothermal fluids are discharged from volcanic vents on the ocean floor, and around these vents there are large quantities of H_2S but virtually no O_2. Hydrogen sulfide is normally considered too toxic to support life, and the lack of O_2 should also pose a critical problem. Nature, however, has adapted to the environment and many new species of animals, including certain clams and mussels, have been discovered only at these volcanic vent sites. Light does not penetrate to such ocean depths, and food chains which depend upon photosynthesis cannot operate. Recent research has shown that the vent communities use geothermal energy (rather than solar energy), and the reaction that starts the foodchain may be represented as:

$$H_2S + \underset{\text{from the volcanic vent}}{CO_2} \rightarrow \underset{\text{from seawater}}{[carbohydrate]} + H_2SO_4$$

This is a case of one animal's poison being the stuff of life to another!

Hydrogen sulfide is slightly soluble in water; a saturated solution at 298 K and atmospheric pressure has a concentration of ≈ 0.1 mol dm^{-3}. Aqueous solutions of H_2S are weakly acidic (equation 11.91) but the extremely small value of the second dissociation constant (equation 11.92) means that this second equilibrium lies to the left-hand side. A consequence of this is that *soluble* metal sulfides (Na_2S, K_2S, $[NH_4]_2S$ and CaS) are readily hydrolysed (equation 11.93).

Only a few metal sulfides are soluble in water: Na_2S, K_2S, $[NH_4]_2S$, CaS.

$$H_2S + H_2O \rightleftharpoons [H_3O]^+ + [SH]^- \qquad pK_a = 7.04 \tag{11.91}$$

$$[SH]^- + H_2O \rightleftharpoons [H_3O]^+ + S^{2-} \qquad pK_a \approx 19 \tag{11.92}$$

$$Na_2S + H_2O \rightleftharpoons NaSH + NaOH \tag{11.93}$$

Hydrogen sulfide burns in air with a blue flame and is oxidized to sulfur dioxide or sulfur depending upon the supply of dioxygen (equations 11.94 and 11.95).

$$\textit{Excess air/dioxygen:} \qquad 2H_2S + 3O_2 \rightarrow 2SO_2 + 2H_2O \tag{11.94}$$

$$\textit{Limited supply of air/dioxygen:} \quad 2H_2S + O_2 \rightarrow 2S + 2H_2O \tag{11.95}$$

In acidic conditions, H_2S is a mild reducing agent and may be oxidized to sulfur(0) (equation 11.96) or to higher oxidation states. We quantify its position as a reducing agent in the next chapter (see Table 12.4).

$$2Fe^{3+} + H_2S \rightarrow 2Fe^{2+} + 2H^+ + \tfrac{1}{8}S_8 \tag{11.96}$$

Hydrogen selenide and hydrogen telluride

Hydrogen selenide (H_2Se) and hydrogen telluride (H_2Te) are colourless, foul smelling and extremely toxic gases. H_2Se is a dangerous fire hazard – on exposure to O_2 it rapidly decomposes to give red selenium. H_2Te decomposes in air or water or when heated. Molecules of H_2Se and H_2Te are bent but, like H_2S, have smaller bond angles than expected from VSEPR theory (∠H–Se–H = 91° and ∠H–Te–H = 89°).

Hydrogen selenide may be prepared from its constituent elements (equation 11.97) but the thermal instability of H_2Te means that it cannot be prepared by an analogous reaction. Both H_2Se and H_2Te can be prepared by hydrolysing the appropriate aluminium chalcogenide (equation 11.98).

$$H_2 + Se \xrightarrow{630\ K} H_2Se \tag{11.97}$$

$$Al_2E_3 + 6H_2O \rightarrow 3H_2E + 2Al(OH)_3 \tag{11.98}$$

E = Se or Te

11.11 BINARY COMPOUNDS CONTAINING HYDROGEN AND GROUP 17 ELEMENTS: HYDROGEN HALIDES

Elements in group 17 form *hydrogen halides* of general formula HX (X = F, Cl, Br or I) which dissolve in water to give acidic solutions called *hydrohalic acids*.

Hydrogen fluoride

Hydrogen fluoride, HF, is produced in an explosive radical chain reaction when H_2 and F_2 are mixed (equation 11.99). Reaction 11.100 is a more convenient method of preparation.

$$H_2(g) + F_2(g) \rightarrow 2HF(g) \qquad (11.99)$$

$$CaF_2(s) + \underset{\text{concentrated}}{2H_2SO_4(aq)} \rightarrow 2HF(g) + Ca(HSO_4)_2(aq) \qquad (11.100)$$

A problem of working in the laboratory with HF is its ability to etch silica glass (equation 11.101) which corrodes glass reaction vessels. On the other hand, the same reaction is used commercially for the etching of patterns on glass. Monel steel (a nickel alloy) or polytetrafluoroethene (PTFE) containers are suitable for storing and handling HF.

$$4HF + SiO_2 \rightarrow SiF_4 + 2H_2O \xrightarrow{2HF(aq)} H_2SiF_6 \qquad (11.101)$$

We have already described the role of hydrogen bonding in the solid state structure of hydrogen fluoride (Figure 11.12). Hydrogen fluoride has a long liquid range (mp 190 K, bp 293 K) and at room temperature is a colourless, fuming and corrosive liquid or a gas depending upon the ambient conditions. Extensive hydrogen bonding exists in the liquid and this contributes to its low volatility; even in the vapour state, some intermolecular interactions persist, giving species of formulae $(HF)_x$ ($x \leq 6$). Although the ability of HF to react with glass colours our view of the compound, it is in many respects a hydrogen-bonded liquid like H_2O. For example, the protein insulin may be recovered unchanged after dissolution in pure liquid HF.

Hydrogen fluoride is completely miscible with water but its solution chemistry is complex. For the dissociation given in equation 11.102, pK_a is 3.45 (at 298 K), and this indicates that HF is a weaker acid than HCl, HBr or HI; a contributing factor is the high bond dissociation enthalpy of the H–F bond (Figure 11.16). Equilibrium 11.102 is complicated by the interaction of fluoride ions and hydrogen fluoride (equation 11.103). By applying Le Chatelier's principle, we can see that if the F^- ions produced in reaction 11.103 are removed from the system by their reaction with HF, more F^- (and, necessarily, more $[H_3O]^+$) will form.

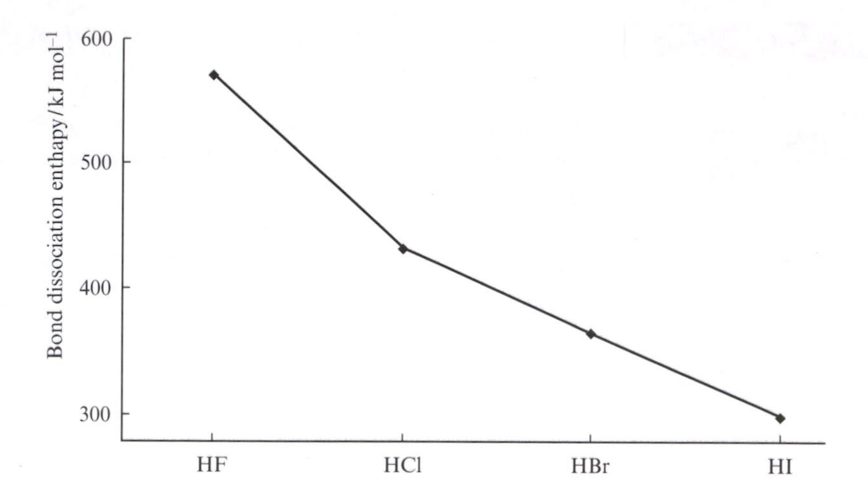

11.16 The trend in bond dissociation enthalpies for the group 17 hydrides.

$$HF(l) + H_2O(l) \rightleftharpoons [H_3O]^+(aq) + F^-(aq) \tag{11.102}$$

$$F^-(aq) + HF(aq) \rightleftharpoons [HF_2]^-(aq) \qquad K = \frac{[HF_2^-]}{[HF][F^-]} = 0.2\ dm^3\ mol^{-1} \tag{11.103}$$

The formation of the $[HF_2]^-$ anion depends on the ability of H and F to be involved in strong hydrogen bonding. Diffraction (X-ray and neutron) studies of $K[HF_2]$ confirm that the $[HF_2]^-$ anion has a linear structure (**11.8**) with an F----F distance of 226 pm. When compared with twice the H–F bond length in HF (2 × 92 pm), it indicates the strength of the hydrogen bonding in $[HF_2]^-$; remember that an H····F hydrogen bond will always be weaker and longer than a two-centre covalent H–F bond.

X-ray and neutron diffraction: see Section 3.2

$$[F—H—F]^-$$

(11.8)

Water-free liquid HF is self-ionizing (equation 11.104) and is highly suitable as a non-aqueous solvent.

$$3HF(l) \rightleftharpoons [H_2F]^+(solv) + [HF_2]^-(solv) \tag{11.104}$$

Hydrogen chloride

Hydrogen chloride, HCl, is a poisonous, colourless gas (bp 188 K) and can be detected by contact with ammonia (equation 11.105). It is conveniently prepared by the action of concentrated H_2SO_4 on solid NaCl (equation 11.106).

$$\underset{\text{colourless gas}}{HCl(g)} + \underset{\text{colourless gas}}{NH_3(g)} \rightarrow \underset{\text{fine, white powder}}{NH_4Cl(s)} \tag{11.105}$$

$$NaCl(s) + H_2SO_4(conc) \rightarrow HCl(g) + NaHSO_4(s) \tag{11.106}$$

Hydrogen chloride is also formed from the reaction of H_2 and Cl_2 but this is by no means as vigorous as that between H_2 and F_2; the standard enthalpies of formation of gaseous HF and HCl are –273 and –92 kJ mol^{-1} respectively. On a commercial scale, reaction 11.106 may be used, but the reaction of H_2 with Cl_2 is preferred. Hydrogen chloride has a wide range of uses, not only in the chemical laboratory as a gas or an aqueous acid (see below), but also in the manufacture of a wide range of inorganic and organic chemicals.

In liquid HCl, the intermolecular association through hydrogen bonding is less than in HF, and the degree of self-ionization is small. The zigzag chain structure of solid HCl is similar to that of HF and is due to hydrogen bonding.

Some reactions involving HCl (e.g. additions to unsaturated hydrocarbons) were described in Chapter 8. Hydrogen chloride oxidizes some metals and the electrochemical series (Table 11.1) can be used to predict which metal chlorides can be formed in this way (equations 11.107 and 11.108).

$$Mg(s) + 2HCl(g) \xrightarrow{\Delta} MgCl_2(s) + H_2(g) \quad (11.107)$$

$$Fe(s) + 2HCl(g) \xrightarrow{\Delta} FeCl_2(s) + H_2(g) \quad (11.108)$$

Hydrogen chloride fumes in moist air and dissolves in water to give a strongly acidic solution (hydrochloric acid) in which it is fully ionized. The chloride ion is precipitated as white AgCl when silver(I) nitrate is added (equation 11.109) and this is a standard qualitative test for the presence of free chloride ions in aqueous solution (equation 11.110).

$$AgNO_3(aq) + HCl(aq) \rightarrow AgCl(s) + HNO_3(aq) \quad (11.109)$$

Qualitative test for Cl^- ions:

$$AgNO_3(aq) + Cl^-(aq) \rightarrow \underset{\text{white precipitate}}{AgCl(s)} + [NO_3]^-(aq) \quad (11.110)$$

Displacement of H_2 from the acid occurs with metals above hydrogen in the electrochemical series – these reactions are analogous to the oxidations shown in equations 11.107 and 11.108 but involve aqueous HCl (equation 11.111).

$$Mg(s) + 2HCl(aq) \rightarrow MgCl_2(aq) + H_2(g) \quad (11.111)$$

Hydrochloric acid neutralizes bases (equations 11.112 and 11.113) and reacts with metal carbonates (equation 11.114) – *these are typical reactions of dilute aqueous acids.*

$$HCl(aq) + NaOH(aq) \rightarrow NaCl(aq) + H_2O(l) \quad (11.112)$$

$$2HCl(aq) + CuO(s) \rightarrow CuCl_2(aq) + H_2O(l) \quad (11.113)$$

$$2HCl(aq) + MgCO_3(s) \rightarrow MgCl_2(aq) + H_2O(l) + CO_2(g) \quad (11.114)$$

The above reactions all occur with *dilute* hydrochloric acid, typically 2 mol dm^{-3}. *Concentrated* hydrochloric acid is oxidized by $KMnO_4$ or MnO_2, to produce dichlorine (equations 11.115 and 11.116) and the former is a convenient synthetic route to Cl_2.

$$16HCl(conc) + 2KMnO_4(s) \rightarrow 5Cl_2(g) + 2KCl(aq) + 2MnCl_2(aq) + 8H_2O(l) \quad (11.115)$$

oxidation

reduction

$$4HCl(conc) + MnO_2(s) \xrightarrow{\Delta} Cl_2(g) + MnCl_2(aq) + 2H_2O(l) \quad (11.116)$$

oxidation

reduction

Hydrogen bromide and hydrogen iodide

Both HBr and HI are choking colourless gases (bp 206 and 238 K respectively). They *cannot* be prepared in reactions analogous to that for HCl in equation 11.106 because both HBr and HI are oxidized by H_2SO_4. Instead, phosphoric acid may be used (equation 11.117) or the reaction of red phosphorus with the respective halogen and water (equation 11.118). H_2 reacts with Br_2 or I_2 to give HBr or HI respectively, but high temperatures and a metal catalyst are needed; compare this with the explosive combination of H_2 and F_2.

$$KX + H_3PO_4 \xrightarrow{\Delta} HX + KH_2PO_4 \quad (11.117)$$

X = Br or I

$$3X_2 + 2P + 6H_2O \rightarrow 6HX + 2H_3PO_3 \quad (11.118)$$

X = Br or I

Both HBr and HI are soluble in water and are fully ionized – hydrobromic hydroiodic acids are strong acids. The bromide or iodide can be precipitated by adding silver nitrate; solid AgBr is cream-coloured and AgI is pale yellow and these precipitations are used in qualitative analysis for Br^- and I^- ions. Both acids undergo similar reactions to hydrochloric acid. As mentioned above, HBr and HI are easily oxidized as exemplified by reactions 11.119–11.121.

$$2Cu^{2+} + 4I^- \rightarrow 2CuI + I_2 \quad (11.119)$$

$$2X^- + Cl_2 \rightarrow X_2 + 2Cl^- \quad (11.120)$$

X = I or Br

$$2I^- + Br_2 \rightarrow I_2 + 2Br^- \quad (11.121)$$

11.12 GROUP 1: THE ALKALI METALS

➤ Alkali metals – physical properties: see Section 2.22; bcc lattices: see Section 7.11

The chemistry of the group 1 metals is dominated by the formation of singly charged cations and the ionization energies relating to equation 11.122 are given in Appendix 8.

$$M(g) \rightarrow M^+(g) + e^- \quad (11.122)$$

Diagonal relationship: see Section 11.14

In many of its properties, lithium is atypical and instead bears a resemblance to magnesium to which it is *diagonally* related in the periodic table.

Appearance, physical properties, sources and uses

Each of the elements lithium, sodium, potassium, rubidium and caesium is a soft, silver-grey solid metal at 298 K; however, the particularly low melting point of caesium (301.5 K, Table 2.7) means that at ambient temperatures, it may be a liquid. Francium (named after France, the country in which it was discovered in 1939) is a radioactive element and only minute quantities of it have ever been handled. Their high reactivities mean that the group 1 metals are not found naturally in the elemental state.

Lithium is produced by electrolysing lithium chloride in a manner analogous to the Downs process (see below). It has the lowest density (0.53 g cm^{-3}) of all the metals in the periodic table.

Sodium is the sixth most abundant element on Earth (in the form of compounds such as NaCl) and the most abundant of the group 1 metals. Compounds of sodium have many applications, including uses in the paper, glass, detergent, chemical and metal industries. Both sodium and potassium are biologically important, being involved, for example, in osmotic control and the body's nervous system. Sodium is manufactured in the Downs process in which molten NaCl is electrolysed (Figure 11.17); $CaCl_2$ is added to reduce the operating temperature to about 870 K; pure NaCl melts at 1073 K. Molten NaCl is composed of free Na^+ and Cl^- ions, and reduction of the Na^+ ions to form sodium (equation 11.123) occurs at the cathode whilst Cl^- ions are oxidized at the anode (equation 11.124). The net result is reaction 11.125, and the molten sodium passes out of the cell where it is cooled and solidifies. The design of the electrolysis cell is critical – NaCl will reform if the metallic sodium and gaseous Cl_2 come into contact with each other.

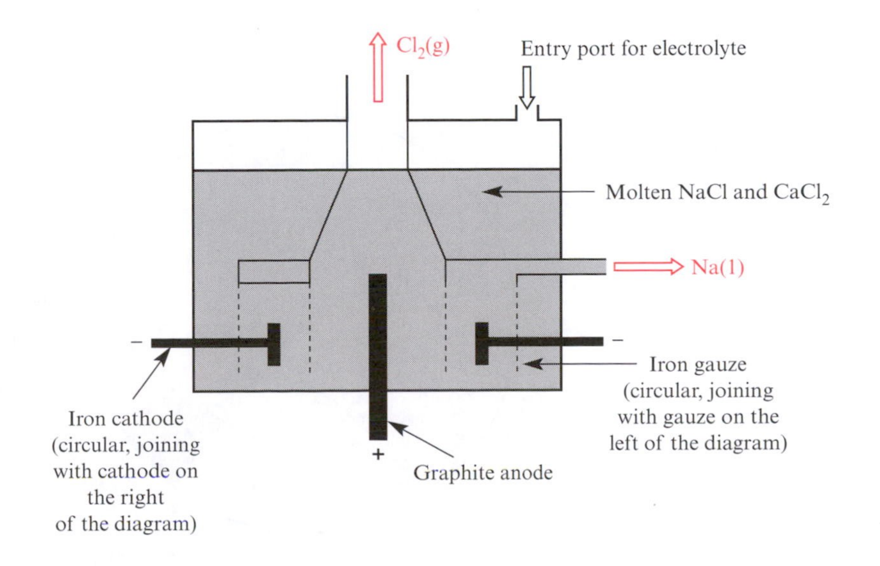

11.17 A schematic representation of the electrolysis cell used in the Downs process to produce sodium commercially from NaCl. The $CaCl_2$ is present to lower the working temperature from 1073 K (the mp of NaCl) to ≈ 870 K. The Na and Cl_2 must be kept separate to prevent reformation of NaCl.

At the cathode: $Na^{+}(l) + e^{-} \rightarrow Na(l)$ (11.123)

At the anode: $2Cl^{-}(l) \rightarrow Cl_2(g) + 2e^{-}$ (11.124)

Overall reaction: $2Na^{+}(l) + 2Cl^{-}(l) \rightarrow 2Na(l) + Cl_2(g)$ (11.125)

reduction

oxidation

Potassium is almost as abundant as sodium in the Earth's crust (2.4% versus 2.6%). Potassium salts are widely used as fertilizers since potassium is an essential plant nutrient. The industrial production of potassium follows reaction 11.126. Rubidium can be prepared similarly by reducing RbCl (equation 11.127).

See problem 11.14

$$KCl(l) + Na(g) \xrightarrow{1000\ K} K(g) + NaCl(s) \qquad (11.126)$$

$$2RbCl + Ca \xrightarrow{\Delta} 2Rb + CaCl_2 \qquad (11.127)$$

Almost all salts of the group 1 metals are soluble in water.

Almost all salts of the group 1 metals are soluble in water. Caesium salts have found some applications in organic syntheses, and the large Cs^{+} cation is often used to give good crystals with large anions.

Alkali metal hydroxides

Sodium hydroxide is an important industrial chemical (Appendix 14) and Figure 11.18 shows some of its uses.

The alkali metals react exothermically with water (equation 11.128) and the reactivity *increases* significantly as the group is descended.

$$2M + 2H_2O \rightarrow 2MOH + H_2 \qquad (11.128)$$

M = Li, Na, K, Rb or Cs

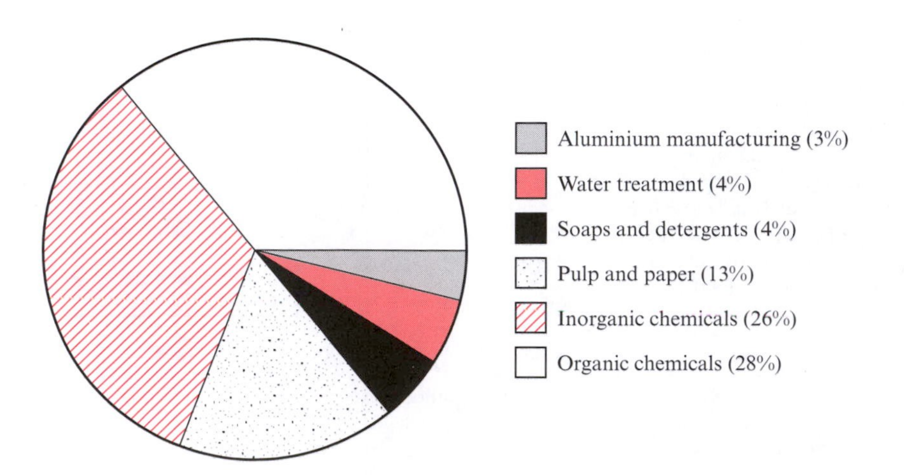

11.18 Industrial uses of sodium hydroxide in Western Europe in 1994. (Data from *Chemistry & Industry*, (1995) p. 832.)

Lithium, sodium and potassium are less dense than water and therefore reaction 11.128 takes place on the surface of the liquid. In the case of potassium, it is violent enough to ignite the H_2 as it is produced. This ignition is an explosive radical chain process, and coupled with the fact that the aqueous KOH formed is caustic, the reaction must be conducted with care. The reactions with rubidium and caesium are even more spectacular. Both of these metals are denser than water (1.53 and 1.84 g cm^{-3} respectively) and so the reaction takes place within the bulk liquid. As ignition of the H_2 occurs, the energy released can cause violent explosions.

Radical chain reaction: see Section 11.2

Lithium, sodium, potassium and rubidium hydroxides are white, crystalline solids, whilst caesium hydroxide is pale yellow. The crystals are *deliquescent,* that is, they absorb water from the surrounding air and finally become liquid. The solubility of the group 1 hydroxides in water tends to increase as the group is descended. The hydroxides of the metals sodium to caesium dissociate fully (equation 11.129) and the presence of free $[OH]^-$ ions means that these aqueous solutions are strongly basic.

A *deliquescent* substance absorbs water from the surrounding air and eventually forms a liquid.

$$MOH(s) \xrightarrow{\text{water}} M^+(aq) + [OH]^-(aq) \qquad (11.129)$$

Alkali metal oxides, peroxides and superoxides

Alkali metals are often stored in paraffin oil to prevent reaction with atmospheric O_2 and water vapour. A piece of freshly cut sodium or potassium is shiny but quickly tarnishes as it comes into contact with the air.

The combustion of the group 1 metals in O_2 leads to various oxides depending on the metal and the conditions. Lithium reacts to give lithium oxide, Li_2O (equation 11.130), but sodium usually forms sodium peroxide, Na_2O_2, which contains the diamagnetic $[O_2]^{2-}$ ion (equation 11.131). By increasing the pressure or temperature, NaO_2 may be produced (equation 11.132) – this contains the paramagnetic superoxide anion, $[O_2]^-$.

$$4Li + O_2 \rightarrow \underset{\text{lithium oxide}}{2Li_2O} \qquad (11.130)$$

$$2Na + O_2 \rightarrow \underset{\text{sodium peroxide}}{Na_2O_2} \qquad (11.131)$$

$$Na + O_2 \rightarrow \underset{\text{sodium superoxide}}{NaO_2} \qquad (11.132)$$

Although under varying conditions all three oxides can be prepared for each of the five metals, the tendency to form the peroxide and the superoxide increases down group 1 with the $[O_2]^{2-}$ and $[O_2]^-$ anions being increasingly stabilized by the larger metal cations K^+, Rb^+ and Cs^+ although all the superoxides are relatively unstable. The colours of the compounds vary from white to orange, following a general trend as the group is descended: Li_2O and Na_2O form white crystals whilst K_2O is pale yellow, Rb_2O is yellow and Cs_2O is orange.

See problem 11.15

Trend in ionic radii: see Figure 6.23

Potassium superoxide is used in breathing masks; KO_2 absorbs water and in so doing produces both O_2 for respiration and KOH which absorbs exhaled CO_2. All the potassium-containing components in equations 11.133 and 11.134 remain in the solid phase and are contained effectively in the breathing mask.

$$4KO_2(s) + 2H_2O(l) \rightarrow 4KOH(s) + 3O_2(g) \quad (11.133)$$

$$KOH(s) + CO_2(g) \rightarrow KHCO_3(s) \quad (11.134)$$

Box 11.9 Suboxides of rubidium and caesium

When oxidized under *controlled* conditions, rubidium and caesium form *suboxides* such as Rb_6O, Cs_7O, Rb_9O_2 and $Cs_{11}O_3$.

An interesting question arises: 'Do the usual oxidation states of +1 for the alkali metal and –2 for the oxygen apply in these compounds?' If we assign an oxidation state of –2 to the oxygen in Rb_6O, then the *formal* oxidation state of rubidium is $+\frac{1}{3}$. However, the formula is more usefully considered in terms of Rb^+ and O^{2-} centres and Rb_6O may thus be written as $(Rb^+)_6(O^{2-})\cdot 4e^-$.

The structures of Rb_6O, Cs_7O, Rb_9O_2 and $Cs_{11}O_3$ all feature octahedral units of metal ions with the oxide ion residing at the centre; the octahedra are fused together by sharing faces (shown in pale red).

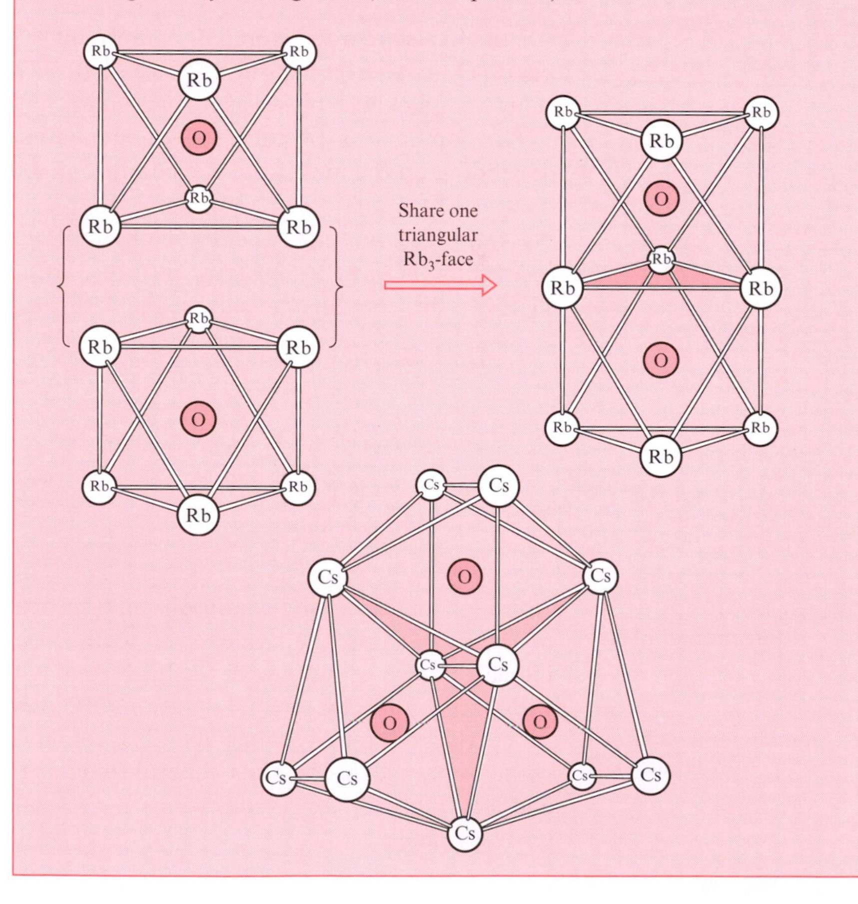

Alkali metal halides

Each group 1 metal reacts with X_2 (X = F, Cl, Br or I) when heated (equation 11.135).

$$2M + X_2 \xrightarrow{\Delta} 2MX \qquad (11.135)$$

The alkali metal halides are white, crystalline high-melting solids with ionic lattices. Sodium chloride, an industrially important chemical, occurs naturally as rock salt and in seawater.

Alkali metals as reducing agents

When an alkali metal reacts with water or dioxygen, it is oxidized. Conversely, the metal behaves as a reducing agent, reducing the hydrogen in H_2O to H_2, or reducing the oxygen in O_2 to O^{2-}, $[O_2]^-$ or $[O_2]^{2-}$. Alkali metals are powerful reducing agents and this property is associated with the ease with which each metal loses an electron: the more thermodynamically favourable is this loss, the more powerful the reducing ability of the metal, and we quantify this process in the next chapter. In Section 8.20 we saw that alkali metals can be used to reduce alkynes, and equations 11.136 to 11.139 further illustrate the role of alkali metals as reducing agents. The reducing power can be increased by dissolving the alkali metal in liquid ammonia as described below.

$$2Na + Cl_2 \rightarrow 2NaCl \qquad (11.136)$$

$$2Cs + F_2 \rightarrow 2CsF \qquad (11.137)$$

$$4Na + ZrCl_4 \rightarrow 4NaCl + Zr \qquad (11.138)$$

$$2Na + 2C_2H_5OH \rightarrow \underset{\text{sodium ethoxide}}{2[C_2H_5O]^- Na^+} + H_2 \qquad (11.139)$$

Alkali metals in liquid ammonia

Alkali metals dissolve in dry liquid ammonia to give coloured solutions. At low concentrations, these are blue and paramagnetic and have high electrical conductivities, but at higher concentrations, bronze coloured 'metallic' solutions are formed in which a variety of aggregates are present. These properties are caused by the ionization of the metal (equation 11.140) and a factor that contributes to this unusual phenomenon is the low value of the first ionization potential of each group 1 metal.

$$\begin{aligned} Na(s) + NH_3(l) \rightarrow Na^+&(\text{solvated by liquid } NH_3) \\ &+ e^-(\text{solvated by liquid } NH_3) \end{aligned} \qquad (11.140)$$

The electron removed from the sodium atom is solvated by the NH_3 molecules but can be considered to be a 'free electron', and therefore the solutions of alkali metals in liquid ammonia are powerful reducing agents (equations

11.141 to 11.143). The alkali metal is oxidized to the corresponding metal ion and this becomes the counter-ion for the reduced partner in the reaction. Thus in equation 11.143, the product is isolated as $Na_2[Fe(CO)_4]$.

$$O_2 \xrightarrow{\text{M in liquid NH}_3} [O_2]^- \quad (11.141)$$

$$[MnO_4]^- \xrightarrow{\text{M in liquid NH}_3} [MnO_4]^{2-} \quad (11.142)$$

$$[Fe(CO)_5] \xrightarrow{\text{Na in liquid NH}_3} [Fe(CO)_4]^{2-} \quad (11.143)$$

Solutions of an alkali metal in ammonia are unstable with respect to the formation of a metal amide and H_2 (equation 11.144). The reaction is slow but is catalysed by some metals (e.g. iron).

$$2Na + 2NH_3 \rightarrow \underset{\text{sodium amide}}{2NaNH_2} + H_2 \quad (11.144)$$

11.13 GROUP 2: THE ALKALINE EARTH METALS

Each of the alkaline earth metals has two electrons in the valence shell and most of their chemistry is that of the M^{2+} ion. Beryllium stands out from the group because it exhibits significant covalency in its compounds.

Appearance, physical properties, sources and uses

Some physical properties of the group 2 metals are listed in Table 11.4; with the exception of barium, these metallic elements have close-packed lattices in the solid state.

Metal lattice: see Section 7.11

Beryllium is relatively rare but it is one of the lightest metals known and possesses one of the highest melting points. It is non-magnetic, has a high thermal conductivity, and is resistant to attack by concentrated nitric acid, as well as to oxidation in air at 298 K. These properties contribute to it being of great industrial importance, and beryllium is used in the manufacture of body-parts in high-speed aircraft and missiles, and in communication satellites; it is also used in nuclear reactors as a moderator and a reflector. Since each beryllium atom has only four electrons, it allows X-rays to pass through virtually unperturbed and is used in the windows of X-ray tubes. Beryllium is found in many natural minerals; emerald and aquamarine, which are two precious forms of the mineral *beryl* (the mixed oxide $3BeO{\cdot}Al_2O_3{\cdot}6SiO_2$), are used in jewellery. Despite the widespread technical uses of beryllium, care must be taken when handling its salts as they are extremely toxic. The element can be prepared by reducing BeF_2 (equation 11.145) or by the electrolysis of molten $BeCl_2$.

X-rays: see Section 3.2

$$BeF_2 + Mg \rightarrow MgF_2 + Be \quad (11.145)$$

Table 11.4 Selected physical properties of the alkaline earth metals.

Element (symbol)	*Physical appearance*	*Melting point / K*	*Boiling point / K*	*Density / g cm^{-3}*	*Notes*
Beryllium (Be)	Steel-grey metal	1560	2744	1.85	Compounds are extremely toxic
Magnesium (Mg)	Silver-white metal	923	1363	1.74	
Calcium (Ca)	Silver coloured metal	1115	1757	1.55	
Strontium (Sr)	Silver coloured metal	1050	1655	2.58	
Barium (Ba)	Silver-white metal	1000	2170	3.50	Soluble compounds are toxic
Radium (Ra)	White metal	973	1413	5.00	Radioactive

Magnesium is the eighth most abundant element in the Earth's crust but it is not present in the elemental state. Two of the major sources are the minerals dolomite (the mixed carbonate $CaCO_3 \cdot MgCO_3$) and magnesite ($MgCO_3$), and it is also present in seawater as magnesium(II) salts. Figure 11.19 summa-

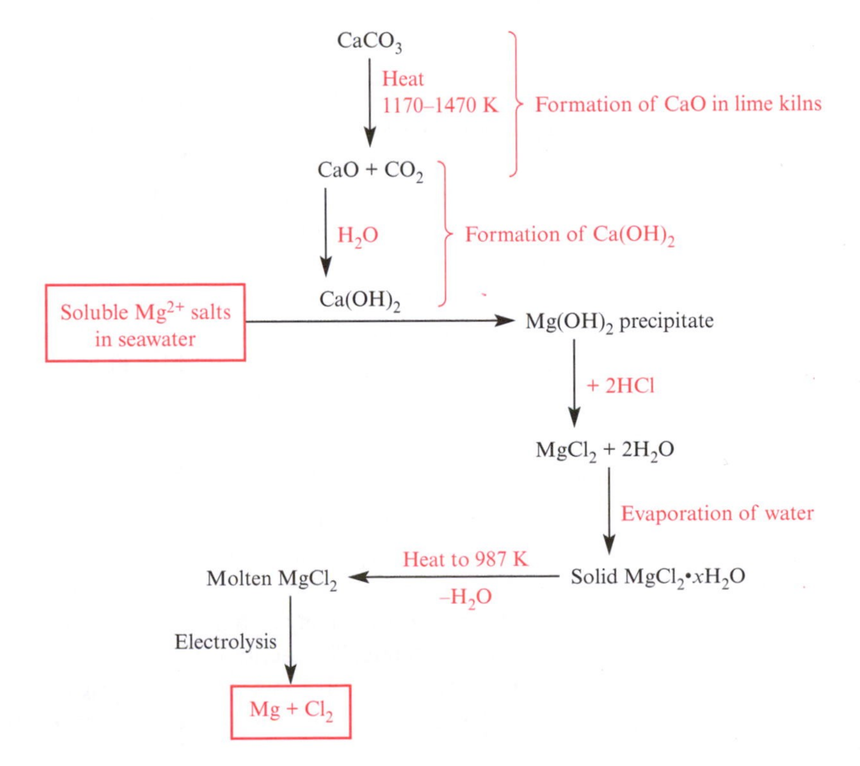

11.19 A summary of the industrial process used to extract magnesium from seawater.

rizes the extraction of magnesium from seawater. It begins with the precipitation of Mg^{2+} ions in the form of magnesium hydroxide, the hydroxide ions being provided by adding $Ca(OH)_2$ (slaked lime) which itself is formed from calcium carbonate which is widely available as limestone, sea shells and other calcareous deposits. The $Mg(OH)_2$ is neutralized with hydrochloric acid (equation 11.146) and the solid magnesium chloride deposited after the water has been evaporated is melted and electrolysed (equations 11.147 and 11.148).

$$2HCl(aq) + Mg(OH)_2(s) \rightarrow MgCl_2(aq) + 2H_2O(l) \qquad (11.146)$$

$$\textit{At the cathode:}\quad Mg^{2+}(l) + 2e^- \rightarrow Mg(l) \qquad (11.147)$$

$$\textit{At the anode:}\quad 2Cl^-(l) \rightarrow Cl_2(g) + 2e^- \qquad (11.148)$$

Large-scale production of magnesium is required to meet the commercial demands of this metal. Its high reactivity is responsible for its uses in photographic flashlights, flares and fireworks, and its low density (Table 11.4) makes it invaluable as a component in alloys. The presence of magnesium in an Mg/Al alloy improves the mechanical strength, welding and fabrication properties of the material, and increases its resistance to corrosion. The properties of a particular alloy depend upon the ratio of Mg : Al and alloys have wide uses in aircraft body-parts, missiles and lightweight tools and containers.

An alloy is an intimate mixture of two or more metals, or metals and non-metals. The aim of alloying is to improve the physical properties and resistance to corrosion, heat, etc. of the material.

Magnesium plays an important biological role: it is involved in phosphate metabolism and is present in chlorophyll. Medical uses of magnesium salts include indigestion powders ('milk of magnesia', $Mg(OH)_2$) and medication for constipation ('Epsom salts', $MgSO_4$).

Calcium is the fifth most abundant element in the Earth's crust and occurs in limestone ($CaCO_3$), gypsum ($CaSO_4{\cdot}2H_2O$) and fluorite (CaF_2), but not in the element state. The metal may be produced by electrolysing molten $CaCl_2$ (reduction of Ca^{2+} occurs at the cathode) or by the reduction of CaO with aluminium (equation 11.149). The uses of calcium metal are not as widespread as those of its compounds.

$$3CaO + 2Al \xrightarrow{\Delta} 3Ca + Al_2O_3 \qquad (11.149)$$

Strontium and barium both occur as the sulfate and carbonate: $SrSO_4$ (celestite), $SrCO_3$ (strontianite), $BaSO_4$ (barite) and $BaCO_3$ (witherite). The oxides of these elements are reduced by aluminium to produce strontium and barium respectively in reactions analogous to equation 11.149. Barium is used as a 'getter' in vacuum tubes; the high reactivity of the metal means that it combines with gaseous impurities such as O_2 and N_2 (equations 11.150 and 11.151) and the effect is to remove residual traces of such gases to give a particularly good vacuum.

$$2Ba + O_2 \rightarrow 2BaO \quad (11.150)$$

$$3Ba + N_2 \rightarrow Ba_3N_2 \quad (11.151)$$

Barium sulfate is a white compound and is insoluble in water; it is used in paints, in X-ray diagnostic work ('barium meals') and in the glass industry. The addition of barium chloride to aqueous solutions containing sulfate ions results in the formation of a white precipitate of $BaSO_4$ (equation 11.152). All water- or acid-soluble compounds of barium are highly toxic and one past use of $BaCO_3$ was as a rat poison.

Qualitative test for $[SO_4]^{2-}$ ions:

$$BaCl_2(aq) + [SO_4]^{2-}(aq) \rightarrow \underset{\text{white precipitate}}{BaSO_4(s)} + 2Cl^-(aq) \quad (11.152)$$

Alkaline earth metal oxides

With the exception of beryllium, the group 2 oxides (MO) are usually formed by the thermal decomposition of the respective carbonate (MCO_3) as shown for calcium in Figure 11.19. In air, the shiny surface of an alkaline earth metal quickly tarnishes duc to the formation of a thin coat of metal oxide. This protects the metal from further reaction under ambient conditions and the metal is said to be *passivated*. The property is particularly noticeable for beryllium which is resistant to reaction with water and acids because of the presence of the protective oxide covering.

$$2Be(s) + O_2 \rightarrow 2BeO(s) \quad (11.153)$$

A metal is passivated if it has a surface coating of the metal oxide which protects it from reaction with e.g. water.

The complete oxidation of beryllium takes place if the powdered metal is ignited in O_2 (equation 11.153) and BeO is a white solid which crystallizes with the wurtzite lattice. When heated in air, magnesium burns with a brilliant white flame, reacting not only with O_2 but also with N_2 in reactions analogous to those for barium. The oxides MgO, CaO, SrO and BaO are white crystalline solids with ionic lattices of the sodium chloride type. Barium oxide reacts further with O_2 when heated at 900 K (equation 11.154), and strontium peroxide can be similarly produced (620 K and under a high pressure of O_2).

➤ Wurtzite lattice: see Section 6.13

➤ NaCl type lattice: see Section 6.8

$$2BaO(s) + O_2(g) \xrightarrow{900\text{ K}} \underset{\text{barium peroxide}}{2BaO_2(s)} \quad (11.154)$$

Calcium oxide is commonly called quicklime or lime, and the addition of a molar equivalent of water to solid CaO produces solid $Ca(OH)_2$ (slaked lime) in a highly exothermic reaction. In contrast, beryllium oxide does not react with water, MgO reacts slowly, and BaO quickly; BaO is used as a drying agent for liquid alcohols and amines. Calcium hydroxide is used as an industrial alkali and has a vital role in the building trade as a component in mortar; dry mixtures of sand and CaO can be stored and transported without problem. Once water is added, the mortar quickly sets, forming solid calcium carbonate as CO_2 is absorbed (equation 11.155). The sand from the mortar acts as a binding agent between the particles of $CaCO_3$.

$$CaO(s) \xrightarrow{H_2O(l)} Ca(OH)_2(s) \xrightarrow{CO_2(g)} CaCO_3(s) + H_2O(l) \qquad (11.155)$$

A solution of calcium hydroxide is called limewater and is used to test for CO_2 – bubbling CO_2 through limewater turns the latter cloudy owing to the precipitation of $CaCO_3$ (equation 11.156).

Qualitative test for CO_2:

$$\underset{\text{limewater}}{Ca(OH)_2(aq)} + CO_2(g) \rightarrow \underset{\text{white precipitate}}{CaCO_3(s)} + H_2O(l) \qquad (11.156)$$

Large amounts of CaO are used in the chemical industry in the manufacture of Na_2CO_3 and NaOH, although the importance of the Solvay process which has been used to produce Na_2CO_3 and $NaHCO_3$ is now significantly less than it was. Calcium oxide is also used in the manufacture of calcium carbide, CaC_2, (equation 11.157) which contains the anion $[C{\equiv}C]^{2-}$.

$$CaO(s) + 3C(s) \xrightarrow{2200\ K} CaC_2(s) + CO(g) \qquad (11.157)$$

The oxides have very high melting points (all > 2000 K) and BeO, MgO and CaO are used as *refractory materials*; such materials are not decomposed at high temperatures. Magnesium oxide is a good thermal conductor but is an electrical insulator, and coupled with its thermal stability, these properties make it suitable for use as an electrical insulator in heating and cooking appliances.

The group 2 hydroxides

Beryllium does not react with water even when heated, since it is passivated (see above). To a certain extent magnesium is also passivated, but reacts with steam or when heated in water (equation 11.158). Calcium, strontium and barium all react with water in a similar manner, with the reactivity increasing down the group. Calcium reacts with cold water and a steady flow of H_2 is produced; with hot water the reaction is rapid. Beryllium hydroxide may be prepared by precipitation from the reaction of $BeCl_2$ and NaOH (but see below).

$$Mg(s) + 2H_2O(l) \xrightarrow{\Delta} Mg(OH)_2(aq) + H_2(g) \qquad (11.158)$$

The properties of beryllium hydroxide differ from those of the later hydroxides. In the presence of an excess of $[OH]^-$, $Be(OH)_2$ forms a tetrahedral *complex anion* **11.9** (equation 11.159) and this exemplifies its behaviour as a Lewis acid.

Box 11.10 Flame tests and fireworks

When the salt of an *s*-block metal is treated with concentrated hydrochloric acid (to produce a volatile metal chloride) and is heated strongly in the non-luminous flame of a Bunsen burner, the flame appears with a characteristic colour – this *flame test* is used in qualitative analysis to determine the identity of the metal ion.

Metal ion	*Colour of flame*	*Additional confirmatory tests*
Lithium	Red	
Sodium	Intense yellow	Cobalt-blue glass completely absorbs sodium light; thus, the flame appears colourless though a piece of blue glass.
Potassium	Lilac (difficult to observe)	The flame appears crimson when viewed through a piece of blue glass.
Calcium	Brick-red	The flame appears pale green when viewed through a piece of blue glass.
Strontium	Crimson	The flame appears purple when viewed through a piece of blue glass.
Barium	Apple-green	

The principal requirement of a firework (apart from the noise!) is to produce bright, coloured lights, and the pyrotechnics industry capitalizes on the emission of coloured light by *s*-block (and other) elements. Typically, white light is the result of the oxidation of magnesium or aluminium (e.g. in sparklers), and yellow, red and green are due to sodium, strontium and barium respectively. Producing fireworks which emit a blue colour is not so easy; copper(I) chloride is used but the intensity of colour is not particularly good. The explosion in a firework is initiated by an oxidizing agent such as potassium chlorate, $KClO_3$, or potassium perchlorate, $KClO_4$. In rockets, an exothermic reaction which will produce the energy to launch the firework is also needed. Fireworks are stored for quite long time periods and must not absorb water – as we saw in the main discussion, many compounds of the *s*-block elements are deliquescent or hygroscopic and thus the component compounds in the fireworks must be carefully chosen.

$$Be(OH)_2(s) + 2[OH]^-(aq) \rightarrow [Be(OH)_4]^{2-}(aq) \quad (11.159)$$

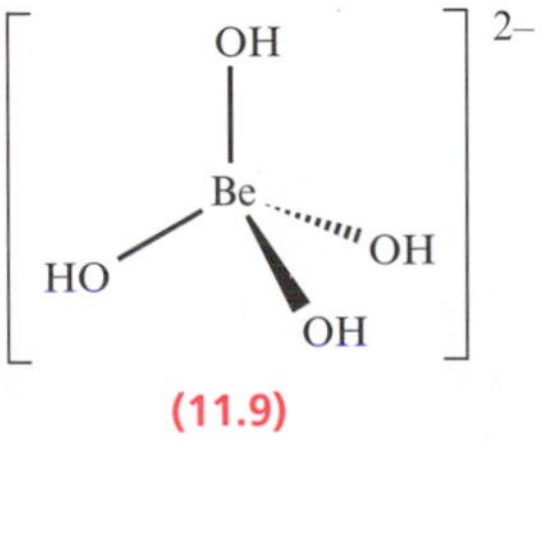

$$[Be(OH)_4]^{2-}$$

(11.9)

Beryllium hydroxide is neutralized by acid and can also function as a Brønsted base (equation 11.160).

$$Be(OH)_2(s) + H_2SO_4(aq) \rightarrow BeSO_4(s) + 2H_2O(l) \qquad (11.160)$$

A compound is *amphoteric* if it can act both as an acid or a base.

When a compound can act both as an acid or a base, it is said to be *amphoteric*. Whereas $Be(OH)_2$ is amphoteric, the later group 2 hydroxides act only as bases with the base strength increasing down the group.

Worked example 11.9 ***pH of a solution of calcium hydroxide***

What is the pH of a 0.001 M solution of calcium hydroxide?

Calcium hydroxide has a fairly low solubility in water but the dissolved salt is fully ionized:

$$Ca(OH)_2(aq) \rightarrow Ca^{2+}(aq) + 2[OH]^-(aq)$$

Each mole of $Ca(OH)_2$ provides two moles of $[OH]^-$ ions.

Therefore, the concentration of hydroxide ions is equal to *twice* the initial concentration of calcium hydroxide: $[OH^-] = 2 \times 0.001 = 0.002$ mol dm^{-3}.

The pH of the solution is given by the equation:

$$pH = -\log[H_3O^+]$$

and $[H_3O^+]$ and $[OH^-]$ are related by the equation:

$$K_w = [H_3O^+][OH^-] = 1 \times 10^{-14} \text{ mol}^2 \text{ dm}^{-6} \text{ (at 298 K)}$$

$$[H_3O^+] = \frac{1 \times 10^{-14}}{0.002} = 5.0 \times 10^{-12} \text{ mol dm}^{-3}$$

$$\begin{aligned} pH &= -\log[H_3O^+] \\ &= -\log(5.0 \times 10^{-12}) \\ &= 11.3 \end{aligned}$$

The group 2 halides

Equation 11.161 shows the general reaction for a group 2 element, M, with a halogen X_2, but this is not always a convenient route to the metal halides.

$$M + X_2 \rightarrow MX_2 \qquad (11.161)$$

$$BeO + Cl_2 + C \xrightarrow{900\text{–}1100 \text{ K}} BeCl_2 + CO \qquad (11.162)$$

Beryllium chloride is prepared from the oxide (equation 11.162) and possesses a polymeric structure similar to that of BeH_2 (Figure 11.4). We described the bonding in BeH_2 in terms of 3c–2e Be–H–Be bridges, but in $BeCl_2$ **11.10**, there are sufficient electrons for all the Be–Cl interactions to be localized 2-centre 2-electron bonds. Each chlorine atom has three lone pairs of electrons and one lone pair per Cl atom is used to form a coordinate bond.

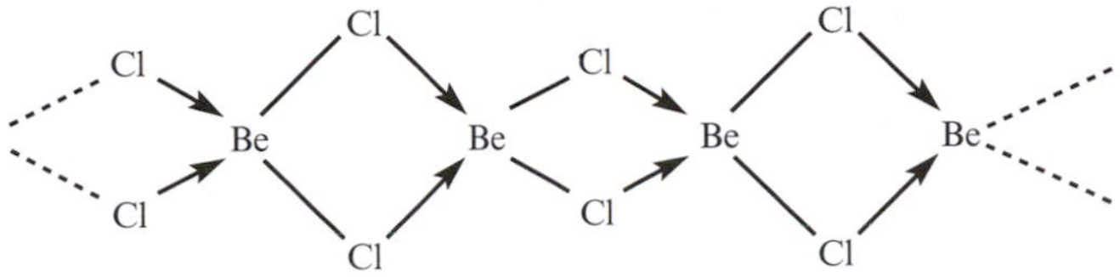

(11.10)

The halides of the later metals form ionic lattices – MgF_2 has a rutile-type lattice and CaF_2, SrF_2 and BaF_2 crystallize with the fluorite structure. Calcium fluoride occurs naturally as the mineral fluorite and is a major source of fluorine. With the exception of BeF_2 which forms a glass, the group 2 halides are white or colourless crystalline solids and the fluorides tend to have considerably higher melting points than the chlorides, bromides and iodides. There is a tendency for crystals of the chlorides, bromides and iodides to be deliquescent, and $SrBr_2$ crystals are *hygroscopic*. A hygroscopic solid absorbs water from the surrounding air but, unlike a deliquescent crystal, does not become a liquid. The deliquescent nature of calcium chloride renders it an excellent drying agent, for example in desiccators.

Fluorite and rutile lattices: see Sections 6.11 and 6.12

A *hygroscopic* solid absorbs water from the surrounding air but does not become a liquid.

Anomalous behaviour of beryllium

After reading this section, you should have gained the impression that beryllium is not a typical alkaline earth metal. Magnesium, calcium, strontium and barium have many properties in common, although the degree of reactivity changes as the group is decended. Beryllium (the first member of the group) shows anomalous chemical behaviour and many of its compounds exhibit covalent character.

The crucial property that makes beryllium behave so distinctly is the high charge density of the Be^{2+} ion and if indeed it existed as a naked ion it would be extremely polarizing. The consequence is that beryllium compounds either exhibit partial ionic character or they contain solvated ions. In water, the beryllium(II) centre is hydrated forming $[Be(H_2O)_4]^{2+}$ in which the positive charge is spread out over the beryllium centre and the four water molecules. The $[Be(H_2O)_4]^{2+}$ ion readily loses a proton because the Be^{2+} centre significantly polarizes the already polar O–H bonds (structure **11.11**).

Charge density: see Section 6.18

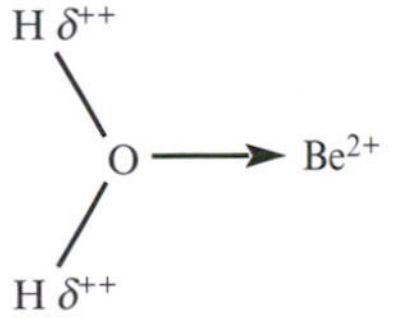

(11.11)

We could envisage that the loss of H^+ (as $[H_3O]^+$) might lead to the formation of $[Be(H_2O)_3(OH)]^+$ but the process does not stop here – various

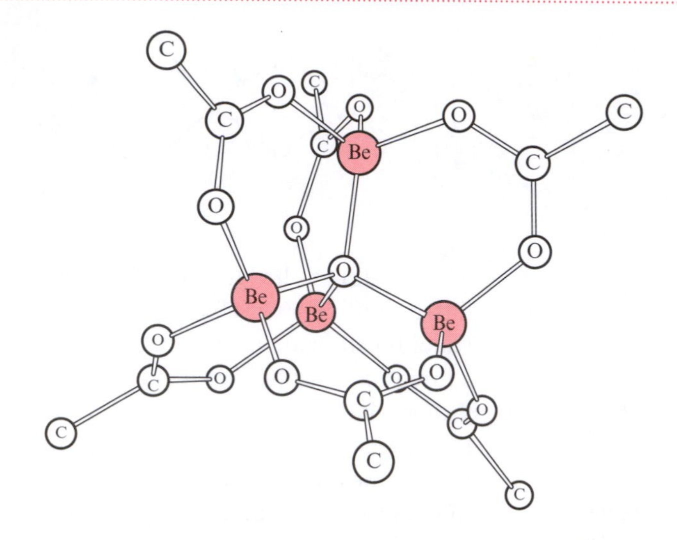

11.20 The structure of $[Be_4O(O_2CMe)_6]$; the central $\{Be_4O\}^{6+}$ unit is a common structural motif in beryllium chemistry. The H atoms of the methyl groups have been omitted.

solution species are present and equation 11.163 gives one example. By applying Le Chatelier's principle to this equilibrium, we can understand why the $[Be(H_2O)_4]^{2+}$ ion only exists in strongly acidic solution; the presence of excess $[H_3O]^+$ ions drives the equilibrium to the left-hand side. A common structural motif is the $\{Be_4O\}^{6+}$ unit and Figure 11.20 shows this in the structure of $[Be_4O(O_2CMe)_6]$.

$$4[Be(H_2O)_4]^{2+}(aq) + 2H_2O(l) \rightleftharpoons 2[(H_2O)_3Be{-}O{-}Be(H_2O)_3]^{2+}(aq) + 4[H_3O]^+(aq) \qquad (11.163)$$

In Chapter 13 we shall see that aluminium behaves in a similar way to beryllium, the two elements being *diagonally related* in the periodic table.

11.14 DIAGONAL RELATIONSHIPS IN THE PERIODIC TABLE: LITHIUM AND MAGNESIUM

Ion size and charge considerations

Although lithium does not stand out so distinctly from group 1 as beryllium does from group 2, there is no doubt that lithium is atypical of the alkali metals in many respects. The Li^+ ion is small (see Figure 6.23) and highly polarizing, and this results in a degree of covalency in some of its compounds. Since the ion is singly charged, the charge density of the Li^+ ion depends only on the surface area (equation 6.42). In group 2, the charge density of a particular ion depends both on the surface area and the fact that the ion is doubly charged. This is usually given as a *charge-to-radius ratio*. The increased charge in going from Li^+ to Mg^{2+} is effectively compensated by the increase in the size of the ion, and as a consequence we observe similar patterns in behaviour between lithium and magnesium despite the fact that they are in different groups. The Li^+ and Mg^{2+} ions are *diagonally related* in the periodic table, as are Be^{2+} and Al^{3+}, and the three ions Na^+, Ca^{2+} and Y^{3+}.

Lithium and magnesium ions are more strongly hydrated in aqueous solution than are the ions of the heavier group 1 and 2 metals. When the nitrates of lithium and magnesium are heated, they decompose to give oxides (equation 11.164) whereas thermal decomposition of the remaining group 1 and 2 nitrates gives a nitrite (equation 11.165) which further decomposes to an oxide.

$$4LiNO_3(s) \xrightarrow{\Delta} 2Li_2O(s) + O_2(g) + 2N_2O_4(g) \qquad (11.164)$$

$$\underset{M = Na, K, Rb, Cs}{2MNO_3(s)} \xrightarrow{\Delta} 2MNO_2(s) + O_2(g) \qquad (11.165)$$

SUMMARY

In this chapter we have discussed some descriptive inorganic chemistry of hydrogen and the *s*-block elements, and in addition we have reviewed acid–base equilibria.

Do you know what the following terms mean?

- kinetic isotope effect
- electropositive
- non-stoichiometric compound
- 3-centre 2-electron interaction
- Lewis acid and base
- coordinate bond
- adduct
- non-aqueous solvent
- self-ionization
- hydrogen bond
- Brønsted acid and base
- conjugate acid and base
- pH
- pK_a
- pK_w
- pK_b
- solvation
- hydration
- disproportionation
- deliquescent
- hygroscopic
- amphoteric
- alloy
- diagonal relationship

You should now be able to discuss aspects of the chemistry, including sources (or syntheses) and uses, of:

- the element hydrogen
- selected hydrides of the *s*-, *p*- and *d*-block elements
- the alkali metals
- halides, oxides and hydroxides of the alkali metals
- the alkaline earth metals
- halides, oxides and hydroxides of the alkaline earth metals

And you should be able:

- to write down expressions for acid or base dissociation constants
- to use a value of K_a to find the concentration of $[H_3O]^+$ ions in solution
- to use a value of K_b to find the concentration of $[OH]^-$ ions in solution
- to inter-relate K_w, K_a and K_b
- to calculate the pH of a solution of a strong acid, a strong base, a weak acid or a weak base

[We consider indicators and buffers in the accompanying workbook.]

Do you know how to test qualitatively for the following?

- $H_2(g)$
- $CO_2(g)$
- $H_2S(g)$
- $HCl(g)$
- Cl^-, Br^- and I^- ions
- $[SO_4]^{2-}$ ions

PROBLEMS

11.1 An absorption in an IR spectrum of an organic compound comes at 3100 cm^{-1} and is assigned to an X–H bond where X = C, O or N. Determine the shift in this band upon deuteration of the X–H bond for each X. (A_r: C = 12, N = 14, O = 16, H = 1, D = 2)

11.2 Why are kinetic isotope effects useful for reactions involving C–H bonds but are not likely to be of significant use in studying reactions involving $^{12}C{=}O$ and $^{13}C{=}O$ labelled compounds?

11.3 Which of the following metal ions might be reduced to the respective metal by H_2: (a) Cu^{2+}; (b) Zn^{2+}; (c) Fe^{2+}; (d) Ni^{2+}; (e) Mn^{2+}? Conversely, which of the following metals would react with dilute sulfuric acid to generate H_2: (a) Mg; (b) Ag; (c) Ca; (d) Zn; (e) Cr?

11.4 The standard enthalpies of formation (298 K) of methanol and octane are –239 and –250 kJ mol^{-1}. Determine the enthalpy change when a mole of each is fully combusted, and assess the relative merits of these compounds as fuels. ($\Delta_f H^o$ $CO_2(g)$ = –393.5, $H_2O(l)$ = –286 kJ mol^{-1}).

11.5 In Section 11.5 we stated that 'the tetrahydroborate anion is *kinetically* stable towards hydrolysis'. What does this mean? How does kinetic stability in this case differ from thermodynamic stability? Suggest why the cobalt is present in the 'cobalt-doped pellets of $Na[BH_4]$' mentioned in Section 11.5.

11.6 Figure 11.21 shows the trends in the boiling points for the group 14 hydrides E_nH_{2n+2} (E = C, Si or Ge) for n = 1, 2 and 3. Comment briefly on the observed trends and the factors that govern the boiling points of these compounds.

11.7 What is the oxidation state of nitrogen in (a) NH_3, (b) $[NH_4]^+$, (c) N_2H_4 and (d) $[NH_2]^-$?

11.8 Use the data in Table 11.5 to verify Trouton's rule.

Table 11.5 Data for problem 11.8.

Compound	*bp / K*	$\Delta_{vap}H$ / kJ mol^{-1}
PCl_3	349	30.5
HI	237	19.8
$CHCl_3$	334	29.2
C_7H_{16} (heptane)	371	31.8

11.9 Discuss what is meant by hydrogen bonding and how it may influence (a) the boiling points and (b) the solid state structures of hydrides of some of the *p*-block elements.

11.10 Given that: $K_a \times K_b = K_w$
show that: $pK_a + pK_b = 14$.

11.11 Determine the pH of a 0.01 M aqueous solution of $MeCO_2H$ ($K_a = 1.7 \times 10^{-5}$ mol dm^{-3}). [Hint: see worked example 11.1.]

11.12 Determine the pH of a 0.05 M aqueous solution of ammonia ($K_b = 1.8 \times 10^{-5}$ mol dm^{-3}). [Hint: see worked example 11.3.]

11.13 Determine the pH of an aqueous solution of 25 cm^3 0.2 M hydrochloric acid to which 18 cm^3 0.25 M NaOH has been added. [Further examples of this type including titrations with weak acids and weak bases are given in the accompanying workbook.]

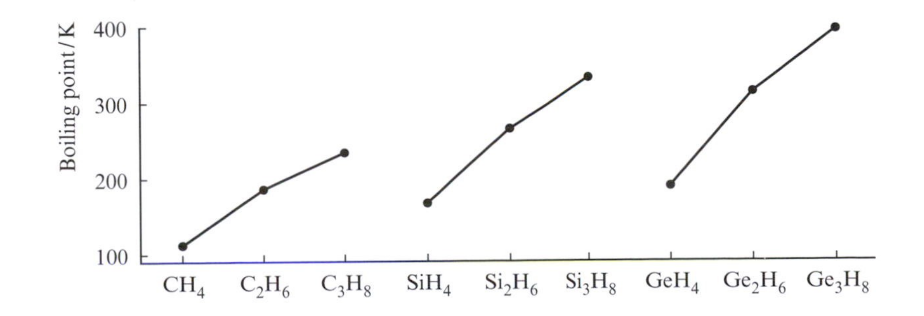

11.21 For Problem 11.6.

11.14 According to the electrochemical series, potassium is a better reducing agent than sodium, and yet in equation 11.126 we showed that sodium reduces K^+ ions in the production of potassium metal. Using Le Chatelier's principle, suggest why the following equilibrium might lie to the right-hand side (at 1120 K) rather than to the left:

$$K^+(l) + Na(g) \rightleftharpoons K(g) + Na^+(l)$$

11.15 Suggest how lattice energy effects might be used to explain why the stabilities of potassium, rubidium and caesium superoxides are greater than that of sodium superoxide, and why lithium usually forms the oxide Li_2O rather than the peroxide or superoxide.

11.16 The first members of periodic groups are often noted for their 'anomalous behaviour'. Discuss some of the chemical properties of (a) lithium and (b) beryllium that would support this statement.

11.17 'Hard water' is a characteristic of areas where limestone ($CaCO_3$) is present and contains Ca^{2+} ions. The following equilibrium describes how calcium ions are freed into the tap-water:

$$CaCO_3(s) + H_2O(l) + CO_2(g) \rightleftharpoons Ca^{2+}(aq) + 2[HCO_3]^-(aq)$$

Carbon dioxide is less soluble in hot water than cold water. Use these data to explain why scales of calcium carbonate form inside a kettle when the tap-water is hard. [Hint: Apply Le Chatelier's principle.]

11.18 Distinguish between the terms *deliquescent* and *hygroscopic*.

11.19 What do you understand by the term *disproportionation*? Give two examples of disproportionation reactions.

11.20 In this chapter we have highlighted several tests used in qualitative inorganic analysis. How would you test for: (a) chloride ions, (b) bromide ions, (c) iodide ions, (d) sulfate ions, (e) carbon dioxide, (f) dihydrogen, (g) gaseous HCl and (h) hydrogen sulfide?

12 Thermodynamics and electrochemistry

Topics

12.1 INTRODUCTION

In Chapter 10 we dealt with kinetics and considered the question: 'How *fast* is a reaction proceeding?' In this chapter we are concerned with chemical *thermodynamics* and ask the questions: 'How *far* will a reaction proceed and how much energy will be consumed or released?' Up to this point, we have dealt with the thermodynamics of reactions in terms of a change in enthalpy, ΔH. This *thermochemical quantity* may provide an *indication* as to whether a reaction is thermodynamically viable or not, *but* some exothermic reactions have a high activation barrier and occur only slowly (or not at all on an observable timescale) and many reactions are endothermic. For example, N_2H_4 is kinetically stable with respect to decomposition into its constituent elements even though this process is exothermic (equation 12.1) and when ammonium chloride dissolves in water, the temperature falls indicating that the reaction is *endothermic* (equation 12.2). Why should reaction 12.2 proceed?

$$N_2H_4(g) \rightarrow N_2(g) + 2H_2(g) \quad \Delta H = -50.6 \text{ kJ per mole of } N_2H_4 \qquad (12.1)$$

$$NH_4Cl(s) \rightarrow [NH_4]^+(aq) + Cl^-(aq) \quad \Delta H = +14.8 \text{ kJ per mole of } NH_4Cl \qquad (12.2)$$

These *thermochemical* data provide only *part* of the story as far as the energetics of a reaction are concerned. The enthalpy change tells us whether the temperature of the *surroundings* or the *system* is raised or lowered as heat is either given out or taken in. Heat is only one form of energy and in Chapter 3 we related the enthalpy change to the change in internal energy, ΔU, of a system – we look again at this relationship in Section 12.2.

In this chapter, we shall see that the *change in free energy*, ΔG, for a reaction is a better guide as to the favourability of the reaction than is ΔH. Notice that, as for ΔU, we are now using the term ***energy*** and not enthalpy. The change in free energy takes into account not only the change in enthalpy, but also the change in the *entropy* of the system. We explore the relationship between changes in free energy, entropy and enthalpy in Section 12.8, but firstly we approach the topic of free energy by considering equilibria, since most readers will already be familiar with the physical significance of large and small values of equilibrium constants. In principle, *all* reactions are equilibria, but a reaction for which the equilibrium constant is very large lies far over towards the right-hand side and to all intents and purposes the reaction has gone to completion. Firstly, we mention some terms which are fundamental to thermodynamics.

The system and the surroundings

Two terms used extensively in thermodynamics are the '*system*' and the '*surroundings*'. The system consists of the reaction components (reactants and products) and is distinguishable from the surroundings which make up everything else in the universe except for the system. The system may be contained within the walls of a reaction vessel (Figure 12.1) or there may be no defined boundary, as for example when a piece of solid CO_2 (dry-ice) sublimes to give gaseous CO_2 – in this case the system may seem to disappear into the surroundings but the molecules of CO_2 still constitute 'the system'. Thermochemical data, ΔH, refer to the 'heat lost by the system to, or gained by the system from, the surroundings'.

If a system is thermally insulated from the surroundings, changes that occur to the system are said to be *adiabatic* and no exchange of heat energy takes place between the system and its surroundings.

Many reactions are carried out at a stated, constant temperature and such a change is termed *isothermal*.

State functions

The state of a system may be described by variable quantities such as temperature, pressure, volume, enthalpy, entropy, free energy and internal energy. These variables are called *state functions* and, provided we know how much material we have in the system, we only need to specify *two* state functions in order to know the other state functions for the system. Consider the ideal gas equation 12.3.

$$PV = nRT \qquad (12.3)$$

where R (molar gas constant) = 8.314×10^{-3} kJ mol^{-1} K^{-1}.

For a particular, fixed system, the amount of material is given by the number of moles, n. The inter-relationship between pressure P, volume V and temperature T means that if we fix *any two* of these variables, then we

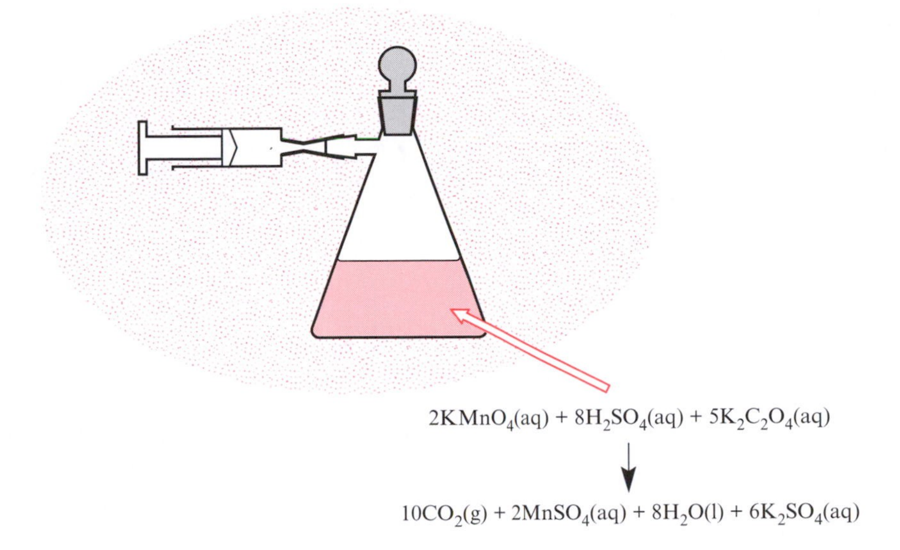

12.1 The distinction between the system and its surroundings may be seen by considering a reaction in which a gas is produced; in this experiment, the gas is collected in a syringe which expands into the surroundings during the reaction.

necessarily know the third; this is the reason why we can specify that the volume of one mole of an ideal gas at 273 K and 10^5 Pa is 22.7 dm^3, and similarly, if we know that at 10^5 Pa the volume of one mole of the gas is 22.7 dm^3, the temperature must be 273 K.

Standard temperature and pressure: see Section 1.9

State functions are inter-related and knowing any *two state functions* automatically fixes *all the others*.

Some state functions are:

Temperature	T	**Pressure**	P
Volume	V	**Internal energy**	U
Enthalpy	H	**Entropy**	S
Free energy	G		

An important property of a state function is that a change in the function is independent of the manner in which the change is made. We have already come across this in the form of Hess's Law of constant heat summation, where ΔH for a reaction is the same irrespective of the thermochemical cycle used (Figure 12.2). Equation 12.4 is a general expression that applies to any function of state:

Σ = summation of

$$\Delta(\text{State function}) = \Sigma(\text{State function})_{\text{products}} - \Sigma(\text{State function})_{\text{reactants}} \quad (12.4)$$

Standard states

Standard states were detailed in Section 1.9.

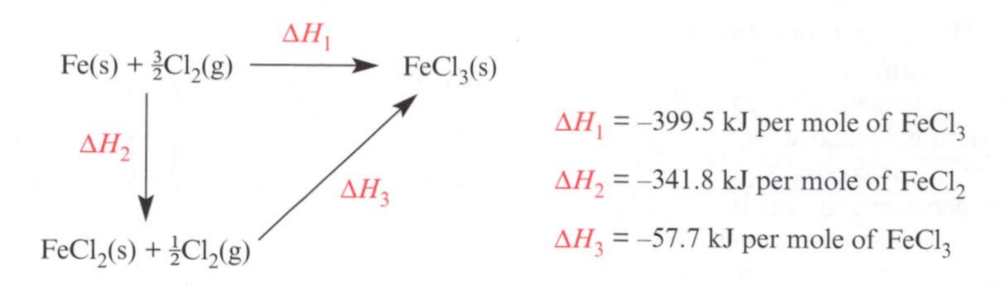

12.2 Iron(III) chloride may be formed directly from iron and dichlorine, or by first forming iron(II) chloride. The enthalpy change in going from $[Fe(s) + \frac{3}{2}Cl_2(g)]$ to $FeCl_3(s)$ is the same, irrespective of the route: $\Delta H_1 = \Delta H_2 + \Delta H_3$.

12.2 INTERNAL ENERGY, *U*

The First Law of Thermodynamics

The First Law of Thermodynamics states that energy cannot be created or destroyed, merely changed from one form to another.

The internal energy, U, of a system is its *total energy*, and is a state function. In a *perfectly isolated* system, the internal energy remains constant and for any change $\Delta U = 0$. Energy can be converted from one form to another, for example into heat or used to do work, but it cannot be created or destroyed. This is the *First Law of Thermodynamics*.

When the internal energy changes, energy *may* be released in the form of heat and the temperature of the surroundings increases (provided that the system is not perfectly insulated). This corresponds to a transfer of energy from the system to the surroundings, and while the total energy of the system *plus* the surroundings remains constant (equation 12.5), there is a change in the internal energy of the system itself.

$$\text{Energy lost by system} = \text{Energy gained by surroundings} \qquad (12.5)$$

Alternatively, if the transfer of heat is in the opposite sense (equation 12.6), the internal energy of the system increases.

$$\text{Energy gained by system} = \text{Energy lost by surroundings} \qquad (12.6)$$

Accompanying the heat transfer, q, may be energy transfer in the form of 'work done', w, and equation 12.7 gives the general relationship for a change in internal energy in the system.

$$\text{Change in internal energy} = \Delta U = q + w \qquad (12.7)$$

where q = change in heat energy (enthalpy) = ΔH w = work done.

If q is positive, heat energy is added to the system and if q is negative, heat is lost from the system to the surroundings.

'Work done' *on* the system includes such events as compressing a gas. Conversely, when a gas expands, work is done *by* the system on the surroundings. At a constant (external) pressure P, the work done in changing the volume by an amount ΔV is given by equation 12.8.

$$\begin{array}{l}\text{Work done at constant pressure} \\ \textbf{\textit{by}} \text{ the system } \textbf{\textit{on}} \text{ the surroundings}\end{array} = -(P \times \Delta V) \qquad (12.8)$$

If the pressure is given in Pa and the volume in m^3, the units of the work done are J (see Table 1.2). Note the sign convention – *work done* ***by*** *the system* ***on*** *the surroundings is defined as negative work*.

$\Delta U \approx \Delta H$: see Sections 3.5, 6.3 and 6.5

The relationship between ΔU and ΔH (equation 12.9) that we introduced earlier now follows by combining equations 12.7 and 12.8. In many reactions, the pressure P corresponds to atmospheric pressure (1 atm = 101 300 Pa, or 1 bar = 10^5 Pa).

$$\Delta U = \Delta H - P\Delta V \tag{12.9}$$

For a system at constant pressure doing work on the surroundings:

(Change in the internal energy of the system) = (Change in enthalpy) – (Pressure x Change in volume)

$$\Delta U = \Delta H - P\Delta V$$

Figure 12.3 summarizes the changes in enthalpy and volume for a system, and emphasizes the sign convention.

How large is the term $P\Delta V$ in relation to ΔH?

The work done by or on the system is typically significantly smaller than the enthalpy change and, in many cases ΔU is very nearly equal to ΔH. This was the approximation that we made in Chapter 3 when we discussed bond

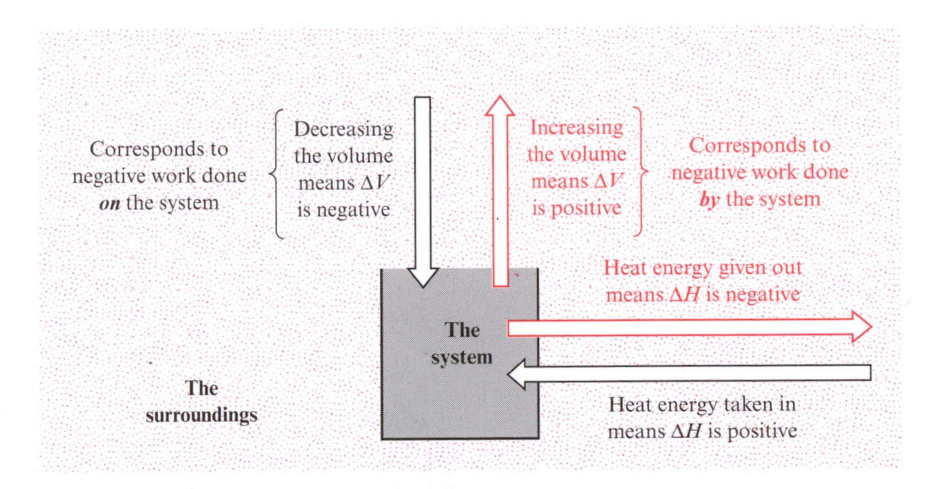

12.3 The system is the term given to the reaction components. The change in enthalpy of a system is given by $\Delta H = H_2 - H_1$ where H_1 and H_2 are the absolute enthalpies of the starting materials and the products respectively. *Absolute enthalpies cannot be measured; only ΔH can be found experimentally*. If $H_1 > H_2$, heat energy has been lost to the surroundings; if $H_2 > H_1$, heat energy has been taken in from the surroundings. The change in volume of the system is given by $\Delta V = V_2 - V_1$ where V_1 and V_2 are the absolute volumes of the starting materials and the products respectively. If the volume of the system increases, for example a gas is evolved in the reaction, then $V_2 > V_1$; if the volume of the system decreases, $V_1 > V_2$.

dissociation energies, and again in Chapter 6 when we took the lattice energy as being close in value to $\Delta_{lattice}H$.

Volume changes associated with reactions involving the formation or consumption of gases are always far larger than ones associated with liquids or solids, and volume changes due to the gaseous components of reactions predominate in the $P\Delta V$ term. Assuming that the ideal gas law (equation 12.3) holds for all gaseous components, we can find the work done by, or on, the system at constant pressure and temperature in terms of the change in the number of moles of gas (equation 12.10).

$$\text{Work done by the system} = -P\Delta V = -(\Delta n)RT \qquad (12.10)$$

If the number of moles of gas increases, Δn is positive and work is done by the system on the surroundings (Figure 12.4). An example is the reaction between a metal carbonate and acid (equation 12.11).

$$\left.\begin{array}{l} CuCO_3(s) + 2HCl(aq) \rightarrow CuCl_2(aq) + H_2O(l) + CO_2(g) \\ \text{For one mole of } CuCO_3\text{:} \\ \Delta n = \text{(moles of gaseous products)} - \text{(moles of gaseous reactants)} \\ \quad = 1 - 0 = 1 \end{array}\right\} \quad (12.11)$$

If the number of moles of gas decreases, Δn is negative and work is done on the system by the surroundings, such as in the oxidation of sulfur dioxide (equation 12.12) or the combustion of an alkane (equation 12.13).

$$\left.\begin{array}{l} 2SO_2(g) + O_2(g) \rightarrow 2SO_3(g) \\ \text{For one mole of } SO_2\text{:} \\ \Delta n = \text{(moles of gaseous products)} - \text{(moles of gaseous reactants)} \\ \quad = 2 - 3 = -1 \end{array}\right\} \quad (12.12)$$

$$\left.\begin{array}{l} C_5H_{12}(l) + 8O_2(g) \rightarrow 5CO_2(g) + 6H_2O(l) \\ \text{For one mole of pentane:} \\ \Delta n = \text{(moles of gaseous products)} - \text{(moles of gaseous reactants)} \\ \quad = 5 - 8 = -3 \end{array}\right\} \quad (12.13)$$

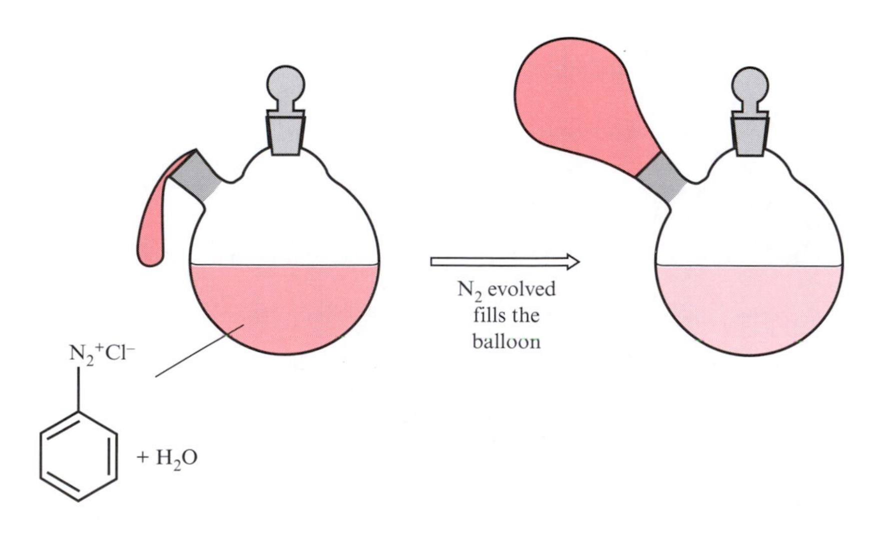

12.4 During the reaction of $[C_6H_5N_2]^+Cl^-$ with water, N_2 is produced which fills the balloon. In doing so, the system does work on the surroundings.

Worked example 12.1 ***Change in internal energy at standard temperature and pressure***

By what amount will the change in internal energy differ from the change in enthalpy for the reaction of 0.05 moles of magnesium carbonate with an excess of dilute hydrochloric acid under conditions of 1 bar pressure and 273 K in the apparatus shown in Figure 12.1? Assume that the syringe is frictionless.

[1 mole of gas at 1 bar pressure and 273 K occupies a volume of 22.7 dm^3; molar gas constant, R = 8.314 J mol^{-1} K^{-1}].

The equation needed is:

$$\Delta U = \Delta H - P\Delta V$$

The reaction is at constant pressure (1 bar) and the difference between ΔU and ΔH is the term $P\Delta V$.
First write a balanced equation for the reaction:

$$MgCO_3(s) + 2HCl(aq) \rightarrow MgCl_2(aq) + H_2O(l) + CO_2(g)$$

0.05 moles of $MgCO_3$ produce 0.05 moles gaseous CO_2 and work is done by the system as the gas produced expands – it 'pushes' against the surrounding atmosphere.
There are two ways of dealing with the problem.

Method 1: The volume occupied by 0.05 moles of gas at 1 bar and 273 K

$$= 0.05 \times 22.7 \text{ dm}^3 = 1.135 \text{ dm}^3$$

Increase in volume, ΔV = Final volume – initial volume = 1.135 dm^3
Note that the change in volume is assumed to be entirely due to the gas released.

Now we must think about the units. By referring to Table 1.2, we see that compatible units of volume, energy and pressure are m^3, J and Pa (see also problem 12.1):

$$\text{Work done by the system} = -P\Delta V$$

With consistent units:

$$P = 10^5 \text{ Pa}$$
$$\Delta V = 1.135 \times 10^{-3} \text{ m}^3$$

$$\text{Work done by the system} = -(10^5 \times 1.135 \times 10^{-3}) = -113.5 \text{ J}$$

Per mole of $MgCO_3$:

$$\text{Difference between } \Delta U \text{ and } \Delta H = \frac{113.5}{0.05}$$
$$= 2270 \text{ J mol}^{-1}$$
$$= 2.27 \text{ kJ mol}^{-1}$$

Method 2: The second (and shorter) method of calculation uses the equation for an ideal gas:

$$PV = nRT$$

At constant pressure (1 bar) and temperature (273 K):

$$P\Delta V = (\Delta n)RT$$

where Δn is the change in number of moles of gas = 0.05 moles

$$P\Delta V = 0.05 \times 8.314 \times 273 = 113.5 \text{ J}$$

and as above:

$$\text{Difference between } \Delta U \text{ and } \Delta H = \frac{113.5}{0.05} \times 10^{-3} = 2.27 \text{ kJ mol}^{-1}$$

Worked example 12.2 ***Change in internal energy at 1 bar pressure and 298 K***

The same reaction as in worked example 12.1 is carried out at 1 bar pressure and 298 K. Calculate the work done by the system on the surroundings.

As in worked example 12.1, there are two methods of calculation. Although both are given here, method 2 is shorter.

Method 1: The only difference from worked example 12.1 is that we have to correct the volume of gas for the change in temperature using the relationship:

$$\frac{P_1V_1}{T_1} = \frac{P_2V_2}{T_2}$$

The pressure is constant, $P_1 = P_2$, and so:

$$\frac{V_1}{T_1} = \frac{V_2}{T_2}$$

$$V_2 = \frac{V_1T_2}{T_1} = \frac{(1.135 \times 10^{-3}) \times 298}{273} = 1.239 \times 10^{-3} \text{ m}^3$$

Work done by the system = $P\Delta V$. With consistent units:

$$\text{Work done} = 10^5 \times 1.239 \times 10^{-3} = 123.9 \text{ J}$$

Per mole of $MgCO_3$:

$$\text{Difference between } \Delta U \text{ and } \Delta H = \frac{123.9}{0.05} \text{ J mol}^{-1} = 2478 \text{ J mol}^{-1} = 2.48 \text{ kJ mol}^{-1}$$

Method 2: Assuming an ideal gas at 10^5 Pa pressure, we can use the equation:

$$PV = nRT$$

and, at constant T and P:

$$P\Delta V = (\Delta n)RT$$

where Δn is the change in number of moles of gas = 0.05 moles.

$$P\Delta V = 0.05 \times 8.314 \times 298 = 123.9 \text{ J}$$

and as above:

$$\text{Difference between } \Delta U \text{ and } \Delta H = \frac{123.9}{0.05} \times 10^{-3} = 2.48 \text{ kJ mol}^{-1}$$

12.3 HEAT CAPACITIES

Definitions

The *molar* heat capacity, C, is a *state function* and is the heat energy required to raise the temperature of one mole of a substance by one kelvin. Equation 12.14 expresses C in a differential form and shows that the heat capacity is defined as the rate of change in q with respect to temperature.

$$C = \frac{\mathrm{d}q}{\mathrm{d}T} \qquad \text{in J mol}^{-1}\text{ K}^{-1} \tag{12.14}$$

However, we must go a step further and define the two distinct *conditions* under which the heat capacity is defined:

- C_P refers to a process at constant pressure
- C_V refers to a process at constant volume.

From equations 12.7 and 12.9, at constant pressure, q is the change in enthalpy, ΔH and so we can write equation 12.15. The notation of the differential is examined in Box 12.1; the subscript P in equation 12.15 refers to the fact that the process is occurring at constant pressure.

$$C_P = \left(\frac{\partial H}{\partial T}\right)_P \tag{12.15}$$

Box 12.1 Partial differentials

We consider partial differentiation with respect to the specific example of internal energy.

The internal energy of a system, U, is a function of temperature, T, and volume, V; that is, U depends upon both these variables. This can be expressed in the form:

$$U = f(T, V)$$

If we differentiate U, we must differentiate with respect to each variable, one at once, with the second variable held constant – this is called *partial differentiation*. The *total differential*, $\mathrm{d}U$, is equal to the sum of the *partial differentials*:

$$\mathrm{d}U = \left(\frac{\partial U}{\partial T}\right)_V \mathrm{d}T + \left(\frac{\partial U}{\partial V}\right)_T \mathrm{d}V$$

A condition of constant pressure refers to experiments such as those carried out at atmospheric pressure where gases are allowed to expand and the pressure remains at atmospheric pressure. In an experiment that is carried out in a *closed container* (for example, in a bomb calorimeter) the pressure may change, but the *volume is constant* and no work is done in expanding the gas. Since the change in volume is zero, equations 12.7 and 12.9 can be simplified to 12.16.

At constant volume: $$\Delta U = q = \Delta H \qquad (12.16)$$

The heat capacity at constant volume is, therefore, defined in terms of the rate of change of internal energy (equation 12.17).

$$C_V = \left(\frac{\partial U}{\partial T}\right)_V \qquad (12.17)$$

The *molar* heat capacity, *C*, is the heat energy required to raise the temperature of one mole of a substance by one kelvin, and the units are J mol^{-1} K^{-1}.

At constant pressure: $C_P = \left(\frac{\partial H}{\partial T}\right)_P$

At constant volume: $C_V = \left(\frac{\partial U}{\partial T}\right)_V$

The relationship between C_P and C_V

For *one mole* of an ideal gas, the enthalpy and internal energy are related according to equation 12.18, which can be rearranged to give equation 12.19.

$$U = H - PV = H - RT \qquad (12.18)$$

$$H = U + RT \qquad (12.19)$$

We can now write the expression for C_P in terms of U (equation 12.20).

$$C_P = \left(\frac{\partial H}{\partial T}\right)_P = \left(\frac{\partial\{U + RT\}}{\partial T}\right)_P \quad \text{for one mole of ideal gas.} \qquad (12.20)$$

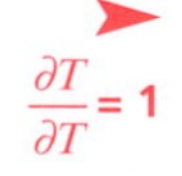

The differential on the right-hand side can be separated into two terms, and, since the differential of T with respect to T is 1, and R is a constant, we can derive a simple relationship between C_P and C_V (equation 12.21).

For one mole of an ideal gas:

$C_P = C_V + R$

$$C_P = \left(\frac{\partial U}{\partial T}\right)_P + \left(R\frac{\partial T}{\partial T}\right)_P = C_V + R \quad \text{for one mole of ideal gas.} \qquad (12.21)$$

Notice that we have taken $\left(\frac{\partial U}{\partial T}\right)_P = \left(\frac{\partial U}{\partial T}\right)_V$. This is true for an *ideal* gas and we discuss this further in the accompanying workbook.

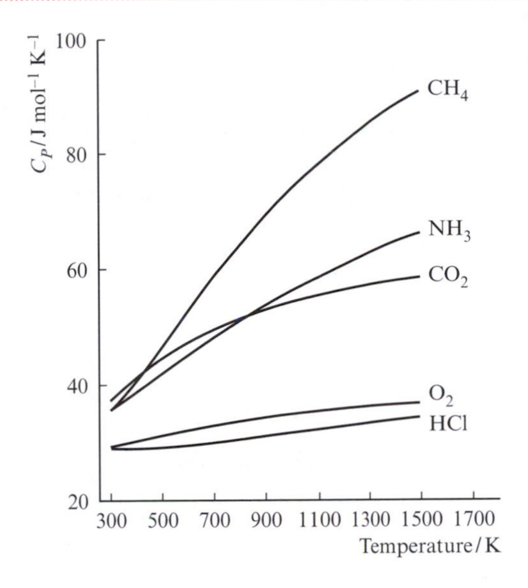

12.5 The variation of C_P with temperature is not very great for diatomic molecules, but is significant for polyatomics.

The variation of C_P with temperature

Most of the reactions with which we are concerned in this book are carried out under conditions of constant pressure and so we must look at C_P in more detail.

Over *small* ranges of temperature, C_P is *approximately* constant although Figure 12.5 illustrates that the variation of C_P with temperature over the range 300–1500 K may be considerable. The variation is not very great for diatomic molecules, but is significant for polyatomic molecules. For larger molecules such as C_2H_6 and C_3H_8, C_P varies from 52.7 to 145.9, and from 73.9 to 205.9 J mol^{-1} K^{-1} respectively between 300 and 1500 K.

12.4 THE VARIATION OF ΔH WITH TEMPERATURE: KIRCHHOFF'S EQUATION

So far we have treated ΔH as being temperature-independent. However, equation 12.15 shows that enthalpy and heat capacity are related and we have just seen that heat capacities *do* depend on temperature. It follows that ΔH is also *temperature-dependent*.

Equation 12.22 restates the definition of C_P. Both H and C_P are state functions and applying equation 12.4 allows us to write equation 12.23 where ΔH is the enthalpy change accompanying a reaction and ΔC_P is the difference in heat capacities (at constant pressure) of the products and the reactants.

$$C_P = \left(\frac{\partial H}{\partial T}\right)_P \tag{12.22}$$

$$\Delta C_P = \left(\frac{\partial \Delta H}{\partial T}\right)_P \qquad \textit{Kirchhoff's equation} \qquad (12.23)$$

This expression is called *Kirchhoff's equation* and in its integrated form (equation 12.24) it can be used to determine how ΔH for a reaction varies with temperature; the integration steps are shown in Box 12.2.

$$\Delta H_{(T_2)} - \Delta H_{(T_1)} = \Delta C_P \times (T_2 - T_1) \qquad (12.24)$$

where $\Delta H_{(T_2)}$ is the enthalpy change at temperature T_2, and $\Delta H_{(T_1)}$ is the enthalpy change at temperature T_1.

Box 12.2 Integration of Kirchhoff's equation

Kirchhoff's equation is in the form of a differential:

$$\Delta C_P = \left(\frac{\partial \Delta H}{\partial T}\right)_P$$

Integrated forms are derived as follows.

Separating the variables in Kirchhoff's equation and integrating between the limits of the temperatures T_1 and T_2 gives:

$$\int_{T_1}^{T_2} \mathrm{d}(\Delta H) = \int_{T_1}^{T_2} \Delta C_P \mathrm{d}T$$

Let ΔH at temperature T_1 be $\Delta H_{(T_1)}$, and at T_2 be $\Delta H_{(T_2)}$. Thus:

$$\Delta H_{(T_2)} - \Delta H_{(T_1)} = \int_{T_1}^{T_2} \Delta C_P \, \mathrm{d}T$$

There are two ways of treating the right-hand side of the equation.

Case 1: Temperature-*independent* ΔC_P. Assume that ΔC_P is a constant; this is often valid over small temperature ranges (see text).

$$\Delta H_{(T_2)} - \Delta H_{(T_1)} = \int_{T_1}^{T_2} \Delta C_P \, \mathrm{d}T = \Delta C_P \int_{T_1}^{T_2} \mathrm{d}T$$

$$\Delta H_{(T_2)} - \Delta H_{(T_1)} = \Delta C_P \times (T_2 - T_1) \qquad \text{(equation 12.24 in text)}$$

Case 2: Temperature-*dependent* ΔC_P. The variation of C_P with T can be empirically expressed as a power series of the type:

$$C_P = a + bT + cT^2 + \ldots$$

where a, b and c are constants. ΔC_P is similarly expressed, and so, if:

$$\Delta C_P = a' + b'T + c'T^2 + \ldots$$

$$\Delta H_{(T_2)} - \Delta H_{(T_1)} = \int_{T_1}^{T_2} \Delta C_P \, \mathrm{d}T = \int_{T_1}^{T_2} \left(a' + b'T + c'T^2\right) \mathrm{d}T$$

$$\Delta H_{(T_2)} - \Delta H_{(T_1)} = \left[a'T + \frac{b'}{2}T^2 + \frac{c'}{3}T^3\right]_{T_1}^{T_2}$$

(You are most likely to encounter Case 1.)

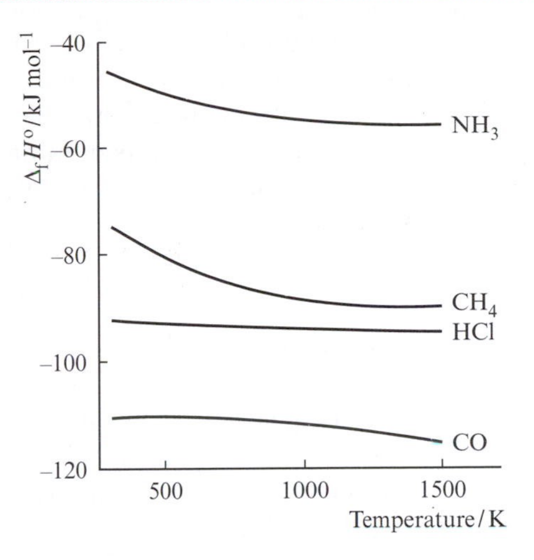

12.6 The variation of $\Delta_f H^o$ with temperature is not very great for diatomic molecules, but is significant for polyatomic; this trend is related to that observed in Figure 12.5 for heat capacities.

Equation 12.24 is only valid if we assume that ΔC_P is independent of temperature – you can test this assumption by determining values of ΔC_P over the ranges 500–600 K and 1000–1100K for each gas shown in Figure 12.5. You should be able to demonstrate that equation 12.24 generally only works over *small* temperature ranges; by studying Figure 12.5, you should deduce how 'small' is defined in this context.

Kirchhoff's equation shows how ΔH for a reaction varies with temperature:

$$\Delta C_P = \left(\frac{\partial \Delta H}{\partial T}\right)_P$$

The temperature dependence of the standard enthalpies of formation of selected compounds is illustrated in Figure 12.6. The temperature range is large (300–1500 K) and although $\Delta_f H^o$ (HCl, g) varies little, $\Delta_f H^o$ (CH_4, g) changes from –74.7 to –90.2 kJ mol^{-1}. As we expect from Kirchhoff's equation, the *trends* in the variation of $\Delta_f H^o$ with temperature (Figure 12.6) are similar to those of C_P with temperature (Figure 12.5).

12.5 EQUILIBRIUM CONSTANTS AND CHANGES IN FREE ENERGY

Equilibrium constants

K_c and K_p: see Section 1.22

Apart from ΔH, the other thermodynamic quantities with which you will be most familiar are equilibrium constants K_p (for a gas phase system) and K_c (for a solution equilibrium). As we saw in Chapter 11, equilibrium constants can be defined for specific types of system including K_a (acid dissociation constant), K_b (base dissociation constant) and K_w (self-ionization constant of water).

The magnitude of K tells us the position of the equilibrium and provides a measure of the *extent of the reaction* in either a forward or backward direction. Consider the general reaction 12.25. The larger the value of K_c (> 1), the more the products dominate over the reactants, and the smaller K_c is (< 1), the more the reactants predominate at the equilibrium state.

$$A + B \rightleftharpoons C + D \qquad K_c = \frac{[C][D]}{[A][B]} \tag{12.25}$$

What do we mean by 'big' and 'small' values of K? The acid dissociation constants in Table 11.3 provide some indication of equilibria which lie towards the left-hand side, in favour of the reactants – the range of K_a values for those selected acids was from 1.5×10^{-2} mol dm^{-3} for the first dissociation step of H_2SO_3, to 2.2×10^{-13} mol dm^{-3} for $[HPO_4]^{2-}$. Very large values of K are exemplified by 3.5×10^{41} bar^{-3} for the formation of water (equation 12.26) or 1.8×10^{48} for the formation of hydrogen fluoride (equation 12.27), both at 298 K. Both these reactions may be explosive and the large values of K reflect the fact that the reactions proceed spontaneously in a forward direction.

$$2H_2(g) + O_2(g) \rightleftharpoons 2H_2O(l) \tag{12.26}$$

$$H_2(g) + F_2(g) \rightleftharpoons 2HF(g) \tag{12.27}$$

The value of K_p for equation 12.28 is 7.5×10^2 bar^{-2} (at 298 K), and the equilibrium lies towards the right-hand side *but* there are significant amounts of N_2 and H_2 present in the reaction mixture at equilibrium.

$$N_2(g) + 3H_2(g) \rightleftharpoons 2NH_3(g) \tag{12.28}$$

Equilibrium constants and ΔH: a lack of consistency?

The standard enthalpies of formation for some selected compounds are given in Table 12.1 and span both exothermic and endothermic reactions. In the third column of the table are values of the equilibrium constants for the corresponding reactions. In most cases, an exothermic reaction is associated with a large value of K, and our previous use of ΔH as an indicator of reaction favourability seems to be fairly well justified. However, the data in the table reveal some inconsistencies – the *trend* in the values of $\Delta_f H^o$ is not fully consistent with the trend in the values of K. One particular inconsistency is with respect to the formation of methylamine where the equilibrium constant, K_p, tells us the extent to which equilibrium 12.29 proceeds to the right-hand side – not very far according to a value of 1.8×10^{-6}, even though the reaction is exothermic.

$$C(gr) + \tfrac{1}{2}N_2(g) + \tfrac{5}{2}H_2(g) \rightleftharpoons CH_3NH_2(g) \tag{12.29}$$

Remember that the *position of equilibrium says nothing about the rate of reaction*. If you mix N_2 and H_2 (equation 12.28) or graphite, N_2 and H_2 (equation 12.29) in a reaction vessel in the laboratory, you should not expect to obtain NH_3 or CH_3NH_2.

Table 12.1 Standard enthalpy and free energy changes, and values of K for the formation of selected compounds at 298 K from their constituent elements in their standard states.[a]

Compound	$\Delta_f H^o$ / $kJ\ mol^{-1}$	K	$\Delta_f G^o$ / $kJ\ mol^{-1}$
$CO_2(g)$	–393.5	1.2×10^{69}	–394.4
$SO_2(g)$	–296.8	4.0×10^{52}	–300.1
$HF(g)$	–273.3	1.8×10^{48}	–275.4
$CO(g)$	–110.5	1.1×10^{24}	–137.2
$HCHO(g)$	–108.7	9.8×10^{17}	–102.7
$HCl(g)$	–92.3	5.0×10^{16}	–95.3
$NH_3(g)$	–45.9	7.5×10^{2}	–16.4
$HBr(g)$	–36.3	2.2×10^{9}	–54.4
$CH_3NH_2(g)$	–22.5	1.8×10^{-6}	+32.7
$BrCl(g)$	+14.6	1.5	–1.0
$HI(g)$	+26.5	0.5	+1.7
$NO_2(g)$	+34.2	6.9×10^{-10}	+52.3
$NO(g)$	+91.3	4.5×10^{-16}	+87.6

[a] Values of K are given as dimensionless but relate to pressures given in bar.

Striking inconsistencies between changes in enthalpy and the extent of reaction are also observed in solubility data. Table 12.2 lists the solubilities in water of selected salts and also the standard enthalpy changes that accompany dissolution. The process to which $\Delta_{sol}H(298\ K)$ refers is given in equation 12.30 for a salt XY; more generally, $\Delta_{sol}H$ refers to dissolution in any solvent (equation 12.31). The *molar enthalpy of solution* refers to the complete dissolution of one mole of XY.

$$XY(s) \xrightarrow{\text{water}} X^+(aq) + Y^-(aq) \tag{12.30}$$

$$XY(s) \xrightarrow{\text{solvent}} X^+(solv) + Y^-(solv) \tag{12.31}$$

It seems quite remarkable that salts such as ammonium nitrate which are very soluble in water dissolve in endothermic processes.

Table 12.2 Solubilities in water of selected salts (at the stated temperature) and molar enthalpies of solution ($\Delta_{sol}H$) at 298 K.

Compound	$\Delta_{sol}H$ / $kJ\ mol^{-1}$	*Solubility (at stated temperature) / mol dm*$^{-3}$	*Solubility (at stated temperature) / g dm*$^{-3}$
KOH	–57.6	19.1 (288 K)	1070 (288 K)
LiCl	–37.0	15.0 (273 K)	637 (273 K)
NaI	–7.5	12.3 (298 K)	1845 (298 K)
NaCl	+3.9	6.1 (273 K)	357 (273 K)
NH_4Cl	+14.8	5.6 (273 K)	295 (273 K)
$NaI{\cdot}2H_2O$	+16.1	17.1 (273 K)	3180 (273 K)
$AgNO_3$	+22.6	7.2 (273 K)	1224 (273 K)
NH_4NO_3	+25.7	14.8 (273 K)	1184 (273 K)
$KMnO_4$	+43.6	0.4 (293 K)	63 (293 K)

Changes in standard free energy, ΔG^o, and the standard free energy of formation, $\Delta_f G^o$

We must now introduce a new state function called the *Gibbs free energy*, G. In Section 12.8 we define G fully, but in this and the next section we study some experimental data and illustrate how the change in standard free energy, ΔG^o, for a reaction is related to the equilibrium constant.

Since G is a state function we can determine changes in free energy for reactions by using equation 12.32.

$$\Delta G = \Sigma G_{products} - \Sigma G_{reactants} \qquad (12.32)$$

Standard states: see Section 1.9

The *standard free energy change* of a reaction, $\Delta_r G^o(T)$ (equation 12.33) refers to a system in which the reactants and products are in their standard states at a specified temperature, T, and at a pressure of 1 bar. As we shall see later, ΔG^o *is temperature-dependent* and it is vital to specify T when quoting values of ΔG^o.

$$\Delta_r G^o(298\ K) = \Sigma[\Delta_f G^o(298\ K)_{products}] - \Sigma[\Delta_f G^o(298\ K)_{reactants}] \qquad (12.33)$$

The *standard free energy of formation* of a compound, $\Delta_f G^o(T)$, is the free energy change that accompanies the formation of one mole of a compound in its standard state from its constituent elements in their standard states. By definition, the standard free energy of formation *and* the standard enthalpy of formation of an element in its standard state are zero, at all temperatures. However, $\Delta_f G^o$ for a *compound* is temperature dependent (see Section 12.6).

The standard free energy of formation of a compound, $\Delta_f G^o(T)$, is the change in free energy that accompanies the formation of one mole of a compound in its standard state from its constituent elements in their standard states.

By definition, $\Delta_f G^o$ and $\Delta_f H^o$ of an *element* in its standard state are zero, at all temperatures.

Worked example 12.3 ***The change in standard free energy***

Determine $\Delta_r G^o(298\ K)$ for the combustion of one mole of ethanol given that $\Delta_f G^o(298\ K)$ $C_2H_5OH(l)$, $H_2O(l)$ and $CO_2(g)$ are –175, –237 and –394 kJ mol^{-1} respectively.

Firstly, write a balanced equation for the combustion of ethanol:

$$C_2H_5OH(l) + 3O_2(g) \xrightarrow{\Delta_r G^o} 2CO_2(g) + 3H_2O(l)$$

The standard free energy of formation of $O_2(g) = 0$ (by definition).

$$\begin{aligned}\Delta_r G^o &= [2 \times \Delta_f G^o(CO_2, g)] + [3 \times \Delta_f G^o(H_2O, l)] - \Delta_f G^o(C_2H_5OH, l)\\ &= [2 \times (-394)] + [3 \times (-237)] - (-175)\\ &= -1324 \text{ kJ per mole of ethanol (at 298 K)}\end{aligned}$$

In order to assess the advantages of using ΔG^o instead of ΔH^o, we return to Table 12.1, where standard free energies of formation are also listed. In contrast to the inconsistencies in the trends of $\Delta_f H^o$ and K that we noted earlier, we now see that the trend in values of $\Delta_f G^o$ *does* follow that in the values of K. A value of $K > 1$ corresponds to a *negative* change in free energy, and if $K < 1$, ΔG is *positive*. The magnitude of the equilibrium constant for the formation of hydrogen iodide (equation 12.34) indicates that the equilibrium lies towards the left-hand side, although there are significant amounts of products in the equilibrium mixture. Although the corresponding change in free energy is positive, it is small.

$$\tfrac{1}{2}H_2(g) + \tfrac{1}{2}I_2(g) \rightleftharpoons HI(g) \qquad K = \frac{(pHI)}{(pH_2)^{\frac{1}{2}}\,(pI_2)^{\frac{1}{2}}} = 0.5 \qquad (12.34)$$

The change in free energy for a reaction is a *reliable* guide to the extent of a reaction *provided that equilibrium is reached* and in this regard, is a far more useful thermodynamic quantity than ΔH. Remember that ΔG and K say nothing about the *rate* of the reaction.

12.6 THE TEMPERATURE DEPENDENCE OF ΔG^o AND K_p: SOME EXPERIMENTAL DATA

In this section we consider some experimental data to assess how K_p and ΔG^o *for a given reaction* vary with temperature.

Case study 1: the formation of ammonia

Haber process: see Box 11.2

ln = $\log_e$

When we discussed the Haber process in Chapter 11, we drew attention to the fact that the equilibrium constant for reaction 12.35 varied with temperature. Over the temperature range 300–1500 K, K_p varies from 6.7×10^2 to 5.8×10^{-4}, and the logarithmic scale ($\ln K_p$) in Figure 12.7a allows this large range of values to be represented more conveniently than plotting K_p itself.

$$N_2(g) + 3H_2(g) \rightleftharpoons 2NH_3(g) \qquad K_p = \frac{(pNH_3)^2}{(pN_2)(pH_2)^3} \qquad (12.35)$$

Figure 12.7b illustrates that $\Delta_f G^o$ per mole of ammonia also varies significantly as a function of temperature; compare this with the much smaller variation in $\Delta_f H^o(NH_3, g)$ shown in Figure 12.6. As the temperature is increased, $\Delta_f G^o$ becomes increasingly *positive* and the decomposition of NH_3 to N_2 and H_2 is the predominant process. This result is mirrored in the trend in the equilibrium constants where at high temperatures, $\ln K_p$ is large and negative, i.e. K_p is *very* small.

Case study 2: metal oxide reduction

In Chapter 11, we saw several examples of the reduction of a metal oxide by a metal, and the electrochemical series provided a means of assessing which

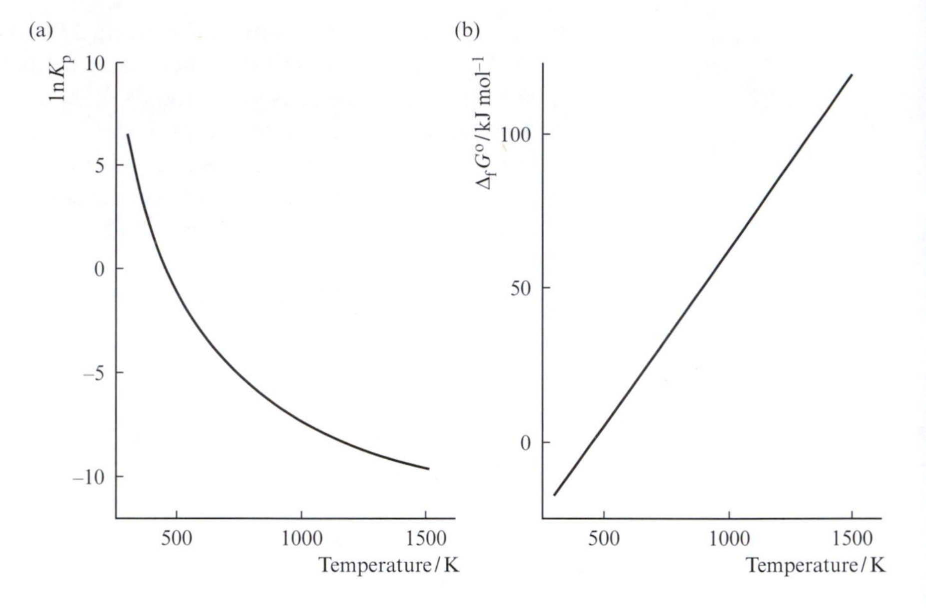

12.7 (a) The variation with temperature of ln K_p for the formation of ammonia (equation 12.28), and (b) the variation of $\Delta_f G^o(NH_3, g)$ with temperature.

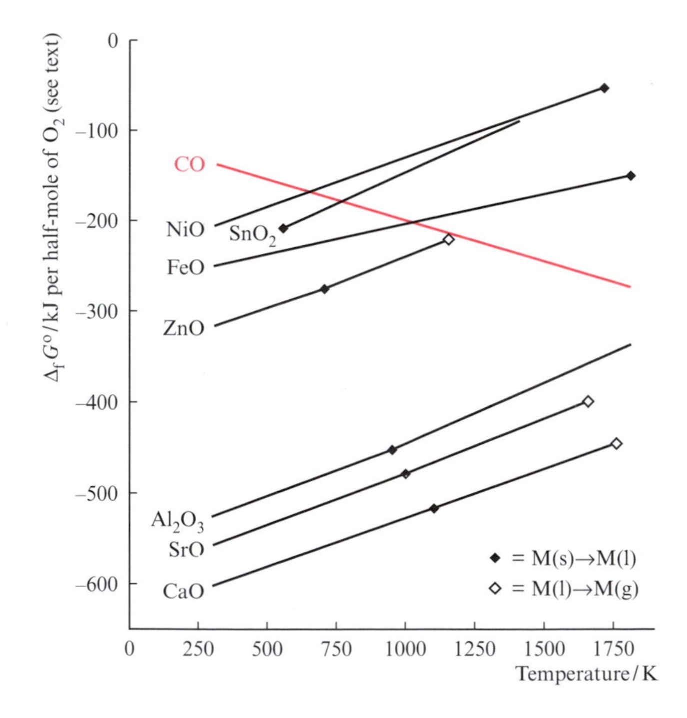

12.8 An Ellingham diagram showing how the standard free energies of formation of several metal oxides and carbon monoxide (the red line) vary as a function of temperature. Values of $\Delta_f G^o$ refer to formation reactions involving a half-mole of O_2:

$M + \frac{1}{2}O_2 \rightarrow MO$,
$\frac{1}{2}M + \frac{1}{2}O_2 \rightarrow \frac{1}{2}MO_2$, or
$\frac{2}{3}M + \frac{1}{2}O_2 \rightarrow \frac{1}{3}M_2O_3$.

The points marked ◆ or ◇ are the melting or boiling points, respectively, of the elemental metal.

metals might reduce which oxides. Free energy data allow us to approach the problem *quantitatively*.

Figure 12.8 shows the temperature variation of $\Delta_f G^o$ for a range of metal oxides. In order that values may be compared to each other, $\Delta_f G^o$ refers to the free energy of formation *per half-mole of* O_2.§

§ We could equally well have plotted the data per mole of O_2.

Box 12.3 The thermite reaction

Aluminium reacts with many metal oxides when heated in the solid state; the aluminium is oxidized and the metal oxide reduced to the respective metal. The general name for this reaction is the *thermite reaction* or *thermite process*, although this name is sometimes used for the specific reaction of iron(III) oxide with aluminium:

$$Fe_2O_3 + 2Al \xrightarrow{\text{heat}} 2Fe + Al_2O_3$$

$$3BaO + 2Al \xrightarrow{\text{heat}} 3Ba + Al_2O_3$$

On a small scale, the two powdered reactants may be mixed and placed in a crucible. A 'fuse' of magnesium ribbon is placed partly into the mixture and the free end is ignited. The reactions between aluminium and most metal oxides are violently exothermic and the temperature is raised such that the metallic product is usually formed in the molten state.

Thus for CaO, $\Delta_f G^o$ refers to reaction 12.36, and for Al_2O_3 it corresponds to reaction 12.37; we show both equations with the metal in the solid state, but this is temperature-dependent (see Figure 12.8).

$$Ca(s) + \tfrac{1}{2}O_2(g) \rightarrow CaO(s) \qquad (12.36)$$

$$\tfrac{2}{3}Al(s) + \tfrac{1}{2}O_2(g) \rightarrow \tfrac{1}{3}Al_2O_3(s) \qquad (12.37)$$

Phase changes: see Section 12.9

Each plot in Figure 12.8 is either linear (e.g. NiO) or has two linear sections (e.g. Al_2O_3). The change in gradient (which may be small) part way along a line occurs at the melting point of the metal. Two observations from Figure 12.8 are:

- each metal oxide is *less* thermodynamically stable (less negative $\Delta_f G^o$) at higher temperatures, and
- the *relative* stabilities of the oxides at any given temperature can be seen directly from the graph.

The second point means that the plots in Figure 12.8 (which constitute an *Ellingham diagram*) can be used to predict which metal(s) will reduce a particular metal oxide, and whether the operating temperature will influence the efficiency of the reduction. The red line in Figure 12.8 shows the way in which $\Delta_f G^o$(CO, g) varies with temperature. This plot contrasts with all the others since CO(g) becomes *more* thermodynamically stable at higher temperatures (more negative $\Delta_f G^o$). As the temperature is raised, carbon becomes more effective at reducing a wide variety of metal oxides.[§] Above 1800 K, a greater range of metal oxides than Figure 12.8 shows can be reduced by carbon, although industrially, this method of obtaining a metal from its oxide is often not commercially viable.

§ The temperature dependence of ΔG and of K_p are quantified in the Gibbs–Helmholz equation and the Van't Hoff isochore. We do not derive these or equation 12.38 in this text, but details may be found in P. W. Atkins, *Physical Chemistry,* 5th edn, Oxford University Press (1994).

Worked example 12.4 ***Using an Ellingham diagram***

(a) At an operating temperature of 800 K, which metal oxides in Figure 12.8 will carbon reduce? (b) Estimate the approximate $\Delta_r G^o$ at 800 K for the reduction of NiO by carbon.

(a) First draw a vertical line on Figure 12.8 at $T = 800$ K. CO is the product when carbon is used as a reducing agent and so as CO is *more* thermodynamically stable than SnO_2 or NiO, carbon should reduce these two oxides:

$$C(s) + NiO(s) \rightarrow CO(g) + Ni(s)$$

$$C(s) + \tfrac{1}{2}SnO_2(s) \rightarrow CO(g) + \tfrac{1}{2}Sn(l)$$

(b) Since free energy is a state function, we can apply a Hess–type cycle to find $\Delta_r G^o$:

$$C(s) + NiO(s) \rightarrow CO(g) + Ni\ (s)$$

$$\Delta_r G^o = [\Delta_f G^o(Ni, s) + \Delta_f G^o(CO, g)] - [\Delta_f G^o(C, s) + \Delta_f G^o(NiO, s)]$$

where all ΔG^o values refer to 800 K. For each of the elements in their standard state

$$\Delta_f G^o(Ni, s) = \Delta_f G^o(C, gr) = 0$$

From the graph, read off values of $\Delta_f G^o$ CO and $\Delta_f G^o$ NiO at 800 K:

$\Delta_f G^o(CO, g) \approx -180$ kJ mol^{-1} (per $\frac{1}{2}$ mole O_2 corresponds to a mole of CO)
$\Delta_f G^o(NiO, s) \approx -150$ kJ mol^{-1} (per $\frac{1}{2}$ mole O_2 corresponds to a mole of NiO)

$$\begin{aligned}\Delta_r G^o &= \Delta_f G^o(CO, g) - \Delta_f G^o(NiO, s)\\ &\approx -180 - (-150)\\ &\approx -30 \text{ kJ per mole of reaction}\end{aligned}$$

The negative value of $\Delta_r G^o$ is in agreement with the prediction that carbon will reduce nickel oxide.

12.7 THE RELATIONSHIP BETWEEN ΔG^o AND K_p: THE REACTION ISOTHERM

In this section we consider how ΔG^o and K_p are related and show how values of ΔG^o and K_p *for different reactions* can be used to find ΔG^o and K_p for other reactions.

The reaction isotherm

Equation 12.38 states *the reaction isotherm* which applies at a pressure of 10^5 kPa (1 bar $\approx$ 1 atm); this is a most important expression with widespread applications in chemical thermodynamics.

$$\Delta G^o = -RT \ln K_p \tag{12.38§}$$

where R (molar gas constant) $= 8.314 \times 10^{-3}$ kJ mol^{-1} K^{-1} and T is the temperature in K.

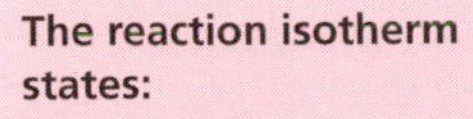

The reaction isotherm states:

$\Delta G^o = -RT \ln K_p$

(at 1 bar)

Figure 12.9 shows the linear relationship between ΔG^o and $\ln K_p$, and by comparison with the general equation for a straight line, we see that the line has a gradient of $-RT$ and passes through the origin (intercept $c = 0$), at which point $K_p = 1$ and $\Delta G^o = 0$.

We are now able to quantify the statements made at the end of Section 12.5. A range of values of ΔG^o and corresponding equilibrium constants at 298 K are shown in Figure 12.10; the pairs of values are related by equation 12.38. If ΔG^o is negative, K_p is greater than 1 and the products predominate over reactants. If ΔG^o is positive, K_p is less than 1 and the reactants are dominant.

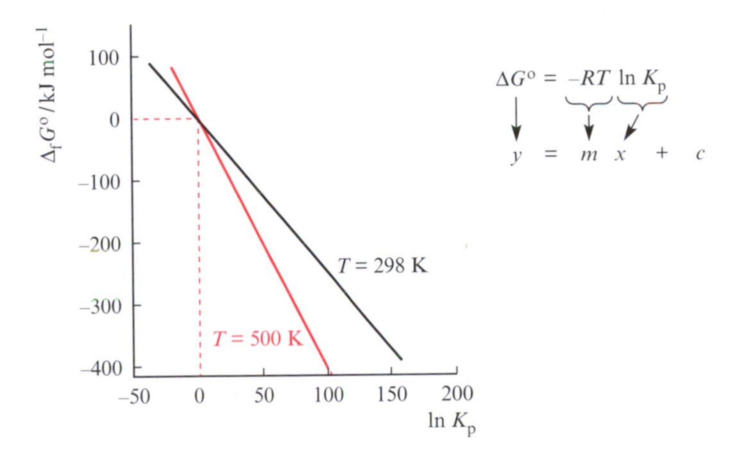

12.9 Plots of ΔG^o against $\ln K_p$ at 298 and 500 K. Both lines pass through the point $\Delta G^o = 0$ when $\ln K_p = 0$ ($K_p = 1$).

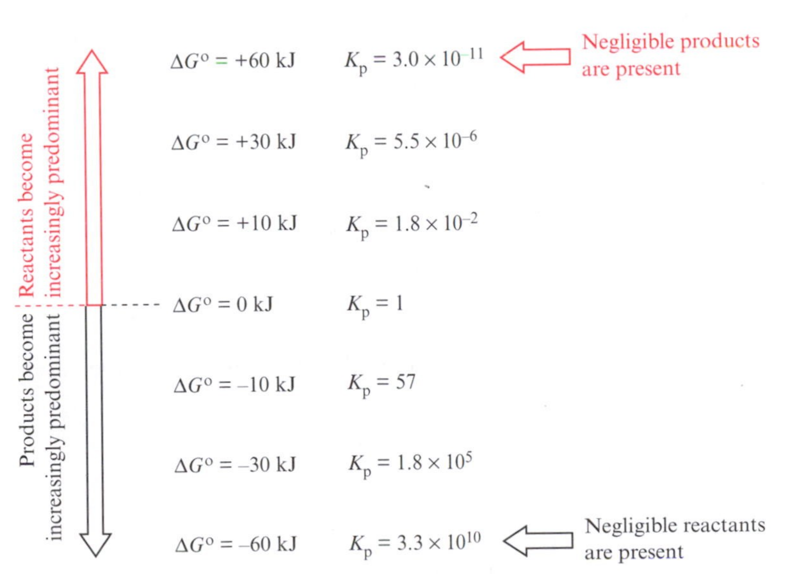

12.10 Values of ΔG^o and corresponding values of K_p at 298 K. An indication is also given of the physical significance of the magnitude of such values.

§ Equation 12.38 is sometimes expressed in terms of $\log K_p$ in which case it becomes: $\Delta G^o = -2.303RT \log K_p$ ($\log = \log_{10}$).

When $\Delta G^{o} = 0$, $K_p = 1$.

If ΔG^{o} is negative, K_p is greater than 1, and the products predominate over the reactants.

If ΔG^{o} is positive, K_p is less than 1, and the reactants predominate over the products.

Worked example 12.5 *Using the reaction isotherm*

Determine K_p at 298 K for the reaction:

$$C_2H_4(g) + H_2(g) \rightleftharpoons C_2H_6(g)$$

if $\Delta_f G^{o}(C_2H_4, g) = +68.4$ kJ mol^{-1} and $\Delta_f G^{o}(C_2H_6, g) = -32.0$ kJ mol^{-1} at 298 K. [$R = 8.314 \times 10^{-3}$ kJ mol^{-1} K^{-1}).

First determine $\Delta_r G^{o}$ for the forward reaction:

$$C_2H_4(g) + H_2(g) \rightleftharpoons C_2H_6(g)$$

$$\Delta_r G^{o} = [\Delta_f G^{o}(C_2H_6, g)] - [\Delta_f G^{o}(C_2H_4, g) + \Delta_f G^{o}(H_2, g)]$$

Since H_2 is in its standard state, $\Delta_f G^{o}(H_2, g) = 0$

$$\begin{aligned} \Delta_r G^{o} &= \Delta_f G^{o}(C_2H_6, g) - \Delta_f G^{o}(C_2H_4, g) \\ &= -32.0 - 68.4 \\ &= -100.4 \text{ kJ per mole of ethene} \end{aligned}$$

The equilibrium constant can be found using the equation:

$$\begin{aligned} \Delta G^{o} &= -RT\ln K_p \\ \ln K_p &= -\frac{\Delta G^{o}}{RT} \\ &= -\frac{-100.4}{8.314 \times 10^{-3} \times 298} \\ &= 40.52 \\ K_p &= 3.96 \times 10^{17} \text{ (at 298 K)} \\ &\approx 4.0 \times 10^{17} \text{ (at 298 K)} \end{aligned}$$

The equilibrium constant is dimensionless because we have determined it from a $\ln K_p$ value.

Does the reaction isotherm only apply to K_p?

It is important to notice that equation 12.38 specifically includes the term K_p and not just K, although it is possible to derive a more general form of the isotherm (equation 12.39).

$$\Delta G^o = -RT \ln K \tag{12.39}$$

The derivation of this equation is treated in detail in more advanced physical chemistry texts, but we need to make some comment about the form of the equilibrium constant K. In principle, K_p, K_c, etc. must be *dimensionless* quantities if we are to take logarithms. However, the form of many expressions for K are not dimensionless since actual concentrations or partial pressures are used, as in equation 12.35. In the case of K_p, we divide each partial pressure (measured in bar) by 1 bar to obtain a dimensionless quantity as shown in equation 12.40.§

$$\left.\begin{aligned} &A + B \rightleftharpoons C + D \\ &K_p = \frac{\left(p_C/P\right)\left(p_D/P\right)}{\left(p_A/P\right)\left(p_B/P\right)} = \frac{(p_C)(p_D)}{(p_A)(p_B)} \qquad \text{(no units) where } P = 1 \text{ bar} \end{aligned}\right\} \tag{12.40}$$

In many systems, the effect of each component on the position of equilibrium is not exactly as expected from its concentration. We get around this problem by defining a new quantity called the *activity*, a, which represents the 'effective' concentration of a component. The difference between the 'real' concentration, x, and the 'effective' concentration, a, arises from interactions *between* the molecules in solution. The quantities x and a are related by the activity coefficient γ (equation 12.41).

$$\gamma = \frac{a}{x} \tag{12.41}$$

The quantity K in equation 12.39 is strictly defined in terms of the *activities* of the components (equation 12.42) rather than their real concentrations.

$$\left.\begin{aligned} &A + B \rightleftharpoons C + D \\ &\text{Strictly, } K_c = \frac{a_C a_D}{a_A a_B} \qquad \text{rather than } K_c = \frac{[C][D]}{[A][B]} \end{aligned}\right\} \tag{12.42}$$

In some cases, a_X *may* be similar to [X], but this is not necessarily so. Very often in the laboratory, we work with dilute solutions for which it is valid to approximate a_X to [X]. A solution containing one mole of X per dm^3 of solution is described as being *one molar*; a solution of X at *unit activity* is described as being *one molal*.

In answer to the question posed in the title of this section – the reaction isotherm applies to K_p and also to solution K_c values *provided the solutions are dilute*. Equation 12.39 is a general statement of the reaction isotherm.

§ If we were working with the SI unit of pascals, our reference pressure would be 10^5 Pa, and we would need to divide by this number.

12.8 GIBBS FREE ENERGY, ENTHALPY AND ENTROPY

Up to this point we have looked at experimental data and said that there is a state function G, changes in which provide a better indicator of reaction spontaneity than do changes in enthalpy. Why is this?

The relationship between G and H

Entropy is defined fully in Section 12.9

Gibbs free energy is defined in equation 12.43 where G, H, T and S are all state functions, and S is the *entropy*. Equation 12.43 is valid under conditions of constant temperature and pressure.

$$G = H - TS \tag{12.43}$$

Gibbs free energy is defined by the equation:

$G = H - TS$

The *change in free energy*, ΔG, is the maximum work that is available in a reversible process at constant temperature and pressure in *addition* to the work done in expansion (equation 12.44); the derivation of equation 12.44 from 12.43 is outlined in Box 12.4.

$$\Delta G = \Delta H - T\Delta S \tag{12.44}$$

For standard state conditions, equation 12.45 is appropriate, where T is often (but not necessarily) 298 K.

$$\Delta G^{o} = \Delta H^{o} - T\Delta S^{o} \tag{12.45}$$

Equations 12.44 and 12.45 show that the change in free energy for a process at a given temperature has contributions both from the change in enthalpy *and* the change in entropy. From our earlier discussion and the values in Table 12.1, we can see that the relationship between the ΔG and ΔH terms for a particular reaction depend critically upon the term $T\Delta S$. Consider the formation of carbon dioxide: $\Delta_f H^{o}$(298 K) CO_2(g) = –393.5 kJ mol^{-1} and $\Delta_f G^{o}$(298 K) CO_2(g) = –394.4 kJ mol^{-1}; the closeness of these values means that the $T\Delta S^{o}$ term is small (equation 12.46). In this instance, both $\Delta_f H^{o}$ and $\Delta_f G^{o}$ are good indicators of reaction spontaneity.

Box 12.4 Deriving $\Delta G = \Delta H - T\Delta S$ from $G = H - TS$

$$G = H - TS$$

$$dG = dH - (TdS + SdT) = dH - TdS - SdT$$

At constant temperature, $dT = 0$, and so:

$$dG = dH - TdS$$

or

$$\Delta G = \Delta H - T\Delta S$$

For the formation of CO_2(g) from carbon and O_2 in their standard states:

$$\left.\begin{aligned} \Delta G^{o} &= \Delta H^{o} - T\Delta S^{o} \\ T\Delta S^{o} &= \Delta H^{o} - \Delta G^{o} \\ &= (-393.5) - (-394.4) = +0.9 \text{ kJ per mole of } CO_2 \end{aligned}\right\} \qquad (12.46)$$

In contrast, for bromine monochloride, $\Delta_f H^o$(298 K) BrCl(g) = +14.6 kJ mol^{-1} and $\Delta_f G^o$(298 K) BrCl(g) = –1.0 kJ mol^{-1}, and the $T\Delta S^o$ term is significant (equation 12.47). Here, the value of $\Delta_f H^o$ would suggest that the reaction between Br_2 and Cl_2 may not proceed, but the magnitude of the $T\Delta S^o$ term compensates for $\Delta_f H^o$ and gives an equilibrium in which there are slightly more products than reactants.

For the formation of BrCl(g) from Br_2 and Cl_2 in their standard states:

$$\left.\begin{aligned} T\Delta S^{o} &= \Delta H^{o} - \Delta G^{o} \\ &= (+14.6) - (-1.0) = +15.6 \text{ kJ per mole of BrCl} \end{aligned}\right\} \qquad (12.47)$$

These results reveal the importance of the $T\Delta S$ term – it is the factor that tips the balance with respect to changes in enthalpy so that some endothermic reactions occur spontaneously. In the next section we look at entropy in some detail.

At constant temperature and pressure, the change in free energy of a reaction is given by:

$$\Delta G = \Delta H - T\Delta S$$

or for standard state conditions (pressure = 1 bar and T = 298 K):

$$\Delta G^o = \Delta H^o - T\Delta S^o$$

12.9 ENTROPY: THE SECOND AND THIRD LAWS OF THERMODYNAMICS

The statistical approach

One way to understand the meaning of the entropy of a system is to use a statistical approach and find the number of possible arrangements of the components (atoms or molecules) of that system. Equation 12.48 gives Boltzmann's definition of entropy, S, where k is the Boltzmann constant, and W is the number of ways of arranging the atoms or molecules in the system whilst maintaining a constant overall energy.§

$$S = k \times \ln W \qquad (12.48)$$

The more arrangements there are, the greater the entropy. Entropy is often expressed in terms of the degree of disorder or randomness of a system: the more randomly arranged are the atoms or molecules in a system, the larger is its entropy.

§ The number of ways of arranging the atoms or molecules in the system does not simply refer to the spatial arrangement of identifiable particles, but also to the distribution of energies within the system.

Standard molar entropy

The standard molar entropy, S^o, of an element or compound is the entropy per mole of a pure substance at 10^5 Pa (1 bar). The units are J mol^{-1} K^{-1}

The Third Law of Thermodynamics states that a pure, perfect crystal at 0 K has zero entropy.

The *standard molar entropy*, S^o, of an element or compound is the entropy per mole of a pure substance at 10^5 Pa (1 bar). Values of S^o(298 K) for some selected elements and compounds are given in Table 12.3. It is tempting to say that there is a general tendency for values of S^o for solid substances to be smaller than those for liquids, whilst gases possess the largest standard entropies. In the broadest sense this statement is true, but we must exercise caution in comparing entropy values across a wide range of substances – a comparison of S^o(298 K) for crystalline KI and liquid mercury bears this out.

In principle, as we approach absolute zero, all systems go to their lowest energy states and the components are perfectly ordered in the solid state. This corresponds to the perfect crystalline state, and the *Third Law of Thermodynamics* states that a pure, perfect crystal at 0 K has zero entropy.

The dependence of entropy on temperature

If we think of entropy as a measure of the degree of randomness of a system, then an increase in temperature is likely to cause an increase in S. Figure 12.11 shows the variation in S^o with temperature for CO_2, CH_4, NH_3, HCl and O_2, all of which are gases over the given temperature range.

The change in entropy as a function of temperature is given by equation 12.49, and the integrated form (equation 12.50) can be used to find the variation in entropy between temperatures T_1 and T_2. The integration steps are set out in Box 12.5.

$$\left(\frac{\partial S}{\partial T}\right)_P = \frac{C_P}{T} \qquad (12.49)$$

$$S_{(T_2)} - S_{(T_1)} = C_P \times \ln\left(\frac{T_2}{T_1}\right) \qquad (12.50)$$

Table 12.3 The standard molar entropies S^o(298 K) for selected elements and compounds.

Substance (standard state at 298 K)	*S^o(298 K) / J mol^{-1} K^{-1}*
Ag(cryst)	42.6
C(graphite)	5.7
C(diamond)	2.4
Fe(cryst)	27.3
I_2(cryst)	116.1
KI(cryst)	106.3
Hg(l)	75.9
PCl_3(l)	217.1
He(g)	126.2
H_2(g)	130.7
HCl(g)	186.9
HBr(g)	198.7
PF_5(g)	300.8

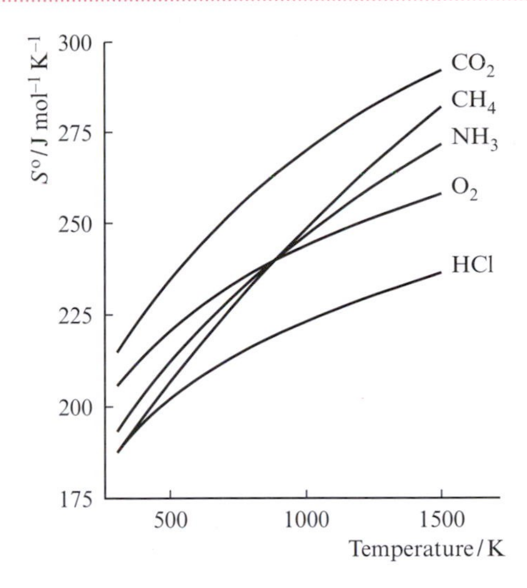

12.11 The variation of S^o with temperature for CO_2, CH_4, NH_3, O_2 and HCl, all of which are gases over the temperature range plotted.

Box 12.5 Integration of $\left(\frac{\partial S}{\partial T}\right)_P = \frac{C_P}{T}$

The integrated form of the equation:

$$\left(\frac{\partial S}{\partial T}\right)_P = \frac{C_P}{T}$$

is derived as follows.

Separating the variables and integrating between the limits of the temperatures T_1 and T_2 gives:

$$\int_{T_1}^{T_2} \mathrm{d}S = \int_{T_1}^{T_2} \frac{C_P}{T} \mathrm{d}T$$

Let S at temperature T_1 be $S_{(T_1)}$, and at T_2 be $S_{(T_2)}$. Integration of the left-hand side of the equation gives:

$$S_{(T_2)} - S_{(T_1)} = \int_{T_1}^{T_2} \frac{C_P}{T.} \mathrm{d}T$$

We now make the assumption that C_P does *not* depend upon temperature; this is *only* valid over small temperature ranges (see Figure 12.6).

With C_P as a constant:

$$S_{(T_2)} - S_{(T_1)} = C_P \int_{T_1}^{T_2} \frac{1}{T} \mathrm{d}T$$

Integrating the right-hand side of the equation between the limits of T_2 and T_1 gives:

$$S_{(T_2)} - S_{(T_1)} = C_P \times \ln\left(\frac{T_2}{T_1}\right)$$

This equation is *only valid* if C_P is approximately constant over the temperature range in question but, as Figure 12.5 showed, this is not always the case. If we wanted to determine the molar entropy of a substance at 298 K, we would need to evaluate the change in S^o between the limits of 0 to 298 K,

12.12 A plot of $\frac{C_P}{T}$ against temperature, T, can be used to find the change in entropy of a substance over a range of temperatures where the heat capacity, C_P, is *not* constant. To find S^o(298 K), determine the area under the graph between 0 and 298 K (the shaded area).

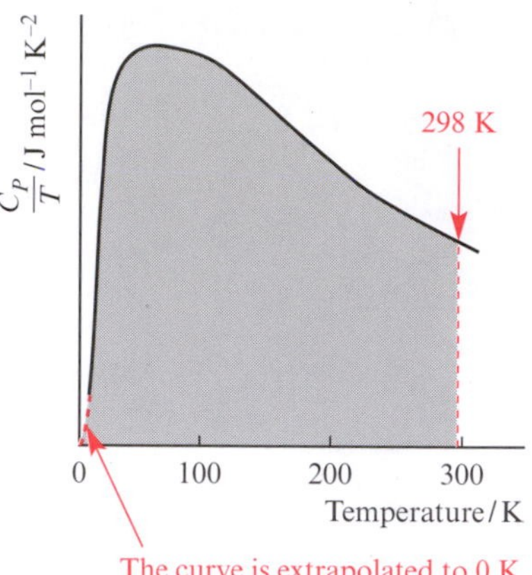

Box 12.6 An integral as an area under a curve

Consider a curve with the general equation $y = f(x)$:

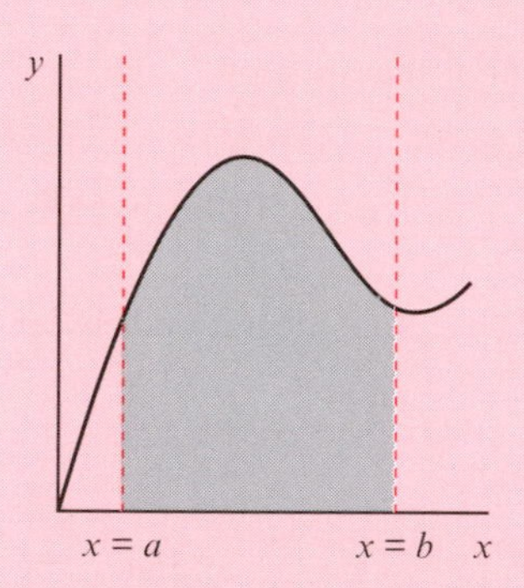

The area under the graph between the limits $x = a$ and $x = b$ is denoted by the definite integral $\int_a^b y\,dx$. We are effectively dividing the shaded area in the graph into an infinite number of vertical strips (drawn parallel to the y axis), finding the area of each strip, and summing the areas together:

$\int_a^b y\,dx$ = area under the curve $y = f(x)$ between the limits $x = a$ and $x = b$

Now apply this to the equation:

$$\left(\frac{\partial S}{\partial T}\right)_P = \frac{C_P}{T}$$

The change in entropy between the limits of T_2 and T_1

$$= \int_{T_1}^{T_2} dS = \int_{T_1}^{T_2} \frac{C_P}{T} dT \quad \text{(from Box 12.4)}$$

or

the change in entropy between the limits of T_2 and T_1

= Area under a plot of $\frac{C_P}{T}$ against T between the limits of T_2 and T_1

i.e. in this case, the function of x is $\frac{C_P}{T}$.

and over this temperature range, C_P is *not* constant. From equation 12.49, it follows that $S^o(298\ K)$ can be determined from the area under a plot of C_P/T against T (Figure 12.12). Measurements of C_P to very low temperatures can be made by calorimetric methods, but the last section of the curve is constructed by an extrapolation to 0 K.

The change in entropy as a function of temperature is given by:

$$\left(\frac{\partial S}{\partial T}\right)_P = \frac{C_P}{T}$$

Assuming that C_P is approximately constant:

$$\Delta S = S_{(T_2)} - S_{(T_1)} = C_P \times \ln\left(\frac{T_2}{T_1}\right)$$

Worked example 12.6 ***Finding the change in entropy of tantalum over a small temperature range***

Determine the change in standard entropy per mole of tantalum between 700 and 800 K at constant pressure, if the average value of C_P over this range is 6.53 J mol^{-1} K^{-1}.

Since we can assume that C_P is constant, the equation needed is:

$$\Delta S^o = S^o_{(T_2)} - S^o_{(T_1)} = C_P \times \ln\left(\frac{T_2}{T_1}\right)$$

$$\Delta S^o = 6.53 \times \ln\left(\frac{800}{700}\right) = 0.87\ \text{J mol}^{-1}\ \text{K}^{-1}.$$

(For comparison, the *actual* values of $S^o(700\ K)$ and $S^o(800\ K)$ are 15.25 and 16.13 J mol^{-1} K^{-1}, giving a value of $\Delta S^o = 0.88$ J mol^{-1} K^{-1}.)

The change of entropy associated with a phase change

If we extend this picture to cover different phases of a given substance, then as Figure 12.13 illustrates for the *d*-block metal tantalum, there are discontinuities in a plot of S^o against temperature which correspond to the solid→ liquid and liquid→gas phase transitions. In terms of the statistical model, this makes good sense; tantalum possesses an ordered crystal structure which will be disrupted upon melting. Similarly, the system becomes more random upon passing from the liquid to the gas phase. Let us consider the phase transitions more carefully. *At the melting point* of tantalum, solid and liquid metal are in equilibrium (equation 12.51).

$$Ta(s) \rightleftharpoons Ta(l) \qquad (12.51)$$

If an infinitesimally small amount of heat energy is given to the system, a small amount of tantalum melts – equilibrium 12.51 moves to the right-hand side. If an infinitesimally small amount of heat energy is lost from the system, a small amount of tantalum solidifies – equilibrium 12.51 moves to

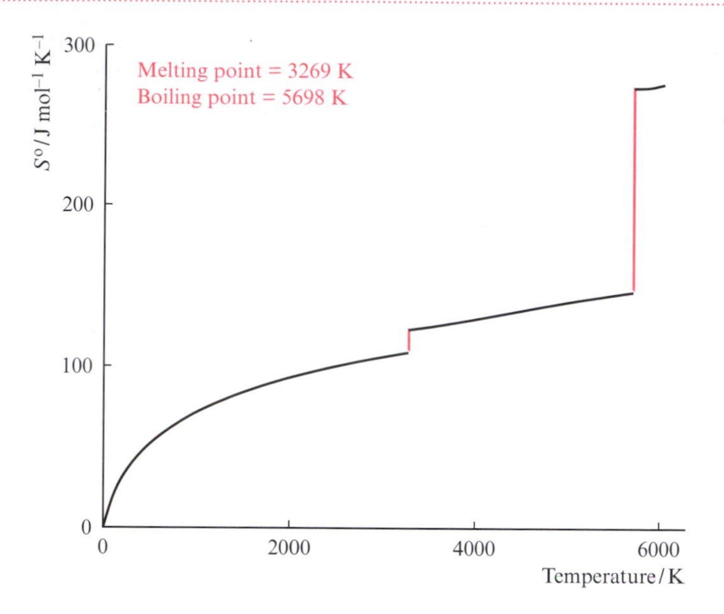

12.13 The variation of S^o with temperature for the metal tantalum; the discontinuities in the graph correspond to phase transitions. (Data from *JANAF Tables*, 1972 update, Dow Chemical Company, Midland, Michigan.)

the left-hand side. However, these changes are *so small* that the system remains virtually at equilibrium, and such a situation is one of *thermodynamic reversibility*. When the solid and liquid tantalum are in equilibrium, their free energies must be equal and ΔG is zero. It is very important to note: it is ΔG that is zero, *not* ΔG^o.

We now apply equation 12.44 to the special case of a phase equilibrium and can write equation 12.52.

$$\text{For an equilibrium: } \Delta G = \Delta H - T\Delta S = 0$$

$$\Delta H = T\Delta S \quad \text{or} \quad \Delta S = \frac{\Delta H}{T} \qquad (12.52)$$

This equation takes two particular forms when we consider the solid ⇌ liquid and liquid ⇌ vapour phase changes. The enthalpy change associated with a solid melting is the enthalpy of fusion ($\Delta_{fus}H$), and when a liquid vaporizes the respective enthalpy change is the enthalpy of vaporization ($\Delta_{vap}H$). The entropy change associated with the solid ⇌ liquid phase change is given by equation 12.53, and equation 12.54 states it for the liquid ⇌ vapour equilibrium.

$$\text{For solid} \rightleftharpoons \text{liquid:} \quad \Delta_{fus}S = \frac{\Delta_{fus}H}{\text{mp}} \quad (\text{mp in K}) \qquad (12.53)$$

$$\text{For liquid} \rightleftharpoons \text{vapour:} \quad \Delta_{vap}S = \frac{\Delta_{vap}H}{\text{bp}} \quad (\text{bp in K}) \qquad (12.54)$$

Worked example 12.7 ***Entropy change of fusion for hexane***

What is the change in entropy of one mole of hexane when it vaporizes at its boiling point under atmospheric pressure? [$\Delta_{vap}H$ = 28.9 kJ mol^{-1}, bp = 342 K].

$$C_6H_{14}(l) \rightleftharpoons C_6H_{14}(g)$$

Firstly, note carefully the sign of $\Delta_{vap}H$. These data (and those of $\Delta_{fus}H$) are given as positive quantities (heat energy taken in by the system), appropriate for liquid → gas (or solid → liquid). If the phase change is in the other direction, the enthalpy change is negative.

$$\begin{aligned}\Delta_{vap}S &= \frac{\Delta_{vap}H}{bp} \\ &= \frac{28.9}{342} \\ &= 0.0845 \text{ kJ mol}^{-1}\text{ K}^{-1} \\ &= 84.5 \text{ J mol}^{-1}\text{ K}^{-1}\end{aligned}$$

Trouton's rule and hydrogen bonding: see Section 11.8

Trouton's rule is a restatement of equation 12.54 but with the message that a great many liquids possess similar values for $\Delta_{vap}S$ ($\approx$ 88 J mol^{-1} K^{-1}). Some anomalies can be explained in terms of the presence of extensive hydrogen bonding in the liquid state – values of $\Delta_{vap}S$ for H_2O, H_2S and H_2Se are 109, 88 and 85 J mol^{-1} K^{-1} respectively. The hydrogen bonding in liquid water produces a relatively ordered system and lowers its entropy; the *change* in the entropy in going from $H_2O(l)$ to $H_2O(g)$ is larger than would have been the case had hydrogen bonding not been an important effect.

Change in entropy for a reaction, ΔS

Since entropy is a state function, we can use equation 12.55 to find the change in entropy, ΔS, for a reaction, or equation 12.56 for the change in standard entropy, ΔS^o.

$$\Delta S = \Sigma S_{products} - \Sigma S_{reactants} \tag{12.55}$$

$$\Delta S^o = \Sigma S^o_{products} - \Sigma S^o_{reactants} \tag{12.56}$$

Values of the standard entropies for elements and compounds are available in standard tables of data, and can be used to calculate $\Sigma S^o_{products}$ and $\Sigma S^o_{reactants}$ for a reaction.

Worked example 12.8 ***Determination of an entropy change for a reaction***

If the standard entropies (at 298 K) of gaseous F_2, Cl_2 and ClF are 203, 223 and 218 J mol^{-1} K^{-1} respectively, determine ΔS^o(298 K) for the formation of ClF from F_2 and Cl_2.

Firstly, write a balanced equation:

$$Cl_2(g) + F_2(g) \rightarrow 2ClF(g)$$

$$\begin{aligned}\Delta_r S^o &= \Sigma S^o_{products} - \Sigma S^o_{reactants} \\ &= (2 \times 218) - (223 + 203) \\ &= 10 \text{ J K}^{-1} \text{ per 2 moles of ClF} \quad \textit{or} \quad 5 \text{ J mol}^{-1}\text{ K}^{-1}\end{aligned}$$

Worked example 12.9 ***Determination of an entropy change for a reaction***

Iridium(VI) fluoride is formed by the direct fluorination of iridium metal. Calculate the standard change in entropy (298 K) during this reaction if S^o(298 K) for Ir(s), F_2(g) and IrF_6(s) are 35.5, 203 and 248 J mol^{-1} K^{-1} respectively.

Firstly, write a balanced equation for the formation of IrF_6.

$$Ir(s) + 3F_2(g) \rightarrow IrF_6(s)$$

$$\begin{aligned}\Delta_r S^o &= \sum S^o_{products} - \sum S^o_{reactants} \\ &= (248) - [35.5 + (3 \times 203)] \\ &= -396.5 \text{ J mol}^{-1} \text{ K}^{-1}\end{aligned}$$

Changes in entropy for reactions can be either positive (an increase in entropy) or negative (a decrease in entropy) as worked examples 12.8 and 12.9 illustrate – but a word of caution: *each value we have calculated is only the entropy change of the system*. Before the story is complete, we must also consider the entropy change of the surroundings during the reaction. If a reaction is exothermic, the heat given out causes an increase in the entropy of the surroundings. Conversely, in an endothermic reaction, heat is taken in by the system causing a decrease in the entropy of the surroundings (Figure 12.14).

The balance between the change in entropy of the system and the change in entropy of the surroundings is sometimes critical in determining whether or not a reaction will be thermodynamically viable (i.e. spontaneous), and in this context we return to equation 12.44: $\Delta G = \Delta H - T\Delta S$. At a constant temperature, the difference between the terms ΔH and $T\Delta S$ determines the sign of ΔG.

Consider worked example 12.8: ΔS^o(298 K) for the formation of ClF(g) is +5 J mol^{-1} K^{-1} meaning that the system becomes *more disordered* as the reaction takes place. Since $\Delta_f H^o$(298 K) ClF(g) is –50.3 kJ mol^{-1}, $\Delta_f G^o$ is also negative (equation 12.57) and the formation of ClF(g) is a thermodynamically favoured process.

$$\left.\begin{aligned} T\Delta S^o &= 298 \times 5 \times 10^{-3} \text{ kJ mol}^{-1} \approx 1.5 \text{ kJ mol}^{-1} \\ \Delta G^o &= \Delta H^o - T\Delta S^o = -50.3 - 1.5 = -51.8 \text{ kJ per mole of ClF(g)} \end{aligned}\right\} \quad (12.57)$$

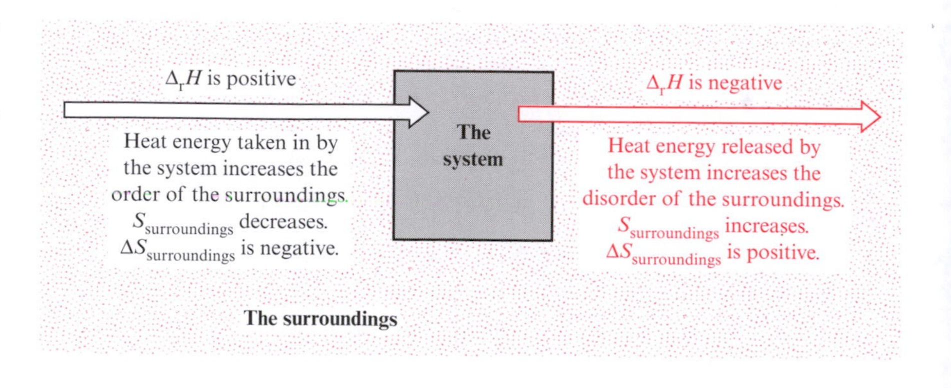

12.14 The effect of enthalpy changes in a system on the entropy of the surroundings, emphasizing sign convention.

Now consider worked example 12.9: $\Delta S^{o}(298\ K)$ for the formation of $IrF_6(s)$ is $-396.5\ J\ mol^{-1}\ K^{-1}$ meaning that the system becomes *more ordered* as the reaction takes place. On the face of it, this does not seem to be a favourable state of affairs, but $\Delta_f H^{o}(298\ K)$ $IrF_6(s)$ is -579.7, and this is more than sufficient to compensate for the negative entropy term (equation 12.58).

$$\left.\begin{aligned} T\Delta S^{o} &= 298 \times (-396.5) \times 10^{-3}\ \text{kJ mol}^{-1} \approx -118.2\ \text{kJ mol}^{-1} \\ \Delta G^{o} &= \Delta H^{o} - T\Delta S^{o} = -579.7 - (-118.2) = -461.5\ \text{kJ per mole of IrF}_6\text{(s)} \end{aligned}\right\} \quad (12.58)$$

In other words, the high bond energy of the Ir–F bonds compensates for the unfavourable entropy change associated with a gaseous reactant giving a solid product.

The Second Law of Thermodynamics can be stated in the form: the *total* entropy always increases during a spontaneous change.

This discussion illustrates the *Second Law of Thermodynamics*. There are various ways in which this law may be stated, but one useful form is that *the total entropy always increases during a spontaneous change*. Note that it is the *total* entropy with which the law is concerned – the sum of the changes in entropies of the surroundings and the system.

12.10 ELECTROCHEMISTRY

In Chapter 11, we saw examples of electrochemical cells in which an electrical current was passed through an electrolyte causing reduction (at the cathode) and oxidation (at the anode) *of the electrolyte*. These were examples of *electrolytic cells* – the decomposition of the electrolyte (for example, molten NaCl) would not have taken place without the passage of an electrical current. In this section we consider another type of electrochemical cell – *the galvanic cell* – in which a spontaneous *redox reaction* occurs and generates an electrical current. In such a cell, *electrical work* is done by the system.

Redox = reduction–oxidation

The Daniell cell and some definitions

Figure 12.15 illustrates a simple electrochemical cell called the *Daniell cell* after its inventor, John Daniell. It consists of two *half-cells*, one containing a strip of copper metal immersed in an aqueous solution of copper(II) sulfate, and the other containing a zinc foil immersed in an aqueous solution of zinc(II) sulfate. The two solutions are connected via a *salt-bridge* – this may consist of gelatine containing potassium chloride or potassium nitrate solution which permits ions to pass between the two half-cells but does not allow the copper(II) and zinc(II) solutions to mix too quickly. The metal strips are connected by an electrically conducting wire, and a voltmeter may be placed in the circuit.

When a metal is placed in a solution containing its ions (e.g. copper in copper(II) sulfate), a state of equilibrium is attained (equation 12.59) and an electrical potential exists which is consequent upon the position of the equilibrium.

$$Cu^{2+}(aq) + 2e^{-} \rightleftharpoons Cu(s) \quad (12.59)$$

The combination of the copper and copper(II) ions is called a *redox couple*; there is the possibility of either the reduction of copper(II) to copper, or the oxidation of copper to copper(II) ions.

The tendency for copper(II) to be reduced is greater than that of zinc(II). This is readily demonstrated by placing a piece of copper metal in a solution containing zinc(II) ions, and in a separate beaker, placing a strip of zinc metal in a solution containing copper(II) ions. In the first beaker, no reaction occurs, but in the second, zinc is oxidized and copper metal is deposited as copper(II) ions are reduced. Reaction 12.60 is the spontaneous process.

$$Zn(s) + Cu^{2+}(aq) \rightarrow Zn^{2+}(aq) + Cu(s) \qquad (12.60)$$

When the two half-cells in the Daniell cell are connected as in Figure 12.15, reaction 12.60 occurs even though the reactants are not in immediate contact. The difference in the electrical potentials of the two half-cells results in a *potential difference* (which can be measured on the voltmeter), and electrons flow *along the wire* from one cell to the other as the reaction proceeds. The direction of electron flow is from the zinc to the copper foil (right-to-left along the wire in Figure 12.15). This corresponds to the production of electrons in the zinc half-cell (equation 12.61) and their consumption in the copper half-cell (equation 12.62).

$$Zn(s) \rightarrow Zn^{2+}(aq) + 2e^- \qquad (12.61)$$

$$Cu^{2+}(aq) + 2e^- \rightarrow Cu(s) \qquad (12.62)$$

The electrical circuit is completed by the passage of ions across the salt-bridge.

The overall electrochemical cell is denoted by the *cell diagram*:

$$Zn(s) \mid Zn^{2+}(aq) \,\vdots\, Cu^{2+}(aq) \mid Cu(s)$$

where the solid vertical lines represent boundaries between phases, and the dotted line stands for the salt-bridge. *The spontaneous process reads from left to right across the cell diagram* – zinc is converted to zinc(II) ions as copper(II) ions transform to copper metal. *The reduction process is always drawn on the right-hand side of the cell diagram.*

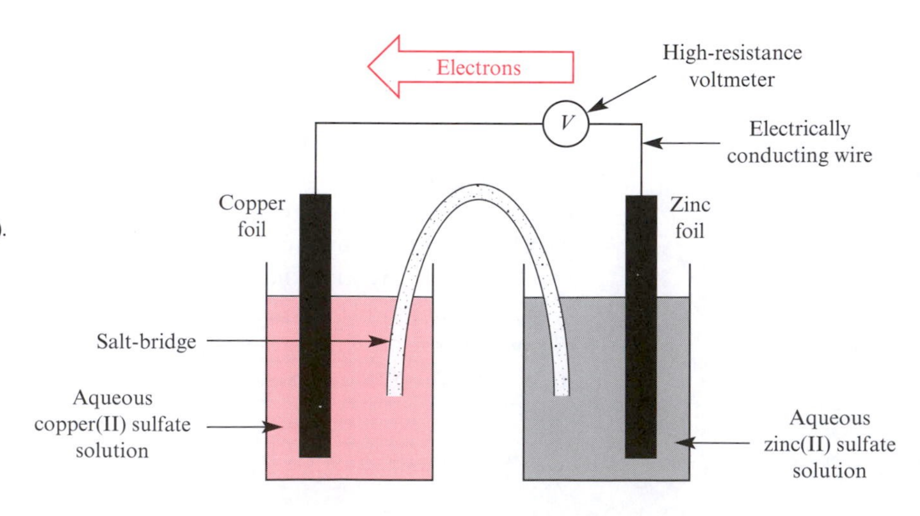

12.15 A representation of the Daniell cell. In the left-hand cell, copper(II) ions are reduced to copper metal, and in the right-hand cell, zinc metal is oxidized to zinc(II) ions. The cell diagram is: $Zn(s) \mid Zn^{2+}(aq) \,\vdots\, Cu^{2+}(aq) \mid Cu(s)$.

The potential difference between the two half-cells is denoted by E_{cell} and its value (in volts) is directly related to the change in free energy for the overall reaction. Equation 12.63 relates E_{cell} to ΔG, and equation 12.64 refers to the *standard cell potential*. The temperature should always be stated.

$$\Delta G = -zFE_{cell} \tag{12.63}$$

$$\Delta G^{o} = -zFE^{o}_{cell} \tag{12.64}$$

where: z = number of moles of electrons transferred *per mole of reaction,*
F = Faraday constant = 96 500 C mol^{-1}.

The *standard cell potential* refers to the conditions:

- the concentration of any solution is 1 mol dm^{-3};§
- the pressure of any gaseous component is 1 bar (10^5 Pa);†
- the temperature is 298 K;
- a solid component is in its standard state.

For a spontaneous reaction, E^{o}_{cell} is always positive, and this corresponds to a negative value of ΔG^{o}.

The change in standard free energy of an electrochemical cell is:

$$\Delta G^{o} = -zFE^{o}_{cell}$$

where: z = number of moles of electrons transferred *per mole of reaction,*

F = Faraday constant = 96 500 C mol^{-1}.

For a spontaneous reaction, E^{o}_{cell} is always positive.

Worked example 12.10 ***The free energy change in a Daniell cell***

If the solutions in the Daniell cell are both of concentration 1 mol dm^{-3}, E^{o}_{cell}(298 K) is 1.1 V. What is ΔG^{o} for the cell reaction? [F = 96 500 C mol^{-1}]

The equation needed is:

$$\Delta G^{o} = -zFE^{o}_{cell}$$

The cell reaction is:

$$Zn(s) + Cu^{2+}(aq) \rightarrow Zn^{2+}(aq) + Cu(s)$$

Per mole of reaction, two moles of electrons are transferred: $z = 2$.

§ Strictly, the standard conditions refer to unit activity, but remember that for a *dilute* solution, the concentration is approximately equal to the activity. Solutions of 1 mol dm^{-3} would not be used in the laboratory; this is discussed later in this section.

† The standard pressure may in some tables of data be taken as 1 atm (101 300 Pa), see Table 12.4.

$$\begin{aligned}\Delta G^{\circ}(298\ \text{K}) &= -zFE^{\circ}_{\text{cell}} \\ &= -(2 \times 96\ 500 \times 1.1)\ \text{J} \\ &= -212\ 300\ \text{J} \\ &= -212.3\ \text{kJ (per mole of reaction)}\end{aligned}$$

This large negative value of ΔG° is consistent with a spontaneous reaction.

The hydrogen electrode and standard reduction potentials

In the Daniell cell, we could measure E°_{cell} directly from the circuit but it is not possible to determine the potentials of the half-cells, although we know qualitatively that the tendency for zinc to lose electrons exceeds that of copper. It would be useful if we knew this information *quantitatively*, and for this purpose, we may use the *standard hydrogen electrode* as a reference electrode for which the cell potential, E°_{cell}, has been assigned 0 V.

Equation 12.65 gives the half-cell reaction that occurs in a standard hydrogen electrode – either H_2 is oxidized or H^+ ions are reduced.

$$2H^+(\text{aq, 1 mol dm}^{-3}) + 2e^- \rightleftharpoons H_2(\text{g, 1 bar}) \tag{12.65}$$

Neither component can be physically connected into the electrical circuit, and when the electrode is constructed, a platinum wire is placed in contact with both the H_2 gas and the aqueous H^+ ions. Platinum metal is inert and is not involved in any redox process. By constructing the electrochemical cell shown in Figure 12.16, the *standard reduction potential* or *standard electrode potential*, E°, for the zinc/zinc(II) couple could, in theory, be determined. In practice, a more dilute solution of aqueous metal ions would be used (see below). The cell diagram for this cell reaction is denoted as follows:

$$\text{Zn(s)} \mid \text{Zn}^{2+}(\text{aq, 1 mol dm}^{-3}) \,\vdots\, 2\text{H}^+(\text{aq, 1 mol dm}^{-3}) \mid [\text{H}_2(\text{g, 1 bar})]\ \text{Pt}$$

The spontaneous reaction, reading from left to right, is the oxidation of zinc (equation 12.66) and the reduction of hydrogen ions (equation 12.67).

$$\text{Zn(s)} \rightarrow \text{Zn}^{2+}(\text{aq}) + 2e^- \tag{12.66}$$

$$2\text{H}^+(\text{aq}) + 2e^- \rightarrow \text{H}_2(\text{g}) \tag{12.67}$$

The cell potential (measured on the voltmeter shown in Figure 12.16) gives a direct measure of the potential of the Zn^{2+}/Zn couple with respect to the standard hydrogen electrode, i.e. E° for the Zn^{2+}/Zn couple. We must, however, pay attention to the sign convention. *After writing the cell diagram such that the spontaneous reaction is read from left to right*, the standard reduction potential of the Zn^{2+}/Zn couple can be determined using equation 12.68. In this instance, the right-hand electrode is the standard hydrogen electrode and the left-hand electrode is the Zn^{2+}/Zn electrode.

$$\begin{aligned}E^{\circ}_{\text{cell}} &= [E^{\circ}_{\text{right-hand electrode}}] - [E^{\circ}_{\text{left-hand electrode}}] \\ &= [E^{\circ}_{\text{reduction process}}] - [E^{\circ}_{\text{oxidation process}}]\end{aligned} \tag{12.68}$$

For the cell in Figure 12.16, E^{o}_{cell} is 0.76 V. Equation 12.69 then shows how the standard reduction potential of the Zn^{2+}/Zn couple is determined; remember that $E^{o} = 0$ for the standard hydrogen electrode (at 298 K).

$$0.76 = 0 - E^{o}_{Zn^{2+}/Zn} \qquad E^{o}_{Zn^{2+}/Zn} = -0.76 \text{ V} \qquad (12.69)$$

Selected standard reduction potentials for half-cells are given in Table 12.4; a wider range is listed in Appendix 12. *The half-cell is always quoted in terms of a reduction reaction.*

Worked example 12.11 ***Standard reduction potential***

The value of E^{o}_{cell} for the cell shown below is 0.80 V.

$$\text{Pt } [H_2(g, 1 \text{ bar})] \mid 2H^+(aq, 1 \text{ mol dm}^{-3}) \;\vdots\; Ag^+(aq, 1 \text{ mol dm}^{-3}) \mid Ag(s)$$

What is the spontaneous cell reaction, and what is the standard reduction potential for the $Ag^+(aq)/Ag(s)$ electrode?

The spontaneous cell reaction is obtained by reading the cell diagram from left to right. The two half-cell reactions are:

$$H_2(g) \rightarrow 2H^+(aq) + 2e^-$$

$$Ag^+(aq) + e^- \rightarrow Ag(s)$$

The full reaction is:

$$H_2(g) + 2Ag^+(aq) \rightarrow 2H^+(aq) + 2Ag(s)$$

For the cell in the diagram above, the standard cell potential is given by the equation:

$$E^{o}_{cell} = [E^{o}_{\text{reduction process}}] - [E^{o}_{\text{oxidation process}}]$$

$$= [E^{o}_{\text{right-hand electrode}}] - [E^{o}_{\text{left-hand electrode}}]$$

$$0.80 = E^{o}_{Ag^+/Ag} - 0$$

$$E^{o}_{Ag^+/Ag} = +0.80 \text{ V}$$

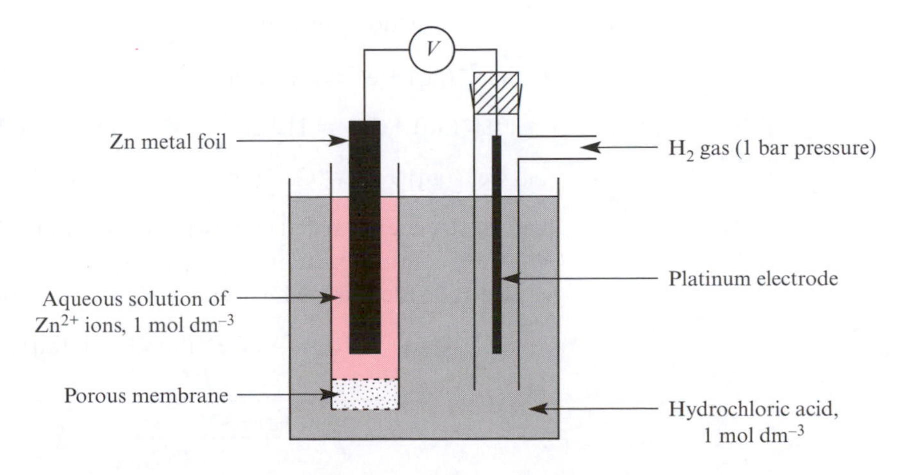

12.16 The standard hydrogen electrode consists of a platinum wire immersed in a 1 mol dm^{-3} solution of aqueous hydrochloric acid through which H_2 (at 1 bar pressure) is bubbled. When the standard hydrogen electrode is combined with a standard Zn^{2+}/Zn couple, the reading on the voltmeter, E^{o}_{cell}, is equal to the standard reduction potential, E^{o}, of the Zn^{2+}/Zn couple.

Other half-cells

By no means all half-cells are of the type M^{n+}/M, i.e. they do not all contain a metal strip immersed in an aqueous solution of the metal ions. For example, half-equations 12.70–12.72 involve aqueous species.

$$Ce^{4+}(aq) + e^- \rightleftharpoons Ce^{3+}(aq) \qquad (12.70)$$

$$Br_2(aq) + 2e^- \rightleftharpoons 2Br^-(aq) \qquad (12.71)$$

$$[MnO_4]^-(aq) + 8H^+(aq) + 5e^- \rightleftharpoons Mn^{2+}(aq) + 4H_2O(l) \qquad (12.72)$$

In these cases, the half-cells are assembled in a similar way to the standard hydrogen electrode, with the reduced and oxidized components in contact with each other and with a platinum wire to allow the flow of electrons. The cell notation for systems involving more complex half-cells follows the same principle as that described above. For the combination of a standard hydrogen electrode and the Fe^{3+}/Fe^{2+} couple, the cell diagram is:

$$Pt\,[H_2(g,\ 1\ bar)]\,|\,2H^+(aq,\ 1\ mol\ dm^{-3})\ \vdots\ Fe^{3+}(aq),\ Fe^{2+}(aq)\,|\,Pt$$

and reaction 12.73 is the spontaneous cell process.

$$H_2(g) + 2Fe^{3+}(aq) \rightarrow 2H^+(aq) + 2Fe^{2+}(aq) \qquad (12.73)$$

Using standard reduction potentials

Standard reduction potentials can be used to determine values of E^o_{cell} for a reaction, and since we can find the related free energy change from E^o_{cell} (equation 12.64), standard reduction potentials are useful means of predicting the thermodynamic viability of a reaction.

The half-cell reactions in Table 12.4 are all written as *reduction reactions* and the name *standard reduction potentials* for the associated values of E^o follows accordingly. A positive E^o means that the reduction is thermodynamically favourable with respect to the reduction of H^+ ions to H_2, while a negative E^o means that the preferential reduction is that of the hydrogen ions. Consider the following three half-equations and their corresponding standard reduction potentials:

$$Cr^{3+}(aq) + e^- \rightleftharpoons Cr^{2+}(aq) \qquad E^o = -0.41\ V$$

$$2H^+(aq) + 2e^- \rightleftharpoons H_2(g) \qquad E^o = 0.00\ V$$

$$Ce^{4+}(aq) + e^- \rightleftharpoons Ce^{3+}(aq) \qquad E^o = +1.72\ V$$

In an electrochemical cell consisting of a standard hydrogen electrode and a Ce^{4+}/Ce^{3+} couple, reaction 12.74 is the spontaneous process, but if the hydrogen electrode is combined with the Cr^{3+}/Cr^{2+} couple, reaction 12.75 occurs.

$$2Ce^{4+}(aq) + H_2(g) \rightarrow 2Ce^{3+}(aq) + 2H^+(aq) \qquad (12.74)$$

reduction

oxidation

Table 12.4 Selected standard reduction potentials (298 K); see Appendix 12 for a more detailed list. The concentration of each aqueous solution is 1 mol dm^{-3}, and the pressure of a gaseous component is 1 bar (10^5 Pa). (Changing the standard pressure to 1 atm (101 300 Pa) makes no difference to the values of E^o at this level of accuracy.)

Reduction half-equation	E^o / V
$Li^+(aq) + e^- \rightleftharpoons Li(s)$	–3.04
$K^+(aq) + e^- \rightleftharpoons K(s)$	–2.93
$Ca^{2+}(aq) + 2e^- \rightleftharpoons Ca(s)$	–2.87
$Na^+(aq) + e^- \rightleftharpoons Na(s)$	–2.71
$Mg^{2+}(aq) + 2e^- \rightleftharpoons Mg(s)$	–2.37
$Al^{3+}(aq) + 3e^- \rightleftharpoons Al(s)$	–1.66
$Zn^{2+}(aq) + 2e^- \rightleftharpoons Zn(s)$	–0.76
$S(s) + 2e^- \rightleftharpoons S^{2-}(aq)$	–0.48
$Fe^{2+}(aq) + 2e^- \rightleftharpoons Fe(s)$	–0.44
$Cr^{3+}(aq) + e^- \rightleftharpoons Cr^{2+}(aq)$	–0.41
$Co^{2+}(aq) + 2e^- \rightleftharpoons Co(s)$	–0.28
$Pb^{2+}(aq) + 2e^- \rightleftharpoons Pb(s)$	–0.13
$Fe^{3+}(aq) + 3e^- \rightleftharpoons Fe(s)$	–0.04
$2H^+(aq) + 2e^- \rightleftharpoons H_2(g, 1\ bar)$	0
$2H^+(aq) + S(s) + 2e^- \rightleftharpoons H_2S(aq)$	+0.14
$Cu^{2+}(aq) + e^- \rightleftharpoons Cu^+(aq)$	+0.15
$Cu^{2+}(aq) + 2e^- \rightleftharpoons Cu(s)$	+0.34
$O_2(g) + 2H_2O(l) + 4e^- \rightleftharpoons 4[OH]^-(aq)$	+0.40
$Cu^+(aq) + e^- \rightleftharpoons Cu(s)$	+0.52
$I_2(aq) + 2e^- \rightleftharpoons 2I^-(aq)$	+0.54
$Fe^{3+}(aq) + e^- \rightleftharpoons Fe^{2+}(aq)$	+0.77
$Ag^+(aq) + e^- \rightleftharpoons Ag(s)$	+0.80
$Br_2(aq) + 2e^- \rightleftharpoons 2Br^-(aq)$	+1.09
$[Cr_2O_7]^{2-}(aq) + 14H^+(aq) + 6e^- \rightleftharpoons 2Cr^{3+}(aq) + 7H_2O(l)$	+1.33
$Cl_2(aq) + 2e^- \rightleftharpoons 2Cl^-(aq)$	+1.36
$[MnO_4]^-(aq) + 8H^+(aq) + 5e^- \rightleftharpoons Mn^{2+}(aq) + 4H_2O(l)$	+1.51
$Ce^{4+}(aq) + e^- \rightleftharpoons Ce^{3+}(aq)$	+1.72
$F_2(aq) + 2e^- \rightleftharpoons 2F^-(aq)$	+2.87

$$2Cr^{2+}(aq) + 2H^+(aq) \rightarrow 2Cr^{3+}(aq) + H_2(g) \quad (12.75)$$

oxidation

reduction

These results show that the cerium(IV) ion is a more powerful oxidizing agent than the hydrogen ion, but that H^+ is a better oxidizing agent than chromium (III) ions. This means that cerium(IV) ions will oxidize chromium(II) ions (equation 12.76).

$$Ce^{4+}(aq) + Cr^{2+}(aq) \rightarrow Ce^{3+}(aq) + Cr^{3+}(aq) \quad (12.76)$$

reduction

oxidation

The thermodynamic favourability of each reaction can be assessed by calculating E^o_{cell} and then ΔG^o(298 K). For reaction 12.74, the cell diagram (showing the spontaneous process from left to right) is:

$$Pt\ [H_2(g, 1\ bar)]\ |\ 2H^+(aq, 1\ mol\ dm^{-3})\ \vdots\ Ce^{4+}(aq), Ce^{3+}(aq)\ |\ Pt$$

and E^o_{cell} (equation 12.77) and ΔG^o (equation 12.78) can be determined.

$$E^o_{cell} = [E^o_{\text{right-hand electrode}}] - [E^o_{\text{left-hand electrode}}] = +1.72 - 0.00 = 1.72\ \text{V} \quad (12.77)$$

reduction process — oxidation process

$$\Delta G^o(298\ \text{K}) = -zFE^o_{cell} = -(2 \times 96\,500 \times 1.72)\ \text{J} \approx -332\ \text{kJ (per mole of reaction)} \quad (12.78)$$

You should be able to show that $\Delta G^o(298\ \text{K})$ for reactions 12.75 and 12.76 are ≈ -79 and -206 kJ (per mole of reaction).

It is important to remember that E^o_{cell} is independent of the amount of material present, whereas ΔG^o is a molar quantity. Whereas E^o_{cell} has identical values for reactions 12.74 *and* 12.79, the value of z (and therefore of ΔG^o) varies.

$$Ce^{4+}(aq) + \tfrac{1}{2}H_2(g) \rightleftharpoons Ce^{3+}(aq) + H^+(aq) \quad (12.79)$$

In Table 12.4, the half-reactions are arranged in order of the values of E^o with the most negative value at the top of the table, and the most positive at the bottom. This is a common format for such tables, and allows you to see immediately the relative oxidizing and reducing abilities of each species. Consider the Ag^+/Ag couple: $E^o = +0.80$ V. Silver(I) ions will oxidize any of the reduced species on the list which lie *above* the Ag^+/Ag couple – for example, copper metal. On the other hand, any oxidized species *below* the Ag^+/Ag couple is able to oxidize Ag to Ag^+. The extent of the reaction depends upon the magnitude of E^o_{cell}. For reaction 12.80, E^o_{cell} is 0.46 V, corresponding to a value of $\Delta G^o(298\ \text{K})$ of -89 kJ (per mole of reaction). This is sufficiently large and negative that very few reactants remain in the system.

$$2Ag^+(aq) + Cu(s) \rightleftharpoons 2Ag(s) + Cu^{2+}(aq)$$
$$E^o_{cell} = +0.80 - (+0.34) = 0.46\ \text{V} \quad (12.80)$$

In contrast, consider the reaction between chromium(III) ions and iron (equation 12.81). The corresponding value of ΔG^o (298 K) is -6 kJ (per mole of reaction), and this indicates that there are only slightly more products than reactants in the system. [*Exercise*: You should confirm this by determining the equilibrium constant.]

$$2Cr^{3+}(aq) + Fe(s) \rightleftharpoons 2Cr^{2+}(aq) + Fe^{2+}(aq)$$
$$E^o_{cell} = -0.41 - (-0.44) = 0.03\ \text{V} \quad (12.81)$$

A useful (if somewhat artificial) 'rule' to remember is the *anticlockwise cycle* which can be used with a table of standard reduction potentials which are arranged as in Table 12.4. Consider any pair of redox couples in the table; each couple is made up of an oxidized and a reduced species. The spontaneous reaction will be between the lower oxidized species and the upper reduced species. For example, as we saw above, Ag^+ (*an oxidized species*) oxidizes Cu (*a reduced species*) and the direction of the spontaneous reaction may be deduced by constructing an anticlockwise cycle:

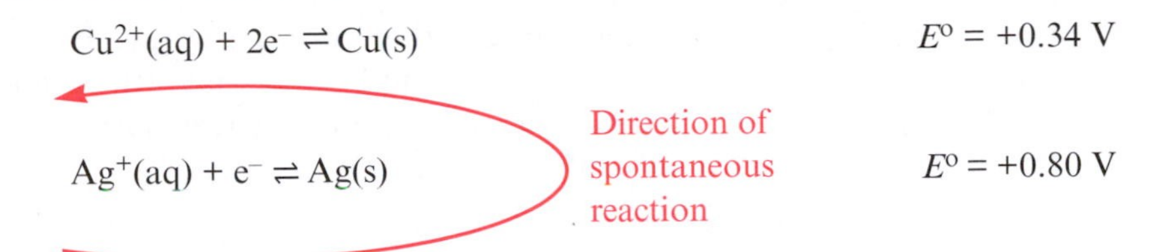

We provide further examples of the uses of standard reduction potentials in the accompanying workbook.

Worked example 12.12 ***Redox equilibria***

Use the standard reduction potentials in Table 12.4 to determine the products of the reaction between iron and aqueous dibromine.

From Table 12.4, there are three redox couples that are relevant to this question:

$$Fe^{2+}(aq) + 2e^- \rightleftharpoons Fe(s) \qquad E^o = -0.44 \text{ V}$$

$$Fe^{3+}(aq) + 3e^- \rightleftharpoons Fe(s) \qquad E^o = -0.04 \text{ V}$$

$$Br_2(aq) + 2e^- \rightleftharpoons 2Br^-(aq) \qquad E^o = +1.09 \text{ V}$$

This suggests two possible reactions:

1. $Fe(s) + Br_2(aq) \rightleftharpoons Fe^{2+}(aq) + 2Br^-(aq)$

or

2. $2Fe(s) + 3Br_2(aq) \rightleftharpoons 2Fe^{3+}(aq) + 6Br^-(aq)$

The value of E^o_{cell} in each case is:

1. $E^o_{cell} = 1.09 - (-0.44) = 1.53$ V

2. $E^o_{cell} = 1.09 - (-0.04) = 1.13$ V

The larger (more positive) E^o_{cell} might initially suggest that reaction **1** is the more thermodynamically favourable process but remember that ΔG^o depends not only on E^o_{cell} but also on z, the number of moles of electrons transferred *per mole of reaction*: in reaction **1**, $z = 2$, but in reaction **2**, $z = 6$.

$\Delta G^o(298 \text{ K})$ in each case is found using the equation:

$$\Delta G^o(298 \text{ K}) = -zFE^o_{cell}$$

1. $\Delta G^o(298 \text{ K}) = -(2 \times 96\,500 \times 1.53) \text{ J} = -295\,290 \text{ J}$

This corresponds to –295 kJ per mole of iron.

2. $\Delta G^o(298 \text{ K}) = -(6 \times 96\,500 \times 1.13) \text{ J} = -\,654\,270 \text{ J}$

This corresponds to –327 kJ per mole of iron.

Reaction **2** is therefore thermodynamically more favourable than **1**. This example emphasizes that although a positive value of E^{o}_{cell} certainly indicates spontaneity, it is ΔG^{o} that is the true indicator of the extent of reaction.

The effect of solution concentration on E^{o}: the Nernst equation

Up to this point, we have considered only *standard* reduction potentials which refer to solution concentrations of 1 mol dm^{-3}. It is common in the laboratory to work with solutions of lower concentrations, and reduction potentials vary with the concentration of the solutions in the electrochemical cells. The dependence is given by the Nernst equation 12.82. Temperature is another variable in the equation, although normally experiments will be carried out at a specified temperature.

$$E = E^{o} + \left\{\frac{RT}{zF} \times \left(\ln \frac{[\text{oxidized form}]}{[\text{reduced form}]}\right)\right\} \qquad \textit{Nernst equation} \qquad (12.82)^{\S}$$

where R = molar gas constant = 8.314 J K^{-1} mol^{-1},
T = temperature in K,
z = number of moles of electrons transferred *per mole of reaction*,
F = Faraday constant = 96 500 C mol^{-1}.

In Table 12.4, the oxidized forms of the half-cells are on the left-hand side and the reduced forms on the right. If the half-cell involves a metal ion/metal couple, the activity of the solid metal is taken to be unity and equation 12.82 simplifies to 12.83.

$$\text{For an } M^{n+}/M \text{ couple:} \qquad E = E^{o} + \left(\frac{RT}{zF} \times \ln[M^{n+}]\right) \qquad (12.83)$$

The Nernst equation states:

$$E = E^{o} + \left\{\frac{RT}{zF} \times \left(\ln \frac{[\text{oxidized form}]}{[\text{reduced form}]}\right)\right\}$$

Worked example 12.13 ***The dependence of E on solution concentration***

The standard reduction potential of the Ag^{+}/Ag couple is +0.80 V. What is the value of E at 298 K when the concentration of silver ions is 0.02 mol dm^{-3}? [R = 8.314 J mol^{-1} K^{-1}, F = 96 500 C mol^{-1}.]

§ The Nernst equation may also be written in the form $E = E^{o} - \left\{\frac{RT}{zF} \times \ln Q\right\}$ where Q is the *reaction quotient*.

The electrode is of the type M^{n+}/M, and so E can be found using the equation:

$$E = E^{o} + \left\{ \frac{RT}{zF} \times \ln [M^{n+}] \right\}$$

The half-equation for the couple is:

$$Ag^{+}(aq) + e^{-} \rightleftharpoons Ag(s)$$

and, one mole of electrons is transferred per mole of reaction: $z = 1$.

$$\begin{aligned} E &= 0.80 + \left\{ \frac{8.314 \times 298 \times \ln(0.02)}{1 \times 96\,500} \right\} \\ &= 0.80 + (-0.10) \\ &= 0.70 \text{ V (at 298 K)} \end{aligned}$$

Equation 12.83 is of the form:

$$y = c + mx$$

and a plot of E against $\ln [M^{n+}]$ is linear. The gradient of the line is equal to $^{RT}/_{zF}$ and the intercept on the vertical axis (when $\ln[M^{n+}] = 0$) is equal to E^{o} for the M^{n+}/M couple. Figure 12.17 shows a plot of $E_{Ag^{+}/Ag}$ against $\ln [Ag^{+}]$ using experimental data collected for dilute solutions; the half-cell potentials may be obtained from the electrochemical cell:

$$\text{Pt } [H_2(g, 1 \text{ bar})] \mid 2H^{+}(aq, 1 \text{ mol dm}^{-3}) \vdots Ag^{+}(aq) \mid Ag(s)$$

for which E_{cell} is equivalent to $E_{Ag^{+}/Ag}$ (equation 12.84).

$$E_{cell} = E_{Ag^{+}/Ag} - E^{o}_{\text{standard hydrogen electrode}} = E_{Ag^{+}/Ag} \qquad (12.84)$$

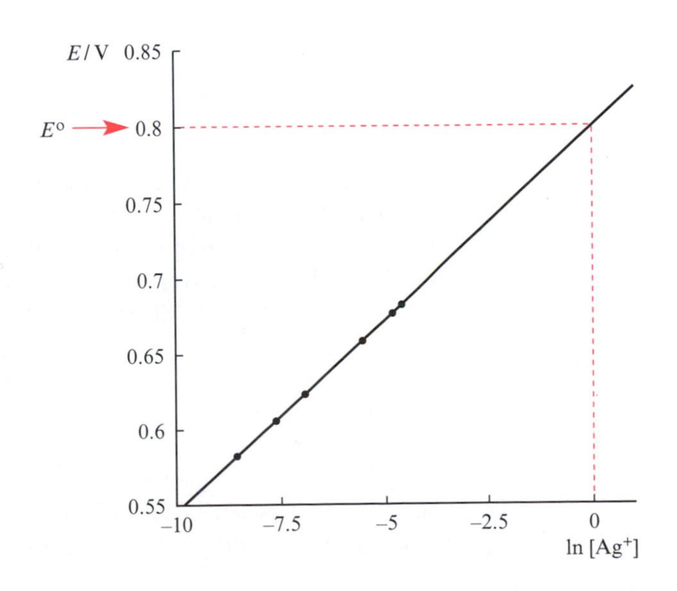

12.17 The dependence of the reduction potential of the Ag^{+}/Ag couple on the concentration of silver(I) ions. When $[Ag^{+}] = 1$ mol dm^{-3}, the cell is under standard conditions; this corresponds to $\ln [Ag^{+}] = 0$, and the reduction potential is the standard value, $E^{o} = +0.80$ V.

In this equation, notice the use of E^o for the hydrogen electrode denoting standard conditions, but E for the Ag^+/Ag electrode because the concentration of the silver(I) ions is variable. The *cell potential* is therefore non-standard (E_{cell} rather than E^o_{cell}). The value of $E^o_{Ag^+/Ag}$ can be found by extrapolation; when $[Ag^+] = 1$ mol dm^{-3} (standard conditions), $\ln[Ag^+] = 0$, and $E^o_{Ag^+/Ag}$ can be determined from Figure 12.17 as shown.

The effect of pH on reduction potential

If the half-cell of interest involves either H^+ or $[OH]^-$ ions, then the value of $E_{half\text{-}cell}$ will be critically dependent upon the pH of the solution. Consider half-cell 12.85.

$$\underbrace{[MnO_4]^-(aq) + 8H^+(aq) + 5e^-}_{\text{oxidized form}} \rightleftharpoons \underbrace{Mn^{2+}(aq) + 4H_2O(l)}_{\text{reduced form}} \qquad E^o = +1.51\text{ V} \qquad (12.85)$$

Using the Nernst equation at 298 K, the value of E can be found (equation 12.86) and we immediately see the dependence of E upon hydrogen ion concentration.

$$E = 1.51 + \left\{\frac{8.314 \times 298}{5 \times 96\,500} \times \left(\ln \frac{[MnO_4^-][H^+]^8}{[Mn^{2+}]}\right)\right\} \qquad (12.86)$$

pH = −log [H⁺]

For a solution in which the ratio $[MnO_4^-] : [Mn^{2+}]$ is 100 : 1, equation 12.86 shows that at pH = 0, $E = +1.53$ V, but at pH = 3, $E = +1.25$ V. You should consider what effect this difference will have on the ability of $KMnO_4$ to oxidize chloride ions to Cl_2 (equation 12.87).

$$Cl_2(aq) + 2e^- \rightleftharpoons 2Cl^-(aq) \qquad E^o = +1.36\text{ V} \qquad (12.87)$$

Reference electrodes

Throughout this section, we have emphasized the use of the standard hydrogen electrode as a reference electrode in electrochemical cells. While this is useful because $E^o_{2H^+/H_2}$ is zero, the cell itself is not very convenient to use – a cylinder of H_2 is needed and the pressure must be maintained at exactly 1 bar. A range of other reference electrodes is available, including the calomel and the silver/silver chloride electrodes.

Figure 12.18 shows the silver/silver chloride electrode. A silver wire coated with silver chloride dips into a saturated aqueous solution of KCl. The half-cell reaction is given in equation 12.88, and the reduction potential depends on the chloride concentration; for $[Cl^-] = 1.0$ mol dm^{-3}, E^o is +0.22 V.

$$AgCl(s) + e^- \rightleftharpoons Ag(s) + Cl^-(aq) \qquad (12.88)$$

Other reduction potentials may be measured *with respect to the silver/silver chloride electrode*, and one use of this reference electrode is in combination

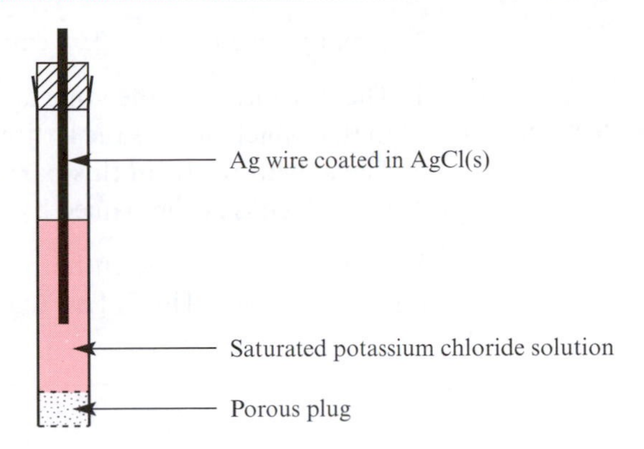

12.18 A representation of the silver/silver chloride reference electrode.

with a pH electrode in a pH meter. The potential of the pH electrode depends upon the $[H^+]$ and use of the electrochemical cell allows pH values to be read directly from a suitably calibrated voltmeter.

Another reference electrode is the dimercury(I) chloride (calomel) electrode, the half-cell reaction for which is given in equation 12.89.

$$Hg_2Cl_2(s) + 2e^- \rightleftharpoons 2Hg(l) + 2Cl^-(aq) \qquad E^o = +0.27\ V \qquad (12.89)$$

Because a variety of reference electrodes are in use, it is always necessary to quote reduction potentials *with respect to a specific reference.*

12.11 ENTHALPY CHANGES AND SOLUBILITY

The data in Table 12.2 illustrated that many salts dissolve in water in endothermic processes. In this section, we show that by considering both the enthalpy *and* entropy changes which accompany the solvation process, it is possible to understand why this phenomenon exists. However, solubility is a complicated topic and our coverage will be relatively superficial.

The standard enthalpy of solution of potassium chloride

NaCl type lattice: see Section 6.8

The K^+ and Cl^- ions in solid KCl are arranged in a sodium chloride type lattice. At 298 K, the solubility of KCl is 0.46 mole (34 g) per 100g water indicating that the process of solvation is thermodynamically favourable.

We can initially approach the problem by using Hess's Law, but first we must define the process that occurs as solid KCl dissolves. An ionic lattice in the solid state is changed into solvated cations and anions in the aqueous solution of KCl (equation 12.90), and the standard enthalpy change (at 298 K) that accompanies the dissolution is $\Delta_{sol}H^o$.

$$KCl(s) \xrightarrow{\text{water}} K^+(aq) + Cl^-(aq) \qquad (12.90)$$

The change in equation 12.90 can be considered in two stages:

$\Delta_{lattice}H^o$: see Section 6.16

1. The disruption of the ionic lattice into gaseous ions – the reverse process to that which defines the lattice energy ($\Delta U(0K) \approx \Delta_{lattice}H^o$)
2. The solvation (or in this particular case, the hydration) of the gaseous ions which is accompanied by a standard enthalpy change $\Delta_{solv}H^o$.

We can now set up the enthalpy cycle shown in equation 12.91 and determine $\Delta_{sol}H^o$ using Hess's Law (equation 12.92).

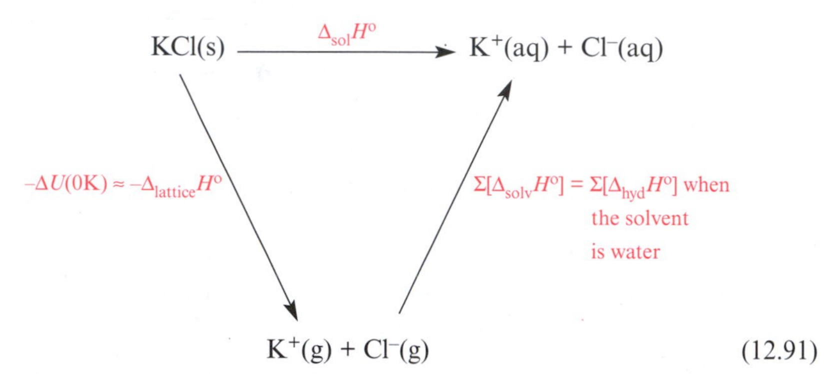

(12.91)

$$\Delta_{sol}H^o(KCl, s) = \Delta_{hyd}H^o(K^+, g) + \Delta_{hyd}H^o(Cl^-, g) + [-\Delta_{lattice}H^o(KCl, s)] \tag{12.92}$$

The lattice energy of KCl is –718 kJ mol^{-1} and so if the dissolution of KCl is to be an exothermic process, the sum of the standard enthalpies of solvation of $K^+(g)$ and $Cl^-(g)$ must be more negative than –718 kJ mol^{-1}. What factors will influence the magnitude of $\Delta_{solv}H^o$ (or $\Delta_{hyd}H^o$) for a particular ion? The solvent molecules must be able to interact strongly with the ion and this depends partly upon the nature of the solvent and partly upon the size and charge of the ion. It is impossible to measure values of $\Delta_{hyd}H^o$(298 K) *directly* for individual ions by calorimetric means since each ion must be present in solution with its counter-ion, but values can be obtained by indirect methods.§ By substituting values for the lattice energy and $\Delta_{hyd}H^o$ into equation 12.92, the standard enthalpy of solution is found to be +20 kJ mol^{-1} (equation 12.93) – the process is *endothermic*.

$$\Delta_{sol}H^o(KCl, s) = (-320) + (-378) + 718 = +20 \text{ kJ mol}^{-1} \tag{12.93}$$

$\Delta_{solv}H^o$(298 K) is the standard enthalpy change that accompanies the solvation of a gaseous ion at 298 K; if the solvent is water, the process of solvation is called hydration.

$\Delta_{sol}H^o$(298 K) is the enthalpy change that accompanies the dissolution of a substance at 298 K.

§ For a more detailed discussion of hydration enthaplies see: W.E. Dasent, *Inorganic Energetics*, 2nd edn, Cambridge University Press, Cambridge (1984).

The change in entropy when potassium chloride dissolves in water

When solid KCl dissolves in water, it is easy to envisage the destruction of the crystal lattice and you might assume that the entropy of the system necessarily increases. However, other changes within the system are occurring:

- the crystal lattice breaks into separate ions;
- hydrogen bonds between water molecules in the bulk solvent are broken as KCl enters the solvent;
- ion–dipole bonds form between the ions and water molecules; and
- the solvated ions are associated with the bulk solvent molecules through hydrogen bonding involving the coordinated water molecules (see Figure 11.13 and accompanying text).

The entropy change for reaction 12.94 is calculated from values of S^o(298 K), available in standard tables.

$$\begin{array}{llll} & KCl(s) & \xrightarrow{\Delta S^o(298\ K)} K^+(aq) + Cl^-(aq) & \\ S^o(298\ K) & +83 & \qquad\qquad +103 \qquad +57 & J\ mol^{-1}\ K^{-1} \end{array} \qquad (12.94)$$

From equation 12.94, ΔS^o(298 K) for the dissolution of KCl is +77 J mol^{-1} K^{-1} and so an increase in entropy accompanies the process.

Why does KCl dissolve in water?

By using the values of $\Delta_{sol}H^o$ and ΔS^o determined above, we can calculate a value for ΔG^o(298 K) for the dissolution in water of one mole of KCl (equation 12.95), making sure that we first convert the value of S^o into appropriate units.

At 298 K:

$$\Delta G^o = \Delta H^o - T\Delta S^o = +20 - (298 \times 77 \times 10^{-3}) \approx -2.9\ kJ\ mol^{-1} \qquad (12.95)$$

The free energy change is negative, and the dissolution of KCl in water is a thermodynamically favourable process – the solubility at 298 K is 4.6 mol dm^{-3}.

Over small temperature ranges, it is often the case that both ΔH and ΔS are approximately constant. For KCl, we have determined that $\Delta_{sol}G^o$(298 K) is –2.9 kJ mol^{-1}. We can estimate the change in free energy for the same process at 320 K from equation 12.96 which assumes that ΔH^o(298 K) ≈ ΔH^o(320 K), and ΔS^o(298 K) ≈ ΔS^o(320 K).

At 320 K:

$$\Delta G^o = \Delta H^o - T\Delta S^o = +20 - (320 \times 77 \times 10^{-3}) \approx -4.6\ kJ\ mol^{-1} \qquad (12.96)$$

The more negative value of ΔG^o at 320 K compared with that at 298 K indicates that the solubility of KCl at 320 K is greater than at 298 K. The actual solubilities of 34 g (298 K) and 40 g (320 K) per 100 g of water confirm our calculations.

Saturated solutions

The solubilities of the salts given in Table 12.2 correspond to the *maximum* amount of compound that will dissolve in a given volume of solvent at a stated temperature. Once this limiting value is exceeded, excess solute remains undissolved and the solution is said to be *saturated.* In a saturated solution, the dissolved and undissolved solutes are in equilibrium with each other, and ΔG for that process is zero.

12.12 SOLUBILITY PRODUCT CONSTANT, K_{sp}

Sparingly soluble salts

Many ionic compounds are only very *sparingly soluble*. This includes many compounds which are commonly described as 'insoluble', e.g. AgCl, $BaSO_4$ and $CaCO_3$. Sparingly soluble means that a saturated solution only contains a small number of ions. When an excess of a sparingly soluble compound is added to a solvent, a system is obtained in which the solid is in equilibrium with ions in the saturated solution. Equation 12.97 exemplifies this for silver chloride, the solubility of which is 1.35×10^{-5} mol dm^{-3} (at 298 K).

$$AgCl(s) \rightleftharpoons Ag^+(aq) + Cl^-(aq) \qquad K_c = \frac{[Ag^+][Cl^-]}{[AgCl]} \tag{12.97}$$

In the expression for the equilibrium constant, $[Ag^+]$ and $[Cl^-]$ are the *equilibrium* concentrations of the aqueous ions and [AgCl] is the equilibrium concentration of the undissolved solid. From Le Chatelier's principle, the addition of more chloride ion to this saturated solution will result in the precipitation of silver chloride. It is more convenient to describe the equilibrium position in terms of the *solubility product* (or *solubility constant*) K_{sp} (equation 12.98).§

$$K_{sp} = [Ag^+][Cl^-] \tag{12.98}$$

The expressions for the solubility products for two other salts which are sparingly soluble in water are given in equations 12.99 and 12.100, and values of K_{sp} for selected salts are listed in Table 12.5.

$$CaF_2(s) \rightleftharpoons Ca^{2+}(aq) + 2F^-(aq) \qquad K_{sp} = [Ca^{2+}][F^-]^2 \tag{12.99}$$

$$Ca_3(PO_4)_2(s) \rightleftharpoons 3Ca^{2+}(aq) + 2[PO_4]^{3-}(aq) \qquad K_{sp} = [Ca^{2+}]^3[PO_4{}^{3-}]^2 \tag{12.100}$$

§ The substitution of [AgCl] = 1 is done on the basis that the activity of a solid is unity and is constant.

Table 12.5 Values of K_{sp}(298 K) for selected sparingly soluble salts.

Compound	*Formula*	K_{sp}*(298 K)*[a]
Barium sulfate	$BaSO_4$	1.1×10^{-10}
Calcium carbonate	$CaCO_3$	4.9×10^{-9}
Calcium hydroxide	$Ca(OH)_2$	4.7×10^{-6}
Calcium phosphate	$Ca_3(PO_4)_2$	2.1×10^{-33}
Copper(I) chloride	$CuCl$	1.7×10^{-7}
Copper(I) sulfide	Cu_2S	2.3×10^{-48}
Copper(II) sulfide	CuS	6.0×10^{-37}
Iron(II) hydroxide	$Fe(OH)_2$	4.9×10^{-17}
Iron(II) sulfide	FeS	6.0×10^{-19}
Iron(III) hydroxide	$Fe(OH)_3$	2.6×10^{-39}
Lead(II) iodide	PbI_2	8.5×10^{-19}
Lead(II) sulfide	PbS	3.0×10^{-28}
Magnesium carbonate	$MgCO_3$	6.8×10^{-6}
Magnesium hydroxide	$Mg(OH)_2$	5.6×10^{-12}
Silver(I) chloride	$AgCl$	1.8×10^{-10}
Silver(I) bromide	$AgBr$	5.4×10^{-13}
Silver(I) iodide	AgI	8.5×10^{-17}
Silver(I) chromate	Ag_2CrO_4	1.1×10^{-12}
Silver(I) sulfate	Ag_2SO_4	1.2×10^{-5}
Zinc(II) sulfide	ZnS	2.0×10^{-25}

[a] The reader should determine the units of each solubility product.

Determining values of K_{sp} from the solubilities of salts

The solubility of silver chloride in water at 298 K is 1.35×10^{-5} mol dm^{-3}. When 1.35×10^{-5} moles of AgCl dissolve in 1dm^3 of solution, 1.35×10^{-5} moles of Ag^+ and 1.35×10^{-5} moles of Cl^- ions are formed, and $[Ag^+]$ and $[Cl^-]$ are both equal to 1.35×10^{-5} mol dm^{-3}. The solubility product constant is given by equation 12.101.

$$K_{sp} = [Ag^+][Cl^-] = (1.35 \times 10^{-5}) \times (1.35 \times 10^{-5}) = 1.82 \times 10^{-10} \text{ mol}^2 \text{ dm}^{-6} \qquad (12.101)$$

Now consider iron(II) hydroxide with a solubility of 2.3×10^{-6} mol dm^{-3} at 298 K. Dissolution of a mole of $Fe(OH)_2$ produces one mole of Fe^{2+} ions and *two* moles of hydroxide ions (equation 12.102).

$$Fe(OH)_2(s) \quad Fe^{2+}(aq) + 2[OH]^-(aq) \qquad (12.102)$$

In a saturated solution of $Fe(OH)_2$ at 298 K, the concentration of Fe^{2+} ions is equal to the solubility of the salt (2.3×10^{-6} mol dm^{-3}), but $[OH^-]$ is double this (4.6×10^{-6} mol dm^{-3}). Equation 12.103 gives the expression for K_{sp}.

$$K_{sp} = [Fe^{2+}][OH^-]^2$$
$$= (2.3 \times 10^{-6}) \times (4.6 \times 10^{-6})^2 = 4.9 \times 10^{-17}\ mol^3\ dm^{-9} \qquad (12.103)$$

Worked example 12.14 ***Determination of the solubility of a salt from K_{sp}***

What is the solubility of calcium hydroxide in water at 298 K if $K_{sp} = 4.7 \times 10^{-6}\ mol^3\ dm^{-9}$?

Firstly, write an equilibrium to show the dissolution of the salt:

$$Ca(OH)_2(s) \rightleftharpoons Ca^{2+}(aq) + 2[OH]^-(aq)$$

The equation for K_{sp} is:

$$K_{sp} = [Ca^{2+}][OH^-]^2$$

In solution, the concentration of hydroxide ions is twice that of the calcium ions, and we can rewrite the above equation in the form:

$$K_{sp} = [Ca^{2+}] \times (2 \times [Ca^{2+}])^2 = 4 \times [Ca^{2+}]^3$$

We choose to write this equation in terms of $[Ca^{2+}]$ rather than the concentration of hydroxide ions because the solubility of the salt (the ultimate goal of the exercise) is equal to $[Ca^{2+}]$.

$$K_{sp} = 4.7 \times 10^{-6} = 4 \times [Ca^{2+}]^3$$

$$[Ca^{2+}] = \sqrt[3]{\frac{4.7 \times 10^{-6}}{4}}$$
$$= 1.05 \times 10^{-2}\ mol\ dm^{-3}$$

The number of moles of $[Ca^{2+}]$ ions in solution is equal to the number of moles of $Ca(OH)_2$ that have dissolved, and so the solubility of $Ca(OH)_2$ is $1.05 \times 10^{-2}\ mol\ dm^{-3}$.

Determining K_{sp} from electrochemical data

The dependence of the reduction potential, E, on solution concentration can be used as a means of measuring very low concentrations of ions, and from these data, values of K_{sp} can be determined. Suppose that we wish to find K_{sp} for AgCl; the electrochemical half-cell must contain a saturated solution of silver chloride (equation 12.104) in addition to a silver electrode which forms a redox couple with the silver(I) ions in solution (equation 12.105).

$$AgCl(s) \rightleftharpoons Ag^+(aq) + Cl^-(aq) \qquad (12.104)$$

$$Ag^+(aq) + e^- \rightleftharpoons Ag(s) \qquad (12.105)$$

By combining this half-cell with a standard hydrogen electrode, we find that $E_{Ag^+/Ag}$ is +0.512 V. Using the Nernst equation, we can now determine the concentration of silver ions present in the saturated solution of silver chloride (equation 12.106). [*Exercise*: What difference to the value of $[Ag^+]$ would it have made if we had measured $E_{Ag^+/Ag}$ less accurately, i.e. +0.51 V?]

$$\left.\begin{aligned}
E &= E^{o} + \left(\frac{RT}{zF} \times \ln[Ag^+]\right) \\
0.512 &= 0.80 + \left(\frac{8.314 \times 298}{1 \times 96\,500} \times \ln[Ag^+]\right) \\
\ln[Ag^+] &= (0.512 - 0.80) \times \left(\frac{96\,500}{8.314 \times 298}\right) = -11.217 \\
[Ag^+] &= e^{-11.217} = 1.34 \times 10^{-5}\ \text{mol dm}^{-3}
\end{aligned}\right\} \quad (12.106)$$

The equilibrium concentrations of Ag^+ and Cl^- ions in the saturated solution of AgCl are equal and K_{sp} is found from equation 12.107.

$$K_{sp} = [Ag^+][Cl^-] = [Ag^+]^2 = (1.34 \times 10^{-5})^2 \approx 1.80 \times 10^{-10}\ \text{mol}^2\ \text{dm}^{-6} \quad (12.107)$$

SUMMARY

Thermodynamics is not the easiest of subjects and in this chapter we have combined experimental observations with theory so that you might see *why* the theory is necessary. Our emphasis has purposely not been on the derivation of equations – this material can be found in more advanced physical chemistry texts.

An understanding of thermodynamics is central to an understanding of why (and in what direction) chemical processes occur. Some of the most important equations that we have encountered in this chapter are listed below. You should be familiar with them all, and *be able to use them.* Further examples of their applications are given in the accompanying workbook.

$$\Delta U = \Delta H - P\Delta V$$

$$C_P = \left(\frac{\partial H}{\partial T}\right)_P$$

$$C_P = \left(\frac{\partial U}{\partial T}\right)_V$$

$$C_P = C_V + R \qquad \text{(for one mole of an ideal gas)}$$

$$\Delta(\text{State function}) = \Sigma(\text{State function})_{\text{products}} - \Sigma(\text{State function})_{\text{reactants}}$$

$$\Delta C_P = \left(\frac{\partial \Delta H}{\partial T}\right)_P \qquad \text{Kirchhoff's equation}$$

$$\Delta G^{\circ} = -RT \ln K_p \qquad \text{Reaction isotherm}$$

$$\Delta G = \Delta H - T\Delta S$$

$$\left(\frac{\partial S}{\partial T}\right)_P = \frac{C_P}{T}$$

$$S_{(T_2)} - S_{(T_1)} = C_P \times \ln\left(\frac{T_2}{T_1}\right) \qquad \text{(for a constant } C_P\text{)}$$

$$\Delta_{\text{fus}}S = \frac{\Delta_{\text{fus}}H}{\text{mp}}$$

$$\Delta_{\text{vap}}S = \frac{\Delta_{\text{vap}}H}{\text{bp}}$$

$$E^{\circ}_{\text{cell}} = E^{\circ}_{\text{reduction process}} - E^{\circ}_{\text{oxidation process}}$$

$$\Delta G^{\circ} = -zFE^{\circ}_{\text{cell}}$$

$$E = E^{\circ} + \left\{\frac{RT}{zF} \times \left(\ln \frac{[\text{oxidized form}]}{[\text{reduced form}]}\right)\right\} \qquad \text{Nernst equation}$$

$$K_{\text{sp}} = [X^+][Y^-] \qquad \text{for a salt XY}$$

Do you know what the following terms mean?

- enthalpy and ΔH
- internal energy
- work done (on or by the system)
- heat capacities (C_P and C_V)
- equilibrium constant (K_c and K_p)
- free energy and ΔG
- spontaneous reaction
- entropy and ΔS
- standard molar entropy
- the First, Second and Third Laws of Thermodynamics
- the reaction isotherm
- Ellingham diagram
- galvanic cell
- standard cell potential
- standard hydrogen electrode
- standard reduction potential
- Nernst equation
- enthalpy of solution
- enthalpy of solvation
- enthalpy of hydration
- saturated solution
- sparingly soluble salt
- solubility product constant

You should now be able:

- to distinguish between the system and the surroundings
- to distinguish between an isothermal and an adiabatic change
- to define and give examples of state functions
- to recognize the limitations of assuming that C_P and ΔH are temperature-invariant
- to discuss the reasons why values of ΔG rather than ΔH are reliable guides to the extent of reactions
- to understand the physical significance of the magnitudes of values of K
- to explain why equilibrium constants may be dimensionless
- to understand the physical significance of the magnitudes and signs of values of ΔG
- to determine $\Delta_r G^{\circ}$ from values of $\Delta_f G^{\circ}$
- to determine ΔS° from values of S°

- to distinguish between a galvanic cell and an electrolytic cell
- to describe the construction of, and reaction within, the Daniell cell, using this as an example of electrochemical processes in a wider sense
- to define a *reference electrode*
- to use tables of standard reduction potentials to determine E^{o}_{cell}
- to determine values of ΔG^{o} from E^{o}_{cell}, and to use the values to judge whether a reaction is spontaneous
- to use tables of standard reduction potentials to predict possible reactions
- to use the Nernst equation to relate E to the concentration of ions in solution
- to justify why some salts dissolve in *endothermic* processes

PROBLEMS

12.1 In equation 12.9, show that units of Pa for pressure and m^3 for volume are compatible with J for ΔH [Hint: refer to Table 1.2.]

12.2 (a) Calculate the work done on the surroundings at 1 bar pressure and 298 K when 0.2 mole H_2O_2 decomposes (volume occupied by 1 mole of gas at 1 bar and 273 K = 22.7 dm^3). (b) If the standard enthalpy of reaction is –98.2 kJ per mole of H_2O_2, what is the corresponding change in the internal energy of the system?

12.3 In equation 12.13, for one mole of pentane, $\Delta n = -3$. What would happen if the reaction were defined so as to produce water in the gas phase?

12.4 How does the *molar* heat capacity of a substance differ from the *specific* heat capacity?

12.5 Estimate the enthalpy change at 400 K that accompanies the reaction:

$$Li(s) + \tfrac{1}{2}Cl_2(g) \rightarrow LiCl(s)$$

if $\Delta_f H^{o}$(298 K) LiCl(s) = –408.6 kJ mol^{-1}, and values of C_P (298 K) are Li 24.8, Cl_2 33.9, and LiCl 48.0 J mol^{-1} K^{-1}.

12.6 For reaction 12.28, what is the effect on the equilibrium concentration of NH_3 of doubling the pressure of (a) N_2 and (b) H_2?

12.7 Determine the free energy change at 298 K that accompanies the reaction:

$$B_2H_6(g) + 6H_2O(l) \rightarrow 2B(OH)_3(s) + 6H_2(g)$$

if values of $\Delta_f G^{o}$(298 K) are B_2H_6(g) +87, $B(OH)_3$(s) –969, H_2O(l) –237 kJ mol^{-1}.

12.8 By using the Ellingham diagram in Figure 12.8, (a) predict which metals will reduce iron(II) oxide at 500 K, (b) estimate ΔG^{o} for the reaction between FeO and carbon at 1200 K, and (c) assess to what extent the reaction between Sn and NiO is affected by changing the temperature from 600 to 1000 K.

12.9 Use the data in Table 12.1 to verify the reaction isotherm and to calculate a value for the molar gas constant.

12.10 The value of $\Delta_f G^{o}$(298 K) C_2H_4(g) is +68 kJ mol^{-1}. What is K_p for this formation reaction? Write an equation for the formation of C_2H_4 from its elements in their standard states and indicate in which direction the equilibrium lies.

12.11 The enthalpies of fusion and vaporization of difluorine are 0.5 and 6.6 kJ mol^{-1}, and the melting and boiling points are 53 and 85 K respectively. Determine values (per mole of F_2) of $\Delta_{fus}S$ and $\Delta_{vap}S$, and comment on their relative magnitudes.

12.12 Calculate the standard change in entropy (298 K) when one mole of sulfur is oxidized to SO_3; S^{o}(298 K) for S(s), O_2(g) and SO_3(g) are 32.1, 205.2, 256.8 J mol^{-1} K^{-1} respectively. Given that the standard enthalpy of formation (298 K) of SO_3(g) is –395.7 kJ mol^{-1}, determine $\Delta_f G^{o}$(298 K).

12.13 Using data in Table 12.4, determine E^{o}_{cell} and ΔG^{o}(298 K) for the following reactions:

(a) $Co^{2+}(aq) + 2Cr^{2+}(aq) \rightleftharpoons Co(s) + 2Cr^{3+}(aq)$

(b) $\tfrac{1}{2}Co^{2+}(aq) + Cr^{2+}(aq) \rightleftharpoons \tfrac{1}{2}Co(s) + Cr^{3+}(aq)$

(c) $[Cr_2O_7]^{2-}(aq) + 14H^{+}(aq) + 6Fe^{2+}(aq) \rightleftharpoons$
$2Cr^{3+}(aq) + 7H_2O(aq) + 6Fe^{3+}(aq)$

[F = 96 500 C mol^{-1}]

12.14 Use Table 12.4 to determine whether hydrogen ions will oxidize iron to iron(II) or iron(III) ions. Rationalize your answer in terms of the thermodynamics of the process.

12.15 Determine the reduction potential for the half-cell:

$$[MnO_4]^-(aq) + 8H^+(aq) + 5e^- \rightleftharpoons Mn^{2+}(aq) + 5H_2O(l)$$

at 298 K, and at pH 5.5 when the ratio $[MnO_4^-]$: $[Mn^{2+}]$ is 75 : 1. [$E^o = +1.51$ V].

12.16 Determine the solubility of $BaSO_4$ in water if $K_{sp} = 1.1 \times 10^{-10}$ mol^2 dm^{-6}.

12.17 (a) If the solubility of silver chromate is 6.5×10^{-5} mol dm^{-3}, determine the solubility product for this salt.

(b) K_{sp} for AgCl is 1.8×10^{-10} mol^2 dm^{-6}. Why can silver chromate (which is red) be used to indicate the end-point in titrations involving the precipitation of AgCl?

13 *p*-Block and high oxidation state *d*-block elements

Topics

1	2	3	4	5	6	7	8	9	10	11	12	13	14	15	16	17	18
H																	He
Li	Be											B	C	N	O	F	Ne
Na	Mg											Al	Si	P	S	Cl	Ar
K	Ca	Sc	Ti	V	Cr	Mn	Fe	Co	Ni	Cu	Zn	Ga	Ge	As	Se	Br	Kr
Rb	Sr	Y	Zr	Nb	Mo	Tc	Ru	Rh	Pd	Ag	Cd	In	Sn	Sb	Te	I	Xe
Cs	Ba	La	Hf	Ta	W	Re	Os	Ir	Pt	Au	Hg	Tl	Pb	Bi	Po	At	Rn
Fr	Ra																

13.1 INTRODUCTION

In this chapter, we continue some descriptive chemistry and apply the thermodynamic principles introduced in Chapter 12 to understand why certain reactions take place.

'Diagonal line': see Section 7.11

Groups 13 to 18 of the periodic table make up the *p*-block and the 'diagonal line' that runs through the block *approximately* partitions the elements into metals and non-metals. In group 13, only the first member, boron, is a non-metal, but by groups 17 and 18, all the members of the group are essentially non-metallic.

The *d*-block§ is composed of ten triads of metallic elements in groups 3 to 12, although lanthanum (La) and actinium (Ac) are the first members of series of elements called the lanthanoids and actinoids, known collectively as the *f*-block elements. The group number gives the total number of valence $(n+1)s$ and nd electrons; for Fe in group 8, the ground state, outer electronic configuration is $4s^23d^6$, and for Ir at the bottom of group 9, the configuration is $6s^25d^7$.

Hydrides of the *p*- and *d*-block: see Chapter 11

The chemistry in this chapter is arranged by group for the *p*-block, with selected *d*-block elements described in Section 13.14. Our purpose in combining some chemistry of the *d*-block elements with that of the *p*-block is to illustrate the extent to which halides, oxides and related anions of the *high oxidation state d*-block elements may be considered as typical covalent species.

13.2 OXIDATION STATES

Inert pair effect: see Section 5.7

By way of introduction, we look at the relationship between the oxidation state of an element in a compound and the periodic group. In going from group 13 to group 18, the number of electrons in the valence shell of an element increases from 3 to 8. For a group 13 element, a maximum oxidation state of +3 is expected, but on going down the group, an oxidation state of +1 is also observed as the inert pair effect becomes significant. In group 14, the maximum oxidation state is +4, but the heavier elements form compounds in both the +2 and +4 oxidation states. The prevalence of *positive* oxidation states for these elements arises from the fact that they are most likely to combine with elements that are more electronegative than themselves (Table 11.2).

Expansion of the octet: see Section 5.16

In group 15, a range of oxidation states from +5 to –3 is observed but it must be stressed that *in its compounds, nitrogen cannot expand its octet.* Thus, the resonance structure **13.1** for the nitrate ion $[NO_3]^-$ ion (in which N is in a formal oxidation state of +5) is correct but resonance form **13.2** is *incorrect*. For later elements in the group, the valence shell can be expanded as in PCl_5 **13.3**.

§ The IUPAC distinction between a transition metal and a *d*-block element is given in Section 1.10.

(13.1) (13.2) (13.3)

$\chi^P(O) = 3.4$

In group 16, oxygen tends to obey the octet rule, but the valence shell may be expanded in sulfur, selenium and tellurium. The electronegativity of oxygen is such that it usually combines with less electronegative elements and is in an oxidation state of –2. Not only can the valence shell in the heavier elements in the group be expanded, but they are also found in oxidation states ranging from –2 to +6.

Each member of group 17 shows an oxidation state of –1, although once again, positive oxidation states are observed for the heavier elements (e.g. in $[ClO_4]^-$, $[BrO_4]^-$ and $[IO_4]^-$). Of the noble gases, xenon (and to a lesser extent krypton) tends to form compounds, but only with very electronegative elements (fluorine and oxygen) and oxidation states of +2, +4, +6 and +8 are observed.

One characteristic of a *d*-block metal is the observation of a range of oxidation states. This is true for most of the metals, but in group 3 (the first triad in the *d*-block), the +3 state predominates and this corresponds to the availability of only three electrons in the valence shell. In group 12, the +2 oxidation state is usual for zinc and cadmium. Figure 13.1 shows the range of oxidation states found for the first row metals. The pattern in the second and third rows is similar, although Ru and Os show higher oxidation states than Fe, as do Rh and Ir compared to Co. Group 11 should be singled out in particular – copper(I) and copper(II) are observed, in contrast to a predominance of the +1 state for silver, and the +1 and +3 states for gold.

It is important to remember that the highest oxidation states in the *d*-block occur when the atom is in a *covalent environment* – structures **13.4** to **13.6** contain manganese(VII), iridium(VI) and gold(III). [*Question*: Why is an Mn^{7+} *ion* not thermodynamically stable?]

Oxidation state	3 Sc	4 Ti	5 V	6 Cr	7 Mn	8 Fe	9 Co	10 Ni	11 Cu	12 Zn
0	*	*	*	*	*	*	*	*	*	*
+1			*	*	*		*	*	*	
+2		*	*	*	*	*	*	*	*	*
+3	*	*	*	*	*	*	*	*	*	
+4		*	*	*	*	*	*	*		
+5			*	*	*	*				
+6				*	*	*				
+7					*					
+8										

13.1 The variation in oxidation states exhibited by the first row *d*-block metals. The most common oxidation states are marked in red. Note the pattern: a single oxidation state (other than zero) is observed for scandium (Sc) and zinc (Zn), and the maximum number of oxidation states is attained for manganese (Mn) in the centre of the row.

(13.4)

(13.5)

(13.6)

13.3 GROUP 13: BORON

1	2		**13**	14	15	16	17	18
H								He
Li	Be		**B**	C	N	O	F	Ne
Na	Mg		**Al**	Si	P	S	Cl	Ar
K	Ca		**Ga**	Ge	As	Se	Br	Kr
Rb	Sr	*d*-block	**In**	Sn	Sb	Te	I	Xe
Cs	Ba		**Tl**	Pb	Bi	Po	At	Rn
Fr	Ra							

In Chapter 7, we discussed the complex structures of the allotropes of boron. Amongst the group 13 elements, boron is unique in forming a wide variety of hydrides, some of which were described in Section 11.5.

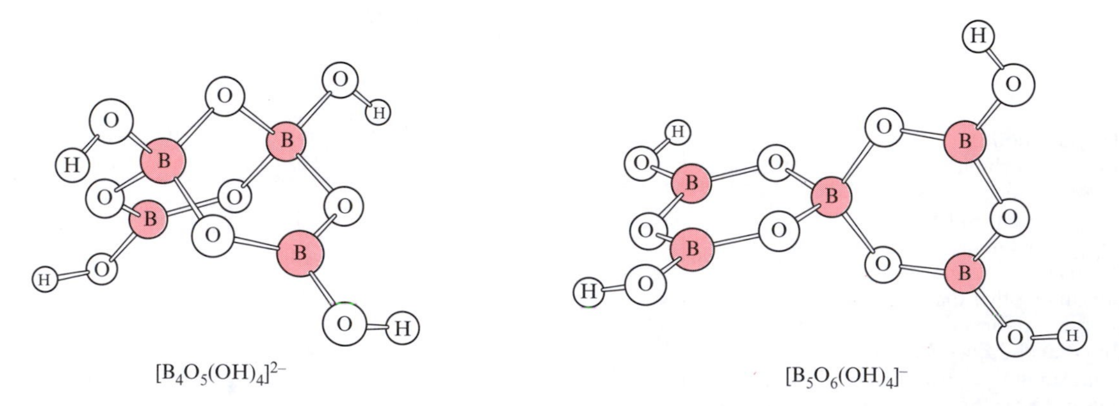

13.2 The structures of the anions $[B_4O_5(OH)_4]^{2-}$ and $[B_5O_6(OH)_4]^-$, determined by X-ray diffraction studies of the salts $[H_3NCH_2CH_2NH_3][B_4O_5(OH)_4]$ and $Cs[B_5O_6(OH)_4]$. Both anions contain trigonal planar and tetrahedral boron centres.

Boron–oxygen compounds

Boron occurs naturally in *borax*, $Na_2[B_4O_5(OH)_4]\cdot10H_2O$, and *kernite*, $Na_2[B_4O_5(OH)_4]\cdot2H_2O$, which are mined in the Mojave Desert, California. The borate anion present in these ores is shown in structure **13.7** and Figure 13.2.

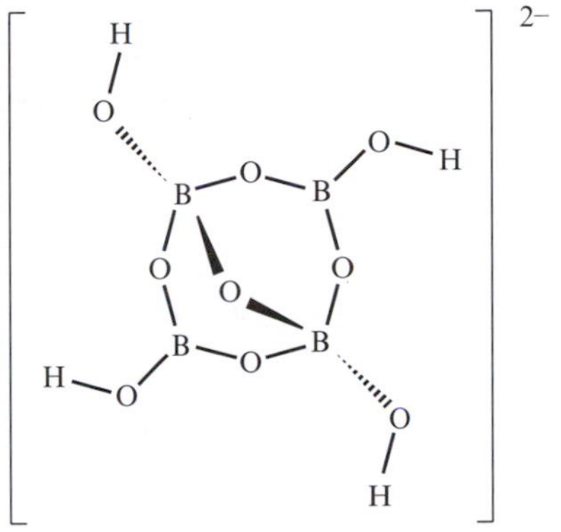

(13.7)

The oxide B_2O_3 is obtained by dehydrating boric acid $B(OH)_3$ (equation 13.1), which is prepared commercially from borax by treatment with aqueous sulfuric acid. After conversion to B_2O_3, boron is produced as a brown, amorphous powder by reaction 13.2.

The formula for boric acid, structure 11.4, can be written as H_3BO_3

$$2B(OH)_3(s) \xrightarrow{\Delta} B_2O_3(s) + 3H_2O(g) \qquad (13.1)$$

$$B_2O_3(s) + 3Mg(s) \xrightarrow{\Delta} 2B(s) + 3MgO(s) \qquad (13.2)$$

At room temperature, B_2O_3 is usually encountered in the form of a vitreous (glassy) solid and is a typical non-metal oxide, reacting slowly with water to give $B(OH)_3$. *Molten* B_2O_3 (> 1270 K) reacts rapidly with steam to give HBO_2 (metaboric acid). The main use of B_2O_3 is in the manufacture of borosilicate glass (including Pyrex), the low thermal expansion of which is a valuable property.

Boric acid has a layered structure in the solid state with trigonal $B(OH)_3$ units linked by hydrogen bonds (Figure 13.3) within each layer. Inter-layer interactions are very weak and as a consequence, boric acid crystals feel slippery and the compound is a good lubricant; this structure-related property is reminiscent of graphite. Boric acid is soluble in water (63.5 g dm^{-3} at 303 K), and aqueous solutions are weakly acidic ($pK_a = 9.1$). In aqueous solution it acts as a *Lewis*, rather than a Brønsted, acid (equation 13.3) – the protons in solution are *not directly* provided by the $B(OH)_3$ molecule.

Graphite: see Section 7.10

$$B(OH)_3(aq) + 2H_2O(l) \rightleftharpoons [B(OH)_4]^-(aq) + [H_3O]^+(aq) \qquad (13.3)$$

In addition to its use as a precursor to B_2O_3 and a range of inorganic borates and related compounds, boric acid is used as flame retardant, a preservative for wood and leather, and as a mild antiseptic. It is also the

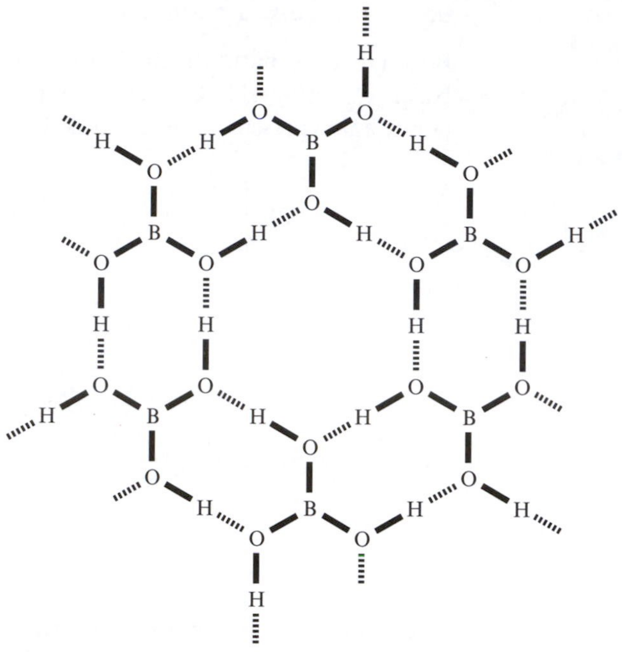

13.3 Boric acid possesses a layer structure. The diagram shows part of one layer in which $B(OH)_3$ molecules are hydrogen bonded together. The bonds within each molecule are highlighted in bold, and the hydrogen bonds are shown by the hashed lines.

Borane hydrolysis: see equation 11.26

product from the hydrolysis of many boron-containing compounds. There are numerous borate anions, some salts of which occur in nature. Examples are $[BO_3]^{3-}$ **13.8**, $[B_3O_6]^{3-}$ **13.9**, $[B_4O_5(OH)_4]^{2-}$ **13.7** (Figure 13.2) and $[B_5O_6(OH)_4]^-$ **13.10** (Figure 13.2). Their structures fall into general types in which the boron is in either a trigonal planar BO_3 or a tetrahedral BO_4 unit. The B–O bond lengths are generally shorter (≈136 pm) in a trigonal planar unit than in a tetrahedral one (≈148 pm). This is due to the donation of π-electrons from an oxygen lone pair to the vacant 2*p* AO on the trigonal planar boron centre (Figure 13.4a). This is called $(p\text{–}p)\pi$-bonding, and can

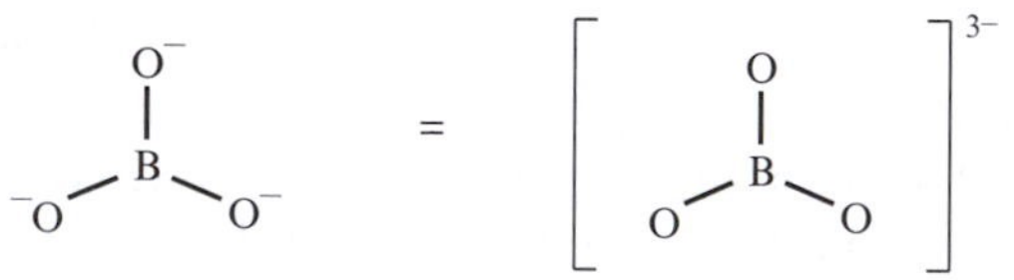

(13.8)

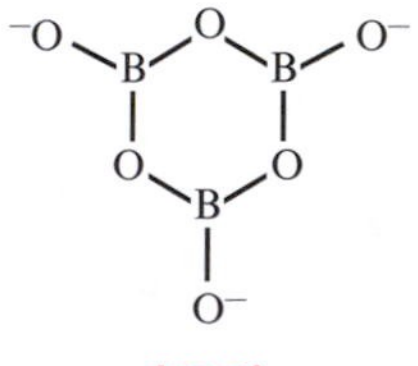

(13.9)

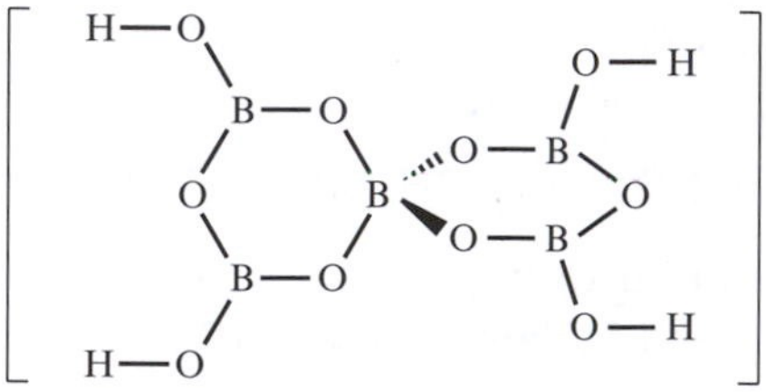

(13.10)

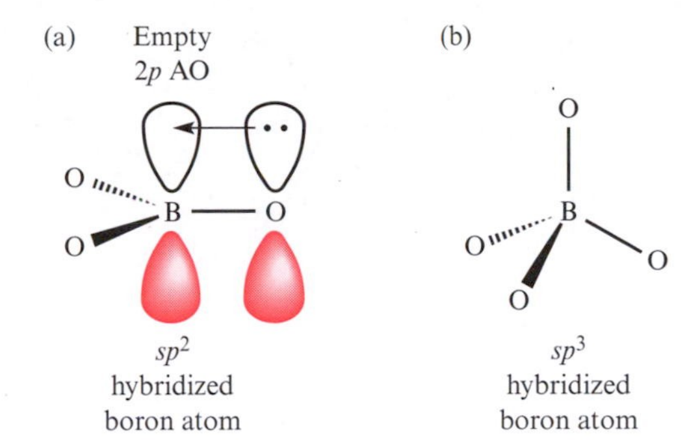

13.4 (a) In a trigonal planar BO_3-unit, the B atom can be considered to be sp^2 hybridized and the vacant $2p$ orbital can accept lone pair electrons from *each* oxygen atom; only one such π-interaction is shown. If all three oxygen atoms are involved, there can only be a *partial* π*-bond* along each B–O vector. The resultant $(p–p)\pi$ interactions strengthen the B–O bonds. (b) In a tetrahedral BO_4 unit, the B atom can be considered to be sp^3 hybridized, and is limited to forming four B–O σ-bonds.

occur in all three B–O interactions, giving *partial* π -character to each. In a tetrahedral geometry (sp^3 hybridized), there is no vacant AO available on the boron atom and B–O π-bonding cannot occur (Figure 13.4b).

$(p–p)\pi$-Bonding arises from the overlap of p-atomic orbitals.

$(p–p)\pi$-Bonding in B_3N_3 rings: see Section 15.13

The cyclic structures that are characteristic of the larger borate anions are also observed in some organic derivatives. The dehydration of boronic acids of the general form $RB(OH)_2$ leads to the formation of cyclic trimers $(RBO)_3$ (equation 13.4). Figure 13.5 shows the structure of $(EtBO)_3$. The boron–oxygen ring is planar by virtue of $(p–p)\pi$ boron–oxygen interactions.

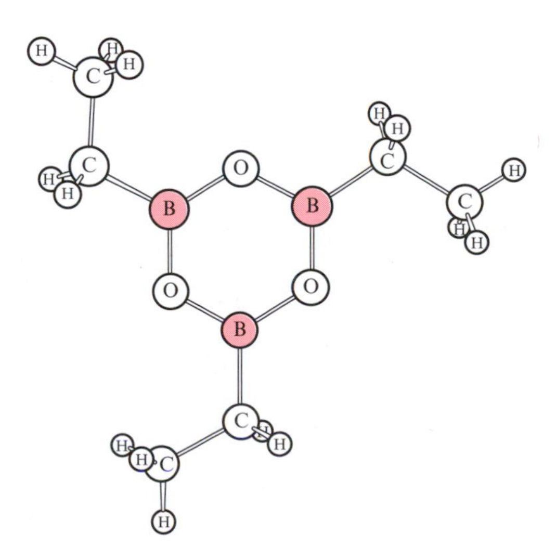

13.5 The structure of the trimer $(EtBO)_3$ as determined by X-ray crystallography. The B_3O_3 ring is planar and each B–O bonding interaction consists of a B–O σ-bond, strengthened with partial π-character.

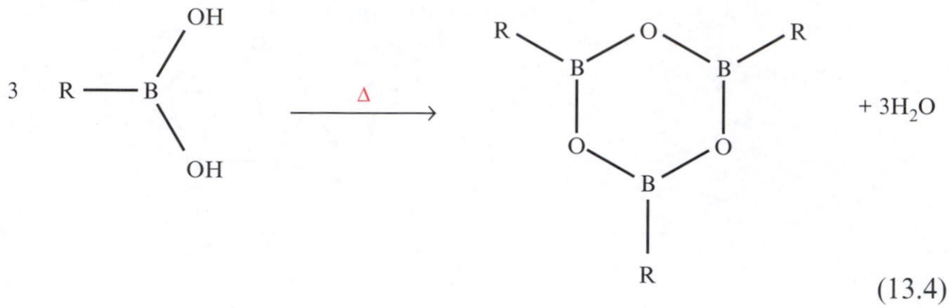

$$3\ R{-}B(OH)_2 \xrightarrow{\Delta} (RBO)_3 + 3H_2O \qquad (13.4)$$

Boron halides

The boron trihalides BF_3, BCl_3, BBr_3 and BI_3 are all monomeric and at room temperature, BF_3 is a colourless gas (bp 172 K), BCl_3 a fuming liquid (bp 285 K), BBr_3 a colourless liquid (mp 227 K, bp 364 K) and BI_3 a colourless solid (mp 323 K). Boron trifluoride may be prepared from B_2O_3 by reaction 13.5, while BCl_3 and BBr_3 are formed by heating boron and dichlorine or dibromine together. Boron triiodide can be prepared from BCl_3 as shown in equation 13.6.

$$B_2O_3 + 3CaF_2 + 3H_2SO_4 \rightarrow 2BF_3 + 3CaSO_4 + 3H_2O \qquad (13.5)$$

$$BCl_3 + 3HI \xrightarrow{\Delta} BI_3 + 3HCl \qquad (13.6)$$

Electron diffraction data for BCl_3: see Section 3.2

Each BX_3 molecule is trigonal planar, consistent with an sp^2 hybridized boron atom. Donation of π-electrons from each halogen atom to the empty $2p$ orbital on the boron atom gives partial B–X π-character and stabilizes the monomer. The importance of B–X $(p–p)\pi$-bonding decreases along the series F > Cl > Br > I; the energy of the $2p$ orbital of boron is poorly matched with that of the $4p$ AO of bromine, and even less well matched with the iodine $5p$ orbital. As the amount of B–X $(p–p)\pi$-bonding *decreases*, the availability of the $2p$ acceptor orbital on the boron centre *increases* and this affects the relative Lewis acid strengths of the trihalides. Consider the forma-

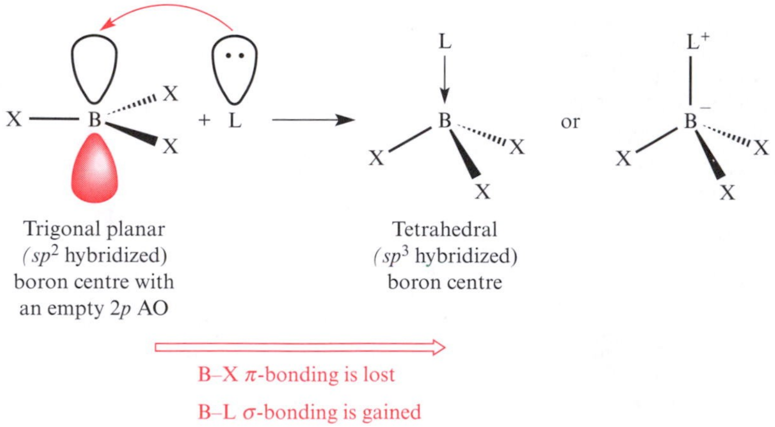

Trigonal planar (sp^2 hybridized) boron centre with an empty $2p$ AO

Tetrahedral (sp^3 hybridized) boron centre

B–X π-bonding is lost

B–L σ-bonding is gained

13.6 When a Lewis base, L, uses a lone pair of electrons to coordinate to the Lewis acid BX_3 (X = F, Cl, Br or I), an adduct is formed. During the reaction, the geometry of the boron centre changes with the loss of B–X $(p–p)$ π-bonding.

tion of a donor–acceptor complex between BX_3 and a Lewis base, L. Figure 13.6 shows the attack by the lone pair of the Lewis base on the boron atom, and the formation of a B–L bond which causes a change from trigonal planar to tetrahedral boron. Concomitant with this change is a *loss* of the bond enthalpy associated with the B–F $(p\text{–}p)\pi$-bonding, but a *gain* of the bond enthalpy due to the formation of the new B–L σ-bond. The thermodynamic stability of the adduct with respect to the separate molecules BX_3 and L depends, in part, upon the difference in these two enthalpy terms. When the Lewis base is pyridine (py), the experimentally determined order of stabilities of the adducts **13.11** is $py{\bullet}BF_3 < py{\bullet}BCl_3 < py{\bullet}BBr_3$, in keeping with the degree of B–X $(p\text{–}p)\pi$-bonding in BX_3 following the order F > Cl > Br.

X = F, Cl or Br

(13.11)

Boron trifluoride plays a part in facilitating some organic transformations by acting as a Lewis acid. It is relatively resistant to hydrolysis, but in an excess of water, reaction 13.7 occurs giving boric acid and aqueous tetrafluoroboric acid.

Lewis acid catalysts: see Chapter 15

$$4BF_3 + 6H_2O \rightarrow \underbrace{3[H_3O]^+ + 3[BF_4]^-}_{\text{aqueous tetrafluoroboric acid}} + B(OH)_3 \qquad (13.7)$$

Pure tetrafluoroboric acid, HBF_4, is not isolable but is commercially available in diethyl ether solution, or as solutions formulated as $[H_3O][BF_4]{\bullet}4H_2O$. The acid is fully ionized in aqueous solution and behaves as a strong 'mineral' acid.§

The halides BCl_3, BBr_3 and BI_3 react rapidly with water (equation 13.8).

$$BX_3 + 3H_2O \rightarrow B(OH)_3 + 3HX \qquad X = Cl, Br, I \qquad (13.8)$$

The reaction of BF_3 with alkali metal fluorides, MF, in a non-aqueous solvent such as HF leads to the formation of the salts $M[BF_4]$. The tetrahedral anion $[BF_4]^-$ can be viewed as a donor–acceptor complex in which the Lewis base is the F^- ion. The B–F bond length increases from 130 pm in BF_3 to 145 pm in $[BF_4]^-$, consistent with the loss of the π-contribution to each B–F bond. All of the B–F bonds are equivalent. The corresponding anions $[BCl_4]^-$, $[BBr_4]^-$ and $[BI_4]^-$ are stabilized only in the presence of large cations such as tetrabutylammonium, $[NBu_4]^+$.

Non-aqueous solvent: see Sections 11.7 and 11.11

Bu = C_4H_9

A second series of boron halides has the formula B_2X_4. B_2Cl_4 is prepared by a radiofrequency discharge of BCl_3 in the presence of mercury, and B_2Br_4 and B_2I_4 can be made similarly. In the solid state, B_2Cl_4 is planar, but in the

§ The $[BF_4]^-$ ion only coordinates (see Chapter 16) very weakly to metal centres and is often used as an 'innocent' anion to precipitate cationic species – similarly $[PF_6]^-$.

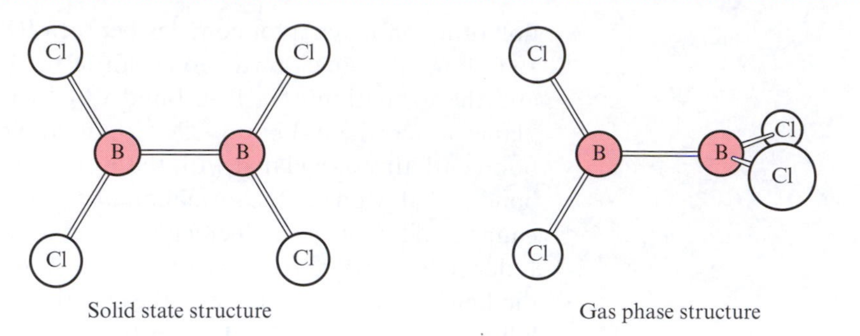

13.7 The solid state structure of B_2Cl_4 is planar, but in the gas phase, a staggered conformation is favoured. In each, the boron atoms are trigonal planar.

gas phase the most stable conformation is a staggered one in which steric interactions between the BCl_2 units are minimized (Figure 13.7). When it reacts with SbF_3, B_2Cl_4 is converted to B_2F_4 which is planar in both the solid state *and* the gas phase. The diboron tetrahalides are inflammable in air. B_2Cl_4 can be reduced to diborane (equation 13.9).

SbF_3 as a fluorinating agent: see Section 13.9

$$B_2Cl_4 \xrightarrow{H_2} B_2H_6 \tag{13.9}$$

13.4 GROUP 13: ALUMINIUM

1	2		13	14	15	16	17	18
H								He
Li	Be		**B**	C	N	O	F	Ne
Na	Mg		**Al**	Si	P	S	Cl	Ar
K	Ca		**Ga**	Ge	As	Se	Br	Kr
Rb	Sr	*d*-block	**In**	Sn	Sb	Te	I	Xe
Cs	Ba		**Tl**	Pb	Bi	Po	At	Rn
Fr	Ra							

Aluminium is the most abundant metal in the Earth's crust and occurs to an extent of 8.1%. The major source is the hydrated oxide *bauxite*, but it is also present as oxides in minerals such as feldspars. Bauxite is refined by the Bayer process to give aluminium oxide Al_2O_3 (alumina), and the metal is then extracted electrolytically. Whereas alumina melts at 2345 K, a mixture of cryolite, $Na_3[AlF_6]$, and alumina melts at 1220 K, making this mixture more convenient to use than pure alumina. Electrolysis of the melt produces aluminium at the cathode (equation 13.10).

In the US and American speaking areas, the element is known as 'aluminum'

At the cathode: $Al^{3+}(l) + 3e^- \rightarrow Al(l)$ (13.10)

Aluminium is a silvery-white, non-magnetic metal (mp 933 K) which is both *ductile* and *malleable*. It has a low density (2.7 g cm^{-3}), a high thermal conductivity and is very resistant to corrosion, being passivated by a coating of oxide. These properties make aluminium industrially important, and its strength can be increased by alloying with copper or magnesium. Thin sheets of aluminium are used as household aluminium foil, confectionery wrappers and emergency (space) blankets.

Passivate/alloy: see Section 11.13

When a metal can be extruded into wires or other shapes without a reduction in strength or the appearance of cracks, the metal is ductile. When a metal can be beaten or rolled into sheets, it is malleable.

Oxides and hydroxides of aluminium

Aluminium oxide can be produced by burning aluminium in O_2 (equation 13.11) or by heating with oxides of metals lower than Al in the electrochemical series (equation 13.12). There are several different structural forms (polymorphs) of Al_2O_3, in addition to a large number of hydrated species.

Electrochemical series: see Table 11.1 and Section 12.6

$$4Al + 3O_2 \xrightarrow{\Delta} 2Al_2O_3 \quad (13.11)$$

$$Fe_2O_3 + 2Al \xrightarrow{\Delta} 2Fe + Al_2O_3 \quad (13.12)$$

Box 13.1 Aluminium oxide and gem stones

The mineral *corundum* is aluminium oxide Al_2O_3; crystals are often barrel-shaped:

The only naturally occurring mineral harder than corundum is diamond. The presence of trace metals leads to the formation of corundum gemstones and these possess characteristic colours: ruby (red), sapphire (blue), oriental amethyst (purple), oriental emerald (green) and oriental topaz (yellow).

The red colour of ruby is due to the presence of chromium(III), while titanium(IV) causes sapphires to be blue, and iron(III) is present in topaz.

Synthetic corundum crystals are made from bauxite by melting it in furnaces. Artificial rubies and sapphires can also be manufactured, and large crystals of these are made by similar methods to corundum. Ruby (and other) crystals are the critical components in many lasers.

Amphoteric: see Section 11.3

The α-form of Al_2O_3, *corundum*, is extremely hard and unreactive and is used industrially as an abrasive. The γ-form (formed by heating α-Al_2O_3 to 2300 K) absorbs water and has amphoteric properties as illustrated in reactions **13.13** and **13.14**.

$$Al_2O_3(s) + 3H_2O(l) + 2[OH]^-(aq) \rightarrow \underset{\mathbf{13.12}}{2[Al(OH)_4]^-} \quad (13.13)$$

$$Al_2O_3(s) + 3H_2O(l) + 6[H_3O]^+(aq) \rightarrow \underset{\mathbf{13.13}}{2[Al(H_2O)_6]^{3+}} \quad (13.14)$$

(13.12)

(13.13)

Ions **13.12** and **13.13** are also formed when aluminium metal reacts with alkali or hot hydrochloric acid, respectively (equations 13.15 and 13.16).

$$2Al(s) + 2KOH(aq) + 6H_2O(l) \rightarrow 2K^+(aq) + 2[Al(OH)_4]^-(aq) + 3H_2(g) \quad (13.15)$$

$$2Al(s) + 6HCl(aq) + 12H_2O(l) \xrightarrow{\Delta} 2[Al(H_2O)_6]^{3+} + 6Cl^-(aq) + 3H_2(g) \quad (13.16)$$

Diagonal relationship: see Section 11.14

The charge density on the aluminium(III) centre is high, making the $[Al(H_2O)_6]^{3+}$ cation acidic (equation 13.17). This behaviour resembles that of beryllium, and exemplifies the periodic *diagonal relationship* between beryllium and aluminium.

$$[Al(H_2O)_6]^{3+}(aq) + H_2O(l) \rightleftharpoons [Al(H_2O)_5(OH)]^{2+}(aq) + [H_3O]^+(aq) \qquad pK_a = 5.0 \quad (13.17)$$

Aluminium hydroxide, $Al(OH)_3$, is a white solid, insoluble in water, and is prepared by the reaction of CO_2 with $[Al(OH)_4]^-$. Reactions 13.18 and 13.19 illustrate the amphoteric behaviour of $Al(OH)_3$ – contrast this with the acidic nature of $B(OH)_3$. One use of $Al(OH)_3$ is as a *mordant* – it absorbs dyes and is used to fix dyes to fabrics.

$$Al(OH)_3(s) + [OH]^-(aq) \rightarrow [Al(OH)_4]^-(aq) \quad (13.18)$$

$$Al(OH)_3(s) + 3[H_3O]^+(aq) \rightarrow [Al(H_2O)_6]^{3+}(aq) \quad (13.19)$$

Aluminium halides

The aluminium trihalides AlX_3 (X = Cl, Br or I) may be prepared by reactions between aluminium and the respective halogen, while reaction 13.20 is a route to AlF_3.

$$Al_2O_3 + 6HF \rightarrow 2AlF_3 + 3H_2O \quad (13.20)$$

Aluminium trifluoride is a colourless solid and sublimes at 1564 K, while $AlCl_3$ is a deliquescent, white solid (sublimation temperature 450 K) which hydrolyses rapidly in moist air releasing HCl. The tribromide (mp 529 K) and triiodide (mp 655 K) are deliquescent solids.

In the solid state, AlF_3 possesses a lattice structure in which each aluminium atom is bonded to six F atoms and each fluorine atom is connected to two Al atoms. An octahedral geometry is also observed in the $[AlF_6]^{3-}$ anion in *cryolite*, $Na_3[AlF_6]$. In crystalline aluminium trichloride, each Al centre is also 6-coordinate but on vaporizing, the lattice collapses to give Al_2Cl_6 molecules (Figure 13.8a). Each aluminium centre is approximately tetrahedral, and the terminal Al–Cl bonds are slightly shorter (206 pm) than the Al–Cl bonds involved in the bridges (220 pm). Although this structure looks reminiscent of that of B_2H_6, the bonding schemes in the two molecules are *not* the same. In B_2H_6, we had to invoke 3c–2e bridge bonds because there were insufficient electrons for localized 2c–2e bonds. In '$AlCl_3$', each chlorine atom has three lone pairs of electrons and, on forming a dimer, each bridging chlorine atom donates a pair of electrons into the empty orbital on an adjacent aluminium atom (Figure 13.8b).

2c–2e = 2-centre 2-electron; 3c–2e = 3-centre 2-electron

At higher temperatures in the gas phase, the dimers dissociate into trigonal planar monomers (equation 13.21). Aluminium tribromide and triiodide are dimeric in the solid and liquid states, and in the gas phase at lower temperatures. Dissociation into monomers occurs at higher temperatures and the associated enthalpy changes are lower than for Al_2Cl_6.

$$Al_2Cl_6(g) \rightleftharpoons 2AlCl_3(g) \quad (13.21)$$

The trihalides are strong Lewis acids because the planar AlX_3 molecule possesses a vacant 3*p* AO on the Al atom. They form adducts such as $Cl_3Al{\bullet}OEt_2$ and $[AlCl_4]^-$ with Lewis bases. Conducting liquids are formed at 298 K when *N*-butylpyridinium chloride **13.14** reacts with $AlCl_3$; the reaction does not yield a simple $[AlCl_4]^-$ salt but instead equilibrium 13.22 is established. These ionic liquids are good solvents for a variety of compounds, although they are very water-sensitive.

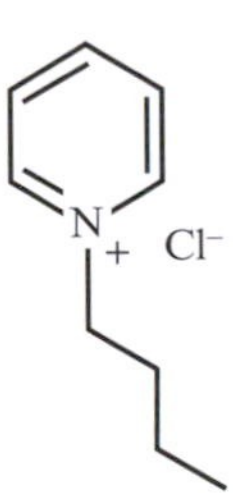

$AlCl_3$ in Friedel–Crafts reactions: see Section 15.6

(13.14)

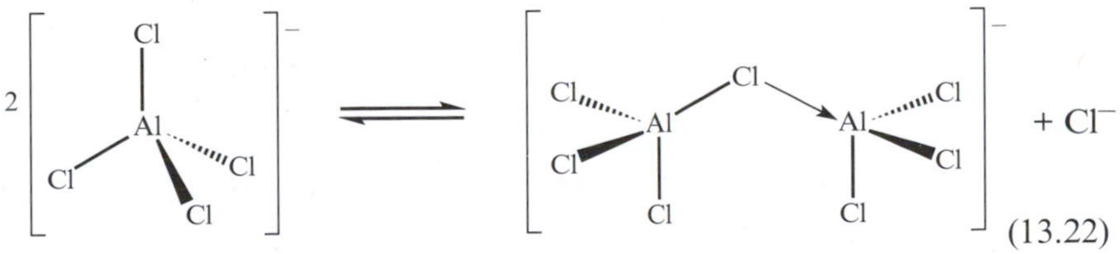

(13.22)

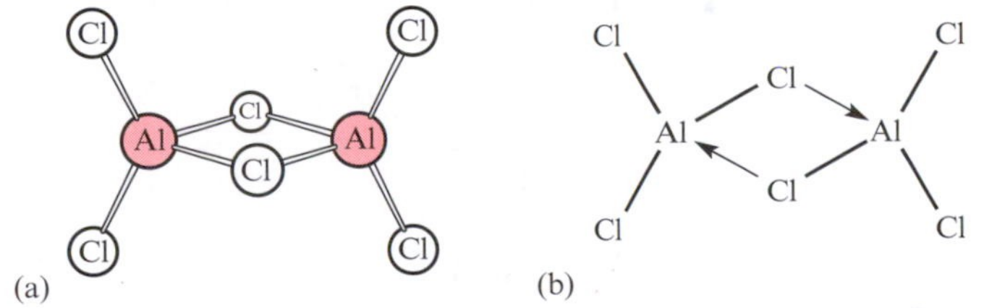

13.8 (a) The gas phase dimeric structure of Al_2Cl_6. (b) The bonding scheme for Al_2Cl_6 involves Cl→Al coordinate bonds within the Al–Cl–Al bridges.

Aluminium sulfate and alums

Aluminium sulfate forms a series of double salts with sulfates of single charged metal and ammonium cations; the general formula of an *alum* is $M_2Al_2(SO_4)_4{\cdot}24H_2O$ where M^+ is an alkali metal cation or $[NH_4]^+$. These include $K_2Al_2(SO_4)_4{\cdot}24H_2O$ (potash alum or just 'alum') which is used in water treatment and as a mordant. Alums are often beautifully crystalline. Members of the series of compounds in which the aluminium is replaced by another M^{3+} ion such as Fe^{3+}, Cr^{3+} or Co^{3+} are also called alums, despite the absence of aluminium – $[NH_4]_2Fe_2(SO_4)_4{\cdot}24H_2O$ (commonly called *ferric or iron(III) alum*) and $KCr(SO_4)_2{\cdot}12H_2O$ (*chrome alum*) are often used in the laboratory for crystal growing experiments.

Aluminium alkyls

Compounds containing Al–C bonds (*organoaluminium compounds*) are important industrial chemicals and are used principally as catalysts. Ziegler–Natta catalysts are mixtures of alkyl aluminium compounds (such as triethylaluminium $AlEt_3$ **13.15**) and titanium(IV) chloride and are used to catalyse the polymerization of alkenes at low temperatures and pressures. The catalysed processes produce stereoregular polymers (isotactic or syndiotactic) which are characteristically high-density materials with crystalline structures.

➤ Addition polymers: see Box 8.13

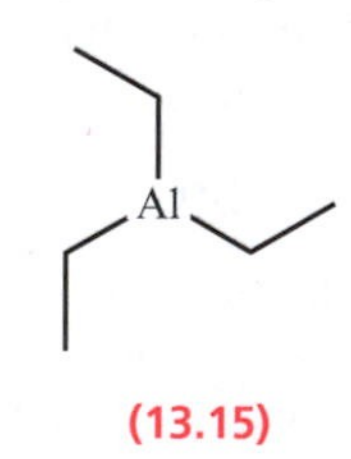

(13.15)

13.5 GROUP 13: GALLIUM, INDIUM AND THALLIUM

1	2		**13**	14	15	16	17	18
H								He
Li	Be		**B**	C	N	O	F	Ne
Na	Mg		**Al**	Si	P	S	Cl	Ar
K	Ca		**Ga**	Ge	As	Se	Br	Kr
Rb	Sr	*d*-block	**In**	Sn	Sb	Te	I	Xe
Cs	Ba		**Tl**	Pb	Bi	Po	At	Rn
Fr	Ra							

Gallium is a silvery-coloured metal (mp 303 K) with an exceptionally long liquid range (303–2477 K). It is a trace element in several naturally occurring minerals, and has important applications in the semiconductor industry and in the manufacture of solid state devices. Indium is a soft, silvery-coloured metal (mp 429 K) and when the pure metal is bent, it emits a high-pitched noise. It occurs naturally in association with zinc-containing minerals. Although gallium and indium compounds are not highly toxic, soluble thallium compounds are extremely harmful – thallium sulfate was used to kill ants and rats, but it is too toxic to other animals (including man) for general use as a method of pest control. Thallium is a soft metal (mp 577 K) and is found as a minor component in some zinc- and lead-containing minerals.

Oxides and hydroxides

Gallium hydroxide, $Ga(OH)_3$, is amphoteric, behaving in a similar manner to $Al(OH)_3$, and forming $[Ga(OH)_4]^-$ in basic solution and $[Ga(H_2O)_6]^{3+}$ in the presence of acid. The $[Ga(H_2O)_6]^{3+}$ ion is a *stronger* acid than $[Al(H_2O)_6]^{3+}$ (compare equations 13.23 and 13.17).

$$[Ga(H_2O)_6]^{3+}(aq) + H_2O(l) \rightleftharpoons [Ga(H_2O)_5(OH)]^{2+}(aq) + [H_3O]^+(aq) \qquad pK_a = 2.6 \qquad (13.23)$$

Indium and thallium(III) oxides and hydroxides are basic (typical of a metal). Thallium(III) oxide loses O_2 on heating (equation 13.24), thallium(III) being reduced to thallium(I).

$$Tl_2O_3(s) \xrightarrow{370\ K} Tl_2O(s) + O_2(g) \qquad (13.24)$$

Halides

Gallium and indium trifluorides are similar to AlF_3 and have high melting points. GaF_3 and InF_3 are prepared by the thermal decomposition of $[NH_4]_3[MF_6]$ (M = Ga or In). $GaCl_3$, $GaBr_3$ and GaI_3 and their indium analogues are formed by reactions of the appropriate metal and halogen. Each gallium compound is a low-melting, deliquescent solid at room temperature; in the solid and liquid states, dimers exist and in the gas phase, dissociation to monomers occurs at higher temperature.

The thallium(III) halides are less stable than those of the earlier group 13 elements – thallium(III) chloride and bromide are thermally unstable with respect to conversion to the respective thallium(I) compounds (equation 13.25).

$$TlBr_3 \rightarrow TlBr + Br_2 \qquad (13.25)$$

In the case of TlI_3, the formula is deceptive – it is *not* a thallium(III) compound but contains thallium(I) and the $[I_3]^-$ anion **13.16**. In the presence of a large excess of iodide ion, the thallium(III) anion $[TlI_4]^-$ is formed.

$Tl^+ [I—I—I]^-$

(13.16)

The Lewis acidity of the MX_3 molecules permits the formation of the ions $[MCl_6]^{3-}$, $[MBr_6]^{3-}$, $[MCl_5]^{2-}$, $[MCl_4]^-$ and $[MBr_4]^-$ (M = Ga or In) and $[TlCl_5]^{2-}$. Reactions occur with neutral Lewis bases, L, to yield adducts of the type $L{\bullet}GaX_3$ or $L_3{\bullet}InX_3$.

See problem 13.5

When $GaCl_3$ is heated with gallium metal, the product appears to be '$GaCl_2$'. However, crystallographic and magnetic data show that it should be formulated as $Ga^I[Ga^{III}Cl_4]$, i.e. a mixed gallium(I)–gallium(III) compound.

13.6 GROUP 14: CARBON AND SILICON

1	2		13	**14**	15	16	17	18
H								He
Li	Be		B	**C**	N	O	F	Ne
Na	Mg		Al	**Si**	P	S	Cl	Ar
K	Ca		Ga	**Ge**	As	Se	Br	Kr
Rb	Sr	*d*-block	In	**Sn**	Sb	Te	I	Xe
Cs	Ba		Tl	**Pb**	Bi	Po	At	Rn
Fr	Ra							

For further chemistry of carbon: see Chapters 8, 14, 15 and 17

The allotropes of carbon were described in Chapter 7. Silicon does not occur naturally in the elemental state, but is still the second most abundant element on Earth – silicon-based minerals include sand, quartz, rock crystal, flint and agate. Silicon may be produced by reducing silica (SiO_2) in an electric furnace, and the Czochralski process (Figure 13.9) is used to draw single crystals of silicon from melts of the element for use in solid state devices or the semiconductor industry.

Oxides

Carbon cycle: see Box 8.1

Carbon dioxide is a minor component of the atmosphere of Earth (Figure 11.1). It is a product of respiration, but is removed from the atmosphere during photosynthesis. When carbon is burned in an excess of O_2, CO_2 forms but when less O_2 is present, carbon monoxide can also be produced

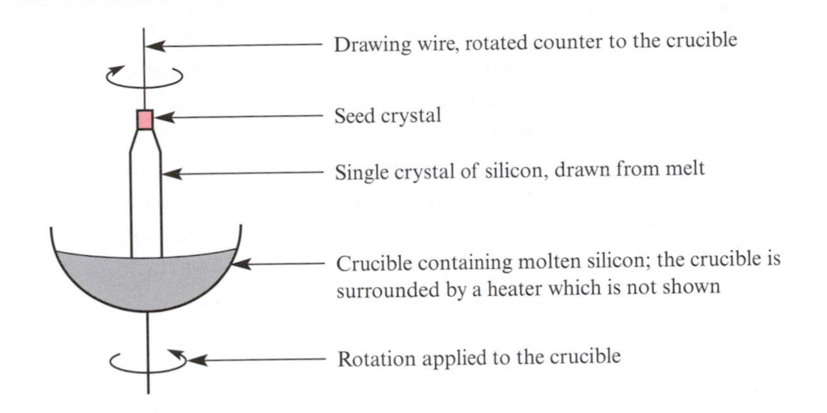

13.9 In the Czochralski process, a seed crystal of silicon is first lowered into molten silicon. It is withdrawn in a specifically controlled manner, and a single crystal of silicon grows. The same technique is used to grow other single crystals, such as gallium arsenide for semiconductor use.

(equations 13.26 and 13.27). Carbon monoxide and carbon dioxide are also produced in the water–gas shift reaction (Section 11.2) and we have already seen the importance of carbon as a reducing agent for metal oxides (Section 12.6). Carbon monoxide is also used industrially to reduce metal and non-metal oxides and is oxidized to CO_2 during the process.

$$\textit{With an excess of } O_2\text{:} \quad C(s) + O_2(g) \xrightarrow{\Delta} CO_2(g) \qquad (13.26)$$

$$\textit{With limited } O_2\text{:} \quad 2C(s) + O_2(g) \xrightarrow{\Delta} 2CO(g) \qquad (13.27)$$

Both CO (bp 82 K) and CO_2 are colourless, odourless gases at room temperature. Solid CO_2 (known as 'dry ice') sublimes at 195 K but may be kept in insulated containers in the laboratory.

Carbon monoxide is almost insoluble in water; it is a very poisonous gas, and its toxicity arises because, acting as a Lewis base, it binds more strongly

Box 13.2 The 'greenhouse effect'

The level of carbon dioxide in the Earth's atmosphere is steadily rising due to the combustion of fossil fuels in industry and the burning of forests (e.g. tropical rain.forests) when land is being cleared. Not only does the latter produce *more* CO_2, but it removes from the Earth vegetation that would otherwise be undergoing photosynthesis and removing CO_2 *from* the atmosphere.

It is possible that the increased amount of CO_2 *may* lead to a rise in the atmospheric temperature. Short-wave radiation from the Sun heats the Earth, but energy that is radiated back from the Earth into the atmosphere has a longer wavelength. This is coincident with the range of the electromagnetic spectrum in which CO_2 absorbs, and the the result would be an increase in the temperature of the atmosphere around the Earth. The phenomenon is known as the 'greenhouse effect'. Consequences of a rise in the Earth's temperature include partial melting of the polar ice caps and of glaciers, and general changes in climate.

Other 'greenhouse gases' include CH_4 (see Box 8.6) and N_2O.

Haemoglobin: see Box 4.4

Metal carbonyls: see Section 16.14

to the iron centres in haemoglobin than does O_2 and prevents the uptake and transport of O_2 in the bloodstream. The ability of CO to coordinate to low oxidation state *d*-block metals leads to the formation of a wide range of metal carbonyl compounds.

The reaction of CO and Cl_2 in the presence of an activated carbon catalyst produces carbonyl dichloride (or phosgene), $COCl_2$ **13.17**. This is a very toxic, colourless gas (bp 281 K) with a choking smell, and was used in the First World War in chemical warfare.

(13.17)

When CO_2 dissolves in water, equilibrium 13.28, which lies to the left-hand side, is established. When we refer to 'carbonic acid', we actually mean a solution of CO_2 in water – pure carbonic acid, H_2CO_3, is not isolable.

$$CO_2(g) + H_2O(l) \rightleftharpoons H_2CO_3(aq) \qquad K \approx 1.7 \times 10^{-3} \tag{13.28}$$

Box 13.3 Low-temperature baths

In the laboratory, it is often necessary to carry out reactions at constant, low temperatures. Dry ice and liquid dinitrogen may be used in conjunction with organic solvents to produce baths covering a range of temperatures. Dry ice (solid CO_2) sublimes at 195 K, and liquid N_2 boils at 77 K.

In the liquid N_2 baths, the cold liquid is poured from a Dewar flask into a solvent that is constantly stirred. As the two are mixed, a *slush* is formed, the temperature of which can be maintained by occasionally adding more liquid N_2.

In a bath that is cooled by dry ice, *small* pieces of solid CO_2 are carefully added to the solvent. Initially, the solid carbon dioxide sublimes, but eventually the temperature of the bath decreases to a point at which solid dry ice is present. The low temperature of the bath is kept constant by adding small pieces of dry ice at intervals.

Typical ranges of temperatures that may be achieved can be seen from the selected bath systems listed below.

Bath contents	*Temperature / K*	*Bath contents*	*Temperature / K*
Liquid N_2 + cyclohexane	279	Dry ice + ethane-1,2-diol	258
Liquid N_2 + cycloheptane	261	Dry ice + heptan-3-one	235
Liquid N_2 + acetonitrile	232	Dry ice + acetonitrile	231
Liquid N_2 + octane	217	Dry ice + ethanol	201
Liquid N_2 + heptane	182	Dry ice + acetone	195
Liquid N_2 + hexa-1,5-diene	132	Dry ice + diethyl ether	173

[ethylene glycol = ethane-1,2-diol]

Carbonic acid is a weak acid, although carbonate and hydrogencarbonate salts are commonly encountered in the laboratory. The pK_a values for equilibria 13.29 and 13.30 indicate that the $[HCO_3]^-$ ion is the dominant species in aqueous solutions of CO_2 and that the $[CO_3]^{2-}$ ion will only form in basic solution. The *rate* of the forward reaction 13.28 is very slow, and fresh solutions contain significant amounts of $CO_2(aq)$ – you should consider how this affects equilibria 13.29 and 13.30.

$$H_2CO_3(aq) + H_2O(l) \rightleftharpoons [H_3O]^+(aq) + [HCO_3]^-(aq) \quad pK_a = 6.4 \quad (13.29)$$

$$[HCO_3]^-(aq) + H_2O(l) \rightleftharpoons [H_3O]^+(aq) + [CO_3]^{2-}(aq) \quad pK_a = 10.3 \quad (13.30)$$

There are remarkable differences between the oxides of carbon and silicon. Silicon monoxide can be prepared by reducing SiO_2 but it is the dioxide that is the more important. Silica, SiO_2, is a crystalline solid with an extended lattice structure. There are many different structural forms of SiO_2, and the most thermodynamically stable is α-quartz (equation 13.31).

$$\alpha\text{-quartz} \underset{}{\overset{845\text{ K}}{\rightleftharpoons}} \beta\text{-quartz} \overset{1145\text{ K}}{\rightleftharpoons} \beta\text{-tridymite} \overset{1745\text{ K}}{\rightleftharpoons} \beta\text{-cristobalite} \quad (13.31)$$

α-Quartz is present in granite and sandstone and occurs as rock crystal and as a range of coloured, crystalline forms such as purple amethyst. The structural building-block is an SiO_4 unit **13.18** which is often represented as a tetrahedron **13.19**. Such units are interconnected through the oxygen atoms to form a three-dimensional lattice, and the zeolite structure described later (Figure 13.13b) is represented in the form of connected tetrahedra.

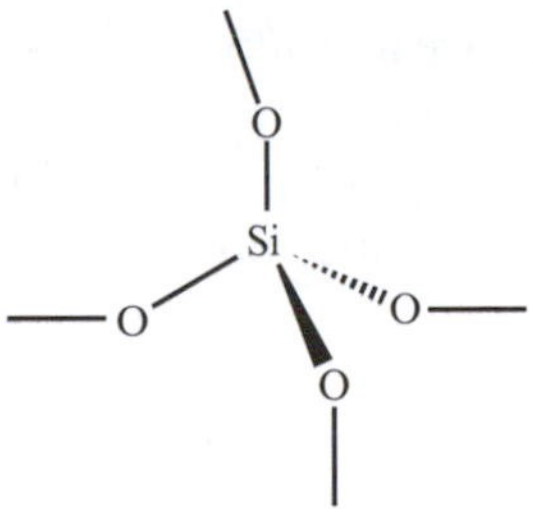

(13.18)

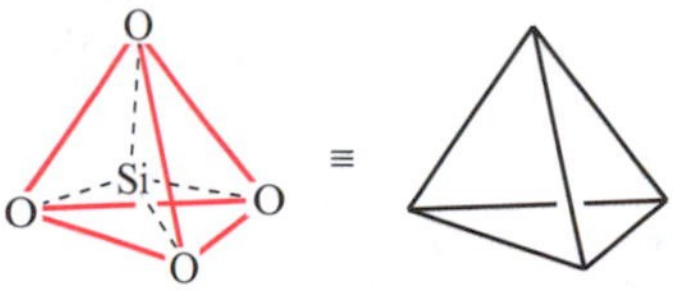

(13.19)

In the various structural modifications of silica, the SiO_4 units are connected in different ways and the simplest to visualize is *cristobalite*. Its structure is related to the diamond-type lattice (Figure 7.22) of elemental silicon but with an oxygen atom bonded between each pair of adjacent silicon atoms giving Si–O–Si bridges.

The critical difference between carbon–oxygen and silicon–oxygen bond formation lies in their relative bond enthalpies. The bond dissociation enthalpy of the C≡O triple bond in CO is 1076 kJ mol^{-1}, and in CO_2, each C=O bond has a dissociation enthalpy of 532 kJ mol^{-1}. In both molecules, the π-contributions to the carbon–oxygen bond are significant. However, if we were to assess the difference in bond enthalpies between two Si=O bonds in a hypothetical SiO_2 *molecule*, O=Si=O, and four Si–O bonds in an extended lattice structure, then the latter are energetically favourable. The opposite is true for carbon.

Etching of silica glass: see Section 11.11

Silica is resistant to attack by acids, with the exception of HF.

$$SiO_2 + 2Na_2CO_3 \xrightarrow{1770\ K} \underset{\text{sodium silicate}}{Na_4[SiO_4]} + 2CO_2 \quad (13.32)$$

Silica reacts slowly with molten Na_2CO_3 (equation 13.32), and after purification, sodium silicate (commonly known as 'water glass') is obtained; it is used commercially in detergents, where it maintains a constant pH and can degrade fats by hydrolysis. Silicon dioxide (and glass) react with concentrated aqueous MOH (M = Na or K) and rapidly with molten hydroxides. *Silica gel* is an amorphous form of SiO_2, produced by treating aqueous sodium silicate with acid. After careful dehydration, silica gel is used as a drying agent; small sachets of silica gel are placed in the cases of optical equipment (e.g. cameras) to keep them moisture-free – the structure of silica gel is an open network into which water penetrates and is trapped. Self-indicating silica gel is doped with cobalt(II) chloride which turns from blue to pink as water is absorbed. Silica gel is used as a stationary phase in chromatography and as a solid state (*heterogeneous*) catalyst.

Sodium silicate, $Na_4[SiO_4]$, is an example of a large family of silicates, the simplest of which (the *orthosilicates*) contain tetrahedral $[SiO_4]^{4-}$ ions. Members of another group of silicates contain *cyclic silicate anions*, such as $[Si_3O_9]^{6-}$ **13.20** and $[Si_6O_{18}]^{12-}$ **13.21**, which occurs in the mineral *beryl*, $Be_3Al_2Si_6O_{18}$. Three-dimensional representations of these silicate anions are shown in Figure 13.10.

The *pyroxenes* are a group of rock-forming silicate minerals in which the silicon and oxygen atoms are arranged in chain structures with a repeat unit of formula $\{Si_2O_6\}_n$ **13.22**. In the mineral *diopside* $CaMgSi_2O_6$, the silicate chains lie parallel to one another and the cations lie between the chains. In

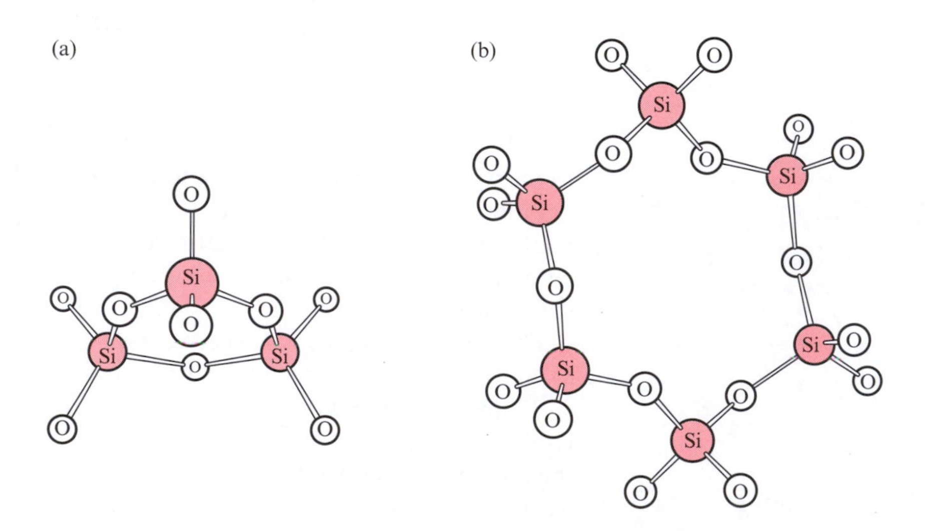

13.10 Three-dimensional representations of the silicate anions $[Si_3O_9]^{6-}$ **13.20** and $[Si_6O_{18}]^{12-}$ **13.21**.

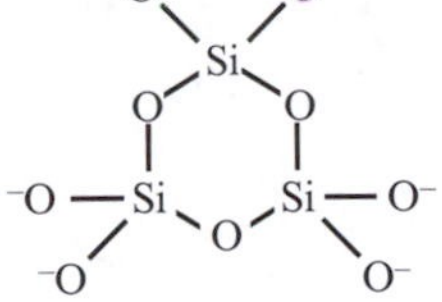

(13.20)

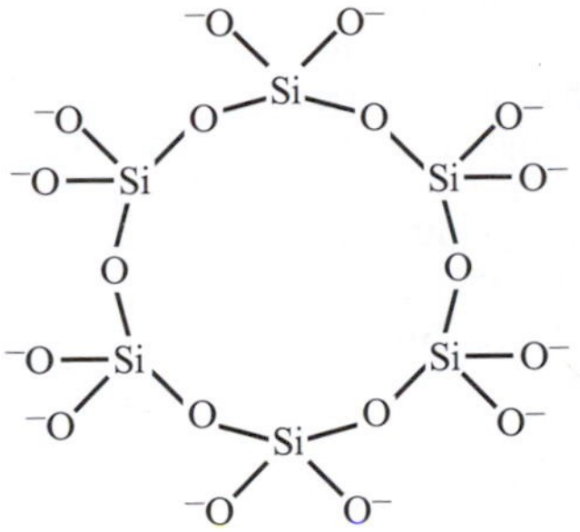

(13.21)

the *amphibole* family of minerals (which includes the fibrous *asbestos*), alternate pairs of adjacent silicate chains are linked together by Si–O–Si bridges and the basic building-block is the $[Si_4O_{11}]^{6-}$ unit.

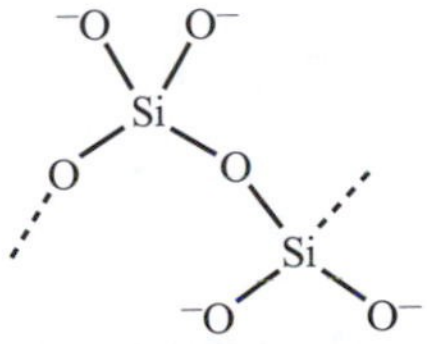

(13.22)

Silicate minerals with sheet structures (the *phyllosilicates*) are produced when SiO_4 units are linked together by Si–O–Si bridges to give a layer of fused Si_4O_4 rings (Figure 13.11). These silicates tend to be flaky, and an example is *talc,* $Mg_3(OH)_2Si_4O_{10}$. *Micas* constitute a related mineral group in which one in four of the silicon atoms is replaced by aluminium; *muscovite mica* has the formula $KAl_2(OH)_2[AlSi_3O_{10}]$. The replacement of silicon by aluminium to give *aluminosilicates* also occurs in some discrete anions and two related species are shown in Figure 13.12. Since silicon and aluminium atoms are about the same size, the substitution of one element for the other results in little perturbation of the structural unit – the four protons in $[H_4Al_4Si_4O_{20}]^{4-}$ compensate for the charge difference between the silicon and aluminium centres.

Box 13.4 Asbestos: fire protection, but a health hazard

In the past, fibrous asbestos found extensive use in the manufacture of heat- and fire-resistant material, and was routinely used in chemical laboratories and in building materials. However, inhalation of asbestos fibres can result in chronic lung disease. Its uses have been drastically curtailed, and it is being removed from many older buildings in which asbestos dust continues to pose a health risk.

13.11 Part of the layer structure of a sheet silicate. Note that the diagram does *not* show a single repeat unit of the layer. Each silicon atom is tetrahedrally sited, and is attached to one terminal and three bridging oxygen atoms. In the three-dimensional structure, the environment at each O atom is bent.

Members of another group of silicate minerals including quartz possess three-dimensional lattices in which each silicon is connected to another silicon atom by an Si–O–Si bridge. The Si : O ratio in these compounds is 1 : 2.

Zeolites represent a large group of natural and synthetic compounds based on aluminosilicate lattices and have a number of unique and useful properties. They contain structured channels (Figure 13.13) of various sizes which small molecules can enter. As different zeolites contain different sized

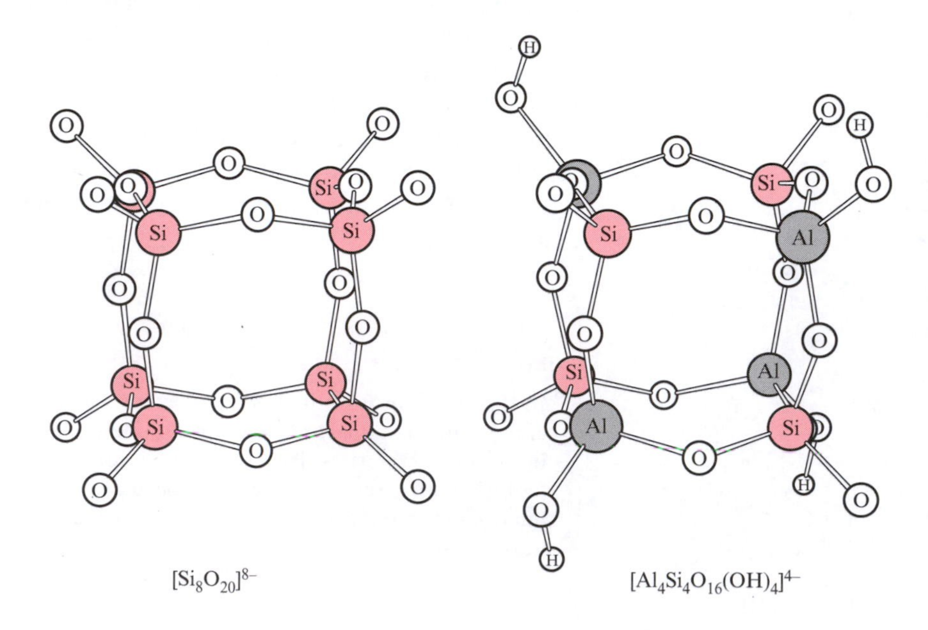

13.12 The structures of the silicate anion $[Si_8O_{20}]^{8-}$ and of the aluminasilicate anion $[Al_4Si_4O_{16}(OH)_4]^{4-}$. The $[Si_8O_{20}]^{8-}$ anion contains a cubic arrangement of silicon atoms supported by bridging oxygen atoms; each Si atom also has a terminal oxygen atom attached to it. Replacement of a tetrahedral silicon centre by aluminium causes little structural perturbation to the structure.

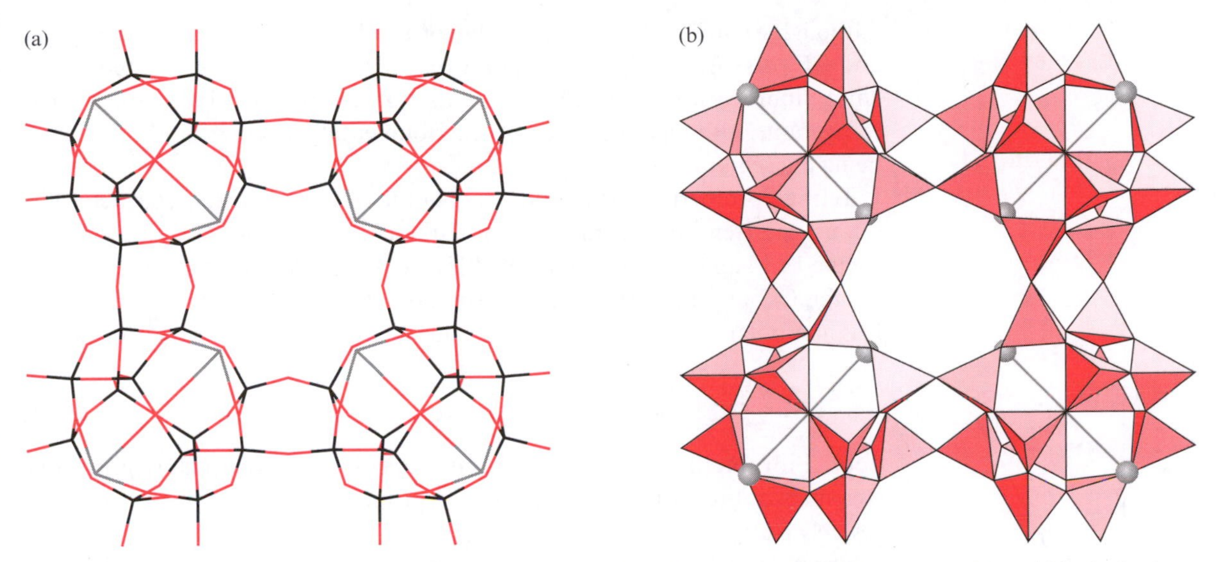

13.13 Two representations of part of the structure of the zeolite ZK5 showing one of the large cavities within the lattice. Diagram (a) shows a conventional 'chemical' representation of the structure whilst diagram (b) shows a representation in which the SiO_4 or $AlO_3(OH)$ units are represented by tetrahedra (see structures **13.18** and **13.19**). The small spheres represent Na^+ ions in the lattice.

channels, it follows that different sizes of molecules can enter them. This observation is the basis for the use of zeolites as catalysts in many industrial processes. An example is the zeolite ZSM-5 with a composition $Na_n[Al_nSi_{96-n}O_{192}]\cdot{\approx}16H_2O$ ($n < 27$) which is used as a catalyst in benzene alkylation, xylene isomerization and methanol to hydrocarbon (for motor fuels) conversions. Another important property of zeolites is related to the OH groups on the aluminium centres; these are strongly acidic and the protons may be replaced by a variety of other cations. In practice, this means that the zeolites are excellent *ion-exchange* materials and find widespread uses in water purification, washing powder manufacture and other applications. In the laboratory, zeolites appear as drying agents in the form of *molecular sieves*.

Benzene and xylene: see Chapter 15

Halides

In Chapter 8 we looked at the halogenation of alkanes, which ultimately gives compounds of the type C_nX_{2n+2} where X = F, Cl, Br or I. In this section we concentrate on CX_4 molecules and on their silicon analogues.

Tetrafluoromethane (or, carbon tetrafluoride) CF_4 is a stable gas (bp 145 K) which may be prepared from silicon carbide (SiC) and F_2, or by reactions 13.33. CF_4 is extremely resistant to chemical attack by acids, alkalis, and oxidizing and reducing agents.

$$CO \xrightarrow{SF_4} CF_4 \xleftarrow{SF_4} CO_2 \qquad (13.33)$$

$\chi^P(C) = 2.6$; $\chi^P(F) = 4.0$

The ozone hole: see Box 8.9

Although each C–F *bond* is polar, the tetrahedral CF_4 *molecule* is non-polar and CF_4 dissolves only in non-polar organic solvents. Fluorocarbons and chlorofluorocarbons (CFCs) are used as high-temperature lubricants and as refrigerants, although they are not 'environmentally friendly' and their use is being phased out.

Tetrachloromethane is a colourless liquid at 298 K with a characteristic odour; the trend in melting points of the carbon halides CX_4 is shown in Figure 13.14 and as expected, the melting point rises with increasing molecular weight. CCl_4 is prepared by the chlorination of methane or by reaction 13.34. It is non-polar, immiscible with water and stable with respect to hydrolysis.

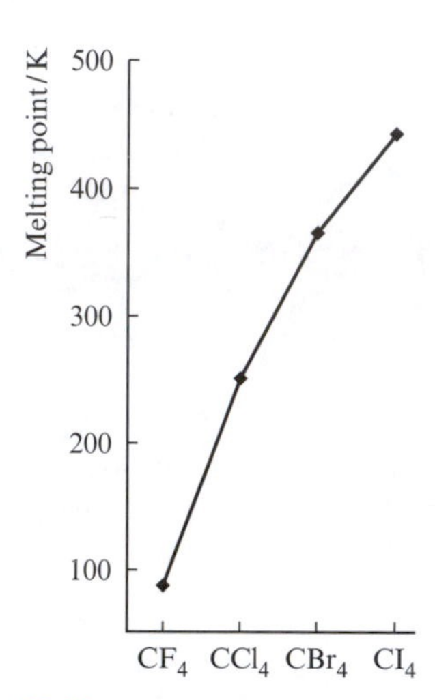

13.14 As the molecular weight of the tetrahalomethane molecules increases, the melting point increases.

$$CS_2 + 3Cl_2 \xrightarrow{\text{Fe catalyst}} CCl_4 + S_2Cl_2 \quad (13.34)$$

Tetrachloromethane was widely used as a solvent in the past (e.g. for dry cleaning) but its potential carcinogenic properties mean that its use is now severely restricted.

The pale yellow compound CBr_4 is a solid at 298 K, and is insoluble in water but soluble in non-polar organic solvents. It is less stable with respect to decomposition than CF_4 and CCl_4, and is usually prepared from CCl_4 (reaction 13.35), or from CH_4 and Br_2.

$$3CCl_4 + 2Al_2Br_6 \xrightarrow{373\text{ K}} 3CBr_4 + 2Al_2Cl_6 \quad (13.35)$$

The C–X bond dissociation enthalpy decreases in the order $D(\text{C–F}) > D(\text{C–Cl}) > D(\text{C–Br}) > D(\text{C–I})$, and CI_4 is even less stable than CBr_4. Red crystals of CI_4 may be prepared from CCl_4 by reaction 13.36 in which $AlCl_3$ functions as a Lewis acid catalyst, but the compound decomposes in the presence of light or when heated (equation 13.37).

$$CCl_4 + 4C_2H_5I \xrightarrow{AlCl_3} CI_4 + 4C_2H_5Cl \quad (13.36)$$

$$2CI_4 \xrightarrow{h\nu \text{ or } \Delta} C_2I_4 + 2I_2 \quad (13.37)$$

The molecular silicon tetrahalides SiX_4 (X = F, Cl, Br or I) can be prepared from their constituent elements (equation 13.38), but whereas the reaction between silicon and F_2 is spontaneous, heat is needed for the other three reactions.

$$Si + 2X_2 \rightarrow SiX_4 \qquad X = \text{F, Cl, Br or I} \quad (13.38)$$

The halides are colourless, and at 298 K, SiF_4 is a gas, $SiCl_4$ and $SiBr_4$ are fuming liquids and SiI_4 is a solid. The fuming of the liquids is due to rapid hydrolysis by moisture in the air and the formation of HCl or HBr (equation 13.39).

$$SiX_4 + 2H_2O \rightarrow SiO_2 + 4HX \qquad X = \text{Cl or Br} \quad (13.39)$$

The ease of hydrolysis of the silicon halides is in marked contrast to the behaviour of the carbon analogues and is due to the availability of low-lying atomic orbitals on silicon, not present in carbon, which facilitate attack by the lone pair of H_2O. SiF_4 undergoes a more complex hydrolysis reaction than $SiCl_4$ and $SiBr_4$, and forms the hexafluorosilicate ion in addition to silica (equation 13.40). Aqueous fluorosilicic acid is a strong acid and is commercially available as an aqueous solution – pure H_2SiF_6 is not isolable.

$$2SiF_4 + 4H_2O \rightarrow SiO_2 + 2HF + \underbrace{2[H_3O]^+ + [SiF_6]^{2-}}_{\text{fluorosilicic acid}} \qquad (13.40)$$

Silicon tetrachloride and related organosilicon halides such as $RSiCl_3$ (R = alkyl) are used widely as synthetic reagents, e.g. in the formation of silicon alkoxides $Si(OR)_4$ (R = alkyl) and siloxanes (equation 13.41). In the siloxane cage produced in this reaction, each silicon atom is tetrahedral, being

Box 13.5 Silicones in personal care products

The manufacture of personal care products (shampoos, conditioners, toothpastes, anti-perspirants, cosmetics, etc.) is a large component of the chemical industry, and *silicone* products play an important role as the ingredients in shampoos and conditioners which improve the softness and silkiness of hair, making it more manageable and easy to style. Silicones are polymeric organosilicon compounds containing Si–O–Si bridges and have many uses other than in the personal care industry – e.g. silicone grease, sealants, varnishes, synthetic rubbers and hydraulic fluids. In the personal product industry in particular, there are manufacturing challenges to overcome – silicones tend to be viscous oils and are immiscible with water. Silicones may be produced by one company for use by another in the manufacture of shampoos and other products – the second manufacturer may require that the silicone product is in a form which is easy to combine with the basic shampoo which is water-based. One way of overcoming the immiscibility problem is for the silicone to be dispersed in water to give an emulsion. In this form, the silicone and water layers do not separate out from each other, and the silicone can be readily incorporated into a shampoo or conditioner.

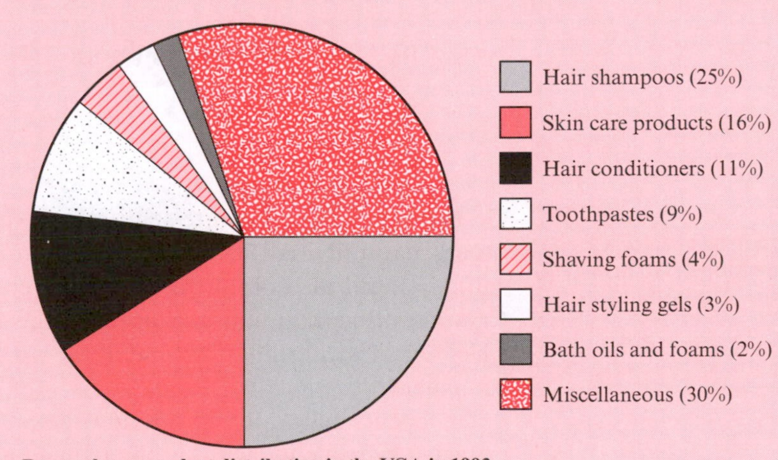

Personal care product distribution in the USA in 1993

(Data from *Chemical & Engineering News*, (1995) p. 34.)

bonded to one terminal R group and three Si–O–Si bridges. Compare this structure with that of the $[Si_8O_{20}]^{4-}$ anion in Figure 13.12.

$$8RSiCl_3 + 24H_2O \xrightarrow{-24HCl} 8RSi(OH)_3 \xrightarrow{\text{dehydrate}} \qquad (13.41)$$

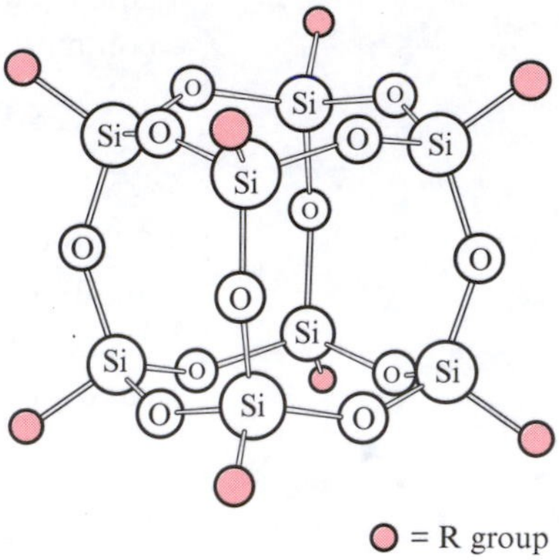

○ = R group

13.7 GROUP 14: GERMANIUM, TIN AND LEAD

1	2		13	**14**	15	16	17	18
H								He
Li	Be		B	**C**	N	O	F	Ne
Na	Mg		Al	**Si**	P	S	Cl	Ar
K	Ca		Ga	**Ge**	As	Se	Br	Kr
Rb	Sr	*d*-block	In	**Sn**	Sb	Te	I	Xe
Cs	Ba		Tl	**Pb**	Bi	Po	At	Rn
Fr	Ra							

Elemental structures: see Sections 7.10 and 7.11

Germanium is a grey-white semi-metal which does not tarnish in air. It occurs naturally as the mineral *argyrodite* (a mixed germanium–silver sulfide) and is also obtained from smelters used in processing zinc ores. Crystals of germanium are used in the semiconductor industry, and when doped with gallium or arsenic, act as transistors for use in electronic equipment.

Tin occurs naturally as *cassiterite* (SnO_2) from which the metal is extracted by reduction with carbon (equation 13.42).

$$SnO_2 + C \xrightarrow{\Delta} Sn + CO_2 \qquad (13.42)$$

Box 13.6 Snails get heavy!

Since Roman times, lead mining has been carried out in parts of Britain. This brings lead to the surface – spoil heaps containing lead-bearing materials are open and may be a health risk. But sometimes, evolution can provide startling solutions. Researchers have found that snails from populations in areas where lead mining has been carried out for many centuries have been able to concentrate lead into their shells where it cannot harm them. Snails from other populations are not able to do this and suffer lead accumulation in body tissue.

One use of tin is in tin-plating metal cans, but because tin is so soft, many applications require that it is alloyed with other elements; alloys include pewter, soldering metal, bronze and die-casting alloy. High-quality window glass is generally manufactured by the Pilkington process which involves floating molten glass on molten tin to generate a flat surface.

The mineral *galena* (PbS) is a major source of lead, and the metal is extracted by converting the sulfide to the oxide, followed by reduction (equation 13.43).

$$PbS \xrightarrow{O_2,\ \Delta} PbO \xrightarrow{\text{C or CO},\ \Delta} Pb \qquad (13.43)$$

Lead is a very soft, malleable and ductile, blue-white metal. Its resistance to corrosion leads to its use for storing corrosive liquids such as sulfuric acid. Major uses of lead (Figure 13.15) include storage batteries, radiation protection (sheet lead) and plumbing materials, although applications for household pipes are now limited owing to the extreme toxicity of this metal.

Oxides and hydroxides

Germanium(IV) oxide can be prepared by hydrolysing $GeCl_4$ and is stable in hot water. It resembles silica, possessing several crystalline forms with simi-

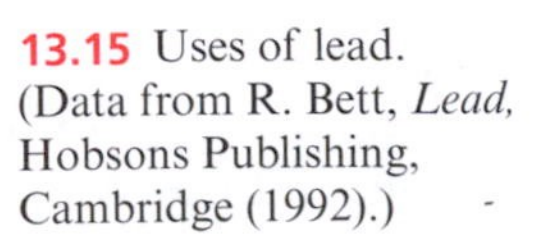

13.15 Uses of lead. (Data from R. Bett, *Lead,* Hobsons Publishing, Cambridge (1992).)

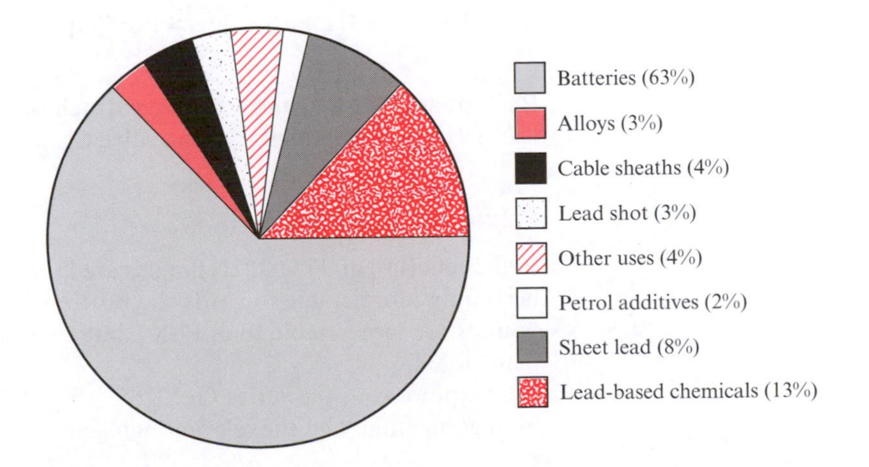

lar lattice types to those of SiO_2; the commercially available form has a β-quartz lattice. When heated with germanium, GeO_2 is reduced to GeO which is stable in air and water at 298 K, but at high temperatures, GeO is oxidized in air back to GeO_2.

There are several oxoanions of germanium(IV) (*germanates*) and reaction 13.44 shows the preparation of the lithium salt of $[Ge_5O_{12}]^{4-}$, and other anions include $[GeO_3]^{2-}$ and $[GeO_4]^{4-}$. Structurally, germanates do *not* resemble silicates.

$$5GeO_2 + 2Li_2O \xrightarrow{\text{molten}} Li_4[Ge_5O_{12}] \qquad (13.44)$$

Rutile lattice: see Section 6.12

Crystalline SnO_2 has a rutile-type lattice; the 6-coordinate tin centres in the lattice illustrate a geometry that is quite common for tin(IV). Tin(IV) oxide is amphoteric and dissolves in acidic and basic solutions. In aqueous hydrochloric acid, the solution species formed is the octahedral $[SnCl_6]^{2-}$ anion (equation 13.45); in contrast, lead(IV) oxide does *not* dissolve in acids. In strongly alkaline solution, $[Sn(OH)_6]^{2-}$ forms (equation 13.46), but in less basic media, $[SnO_3]^{2-}$ is present.

$$SnO_2(s) + 6HCl(aq) \rightarrow 2[H_3O]^+(aq) + [SnCl_6]^{2-}(aq) \qquad (13.45)$$

$$SnO_2(s) + 2KOH(aq) + 2H_2O(l) \rightarrow 2K^+(aq) + [Sn(OH)_6]^{2-}(aq) \qquad (13.46)$$

Tin(II) oxide is prepared by the reaction of any tin(II) salt with dilute alkali. The oxide is amphoteric, dissolving in both acids and alkalis. In strongly alkaline solution, the anion $[Sn(OH)_3]^-$ is formed but more complex species may exist, depending upon the pH.

The red form of lead(II) oxide is known as *litharge*, but above 760 K, a yellow form exists. Lead(II) oxide has wide application in lead-acid batteries. It is prepared by heating lead and O_2, and further reaction in air at 730 K yields Pb_3O_4 (*red lead*). This diamagnetic oxide is a mixed lead(II)–lead(IV) compound ($2PbO{\bullet}PbO_2$), and is used as a pigment and a corrosion-resistant coating for steel and iron. Lead oxides have important uses in the manufacture of 'lead crystal' glass. Strong oxidizing agents are needed to prepare lead(IV) oxide because PbO_2 is readily reduced as the standard reduction potential for half-equation 13.47 indicates.

$$PbO_2(s) + 4H^+(aq) + 2e^- \rightleftharpoons Pb^{2+}(aq) + 2H_2O(l) \qquad E^o = +1.46\ V \qquad (13.47)$$

PbO_2 reacts with concentrated hydrochloric acid to give solutions of lead(IV) halo-compounds which evolve Cl_2.

Halides

Both metal(IV) and metal(II) halides are known, with the +2 oxidation state becoming more stable towards the bottom of group 14. Thus PbX_2 compounds are more stable than PbX_4, but GeX_4 compounds are more stable than GeX_2.

Compounds in the series GeX_4 (X = F, Cl, Br or I) can all be prepared from germanium and the relevant halogen. Under ambient conditions, GeF_4

Box 13.7 Lead storage batteries

A lead-acid battery in a motor vehicle consists of a series of electrochemical cells. Each cell contains several lead–antimony electrodes in the forms of grid-like plates, the design of which ensures a large surface area. Alternate electrode-plates possess a coating of lead(IV) oxide or lead(II) sulfate which renders them as either a cathode or an anode respectively. Each electrochemical cell is filled with sulfuric acid which functions as the electrolyte.

The two half reactions that are involved in the lead-acid cell are:

$$PbSO_4(s) + 2e^- \rightleftharpoons Pb(s) + [SO_4]^{2-}(aq) \qquad E^o = -0.36\ V$$

$$PbO_2(s) + 4H^+(aq) + [SO_4]^{2-}(aq) + 2e^- \rightleftharpoons PbSO_4(s) + 2H_2O(l) \qquad E^o = +1.69\ V$$

giving (under standard conditions) a value of E^o_{cell} of +2.05 V.

A normal automobile battery runs at 12 V and contains six cells connected *in series*; when cells are connected in a sequential manner, the overall cell potential equals the sum of the separate cell potentials.

By combining the two half-cells above (using the *anti-clockwise cycle* detailed in Chapter 12), the overall cell reaction can be written as:

$$PbO_2(s) + 4H^+(aq) + 2[SO_4]^{2-}(aq) + Pb(s) \rightleftharpoons 2PbSO_4(s) + 2H_2O(l)$$

and it is apparent that sulfuric acid is constantly consumed by the electrolysis cell. Recharging a cell drives the cell-reaction in the opposite direction to its spontaneous route, thereby regenerating aqueous H_2SO_4.

is a colourless gas, $GeCl_4$ a colourless liquid, while $GeBr_4$ melts at 299 K; GeI_4 is a red-orange solid (mp 417 K). The tetrahalides are tetrahedral monomers. Hydrolysis occurs with the formation of HX, and the Lewis acidity of the compounds is illustrated in their ability to form ions such as $[GeCl_6]^{2-}$ (from $GeCl_4$ and Cl^-) and adducts including *trans*-$[Ge(py)_2Cl_4]$ **13.23** (from $GeCl_4$ and pyridine).

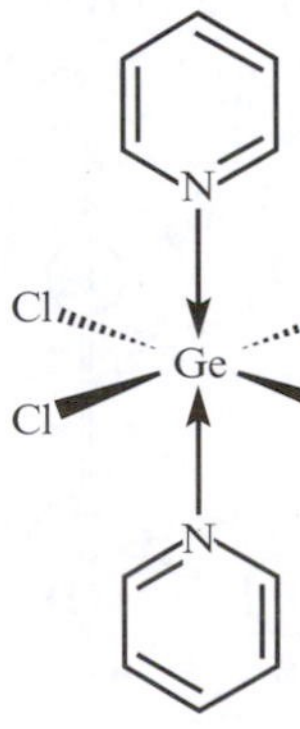

(13.23)

Germanium reduces GeF_4 in a *conproportionation* reaction (equation 13.48) to GeF_2 (a white solid), and $GeCl_2$ can similarly be prepared. The reduction of $GeBr_4$ with zinc gives $GeBr_2$. Heating GeX_2 (X = Cl, Br) causes disproportionation (equation 13.49).

$$\mathrm{Ge} + \mathrm{GeF_4} \longrightarrow 2\mathrm{GeF_2} \quad \text{(oxidation: Ge} \rightarrow \mathrm{GeF_2}\text{; reduction: } \mathrm{GeF_4} \rightarrow \mathrm{GeF_2}\text{)} \tag{13.48}$$

$$2\mathrm{GeX_2} \xrightarrow{\Delta} \mathrm{GeX_4} + \mathrm{Ge} \qquad \begin{array}{l} X = \mathrm{Cl},\ T = 1270\ \mathrm{K\ in\ a\ vacuum} \\ X = \mathrm{Br},\ T = 420\ \mathrm{K} \end{array} \tag{13.49}$$

In a conproportionation reaction, oxidation and reduction occur to give the same species. It is the reverse of a disproportionation reaction.

The Lewis acidity of GeF_2 is illustrated in its reaction with F^- to give $[GeF_3]^-$ (equation 13.50), and a similar reaction can occur between $GeCl_2$ and Cl^-. The solid state structure of the products may contain discrete pyramidal $[GeX_3]^-$ anions (as in $Rb[GeCl_3]$), but the nature of the cation has a significant influence on the structure and the presence of separated $[GeX_3]^-$ anions cannot be assumed.

$$\underset{\text{Lewis acid}}{\mathrm{GeF_2}} + \mathrm{CsF} \rightarrow \underset{\text{fluoride ion donor}}{\mathrm{Cs[GeF_3]}} \tag{13.50}$$

Hygroscopic: see Section 11.13

Tin(IV) fluoride may be prepared from $SnCl_4$ and HF; crystals are hygroscopic, and SnF_4 sublimes at 978 K to give a vapour containing tetrahedral molecules. At 298 K, SnF_4 is a white solid and possesses an extended lattice consisting of sheets containing 6-coordinate Sn atoms (Figure 13.16). These physical and structural properties contrast greatly with those of CF_4, SiF_4 and GeF_4.

Tin(IV) chloride, bromide and iodide are prepared from the constituent elements and resemble their silicon and germanium analogues. Hydrolysis occurs with loss of HX, but hydrates can also be isolated, for example

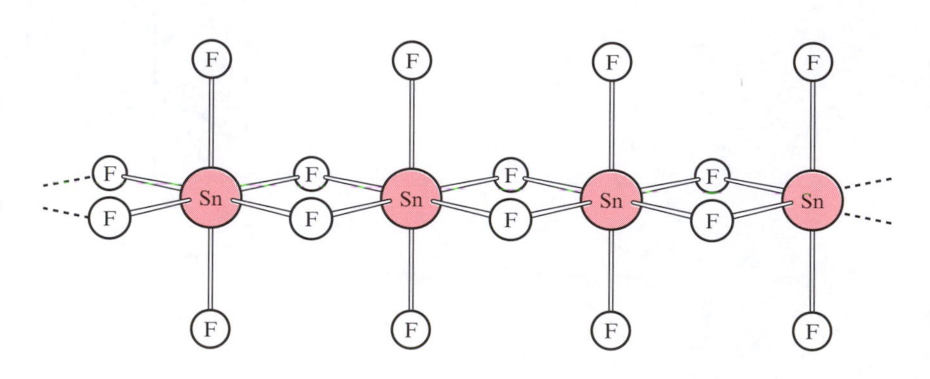

13.16 In SnF_4, bridging fluorine atoms between pairs of tin centres results in the formation of a 'sheet' structure in the solid state. The tin atom is 6-coordinate but the stoichiometry remains Sn : F = 1 : 4.

$SnCl_4{\cdot}4H_2O$ which forms opaque crystals. The tin(IV) halides are Lewis acids; $SnCl_4$ reacts with chloride ion to give $[SnCl_6]^{2-}$ **13.24** (equation 13.51) or $[SnCl_5]^-$ **13.25**, and adducts with other Lewis bases include $THF{\cdot}SnF_4$ and *cis*-$[Sn(MeCN)_2Cl_4]$ **13.26**. The Lewis acid strengths of the SnX_4 molecules decrease in the order $SnF_4 > SnCl_4 > SnBr_4 > SnI_4$.

$$2KCl + SnCl_4 \xrightarrow{\text{in presence of HCl(aq)}} K_2[SnCl_6] \qquad (13.51)$$

(13.24) **(13.25)**

(13.26)

The reaction of tin and hydrogen chloride gives tin(II) chloride, a white solid at room temperature. $SnCl_2$ reacts slowly in alkaline solutions to give a range of oxo and hydroxy species, but the white hydrate $SnCl_2{\cdot}2H_2O$, used as a reducing agent, can be isolated and is commercially available. $SnCl_2$ is a Lewis acid and reacts with chloride ion to give $[SnCl_3]^-$ **13.27**.

(13.27)

The lead(IV) halides are far less stable than the lead(II) compounds, all of which are crystalline solids at 298 K, and can be precipitated by mixing aqueous solutions of a soluble halide salt and a soluble lead(II) salt (equation 13.52). The choice of lead(II) salt is critical since few lead(II) compounds are very soluble in water.

$$Pb(NO_3)_2(aq) + 2NaCl(aq) \rightarrow PbCl_2(s) + 2NaNO_3(aq) \qquad (13.52)$$

Lead(II) chloride is sparingly soluble in water ($K_{sp} = 1.2 \times 10^{-5}$ mol^3 dm^{-9}) but in the presence of hydrochloric acid, the solubility increases as $PbCl_2$ acts as a Lewis acid and forms $[PbCl_4]^{2-}$ **13.28**.

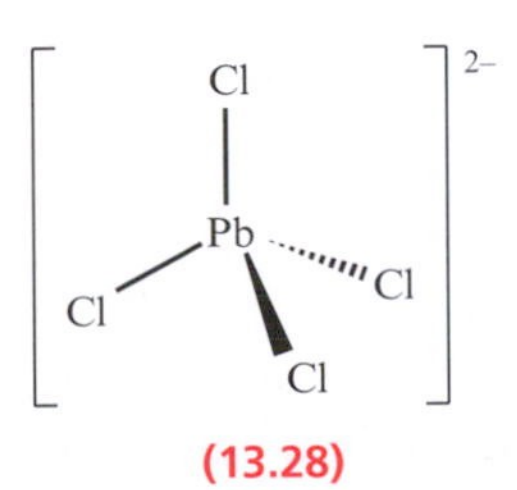

(13.28)

13.8 GROUP 15: NITROGEN AND PHOSPHORUS

1	2		13	14	**15**	16	17	18
H								He
Li	Be		B	C	**N**	O	F	Ne
Na	Mg		Al	Si	**P**	S	Cl	Ar
K	Ca		Ga	Ge	**As**	Se	Br	Kr
Rb	Sr	*d*-block	In	Sn	**Sb**	Te	I	Xe
Cs	Ba		Tl	Pb	**Bi**	Po	At	Rn
Fr	Ra							

Dinitrogen makes up 78% of the Earth's atmosphere (Figure 11.1) and can be obtained by liquefaction and fractional distillation of air. Reaction 13.53 is a convenient laboratory synthesis, but since ammonium nitrite is potentially explosive, N_2 is prepared as required by mixing aqueous solutions of sodium nitrite and an ammonium salt. N_2 is also formed when NH_4NO_3 is heated *strongly* (equation 13.54) but this can be explosive – ammonium nitrate is a powerful oxidizing agent and a component of dynamite.

$$\underset{\text{ammonium nitrite}}{NH_4NO_2(aq)} \xrightarrow{\Delta} N_2(g) + 2H_2O(l) \qquad (13.53)$$

$$\underset{\text{ammonium nitrate}}{2NH_4NO_3(s)} \xrightarrow{>570\text{ K}} 2N_2(g) + O_2(g) + 4H_2O(g) \qquad (13.54)$$

Dinitrogen is a colourless, odourless gas, and is generally unreactive; the N≡N bond is strong ($D = 945$ kJ mol^{-1}) due to the effective overlap of nitrogen $2p$ AOs. Nitrogen is the only member of group 15 to be able to form effective π-interactions in a homonuclear bond; for the heavier elements, the np AOs are more diffuse. The lack of reactivity means that N_2 is used to provide inert atmospheres in the laboratory and in the electronics industry during the production of transistors and other components. Liquid N_2 (bp 77 K) is used widely as a coolant.

Sizes of orbitals: see Section 2.11

N_2 as a coolant: see Box 13.3

When heated with lithium, magnesium or calcium, N_2 reacts to give a nitride (equation 13.55).

Reaction of N_2 with H_2: see Section 11.2

$$N_2(g) + 3Mg(s) \xrightarrow{\Delta} Mg_3N_2(s) \qquad (13.55)$$

In contrast to nitrogen, phosphorus does not occur in the elemental state, but is found in minerals such as *apatite*, an impure form of calcium phos-

Box 13.8 The nitrogen cycle

The nitrogen cycle illustrates how N_2 is removed from the atmosphere and converted into useful sources. Ultimately, the element is returned to the atmosphere. The action of lightning during thunderstorms converts N_2 to nitrogen monoxide:

$$N_2(g) + O_2(g) \rightarrow 2NO(g)$$

Subsequent reaction with O_2 occurs:

$$2NO(g) + O_2(g) \rightarrow 2NO_2(g)$$

Nitric acid is produced by the action of rain:

$$4NO_2(g) + 2H_2O(l) + O_2(g) \rightarrow 4HNO_3(aq)$$

Soluble nitrates are formed when the nitric acid reacts with, for example, calcium carbonate:

$$2HNO_3(aq) + CaCO_3(s) \rightarrow Ca(NO_3)_2(aq) + H_2O(l) + CO_2(g)$$

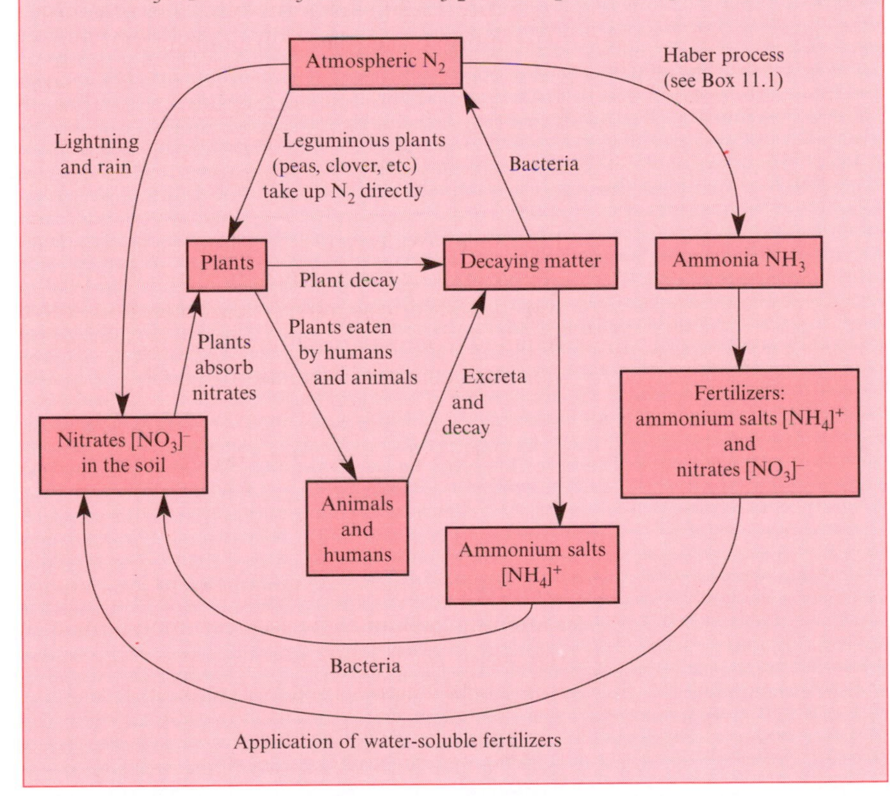

phate, $Ca_3(PO_4)_2$. One method of extraction involves heating the mineral with silica and carbon in a furnace (equation 13.56); phosphorus is formed as a vapour and can be collected in water or oxidized to phosphoric acid which is used for the manufacture of fertilizers.

$$2Ca_3(PO_4)_2 + 10C + 6SiO_2 \xrightarrow{\Delta} P_4 + 6CaSiO_3 + 10CO \qquad (13.56)$$

In addition to its vital role in the fertilizer industry, phosphorus finds application in the production of steels and phosphor-bronze, and the salt Na_3PO_4 is used as a water softener (see later).

Allotropes of phosphorus: see Sections 7.8 and 7.10

In sharp contrast to the inactivity of N_2 at 298 K, P_4 is stored under water because it ignites spontaneously in air (equation 13.57). The product, P_4O_{10}, is often referred to as phosphorus pentoxide, 'P_2O_5' (see later).

$$P_4(\text{white}) + 5O_2(g) \xrightarrow{\Delta} P_4O_{10}(s) \quad (13.57)$$

Oxides of nitrogen

Some common oxides of nitrogen are listed in Table 13.1. In N_2O and NO, the nitrogen atom is in a formal oxidation state of +1 and +2 respectively; N_2O is linear and structure **13.29** gives a valence bond representation of the molecule. Nitrogen monoxide is a radical, and although the valence bond structure **13.30** shows the odd electron on nitrogen, a molecular orbital description places the electron in an NO π^* orbital.

Bonding in NO: see Section 4.16

$$\overset{-}{N}{\equiv}\overset{+}{N}{=}O \qquad \dot{N}{=}O$$

113 pm 119 pm — 115 pm

(13.29) **(13.30)**

Dinitrogen oxide N_2O is a colourless gas at 298 K, and is quite unreactive. It is prepared by the potentially explosive conproportionation reaction 13.58 but the conditions must be carefully controlled – relatively mild heating is required (compare reactions 13.54 and 13.58). [*Question*: what are the oxidation state changes in equation 13.58?]

$$NH_4NO_3 \xrightarrow{520\text{ K}} N_2O + 2H_2O \quad (13.58)$$

oxidation

reduction

Dinitrogen oxide is probably best known for its use as a general anaesthetic (laughing gas), but it is used commercially in the synthesis of sodium azide (equation 13.59), and as a precursor to other azides such as $Pb(N_3)_2$ (equation 13.60) which is used as a detonator.

$$N_2O + NaNH_2 \xrightarrow{470\text{ K}} \underset{\text{sodium azide}}{NaN_3} + H_2O \quad (13.59)$$

$$2NaN_3 + Pb(NO_3)_2 \rightarrow Pb(N_3)_2 + 2NaNO_3 \quad (13.60)$$

Nitrogen monoxide NO is prepared on an industrial scale from ammonia (equation 13.61), but is also one of several oxides formed when N_2 is oxidized; generally, these combustion products are written as NO_x (pronounced 'NOX').

Table 13.1 Oxides of nitrogen. What is the formal oxidation state of nitrogen in each compound?

Formula	*Name*
N_2O	Dinitrogen oxide; nitrous oxide
NO	Nitrogen oxide; nitric oxide
N_2O_3	Dinitrogen trioxide
N_2O_4	Dinitrogen tetraoxide
NO_2	Nitrogen dioxide
N_2O_5	Dinitrogen pentaoxide

$$4NH_3 + 5O_2 \xrightarrow{\text{1300 K, Pt catalyst}} 4NO + 6H_2O \quad (13.61)$$

$$\tfrac{1}{2}N_2(g) + \tfrac{1}{2}O_2(g) \rightleftharpoons NO(g) \quad \Delta H^o = +91 \text{ kJ per mole of NO} \quad (13.62)$$

Equilibrium 13.62 shows the formation of gaseous NO, and Figure 13.17 illustrates how this endothermic reaction shifts to the right-hand side as the temperature is raised – a situation relevant to the combustion of motor and aircraft fuels. *Catalytic converters* in motor vehicles decrease NO_x, CO and hydrocarbon emissions. NO_x contributes to the formation of smogs over major cities, notably in Los Angeles, but its role as a pollutant is in sharp contrast to the essential role of NO in biological systems.

NO in biology: see Box 4.5

Even though NO is a radical, it does not dimerize unless it is cooled to low temperatures. The reactions of NO with halogens leads to FNO, ClNO and BrNO (equation 13.63). With respect to decomposition into NO and halogen, FNO is the most stable of the three compounds. The halides FNO and ClNO are used widely as sources of the nitrosyl cation, $[NO]^+$, and reactions with suitable Lewis acids yield nitrosyl salts (equations 13.64 and 13.65).

FNO, ClNO and BrNO are bent molecules: see Figure 5.6

$$2NO + Cl_2 \xrightarrow{\text{190 K}} 2ClNO \quad (13.63)$$

$$ClNO \xrightarrow{AlCl_3} [NO]^+[AlCl_4]^- \quad (13.64)$$

$$FNO \xrightarrow{PF_5} [NO]^+[PF_6]^- \quad (13.65)$$

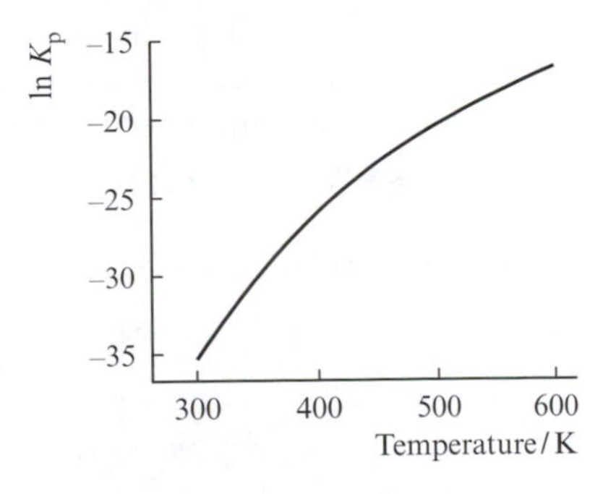

13.17 A graph showing the temperature-dependence of ln K_p for the equilibrium: $\tfrac{1}{2}N_2(g) + \tfrac{1}{2}O_2(g) \rightleftharpoons NO(g)$

At the high temperatures of fuel combustion in aviation and motor engines, the equilibrium is shifted towards the formation of NO. Nitrogen oxides (NO_x) are major pollutants.

Box 13.9 Catalytic converters

A catalytic converter consists of a honeycomb ceramic structure coated in finely divided aluminium oxide in which particles of catalytically active metals including platinum, palladium and rhodium are dispersed. The assembly is enclosed in a stainless steel vessel. Exhaust gases from a motor-vehicle pass through the heated catalytic chamber in which CO and hydrocarbons are oxidized and NO_x is reduced:

$$2CO(g) + O_2(g) \rightarrow 2CO_2(g)$$
$$2C_xH_y(g) + zO_2(g) \rightarrow 2xCO_2(g) + yH_2O(g)$$
$$2NO + 2CO \rightarrow N_2 + 2CO_2$$
$$2NO(g) + 2H_2(g) \rightarrow N_2(g) + 2H_2O(g)$$

Originally, catalytic converters concentrated on the oxidation processes, but are now designed to catalyse both oxidation and reduction reactions, so enhancing their environmental impact. The design of the converter provides a large surface area and efficient conversion of the toxic gases into non-pollutants.

A problem, however, is that the catalytic converter must be hot (≥ 600 K) before it becomes effective, and this precludes it being effective during the starting of a car engine and its initial cold-running. Electrically heated converters have been designed but they can drain car-battery resources. In 1994, researchers at Tufts University in the USA came up with a novel idea which involves the reversible dehydration of lithium bromide. While the vehicle is running and the engine hot, hydrated LiBr is dehydrated. As it is re-hydrated, the energy released can be used to heat the catalytic converter just before ignition. This process has not yet been adopted commercially. (Refer to Section 12.11 for enthalpies of hydration.)

Oxidation of NO with O_2 gives NO_2 (equation 13.66), and NO_2, like NO, has an unpaired electron.

$$\underset{\text{colourless gas}}{2NO(g)} + O_2(g) \rightleftharpoons \underset{\text{brown gas}}{2NO_2(g)} \qquad (13.66)$$

$$2NO_2(g) \rightleftharpoons N_2O_4(g) \qquad K_p\,(298\text{ K}) = 8.7 \qquad (13.67)$$

We consider equilibrium 13.67 in the accompanying workbook

Nitrogen dioxide (Figure 13.18) readily dimerizes and exists in the gas phase in equilibrium with N_2O_4 (equation 13.67), although above 420 K the system essentially contains only NO_2. In the liquid phase, the equilibrium lies well over to the right-hand side, and in the solid state (mp 262 K), only N_2O_4 molecules are present. The N_2O_4 molecule is planar (Figure 13.18) with a particularly long N–N bond (175 pm). The N–O bond distances in NO_2 and N_2O_4 are consistent with the presence of some nitrogen–oxygen $(p–p)\pi$-bonding (Figure 13.19).

A convenient synthesis of NO_2 or N_2O_4 is the thermal decomposition of *dry* lead(II) nitrate (equation 13.68); in the presence of water, NO_2 is hydrolysed to nitrous and nitric acids (see below). By cooling the brown gaseous NO_2 to about 273 K, the equilibrium in equation 13.68 is shifted towards the dimer and N_2O_4 condenses as a yellow liquid.

$$2Pb(NO_3)_2(s) \xrightarrow{\Delta} 2PbO(s) + 4NO_2(g) + O_2(g)$$
$$\Big\updownarrow \text{cool to} \approx 273\text{ K} \qquad (13.68)$$
$$2N_2O_4(l) \text{ as the major component}$$

13.18 The molecular structures of NO_2, N_2O_4, N_2O_3 and N_2O_5, and resonance structures for N_2O_3 indicating that two of the N–O bonds have partial π-character. What resonance structures contribute towards the bonding in NO_2, N_2O_4 and N_2O_5?

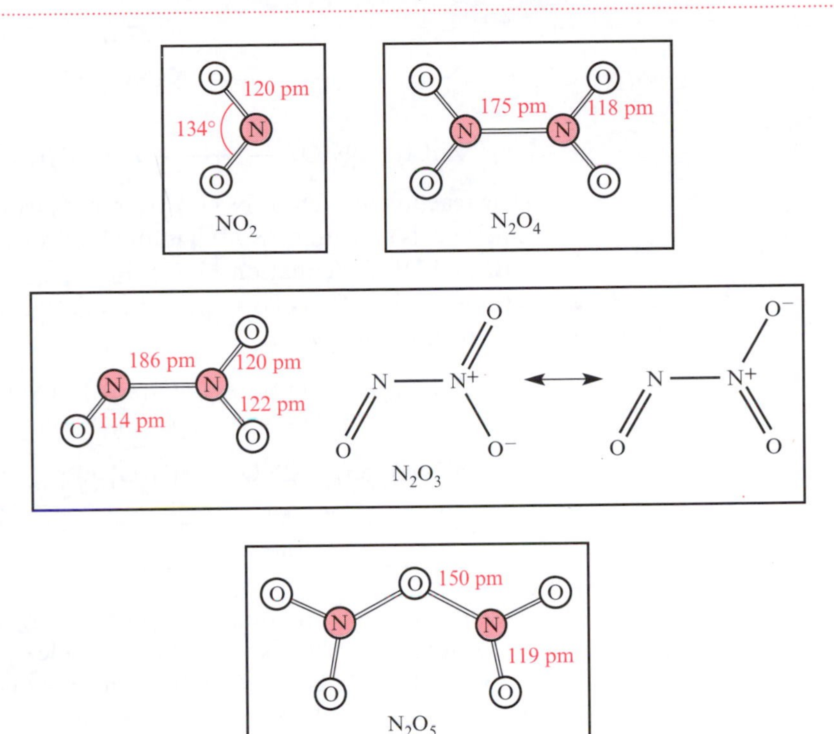

Hydrazine: see Section 11.7

Dinitrogen tetraoxide is a powerful oxidizing agent and the reaction with hydrazine has been used to fuel rockets. Pure liquid N_2O_4 is self-ionizing (equation 13.69) and is used as a non-aqueous solvent. The equilibrium lies towards the left-hand side but can be pushed to the right if nitromethane ($MeNO_2$) or dimethyl sulfoxide (Me_2SO) is added as a co-solvent.

$$N_2O_4(l) \rightleftharpoons [NO]^+(solv) + [NO_3]^-(solv) \quad (13.69)$$

Dry liquid N_2O_4 is used to prepare anhydrous metal nitrates such as $Cu(NO_3)_2$ which cannot be obtained by the evaporation of an aqueous solution or by heating hydrated copper(II) nitrate. The compounds $NaNO_3$ and $Zn(NO_3)_2$ may be prepared by reactions 13.70 and 13.71. As the sodium is oxidized in reaction 13.70, the $[NO]^+$ ion undergoes a one-electron reduction.

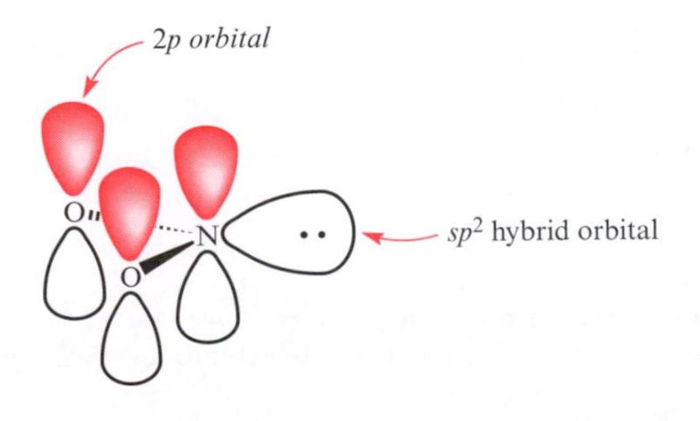

13.19 In NO_2, the nitrogen atom can be considered to be sp^2 hybridized with a lone pair of electrons occupying one hybrid orbital. Overlap can occur betwen the nitrogen $2p$ orbital (which is singly occupied) and the two oxygen $2p$ AOs, giving *partial* π-character to each N–O bond.

$$\mathrm{Na + N_2O_4 \xrightarrow{N_2O_4\ solvent} NaNO_3 + NO} \qquad (13.70)$$

$$\mathrm{ZnCl_2 + 2N_2O_4 \xrightarrow{N_2O_4\ solvent} Zn(NO_3)_2 + 2ClNO} \qquad (13.71)$$

The reaction of NO_2 or N_2O_4 with water gives a 1 : 1 mixture of nitrous and nitric acids (equation 13.72), although nitrous acid disproportionates to give NO and HNO_3 (equation 13.73). Because of the formation of these acids, atmospheric NO_2 is corrosive and contributes to the pollution problem called 'acid rain'.

Acid rain: see Box 13.14

$$\mathrm{2NO_2(g) + H_2O(l) \rightarrow \underset{nitric\ acid}{HNO_3(aq)} + \underset{nitrous\ acid}{HNO_2(aq)}} \qquad (13.72)$$

$$\mathrm{3HNO_2(aq) \rightarrow 2NO(g) + [H_3O]^+(aq) + [NO_3]^-(aq)} \qquad (13.73)$$

reduction (HNO₂ → NO)

oxidation (HNO₂ → [NO₃]⁻)

When equal molar amounts of NO and NO_2 are mixed at low temperature, N_2O_3 forms as a blue solid. The molecule is planar (Figure 13.18) with a long N–N single bond. The different N–O bond lengths suggest that the unique bond is of bond order 2, but the other two nitrogen–oxygen interactions possess *partial* π-character. Above its melting point (173 K), N_2O_3 dissociates into NO and NO_2. N_2O_3 dissolves in water to give nitrous acid (equation 13.74) which may then disproportionate (as shown above).

$$\mathrm{N_2O_3 + H_2O \rightarrow 2HNO_2} \qquad (13.74)$$

Dinitrogen pentaoxide is prepared by dehydrating nitric acid (equation 13.75) and forms colourless deliquescent crystals.

Deliquescent: see Section 11.12

$$\mathrm{2HNO_3 \xrightarrow[(dehydrating\ agent)]{P_4O_{10}} N_2O_5 + H_2O} \qquad (13.75)$$

In the solid state, crystalline N_2O_5 is composed of $[NO_2]^+$ and $[NO_3]^-$ ions, but it sublimes at 305 K to a give a vapour containing planar molecules (Figure 13.18). N_2O_5 is a powerful oxidizing agent (reaction 13.76).

$$\mathrm{N_2O_5 + I_2 \rightarrow I_2O_5 + N_2} \qquad (13.76)$$

Oxoacids of nitrogen

The two most important *oxoacids* of nitrogen are HNO_3 (nitric acid) and HNO_2 (nitrous acid). Nitric acid is a strong acid, but for HNO_2, $pK_a = 3.37$. The conjugate bases are the nitrate and nitrite ions respectively.

An *oxoacid* is a compound that contains oxygen, and at least one hydrogen atom attached to oxygen, and which produces a conjugate base by proton loss.

In the gas phase, molecules of HNO_2 **13.31** are planar with a *trans*-configuration.

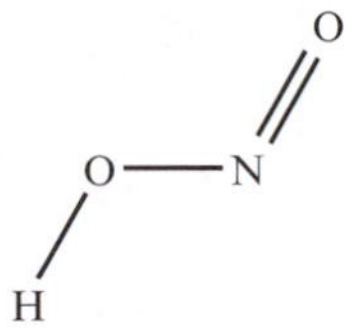

O—N=O with H on O (trans)

(13.31)

Pure HNO_2 is unstable, but dilute aqueous solutions can be prepared from *s*-block metal nitrites (equation 13.77). On standing in aqueous solution, HNO_2 disproportionates (equation 13.73).

$$NaNO_2(aq) + HCl(aq) + H_2O(l) \xrightarrow{< 273\ K} NaCl(aq) + [H_3O]^+(aq) + [NO_2]^-(aq) \qquad (13.77)$$

Diazo compounds: Section 15.11

Reaction 13.77 is used to prepare HNO_2 *in situ* for the formation of diazo derivatives (reaction 13.78).

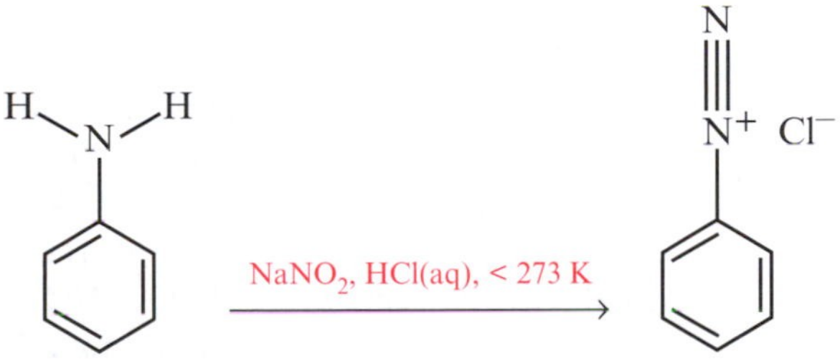

$$C_6H_5NH_2 \xrightarrow{NaNO_2,\ HCl(aq),\ < 273\ K} [C_6H_5N\equiv N]^+ Cl^- \qquad (13.78)$$

Most nitrite salts are water-soluble and are slightly toxic, although they are used in the curing of meat because they inhibit the oxidation of blood and prevent the meat from turning brown.

Nitric acid is manufactured on a large scale by the Ostwald process (equations 13.79 and 13.80).

$$4NH_3(g) + 5O_2(g) \xrightarrow[\text{Platinum/rhodium catalyst}]{1120\ K,\ 500\ kPa} 4NO(g) + 6H_2O(g) \qquad (13.79)$$

$$2NO(g) \xrightarrow{O_2} 2NO_2(g) \xrightarrow{H_2O} HNO_3(aq) + NO(g) \qquad (13.80)$$

Pure nitric acid is a colourless liquid (bp 356 K) and the brown-yellow coloration that concentrated aqueous solutions often develop is due to the decomposition reaction 13.81.

$$\underset{\text{colourless}}{4HNO_3(aq)} \rightleftharpoons \underset{\text{brown}}{4NO_2(aq)} + 2H_2O(l) + O_2(g) \qquad (13.81)$$

In the gas phase, HNO_3 is monomeric with a planar NO_3 framework (Figure 13.20a). The resonance structures shown account for the observed equivalence of two of the N–O bonds (121 pm) and the distance of 141 pm for the third N–O bond indicates a bond order of 1. In the planar nitrate ion, all the

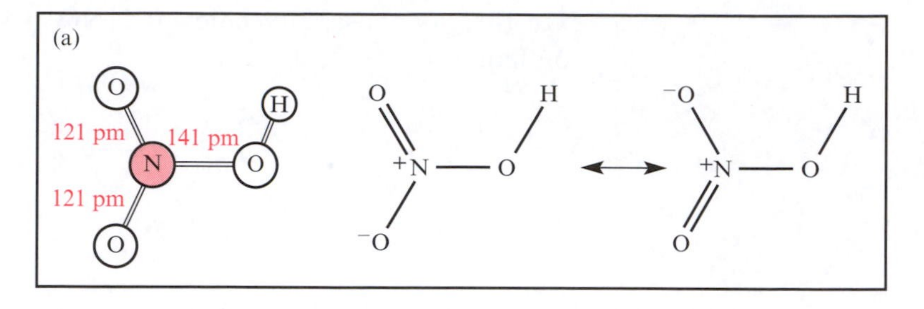

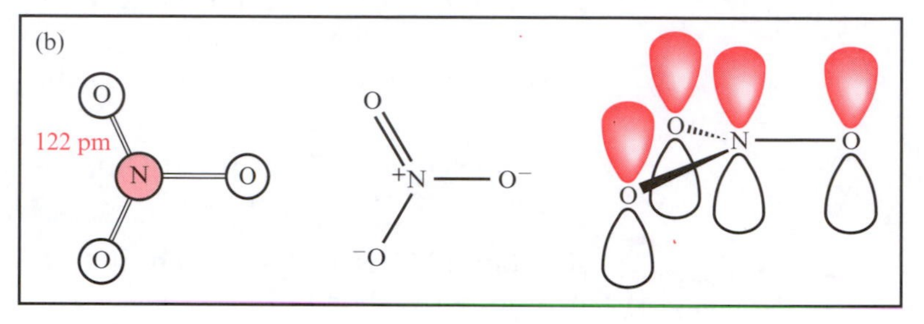

13.20 (a) The molecular and resonance structures of HNO_3. (b) The molecular structure of the planar $[NO_3]^-$ anion; the equivalence of the three N–O bonds can be rationalized by valence bond theory (one of three resonance structures is shown) or by MO theory (partial π-bonds are formed by overlap of N and O 2*p* AOs and the π-bonding is delocalized over the NO_3 framework).

N–O bond distances are equal (Figure 13.20b). One of *three* contributing resonance structures is shown in the figure, but the bonding can also be described in terms of three N–O σ-bonds and additional $(p\text{–}p)\pi$-bonding delocalized over the NO_3 framework.

As well as being a strong acid, HNO_3 is a good oxidizing agent and some half-equations involving HNO_2, $[NO_2]^-$, HNO_3 and $[NO_3]^-$ are listed in Table 13.2 along with the corresponding E^o values. By combining these data with those in Table 12.4, you should be able to suggest redox reactions in which HNO_2, $[NO_2]^-$, HNO_3 and $[NO_3]^-$ are involved. Nitric acid oxidizes most metals, but gold, rhodium, platinum and iridium are exceptions. With *very* dilute acid, the oxidations may proceed as in reaction 13.82, but at higher concentrations, the reactions become more complex (equations 13.83 and 13.84). [*Question*: What are the reduction and oxidation steps in each reaction?]

Table 13.2 Standard reduction potentials involving the nitrate and nitrite ions. Some reactions occur in acidic solution, and some in basic medium.

Half-equation	E^o/V
$2[NO_3]^-(aq) + 2H_2O(l) + 2e^- \rightleftharpoons N_2O_4(g) + 4[OH]^-(aq)$	–0.85
$[NO_2]^-(aq) + H_2O(l) + 3e^- \rightleftharpoons NO(g) + 2[OH]^-(aq)$	–0.46
$[NO_3]^-(aq) + H_2O(l) + 2e^- \rightleftharpoons [NO_2]^-(aq) + 2[OH]^-(aq)$	+0.01
$2[NO_2]^-(aq) + 3H_2O(l) + 4e^- \rightleftharpoons N_2O(g) + 6[OH]^-(aq)$	+0.15
$2HNO_2(aq) + 4H^+(aq) + 4e^- \rightleftharpoons H_2N_2O_2(aq) + 2H_2O(l)$	+0.86
$[NO_3]^-(aq) + 3H^+(aq) + 2e^- \rightleftharpoons HNO_2(aq) + H_2O(l)$	+0.93
$[NO_3]^-(aq) + 4H^+(aq) + 3e^- \rightleftharpoons NO(g) + 2H_2O(l)$	+0.96
$HNO_2(aq) + H^+(aq) + e^- \rightleftharpoons NO(g) + H_2O(l)$	+0.98
$2HNO_2(aq) + 4H^+(aq) + 4e^- \rightleftharpoons N_2O(g) + 3H_2O(l)$	+1.30

$$Mg(s) + 2HNO_3(aq) \rightarrow Mg(NO_3)_2(aq) + H_2(g) \quad (13.82)$$
(*very* dilute HNO_3)

$$3Cu(s) + 8HNO_3(aq) \rightarrow 3Cu(NO_3)_2(aq) + 4H_2O(l) + 2NO(g) \quad (13.83)$$
(dilute HNO_3)

$$Cu(s) + 4HNO_3(aq) \rightarrow Cu(NO_3)_2(aq) + 2H_2O(l) + 2NO_2(g) \quad (13.84)$$
(conc. HNO_3)

Passivate: see Section 11.13

Some metals such as aluminium, iron and chromium are passivated when placed in concentrated HNO_3. Although concentrated HNO_3 does not attack gold, a 3 : 1 (by volume) combination of concentrated HCl and HNO_3 solutions oxidizes gold to gold(III), the latter being formed as $[AuCl_4]^-$. This mixture of acids is called *aqua regia* and is used in cleaning baths for laboratory glassware.

Nitryl ion = nitronium ion; Nitration: see Section 15.6

A mixture of concentrated sulfuric and nitric acids produces the nitryl ion, $[NO_2]^+$ (equation 13.85), and is used widely as a nitrating agent in organic chemistry. In reaction 13.85, HNO_3 acts *as a Brønsted base*, since H_2SO_4 is a stronger acid than HNO_3.

$$\underset{\text{conc.}}{2H_2SO_4(aq)} + \underset{\text{conc.}}{HNO_3(aq)} \rightarrow 2[HSO_4]^-(aq) + [NO_2]^+(aq) + [H_3O]^+(aq) \quad (13.85)$$

Most nitrate salts are water-soluble and a qualitative test for the $[NO_3]^-$ ion is the *brown ring test*. This is carried out by acidifying the test solution with a few drops of *dilute* sulfuric acid, boiling the solution and allowing it to cool. After the addition of an equal volume of aqueous iron(II) sulfate solution, cold, concentrated sulfuric acid is added *carefully* (Figure 13.21). If $[NO_3]^-$ is present, a brown ring forms at the interface of the concentrated acid and aqueous layers. The origin of the brown colour is the formation of an octahedral ion containing iron and a nitrosyl ligand (equation 13.86).

Complex ions: see Chapter 16

$$Fe^{2+}(aq) + [NO_3]^- \xrightarrow{\text{conc. } H_2SO_4} [Fe(NO)(H_2O)_5]^{2+} \quad (13.86)$$

We conclude this section on oxoacids of nitrogen by comparing the results of heating some nitrite and nitrate salts (equations 13.87 and 13.88).

$$\left.\begin{array}{l} 2LiNO_2 \xrightarrow{\Delta} Li_2O + NO + NO_2 \\ 4LiNO_3 \xrightarrow{\Delta} 2Li_2O + 4NO_2 + O_2 \end{array}\right\} \quad (13.87)$$

$$\left.\begin{array}{l} Pb(NO_2)_2 \xrightarrow{\Delta} PbO + NO + NO_2 \\ 2Pb(NO_3)_2 \xrightarrow{\Delta} 2PbO + 4NO_2 + O_2 \end{array}\right\} \quad (13.88)$$

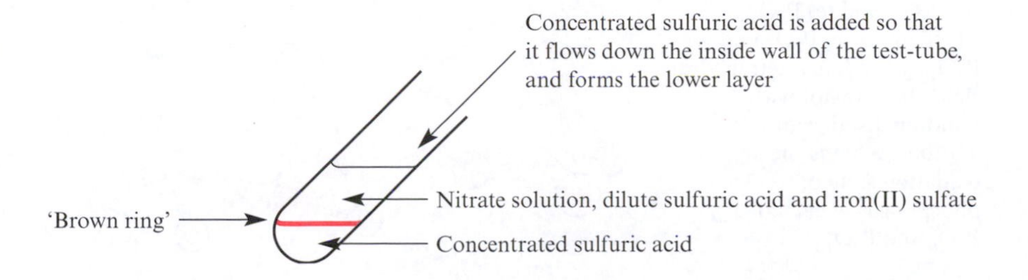

13.21 The 'brown-ring' test is used in qualitative analysis to confirm the presence of nitrate ions.

Oxides of phosphorus

When white phosphorus is exposed to air, it burns spontaneously to give phosphorus(V) oxide P_4O_{10} **13.32**, but if the supply of O_2 is limited, phosphorus(III) oxide, P_4O_6 **13.33** can be isolated.

(13.32)

(13.33)

Figure 13.22 illustrates the relationship between the structures of P_4, P_4O_6 and P_4O_{10}. The bonding in P_4 is described in terms of six localized P–P single bonds, and in the first step of oxidation, each P–P bond is converted into a P–O–P unit; each P–O bond distance of 165 pm is consistent with a single bond. Although a tetrahedral arrangement of phosphorus atoms is retained in P_4O_6, the P–P distances increase from 221 pm (bonding) in P_4 to 295 pm in P_4O_6 (non-bonding). Further oxidation with the formation of P=O double bonds occurs to give P_4O_{10}. The molecular units are retained in the gas phase.

Hydrolysis of P_4O_6 yields H_3PO_3, whilst P_4O_{10} gives H_3PO_4 (see below). P_4O_{10} is available commercially as a white powder but is extremely hygroscopic. It is used as a drying and dehydrating agent – if a dish containing P_4O_{10} is exposed to moist air, a brown, viscous layer forms on the surface of the white powder preventing it from absorbing more water.

Oxoacids of phosphorus

Phosphorus possesses far more oxoacids than does nitrogen, and Table 13.3 lists selected oxoacids and gives their structures – each phosphorus atom is

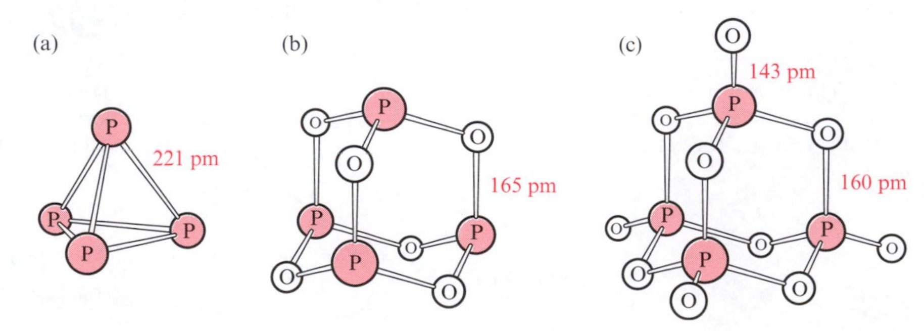

13.22 The molecular structures of (a) P_4, (b) P_4O_6, and (c) P_4O_{10}. The P atoms in P_4O_6 and P_4O_{10} are still in a tetrahedral shape but are not within bonding distance of one another. What is the oxidation state of phosphorus in each of P_4, P_4O_6 and P_4O_{10}?

Table 13.3 Selected oxoacids of phosphorus. The names in parentheses are older names, still in use. The number of OH groups determines the basicity of the acid; this is *not necessarily* the same as the total number of hydrogen atoms – for example in H_3PO_3.

Formula	*Name*	*Structure*
H_3PO_4	Phosphoric acid (orthophosphoric acid)	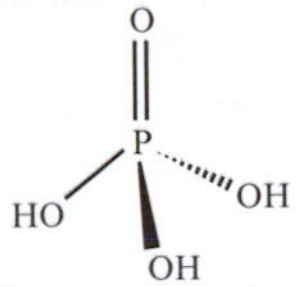
H_3PO_3 or $HPO(OH)_2$	Phosphonic acid (phosphorous acid)	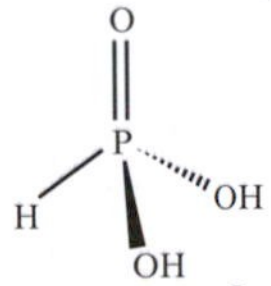
H_3PO_2 or $H_2PO(OH)$	Phosphinic acid	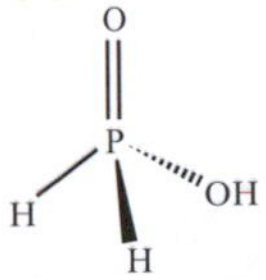
$H_4P_2O_7$	Diphosphoric acid	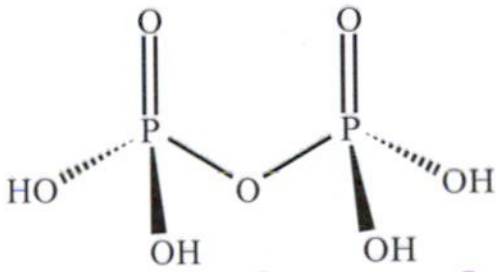
$H_4P_2O_5$	Diphosphonic acid (diphosphorous acid)	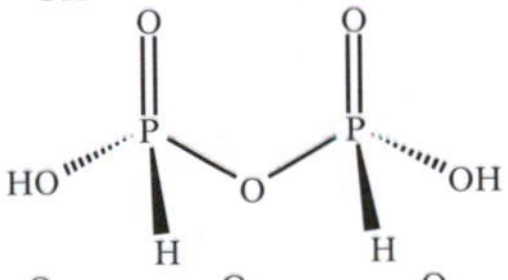
$H_5P_3O_{10}$	Triphosphoric acid	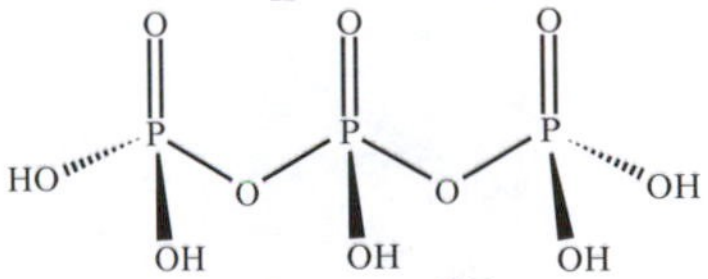
$H_4P_2O_6$	Hypophosphoric acid	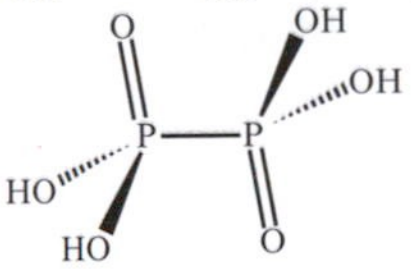
$(HPO_3)_n$	Polymetaphosphoric acid	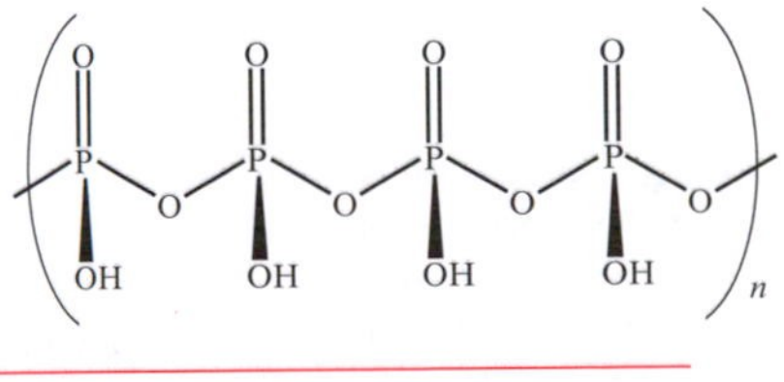

tetrahedrally bonded. The basicity of each acid corresponds to the number of OH groups, *and not simply to the number of hydrogen atoms*. Phosphinic acid, H_3PO_2, is monobasic (equation 13.89) because the two *hydrogen atoms which are bonded directly to the phosphorus atom are not lost as protons*.

$$H_3PO_2(aq) + H_2O(l) \rightleftharpoons [H_3O]^+(aq) + [H_2PO_2]^-(aq) \qquad (13.89)$$

Box 13.10 Phosphates in lakes: friend or foe?

Phosphates from detergents and fertilizers are a factor in the increase of algae in lakes – excessive growth of algae is called *eutrophication* and its presence depletes the lakes of a supply of O_2 and affects fish and other water-life. The algae population can increase as zooplankton (algae are their natural food) are killed off. But the phosphate problem is not clear cut.

Firstly, it has been shown that phosphates are not the only pollutant to blame for this upset in the natural balance – zooplankton are also affected by heavy metal wastes, oil and insecticides.

Secondly, the presence of phosphates in a lake may be able to offset some of the problems caused by acid rain (see Box 13.14). Acid rain lowers the pH of lakes (i.e. the water becomes more acidic). Although neutralization can be achieved by adding lime, this leads to higher levels of Ca^{2+} ions which in turn affects the variety of life supported by the lake's waters. A study carried out in the Lake District of Britain has shown the presence of phosphates encourages water-plant growth (i.e. the phosphate acts as a fertilizer) and the plants absorb nitrates (a natural part of the N_2 cycle, Box 13.8). Hydroxide ions are produced as nitrate ions are taken up, and neutralization of the lake's water can occur. The report of this research (*Nature*, (1994) vol. 377, p. 504) brings into question whether phosphates are as bad for our water-courses as other reports may suggest.

Eutrophic lake in Lithuania (courtesy Chris Mason)

An acid anhydride is formed when one or more molecules of acid lose one or more molecules of water.

Phosphoric acid, H_3PO_4, is formed by reaction 13.90 – the oxide is the *anhydride* of the acid – or when calcium phosphate ('phosphate rock') reacts with concentrated sulphuric acid (equation 13.91). In its pure state, H_3PO_4 forms deliquescent, colourless crystals (mp 315 K) which quickly turn into a viscous liquid. The acid is available commercially as an 85% aqueous solution; even this is viscous, owing to extensive hydrogen bonding.

$$P_4O_{10}(s) + 6H_2O(l) \rightarrow 4H_3PO_4(aq) \qquad (13.90)$$

$$Ca_3(PO_4)_2(s) + \underset{\text{concentrated}}{3H_2SO_4(aq)} \rightarrow 2H_3PO_4(aq) + 3CaSO_4(s) \qquad (13.91)$$

Industrially, phosphoric acid is very important and is used on a large scale in the production of fertilizers, detergents and food additives, being responsible for the sharp taste of many soft drinks. It is also used to remove oxide and scale from the surfaces of iron and steel.

Aqueous H_3PO_4 is a tribasic acid, but the values of pK_a for equilibria 13.92 to 13.94 show that only the first proton is readily lost. With NaOH, H_3PO_4 can react to give three salts NaH_2PO_4, Na_2HPO_4 and Na_3PO_4.

$$H_3PO_4(aq) + H_2O(l) \rightleftharpoons [H_3O]^+(aq) + [H_2PO_4]^-(aq) \quad pK_a = 2.12 \qquad (13.92)$$

$$[H_2PO_4]^-(aq) + H_2O(l) \rightleftharpoons [H_3O]^+(aq) + [HPO_4]^{2-}(aq) \quad pK_a = 7.21 \qquad (13.93)$$

$$[HPO_4]^{2-}(aq) + H_2O(l) \rightleftharpoons [H_3O]^+(aq) + [PO_4]^{3-}(aq) \quad pK_a = 12.67 \qquad (13.94)$$

Phosphoric acid is not a strong oxidizing agent, as indicated by the value of E^o for half-equation 13.95 – compare this with the values listed in Table 3.2 for nitric and nitrous acids.

$$H_3PO_4(aq) + 2H^+(aq) + 2e^- \rightleftharpoons H_3PO_3(aq) + H_2O(l) \quad E^o = -0.28 \text{ V} \qquad (13.95)$$

When H_3PO_4 is heated above 480 K, it is dehydrated to diphosphoric acid (equation 13.96).

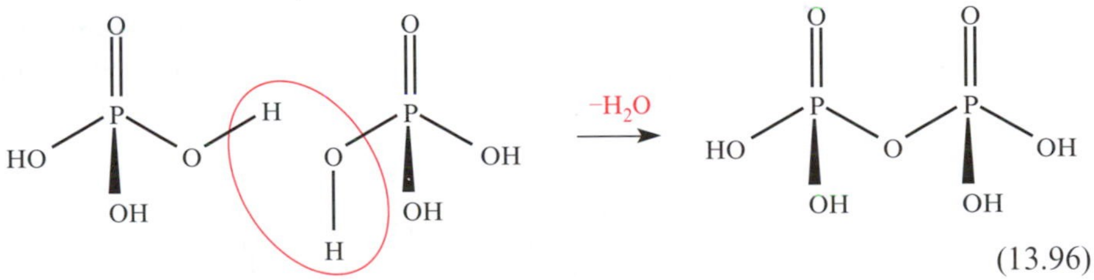

$-H_2O$

(13.96)

Further heating results in further H_2O elimination and the formation of polyphosphoric acids such as triphosphoric acid. The sodium salt of this pentabasic acid (Table 3.3) $Na_5P_3O_{10}$ is used in detergents where it acts as a water softener. 'Hard water' is due to the presence of Ca^{2+} and Mg^{2+} ions, and the $[P_3O_{10}]^{3-}$ anion **13.34** is a *sequestering agent* which forms stable water-soluble complexes with the group 2 ions, precluding them from forming insoluble complexes with the stearate ion in soaps. Such precipitates appear as 'scum' in household sinks and baths.

Stearic acid and soap: see Section 17.8

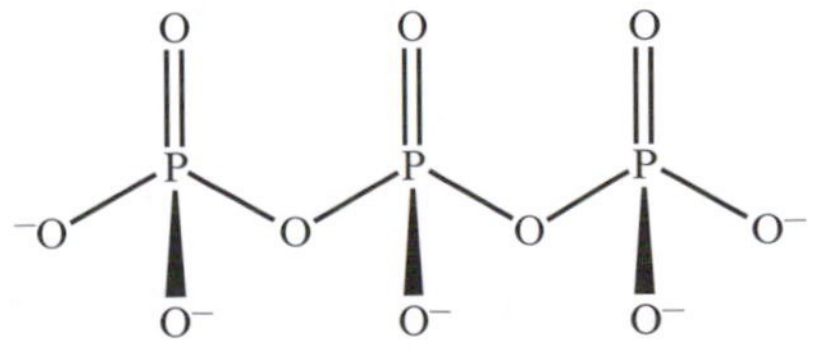

(13.34)

Box 13.11 Artificial bone injected as a liquid

Recent research in the USA has led to the possibility that bone can be formed *in situ*, thereby bringing a new dimension to the surgeries involved in bone repair. Solid $Ca(H_2PO_4)_2 \cdot H_2O$, $Ca_3(PO_4)_2$ and $CaCO_3$ are combined with a solution of sodium phosphate to form a paste which can be injected into a patient at the site of a bone fracture. Within minutes, the paste hardens and forms artificial bone. In time, the body replaces this material with natural bone.

Phosphates are biologically important, and the structural material of bones and teeth is *apatite* $Ca_5(OH)(PO_4)_3$. Tooth decay involves acid attack on the phosphate, but reaction with fluoride ion (which is commonly added to water supplies) can lead to the formation of fluoroapatite as a coating on teeth (equation 13.97) making them more resistant to decay.

$$\underset{\text{apatite}}{Ca_5(OH)(PO_4)_3} + F^- \rightarrow \underset{\text{fluoroapatite}}{Ca_5F(PO_4)_3} + [OH]^- \qquad (13.97)$$

All living cells contain *adenosine triphosphate* (ATP) **13.35** and its conversion to adenosine diphosphate (ADP) by the cleavage of one phosphate group during hydrolysis (equation 13.98) releases energy used for functions such as cell growth and muscle movement. The conversion of ATP to ADP occurs at pH 7.4 (biological pH) and ΔG for reaction 13.98 (which is an over-simplification of this essential biological process) is approximately –40 kJ per mole of reaction. ATP is continually being reformed using energy from the burning of organic compounds, so that there is a constant supply of stored energy.

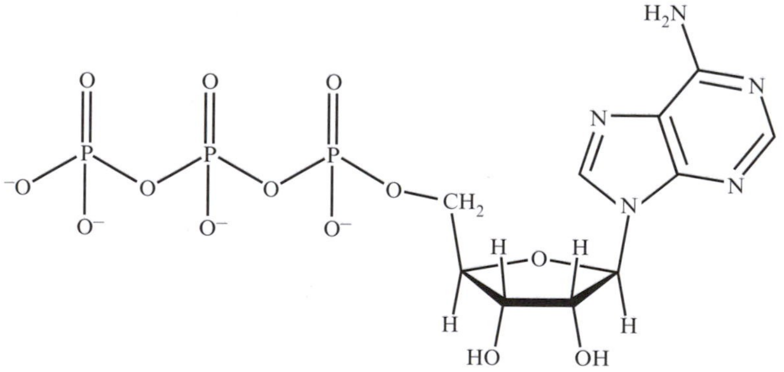

(13.35)

$$[ATP]^{4-} + 2H_2O \rightarrow [ADP]^{3-} + [HPO_4]^{2-} + [H_3O]^+ \qquad (13.98)$$

Phosphonic acid is formed when P_4O_6 or PCl_3 is hydrolysed (equations 13.99 and 13.100) and the latter method is used industrially.

$$P_4O_6(s) + 6H_2O(l) \rightarrow 4H_3PO_3(aq) \qquad (13.99)$$

$$PCl_3(l) + 3H_2O(l) \rightarrow H_3PO_3(aq) + 3HCl(aq) \qquad (13.100)$$

Box 13.12 ATP and fireflies: bioluminescence

The firefly is a common sight at dusk in North America. This remarkable insect attracts its mate by emitting flashes of light from its rear body. This is called *bioluminescence* and the basis is a light-emitting chemical reaction. A compound called luciferin is oxidized in this process. The overall reaction uses an enzyme catalyst (luciferase) and the biological fuel ATP which is reduced to adenosine monophosphate (AMP) and inorganic phosphate. Many bioluminescent reactions are known and a walk through the fields or woods at night in summer or autumn may reveal many spectacular light-emitting insects and fungi.

Luciferin

Pure H_3PO_3 forms colourless, deliquescent crystals (mp 343 K). It contains *two* OH groups (Table 13.3) and in aqueous solution acts as a dibasic acid (equations 13.101 and 13.102) forming salts such as NaH_2PO_3 and Na_2HPO_3.

$$H_3PO_3(aq) + H_2O(l) \rightleftharpoons [H_3O]^+(aq) + [H_2PO_3]^-(aq) \quad pK_a = 2.00 \qquad (13.101)$$

$$[H_2PO_3]^-(aq) + H_2O(l) \rightleftharpoons [H_3O]^+(aq) + [HPO_3]^{2-}(aq) \quad pK_a = 6.59 \qquad (13.102)$$

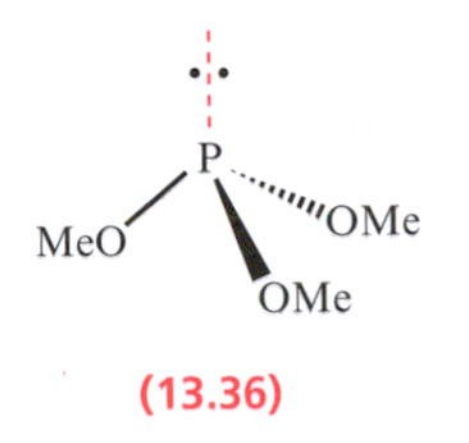

(13.36)

Salts containing the $[HPO_3]^{2-}$ ion are called *phosphonates* but the name 'phosphite' is still commonly used. This can cause confusion since 'phosphite' is also used for organic compounds of formula $P(OR)_3$; e.g. $P(OMe)_3$ **13.36** is trimethylphosphite. There is *no inorganic acid* of formula $P(OH)_3$ – it is a common mistake to think that the formula H_3PO_3 refers to a tribasic acid of this structural type.

Pure phosphinic acid H_3PO_2 is difficult to isolate as it readily decomposes (reaction 13.103). It forms deliquescent, white crystals which rapidly absorb water to give an oily liquid, and in aqueous solution, the acid is monobasic ($pK_a = 1.1$) (equation 13.104).

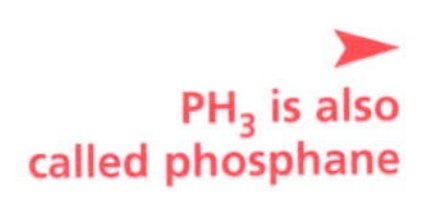

$$2H_3PO_2 \rightarrow \underset{\text{phosphine}}{PH_3} + H_3PO_4 \qquad (13.103)$$

$$H_3PO_2(aq) + H_2O(l) \rightleftharpoons [H_3O]^+(aq) + \underset{\text{phosphinate ion}}{[H_2PO_2]^-(aq)} \qquad (13.104)$$

Phosphinate salts are produced when white phosphorus reacts with aqueous alkali (equation 13.105).

$$P_4 + 4[OH]^- + 4H_2O \rightarrow 4[H_2PO_2]^- + 2H_2 \qquad (13.105)$$

Halides of nitrogen and phosphorus

Dipole moment of NX_3: see Section 5.13

Nitrogen forms trihalides of formula NX_3 (X = F, Cl or Br), but NBr_3 is unstable. They are molecular compounds with trigonal pyramidal structures. At room temperature, NCl_3 is an explosive liquid, but is used industrially as a bleach. It hydrolyses rapidly in water (equation 13.106).

$$NCl_3 + 3H_2O \rightarrow NH_3 + \underset{\text{hypochlorous acid}}{3HOCl} \quad (13.106)$$

Hypochlorous acid: see Section 13.12

Nitrogen trifluoride can be produced by electrolysing molten NH_4F and HF, or by reaction 13.107. It is a colourless gas at 298 K and is thermodynamically more stable than NCl_3 or NBr_3. NF_3 resists attack by water and dilute acids and alkalis, but is an effective fluorinating agent converting some elements into their respective fluorides.

$$4NH_3 + 3F_2 \xrightarrow{\text{copper catalyst}} NF_3 + 3NH_4F \quad (13.107)$$

When NF_3 is heated with copper (equation 13.108), dinitrogen tetrafluoride is formed. This colourless gas exists in equilibrium with NF_2 radicals (equation 13.109).

$$2NF_3 + Cu \xrightarrow{\Delta} N_2F_4 + CuF_2 \quad (13.108)$$

$$\underset{\text{colourless}}{N_2F_4(g)} \rightleftharpoons \underset{\text{blue}}{2NF_2(g)} \qquad K_p\ (420\ K) = 0.03 \quad (13.109)$$

Dinitrogen tetrafluoride (Figure 13.23) adopts both staggered and *gauche* conformers at 298 K – compare this with hydrazine **11.5**. N_2F_4 is very reactive and is a powerful fluorinating agent (reactions 13.110 and 13.111). It reacts with *fluoride acceptors* such as AsF_5 to give salts (equation 13.112).

Fluoride acceptors: see Section 13.9

$$N_2F_4 + 10Li \xrightarrow{\Delta} 4LiF + 2Li_3N \quad (13.110)$$

$$N_2F_4 + SiH_4 \rightarrow SiF_4 + N_2 + 2H_2 \quad (13.111)$$

$$N_2F_4 + AsF_5 \rightarrow [N_2F_3]^+[AsF_6]^- \quad (13.112)$$

The difluoride N_2F_2 is planar and exists as both the *cis* and *trans*-isomers (**13.37** and **13.38**), *cis*-N_2F_2 being more thermodynamically stable, but also more reactive.§ The *trans*-isomer can be selectively prepared by reaction 13.113, but on heating to 370 K, it isomerizes to give an equilibrium mixture of isomers in which *cis*-N_2F_2 predominates.

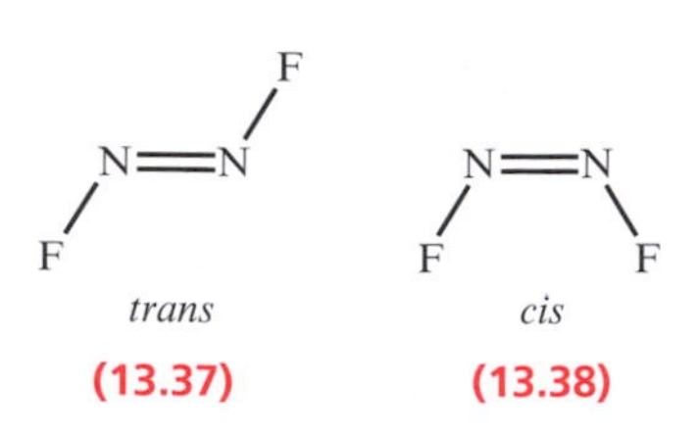

(13.37) (13.38)

§ Inorganic nomenclature generally retains *cis*- and *trans*- rather then (*Z*)- and (*E*)-isomers.

$$2N_2F_4 + 2AlCl_3 \xrightarrow{200\ K} trans\text{-}N_2F_2 + N_2 + 3Cl_2 + 2AlF_3 \quad (13.113)$$

In comparing the halides of nitrogen with those of phosphorus, we must bear in mind that:

- N–N π-interactions are stronger than P–P π-interactions, and as a consequence, there is no phosphorus analogue of N_2F_2; and
- phosphorus can expand its valence shell and so can form halides of the type PX_5.

The phosphorus trihalides PX_3 (X = F, Cl, Br or I), all of which are toxic, have pyramidal structures; at 298 K, PF_3 is a colourless gas, PCl_3 and PBr_3 are colourless liquids and PI_3 is a red crystalline solid. PF_3 is highly toxic – like CO, PF_3 reacts with the iron centre in blood haemoglobin. In contrast to the later trihalides, PF_3 hydrolyses slowly in moist air (equation 13.114).

Haemoglobin: see Box 4.4

$$PF_3 + 3H_2O \rightarrow 3HF + H_3PO_3 \quad (13.114)$$

The compounds PCl_3, PBr_3 and PI_3 may be prepared from their respective elements, and PF_3 is made from PCl_3 using CaF_2 as the fluorinating agent. PCl_3 is manufactured on a large scale, being a precursor to many *organophosphorus compounds* (see below), used as flame retardants and fuel additives, and in the formation of insecticides and nerve gases for chemical warfare.

The reaction between water and PCl_3 is vigorous (equation 13.115), while oxidation with O_2 gives the oxychloride $POCl_3$ **13.39** (equation 13.116).

$$PCl_3 + 3H_2O \rightarrow 3HCl + H_3PO_3 \quad (13.115)$$

$$2PCl_3 + O_2 \rightarrow 2POCl_3 \quad (13.116)$$

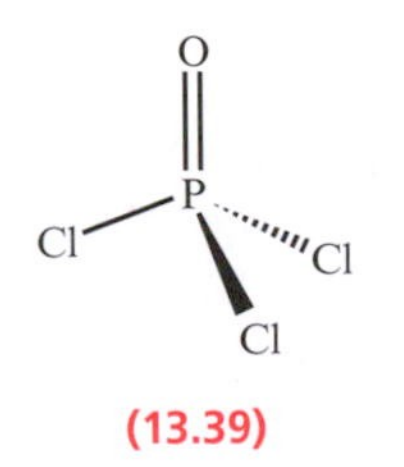

(13.39)

All four P_2X_4 compounds (X = F, Cl, Br or I) are known, but P_2I_4 is the most stable member of the series. It can be prepared by heating red phosphorus with I_2, and is a red crystalline solid at 298 K, consisting of P_2I_4 molecules with a staggered conformation like that of one conformer of N_2F_4 (Figure 13.23). Reactions of P_2I_4 often involve P–P bond cleavage (equation 13.117).

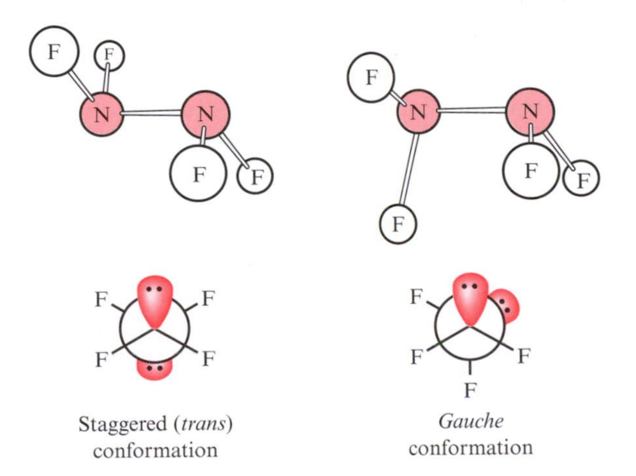

13.23 N_2F_4 exists in both the staggered and *gauche* conformers, shown here as molecular structures and Newman projections.

Box 13.13 Chemical weapons

In the First and Second World Wars, chemical weapons based on simple toxic chemicals such as phosgene and mustard gas were available for active use and also as a deterrent.

Phosgene

Mustard gas

During and after the Second World War, the development of chemical weapons moved to nerve gases, for example Sarin and Soman. These agents work by enzyme inhibition in the nervous system and inhalation of as little as 1 mg is fatal. Sarin was the nerve gas used in the attacks in Japan that brought chaos to commuters in March 1995.

Sarin

Soman

Ways must be found to destroy chemical weapons that are stockpiled, since current policy in most countries is towards chemical weapon disarmament. For example, Sarin may be chemically destroyed by hydrolysis:

$$(Me_2HCO)P(O)(Me)F + H_2O \rightarrow (Me_2HCO)P(O)(Me)OH + HF$$

$$\downarrow H_2O$$

$$Me_2HCOH + MeP(O)(OH)_2$$

and if aqueous sodium hydroxide is used for the reaction, effectively harmless sodium salts are formed.

$$P_2I_4 + Br_2 \rightarrow 2PI_2Br \qquad (13.117)$$

The phosphorus(V) halides exist as trigonal bipyramidal molecules in the gas phase, consistent with VSEPR theory, but only PF_5 is a gas at 298 K – PCl_5, PBr_5 and PI_5 are *ionic* crystalline solids. For PCl_5, chloride transfer occurs to give a solid consisting of tetrahedral $[PCl_4]^+$ cations and octahedral $[PCl_6]^-$ anions. In the solid bromide and iodide, free halide ions are present and the compounds are formulated as $[PBr_4]^+Br^-$ and $[PI_4]^+I^-$. Although the molecular structure of PF_5 **13.40** indicates two distinct fluorine sites (axial and equatorial), the ^{19}F NMR spectrum of a solution of PF_5

Stereochemically non-rigid: see Section 5.12

shows only one signal, because PF_5 is stereochemically non-rigid and the fluorine nuclei are equivalent on the NMR spectroscopic timescale. We consider the NMR spectroscopic data in the accompanying workbook.

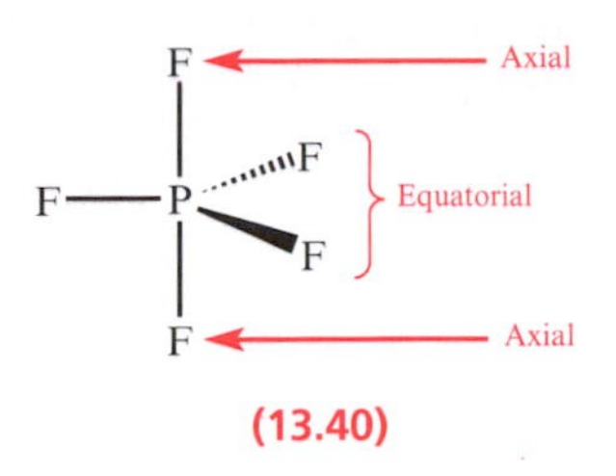

(13.40)

Reactions of PCl_5 and alcohols: see Section 14.10

Of the phosphorus(V) halides, PCl_5 is the most important, and its synthetic use is widespread. It is made industrially from PCl_3 (equation 13.118) and may be converted to PF_5 by reaction with a fluorinating agent such as CaF_2. However, PF_5 itself is a good fluoride acceptor and fluorination of PCl_5 with KF gives the salt $K[PF_6]$. PF_5 and PCl_5 are moisture-sensitive (equation 13.119) but if the conditions of PCl_5 hydrolysis are controlled, $POCl_3$ can be produced (equation 13.120).

$$PCl_3 + Cl_2 \rightarrow PCl_5 \qquad (13.118)$$

$$PCl_5 + 4H_2O \rightarrow H_3PO_4 + 5HCl \qquad (13.119)$$

$$PCl_5 + H_2O \rightarrow POCl_3 + 2HCl \qquad (13.120)$$

13.9 GROUP 15: ARSENIC, ANTIMONY AND BISMUTH

1	2		13	14	**15**	16	17	18
H								He
Li	Be		B	C	N	O	F	Ne
Na	Mg		Al	Si	**P**	S	Cl	Ar
K	Ca		Ga	Ge	**As**	Se	Br	Kr
Rb	Sr	*d*-block	In	Sn	**Sb**	Te	I	Xe
Cs	Ba		Tl	Pb	**Bi**	Po	At	Rn
Fr	Ra							

Arsenic is a steel-grey, brittle semi-metal and occurs in the elemental state and in the sulfide ores *arsenopyrite* (FeAsS), *realgar* (As_4S_4) and *orpiment* (As_2S_3). When arsenopyrite is heated, it decomposes to give arsenic and FeS.

Compounds of arsenic are toxic – much beloved by crime-writers! Gallium arsenide has widespread uses in solid state devices and semiconductors.

Antimony is not an abundant element and is found in various minerals of which *stibnite* Sb_2S_3 is the most important and from which it can be extracted by reduction using scrap iron (equation 13.121).

$$Sb_2S_3 + 3Fe \rightarrow 2Sb + 3FeS \qquad (13.121)$$

reduction

oxidation

Antimony is a toxic, brittle, blue-white metal, which is a poor electrical and thermal conductor. It resists oxidation in air at 298 K, but is converted to oxides at higher temperatures (see below). Some of its uses are in the semiconductor industry, in alloys (e.g. it increases the strength of lead) and in batteries.

Bismuth is a soft, grey metal, occurring in its elemental state and in *bismite* (Bi_2O_3), *bismuthinite* (Bi_2S_3) and in some copper-, lead- and silver-containing ores. It is a component of alloys (e.g. with tin) and compounds of bismuth are used in the cosmetics industry, as oxidation catalysts and in high-temperature superconductors.

Oxides of arsenic, antimony and bismuth

When heated with O_2, arsenic forms the highly toxic As_2O_3 (equation 13.122) which has a smell resembling garlic. This oxide is an important precursor to many arsenic compounds and is manufactured by several routes, including roasting the sulfide ores realgar or orpiment in air.

$$4As + 3O_2 \rightarrow 2As_2O_3 \qquad (13.122)$$

At 298 K, As_2O_3 is a crystalline solid containing As_4O_6 units, and in the vapour state, As_4O_6 molecules, similar to P_4O_6, are present. Unlike P_4O_6, As_4O_6 does not readily oxidize to arsenic(V) oxide. Instead, As_4O_{10} can be formed by first making H_3AsO_4 (equation 13.123). As this reaction suggests, As_4O_{10} dissolves in water to give arsenic acid, H_3AsO_4.

$$2As_2O_3 \xrightarrow{\text{conc. } HNO_3} 4H_3AsO_4 \xrightarrow{\text{dehydration}} As_4O_{10} + 6H_2O \qquad (13.123)$$

Antimony(III) oxide, Sb_2O_3, forms when antimony burns in O_2 and, structurally, it is similar to As_2O_3. It is used as a flame retardant and in paints, adhesives and plastics. Oxidation to antimony(V) oxide occurs when Sb_2O_3 reacts with high pressures of O_2 at elevated temperatures. Neither arsenic(V) nor antimony(V) oxide resembles P_4O_{10} in structure. Bismuth(III) oxide (the mineral *bismite*) has many uses in the glass and ceramic industry, and for catalysts and magnets. Bismuth(V) oxide is very unstable.

Oxoacids and related salts of arsenic, antimony and bismuth

The later members of group 15 form fewer oxoacids than phosphorus. Arsenic acid, H_3AsO_4 is tribasic (equations 13.124–13.126).

$$H_3AsO_4(aq) + H_2O(l) \rightleftharpoons [H_3O]^+(aq) + [H_2AsO_4]^-(aq) \qquad pK_a = 2.25 \qquad (13.124)$$

$$[H_2AsO_4]^-(aq) + H_2O(l) \rightleftharpoons [H_3O]^+(aq) + [HAsO_4]^{2-}(aq) \qquad pK_a = 6.77 \qquad (13.125)$$

$$[HAsO_4]^{2-}(aq) + H_2O(l) \rightleftharpoons [H_3O]^+(aq) + [AsO_4]^{3-}(aq) \qquad pK_a = 11.60 \qquad (13.126)$$

In contrast to H_3PO_4, arsenic acid is a good oxidizing agent as the E^o value for half-reaction 13.127 indicates.

$$H_3AsO_4(aq) + 2H^+(aq) + 2e^- \rightleftharpoons HAsO_2(aq) + 2H_2O(l) \qquad E^o = +0.56\ V \qquad (13.127)$$

The acid 'H_3AsO_3' has not been isolated in the pure state, but *arsenite* salts containing the $[AsO_3]^{3-}$ anion are known. Although antimonites are well-characterized salts, oxoacids of antimony(III) are not stable. The trend continues for bismuth, and no oxoacids are known, although some bismuthate salts are, and sodium bismuthate is an orange solid which is insoluble in water and a powerful oxidizing agent.

Halides of arsenic, antimony and bismuth

Each of arsenic, antimony and bismuth forms a halide of formula EX_3 (X = F, Cl, Br or I) and we focus here on the fluorides and chlorides. Arsenic(III) fluoride is a colourless liquid (mp 267 K, bp 330 K) and may be prepared from arsenic and F_2, but glass containers should be avoided for storage, because, in the presence of moisture, AsF_3 reacts with silica. The trichloride can be prepared by the reaction of As_2O_3 with HCl (equation 13.128) and $AsCl_3$ is a colourless liquid at 298 K. Both AsF_3 and $AsCl_3$ have molecular, trigonal pyramidal structures in the solid, liquid and gas phases. Each of the arsenic trihalides is hydrolysed by water.

$$As_2O_3 + 6HCl \rightarrow 2AsCl_3 + 3H_2O \qquad (13.128)$$

Arsenic(V) fluoride is the only halide of arsenic to be stable in the +5 oxidation state. It is a colourless gas at 298 K, with a trigonal bipyramidal structure, and can be made by reaction 13.129 in which Br_2 acts as an oxidizing agent. AsF_5 is an excellent *fluoride acceptor* as we discuss below.

$$AsF_3 + 2SbF_5 + Br_2 \rightarrow AsF_5 + 2SbBrF_4 \qquad (13.129)$$

In contrast to gaseous AsF_3, SbF_3 is a colourless crystalline solid (mp 563 K). Trigonal pyramidal molecular units are present in the crystal lattice, but fluorine atoms form bridges between antimony centres such that each antimony is six (rather than three) coordinate although the geometry is very distorted. SbF_3 is prepared by reacting Sb_2O_3 with HF, and is widely used as a fluorinating agent (equation 13.130) although reactions may be complicated by SbF_3 acting as an oxidizing agent *in addition to* donating fluorine (equation 13.131).

$$SiCl_4 \xrightarrow{SbF_3} SiCl_3F + SiCl_2F_2 + SiClF_3 + SiF_4 \qquad (13.131)$$

$$3\ PhPCl_2 + 4SbF_3 \rightarrow 3\ PhPF_4 + 2SbCl_3 + 2Sb \qquad (13.131)$$

Phosphorus(III) Phosphorus(V)

= phenyl group: see Chapter 15

The reaction of Sb_2O_3 with HCl gives $SbCl_3$ which is a crystalline solid at 298 K forming white, deliquescent crystals:

$$SbF_3 + F_2 \rightarrow SbF_5 \qquad (13.132)$$

Difluorine oxidizes SbF_3 to SbF_5 (equation 13.132). This is a viscous liquid (mp 281 K) due to the formation of Sb–F–Sb bridges between SbF_5 units, and in the solid state, the compound is tetrameric (Figure 13.24a). The oxidation of $SbCl_3$ with Cl_2 gives $SbCl_5$ which in the liquid state consists of trigonal bipyramidal molecules but in the solid (mp 276 K) consists of dimers with Sb–Cl–Sb bridges (Figure 13.24b).

The Lewis acidity of arsenic and antimony halides

An important property of arsenic and antimony trihalides and pentahalides is their Lewis acidity. The solid state structures of SbF_5 and $SbCl_5$ illustrate the tendency of antimony to accept a lone pair of electrons from a halide

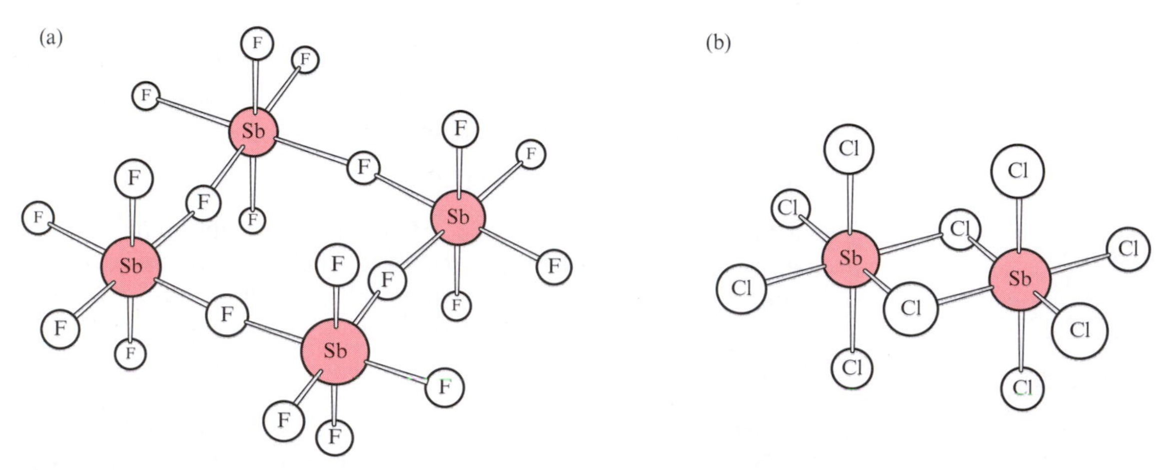

13.24 (a) The solid state structure of antimony(V) fluoride contains tetrameric molecules. Although each Sb atom is 6-coordinate, the stoichiometry remains Sb : F = 1 : 5. (b) Solid antimony(V) chloride consists of dimers Sb_2Cl_{10}, and again the coordination number of the antimony centre is 6 but the stoichiometry is Sb : Cl = 1 : 5.

and the formation of $\{SbF_5\}_4$ and $\{SbCl_5\}_2$ (Figure 13.24) is similar to that of Al_2Cl_6 illustrated in Figure 13.8. Halide bridge formation involves the donation of a lone pair of electrons, but *not* the complete transfer of a halide ion. When complete halide *transfer* occurs, ionic products are obtained as in reactions 13.133 and 13.134 – $SbCl_5$ is a particularly strong chloride acceptor. [*Question*: What shapes does the VSEPR theory predict for the ions formed in these reactions?]

$$\underset{\text{chloride acceptor}}{SbCl_5} + \underset{\text{chloride donor}}{AlCl_3} \rightarrow [AlCl_2]^+[SbCl_6]^- \quad (13.133)$$

$$\underset{\text{chloride acceptor}}{SbCl_5} + \underset{\text{chloride donor}}{PCl_5} \rightarrow [PCl_4]^+[SbCl_6]^- \quad (13.134)$$

The Lewis acidity of SbF_5 means that in anhydrous hydrogen fluoride, it functions as an acid (equation 13.135).

$$2HF + SbF_5 \rightleftharpoons [H_2F]^+ + [SbF_6]^- \quad (13.135)$$

Antimony(V) and arsenic(V) fluorides react with alkali metal fluorides or with FNO to give salts (equation 13.136).

$$\left.\begin{matrix} MF \\ \text{e.g. } M = Na, K, Cs \\ \text{or} \\ FNO \end{matrix}\right\} \begin{matrix} \xrightarrow{AsF_5} M^+[AsF_6]^- \text{ or } [NO]^+[AsF_6]^- \\ \xrightarrow{SbF_5} M^+[SbF_6]^- \text{ or } [NO]^+[SbF_6]^- \end{matrix} \quad (13.136)$$

The $[AsF_6]^-$ anion is octahedral, but $[SbF_6]^-$ shows a tendency to associate with Lewis bases in solid state structures; for example, with SbF_5, F→Sb coordinate bond formation leads to the formation of $[Sb_2F_{11}]^-$ (Figure 13.25). Further association may take place to give $[Sb_3F_{16}]^-$. In $[SbF_6]^-$, $[Sb_2F_{11}]^-$ and $[Sb_3F_{16}]^-$, each antimony(V) centre is octahedral.

Coordinate bond formation is not restricted to halides, and adducts with oxygen and nitrogen donors are known. When SbF_5 reacts with anhydrous HSO_3F (equation 13.137), the product is known as *super-acid* because it is capable of protonating exceptionally weak bases. Figure 13.26 shows the solid state structure of a related acid, $SbF_5OSO(OH)CF_3$.

HSO_3F: see Section 13.10

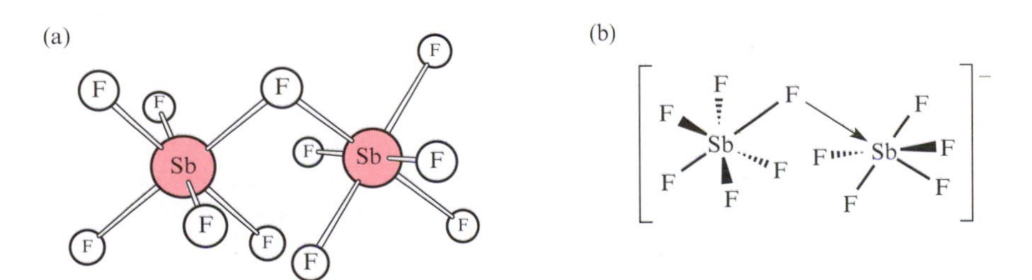

13.25 (a) The structure of the $[Sb_2F_{11}]^-$ anion, and (b) a representation of the structure indicating coordinate bond formation between $[SbF_6]^-$ and $[SbF_5]$.

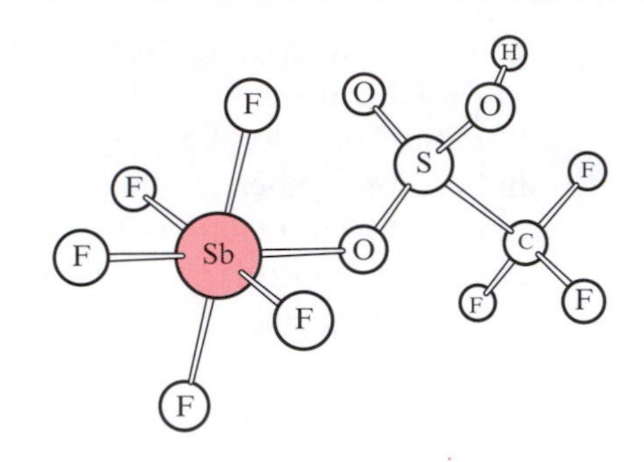

13.26 The solid state structure of the adduct $SbF_5OSO(OH)CF_3$.

$$2HSO_3F + SbF_5 \rightleftharpoons [H_2SO_3F]^+ + [SbF_5OSO_2F]^- \quad (13.137)$$

(13.41)

The trihalides of arsenic and antimony are also strong Lewis acids, and accept halide ions to give anions such as $[SbF_4]^-$ **13.42** and $[SbF_5]^{2-}$ **13.43**.

(13.42) **(13.43)**

13.10 GROUP 16: OXYGEN AND SULFUR

1	2		13	14	15	**16**	17	18
H								He
Li	Be		B	C	N	**O**	F	Ne
Na	Mg		Al	Si	P	**S**	Cl	Ar
K	Ca		Ga	Ge	As	**Se**	Br	Kr
Rb	Sr	*d*-block	In	Sn	Sb	**Te**	I	Xe
Cs	Ba		Tl	Pb	Bi	**Po**	At	Rn
Fr	Ra							

The group 16 elements are called the *chalcogens*. Dioxygen constitutes 21% of the Earth's atmosphere (see Figure 11.1), and oxygen-containing compounds (e.g. water, sand (SiO_2), limestone, silicates, bauxite and haematite

(Fe_2O_3)) constitute 49% of the Earth's crust. It is a component of many hundreds of thousands of compounds and is essential to life, being taken in and converted to CO_2 during respiration. Reactions 13.138 and 13.139 are convenient laboratory preparations of O_2, and a mixture of $KClO_3$ and MnO_2 used to be sold as 'oxygen mixture'. It can be produced by the electrolysis of water (equation 13.140), and on an industrial scale, O_2 is obtained by the liquefaction and fractional distillation of air.

$$2H_2O_2 \xrightarrow{MnO_2 \text{ or Pt catalyst}} O_2 + 2H_2O \qquad (13.138)$$

$$\underset{\text{potassium chlorate}}{2KClO_3} \xrightarrow{\Delta,\ MnO_2 \text{ catalyst}} 3O_2 + 2KCl \qquad (13.139)$$

Caution! chlorates are potentially explosive

$$\left.\begin{array}{ll} \textit{At the anode:} & 4[OH]^-(aq) \rightarrow O_2(g) + 2H_2O(l) + 4e^- \\ \textit{At the cathode:} & 2[H_3O]^+ + 2e^- \rightarrow H_2(g) + 2H_2O(l) \end{array}\right\} \qquad (13.140)$$

The diradical O_2: see Section 3.18

The ozone layer: see Box 8.9

Dioxygen is a colourless, odourless gas (bp 90 K), but liquid O_2 is blue. It is paramagnetic owing to the presence of two unpaired electrons and is very reactive – most elements combine with O_2 to form oxides. Electric discharges (for example in thunderstorms) or UV light convert O_2 into ozone (equation 13.141). Ozone is a colourless, toxic gas with a characteristic 'electric' smell. Molecules of O_3 are bent (Figure 13.27) with an O–O bond distance (128 pm) longer than in O_2 (121 pm) consistent with a lower bond order. The layer of ozone in the Earth's atmosphere is essential in preventing harmful UV radiation reaching the planet.

$$3O_2 \xrightarrow[\text{or } h\nu]{\text{electrical discharge}} \underset{\text{ozone}}{2O_3} \qquad (13.141)$$

Allotropes of sulfur: see Section 7.8

Elemental sulfur occurs in deposits around volcanoes and hot springs, as well as in minerals including *iron pyrites* (FeS_2, also called *fool's gold*, and which contains the $[S_2]^{2-}$ anion), *galena* (PbS), *sphalerite* (or zinc blende ZnS), *cinnabar* (HgS), *realgar* (As_4S_4) and *stibnite* (Sb_2S_3). Sulfur is obtained commercially from natural underground deposits by the Frasch process in which superheated water (440 K under pressure) is used to melt sulfur, and compressed air is then applied to force the sulfur to the surface (Figure 13.28). Sulfur produced in this way is ≈99% pure and is called *roll sulfur*.

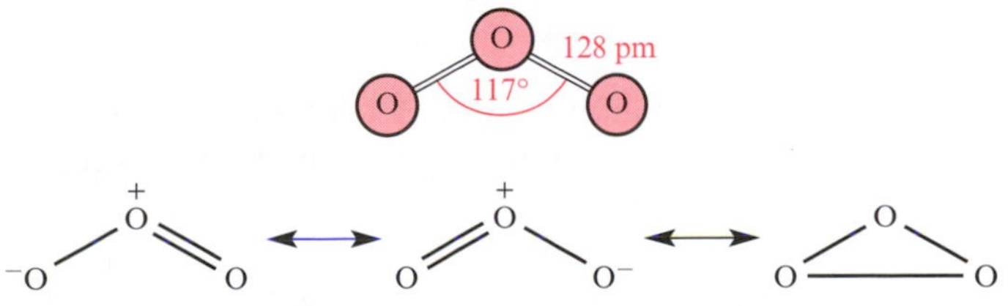

13.27 The structure of the ozone molecule O_3, and valence bond representations of the molecule. The contribution made by the cyclic resonance structure means that the average O–O bond order will be less than 1.5.

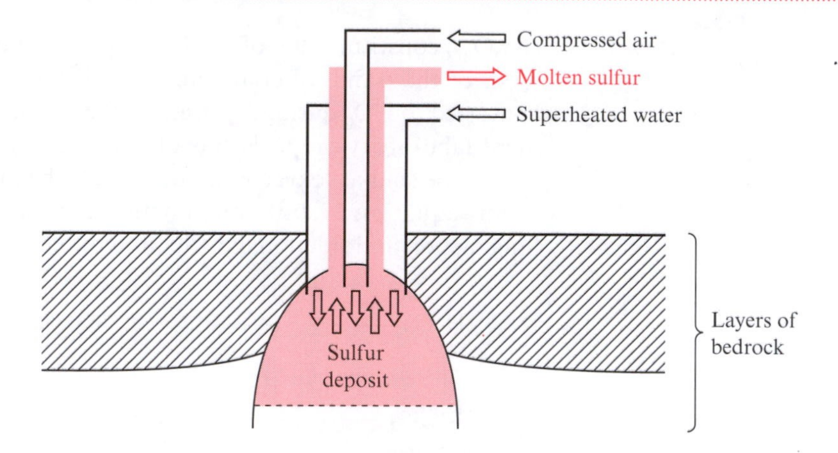

13.28 A schematic representation of the Frasch process, used to extract native sulfur from underground deposits. Superheated water melts the sulfur which is then driven to the surface under pressure.

Compounds of sulfur occur in natural gas and crude petroleum, and whereas once the sulfur content was wasted, techniques are now available to recover the element.

A major use of sulfur is in the production of sulfuric acid, ranked number one in manufactured chemical products both in 1993 and 1994 in the USA (Appendix 14). On an industrial level, it is also incorporated into a wide range of sulfur-containing organic compounds and inorganic compounds including $CaSO_4$, $[NH_4]_2SO_4$, CS_2, H_2S and SO_2.

Oxides of sulfur

The two most important oxides of sulfur are sulfur dioxide and sulfur trioxide. SO_2 is formed when sulfur burns in air or O_2 (equation 13.142), or by roasting sulfide ores with O_2 (equation 13.143) and this is the first step in the manufacture of sulfuric acid (see below).§

$$S(s) + O_2(g) \xrightarrow{\Delta} SO_2(g) \qquad (13.142)$$

$$4FeS_2 + 11O_2 \xrightarrow{\Delta} 2Fe_2O_3 + 8SO_2 \qquad (13.143)$$

Reactions 13.144 or 13.145 are convenient laboratory preparations of SO_2 and the gas is best dried by passage through concentrated sulfuric acid.

$$Na_2SO_3(s) + \underset{\text{conc.}}{2HCl(aq)} \rightarrow SO_2(g) + 2NaCl(aq) + H_2O(l) \qquad (13.144)$$

$$Cu(s) + \underset{\text{conc.}}{2H_2SO_4(aq)} \rightarrow SO_2(g) + CuSO_4(aq) + 2H_2O(l) \qquad (13.145)$$

Sulfur dioxide is a colourless, dense and toxic gas with a choking smell. It is easily liquefied (bp 263 K) and liquid SO_2 is widely used as a non-aqueous

§ In equations, we represent elemental sulfur simply as 'S' and do not specify whether S_8 rings or other allotropes are present.

solvent. It dissolves in water to give sulfurous acid, but equilibrium 13.146 lies well to the left-hand side and solutions contain significant amounts of dissolved SO_2. This situation parallels that of CO_2 in water.

$$SO_2(g) + H_2O(l) \rightleftharpoons H_2SO_3(aq) \qquad K = 1.3 \times 10^{-2} \qquad (13.146)$$

The SO_2 molecule is bent **13.44**. This structure may look similar to that of O_3 (Figure 13.27), but, whereas oxygen must obey the octet rule, sulfur can expand its octet. In SO_2 the sulfur–oxygen bond order is 2, but in O_3 the oxygen-oxygen bond order is less than 1.5.

(13.44)

Sulfur dioxide reacts with $COCl_2$ to give *thionyl dichloride* (equation 13.147) – commonly called 'thionyl chloride' – and is prepared industrially by reaction 13.148. It is used commercially in the synthesis of anhydrous metal chlorides and for the conversion of organic –OH to –Cl groups. It is a colourless liquid at 298 K, consisting of trigonal pyramidal molecules **13.45**, and is rapidly hydrolysed by water (equation 13.149).

Uses of $SOCl_2$: see Section 14.10

$$SO_2 + COCl_2 \rightarrow SOCl_2 + CO_2 \qquad (13.147)$$

$$2SO_2 + S_2Cl_2 + 3Cl_2 \rightarrow 4SOCl_2 \qquad (13.148)$$

$$SOCl_2 + H_2O \rightarrow SO_2 + 2HCl \qquad (13.149)$$

(13.45)

Sulfur dioxide is oxidized to SO_3 by O_2 in the presence of a catalyst such as vanadium(V) oxide. Figure 13.29 shows that equilibrium 13.150 shifts further towards the left-hand side as the temperature is raised, and industrial

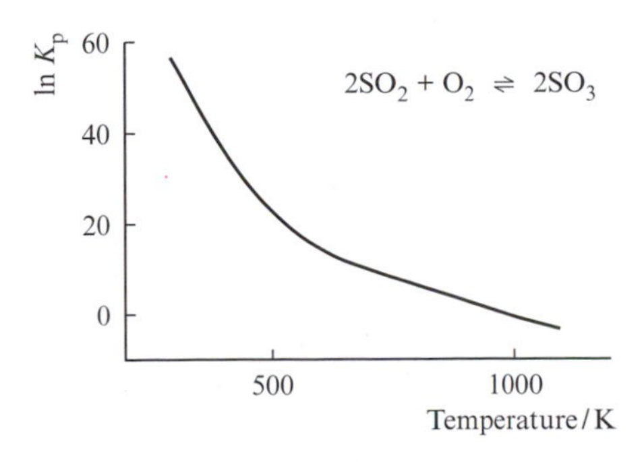

13.29 A graph illustrating the temperature-dependence of the equilibrium:

$$2SO_2(g) + O_2(g) \rightleftharpoons 2SO_3(g)$$

As the temperature is raised, ln K_p *decreases*, meaning that there is less SO_3 in the equilibrium mixture. In industrial processes in which SO_3 is produced, a temperature of about 700 K is used.

Box 13.14 Acid rain

Sulfur dioxide is produced naturally when volcanoes erupt, but the combustion of fossil fuels (e.g. coal and oil) releases large quantities of SO_2 into the atmosphere and these emissions are concentrated in the industrial regions of the world. The SO_2 dissolves in atmospheric water vapour, and forms H_2SO_3 and H_2SO_4. The formation of the acids takes time – the reactions leading to the formation of H_2SO_4 involve radicals:

$$SO_2 + O_2 + OH^{\bullet} \rightarrow SO_3 + HO_2^{\bullet}$$
$$SO_3 + H_2O \rightleftharpoons H_2SO_4$$

The acids are transported in the atmosphere in directions that are dictated by the prevailing winds. In Europe, SO_2 from the industrialized areas of the United Kingdom, France and Germany finds its way to Scandinavia and is deposited on the landscape as *acid rain*. One of the consequences of acid rain is that trees die, and when the acid rain enters natural water courses, fish die and this in turn affects other wildlife which depend upon the fish for food. Acid rain falling on stone buildings causes corrosion – the ever-disappearing faces of gargoyles on ancient churches are reminders of the problems of pollution.

NO_x emissions, discussed in Section 13.8, also contribute to the problem of acid rain. It is possible to 'scrub' industrial waste gases that are emitted from industry but the processes are expensive and new methods such as bacterial control are being sought.

processes in which SO_3 is produced use a relatively low operating temperature (≈700 K). The importance of this equilibrium lies in its role in the formation of sulfuric acid (see below).

$$2SO_2(g) + O_2(g) \rightleftharpoons 2SO_3(g) \qquad (13.150)$$

Sulfur dioxide is a weak reducing agent in acidic solution, and a rather stronger one in basic aqueous solution (when $[SO_3]^{2-}$ and $[HSO_3]^-$ ions are formed) as indicated by the E^o values for half-equations 13.151 and 13.152. The *oxidation* product in each case is the *sulfate ion*. Although we are considering the reactivity of SO_2, these equations show the presence of H_2SO_3 or $[SO_3]^{2-}$ in acidic or alkaline aqueous conditions rather than indicating a direct involvement by SO_2. This is because the aqueous solutions of SO_2 behave as though they contain these species as we discuss later in this section.

$$[SO_4]^{2-}(aq) + H^+(aq) + 2e^- \rightleftharpoons H_2SO_3(aq) + H_2O(l) \qquad E^o = +0.17\ V \qquad (13.151)$$

$$[SO_4]^{2-}(aq) + H_2O(l) + 2e^- \rightleftharpoons [SO_3]^{2-}(aq) + 2[OH]^-(aq) \qquad E^o = -0.93\ V \qquad (13.152)$$

At 298 K, SO_3 is a volatile white solid which is usually stored in sealed vessels. There are three modifications of the oxide of which α-SO_3 (mp 290 K) and β-SO_3 (mp 335 K) crystallize as silky, fibrous needles. In these forms, the sulfur and oxygen atoms are bonded in infinite arrays, but in the gas phase, trigonal planar SO_3 molecules **13.46** exist. In the liquid state, monomers and some trimers (Figure 13.30) are present.

(13.46)

SO_3 reacts extremely vigorously with water (equation 13.153) but concentrated H_2SO_4 dissolves SO_3 to give *oleum* consisting of a mixture of polysulfuric acids (see below). The reaction of oleum with water gives concentrated H_2SO_4 in a less vigorous reaction.

$$SO_3(g) + H_2O(l) \rightarrow H_2SO_4(aq) \qquad (13.153)$$

With HF and HCl, SO_3 reacts to give fluoro- or chlorosulfuric acid (equation 13.154). Anhydrous HSO_3F undergoes self-ionization (equation 13.155) – the equilibrium lies well over to the left-hand side, but is pushed towards the right in the presence of SbF_5 (equation 13.137). The combination of HSO_3F and SbF_5 generates one of the strongest acids known.

(13.47)

$$SO_3 + HF \rightarrow \underset{\mathbf{13.47}}{HSO_3F} \qquad (13.154)$$

$$2HSO_3F \rightleftharpoons [H_2SO_3F]^+ + [SO_3F]^- \qquad K \approx 10^{-8} \qquad (13.155)$$

Oxoacids of sulfur and their salts

Selected oxoacids of sulfur, of which sulfuric acid is by far the most important, are listed in Table 13.4. It is manufactured from SO_2 by the *Contact process* – SO_2 is first converted to SO_3 (equilibrium 13.150), the SO_3 is dissolved in concentrated H_2SO_4 to give oleum, dilution of which with water yields H_2SO_4. The major uses of sulfuric acid include the manufacture of inorganic sulfate salts, and the production of phosphorus and nitrogen-containing fertilizers (e.g. $[NH_4]_2SO_4$), as well as applications in the detergents and synthetic fibres industries (Figure 13.31).

Industrial ranking of H_2SO_4: see Appendix 14

Pure H_2SO_4 is a viscous liquid due to extensive intermolecular hydrogen bonding, and self-ionization occurs (equation 13.156), but is complicated by dehydration steps such as that in equation 13.157.

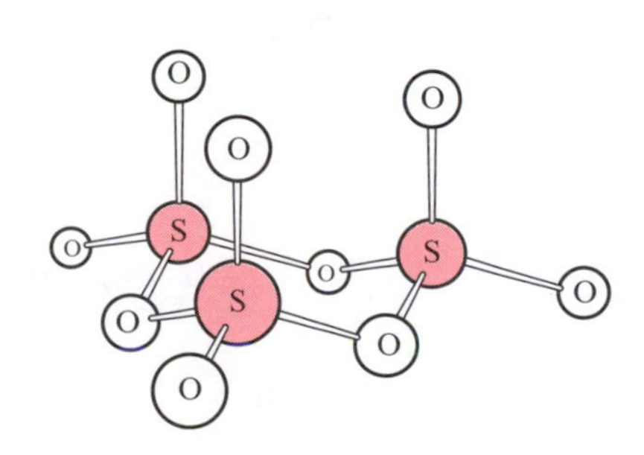

13.30 The structure of the trimer S_3O_9, present (along with monomeric SO_3) in liquid sulfur trioxide. In the monomer, each sulfur atom is in a trigonal planar environment, but in the trimer, each is tetrahedrally sited.

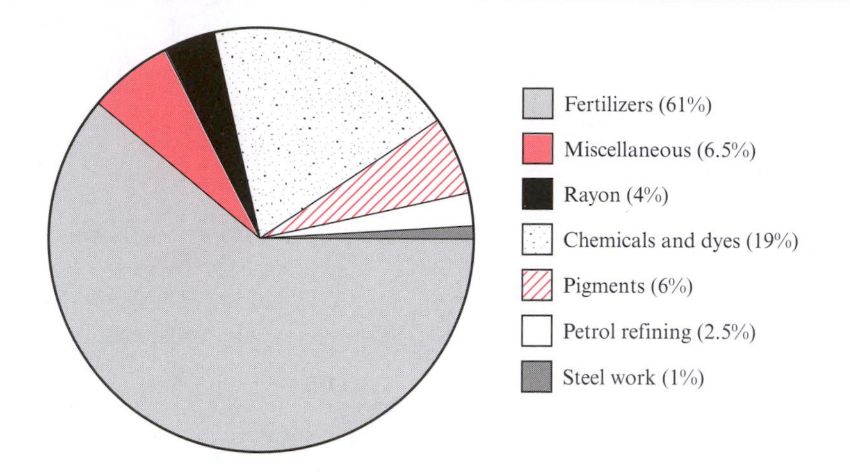

13.31 Applications of sulfuric acid. (Data from R. B. King, *Encyclopedia of Inorganic Chemistry*, (1994) vol. 7, Wiley, New York, p. 3968.)

Table 13.4 Selected oxoacids of sulfur. The number of OH groups determines the basicity of the acid; notice that *unlike* the case for phosphorus where some acids contain direct P–H bonds (Table 13.3), there are no oxoacids of sulfur containing direct S–H bonds.

Formula	*Name*	*Structure*
H_2SO_4	Sulfuric acid	
$H_2S_2O_3$	Thiosulfuric	
$H_2S_2O_7$	Disulfuric acid	
$H_2S_2O_8$	Peroxydisulfuric acid	
H_2SO_3	Sulfurous acid	
$H_2S_2O_4$	Dithionous acid	

$$\underset{\text{pure acid}}{2H_2SO_4(l)} \rightleftharpoons [H_2SO_4]^+(solv) + [HSO_4]^-(solv) \quad K \approx 1.7 \times 10^{-4} \tag{13.156}$$

$$H_2SO_4 + [HSO_4]^- \rightleftharpoons H_2O + [HS_2O_7]^- \tag{13.157}$$

Concentrated H_2SO_4 is a powerful dehydrating agent. If it is poured carefully onto a pile of sugar crystals, hydrogen and oxygen are removed from the sugar as water (equation 13.158) and a sticky, black deposit of carbon is left; the black product appears to 'grow' rather like a volcano erupting as gaseous CO and CO_2 are produced by oxidation processes that accompany the reaction.

$$\underset{\text{sucrose}}{C_{12}H_{22}O_{11}(s)} \xrightarrow{\text{concentrated } H_2SO_4} 11H_2O(l) + 12C(s) \qquad (13.158)$$

In aqueous solution, H_2SO_4 acts as a strong acid (equation 13.159) but the $[HSO_4]^-$ ion is a fairly weak acid (equation 13.160).

$$H_2SO_4(aq) + H_2O(l) \rightarrow [H_3O]^+ + \underset{\text{hydrogensulfate(1–) ion}}{[HSO_4]^-} \quad \textit{fully dissociated} \qquad (13.159)$$

$$[HSO_4]^-(aq) + H_2O(l) \rightleftharpoons [H_3O]^+ + \underset{\text{sulfate ion}}{[SO_4]^{2-}} \quad pK_a = 1.92 \qquad (13.160)$$

Sulfate(2–) is usually known simply as *sulfate*

Two series of salts are formed, for example $NaHSO_4$ and Na_2SO_4. *Dilute* aqueous sulfuric acid (typically 2 M) neutralizes bases (equations 13.161 and 13.162), and reacts with metal carbonates (equation 13.163) and some metals (equation 13.164). By looking at the standard reduction potentials in Table 12.4, you should be able to predict which metals will react with 0.5 M H_2SO_4; remember that the standard conditions for the redox cells is $[H^+] = 1$ mol dm^{-3}, and one mole of sulfuric acid provides two moles of H^+ ions.

$$H_2SO_4(aq) + NaOH(aq) \rightarrow NaHSO_4(aq) + H_2O(l) \qquad (13.161)$$

$$H_2SO_4(aq) + 2NaOH(aq) \rightarrow Na_2SO_4(aq) + H_2O(l) \qquad (13.162)$$

$$H_2SO_4(aq) + CuCO_3(s) \rightarrow CuSO_4(aq) + H_2O(l) + CO_2(g) \qquad (13.163)$$

$$H_2SO_4(aq) + Mg(s) \rightarrow MgSO_4(aq) + H_2(g) \qquad (13.164)$$

Many sulfate salts have commercial applications and an important one is in the fertilizer industry. Other uses include lead(II) sulfate in lead storage batteries, copper(II) sulfate in fungicides, magnesium sulfate as a laxative, and hydrated calcium sulfate ('plaster of Paris') in casts for broken bones.

Concentrated sulfuric acid is a good oxidizing agent (reaction 13.165). [*Question*: Why does *dilute* aqueous H_2SO_4 *not* react with copper to give $CuSO_4$?]

$$Cu(s) + \underset{\text{conc.}}{2H_2SO_4(aq)} \xrightarrow{\Delta} CuSO_4(aq) + SO_2(g) + 2H_2O(l) \qquad (13.165)$$

When we refer to 'sulfurous acid' (equations 13.166 and 13.167) we mean a *solution of sulfur dioxide in water* (equilibrium 13.146) – pure H_2SO_3 is *not* isolable. This exactly parallels the situation for 'carbonic acid'. Nonetheless, salts containing the $[HSO_3]^-$ and $[SO_3]^{2-}$ ions are common. Equilibria 13.166 and 13.167 are complicated by the fact that any aqueous solution of 'H_2SO_3' contains significant amounts of SO_2. Earlier in this section, we described the use of SO_2 as a reducing agent, and it should now be clear why half-equations 13.151 and 13.152 showed the presence of H_2SO_3 or $[SO_3]^{2-}$, rather than SO_2, in acidic or alkaline aqueous conditions respectively. Sodium sulfite is prepared by dissolving SO_2 in aqueous sodium hydroxide solution (reaction 13.168).

$$H_2SO_3(aq) + H_2O(l) \rightleftharpoons [H_3O]^+(aq) + [HSO_3]^-(aq) \qquad pK_a = 1.82 \tag{13.166}$$

$$[HSO_3]^-(aq) + H_2O(l) \rightleftharpoons [H_3O]^+(aq) + [SO_3]^{2-}(aq) \qquad pK_a = 6.92 \tag{13.167}$$

$$SO_2(aq) + 2NaOH(aq) \rightarrow Na_2SO_3(aq) + H_2O\,(l) \tag{13.168}$$

Although the anhydrous thiosulfuric acid $H_2S_2O_3$ exists, it decomposes in aqueous solutions. However, salts containing the thiosulfate anion, $[S_2O_3]^{2-}$, are stable and sodium thiosulfate is commercially important. It is used in the photographic industry as a reducing agent and to remove excess Cl_2 in bleaching processes. It is also commonly encountered in the laboratory as a reagent in I_2 analysis (reactions 13.169 and 13.170).

I_2 is insoluble in water: see Section 13.12

$$I_2(s) + \underset{\text{aqueous KI}}{I^-(aq)} \rightarrow [I_3]^-(aq) \tag{13.169}$$

$$2Na_2S_2O_3(aq) + [I_3]^-(aq) \rightarrow 2NaI(aq) + Na_2S_4O_6(aq) + I^-(aq) \tag{13.170}$$

Fluorides and chlorides of oxygen and sulfur

H_2O_2: see Section 11.10

Oxygen forms halides of the type OX_2 and O_2X_2 (X = halogen)§ in which it obeys the octet rule, and these compounds are structurally similar to H_2O and H_2O_2. At 298 K, oxygen difluoride, OF_2, is a highly toxic, pale yellow gas. It is formed by reaction 13.171, although this preparative route may be complicated by the fact that OF_2 reacts with hydroxide ion.

$$2NaOH(aq) + 2F_2(g) \rightarrow OF_2(g) + 2NaF(aq) + H_2O(l) \tag{13.171}$$

The role of the $ClO^{\bullet}$ radical in the depletion of the ozone layer: see Box 8.9

Oxygen difluoride is hydrolysed by water only slowly, but with steam, it explodes. When irradiated, OF_2 decomposes and the $FO^{\bullet}$ radicals combine to give O_2F_2 (equation 13.172). Above 220 K, this yellow gas is unstable with respect to decomposition into O_2 and F_2. When O_2F_2 combines with BF_3 or SbF_5, fluoride transfer occurs to give salts of the $[O_2]^+$ ion (reaction 13.173).

$$2OF_2 \xrightarrow{h\nu} 2FO^{\bullet} + 2F^{\bullet}; \quad 2FO^{\bullet} \rightarrow O_2F_2 \tag{13.172}$$

$$2O_2F_2 + 2SbF_5 \rightarrow 2[O_2][SbF_6] + F_2 \tag{13.173}$$

§ In formulae for binary compounds between oxygen, fluorine and chlorine, the symbols should be placed in the order Cl, O, F. Thus OF_2 and O_2F_2, but Cl_2O and ClO_2, are correct.

Oxygen dichloride Cl_2O is explosively unstable, and reacts with water to give hypochlorous acid (equation 13.174).

$$Cl_2O + H_2O \rightarrow 2HOCl \quad (13.174)$$

Chlorine dioxide ClO_2 is a paramagnetic, yellow gas at room temperature and is prepared in a hazardous reaction between concentrated H_2SO_4 and potassium chlorate ($KClO_3$), or by reaction 13.175 which is marginally less hazardous. An industrial application of ClO_2 is as a bleaching agent.

$$\underset{\text{potassium chlorate}}{2KClO_3} + \underset{\text{ethanedioic acid (oxalic acid)}}{2H_2C_2O_4} \rightarrow 2ClO_2 + K_2C_2O_4 + 2CO_2 + 2H_2O \quad (13.175)$$

We consider some further compounds formed between oxygen and the halogens in Section 13.12.

The number of sulfur fluorides exceeds the number of other halides of sulfur. Two structural isomers of S_2F_2 exist. In isomer **13.48**, the sulfur atom is in a formal oxidation state of +1, whilst in **13.49**, the *central* sulfur is in oxidation state +4. Structure **13.48** has a *gauche* conformation similar to that of H_2O_2 (Figure 11.15).

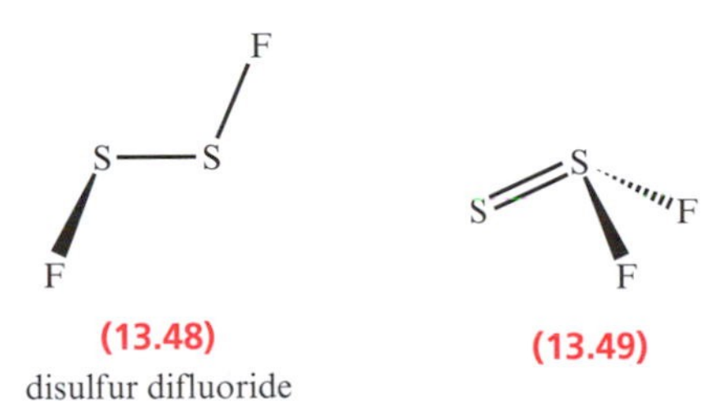

(13.48) disulfur difluoride

(13.49)

Sulfur difluoride SF_2 is not a well-known species, unlike its chloride analogue SCl_2, but the higher fluorides SF_4 and SF_6 are stable gases at room temperature.

$$3SCl_2 + 4NaF \rightarrow S_2Cl_2 + SF_4 + 4NaCl \quad (13.176)$$

Sulfur tetrafluoride can be prepared from SCl_2 (equation 13.176) and is useful as a selective fluorinating agent. SF_4 acts both as a Lewis acid (reaction 13.177) and as a Lewis base (reactions 13.178 and 13.179). Anhydrous conditions are essential in reactions involving SF_4, because it rapidly hydrolyses to give HF and SOF_2. [*Question*: in reaction 13.178, why are the products not $[BF_2]^+$ and $[SF_5]^-$?]

$$SF_4 + KF \rightarrow K[SF_5] \quad (13.177)$$

$$SF_4 + BF_3 \rightarrow [SF_3]^+ + [BF_4]^- \quad (13.178)$$

$$SF_4 + HF \rightleftharpoons [SF_3]^+ + [HF_2]^- \qquad K = 4 \times 10^{-2} \quad (13.179)$$

Although SF_4 can be formed when sulfur reacts with F_2, SF_6 is the dominant product when an excess of F_2 is used. Sulfur hexafluoride consists of octahedral molecules and is *kinetically* stable – even though reaction 13.180 is *thermodynamically* favoured, SF_6 is resistant to attack even by steam.

$$SF_6(g) + 3H_2O(g) \rightarrow SO_3(g) + 6HF(g)$$
$$\Delta G^o = -221 \text{ kJ per mole of } SF_6 \qquad (13.180)$$

The lower chlorides of sulfur are S_2Cl_2 and SCl_2; S_2Cl_2 is a toxic, foul-smelling yellow liquid and is prepared by reacting molten sulfur with Cl_2. Structurally it resembles disulfur difluoride and has a *gauche* conformation. S_2Cl_2 reacts further with Cl_2 to give SCl_2 which is a poisonous, red liquid at 298 K, but SCl_2 tends to decompose according to equilibrium 13.181.

$$2SCl_2 \rightleftharpoons S_2Cl_2 + Cl_2 \qquad (13.181)$$

Both SCl_2 and S_2Cl_2 are reagents in synthetic chemistry for the preparation of organosulfur compounds.

13.11 GROUP 16: SELENIUM AND TELLURIUM

1	2		13	14	15	**16**	17	18
H								He
Li	Be		B	C	N	**O**	F	Ne
Na	Mg		Al	Si	P	**S**	Cl	Ar
K	Ca		Ga	Ge	As	**Se**	Br	Kr
Rb	Sr	*d*-block	In	Sn	Sb	**Te**	I	Xe
Cs	Ba		Tl	Pb	Bi	**Po**	At	Rn
Fr	Ra							

Allotropes of selenium: see Section 7.8

Selenium is found in only a few minerals, and one commercial source is flue dusts deposited during the refining of copper sulfide ores. An important property of selenium is its ability to convert light into electricity, and the element is used in photoelectric cells and in photographic exposure meters. Below its melting point (490 K), selenium is a semiconductor.

Tellurium is usually found combined with other metals, e.g. with gold in the mineral *calaverite*. Pure metallic tellurium has a silver-white appearance and both it and its compounds are toxic; if you are exposed to tellurium, your breath develops a garlic-like smell. Some of the applications of the element arise from its semiconducting properties.

The heaviest element in group 16 is polonium, but only milligram quantities of this radioactive element are available.

Compounds with selenium–oxygen or tellurium–oxygen bonds

In contrast to SO_2, SeO_2 and TeO_2 (formed from the element and O_2) are crystalline solids at 298 K with lattice structures. Selenium dioxide dissolves in water to give selenous acid, H_2SeO_3, and its ease of reduction makes it a suitable oxidizing agent for some organic reactions. There are two structural forms of solid TeO_2, of which white crystalline α-TeO_2 dissolves in water to give H_2TeO_3. Whereas both SO_2 and TeO_2 are readily oxidized to trioxides, SeO_3 is thermodynamically *unstable* with respect to SeO_2 and O_2.

In contrast to sulfur, selenium and tellurium do not form a wide range of oxoacids. Both H_2SeO_3 and H_2TeO_3 are white solids at 298 K, and can be dehydrated to the respective metal dioxide. The acids are dibasic, and two sets of salts containing the $[HSeO_3]^-$ and $[SeO_3]^{2-}$, or $[HTeO_3]^-$ and $[TeO_3]^{2-}$ anions are formed.

Fluorides and chlorides of selenium and tellurium

A mixture of SeF_2, Se_2F_2 and SeF_4 is obtained when selenium vapour and F_2 react. Selenium tetrafluoride is a good fluorinating agent, and since it is a liquid at room temperature (mp 259 K), it is more convenient to use than gaseous SF_4. Liquid SeF_4 contains discrete molecules (Figure 13.32a) but in the solid state, they form an ordered array with significant intermolecular interactions. In solid TeF_4, association between the TeF_4 units is more extensive owing to the formation of Te–F–Te bridges (Figure 13.32b).

Like SF_4, SeF_4 acts as a *fluoride donor* but equilibrium 13.182 lies further to the left than the corresponding equilibrium involving SF_4 (equation 13.179). SeF_4 can also act as a *fluoride acceptor* as in reaction 13.183. Similarly, TeF_4 behaves both as a fluoride donor and acceptor.

$$SeF_4 + HF \rightleftharpoons [SeF_3]^+ + [HF_2]^- \qquad K = 4 \times 10^{-4} \qquad (13.182)$$

$$SeF_4 + CsF \rightleftharpoons Cs^+[SeF_5]^- \qquad (13.183)$$

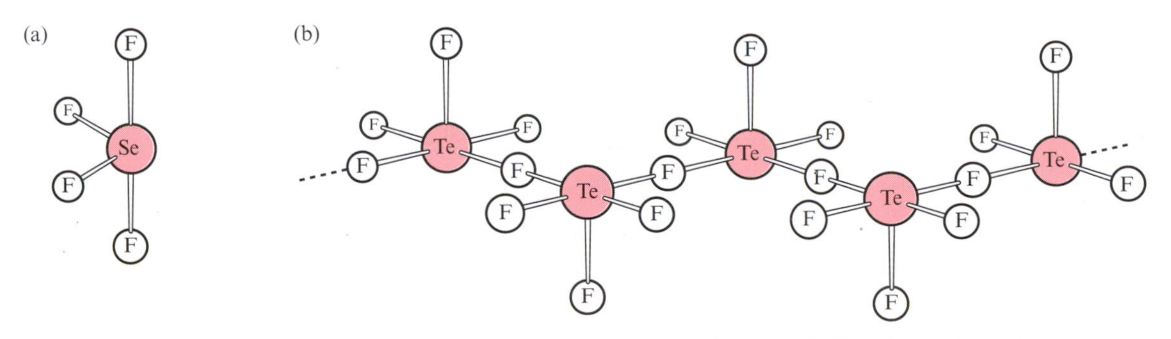

13.32 The structures of (a) SeF_4 and (b) TeF_4 in the solid state. Each tellurium atom in (b) is 5-coordinate but the stoichiometry remains Te : F = 1 : 4. Why is the geometry of SeF_4 not tetrahedral? Why is the geometry of each 5-coordinate tellurium atom square pyramidal and not trigonal bipyramidal?

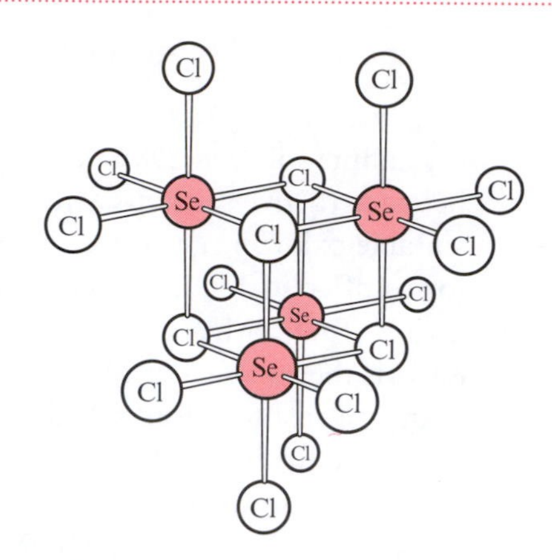

13.33 The structure of the molecular Se_4Cl_{16} unit present in the crystal lattice of $SeCl_4$. Similar molecular units of formula Te_4Cl_{16} are present in $TeCl_4$.

Whereas SCl_4 is unstable, $SeCl_4$ is produced in the reaction of selenium with Cl_2 although it decomposes when heated (equation 13.184). $TeCl_4$ is prepared similarly and is the only stable chloride that tellurium forms. Both $SeCl_4$ and $TeCl_4$ are crystalline solids at 298 K, and their solid state lattices contain tetrameric Se_4Cl_{16} or Te_4Cl_{16} units (Figure 13.33).

$$SeCl_4 \xrightarrow{\Delta} SeCl_2 + Cl_2 \qquad (13.184)$$

Selenium and tellurium tetrachlorides *accept* chloride ions to give the anions $[SeCl_6]^{2-}$ and $[TeCl_6]^{2-}$ (equation 13.185). Each possesses a *regular octahedral* structure, despite having 14 electrons in the valence shell – the lone pair is stereochemically inactive.

Stereochemically inactive lone pair: see Section 5.7

$$SeCl_4 + 2KCl \rightarrow K_2[SeCl_6] \qquad (13.185)$$

Donation of chloride ion by $SeCl_4$ is observed in its reaction with $AlCl_3$, and an X-ray diffraction study of the crystalline product confirms the presence of discrete $[SeCl_3]^+$ and $[AlCl_4]^-$ ions.

Selenium hexafluoride (produced from selenium and F_2) is a gas at 298 K and is quite stable, resisting hydrolysis even in alkaline aqueous solution. TeF_6 is well characterized, but binary *chlorides* of selenium(VI) and tellurium(VI) are not known. It is the oxidizing ability of F_2 (emphasized by the value of E^o for half-equation 13.186) and the small size of fluorine that makes the formation of SF_6, SeF_6 and TeF_6 feasible.

$$F_2(g) + 2e^- \rightleftharpoons 2F^-(aq) \qquad E^o = +2.87 \text{ V} \qquad (13.186)$$

13.12 GROUP 17: THE HALOGENS

1	2		13	14	15	16	**17**	18
H								He
Li	Be		B	C	N	O	**F**	Ne
Na	Mg		Al	Si	P	S	**Cl**	Ar
K	Ca		Ga	Ge	As	Se	**Br**	Kr
Rb	Sr	*d*-block	In	Sn	Sb	Te	**I**	Xe
Cs	Ba		Tl	Pb	Bi	Po	**At**	Rn
Fr	Ra							

Fluorine is not found in the elemental state of F_2, but occurs as the fluoride ion in *fluorite* CaF_2, *cryolite* $Na_3[AlF_6]$ and a range of other minerals. Because of its strongly oxidizing nature, F_2 must be prepared by electrolytic oxidation of F^-, and in the industrial route, a mixture of anhydrous molten KF and HF is electrolysed (equation 13.187).

$$\textit{At a carbon anode: } 2F^-(l) \rightarrow F_2(g) + 2e^- \qquad (13.187)$$

Difluorine is a pale-yellow gas (bp 85 K) consisting of discrete F_2 molecules. Fluorine is the most electronegative of all the elements, and F_2 is corrosive, reacting with most elements and compounds. It is a dangerous gas to use, being extremely corrosive and must be handled in special steel containers. The developments of the atomic bomb and the nuclear energy industry have demanded large quantities of F_2 – a major use is in the production of uranium hexafluoride during nuclear fuel enrichment processes.

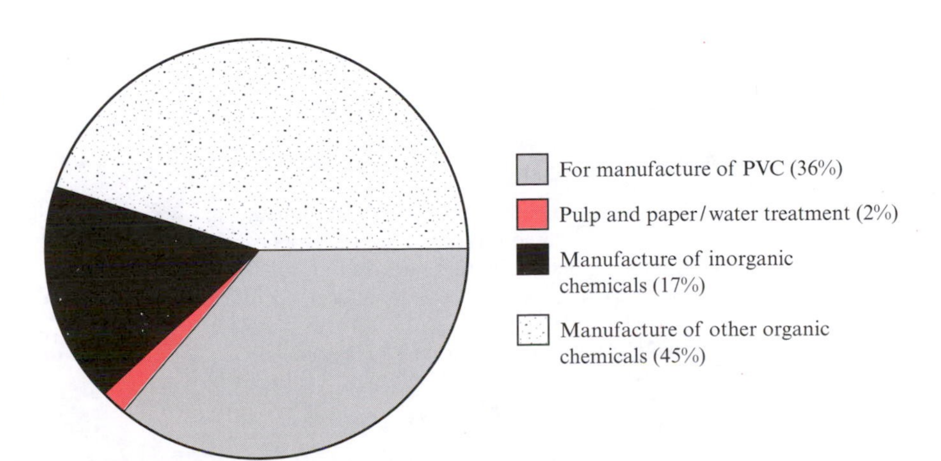

13.34 Uses of Cl_2 in Western Europe in 1994. The percentage used for pulp and paper bleaching and water chlorination has dropped in recent years owing to environmental concern. Applications for CFCs have also been phased out. (Data from *Chemistry & Industry*, (1995) p. 832.)

The structures of the halogens: see Section 7.7

Dichlorine is amongst the most important industrial chemicals (Appendix 14) and is extracted from NaCl using the Downs process (Figure 11.17). On a laboratory scale, reaction 13.188 is a convenient synthesis of Cl_2.

$$MnO_2(s) + 4HCl(aq) \rightarrow MnCl_2(aq) + Cl_2(g) + 2H_2O(l) \qquad (13.188)$$

Dichlorine is a pale green-yellow gas (bp 239 K) with a characteristic smell. When inhaled, it causes irritation of the respiratory system and liquid Cl_2 burns the skin. Drinking-water supplies are often chlorinated in order to make the water safer for human consumption, and Cl_2 is used widely as a disinfectant and a bleach. Its conversion into chlorine-containing compounds, including HCl, is a major industrial use (Figure 13.34). It reacts with many elements to give the respective chlorides.

Box 13.15 Nuclear reactors and the reprocessing of nuclear fuels

Uranium is a nuclear fuel, and consists naturally of the isotopes $^{234}_{92}U$ (< 0.01%), $^{235}_{92}U$ (≈0.72%) and $^{238}_{92}U$ (≈99.28%) – of these, $^{235}_{92}U$ has the ability to *capture a thermal neutron* to yield the unstable isotope $^{236}_{92}U$. This is followed by nuclear fission:

$$^{235}_{92}U + ^{1}_{0}n \rightarrow ^{236}_{92}U \xrightarrow{\text{nuclear fusion}} ^{141}_{56}Ba + ^{92}_{36}Kr + 3\,^{1}_{0}n$$

and release of about 2×10^{10} kJ of energy per mole of reaction. The production of extra neutrons means that there is the potential for a *chain reaction* and this is the basis for the atomic bomb. In a nuclear reactor, nuclear energy is harnessed for electricity among other uses, but the nuclear reaction must be efficiently *controlled*. This is done by inserting boron or cadmium rods (which absorb neutrons) between the uranium fuel rods.

As $^{235}_{92}U$ is used up in the nuclear reactor, the fuel becomes spent. Spent nuclear fuels are not disposed of but are *reprocessed* – this recovers uranium and separates $^{235}_{92}U$ from fission products. After preliminary processes including *pond storage* (which allows time for short-lived radioactive products to decay), reprocessing begins by converting uranium metal to soluble uranyl nitrate:

$$U(s) + 8HNO_3(aq) \rightarrow [UO_2][NO_3]_2(aq) + 4H_2O(l) + 6NO_2(g)$$

This is followed by conversion to uranium(VI) oxide, reduction to uranium(IV) oxide, and fluorination to give UF_6:

$$[UO_2][NO_3]_2(s) \xrightarrow{570\text{ K}} UO_3(s) + NO(g) + NO_2(g) + O_2(g)$$

$$UO_3(s) + H_2(g) \xrightarrow{970\text{ K}} UO_2(s) + H_2O(g)$$

$$UO_2(s) + 4HF(aq) \rightarrow UF_4(s) + 2H_2O(l)$$

$$UF_4(s) + F_2(g) \xrightarrow{720\text{ K}} UF_6(g)$$

The final stage of reprocessing is the separation of $^{235}_{92}UF_6$ from $^{238}_{92}UF_6$ and is achieved by using a special centrifuge. *Graham's Law* states that the rate of gas effusion (or diffusion) depends upon the molecular mass:

$$\text{Rate of effusion} \propto \frac{1}{\sqrt{\text{Molecular mass}}}$$

Graham's Law is discussed further in Chapter 1 of the accompanying workbook. Application of this law shows that $^{235}_{92}UF_6$ and $^{238}_{92}UF_6$ can be separated by subjecting them to a centrifugal force which moves the molecules to the outer wall of their container, but at different rates. In this way it is possible to obtain $^{235}_{92}U$ *enriched* UF_6. After this process, the hexafluoride is converted back to uranium metal before the fuel is ready for reuse in the nuclear reactor:

$$UF_6(g) + H_2(g) + 2H_2O(g) \xrightarrow{870\text{ K}} UO_2(s) + 6HF(g)$$

$$UO_2(s) + 2Mg(s) \xrightarrow{\Delta} U(s) + 2MgO(s)$$

Dibromine is a dark-orange, volatile liquid (mp 266 K, bp 332 K) and is the only non-metallic element which is a liquid at room temperature. In the laboratory it is often encountered as the aqueous solution 'bromine water'. Br_2 is toxic – the liquid produces skin burns and the vapour has an unpleasant odour and causes eye and respiratory irritations. Bromine occurs as the bromide ion in brine wells in parts of the USA, and can be extracted using reaction 13.189. Some uses of Br_2 include its role in flame-proofing materials, dyes, bromide compounds for the photographic industry and organic bromides.

$$2Br^-(aq) + Cl_2(g) \rightarrow Br_2(aq) + 2Cl^-(aq) \qquad (13.189)$$

(oxidation: $Br^- \rightarrow Br_2$; reduction: $Cl_2 \rightarrow Cl^-$)

At 298 K, diiodine forms dark purple crystals containing I_2 molecules (mp 387 K, bp 458 K) and the solid sublimes readily at atmospheric pressure into a purple vapour which irritates the eyes. Iodine occurs as the I^- ion in seawater and is taken up by seaweed from which it is possible to extract I_2. On an industrial scale, it is extracted from oil and brine wells by using Cl_2 to oxidize I^- to I_2 (analogous to reaction 13.189). Unlike F_2, Cl_2 and Br_2, I_2 is only very sparingly soluble in water, and laboratory solutions are usually made up in aqueous potassium iodide in which I_2 is soluble giving $[I_3]^-$ salts (equation 13.169). Iodine is an essential element for life, and a deficiency results in a swollen thyroid gland; 'iodized salt' (NaCl with added I^-) provides supplemental iodine to the body. Iodide salts such as KI have applications in the photographic industry, and solutions of I_2 in aqueous KI are used as disinfectants for wounds.

The trend in reactivities of the halogens reflects the decrease in electronegativity values from F ($\chi^P = 4.0$) to I ($\chi^P = 2.7$) and the decreasing oxidizing

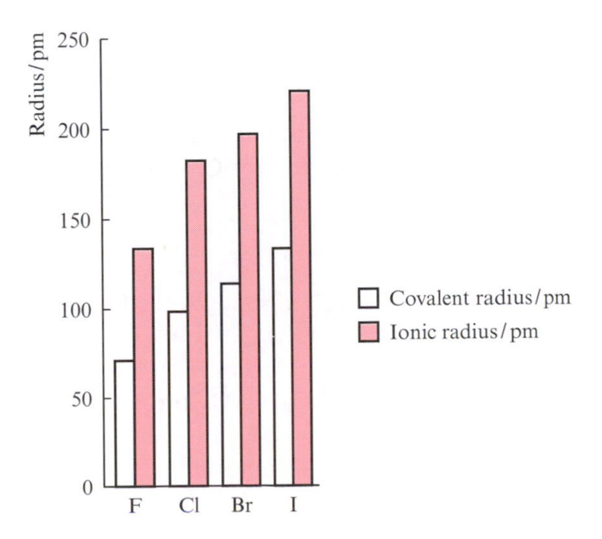

13.35 The trend in the covalent and ionic radii of the halogens.

Table 13.5 Standard reduction potentials for the reduction of the halogens. F_2 is the most readily reduced and is the most powerful oxidizing agent in the series.

Reduction half-reaction	E^o/V
$I_2(aq) + 2e^- \rightleftharpoons 2I^-(aq)$	+0.54
$Br_2(aq) + 2e^- \rightleftharpoons 2Br^-(aq)$	+1.09
$Cl_2(aq) + 2e^- \rightleftharpoons 2Cl^-(aq)$	+1.36
$F_2(aq) + 2e^- \rightleftharpoons 2F^-(aq)$	+2.87

strength of the element (Table 13.5). Fluorine is the most electronegative element in the periodic table, and in its compounds it always has a formal oxidation state of –1. The lower halogens form compounds with a range of oxidation states (see below). The relative sizes of the halogen atoms and ions (Figure 13.35) also contribute to some of the differences between the chemistries of the group 17 elements. The small covalent radius of fluorine permits elements with which fluorine combines to attain high coordination numbers. For example, six fluorine atoms will fit around a sulfur atom, but six iodine atoms will not. In ionic metal fluorides, the small size of the F^- ion contributes to the high values of lattice enthalpies and to the thermodynamic stability of the compounds.

Dependence of lattice enthalpy on ion size: see Section 6.15

The relative strength of F_2 as an oxidizing agent with respect to the other halogens is seen in the formation of *interhalogen compounds*. These are compounds formed by combining two *different* halogens. In each of ClF, ClF_3, BrF_5 and IF_7, the formal oxidation state of fluorine is –1, while the oxidation states of the other halogens range from +1 to +7. In *none* of the interhalogen compounds is fluorine in a positive oxidation state — F_2 is able to oxidize the lower halogens (equation 13.190) but Cl_2, Br_2 and I_2 are *not* able to oxidize F_2.

$$F_2 \xrightarrow{Cl_2,\ 470\ K} ClF_3 \qquad F_2 \xrightarrow{Br_2,\ 420\ K} BrF_5 \qquad F_2 \xrightarrow{I_2,\ 530\ K} IF_7 \tag{13.190}$$

Hydrogen halides: see Section 11.11

Examples of reactions of halogens with metals and non-metals are considered in the relevant sections of Chapters 11 and 13.

Oxides of the halogens

We have already described some of the lower oxides of fluorine and chlorine (OF_2, O_2F_2, Cl_2O and ClO_2). Fluorine does not form higher oxides because it is a stronger oxidizing agent than oxygen, but the other halogens form compounds such as Cl_2O_6, Cl_2O_7 and I_2O_5.

Dichlorine hexaoxide Cl_2O_6 is a dark red liquid at 298 K and is prepared by the reaction between ozone and ClO_2 (equation 13.191). It must be handled with extreme care as it explodes when in contact with organic material.

$$2ClO_2 + 2O_3 \rightarrow Cl_2O_6 + 2O_2 \tag{13.191}$$

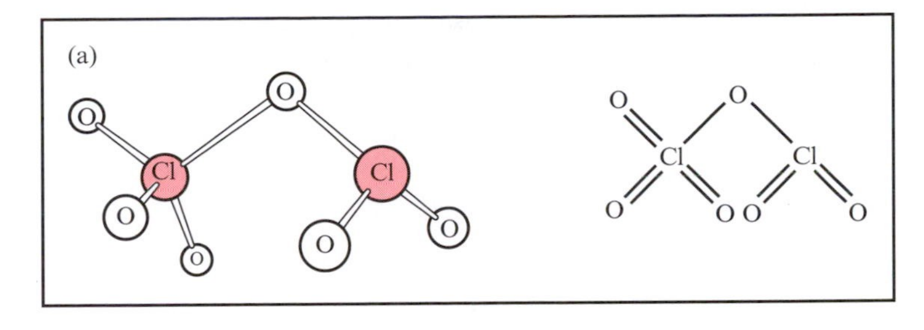

(b)

141 pm

171 pm

13.36 (a) The proposed structure of Cl_2O_6 and (b) the confirmed structure of Cl_2O_7.

There has been much speculation as to the exact structure of Cl_2O_6, but the probable structure (Figure 13.36a) has one chlorine atom in a trigonal pyramidal site (with a lone pair of electrons on chlorine) and one tetrahedrally sited. In the solid state (mp 277 K), it appears that $[ClO_2]^+$ and $[ClO_4]^-$ ions are present. The oxide is unstable with respect to decomposition into ClO_2 and O_2, and with water, reaction 13.192 occurs.

$$\underset{}{Cl_2O_6} + H_2O \rightarrow \underset{\text{perchloric acid}}{HClO_4} + \underset{\text{chloric acid}}{HClO_3} \tag{13.192}$$

The oxide Cl_2O_7 is a colourless oily liquid (mp 182 K, bp 355 K) which is shock-sensitive and potentially explosive. The differences in the Cl–O bond distances (Figure 13.36b) are consistent with the presence of single and double bonds. It is formed by dehydrating perchloric acid (equation 13.193), but the reaction is reversed when Cl_2O_7 contacts water.

$$2HClO_4 \xrightarrow{H_3PO_4,\ 263\ K} H_2O + Cl_2O_7 \tag{13.193}$$

Bromine oxides are all unstable, but iodine forms I_2O_5 as white, hygroscopic crystals which decompose above 570 K. The solid state structure consists of an extended array. The dehydration of HIO_3 is one method of preparing I_2O_5, but the compound readily takes up water again to regenerate the acid. I_2O_5 is a powerful oxidizing agent and oxidizes CO to CO_2. This reaction may be used in the quantitative analysis of carbon monoxide, the I_2 produced being analysed by titration with sodium thiosulfate using a starch indicator (equation 13.194). The I_2 in equation 13.194 is in the form of $[I_3]^-$ (see equation 13.169).

Quantitative analysis for CO:

$$\left.\begin{array}{l}5CO + I_2O_5 \rightarrow 5CO_2 + I_2 \\ I_2 + 2Na_2S_2O_3 \rightarrow 2NaI + Na_2S_4O_6\end{array}\right\} \quad (13.194)$$

Box 13.16 Starch test for I_2

Starch is a widespread component of plant material and is obtained from potatoes, wheat and rice. One of the components of starch is amylose which consists of α-D(+)glucopyranose units linked together to produce helical coils.

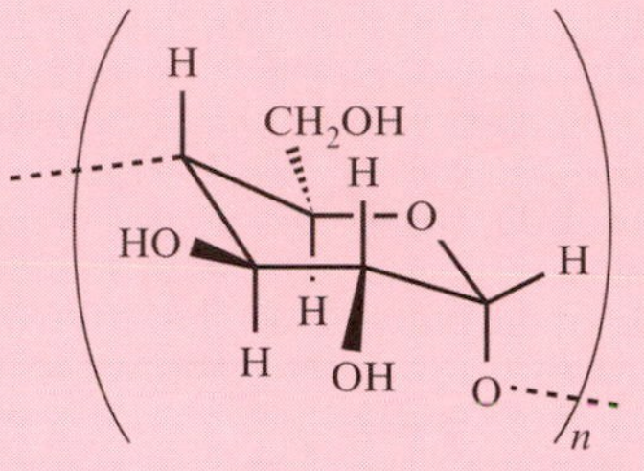

α-D(+)glucopyranose unit

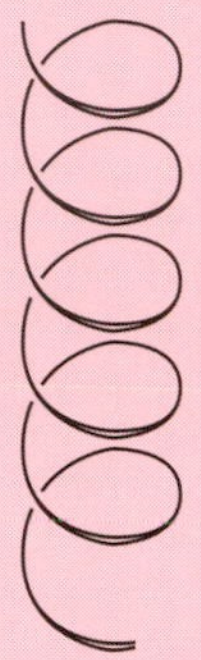

Helical coil

Solutions of I_2 in aqueous KI contain a series of polyiodide ions including $[I_3]^-$ and $[I_5]^-$:

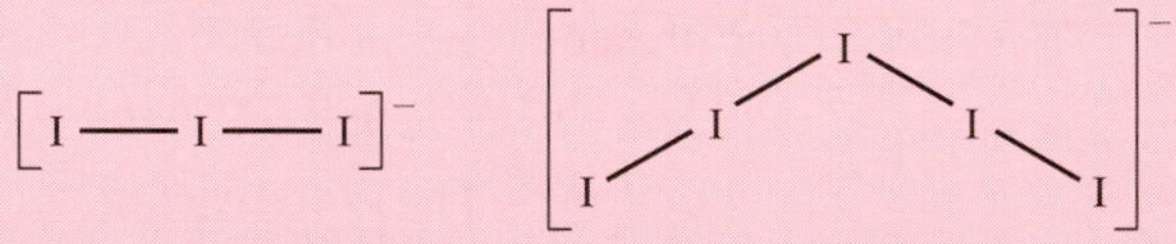

$$[I—I—I]^- \qquad [I_5]^-$$

The $[I_5]^-$ ion is trapped within the channels that run through the helical structures of α-D(+)glucopyranose units. The adduct formed has a characteristic blue-black colour. It is the formation of this coloured adduct that accounts for the use of starch as an 'indicator' in titrations involving iodine.

Oxoacids of the halogens and the oxoanions

The oxides Cl_2O, Cl_2O_7 and I_2O_5 are the *anhydrides* of HOCl, $HClO_4$ and HIO_3 respectively. The acids and their oxoanions are all good oxidizing agents, and Table 13.6 lists selected half-equations and the corresponding standard reduction potentials – remember that the more positive the value of E^o, the more readily the reduction process occurs.

The only oxoacid of fluorine is HOF, and it spontaneously decomposes to HF and O_2. Hypochlorous acid, HOCl, and its heavier congeners HOBr and HOI, cannot be isolated as pure compounds, but are encountered as aqueous solutions. They are weak, monobasic acids (equilibrium 13.195), but the $[OX]^-$ anions are unstable with respect to disproportionation (equation 13.196), the reaction being slow for $[OCl]^-$, fast for $[OBr]^-$ and very rapid for $[OI]^-$.

Table 13.6 Standard reduction potentials for selected reduction reactions involving oxoacids of chlorine, bromine and iodine.

Reduction half-reaction	E^o/V
$[ClO_4]^-(aq) + H_2O(l) + 2e^- \rightleftharpoons [ClO_3]^-(aq) + 2[OH]^-(aq)$	+0.36
$[BrO_3]^-(aq) + 3H_2O(l) + 6e^- \rightleftharpoons Br^-(aq) + 6[OH]^-(aq)$	+0.61
$[BrO]^-(aq) + H_2O(l) + 2e^- \rightleftharpoons Br^-(aq) + 2[OH]^-(aq)$	+0.76
$[ClO]^-(aq) + H_2O(l) + 2e^- \rightleftharpoons Cl^-(aq) + 2[OH]^-(aq)$	+0.84
$[IO_3]^-(aq) + 6H^+(aq) + 6e^- \rightleftharpoons I^-(aq) + 3H_2O(l)$	+1.09
$2[IO_3]^-(aq) + 12H^+(aq) + 10e^- \rightleftharpoons I_2(aq) + 6H_2O(l)$	+1.20
$[ClO_4]^-(aq) + 8H^+(aq) + 8e^- \rightleftharpoons Cl^-(aq) + 4H_2O(l)$	+1.39
$2[ClO_4]^-(aq) + 16H^+(aq) + 14e^- \rightleftharpoons Cl_2(aq) + 8H_2O(l)$	+1.39
$[BrO_3]^-(aq) + 6H^+(aq) + 6e^- \rightleftharpoons Br^-(aq) + 3H_2O(l)$	+1.42
$[ClO_3]^-(aq) + 6H^+(aq) + 6e^- \rightleftharpoons Cl^-(aq) + 3H_2O(l)$	+1.45
$2[ClO_3]^-(aq) + 12H^+(aq) + 10e^- \rightleftharpoons Cl_2(aq) + 6H_2O(l)$	+1.47
$2[BrO_3]^-(aq) + 12H^+(aq) + 10e^- \rightleftharpoons Br_2(aq) + 6H_2O(l)$	+1.48
$HOCl(aq) + H^+(aq) + 2e^- \rightleftharpoons Cl^-(aq) + H_2O(l)$	+1.48
$2HOCl(aq) + 2H^+(aq) + 2e^- \rightleftharpoons Cl_2(aq) + 2H_2O(l)$	+1.61

$$\left.\begin{array}{l} HOX(aq) + H_2O(l) \rightleftharpoons [H_3O]^+(aq) + [OX]^-(aq) \\ \qquad X = Cl \quad pK_a = 4.53 \\ \qquad X = Br \quad pK_a = 8.69 \\ \qquad X = I \quad pK_a = 10.64 \end{array}\right\} \quad (13.195)$$

$$3[OX]^-(aq) \rightarrow [XO_3]^-(aq) + 2X^-(aq) \quad (13.196)$$

Salts such as NaOCl, KOCl and $Ca(OCl)_2$ can be isolated (e.g. reaction 13.197) and are valuable oxidizing agents; $Ca(OCl)_2$ is a component of bleaching powder, and NaOCl is a bleaching agent and disinfectant.

$$2CaO(s) + 2Cl_2(g) \rightarrow Ca(OCl)_2(s) + CaCl_2(s) \quad (13.197)$$

Chloric and bromic acids, $HClO_3$ and $HBrO_3$, are both strong acids but cannot be isolated as pure compounds. Aqueous solutions of $HClO_3$ can be made by reaction 13.198, and bromic acid is prepared similarly.

$$Ba(ClO_3)_2(aq) + H_2SO_4(aq) \rightarrow BaSO_4(s) + 2HClO_3(aq) \quad (13.198)$$

Iodic acid HIO_3 is a stable, white solid at room temperature, and is produced by reacting I_2O_5 with water. In the solid state, trigonal HIO_3 molecules are present with extensive intermolecular hydrogen bonding. In aqueous solution it behaves as a monobasic acid ($pK_a = 0.77$).

The three halic acids and the oxoanions derived from them are strong oxidizing agents (see Table 13.6), and a conproportionation reaction used in volumetric analysis combines the reduction of the $[XO_3]^-$ ion to X_2 with the oxidation of X^- to X_2 (e.g. reaction 13.199). Reactions 13.200 and 13.201 further illustrate the uses of $[ClO_3]^-$ and $[IO_3]^-$ as oxidizing agents.

$$[IO_3]^-(aq) + 5I^-(aq) + 6H^+(aq) \rightarrow 3I_2(aq) + 3H_2O(l) \quad (13.199)$$

$$[ClO_3]^-(aq) + 6Fe^{2+}(aq) + 6H^+(aq) \rightarrow Cl^-(aq) + 6Fe^{3+}(aq) + 3H_2O(l) \quad (13.200)$$

$$[IO_3]^-(aq) + 3[SO_3]^{2-}(aq) \rightarrow I^-(aq) + 3[SO_4]^{2-}(aq) \quad (13.201)$$

The disproportionation of the chlorate ion (reaction 13.202) is thermodynamically favourable but takes place only slowly. Alkali metal chlorates decompose by this route when they are heated above their melting points, but in the presence of a suitable catalyst, $KClO_3$ decomposes with loss of O_2 (equation 13.203) and, as we saw earlier, this method can be used for small-scale preparations of O_2. *Caution is required as there is a risk of explosion*. Potassium chlorate is used as a component of fireworks and in safety matches where its ready reduction by, for example, sulfur provides a source of O_2 for the combustion of the match.

$$\underset{\text{chlorate ion}}{4[ClO_3]^-} \rightarrow \underset{\text{perchlorate ion}}{3[ClO_4]^-} + \underset{\text{chloride ion}}{Cl^-} \quad (13.202)$$

$$2KClO_3(s) \xrightarrow{\Delta,\ MnO_2\ \text{catalyst}} 2KCl(s) + 3O_2(g) \quad (13.203)$$

Of the perhalate salts (salts containing $[XO_4]^-$), perchlorates are most commonly encountered in the laboratory, *but they should always be regarded as hazardous explosives and avoided whenever possible* – mixtures of ammonium perchlorate and aluminium are used in missile propellants. The parent acid $HClO_4$ (perchloric acid) can be produced by reacting $NaClO_4$ with concentrated HCl, and is a colourless, hygroscopic liquid when pure. The chlorine atom is tetrahedrally sited (Figure 13.37) and is in its highest oxidation state of +7, using all of its valence electrons to form covalent bonds. The difference in the Cl–O bond lengths reflects the differing bond orders, and in the tetrahedral $[ClO_4]^-$ anion, all the Cl–O bond distances are equal (144 pm). [*Question*: How would you describe the bonding in this ion?]

The space shuttle: see Box 11.2

Anhydrous $HClO_4$ is a strong oxidizing agent, but aqueous solutions of the acid or its salts are *kinetically* stable; remember that the values of E^o in Table 3.6 for reductions involving $[ClO_4]^-$ provide *thermodynamic*, but not kinetic, information. In aqueous solution, $HClO_4$ behaves as an oxidizing agent and a strong, monobasic acid and most metals form perchlorate salts. Alkali metal perchlorates are usually prepared by the disproportionation of the corresponding chlorate salts (reaction 13.202).

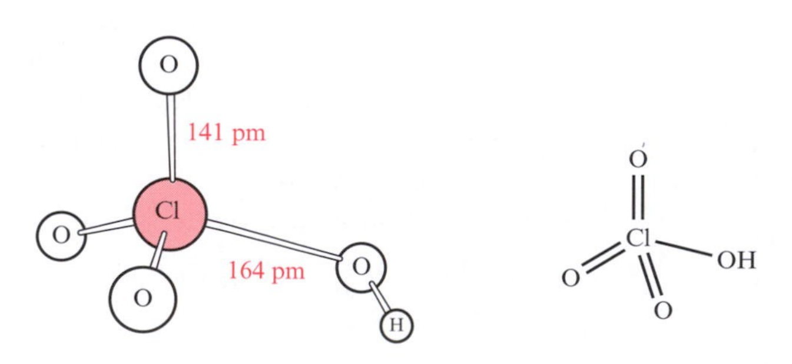

13.37 The structure of perchloric acid, $HClO_4$.

13.13 GROUP 18: THE NOBLE GASES

1	2		13	14	15	16	17	**18**
H								**He**
Li	Be		B	C	N	O	F	**Ne**
Na	Mg		Al	Si	P	S	Cl	**Ar**
K	Ca		Ga	Ge	As	Se	Br	**Kr**
Rb	Sr	*d*-block	In	Sn	Sb	Te	I	**Xe**
Cs	Ba		Tl	Pb	Bi	Po	At	**Rn**
Fr	Ra							

Physical and structural properties of noble gases: see Sections 2.21 and 7.6

Helium is the second most abundant element in the universe (after hydrogen) and is used widely wherever inert atmospheres are needed (e.g. as a protective environment during the growth of single silicon or germanium crystals for the semiconductor industry) and in balloons and advertising 'blimps'. Liquid helium is an important coolant and is used in highfield NMR spectrometers including those used for medical imaging.

Neon is a rare element obtained by the liquefaction of air followed by distillation. It is used in electric discharge tubes in which it produces a red glow, and a major application of this is in advertising signs. Argon constitutes 0.94% of the atmosphere and is obtained by the liquefaction of air. It is used to provide inert environments including those in glove or dry boxes in the laboratory, and as an atmosphere in fluorescent lighting tubes. Krypton too is rare, being present in the Earth's atmosphere to an extent of only one part per million (1 ppm), and xenon is even less abundant. Nonetheless, it is xenon that is the focus of attention of this section, because it is the only noble gas for which a reasonably extensive chemistry is known. There is little evidence that helium, neon and argon form compounds, and the chemistry of krypton is essentially limited to that of KrF_2. The heaviest noble gas is radon; it is radioactive and has been a serious health hazard in uranium mines where people exposed to it have developed lung cancer.

Bartlett prepared the first compound of xenon in 1962 by reacting Xe with PtF_6 which is a very strong oxidizing agent (equation 13.204). Even after 30 years, the *exact* nature of this product is unclear.

$$Xe + PtF_6 \rightarrow \text{'}XePtF_6\text{'} \qquad (13.204)$$

A noble gas is characterized by having a *complete* valence shell of electrons, and it is reasonable to assume that their reactivity will, as a consequence, be

Box 13.17 Radon levels in our homes

The radioactive decay of uranium leads to the formation of radon. Uranium occurs naturally in low abundance in many rock deposits, and the radon gas released may be trapped under the Earth's surface or may find a route to above the ground – much depends upon the geological rock formations. In some areas of Britain, levels of radon in houses are higher than in surrounding areas, and this appears to be related to the bedrock formations in each region. Research points towards a link between exposure to radon and lung cancer, and there is some concern over the health risks that may be involved in homes in which radon levels are relatively high.

low or even non-existent. Xenon, however, combines with fluorine and oxygen to form a wide range of compounds and this restriction is because:

- F_2 and O_2 are strong oxidizing agents; and
- F and O are highly electronegative elements.

Fluorides of xenon

Xenon forms the neutral fluorides XeF_2, XeF_4 and XeF_6 which are prepared from Xe and F_2 under different conditions (equation 13.205); all are crystalline solids at room temperature.

$$Xe + F_2 \begin{cases} \xrightarrow{\text{670 K, 2 : 1 ratio}} XeF_2 \\ \xrightarrow{\text{670 K, 6 bar, 1 : 5 ratio}} XeF_4 \\ \xrightarrow{\text{570 K, 50 bar, 1 : 20 ratio}} XeF_6 \end{cases} \quad (13.205)$$

The molecular structures of XeF_2 (linear) and XeF_4 (square planar) are as expected by VSEPR theory. Xenon hexafluoride is isoelectronic with $[TeCl_6]^{2-}$ but they are *not* isostructural. The *regular* octahedral structure of $[TeCl_6]^{2-}$ is explained in terms of a stereochemically inactive lone pair but it appears that in the vapour phase, XeF_6 is stereochemically non-rigid with the lone pair of electrons occupying different sites and causing the overall structure to be *distorted octahedral* (Figure 13.38). In the solid state, $[XeF_5]^+$ and F^- ions are present, and the fluoride ions bridge between the square pyramidal cations.

$[TeCl_6]^{2-}$: see Section 5.7

Xenon(II) fluoride is commercially available and is used widely as a fluorinating and oxidizing agent; often the solvent for the reaction is anhydrous hydrogen fluoride (equations 13.206 and 13.207). Using XeF_2 as a fluorinating agent is advantageous because the only other product is inert, gaseous xenon.

$$S + 3XeF_2 \xrightarrow{\text{anhydrous HF}} SF_6 + 3Xe \quad (13.206)$$

$$2Ir + 5XeF_2 \xrightarrow{\text{anhydrous HF}} 2IrF_5 + 5Xe \quad (13.207)$$

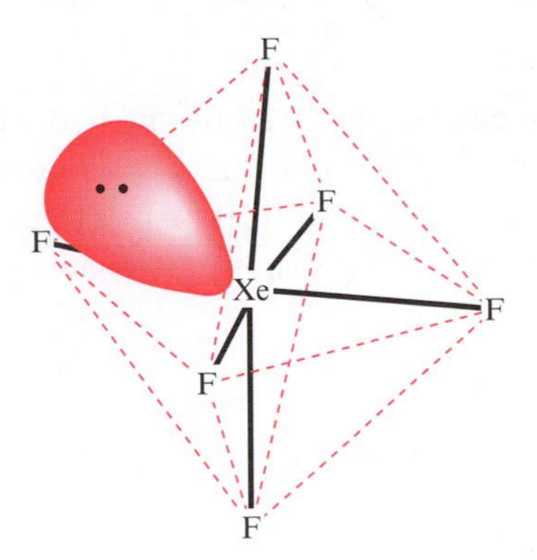

13.38 The distorted octahedral structure of XeF_6 in the gas phase is caused by the presence of a lone pair of electrons, but the molecule is stereochemically non-rigid and exchanges between several possible arrangements.

Xenon(II) fluoride is reduced by water to xenon (equation 13.208) but the reaction is slow unless a base is present. In the presence of water, xenon(IV) fluoride disproportionates (equation 13.209), whilst XeF_6 may be partially (equation 13.210) or completely (equation 13.211) hydrolysed without a change in oxidation state. In each hydrolysis, HF is produced.

$$2XeF_2 + 2H_2O \rightarrow 2Xe + 4HF + O_2 \quad (13.208)$$

$$6XeF_4 + 12H_2O \rightarrow 4Xe + 2XeO_3 + 24HF + 3O_2 \quad (13.209)$$

$$XeF_6 + H_2O \rightarrow XeOF_4 + 2HF \quad (13.210)$$

(13.50)

$$XeF_6 + 3H_2O \rightarrow XeO_3 + 6HF \quad (13.211)$$

(13.50)

A problem of working with XeF_6 is that it attacks silica glass. Initially, $XeOF_4$ is formed (equation 13.212) but further reaction leads to XeO_3. Copper or platinum apparatus is commonly used in place of glass.

$$2XeF_6 + SiO_2 \rightarrow 2XeOF_4 + SiF_4 \quad (13.212)$$

The donation and acceptance of fluoride ions by XeF_2, XeF_4 and XeF_6 is illustrated by the formation of the ions $[XeF]^+$, $[XeF_5]^+$, $[XeF_5]^-$ and $[XeF_8]^{2-}$. The $[XeF]^+$ ion acts as a Lewis acid and forms an adduct with XeF_2 (Figure 13.39).

13.39 Adduct formation between XeF_2 and $[XeF]^+$ leads to the formation of $[Xe_2F_3]^+$.

Oxides of xenon

Xenon(VI) oxide can be prepared by the hydrolysis of XeF_4 or XeF_6 (see above) but is highly explosive. It dissolves in water without change, but in strongly alkaline solution, equilibrium 13.213 is set up involving the xenate anion $[HXeO_4]^-$ which slowly disproportionates giving the perxenate ion and xenon (equation 13.214).

$$Xe^{VI}O_3 + [OH]^- \rightleftharpoons [HXe^{VI}O_4]^- \quad (13.213)$$

$$2[HXe^{VI}O_4]^- + 2[OH]^- \rightarrow [Xe^{VIII}O_6]^{4-} + Xe^0 + 2H_2O + O_2 \quad (13.214)$$

oxidation

reduction

In acidic solution, both XeO_3 and $[XeO_6]^{4-}$ are very powerful oxidizing agents as the standard reduction potentials for equilibria 13.215 and 13.216 show. By comparing these potentials with those in Appendix 12, we can see that $[XeO_6]^{4-}$ is capable of oxidizing aqueous Cr^{3+} to $[Cr_2O_7]^{2-}$, Br_2 to $[BrO_3]^-$ and Mn^{2+} to $[MnO_4]^-$.

$$XeO_3(aq) + 6H^+(aq) + 6e^- \rightleftharpoons Xe(g) + 3H_2O(l) \qquad E^o = +2.10\ V \quad (13.215)$$

$$H_4XeO_6(aq) + 2H^+(aq) + 2e^- \rightleftharpoons XeO_3(aq) + 3H_2O(l) \qquad E^o = +2.42\ V \quad (13.216)$$

13.14 SOME COMPOUNDS OF HIGH OXIDATION STATE *d*-BLOCK ELEMENTS

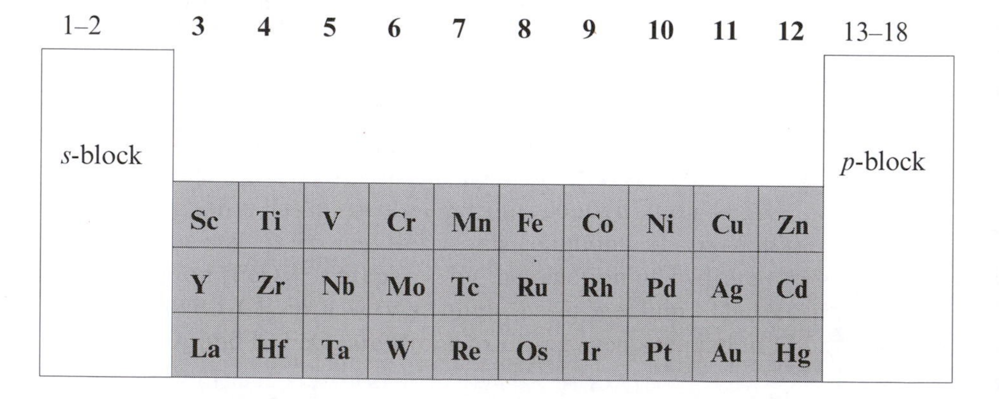

Older nomenclature labelled the groups in the *d*-block so that each was related to a group in the *s*- or *p*-block (Figure 13.40). Although this notation has been largely superseded by current IUPAC recommendations, the reasons behind the choice of labels reminds us that, in their highest oxidation states, the chemistry of the transition metals is reminiscent of that of related elements in the *s*- and *p*-blocks. We illustrate this point by briefly considering some oxides, oxoanions and halides of the *d*-block metals.

IUPAC group number	3	4	5	6	7	8	9	10	11	12
Older group number	IIIA	IVA	VA	VIA	VIIA	←	VIII	→	IB	IIB
Related *s*- or *p*-block group	13	14	15	16	17	←	18	→	1	2

13.40 The IUPAC group numbering for the *d*-block elements, and the corresponding older numbering system which emphasized the relationship between these metals and elements in the *s*- and *p*-blocks.

The highest oxidation state of titanium (group 4) is +4, and the properties of the titanium(IV) halides are, in many respects, similar to those of the heavier group 14 halides, but are *not* similar to those of CX_4 (X = F, Cl, Br or I). For example, TiX_4 compounds are all readily hydrolysed by water (equation 13.217). Titanium(IV) halides are Lewis acids and in concentrated hydrochloric acid, $TiCl_4$ forms the octahedral anion $[TiCl_6]^{2-}$. The $[TiCl_5]^-$ anion can also be formed and this is isostructural with $[SnCl_5]^-$. $TiCl_4$ also forms a complex with pyridine (equation 13.218) with a structure reminiscent of that of $[Ge(py)_2Cl_4]$ **13.23**. The oxide TiO_2 possesses the rutile lattice, as does SnO_2.

$$TiX_4 + 2H_2O \rightarrow TiO_2 + 4HX \qquad X = F, Cl, Br \text{ or } I \tag{13.217}$$

$$TiCl_4 + 2py \rightarrow [Ti(py)_2Cl_4] \tag{13.218}$$

py = (pyridine ring, N)

In group 5, vanadium forms compounds in the +5 oxidation state that find analogues in phosphorus chemistry. For example, vanadium reacts with O_2 to give V_2O_5 (as well as lower oxides) and with F_2 to yield VF_5. However, the product of heating vanadium metal with Cl_2 is VCl_4, and Br_2 and I_2 only oxidize the metal as far as the +3 state giving VBr_3 and VI_3, respectively. The structure of V_2O_5 is not like that of phosphorus(V) oxide, and in the solid state it consists of a complicated lattice structure. In strongly basic solution, V_2O_5 reacts to give oxoanions including $[VO_4]^{3-}$ which is tetrahedral like $[PO_4]^{3-}$. If the pH of the solution is lowered (i.e. the solution becomes less basic), condensation of the $[VO_4]^{3-}$ anions occurs, which is reminiscent of (but more complicated than) the condensation of phosphate anions.

The highest oxidation state of chromium (group 6) is +6 as observed in the oxide CrO_3, the unstable compound CrF_6, and the oxoanions $[CrO_4]^{2-}$ (chromate) and $[Cr_2O_7]^{2-}$ (dichromate). Solutions of chromium(VI) oxide are acidic like those of SO_3, and in basic solution it forms the tetrahedral chromate ion **13.51**. Solutions of chromate salts are yellow, but turn orange as the pH is lowered, because in acidic solution, the dichromate ion **13.52** is favoured. Equilibrium 13.219 can be pushed to the left- or the right-hand side by altering the pH of the solution; chromate is favoured in basic solution and dichromate in acidic solution.

$$[Cr_2O_7]^{2-}(aq) + 2[OH]^-(aq) \rightleftharpoons 2[CrO_4]^{2-}(aq) + H_2O(l) \quad (13.219)$$

orange yellow

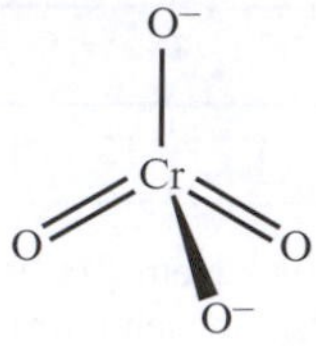

(13.51)

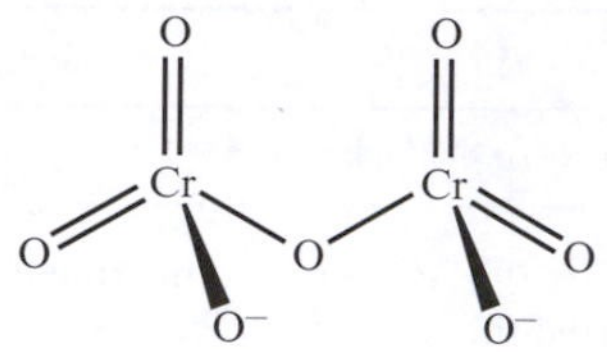

(13.52)

Acidic solutions of dichromate salts are strongly oxidizing as the standard reduction potential in equation 13.220 indicates.

$$[Cr_2O_7]^{2-}(aq) + 14H^+(aq) + 6e^- \rightleftharpoons 2Cr^{3+}(aq) + 7H_2O(aq) \qquad E^o = +1.33\ V \quad (13.220)$$

The manganese atom has seven electrons in its valence shell and the maximum oxidation state is +7. The oxide Mn_2O_7 is unstable and explosive; Cl_2O_7 has similar properties. The manganate(VII) anion $[MnO_4]^-$ (commonly known as permanganate) is most usually encountered as the purple potassium salt and is a powerful oxidizing agent as indicated by the standard reduction potential for equation 13.221. Compare this with the reduction potentials for the perchlorate ion given in Table 13.6. Equation 13.221 refers to an acidic solution; in basic aqueous solution, $[MnO_4]^-$ is reduced to manganese(IV) oxide (equation 13.222). The reduction potential of manganate(VII) is such that it can oxidize water; solutions of $KMnO_4$ are *thermodynamically* unstable with respect to equation 13.222, but are *kinetically* stable and may be kept for short periods in the dark.

$$[MnO_4]^-(aq) + 8H^+(aq) + 5e^- \rightleftharpoons Mn^{2+}(aq) + 4H_2O(aq) \qquad E^o = +1.51\ V \quad (13.221)$$

$$[MnO_4]^-(aq) + 2H_2O(aq) + 3e^- \rightleftharpoons MnO_2(s) + 4[OH]^-(aq) \qquad E^o = +\ 0.59\ V \quad (13.222)$$

Although iron has a valence electronic configuration $4s^23d^6$, the maximum oxidation state is +6, and this is only commonly observed in the tetrahedral ferrate ion $[FeO_4]^{2-}$, prepared by the reaction of hydrated iron(III) oxide Fe_2O_3 with Cl_2 in a very strongly basic medium. The red-purple salts Na_2FeO_4 and K_2FeO_4 are strong oxidizing agents (equation 13.223).

$$[FeO_4]^{2-}(aq) + 8H^+(aq) + 3e^- \rightleftharpoons Fe^{3+}(aq) + 4H_2O(l) \qquad E^o = +2.20\ V \quad (13.223)$$

As with the noble gases, the highest oxidation state for the group 8 metals is only observed for the heaviest elements in the group. Osmium forms the white crystalline oxide OsO_4 and the hexafluoride OsF_6, neither of which has an iron analogue. Similarly, in group 9, cobalt exhibits a maximum oxidation state of +4, whereas iridium forms iridium(VI) fluoride. A similar trend is seen in group 10, where nickel forms compounds in highest oxidation state +4, but platinum, at the bottom of the triad, forms the octahedral PtF_6. As expected, PtF_6 is an extremely powerful oxidizing agent, and the first example of a noble gas compound was obtained from the reaction of PtF_6 with xenon (equation 13.204). Bartlett was encouraged to perform this reaction after his observation that PtF_6 oxidized O_2 to $[O_2]^+[PtF_6]^-$; the ionization potential of O_2 is similar to that of Xe.

SUMMARY

This chapter has been concerned with the descriptive chemistry of elements in the *p*-block and of higher oxidation state compounds of the *d*-block metals.

Do you know what the following terms mean?

- $(p\text{–}p)\pi$-bonding
- conproportionation
- ductile
- malleable
- oxoacid
- acid anhydride

You should now be able

- to discuss trends in chemical reactivities of elements within the *p*-block groups
- to comment on the differences between the first and second members of a group
- to state the characteristic oxidation states of *p*-block elements
- to give selected and characteristic reactions of *p*-block elements
- to illustrate the diversity of oxoacids and oxoanions formed by nitrogen, phosphorus and sulfur and discuss some of their reactions
- to discuss some of the properties of halides of the *p*-block elements
- to understand the reasons why compounds of fluorine contain the element in the –1 oxidation state, but the other group 17 elements can attain positive oxidation states
- to discuss the reactivity of xenon
- to give representative chemistry of the high oxidation state *d*-block elements

PROBLEMS

13.1 What is the oxidation state of the central atom (in red) in each of the following compounds or ions: (a) SCl_2; (b) $POCl_3$; (c) $[XeF_5]^-$; (d) $[Sb_2F_{11}]^-$; (e) $[HPO_4]^{2-}$; (f) Al_2O_3; (g) Me_2SnF_2; (h) $[Cr_2O_7]^{2-}$; (i) $[MnO_4]^-$; (j) $[IO_4]^-$, (k) $[S_2]^{2-}$

13.2 Calculate ΔG^o(298 K) for the reaction given in equation 13.2 if (at 298 K) $\Delta_f G^o$ B_2O_3(s) and MgO(s) = –1194 and –569 kJ mol^{-1} respectively.

13.3 The Pauling electronegativities of B, F, Cl and Br are listed in Table 4.2. (a) What are the *relative* dipole moments of B–F, B–Cl and B–Br bonds? (b) When pyridine reacts with BF_3, BCl_3 and BBr_3, which trihalide might you expect to attract the Lewis acid the most, based solely on your answer to part (a)? (c) Rationalize why the formation of Br_3B•py (from pyridine and BBr_3) is more favourable than the formation of Cl_3B•py, which in turn is more favourable than the formation of F_3B•py. Does this ordering support or contradict your answer to (b)? What conclusions can you draw?

13.4 How do magnetic data assist in the formulation of '$GaCl_2$' as $Ga[GaCl_4]$?

13.5 Use VSEPR theory to predict the structures of (a) $[InBr_6]^{3-}$, (b) $[GaCl_5]^{2-}$ and (c) $[GaCl_4]^-$. In the salt $[(C_2H_5)_4N]_2[InCl_5]$, the dianion has a square-based pyramidal structure. Suggest possible reasons for this geometry. [Hint: Look back at Section 5.12.]

13.6 By considering the atomic orbitals that are available on carbon and silicon, suggest why two C=O double bonds are energetically preferred to four C–O single bonds, but four Si–O single bonds are energetically preferred to two Si=O double bonds.

13.7 Describe, with structural diagrams, representative groups of silicates.

13.8 In equations 13.53 and 13.54, what are the oxidation states of the N atoms in the reactants and products? What general class of reaction is each?

13.9 Draw the structure of the azide ion. With which oxide of nitrogen is it isoelectronic?

13.10 Which of the following oxides of nitrogen are paramagnetic: (a) NO; (b) N_2O; (c) NO_2; (d) N_2O_4; (e) N_2O_3?

13.11 Write down an expression for the equilibrium constant for reaction 13.62 and, using Figure 13.17, estimate the value of K at 400 K.

13.12 Draw resonance structures to describe the bonding in N_2O_4, bearing in mind the bond lengths shown in Figure 13.18.

13.13 Suggest likely products for the following reactions which *are balanced* on the left-hand side:

$SbF_5 + BrF_3 \rightarrow$

$KF + AsF_5 \rightarrow$

$PF_5 + 2SbF_5 \rightarrow$

13.14 What other valence bond representations are possible for structure **13.41**?

13.15 The structure of molecular SeF_4 is shown in Figure 13.32. How many fluorine environments are there? The ^{19}F NMR spectrum liquid SeF_4 shows only one signal at 298 K. Suggest a reason for this observation.

13.16 Suggest why some *p*-block elements occur in the elemental state while others occur only as compounds. Give examples.

13.17 Write out possible products from the reaction of (a) I_2 with F_2, and (b) Cl_2 with Br_2. In each case, also give the expected reactions of the products you have written with water.

13.18 Suggest why compounds of the noble gases appear to be essentially restricted to xenon? Your arguments might lead you to suggest that radon should form compounds similar to those of xenon – why do you think such compounds have not been included in our discussions?

13.19 If xenon oxo-compounds are such good oxidizing agents, and only give gaseous xenon as a product, why do you think they are not more commonly used in the laboratory?

13.20 What do you understand by the term *acid anhydride*? What are the anhydrides of (a) HNO_2, (b) HNO_3, (c) H_3PO_4, (d) H_3PO_3, (e) HOCl and (f) HIO_3?

14 Polar organic compounds

Topics

14.1 INTRODUCTION

Although we concentrated on hydrocarbons in Chapter 8, the vast majority of organic compounds contain substituents other than hydrogen; examples include chloroethane **14.1**, propan-1-ol **14.2** and 1-butylamine **14.3**. The chemistry of these compounds often depends primarily upon reactions of the functional group – the C–Cl bond in **14.1**, the –OH group in **14.2** or the $-NH_2$ group in **14.3**.

CH_3CH_2Cl **(14.1)** $CH_3CH_2CH_2OH$ **(14.2)** $CH_3CH_2CH_2CH_2NH_2$ **(14.3)**

Numbering an aliphatic carbon chain: see Section 8.2

In this chapter we explore the structure and reactivity of four particular groups of aliphatic compounds:

- halogenoalkanes in which the functional group is a carbon–halogen bond,
- ethers in which the characteristic unit is a C–O–C group;
- alcohols which contain the –OH group; and
- amines and alkyl ammonium salts with the general formulae RNH_2, R_2NH, R_3N and $[R_4N]^+$.

These compounds have a feature in common – each possesses a polar C–X bond where X is a halogen, oxygen or nitrogen.

14.2 POLAR C–X BONDS

Polar bonds and molecular dipole moments

Molecular dipole moments: see Section 5.13

The presence of polar bonds in a molecule does not necessarily imply that the *molecule* has a dipole moment – we have seen that tetrahedral molecules such as CF_4 **14.4** are non-polar because the four C–F bond dipole moments cancel each other out. On the other hand, in H_2O **14.5**, the two O–H bond dipole moments combine to give an overall *molecular dipole moment*.

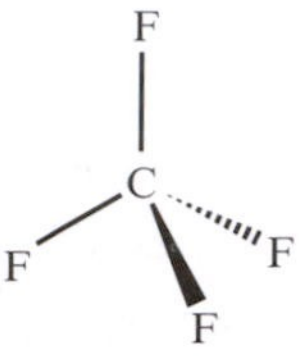

(14.4)

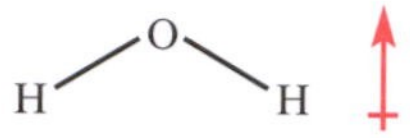

(14.5)

In a polyatomic molecule, the presence of a molecular dipole moment depends on *both* the presence of bond dipole moments *and* the shape of the molecule.

Of course, many molecules contain more than one type of substituent, and the various bond dipole moments act in different directions. Whether or not there is a resultant *molecular* dipole moment depends on the geometry of the molecular *and* the relative magnitudes of the individual bond dipoles. Consider fluoromethane **14.6** – the electronegativity values are $\chi^P(C) = 2.6$, $\chi^P(H) = 2.2$ and $\chi^P(F) = 4.0$. Each carbon–hydrogen bond has a small dipole moment in the sense $C^{\delta-}–H^{\delta+}$, and the carbon–fluorine bond is polar in the sense $C^{\delta+}–F^{\delta-}$. There is a resultant molecular dipole moment (1.86 D) as indicated in structure **14.6**.

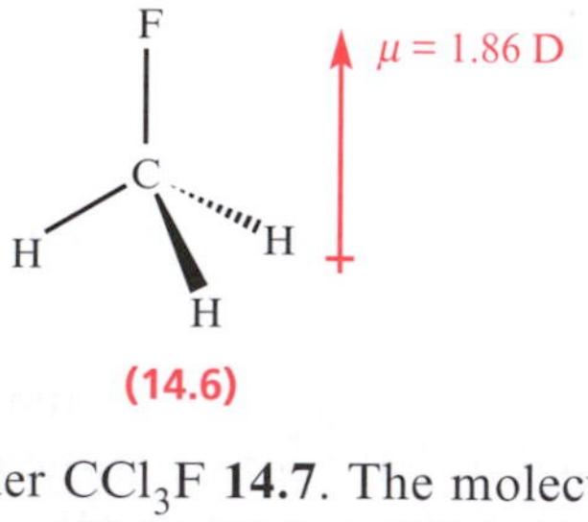

$\mu = 1.86$ D

(14.6)

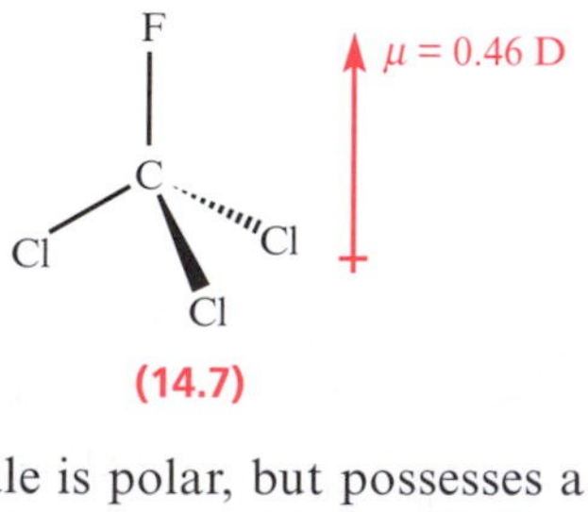

$\mu = 0.46$ D

(14.7)

$\chi^P(Cl) = 3.2$

Now consider CCl_3F **14.7**. The molecule is polar, but possesses a smaller dipole moment (0.46 D) than CH_3F. Each carbon–chlorine bond is polar in the sense $C^{\delta+}–Cl^{\delta-}$ and the resultant of these three dipole moments opposes the moment of the C–F bond in CCl_3F. It can be dangerous to draw rapid conclusions about the magnitude and direction of a molecular dipole moment and we explore this further in the accompanying workbook. Although we may be able to predict or rationalize the presence of a molecular dipole moment in terms of electronegativities and molecular shape, other

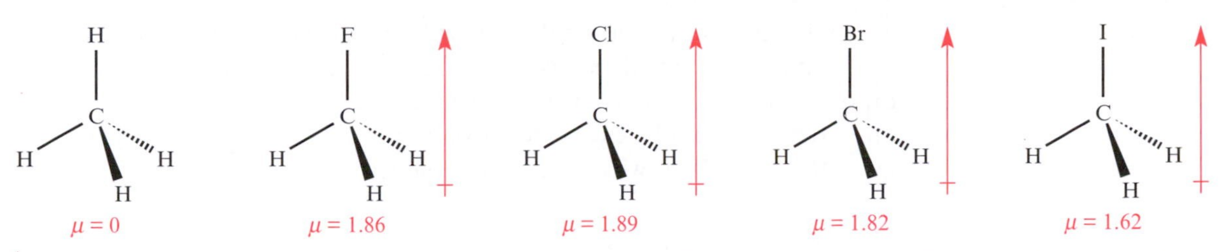

14.1 Values of the molecular dipole moments (μ) in debyes for CH_3X (X = H, F, Cl, Br and I) measured in the gas phase.

factors may also be important. Figure 14.1 uses halogenoalkane derivatives of methane to exemplify an anomaly – each molecule has a tetrahedral carbon centre and on the basis of electronegativity values, we expect the molecular dipoles to be in the order $CH_3F > CH_3Cl > CH_3Br > CH_3I > CH_4$. The origins of the observed effects are subtle and by no means predictable.

The effect of a polar bond on the reactivity of a molecule

The presence of the polar C–X bond can influence the chemistry of an organic molecule in many ways, and we concentrate upon two:

- the charge distribution in the molecule is altered and the carbon atom to which X is attached becomes a centre of reactivity; and
- the charge distribution in the molecule is altered and the reactivity of carbon centres other than that attached directly to X is modified.

Consider the first of these points. In C_2H_6, all the hydrogen atoms are equivalent and ethane undergoes free radical substitution reactions such as that with $Cl^{\bullet}$ to give C_2H_5Cl as one product. In C_2H_5Cl, the C–Cl bond is polar (structure **14.8**) and the carbon centre which carries the δ^+ charge attracts *nucleophiles* – species that *donate electrons*.

Nucleophile and electrophile: see Section 8.7

$$CH_3-\overset{\displaystyle H}{\underset{\displaystyle H}{\overset{|}{\underset{|}{\overset{\delta^+}{C}}}}}-\overset{\delta^-}{Cl}$$

(14.8)

The second point concerns longer-range effects in charge distribution brought about by the presence of an electronegative atom. Consider chloroethane again. In the C–Cl bond, the electronegative Cl atom draws electron density towards itself, making the carbon centre partially positively charged. This is not the whole story as structure **14.9** shows because atom C(1) can partly compensate for the movement of electron density towards the halogen atom by withdrawing electron density from atom C(2). This is called the *inductive effect*.

$$\underset{2}{\overset{\delta\delta^+}{CH_3}} \rightarrow \underset{1}{\overset{\delta^+}{CH_2}} \rightarrow \overset{\delta^-}{Cl}$$

(14.9)

The inductive effect operates through σ-bonds and is also referred to as a *σ-effect*. Its magnitude depends upon distance, and the size of the induced dipole at a carbon centre usually decreases as the carbon atom becomes more remote from the electronegative atom. This is represented in **14.9** by using the symbol $\delta\delta^+$ which signifies a *smaller* positive charge than does δ^+.

The inductive effect operates mainly through σ-bonds, and is the redistribution of charge in a molecule caused by atomic electronegativity differences.

An additional effect occurs through space and this is called the *field effect*. Since experimental observables are usually the net result of both the inductive *and* field effects, it is generally difficult to determine the magnitude of each separate phenomenon.

The field effect is similar to the inductive effect *but operates through space*.

The inductive effect may be used to partially rationalize the increased strength of haloacetic acids over the parent acid. For example, the pK_a values of CH_3CO_2H and CCl_3CO_2H are 4.75 and 0.70 respectively; the greater inductive effect of the CCl_3 group with respect to the CH_3 group encourages dissociation of the O–H bond (structure **14.10**) and stabilizes the anion $[CCl_3CO_2]^-$.

(14.10)

If a functional group attracts electrons towards itself, it is said to be *electron-withdrawing* and examples of such groups are –F, –Cl, –Br and –CN. This is the opposite of the electron-releasing effect that we described for alkyl groups. In the same way that electron-releasing groups can stabilize radicals and carbenium ions, electron-withdrawing groups are able to stabilize carbanion centres as shown in structure **14.11** – the carbanion $[CCl_3]^-$ is easily formed by deprotonating $CHCl_3$. The negative charge is spread out amongst the electron-withdrawing substituents.

Carbenium ion and electron-releasing alkyl groups: see Section 8.14

Carbanion: see Section 8.17

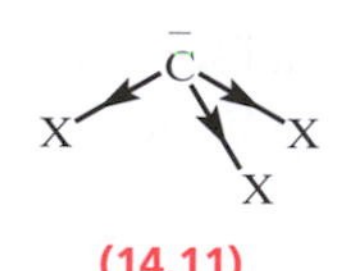

(14.11)

14.3 HALOGENOALKANES 1: STRUCTURE, SYNTHESIS AND PHYSICAL PROPERTIES

Structure and nomenclature

Nomenclature for hydrocarbons: see Section 8.2

In this section, we expand the basic rules of organic nomenclature to incorporate halogen functional groups. The general formula for a *halogenoalkane* is RX where R is the alkyl group and X is F, Cl, Br or I; these compounds are also referred to as *alkyl halides*.

Consider a halogenoalkane containing one halogen atom. The alkyl group may be a straight or branched chain, and in naming a relatively simple compound,§ the same basic rules that we outlined in Section 8.2 apply, but with the caveat that the principal chain should be the *longest that contains the functional group*:

1. **Find the longest straight chain in the molecule *containing the C atom attached to the functional group;* this gives the root-name and the chain is called the *principal chain.***
2. **The principal chain is numbered from one end, this being chosen to give the lowest position-numbers for the substituents.**
3. **Prefix the root-name with the substituent descriptors, and if more than one is present, give them in alphabetical order.**

The name of a halogenoalkane is formed by prefixing the alkane stem with *fluoro-*, *chloro-*, *bromo-* or *iodo-*. Compounds **14.12** and **14.13** are straight chain, bromo-derivatives of pentane. In each case, the principal carbon chain is numbered so that the bromine substituents have the lowest possible position-numbers, meaning that in **14.12** and **14.13** the chains are numbered from the right-hand end.

(14.12) (14.13)

Figure 14.2 shows four isomers of chloromethylpentane. The principal chain *must contain the carbon atom attached to the chloro-substituent*. The principal chain for each of isomers (a)–(c) contains five carbon atoms, and alphabetically, 'chloro' comes before 'methyl' – hence the ordering of the substituent labels in the name for each isomer. However, in isomer (d), the chloro-group is located in a four-carbon chain and the principal chain is based on 'butane'. The name 1-chloro-2-ethylbutane is correct; the name 3-(chloromethyl)pen-

§ As the complexity of the molecule increases, the rules for giving it a systematic name become more complicated.

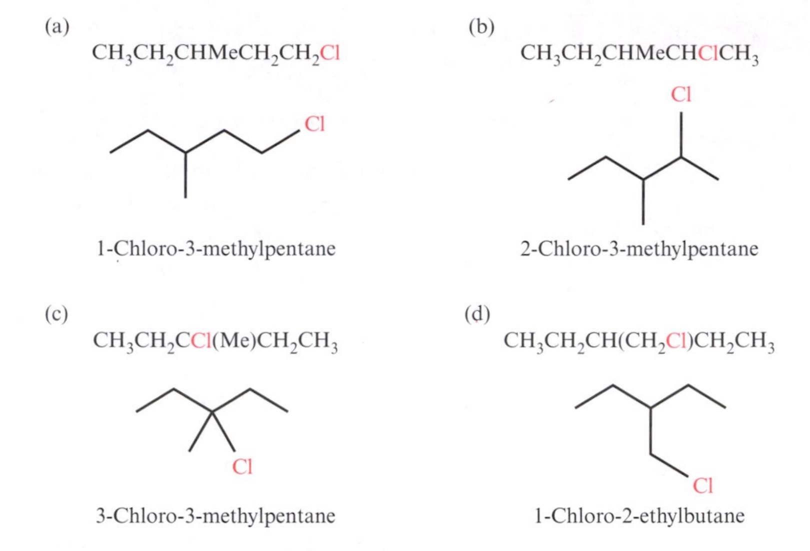

14.2 Structural formulae and systematic names for four isomers of chloromethylpentane.

tane is incorrect. [*Question*: How many other isomers are there with the formula $C_6H_{13}Cl$ which retain the root-name 'pentane'? And how many isomers have 'butane' as the principal chain?]

We now consider compounds with two halogen substituents, and begin with straight chain halogenoalkanes. Structure **14.14** is 1,4-dibromopentane; when different halogens are present, the relevant prefixes are ordered alphabetically, e.g. 1-bromo-2-chlorobutane and 2-chloro-2-fluoropentane.

$$\underset{1}{CH_2Br}\underset{2}{CH_2}\underset{3}{CH_2}\underset{4}{CHBr}\underset{5}{CH_3}$$

(14.14)

When the halogen atom is attached to a terminal carbon atom as in **14.12**, the compound is called a *primary* halogenoalkane. In a *secondary* (or *sec*-) halogenoalkane, *two* R groups are attached to the same carbon centre as the halogen atom as in **14.13**. 3-Chloro-3-methyl pentane (Figure 14.2) is an example of a *tertiary* (or *tert*-) halogenoalkane.

A *primary* halogenoalkane has the general formula RCH_2X.

A *secondary* halogenoalkane has the general formula R_2CHX.

A *tertiary* halogenoalkane has the general formula R_3CX.

The R groups do *not* have to be the same as each other.

Synthesis

Halogenoalkanes can be prepared from alkanes by free radical substitution, or from alkenes by addition reactions as exemplified in equations 14.1 and

Halogenation of alkanes: see Section 8.10; Addition of halogens to alkenes: see Section 8.13

14.2. The disadvantage of the reaction between an alkane and a halogen X_2 is that it is non-specific, giving a mixture of products – what other derivatives might be formed in reaction 14.1? On the other hand, reaction 14.2 gives good yields of 1,2-dibromoethane.

$$C_2H_6 + Cl_2 \xrightarrow{h\nu} \underset{\text{chloroethane}}{C_2H_5Cl} + HCl \quad \textit{Free radical substitution} \tag{14.1}$$

$$C_2H_4 + Br_2 \rightarrow \underset{\text{1,2-dibromoethane}}{CH_2BrCH_2Br} \quad \textit{Electrophilic addition} \tag{14.2}$$

Since the reactions between alkanes or alkenes and F_2 are explosive, fluoroalkanes are usually prepared using an alternative fluorinating agent. Equation 14.3 illustrates the use of antimony trifluoride – the fluorination takes place by partial *halogen exchange*.

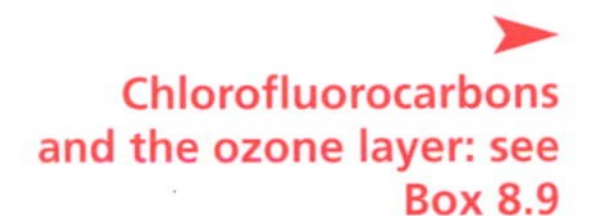
Chlorofluorocarbons and the ozone layer: see Box 8.9

$$C_2Cl_6 \xrightarrow{SbF_3} \underset{\text{tetrachloro-1,2-difluoroethane}}{CCl_2FCCl_2F} \tag{14.3}$$

A convenient and commonly used route for the preparation of halogenoalkanes is from the corresponding alcohols, and we discuss these reactions in Section 14.10.

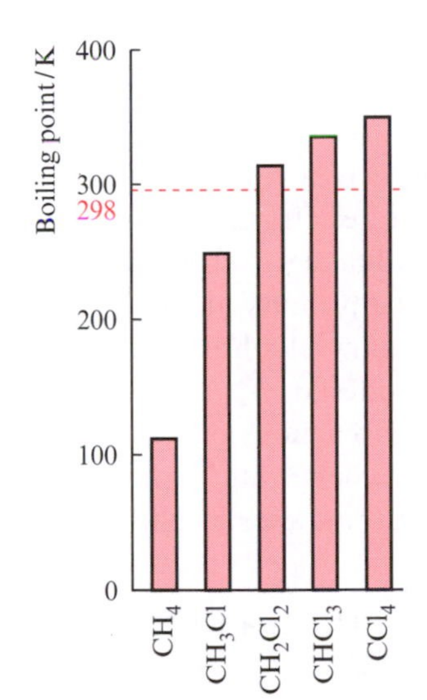

14.3 The trend in the boiling points of members of the family of compounds $CH_{4-x}Cl_x$ for $x = 0$ to 4.

Physical properties

A halogenoalkane usually has a higher boiling point than the corresponding alkane – trichloromethane and tetrachloromethane are liquids at room temperature while methane and chloromethane are gases (Figure 14.3). Dichloromethane is usually a liquid at room temperature, although this is not true in every part of the world! The trend in boiling points follows the increase in molecular weight.

The liquid chloro-, bromo- and iodoalkanes are usually denser than water, and the density increases with higher degrees of halogen substitution – the densities of CH_2Cl_2, $CHCl_3$ and CCl_4 are 1.33, 1.48 and 1.59 g cm^{-3} respectively. Liquid halogenoalkanes are usually immiscible with water. Since CCl_4 is non-polar, this is expected, but even though CH_2Cl_2 and $CHCl_3$ are polar, they too are immiscible.

Although liquid halogenoalkanes have been widely used as solvents in the laboratory solvents and for dry-cleaning, it is now known that some of them may cause liver damage and others have been implicated as cancer-causing agents. Consequently, the use of these valuable solvents is becoming increasingly restricted. The search for safe solvents continues!

14.4 HALOGENOALKANES 2: REACTIVITY

Halogenoalkanes usually undergo the following general types of reactions:

- **nucleophilic substitution of halide ion;**
- **β-elimination of HX to give an unsaturated hydrocarbon;**
- **formation of Grignard reagents;**
- **formation of organolithium reagents.**

Nucleophilic substitution reactions

In this section, we give examples of reactions in which the halo group in a halogenoalkane is replaced by a nucleophile, and in Section 14.5 we discuss the mechanisms of these *nucleophilic substitutions*. Equations 14.4 and 14.5 give general representations of the reaction. The nucleophile Y or Y^- is a Lewis base and donates a pair of electrons to the carbon atom to which halogen atom X is attached; a C–Y covalent single bond forms and the C–X bond is broken. The halide ion X^- is called the *leaving group*.

$$Y^- + RCH_2—X \longrightarrow RCH_2—Y + X^- \qquad (14.4)$$

$$Y\!: + RCH_2—X \longrightarrow RCH_2—\overset{+}{Y} + X^- \qquad (14.5)$$

The nucleophile may be neutral or an anion, but in each case it must provide a pair of electrons, and the leaving group takes a pair of electrons away with it.

When a halogenoalkane RX undergoes a *nucleophilic substitution* reaction, a Lewis base displaces the halo group, and the *leaving group* is the halide ion X^-.

The simple displacement reaction of a halogenoalkane with hydroxide ion gives an alcohol – a primary halogenoalkane gives a primary alcohol, a secondary halogenoalkane is converted to a secondary alcohol, and so on although tertiary halogenoalkanes often tend to undergo *elimination* rather than substitution reactions as we describe later on. Reactions proceed faster with the negatively charged hydroxide ion which is a better nucleophile than water. Equations 14.6–14.8 suggest that the substitutions are straightforward but as we shall see, this tends not to be the case.

$$\underset{\text{1-chlorobutane}}{CH_2CH_2CH_2CH_2Cl} + [OH]^- \rightarrow \underset{\text{butan-1-ol}}{CH_2CH_2CH_2CH_2OH} + Cl^- \qquad (14.6)$$

$$\underset{\text{2-chlorobutane}}{CH_2CH_2CHClCH_3} + [OH]^- \rightarrow \underset{\text{butan-2-ol}}{CH_2CH_2CH(OH)CH_3} + Cl^- \qquad (14.7)$$

$$\underset{\text{2-chloro-2-methylpropane}}{CH_3CMeClCH_3} + [OH]^- \rightarrow \underset{\text{2-methylpropan-2-ol}}{CH_3CMe(OH)CH_3} + Cl^- \qquad (14.8)$$

Formation of alkoxide ions: see Section 14.10

Similar reactions occur with *alkoxide ions* – an alkoxide ion has the form $[RO]^-$ and is derived from an alcohol ROH ($pK_a \approx 16$). The reaction of an alkoxide ion with a halogenoalkane is a method for preparing *ethers* and is known as the *Williamson synthesis*. Equation 14.9 shows the formation of diethyl ether **14.15** from the reaction of bromoethane and sodium ethoxide. Diethyl ether is a symmetrical ether of the general type ROR.

Aliphatic ethers: see Sections 14.7 and 14.8

$$C_2H_5Br + Na^+[^-OC_2H_5] \rightarrow C_2H_5OC_2H_5 + NaBr \qquad (14.9)$$

sodium ethoxide (14.15)

or Et–O–Et

(14.15)

The Williamson synthesis is the reaction of a halogenoalkane and an alkoxide ion to give an ether:

$RX + [R'O]^- \rightarrow ROR' + X^-$

The ability of RX compounds to react with nucleophiles means that many are potent *alkylating agents*. The alkylation of the organic components of DNA (the material that carries genetic information) is strongly implicated in carcinogenic (cancer-forming) and mutagenic (mutation-forming) processes.

An alkylating agent provides an alkyl group in a reaction. The process is called alkylation. Alkylating agents may be nucleophiles or electrophiles.

Elimination of HX to give an alkene

There are several types of *elimination* reaction involving the loss of two atoms or groups, and in a β-*elimination*, the groups are lost from *adjacent* atoms. Halogenoalkanes undergo β-eliminations to give alkenes by the general reaction 14.10 which is also described as a *dehydrohalogenation*.

$$\text{H–C–C–X} \longrightarrow \text{C=C} + \text{HX} \qquad (14.10)$$

In a β-elimination, two atoms groups are lost from adjacent atoms. The β-elimination of HX from a halogenoalkane gives an alkene.

The elimination of HX from a halogenoalkane requires a strong base such as a solution of potassium hydroxide in ethanol – this combination is called *alcoholic KOH*. The reaction of 1-chloropropane with alcoholic KOH gives propene (equation 14.11). In general, *if the halogenoalkane possesses a terminal halo-group, the double bond in the product is in the 1-position* (e.g. 1-chlorobutane gives but-1-ene, and 1-bromohexene gives hex-1-ene).

$$\underset{\text{1-chloropropane}}{CH_3CH_2CH_2Cl} \xrightarrow{\text{alcoholic KOH}} \underset{\text{propene}}{CH_3CH{=}CH_2} + HCl \qquad (14.11)$$

For a halo-compound such as 2-bromobutane, elimination of HBr can occur in one of two ways as shown in equation 14.12. Elimination to produce the more highly substituted alkene is the favoured pathway. We examine the reasons for this preference in Section 14.6.

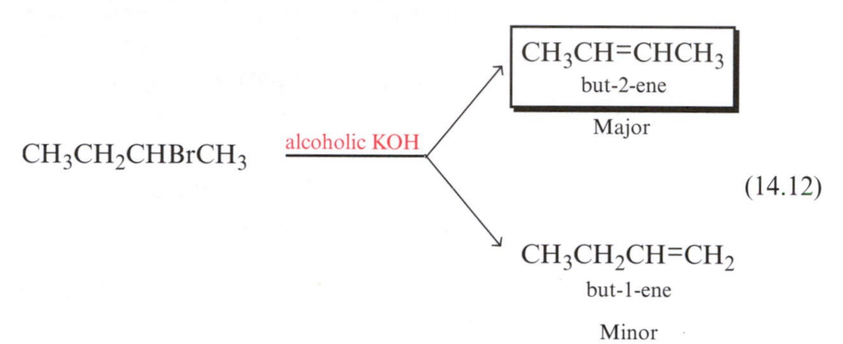

Reaction with alcoholic silver nitrate: qualitative analysis

One simple way to tell if a compound contains a halogen atom involves reaction with an alcoholic solution of silver nitrate. The C–Cl (or C–X) bonds break and the halide is precipitated as silver halide (equation 14.13).

$$\underset{R = \text{alkyl}}{RX + AgNO_3} \xrightarrow{\text{aqueous alcohol}} ROH + AgX(s) + [NO_3]^- \qquad (14.13)$$

The formation of white silver chloride, cream silver bromide or pale-yellow silver iodide precipitates is often used as a qualitative test for a halogenoalkane, and readily distinguishes the compound from the parent alkane. It must be stressed that *not all organohalogen compounds undergo this reaction*. In particular, some halogenoalkanes such as Me_3CH_2Cl, CCl_4 and RCH=CHCl are unreactive, as are aromatic halogen compounds. Furthermore, no precipitate is obtained when fluoroalkanes are treated with $AgNO_3$. [*Question*: Why not? *Hint*: Can you find a value of K_{sp} for AgF?]

Aromatic compounds: see Chapter 15

Two further features of reaction 14.13 should be noted:

- the order of reactivity of the halo-compounds is RI > RBr > RCl;
- the order of reactivity of primary, secondary and tertiary halogenoalkanes is $R_3CX > R_2CHX > RCH_2X$.

These observations may be rationalized in terms of the mechanism of the reactions (see problem 14.9).

Formation of Grignard reagents

Grignard reagents are *organometallic* compounds of the form RMgX where R is an alkyl group and X is a halide. The Mg–C bond is highly polar ($Mg^{\delta+}$–$C^{\delta-}$) and this contributes to the reactivity of Grignard reagents, making them extremely important synthetic reagents. They are air- and moisture-sensitive and are usually prepared *in situ*, i.e. the compounds are not isolated but are prepared in solution as required by the reaction of magnesium with a halogenoalkane under moisture- and oxygen-free conditions (equation 14.14). Common solvents are dry diethyl ether (Et_2O, **14.15**) or tetrahydrofuran (THF, **14.16**). The reaction can be initiated by adding a crystal of I_2 which reacts to give a clean and highly reactive magnesium surface. The rate at which reaction 14.14 proceeds follows the order RI > RBr > RCl. Equations 14.15 and 14.16 show the formation of Grignard reagents with primary and secondary carbon centres.

Cyclic ethers: see Section 15.3

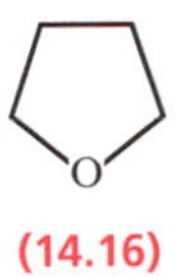

(14.16)

$$RX + Mg \xrightarrow{\text{anhydrous } Et_2O} RMgX \quad (14.14)$$

$$CH_3CH_2CH_2CH_2Br + Mg \xrightarrow{\text{anhydrous } Et_2O} CH_3CH_2CH_2CH_2MgBr \quad (14.15)$$

$$CH_3CH_2CHClCH_3 + Mg \xrightarrow{\text{anhydrous } Et_2O} CH_3CH_2CH(MgCl)CH_3 \quad (14.16)$$

A Grignard reagent is a compound of formula RMgX (X = halogen) used in alkylation reactions.

Formation of organolithium reagents

Organolithium compounds, RLi, are a second group of organometallic compounds with widespread synthetic uses, in particular as alkylating agents (i.e. they behave as a source of a carbanion R^-). Alkyllithium compounds may be prepared by reacting lithium with an appropriate halogenoalkane (equation 14.17), although air- and moisture-free conditions are essential because both lithium metal and the organolithium derivatives react with H_2O and O_2. A particularly important organolithium reagent is butyllithium, BuLi, which is commercially available as a solution in hexane, sealed under an atmosphere of N_2.

$$RX + 2Li \xrightarrow{\text{anhydrous } Et_2O} RLi + LiCl \quad (14.17)$$

An organolithium compound is of the general type RLi (R = alkyl) and is an alkylating reagent

Box 14.1 Selected synthetic uses of alkyl Grignard reagents, RMgX

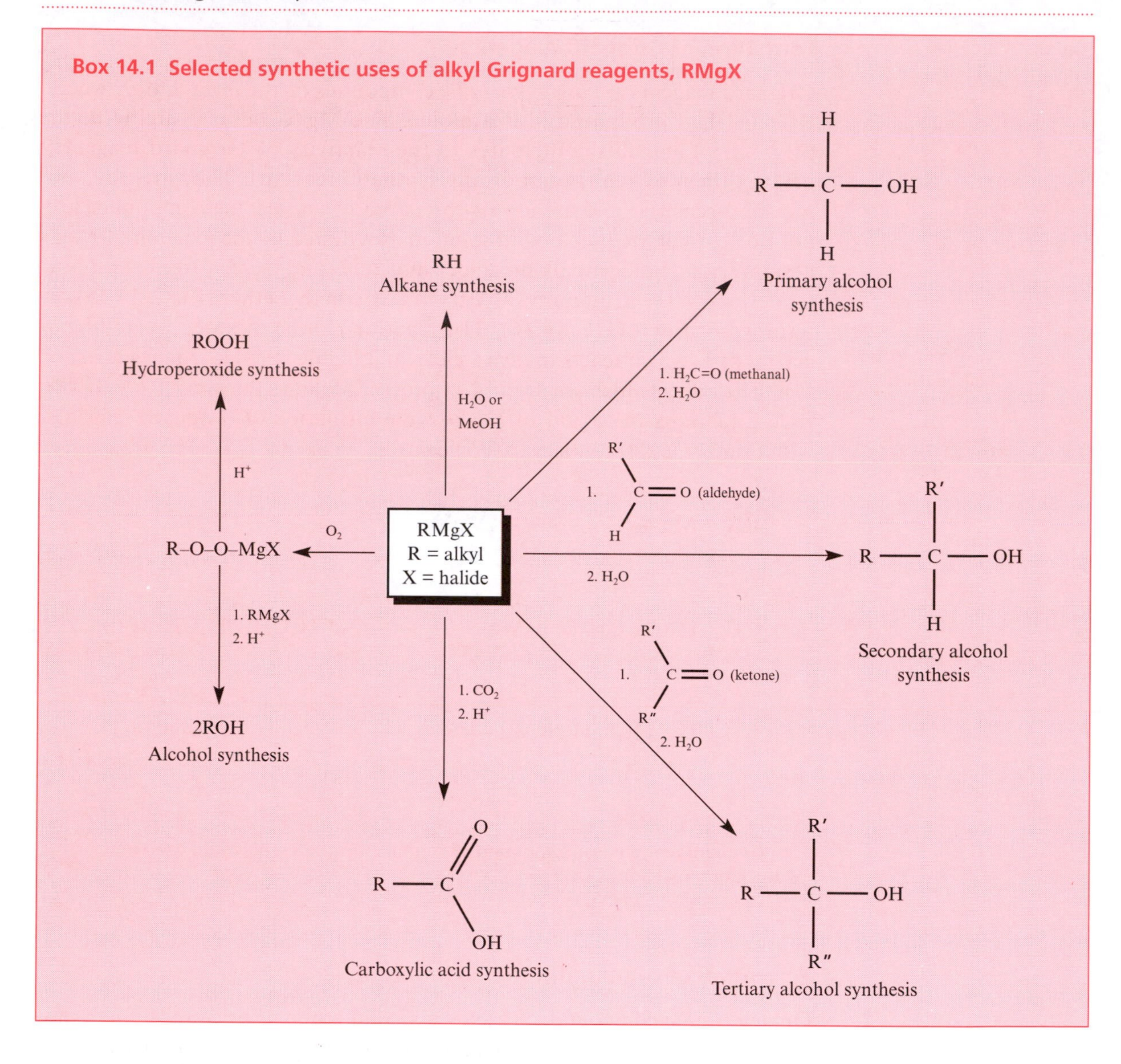

14.5 THE MECHANISM OF NUCLEOPHILIC SUBSTITUTION

Reaction kinetics: some experimental data

See problem 10.6

In Chapter 10, problems dealing with the kinetics of reactions included the reaction between 2-chloro-2-methylpropane and water (equation 14.18). These data are presented in Figure 14.4, and show that the reaction is first order with respect to the halogenoalkane. Further data provide evidence that the reaction is zero order with respect to water. We therefore write a rate equation for the reaction that illustrates that Me_3CCl is the only component in the rate determining step (equation 14.19).

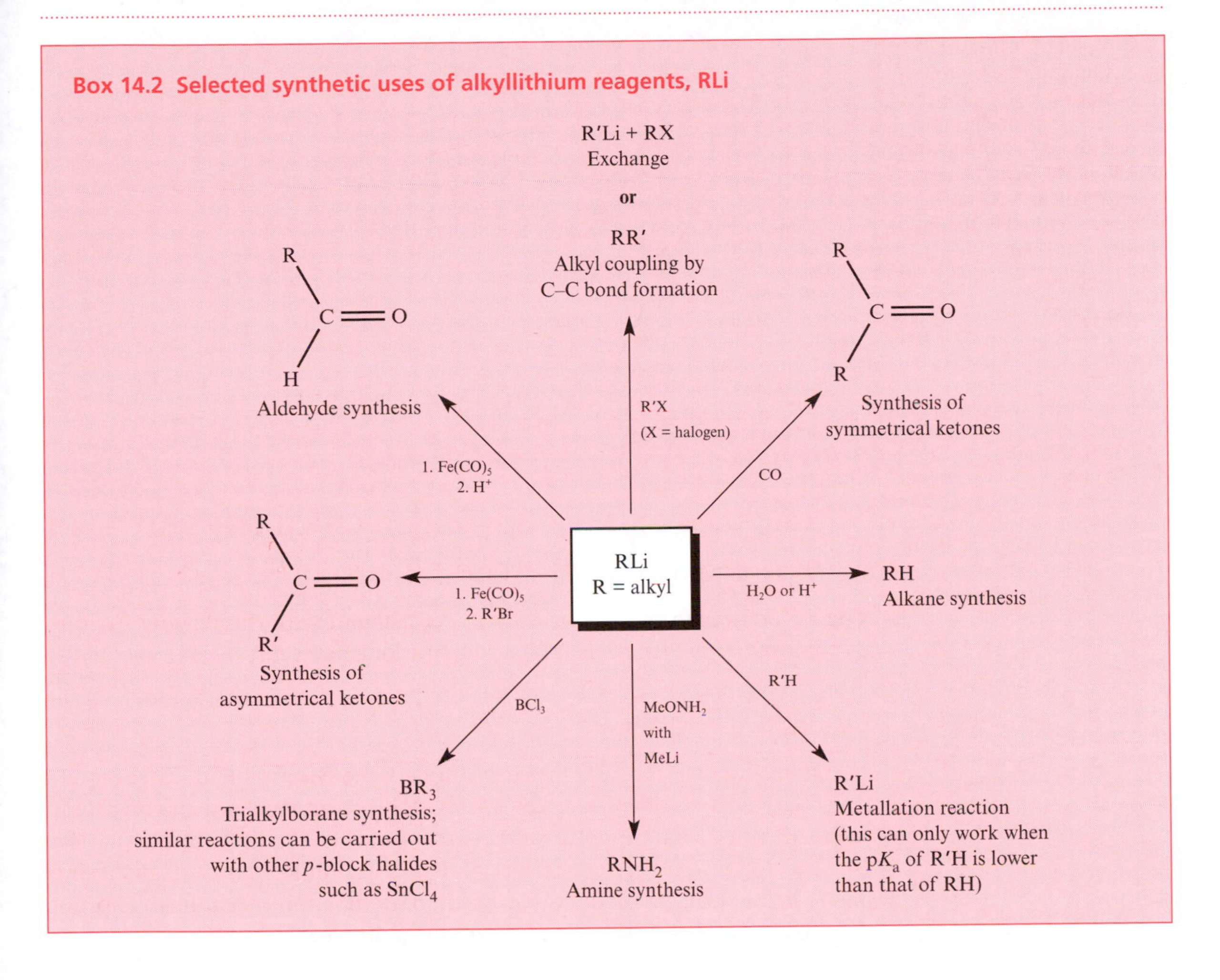

$$\mathrm{Me_3CCl} + \mathrm{H_2O} \xrightarrow{\text{solvent}} \mathrm{Me_3COH} + \mathrm{HCl} \qquad (14.18)$$

$$\text{Rate of reaction} = -\frac{d[\mathrm{Me_3CCl}]}{dt} = k[\mathrm{Me_3CCl}] \qquad (14.19)$$

Reaction 14.18 is an example of a nucleophilic substitution reaction – the nucleophile is water, and the net result of the attack is an exchange of the chloro-group for an –OH substituent. However, kinetic data indicate that the water is *not* involved in the rate determining step of the reaction. This is not a general result for all halogenoalkanes – if we consider the kinetics of the nucleophilic substitution reaction between hydroxide ion and 1-bromobutane (equation 14.20), we find that the reaction is first order with respect to *each* reactant (equation 14.21).

14.4 A plot of ln[Me_3CCl] against time for the hydrolysis of 2-chloro-2-methylpropane. The linear graph means that the reaction is first order with respect to Me_3CCl. (Data from: A. Allen, A.J. Haughey, Y. Hernandez and S. Ireton, *Journal of Chemical Education*, (1991) vol. 68, p. 609.)

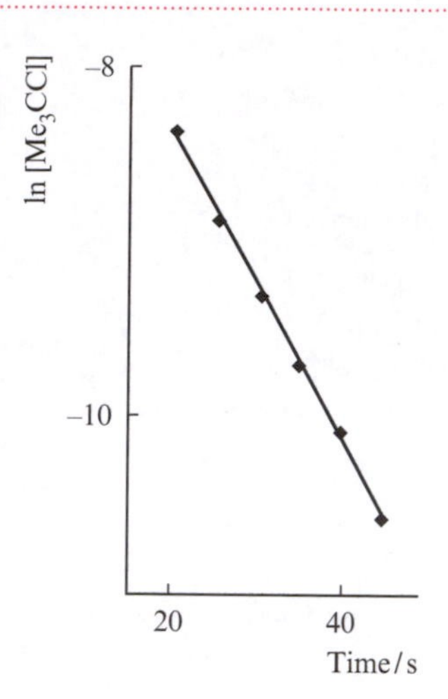

$$CH_3CH_2CH_2CH_2Br + [OH]^- \rightarrow CH_3CH_2CH_2CH_2OH + Br^- \quad (14.20)$$

$$\text{Rate of reaction} = k[CH_3CH_2CH_2CH_2Br][OH^-] \quad (14.21)$$

The differences in the kinetic data are consistent with there being various mechanisms by which nucleophilic substitution can occur.

Dissociative mechanism

In this section, we consider the reaction between Me_3CCl and water (equation 14.18) which obeys rate law 14.19. A reaction mechanism consistent with the kinetic data involves a *two-step process*. The first step (equation 14.22) is slow (rate-determining) and is a unimolecular step involving *dissociative* cleavage of the C–Cl bond. The reaction mechanism is often described as S_N1 (substitution nucleophilic unimolecular).

Rate-determining step is abbreviated to RDS

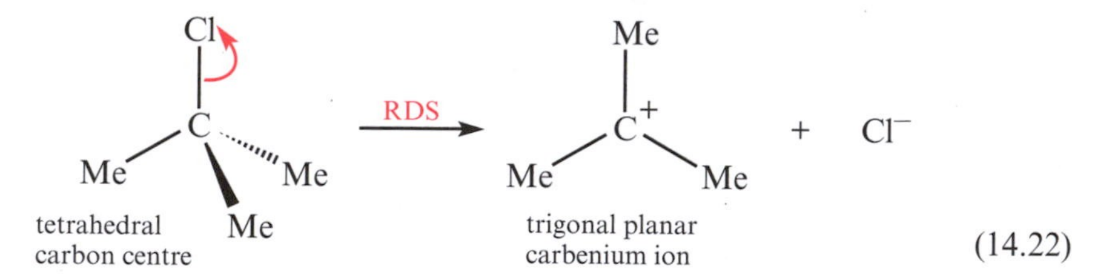

(14.22)

The carbenium ion formed in reaction 14.22 is a reaction *intermediate* which reacts in a *fast step* with any available nucleophile – this could be Cl^- (the outcome of which is to reverse reaction 14.22) or water as shown in equation 14.23.

$H_2\ddot{O}$ + [trigonal planar carbenium ion Me_3C^+] —fast step→ [tetrahedral carbon centre $Me_3C–OH_2^+$] —$-H^+$→ $Me_3C–OH$ (14.23)

nucleophile — trigonal planar carbenium ion — tetrahedral carbon centre

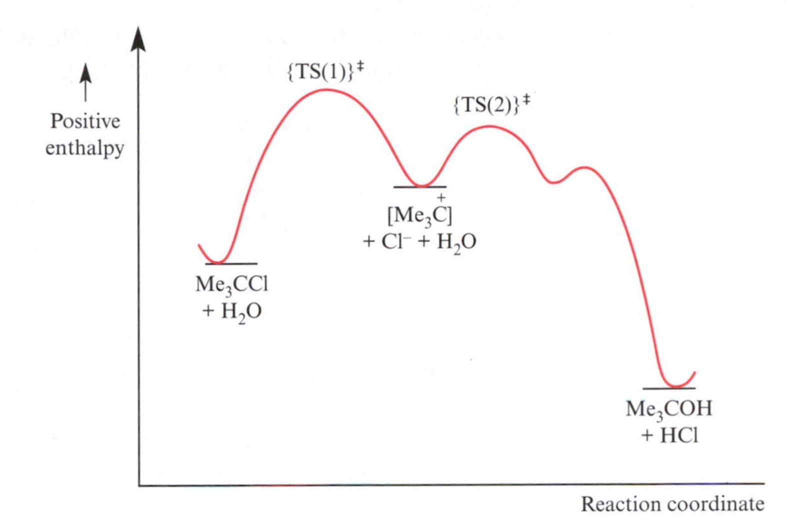

14.5 Reaction profile for the S_N1 reaction of 2-chloro-2-methylpropane with water. The intermediate is a carbenium ion, and $\{TS(1)\}^{\ddagger}$ and $\{TS(2)\}^{\ddagger}$ are the transition states involving C–Cl bond cleavage and C–OH_2 bond formation respectively. In the last step in the reaction profile, an O–H bond breaks, releasing H^+.

The first step in reaction 14.23 shows the attack of H_2O at the positive carbon centre of the $[Me_3C]^+$ ion. As the C–O bond forms, the positive charge is transferred to the oxygen centre. This is followed by the loss of a proton, and formation of 2-methylpropan-2-ol.

The reaction profile for the reaction of 2-chloro-2-methylpropane with water is shown in Figure 14.5. The carbenium ion intermediate occurs at a local energy minimum. Transition state $\{TS(1)\}^{\ddagger}$ may be described in terms of a species which is part-way between Me_3CCl and $[Me_3C]^+$ and can be represented as having the 'stretched bond' shown in **14.17**. The second transition state, $\{TS(2)\}^{\ddagger}$, may similarly be considered as being a species in which a water molecule has begun to form a bond with the carbenium ion.

$$\left\{ Cl^- \cdots \overset{+}{C}(Me)_3 \right\}^{\ddagger}$$

(14.17)

As the intermediate carbenium ion is planar, the incoming nucleophile can attack from above or below the plane – the consequences of this are discussed later in this section.

Exchange mechanism

The rate equation for the reaction between 1-bromobutane and hydroxide ion is second order overall and depends on *both* the concentration of the halogenoalkane and the nucleophile (equations 14.20 and 14.21). A mechanism that is consistent with these data involves a *one-step process* and is shown in equation 14.24 – it is a bimolecular exchange mechanism, often abbreviated to S_N2 (substitution nucleophilic bimolecular).

Pr = propyl = C_3H_7

$$[OH]^- + \mathrm{C(H)(H)(Pr)Br} \longrightarrow \mathrm{C(OH)(H)(H)(Pr)} + Br^- \qquad (14.24)$$

In contrast to the sequence in a dissociative process, the formation of the C–OH bond and cleavage of the carbon–halogen bond take place in a single step in an exchange mechanism – it is a *synchronous* process. The reaction profile is shown in Figure 14.6, and the transition state can be represented in terms of a carbon centre involving two 'stretched bonds' – one in the middle of being formed and the other in the process of being broken. The stereochemical consequences of this transition state are discussed below.

Stereochemistry of the S_N1 and S_N2 reactions

We have now outlined two *limiting* mechanisms whereby nucleophilic substitution may occur, and we have seen that experimentally observable differences in the kinetics of these reactions provide some evidence for the mechanisms. These two descriptions represent *extremes of behaviour* shown exclusively by few, if any, molecules. A second observable that distinguishes them is the *stereochemistry of the product in relation to the reactant*.

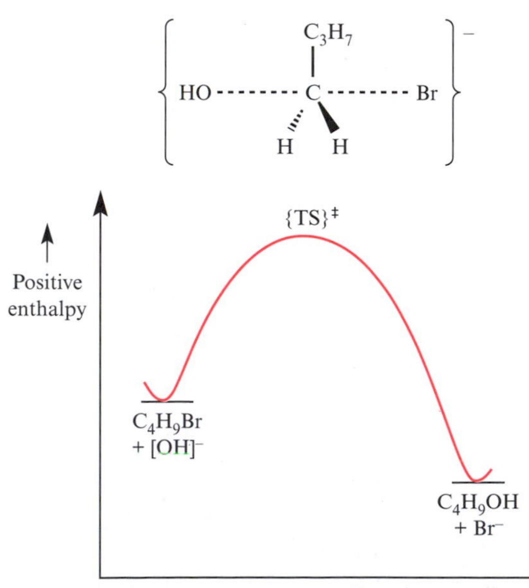

14.6 Reaction profile for the reaction of 1-bromobutane with hydroxide ion which proceeds by an S_N2 mechanism. In the transition state, the central carbon atom has 'stretched bonds' to the nucleophile and the leaving group. The C–C(butyl) and the two C–H bonds are coplanar.

Chiral compounds: see Section 8.6

When the chiral compound **14.18** undergoes nucleophilic substitution with hydroxide ion, the product has the structure shown in **14.19**. The important point is that the *absolute configuration has changed (inverted)* – that is, the methyl and hydrogen groups have changed their positions *relative* to the Br or OH substituent.

(14.18) (14.19)

This result is observed only for the nucleophilic substitution reactions that obey overall second order kinetics (i.e. exchange reactions) and can be rationalized in terms of the transition state shown in Figure 14.6. The attack of the nucleophile and the release of the leaving group take place along a linear path, and the three remaining substituents attached to the central carbon atom are forced into a plane in the transition state. Once the bond to the leaving group has been ruptured, a tetrahedral geometry at the central carbon atom is re-established but with a configuration that is the inverse of that in the starting compound. This is demonstrated in Figure 14.7 for the reaction between 2-bromobutane and hydroxide ion. An S_N2 reaction is accompanied by *complete inversion* at the central carbon atom. This process is known as *Walden inversion*.

In a dissociative reaction, the formation of a planar carbenium ion $[R_3C]^+$ as an *intermediate* means that the nucleophile can attack from either side of the plane. In Figure 14.8a, attack occurs from the top-side and the stereochemistry of the product (which contains an asymmetric carbon atom) is specific. In Figure 14.8b, water attacks the carbenium ion from below – once again the formation of the product is stereochemically specific, *but* the

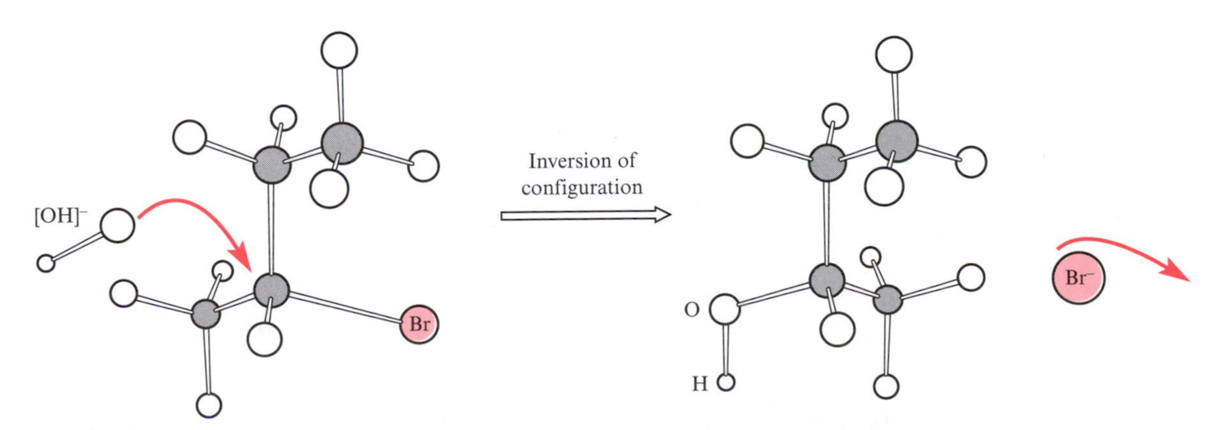

14.7 Inversion of configuration occurs during an S_N2 reaction. Here it is illustrated by the reaction of hydroxide ion with 2-bromobutane; the presence of the asymmetric carbon centre allows a particular optical isomer to be recognized.

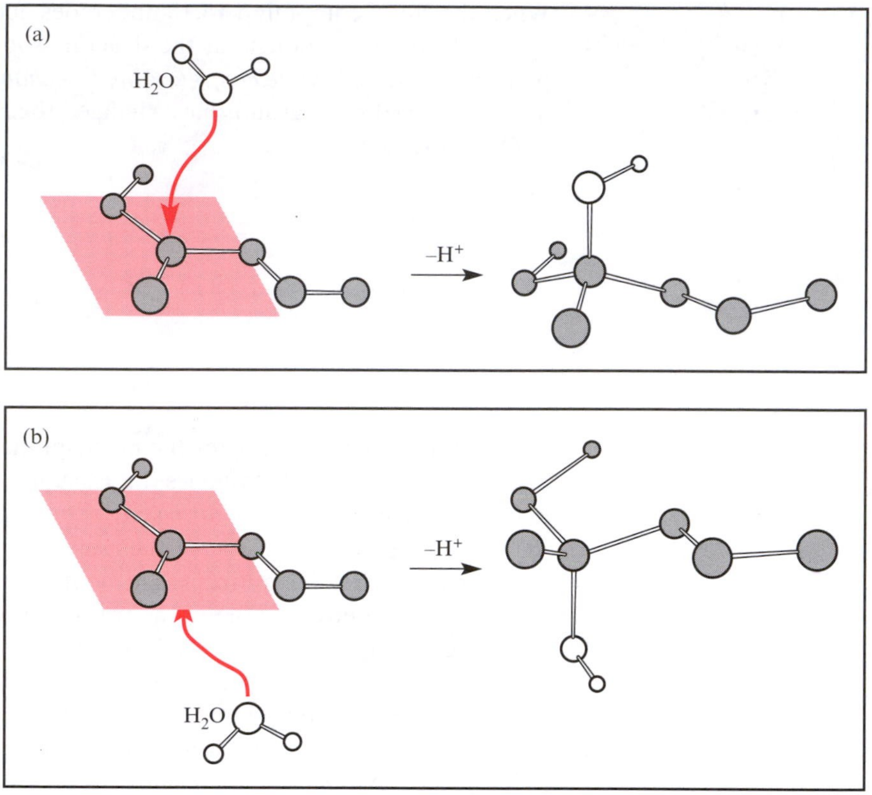

14.8 In an S_N1 reaction, the configuration at the central carbon atom may be retained or inverted. (a) Attack by water from the top-side of the planar carbenium ion $[(CH_3)(C_2H_5)(C_3H_7)C]^+$ and (b) attack from below, respectively produce optical isomers of $(CH_3)(C_2H_5)(C_3H_7)COH$. Individually, each of paths (a) and (b) is stereochemically specific, but the overall formation of enantiomers makes the S_N1 mechanism stereochemically *non-specific*. Alkyl hydrogen atoms have been omitted.

(a) H_2O $\xrightarrow{-H^+}$

(b) H_2O $\xrightarrow{-H^+}$

Enantiomer: see Section 8.6

product is the enantiomer of the product of the reaction shown in Figure 14.8a. Both the top and bottom faces of the carbenium ion are identical and so equal quantities of the two enantiomers *must* be formed. A reaction that proceeds by an S_N1 mechanism therefore gives rise to a mixture of equal amounts of enantiomers (equation 14.25) and this is called a *racemic mixture*. The S_N1 mechanism is stereochemically *non-specific*.

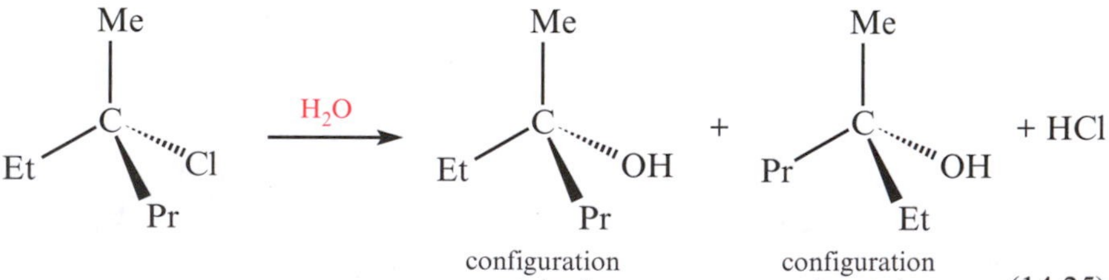

$$\text{C(Me)(Et)(Pr)Cl} \xrightarrow{H_2O} \underset{\text{configuration retained}}{\text{C(Me)(Et)(Pr)OH}} + \underset{\text{configuration inverted}}{\text{C(Me)(Pr)(Et)OH}} + \text{HCl} \quad (14.25)$$

A *racemic mixture* contains equal amounts of enantiomers of a chiral compound.

In reality, the situation is more complex. In some cases, *partial inversion* occurs because the 'real' mechanism might involve a transition state which lies somewhere between the two extremes we have described.

Factors influencing mechanism

Several factors influence the mechanism of a nucleophilic substitution reaction in halogenoalkanes, but a crucial point to bear in mind is that *for the dissociative pathway to be preferred, heterolytic cleavage of the C–X bond must readily occur.*

1. Nature of the carbenium ion

A carbenium ion is stabilized by electron-releasing alkyl groups directly attached to the positively charged carbon centre, and we have already seen that the order of stabilities of carbenium ions is:

Carbenium ions: see Section 8.14

$$[R_3C]^+ > [R_2CH]^+ > [RCH_2]^+ > [CH_3]^+$$

A *dissociative mechanism* (S_N1) is facilitated when the intermediate is a tertiary carbenium ion, and the preference for the unimolecular pathway follows the sequence:

$$R_3CX > R_2CHX > RCH_2X > CH_3X$$

Conversely, the preference for an *exchange mechanism* (S_N2) follows the order:

$$CH_3X > RCH_2X > R_2CHX > R_3CX$$

For secondary halogenoalkanes in particular the mechanism is often in between the extremes of S_N1 and S_N2.

2. Concentration of the nucleophile

If the nucleophile is present in a high concentration, the chances of a collision with a halogenoalkane molecule are increased, favouring the bimolecular exchange mechanism. Conversely, low concentrations of the nucleophile lead to a preference for a unimolecular dissociative mechanism.

3. Lewis basicity of the nucleophile

A strong Lewis base will attack a $C^{\delta+}$ carbon centre more readily than a weak base. This will facilitate an exchange mechanism – the strong Lewis base will help to drive out the leaving group. On the other hand, a weaker Lewis base will be more likely to attack a carbenium ion than a carbon centre with a smaller positive charge.

4. Solvent polarity

The tendency for the heterolytic cleavage of a C–X bond to occur increases as the polarity of the solvent increases, and in a polar solvent, we observe a preference for a dissociative mechanism.

Clearly, the overall situation is complicated! With so many contributing factors, it is often difficult to predict with any certainty to what extent either mechanism will be favoured.

Rearrangements accompanying dissociative reactions

One complicating factor in a dissociative mechanism is that there may be a tendency for the carbenium ion to rearrange *before* the nucleophile has time to attack. For example, the primary carbenium ion **14.20** may undergo

Relative stabilities of carbenium ions: see Section 8.14

methyl migration in order to form the more stable tertiary carbenium ion **14.21** (equation 14.26).

Me, Me, Me–C–C^+H$_2$ (primary carbenium ion **(14.20)**) —methyl migration→ Me_2C^+–CH(H)Me (tertiary carbenium ion **(14.21)**) (14.26)

Carbenium ion **14.21** may now be attacked by a nucleophile, or it may eliminate H^+ to give an alkene. This competitive reaction pathway is shown in equation 14.27, and such elimination reactions are discussed in more detail in the next section. Of course, there is also the possibility that the conversion of **14.20** to **14.21** was not complete, leaving some of the primary carbenium ion to react with the nucleophile – reaction with $[OH]^-$ would give 2,2-dimethylpropan-1-ol.

Me_2C^+–CH(H)Me —$[OH]^-$→ $Me_2C(OH)CH_2Me$ (2-methylbutan-2-ol)

Me_2C^+–CH(H)Me —$-H^+$→ $Me_2C{=}CHMe$ (2-methylbut-2-ene) (14.27)

Evidence for the two mechanisms for nucleophilic substitutions:

Unimolecular (dissociative, S_N1)

- **shows first order kinetics overall with no dependence on the concentration of the nucleophile;**
- **is stereochemically non-specific, and with chiral compounds, results in a partial or complete racemization; and**
- **is accompanied by the formation of additional products in cases when the carbenium ion shows a tendency to rearrange.**

Bimolecular (exchange, S_N2)

- **shows second order kinetics overall with a dependence on both the concentration of the nucleophile and the halogenoalkane, and**
- **is accompanied by inversion of configuration (stereochemically specific).**

14.6 THE MECHANISM OF β-ELIMINATION TO FORM ALKENES, AND ITS COMPETITION WITH NUCLEOPHILIC SUBSTITUTION

In equation 14.10, we showed a general β-elimination reaction in which a halogenoalkane loses HX to give an alkene. We can think of this in terms of the elimination of H^+ and X^- which combine to give a hydrogen halide. In the last section, we mentioned that a carbenium ion may eliminate H^+ to give an alkene. Remember that the carbenium ion was initially formed by the cleavage of a carbon–halogen bond and so the *overall* reaction was actually that shown in equation 14.10. This latter description of a β-elimination reaction indicates a *two-step* process, but is this always the case? The answer is 'no', and in this section we describe two mechanisms by which a halogenoalkane may undergo the elimination of HX – one is unimolecular and the other bimolecular. As you work through this section, keep in mind the close relationships between the unimolecular or bimolecular elimination and substitution mechanisms. *The crucial difference between E2 and S_N2 reactions is whether the Lewis base acts as a base or a nucleophile.*

The E1 mechanism: unimolecular elimination

The reaction of 2-chloro-2-methylpropane with H_2O gave 2-methylpropan-2-ol (equation 14.18). However, treatment with *hydroxide ion* leads to the formation of 2-methylpropene. This reaction shows a first order dependence overall with *no* dependence on $[OH^-]$. The reaction is a β-elimination, and a mechanism consistent with the kinetic data involves the formation of a carbenium ion in the rate-determining step (i.e. the same pathway as shown in equation 14.22), followed by the loss of a proton in a *fast step* (equation 14.28). This process is called an E1 mechanism (elimination unimolecular).

$$Me_3CCl \xrightarrow[-Cl^-]{RDS} Me_2C^+CH_2{-}H \xrightarrow[-H_2O]{fast} Me_2C{=}CH_2 \qquad (14.28)$$

tetrahedral carbon centre; trigonal planar carbenium ion ($[OH]^-$ removes H)

As we saw above, carbenium ions may rearrange before undergoing further reaction, and the nature of the elimination products reflects this as reactions 14.26 and 14.27 showed.

The competition between an E1 and S_N1 reaction depends on a range of factors, but it should reflect the relative stabilities of the elimination and substitution products. The loss of the proton can be facilitated by the presence of a strong base which is a poor nucleophile, for example, $Li\{N(CHMe_2)_2\}$. Steric factors may also swing the pathway in favour of elimination – it may help to relieve steric strain by creating two trigonal planar carbons in preference to tetrahedral centres. For example, the tertiary carbenium ion $[Bu_3C]^+$ **14.22** should prefer to eliminate H^+ (equation 14.29) rather than react with a nucleophile; the latter would force the three butyl substituents closer together than in the alkene product.

$-H^+$

(14.22)

(14.29)

The E2 mechanism: bimolecular elimination

In the presence of base, the β-elimination of HX from a halogenoalkane may obey second order kinetics with the rate equation being of the form shown in equation 14.30.

$$\text{Rate of reaction} = k[\text{halogenoalkane}][\text{base}] \quad (14.30)$$

A mechanism that is consistent with these experimental results involves a *one-step* process (equation 14.31) – it is a bimolecular mechanism, abbreviated to E2 (elimination bimolecular). The base is shown as a neutral molecule B which carries a lone pair of electrons, but it could be negatively charged, for example ethoxide ion $[EtO]^-$. The process is a *base-promoted elimination reaction.*

$+ [BH]^+ + X^-$

B:

(14.31)

The reaction profile for the base-promoted elimination of HBr from 2-bromobutane is shown in Figure 14.9. In the transition state, base B forms a bond with the proton that is about to be eliminated, and *simultaneously*, the C–H bond begins to break, the π-component of the C=C double bond begins to form, and the C–Br bond starts to cleave. For this synchronous process to occur, all five atoms involved in the making and breaking of bonds should be coplanar. There are only two conformations of the halogenoalkane that allow the planar transition state, and these are shown in the sawhorse drawings **14.23** and **14.24**.

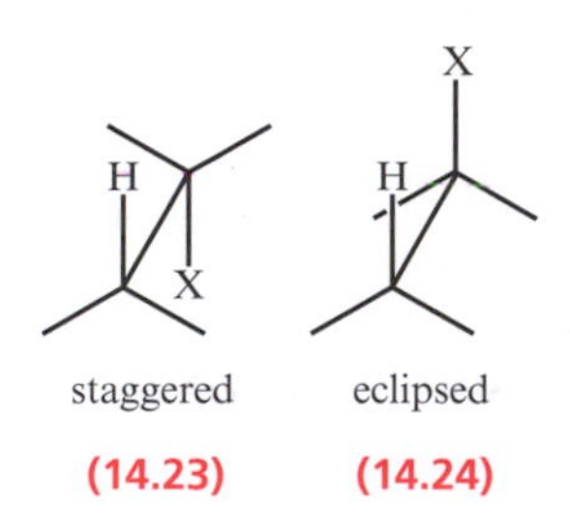

staggered **(14.23)** eclipsed **(14.24)**

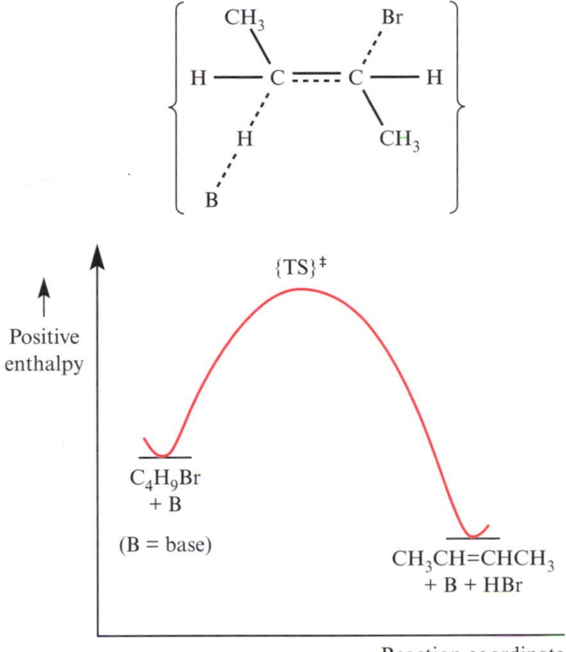

14.9 Reaction profile for the reaction of 2-bromobutane in alcoholic KOH which proceeds by an E2 mechanism. In the transition state, the base B (which is hydroxide ion), the hydrogen atom that is attacked, the central two carbon atoms, and the halogen atom all lie in one plane. The dotted lines represent 'stretched bonds'.

Elimination of HX from a halogenoalkane in a staggered conformation is called *anti*-elimination, and from an eclipsed conformer, is *syn*-elimination. The synchronous nature of the E2 process means that free rotation about the central C–C bond *does not* occur during the elimination of HX, and *anti*- and

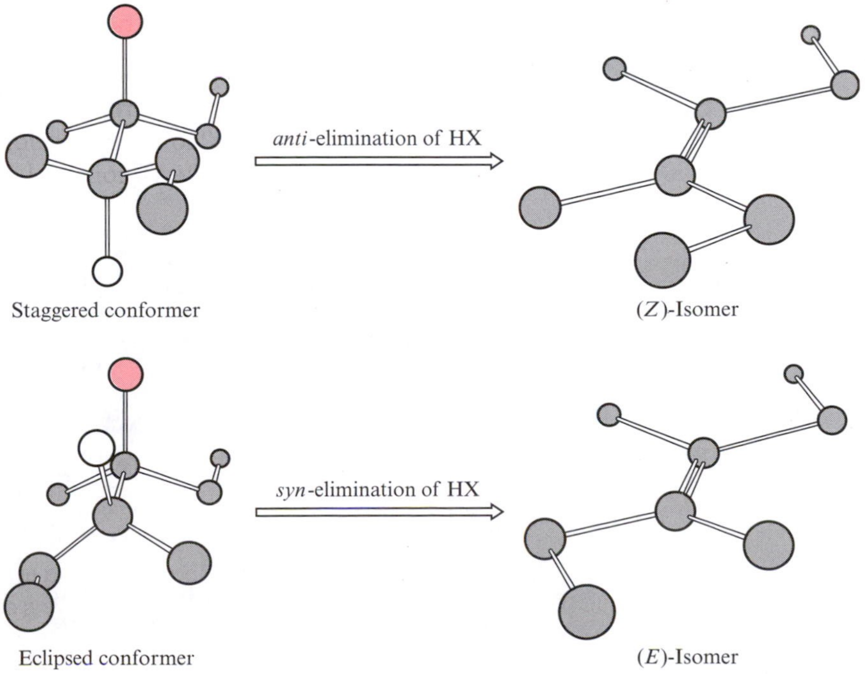

anti-elimination of HX

Staggered conformer

(Z)-Isomer

syn-elimination of HX

Eclipsed conformer

(E)-Isomer

14.10 *Anti*- and *syn*-elimination of HCl from one stereoisomer of 3-chloro-3,4-dimethylhexane. See also problem 14.11.

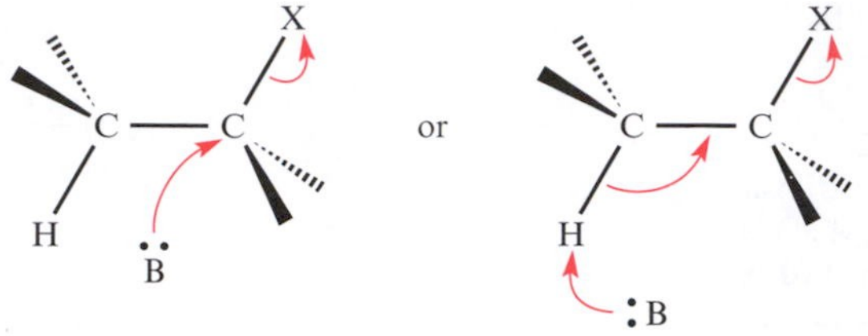

14.11 When a base B attacks a halogenoalkane, elimination of HX by the E2 process will compete with nucleophilic substitution by the S_N2 mechanism.

syn-eliminations lead exclusively to either (*Z*)- or (*E*)-isomers of the alkene (if the substitution pattern is such that the isomers can be recognized). This is illustrated in Figure 14.10 with the elimination of HCl from one stereoisomer of 3-chloro-3,4-dimethylhexane. *anti*-Elimination leads to the formation of (*Z*)-3,4-dimethylhex-3-ene, and *syn*-elimination gives (*E*)-3,4-dimethylhex-3-ene. There are three other stereoisomers of 3-chloro-3,4-dimethylhexane, and in problem 14.11, you are asked to consider the consequences of starting with these isomers. Although these reactions are complicated, it is most commonly found that halogenoalkanes undergo *anti*-elimination.

Clearly, there will always be a balance between the reaction of a nucleophile leading to an S_N2 process or the nucleophile acting as a base and leading to an E2 process (Figure 14.11).

Box 14.3 Diastereomers

We have already encountered optical isomers (or enantiomers) containing one asymmetric carbon atom; for example, 2-chlorobutane is chiral:

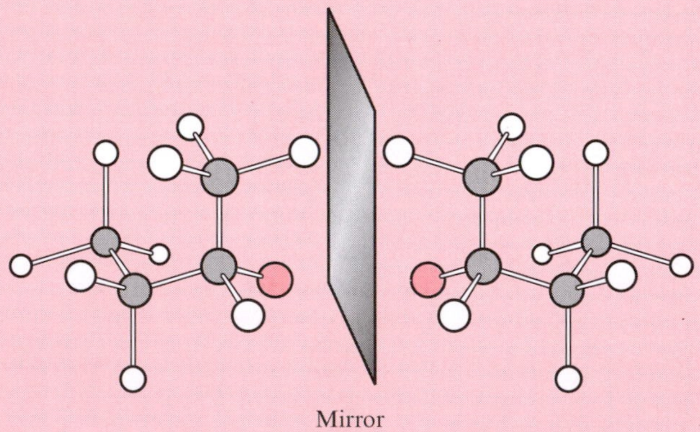

Mirror plane

If a molecule contains more than one asymmetric centre, the situation becomes more complicated. Consider 3-chloro-3,4-dimethylhexane. We can draw structure **A** (in which all but one hydrogen atom have been omitted for clarity) and its mirror image **B**. **A** and **B** are enantiomers (optical isomers), and they are also referred to as *stereoisomers*.

continues ▶

A B

Mirror plane

We can draw another stereoisomer of 3-chloro-3,4-dimethylhexane, structure **C**, and it too has an enantiomer, structure **D**.

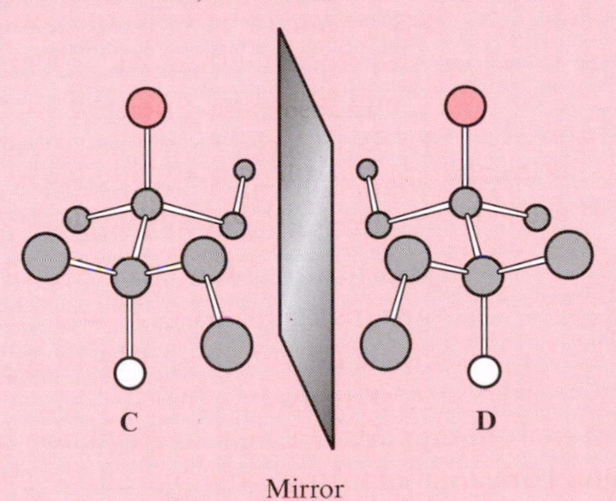

We now have four stereoisomers which can be related to one another in the form of two pairs of enantiomers – **A** and **B**, and **C** and **D**. However, while **A** and **C** are non-superimposable, they are *not* mirror images of each other and so they are not enantiomers. These stereoisomers are called *diastereomers.* Similarly, the pairs **A** and **D**, **B** and **C**, and **B** and **D** are diastereomers.

14.7 ALIPHATIC ETHERS 1: STRUCTURE, SYNTHESIS AND PHYSICAL PROPERTIES

Structure and nomenclature

An aliphatic ether contains an oxygen atom that is covalently bonded to two alkyl groups, and the two classes are:

- symmetrical ethers of general formula ROR in which the R groups are the same, and
- asymmetrical ethers of general formula ROR′ in which R and R″ are different.

An ether may be named by prefixing the word 'ether' with the alkyl substituents – the symmetrical ethers **14.25** and **14.26** are called dimethyl ether and diethyl ether respectively. In an asymmetrical ether, the two alkyl substituent names are placed in alphabetical order, so **14.27** is called ethyl methyl ether. Notice the use of spaces between the component names.

Me–O–Me **(14.25)** Et–O–Et **(14.26)** Me–O–Et **(14.27)**

An alternative nomenclature (also recommended by IUPAC) is to recognize the –OR group as an *alkoxy* substituent. By this nomenclature, **14.25** is methoxymethane, and **14.26** is ethoxyethane. For an asymmetrical ether, the parent compound is established by finding the longest carbon chain; thus, **14.27** is methoxyethane, and *not* ethoxymethane. In this nomenclature there are *no spaces* between the component names.

A third method of nomenclature is useful when more than one ether group is present. First we must recognize that an –O– group in an ether can be considered to replace an isoelectronic $-CH_2$ unit in the parent aliphatic chain. The name of the ether is derived by prefixing the name of the parent compound with *oxa*, and by numerical descriptors to show the positions of substitution. For structure **14.28**, the parent chain is heptane, but the CH_2 groups at atoms C(2) and C(5) have been replaced by oxygen atoms. The name for compound **14.28** is 2,5-dioxaheptane and, similarly, **14.29** is called 2,5,8-trioxanonane.

(14.28) **(14.29)**

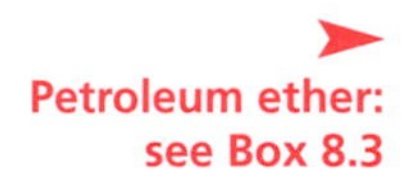
Petroleum ether: see Box 8.3

Common names are still in everyday use for some ethers, and Table 14.1 lists several that you may encounter. Note that *petroleum ether* is a hydrocarbon, and not an ether at all!

Table 14.1 Selected non-cyclic ethers for which common names are frequently used.

Trivial name	*Systematic name*	*Structure*
Ether or ethyl ether	Diethyl ether	
Monoglyme	1,2-Dimethoxyethane, or 2,5-dioxahexane	
Diglyme	Bis(2-methoxyethyl) ether, or 2,5,8-trioxanonane	
Triglyme	1,2-Bis(2-methoxyethoxy)ethane, or 2,5,8,11-tetraoxadodecane	

Synthesis

Dehydration of alcohols: see Section 14.10

Symmetrical ethers can be prepared from alcohols by heating them in the presence of acid (equation 14.32). This method is successful *under carefully controlled conditions*, because there is potential competition between this reaction and intramolecular dehydration of the alcohol to give an alkene.

$$2C_2H_5OH \xrightarrow{H_2SO_4,\ \Delta} C_2H_5OC_2H_5 + H_2O \qquad (14.32)$$

Williamson synthesis: see Section 14.4

The Williamson synthesis is widely used to prepare both symmetrical and asymmetrical ethers (equations 14.33 and 14.34).

$$RX + [OR]^- \rightarrow ROR + X^- \qquad \textit{Symmetrical ether} \qquad (14.33)$$

$$RX + [OR']^- \rightarrow ROR' + X^- \qquad \textit{Asymmetrical ether} \qquad (14.34)$$

An alkoxide ion is both a good nucleophile and a strong base, and so reactions 14.33 and 14.34 run the risk of being complicated by competing elimination reactions. In the previous sections, we saw that tertiary halogenoalkanes (which tend to react via a dissociative pathway) are prone to giving elimination and rearrangement products when they undergo nucleophilic substitution. On the other hand, the nucleophilic substitution of primary halogenoalkanes is less likely to have a more favourable competing elimination reaction. It is important to consider these observations if you are preparing an *asymmetrical* ether by the Williamson method. There will always be two possible combinations of reagents and one reaction route should be better than the other in terms of maximizing the yield of the ether. In equation 14.35, the reaction of 2-chloro-2-methylpropane with methoxide ion will give rise to elimination and rearrangement products. On the other hand, the reaction between chloromethane and $K[OCMe_3]$ should produce the ether in good yield.

$CH_3Cl + K[OCMe_3]$ — good choice → CH_3–O–$C(Me)_3$

$Me_3CCl + K[OCH_3]$ — very poor choice (dashed arrow) → CH_3–O–$C(Me)_3$

(14.35)

Physical properties

An ether can be thought of in terms of the exchange of a CH_2 unit in an alkane for an oxygen atom, and the introduction of this atom leads to the molecule having a small dipole moment as shown for Me_2O **14.30**. Values of dipole moments for selected ethers are listed in Table 14.2. Despite being polar, the boiling points given in Table 14.2 indicate that the intermolecular interactions between ether molecules are similar to those between their parent alkanes.

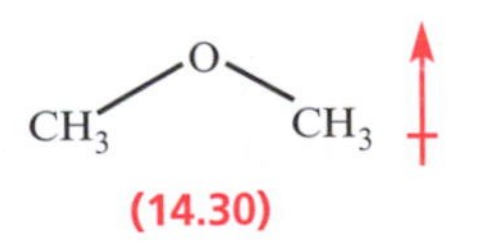

(14.30)

Table 14.2 Dipole moments, boiling points, and enthalpies of vaporization of selected ethers and parent alkanes. The 'parent' alkane is defined as one in which an exchange of a CH_2 unit for an O atom would yield the corresponding ether, for example, $CH_3CH_2CH_3$ is the parent ether of CH_3OCH_3.

Ether	*Relative molecular mass*	*Dipole moment / D*	*Boiling point / K*	*Enthalpy of vaporization / kJ mol^{-1}*	*Parent alkane*	*Relative molecular mass*	*Dipole moment*	*Boiling point / K*	*Enthalpy of vaporization / kJ mol^{-1}*
Dimethyl ether	46	1.30	248	21.5	Propane	44	0	231	19
Diethyl ether	74	1.15	308	26.5	Pentane	72	0	309	26
Dipropyl ether	102	1.21	364	31	Heptane	100	0	371	32
Dibutyl ether	130	1.17	413	36.5	Nonane	128	0	424	37

The low enthalpies of vaporization (Table 14.2) mean that the low molecular weight ethers are volatile; diethyl ether spilt on your hand will quickly evaporate as it absorbs heat from your body. Compare the value of $\Delta_{vap}H$ for Me_2O (21.5 kJ mol^{-1}) with that for ethanol (38.5 kJ mol^{-1}) – both compounds have the same molecular mass but the molecules in liquid ethanol are associated by hydrogen bonds and those of dimethyl ether are not.

Hydrogen bonding: see Section 11.8

Although diethyl ether is widely used for the extraction of organic compounds from aqueous solutions, it is not completely insoluble in water (a saturated solution contains about 8 g of Et_2O per 100 cm^3). After an 'ether extraction' (Figure 14.12), the aqueous layer consists of a saturated solution

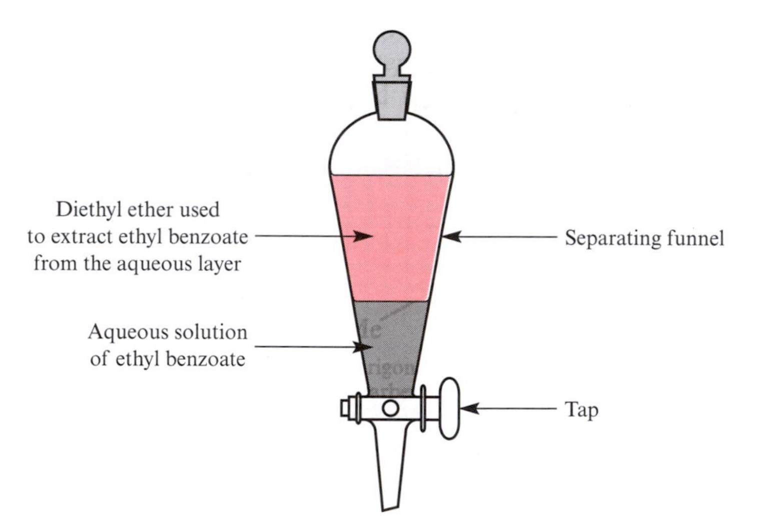

14.12 Ethyl benzoate, an ester, can be prepared from benzoic acid and ethanol (see Section 17.8). The product is contained in an aqueous solution and may be extracted by adding diethyl ether in a separating funnel. The funnel is stoppered and shaken; the pressure which builds up in the funnel must be released by inverting the funnel and opening the tap. After closing the tap again, the funnel is left as shown so that the two liquid layers separate out. The stopper is taken out, and the lower layer is removed through the tap. The ether layer is then poured out into a separate flask and the aqueous layer is returned to the funnel. This process is repeated until no more ethyl benzoate is extracted. The combined ether extracts are dried by adding some powdered anhydrous $MgSO_4$ or $CaCl_2$.

of Et_2O in water, and the ether layer consists of a saturated solution of water in ether. This is why the organic phase from an ether extraction must be dried before the solvent is removed. It is usually found that the solubility of diethyl ether in water decreases when ionic salts are dissolved in the aqueous phase. This is readily understood in terms of a competition between the Et_2O molecules for hydrogen bonding with the water molecules, and the requirement of the ions for solvation by the water molecules. The enthalpies of solvation of the ions are significantly greater than those of hydrogen bonding. In the laboratory this means that the addition of sodium chloride to the aqueous phase drastically reduces the solubility of the diethyl ether – the process is known as *salting out*.

Solvation of ions: see Section 12.11

Both the dipole moment and the ability to form hydrogen bonds with compounds containing an XH group (X = electronegative atom) contribute towards making ethers useful solvents. As most of the lower *alkyl ethers are highly volatile and inflammable, their vapours present a significant explosion hazard!*

Identification of ethers by infrared spectroscopy

The infrared spectrum of a dialkyl ether exhibits a strong absorbance in the range 1080–1150 cm^{-1}, assigned to the C–O stretching mode. This is illustrated in Figure 14.13 with the IR spectra of two symmetrical ethers, which are compared with that of an alkane of similar chain length.

IR spectra for organic compounds: see Section 9.8

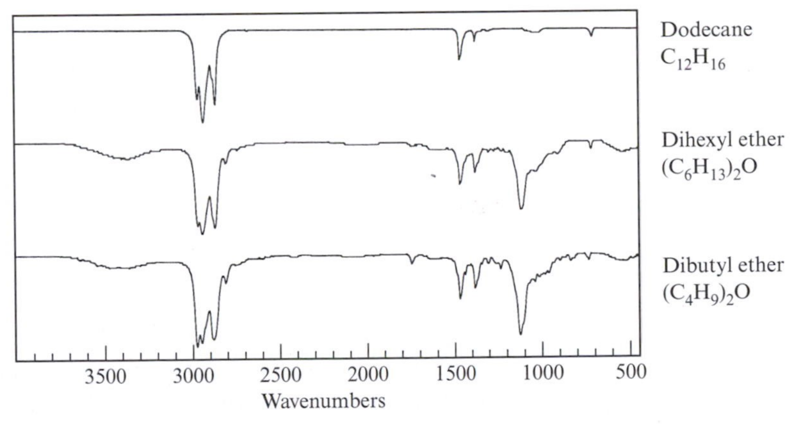

14.13 The IR spectra of dibutyl ether, $(C_4H_9)_2O$, and dihexyl ether, $(C_6H_{13})_2O$, both exhibit a strong absorbance at 1120 cm^{-1} assigned to the C–O stretching mode. Compare these spectra with that of an alkane of similar chain length – dodecane $C_{12}H_{26}$. Common to all three spectra are absorbances due to the aliphatic C–H stretches, and characteristic bands in the fingerprint region around 1400 cm^{-1}. To what do you assign the broad band near 3400 cm^{-1} in the spectra of the ethers?

14.8 ALIPHATIC ETHERS 2: REACTIVITY

Ethers are not particularly reactive compounds. They can cleaved by using concentrated HBr or HI (equation 14.36).

$$\text{ROR} + \text{HX} \xrightarrow{\Delta} \text{ROH} + \text{RX} \qquad X = \text{Br or I} \qquad (14.36)$$

The first step in the reaction is protonation of the oxygen atom (equation 14.37), followed by nucleophilic attack by bromide or iodide ion, with the substitution occurring by either an exchange or dissociative mechanism (equations 14.38 and 14.39). The *neutral alcohol* is a good leaving group. Attack by X^- *before* the protonation step would require the loss of an alkoxide ion, but this process is unfavourable because a *strong base is a poor leaving group*.

$$\text{R–}\ddot{\text{O}}\text{–R}' + \text{H}^+ \longrightarrow \text{R–O}^+\text{(H)–R}' \qquad (14.37)$$

$$\text{R–O}^+\text{(H)–R}' + X^- \longrightarrow \text{ROH} + \text{R}'\text{X} \qquad \textit{Exchange} \qquad (14.38)$$

$$\text{R–O}^+\text{(H)–R}' \longrightarrow \text{R–O–H} + \text{R}'^+ \xrightarrow{X^-} \text{ROH} + \text{R}'\text{X} \qquad X = \text{Br or I} \qquad \textit{Dissociative} \qquad (14.39)$$

The products could be ROH and R′X (as shown) or R′OH and RX. Mixtures of products are often obtained, but if either R or R′ is a methyl group, then the formation of CH_3X is generally favoured over CH_3OH (equation 14.40).

$$CH_3OCH_2CH(Me)CH_3 + HI \xrightarrow{\Delta} CH_3I + CH_3CH(Me)CH_2OH \qquad (14.40)$$

14.9 ALIPHATIC ALCOHOLS 1: STRUCTURE, SYNTHESIS AND PHYSICAL PROPERTIES

Structure and nomenclature

The functional group of an alcohol is an –OH group, and the general formulae for primary, secondary and tertiary alcohols are RCH_2OH, R_2CHOH and R_3COH respectively. For a straight chain primary alcohol, the name is derived by attaching the ending -ol to the root name of the alkane – the principal chain is found as described in Section 14.3. Compound **14.31** is butanol and **14.32** is hexanol, but the more complete names of butan-1-ol and hexan-1-ol leave no ambiguity as to the position of the –OH group.

$CH_3CH_2CH_2CH_2OH$ or OH

(14.31)

$CH_3CH_2CH_2CH_2CH_2CH_2OH$ or OH

(14.32)

For secondary and tertiary alcohols, the name is derived by recognizing the –OH group as a substituent in a straight or branched alkane chain. Compound **14.33** is butan-2-ol, and **14.34** is 2-methylpentan-2-ol.

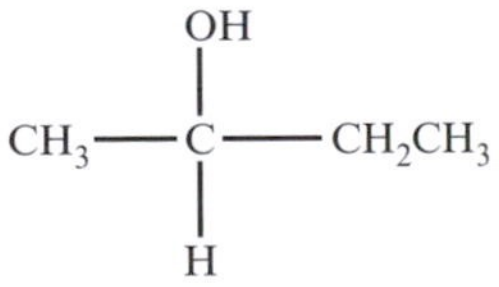

CH_3–C(OH)(H)–CH_2CH_3 or OH

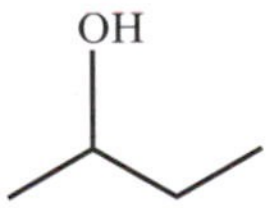

(14.33)

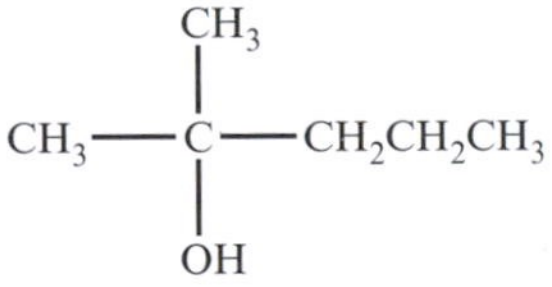

CH_3–C(CH_3)(OH)–$CH_2CH_2CH_3$ or OH

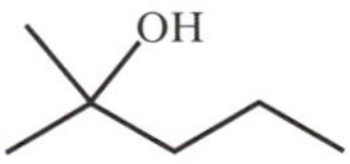

(14.34)

HO OH

(14.35)

OH OH OH

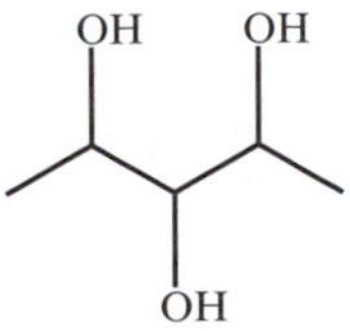

(14.36)

When a compound possesses more than one alcohol functionality, the name must specify the total number of, as well as the position-numbers for, the –OH groups. Compound **14.35** is ethane-1,2-diol. Notice that for a diol (or triol, etc.) the 'e' on the end of the alkane root-name is retained. The systematic name for compound **14.36** is pentane-2,3,4-triol.

Many alcohols are referred to by common names, and some common laboratory alcohols are listed in Table 14.3.

Table 14.3 Selected alcohols for which common names are frequently used.

Trivial name	*Systematic name*	*Structure*
Ethylene glycol	Ethane-1,2-diol	$HOCH_2CH_2OH$
Glycerol (glycerine)	Propane-1,2,3-triol	$HOCH_2CH(OH)CH_2OH$
tert-Butyl alcohol	2-Methylpropan-2-ol	$(CH_3)_3COH$
Isopropyl alcohol	Propan-2-ol	$(CH_3)_2CHOH$

Synthesis and industrial manufacture

Earlier in this chapter we saw that an alcohol may be prepared by hydrolysing the corresponding halogenoalkane. Other methods of preparing alcohols that we have previously discussed include alkene addition reactions – the acid-catalysed addition of water gives Markovnikov addition products in the case of asymmetric alkenes (equation 14.41), and the conversion of alkenes to organoboranes followed by reaction with H_2O_2 and NaOH gives alcohols by anti-Markovnikov addition (equation 14.42).

Conversion of alkenes to alcohols: see Sections 8.13, 8.14 and 8.18

$$CH_3CH{=}CH_2 + H_2O \xrightarrow{H^+ \text{ (catalyst)}} CH_3CH(OH)CH_3 \qquad (14.41)$$

$$CH_3CH{=}CH_2 \xrightarrow{\text{hydroboration}} (CH_3CH_2CH_2)_3B \xrightarrow{H_2O_2,\ [OH]^-} 3CH_3CH_2CH_2OH \qquad (14.42)$$

Grignard reagents RMgX (X = Cl or Br) react with carbonyl compounds (compounds containing a C=O functional group) to give primary, secondary or tertiary alcohols. The product depends upon the substituents attached to the carbonyl group – if the precursor is methanal, the product is a primary alcohol (equation 14.43), if the starting compound is a higher aldehyde, a secondary alcohol is formed (equation 14.44), and if we begin with a ketone, the result is a tertiary alcohol (equation 14.45).

Carbonyl compounds: see Chapter 17

$$\underset{\text{methanal}}{H_2C{=}O} \xrightarrow[\text{2. } H_2O]{\text{1. RMgX}} \underset{\text{primary alcohol}}{RCH_2OH} \qquad (14.43)$$

$$R'(H)C{=}O \xrightarrow[\text{2. } H_2O]{\text{1. RMgX}} R{-}C(R')(H){-}OH \qquad (14.44)$$

aldehyde $R' \neq H$ → secondary alcohol

$$R'(R'')C{=}O \xrightarrow[\text{2. } H_2O]{\text{1. RMgX}} R{-}C(R')(R''){-}OH \qquad (14.45)$$

ketone → tertiary alcohol

The industrial preparations of methanol and ethanol deserve a special mention. In 1994, methanol ranked as 22nd among chemicals manufactured in the United States of America. It has been referred to in the past as *wood alcohol* because of a method of manufacture which involved the heating of wood in the absence of air. It is now manufactured from carbon monoxide by reaction 14.46. Methanol is a toxic liquid, the drinking or inhalation of which may cause blindness or death.

Industrial rankings of chemicals: see Appendix 14

$$CO + 2H_2 \xrightarrow[\text{ZnO catalyst}]{\text{150 bar, 670 K}} CH_3OH \qquad (14.46)$$

Ethanol is the alcohol present in alcoholic drinks, and for this purpose it is manufactured by the fermentation of glucose contained in, for example, grapes (for wine production), barley (for whisky) and cherries (for kirsch). The reaction is catalysed by a series of enzymes in yeast (equation 14.47).

$$\text{glucose} \xrightarrow{\text{yeast}} 2C_2H_5OH + 2CO_2 \qquad (14.47)$$

Ethanol is widely used as a solvent and a precursor for other chemicals including acetic acid, and in this context it is manufactured by the acid-catalysed addition of water to ethene. Pure C_2H_5OH is often called *absolute alcohol*, but it is often used as a solution containing 5% water (and drunk in more dilute solutions!).

Ethylene glycol: see Section 8.13

Ethane-1,2-diol is also of industrial importance and is a component of 'anti-freeze' for motor-vehicle radiator water. It is manufactured by the oxidation of ethene (equation 14.48) rather than the smaller scale synthesis described in Section 8.13.

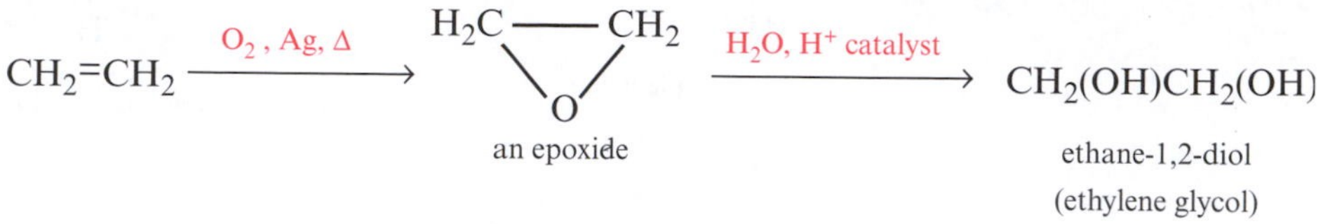

$$CH_2{=}CH_2 \xrightarrow{O_2,\ Ag,\ \Delta} \underset{\text{an epoxide}}{H_2C{-}CH_2\ (\text{bridged by O})} \xrightarrow{H_2O,\ H^+\ \text{catalyst}} \underset{\text{ethane-1,2-diol (ethylene glycol)}}{CH_2(OH)CH_2(OH)} \qquad (14.48)$$

OCOR

R″OCO OCOR′

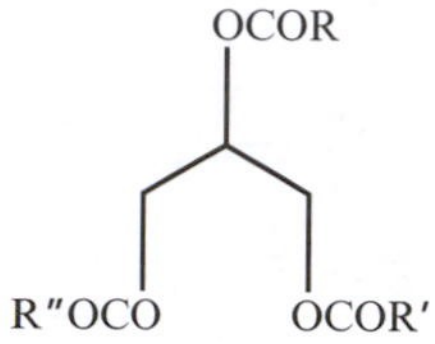

(14.37)

Propane-1,2,3-triol (glycerol or glycerine) is the critical component of all fats which are of formula **14.37**. It is used as a sweetener, and as a softener in soap. Its reaction with nitric acid produces nitroglycerine (equation 14.49) which is used for the treatment of angina, but is also the explosive constituent of dynamite.

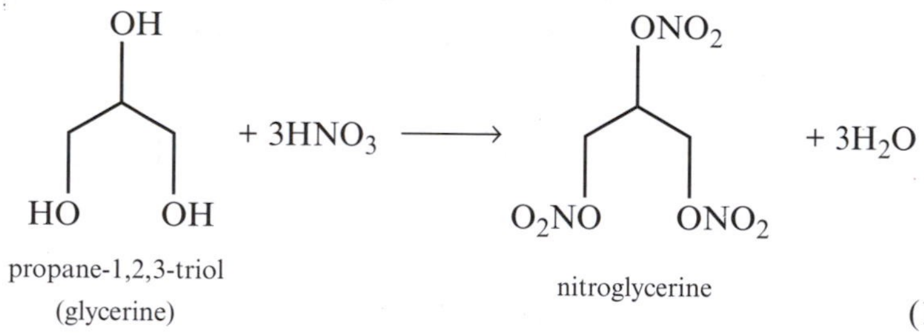

$$\underset{\text{propane-1,2,3-triol (glycerine)}}{HOCH_2CH(OH)CH_2OH} + 3HNO_3 \longrightarrow \underset{\text{nitroglycerine}}{O_2NOCH_2CH(ONO_2)CH_2ONO_2} + 3H_2O \qquad (14.49)$$

Physical properties

The –OH group in alcohols leads to intermolecular hydrogen bonding in the solid and liquid states. The boiling points and enthalpies of vaporization of

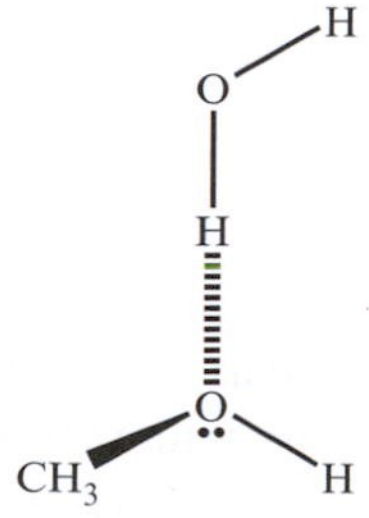

(14.38)

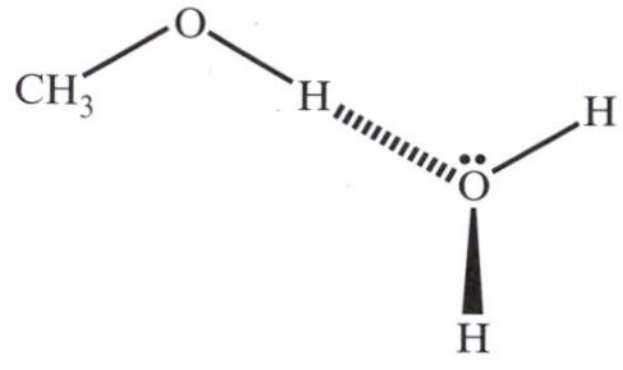

(14.39)

selected alcohols are listed in Table 14.4, and their relatively high values reflect the intermolecular association in the liquid state. The lower liquid alcohols (methanol, ethanol and propanol) are completely miscible with water owing to the formation of hydrogen bonds between alcohol and water molecules. Structures **14.38** and **14.39** show two possible hydrogen-bonded interactions between methanol and water.

Homologue: see Section 8.2

Box 14.4 Explosives and rocket propellants

Both nitroglycerine and TNT (trinitrotoluene) are well-known explosives; TNT is used in a variety of bombs. The common structural feature of these compounds is the nitro group (NO_2). Upon explosion, the molecules decompose – nitroglycerine (mp 286 K) forms *only* gaseous products (29 moles per 4 moles of explosive):

$$4O_2NOCH(CH_2ONO_2)_2(l) \rightarrow 6N_2(g) + 12CO_2(g) + 10H_2O(g) + O_2(g)$$

Each reaction is therefore accompanied by a large *positive* change in entropy, making the reaction entropically *very* favourable. The detonation releases 1415 kJ of energy per mole of nitroglycerine.

Other nitro-derivatives have been used (or studied for potential use) by the military as explosives and rocket propellants. With the end of the Cold War, more information about these compounds has become available to the public. Although an explosive and a rocket propellant may appear to do the same job, the design requirements are different. For example, one aim of a propellant is to fire a missile over a given range in a controlled manner, whereas an explosive releases all of its energy rapidly. A further point to consider is the thermal and shock sensitivities of the materials – hazards prior to the point of ignition must be minimized. Picric acid is explosive but its high shock sensitivity can result in premature detonation, and it has generally been replaced by other (slightly!) less hazardous materials.

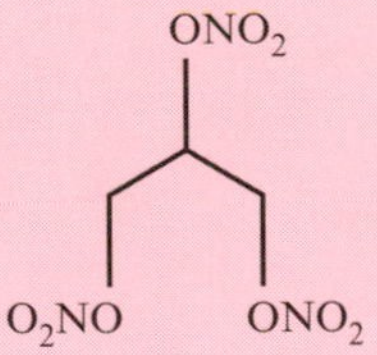

nitroglycerine (nitroglycerin, trinitroglycerol)

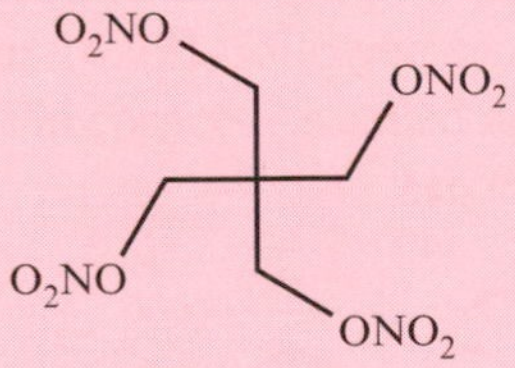

pentaerythritol tetranitrate

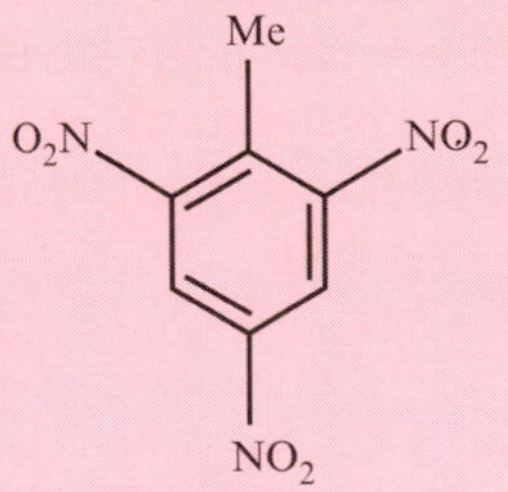

TNT

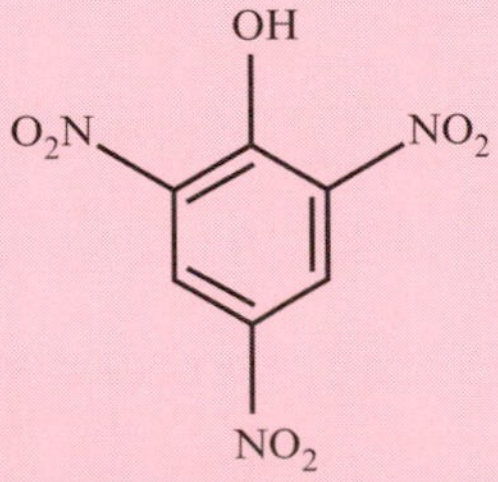

picric acid (2,4,6-trinitrophenol)

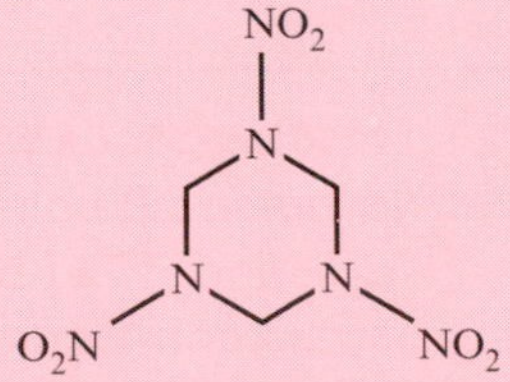

RDX

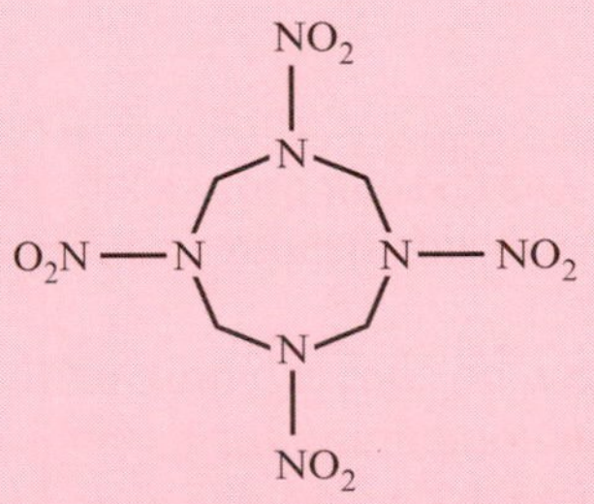

HMX

Table 14.4 Boiling points and enthalpies of vaporization of selected alcohols. Values for water are included for comparison.

Alcohol	*Relative molecular mass*	*Boiling point /K*	$\Delta_{vap}H$ / $kJ\ mol^{-1}$
Water	18	373	41
Methanol	32	338	35
Ethanol	46	351	38.5
Propan-1-ol	60	370	41
Propan-2-ol	60	355	40
Butan-1-ol	74	391	43
Butan-2-ol	74	372.5	41
2-Methylpropan-1-ol	74	381	42
2-Methylpropan-2-ol	74	355	39
Pentan-1-ol	88	411	44
Hexan-1-ol	102	431	44.5

The solubilities in water of butanol, pentanol, hexanol and heptanol are dramatically lower than those of their lighter homologues, and the higher alcohols are essentially immiscible with water. This can be attributed to the increased aliphatic chain length of the molecule. The –OH group is *hydrophilic* ('water-loving'), while the hydrocarbon portion of an alcohol molecule is *hydrophobic* ('water-hating') (Figure 14.14). As the molecular mass of the alcohol increases, the inability of the hydrocarbon chain to form hydrogen bonds with the water predominates over the effects of the hydrophilic group – as a result, the solubility of the alcohol in water decreases dramatically.

Hydrophilic means 'water-loving' and hydrophobic means 'water-hating'.

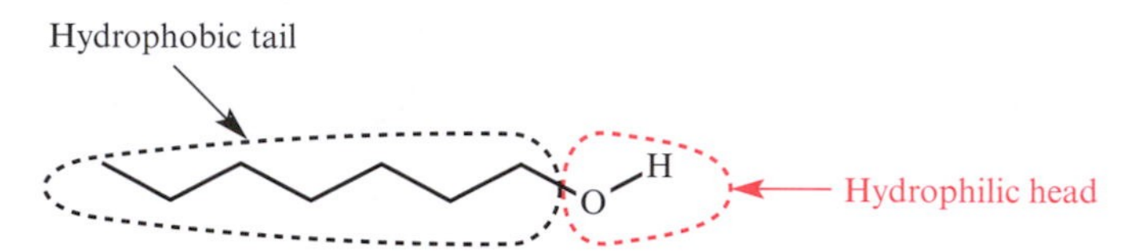

14.14 The –OH group of an alcohol can form hydrogen bonds with water and is called a *hydrophilic group*. In a long chain alcohol such as heptan-1-ol, the OH functionality forms a hydrophilic head, while the aliphatic chain is a *hydrophobic* tail.

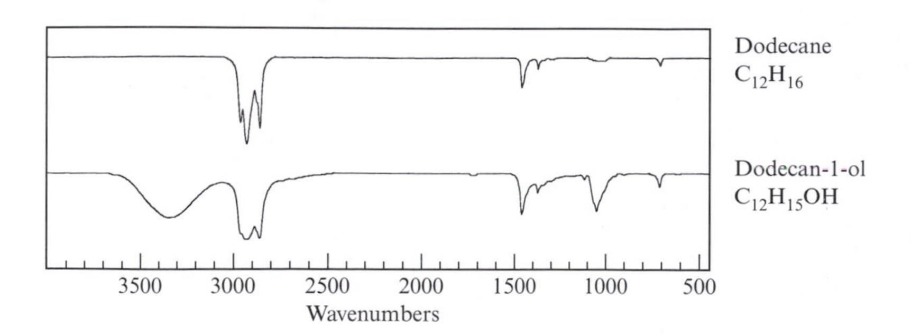

14.15 The IR spectra of dodecane and dodecan-1-ol. Common to both spectra are absorbances due to the aliphatic C–H stretches, and characteristic bands in the fingerprint region. The broad band centred at 3330 cm^{-1} in the spectrum of dodecan-1-ol is assigned to the O–H stretch.

Box 14.5 Pheromones 1: a female moth attracts her mate

Pheromones are chemical messengers and nature has provided female moths with the ability to secrete pheromones to act as sexual attractants. But a pheromone is more than a perfume! A particular species of moth secretes its own particular pheromone or group of pheromones and so attracts males of its own species. Some female moths such as the vapourer and gypsy moths are wingless. It is a remarkable sight to watch a newly emerged wingless female vapourer moth – her life is restricted to sitting on the pupal cocoon and from there she must play her part to ensure the next generation of her species. She pumps out pheromones which can be sensed by male moths some distance away. Large antennae for sensing pheromones are characteristic features of many male moths, and it is not long before the female entices her mate! Many moth pheromones consist of linear C_n chains with a terminal functional group (e.g. an alcohol or aldehyde group) – the two examples shown below may appear to be deceptively simple molecules.

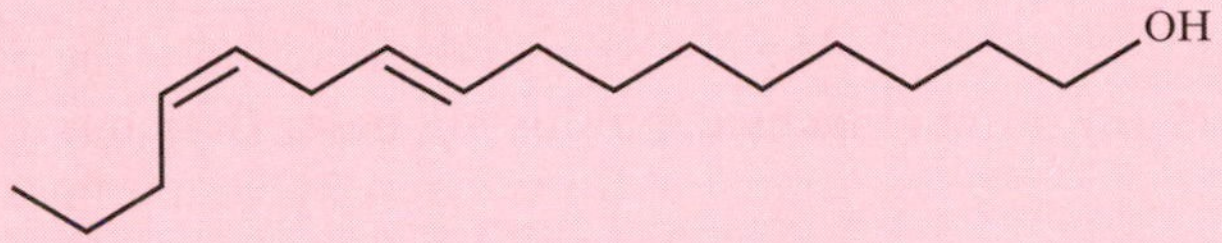

OH

Pheromone of the female silkworm moth

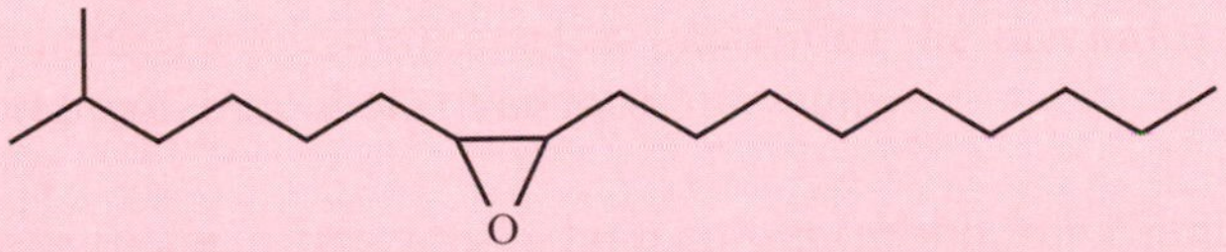

O

Pheromone of the female gypsy moth

(See also Box 15.12: Pheromones 2: ant trails)

Identification of alcohols by infrared spectroscopy

IR and NMR spectra of organic compounds: see Section 9.8

The appearance in an IR spectrum of a broad and strong absorbance near to 3300 cm^{-1} is characteristic of an alcohol functional group. In Figure 14.15 we compare the IR spectra of dodecane and dodecan-1-ol. Apart from the broad band centred at 3330 cm^{-1}, the two spectra look quite similar.

14.10 ALIPHATIC ALCOHOLS 2: REACTIVITY

Reactions accompanied by complete degradation: combustion

The combustion of an alcohol gives carbon dioxide and water (equation 14.50); ethanol burns with a blue flame.

$$C_2H_5OH + 3O_2 \rightarrow 2CO_2 + 3H_2O \tag{14.50}$$

At present, methanol ($\Delta_c H = -726$ kJ mol^{-1}) is being considered as an alternative to petroleum hydrocarbon as a fuel. It is particularly attractive in countries where it is available in bulk from biomass or from catalytic processes. We single out alcohols for a mention of combustion on the grounds of their potential as fuels, but note that most other organic compounds also burn well.

Reactions that cleave the R–OH bond

1. Dehydration

We saw in equation 14.32 that, under *carefully controlled conditions*, the *intermolecular* elimination of water from an alcohol could lead to a symmetrical ether. The *intramolecular* elimination of H_2O occurs when an alcohol is heated with concentrated sulfuric acid (equation 14.51) or is passed over hot aluminium oxide.

$$C_2H_5OH \xrightarrow{\text{conc. } H_2SO_4\text{, 450 K}} CH_2{=}CH_2 + H_2O \tag{14.51}$$

The first step in the reaction is the protonation of the –OH group, and the subsequent loss of water (which is a *good leaving group*) generates a carbenium ion. In reaction 14.52, a primary carbenium ion is formed. This could eliminate H^+ in an E1 process to yield but-1-ene (equation 14.53), but the observed major product is actually but-2-ene. This can be explained in terms of the rearrangement of the intermediate to a more stable secondary carbenium ion prior to the loss of H^+ (equation 14.54).

$$CH_3CH_2CH_2CH_2OH \xrightarrow{H^+} CH_3CH_2CH_2CH_2{-}\overset{+}{O}H_2 \xrightarrow{-H_2O} CH_3CH_2CH_2\overset{+}{C}H_2 \tag{14.52}$$

$$CH_3CH_2CH_2\overset{+}{C}H_2 \xrightarrow{-H^+} CH_3CH_2CH{=}CH_2 \quad \text{(but-1-ene)} \quad \textit{Not favourable} \tag{14.53}$$

$$CH_3CH_2CH_2\overset{+}{C}H_2 \xrightarrow{\text{rearrangement}} CH_3CH_2\overset{+}{C}HCH_3 \xrightarrow{-H^+} CH_3CH{=}CHCH_3 \quad \text{(but-2-ene)} \quad \textit{Favourable} \tag{14.54}$$

Box 14.6 Methanol and ethanol: alternative motor fuels?

Methanol can be produced from natural gas and from biomass (i.e. plant and animal material) and is considered to be a possible alternative fuel for motor-vehicles. Amongst its advantages are the lower levels of NO_x emissions (see Section 13.8) but disadvantages include fire and explosion hazards and high toxicity. Mixed with hydrocarbon-based fuels, methanol may be more viable.

Ethanol can also be manufactured from biomass and it possesses a high octane number (see Section 8.9). 'Gasohol' (a petrol mixture containing 10% ethanol) is becoming available in the American continents, and Brazil uses large amounts of ethanol-containing fuels. Cars do not have to be specially adapted, although they may suffer from reduced performance. Within many European countries, the EU restricts the ethanol content of motor fuels to 5%.

Primary or secondary carbenium ions may undergo rearrangement to form more stable intermediates either by *proton* or *methyl migration* (for example as in reaction 14.26). Problem 14.18 asks you to predict the major products of some alcohol dehydration reactions, bearing in mind the possibility of such rearrangements.

2. Conversion of an alcohol to a halogenoalkane

Alcohols react with phosphorus(III) or phosphorus(V) halides (PCl_3, PBr_3, PI_3 or PCl_5) to give halogenoalkanes (equations 14.55 and 14.56). Thionyl chloride, $SOCl_2$, can also be used to convert primary, secondary or tertiary alcohols to halogenoalkanes (equation 14.57) and the beauty of this latter method is that the unwanted side-products are gaseous, making isolation of the product facile.

$$3ROH + PX_3 \rightarrow 3RX + H_3PO_3 \qquad X = Br \text{ or } I \tag{14.55}$$

$$ROH + PCl_5 \rightarrow RCl + POCl_3 + HCl \tag{14.56}$$

$$ROH + SOCl_2 \rightarrow RCl + SO_2(g) + HCl(g) \tag{14.57}$$

Reactions that cleave the RO–H bond

1. Reaction with alkali metals

When a piece of sodium is placed in ethanol, H_2 is evolved in a reaction that is reminiscent of, but less violent than, that of sodium with water. The alcohol acts as a weak acid and forms a *sodium alkoxide* salt (equation 14.58). The pK_a value of ethanol is ≈16, slightly higher than pK_w.

$pK_w = 14.00$

$$2Na + 2C_2H_5OH \rightarrow \underset{\text{sodium ethoxide}}{2Na^+[C_2H_5O]^-} + H_2 \tag{14.58}$$

Reactions between sodium and alcohols are often used to destroy excess sodium (e.g. after the sodium has been used as a drying agent for a hydrocarbon solvent), but there is a risk of fire owing to the production of the H_2 in an inflammable alcohol medium. Choosing higher molecular weight alcohols reduces the fire hazard – however, if the alcohol is too bulky, the reaction is very slow. Propan-2-ol is a suitable compromise (equation 14.59).

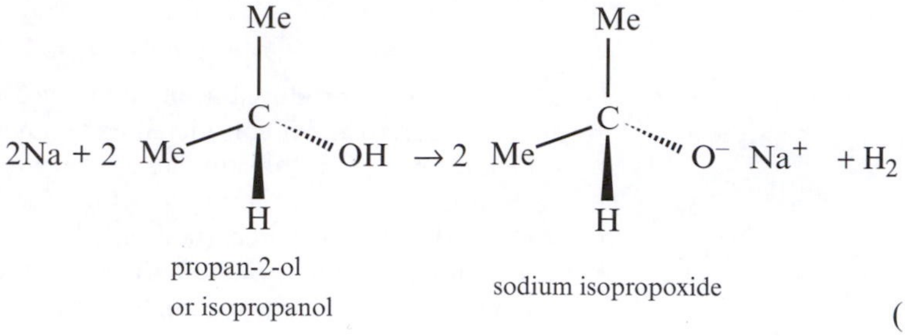

$$2Na + 2\ Me_2CHOH \rightarrow 2\ Me_2CHO^-\ Na^+ + H_2 \qquad (14.59)$$

propan-2-ol or isopropanol → sodium isopropoxide

2. Oxidation

Aldehydes, ketones and carboxylic acids: see Chapter 17

Although combustion is an oxidation process, in this section we are concerned with the selective oxidation of alcohols to aldehydes, ketones and carboxylic acids.

The general reactions are summarized in Figure 14.16 – only primary and secondary alcohols are readily oxidized. The oxidation of a secondary alcohol to a ketone can be achieved using acidified potassium dichromate at room temperature (equation 14.60), and the ketone that is produced cannot easily be oxidized further.

$$\underset{\text{butan-2-ol}}{CH_3CH_2CH(OH)CH_3} \xrightarrow{K_2Cr_2O_7,\ H_2SO_4} \underset{\text{butanone}}{CH_3CH_2C(O)CH_3} \qquad (14.60)$$

Using the same conditions for the oxidation of a primary alcohol leads to complications – the initial product is an aldehyde, but further oxidation takes place to give a carboxylic acid (equation 14.61).

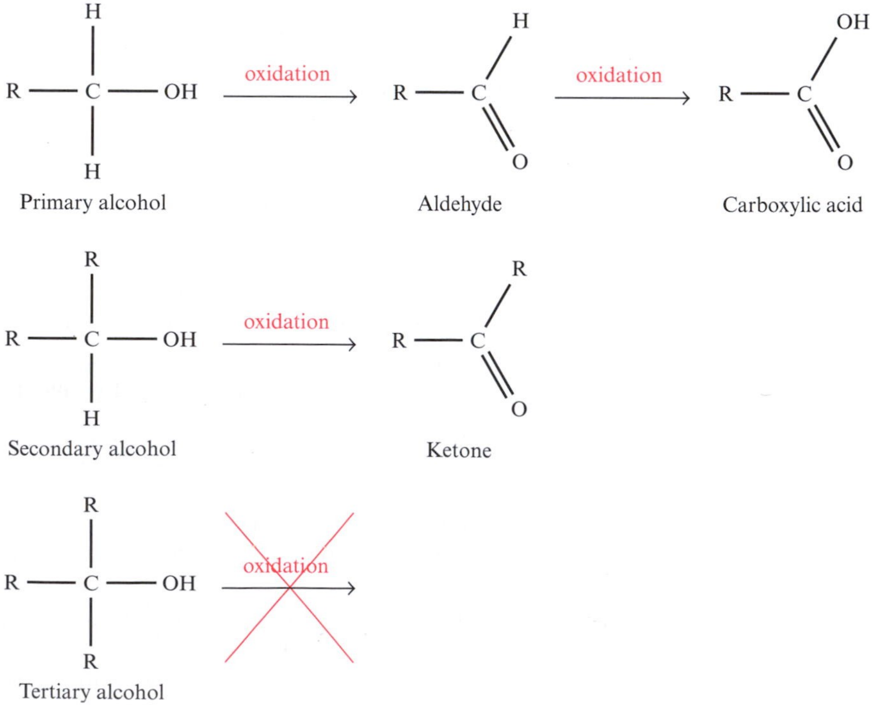

$$RCH_2OH \xrightarrow{\text{oxidation}} RC(H)=O \xrightarrow{\text{oxidation}} RC(OH)=O$$

Primary alcohol → Aldehyde → Carboxylic acid

$$R_2CHOH \xrightarrow{\text{oxidation}} R_2C=O$$

Secondary alcohol → Ketone

$$R_3COH \not\xrightarrow{\text{oxidation}}$$

Tertiary alcohol

14.16 The oxidation of primary alcohols leads to aldehydes and, ultimately, to carboxylic acids. Secondary alcohols are oxidized to ketones. Tertiary alcohols cannot be oxidized to give similar products, but remember that combustion is also oxidation, and all three classes of alcohols burn to give carbon dioxide and water.

$$\underset{\text{butan-1-ol}}{CH_3CH_2CH_2CH_2OH} \xrightarrow{K_2Cr_2O_7,\ H_2SO_4} \underset{\text{butanoic acid}}{CH_3CH_2CH_2CO_2H} \qquad (14.61)$$

Several reagents oxidize primary alcohols only as far as the corresponding aldehyde (equation 14.62) and these include Corey's reagent (pyridinium chlorochromate) **14.40**, and pyridinium dichromate **14.41**.

$$\underset{\text{butan-1-ol}}{CH_3CH_2CH_2CH_2OH} \xrightarrow{\text{Corey's reagent}} \underset{\text{butanal}}{CH_3CH_2CH_2CHO} \qquad (14.62)$$

(14.40) **(14.41)**

3. Esterification

The reaction between an alcohol and a carboxylic acid produces an *ester* (equation 14.63) and these compounds are characterized by their characteristic, often fruity, odours.

Esters: see Chapter 17

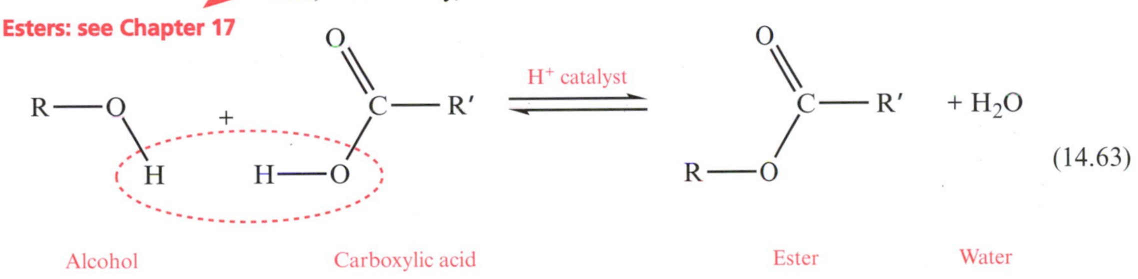

14.11 ALIPHATIC AMINES AND AMMONIUM SALTS 1: STRUCTURE, SYNTHESIS AND PHYSICAL PROPERTIES

Structure and nomenclature

The functional group of a *primary amine* is an $-NH_2$ group, and the general formula of an aliphatic primary amine is RNH_2. We can think in terms of this compound being derived from ammonia by the replacement of one hydrogen atom by an alkyl group. Further replacements lead to secondary (R_2NH) and tertiary (R_3N) amines.

For a straight chain primary amine, the name is derived by suffixing the root-name of the alkane with *amine* in addition to numerical descriptors that indicate the position of the NH_2 group. In a primary amine, *the NH_2 group is not necessarily at the end of the carbon chain*. For example **14.42** is 1-butylamine, **14.43** is 2-butylamine, and **14.44** is 4-methyl-2-pentylamine.

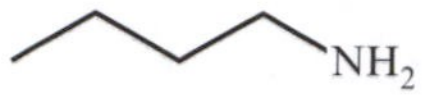

NH_2 or $CH_3CH_2CH_2CH_2NH_2$

(14.42)

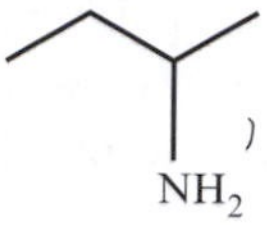

NH_2 or $CH_3CH_2CH(NH_2)CH_3$

(14.43)

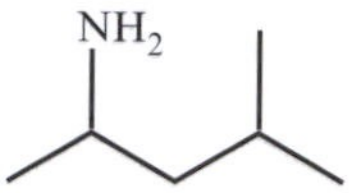

NH_2

or $CH_3CH(NH_2)CH_2CH(Me)CH_3$

(14.44)

Primary amines containing more than one NH_2 group are named using the prefix *amino* and by specifying the total number of, as well as the position-numbers for, the functional groups. Compound **14.45** is 1,2-diaminoethane and **14.46** is 1,2-diaminopropane. An older name for 1,2-diaminoethane is ethylenediamine.

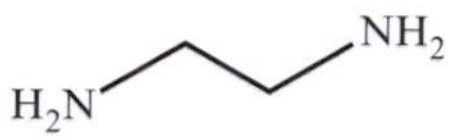

1, 2-Diaminoethane: see Section 16.3

H_2N NH_2 or $H_2NCH_2CH_2NH_2$

(14.45)

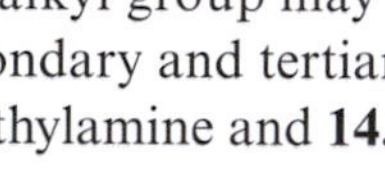

NH_2

NH_2 or $CH_3CH(NH_2)CH_2NH_2$

(14.46)

More than one alkyl group may be attached to the nitrogen atom and this gives rise to secondary and tertiary amines – **14.47** is diethylamine (Et_2NH), **14.48** is ethylmethylamine and **14.49** is triethylamine (Et_3N).

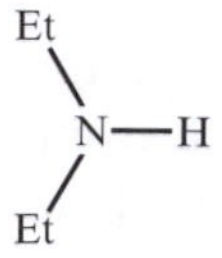

Et, N—H, Et (14.47) | Et, N—H, Me (14.48) | Et, N—Et, Et (14.49)

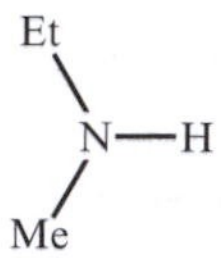

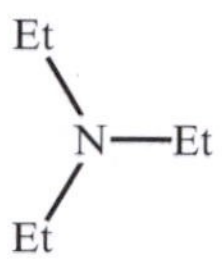

Just as we considered an aliphatic ether to be derived from an alkane with a CH_2 replaced by an isoelectronic O atom, so we can consider a secondary amine to be derived from an alkane by replacing a CH_2 group by an NH unit. A secondary amine can be named by prefixing the alkane root-name

Isoelectronic: see Section 4.12

with *aza* accompanied by numerical descriptors to show the position of the NH group or groups. An alternative name for **14.47** is 3-azapentane and for **14.48** is 2-azabutane. Compound **14.50** is 2,5-diazanonane.

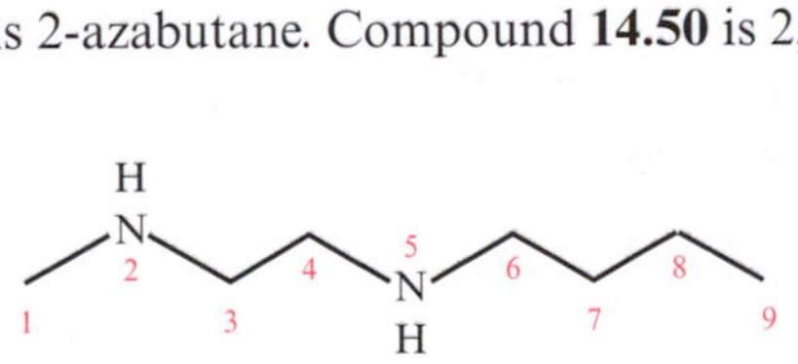

(14.50)

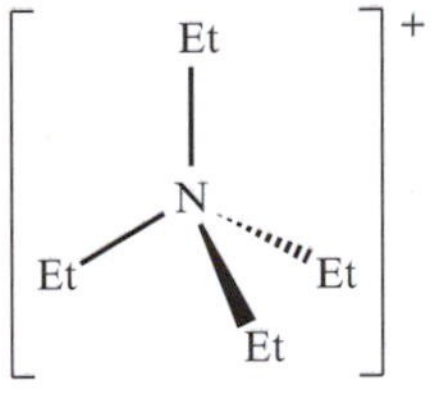

(14.51)

You will often encounter *tetraalkylammonium ions*, particularly in inorganic chemistry. The general formula of a symmetrical *quaternary ammonium ion* is $[R_4N]^+$ and it is named by using the name of the appropriate alkyl group: $[Et_4N]^+$ **14.51** is tetraethylammonium ion, and $[Bu_4N]^+$ is tetrabutylammonium ion. Other ammonium salts are similarly named: $[Et_3NH]^+$ is triethylammonium ion, $[Et_2NH_2]^+$ is diethylammonium ion and $[EtNH_3]^+$ is ethylammonium ion. These ions are useful as their salts are often soluble in 'organic' solvents such as CH_2Cl_2.

Synthesis

Lithium aluminium hydride as a reducing agent: see Box 11.5

Primary amines can be prepared by the reduction of nitriles (equation 14.64); for example, acetonitrile can be converted into ethylamine (equation 14.65). Other reducing agents such as H_2 in the presence of a catalyst can also be used.

$$R{-}C{\equiv}N \xrightarrow{Li[AlH_4]} RCH_2NH_2 \qquad (14.64)$$

$$\underset{\text{acetonitrile}}{CH_3C{\equiv}N} \xrightarrow{Li[AlH_4]} \underset{\text{ethylamine}}{CH_3CH_2NH_2} \qquad (14.65)$$

An alternative method which *specifically gives a primary amine* is the *Gabriel synthesis* in which an alkyl halide is treated with potassium phthalimide and the product then reacted with either water or hydrazine (equation 14.66).

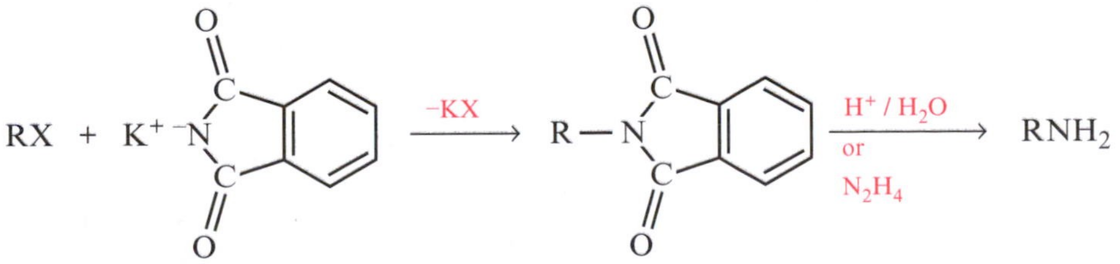

$$RX + K^+\,{}^-N(\text{phthalimide}) \xrightarrow{-KX} R{-}N(\text{phthalimide}) \xrightarrow[\text{or } N_2H_4]{H^+/H_2O} RNH_2 \qquad (14.66)$$

potassium phthalimide

Alkyl halides can be converted to amines by reaction with ammonia, but this method is not specific and yields primary, secondary and tertiary amines, in addition to tetraalkylammonium ions if the alkyl halide is in excess. Equations 14.67–14.70 show the sequence of reactions that can occur when chloroethane reacts with ammonia.

$$CH_3CH_2Cl + NH_3 \rightarrow \underset{\text{ethylamine}}{CH_3CH_2NH_2} + HCl \quad (14.67)$$

$$CH_3CH_2Cl + CH_3CH_2NH_2 \rightarrow \underset{\text{diethylamine}}{(CH_3CH_2)_2NH} + HCl \quad (14.68)$$

$$CH_3CH_2Cl + (CH_3CH_2)_2NH \rightarrow \underset{\text{triethylamine}}{(CH_3CH_2)_3N} + HCl \quad (14.69)$$

$$CH_3CH_2Cl + (CH_3CH_2)_3N \rightarrow \underset{\text{tetraethylammonium chloride}}{[(CH_3CH_2)_4N]^+Cl^-} \quad (14.70)$$

Aldehydes and ketones: see Chapter 17

Primary amines in which the $-NH_2$ group is attached to the terminal carbon atom of an aliphatic chain can also be prepared by the *reductive amination of aldehydes*. In this case the reaction is carried out using H_2 (or other reducing agent) and ammonia (equation 14.71).

$$\underset{\text{pentanal}}{CH_3CH_2CH_2CH_2C(=O)H} \xrightarrow{H_2 \text{ with Ni catalyst, } NH_3} \underset{\text{1-pentylamine}}{CH_3CH_2CH_2CH_2CH_2NH_2} \quad (14.71)$$

Note the difference between a *secondary amine*, RR′NH, and an amino-derivative of a *secondary alkyl group*, RR′CHNH$_2$

Primary amines in which the $-NH_2$ group is attached to a *secondary alkyl group* (i.e. of the general type shown in structure **14.52** in which R and R′ may be identical or different) are produced by the *reductive amination of ketones*. The $-NH_2$ group will always end up in the position originally occupied by the carbonyl group (reaction 14.72).

$$RR'C(H)NH_2$$

(14.52)

$$\underset{\text{hexan-3-one}}{CH_3CH_2CH_2C(=O)CH_2CH_3} \xrightarrow{H_2 \text{ with Ni catalyst, } NH_3} \underset{\text{3-hexylamine}}{CH_3CH_2CH_2CH(NH_2)CH_2CH_3} \quad (14.72)$$

A secondary amine was produced in reaction 14.68, but we noted that this methodology is *not* specific to a particular class of amine. A specific synthesis of a secondary amine is the reductive amination of an aldehyde or ketone using a *primary amine as the aminating agent*. For example, ethanal reacts with ethylamine under reducing conditions to give diethylamine (equation 14.73). [*Question*: What is the product when acetone undergoes reductive amination with ethylamine?]

$$\underset{\text{ethanal}}{CH_3-C(=O)H} + CH_3CH_2NH_2 \xrightarrow{H_2 \text{ with Ni catalyst}} CH_3-CH_2-NHCH_2CH_3 \equiv \underset{\text{diethylamine}}{Et_2NH} \qquad (14.73)$$

Physical properties

Like ammonia, amines possess characteristic smells, often described as 'fishy'. Amines are polar molecules, but the presence of N–H bonds in primary and secondary amines means that some of their physical properties contrast with those of tertiary amines. In the neat liquid and in the solid state, primary and secondary amines can form intermolecular hydrogen bonds, but this is not generally true for tertiary amines. Table 14.5 lists boiling points and enthalpies of vaporization for selected amines; notice the general increase with increasing molecular mass for the series of straight chain primary amines, and the significant difference between the values for 1-hexylamine (a primary amine) and triethylamine (a tertiary amine), both of which have the same molecular mass.

Table 14.5 Boiling points and enthalpies of vaporization of selected amines. Values for ammonia are included for comparison.

Amine	*Relative molecular mass*	*Boiling point* / K	$\Delta_{vap}H$ / kJ mol^{-1}
Ammonia	17	240	23
Methylamine	31	267	26
Dimethylamine	45	280	26
1-Propylamine	59	320	29.5
2-Propylamine	59	305	28
1-Butylamine	73	350	32
2-Butylamine (*sec*-butylamine)	73	336	30
Diethylamine	73	328.5	29
Isobutylamine	73	341	31
tert-Butylamine	73	317	28
1-Hexylamine	101	406	36.5
Triethylamine	101	362	31

Amines form hydrogen bonds with water. Primary, secondary and tertiary amines may interact as shown in structure **14.53**, and the nitrogen-bound hydrogen atoms in primary and secondary amines may form hydrogen bonds with oxygen as in structure **14.54**. Low molecular weight amines are water-soluble and miscible with polar solvents such as ethers and alcohols.

X = H or R
(14.53)

X = H or R
(14.54)

Identification of amines by infrared spectroscopy

Interpretation of IR and NMR spectra for organic compounds: see Section 9.8

The IR spectrum of a primary or secondary amine exhibits a broad absorbance in the region 3000–3500 cm^{-1} assigned to the N–H stretching mode. Since a tertiary amine does not possess an N–H bond, its IR spectrum does not exhibit this characteristic absorption. Figure 14.17 shows the IR spectra of 1-butylamine (a primary amine) and triethylamine (a tertiary amine).

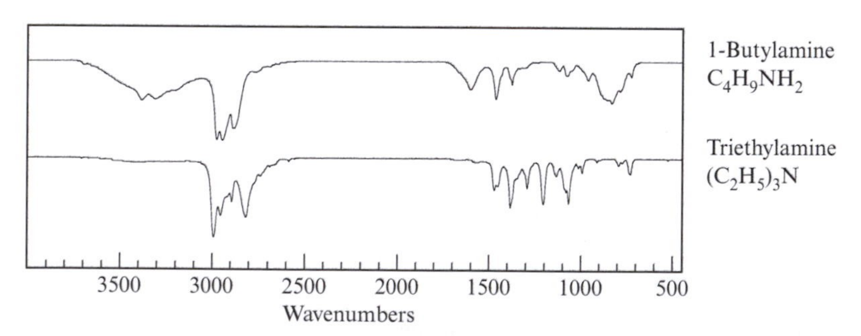

14.17 The IR spectra of 1-butylamine and triethylamine. The spectra show absorbances due to the aliphatic C–H stretches, and characteristic bands in the fingerprint region. The broad band centred around 3330 cm^{-1} in the spectrum of 1-butylamine is assigned to the N–H stretch; similar bands are characteristic of primary and secondary amines but are absent from the spectra of tertiary amines such as Et_3N.

14.12 ALIPHATIC AMINES AND AMMONIUM SALTS 2: REACTIVITY

Amines as Brønsted bases

Ammonia: see Section 11.7; K_b and pK_b: see Section 11.9

In Chapter 11 we discussed the basicity of ammonia. Similarly, amines also behave as Brønsted bases and values of pK_b for selected amines are listed in Table 14.6. For an amine RNH_2, R_2NH or R_3N, the value of pK_b refers to the processes shown in equations 14.74–14.76 respectively, where the products are alkylammonium, dialkylammonium and trialkylammonium ions.

$$\underset{\text{primary amine}}{RNH_2(aq)} + H_2O(l) \rightleftharpoons \underset{\text{alkylammonium ion}}{[RNH_3]^+(aq)} + [OH]^-(aq) \quad (14.74)$$

$$\underset{\text{secondary amine}}{R_2NH(aq)} + H_2O(l) \rightleftharpoons \underset{\text{dialkylammonium ion}}{[R_2NH_2]^+(aq)} + [OH]^-(aq) \quad (14.75)$$

$$\underset{\text{tertiary amine}}{R_3N(aq)} + H_2O(l) \rightleftharpoons \underset{\text{trialkylammonium ion}}{[R_3NH]^+(aq)} + [OH]^-(aq) \quad (14.76)$$

For amines with two functional groups such as 1,2-diaminoethane, the two values of pK_b refer to the proton transfers shown in equations 14.77 and 14.78.

$$H_2NCH_2CH_2NH_2(aq) + H_2O(l) \rightleftharpoons [H_2NCH_2CH_2\overset{+}{N}H_3](aq) + [OH]^-(aq) \qquad pK_b = 3.29 \quad (14.77)$$

$$[H_2NCH_2CH_2\overset{+}{N}H_3](aq) + H_2O(l) \rightleftharpoons [H_3\overset{+}{N}CH_2CH_2\overset{+}{N}H_3](aq) + [OH]^-(aq) \qquad pK_b = 6.44 \quad (14.78)$$

The values of pK_b in Table 4.6 illustrate that aliphatic amines are stronger bases than water or alcohols, but weaker bases than hydroxide or alkoxide

Table 14.6 Values of pK_b for selected amines. Ammonia is included for comparison. For amines with two NH_2 groups, the values of pK_b refer to the gain of the first and second protons respectively.

Amine	*Formula*	pK_b
Ammonia	NH_3	4.75
Methylamine	CH_3NH_2	3.34
Ethylamine	$CH_3CH_2NH_2$	3.19
1-Propylamine	$CH_3CH_2CH_2NH_2$	3.29
1-Butylamine	$CH_3CH_2CH_2CH_2NH_2$	3.23
tert-Butylamine	$CH_3C(NH_2)(CH_3)CH_3$	3.17
Diethylamine	$(CH_3CH_2)_2NH$	3.51
Triethylamine	$(CH_3CH_2)_3N$	2.99
1,2-Diaminoethane	$H_2NCH_2CH_2NH_2$	3.29, 6.44
1,3-Diaminopropane	$H_2NCH_2CH_2CH_2NH_2$	3.06, 4.97
1,2-Diaminopropane	$CH_3CH(NH_2)CH_2NH_2$	4.18, 7.39

ions. The reactions of amines with Brønsted acids result in the formation of salts (equations 14.79–14.81).

$$\underset{\text{triethylamine}}{Et_3N} + HCl \rightarrow \underset{\text{triethylammonium chloride}}{[Et_3NH]Cl(aq)} \quad (14.79)$$

$$\underset{\text{ethylamine}}{EtNH_2} + HNO_3 \rightarrow \underset{\text{ethylammonium nitrate}}{[EtNH_3][NO_3]} \quad (14.80)$$

$$\underset{\textit{tert}\text{-butylamine}}{2Me_3CNH_2} + H_2SO_4 \rightarrow \underset{\textit{tert}\text{-butylammonium sulfate}}{[Me_3CNH_3]_2[SO_4]} \quad (14.81)$$

tert-Butyl: see Figure 8.7

Alkene elimination from a quaternary ammonium ion

A quaternary ammonium ion is susceptible to attack by strong base (e.g. $[OH]^-$ or $[NH_2]^-$) and undergoes an elimination reaction to yield a tertiary amine and an alkene (equation 14.82).

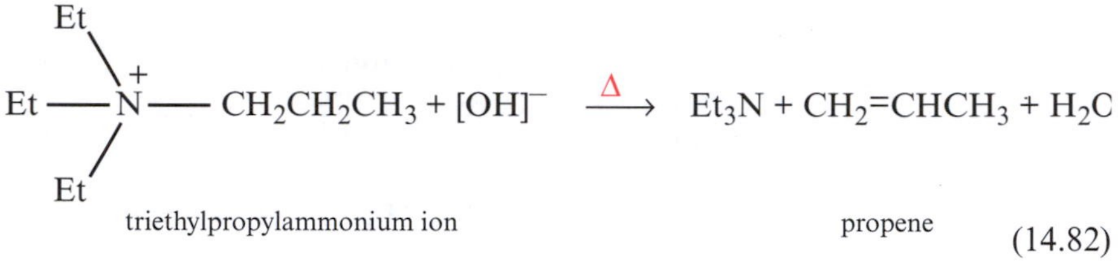

$$\underset{\text{triethylpropylammonium ion}}{Et_3\overset{+}{N}{-}CH_2CH_2CH_3} + [OH]^- \xrightarrow{\Delta} Et_3N + \underset{\text{propene}}{CH_2{=}CHCH_3} + H_2C \quad (14.82)$$

The mechanism is usually E2, and *syn*-elimination is favoured.

14.13 AN OVERVIEW: COMBINING REACTIONS INTO A SYNTHETIC STRATEGY

Although we have presented the reactivity of the halo-, ether, alcohol and amine functional groups in terms of individual reactions, it is often the case in the laboratory that you will wish to plan a synthetic route that involves a series of reactions – *a multi-step synthesis*. Figure 14.18 puts together some of the reactions from this chapter, and illustrates how to convert, for example, a primary amine into an ether, and an alcohol into an alkane; see also problem 14.24.

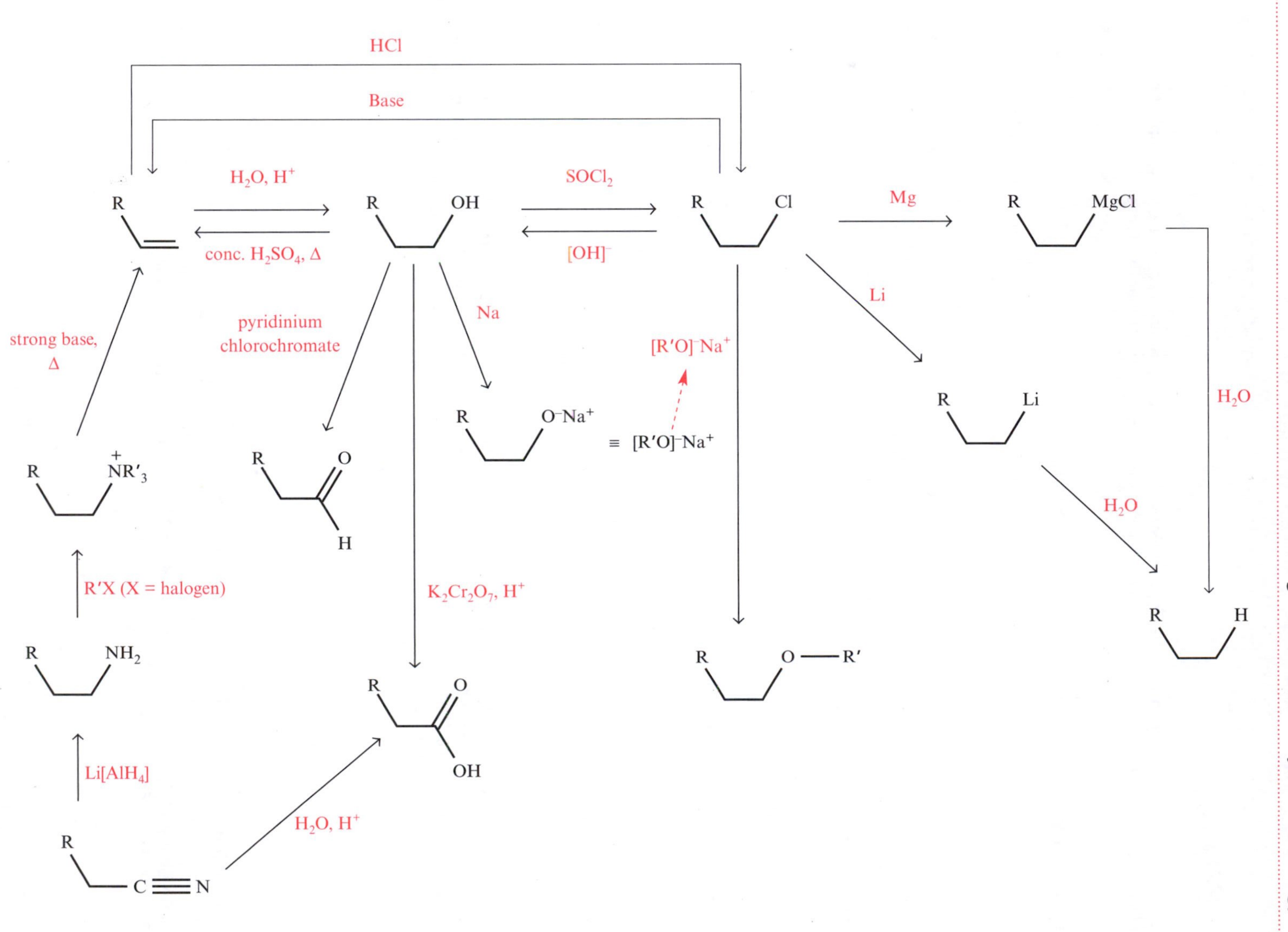

14.18 A summary of some of the synthetic interconversions discussed in this chapter.

SUMMARY

This chapter has been concerned with the chemistry of acyclic halogenoalkanes, and acyclic, aliphatic ethers, alcohols and amines, and the discussions have also introduced some new mechanistic ideas.

Do you know what the following terms mean?

- inductive effect
- nucleophilic substitution
- β-elimination (leading to alkene formation)
- *syn*-elimination
- *anti*-elimination
- leaving group
- dehydrohalogenation
- alkoxide ion
- alkylation
- alcoholic KOH
- organolithium compound
- Grignard reagent
- retention of configuration
- inversion of configuration
- stereoisomer
- racemic mixture
- methyl migration
- hydrophobic
- hydrophilic

You should also be able:

- systematically to name straight and branched chain halogenoalkanes containing one or more halo-substituents
- systematically to name symmetrical and asymmetrical aliphatic ethers, and recognize the advantages of using different systems of nomenclature in different situations
- systematically to name straight and branched chain aliphatic alcohols containing one or more –OH functionalities
- systematically to name acyclic, aliphatic primary, secondary and tertiary amines, and recognize the advantages of using different systems of nomenclature in different situations
- to outline methods of preparation of halogenoalkanes, ethers, alcohols and amines
- to describe typical physical properties of acyclic halogenoalkanes, ethers, alcohols and amines, and indicate how these differ from those of the parent alkanes
- to recognize characteristic IR spectroscopic absorptions of functional groups in alcohols, and straight chain ethers and amines
- to give important uses of some ethers and alcohols, and appreciate any hazards of these compounds
- to discuss the contributions that intermolecular interactions and molecular mass make to trends in boiling points of halogenoalkanes, ethers, alcohols and amines.
- to give examples of nucleophilic substitution reactions
- to discuss the S_N1 and S_N2 mechanisms, and to appreciate the factors that might favour one mechanism over the other, and the competitive elimination reactions that may occur
- to give examples of β-elimination reactions
- to discuss the E1 and E2 mechanisms, and appreciate the factors that might favour one mechanism over the other
- to comment on the kinetic data that would support the proposal of either an S_N1, S_N2, E1 or E2 mechanism
- to discuss the stereochemical consequences of *syn*- versus *anti*-elimination pathways
- to outline the different ways in which primary, secondary and tertiary alcohols are oxidized
- to be able to plan simple, multi-step syntheses which combine the types of reaction outlined in this chapter and Chapter 8.

PROBLEMS

14.1 In which directions are the following single bonds polarized, if at all: (a) C–F; (b) N–O; (c) C–O; (d) O–H; (e) N–H; (f) C–S? (Electronegativity values are given in Table 4.2.)

14.2 The molecular dipole moments, measured in the gas phase, of CH_2Br_2, $CClF_3$ and HCN are 1.43, 0.5 and 2.98 respectively. Draw the structure of each molecule and indicate the direction in which the resultant dipole moment acts. Draw the structure methanol, and show the *approximate* direction in which you expect the dipole moment (1.70 D) to act.

14.3 Draw the structures of *cis*- and *trans*-1,2-dichloroethene. How can values of the molecular dipole moments help to distinguish between these isomers?

14.4 Figure 14.19 shows the variation in boiling points along the series of compounds C_2H_5X (X = H, F, Cl, Br and I). What factors contribute to this trend?

14.5 Give systematic names for the following compounds:

(a) $CH_3CHICHICH_3$;

(b) $CCl_3CH_2CH_3$;

(c) $CH_3CHBrCH_2CH_2CMe_2CH_3$;

(d) $CH_2FCH_2CHFCHFCH_3$;

(e) $CH_3CH(CH_2Br)CHBrCH_3$.

14.6 How might you prepare 2-bromopropane in as specific a manner as possible? [*Hint*: refer to Section 8.15.]

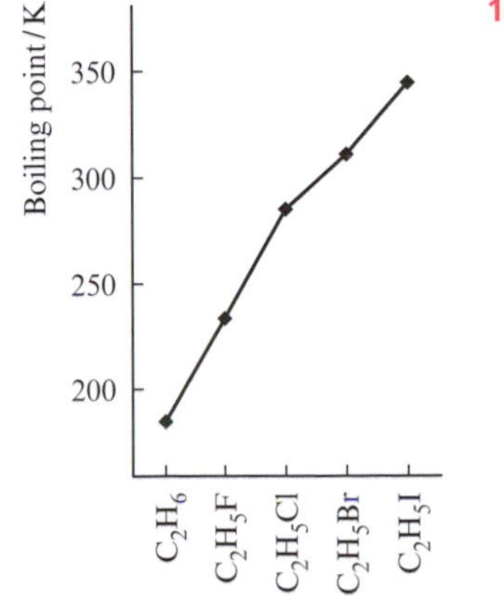

14.19 For problem 14.4.

14.7 Outline the mechanisms by which 2-chloro-2-methylbutane might be expected to react with water and hydroxide ion, paying particular attention to the stereochemistry of the reactions and any competing reaction pathways. What would be the expected rate equations for the reactions? Would your answer be equally valid for the reactions with 1-chloro-2-methylbutane?

14.8 What do you understand by the terms (a) nucleophile, (b) β-elimination, (c) methyl migration, (d) inversion of configuration, (e) racemic mixture, (f) partial racemization, (g) stereoisomer and (h) diastereomer? [Part (h) requires that you have worked through Box 14.3.]

14.9 By considering the mechanisms of nucleophilic substitutions, suggest why the reaction of aqueous alcoholic $AgNO_3$ with 2-bromo-2-methylpropane proceeds more rapidly than that with 1-bromobutane. Suggest why the rates of the reactions between aqueous alcoholic $AgNO_3$ and $CH_3C(Me)XCH_3$ (X = Cl, Br, I) follow the sequence I > Br > Cl.

14.10 A primary carbenium ion of the particular type shown in structure **14.20** is unable to undergo proton elimination to form an alkene. What specific property of this carbenium ion makes the elimination impossible?

14.11 The elimination of HCl from one stereoisomer from 3-chloro-3,4-dimethylhexane by an E2 mechanism was discussed in Section 14.6. From this stereoisomer, which we label here as **A**, we saw that *syn*-elimination gives exclusively *E*-3,4-dimethylhex-3-ene, and *anti*-elimination gives the (*Z*)-isomer. (a) Draw the enantiomer of **A**, and label this **B**. (b) What are the products of the *syn*- and *anti*-elimination of HCl from **B**? (c) Draw a third stereoisomer (a diastereomer, see Box 14.3) of **A** and label this **C**. (d) What are the products of the *syn*- and *anti*-elimination of HCl from **C**? (e) Draw the enantiomer of **C**, and label this **D**. (f) Are the products of *syn*- and *anti*-elimination of HCl from **D** the same or different from those formed from **C**? (g) What conclusions can you draw about the specificity of the elimination of HCl from 3-chloro-3,4-dimethylhexane by the E2 mechanism?

14.12 Give systematic names for the following ethers:

(a) $CH_3OCH_2CH_2CH_3$;
(b) $CH_3CH_2CH(OMe)CH_2CH_3$;
(c) $CH_3CH_2OCH(Me)CH_3$.

14.13 Why would it not be efficient to attempt to prepare ethyl methyl ether by a method analogous to that shown in equation 14.32? What reaction(s) could you employ to prepare this ether?

14.14 Draw structural formulae for the following alcohols:

(a) pentan-2-ol;
(b) 2-chlorobutan-1-ol;
(c) 2-chlorobutan-2-ol;
(d) propane-1,3-diol;
(e) butane-1,1,3-triol.

14.15 Suggest reasons for the trends in boiling points among the alcohols listed in Table 14.4. Why is the boiling point of water higher than that of methanol even though the relative molecular mass is lower?

14.16 A saturated compound **X** contains carbon, hydrogen and 30.8% oxygen by weight, and has a relative molecular mass of 104. The IR spectrum of **X** is shown in Figure 14.20. What is the molecular formula of **X**? What functional groups does **X** contain? Suggest a possible structure for the compound.

14.17 Write a balanced equation for the reaction of ethanol with sodium. What happens when sodium ethoxide is treated with water? (pK_a $C_2H_5OH \approx 16$.) How would you expect aluminium to react with propan-2-ol?

14.18 What would be the major products of the acid-catalysed dehydrations of (a) butan-2-ol, (b) 2-methylbutan-1-ol, (c) pentan-1-ol and (d) 3,3-dimethylbutan-2-ol? [*Hint*: Remember the possibilities of rearrangement products.]

14.19 Suggest a mechanism by which 3-methylbutan-2-ol reacts with chloride ion in the presence of acid. What is likely to be the major product? [*Hint*: What is the most stable carbenium ion that can be formed?]

14.20 (a) What is obtained when propan-2-ol reacts with acidified potassium dichromate?
(b) If the same reaction were carried out using propan-1-ol, what products would be formed?

14.21 Draw structural formulae for:

(a) 1-propylamine,
(b) 2-propylamine,
(c) tributylamine,
(d) dipropylamine,
(e) ethylmethylamine,
(f) 2-azahexane and
(g) 2,5,8-triazanonane.

14.22 The values of pK_b for NH_3, $EtNH_2$, Et_2NH and Et_3N are 4.75, 3.19, 3.51 and 2.99 respectively. (a) What do these values tell you about the relative base strengths of NH_3 and the organic amines? (b) Suggest reasons for the trend in values. (c) Calculate the pH of a 0.1 M aqueous solution of triethylamine.

14.23 What would be the product of the reaction of an excess of HBr with 1,3-diaminopropane?

14.24 Suggest a multi-step route by which ethene might be converted to 1-propylamine. [*Hint*: Cyanide ion is a nucleophile.]

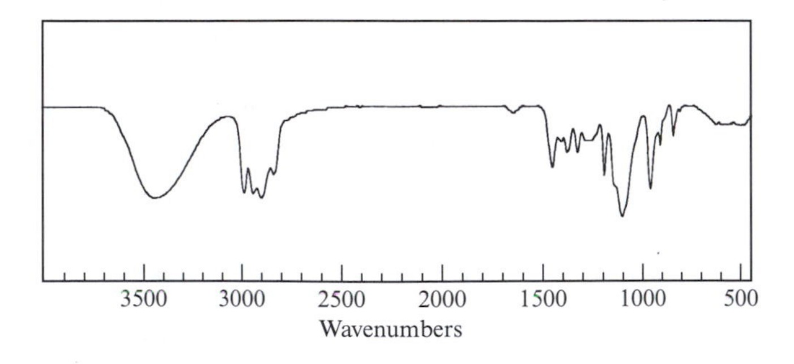

14.20 For problem 14.16.

15 Rings

Topics

15.1 INTRODUCTION

See Sections 7.8, 11.5, 13.4, 13.9 and 13.10

In this chapter we discuss the chemistry of molecules with cyclic structures. We have already seen several compounds which undergo dimerization, trimerization or tetramerization reactions to give cyclic molecules, e.g. EH_3 to E_2H_6 (E = B or Ga) **15.1**, $AlCl_3$ to Al_2Cl_6 **15.2**, $SbCl_5$ to Sb_2Cl_{10} **15.3**, and SO_3 to S_3O_9 **15.4**, and we have described some allotropes with cyclic molecular structures, for example S_6, S_8 and Se_8.

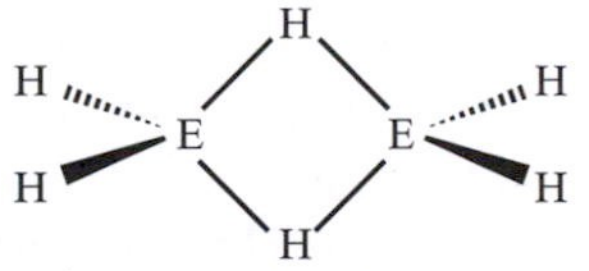

E = B or Ga

(15.1)

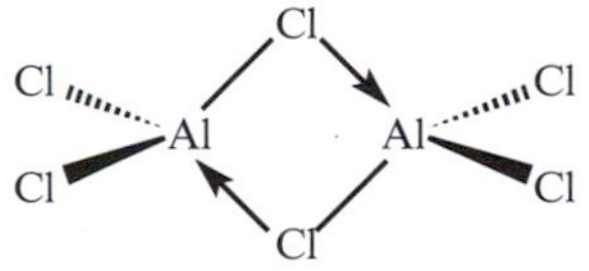

(15.2)

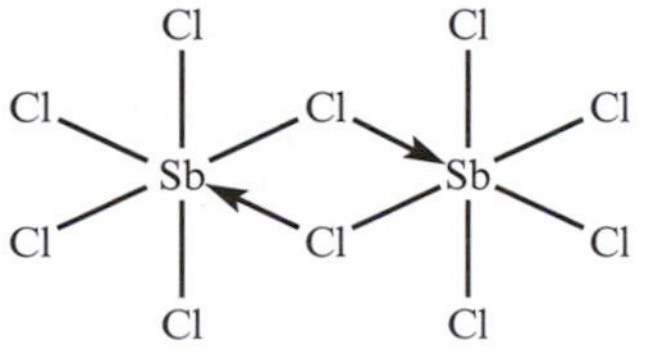

(15.3)

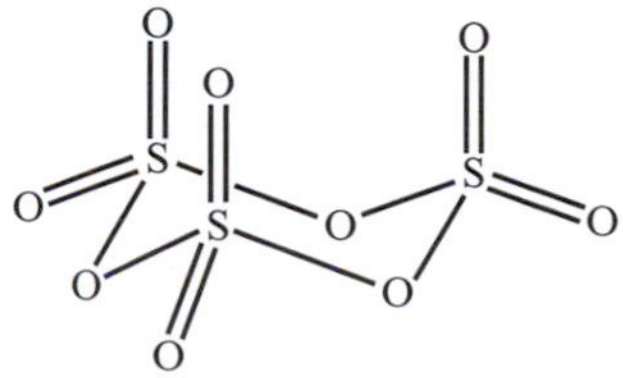

(15.4)

We have also illustrated that the dehydration of boronic acids $RB(OH)_2$ leads to the formation of cyclic trimers $(RBO)_3$ (equation 13.4).

We begin this chapter by considering some saturated rings, and then move on to unsaturated systems and compounds in which the bonding is delocalized.

15.2 SATURATED RINGS 1: CYCLOALKANES

The discussion of saturated hydrocarbons in Chapter 8 was restricted to compounds in which the carbon atoms formed chains – aliphatic, acyclic alkanes. In this section we consider related compounds in which some or all of the carbon atoms are joined into rings – aliphatic, cycloalkanes.

Structure and nomenclature

An *unsubstituted* cycloalkane with a single ring (monocyclic) has the generic formula C_nH_{2n}, and the compound name indicates the number of carbon atoms in the cyclic portion of the molecule. Compound **15.5** is cyclohexane and **15.6** is cyclopropane.

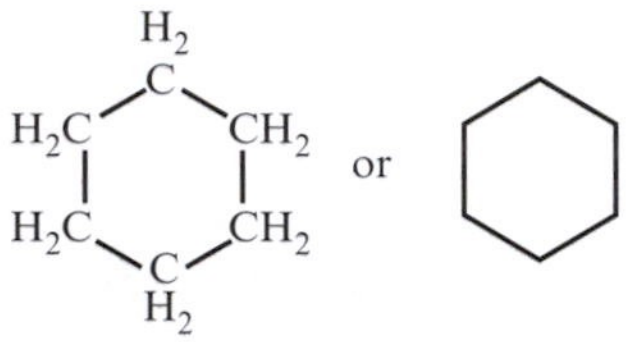

(15.5)

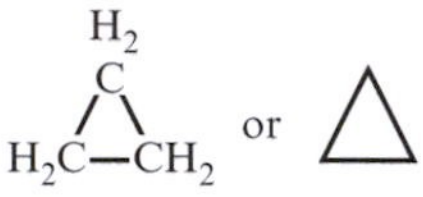

(15.6)

A monocyclic compound contains a single ring.

In an *unsubstituted* cycloalkane (C_nH_{2n}), all the carbon atoms are equivalent. This contrasts with their acyclic counterparts, and ^{1}H or ^{13}C NMR spectroscopies can be used to distinguish between them; Figure 15.1 shows the ^{13}C NMR spectra of cyclopentane and pentane.

A *substituted* cycloalkane is named so that the position numbers for the substituents are as *low as possible*. Compound **15.7** is methylcyclohexane (no position number is needed – why?) and **15.8** is 1,3-dimethylcyclohexane; the numbering begins with C(1) as the position of first substitution.

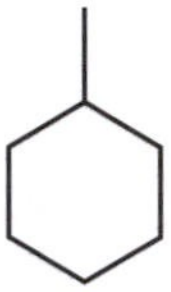

(15.7)

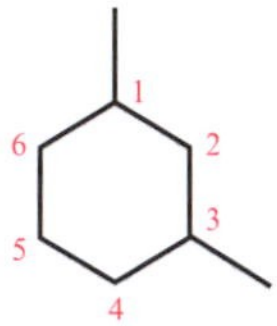

(15.8)

Ring conformation

Whereas there is usually free rotation about the C–C bonds in an acyclic alkane, such rotation is restricted in the smaller cycloalkanes. The C_3 ring in

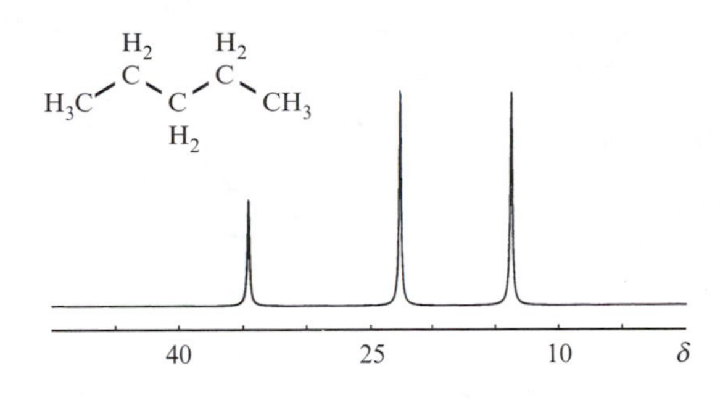

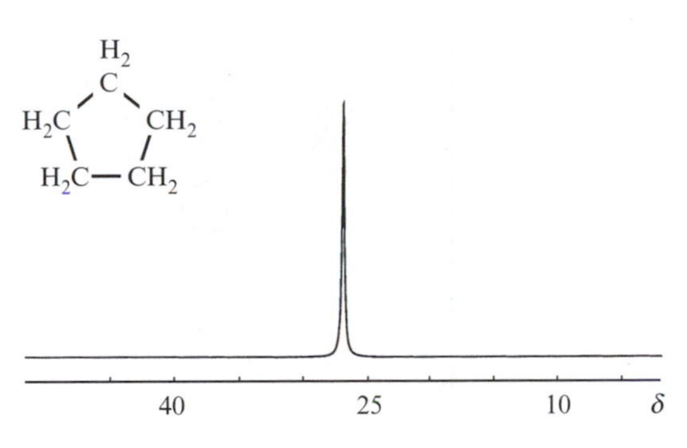

15.1 The ^{13}C NMR spectrum of pentane shows three signals but that of cyclopentane shows only one.

cyclopropane (Figure 15.2) is constrained to being planar, and this imposes eclipsed conformations along each C–C bond. Cyclopropane is termed a *strained ring* and its chemistry reflects this (see below).

Staggered and eclipsed conformers: see Section 8.5

As the C_n ring size increases, *partial* rotation about the C–C bonds becomes possible. If the C_n rings of cyclobutane and cyclopentane adopt planar conformations, adjacent CH_2 groups are eclipsed. However, if partial rotation occurs about one or more of the C–C bonds, the ring becomes non-planar, non-bonded H······H interactions are reduced, and the energy of the system is lowered. In Figure 15.3 we compare planar and folded structures for cyclobutane and cyclopentane; the folded conformations are close to those observed experimentally. Despite possessing some degree of flexibility, the ring in cyclobutane is strained. Larger rings have greater degrees of flexibility – the partial rotation about the C–C bonds in cyclohexane results in 'flipping' between the chair and boat conformations with the twist-boat as an intermediate structure (Figure 15.4a). Cyclohexane is a liquid at 298 K,

Boat and chair conformers: see Section 7.8. Conformations and steric energies: see Section 8.5

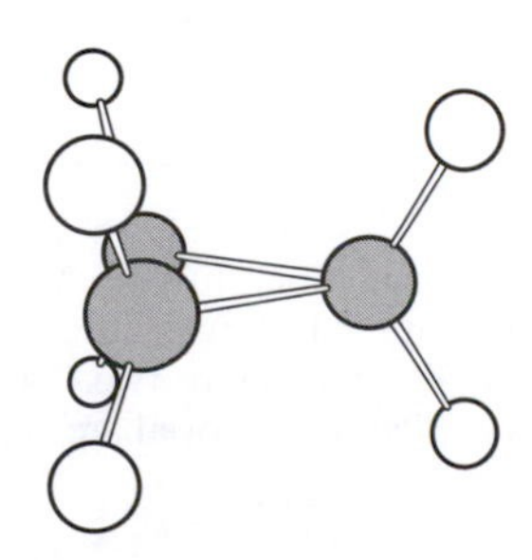

15.2 The structure of cyclopropane. The ring is planar and the conformation along each C–C single bond is eclipsed.

15.3 In planar cyclobutane (a) and cyclopentane (b) the conformation along each C–C single bond is eclipsed. An energetically more favourable conformation is achieved if the ring is folded; this is the result of partial rotation about the C–C single bonds.

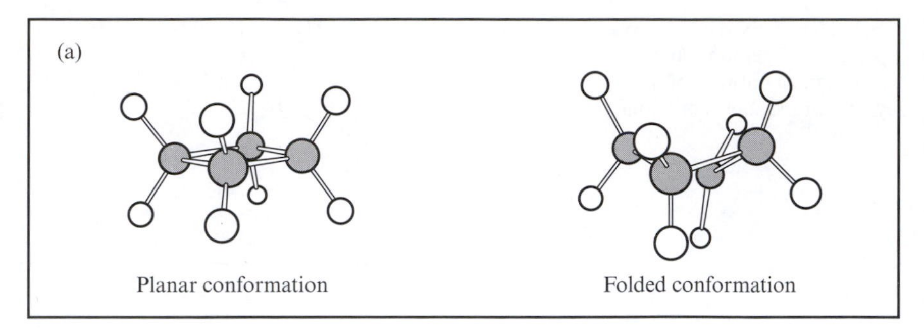

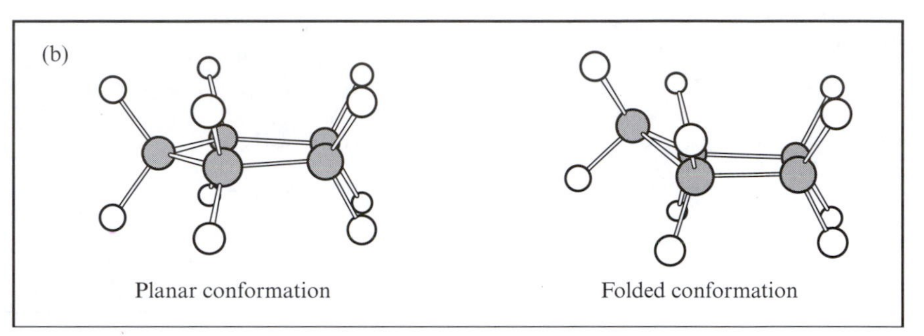

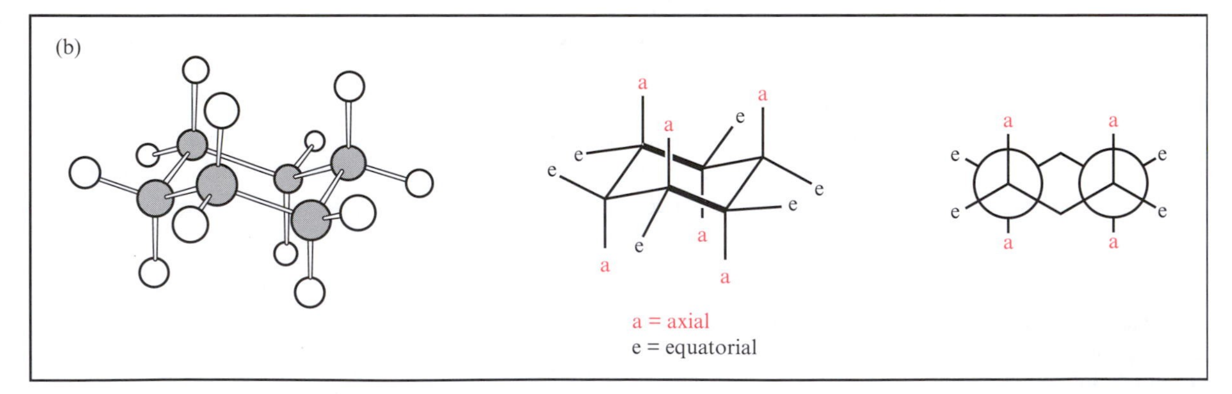

15.4 (a) The chair and boat forms of cyclohexane are the two extreme conformations; the chair is energetically favoured. The twist-boat is an intermediate conformation. (b) The chair conformation of C_6H_{12} is shown as a ball-and-stick structure and in a schematic form that shows that the hydrogen atoms are either in axial or equatorial positions. The Newman projection on the right-hand side is viewed along the two C–C bonds which are highlighted as bold lines in the middle diagram. The origins of the terms axial and equatorial are apparent.

15.5 (a) There are two conformations of the chair form of methylcyclohexane depending upon the site of substitution. The space-filling diagrams illustrate that the methyl group experiences greater steric crowding when it occupies the axial site.
(b) There are two *isomers* of 1,3-dimethylcyclohexane in which the methyl groups are either *cis* (same side of the ring) or *trans* (opposite sides of the ring); there are two *conformers* of the *cis*-isomer. The *equatorial,equatorial* conformer is sterically favoured. Note that *cis* and *trans* is recommended in preference to the *Z* and *E* nomenclature that is used for alkenes.

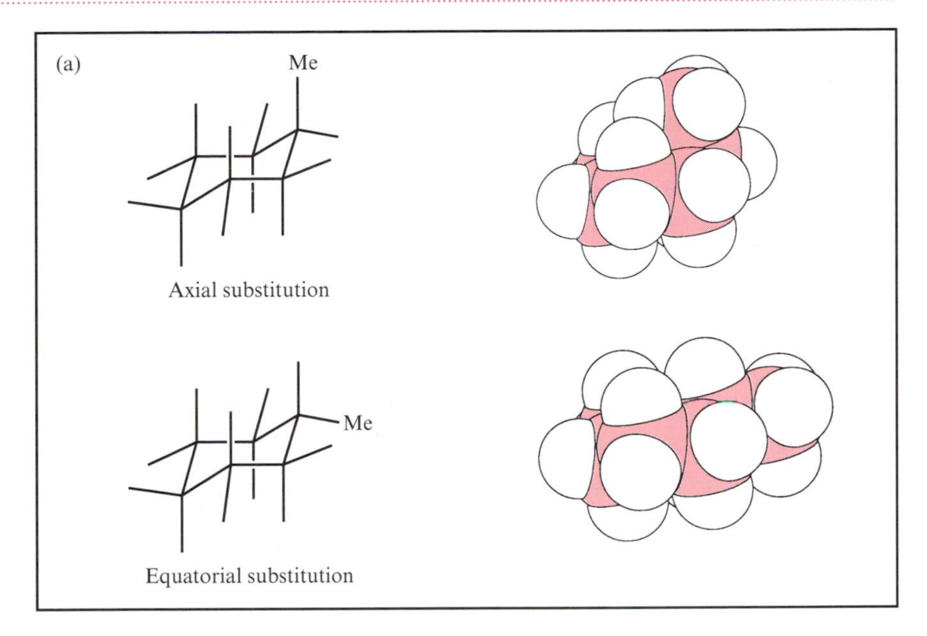

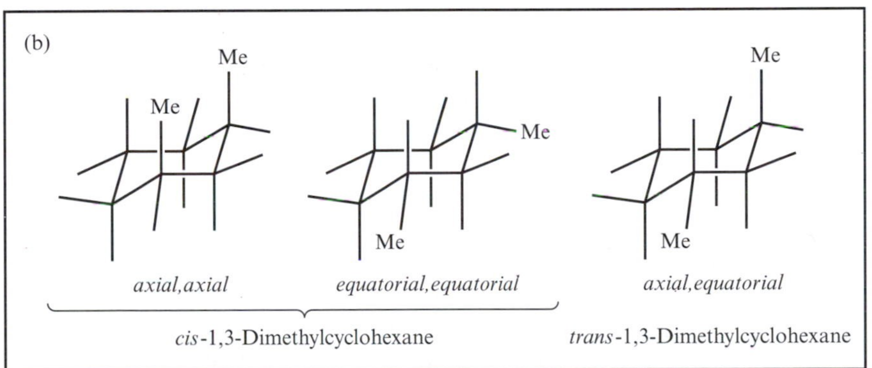

and although the boat and chair forms are in equilibrium, the chair form is favoured to an extent of $\approx 10^4 : 1$; H······H interactions are greater in the boat form than in the chair conformer.

In the chair form of cyclohexane (Figure 15.4b), there are *two different sites* for the hydrogen atoms, termed axial and equatorial. Their origins can be appreciated by considering a Newman projection of the molecule, viewed along two of the six C–C bonds. The presence of different sites leads to the possibility of structural isomerism and different conformations. There are two *conformers* of methylcyclohexane depending whether the CH_3 substituent occupies an axial or an equatorial position (Figure 15.5a). The structures are *conformers* because the interconversion between the chair and boat forms of the ring leads to an interconversion between the axial and equatorial sites (Figure 15.6). Figure 15.5b shows the three conformers of the two isomers of 1,3-dimethylcyclohexane. The space-filling diagrams of the conformers of methylcyclohexane illustrate that the equatorial site is less

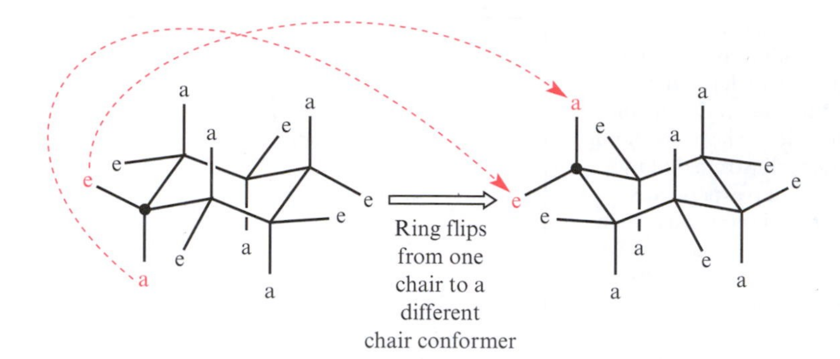

15.6 When the cyclohexane ring 'flips' between two different chair conformations, axial and equatorial sites are interconverted.

sterically crowded than the axial site; the equatorially substituted conformer is energetically favoured, although the energy difference is small, about 7 kJ mol^{-1}. Similarly, the *equatorial,equatorial* conformer of 1,3-dimethylcyclohexane is favoured.

The chair conformation of cyclohexane is preferred over the twist-boat or boat:

E(chair) << E(twist-boat) < E(boat)

Equatorial sites in the chair conformer of cyclohexane are less sterically crowded than axial sites.

Synthesis

Cycloalkanes are present in crude petroleum, but the relative amounts depend on the source of the oil deposit, and it is necessary to develop synthetic methods which are specific to particular ring sizes. Two general strategies are:

- to use addition reactions of alkenes;
- to cyclize an acyclic compound.

Cyclopropane can be prepared by the addition of a methylene (CH_2) unit to ethene. Although we have seen methylene groups as components of alkane chains, we have not considered them as independent molecules, but when, for example, diazomethane is photolysed, N_2 is released and carbene $:CH_2$ is formed (equation 15.1). Carbene is electron deficient and adds across the C=C double bond in ethene to give cyclopropane as (equation 15.2).§

$$CH_2N_2 \xrightarrow{h\nu} :CH_2 + N_2 \quad (15.1)$$

$$CH_2{=}CH_2 + :CH_2 \rightarrow \bigtriangledown \quad (15.2)$$

§ The reactions of carbenes are complex; for further details, see J. March, *Advanced Organic Chemistry: Reactions, Mechanisms and Structure,* 4th edn, Wiley, New York (1992).

Substituted cyclopropanes can be prepared using other alkenes (equation 15.3).

$$CH_2CH{=}CH_2 + :CH_2 \rightarrow \text{Me-cyclopropane} \tag{15.3}$$

propene — methylcyclopropane

We considered the radical polymerization of ethene in Section 8.16, but the *cyclo*addition of alkenes may also occur. The photochemical reaction of two molecules of ethene gives cyclobutane. This reaction is a dimerization (equation 15.4); similar reactions occur between a range of alkenes and in some cases it is possible to react one alkene specifically with a different one. The process is termed *cycloaddition*.

$$\left.\begin{array}{l} C_2H_4 + C_2H_4 \xrightarrow{h\nu} C_4H_8 \\ \| + \| \xrightarrow{h\nu} \square \end{array}\right\} \tag{15.4}$$

The cycloaddition of unsaturated molecules such as alkenes or alkynes is their reaction to form cyclic molecules.

The second synthetic strategy for cycloalkanes involves the cyclization of *functionalized* alkanes. For example, 1,3-dibromopropane undergoes dehalogenation and ring-closure when treated with zinc and sodium iodide (equation 15.5); this is an example of a *Wurtz reaction* in which a halogenoalkane is used in a carbon–carbon bond-forming process. Often, the reagents used for a Wurtz reaction are a halogenoalkane and sodium in an ether solvent.

$$BrCH_2CH_2CH_2Br \xrightarrow{Zn,\ NaI} \triangledown \tag{15.5}$$

In a Wurtz reaction, a halogenalkane is used in a C–C bond-forming reaction. Often it is carried out by the reaction of the halogenoalkane with sodium in an ether solvent.

Reactions

Cycloalkanes undergo similar radical substitution reactions to acyclic alkanes, and the chlorination of cyclopentane is shown in equation 15.6. The mechanism is similar to that outlined in Section 8.11 – initiation, propagation and termination steps make up the radical chain reaction, and the formation of chlorocyclopentane in reaction 15.6 is accompanied by polychlorinated products.

$$\text{cyclopentane} + Cl_2 \xrightarrow{h\nu} \text{chlorocyclopentane} + HCl \qquad (15.6)$$

chlorocyclopentane

Competitive pathways in the reaction of propane and Cl_2: see Section 8.12

In an *unsubstituted* cycloalkane, all of the carbon centres are identical and the abstraction of an $H^{\bullet}$ radical from the cycloalkane in a propagation step can only form an alkyl radical with a secondary carbon atom. Thus, there are no competitive reactions as when, for example, Cl_2 reacts with propane.

Cyclopropane in particular undergoes *ring-opening reactions*. These are *additions* which result in the relief of ring-strain, and the reactivity of cyclopropane is often similar to that of an alkene. For example, cyclopropane reacts with H_2 to give propane (equation 15.7), with water in the presence of acid to yield propan-1-ol (equation 15.8), and with hydrogen halides to give 1-halopropanes (equation 15.9).

$$\text{cyclopropane} + H_2 \xrightarrow{\Delta,\ \text{nickel catalyst}} CH_3CH_2CH_3 \qquad (15.7)$$

$$\text{cyclopropane} + H_2O \xrightarrow{\text{conc. } H_2SO_4} CH_3CH_2CH_2OH \qquad (15.8)$$

$$\text{cyclopropane} + HBr \rightarrow CH_3CH_2CH_2Br \qquad (15.9)$$

Although we have likened these reactions to those of alkenes, the three-membered ring does *not* undergo additions as readily as a C=C double bond. Furthermore, the mechanisms of the reactions may differ from those observed for alkenes.

15.3 SATURATED RINGS 2: CYCLIC ETHERS

Structure and nomenclature

The replacement of a methylene unit in a cycloalkane by an oxygen atom produces a cyclic ether: *note that this is a 'paper exercise', and not a synthetic method!* Cyclic ethers are examples of *heterocyclic molecules*, carbon-based rings containing one or more *heteroatoms* (atoms other than carbon).

In Section 14.7, we described three methods of naming aliphatic acyclic ethers, and of these, the use of the prefix *oxa* is the preferred choice for larger cyclic ethers. Compound **15.9** is 1,4,7-trioxacyclononane, and the name is derived by prefixing the root-name 'cyclononane' with *oxa* and appropriate numeric descriptors, with the ring numbering beginning so as to place one of the O atoms at position 1.

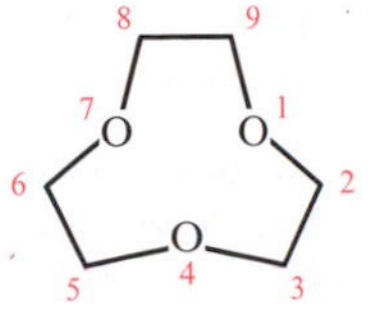

(15.9)

Small rings each have a particular name and we shall only consider the three- and five-membered rings. Structure **15.10** is called *oxirane*, the -irane ending being characteristic of a *saturated* three-atom ring and the prefix ox- indicating that the heteroatom is oxygen. An older name for oxirane is ethylene oxide, and a general name used for compounds of the type shown in structure **15.11** is *epoxides*.

(15.10)  (15.11)

One of the commonest cyclic ethers that you are likely to encounter in the laboratory is *tetrahydrofuran* (abbreviated to THF), **15.12**. The IUPAC-recommended name for **15.12** is *oxolane*; the -olane signifies the presence of a *saturated* five-membered ring, but the name tetrahydrofuran is almost always used. If two oxygen atoms are present, the systematic name is used, and **15.13** is 1,3-dioxolane.

(15.12) 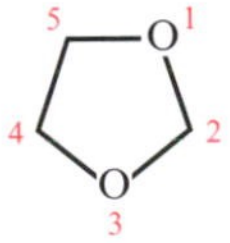 (15.13)

Synthesis

Many cyclic ethers are available commercially, and here we focus on methods by which oxirane and THF can be prepared.

Ethane-1,2-diol: see Section 14.9

The catalytic oxidation of ethene leads to oxirane (equation 15.10) and this reaction is the first step in the industrial preparation of ethane-1,2-diol.

$$CH_2{=}CH_2 \xrightarrow{O_2,\ Ag,\ \Delta} \text{oxirane (ethylene oxide)} \qquad (15.10)$$

Preparation of 1-chloropropan-2-ol: see Section 8.13

On a smaller scale, oxiranes can be prepared from chloro-substituted alcohols; the –OH and –Cl substituents must be on adjacent carbon atoms as in reaction 15.11. The normal laboratory preparation is by the reaction of an

Box 15.1 Epoxy resins

The glue *Araldite* is sold in two separate tubes, which, when mixed, harden rapidly. One tube contains an *epoxy-resin* and the second, a diamine. The reaction that occurs on mixing the two components can be represented as follows:

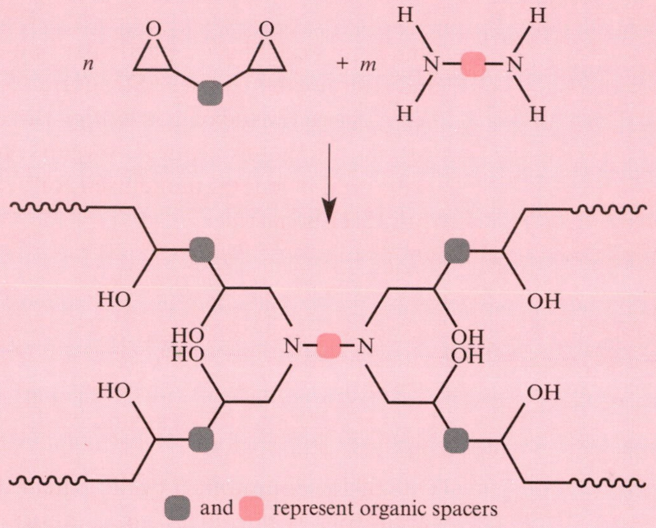

■ and ■ represent organic spacers

The strained epoxide ring undergoes an instantaneous ring-opening reaction as soon as the amine is made available as a reaction partner. The four H atoms in the amine are points from which the polymer can 'grow' and a *cross-linked* (rather than a single chain) structure results.

An organic peracid has the general formula

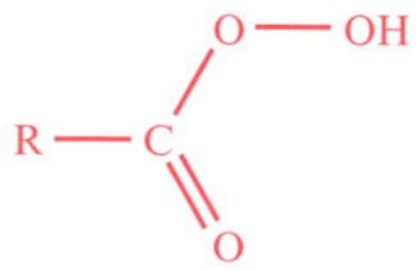

R—C(=O)—O—OH

alkene with a *peracid* such as perbenzoic acid. Reaction 15.12 shows the conversion of 2,3-dimethylbut-2-ene to tetramethyloxirane.

$$CH_3CH(OH)CH_2Cl \xrightarrow{\text{conc. aq. } [OH]^-} \text{methyloxirane} \quad (15.11)$$

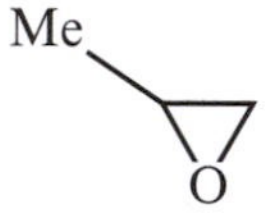

$$Me_2C{=}CMe_2 \xrightarrow{PhC(O)O{-}OH} \text{tetramethyloxirane} \quad (15.12)$$

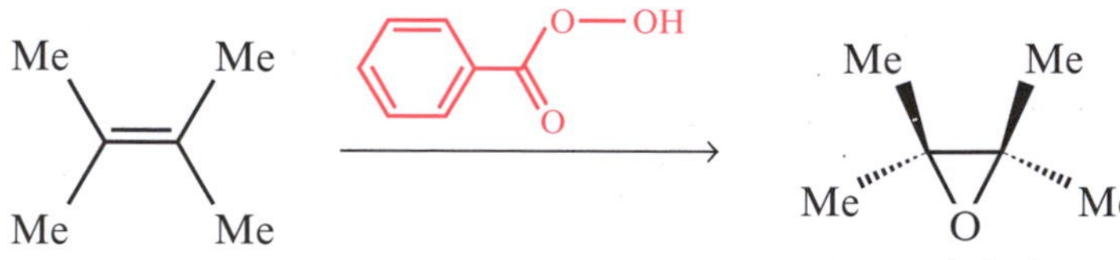

An alcohol with the arrangement of –OH and –H atoms shown in structure **15.14** undergoes ring-closure when treated with lead(IV) acetate; with butan-1-ol, THF is formed (equation 15.13).

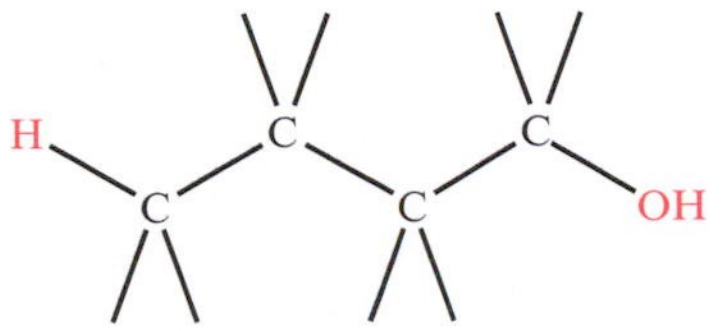

H, C, C, C, C, OH

(15.14)

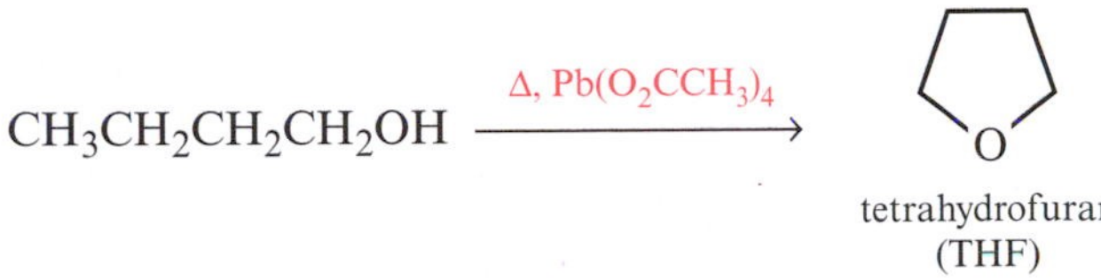

$$CH_3CH_2CH_2CH_2OH \xrightarrow{\Delta,\ Pb(O_2CCH_3)_4} \text{tetrahydrofuran (THF)} \quad (15.13)$$

Tetrahydrofuran may also be prepared by the catalytic hydrogenation of furan (equation 15.14).

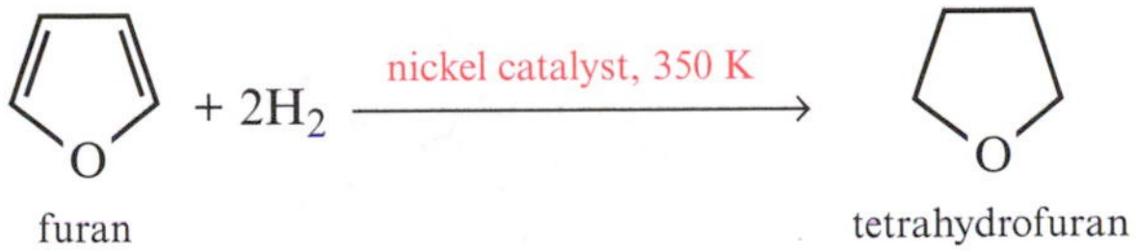

$$\text{furan} + 2H_2 \xrightarrow{\text{nickel catalyst, 350 K}} \text{tetrahydrofuran} \quad (15.14)$$

Physical properties and uses

The physical properties of cyclic ethers follow similar trends to those of their acyclic analogues, although as the data in Table 15.1 illustrate, the dipole moments (measured in the gas phase) of oxirane and THF are higher than those of the corresponding acyclic ethers, and both cyclic ethers possess higher boiling points than acyclic ethers of similar molecular mass.

Like acyclic ethers, cyclic ethers show a strong absorption in their infrared spectra in the range 1080–1150 cm^{-1} (Figure 15.7).

Table 15.1 Dipole moments, boiling points and enthalpies of vaporization for oxirane and THF, and for acyclic ethers of comparable molecular mass. Compounds are arranged according to increasing molecular mass.

Ether	*Structure*	*Dipole moment for gas phase molecule / D*	*Relative molecular mass*	*Boiling point / K*	$\Delta_{vap}H$ / $kJ\ mol^{-1}$
Oxirane		1.89	44	284	25.5
Dimethyl ether	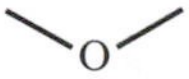	1.30	46	248	21.5
Tetrahydrofuran		1.75	72	338	30
Diethyl ether	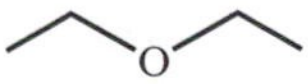	1.15	74	308	26.5
Methyl propyl ether		1.11	74	312	27

Box 15.2 Crown ethers: large cyclic ethers that encapsulate alkali metal ions

In Figure 11.13, we showed a hydrated potassium ion. The oxygen atoms of large cyclic polyethers are able to form similar coordinate bonds with alkali metal ions such that the metal ion lies within the organic molecule. Since the ionic radii of the alkali metals vary, a cyclic ether of a particular ring size will be better suited to a particular metal ion.

The structure of 1,4,7,10,13,16-hexaoxacyclooctadecane is shown in diagrams (a)–(c). This is one of the so-called *crown ethers* (the name arises from the crown-shaped ring) and is usually called 18-crown-6. Its size is compatible with that of the K^+ ion. The 18-crown-6 ring is flexible (by restricted rotation about the C–O and C–C single bonds) and can alter its conformation so that the six oxygen atoms are arranged about the K^+ ion as shown in diagram (d).

(a)

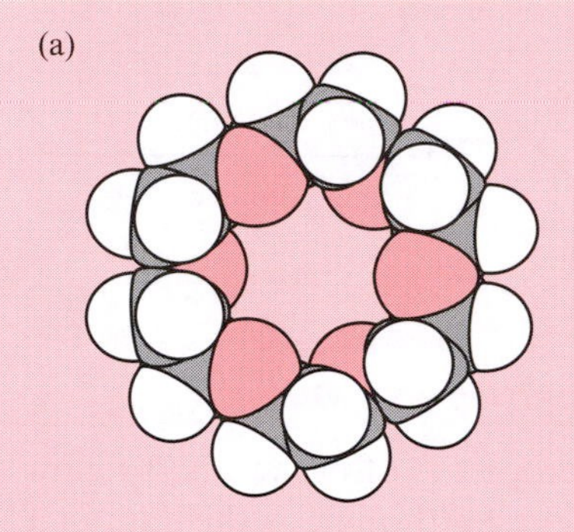

(b)

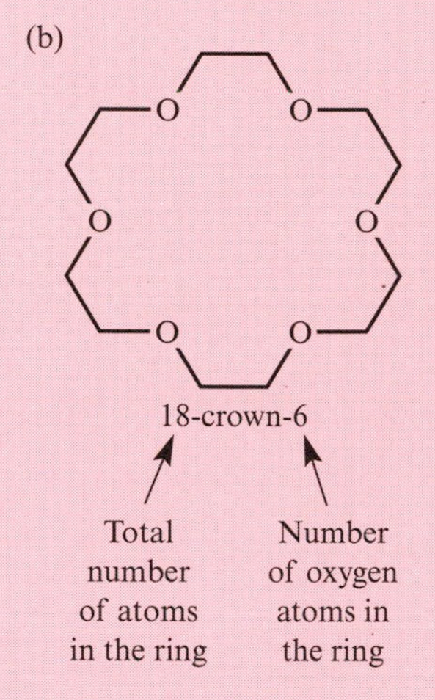

(c)

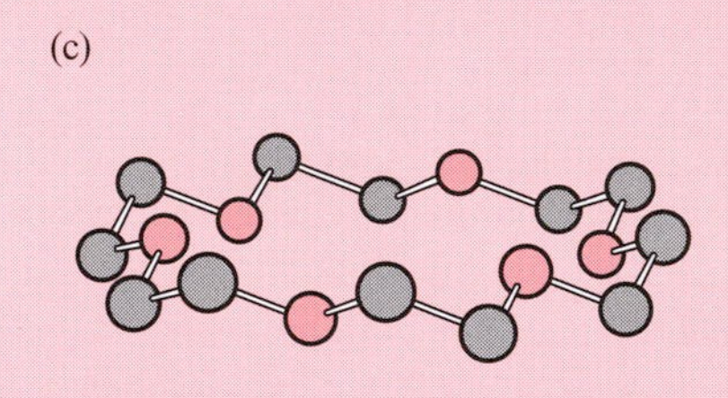

(d)

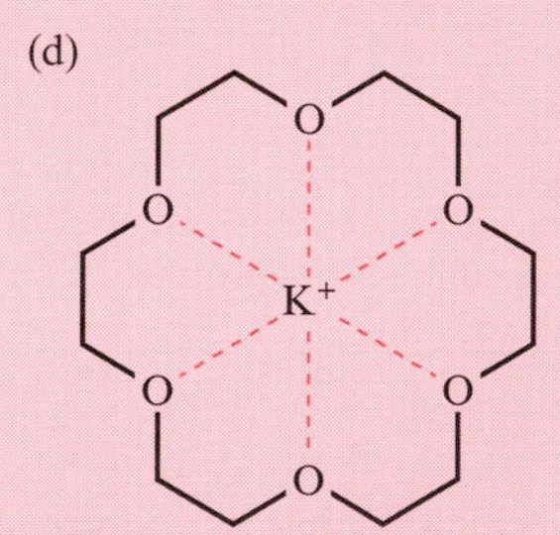

The advantage of the [K(18-crown-6)]$^+$ and similar cations is that they increase the solubility of salts. For example, $KMnO_4$ is insoluble in benzene, but when 18-crown-6 is added, it becomes soluble, giving a purple benzene solution; the compound that dissolves is [K(18-crown-6)][MnO_4].

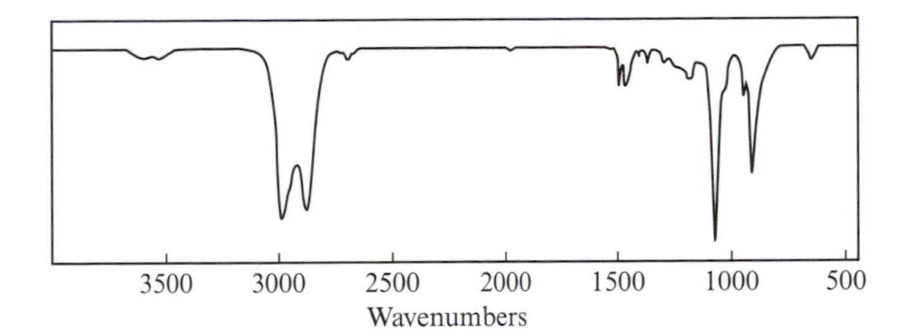

15.7 The IR spectrum of tetrahydrofuran. The absorptions close to 3000 cm^{-1} are due to the C–H stretches. In the fingerprint region, the strong absorption at 1070 cm^{-1} is assigned to the C–O stretch. Compare this spectrum with those of dibutyl ether and dihexyl ether (Figure 14.13).

Cyclic ethers are widely used as solvents in synthetic chemistry. For example, THF may be preferred over diethyl ether on the grounds of its higher boiling point and greater polarity, and is commonly used in reactions involving organolithium compounds and Grignard reagents.

15.4 UNSATURATED RINGS: CYCLOALKENES

Structure and nomenclature

The general formula for an unsubstituted monocyclic alkene containing one C=C double bond is C_nH_{2n-2}. Its name must indicate both the number of carbon atoms, the fact that the structure is cyclic, and the presence of the C=C functional group. Compound **15.15** is cyclohexene and **15.16** is cyclobutene. No position-numbers are required – why not?

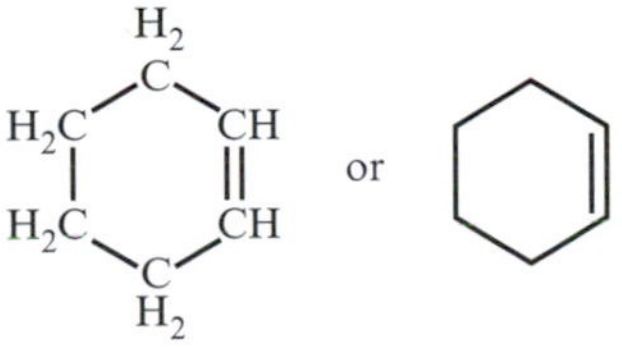

(15.15) 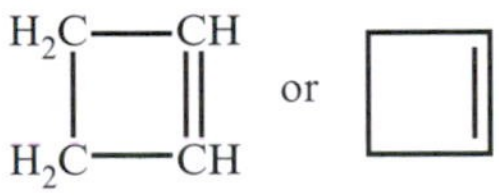 (15.16)

The structure of cyclopentene is shown in Figure 15.8a. The presence of the two trigonal planar carbon atoms results in the ring being less flexible than that of cyclopentane (Figure 15.3b). The C_5 ring in cyclopentadiene (Figure 15.8b) is even less flexible.

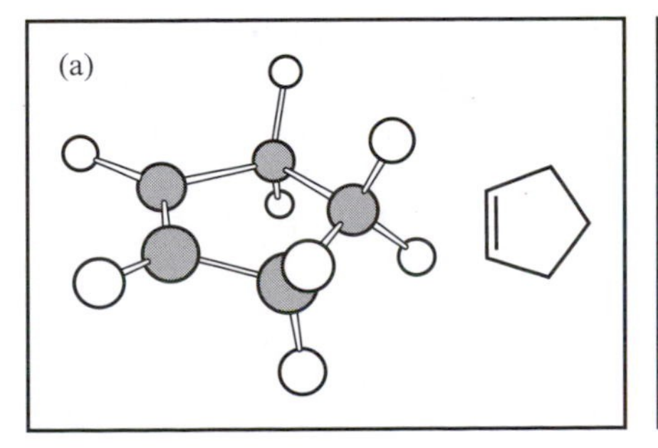

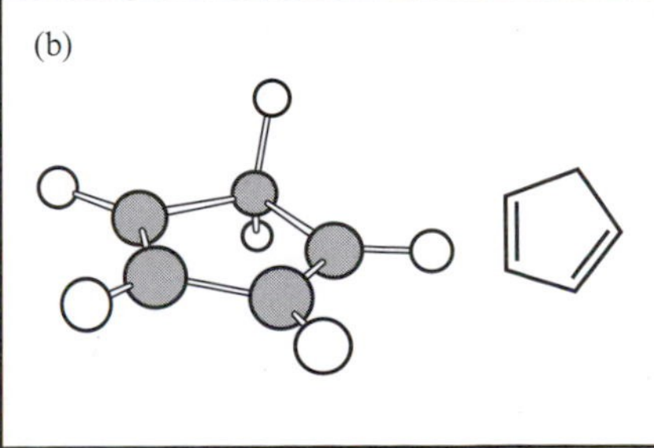

15.8 (a) The structure and a schematic representation of cyclopentene. There are three tetrahedral and two trigonal planar carbon centres.
(b) The structure and a schematic representation of cyclopentadiene, in which there are four trigonal planar and one tetrahedral carbon environments.

A diene contains two C=C functional groups.

Cyclic *dienes* (with the exceptions of cyclobutadiene and cyclopentadiene – why?) require position-numbers to define the relative positions of the two double bonds. The two isomers of cyclohexadiene are shown in structures **15.17** and **15.18**; as previously, the ring numbering begins so that one double bond is designated as being at position C(1). The geometry of the ring precludes the double bonds from sharing a common carbon atom. [*Question*: Why is this? Think about atom C(3) in penta-2,3-diene.]

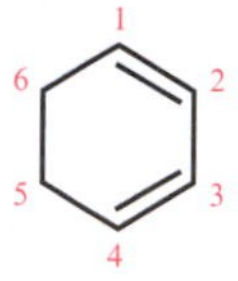

cyclohexa-1,3-diene
(15.17)

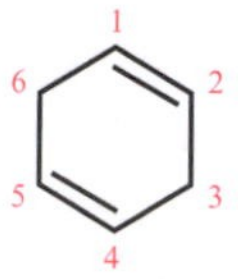

cyclohexa-1,4-diene
(15.18)

When numbering a *substituted* ring, the position of the C=C double bond takes priority over the positions of the substituents. Thus, **15.19** is 3,5-dichlorocyclohexene; the position of the double bond is assumed to be at carbon atom C(1), and need not be specified. In a diene, the double bonds should take the lowest position descriptors possible, and compound **15.20** is named 5-ethylcyclohexa-1,3-diene.

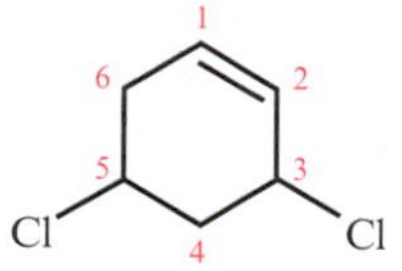

Cl Cl
(15.19)

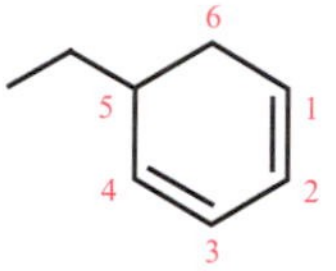

(15.20)

Synthesis

The dehydrogenation of a cycloalkane to give a specific cycloalkene or a diene with the double bonds in particular positions is synthetically difficult and is not usually a viable synthetic strategy.

Cyclohexenes may be prepared by *cycloaddition* of ethenes and buta-1,3-dienes – an example of a *Diels–Alder reaction*. In such a reaction, a double bond adds across the 1,4-positions of a 1,3-conjugated diene to give a cycloalkene (equation 15.15). The precursor containing *one* double bond (CH_2=CHX in equation 15.15) is called a *dienophile* (it 'loves dienes'). Each full-headed arrow in reaction 15.15 represents a two-electron transfer; two new C–C σ-bonds and one new π-bond are formed.§

Ethene (X = H in equation 15.15) is a poor dienophile, and for a successful reaction, the dienophile should bear at least one electron-withdrawing substituent X such as Cl, Br, CN, CH_2Cl, CH_2OH, CHO or CO_2H.

§ The mechanisms of cycloaddition reactions are complex. Detailed accounts are given in J. March, *Advanced Organic Chemistry: Reactions, Mechanisms and Structure*, 4th edn. Wiley, New York (1992).

(15.15)

see text for identity of X

In a Diels–Alder reaction, a double bond adds across the 1,4-positions of a 1,3-conjugated diene.

In related reactions, acetylenes with electron-withdrawing substituents behave as dienophiles, and reaction 15.16 shows the formation of a substituted hexa-1,4-diene.

(15.16)

The trimerization and tetramerization of acetylene are also cycloadditions, and occur when HC≡CH is heated in the presence of a nickel(0) or nickel(II) catalyst (equation 15.17).

$$\text{HC}{\equiv}\text{CH} \xrightarrow{\text{nickel catalyst}} \text{benzene} + \text{cyclooctatetraene} \qquad (15.17)$$

The cycloaddition of buta-1,3-diene is also catalysed by nickel and other transition metal compounds. This reaction leads to the formation of dimers, trimers and tetramers amongst other products. Equation 15.18 gives a simplified picture of the reaction.

(15.18)

Reactions

Cycloalkenes undergo electrophilic and radical *addition reactions* with mechanisms similar to those of acyclic alkenes, and equations 15.19–15.22 show representative reactions of cyclohexene. [*Question*: Suggest why the product of the addition of Br_2 is shown as having a particular stereochemistry.]

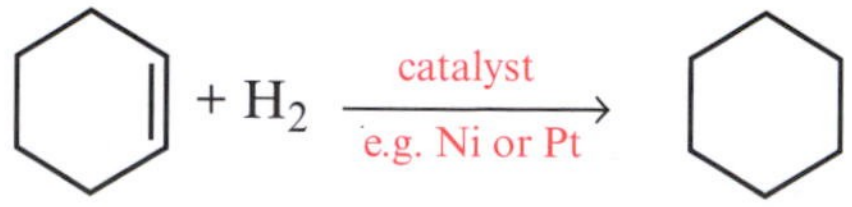

+ H_2 —catalyst e.g. Ni or Pt→ cyclohexane (15.19)

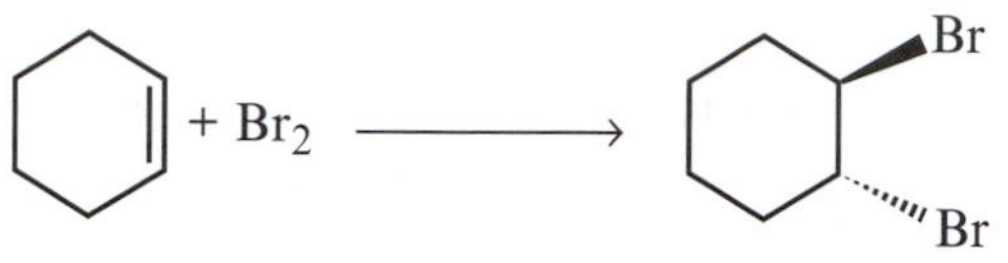

+ Br_2 → 1,2-dibromocyclohexane (15.20)

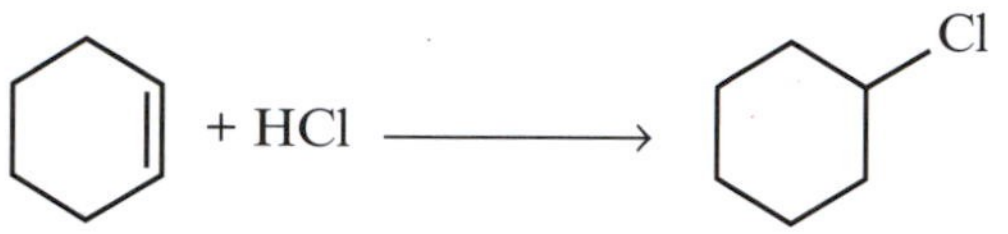

+ HCl → chlorocyclohexane (15.21)

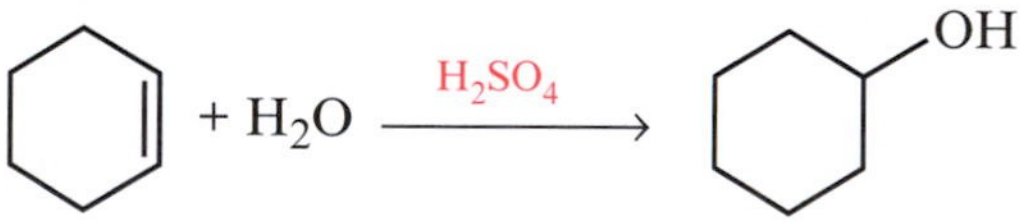

+ H_2O —H_2SO_4→ cyclohexanol (15.22)

Since these additions are to a *symmetrical* alkene, there are no other products in reactions 15.21 and 15.22 arising from competing reactions; contrast this with the additions of HCl and H_2O to hex-1-ene or hex-2-ene. If on the other hand the substrate is 1-methylcyclohexene, then competition arises during the electrophilic addition of a reagent such as HBr due to the formation of either a tertiary or secondary carbenium intermediate (equation 15.23).

Additions to asymmetrical alkenes: see equations 8.36 and 8.48

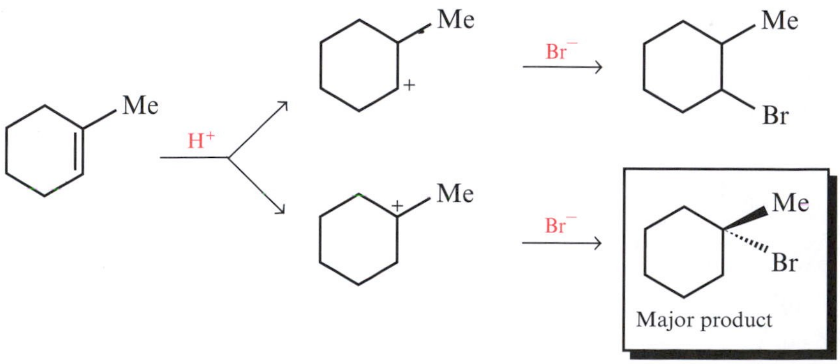

Me, H^+, Me, +, Br^-, Me, Br

Me, +, Br^-, Me, Br

Major product

(15.23)

The oxidation of the C=C double bond in a cycloalkene leads to the formation of an epoxide (equation 15.24) in a reaction similar to that in equations 15.12. The product is a *bicyclic* molecule which contains two rings sharing a common C–C bond.

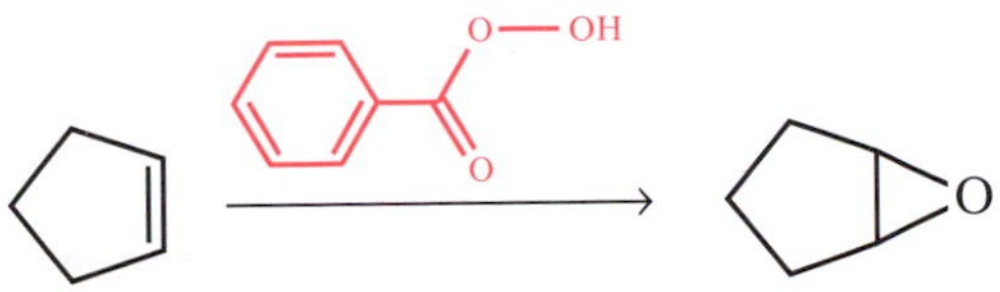

(15.24)

A bicyclic molecule consists of two rings which have ring-atoms in common:

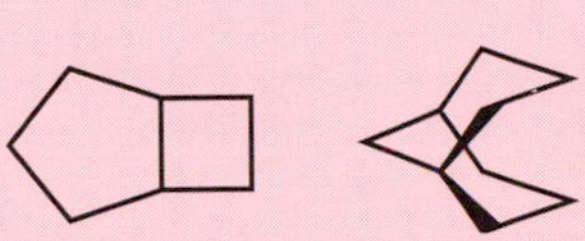

Molecules with a single ring-*atom* in common are called *spirocyclic*:

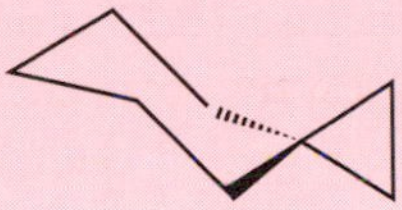

15.5 BENZENE 1: STRUCTURE, BONDING AND SPECTROSCOPY

Benzene, C_6H_6, is the simplest member of a group of compounds called *aromatic hydrocarbons* – we define what is meant by 'aromatic' later in this section.

At room temperature, benzene is a colourless liquid (mp 278.5 K, bp 353 K) with a characteristic odour. It is non-polar, immiscible with water, but miscible with other non-polar organic solvents. At one time, it was widely used as a solvent, but now that its level of toxicity has been recognized, its use in the laboratory has become restricted; benzene is a carcinogen and prolonged exposure may cause leukaemia.

Structure and the Kekulé model

Benzene is planar (Figure 15.9) and all the C–C bonds are of *equal length*. The value of 140 pm is in between typical values for C–C single (154 pm)

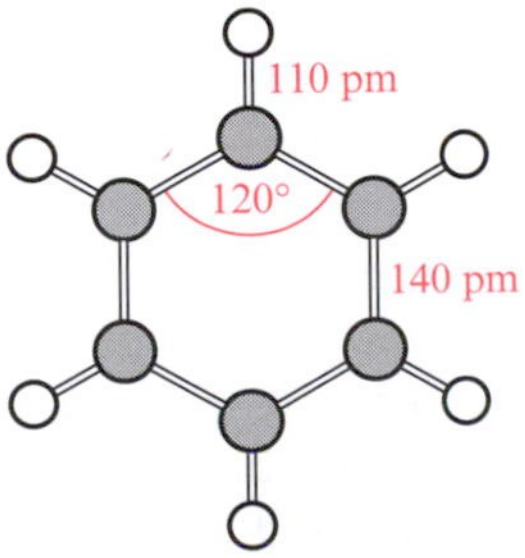

15.9 Benzene consists of a planar C_6 ring, and each carbon atom is in a trigonal planar environment. The carbon–carbon bond distances are all equal and the value of 140 pm is in between typical values for C–C single and C=C double bonds.

and C=C double bonds (134 pm). The valence bond model, based on the ideas of Kekulé, explains this observation in terms of the two resonance structures **15.21**. If the structures contribute *equally* to the bonding, then the average bond order of each C–C bond is 1.5.

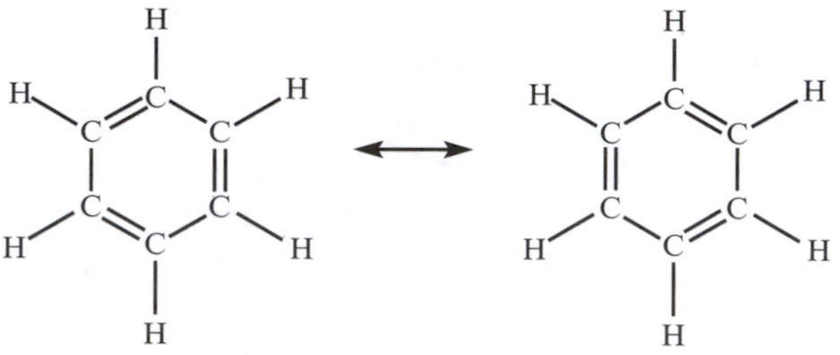

(15.21)

Is benzene a triene?

Structure **15.22** is one of two representations used for benzene, and it may suggest that benzene could be described as a cyclic triene.

(15.22)

This model implies the presence of localized double and single carbon–carbon bonds alternating around the ring, but the equivalence of the C–C bond distances is not consistent with this picture. Thermochemical data also indicate that benzene should *not* be described as a triene with localized carbon–carbon bonds. When cyclohexene is hydrogenated (equation 15.25), the enthalpy change accompanying the reaction is –118 kJ per mole of cyclohexene.

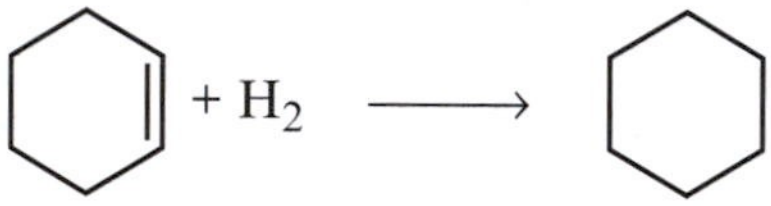

$$\text{cyclohexene} + H_2 \longrightarrow \text{cyclohexane} \qquad (15.25)$$

$$\text{benzene} + 3H_2 \longrightarrow \text{cyclohexane} \qquad (15.26)$$

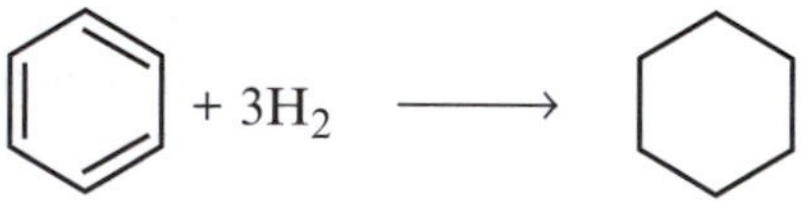

The observed value of $\Delta_r H$ for the hydrogenation of one mole of cyclohexadiene is approximately twice that of cyclohexene, and so we appear to be justified in using the value of –118 kJ as a measure of the enthalpy change associated with the *addition of one mole of H_2 to one mole of C=C bonds*. We predict that when a *model* compound 'cyclohexatriene' is hydrogenated (equation 15.26), the change in enthalpy is $-(3 \times 118)$ kJ per mole of triene as shown on the left-hand side of Figure 15.10. The *observed* value is significantly less than this as the right-hand of Figure 15.10 indicates, meaning that

15.10 The experimental molar enthalpy of hydrogenation ($\Delta_r H$) of benzene is smaller than that predicted by assuming a cyclic triene model.

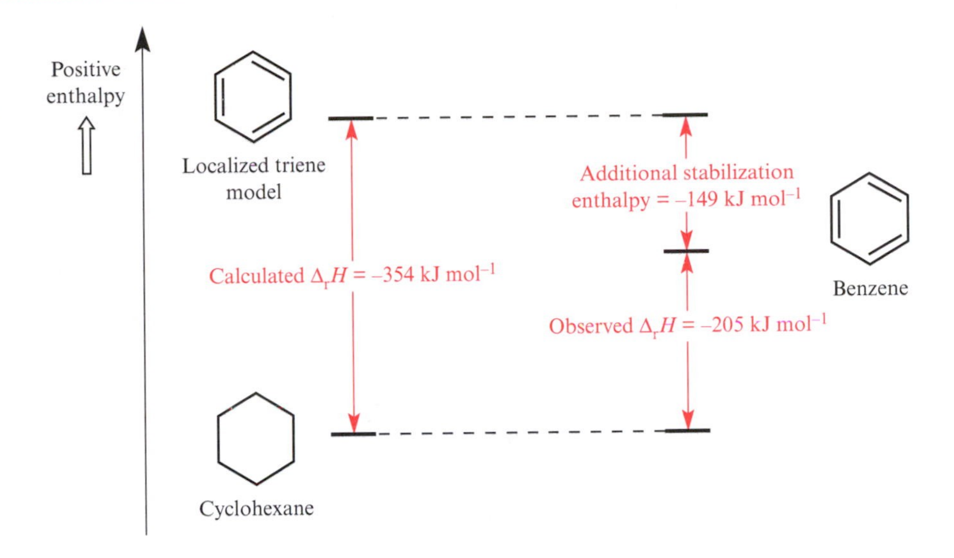

benzene is thermodynamically *more stable* than the triene model suggests.

Further evidence for the failure of the triene model comes from the reactivity of benzene. For example, whereas an aqueous solution of Br_2 is decolorized when ethene is bubbled through it (equation 15.27), benzene reacts with Br_2 only in the presence of a catalyst, and the reaction is a *substitution*, not an addition (equation 15.28).

$$CH_2{=}CH_2 + Br_2 \rightarrow CH_2BrCH_2Br \qquad (15.27)$$

$$C_6H_6 + Br_2 \xrightarrow{FeBr_3 \text{ catalyst}} C_6H_5Br + HBr \qquad (15.28)$$

A delocalized bonding model

The resonance pair **15.21** can also be represented by a delocalized bonding approach. An sp^2 hybridization scheme is consistent with each trigonal planar carbon centre, and Figure 15.11 illustrates that the carbon sp^2 hybrid orbitals and the hydrogen 1*s* AOs overlap to form a σ-bonding framework consisting of localized C–C and C–H interactions. Each carbon atom provides three electrons and each hydrogen atom contributes one electron to the σ-bonding interactions. The six remaining carbon 2*p* AOs can overlap to give a delocalized π-molecular orbital (Figure 15.11b) which is continuous around the C_6 ring. Each carbon atom provides one electron to the π-interactions making this a *six-electron delocalized system*.

The relationship between the Kekulé and delocalized models of bonding in benzene is similar to that which we described for the allyl anion in structures **8.17** and **8.18**, and **15.23** shows a representation of the benzene ring

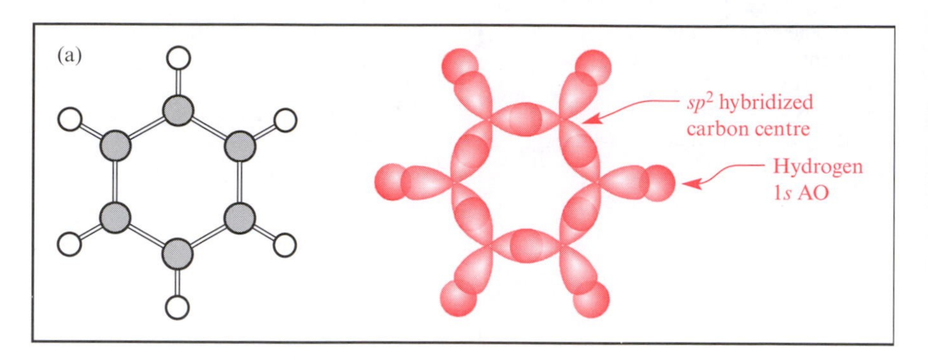

15.11 In benzene, each carbon atom provides four electrons and each hydrogen atom provides one. (a) The σ-bonding framework in benzene can be described in terms of sp^2 hybridized carbon centres. Each C–H interaction is a localized, 2-centre 2-electron σ-bond formed by the overlap of a carbon sp^2 hybrid orbital with a hydrogen 1*s* AO. Each C–C σ-interaction is a localized single bond. (b) After forming three σ-bonds, each carbon atom has one 2*p* AO remaining and overlap can occur between *all six* AOs to give a delocalized π-system. Each carbon atom provides one electron, and a six-electron π-system is therefore formed.

Delocalized π-bonding: see Section 9.10

which parallels structure **8.18**. Both bonding models arise from a *valence bond approach*; a full molecular orbital treatment is beyond the scope of this discussion, although we provide an insight into this method in Box 15.3.

(15.23)

It is important to remember that the resonance pair **15.21** *is exactly equivalent* to structure **15.23**.

Aromaticity and the Hückel (4*n*+2) rule

At the beginning of this section, we stated that benzene is an *aromatic* compound. By aromatic we mean that there is some special character associated with the ring system. While this is difficult to quantify in a simple definition, an aromatic species is typified by a planar ring with a delocalized π-system; diagrammatic representations of the ring may show alternating single and double carbon–carbon bonds. In contrast to benzene, cyclobutadiene (although planar) exhibits C–C and C=C bond lengths of about 159 and 134

Box 15.3 A molecular orbital approach to the π-bonding in benzene

[Before working through this box, you may wish to review the MO approach to the π-bonding in the allyl anion in Section 9.10.]

The π-bonding in benzene arises from the overlap of carbon $2p$ atomic orbitals and since there are six AOs, we must form six π-MOs as shown below.

An in-phase overlap of all six carbon $2p$ AOs generates the lowest energy π-bonding MO and this possesses π-bonding character which extends continuously (no phase change) over all the carbon centres. The highest energy MO results if adjacent $2p$ AOs are out-of-phase with each other; this orbital is antibonding between each pair of adjacent carbon atoms, and contains three nodal planes which lie perpendicular to the plane of the benzene ring. The remaining four MOs consist of two degenerate pairs containing either one or two nodal planes. In two of these MOs, the nodal planes pass through two carbon atoms and this means that there is no contribution from these $2p$ AOs; compare this with the description of the allyl anion in Figure 9.25.

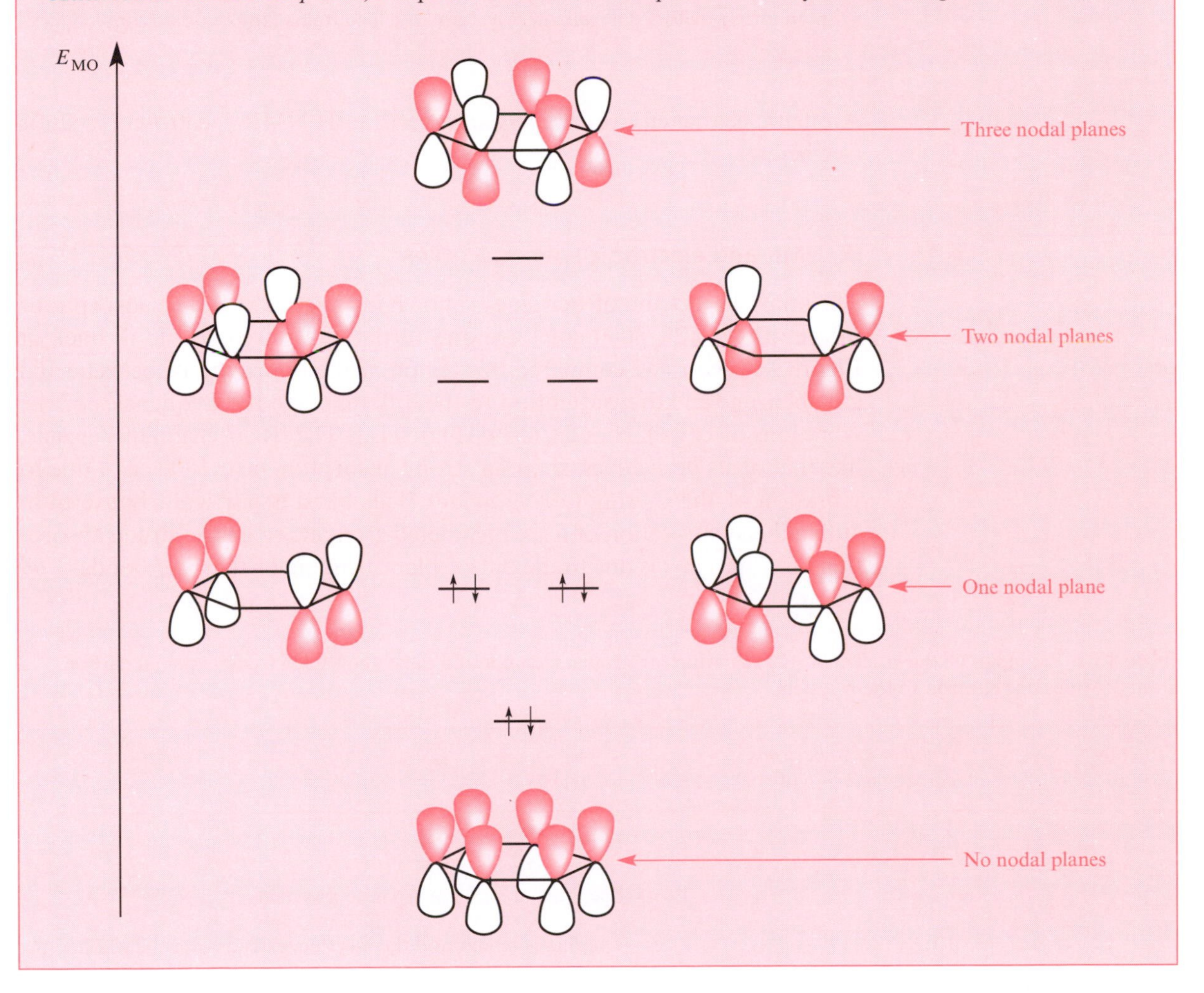

pm respectively[§] and is *not* aromatic. Most cyclobutadienes are only stable for very short periods, and C_4H_4 has a lifetime of less than five seconds.

§ These bond lengths are for a derivative of cyclobutadiene.

We have already seen that benzene possesses a delocalized π-system extending around the ring, but this in itself does not make the compound aromatic. The *number* of π-electrons in the ring is critical, and the *Hückel rule* states that there must be $(4n+2)$ π-electrons, where n is an integer. Benzene has six π-electrons and obeys the Hückel rule ($n = 1$), but cyclobutadiene with four π-electrons does not – an integral value of n cannot be found that satisfies the equation: $(4n+2) = 4$. Cyclobutadiene is an *anti-aromatic* species. Table 15.2 lists several aromatic hydrocarbons and shows how the $(4n+2)$ rule is applied. Note that *n is not equal to the number of carbon atoms.*

The Hückel $(4n+2)$ rule states that an aromatic compound must possess a planar ring, with a delocalized π-system which contains $(4n+2)$ π-electrons.

Aromatic character gives rise to characteristic ^{1}H NMR spectroscopic shifts (see below).

IR, NMR and electronic spectroscopies

The infared spectrum of benzene is shown in Figure 15.12. The absorptions in the range 3120–3000 cm^{-1} are due to the C–H stretches; look back at Figure 9.19 where we compared this region of the IR spectra of benzene and ethylbenzene and drew attention to the differences in the frequencies of the vibrations of C(sp^2)–H and C(sp^3)–H stretches. The IR spectrum of benzene, like that of its derivatives, shows a strong absorption near 1500 cm^{-1} due to vibration of the C_6 ring (a *ring mode*). This band is particularly useful in aiding the identification of a six-membered aromatic ring. The strong absorption near 700 cm^{-1} is due to the out-of-plane bending of the C–H bonds.

Table 15.2 Examples of the Hückel $(4n+2)$ rule which an aromatic compound must obey. Each trigonal planar carbon atom provides one electron to the π-system.

Structure	*Number of π-electrons corresponding to* $(4n+2)$	n	*Comments*
	6	1	
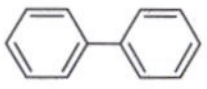	6 (twice)	1	Each planar ring is a separate π-system
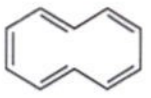	10	2	The ring is planar and the π-system extends over the whole ring
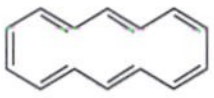	14	3	The ring is planar and the π-system extends over the whole ring
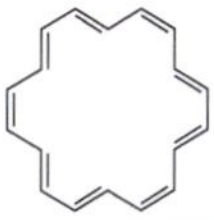	18	4	The ring is planar and the π-system extends over the whole ring

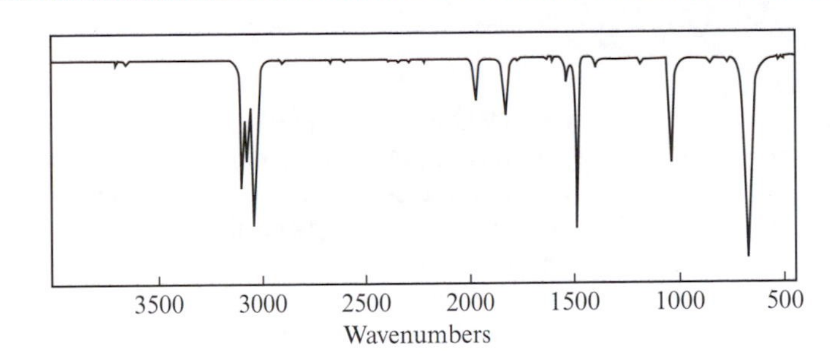

15.12 The IR spectrum of benzene. The absorptions near to 3000 cm^{-1} are assigned to the C–H stretches, and the strong absorption at 1480 cm^{-1} arises from the vibration of the C_6 ring (a *ring mode*). The strong band at 680 cm^{-1} is due to the out-of-plane C–H bending vibrations.

The ^1H NMR spectrum of benzene is shown in Figure 15.13a. The six protons are equivalent and the spectrum consists of a singlet at δ +7.2. This relatively downfield shift is characteristic of the H atoms in benzene and those attached *directly* to the C_6 ring in derivatives of benzene and related six-membered ring aromatic compounds. It distinguishes benzene-like protons from those in alkenes; the typical shift range for aromatic protons is δ +6 to +10, and this is characteristically further downfield than signals assigned to alkene protons (see Table 9.12).

Chemical shift values: see Section 9.12

Figure 15.13b illustrates the ^{13}C NMR spectrum of benzene. The carbon centres are equivalent and there is a single resonance at δ +128. This is typical of an sp^2 hybridized carbon atom (see Table 9.11) and the carbon nuclei in derivatives of benzene resonate at similar frequencies. Practice in the interpretation of some ^{13}C NMR spectra is given in the accompanying workbook.

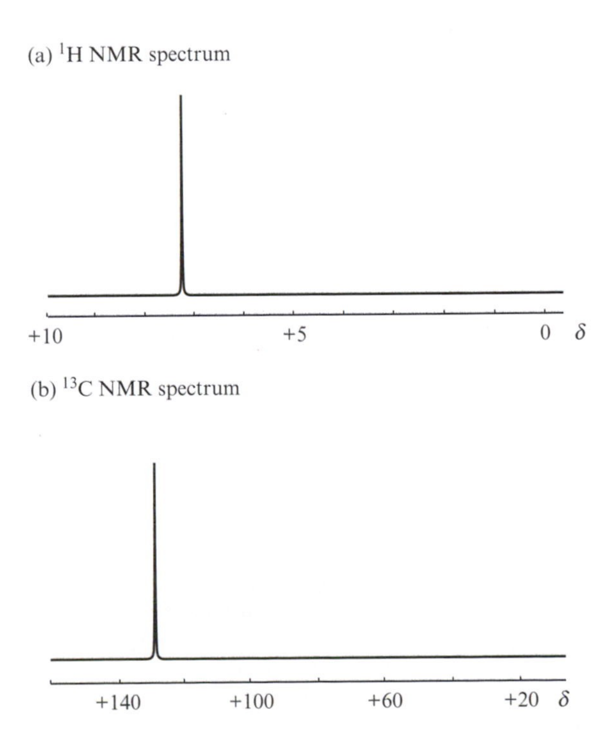

15.13 (a) The ^1H NMR spectrum of benzene consists of a single resonance at δ +7.2. (b) The ^{13}C NMR spectrum of benzene confirms the equivalence of the carbon centres; the resonance is at δ +128. The effects of ^{13}C–^1H spin–spin coupling have been removed from the spectrum.

Box 15.4 Deshielding of aromatic protons

Compared to typical alkene protons, those in benzene give rise to a signal in the ^{1}H NMR spectrum which is further downfield (see Table 9.12). The protons attached directly to the C_6 ring in derivatives of benzene also give signals close to that of benzene (δ +7.2). Resonances in the region δ +6 to +10 are characteristic of many aromatic compounds. What is the origin of this effect?

When a benzene molecule is placed in the magnetic field of an NMR spectrometer, the π-electrons undergo a circular motion (a *ring current*) which generates a secondary (induced) magnetic field.

Applied magnetic field, $\boldsymbol{B}_0$

Secondary (induced) field

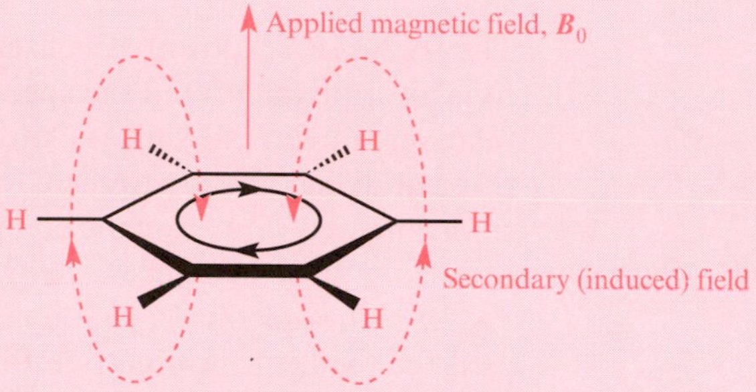

In benzene, the induced field reinforces the applied field and the protons experience a greater magnetic field than do protons not affected by the induced field; they are said to be *deshielded*. In the ^{1}H NMR spectrum, deshielded protons appear at more positive chemical shift values.

In some aromatic systems, the induced magnetic field *opposes* the applied field and the aromatic protons are *shielded*, and in the ^{1}H NMR spectrum, signals are at lower chemical shift. The classic example of this effect is observed in the spectrum of 18-annulene, the structure of which is shown below. This is a planar, aromatic molecule – the Hückel ($4n+2$) rule is satisfied with $n = 4$. The ring current *shields* the protons shown in red (these lie *within* the ring), but *deshields* the protons shown in black (these lie *outside* the ring).

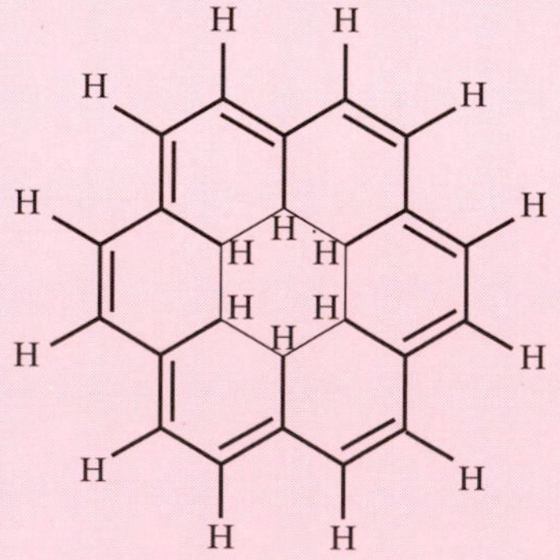

At low temperatures, the ^{1}H NMR spectrum of 18-annulene contains two signals close to δ +9 and –3.

Table 15.3 Observed values of λ_{max} for benzene and some related aromatic molecules. Only the most intense band in each spectrum is detailed.

Compound	*Structure*	λ_{max} / nm	ε_{max} / $dm^3\ mol^{-1}\ cm^{-1}$
Benzene		183	46 000
Biphenyl		246	20 000
Naphthalene		220	132 000
Anthracene		256	182 000
Styrene		244	12 000
Diphenylacetylene		236	12 500

π-Conjugation and electronic spectra: see Section 9.10

We have already seen how the effects of π-conjugation permit the $\pi \rightarrow \pi^*$ transition to be observed in the near-UV part of the electronic spectrum of conjugated polyenes. Similarly, the presence of the delocalized π-system in benzene reduces the energy difference between the highest-lying π MO and the lowest-lying π^* MO, and the associated $\pi \rightarrow \pi^*$ transition is observed in the near-UV. The electronic spectrum of benzene has three absorptions, the most intense of which occurs at λ_{max} = 183 nm (ε_{max} = 46 000 $dm^3\ mol^{-1}\ cm^{-1}$). Selected electronic spectroscopic data for benzene and some of its derivatives are given in Table 15.3.

15.6 BENZENE 2: SYNTHESIS, NOMENCLATURE AND REACTIVITY

Synthesis

Industrial chemical rankings: see Appendix 14

Benzene is commercially available, and it lies seventeenth in the 1994 ranking of chemicals manufactured in the United States of America. Benzene may be prepared by the catalytic dehydrogenation of cyclohexane (equation 15.29). The general process for converting an alkane into an aromatic compound is *catalytic reforming* and the method is used to produce benzene on an industrial scale.

$$\text{cyclohexane} \xrightarrow[-3H_2]{\text{platinum catalyst, 380 K}} \text{benzene} \qquad (15.29)$$

Equation 15.17 showed the cycloaddition reaction of acetylene and by altering the reaction conditions, it is possible to swing the pathway in favour of benzene and away from other products.

Nomenclature

The usual abbreviation for a general *aryl* substituent is Ar.

Before discussing some of the reactions of benzene, we must define the basic rules for naming derivatives of benzene. An aromatic radical is called an *aryl* group; whereas R stands for a general alkyl group, Ar is usually used to represent a general aryl substituent.

Consider a compound of formula C_6H_5X with the structure **15.24**.

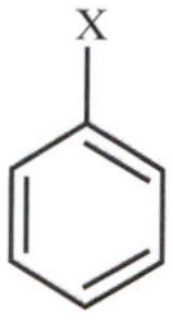

X

(15.24)

In some compounds of this type, the group X is considered to be a substituent of the benzene ring and the compound is named by prefixing the word 'benzene' with the name of the functional group. This nomenclature is used for halogeno derivatives such as **15.25**, alkyl derivatives such as **15.26** and nitro derivatives, **15.27**. An exception is benzenesulfonic acid, **15.28**.

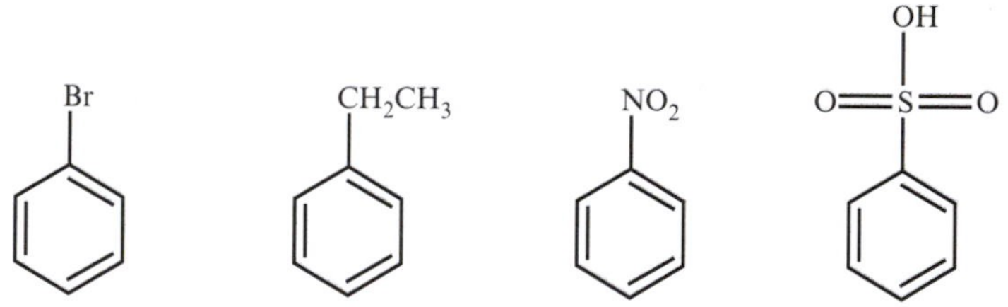

Br — bromobenzene (15.25)

CH_2CH_3 — ethylbenzene (15.26)

NO_2 — nitrobenzene (15.27)

OH, O=S=O — benzenesulfonic acid (15.28)

In other compounds, the C_6H_5 group, the *phenyl* group abbreviated to Ph, is considered to be a substituent of the X group, and the compound is named by prefixing the name of the X group with the word 'phenyl'. This nomenclature is used for amino-derivatives such as **15.29**, and for phosphines, e.g. **15.30**. Many trivial names for aromatic compounds remain in use, and phenylamine **15.31** is commonly known as aniline.

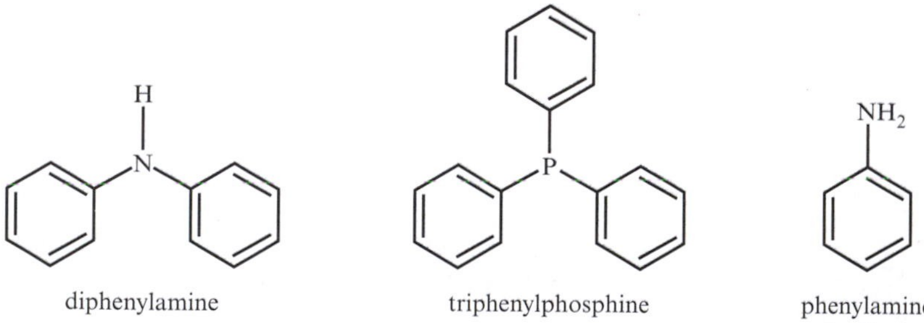

H, N — diphenylamine (15.29)

P — triphenylphosphine (15.30)

NH_2 — phenylamine (aniline) (15.31)

Structures **15.32–15.38** show some derivatives of benzene for which you are likely to encounter common, rather than systematic, names. We return to the derivatives containing C=O groups in Chapter 17.

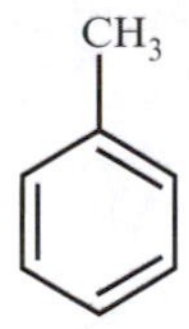

toluene **(15.32)** phenol **(15.33)** benzoic acid **(15.34)** benzaldehyde **(15.35)**

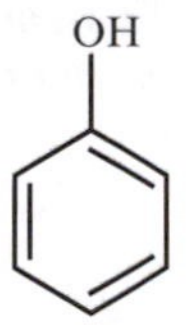

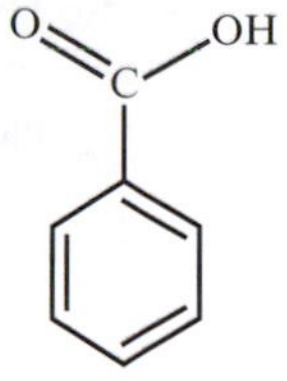

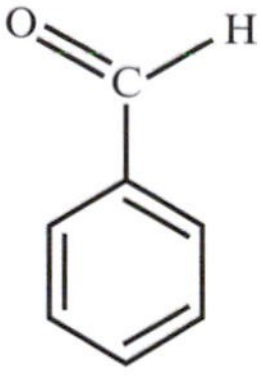

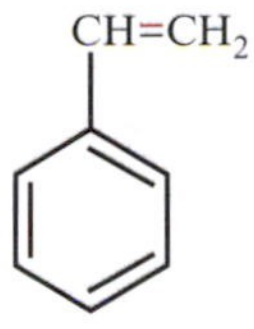

styrene **(15.36)** benzoyl chloride **(15.37)** anisole **(15.38)**

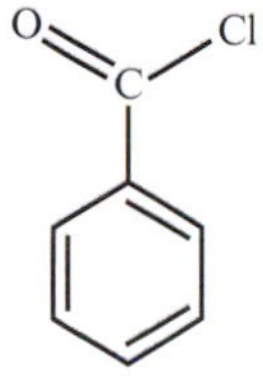

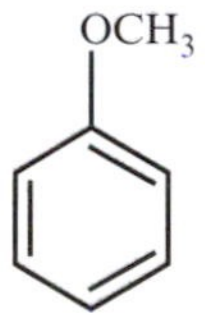

When there is more than one substituent on the benzene ring, numerical descriptors are used so as to place one substituent at position C(1). Compound **15.39** is 1,4-dibromobenzene, and **15.40** is 1,3,5-trimethylbenzene (commonly called mesitylene). Using the position numbers distinguishes between structural isomers. Dibromobenzene has three isomers – 1,2-, 1,3- and 1,4-dibromobenzene.

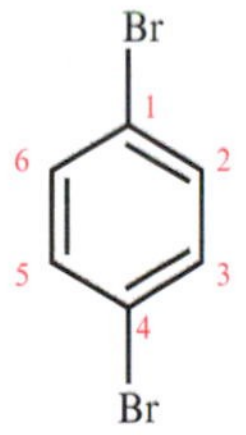

(15.39) 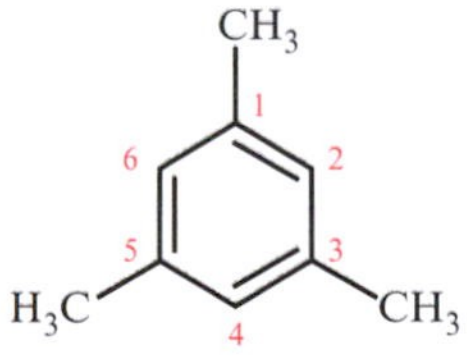 **(15.40)**

It is important to distinguish between a *phenyl* and a *benzyl* group. A benzyl substituent, has the formula $C_6H_5CH_2$ and benzyl chloride **15.41** behaves as a halogenoalkane (not a halobenzene) since the chloro-substituent is part of the alkyl group and is not attached directly to the benzene ring.

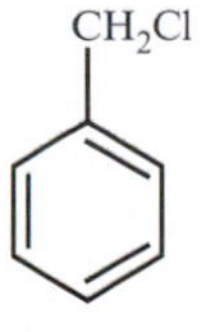

(15.41)

Box 15.5 *Ortho-*, *meta-* and *para-*substitution patterns

The 1,2-, 1,3- and 1,4-isomers of compounds such as dibromobenzene may also be named by using the prefixes *ortho-*, *meta-* and *para-* respectively. They are abbreviated to *o-*, *m-* and *p-*. Isomers of dimethylbenzene, the trivial name of which is *xylene*, are shown below.

1,2-dimethylbenzene
ortho-xylene or *o*-xylene

1,3-dimethylbenzene
meta-xylene or *m*-xylene

1,4-dimethylbenzene
para-xylene or *p*-xylene

Reactions of benzene

Like other hydrocarbons, benzene can be burnt to give carbon dioxide and water (equation 15.30).

$$2C_6H_6(l) + 15O_2(g) \rightarrow 12CO_2(g) + 6H_2O(l) \qquad (15.30)$$

$$\Delta_c H(298\ K) = -3268\ kJ\ \text{per mole of benzene}$$

The hydrogenation of benzene to give cyclohexane (the reverse of reaction 15.29) can be achieved by heating C_6H_6 and H_2 in the presence of a suitable catalyst. Under these conditions, it is difficult to isolate the intermediate unsaturated products (cyclohexadiene and cyclohexene), but equation 15.31 shows that reduction using sodium in liquid ammonia in the presence of an alcohol gives cyclohexa-1,4-diene. This is called a *Birch reduction*.

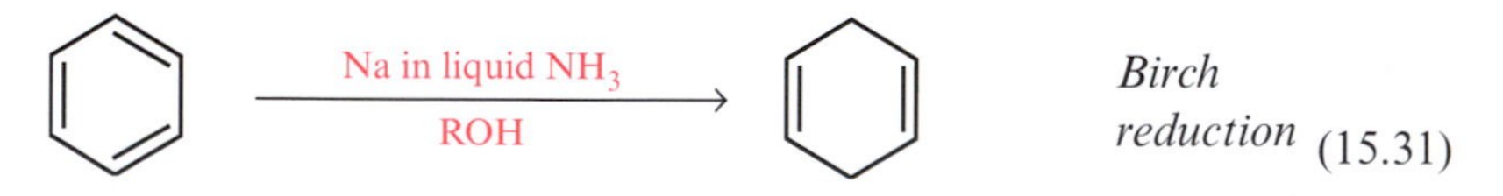

(15.31)

The characteristic reactions of benzene are *substitution reactions* rather than additions, and most require a catalyst. The chlorination of benzene to give chlorobenzene occurs when C_6H_6 reacts with Cl_2 in the presence of aluminium chloride (equation 15.32). This is an example of a *Lewis acid catalysed* reaction and in the next section we look more closely at the mechanism of this important synthetic method. Bromination takes place in a similar reaction, and other catalysts that are suitable include $FeCl_3$ (for chlorination), and $FeBr_3$ (for bromination). Iodination is more difficult, since I_2 is less reactive than Cl_2 or Br_2.

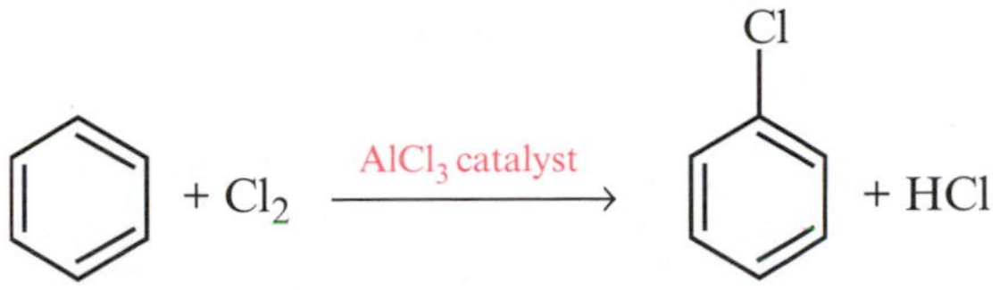

$$\text{C}_6\text{H}_6 + \text{Cl}_2 \xrightarrow{\text{AlCl}_3\text{ catalyst}} \text{C}_6\text{H}_5\text{Cl} + \text{HCl} \qquad (15.32)$$

Alkyl derivatives of benzene may be prepared by Friedel–Crafts *alkylation* reactions. Benzene reacts with chloro- or bromoalkanes in the presence of a Lewis acid catalyst such as aluminium chloride or iron(III) bromide (equation 15.33). However, *polyalkylation* is usually a complicating factor (see Section 15.9), and additionally, rearrangements may occur as reaction 15.34 illustrates.

Lewis acid: see Section 11.5

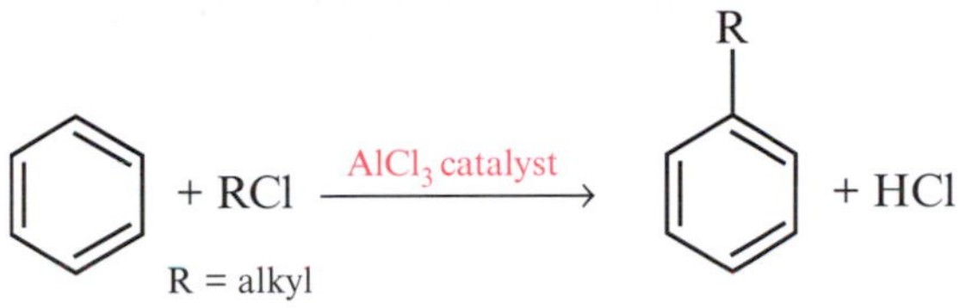

$$\text{C}_6\text{H}_6 + \text{RCl} \xrightarrow{\text{AlCl}_3\text{ catalyst}} \text{C}_6\text{H}_5\text{R} + \text{HCl} \qquad (15.33)$$

R = alkyl

$$\text{C}_6\text{H}_6 + \text{CH}_3\text{CH}_2\text{CH}_2\text{Cl} \xrightarrow{\text{AlCl}_3\text{ catalyst}} \text{C}_6\text{H}_5\text{CH}_2\text{CH}_2\text{CH}_3 + \text{HCl} \;\text{ or }\; \text{C}_6\text{H}_5\text{CH}(\text{CH}_3)_2 + \text{HCl} \qquad (15.34)$$

cumene
Major product

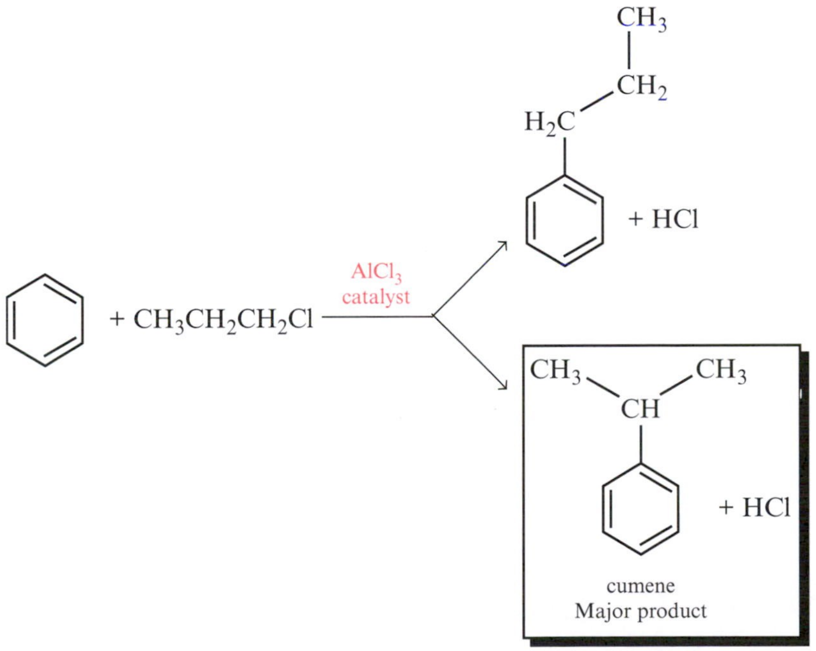

The common name for isopropylbenzene is *cumene*

Friedel–Crafts *acylation* of benzene is the reaction with an acyl chloride, RC(O)Cl, to give a ketone, and equation 15.35 shows the formation of ethyl phenyl ketone. We consider ketones in Chapter 17, but note here that the form of the reaction (i.e. an $AlCl_3$ catalyst and the use of a chloride precursor) is the same as those for the Friedel–Crafts halogenation and alkylation of benzene. The method does *not* suffer from the rearrangements and multisubstitutions of the alkylations described above.

$$\text{C}_6\text{H}_6 + \text{Et—C(=O)Cl} \xrightarrow{\text{AlCl}_3\text{ catalyst}} \text{C}_6\text{H}_5\text{C(=O)Et} + \text{HCl} \quad (15.35)$$

In a Lewis acid-catalysed reaction, a halide YX reacts with an aromatic compound ArH in the presence of a catalyst (e.g. $AlCl_3$ or $FeBr_3$) to give ArY and HX, where X may be Cl or Br. Examples are chlorination (Y = Cl), bromination (Y = Br), alkylation (Y = alkyl) and acylation (Y = RCO).

Nitrobenzene is a precursor to aromatic amines: see Section 15.11

The nitration of benzene to give nitrobenzene is carried out using a mixture of concentrated nitric and sulfuric acids (equation 15.36).

$$\text{C}_6\text{H}_6 + \text{HNO}_3 \xrightarrow{\text{H}_2\text{SO}_4} \text{C}_6\text{H}_5\text{NO}_2 + \text{H}_2\text{O} \quad (15.36)$$

A solution of SO_3 in concentrated H_2SO_4 and SO_3 is called *oleum*

The reaction between benzene and sulfuric acid in the presence of sulfur trioxide leads to *sulfonation* of the ring (equation 15.37). One use of benzenesulfonic acid is as a precursor to phenol (see equation 15.59).

$$\text{C}_6\text{H}_6 + \text{H}_2\text{SO}_4 \xrightarrow{\text{SO}_3} \text{C}_6\text{H}_5\text{SO}_3\text{H} + \text{H}_2\text{O} \quad (15.37)$$

15.7 THE MECHANISM OF ELECTROPHILIC SUBSTITUTION

The general mechanism

Most of the reactions described in the previous section are *electrophilic substitutions*. The mechanism involves two steps:

- the reaction between the aromatic ring and an electrophile to give an *intermediate carbenium ion*; and
- loss of the leaving group.

The first step in the reaction of benzene with an electrophile X^+ is shown in equation 15.38. The direction of movement of the electron pair shows that the *aromatic ring behaves as a nucleophile*, donating a pair of electrons to the electrophile.

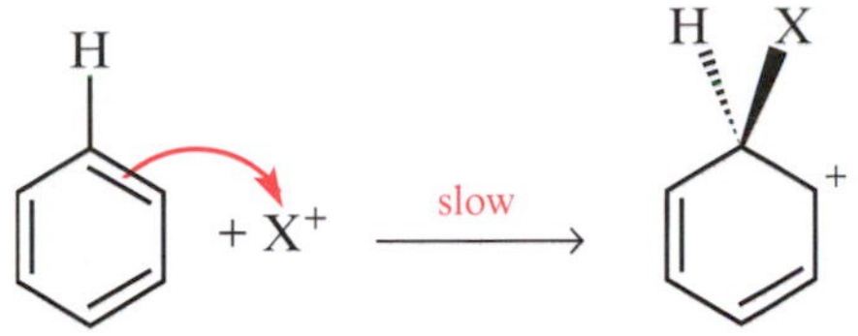

(15.38)

The carbenium ion formed by the *addition* of the electrophile to the aromatic ring is a called a *Wheland intermediate* and possesses a 4-coordinate carbon atom (Figure 15.14). The positive charge on the carbenium ion in equation 15.38 is shown as being localized on one carbon centre. This structure is only one of three contributing resonance structures which together can be represented by structure **15.42** in which the charge is delocalized (Figure 15.14). Throughout the remainder of the chapter, we shall use structure **15.42** to represent the Wheland intermediate. The C_6 ring in the Wheland intermediate contains one sp^3 and five sp^2 carbon atoms and the π-system, which was originally fully delocalized around the benzene ring, has been disrupted. The ring has lost its aromaticity because the π-system contains only five electrons and does not obey the Hückel ($4n+2$) rule.

In the second step of the electrophilic substitution reaction, a proton is lost from the intermediate by heterolytic cleavage of the C–H bond (equation 15.39) and the aromatic six π-electron system reforms. The reaction profile is shown in Figure 15.15.

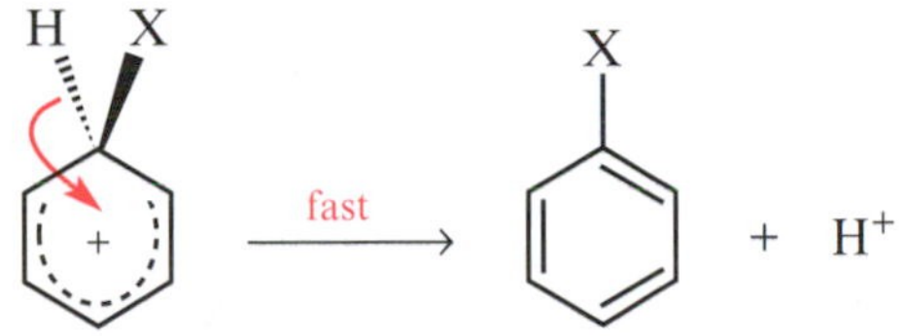

(15.39)

We now consider individual reactions and see how their pathways conform to this general mechanism.

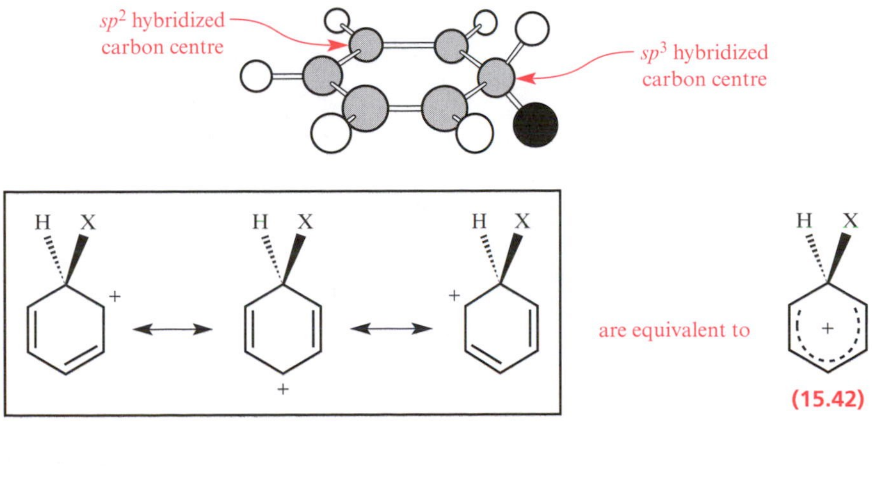

15.14 The Wheland intermediate possesses one tetrahedral carbon atom; the black atom represents the electrophile, X. The remaining part of the ring is planar. The bonding in the Wheland intermediate can be described in terms of a set of three resonance structures, which together are equivalent to a single structure with delocalized bonding.

(15.42)

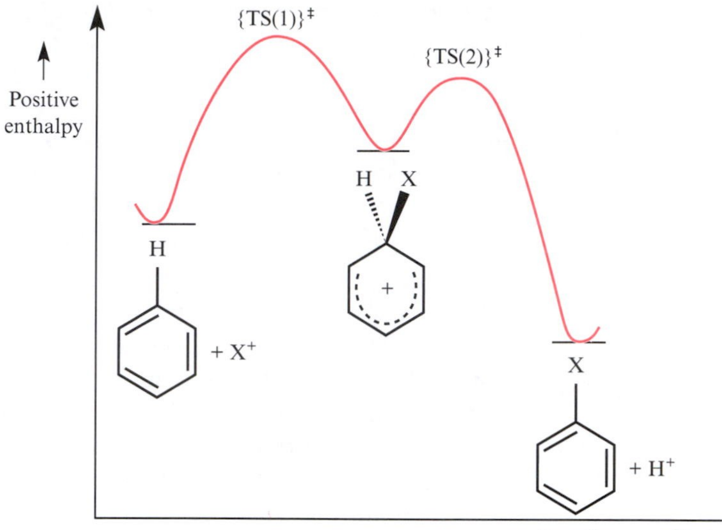

15.15 A representation of the reaction profile for the electrophilic substitution of H^+ in benzene by X^+.

Chlorination and bromination

Halogenation reactions require a catalyst to encourage the formation of the electrophile, and in the reaction between benzene and Cl_2, $AlCl_3$ is the catalyst. Aluminium chloride is a Lewis acid and accepts a chloride ion to give $[AlCl_4]^-$,

$$AlCl_3 + Cl_2 \rightarrow Cl{-}Cl^+{-}Al^-Cl_3 \qquad (15.40)$$

$$C_6H_6 + Cl{-}Cl^+{-}Al^-Cl_3 \longrightarrow [AlCl_4]^- + [C_6H_6Cl]^+ \qquad (15.41)$$

$$[C_6H_6Cl]^+ + [AlCl_4]^- \longrightarrow C_6H_5Cl + HCl + AlCl_3 \qquad (15.42)$$

catalyst is regenerated

thereby facilitating the heterolytic cleavage of the Cl–Cl bond in Cl_2, effectively releasing Cl^+ which reacts with the aromatic ring (equations 15.40–15.42). A similar reaction occurs between benzene and Br_2 in the presence of $FeBr_3$.

Friedel–Crafts alkylation and acylation

In a Friedel–Crafts alkylation reaction, the Lewis acid catalyst functions in a similar way to that in the halogenations described above. The reaction of an alkyl chloride RCl with $AlCl_3$ produces the R^+ carbenium ion which then acts as an electrophile towards benzene (equations 15.43 and 15.44).

$$RCl + AlCl_3 \rightarrow R^+ + [AlCl_4]^- \qquad (15.43)$$

H R

R

+ R^+ ⟶ ⟶ + H^+

(15.44)

Although we show the carbenium ion R^+ in equation 15.43 as a free entity, this is probably not the case, and the reaction may proceed in a similar manner to that shown in equation 15.40. The rearrangement products that we described in equation 15.34 can be understood in terms of the relative stabilities of carbenium ions. The order of stabilities:

Relative stabilities of carbenium ions: see Section 8.14

$$[R_3C]^+ > [R_2CH]^+ > [RCH_2]^+$$

means that when the alkylating reagent is a primary halogenoalkane, the favoured product does *not* contain a straight chain alkyl substituent. Rearrangement of the carbenium ion occurs to give a more stable species, and so the reaction of 1-chloropropane with benzene leads predominantly to isopropylbenzene and not propylbenzene (equations 15.34 and 15.45).

$$CH_3CH_2CH_2Cl + AlCl_3 \rightarrow CH_3CH_2\overset{+}{C}H_2 + [AlCl_4]^-$$

primary carbenium ion

↓ rearrange

$$CH_3\overset{+}{C}HCH_3 \qquad (15.45)$$

secondary carbenium ion

Another complicating factor in the alkylation of aromatic rings is that the introduction of the first alkyl substituent *activates* the ring towards further substitution, and we illustrate this problem in Section 15.9 with the preparation of toluene.

Acylation occurs in a similar manner to alkylation. The precursor is an *acyl chloride* RC(O)Cl which reacts with the Lewis base catalyst (equation

15.46) to give an electrophile which can react with benzene to produce a ketone (equation 15.47). The acyl group *deactivates* the ring and multiple substitution (see above) is not a problem.

$$R{-}C(Cl){=}O + AlCl_3 \longrightarrow R{-}C{\equiv}\overset{+}{O} + [AlCl_4]^-$$ (15.46)

$$C_6H_6 + R{-}C{\equiv}\overset{+}{O} \longrightarrow [C_6H_6(COR)]^+ \longrightarrow C_6H_5COR + H^+$$ (15.47)

Nitration

Nitration of benzene cannot be carried out using nitric acid alone, but requires a mixture of concentrated nitric and sulfuric acids. Sulfuric acid is a stronger acid than HNO_3 and an acid–base reaction occurs to generate $[NO_2]^+$, the nitryl ion (equation 15.48). This electrophile reacts with benzene to give nitrobenzene (equation 15.49).

The older name for $[NO_2]^+$ is the nitronium ion

$$2H_2SO_4 + HNO_3 \rightarrow [NO_2]^+ + [H_3O]^+ + 2[HSO_4]^-$$ (15.48)

$$C_6H_6 + [NO_2]^+ \longrightarrow [C_6H_6(NO_2)]^+ \longrightarrow C_6H_5NO_2 + H^+$$ (15.49)

15.8 ORIENTATION OF ELECTROPHILIC SUBSTITUTIONS IN MONOSUBSTITUTED AROMATIC RINGS

All the potential sites of attack in benzene are equivalent, and so when *one* electrophile X^+ substitutes for H^+ in benzene, only a single product can be obtained (equation 15.50).

$$\underset{C_6H_6}{\text{benzene}} \xrightarrow[-H^+]{X^+} \underset{C_6H_5X}{\text{PhX}}$$ (15.50)

However, if the substrate for electrophilic substitution is C_6H_5X, there are *three different sites* that a second electrophile Y^+ could attack (equation 15.51).

$$C_6H_5X \xrightarrow[-H^+]{Y^+} 2\text{-}YC_6H_4X \text{ or } 3\text{-}YC_6H_4X \text{ or } 4\text{-}YC_6H_4X \quad (15.51)$$

Ortho, meta and para positions: see Box 15.5

The position of attack of Y^+ is determined by the identity of substituent X. If the predominant product is 2-YC_6H_4X, X is said to be *ortho*-directing. If it is 3-YC_6H_4X, X is *meta*-directing, and if the major product is 4-YC_6H_4X, X is *para*-directing. Note that product selectivity is rarely exclusive. Table 15.4 lists orientation effects for some substituents and shows that an *ortho*-directing group is generally also *para*-directing. This leads to a mixture of isomers in reaction 15.51, and the ratio of *ortho* : *para* products varies with X and Y. Examples of these effects are given in the following sections where we explore some of the chemistry of toluene (X = Me), phenol (X = OH), nitrobenzene (X = NO_2) and aniline (X = NH_2).

The nature of X also determines whether the *rate of substitution* of Y^+ for H^+ in C_6H_5X is faster or slower that the rate of substitution of X^+ for H^+ in benzene. If the rate is *faster*, then X is said to *activate* the ring towards electrophilic substitution, and if it is *slower*, then the ring has been *deactivated* (Table 15.4). We examine the reasons behind the different orientation effects in the accompanying workbook.

Table 15.4 The orientation and activation effects of substituents in monosubstituted aromatic rings.

Substituent X	*Ortho-, meta- or para-directing*	*Activating or deactivating*
CH_3	*ortho-* and *para-*	Activating
C_2H_5	*ortho-* and *para-*	Activating (less than CH_3)
OH	*ortho-* and *para-*	Activating
NO_2	*meta-*	Deactivating
NH_2	*ortho-* and *para-*	Activating
SO_3H	*meta-*	Deactivating
Cl	*ortho-* and *para-*	Deactivating
Br	*ortho-* and *para-*	Deactivating

15.9 SOME DERIVATIVES OF BENZENE 1: TOLUENE

The systematic name for toluene is methylbenzene

In this and the following three sections, we look at some important derivatives of benzene. A further derivative, benzoyl chloride $C_6H_5C(O)Cl$, is included in discussions of acyl chlorides in Chapter 17. The first compound is toluene **15.43**.

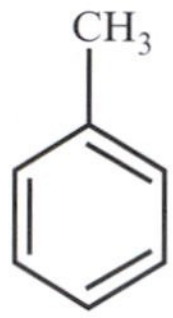

CH_3

(15.43)

Physical properties and synthesis

Toluene is a colourless liquid (mp 178 K, bp 384 K). It is less toxic than benzene and has, as far as possible, replaced benzene as a laboratory solvent. Like benzene, it is non-polar and immiscible with water.

Toluene may be prepared by a Friedel–Crafts methylation of benzene (equation 15.52) but the reaction is complicated by the fact that the methyl group activates the ring towards further attack by Me^+ at the *ortho* and *para* positions. Thus, 1,2-dimethylbenzene (commonly called 1,2-xylene) **15.44** and 1,4-dimethylbenzene (or 1,4-xylene) **15.45**, with smaller amounts of mesitylene **15.40**, are by-products. On an industrial scale, the dehydrogenation of methylcyclohexane is a convenient route (equation 15.53), and is an example of a catalytic reforming process.

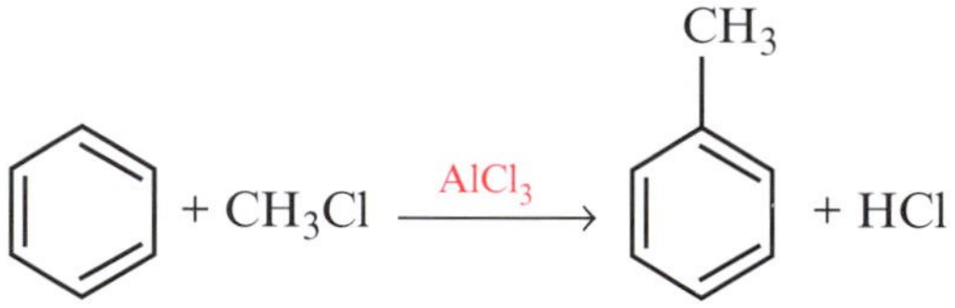

$$C_6H_6 + CH_3Cl \xrightarrow{AlCl_3} C_6H_5CH_3 + HCl \tag{15.52}$$

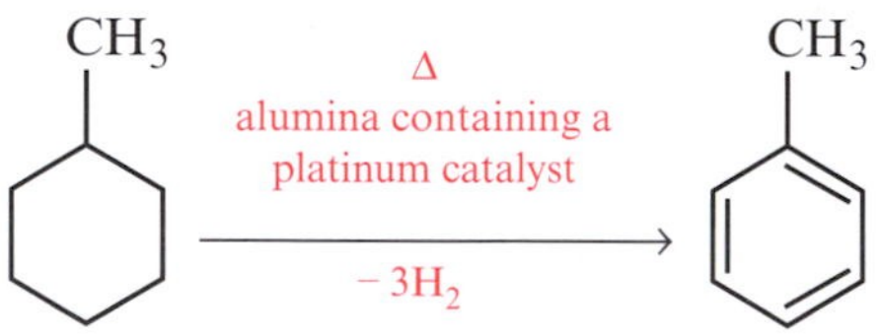

$$\text{methylcyclohexane} \xrightarrow[-\,3H_2]{\Delta,\ \text{alumina containing a platinum catalyst}} \text{toluene} \tag{15.53}$$

CH_3 CH_3 CH_3 CH_3

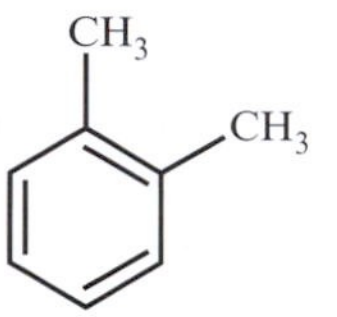

(15.44) **(15.45)**

NMR spectroscopy

Typical ^{13}C and ^{1}H NMR shift ranges: see Tables 9.11 and 9.12

The ^{13}C NMR spectrum of toluene is shown in Figure 15.16 and illustrates the distinction in chemical shifts between alkyl and aryl carbon atoms, as well as the effects of the loss of symmetry in going from benzene to a monosubstituted derivative. The spectrum has five resonances. The signal at δ +20.4 is typical of an sp^3 hybridized carbon centre, and can be assigned to the carbon atom of the methyl substituent. The remaining signals are close to that of benzene (δ +128, Figure 15.13b) but the presence of the substituent reduces the symmetry of the ring resulting in four carbon environments in a ratio 1 : 2 : 2 : 1.

In the ^{1}H NMR spectrum of toluene, a singlet resonance at δ +2.3 is assigned to the methyl protons. There is a complex pattern of multiplets around δ +7.1. Compare this with the ^{1}H NMR spectrum of benzene in Figure 15.13a.

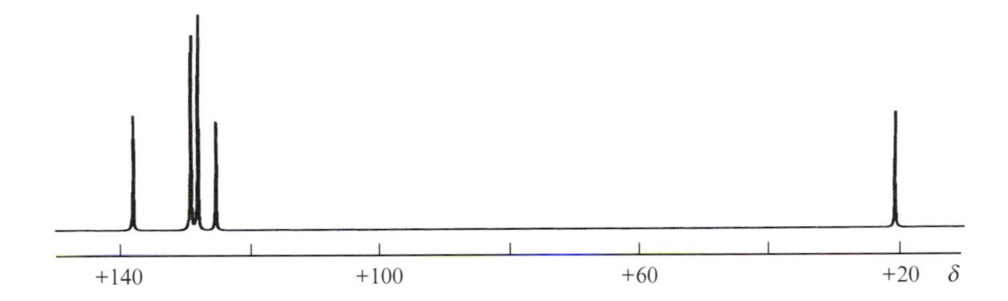

15.16 The ^{13}C NMR spectrum of toluene.

Reactions of toluene

We have already seen that the methyl group in toluene activates the aromatic ring towards electrophilic substitution at the 2-, 4- and 6-positions (structure **15.46**).

CH_3 1 2 3 4 5 6

(15.46)

Nitration proceeds as shown in equation 15.54 – note that substitution at the 6-position is identical to attack at the 2-position.

CH_3 $\xrightarrow[-H_2O]{HNO_3,\ H_2SO_4}$ CH_3, NO_2 + CH_3, NO_2

2-nitrotoluene 4-nitrotoluene (15.54)

Radical substitution of alkanes: see Sections 8.10 to 8.12

In the presence of a Lewis acid catalyst, toluene can be converted into 2-chlorotoluene, 4-chlorotoluene, 2-bromotoluene or 4-bromotoluene and reaction 15.55 illustrates bromination. In contrast to benzene, toluene possesses both aryl *and* alkyl hydrogen atoms, and halogenation of the methyl group occurs when toluene reacts with Cl_2 or Br_2 under *thermal* or *photolytic conditions* (equation 15.56).

CH$_3$ — Br_2, $FeBr_3$ / − HBr → CH$_3$, Br + CH$_3$, Br

2-bromotoluene 4-bromotoluene (15.55)

CH$_3$ — Cl_2, $h\nu$ / − HCl → CH_2Cl + $CHCl_2$ + CCl_3 (15.56)

In toluene, the aryl ring is susceptible to *electrophilic substitution* by Cl^+ and Br^+, and the alkyl group undergoes *radical substitution* by $Cl^{\bullet}$ or $Br^{\bullet}$.

Neither benzene nor alkanes are oxidized by potassium permanganate or potassium dichromate, but the methyl group in toluene can be oxidized to give a carboxylic acid. Oxidation is usually carried out using hot alkaline $KMnO_4$ (equation 15.57). A similar reaction occurs with ethylbenzene, but the product still has the CO_2H group attached *directly to the ring* – $PhCH_2CH_3$ is oxidized to $PhCO_2H$.

CH$_3$ — hot alkaline $KMnO_4$ → CO_2H

benzoic acid (15.57)

15.10 SOME DERIVATIVES OF BENZENE 2: PHENOL

Phenols are compounds in which an –OH group is attached *directly to an aromatic ring*. The archetypal molecule in this family is called phenol **15.47**.

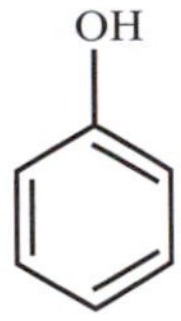

(15.47)

Physical properties and synthesis

Phenol is a colourless, hygroscopic crystalline solid at room temperature (mp 316 K) and is soluble in water and a range of organic solvents. An old name for phenol is 'carbolic acid' and it is a strong disinfectant; J J. Lister recognized this property and initiated the use of antiseptics in surgery in 1865.

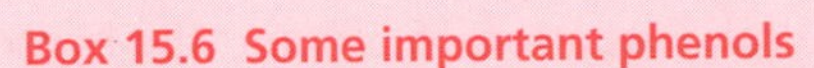

Box 15.6 Some important phenols

Many phenols are in common use in the laboratory and trivial names remain in use. Some of those you are most likely to encounter are shown below:

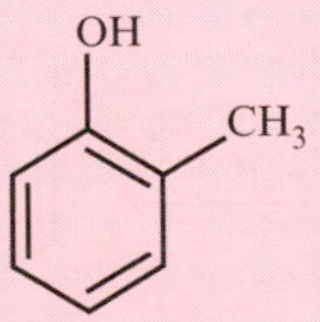

o-cresol
2-cresol
(also 3- and 4-derivatives)
(used in syntheses of azo-dyes)

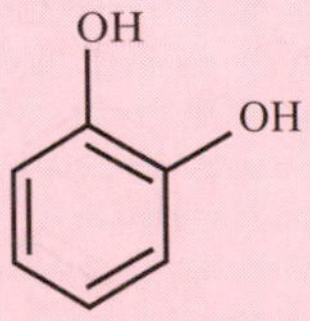

catechol
(reducing agent; used in photographic developing)

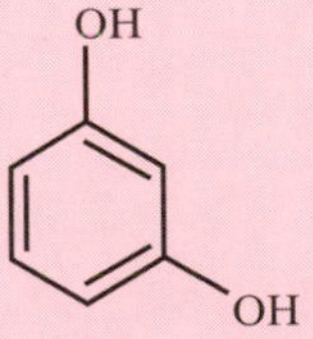

resorcinol
(used in syntheses of dyes, e.g. in fluorescein dyes, and in resins and adhesives)

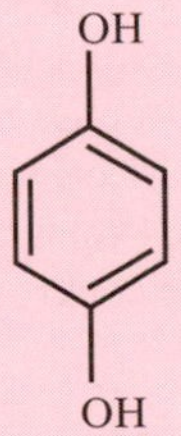

hydroquinone
(reducing agent; used in photographic development processes)

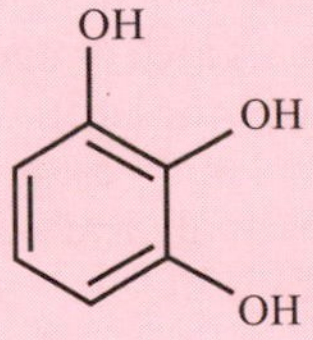

pyrogallol
(alkaline solutions are used to remove O_2 from gas mixtures)

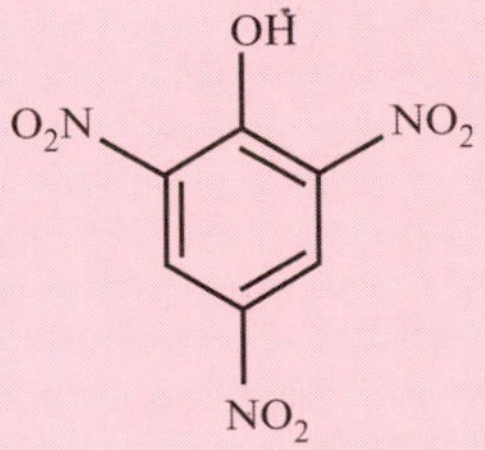

picric acid
(used as a dye, and metal salts have applications as explosives)

Phenol is used industrially on a large scale, and in 1994 it was ranked as the thirty-third manufactured chemical in the USA. Some of its applications lie in the formation of phenolic resins, bisphenol A (a precursor in the manufacture of epoxy resins), and caprolactam (used in the manufacture of Nylon 6).

Phenol cannot be conveniently made from benzene in a single step, and an industrial method uses cumene as the precursor (equation 15.58).

Preparation of cumene: see equation 15.34

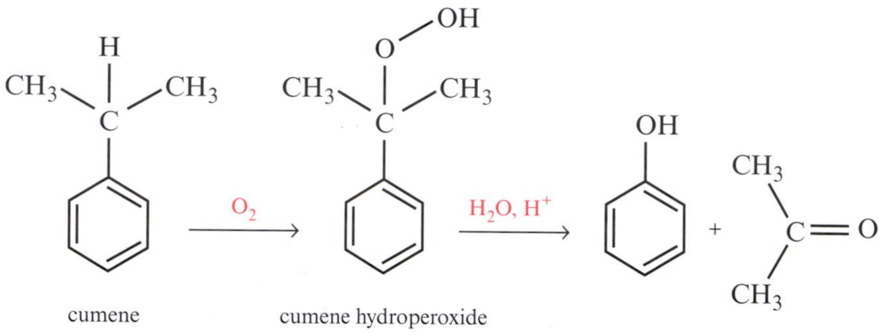

$$\text{cumene } (C_6H_5CH(CH_3)_2) \xrightarrow{O_2} \text{cumene hydroperoxide } (C_6H_5C(CH_3)_2OOH) \xrightarrow{H_2O,\ H^+} C_6H_5OH + (CH_3)_2C{=}O \qquad (15.58)$$

Two other methods of preparing phenol are the fusion of a benzenesulfonate salt with alkali (equation 15.59) and the hydrolysis of a diazonium salt (see Section 15.11).

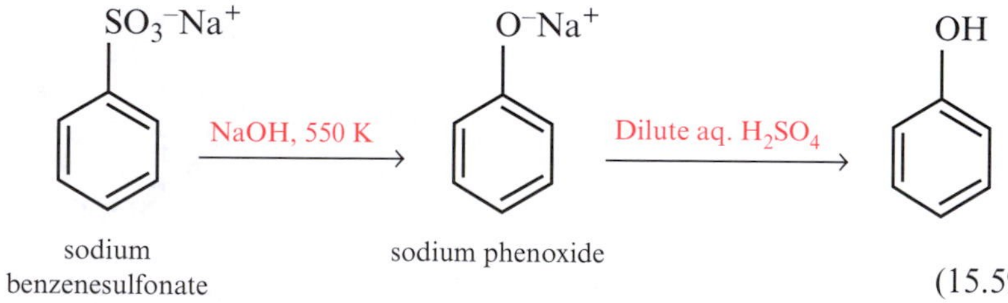

$$\text{sodium benzenesulfonate } (C_6H_5SO_3^-Na^+) \xrightarrow{NaOH,\ 550\ K} \text{sodium phenoxide } (C_6H_5O^-Na^+) \xrightarrow{\text{Dilute aq. } H_2SO_4} C_6H_5OH \qquad (15.59)$$

Box 15.7 Aromatics attract cockroaches

Cockroaches are encountered as pests in households world wide and chemical baits are commonly used to attract the insects to traps or poisons. Phenol and naphthol attract male cockroaches but only those of one particular species, making these aromatics too selective to be useful! However, derivatives of the partially hydrogenated 1,2,3,4-tetrahydronaphthalene have been found to be irresistible to male and female cockroaches of all common species.

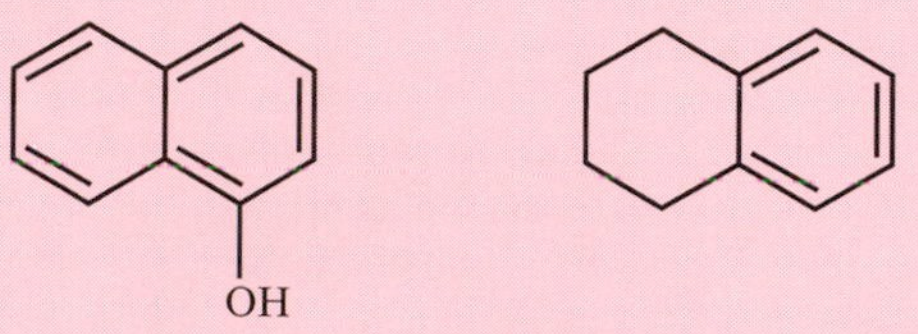

1-naphthol (OH)

1,2,3,4-tetrahydronaphthalene

Acidity

Carboxylic acids: see Chapter 17

Phenol is a weak acid ($pK_a = 9.89$). In aqueous solution, other phenols behave similarly and this property makes them stand apart from aliphatic alcohols (e.g. ethanol has a pK_a value of ≈16). However, phenols are typically *weaker acids* than carboxylic acids. What is so special about the direct attachment of an OH group to an aromatic ring?

In equilibrium 15.60, the factor that determines the acidity of phenol is the stability of the *phenoxide ion*. The negative charge on the anion may be spread out over the aromatic ring and this can be represented by a series of resonance structures (Figure 15.17a). Remember that, in general, the more resonance structures you can draw for a system, the more stable it is likely to be. Alternatively, the bonding can be described in terms of overlap of an oxygen $2p$ atomic orbital and the carbon $2p$ orbitals that comprise the π-system of the C_6 ring (Figure 15.17b).

Resonance stabilization: see Sections 5.17 and 8.15

$$\text{phenol (C}_6\text{H}_5\text{OH)} + H_2O \rightleftharpoons \text{phenoxide ion (C}_6\text{H}_5\text{O}^-) + [H_3O]^+ \qquad (15.60)$$

Reactions of phenol

Phenol reacts with alkali metal hydroxides to give phenoxide salts (equation 15.61) but is *not* a strong enough acid to neutralize sodium hydrogencarbonate. This observation can be used to distinguish phenol from a carboxylic acid which *does* react with $NaHCO_3$.

(a) O⁻ ⟷ O ⟷ O ⟷ O

(b)

15.17 (a) Resonance structures for the phenoxide ion showing the localization of the negative charge at different sites, (b) The delocalized π-system of the aryl ring can be extended to include a $2p$ AO of oxygen, thereby allowing the negative charge in the phenoxide ion to be spread out over C_6O framework. This helps to stabilize the phenoxide ion.

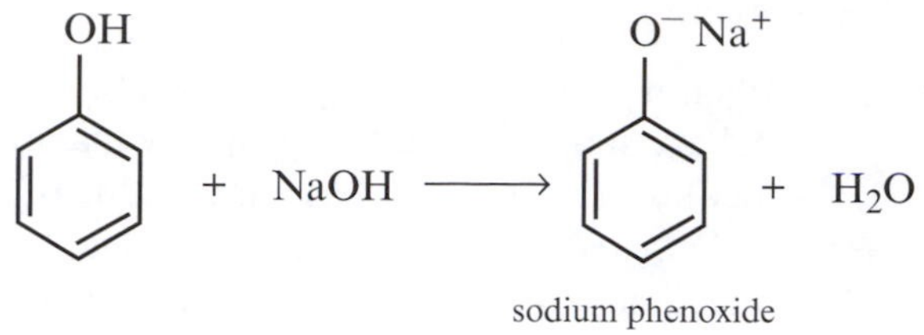

$$C_6H_5OH + NaOH \longrightarrow C_6H_5O^-Na^+ + H_2O$$

sodium phenoxide (15.61)

Table 15.4 lists the –OH substituent as being *ortho-/para*-directing and capable of activating the phenyl ring towards further electrophilic substitution. The OH group is *strongly activating*, and the conditions under which phenol undergoes electrophilic substitutions are less harsh than those required for the corresponding reactions of benzene. Nitration of phenol can be carried out using HNO_3 alone (equation 15.62). Since the NO_2 group is *deactivating*, the reaction may be halted after the introduction of one nitro-substituent. However, activation of the ring by the OH group means that it is possible to introduce *more than one* NO_2 group and nitration of phenol can lead to the formation of 2,4,6-trinitrophenol (picric acid).

Picric acid: see Box 14.4

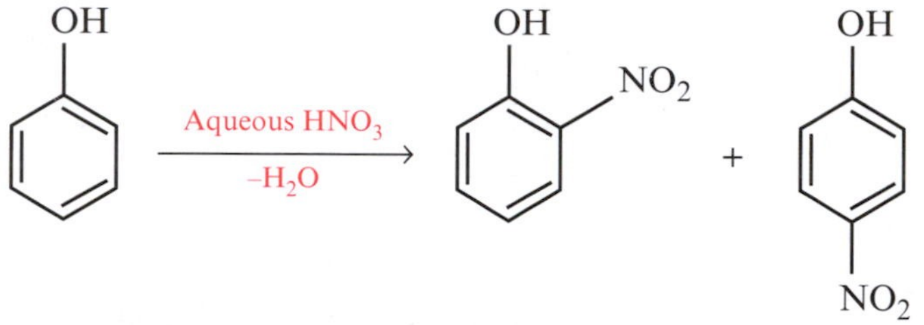

$$C_6H_5OH \xrightarrow[-H_2O]{\text{Aqueous } HNO_3} \textit{o-}O_2NC_6H_4OH + \textit{p-}O_2NC_6H_4OH \qquad (15.62)$$

Sulfonation requires only sulfuric acid (equation 15.63), and bromination or chlorination is effected using aqueous dibromine or dichlorine; compare these conditions with those needed to form the analogous derivatives of benzene. The halogenations are rapid, and multi-substitution is common (equation 15.64) – 2,4,6-trichlorophenol is a disinfectant, marketed commercially in a mixture with other compounds as TCP.

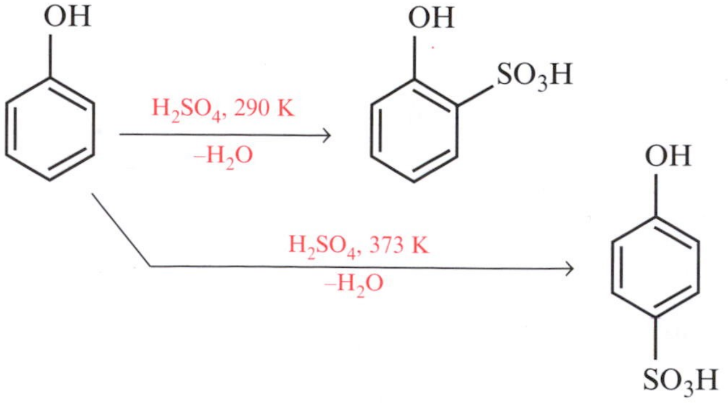

$$C_6H_5OH \xrightarrow[-H_2O]{H_2SO_4,\ 290\ K} \textit{o-}HO_3SC_6H_4OH$$

$$C_6H_5OH \xrightarrow[-H_2O]{H_2SO_4,\ 373\ K} \textit{p-}HO_3SC_6H_4OH \qquad (15.63)$$

Box 15.8 Timber treatment: environmental side-effects

2,3,4,5,6-Pentachlorophenol (PCP) is a fungicide used to treat wood, but there are environmental side-effects. Field research has shown that pipistrelle bats roosting in wooden boxes treated with PCP died within a short period after exposure to the chemical in the concentrations used in the study. Bats often roost in the roof spaces of older houses and barns, and may be in close proximity of timbers treated with PCP. Not all chemicals used as wood preservatives are as toxic with respect to bats and other mammals, but we must be aware of and assess the environmental effects that such treatments may have.

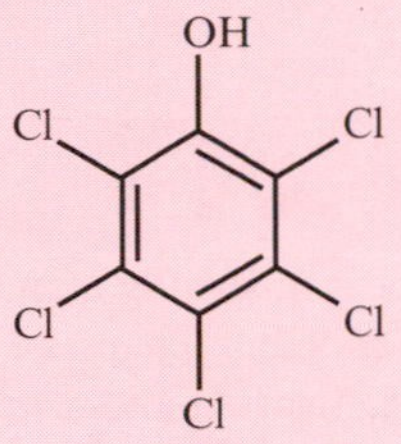

OH
Cl Cl
Cl Cl
Cl
PCP

PCP is not the only chloro-substituted aromatic compound to have environmental side effects; further examples include DDT (dichlorodiphenyltrichloroethane) and the family of PCB (polychlorinated biphenyl) derivatives.

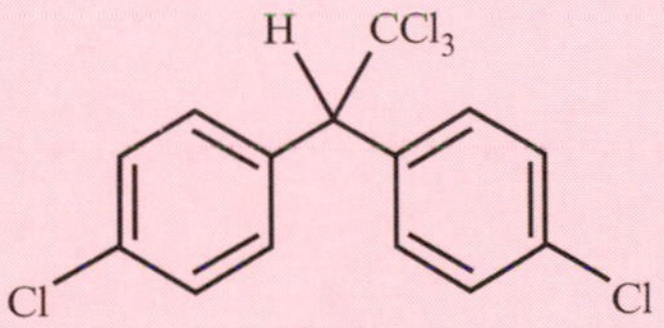

H CCl_3
Cl Cl
DDT

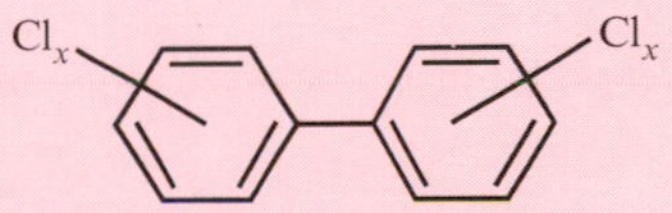

Cl_x Cl_x
PCB

DDT has been used as a pesticide for over 50 years, while PCBs have applications in, for example, paints and adhesives. Today, legislative measures have been taken to control the use of DDT and to restrict the quantities being discharged into water-courses.

An article dealing with the environmental problems of these and related chemicals and methods of disposing of them has been written by M L. Hitchman *et al*, *Chemical Society Reviews,* (1995) vol. 24, p. 423.

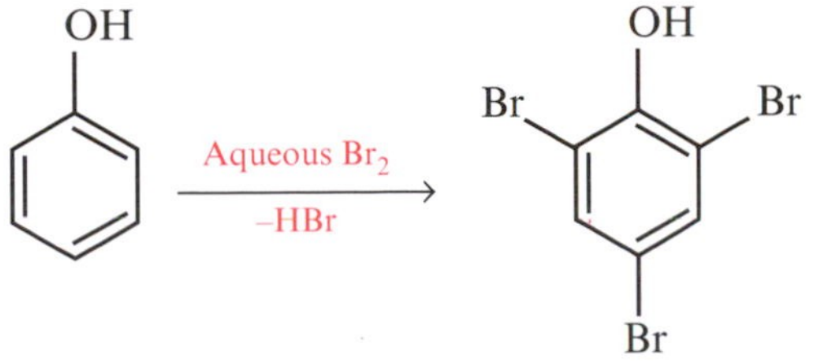

OH → (Aqueous Br_2, $-HBr$) → OH, Br, Br, Br

(15.64)

Williamson synthesis: see Section 14.4

The ready formation of the phenoxide ion is utilized in the Williamson synthesis to form ethers containing the PhO group. Equation 15.65 shows the stepwise reaction of phenol with iodoethane in the presence of alkali; sodium hydroxide reacts with phenol to generate the phenoxide ion.

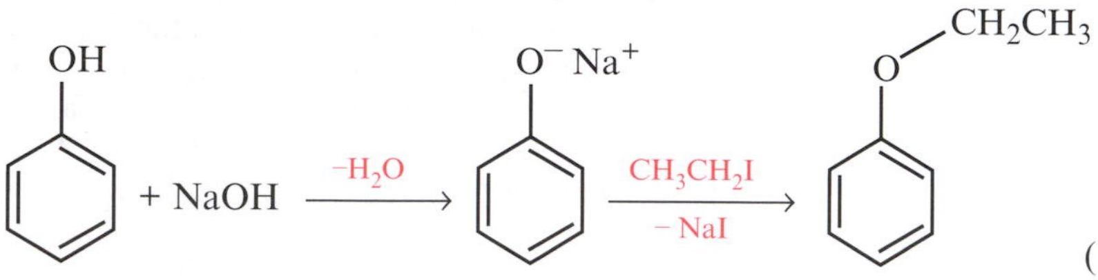

$$\text{PhOH} + \text{NaOH} \xrightarrow{-H_2O} \text{PhO}^- \text{Na}^+ \xrightarrow[-\text{NaI}]{CH_3CH_2I} \text{PhOCH}_2\text{CH}_3 \qquad (15.65)$$

Box 15.9 Thermosetting polymers

One of the early polymers to be industrially manufactured was Bakelite, prepared from phenol and methanal (reactions of aldehydes are described in Chapter 17).

The –OH group is strongly activating and *ortho/para*-directing, and in the first step, electrophilic substitution of the hydrogen atom at the 2-position of the phenyl ring in PhOH occurs:

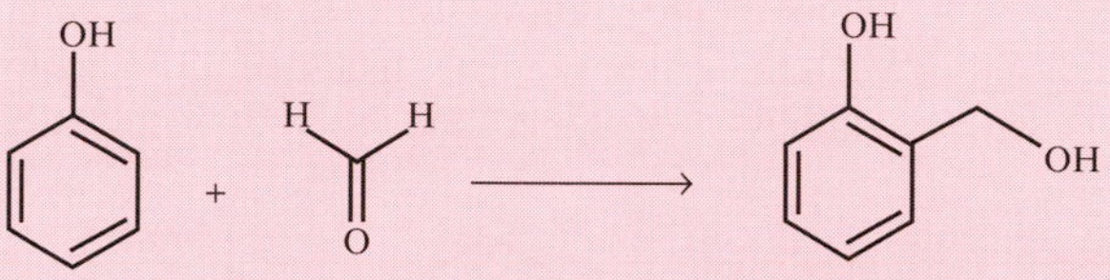

$$\text{PhOH} + \text{HCHO} \longrightarrow 2\text{-HOCH}_2C_6H_4OH$$

Water is eliminated between adjacent monomer units in a *condensation reaction* to give a polymer:

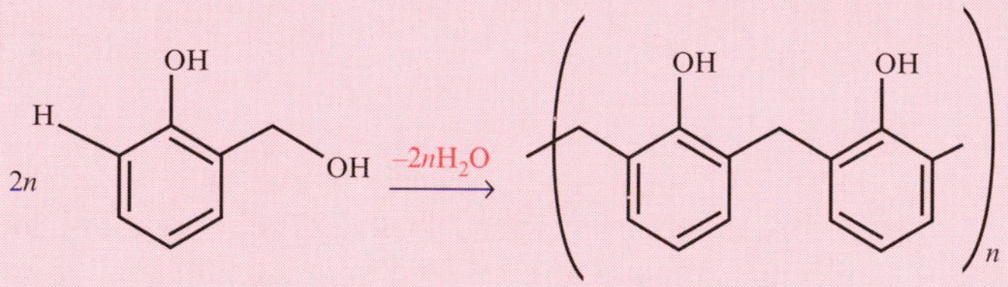

$$2n \; 2\text{-HOCH}_2C_6H_4OH \xrightarrow{-2nH_2O} [\text{polymer}]_n$$

This polymer is a chain, but a *cross-linked* polymer can be obtained by controlling the stoichiometry of the reaction between phenol and methanal:

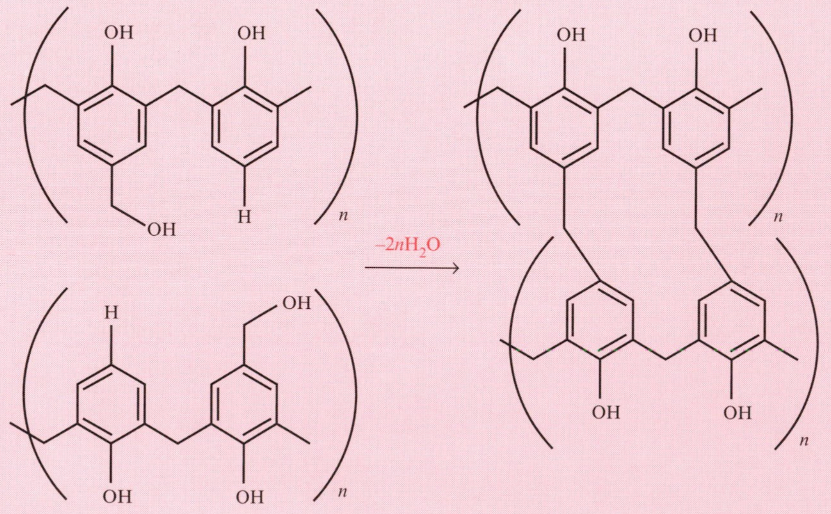

$$\xrightarrow{-2nH_2O}$$

Cross-linking provides the polymer with added structural rigidity, making Bakelite a hard material that does not soften on heating.

15.11 SOME DERIVATIVES OF BENZENE 3: NITROBENZENE, ANILINE AND DIAZONIUM SALTS

Physical properties, syntheses and uses

Nitrobenzene is a colourless liquid with a relatively long liquid range (mp 279 K, bp 484 K). It possesses a characteristic smell (rather like marzipan) and is extremely toxic, although it has been used in the cosmetics industry. It is manufactured from benzene (equation 15.49), is an industrial precursor to aniline, and has widespread applications in the dye-stuffs industry (see below). The reduction of nitrobenzene to aniline can be carried out by several methods, for example by reaction 15.66. In smaller-scale preparations, the reducing agent is often granulated tin in hydrochloric acid, and the acidic conditions mean that the initial product is an anilinium salt which can be neutralized by base to give aniline (equation 15.67).

$$C_6H_5NO_2 \xrightarrow[\text{or Fe, aqueous HCl}]{H_2\text{, copper catalyst}} C_6H_5NH_2 \qquad (15.66)$$

$$C_6H_5NO_2 \xrightarrow{\text{Sn, aqueous HCl}} \underset{\text{anilinium chloride}}{C_6H_5NH_3^+Cl^-} \xrightarrow[-NaCl,\ -H_2O]{\text{aqueous NaOH}} C_6H_5NH_2 \qquad (15.67)$$

Aniline is a colourless, oily and toxic liquid (bp 457 K) but often appears brownish owing to the presence of oxidation products. It is slightly soluble in water; at 298 K, 100 cm^3 of water dissolves 3.7 g of aniline. Its uses include those in the dyes and pharmaceutical industries.

Basicity of aniline

pK_b and pK_a values: see Section 11.9

Aniline is an *aromatic amine* and behaves as base (equations 15.68 and 15.69) *but only a very weak one* – compare the pK_b values for aniline and some of its derivatives (Table 15.5) with those for selected aliphatic amines (Table 14.6).

$$C_6H_5NH_2(aq) + H_2O(l) \rightleftharpoons \underset{\text{anilinium ion}}{C_6H_5NH_3^+}(aq) + [OH]^-(aq) \qquad pK_b = 9.37 \qquad (15.68)$$

Table 15.5 Values of pK_b for aniline and selected derivatives. For comparison, the value of pK_b for NH_3 is 4.75. The relationship between pK_b and pK_a is discussed in Section 11.9.

Amine	*Structure*	pK_b
Aniline	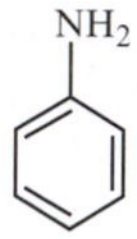 NH_2 (benzene ring)	9.37
N-Methylaniline[a]	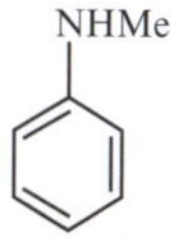 NHMe (benzene ring)	9.15
N,N-Dimethylaniline[a]	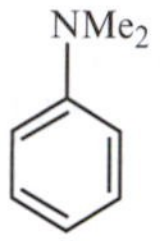 NMe_2 (benzene ring)	8.85
2-Chloroaniline	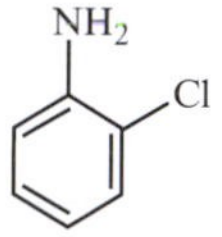 NH_2, Cl (benzene ring)	11.35
4-Nitroaniline	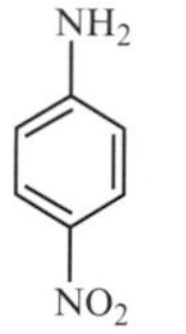 NH_2, NO_2 (benzene ring)	13.00

[a] The prefix *N*- denotes that the methyl group is attached to the nitrogen atom and not the aromatic ring.

$$2PhNH_2(aq) + H_2SO_4(aq) \rightarrow [PhNH_3]_2[SO_4](aq) \quad (15.69)$$

The weak basicity of aniline can be understood in terms of resonance stabilization. Whereas a series of resonance structures can be drawn for $PhNH_2$ (Figure 15.18), the same is not true of the anilinium ion, and so equilibrium 15.68 lies to the left-hand side. Table 15.5 includes chloro and nitro derivatives of aniline to illustrate the effects that substituents can have on protonation of the amine group. Both Cl and NO_2 are electron-withdrawing and *destabilize* the positively charged nitrogen centre in the corresponding anilinium ions, making 2-chloroaniline and 4-nitroaniline *weaker* bases than aniline. Conversely, substituents such as alkyl groups which are electron-releasing *stabilize* the corresponding anilinium ions and make these substituted anilines *more basic* than $PhNH_2$. This effect is also observed in the *N*-methyl derivatives listed in Table 15.5.

The prefix *N*- before a substituent descriptor in the name of an amine or other nitrogen-containing compound signifies that the substituent is attached to the nitrogen atom.

15.18 Resonance structures for aniline.

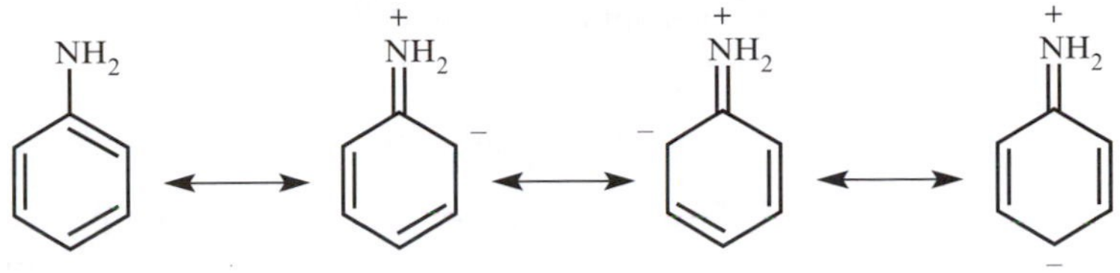

NH_2 ⟷ $\overset{+}{N}H_2$ ⟷ $\overset{+}{N}H_2$ ⟷ $\overset{+}{N}H_2$

Reactions

The conversion of nitrobenzene to aniline (equation 15.66) is one possible reduction reaction, but when nitrobenzene reacts with zinc and water, *phenyl hydroxylamine* is produced (equation 15.70). This is also a reduction reaction.

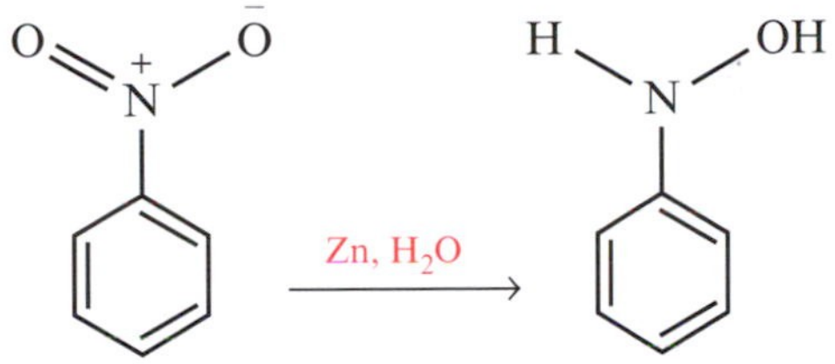

$$C_6H_5NO_2 \xrightarrow{Zn,\ H_2O} C_6H_5N(H)OH \tag{15.70}$$

Aniline reacts with alkyl halides to give *N*-alkyl derivatives, and equation 15.71 illustrates the conversion of aniline to *N*,*N*-dimethylaniline.

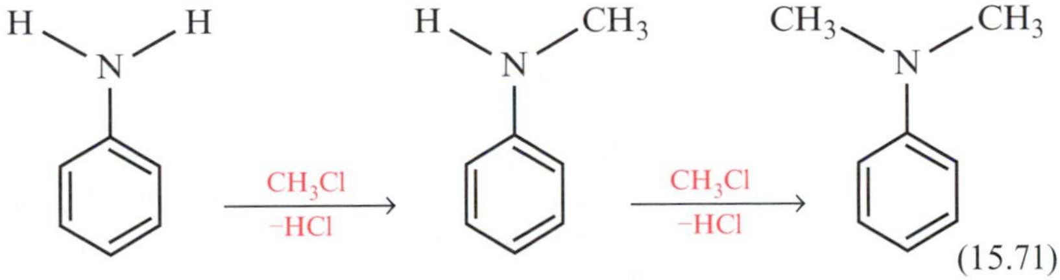

$$C_6H_5NH_2 \xrightarrow[-HCl]{CH_3Cl} C_6H_5N(H)CH_3 \xrightarrow[-HCl]{CH_3Cl} C_6H_5N(CH_3)_2 \tag{15.71}$$

One of the most important reactions of anilines is their conversion to *diazonium salts*. A diazonium cation has the general structure **15.48** and the central nitrogen atom is in a linear environment. Diazonium salts are made by treating aniline with nitrous acid, prepared *in situ* by the action of an acid such as HCl on sodium nitrite (equation 15.72).

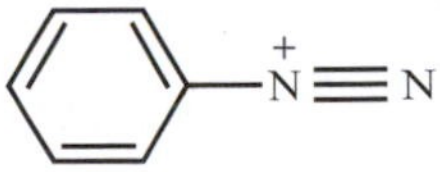

Nitrous acid: see Oxyacids of nitrogen in Section 13.8

$$C_6H_5-\overset{+}{N}\equiv N$$

(15.48)

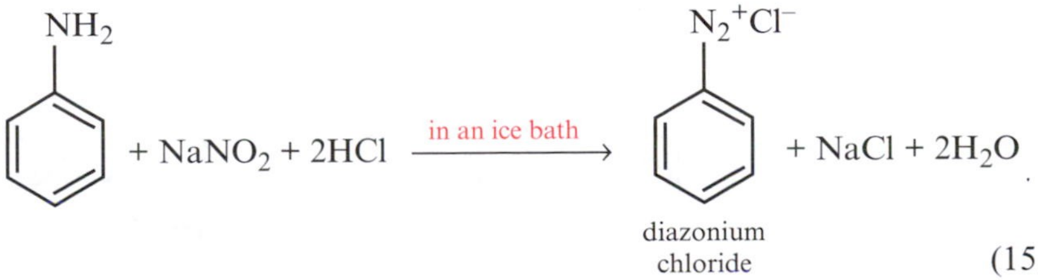

$$C_6H_5NH_2 + NaNO_2 + 2HCl \xrightarrow{\text{in an ice bath}} C_6H_5N_2^+Cl^- + NaCl + 2H_2O \tag{15.72}$$

diazonium chloride

Diazonium salts undergo two general types of reaction in which:

- N_2 is eliminated; or
- an –N=N– unit (an *azo* group) is formed.

The first group of reactions is exemplified by hydrolysis (equation 15.73). This occurs slowly in the aqueous medium of reaction 15.72, but the designed hydrolysis of diazonium salts is a convenient route to phenols.

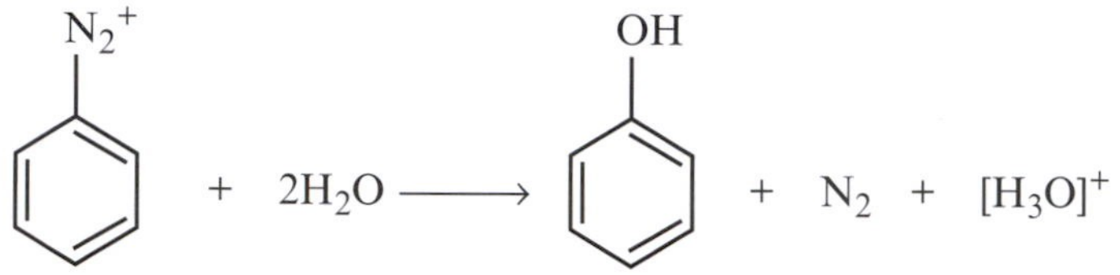

$$PhN_2^+ + 2H_2O \longrightarrow PhOH + N_2 + [H_3O]^+ \tag{15.73}$$

The conversion of $[PhN_2]^+$ to chloro- or bromobenzene in the presence of a copper(I) halide catalyst is called the *Sandmeyer reaction* and is a historically important method of preparing halobenzenes (equation 15.74).

In the *Sandmeyer reaction*, a diazonium ion may be converted to a chloro- or bromobenzene.

$$PhN_2^+X^- \xrightarrow[X = Cl \text{ or } Br]{CuX \text{ catalyst}} PhX + N_2 \quad \textit{Sandmeyer reaction}$$

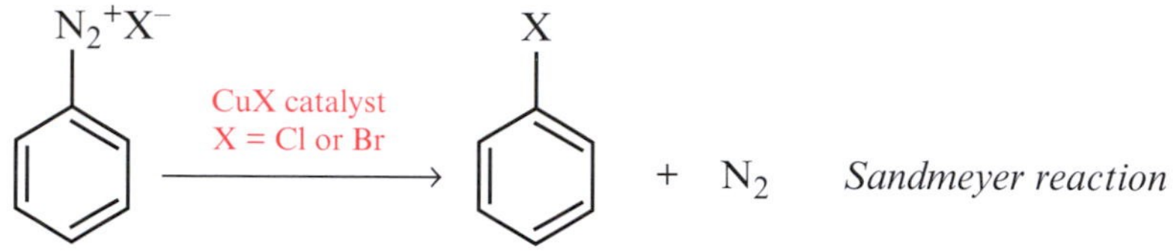

X = Cl or Br (15.74)

The coupling of a diazonium ion with phenols and amines takes place *without* the elimination of N_2 and gives an *azo*-compound of the general type **15.49** – the –N=N– unit is an *azo*-group. The need for a phenol or amine precursor arises from the requirement that the compound coupling with the diazonium ion must possess very electron-releasing substituents.

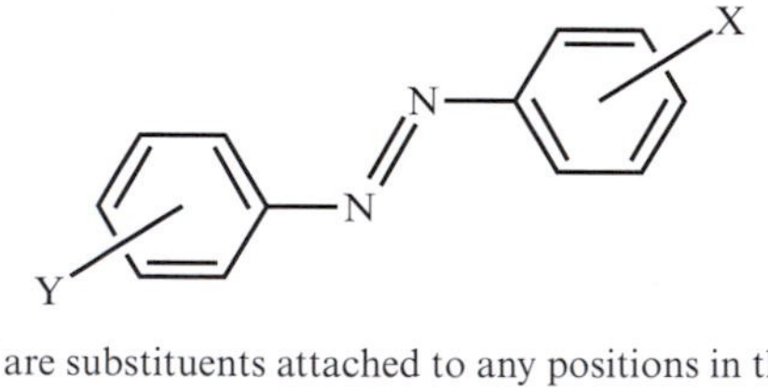

$$Y\text{-}C_6H_4\text{-}N{=}N\text{-}C_6H_4\text{-}X$$

X and Y are substituents attached to any positions in the rings

(15.49)

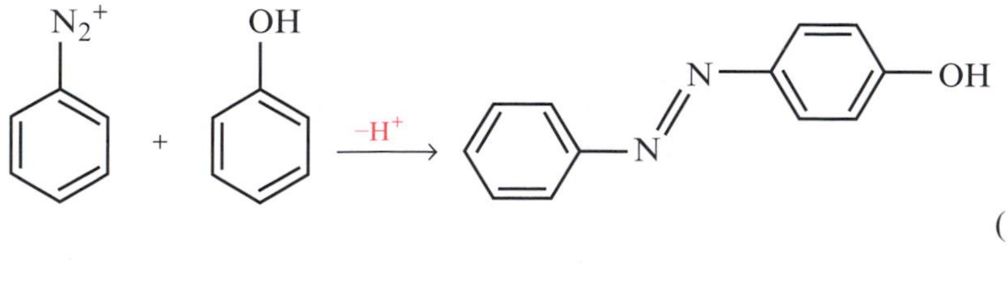

$$PhN_2^+ + PhOH \xrightarrow{-H^+} Ph\text{-}N{=}N\text{-}C_6H_4\text{-}OH \tag{15.75}$$

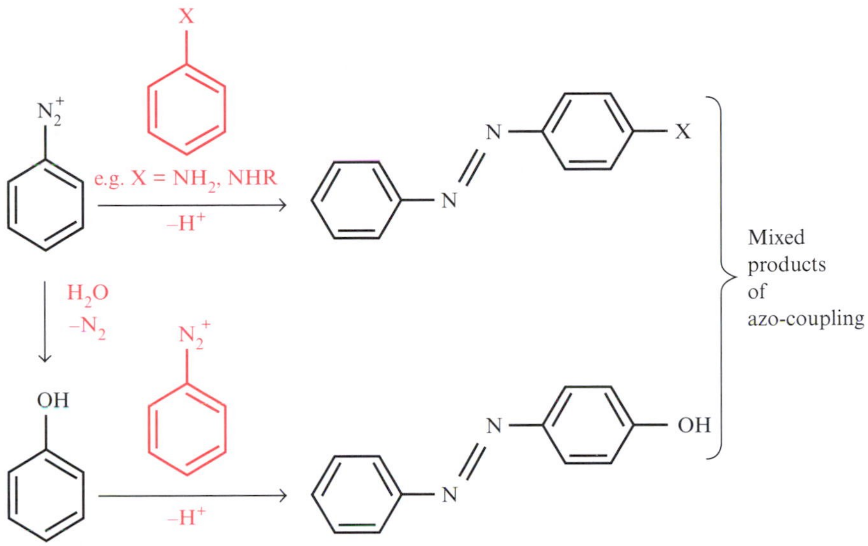

15.19 When an amine reacts with a diazonium ion in aqueous solution, competitive *azo*-coupling reactions occur, the second pathway arising from the hydrolysis of $[PhN_2]^+$.

Equation 15.75 shows the reaction of phenol with $[PhN_2]^+$ – notice that the C–N bond formation usually takes place at the carbon atom at the 4-position, opposite to the electron-releasing substituent. If the 4-position is blocked by another substituent, C–N bond formation occurs at the 2-position. The reaction can be carried out in aqueous solution but hydrolysis of the diazonium cation (equation 15.73) competes with the azo-coupling. In the reaction shown, the hydrolysis of $[PhN_2]^+$ to PhOH would only reduce the yield of the desired product, but if the substrate were an amine, then as Figure 15.19 illustrates, hydrolysis of the diazonium ion produces phenol which can then compete with the amine in *azo*-coupling reactions. The problem can be partially solved by adjusting the pH of the reaction solution – amines undergo azo-coupling relatively rapidly in moderately acidic solution, whereas the coupling of phenols is favoured under moderately alkaline conditions.

The –N=N– group is a chromophore and some azo-compounds absorb light in the visible region making them coloured and of commercial use as biological stains and dyes (Figure 9.27).

–N=N– chromophore: see Section 9.11

15.12 PYRIDINE: AN AROMATIC HETEROCYCLIC COMPOUND

In this section we consider the *aromatic heterocyclic* compound *pyridine* **15.50** which is related to benzene by the replacement of one CH group by a nitrogen atom. A common abbreviation for pyridine is py. Ring numbering begins at the nitrogen atom as shown in structure **15.50**.

CH and N are isoelectronic

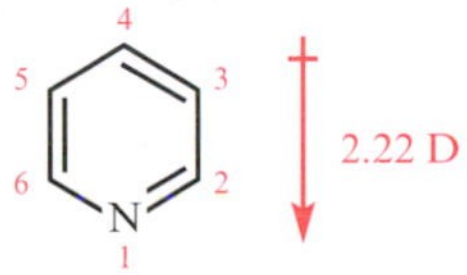

(15.50)

Physical properties and uses

Pyridine is a colourless liquid at room temperature (bp 388 K) with a characteristic odour; it is toxic, and prolonged exposure may cause sterility in men. It is a polar molecule and the dipole moment given in structure **15.50** is for the gas phase species. Pyridine has a similar density to water (0.98 g cm^{-3}) and is miscible with water and with a range of organic solvents including alcohols, ethers and benzene.

Pyridine is used widely as a solvent and has application in the plastics industry and in the manufacture of some pharmaceuticals. A derivative is nicotinic acid **15.51**, which is a vitamin and an essential component in the diet of mammals. The name derives from the *alkaloid* nicotine **15.52** which may be oxidized to nicotinic acid.

(15.51) (15.52)

Structure and bonding

Pyridine is planar (Figure 15.20a) with bond lengths and angles similar to those in benzene (Figure 15.9); the lowering in symmetry in going from benzene to pyridine means that the bond lengths in the ring are not equal. The bonding description that we gave for benzene is equally valid for pyridine –

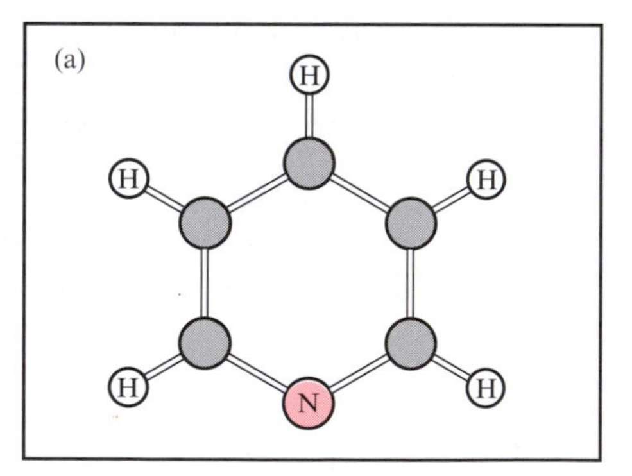

(b)

Nitrogen 2*p* AO

Nitrogen sp^2 hybrid orbital

15.20 (a) The planar structure of pyridine. (b) Pyridine possesses a delocalized π-system, and the lone pair on the nitrogen atom may be considered to be accommodated in an outward-pointing sp^2 hybrid orbital.

Box 15.10 Alkaloids

The name *alkaloid* means 'alkali-like' and is a term used to encompass a wide range of naturally occurring organic bases. Most alkaloids are optically active, and are produced naturally as one particular enantiomer. Many alkaloids are toxic, but at the same time are the active ingredients in a large number of drugs. The pyridine ring is a structural unit in many alkaloids as are quinoline, isoquinoline and pyrimidine, all of which are related to pyridine.

quinoline 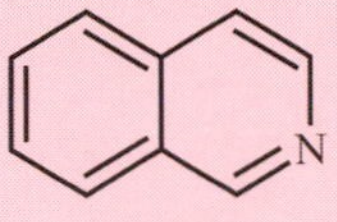   isoquinoline 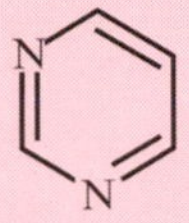   pyrimidine

Alkaloid drugs, many of which are addictive, include nicotine (present in tobacco), morphine (from opium, highly addictive), caffeine (a stimulant in tea and coffee, and an additive in cola drinks), quinine (from cinchoa bark, used in the treatment of malaria), cocaine (stimulates the nervous system and is found in coca) and strychnine (stimulates the nervous system, causing convulsions when taken in large doses).

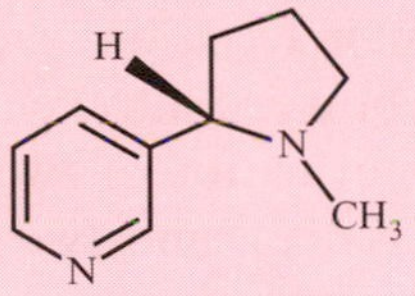

nicotine

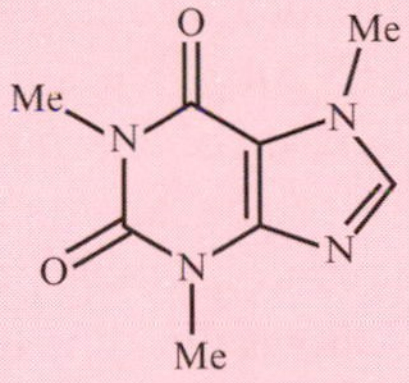

caffeine

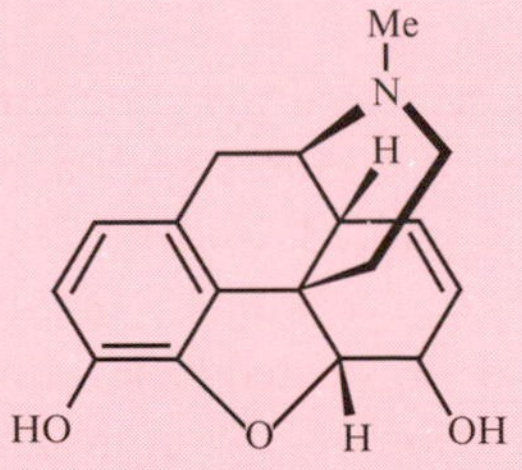

morphine

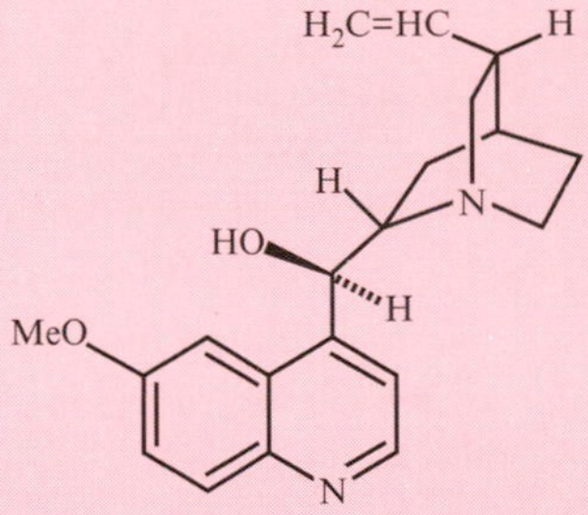

quinine

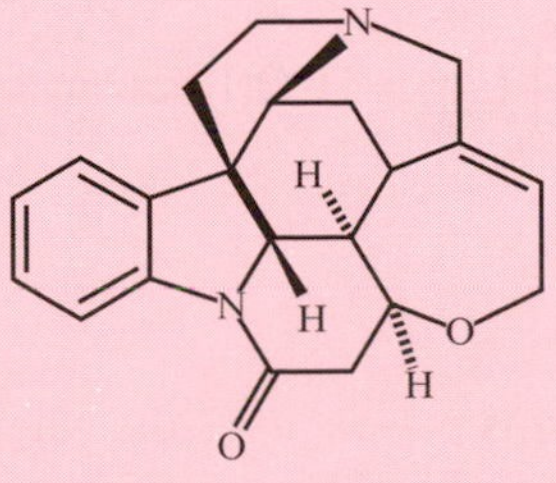

strychnine

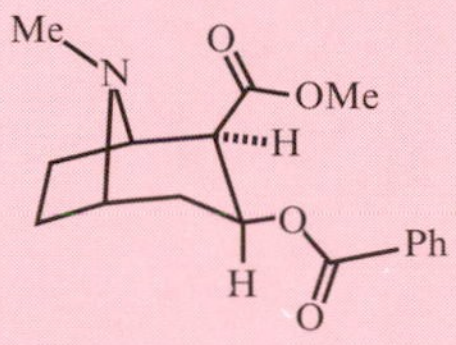

cocaine

Box 15.11 4-Aminopyridine: a drug in the treatment of multiple sclerosis

Multiple sclerosis (MS) is a debilitating disease that affects the nervous system. The search for drugs to treat MS has led some researchers to investigate 4-aminopyridine. In a patient suffering from MS, the nerve fibres and potassium channels are exposed and this impairs the workings of the nervous system. It has been found that 4-aminopyridine functions as a potassium channel blocker and by taking this drug, MS patients regain some strength.

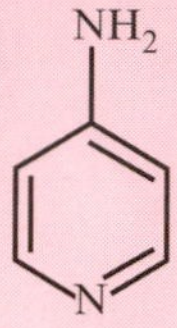

NH_2

N

4-aminopyridine

two Kekulé-like resonance structures may be drawn, or we can represent the delocalized π-system in terms of six overlapping $2p$ atomic orbitals (Figure 15.20b). The sp^2 hybridized nitrogen atom has five valence electrons – two are involved in the formation of two N–C σ-bonds, one is used in the 6π-electron aromatic system, and two are accommodated in an sp^2 hybrid orbital which points *outside* and *in-the-plane* of the ring. The lone pair allows pyridine to function as a base.

Reactivity

We have already seen that pyridine can behave as a Lewis base, for example in the formation of adducts such as $py{\bullet}BX_3$ (X = F, Cl or Br) **13.11**, and *trans*-$[Ge(py)_2Cl_4]$ **13.23**, and in Chapter 16 we show how pyridine and related molecules can coordinate to *d*-block metal centres.

pK_b and pK_a values: see Section 11.9

Pyridine is also a Brønsted base – it is, however, *not an amine*. The value of pK_b is 8.75 (equation 15.76), compared with values of 4.75 for NH_3, and 9.37 for aniline.

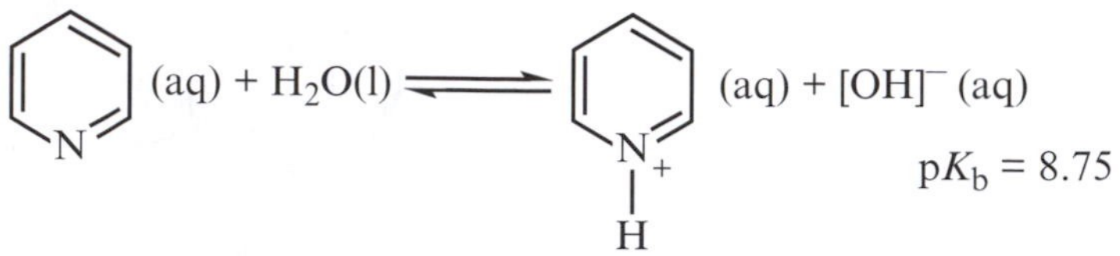

$$C_5H_5N(aq) + H_2O(l) \rightleftharpoons [C_5H_5NH]^+(aq) + [OH]^-(aq) \qquad pK_b = 8.75 \qquad (15.76)$$

Pyridine forms *pyridinium salts* when it reacts with acids, and the formation of pyridinium chloride is shown in equation 15.77. The salt may be abbreviated to [pyH]Cl.

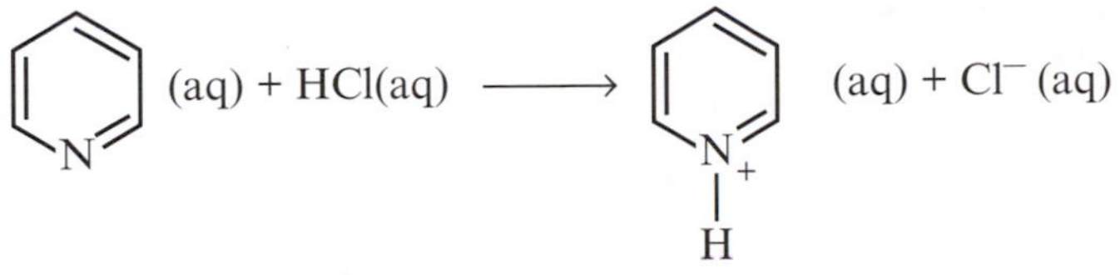

N (aq) + HCl(aq) ⟶ N+ H (aq) + Cl^- (aq)

pyridinium chloride (15.77)

Quaternary pyridinium salts containing cations of the general type **15.53** can be formed by treating pyridine with the appropriate alkyl iodide, and equation 15.78 shows the formation of *N*-methylpyridinium iodide. The use of the prefix *N*- distinguishes the *N*-methylpyridinium cation from methylpyridinium ions in which the methyl group is attached to the ring. [*Question*: What is the structure of the 3-methylpyridinium ion?]

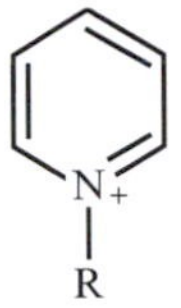

N+ R

R = alkyl

(15.53)

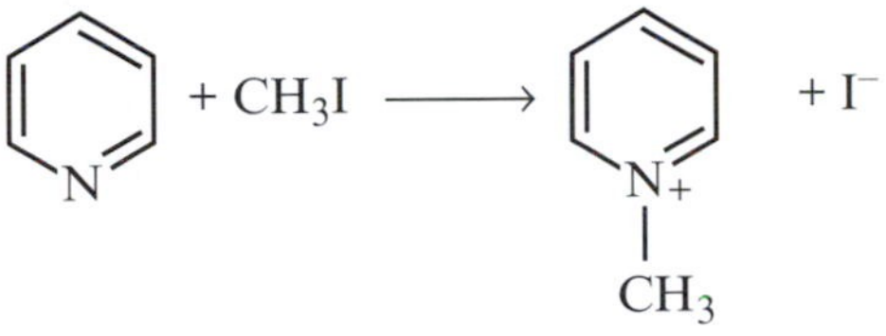

N + CH_3I ⟶ N+ CH_3 + I^-

N-methylpyridinium iodide (15.78)

➤ *N*-Butylpyridinium chloride: see structure 13.14

Compared with benzene, pyridine is extremely deactivated towards electrophilic substitution reactions and does not undergo, for example, normal Friedel–Crafts reactions.

Box 15.12 Pheromones 2: ant trails

As we saw in Box 14.5, pheromones are chemical messengers. When walking through woodland, the sight of a trail of ants is common. When worker ants discover food, they secrete pheromones to mark a trail allowing ants from the same nest to find the food supply. Some of the ant pheromones are heterocyclic compounds and include those shown below.

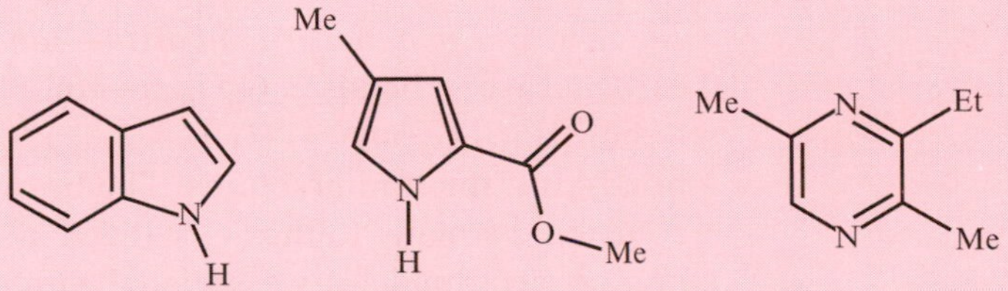

N H　　Me N H O O Me　　Me N Et N Me

[See also Box 14.5 Pheromones 1: A female moth attracts her mate]

15.13 BORAZINE: A COMPOUND WHICH IS ISOELECTRONIC WITH BENZENE

Physical properties, structure and bonding

Borazine $(HBNH)_3$, **15.54**, is sometimes referred to as an 'inorganic analogue of benzene', although this description should be approached *with caution*. By using the isoelectronic principle, we can see that the relationship lies in the fact that $[BH]^-$, $[NH]^+$ and CH are isoelectronic units, and replacing each unit in borazine by an isoelectronic CH generates benzene.

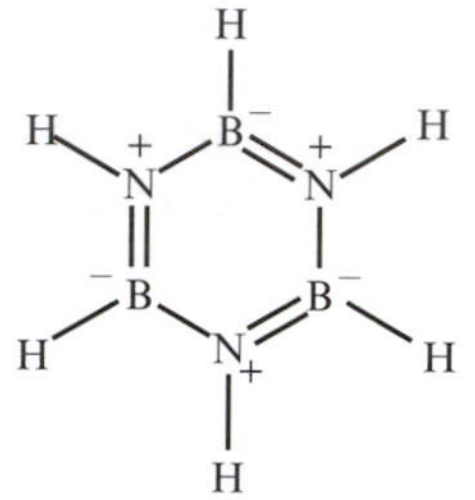

(15.54)

Borazine is a colourless liquid at room temperature (mp 215 K, bp 328 K) consisting of discrete molecules. All the B–N bond distances in the planar ring are equal (144 pm), indicating that structure **15.54** should really be represented either by the resonance pair **15.55** or by a delocalized bonding scheme such as we showed for benzene in Figure 15.11.

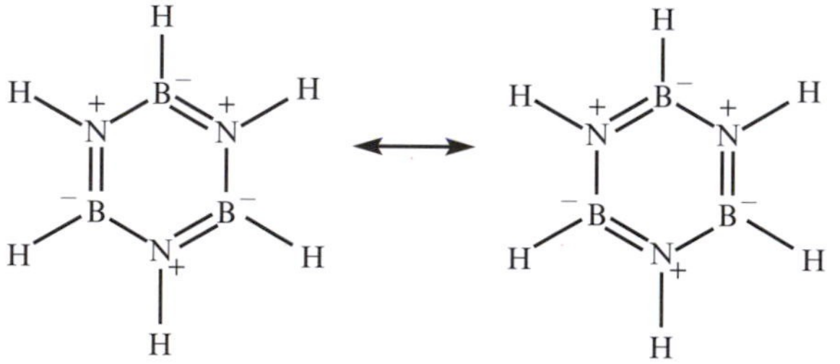

(15.55)

But now we come to a critical distinction between benzene and borazine: the carbon–carbon bonds in C_6H_6 are non-polar, but the boron–nitrogen bonds in borazine are polar – $\chi^P(B) = 2.0$, $\chi^P(N) = 3.0$. This affects both the σ-bonds and the π-interactions. The boron and nitrogen $2p$ orbitals do *not* contribute equally to the delocalized π-system with the result that there is more π-electron density associated with the nitrogen than the boron atoms. This results in the reactivity pattern of the boron–nitrogen ring being significantly different from that of benzene (see below). *You should notice that the charges in the resonance structures do not correspond to the observed charge separations in the B–N bonds: boron is δ^+ and nitrogen is δ^-.* Why is this?

Synthesis

The synthesis of borazine was first carried out by Stock in the 1920s by heating the adduct formed from ammonia and diborane (equation 15.79), but it is more effectively prepared by reaction 15.80.

$$3B_2H_6 + 6NH_3 \rightarrow 3[B_2H_6{\bullet}2NH_3] \xrightarrow{\Delta} 2\,(HBNH)_3 + 12H_2 \quad (15.79)$$

$$3[NH_4]Cl + 3BCl_3 \xrightarrow[-9HCl]{\Delta} (ClBNH)_3 \xrightarrow[-3NaCl,\ -\frac{3}{2}B_2H_6]{Na[BH_4]} (HBNH)_3 \quad (15.80)$$

Reactions

The charge distribution around the B_3N_3 ring in borazine is shown in Figure 15.21. Unlike benzene, borazine is susceptible to attack by both electrophiles (at nitrogen) and nucleophiles (at boron) and undergoes addition reactions with HCl, H_2O and CH_3OH (equation 15.81). On heating, the saturated

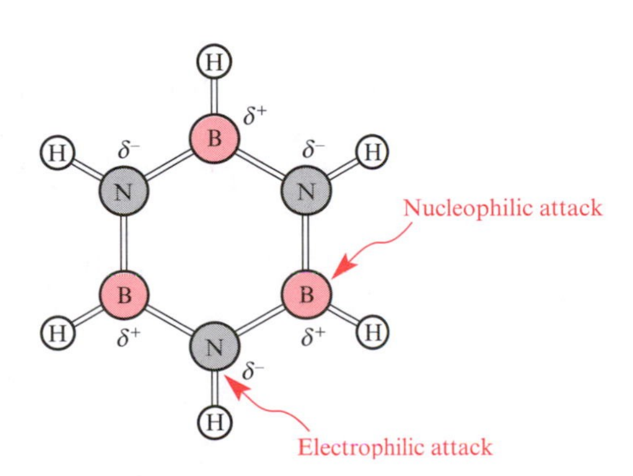

15.21 The charge distribution in borazine. Electrophiles attack nitrogen and nucleophiles attack boron atoms.

compounds eliminate H_2 to give a *B*-substituted product $(XBNH)_3$ (equation 15.82).

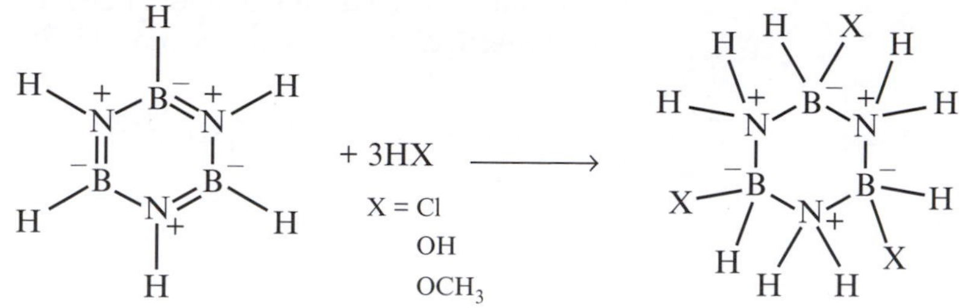

$$(HBNH)_3 + 3HX \longrightarrow (XHBNH_2)_3 \qquad X = Cl,\ OH,\ OCH_3$$

(15.81)

$$(XHBNH_2)_3 \xrightarrow{\Delta} (XBNH)_3 + 3H_2$$

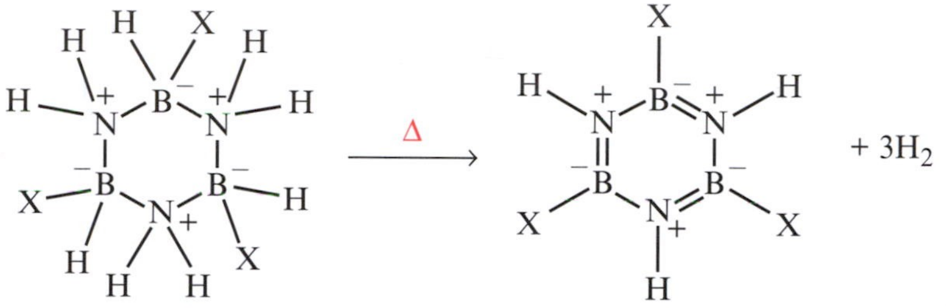

(15.82)

The ready *addition* of Cl_2 or Br_2 to borazine (equation 15.83) contrasts with *substitution* of Cl or Br for H when benzene undergoes halogenation.

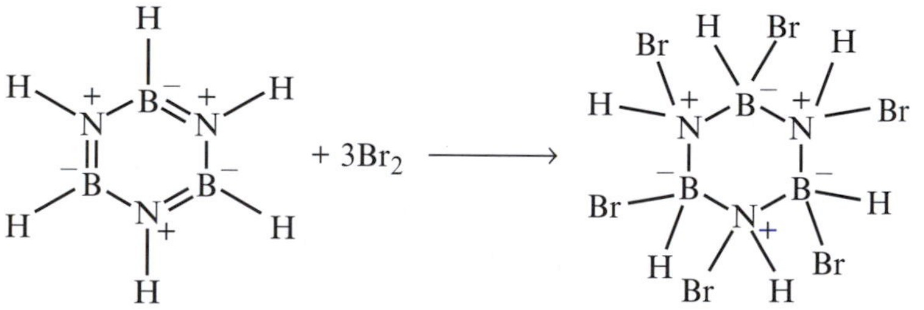

$$(HBNH)_3 + 3Br_2 \longrightarrow (BrHBNHBr)_3$$

(15.83)

A boron–nitrogen analogue of cyclohexane can be prepared by reducing $(HClBNH_2)$ (reaction 15.84) and the product **15.56** has a chair conformation.

$$(HBNH)_3 + 3HCl \rightarrow (ClHBNH_2)_3 \xrightarrow{Na[BH_4]} (H_2BNH_2)_3 \qquad (15.84)$$

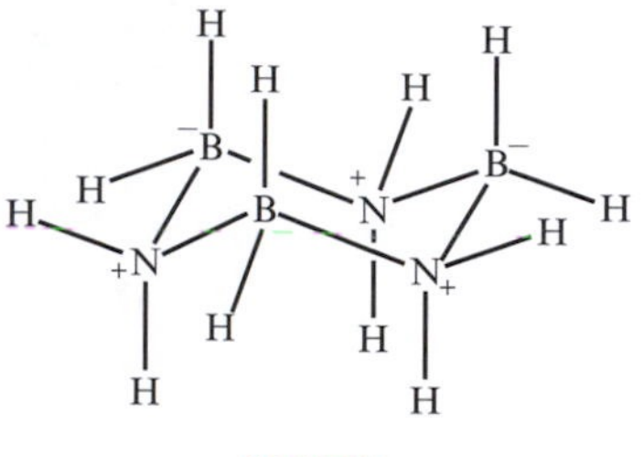

(15.56)

Box 15.13 Benzene and graphite *versus* borazine and boron nitride

The structural and physical properties of graphite were described in Section 7.10. Each hexagonal unit present in a layer of α-graphite (Figure 7.23) is similar in dimensions to the C_6 ring of benzene; the C–C bond distances in α-graphite are 142 pm compared with 140 pm in benzene. The six π-electrons in benzene form a discrete system, delocalized over the ring. In graphite, the π-electrons are involved in an extended delocalized system and this gives rise to a high electrical conductivity. The distance between two adjacent layers in α-graphite is 335 pm.

Just as benzene and graphite are structurally related, so too are borazine and one structural modification of boron nitride $(BN)_x$. The structure is composed of layers of hexagonal B_3N_3 rings which share common edges as shown below. The B–N distances within a layer are 145 pm, close to the 144 pm found in borazine. The layers lie directly over one another and are arranged so that boron and nitrogen atoms lie over each other. This contrasts with the staggered pattern found in graphite (Figure 7.23). The inter-layer separation in boron nitride is 330 pm, similar to that in α-graphite.

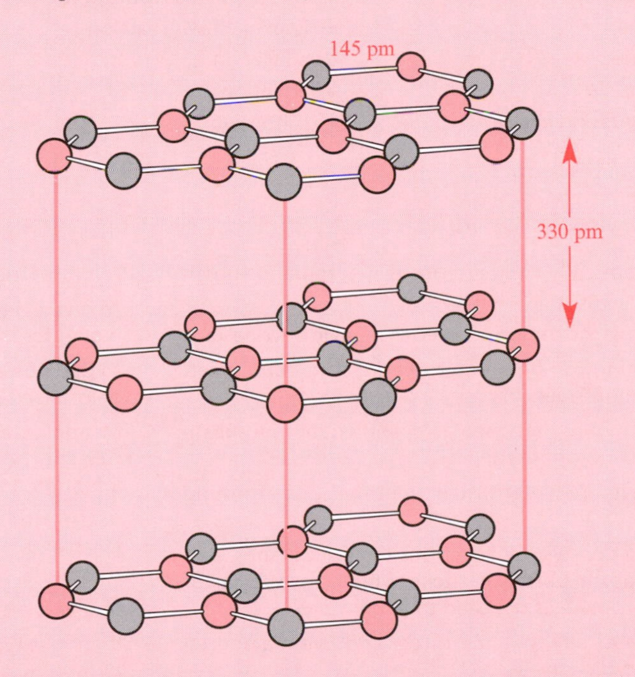

When this form of boron nitride is heated above 1770 K in the presence of lithium nitride, it transforms to a diamond-type lattice in which the B–N bond distances are 157 pm.

SUMMARY

This chapter has been concerned with the chemistry of cyclic molecules — saturated cycloalkanes and cyclic ethers, unsaturated cycloalkenes with localized C=C bonds, aromatic carbon compounds exemplified by benzene, toluene, phenol, nitrobenzene, aniline and diazonium salts, heteroaromatic compounds exemplified by pyridine, and an inorganic compound, borazine, that is structurally related to benzene. We have also introduced electrophilic substitution as a new type of reaction mechanism.

Do you know what the following terms mean?

- monocyclic
- bicyclic
- chair and boat conformers
- cycloaddition
- ring-opening reaction
- epoxide
- oxirane
- oxolane
- dienophile
- Diels–Alder reaction
- aromaticity
- Hückel ($4n+2$) rule
- aryl
- phenyl
- benzyl
- electrophilic substitution
- Friedel–Crafts reaction
- alkylation
- acylation
- sulfonation
- nitration
- Wheland intermediate
- ring activation
- orientation of electrophilic substitutions (specifically directing groups)
- phenol and phenoxide ion
- aniline and anilinium ion
- pyridine and pyridinium ion
- borazine

You should also be able:

- systematically to name unsubstituted and substituted monocyclic alkanes
- to discuss the conformational changes in a cycloalkane
- to distinguish between the boat, chair and twist-boat conformers of cyclohexane
- to distinguish between axial and equatorial substitution in cyclohexane
- to give representative methods of forming cycloalkanes
- to describe typical reactions of cycloalkanes, and distinguish between those which are typical of cyclopropane as opposed to larger rings
- systematically to name three- and five-membered monocyclic ethers
- to draw the structure of tetrahydrofuran (THF)
- to give synthetic routes to tetrahydrofuran and oxiranes
- systematically to name monocyclic alkenes and dienes
- to give examples of Diels–Alder cycloadditions, noting the substituent types required
- to discuss typical reactions of cycloalkenes and compare these with corresponding reactions of acyclic alkenes
- to draw Kekulé structures for benzene
- to give evidence for the failure of the cyclic triene model for benzene
- to describe a bonding model for benzene involving delocalized π-electrons
- to appreciate how and when the Hückel ($4n+2$) rule may be applied as a test for aromaticity (is this a fail-safe test?)
- to describe characteristic ^{1}H and ^{13}C NMR, IR and electronic properties of benzene, and relate these to other aromatic compounds
- to give methods of preparing benzene
- to discuss typical reactions of benzene
- to describe (with examples) the mechanism of electrophilic substitution in benzene
- to compare and contrast typical reactions of toluene, phenol, nitrobenzene and aniline with those of benzene
- to compare the structures and bonding in benzene, pyridine and borazine
- to give typical reactions of pyridine and borazine, comparing them with those of benzene

PROBLEMS

Problems 15.1, 15.2 and 15.6 include revision material based on earlier chapters.

15.1 Describe the bonding in the molecules shown in structures **15.1–15.4**, paying particular attention to similarities and differences between the roles of the bridging H, Cl or O atoms.

15.2 Classify each of the following allotropes according to whether they possess monocyclic or cage structures: (a) S_6; (b) S_8; (c) Se_8; (d) P_4; (e) C_{60}; (f) C_{70}. Comment upon the bonding in each molecule. Suggest why nitrogen forms diatomic molecules at 298 K, but white phosphorus does not?

15.3 Draw the structures of cyclobutane, 1,1-dimethylcyclooctane and 1,2-dibromocyclohexane. How many isomers of dibromocyclohexane are possible?

15.4 Give names for the unsubstituted cyclic alkanes C_5H_{10} and C_9H_{18}.

15.5 Name each of the following cycloalkanes:

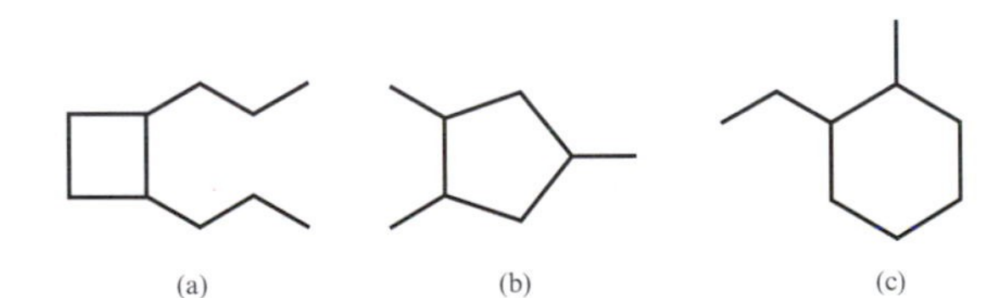

15.6 Figure 15.22 shows the ^{13}C NMR spectra of methylcyclopentane, ethylcyclopentane and 1,2-dimethylcyclopentane but each spectrum is unassigned. Assign each spectrum to a particular compound, and suggest possible assignments for the signals in each spectrum. No ^{13}C–^{1}H spin–spin coupling is shown in the spectra. [*Hint*: Refer to Table 9.11.]

15.7 Name each of the following cycloalkenes:

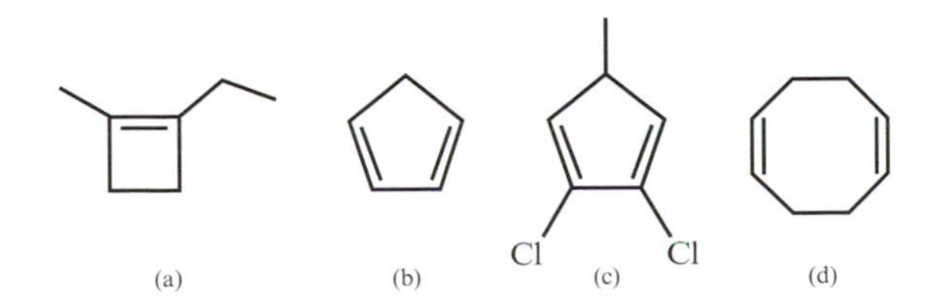

15.8 By drawing a simple mechanism similar to that in equation 15.15, show how the cycloaddition products in equation 15.18 arise.

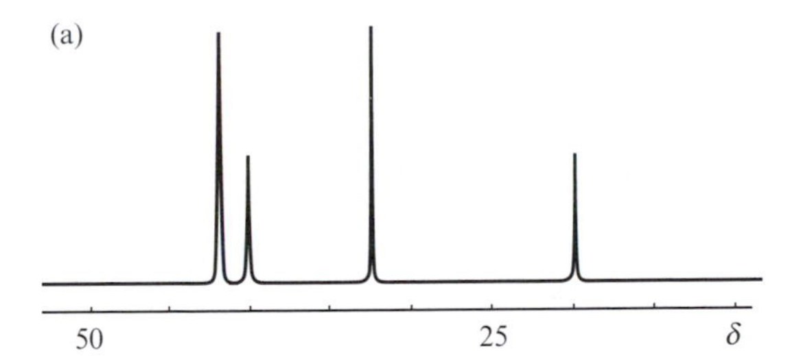

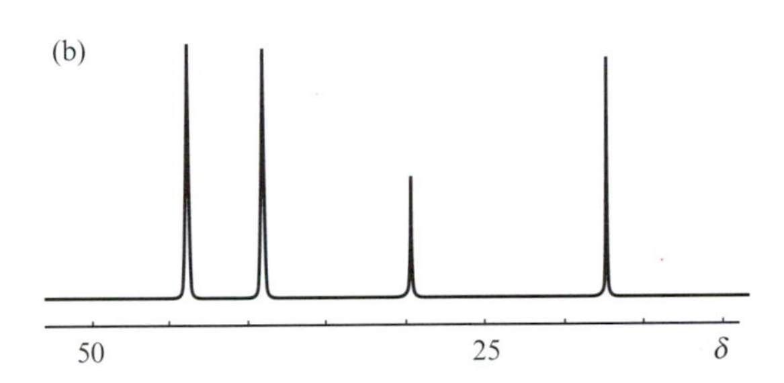

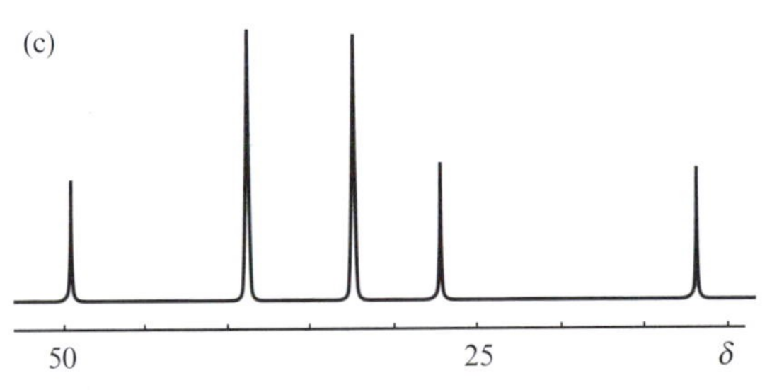

15.22 For problem 15.6.

15.9 How many π-electrons are there in each of the following rings? Which of the systems are delocalized? Which obey the Hückel rule? Which compounds are aromatic?

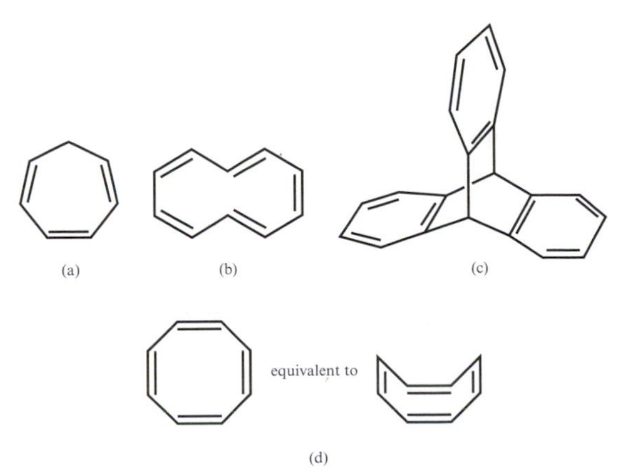

15.10 Draw structural formulae for the following compounds:

(a) iodobenzene;
(b) (*E*)-1,2-diphenylethene;
(c) 1,4-dichlorobenzene;
(d) 1,3,5-triethylbenzene;
(e) 3-chloroaniline,
(f) 2-bromo-4-phenylheptane.

15.11 Suggest possible products from the following reactions:

(a) + Br_2 $\xrightarrow{FeBr_3}$

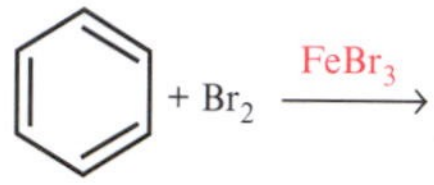

(b) + Me_3CCl $\xrightarrow{AlCl_3}$

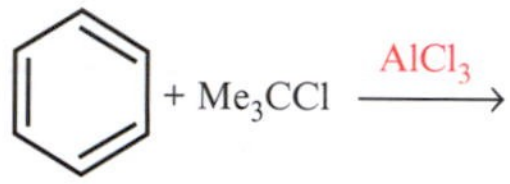

(c) + $CH_3CH_2CH_2Cl$ $\xrightarrow{AlCl_3}$

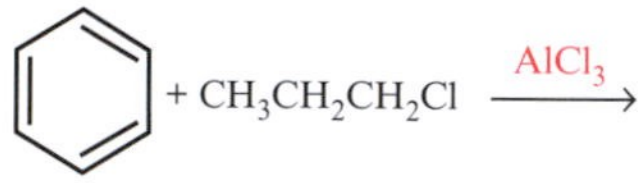

(d) + $CH_3CH_2CH_2COCl$ $\xrightarrow{AlCl_3}$

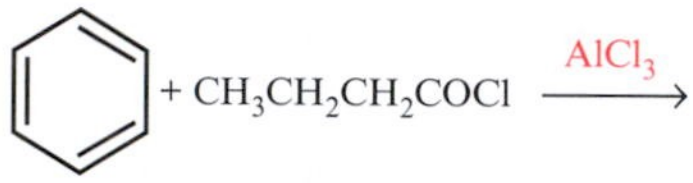

15.12 Which of the following compounds (and why) might be suitable catalysts in a Friedel–Crafts reaction: (a) BF_3; (b) NH_3; (c) $GaCl_3$; (d) $SbCl_5$?

15.13 Suggest likely products from the reaction of excess benzene with CH_2Cl_2 in the presence of $AlCl_3$.

15.14 Suggest possible products in the following reactions:

(a) CH_3 $\xrightarrow{\text{concentrated } HNO_3/H_2SO_4}$

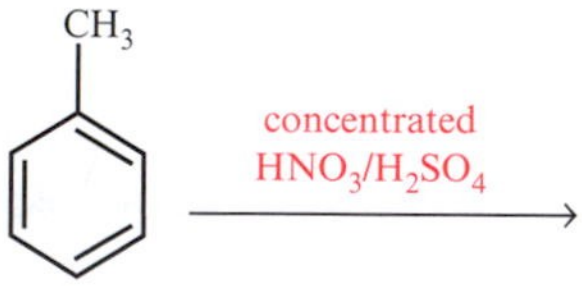

(b) CH_3 CH_3 $\xrightarrow{\text{hot alkaline } KMnO_4}$

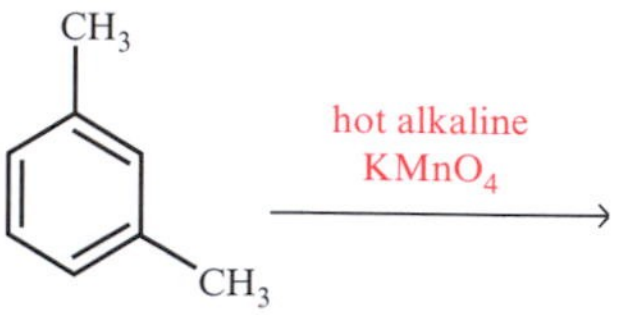

(c) OH + KOH ⟶

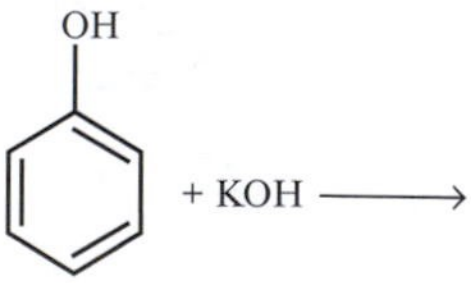

(d) O^-Na^+ Me Me + $PhCH_2Br$ ⟶

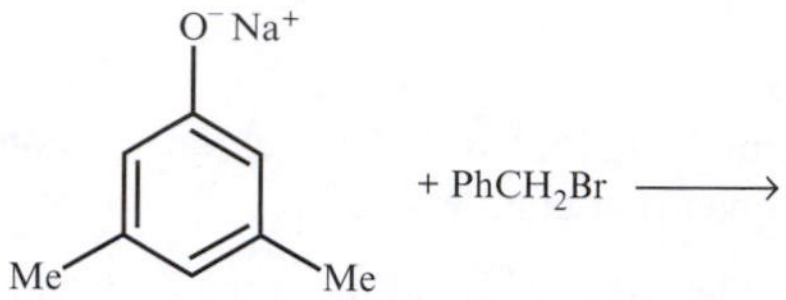

15.15 Calculate the pH of a 0.1 M aqueous solution of phenol ($pK_a = 9.89$).

15.16 Classify each of the following as either an aromatic or aliphatic amine: (a) diphenylamine; (b) benzylamine; (c) 2-bromoaniline; (d) 2-phenylethylamine.

15.17 Suggest likely products for the following reactions:

(a) NO_2 $\xrightarrow{Sn/HCl(aq)}$

(b) N_2^+ $\xrightarrow{H_2O}$ $\xrightarrow{\text{excess } PhN_2^+}$

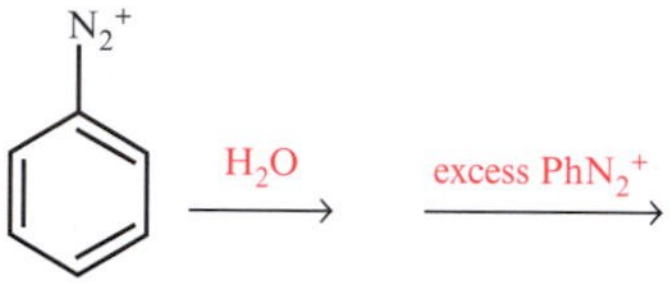

(c) N_2^+ $\xrightarrow{\text{Boil with aq. KI}}$

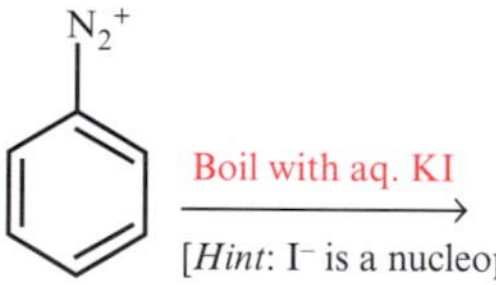

[*Hint*: I^- is a nucleophile.]

15.18 Estimate the pH of 0.2 M aqueous solutions of (a) aniline and (b) pyridine. [pK_b aniline = 9.37, pK_b pyridine =8.75.]

15.19 Suggest likely products from the following reactions of pyridine:

(a) N $\xrightarrow{HClO_4(aq)}$

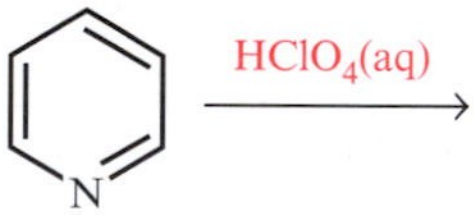

(b) CH_3 N $\xrightarrow{\text{hot alkaline } KMnO_4}$

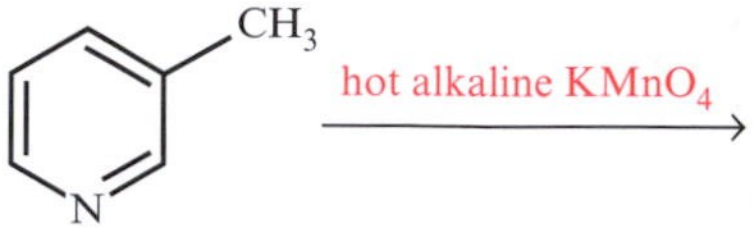

(c) N $\xrightarrow{H_2 \text{ 3 bar, Pt catalyst}}$

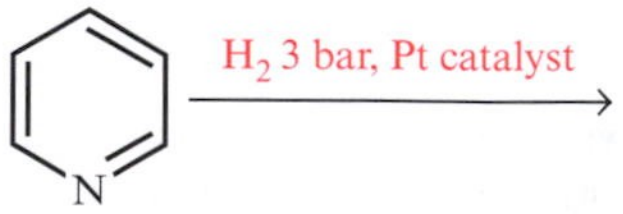

(d)

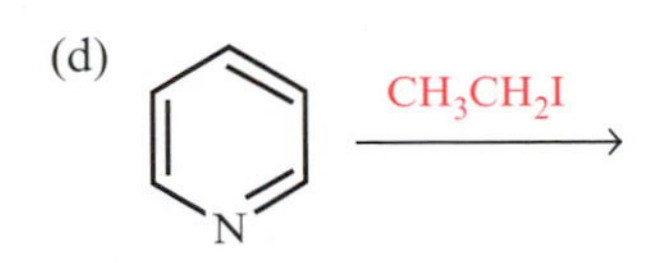

15.20 Make a comparative list of the properties of benzene and borazine that illustrate similarities *and* differences between these compounds.

15.21 Figure 15.23 shows the ^{13}C NMR spectra of THF, cyclohexa-1,4-diene, cyclohexa-1,3-diene, cycloheptatriene and pyridine. Assign each spectrum to the correct compound, giving reasons for your choices.

(a)

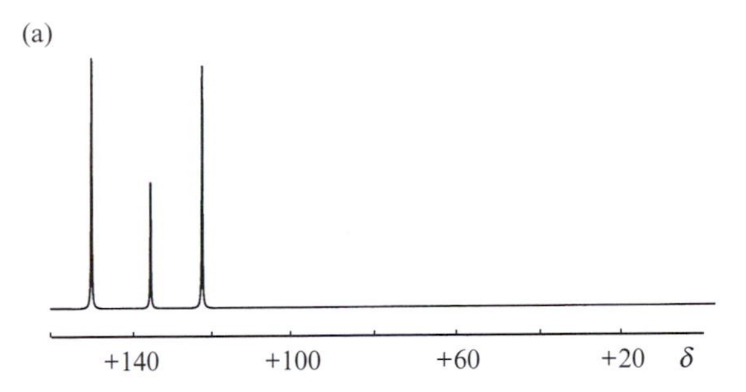

(b)

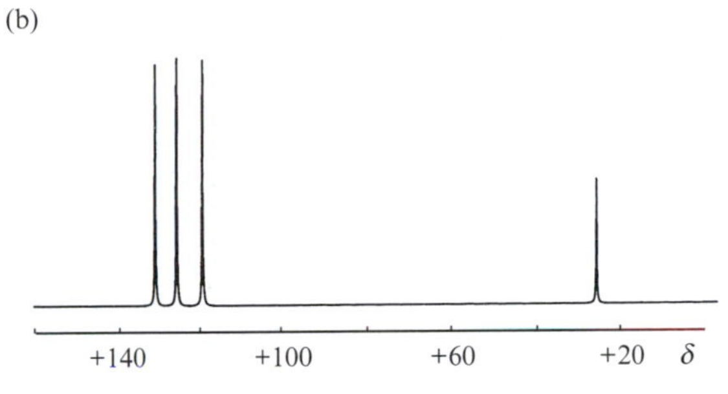

(c)

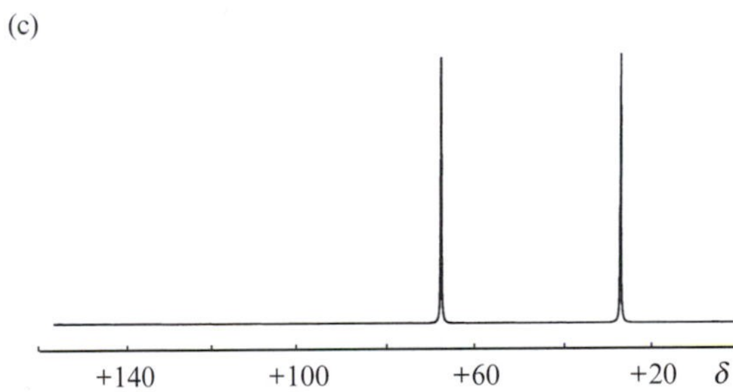

(d)

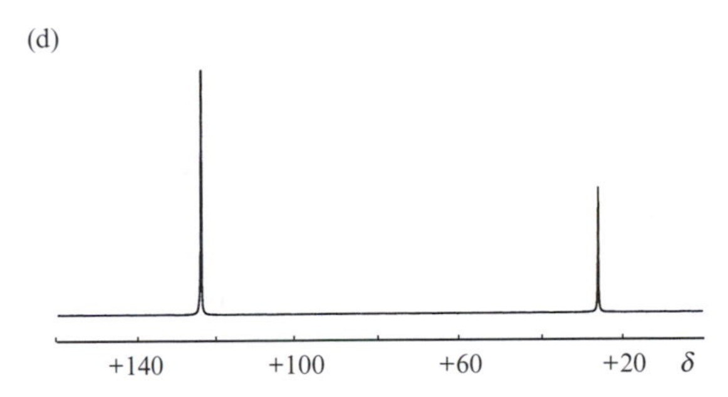

(e)

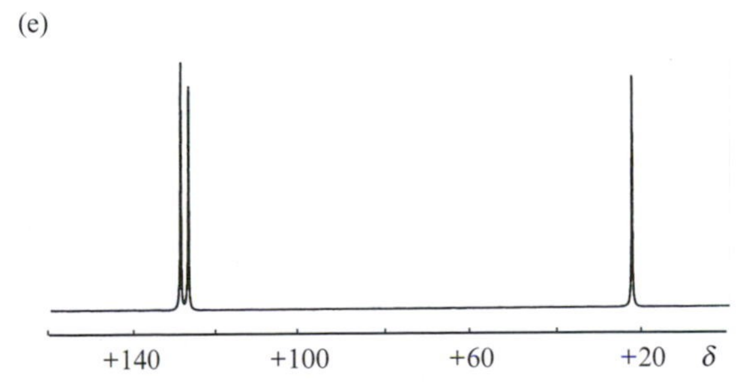

15.23 For problem 15.21

16 Coordination complexes of the *d*-block metals

Topics

1	2	**3**	**4**	**5**	**6**	**7**	**8**	**9**	**10**	**11**	**12**	13	14	15	16	17	18
H																	He
Li	Be											B	C	N	O	F	Ne
Na	Mg											Al	Si	P	S	Cl	Ar
K	Ca	**Sc**	**Ti**	**V**	**Cr**	**Mn**	**Fe**	**Co**	**Ni**	**Cu**	**Zn**	Ga	Ge	As	Se	Br	Kr
Rb	Sr	**Y**	**Zr**	**Nb**	**Mo**	**Tc**	**Ru**	**Rh**	**Pd**	**Ag**	**Cd**	In	Sn	Sb	Te	I	Xe
Cs	Ba	**La**	**Hf**	**Ta**	**W**	**Re**	**Os**	**Ir**	**Pt**	**Au**	**Hg**	Tl	Pb	Bi	Po	At	Rn
Fr	Ra																

16.1 INTRODUCTION: SOME TERMINOLOGY

Coordination complexes and coordinate bonds

This chapter is concerned with some aspects of the chemistry of the *d*-block metals. In Section 13.14, we considered some *d*-block metal halides, oxides and oxyanions (e.g. $[VO_4]^{3-}$, CrO_3, $[MnO_4]^-$ and OsF_6) which contained metals in high oxidation states. In each species, the metal was essentially *covalently bonded* to the surrounding atoms. Now we turn our attention to *metal complexes* in which the *d*-block elements are in lower oxidation states and are surrounded by molecules or ions which function as Lewis bases and form coordinate bonds with the central metal ion. We have already discussed related complexes when we described the formation of *hydrated ions* in Section 11.8, but here, *ion–dipole interactions* were primarily responsible for the formation of the first hydration shell (Figure 11.13).

Coordinate bond: see Section 11.5

A great deal of the chemistry of the *d*-block metals is concerned with *coordination compounds or complexes* and these include species such as $[Ni(NH_3)_6]^{2+}$ **16.1**, and $[PtCl_4]^{2-}$ **16.2** in which the *coordination numbers* of the nickel(II) and platinum(II) centres are six and four respectively.

Coordination number: see Section 5.1

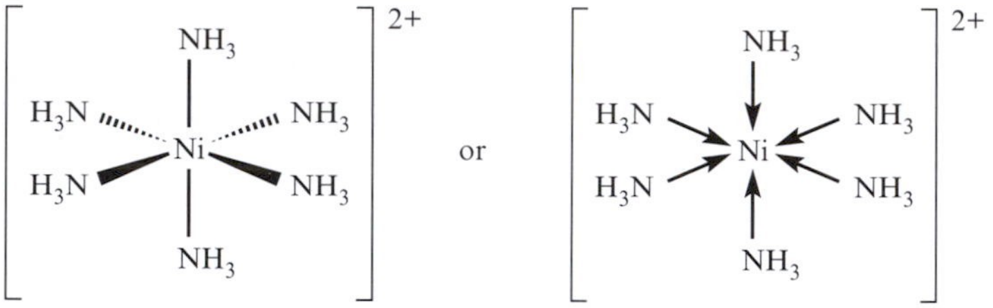

(16.1)

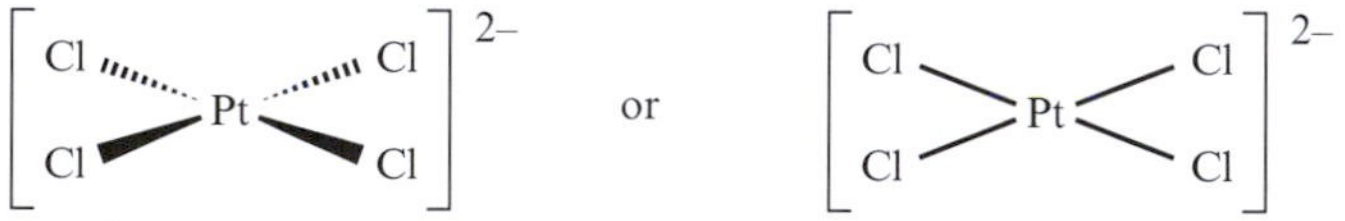

(16.2)

Two representations of each of complexes **16.1** and **16.2** are shown – the left-hand diagram emphasizes the *stereochemistry* while the right-hand structure emphasizes the nature of the *bonding*. Each of the chloride ions or ammonia molecules forms a bond with the ionic metal centre by donating a lone pair of electrons. The molecules or ions surrounding the metal ions are called *ligands*; this term is derived from the Latin verb '*ligare*' meaning 'to bind'. As an extension of the formalisms introduced in structures **11.1** and **11.2**, we denote a bond between a metal ion and a *neutral ligand* by an arrow (as in **16.1**), and between a metal ion and an *anionic ligand* by a line (as in **16.2**). In a coordination compound, the *ligands coordinate to the metal ion*.

Ligand: see Section 16.3

The existence of coordination complexes was first recognized by Alfred Werner, who was awarded a Nobel prize for chemistry in 1913 for his pioneering studies. A typical example of one of the problems encountered by Werner was in the structural formulation of a compound with the molecular formula $CoCl_3(NH_3)_6$. This compound was found to be ionic and it possesses a *complex cation* $[Co(NH_3)_6]^{3+}$ in which the cobalt(III) ion is sited at the centre of an octahedral arrangement of six ammonia molecules (structure **16.3**); the three chlorides are present as free ions. Further members of this group of compounds include $CoCl_3(NH_3)_5$ and $CoCl_3(NH_3)_4$, and in each, the cobalt(III) centre is octahedrally sited with one or two chloride ions directly bonded to the metal ion (structures **16.4** and **16.5**); the remaining chloride ions are non-coordinated.

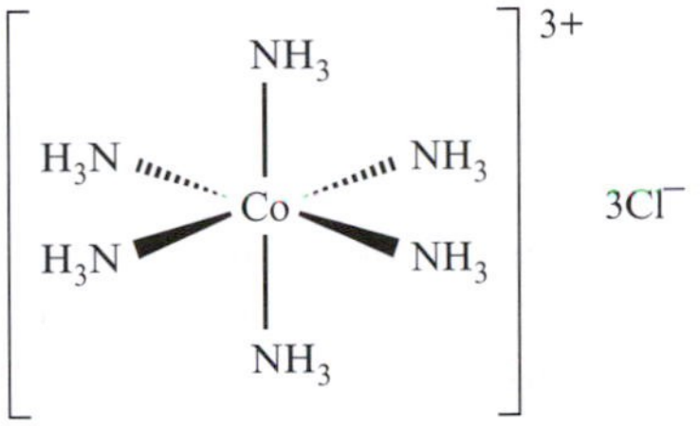

$[Co(NH_3)_6]^{3+}$ $3Cl^-$

(16.3)

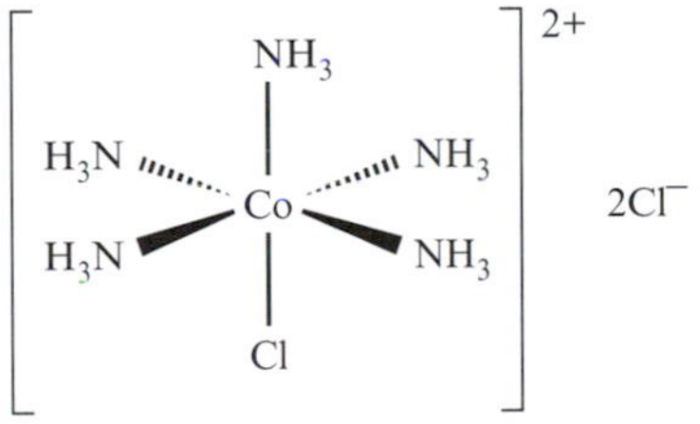

$[Co(NH_3)_5Cl]^{2+}$ $2Cl^-$

(16.4)

$[Co(NH_3)_4Cl_2]^{+}$ Cl^-

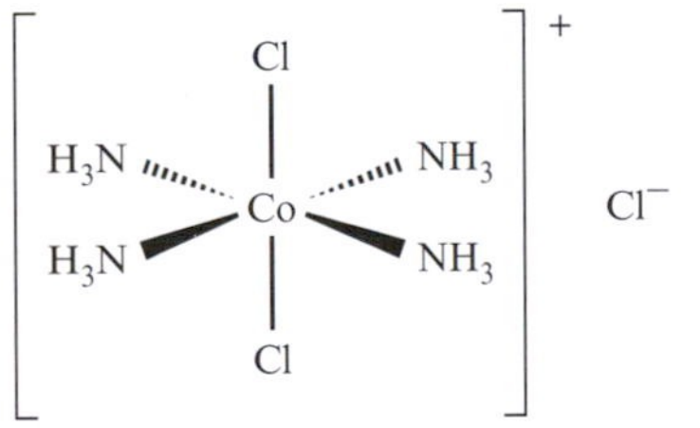

(16.5)

Structure **16.5** shows one of two possible *stereoisomers* of the $[Co(NH_3)_4Cl_2]^+$ cation. The chloride ions are drawn in a *trans*-arrangement, but a *cis*-isomer is also possible. [*Question*: What is the structure of *cis*-$[Co(NH_3)_4Cl_2]^+$?]

trans- and *cis*-isomers: see Section 5.11

Primary valence and coordination number

In each of $[Co(NH_3)_6]Cl_3$, $[Co(NH_3)_5Cl]Cl_2$, and $[Co(NH_3)_4Cl_2]Cl$, the *primary valence* or *oxidation state* of the cobalt centre is +3, but the coordination number is six. Once recognized, the concept of complex ion formation opened up a new and exciting area of chemistry. A crucial principle is that *the coordination number of the metal centre is not related to the primary valence (the oxidation state) of the metal ion.*

In a complex, the coordination number of the metal centre is *not* related to the oxidation state of the metal ion.

Atomic *d* orbitals

We have already mentioned *d* orbitals in discussions of atomic structure (Chapter 2), molecular shapes (Chapter 5) and the chemistry of the *p*-block and high oxidation states *d*-block elements (Chapter 13), but we have not drawn attention to the spatial properties of these atomic orbitals.

Quantum numbers: see Section 2.10

For the principal quantum number $n = 3$, there are three possible values for the quantum number l ($l = 0, 1, 2$). An orbital quantum number $l = 2$ is associated with a *d* orbital. For $l = 2$, there are five possible values of m_l (for $l = 2$, $m_l = +2, +1, 0, -1, -2$) and this means that there are five real solutions to the Schrödinger equation – a set of *d* orbitals is *five-fold degenerate*. The labels given to these five AOs are d_{xy}, d_{xz}, d_{yz}, d_{z^2} and $d_{x^2-y^2}$ and their spatial

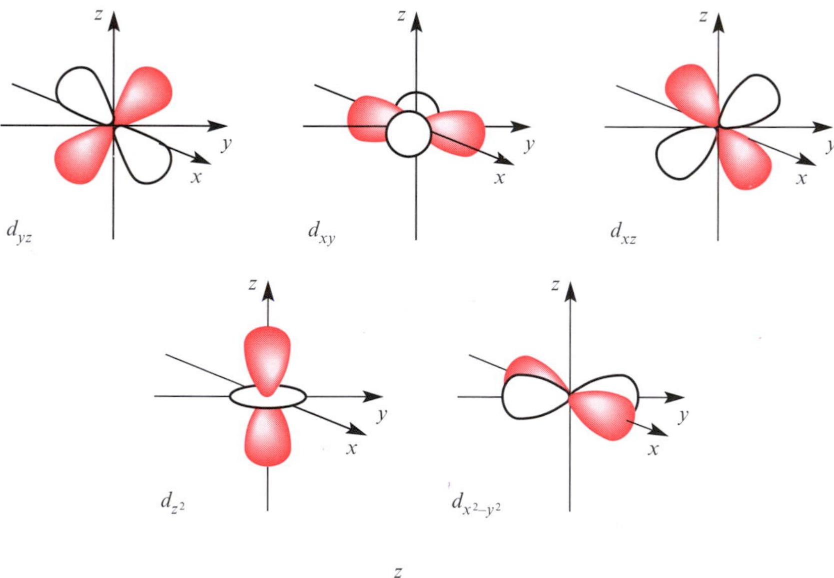

16.1 The five 3*d* AOs. Each has two nodal planes.

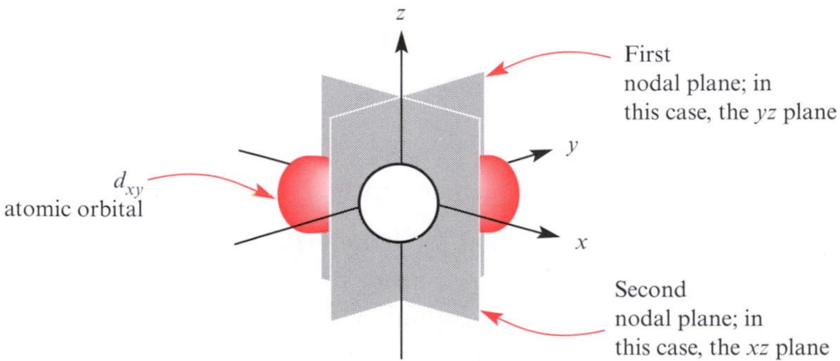

16.2 The two sign changes in an atomic d_{xy} orbital. The nodal planes lie in the *xz* and *yz* planes

Box 16.1 The shapes of *d* atomic orbitals

Figure 16.1 shows that the lobes of the d_{xy}, d_{xz} and d_{yz} orbitals point *between* the Cartesian axes and each orbital lies in one of the three planes defined by these axes.

The $d_{x^2-y^2}$ orbital is related to d_{xy} — its lobes point *along* (rather than between) the *x* and *y* axes.

We could envisage being able to draw two more atomic orbitals which are related to the $d_{x^2-y^2}$ orbital, namely the $d_{z^2-x^2}$ and $d_{z^2-y^2}$ orbitals. This would give a total of *six* orbitals. But, only *five* real solutions (i.e. five values of m_l) of the Schrödinger equation are allowed for $l = 2$.

The problem is solved by taking a *linear combination* (see Section 3.12) of the $d_{z^2-x^2}$ and $d_{z^2-y^2}$ orbitals. This means that the two are combined with the result that the fifth real solution to the Schrödinger equation corresponds to the d_{z^2} orbital:

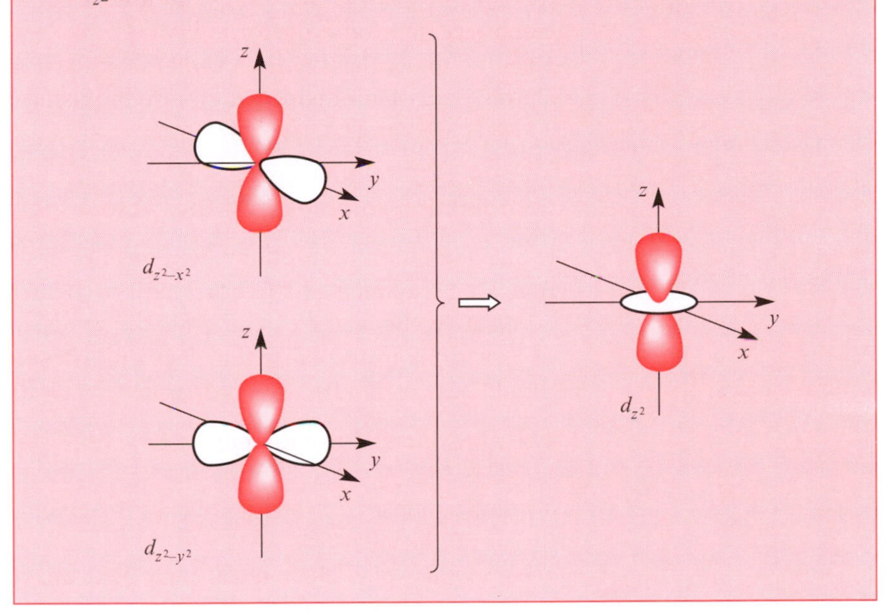

properties (shapes) are shown in Figure 16.1. Notice that, like *p* orbitals, *d* orbitals are *directional*. Each *d* orbital possesses *two nodal planes* and this is shown for the d_{xy} orbital in Figure 16.2. [*Question*: where are the nodal planes in each of the other *d* orbitals?]

The radial distribution function of a 3*d* atomic orbital possesses *no radial node* (Figure 2.12) and we look further at the dependence of *d* orbital energy on effective nuclear charge in Section 16.2.

Geometries of complexes: revision of the Kepert theory

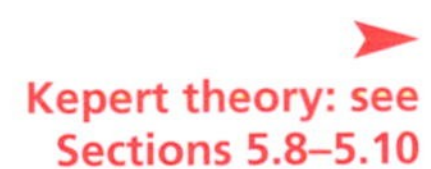
Kepert theory: see Sections 5.8–5.10

Kepert theory allows us to predict the shape of a complex by considering the repulsive forces between ligands. To recap, the donor atoms lie at the vertices of a polyhedral shell that surrounds the metal ion and the basic structural types are based upon:

- a linear arrangement for two donor atoms;
- a trigonal planar arrangement for three donor atoms;
- a tetrahedral arrangement for four donor atoms;
- a trigonal bipyramidal or square-based pyramidal arrangement for five donor atoms;
- an octahedral arrangement for six donor atoms.

Shapes and orbital hybridization: see Section 5.18

An exception to these rules is that some 4-coordinate complexes (e.g. $[PtCl_4]^{2-}$ and $[AuCl_4]^-$) are square-planar.

Nomenclature

The nomenclature of coordination complexes is beyond the scope of this text and we shall refer to these complexes only by formulae. However, we emphasize one point: the formula of a coordination complex, even when it is a neutral molecule, is written *within square brackets*. For an ionic species, the charge is written outside the square brackets as usual, e.g. $[PtCl_2(NH_3)_2]$, $[Co(NH_3)_6]^{2+}$ and $[ZnCl_4]^{2-}$.

Coordination complexes of the *d*-block metals: chapter aims

In the discussion that follows, we consider first some terms and definitions that are fundamental to coordination chemistry; then we look at some descriptive chemistry of the *d*-block metals, and rationalize some of the observations in terms of the electronic structure of the metal ions. You should bear in mind several properties that typify transition metal ions:

- a *d*-block metal usually possesses several stable oxidation states (Figure 13.1);
- compounds of *d*-block metals are often coloured, although the colour may not be intense;
- compounds of *d*-block metals may be diamagnetic or paramagnetic;
- the formation of complexes is common, but not all *d*-block compounds are complexes.

Dia- and paramagnetism: see Section 3.18

16.2 ELECTRONIC CONFIGURATIONS OF THE *d*-BLOCK METALS AND THEIR IONS

Figure 16.3 shows the *d*-block metals and Table 16.1 lists the names and ground state electronic configurations of each element. The group number gives the *total number of valence electrons* for each member of a particular triad. For example, in group 4, each of titanium, zirconium and hafnium possesses a configuration $(n+1)s^2nd^2$. In Section 2.19, we stated that the *approximate* order in which atomic orbitals are occupied in the ground state of an atom follows the sequence:

$$1s < 2s < 2p < 3s < 3p < 4s < 3d < 4p < 5s < 4d < 5p < 6s < 5d \approx 4f < 6p < 7s < 6d \approx 5f$$

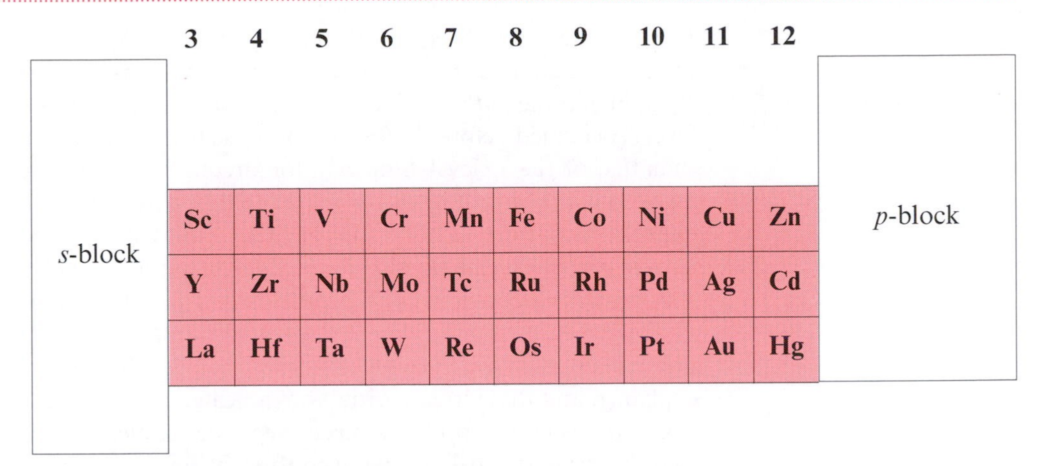

16.3 The periodic table highlighting the *d*-block elements. Each group of three metals is called a *triad*. Lanthanum (La) is the first metal in the series of the *f*-block *lanthanoid metals* which are omitted from the table.

Table 16.1 The *d*-block metals and their ground state electronic configurations.

Element symbol	*Element name*	*Group number*	*Ground state electronic configuration*	*Comment*
Sc	Scandium	3	$[Ar]4s^23d^1$	
Ti	Titanium	4	$[Ar]4s^23d^2$	
V	Vanadium	5	$[Ar]4s^23d^3$	
Cr	Chromium	6	$[Ar]4s^13d^5$	Half-filled *d*-level
Mn	Manganese	7	$[Ar]4s^23d^5$	Half-filled *d*-level
Fe	Iron	8	$[Ar]4s^23d^6$	
Co	Cobalt	9	$[Ar]4s^23d^7$	
Ni	Nickel	10	$[Ar]4s^23d^8$	
Cu	Copper	11	$[Ar]4s^13d^{10}$	Filled *d*-level
Zn	Zinc	12	$[Ar]4s^23d^{10}$	Filled *d*-level
Y	Yttrium	3	$[Kr]5s^24d^1$	
Zr	Zirconium	4	$[Kr]5s^24d^2$	
Nb	Niobium	5	$[Kr]5s^14d^4$	Irregular configuration
Mo	Molybdenum	6	$[Kr]5s^14d^5$	Half-filled *d*-level
Tc	Technetium	7	$[Kr]5s^24d^5$	Half-filled *d*-level
Ru	Ruthenium	8	$[Kr]5s^14d^7$	Irregular configuration
Rh	Rhodium	9	$[Kr]5s^14d^8$	Irregular configuration
Pd	Palladium	10	$[Kr]5s^04d^{10}$	Filled *d*-level; irregular
Ag	Silver	11	$[Kr]5s^14d^{10}$	Filled *d*-level
Cd	Cadmium	12	$[Kr]5s^24d^{10}$	Filled *d*-level
La	Lanthanum	3	$[Xe]6s^25d^1$	
Hf	Hafnium	4	$[Xe]4f^{14}6s^25d^2$	
Ta	Tantalum	5	$[Xe]4f^{14}6s^25d^3$	
W	Tungsten	6	$[Xe]4f^{14}6s^25d^4$	
Re	Rhenium	7	$[Xe]4f^{14}6s^25d^5$	Half-filled level
Os	Osmium	8	$[Xe]4f^{14}6s^25d^6$	
Ir	Iridium	9	$[Xe]4f^{14}6s^25d^7$	
Pt	Platinum	10	$[Xe]4f^{14}6s^15d^9$	Irregular configuration
Au	Gold	11	$[Xe]4f^{14}6s^15d^{10}$	Filled *d*-level
Hg	Mercury	12	$[Xe]4f^{14}6s^25d^{10}$	Filled *d*-level

aufbau principle: see Sections 2.18 and 2.19.

However we emphasized that the energy of electrons in orbitals is affected by the nuclear charge, the presence of other electrons, and the overall charge. By applying the *aufbau* principle, we predict that for a titanium atom, the 4*s* level is occupied before the 3*d* level because the energy of the 4*s* level is lower than that of the 3*d* level. Similarly, for zirconium, the 5*s* level is before the 4*d* level, and for hafnium, the 6*s* level is filled before the 5*d* level. We might expect this pattern to continue for other metal atoms, and that the $(n+1)s$ orbital will always be filled before the corresponding *nd* level. However, the ground state electronic configurations in Table 16.1 show that this is not always the case. Some apparent irregularities may be rationalized (for example, Cr and Cu which are considered below), but others are not easily explained and these irregularities are indicated in the table.

Copper is in group 11 and has eleven valence electrons. We might predict that the ground state electronic configuration is $[Ar]4s^2 3d^9$ but in practice the electrons occupy the orbitals so as to give a configuration $[Ar]4s^1 3d^{10}$. This provides the copper atom with a completely filled 3*d* shell, a situation that is energetically favourable. Similarly, chromium with six valence electrons possesses a ground state electronic configuration of $[Ar]4s^1 3d^5$ in preference to $[Ar]4s^2 3d^4$, and has a half-filled 3*d* level.§

In *d*-block *metal ions*, the configurations are $(n+1)s^0 nd^m$. For example, the electronic configuration of an Fe^{2+} ion is $[Ar]3d^6$, and that of a Ti^{3+} ion is $[Ar]3d^1$. The origins of this effect are rather complex, but the end result is that the energies of the *nd* orbitals drop *below* those of the $(n+1)$s orbital (3*d* is lower than 4*s*). In part this is due to the different ways in which these orbitals behave as the effective nuclear charge increases as we move from the metal atom to metal ion. The effective nuclear charge experienced by an electron is dependent upon the orbital in which it finds itself. The overall effect is a *contraction* of the *d*-orbitals in the metal *ions* such that the electrons are closer to the nucleus. This means that the electrostatic attraction between the negatively charged electrons and the positively charged nucleus increases and the *nd* orbitals are stabilized with respect to the $(n+1)s$ orbital.†

Effective nuclear charge, penetration and shielding: see Section 2.15

16.3 LIGANDS

Ligand denticity and monodentate ligands

A ligand is a Lewis base and has at least one pair of electrons with which it can form a coordinate bond with a metal centre. Most ligands are neutral or anionic, and all can donate one or more electrons through one or more atoms to a metal ion. An atom on which the electrons reside is called a *donor atom* of the ligand – when ammonia acts as a ligand as in structure **16.1**, the donor atom is nitrogen.

§ It is a general phenomenon that an electronic arrangement in which a set of orbitals is half or fully occupied is particularly favoured. The origins of this effect are beyond this text, but examples may be found in the *p*- and *f*-blocks as well as in the *d*-block.

† For a more in-depth discussion of this topic, see M. Gerloch and E.C. Constable, *Transition Metal Chemistry: The Valence Shell in d-Block Chemistry*, VCH, Weinheim (1994).

The number of donor atoms through which a ligand coordinates to a metal ion is defined as the *denticity of the ligand*, and a ligand that coordinates to a metal centre through one donor atom is *monodentate*. Examples of monodenate ligands are given in Table 16.2.

In a coordination complex, a metal ion is surrounded by molecules or ions (the ligands) which act as Lewis bases and form coordinate bonds with the metal centre which acts as a Lewis acid. The atoms in the ligands that are bonded to the metal centre are the donor atoms.

Polydentate ligands and chelate rings

If a ligand binds to a metal ion through two donor atoms, the ligand is termed *didentate* (Table 16.2). 1,2-Diaminoethane **16.6** (usually abbreviated to en) is capable of behaving as a didentate ligand as in the complex ion $[Co(en)_3]^{3+}$ (Figure 16.4a).§ However, en could also function as a monodentate ligand as is shown in Figure 16.4b.

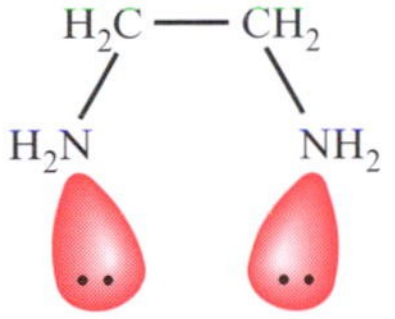

(16.6)

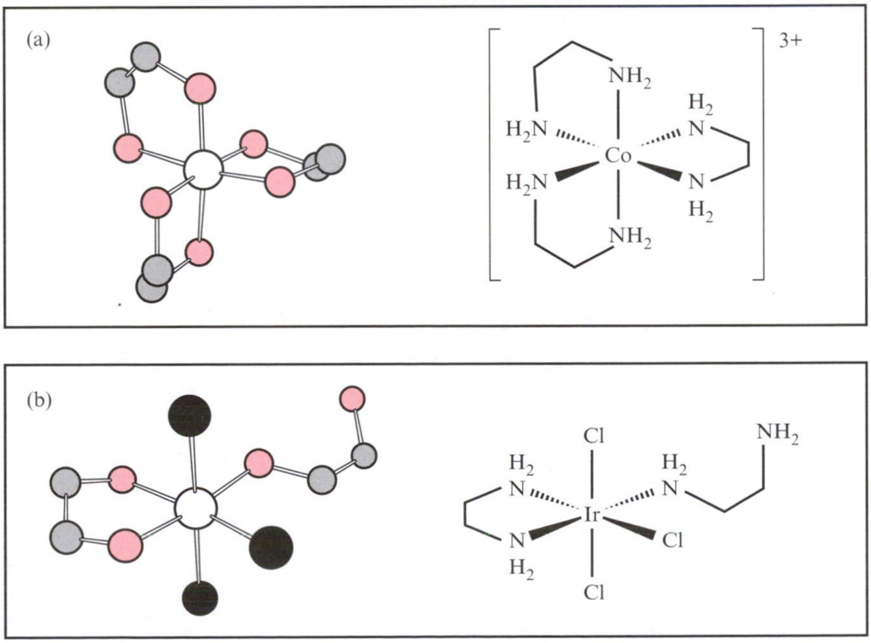

16.4 The solid state structures (determined by X-ray diffraction methods) and diagrammatic representations of (a) $[Co(en)_3]^{3+}$ (en = 1, 2-diaminoethane) determined for the salt $[Co(en)_3]Br_3{\cdot}H_2O$ and (b) *mer*-$[IrCl_3(\text{en-}N)(\text{en-}N,N')]$ in which one en ligand is monodentate and one is didentate. Hydrogen atoms in the structural figures have been omitted for simplicity.

§ The 1990 Recommendations of the IUPAC replace the term *bidentate* by *didentate*.

Table 16.2 Some typical monodentate and polydentate ligands and representative *d*-block metal complexes.

Name of free ligand	*Name of coordinated ligand*	*Formula or structure of ligand*	*Common abbreviation*	*Donor atoms*	*Representative complex*
Monodentate ligands					
Water	Aqua	H_2O		*O*	$[Mn(H_2O)_6]^{2+}$
Hydroxide	Hydroxo	$[OH]^-$		*O*	$[Zn(OH)_4]^{2-}$
Ammonia	Ammine	NH_3		*N*	$[Co(NH_3)_6]^{3+}$
Fluoride	Fluoro	F^-		*F*	$[CrF_6]^{3-}$
Chloride	Chloro	Cl^-		*Cl*	$[ZnCl_4]^{2-}$
Bromide	Bromo	Br^-		*Br*	$[FeBr_4]^{2-}$
Iodide	Iodo	I^-		*I*	$[AuI_2]^-$
Acetonitrile	Acetonitrile	MeCN		*N*	$[Fe(NCMe)_6]^{2+}$
Cyanide	Cyano	$[CN]^-$		*C*	$[Fe(CN)_6]^{3-}$
Thiocyanate	Thiocyanate	$[NCS]^-$		*N* or *S*	$[Co(CN)_5(SCN)]^{3-}$ In solution, linkage isomers are present, $[Co(CN)_5(\boldsymbol{S}CN)]^{3-}$ and $[Co(CN)_5(SC\boldsymbol{N})]^{3-}$
Pyridine	Pyridine	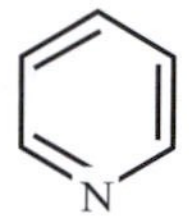 N	py	*N*	$[Fe(py)_6]^{2+}$
Tetrahydrofuran	Tetrahydrofuran	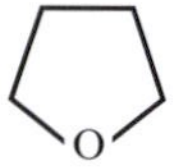 O	THF or thf	*O*	*trans*–$[YCl_4(THF)_2]^-$
Didentate ligands					
1,2-Diaminoethane[a]	1,2-Diaminoethane	 H_2N NH_2	en	*N,N'*	$[Co(en)_3]^{3+}$
Acetylacetonate ion	Acetylacetonato	$[MeC(O)CHC(O)Me]^-$ 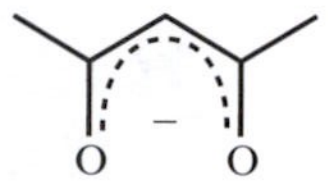 O – O	$[acac]^-$	*O,O'*	$[Cr(acac)_3]$

Oxalate or ethanedioate ion	Oxalato	$[C_2O_4]^{2-}$ 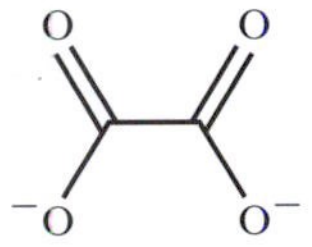	$[ox]^{2-}$	*O*,*O′*	$[Fe(ox)_3]^{3-}$
Glycinate ion	Glycinato	$[H_2NCH_2CO_2]^-$ 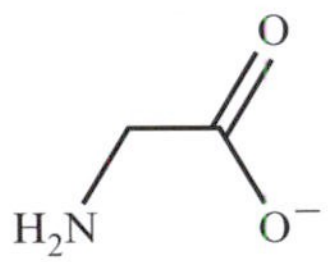	$[gly]^-$	*N*,*O*	$[Ni(gly)_3]^-$
2,2′-Bipyridine	2,2′-Bipyridine	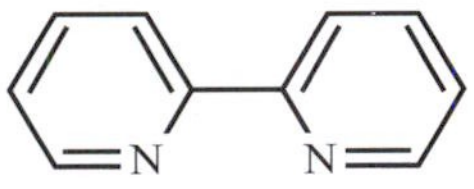	bpy (or bipy)	*N*,*N′*	$[Ru(bpy)_3]^{2+}$
1,10-Phenanthroline	1,10-Phenanthroline	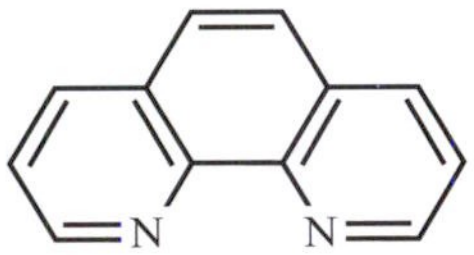	phen	*N*,*N′*	$[Fe(phen)_3]^{2+}$
Tridentate ligands					
1,4,7-Triazaheptane[a]	1,4,7-Triazaheptane		dien	*N*,*N′*,*N″*	$[Co(dien)_2]^{3+}$
2,2′:6′,2″-Terpyridine	2,2′:6′,2″-Terpyridine	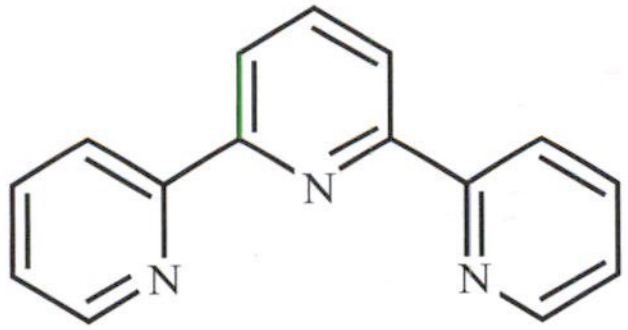	tpy	*N*,*N′*,*N″*	$[Ru(tpy)_2]^{2+}$
Hexadentate ligands					
N,*N*,*N′*,*N′*–Ethylenediaminetetraacetate ion[b]	*N*,*N*,*N′*,*N′*–Ethylenediaminetetraacetato	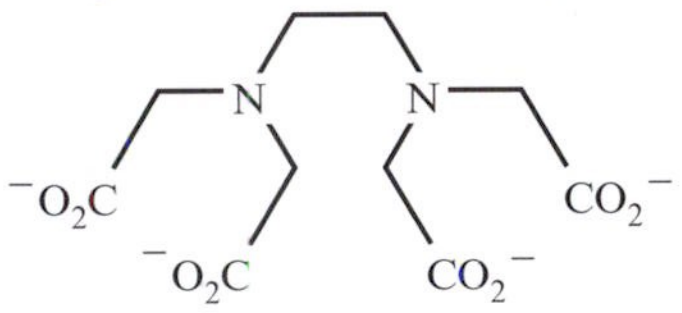	$[edta]^{4-}$	*N*,*N′*,*O*,*O′*,*O″*,*O‴*	$[Co(edta)]^-$

[a]The older names for 1,2-diaminoethane and 1,4,7-triazaheptane (still in common use) are ethylenediamine and diethylenetriamine.
[b]Although not systematic by IUPAC, this is the commonly accepted name for this anion. The prefix *N*,*N*,*N′*,*N′*– indicates the positions of substitution of the $CH_2CO_2^-$ groups (see Section 15.11).

The denticity of a ligand is the number of donor atoms which bind to the metal centre in a complex. Collectively, ligands which bond through more than one donor atom are called polydentate.

Number of donor atoms bonded to the metal centre.	Denticity
1	Monodentate
2	Didentate
3	Tridentate
4	Tetradentate
5	Pentadentate
6	Hexadentate

Bonding in $[acac]^-$: see Section 17.11

The coordination to a metal ion of a *didentate* $H_2NCH_2CH_2NH_2$ ligand results in the formation of a *non-planar*, *five-membered chelate ring* **16.7** (Figure 16.4a). When it adopts this type of bonding mode, a ligand is said to be *chelating*, and it is usually the case that complexes containing chelating ligands possess an enhanced stability over complexes with similar monodentate ligands. Five- and six-membered chelate rings are particularly common. The $[acac]^-$ ligand forms a *planar six-membered* chelate ring **16.8** as in $[Cr(acac)_3]$ (Figure 16.5).

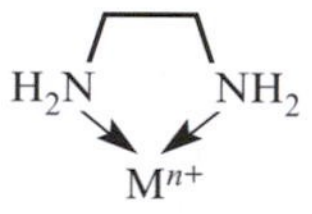

H_2N NH_2 M^{n+}

(16.7)

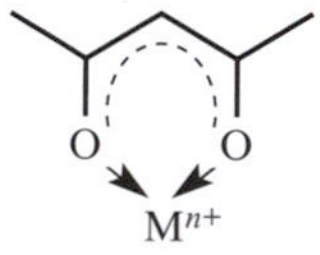

O O M^{n+}

(16.8)

The attachment of two donor atoms from one ligand to a single metal ion forms a chelate ring.

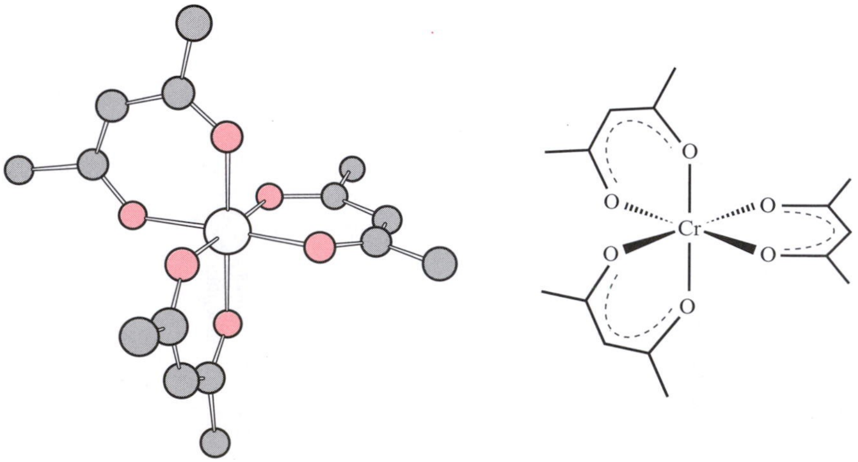

16.5 The solid state structure of $[Cr(acac)_3]$ and a schematic representation of the complex. Each of the chelate rings is planar.

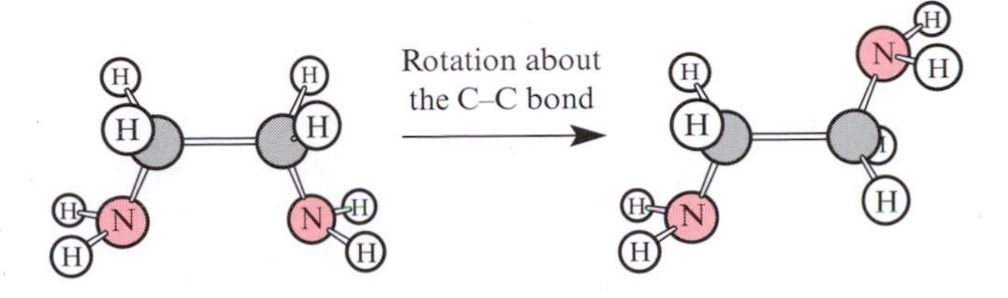

16.6 The ligand 1,2-diaminoethane (en) can undergo free bond rotation about C–C and C–N single bonds, and the results is a change in conformation. The diagram illustrates bond rotation about the C–C bond.

Conformers: see Section 8.5

The ability of 1,2-diaminoethane to function as either a didentate or a monodentate ligand arises because the N–C–C–N chain can alter its conformation by rotation about the C–C bond (Figure 16.6) or the C–N bonds. Some didentate ligands such as 1,10-phenanthroline **16.9** (abbreviated to phen) have a rigid framework and a chelated mode is usually forced upon the ligand. In Section 9.5, we saw how a related ligand 4,7-diphenyl-1,10-phenanthroline **9.1** was used to analyse for iron(II) ions. The complex formed in this case is the octahedral dication $[FeL_3]^{2+}$, where L is ligand **9.1**. The ligand phen forms a similar complex $[Fe(phen)_3]^{2+}$.

(16.9) (16.10)

Pyridine: see Section 15.12

A ligand such as 2,2′-bipyridine (bpy) **16.10** may undergo free bond rotation about the C–C bond connecting the two pyridine rings. However, in most of its complexes, this ligand acts as a didentate ligand and forms a planar, five-membered chelate ring (Figure 16.7).

A tridentate ligand possesses three donor atoms. An example is 2,2′:6′,2″-terpyridine (abbreviated to tpy) and Figure 16.8 shows its structure and that of the complex $[Fe(tpy)_2]^{2+}$. Ligands with four, five and six-donor atoms are tetradentate, pentadentate and hexadentate, although collectively, any ligand

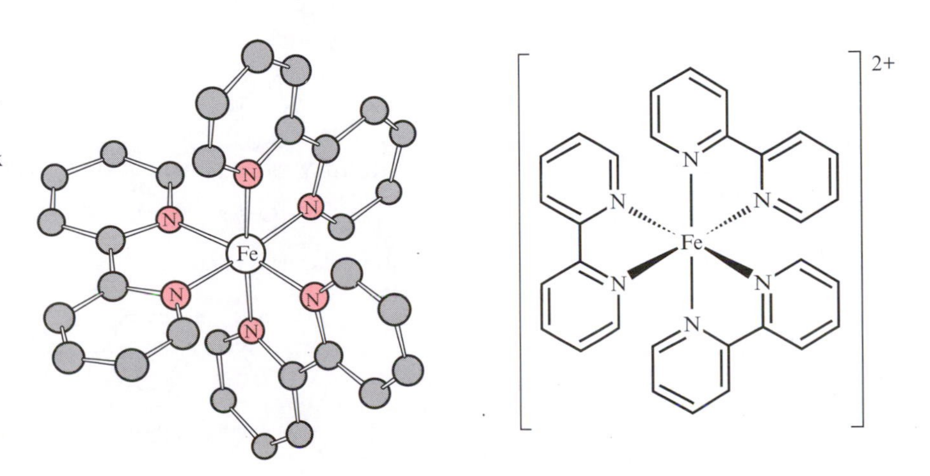

16.7 The stucture of $[Fe(bpy)_3]^{2+}$, determined by X-ray diffraction methods for the tartrate salt. A schematic representation of the complex cation is also shown.

Box 16.2 Molecular architecture by ligand design

The emphasis in this chapter is on coordination compounds which contain a single *d*-block metal centre – *mononuclear* complexes. Ligands such as 2,2′-bipyridine and phenanthroline usually act in a didentate mode, binding to one metal centre, although bpy is more flexible than phen. By designing ligands with donor atoms in particular sites, it is possible to encourage the formation of *polynuclear* complexes. An example is the ligand 4,4′-bipyridine, in which the two nitrogen donor atoms are remote from each other. The only flexibility that the ligand has is rotation about the inter-ring C–C bond.

4,4′-Bipyridine

4,4′-Bipyridine cannot act as a chelating ligand *but* it can coordinate to two different metal atoms and has the potential to be a building-block in a polymeric structure. This is seen in the silver(I) complex $[Ag(4,4'\text{-bpy})_2][CF_3SO_3]$ which (in the solid state) possesses a network structure with the 4,4′-bipyridine ligands connecting pairs of silver(I) centres; the $[CF_3SO_3]^-$ anions are distributed within the network and are not shown in the structural diagram below:

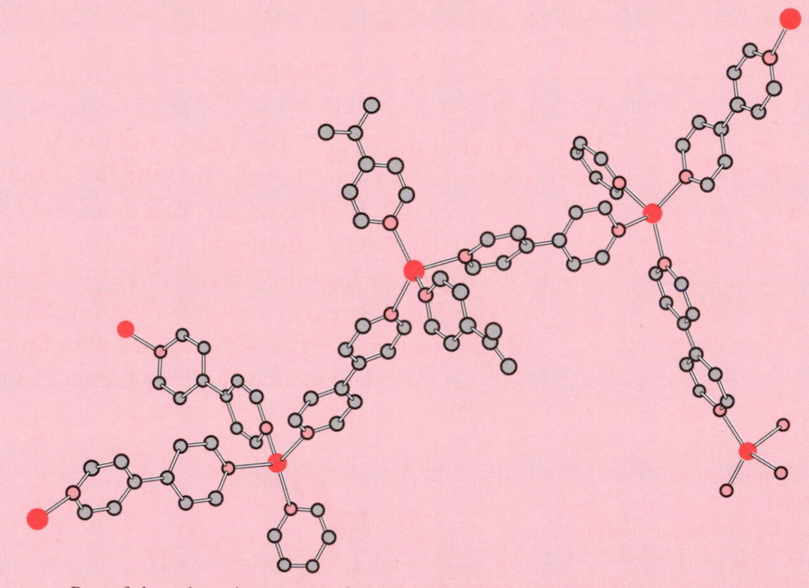

Part of the polymeric structure of $[Ag(4,4'\text{-bpy})_2][CF_3SO_3]$ with the anions omitted

○ = N ○ = C ● = Ag

3,6-Di(2-pyridyl)pyridazine is related to 2,2′-bipyridine and is tetradentate:

3,6-di(2-pyridyl)pyridazine

With copper(I) ions, it forms a tetranuclear complex, the structure of which is a consequence of the arrangement of the nitrogen donor atoms *and* the tetrahedral geometry of the copper(I) centres.

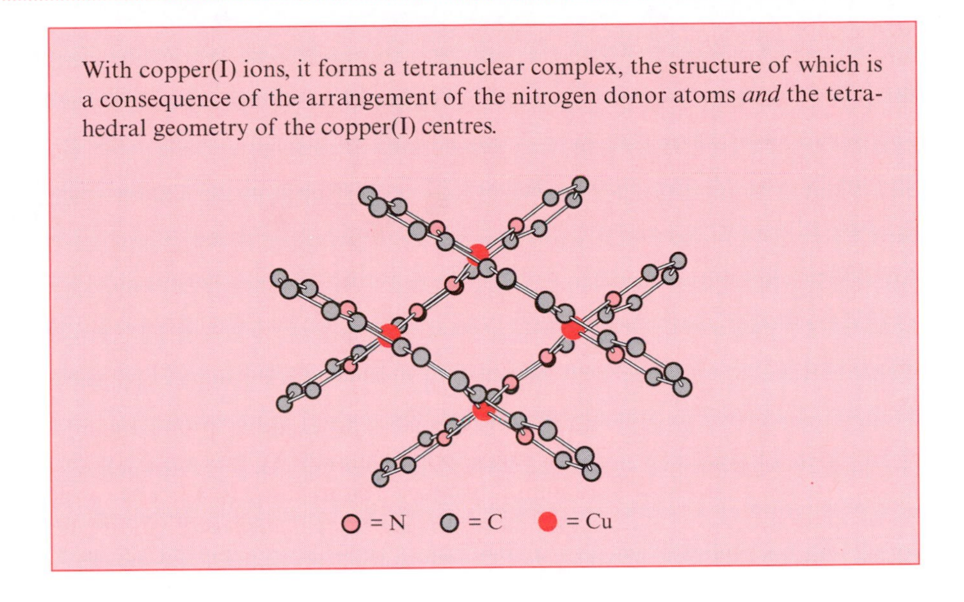

with more than one donor atom may be referred to as being *polydentate*. Some common ligands and examples of complexes involving them are listed in Table 16.2.

Cyclic ethers and thioethers

Cyclic ethers and crowns: see Section 15.3 and Box 15.2

The oxygen atoms in a cyclic ether are potential donor atoms. An ether such as THF (Table 16.2) can only function as a monodentate ligand as in the complex anion *trans*-$[YCl_4(THF)_2]^-$ (Figure 16.9). Larger cyclic ethers with nine or more atoms and three or more oxygen atoms are called *macrocyclic ligands*

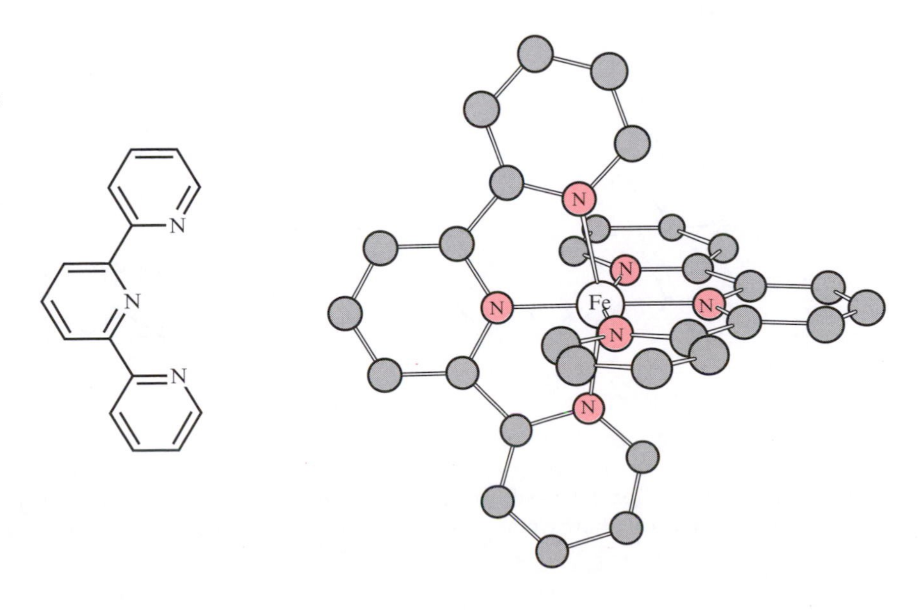

16.8 The structures of the terpyridine ligand (tpy) and the complex $[Fe(tpy)_2]^{2+}$. This structure was determined by X-ray diffraction methods for the perchlorate salt $[Fe(tpy)_2][ClO_4]_2$.

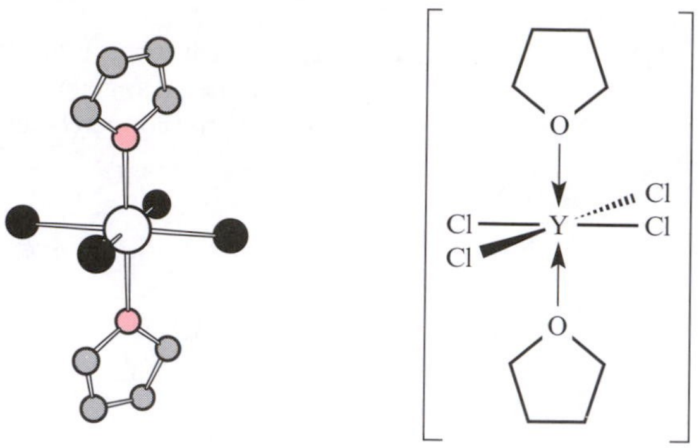

16.9 Cyclic ethers form some complexes with *early d*-block metals. Tetrahydrofuran (THF) can act as a monodentate ligand as in the yttrium(III) complex *trans*-$[YCl_4(THF)_2]^-$.

and are a special type of polydentate ligand; examples include 1,4,7,10-tetraoxacyclododecane **16.11** and 1,4,7,10,13,16-hexaoxacyclooctadecane (commonly called 18-crown-6) **16.12**. The oxygen atoms can be replaced by sulfur atoms to give *cyclic thioethers*; ligand **16.13** is 1,4,7-trithiacyclononane.

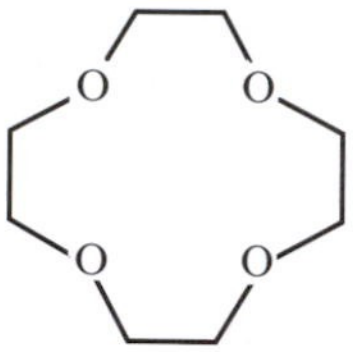

(16.11) 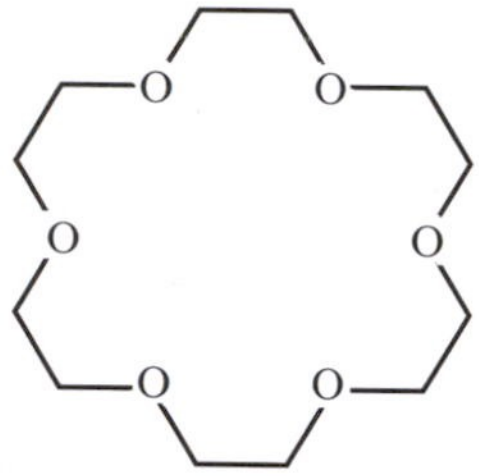 (16.12) 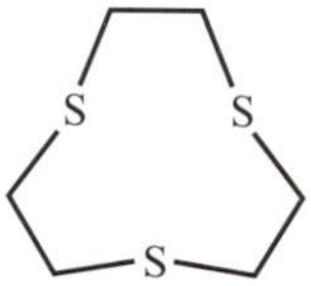 (16.13)

Box 16.3 Porphyrin ligands

Some biologically important macrocyclic ligands contain a porphyrin ring. The parent porphyrin is of the type shown on the left-hand side below – the R groups may be, for example, Ph in an isolated porphyrin or may be linkages into protein chains as in haemoglobin (Box 4.4).

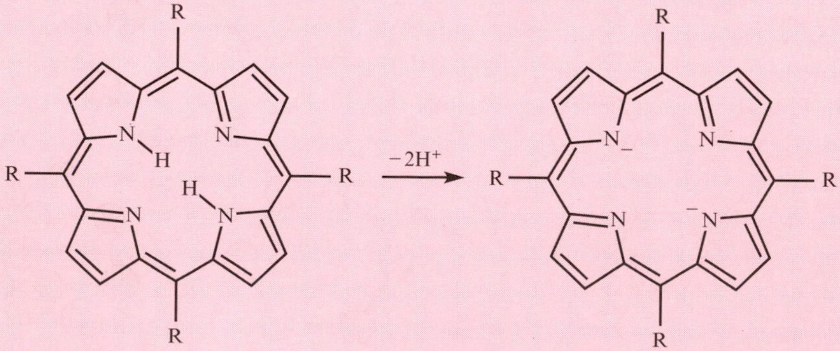

$-2H^+$

The central cavity of a porphyrin molecule contains two N-donor atoms and two NH groups. When these are deprotonated, the *porphyrinato ligand* is formed and the four nitrogen donor atoms can coordinate to a variety of metal centres. In Box 4.4, we illustrated coordination to iron. Chlorophyll (the green pigment in plants which is responsible for photosynthesis) contains a porphyrin coordinated to Mg^{2+}.

A macrocyclic ligand possesses nine more atoms and three or more donor atoms. In the case of a cyclic ether, the donor atoms are oxygens.

In Box 15.2 we described how crown ethers can interact with alkali metal ions, and a few similar complexes are known to form between cyclic ethers and *early* metals in the *d*-block, for example scandium(III). Cyclic *thioethers* form complexes with a range of *d*-block metal ions. When ligand **16.13** coordinates to a metal centre, the ion resides out of the plane containing the three sulfur atoms (Figure 16.10a) but larger cyclic ethers provide a cavity that can accommodate the metal ion within the ring as the structure of [Sc(benzo-15-crown-5)Cl_2]$^+$ illustrates (Figure 16.10b).

Donor atom notation

When we write the formula of a complex ion such as [Co(en)$_3$]$^{2+}$, there is an in-built assumption that the en ligand coordinates through both the nitrogen donor atoms. However, we have also seen that the en ligand can coordinate in a monodentate fashion. In order to distinguish between these two situations, we can add suffixes to the ligand name which denote the donor atoms involved in coordination. In a monodentate mode, the ligand may be specified as en-*N*, and in a didentate mode, as en-*N*,*N'*. Notice two features of this notation:

- the symbols for the donor atoms are written in italic;
- the two *different* nitrogen donor atoms are distinguished by using a prime mark on the second atom.

The structure shown in Figure 16.4b is of the compound *mer*-[IrCl_3(en-*N*)(en-*N*,*N'*)]. Further examples of the use of this notation are shown in the donor atom list in Table 16.2; additional prime marks are added for additional like-atoms, for example in the [edta]$^{4-}$ ligand.

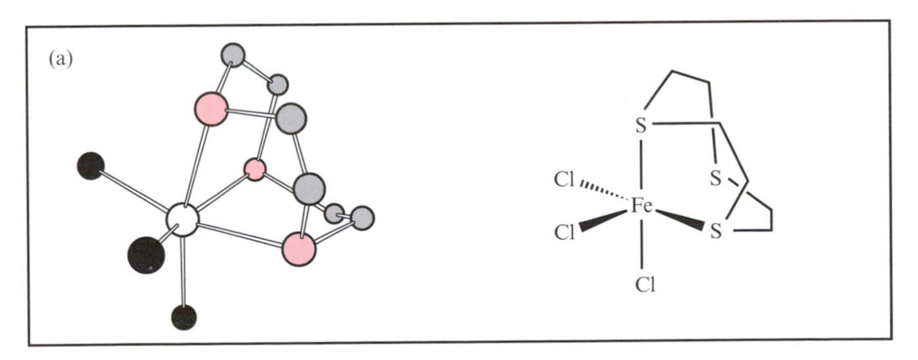

16.10 (a) When a macrocyclic ligand coordinates to a metal ion, the metal may lie above the plane containing the donor atoms as in the iron (III) complex [Fe(L)Cl_3] where L is 1,4,7-trithiacyclononane (ligand **16.13**). (b) Larger macrocyclic ligands may be able to accommodate the metal ion within the ring as in the scandium (III) complex *trans*-[ScCl_2(benzo-15-crown-5)]$^+$; the geometry of the coordination sphere is a pentagonal bipyramid.

16.4 THE ELECTRONEUTRALITY PRINCIPLE

In Chapter 4 we discussed polar covalent bonds, and in Chapter 6 we considered ionic compounds but emphasized that the ionic model may be adjusted to take account of some covalent character. The coordinate bond formed between the donor atom of a ligand and a metal ion is a particular type of bond that requires some thought. What we are implying by the representation shown in Figure 16.11a for $[Co(NH_3)_6]^{3+}$ is that a total of six *pairs of electrons* are donated from the ligands to the metal centre. In the *covalently* bonded representation, a single net unit of charge is transferred from each ligand to the cobalt(III) centre and Figure 16.11b shows the resulting charge distribution. The differences between Figures 16.11a and 16.11b are equivalent to the differences between the two representations of the bonding in $THF{\bullet}BH_3$ in structures **11.1** and **11.2**.

Now consider the consequence of assuming a wholly *ionic* model for the bonding in $[Co(NH_3)_6]^{3+}$. The tripositive charge remains localized on the cobalt ion and the six ammonia molecules remain neutral (Figure 16.11c). Which of these two models is correct, or is an intermediate situation realistic?

The solution and solid state properties of coordination compounds suggest that the ionic model is not satisfactory; for example, the compound $[Co(NH_3)_6]Cl_3$ dissolves in water to give $[Co(NH_3)_6]^{3+}$ and Cl^- ions, rather than Co^{3+} ions, Cl^- ions and NH_3 molecules. Simple electrostatic interactions as indicated by the ionic model are not likely to be sufficiently strong to allow the $[Co(NH_3)_6]^{3+}$ complex ion to be maintained in aqueous solution. If we return to the covalent model for $[Co(NH_3)_6]^{3+}$, the charge distribution in Figure 16.11b causes a problem in terms of the electron-withdrawing capacities of the cobalt(III) and nitrogen centres. The metal centre is

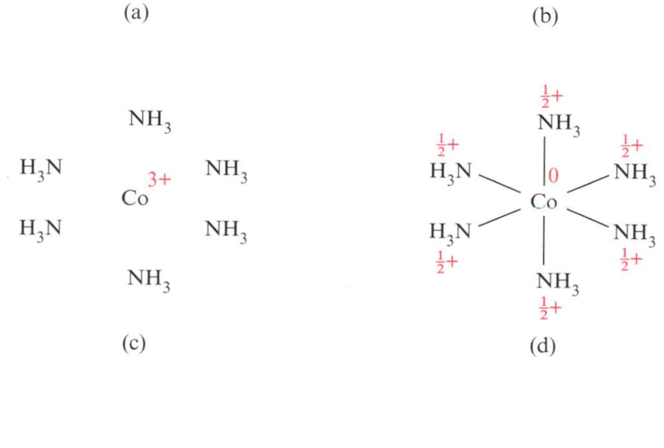

16.11 The complex cation $[Co(NH_3)_6]^{3+}$ (a) showing the presence of coordinate bonds, (b) showing the charge distribution that results from a 100% covalent model, (c) showing the charge distribution that results from a 100% ionic model, and (d) showing the charge distribution that results by applying the electroneutrality principle.

electropositive whereas the nitrogen is *electronegative* and we expect the cobalt–nitrogen bond to be polarized in the sense $Co^{\delta+}$–$N^{\delta-}$. This suggests that the charge distribution in Figure 16.11b is unrealistic.

Neither a wholly ionic nor a wholly covalent model is appropriate to describe the interactions between ligands and metal ions. The problem was addressed by Pauling who put forward the *electroneutrality principle*. This states that the distribution of charge in a molecule or ion is such that the charge on any single atom is within the range +1 to –1, and ideally should be close to zero.

Linus Pauling: see Box 4.1

Pauling's *electroneutrality principle* states that the distribution of charge in a molecule or ion is such that the charge on any single atom is within the range +1 to –1 (ideally close to zero).

If we apply the electroneutrality principle to the $[Co(NH_3)_6]^{3+}$ ion, then, ideally, the net charge on the cobalt centre should be zero. That is, the Co^{3+} ion may accept a total of only three electrons from the six NH_3 ligands, and this leads to the charge distribution shown in Figure 16.11d. The charge transferred from each N to Co is only half an electron and this is midway between the limiting covalent (one electron transferred) and ionic (no electrons transferred) models. The electroneutrality principle results in a bonding description for the $[Co(NH_3)_6]^{3+}$ ion which is 50% ionic (or, 50% covalent). **Note:** The electroneutrality principle is only an *approximate method* of estimating the charge distribution in molecules and complex ions.

Worked example 16.1 ***The electroneutrality principle***

Estimate the charge distriubtion on the metal and each donor atom in the octahedral ion $[Fe(CN)_6]^{4-}$.

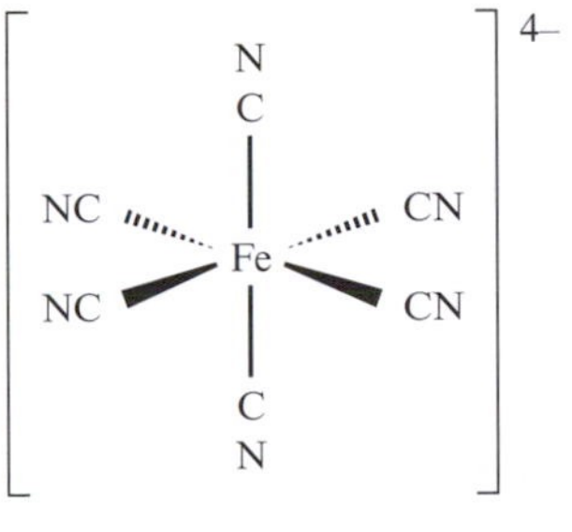

The ion $Fe(CN)_6]^{4-}$ is an iron(II) complex, and the ionic model would imply a charge distribution of:

N
C$^-$

$^-$NC $\quad$ Fe^{2+} $\quad$ CN$^-$

$^-$NC $\quad\quad\quad$ CN$^-$

C$^-$
N

while 100% covalent character leads to a charge distribution of:

N
C ^{0}

NC ^{0} $\quad$ Fe $^{4-}$ $\quad$ CN ^{0}

NC ^{0} $\quad\quad\quad$ CN ^{0}

C ^{0}
N

By the electroneutrality principle, the charge distribution should ideally be such that the iron centre bears no net charge. It must therefore gain two electrons (to counter the Fe^{2+} charge) from the six $[CN]^-$ ligands.

Each $[CN]^-$ therefore loses $\frac{1}{3}$ of an electron. The charge per cyano ligand therefore changes from 1– to $\frac{2}{3}$–.

The final charge distribution is approximately:

N
C $\frac{2}{3}-$

NC $\frac{2}{3}-$ $\quad$ Fe ^{0} $\quad$ CN $\frac{2}{3}-$

NC $\frac{2}{3}-$ $\quad\quad\quad$ CN $\frac{2}{3}-$

C $\frac{2}{3}-$
N

16.5 ISOMERISM

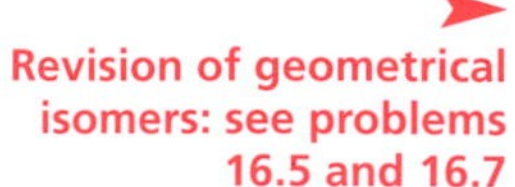

Revision of geometrical isomers: see problems 16.5 and 16.7

The discussion of geometrical isomerism in Section 5.11 included examples of mononuclear complexes of the *d*-block metals. We have also discussed optical isomers in the context of chiral organic molecules (Section 8.6) – the real test for chirality is the non-superimposability of the mirror image of an object on the original object. Both optical and geometrical isomers are types of *stereoisomers* in which the molecules are *spatially different* but pairs of atoms in a given pair of isomers remain bonded to one another. Look back at Figures 8.10 and 8.11 and at the structures in Section 5.11 to confirm this statement.

Geometrical isomers include *cis*- and *trans*-isomers, and *mer*- and *fac*-isomers.

In this section we consider types of *structural isomerism* that coordination complexes may exhibit, and extend our discussion of optical isomers. Structural isomers include hydration, ionization, linkage and coordination isomers and differ from stereoisomers in that one or more pairs of atoms in a given pair of isomers do *not* remain bonded to one another.

Hydration and ionization isomers

Hydration and ionization isomers result from the interchange of ligands within the first coordination sphere with those outside it. The compound $CrCl_3(H_2O)_6$ is obtained commercially as a green solid and has the structural formula $[CrCl_2(H_2O)_4]Cl{\bullet}2H_2O$ – two chloride ions are bonded to the chromium(II) centre and there is one non-coordinated chloride ion and two water molecules of crystallization. When this solid is dissolved in water, the solution slowly turns blue-green and this transformation is the result of an isomerization in which one chloride ligand is replaced by a water molecule; crystallization yields $[CrCl(H_2O)_5]Cl_2{\bullet}H_2O$. The compound consists of the complex cation $[CrCl(H_2O)_5]^{2+}$ **16.14**, two Cl^- ions and a water molecule of crystallization.

(16.14)

Ionization isomers result from an exchange involving an anionic ligand. The purple complex $[Co(NH_3)_5(SO_4)]Br$ and the pink complex $[Co(NH_3)_5Br][SO_4]$ are ionization isomers, and differ in having either a Br^- or $[SO_4]^{2-}$ ion coordinated (or not) to the cobalt(III) centre.

Hydration and ionization isomers of a compound are the result of the exchange of ligands within the first coordination sphere with those outside it. Hydration isomers involve the exchange of water, and ionization isomers involve the exchange of anionic ligands.

Linkage isomerism

$\bar{N}{=}C{=}S$

(16.15)

If a ligand possesses two or more different donor sites, there is more than one way in which it can bond to a metal ion. One resonance structure of the thiocyanate ion $[NCS]^-$ is shown in **16.15**; what other resonance structures may make significant contributions? Both the nitrogen and sulfur atoms are potential donor atoms, and *linkage isomers* of a complex may exist.

Structures **16.16** and **16.17** illustrate linkage isomers of the complex cation $[Co(NH_3)_5(NCS)]^{2+}$ in which the thiocyanate ligand is either *S*- or *N*-bonded.

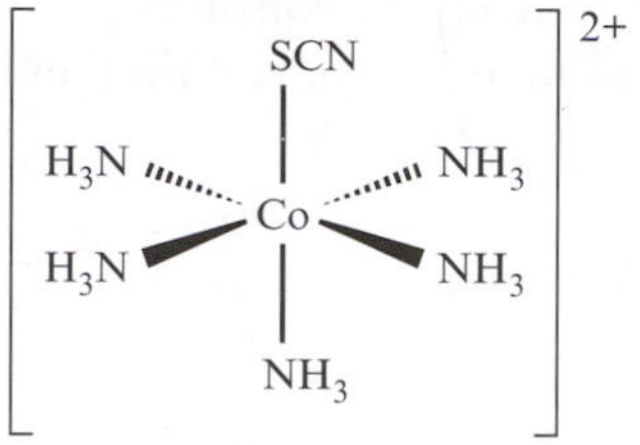

(16.16)

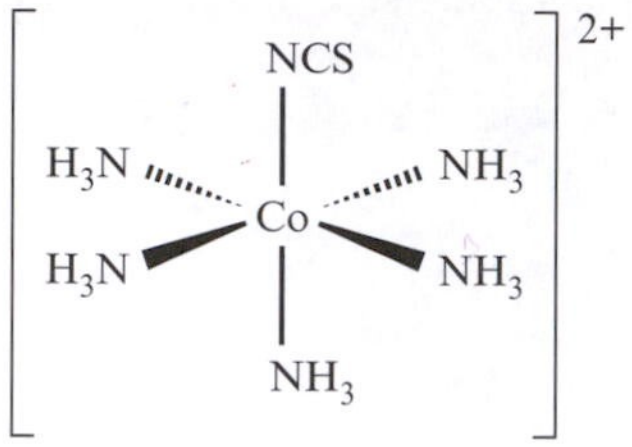

(16.17)

If a ligand possesses two or more different donor sites, linkage isomers may be formed because the ligand can attach to the metal centre in more than one way.

Coordination isomerism

If a coordination compound contains both a complex cation *and* a complex anion, it may be possible to isolate isomers of the compound in which ligands are exchanged between the two metal centres. Such complexes are examples of *coordination isomers*. Consider $[Co(bpy)_3][Fe(CN)_6]$ which contains the cation $[Co(bpy)_3]^{3+}$ and the anion $[Fe(CN)_6]^{3-}$. This is not the only arrangement of the ligands, and other possibilities are:

$[Co(bpy)_2(CN)_2]^+$ $[Fe(bpy)(CN)_4]^-$
$[Fe(bpy)_2(CN)_2]^+$ $[Co(bpy)(CN)_4]^-$
$[Fe(bpy)_3]^{3+}$ $[Co(CN)_6]^{3-}$

Each of these is a distinct chemical compound and may be isolated and characterized. For example, $[Co(bpy)_3][Fe(CN)_6]$ is prepared from the reaction of $K_3[Fe(CN)_6]$ and $[Co(bpy)_3]Cl_3$.

If a coordination compound contains both a complex cation *and* a complex anion, it may be possible to isolate coordination isomers in which ligands are exchanged between the two different metal centres.

Optical isomerism

Optically active isomers (*enantiomers*) of a coordination compound most often occur when chelating ligands are involved. The cation *cis*-$[Co(en)_2Cl_2]^+$ and its mirror image are shown in Figure 16.12a. Figure 16.12b shows that the mirror image (labelled isomer B) is non-superimposable upon the original

16.12 The complex cation $[Co(en)_2Cl_2]^+$ is optically active. (a) Isomers A and B are related by a reflection through a mirror plane. (b) Rotating isomer B shows that it is the *non-superimposable mirror image* of isomer A. A and B are therefore enantiomers.

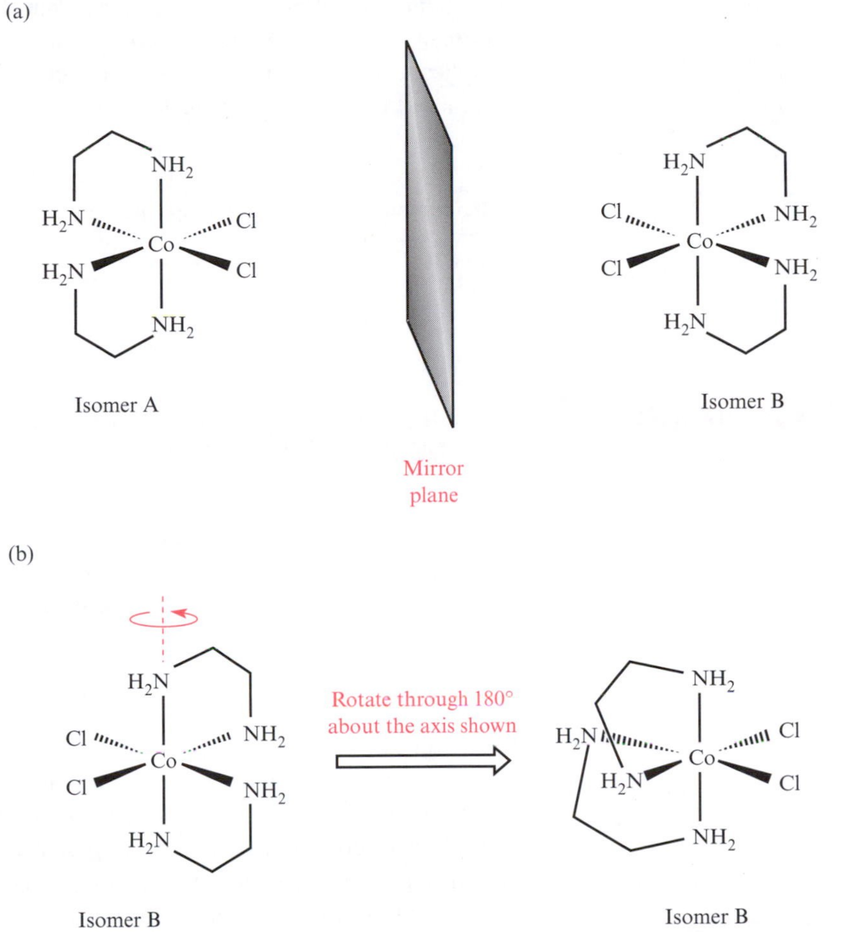

(a)

Isomer A

Mirror plane

Isomer B

(b)

Rotate through 180° about the axis shown

Isomer B

Isomer B

structure (labelled isomer A). Isomers A and B are optical isomers and *cis*-$[Co(en)_2Cl_2]^+$ is a chiral cation. Similarly, $[Cr(acac)_3]$ is optically active because it has a non-superimposable mirror image (Figure 16.13). On the

16.13 Optical isomers of the coordination complex $[Cr(acac)_3]$.

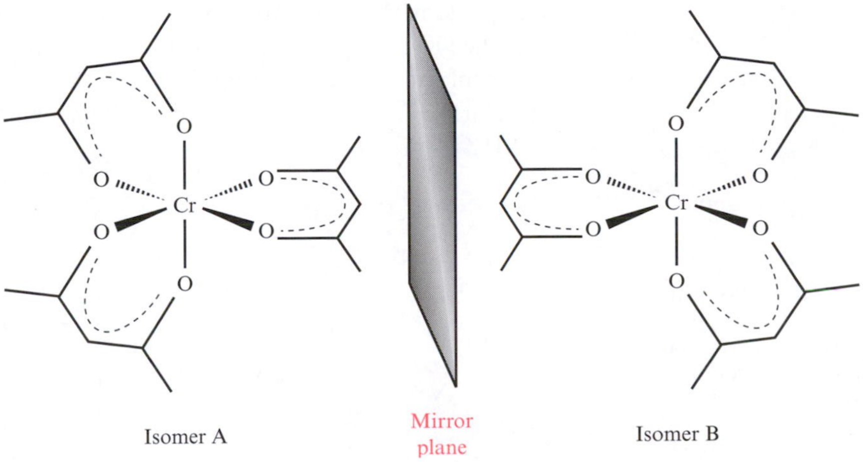

Isomer A

Mirror plane

Isomer B

other hand, you should be able to show that the octahedral complexes *trans*-$[Co(en)_2Cl_2]^+$ and $[Mn(en)(H_2O)_4]^{2+}$ are *not* optically active. Further complications arise if the ligand itself is chiral, and this and additional questions about optical activity are addressed in the accompanying workbook.

If a complex and its mirror image are non-superimposable, the compound is optically active.

16.6 THE FORMATION OF FIRST ROW *d*-BLOCK METAL COORDINATION COMPLEXES

Group number	3	4	5	6	7	8	9	10	11	12
First row *d*-block metal	Sc	Ti	V	Cr	Mn	Fe	Co	Ni	Cu	Zn

Aqua complexes and complexes with hydroxide ligands

The simplest method of forming a complex ion is to dissolve a *d*-block metal salt such as anhydrous cobalt(II) chloride in water. On dissolution, the cobalt(II) ion is present as the *hexaaqua ion* $[Co(H_2O)_6]^{2+}$ (Figure 16.14), and solutions containing this ion are pale pink. Although 'naked' Co^{2+} ions (and other *d*-block metal ions) are *not present in aqueous solution*, you will often find $Co^{2+}(aq)$ written in equations. This is strictly incorrect – water molecules are present in the first coordination sphere of the metal ion.

Hexaaqua ions such as $[Co(H_2O)_6]^{2+}$ are found for all of the first row *d*-block *metal(II) ions*, although the $[M(H_2O)_6]^{2+}$ ions are not always stable in aqueous solution. For example, $[Ti(H_2O)_6]^{2+}$ is not stable and solutions readily oxidize to the purple hexaaqua ion $[Ti(H_2O)_6]^{3+}$. However, since the charge density on the titanium(III) centre is high, it is strongly polarizing and the $[Ti(H_2O)_6]^{3+}$ cation readily loses a proton (equation 16.1). In dilute hydrochloric acid, equilibrium 16.1 is pushed towards the left-hand side, and the $[Ti(H_2O)_6]^{3+}$ cation is stabilized. This behaviour resembles that of the $[Al(H_2O)_6]^{3+}$ ion.

Polarizing power: see Section 6.18

$[Al(H_2O)_6]^{3+}$: see equation 13.17

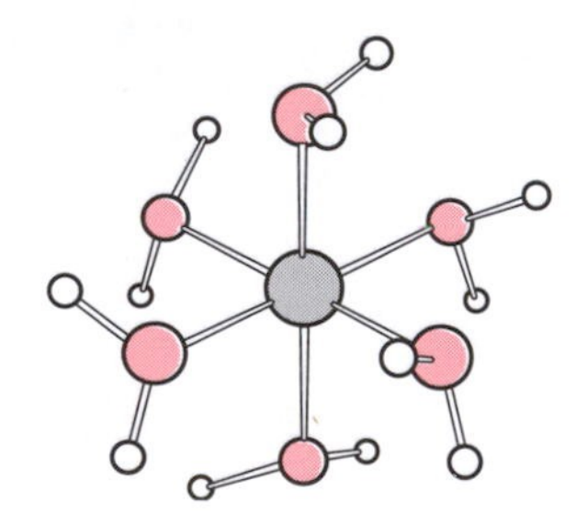

$$\left[\begin{array}{ccc} & OH_2 & \\ H_2O & \rightarrow Co \leftarrow & OH_2 \\ H_2O & & OH_2 \\ & OH_2 & \end{array} \right]^{2+}$$

16.14 The octahedral structure of the complex cation $[Co(H_2O_6]^{2+}$ and a schematic representation showing the coordinate bonds.

$$[Ti(H_2O)_6]^{3+}(aq) + H_2O(l) \rightleftharpoons [Ti(H_2O)_5(OH)]^{2+}(aq) + [H_3O]^+(aq) \quad pK_a = 3.9 \qquad (16.1)$$

The next metal in the first row of the *d*-block is vanadium. Pale purple solutions containing the $[V(H_2O)_6]^{2+}$ ion are readily oxidized and solutions turn green as the $[V(H_2O)_6]^{3+}$ ion is formed. (This oxidation process is not simple and we consider it further in the accompanying workbook.) Although $[V(H_2O)_6]^{2+}$ is the main species present in an aqueous solution of a vanadium(II) salt, the vanadium(III) ion is strongly polarizing and $[V(H_2O)_6]^{3+}$ is acidic (equation 16.2); the pK_a value shows it to be a stronger acid than $[Ti(H_2O)_6]^{3+}$, and $[V(H_2O)_6]^{3+}$ is only stabilized in acidic solutions.

$$[V(H_2O)_6]^{3+}(aq) + H_2O(l) \rightleftharpoons [V(H_2O)_5(OH)]^{2+}(aq) + [H_3O]^+(aq) \quad pK_a = 2.9 \qquad (16.2)$$

The formation of a *hydroxo-ligand*, $[OH]^-$, is often associated with a change in the *nuclearity* of the solution species. Equation 16.2 shows the formation of a *mononuclear* complex cation (i.e. one vanadium centre), but a *dinuclear* solution species is also present and this is possible because of the formation of *hydroxo-bridges* between two vanadium centres. The presence of a ligand in a bridging mode is denoted in the formula by the Greek letter μ (pronounced 'mu'), written as a prefix – the structure of $[(H_2O)_4V(\mu\text{-}OH)_2V(H_2O)_4]^{4+}$ is shown in structure **16.18**. It is the hydroxo and not the aqua ligands that adopt the bridging sites.

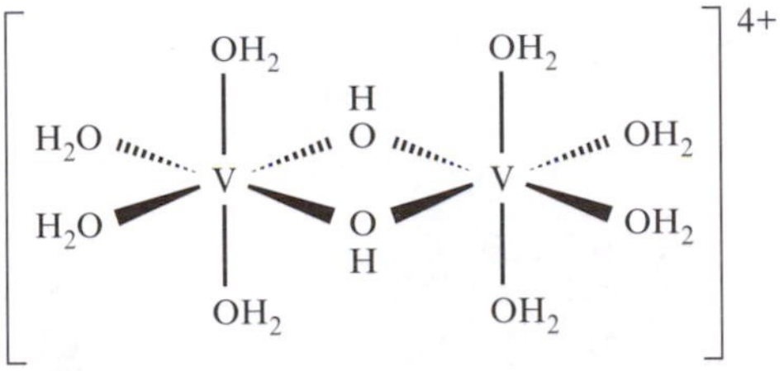

(16.18)

The nuclearity of a metal complex refers to the number of metal centres. A mononuclear complex has one metal centre, a dinuclear complex contains two and a trinuclear complex possesses three.

In a polynuclear coordination compound, the use of the prefix μ- means that the ligand is in a bridging position between two metal centres.

Vanadium(IV) is even more polarizing than vanadium(III) and the $[V(H_2O)_6]^{4+}$ ion does not exist because it readily loses two protons *from the same aqua ligand* to give $[V(O)(H_2O)_5]^{2+}$ **16.19** which contains a V=O double bond. Solutions containing this ion are blue. You will often find the formula

Box 16.4 Layered structures: a vanadium(IV) oxoorganophosphate

In the design of new materials, an aim is to achieve compounds with structures that possess particular features – for example, layered structures with channels that might be occupied by particular types of small molecules. In the solid state, the salt $[Et_2NH_2][Me_2NH_2][V_4O_4(OH)_2(PhPO_3)_4]$ has a structure consisting of layers of hydroxo-bridged vanadium atoms, each of which is also bonded to an oxo group and to the oxygen atoms of an organophosphate ligand. Each vanadium atom is 5-coordinate and in a single layer, all the V=O bonds are parallel to each other and lie on the same side of the layer. The first structure below views one layer looking down the O=V bonds:

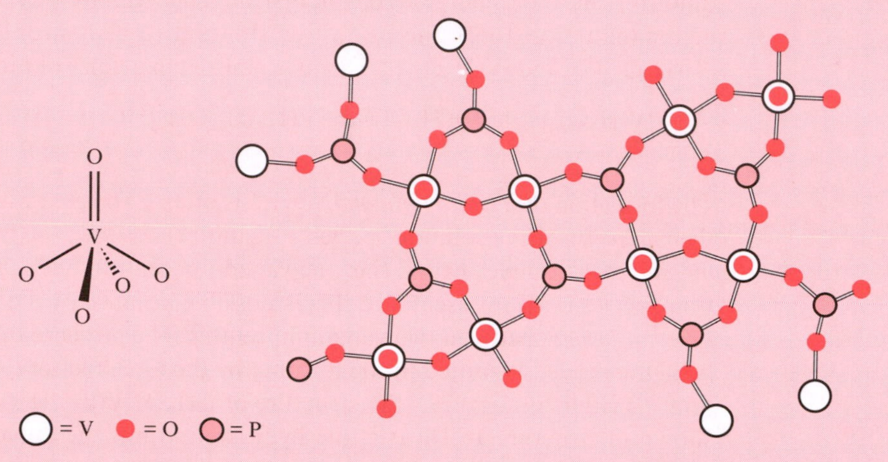

In the crystal lattice, alternate layers face in opposite directions, providing the overall structure with hydrophobic and hydrophilic channels (see Figure 14.14) which are occupied by the $[Et_2NH_2]^+$ and $[Me_2NH_2]^+$ cations. The diagram below shows a space-filling diagram of two layers with the cations in one of the channels; hydrogen atoms have been omitted from the diagram.

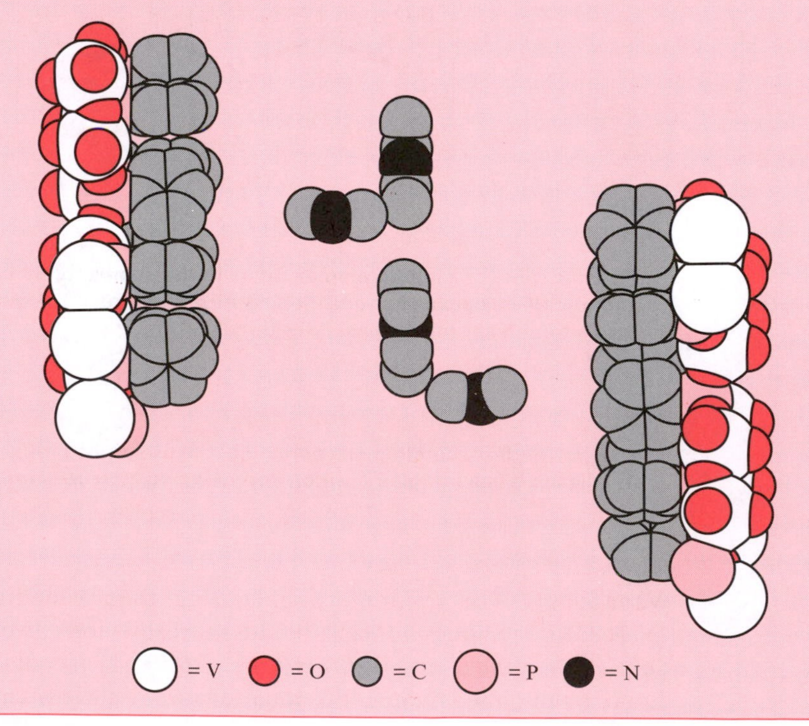

of this ion written simply as $[VO]^{2+}$; it is known as the oxovanadium(IV) or vanadyl ion.

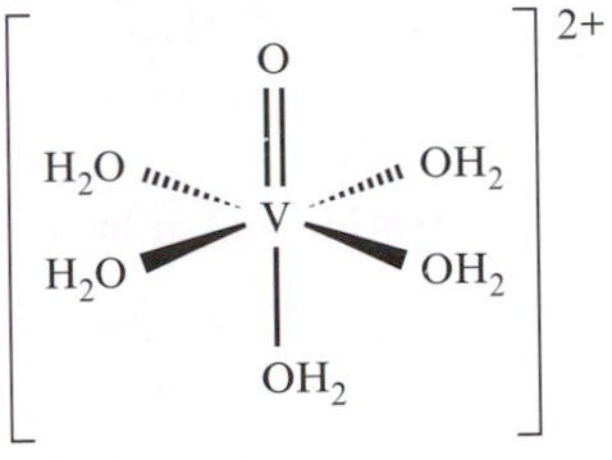

(16.19)

Chromium(II) is also readily oxidized in aqueous solution; solutions containing the $[Cr(H_2O)_6]^{2+}$ ion are sky-blue and oxidation gives the violet $[Cr(H_2O)_6]^{3+}$ ion. Once again, the polarizing metal(III) ion leads to easy proton loss (equation 16.3) and the formation of a hydroxo-complex which condenses to a dinuclear complex (equation 16.4) with a structure similar to **16.18**.

$$[Cr(H_2O)_6]^{3+}(aq) + H_2O(l) \rightleftharpoons [Cr(H_2O)_5(OH)]^{2+}(aq) + [H_3O]^+(aq) \quad pK_a = 4.0 \tag{16.3}$$

$$2[Cr(H_2O)_5(OH)]^{2+}(aq) \rightleftharpoons [(H_2O)_4Cr(\mu\text{-}OH)_2Cr(H_2O)_4]^{4+} + 2H_2O \tag{16.4}$$

Manganese follows chromium in the first row of the *d*-block, and manganese(II) salts such as $MnCl_2{\bullet}4H_2O$, $Mn(NO_3)_2{\bullet}6H_2O$ and $MnSO_4{\bullet}H_2O$ dissolve in water to give very pale pink solutions containing the octahedral $[Mn(H_2O)_6]^{2+}$ ion. Iron(II) salts such as $FeSO_4{\bullet}7H_2O$ dissolve to give pale green aqueous solutions, although atmospheric O_2 oxidizes iron(II) to iron(III). Solutions containing the $[Fe(H_2O)_6]^{3+}$ hexaaqua ion are purple but when it is prepared, solutions usually appear yellow due to the formation of the hydroxo species $[Fe(H_2O)_5(OH)]^{2+}$ and $[Fe(H_2O)_4(OH)_2]^+$ (equations 16.5 and 16.6). These observations emphasize the polarizing power of the Fe^{3+} ion. In equilibria 16.5 and 16.6, *successive* aqua ligands are deprotonated. The purple colour of $[Fe(H_2O)_6]^{3+}$ can be seen in the alum $KFe(SO_4)_2{\bullet}12H_2O$.

Alums: see Section 13.4

$$[Fe(H_2O)_6]^{3+}(aq) + H_2O(l) \rightleftharpoons [Fe(H_2O)_5(OH)]^{2+}(aq) + [H_3O]^+(aq) \quad pK_a = 2.0 \tag{16.5}$$

$$[Fe(H_2O)_5(OH)]^{2+}(aq) + H_2O(l) \rightleftharpoons [Fe(H_2O)_4(OH)_2]^+(aq) + [H_3O]^+(aq) \quad pK_a = 3.3 \tag{16.6}$$

As with the hydroxo complexes of vanadium and chromium described above, those of iron(III) undergo condensation reactions to form dinuclear compounds such as $[(H_2O)_4Fe(\mu\text{-}OH)_2Fe(H_2O)_4]^{4+}$. This type of process continues until, finally, solid state iron(III) hydroxides are produced.

The later first row *d*-block elements all form metal(II) hexaaqua ions; solutions containing the $[Co(H_2O)_6]^{2+}$ ion are pale pink, those of $[Ni(H_2O)_6]^{2+}$ are pale green, and solutions with the $[Cu(H_2O)_6]^{2+}$ cation are pale blue. From titanium to nickel, all the solution species that we have described have been coloured, but when zinc(II) salts dissolve in water, colourless solutions are formed even though they contain an aqua ion of zinc(II), $[Zn(H_2O)_6]^{2+}$. We return to the question of colour in Section 16.10.

Ammine complexes

The addition of ammonia to an aqueous solution of *d*-block metal(II) ions *may* result in the *displacement* of aqua by ammine ligands.§ The number of ligands that are displaced depends upon the metal and the concentrations of the aqua complex and NH_3 in solution. Equation 16.7 shows that the reaction between $[Ni(H_2O)_6]^{2+}$ and NH_3 leads to the displacement of all six ligands, while equilibrium 16.8 shows that only four of the six aqua ligands in $[Cu(H_2O)_6]^{2+}$ are readily displaced.

$$[Ni(H_2O)_6]^{2+}(aq) + 6NH_3(aq) \rightleftharpoons [Ni(NH_3)_6]^{2+}(aq) + 6H_2O(l) \qquad (16.7)$$

$$[Cu(H_2O)_6]^{2+}(aq) + 4NH_3(aq) \rightleftharpoons [Cu(H_2O)_2(NH_3)_4]^{2+}(aq) + 4H_2O(l) \qquad (16.8)$$

In a ligand displacement reaction, one ligand in the coordination sphere of the metal ion is replaced by another.

In Section 16.7 we consider the equilibrium constants for reactions of this type. Displacement reactions are not always the best methods of preparing ammine complexes and the reactions of metal salts with gaseous ammonia may provide alternative routes. Solid iron(II) halides, FeX_2, react with gaseous NH_3 to give $[Fe(NH_3)_6]X_2$ in addition to $[Fe(NH_3)_5X]X$ and $[Fe(NH_3)_4X_2]$. The reaction must be carried out in the solid state because iron(II) ammine complexes are not stable in aqueous solution. Iron(III) does not form simple ammine complexes.

The complex cation $[Co(NH_3)_6]^{3+}$ is usually prepared by oxidizing the corresponding cobalt(II) complex (equation 16.9) which may itself be prepared by the displacement of aqua ligands in $[Co(H_2O)_6]^{2+}$ by ammonia ligands. (What is the role of the ammonium ions in reaction 16.9?) Alternatively, oxidation and complex formation may be carried out in one step as in reaction 16.10.

$$4[Co(NH_3)_6]^{2+} + O_2 + 4[NH_4]^+ \rightarrow 4[Co(NH_3)_6]^{3+} + 2H_2O + 4NH_3 \qquad (16.9)$$

oxidation

reduction

$$4CoCl_2 + O_2 + 4[NH_4]Cl + 20NH_3 \xrightarrow{\text{activated charcoal}} 4[Co(NH_3)_6]Cl_3 + 2H_2O \qquad (16.10)$$

§ Do not confuse the name *ammine* for a coordinated ammonia ligand with *amine*, a molecule of type RNH_2, R_2NH or R_3N.

Chloro ligands

In Chapter 13, we illustrated many examples of reactions between a *p*-block metal halide and excess halide ion in which the metal halide acted as a Lewis acid, for example the formation of $[AlCl_4]^-$ and $[SnCl_6]^{2-}$. Similar reactions occur with some of the *d*-block metal halides and equations 16.11 to 16.13 show representative reactions; each chlorocomplex has a tetrahedral structure **16.20**. The nickel(II) complex anion $[NiCl_4]^{2-}$ is also tetrahedral.

$$FeCl_3(aq) + Cl^-(aq) \xrightleftharpoons{\text{in concentrated HCl}} [FeCl_4]^-(aq) \qquad (16.11)$$

$$ZnCl_2(aq) + 2Cl^-(aq) \xrightleftharpoons{\text{in concentrated HCl}} [ZnCl_4]^{2-}(aq) \qquad (16.12)$$

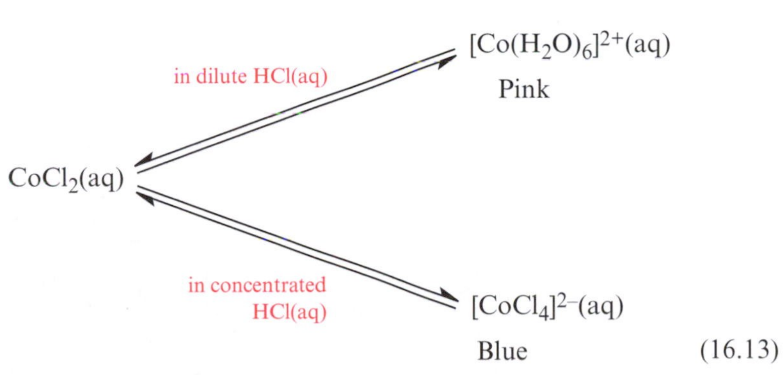

(16.13)

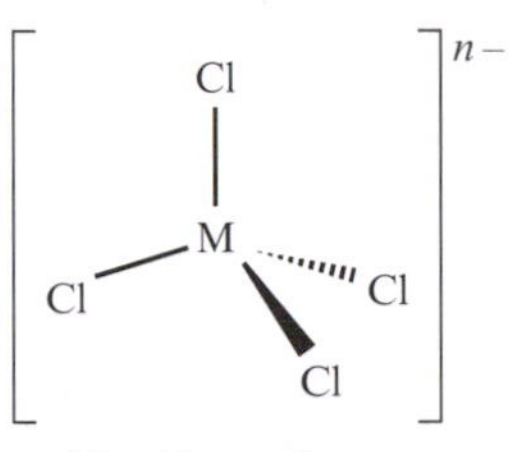

M = Fe; n = 1
M = Co, Zn; n = 2

(16.20)

The chloro complexes may be isolated from solution by exchanging the cation (which is $[H_3O]^+$ in aqueous medium) for a much larger counter-ion such as tetraethylammonium $[Et_4N]^+$. After adding $[Et_4N]Cl$, the desired salt precipitates out of solution (equation 16.14).

$$2[Et_4N]Cl(s) + [ZnCl_4]^{2-}(aq) \rightarrow [Et_4N]_2[ZnCl_4](s) + 2Cl^-(aq) \qquad (16.14)$$

In equation 16.13, the effects of varying the concentration of chloride ion are apparent. If a pink solution of cobalt(II) chloride in dilute hydrochloric acid is heated, the concentration of chloride ion increases as water evaporates, and the solution turns blue as $[Co(H_2O)_6]^{2+}$ is converted to $[CoCl_4]^{2-}$ (equation 16.15). This process is used in experiments involving 'invisible ink' and is reversible; solutions containing the blue $[CoCl_4]^{2-}$ ion turn pale pink when diluted with water.

$$[Co(H_2O)_6]^{2+}(aq) + 4Cl^-(aq) \rightleftharpoons [CoCl_4]^{2-}(aq) + 6H_2O(l) \qquad (16.15)$$

Similar effects are observed with other metal ions, for example, $[Ti(H_2O)_6]^{3+}$ is stabilized in *dilute* hydrochloric acid (equation 16.1), but in *concentrated* acid, chloride ion displaces an aqua ligand and $[TiCl(H_2O)_5]^{2+}$ **16.21** is formed.

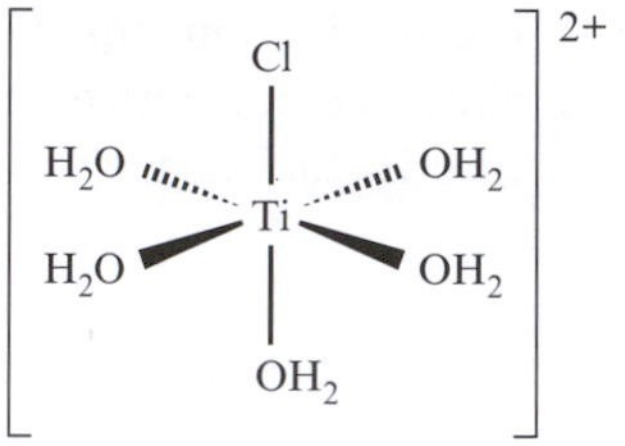

(16.21)

Cyano complexes

The cyano ligand $[CN]^-$ usually bonds to metal ions through the *carbon atom* and each M–C≡N unit is linear. With chromium(II), it forms the octahedral complex anion $[Cr(CN)_6]^{4-}$, and octahedral complexes are also formed with iron(II) and iron(III). Many stable salts of $[Fe(CN)_6]^{4-}$ exist, for

Box 16.5 Prussian blue and Turnbull's blue

When $[Fe(CN)_6]^{4-}$ is added to an aqueous solution containing iron(III) ions, a deep blue compound called Prussian blue is formed. The same blue colour is observed when $[Fe(CN)_6]^{3-}$ is added to an aqueous solution containing iron(II) ions; this blue complex is referred to as Turnbull's blue. Both Prussian blue and Turnbull's blue are hydrated salts and possess the formula $Fe^{III}{}_4[Fe^{II}(CN)_6]_3 \cdot xH_2O$ (where x is about 14).

A related compound is the potassium salt $KFe[Fe(CN)_6]$ and this is referred to as 'soluble Prussian blue'.

The solid state structures of these compounds contain cubic arrangements of iron centres with cyano groups bridging between the metals. One-eighth of the unit cell of $KFe[Fe(CN)_6]$ is shown below, with the K^+ ions removed. Each iron centre is octahedrally sited and is in either an FeN_6 or an FeC_6 environment; the lattice must be extended to see this feature. Half of the iron sites are Fe(II) and half are Fe(III), and the K^+ ions occupy cavities within the cubic lattice.

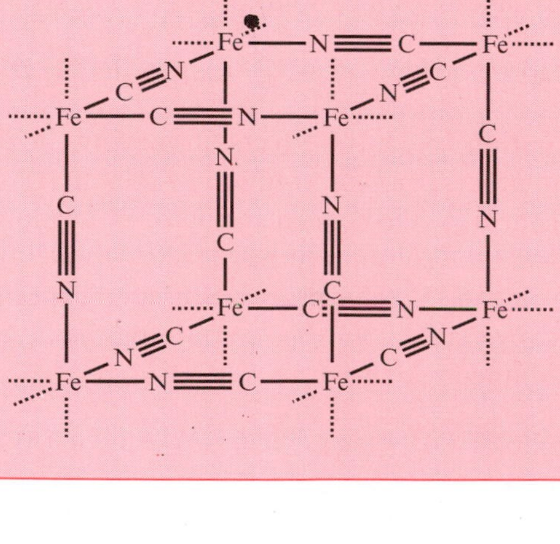

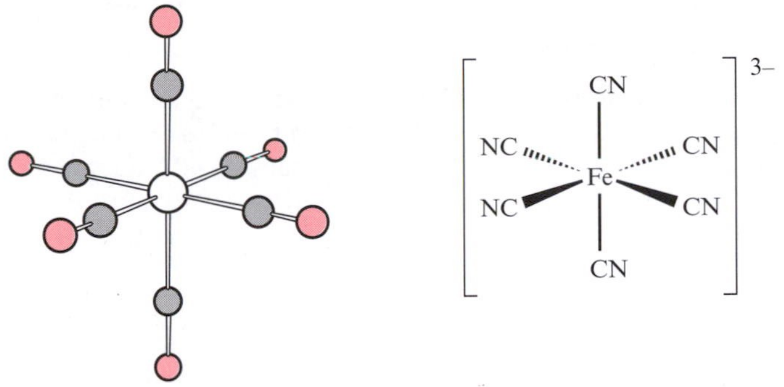

16.15 The octahedral structure of $[Fe(CN)_6]^{3-}$ as determined for the salt $Cs_2K[Fe(CN)_6]$. Each of the Fe–C–N bond angles is approximately 180°.

example $K_4[Fe(CN)_6]$; the complex anion is stable in aqueous solution. The iron(III) containing complex $[Fe(CN)_6]^{3-}$ (Figure 16.15) is more toxic than $[Fe(CN)_6]^{4-}$, because it releases cyanide ion more readily than does the iron(II) species. When $[Fe(CN)_6]^{4-}$ is added to an aqueous solution containing iron(III) ions, a deep blue compound is formed and this is a qualitative test for Fe^{3+}.

Qualitative test for iron(III) ions: **When $[Fe(CN)_6]^{4-}$ is added to an aqueous solution containing iron(III) ions, a deep blue complex called Prussian blue is formed.**

Qualitative test for iron(II) ions: **When $[Fe(CN)_6]^{3-}$ is added to an aqueous solution containing iron(II) ions, a deep blue complex called Turnbull's blue is formed.**

Although most 4-coordinate complexes of the first row *d*-block metals are tetrahedral, *those containing nickel(II) centres are* ***sometimes*** *square planar*. For example, $[NiCl_4]^{2-}$ and $[NiBr_4]^{2-}$ are tetrahedral, but $[Ni(CN)_4]^{2-}$ **16.22** is square planar.

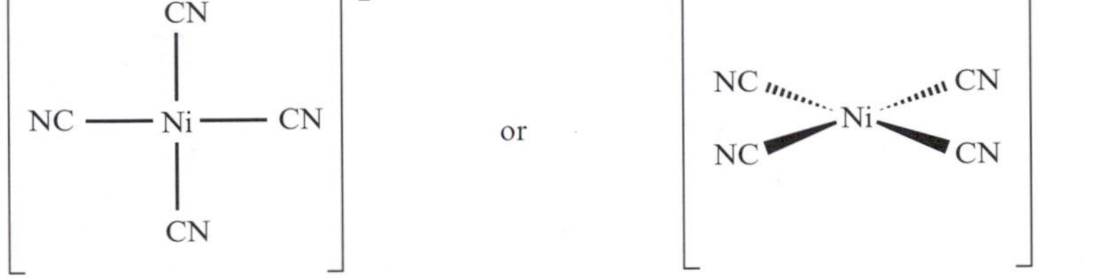

(16.22)

Didentate and polydentate ligands

Table 6.2 lists examples of both symmetrical and asymmetrical didentate ligands. The presence of three *symmetrical* didentate ligands in an octahedral coordination sphere leads to the formation of enantiomers and, if the precursor is achiral (i.e. non-chiral) such as $[Co(H_2O)_6]^{2+}$ (equation 16.16), the

Box 16.6 Anti-cancer drugs

The square planar complex *cis*-$[PtCl_2(NH_3)_2]$ (cisplatin) is an antitumour drug and has been recognized in this capacity since the 1960s. It is used in the treatment of bladder and cervical tumours and ovarian and testicular cancers, but, unfortunately, side-effects include nausea and kidney damage. An alternative drug which is clinically active but has fewer side-effects than cisplatin is carboplatin.

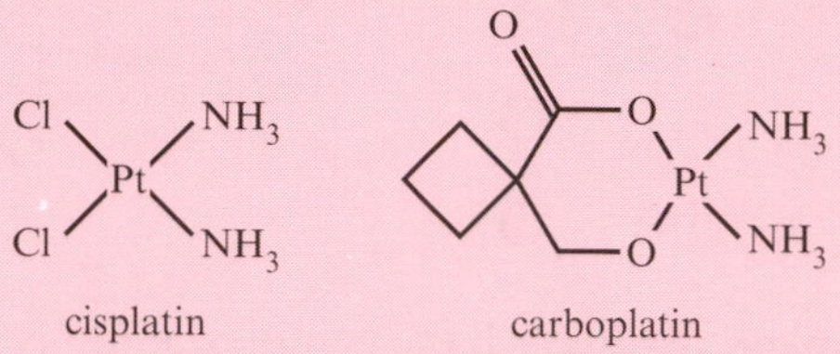

cisplatin carboplatin

Research to find other active platinum-containing anti-cancer drugs is actively being pursued, and an understanding of how cisplatin and related active compounds work is an important aspect of drug development.

DNA (see Box 11.6) is composed of a sequence of nucleotides bases including guanine. It appears that when cisplatin is administered to a patient, the *cis*-$Pt(NH_3)_2$ units are coordinated by nitrogen donor atoms in the guanine bases of DNA.

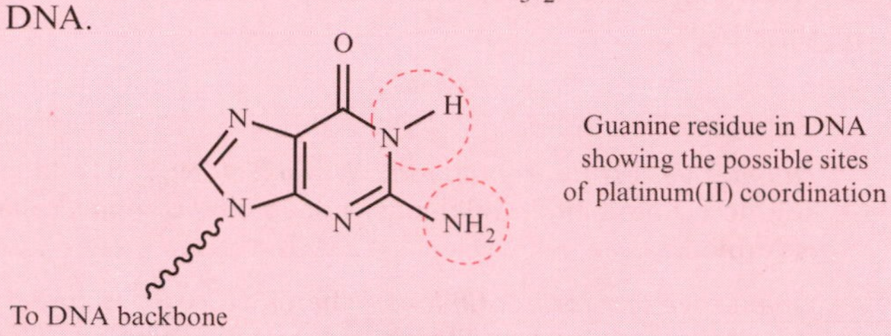

To DNA backbone

Guanine residue in DNA showing the possible sites of platinum(II) coordination

Designing 'model ligands' which might mimic biological systems is commonly employed as a method of investigating otherwise complex systems. For example, a ligand incorporating a cytosine and two guanine units, connected to a phosphate-containing chain similar to the backbone of DNA has been shown to react with cisplatin to give the complex in which two nitrogen atoms, one from each of the guanine residues, coordinate to the platinum(II) centre. The structure of the complex is shown below; hydrogen atoms, except those in the NH_3 ligands, are omitted from the diagram.

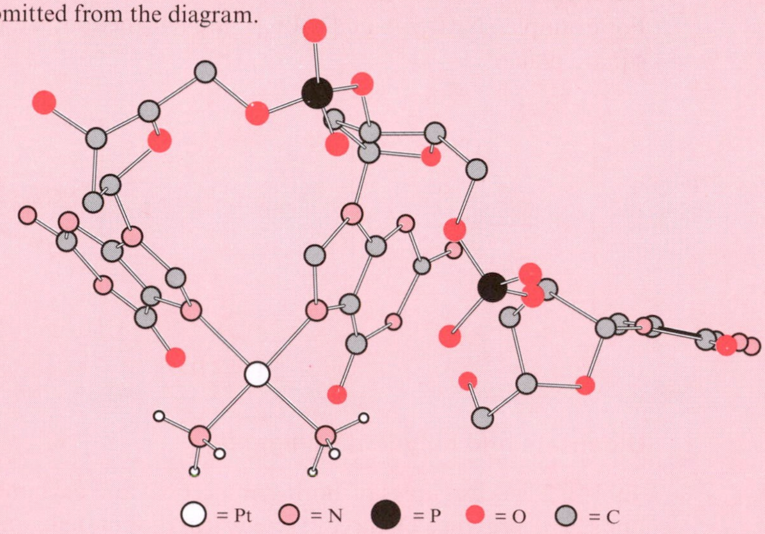

= Pt = N = P = O = C

Research in the area of platinum anticancer drugs has been described by Jan Reedijk, *Chemical Communications*, (1996) p. 801.

Racemic mixture: see Section 14.5

product is formed as a racemic mixture. These cobalt(II) complexes readily oxidize to the corresponding cobalt(III) compounds as we showed for the ammine complexes in equation 16.9.

$$[Co(H_2O)_6]^{2+}(aq) + 3\,en(aq) \rightleftharpoons [Co(en)_3]^{2+}(aq) + [Co(en)_3]^{2+}(aq) + 6H_2O(l) \quad (16.16)$$

N⌒N = 1,2-diaminoethane (en)

$^{-}O\text{–}CH_2CH_2\text{–}NH_2$

(16.23)

If the ligands are asymmetrical, geometrical isomers are also possible. For example, the reaction of $CrCl_3$ with $HOCH_2CH_2NH_2$ in water may give a mixture of isomers (Figure 16.16) of the octahedral complex $[Cr(OCH_2CH_2NH_2)_3]$ where the ligand is **16.23**.

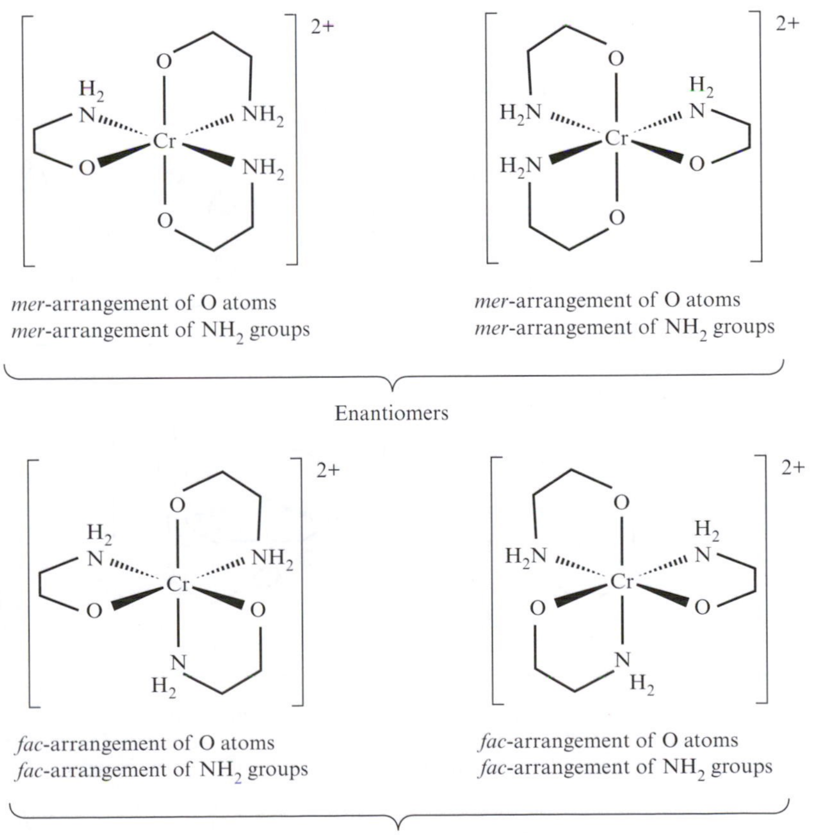

16.16 The complex cation $[Cr(OCH_2CH_2NH_2)_3]^{2+}$ possesses *mer*- and *fac*-geometrical isomers, and each forms as an enantiomeric pair. Notice that a *fac*- arrangement of NH_2 groups *necessarily* means a *fac*-arrangement of O atoms, and similarly for the *mer*-arrangements.

Related complexes which can be formed by adding the ligand en to aqueous solutions of metal(II) chlorides or nitrates are $[Cr(en)_3]^{2+}$, $[Mn(en)_3]^{2+}$, $[Fe(en)_3]^{2+}$ and $[Ni(en)_3]^{2+}$, although the en ligand readily dissociates from the chromium(II) centre in aqueous solution. 1,2-Diaminoethane may also displace ammonia from ammine complexes (equation 16.17).

$$[Ni(NH_3)_6]^{2+}(aq) + 3en \rightleftharpoons [Ni(en)_3]^{2+}(aq) + 6NH_3(aq) \qquad (16.17)$$

Entropy: see Sections 12.8 and 12.9

The driving force behind many of the displacements of two monodentate for one didentate ligand may be attributed to a change in entropy and is referred to as *the chelate effect*. Similarly, the replacement of three monodentate ligands by one tridentate ligand is favourable. When a dien ligand (Table 16.2) coordinates to a metal ion in a tridentate manner, *two chelate rings* are formed (structure **16.24**). This makes dien complexes more stable than corresponding en or ammine complexes; $[Ni(dien)_2]^{2+}$ is thermodynamically more stable than $[Ni(en)_3]^{2+}$ which is thermodynamically more stable than $[Ni(NH_3)_6]^{2+}$.

(16.24)

Conformation: see Section 8.5

Figure 16.17a shows the structure of the dien ligand in an *extended conformation*; rotation can occur around any of the single bonds and this makes dien flexible enough to adopt either a *mer*- or *fac*-configuration in an octahedral complex. When dien reacts with cobalt(II) ions in air (i.e. an oxidizing environment) in aqueous solution, the complex cation $[Co(dien)_2]^{3+}$ is formed (equation 16.18). A mixture of the three isomers shown in Figure 16.17b is formed; one *mer*-arrangement of the two dien ligands is possible,

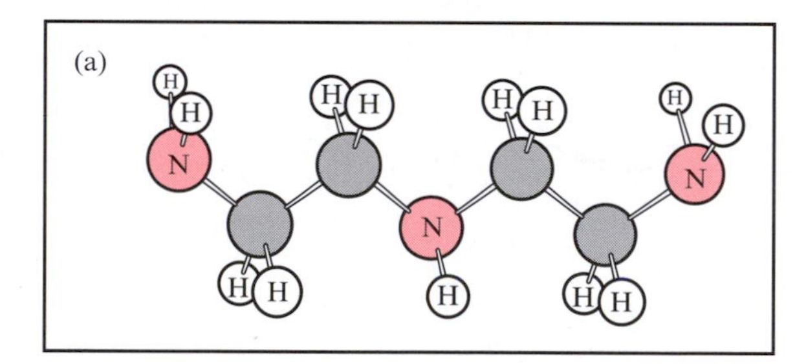

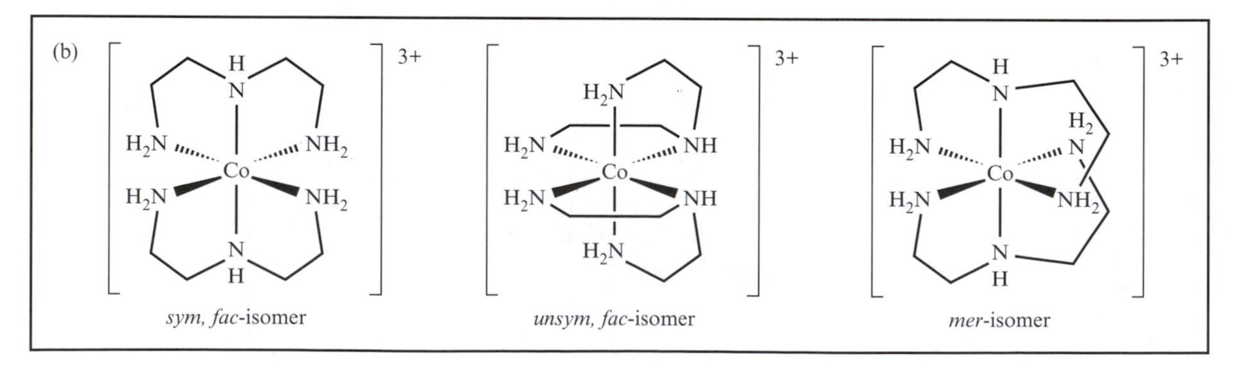

16.17 (a) The dien ligand ($H_2NCH_2CH_2NHCH_2CH_2NH_2$) in an extended conformation. Free rotation can occur about any of the single bonds. (b) The complex cation $[Co(dien)_2]^{3+}$ is produced as a mixture of isomers; the *mer*- and *unsym, fac*-isomers are present as racemic mixtures (*sym*- = symmetrical; *unsym*- = unsymmetrical.)

$$[Co(H_2O)_6]^{2+}(aq) + 2\,dien(aq) \xrightleftharpoons[]{\text{in the presence of air and activated charcoal}} [Co(dien)_2]^{3+}(aq) + 6H_2O(l) \quad (16.18)$$

Co(II) Co(III)

but there are *two different fac*-arrangements. In the *sym,fac*-isomer, the ligands are positioned with the central NH groups *trans*- to one another, whilst in the *unsym,fac*-isomer, the NH groups are *cis* to each other. These three isomers can be separated by using column chromatography, but additionally, each of the *mer*- and *unsym,fac*-isomers is optically active.

If the tridentate ligand is *not* as flexible as dien, the number of isomers of an octahedral complex may be reduced. For example, although free rotation about the C–C bonds *between* the aromatic rings in the terpyridine ligand is possible, the ligands in $[Fe(tpy)_2]^{2+}$ can only adopt a *mer*-arrangement (Figure 16.8).

Flexibility plays a vital role in the coordination behaviour of polydentate ligands, and an example is $[edta]^{4-}$ **16.25**; the parent carboxylic acid is tetrabasic and is represented by the abbreviation H_4edta. The two nitrogen atoms and the four O^- groups are donor atoms and $[edta]^{4-}$ is a potentially hexadentate ligand. The conformation of the ligand can change by free rotation about the single C–C and C–N bonds; compare the conformation shown in structure **16.25** with that in Table 16.2. When $[edta]^{4-}$ coordinates to a metal ion, it may wrap around the metal centre to give an octahedral complex. Equations 16.19 and 16.20 show a two-step preparation of the ammonium salt of $[Co(edta)]^-$, the structure of which is shown in Figure 16.18.

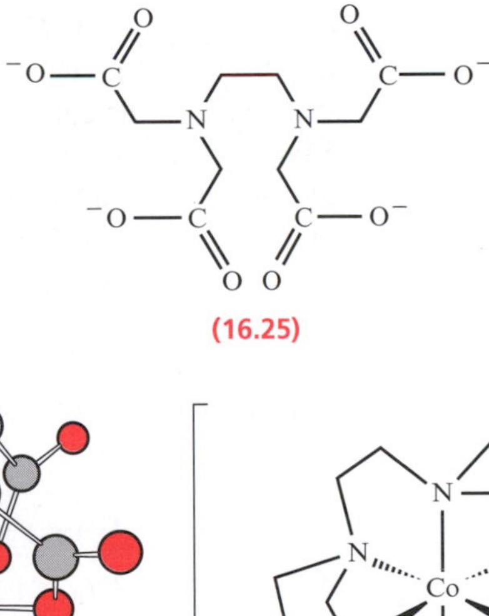

(16.25)

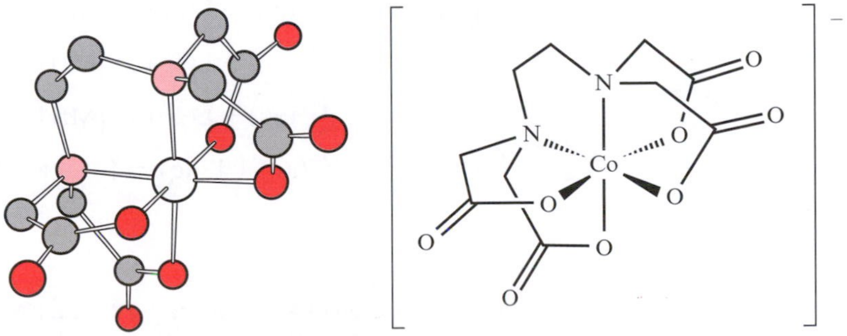

16.18 The solid state structure of $[Co(edta)]^-$ as determined by X-ray diffraction methods for the salt $[NH_4][Co(edta)\cdot 2H_2O$. The two nitrogen donor atoms must be mutally *cis*, and the flexibility of the ligand allows the oxygen donor atoms to occupy the remaining four coordination sites so as to complete the octahedral coordination shell.

$$Co(OH)_2 + H_4edta + 2NH_3 \rightleftharpoons \underset{\text{cobalt(II)}}{[NH_4]_2[Co(edta)]} + 2H_2O \qquad (16.19)$$

$$2[NH_4]_2[Co(edta)] + H_2O_2 \rightarrow \underset{\text{cobalt(III)}}{2[NH_4][Co(edta)]} + 2H_2O + 2NH_3 \qquad (16.20)$$

16.7 LIGAND EXCHANGE AND COMPLEX STABILITY CONSTANTS

Many of the complex formation reactions shown in Section 16.6 were written as *equilibria* and in this section we consider the thermodynamic stability of coordination complexes and values of their *stability constants*.

Stepwise stability constants, *K*

The replacement of one aqua ligand in a hexaaqua ion by another ligand L is shown in equation 16.21. The relative stabilities of the two complex ions are given by the equilibrium constant, *K*, and this particular type of equilibrium constant is called the *stability constant of the product complex* (equation 16.22). The concentration of water is taken to be unity.

$$[M(H_2O)_6]^{2+}(aq) + L(aq) \rightleftharpoons [M(H_2O)_5L]^{2+}(aq) + H_2O(l) \qquad (16.21)$$

$$K_1 = \frac{[M(H_2O)_5L^{2+}][H_2O]}{[M(H_2O)_6{}^{2+}][L]} = \frac{[M(H_2O)_5L^{2+}]}{[M(H_2O)_6{}^{2+}][L]} \qquad (16.22)$$

The stability constant K_1 refers to the stepwise replacement of the first aqua ligand by L, and there are six stability constants (K_1, K_2, K_3, K_4, K_5 and K_6) associated with the conversion of $[M(H_2O)_6]^{2+}$ to $[ML_6]^{2+}$. (This is similar to the manner in which we consider the stepwise dissociation of polybasic acids; we define a value of K_a for the stepwise loss of *each* proton.) Equations 16.23–16.27 are the five steps which follow equilibrium 16.21 in the formation of $[ML_6]^{2+}$.

➤ Polybasic acids: see Section 11.9

➤ See problem 16.12

$$[M(H_2O)_5L]^{2+}(aq) + L(aq) \rightleftharpoons [M(H_2O)_4L_2]^{2+}(aq) + H_2O(l) \qquad (16.23)$$

$$[M(H_2O)_4L_2]^{2+}(aq) + L(aq) \rightleftharpoons [M(H_2O)_3L_3]^{2+}(aq) + H_2O(l) \qquad (16.24)$$

$$[M(H_2O)_3L_3]^{2+}(aq) + L(aq) \rightleftharpoons [M(H_2O)_2L_4]^{2+}(aq) + H_2O(l) \qquad (16.25)$$

$$[M(H_2O)_2L_4]^{2+}(aq) + L(aq) \rightleftharpoons [M(H_2O)L_5]^{2+}(aq) + H_2O(l) \qquad (16.26)$$

$$[M(H_2O)L_5]^{2+}(aq) + L(aq) \rightleftharpoons [ML_6]^{2+}(aq) + H_2O(l) \qquad (16.27)$$

In the formation of a complex $[ML_6]^{2+}$ from $[M(H_2O)_6]^{2+}$, the stability constant K_1 refers to the stepwise replacement of the first aqua ligand by L. There are six stability constants (K_1, K_2, K_3, K_4, K_5 and K_6) associated with steps in the overall process.

The stepwise displacements of the aqua ligands in $[Ni(H_2O)_6]^{2+}$ by the didentate ligands en are shown in equations 16.28 to 16.30; the stability constants show that the first displacement (equation 16.28) of two H_2O molecules by one 1,2-diaminoethane ligand is more favourable than the second step, and this is more favourable than the third step.

$$[Ni(H_2O)_6]^{2+}(aq) + en(aq) \rightleftharpoons [Ni(H_2O)_4(en)]^{2+}(aq) + 2H_2O(l) \quad K_1 = 2.81 \times 10^7 \text{ mol}^{-1}\text{ dm}^3 \quad (16.28)$$

$$[Ni(H_2O)_4(en)]^{2+}(aq) + en(aq) \rightleftharpoons [Ni(H_2O)_2(en)_2]^{2+}(aq) + 2H_2O(l) \quad K_2 = 1.70 \times 10^6 \text{ mol}^{-1}\text{ dm}^3 \quad (16.29)$$

$$[Ni(H_2O)_2(en)_2]^{2+}(aq) + en(aq) \rightleftharpoons [Ni(en)_3]^{2+}(aq) + 2H_2O(l) \quad K_3 = 2.19 \times 10^4 \text{ mol}^{-1}\text{ dm}^3 \quad (16.30)$$

Worked example 16.2 ***Stepwise formation of $[Ni(NH_3)_6]^{2+}$***

The following graph shows a plot of log K (K is the stability constant) against the number, n, of ammonia ligands present in the complex $[Ni(H_2O)_{6-n}(NH_3)_n]^{2+}$:

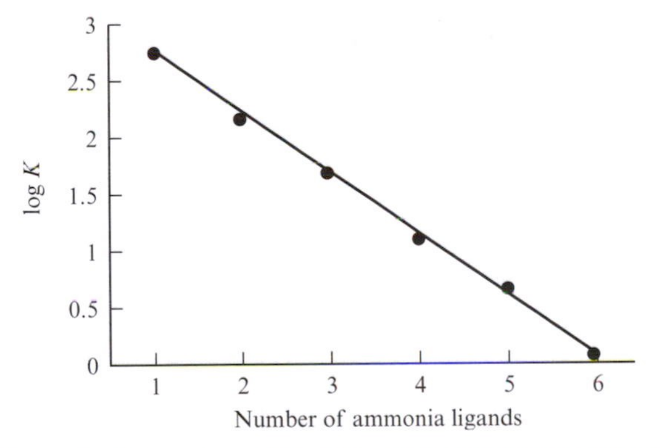

To what reactions do the values of log K = 2.70 and 0.65 correspond? What does the graph tell you about the displacement of successive aqua by ammonia ligands?

The stability constants correspond to the stepwise displacement of aqua by ammonia ligands in going from $[Ni(H_2O)_6]^{2+}$ to $[Ni(NH_3)_6]^{2+}$.

The value of log K = 2.70 corresponds to the first ligand exchange and to the equilibrium:

$$[Ni(H_2O)_6]^{2+}(aq) + NH_3(aq) \rightleftharpoons [Ni(H_2O)_5(NH_3)]^{2+}(aq) + H_2O(l)$$

The value of log K = 0.65 corresponds to $n = 5$, and therefore to the reaction:

$$[Ni(H_2O)_2(NH_3)_4]^{2+}(aq) + NH_3(aq) \rightleftharpoons [Ni(H_2O)(NH_3)_5]^{2+}(aq) + H_2O(l)$$

The graph shows a decrease in the value of log K as successive aqua ligands are displaced by ammonia ligands. As log K decreases, K decreases, and this means that successive equilibria (corresponding to successive ligand replacements) lie further to the left-hand side. It is less favourable to displace an aqua ligand by ammonia in $[Ni(H_2O)_5(NH_3)]^{2+}$ than in $[Ni(H_2O)_6]^{2+}$, and less favourable still in later complexes in the series.

Overall stability constants, β

Instead of considering the *stepwise* formation of a complex such as $[ML_6]^{2+}$, it may be more appropriate or convenient to consider the formation in a single step such as in reaction 16.31.

$$[M(H_2O)_6]^{2+}(aq) + 6L(aq) \rightleftharpoons [ML_6]^{2+}(aq) + 6H_2O(l) \qquad (16.31)$$

We can write down an expression for the equilibrium constant for this reaction but instead of calling the equilibrium constant K, the symbol β is used – β_n is the *overall stability constant*, whereas values of K_n represent *stepwise stability constants* for a series of n equilibria. Equation 16.32 gives the expression for the overall stability constant for the formation of $[ML_6]^{2+}$ from $[M(H_2O)_6]^{2+}$.

$$\beta_6 = \frac{[ML_6{}^{2+}]}{[M(H_2O)_6{}^{2+}][L]^6} \qquad (16.32)$$

For the formation of a complex $[ML_6]^{2+}$ in the reaction:

$$[M(H_2O)_6]^{2+}(aq) + 6L(aq) \rightleftharpoons [ML_6]^{2+}(aq) + 6H_2O(l)$$

the overall stability constant is β_6 and is given by the equation:

$$\beta_6 = \frac{[ML_6{}^{2+}]}{[M(H_2O)_6{}^{2+}][L]^6}$$

For an *n*-step process, the overall stability constant is β_n.

Free energy changes for the formation of complexes

The reaction isotherm (equation 16.33) gives the relationship between a stability constant and the free energy of formation of a complex and is used to determine thermodynamic data for the formation of complexes.

$R = 8.314 \text{ J mol}^{-1} \text{ K}^{-1}$

$$\Delta G^o = -RT \ln K \qquad (16.33)$$

By applying this equation to the series of reactions given in equations 16.28 to 16.30 (for which values of K are measured at 298 K), values of ΔG^o of –42.5, –35.5 and –24.8 kJ mol^{-1} are obtained for the successive displacement of two aqua ligands by a didentate 1,2-diaminoethane ligand. The more en ligands that are added, the less favourable the reaction becomes. This trend is seen in many complex-forming reactions and another example was illustrated in worked example 16.2.

Worked example 16.3 ***The thermodynamics of complex formation***

The first five stepwise stability constants (at 298 K) corresponding to the reaction of $[Cu(H_2O)_6]^{2+}$ with ammonia are 1.58×10^4, 3.16×10^3, 7.94×10^2, 1.25×10^2 and 0.32. Write down the equilibria to which these values correspond and determine values of ΔG^o for each step. What can you conclude about the overall reaction between $[Cu(H_2O)_6]^{2+}$ and ammonia? [$R = 8.314$ J mol^{-1} K^{-1}]

The relationship between K and ΔG^o is:

$$\Delta G^o = -RT \ln K$$

By applying this equation, we can determine the change in standard free energy for the displacements of the first five aqua ligands in $[Cu(H_2O)_6]^{2+}$ by ammine ligands.

For the first step:

$$[Cu(H_2O)_6]^{2+}(aq) + NH_3(aq) \rightleftharpoons [Cu(H_2O)_5(NH_3)]^{2+}(aq) + H_2O(l)$$

$$K = 1.58 \times 10^4 \qquad \ln K = 9.67$$

$$\Delta G^o = -(8.314 \times 10^{-3} \times 298 \times 9.67) = -23.9 \text{ kJ per mole of reaction}$$

Similarly for the next four steps:

$$[Cu(H_2O)_5(NH_3)]^{2+}(aq) + NH_3(aq) \rightleftharpoons [Cu(H_2O)_4(NH_3)_2]^{2+}(aq) + H_2O(l)$$
$$\Delta G^o = -20.0 \text{ kJ per mole of reaction}$$

$$[Cu(H_2O)_4(NH_3)_2]^{2+}(aq) + NH_3(aq) \rightleftharpoons [Cu(H_2O)_3(NH_3)_3]^{2+}(aq) + H_2O(l)$$
$$\Delta G^o = -16.5 \text{ kJ per mole of reaction}$$

$$[Cu(H_2O)_3(NH_3)_3]^{2+}(aq) + NH_3(aq) \rightleftharpoons [Cu(H_2O)_2(NH_3)_4]^{2+}(aq) + H_2O(l)$$
$$\Delta G^o = -12.0 \text{ kJ per mole of reaction}$$

$$[Cu(H_2O)_2(NH_3)_4]^{2+}(aq) + NH_3(aq) \rightleftharpoons [Cu(H_2O)(NH_3)_5]^{2+}(aq) + H_2O(l)$$
$$\Delta G^o = +2.8 \text{ kJ per mole of reaction}$$

These data illustrate two points:

- it becomes increasingly less thermodynamically favourable to replace successive aqua by ammine ligands in $[Cu(H_2O)_6]^{2+}$; and
- the displacement of the fifth ligand is thermodynamically *unfavourable* because ΔG^o for this step is positive.

16.8 HARD AND SOFT METALS AND DONOR ATOMS

We have already seen that some metal ions show a preference for particular types of donor atoms; for example, early metals in the *d*-block form complexes with cyclic ethers (oxygen donor atoms) but later metals prefer thioether ligands (sulfur donor ligands). In this section we describe the concept of *hard* and *soft* metals and donor atoms – this provides us with a method of 'matching' metal centres with appropriate ligand types.

Case study 1: iron(III) halide complexes

The displacement of an aqua ligand by a halide X^- in $[Fe(H_2O)_6]^{3+}$ is shown in equation 16.34; the solution must be acidic in order to stabilize the hexaaquairon(III) ion. Equation 16.35 gives an alternative form of this equation, where the $Fe^{3+}(aq)$ and $[FeX]^{2+}(aq)$ signify hydrated ions. We include this because the shorthand method of writing hydrated species is so widely adopted, even though it provides less information about the identity of the aqua-species.

$$[Fe(H_2O)_6]^{3+}(aq) + X^-(aq) \rightleftharpoons [Fe(H_2O)_5X]^{2+}(aq) + H_2O(l) \quad (16.34)$$

$$Fe^{3+}(aq) + X^-(aq) \rightleftharpoons [FeX]^{2+}(aq) \quad (16.35)$$

Table 16.3 lists stability constants for the formation of $[FeX]^{2+}(aq)$ for fluoride, chloride and bromide complexes, and the data indicate that formation of the fluoride complex is highly favourable – products predominate in the equilibrium. Although the stability constants for $[FeCl]^{2+}(aq)$ and $[FeBr]^{2+}$ are much smaller than for $[FeF]^{2+}$, the formation of the chloro and bromo complexes is still favourable, but $[FeBr]^{2+}$ is less stable than $[FeCl]^{2+}$ with respect to the hexaaqua ion.

The trend in $[FeX]^{2+}(aq)$ complex stability is:

F >> Cl > Br

Case study 2: mercury(II) halide complexes

Equation 16.36 shows the reaction between a halide ion and a mercury(II) aqua ion, and Table 16.3 gives stability constants for the $[HgX]^+$ complexes with different halo ligands.

$$Hg^{2+}(aq) + X^-(aq) \rightleftharpoons [HgX]^+(aq) \quad (16.36)$$

For each halide, equilibrium 16.36 lies over to the right-hand side, and the coordination of an iodo ligand to Hg^{2+} is the most favourable. The trend in $[HgX]^+(aq)$ complex stability is:

I > Br > Cl >> F

Hard and soft metal ions and ligands

The two case studies above illustrate that while fluoride ion is the halide preferred by an aqueous iron(III) ion, iodide is the preferred ligand for mercury(II) ions. The fact that Fe^{3+} and Hg^{2+} exhibit opposite trends in

Table 16.3 Stability constants for the formation of iron(III) and mercury(II) halides $[FeX]^{2+}(aq)$ and $[HgX]^+(aq)$. The equilibria are defined in equations 16.35 and 16.36.

Metal ion	*Stability constant K_1*			
	X = F	*X = Cl*	*X = Br*	*X = I*
$Fe^{3+}(aq)$	1×10^6	25.1	3.2	–
$Hg^{2+}(aq)$	10.0	5.0×10^6	7.9×10^8	7.9×10^{12}

their preferences for halo ligands is not unique to this pair of metal centres. The ions Sc^{3+}, Cr^{3+}, Co^{3+} and Zn^{2+} behave in a similar manner to Fe^{3+}, while Cu^{+}, Ag^{+} and Pt^{2+} behave similarly to Hg^{2+}.

Metal ions are classified as being either *hard* (*class A*) or *soft* (*class B*) – Fe^{3+} is hard while Hg^{2+} is soft.

A hard metal ion carries a high positive charge (often +3 or +4) and/or has a high charge density. The consequence of this is that the valence electrons of the metal ion are tightly held and the metal–ligand bonding possesses a high degree of ionic character. This in turn has an effect on the type of ligands with which the metal ion prefers to interact – a hard metal ion will favour ligands containing donor atoms which are highly electronegative. Such ligands are termed *hard ligands* and include fluoride and chloride ions and oxygen or nitrogen donor atoms.

Charge density and polarization of ions: see Section 6.18

A soft metal ion carries a low positive charge (often 1+ to 2+) and/or has a low charge density. Such metal ions are easily polarized and the metal–ligand bonding has a high covalent character. In contrast to a hard metal, a soft metal ion favours ligands containing less electronegative donor atoms and these include iodide and sulfur donor atoms.

Many metal–ligand interactions can be classified as being either hard–hard (for example, iron(III)–fluoride ion) or soft–soft (for example, mercury(II)–iodide ion). However, some metal ions and ligands are intermediate in character and this permits them to be more versatile in their complex formation than the limiting hard and soft metal ions and ligands. Table 16.4 lists representative hard, soft and intermediate metal ions and ligands; the donor atoms in the ligands are highlighted in red.

Table 16.4 Representative hard, soft and intermediate *d*-block metal centres and ligands. Donor atoms in the ligands are indicated in red; ligands such as nitrate and sulfate ions are oxygen donors and may be mono- or didentate. Notice that higher oxidation state centres tend to be hard, and lower oxidation state centres tend to be soft.

Hard metal ions	Mn(II), Zn(II)
	Sc(III), Y(III), Cr(III), Fe(III), Co(III)
	Ti(IV), Hf(IV), V(IV)
Hard ligands	F^-, Cl^-
	H_2O, $[OH]^-$, $[OR]^-$, ROH, R_2O, $[RCO_2]^-$, $[SO_4]^{2-}$, $[NO_3]^-$
	NH_3, RNH_2
Intermediate metal ions	Fe(II), Co(II), Ni(II), Cu(II)
	Rh(III), Ir(III), Ru(III), Os(III)
Intermediate ligands	Br^-
	C_5H_5N (pyridine), $[NCS]^-$
Soft metal ions	Cu(I), Ag(I), Au(I)
	Hg(II), Cd(II), Pd(II), Pt(II)
Soft ligands	I^-
	$[CN]^-$, CO
	PR_3
	$[NCS]^-$, $[RS]^-$, RSH, R_2S

We can now appreciate why scandium(III) and yttrium(III) ions form complexes with the oxygen donor atoms in ethers (Figures 16.9 and 16.10b), why fluoride and chloride ions form more stable complexes with iron(III) ions than does iodide ion, and why iodo-mercury(II) complexes are more stable than fluoro-mercury(II) compounds.

Problem 16.15 asks you to use the data in Table 16.4 to explore this concept further. However, it must be noted that while the concept of 'matching' hard metal ions with hard ligands and soft metal ions with soft ligands provides us with a means of understanding why certain complexes are more favourable than others, it is *not* a foolproof method of predicting whether a complex will form at all. For example, the reaction of aqueous chromium(III) acetate with potassium cyanide gives the compound $K_3[Cr(CN)_6]$ which contains the ion $[Cr(CN)_6]^{3-}$ **16.26**. This is composed of a hard Cr(III) centre and soft cyano ligands.

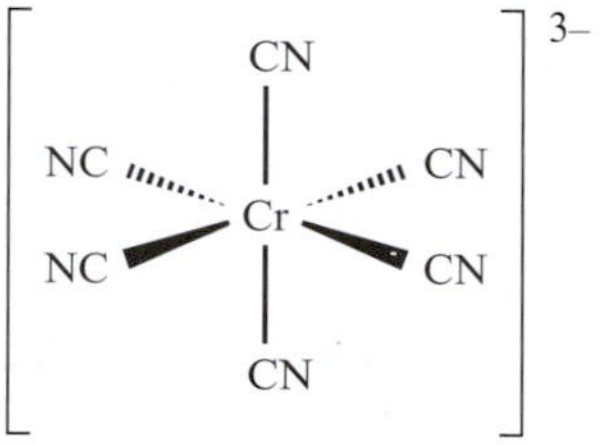

(16.26)

16.9 THE THERMODYNAMIC STABILITY OF HEXAAQUA METAL IONS

One of the usual characteristics of a *d*-block metal is that it possesses several stable oxidation states. Exceptions include the group 12 metals zinc and cadmium which form compounds almost exclusively in the +2 oxidation state, and the group 3 metals (scandium, yttrium and lanthanum) which predominantly form compounds in the +3 oxidation state. In this section we consider the reduction potentials of couples containing *d*-block metal ions, and quantify some of the reduction–oxidation (*redox*) reactions that are observed in aqueous solution.

Reduction potentials for hexaaqua metal ions

Table of reduction potentials: see Appendix 12

Table 16.5 lists values of E^o for selected couples involving first row *d*-block metal ions. A very important point to remember is that each metal ion is of the type $M^{n+}(aq)$ and is an *aqua ion*. For example, $Co^{2+}(aq)$ tells us we have an aqueous solution of cobalt(II) ions, but also indicates the presence of the complex ion $[Co(H_2O)_6]^{2+}$. Although we have made this point before, we reiterate it here because the presence of ligands other than water can significantly influence the reduction potential of a metal–ion couple. For example, the value for E^o $Fe^{3+}(aq)/Fe^{2+}(aq)$ is +0.77 V and refers to half-reaction 16.37.

Table 16.5 Selected standard reduction potentials (298 K). The concentration of each aqueous solution is 1 mol dm^{-3}, and the pressure of a gaseous component is 1 bar.

Reduction half-equation	E^o/V
$Ti^{2+}(aq) + 2e^- \rightleftharpoons Ti(s)$	–1.63
$Mn^{2+}(aq) + 2e^- \rightleftharpoons Mn(s)$	–1.19
$Zn^{2+}(aq) + 2e^- \rightleftharpoons Zn(s)$	–0.76
$Fe^{2+}(aq) + 2e^- \rightleftharpoons Fe(s)$	–0.44
$Cr^{3+}(aq) + e^- \rightleftharpoons Cr^{2+}(aq)$	–0.41
$Ti^{3+}(aq) + e^- \rightleftharpoons Ti^{2+}(aq)$	–0.37
$Co^{2+}(aq) + 2e^- \rightleftharpoons Co(s)$	–0.28
$V^{3+}(aq) + e^- \rightleftharpoons V^{2+}(aq)$	–0.26
$Ni^{2+}(aq) + 2e^- \rightleftharpoons Ni(s)$	–0.25
$Fe^{3+}(aq) + 3e^- \rightleftharpoons Fe(s)$	–0.04
$2H^+(aq) + 2e^- \rightleftharpoons H_2(g, 1 bar)$	0
$Cu^{2+}(aq) + e^- \rightleftharpoons Cu^+(aq)$	+0.15
$Cu^{2+}(aq) + 2e^- \rightleftharpoons Cu(s)$	+0.34
$O_2(g) + 2H_2O(l) + 4e^- \rightleftharpoons 4OH^-(aq)$	+0.40
$Cu^+(aq) + e^- \rightleftharpoons Cu(s)$	+0.52
$O_2(g) + 2H^+(aq) + 2e^- \rightleftharpoons H_2O_2(aq)$	+0.70
$Fe^{3+}(aq) + e^- \rightleftharpoons Fe^{2+}(aq)$	+0.77
$Ag^+(aq) + e^- \rightleftharpoons Ag(s)$	+0.80
$O_2(g) + 4H^+(aq) + 4e^- \rightleftharpoons 2H_2O(l)$	+1.23
$[Cr_2O_7]^{2-}(aq) + 14H^+(aq) + 6e^- \rightleftharpoons 2Cr^{3+}(aq) + 7H_2O(l)$	+1.33
$[MnO_4]^-(aq) + 8H^+(aq) + 5e^- \rightleftharpoons Mn^{2+}(aq) + 4H_2O(l)$	+1.51
$H_2O_2(aq) + 2H^+(aq) + 2e^- \rightleftharpoons 2H_2O(l)$	+1.78
$Co^{3+}(aq) + e^- \rightleftharpoons Co^{2+}(aq)$	+1.92

$$[Fe(H_2O)_6]^{3+} + e^- \rightleftharpoons [Fe(H_2O)_6]^{2+} \qquad E^o = +0.77\ V \qquad (16.37)$$

$$[Fe(CN)_6]^{3-} + e^- \rightleftharpoons [Fe(CN)_6]^{4-} \qquad E^o = +0.36\ V \qquad (16.38)$$

$$[Fe(bpy)_3]^{3+} + e^- \rightleftharpoons [Fe(bpy)_3]^{2+} \qquad E^o = +1.03\ V \qquad (16.39)$$

When the iron(III) centre is surrounded by cyano or 2,2′-bipyridine ligands, their presence markedly alters the ease with which the reduction of iron(III) to iron(II) takes place (equations 16.38 and 16.39). The reasons for this difference are beyond the scope of our discussion, but it is important to remember that *you can only apply the reduction potentials listed in Table 16.5 to aqua ions*, and that more detailed tables should be consulted if other ligands are involved.

The reduction potential of a couple such as M^{n+}/M^{m+} is dependent upon the ligands present in the coordination shell.

The stability of $M^{2+}(aq)$ ions in aqueous solutions

In Section 16.6 we considered some hexaaqua ions of the first row *d*-block metals. We stated that in aqueous solution, titanium(II), vanadium(II) and chromium(II) are readily oxidized to titanium(III), vanadium(III) and chromium(III) respectively. We also saw that atmospheric O_2 can oxidize iron(II) to iron(III). By considering the reduction potentials in Table 16.5, we can quantify these observations.

Predicting reactions for values of E^o_{cell}: see Section 12.10

Each of the $Cr^{3+}(aq)/Cr^{2+}(aq)$, $Ti^{3+}(aq)/Ti^{2+}(aq)$ and $V^{3+}(aq)/V^{2+}(aq)$ standard reduction potentials is *negative* and this means that the $M^{2+}(aq)$ ions liberate H_2 from *acidic solution*; standard conditions mean that the concentration of the acid is 1 mol dm^{-3}. Equations 16.40–16.42 illustrate this for chromium(II).

$$Cr^{3+}(aq) + e^- \rightleftharpoons Cr^{2+}(aq) \qquad E^o = -0.41\ V \qquad (16.40)$$

$$2H^+(aq) + 2e^- \rightleftharpoons H_2(g) \qquad E^o = 0\ V \qquad (16.41)$$

$$2Cr^{2+}(aq) + 2H^+(aq) \rightarrow 2Cr^{3+}(aq) + H_2(g) \qquad (16.42)$$

oxidation

reduction

The value of E^o_{cell} for reaction 16.42 is 0.41 V and from this, the change in standard free energy can be determined (equation 16.43).

$$\begin{aligned}\Delta G^o(298\ K) = -zFE^o_{cell} &= -(2 \times 96\,500 \times 0.41)\\ &= -79.1 \times 10^3\ J\ \text{per mole of reaction}\\ &= -79.1\ kJ\ \text{per mole of reaction}\end{aligned} \qquad (16.43)$$

The negative value of ΔG^o indicates that hydrogen ions will oxidize acidic aqueous chromium(II) to chromium(III) ions, i.e. the $[Cr(H_2O)_6]^{2+}$ ion is not stable in acidic aqueous solution. This means that even in the total absence of air, chromium(II) solutions (acidic, aqueous) are unstable with respect to the evolution of H_2 and formation of Cr(III). We know nothing *per se* about the *rate* of this reaction which is actually quite slow. In the presence of air, oxidation by O_2 occurs. Once again, this makes chromium(II) unstable with respect to chromium(III), but this time half-reactions 16.44 and 16.45 are relevant. The value of E^o_{cell} for the overall reaction 16.46 is +1.64 V, and the change in standard free energy for the overall reaction is calculated in equation 16.47.

$$Cr^{3+}(aq) + e^- \rightleftharpoons Cr^{2+}(aq) \qquad E^o = -0.41\ V \qquad (16.44)$$

$$O_2(aq) + 4H^+(aq) + 4e^- \rightleftharpoons 2H_2O(l) \qquad E^o = +1.23\ V \qquad (16.45)$$

$$4Cr^{2+}(aq) + O_2(g) + 4H^+(aq) \rightarrow 4Cr^{3+}(aq) + 2H_2O(l) \qquad (16.46)$$

oxidation

reduction

$$\begin{aligned}\Delta G^o(298\ K) = -zFE^o_{cell} &= -(4 \times 96\,500 \times 1.64)\\ &= -63.3 \times 10^4\ J\ \text{per mole of reaction}\\ &= -633\ kJ\ \text{per mole of reaction}\end{aligned} \qquad (16.47)$$

As we saw in Section 12.10, reduction potentials which involve H^+ or $[OH]^-$ ions are dependent upon the pH of the solution and the values of ΔG^o determined above refer to *standard conditions*. We leave it to the reader to consider the effects on ΔG^o of reducing the concentration of H^+ ions for either of reactions 16.42 and 16.46.

16.10 COLOURS

A characteristic feature of compounds of *d*-block metals is that they are often coloured, although the colour may not be intense. Although *dilute* aqueous solutions of potassium permanganate $K[MnO_4]$ are intense purple, this depth of colour should not be considered 'normal' for a *d*-block metal compound. Blue crystals of copper(II) sulfate dissolve in water to give a *pale blue* solution; the solution must be very concentrated before the colour can be described as being intense. 'Weakly' and 'intensely' coloured solutions are quantified by molar extinction coefficients.

Molar extinction coefficient: see Section 9.4

The colours of the hexaaqua ions of the first *d*-block row metals are listed in Table 16.6, and the complex ions included in the right-hand side of the table illustrate that the colour is dependent not only on the metal ion but also on the ligands present in the coordination sphere.

The fact that solutions containing complexes of the *d*-block metals are coloured means that these species must absorb visible light. In Section 9.11, we introduced this idea by considering the electronic spectrum of an aqueous solution containing the $[Ti(H_2O)_6]^{3+}$ ion. The spectrum exhibits a band with λ_{max} = 510 nm (corresponding to the absorbed light), and the observed colour of aqueous titanium(III) ions is purple.

Complementary colours: see Section 9.5

The electronic transitions that occur and give rise to the characteristic colours of many *d*-block metal complexes are known as '*d–d*' transitions. When an ion of a *d-block element possesses partially filled d orbitals*, electronic transitions between *d* orbitals may occur, but the zinc(II) ion has a filled 3*d* shell, and no electronic transitions are possible. This explains why solution of zinc(II) compounds are usually colourless.

We now come to two major problems:

- Are not *d* orbitals with the same principal quantum number degenerate? If so, transitions between them may not be accompanied by a change in electronic energy. The consequence of this is that no electronic spectrum should be visible.

Table 16.6 Colours of some hexaaqua ions and other complex ions of the *d*-block metals. Ligand abbreviations are given in Table 16.2.

Hexaaqua metal ion	*Observed colour in aqueous solution*	*Complex ion*	*Observed colour in aqueous solution*
$[Ti(H_2O)_6]^{3+}$	Violet	$[Co(CN)_6]^{3-}$	Yellow
$[V(H_2O)_6]^{3+}$	Green	$[Co(en)_3]^{3+}$	Yellow
$[Cr(H_2O)_6]^{3+}$	Green	$[Co(ox)_3]^{3-}$	Dark green
$[Cr(H_2O)_6]^{2+}$	Blue	$[Co(NH_3)_6]^{3+}$	Orange
$[Fe(H_2O)_6]^{2+}$	Very pale green	$[Cu(en)_3]^{2+}$	Blue
$[Co(H_2O)_6]^{2+}$	Pink	$[CoCl_4]^{2-}$	Blue
$[Ni(H_2O)_6]^{2+}$	Green		
$[Cu(H_2O)_6]^{2+}$	Blue		
$[Zn(H_2O)_6]^{2+}$	Colourless		

Laporte selection rule: see Section 2.16

- Are not 'd–d' transitions disallowed by the *Laporte selection rule*? This rule states that Δl must change by +1 or –1, and therefore a $3d \rightarrow 3d$ transition is disallowed because $\Delta l = 0$.

In the next section, we introduce a method of approaching the bonding in d-block metal complexes and it provides us with a partial answer to the above questions.

16.11 CRYSTAL FIELD THEORY

In Section 16.4, we considered the electroneutrality principle and saw that neither a wholly ionic nor a wholly covalent bonding picture provided a realistic charge distribution in the complex. Nonetheless, the assumptions that ligands may be considered to be negative point charges and that there is *no* metal–ligand covalent bonding are the bases for a model put forward to explain, among other things, the electronic spectroscopic characteristics of d-block metal complexes. The model is called *crystal field theory* and in this section we apply it to octahedral complexes.

Crystal field splitting

Point charges and electrostatic interactions: see Section 6.6.

Consider an octahedral metal complex $[ML_6]^{n+}$ (Figure 16.19a) in which a metal ion M^{n+} is surrounded by six ligands. We can view this complex in terms of a positively charged metal ion surrounded by six point negative charges (the electrons on the ligands) as shown in Figure 16.19b. We must remember, however, that the metal ion consists of a positively charged nucleus surrounded by negatively charged electrons.

Although there are electrostatic attractions between the metal ion and the ligands, as the ligands approach the d-block M^{n+} ion, the electrons in the d orbitals will be *repelled* by the point charges (the ligands). This has a *destabilizing effect* upon the metal electrons. The valence electrons of a first row d-block metal ion occupy the $3d$ atomic orbitals. If the electrostatic field

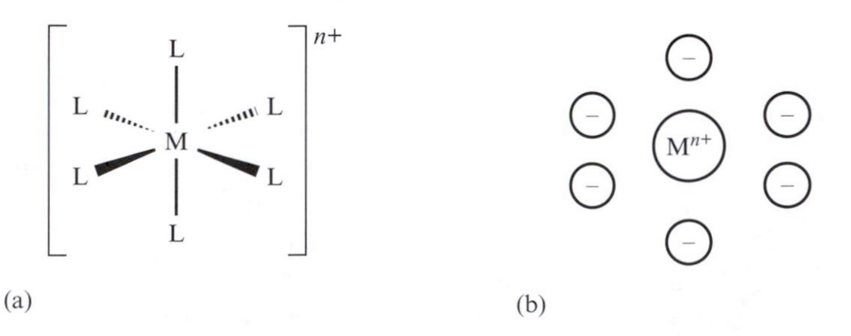

16.19 (a) A structural representation of an octahedral complex $[ML_6]^{n+}$ in which a metal ion M^{n+} is surrounded by six neutral ligands L. (b) The complex can be considered in terms of six negative point charges surrounding the positively charged metal ion. *Remember that the metal ion M^{n+} contains a positively charged nucleus surrounded by negatively charged electrons.*

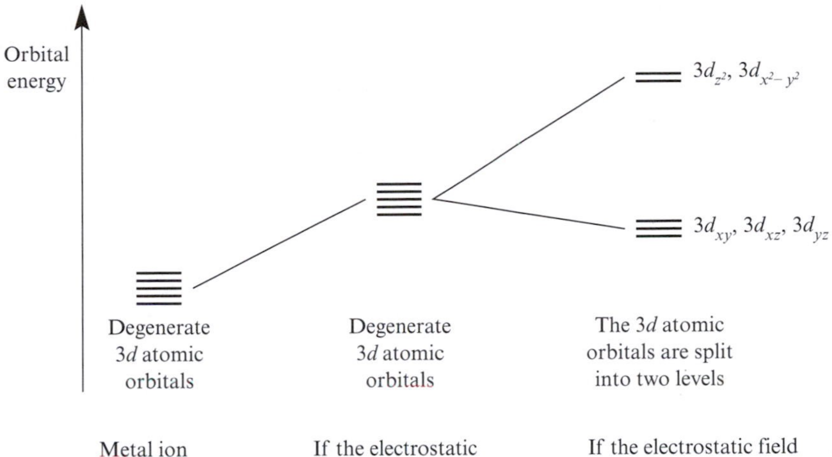

16.20 The charges in the energies of electrons occupying the 3*d* atomic orbitals of an ion M^{n+} when the metal ion is surrounded by a spherical crystal field and an octahedral crystal field. The energy changes are shown in terms of the orbital energies.

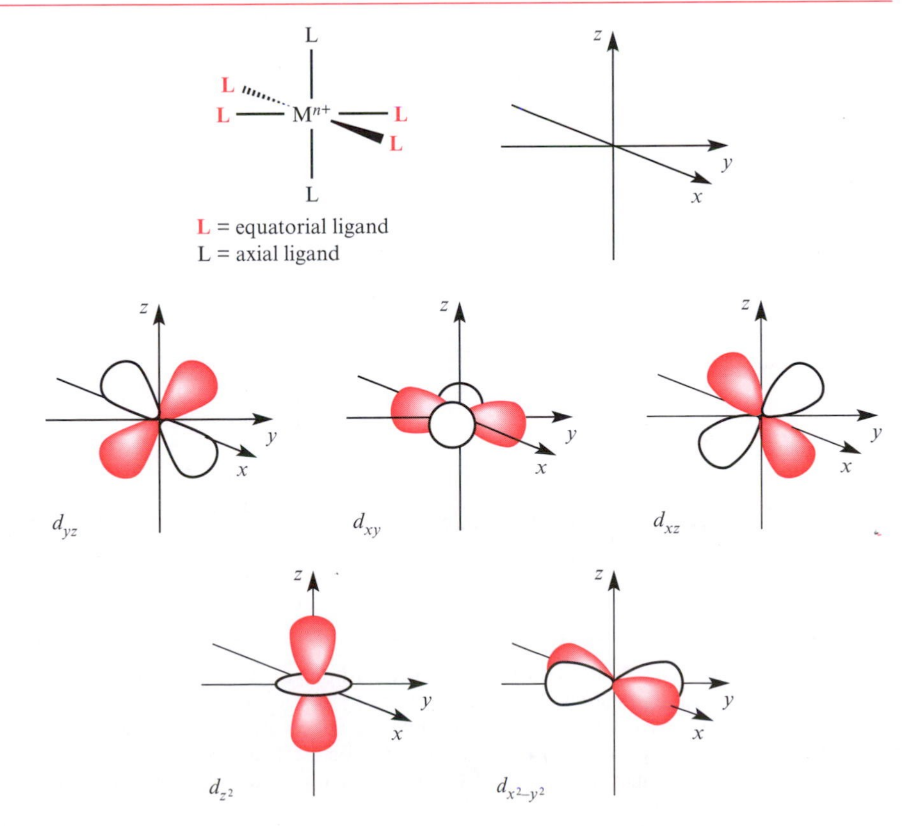

16.21 The six M–L bonds in an octahedral complex $[ML_6]^{n+}$ may be defined as lying along Cartesian axes. The directionalities of the five *d* orbitals can therefore be related to the directions of the M–L bond vectors.

(termed the *crystal field*) were *spherical*, then all the metal ion $3d$ electrons would be destabilized to the same extent as illustrated in Figure 16.20. In an octahedral complex $[ML_6]^{n+}$, the electrostatic field is not spherical but octahedral. Before we look at how this changes the picture, we must recognize the relationship between the shape of the $[ML_6]^{n+}$ cation and the directionalities of the five $3d$ orbitals.

Figure 16.21 shows that the M–L vectors may be defined so as to lie along a set of Cartesian axes. This means that the $3d_{z^2}$ orbital points *directly* at two of the ligands (two axial L in Figure 16.21) and the $3d_{x^2-y^2}$ orbital points *directly* at four of the ligands (the equatorial L in Figure 16.21). The three orbitals d_{xy}, d_{xz} and d_{yz} are related to one another and their lobes point *between* the Cartesian axes, that is, *between* the ligands. Now consider the effect of placing the metal ion at the centre of an octahedral crystal field. The electrons in the $3d_{z^2}$ and $3d_{x^2-y^2}$ orbitals are repelled by the point charge ligands. The electrons in the $3d_{xy}$, $3d_{xz}$ and $3d_{yz}$ orbitals are also repelled by the point charges, but since these orbitals do *not* point directly at the ligands, the repulsion is less than that experienced by the electrons in the $3d_{z^2}$ and $3d_{x^2-y^2}$ orbitals. The net effect is that the electrons in the $3d_{z^2}$ and $3d_{x^2-y^2}$

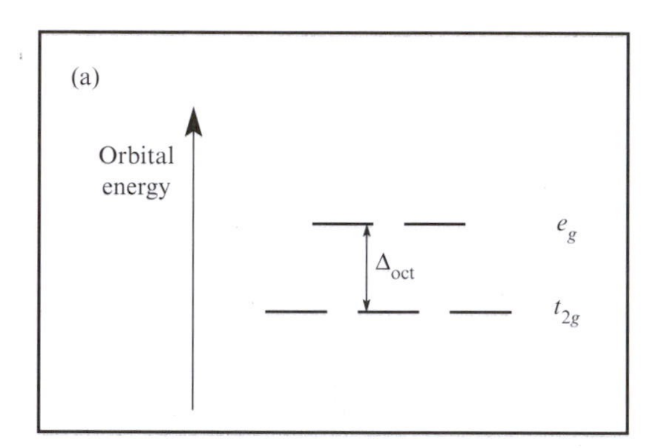

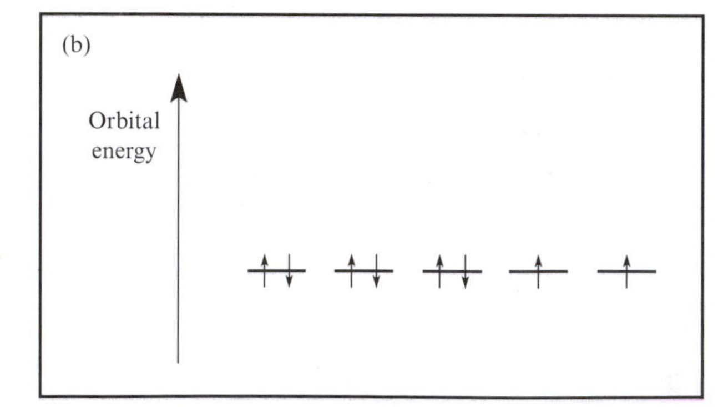

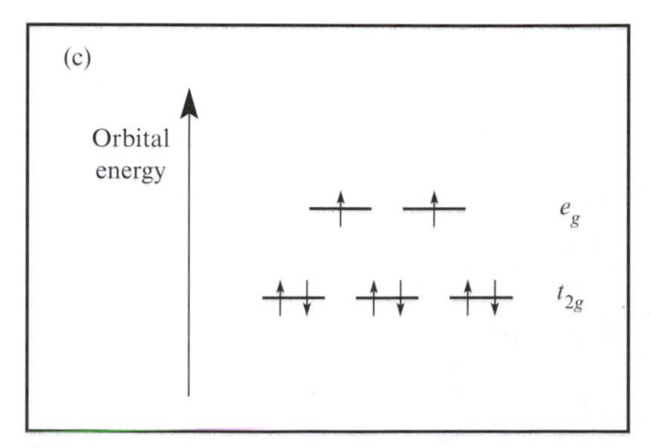

16.22 (a) In an octahedral field, the five d orbitals split into two sets: the higher energy e_g set is doubly degenerate and consists of the $3d_{z^2}$ and $3d_{x^2-y^2}$ orbitals, while the lower energy t_{2g} set is triply degenerate and consists of the $3d_{xy}$, $3d_{xz}$ and $3d_{yz}$ orbitals. (b) In a free, gaseous Ni^{2+} ion, the $3d$ orbitals are degenerate and the eight valence eletrons occupy them according to the Pauli exclusion principle. (c) In an octahedral complex such as $[Ni(H_2O_6]^{2+}$, the $3d$ orbitals are split into the t_{2g} and e_g sets and the eight valence electrons of the nickel(II) ion do not all possess the same orbital energy.

orbitals are destabilized *with respect to a spherical field* while those in the $3d_{xy}$, $3d_{xz}$ and $3d_{yz}$ orbitals are stabilized. This is represented on the right-hand side of Figure 16.20. The final result is that the 3*d* orbitals (which were originally degenerate) are split into two sets – the $3d_{z^2}$ and $3d_{x^2-y^2}$ orbitals form a higher energy set, and the $3d_{xy}$, $3d_{xz}$ and $3d_{yz}$ orbitals make up a lower energy set. This is called the *crystal field splitting*.

The two sets of *d* orbitals of the metal ion in an octahedral complex are conveniently labelled as the e_g and the t_{2g} sets. The e_g set is doubly degenerate and consists of the $3d_{z^2}$ and $3d_{x^2-y^2}$ orbitals, while the t_{2g} set is triply degenerate and consists of the $3d_{xy}$, $3d_{xz}$ and $3d_{yz}$ orbitals (Figure 16.22a). The energy difference between the e_g and the t_{2g} orbitals is called Δ_{oct} (pronounced 'delta oct').

Crystal field theory is a purely electrostatic model which may be used to show that the *d* orbitals in a *d*-metal ion complex are no longer degenerate. In an octahedral complex, crystal field splitting results in there being two sets of *d* orbitals:

Orbital energy

d_{z^2}, $d_{x^2-y^2}$

Δ_{oct}

d_{xy}, d_{xz}, d_{yz}

Box 16.7 The e_g and t_{2g} labels

We will not delve deeply into the origins of the t_{2g} and e_g labels but it is useful to be aware of the following:

- A triply degenerate level is labelled *t*.
- A doubly degenerate level is labelled *e*.
- The subscript $_g$ means *gerade*, German for 'even'.
- The subscript $_u$ means *ungerade*, German for 'odd'.
- *Gerade* and *ungerade* designate the behaviour of the wavefunction under the operation of inversion.

Electronic configurations

So far, we have written the ground state electronic configuration of a *d*-block metal ion in terms of occupying a *degenerate* set of five *d* orbitals. Indeed, this is correct for the free, gaseous ion such as Ni^{2+} (Figure 16.22b), but what happens when we form $[Ni(H_2O)_6]^{2+}$?

In an octahedral complex, the metal *d* orbitals can be considered in terms of the t_{2g} and e_g sets (Figure 16.22a) and when the eight valence electrons of

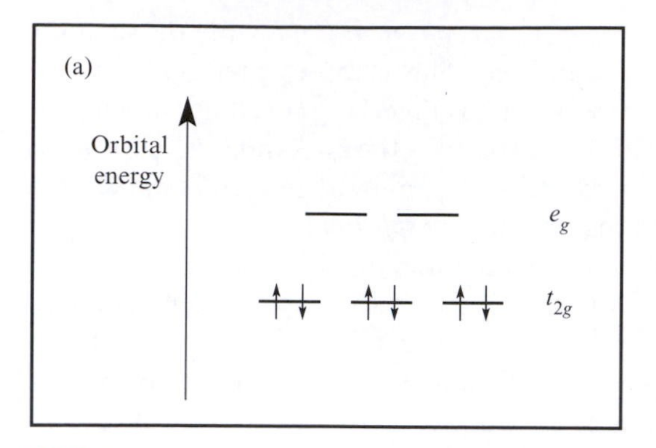

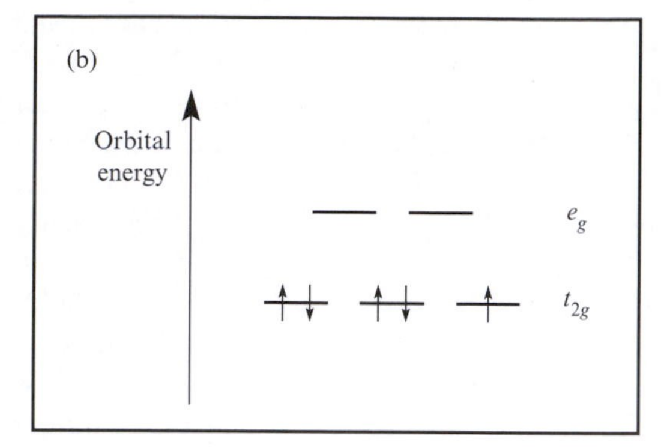

16.23 The arrangements of the 3d electrons in the octahedral complexes (a) $[Fe(CN)_6]^{4-}$ and (b) $[Mn(CN)_6]^{4-}$.

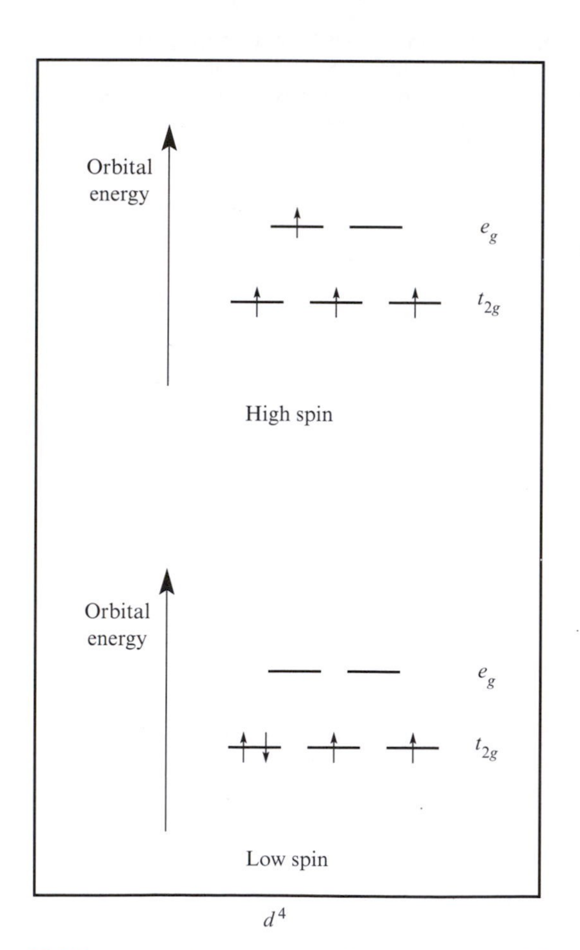

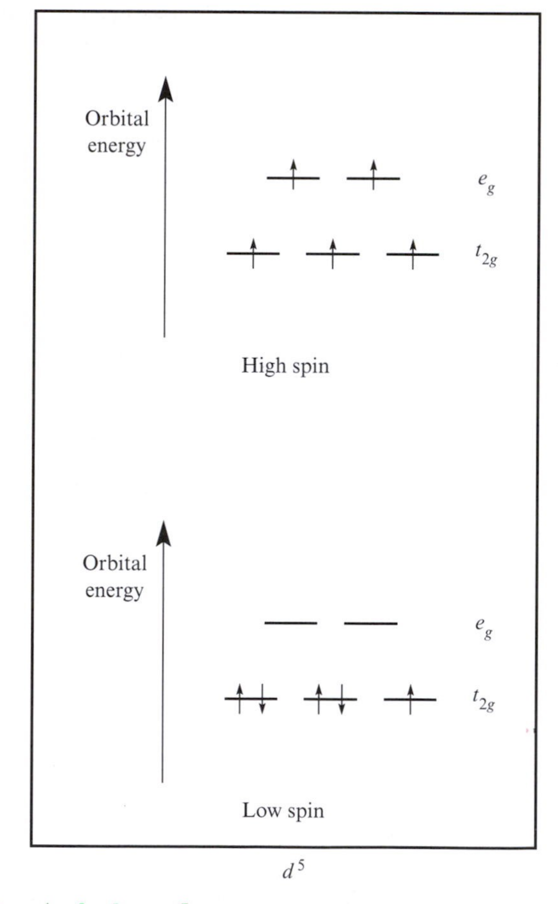

16.24 (Part A) High- and low-spin arrangements of electrons for the d^4, d^5, d^6 and d^7 configurations. Only one electron can be promoted from the t_{2g} to e_g level in the d^4 and d^7 cases, but two can be promoted for d^5 and d^6. (Why is this?)

the Ni(II) ion in $[Ni(H_2O)_6]^{2+}$ occupy the five 3*d* orbitals, they do so such that two of the electrons are at a higher energy than the other six (Figure 16.22c). The d^8 configuration can be more informatively written as $t_{2g}{}^6e_g{}^2$. Figure 16.23 shows the occupancy of the 3*d* orbitals of the Fe(II) ion in $[Fe(CN)_6]^{4-}$ and the Mn(II) ion in $[Mn(CN)_6]^{4-}$; the configurations are $t_{2g}{}^6e_g{}^0$ and $t_{2g}{}^5e_g{}^0$ respectively.

Strong and weak field ligands: the spectrochemical series

The energy gap Δ_{oct} is a measure of the electrostatic repulsions between the metal *d* electrons and the point charge ligands, and some ligands cause a greater splitting of the *d* orbitals than others. Ligands that cause a large crystal field splitting are called *strong field ligands* and $[CN]^-$ is an example. Conversely, the interaction of a metal ion with *weak field ligands* results in a

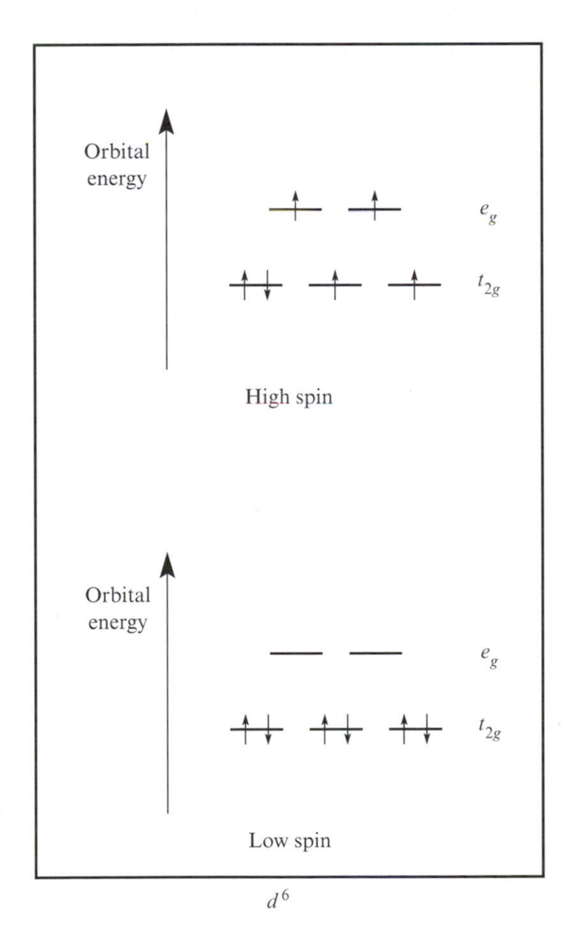

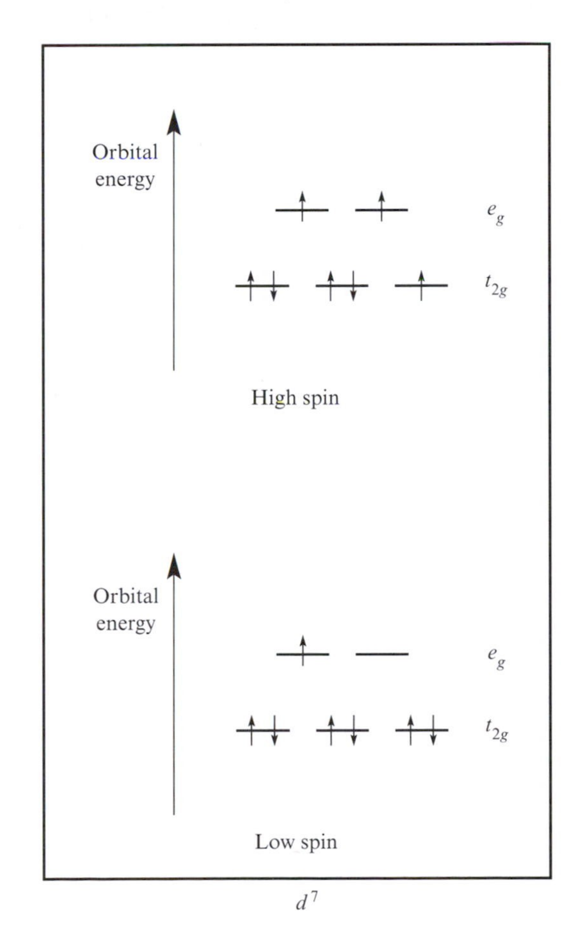

16.24 Part B

smaller splitting of the *d* orbitals. The ordering of ligands according to their ability with respect to crystal field splitting is given by the *spectrochemical series* in which I^- is the weakest field ligand:

$$I^- < Br^- < [NCS]^- < Cl^- < F^- < [OH]^- < [ox]^{2-} \approx H_2O < [NCS]^- < NH_3 < en < [CN]^- < PR_3$$

The spectrochemical series orders ligands according to their ability with respect to crystal field splitting:

$$I^- < Br^- < [NCS]^- < Cl^- < F^- < [OH]^- < [ox]^{2-} \approx H_2O < [NCS]^- < NH_3 < en < [CN]^- < PR_3$$

weak field ligands ... **strong field ligands**

High- and low-spin electronic configurations

For metal ions with d^4, d^5, d^6 or d^7 configurations, it is possible for the electrons to be arranged in the five *d* orbitals in two different ways (Figure 16.24). Consider the d^4 case. If the value of Δ_{oct} is relatively small, it may be energetically favourable to adopt the *high-spin* $t_{2g}{}^3e_g{}^1$ configuration rather than the *low-spin* $t_{2g}{}^4$ in which there are significant electrostatic replusions between the electrons paired in the same t_{2g} orbital. The adoption of a high- or low-spin metal ion will depend upon the relative magnitudes of the *spin-pairing energies* and Δ_{oct}.

It follows that strong field ligands which cause the greatest splitting of the *d* orbitals will tend to give rise to low-spin electronic configurations. Similarly, weak field ligands may lead to high-spin configurations. In Figure 16.23, the configurations for $[Fe(CN)_6]^{4-}$ and $[Mn(CN)_6]^{4-}$ are low spin, consistent with the fact that $[CN]^-$ is a strong field ligand. We emphasize that it is not possible to predict with complete certainty which of the high- or low-spin configurations will be observed in practice, and in Section 16.13 we look at experimental data that can distinguish between them.

Worked example 16.4 ***High- and low-spin configurations***

In an octahedral complex, why can the Ti(III) ion only have one electronic configuration?

Titanium is in group 4 and the electronic configuration of Ti^{3+} is $[Ar]3d^1$.

In a octahedral field, the 3*d* orbitals split and the one electron occupies the t_{2g} level.

If this electron is promoted to the e_g level, it would be at the expense of completely vacating the t_{2g} orbitals and therefore the configuration shown is the only one available for the Ti(III) ion.

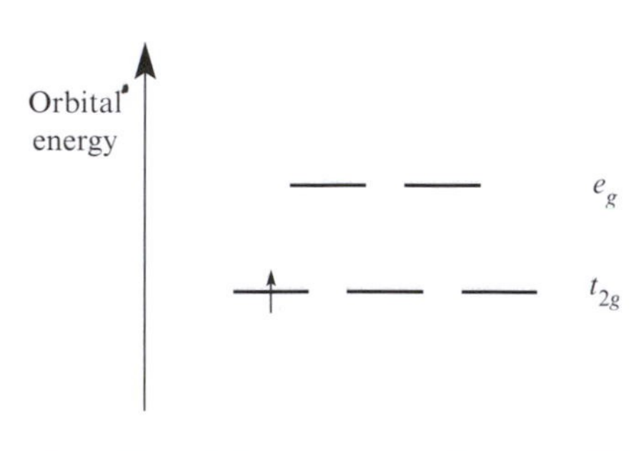

A similar argument can be applied to show that d^2 and d^3 configurations can each only have one arrangement.

16.12 ELECTRONIC SPECTRA

'*d–d*' Transitions

The crystal field splitting of the *d* orbitals in an octahedral complex leads us to a qualitative understanding as to why such complexes are coloured. If the t_{2g} and e_g sets of five *d* orbitals are *partially occupied* as is the case for $[Ni(H_2O)_6]^{2+}$ (Figure 16.22c), it would appear to be possible for an electron from the lower t_{2g} level to undergo a transition to the e_g. The energy absorbed is equal to Δ_{oct}.

Laporte selection rule: see Section 2.16

There is, however, a problem – a '*d–d*' transition is *Laporte forbidden*. The observed electronic spectra of *d*-block complexes are consistent with the fact that '*d–d*' transitions *appear to occur*. In fact the selection rule is absolutely obeyed and '*d–d*' transitions are strictly forbidden. However, the picture of the interactions that we developed in Section 16.11 is overly simplistic, and in this text we shall simply relate Δ_{oct} to the spectroscopic transitions that are observed.§

The relationship between λ_{max} and Δ_{oct}

In this section, we look more closely at how the energy gap between the t_{2g} and e_g levels influences the value of λ_{max} in the electronic spectrum of a complex.

The $[Ti(H_2O)_6]^{3+}$ ion absorb lights at 510 nm. The aqua ligand is relatively weak field and Δ_{oct} is correspondingly small. Exchanging the aqua for even weaker field chloro or bromo ligands gives $[TiCl_6]^{3-}$ and $[TiBr_6]^{3-}$ respectively and in these anions the energy gap between the t_{2g} and e_g levels is *smaller* than in $[Ti(H_2O)_6]^{3+}$. We expect therefore that the light absorbed by these halo-complexes should be of a *longer wavelength* (equation 16.48) than that absorbed by $[Ti(H_2O)_6]^{3+}$, and indeed the observed values of λ_{max} for $[TiCl_6]^{3-}$ and $[TiBr_6]^{3-}$ are 780 and 850 nm respectively.

$$E = h\nu \qquad \text{or} \qquad E \propto \frac{1}{\lambda} \tag{16.48}$$

Although we have talked in terms of the field strength of the ligands to gain insight into the spectroscopic properties of these complexes, it is in fact the opposite arguments that allow us to construct the spectrochemical series in the first place!

§ For a more detailed discussion of how '*d–d*' electronic spectra arise, see M. Gerloch and E. C. Constable, *Transition Metal Chemistry: The Valence Shell in d-Block Chemistry*, VCH, Weinheim (1994).

16.13 MAGNETISM: THE SPIN-ONLY FORMULA

High- and low-spin configurations: how many unpaired electrons?

We have already seen that the presence of strong or weak field ligands in an octahedral complex may result in a change from low- to high-spin electronic arrangements for the d^4, d^5, d^6 and d^7 configurations. If we consider the high- and low-spin configurations of the d^4 ion chromium(II), Figure 16.24 reveals that the difference between them is *the number of unpaired electrons*. A low-spin Cr(II) ion possesses two unpaired electrons but a high-spin ion has four. This distinction gives rise to differences in the magnetic properties of the two chromium(II) centres, and in this section we look at the relationship between the *effective magnetic moment* of a complex and the number of unpaired electrons present.

The Gouy balance

The Gouy method is used to measure the effective magnetic moment of a compound, and the basis of the method is the interaction of unpaired electrons with a magnetic field. If a material is *diamagnetic* (all electrons paired) it is repelled by a magnetic field, but a *paramagnetic* material (possessing one or more unpaired electrons) is attracted by it.

Dia- and paramagnetism: see Section 3.18.

Figure 16.25 schematically illustrates the apparatus used in the Gouy method. The compound under study is contained in a glass tube which is suspended from a balance on which the weight of the sample can be recorded. The glass tube is positioned so that one end of the sample lies at the point of maximum magnetic flux in an electromagnetic field – this is initially switched off. (An electromagnet is more convenient to use than a permanent magnet

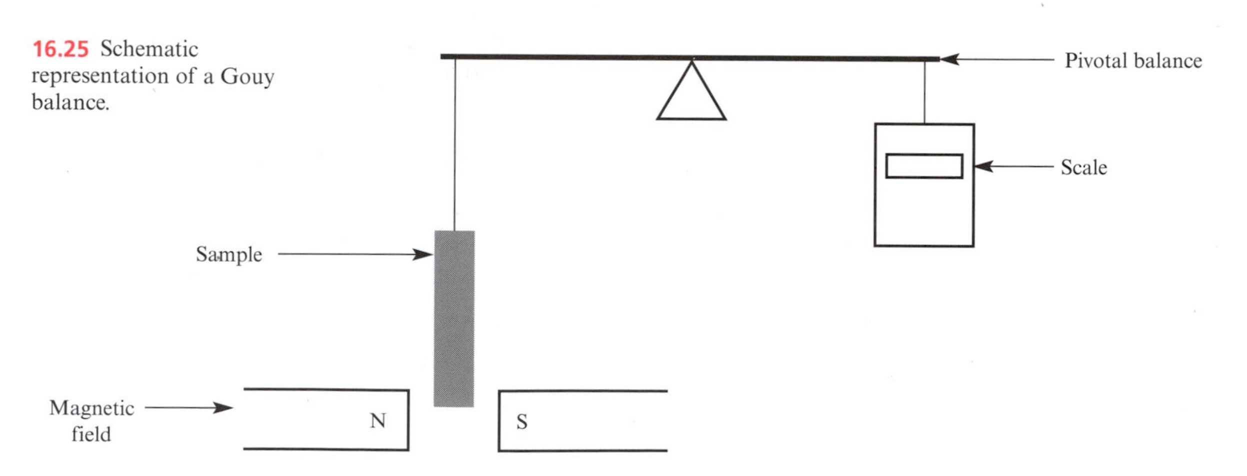

16.25 Schematic representation of a Gouy balance.

because the field can be switched on and off.) The other end of the sample lies at a point of low magnetic flux. When the magnetic field is applied, paramagnetic compounds are attracted into it, and the extent of attraction depends upon the number of unpaired electrons. The attraction causes a change in the weight of the sample and from these readings, it is possible to determine the *magnetic susceptibility*, χ, of the compound. (The details of the derivation are beyond the scope of this text.) Usually, values of the *effective magnetic moment* are reported in the chemical literature rather than values of χ, and the two quantities are related by equation 16.49.

$$\mu_{\text{eff}} = 2.828 \sqrt{\chi \times T} \tag{16.49}$$

where μ_{eff} = effective magnetic moment in units of bohr magnetons (BM),
χ = magnetic susceptibility,
T = temperature.

Box 6.8 Ferromagnetic chains

The magnetic behaviour of a substance is usually described in terms of the Curie–Weiss Law which states that the *molar magnetic susceptibility* of a substance is inversely proportional to the temperature. *Ferromagnetic materials* do not conform to the Curie–Weiss Law until temperatures above the Curie temperature. *Ferromagnetism* arises through communication between metal centres, each of which has one or more unpaired electrons. Interaction between electrons on different metal centres causes their spins to become aligned but the electrons do *not* pair up so as to form a covalent bond.

The salt $[Ni(en)_2]_3[Fe(CN)_6]_2{\cdot}2H_2O$ is a beautiful example of a ferromagnetic material. In the solid state, X-ray diffraction studies have shown that the nickel(II) and iron(III) centres are connected by bridging cyano ligands and each metal is 6-coordinate. The structure consists of interconnected helical chains. The ferromagnetism arises because the cyano-bridges allow the iron and nickel centres to communicate electronically with one another. This is just one example of the process of self-assembly of magnificent structures in the solid state.

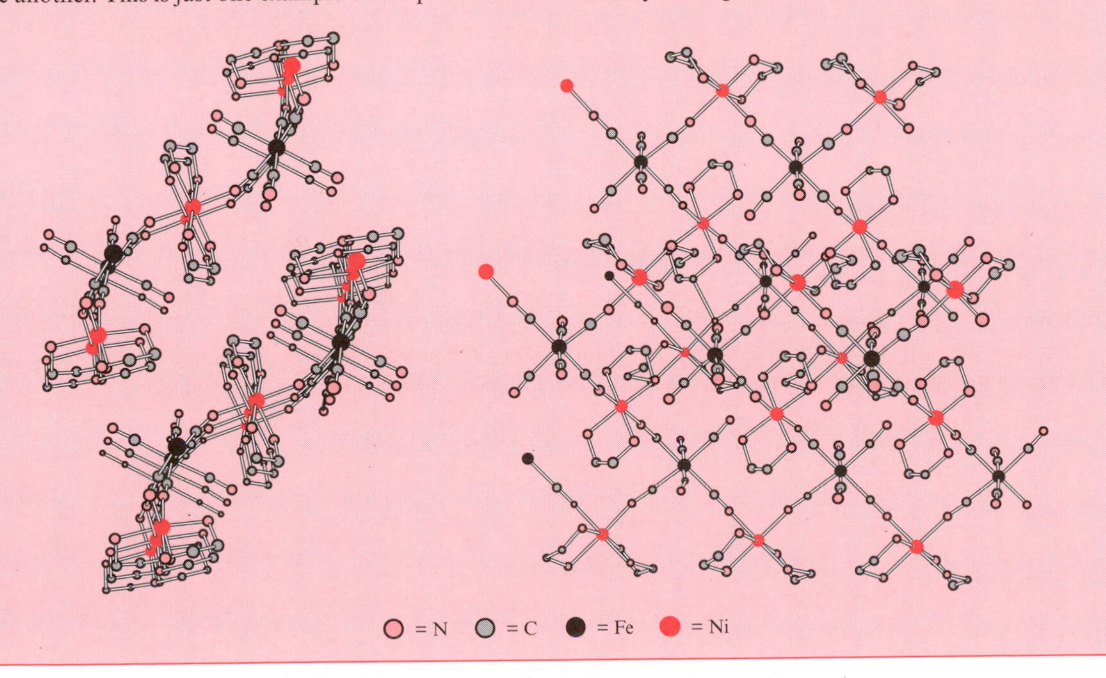

Table 16.7 Experimentally determined values of the effective magnetic moments of selected octahedral complexes and values of μ calculated using the spin only formula (equation 16.50). Ligand abbreviations are listed in Table 16.2.

Compound	*Experimental* μ_{eff} */ BM*	*Spin only* μ */ BM*
$K_4[Mn(CN)_6]$	2.18	1.73
$[VCl_4(bpy)]$	1.77	1.73
$K_3[V(ox)_3]$	2.80	2.83
$K_2[Mn(CN)_6]$	3.94	3.87
$[MnCl_4(bpy)]$	3.82	3.87
$[Cr(en)_3]Br_2$	4.75	4.90
$[Cr(dien)_2]Cl_2$	4.81	4.90
$K_3[Fe(ox)_3]$	5.95	5.92

We shall only be concerned with values of μ_{eff}, the units of which are bohr magnetons (BM). Experimental values of the effective magnetic moments of selected octahedral complexes are listed in Table 16.7 and are compared with calculated values which we discuss below.

The spin-only formula

There are two effects that contribute towards the paramagnetism: the electron spin and the orbital angular momentum. For octahedral complexes involving first row *d*-block metals, it is *approximately true* that the value of μ_{eff} can be estimated by assuming that the contribution made by the orbital angular momentum is negligible. Equation 16.50 is the *spin-only formula* and shows a simple dependence of the spin-only magnetic moment on the number of unpaired electrons, n.

$$\text{Spin-only magnetic moment} = \mu(\text{spin-only}) = \sqrt{n(n+2)} \qquad (16.50)$$

where n = number of unpaired electrons

Box 16.9 An alternative form of the spin-only formula

In equation 16.50, we write the spin-only formula in terms of the number of unpaired electrons, n. An alternative form of the equation is in terms of the *total spin quantum number, S*.

In Section 2.10, we saw that the spin quantum number s determines the magnitude of the spin angular momentum of an electron and has a value of $\frac{1}{2}$. The *total spin quantum number S* is given by:

Number of unpaired electrons:	1	2	3	4	5
Total spin quantum number, S:	$\frac{1}{2}$	1	$\frac{3}{2}$	2	$\frac{5}{2}$

Since:

$$S = n \times \tfrac{1}{2} = \tfrac{n}{2} \quad \text{or} \quad n = 2 \times S$$

we can rewrite the spin-only formula in the form: $\mu(\text{spin-only}) = \sqrt{4S(S+1)}$

For the chromium(II) ion in a high-spin configuration, there are four unpaired electrons (Figure 16.24) and the spin-only value of μ is calculated to be 4.90 BM (equation 16.51).

$$\mu(\text{spin-only}) = \sqrt{n(n+2)} = \sqrt{4 \times 6} = \sqrt{24} = 4.90 \text{ BM} \qquad (16.51)$$

Measured effective magnetic moments are used to distinguish between high- and low-spin configurations. In a low-spin configuration, chromium(II) has only two unpaired electrons and the observed value of μ_{eff} is expected to be close to 2.83 BM.

Worked example 16.5 ***Effective magnetic moments***

The experimentally determined magnetic moment of the compound $[V(en)_3]Cl_2$ is 3.84 BM. How many unpaired electrons are present?

The effective magnetic moment can be estimated by using the spin-only formula:

$$\mu(\text{spin-only}) = \sqrt{n(n+2)}$$

Rather than substituting in the experimental value of μ_{eff}, it is quicker to test values of n (which can only be 1, 2, 3, 4 or 5) in the spin-only formula.

If $n = 3$:

$$\mu(\text{spin-only}) = \sqrt{3(3+2)} = \sqrt{15} = 3.87$$

This value is close to the experimental value of μ_{eff} and so there are three unpaired electrons on the vanadium centre in $[V(en)_3]^{2+}$.

This is consistent with a vanadium(II) ion with electronic configuration $[Ar]3d^3$.

As the data in Table 16.7 illustrate, the spin-only formula is a fairly good approximation for octahedral complexes containing first row *d*-block metal ions. However, you should appreciate that the fit between the spin-only and measured effective magnetic moments varies and the values for $K_4[Mn(CN)_6]$ and $[VCl_4(bpy)]$ illustrate this. Complexes with other geometries or containing second and third row *d*-block metals show much larger deviations. Second and third row complexes are generally low spin but with values significantly lower than the spin-only formula predicts.

16.14 METAL CARBONYL COMPOUNDS

The discussion in this chapter has focused on complexes of M^{2+} and M^{3+} ions with ligands that donate electrons, and we have seen that the electroneutrality principle can be used to assess reasonable charge distributions. In this final section, we introduce another group of complexes called *metal carbonyl compounds* which are representative of the much larger class of *organometallic compounds.*

The carbon monoxide ligand

Haemoglobin: see Box 4.4

Bonding in CO: see Section 4.14

Carbon monoxide is a Lewis base and we have already seen that its toxicity is due to the fact that it binds more strongly to the iron centres in haemoglobin than does O_2. Carbon monoxide also forms complexes with *low oxidation state* metals from the *d*-block. Its Lewis basicity arises from the properties of the highest *occupied* molecular orbital (HOMO) in the CO molecule (Figure 16.26). This MO consists largely of carbon character and may be considered to be a carbon-centred lone pair. The lowest *unoccupied* molecular orbitals (LUMO) in CO are a degenerate set of π^* orbitals, again with more carbon than oxygen character, and one of the orbitals is shown in Figure 16.26.

The low oxidation state metal centre

By *low oxidation state*, we usually think in terms of a metal(0) centre but also include metals in oxidation states +1, –1 and –2. In iron pentacarbonyl,

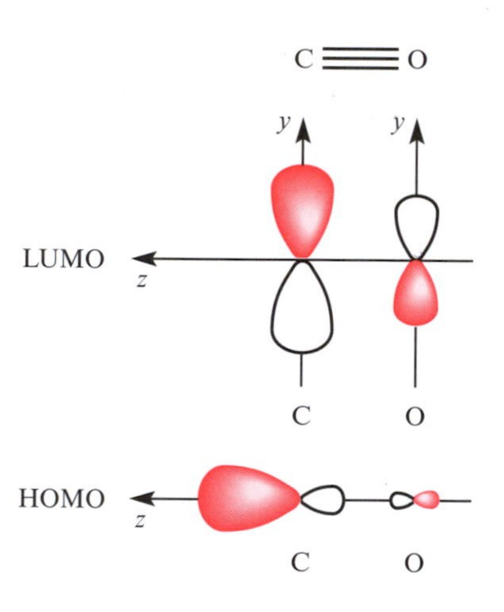

16.26 The highest occupied molecular orbital (HOMO) and one of the two degenerate lowest unoccupied molecular obitals (LUMO) of carbon monoxide. More detailed diagrams are given in Section 4.14.

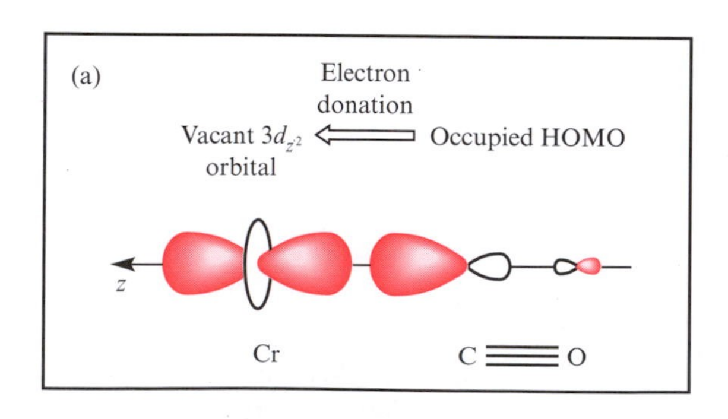

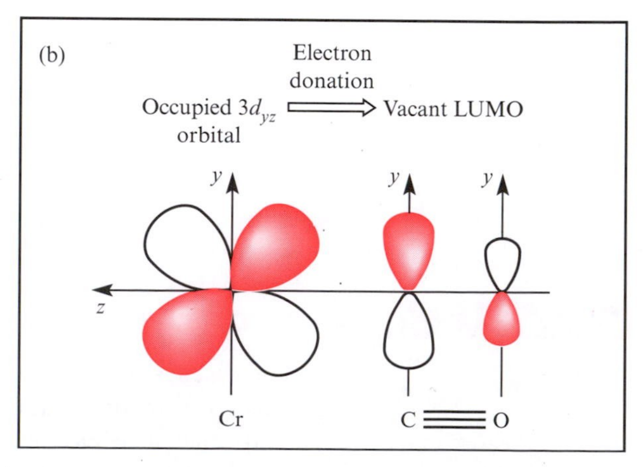

16.27 (a) The formation of one chromium–carbon σ-bond, and (b) the formation of a chromium–carbon π-bond in $[Cr(CO)_6]$. The electronic charge is transferred in *opposite* directions in the two interactions, and the overall effect is a *synergic interaction.*

$Fe(CO)_5$, the iron centre is in an oxidation state of zero, and in $[Fe(CO)_4]^{2-}$, the oxidation state is –2.

In a metal carbonyl compound such as $[Cr(CO)_6]$ **16.27**, each CO ligand can donate a pair of electrons to the metal centre but this greatly increases the negative charge at the metal. This σ-interaction is shown for one Cr–CO interaction in Figure 16.27a; the metal orbital involved is one of the e_g orbitals, exemplified here by the d_{z^2} orbital.

CO
OC CO
Cr
OC CO
CO

(16.27)

By Pauling's electroneutrality principle, the distribution of charge in a molecule or ion is such that the charge on any single atom is ideally close to zero, and since the chromium centre in $[Cr(CO)_6]$ bears no formal charge, we appear to be saying that it is unfavourable for the chromium atom to form coordinate bonds with the six carbonyl ligands which would give a formal charge at the Cr(0) centre of –6. How does the molecule cope with this situation?

CO: a π-acceptor ligand

We used the electroneutrality principle to introduce partially ionic metal–ligand bonding with higher oxidation state metal ions. However, with the $[Cr(CO)_6]$ molecule, no degree of ionic character will remove the negative charge that appears to have built up on the electropositive Cr(0) centre, and we need another means of removing this negative charge from the metal.

The key to the success of carbon monoxide in stabilizing low oxidation state metal centres is its ability to act as an *acceptor ligand.* The low-lying π^* orbitals can overlap with the occupied t_{2g} orbitals of the metal and the result is the formation of a coordinate π-bond in which electrons *from the metal are donated to the ligand* (Figure 16.27b).

The π-interaction operates in addition to the σ-interaction shown in Figure 16.27a and the overall effect is that the negative charge that builds up on the chromium centre as a result of the formation of six metal–carbon σ-bonds is redistributed *back to the ligands* as a consequence of the metal–carbon π-bonds. This is called a *synergic effect* and the bonding theory summarized in Figure 16.27 is called the Dewar–Chatt–Duncanson model.

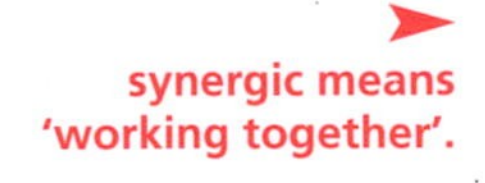

synergic means 'working together'.

Infrared spectroscopic evidence for π-back donation

The back donation of electrons from a metal to CO ligand involves the occupancy of a carbon–oxygen π^* MO; that is, an orbital in which there is carbon–oxygen *antibonding character*. This leads to a weakening of the carbon–oxygen bond and the result is readily observed using IR spectroscopy because it gives a *reduction* in the value of the C–O stretching

Table 16.8 Infrared spectroscopic data for two series of isoelectronic metal carbonyls.

Octahedral complex	ν_{CO}/ cm^{-1}	*Tetrahedral complex*	ν_{CO}/ cm^{-1}
$Cr(CO)_6$	2000	$Ni(CO)_4$	2060
$[Mn(CO)_6]^+$	2090	$[Co(CO)_4]^-$	1890
$[V(CO)_6]^-$	1860	$[Fe(CO)_4]^{2-}$	1790

frequency in going from free CO to coordinated CO. Table 16.8 lists values of ν_{CO} for a series of isoelectronic complexes, each of which possesses an octahedral structure. In the spectrum of carbon monoxide, an absorption is observed at 2143 cm^{-1}.

Shapes of mononuclear metal carbonyl complexes

Kepert theory: see Sections 5.8–5.10

The theory of Kepert can successfully be applied to mononuclear metal carbonyl complexes. Binary carbonyl complexes with four ligands are tetrahedral, with five ligands are trigonal bipyramidal and with six ligands are octahedral, and the structures of the neutral compounds $[Ni(CO)_4]$, $[Fe(CO)_5]$ and $[Cr(CO)_6]$ are shown in Figure 16.28. The anions $[Co(CO)_4]^-$ and $[Fe(CO)_4]^{2-}$ are both isoelectronic and isostructural with $Ni(CO)_4$. Similarly $[Mn(CO)_5]^-$ is related to $[Fe(CO)_5]$, and $[Mn(CO)_6]^+$ and $[V(CO)_6]^-$ are isostructural with $[Cr(CO)_6]$.

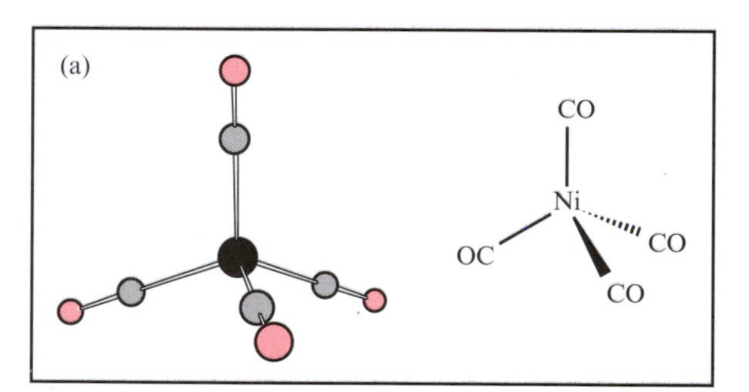

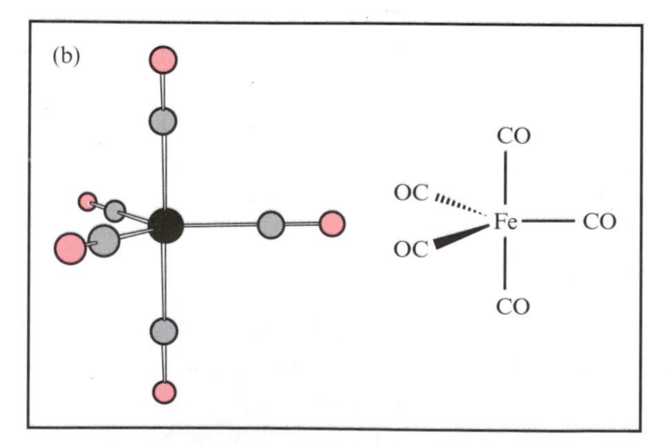

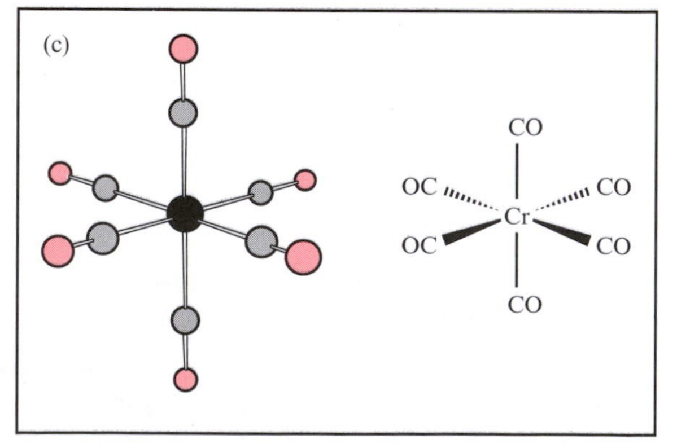

16.28 The structures of the binary metal carbonyl compounds (a) $Ni(CO)_4$, (b) $Fe(CO)_5$ and (c) $Cr(CO)_6$.

SUMMARY

This chapter has introduced some of the basic concepts of the coordination chemistry of the *d*-block metals, with an emphasis on the elements in the first row. We have also looked briefly at the special properties of the CO ligand that make it able to stabilize *low* oxidation state metal centres.

Do you know what the following terms mean?

- coordinate bond
- coordination number
- ligand
- monodentate
- didentate
- polydentate
- donor atom
- chelate
- geometrical isomers
- structural isomers
- hexaaqua ion
- ammine complex
- nuclearity of a complex
- hard and soft donor atoms
- hard and soft metal centres
- intensity of colour
- crystal field splitting
- Δ_{oct}
- strong and weak field ligand
- spectrochemical series
- high-spin and low-spin configurations
- spin-only formula
- π-acceptor ligand
- synergic effect
- Dewar–Chatt–Duncanson model

You should now be able to:

- to know the symbols and names of the first row *d*-block metals
- to write down the elements in sequence in the first row of the *d*-block
- to write down the ground state electronic configuration of metal atoms and ions in the first row of the *d*-block
- to recognize potential donor atoms in a ligand
- to recognize how flexible a ligand may be in terms of its ability to adopt different conformations
- to draw the abbreviations and structures of representative ligands including py, $[NCS]^-$, en, dien, bpy, phen, $[acac]^-$ and $[edta]^{4-}$.
- to estimate the charge distribution in a complex of the type $[ML_6]^{n+}$
- to work out the possible isomers of a given complex
- to explain why some aqua ions are acidic in aqueous solution
- to give representative preparative routes to selected complexes
- to suggest *possible* products from complex-forming reactions, including isomers
- to write down an expression for the stability constant (stepwise or overall) of a complex
- to obtain values of ΔG° from stability constant data
- to use the concept of hard and soft metal ions and ligands to suggest reasons for the preferences shown in the formation of some complexes
- to use values of E° to assess the stability of an aqueous ion in solution with respect to oxidation by, for example, H^+ or O_2, recognizing the restrictions that working under standard conditions imposes
- to explain why the *d*-orbitals are not degenerate in an octahedral complex
- to write down the electronic configuration of a given metal ion in an octahedral environment
- to understand the effects that the presence of strong and weak field ligands have on the electronic configuration of a metal ion in an octahedral environment
- to describe a method of *measuring* the effective magnetic moment of a complex
- to estimate the magnetic moment of complex containing a first row *d*-block metal ion
- to estimate the number of unpaired electrons present given a value of μ_{eff}
- to discuss the molecular orbital properties of CO that lead to its ability to act as a π-acceptor ligand
- to outline the bonding in a mononuclear metal carbonyl compound in terms of the interactions in a single M–CO unit.
- to explain what spectroscopic evidence supports the Dewar–Chatt–Duncanson model

PROBLEMS

Refer to Table 16.2 for ligand abbreviations

16.1 Draw radial distribution curves for the 4*s* and 3*d* atomic orbitals. How is an electron in a 4*s* orbital affected by an increase in nuclear charge? How is an electron in a 3*d* orbital affected by an increase in nuclear charge?

16.2 Write down the ground state electronic configurations of the following atoms and ions: (a) Fe; (b) Fe^{3+}; (c) Cu; (d) Zn^{2+}; (e) V^{3+}; (f) Ni; (g) Cr^{3+}; (h) Cu^+.

16.3 Draw the structures of the following ligands and mark the potential donor atoms:

(a) 2-aminopyridine
(b) oxalate ion
(c) phenoxide ion
(d) 1,3-diaminopropane
(e) 3,6-diazaoctane
(f) trimethylphosphine
(g) THF
(h) 2,5-dioxaheptane

Which ligands could be chelating?

16.4 What is the oxidation state of the metal in following complexes?

(a) $[CoCl_4]^{2-}$
(b) $[Fe(CN)_6]^{3-}$
(c) $[Ni(gly\text{–}N,O)_3]^-$
(d) $[Fe(bpy)_3]^{3+}$
(e) $[Cr(acac)_3]$
(f) $[Co(H_2O)_6]^{2+}$
(g) $[Ni(en)_3]^{2+}$
(h) $[Ni(edta)]^{2-}$
(i) $[Cu(ox)_2]^{2-}$
(j) $[Cr(en)_2F_2]^+$

16.5 Use the Kepert theory to predict the structures of the following complexes. Could any of these complexes possess geometrical isomers? If so, suggest which isomer might be favoured, (none of the 4-coordinate complexes is square planar):

(a) $[Zn(OH)_4]^{2-}$
(b) $[Ru(NH_3)_5(NCMe)]^{2+}$
(c) $[CoBr_2(PPh_3)_2]$
(d) $[WBr_4(NCMe)_2]$
(e) $[ReCl(CO)_3(py)_2]$
(f) $[Cr(NPr_2)_3]$
(g) $[CrCl_3(NMe_2)_2]$

[*Hint*: $[NPr_2]^-$ and $[NMe_2]^-$ are *amido* ligands, isoelectronic with the corresponding ether R_2O.]

16.6 Estimate the charge distribution over the metal and donor atom centres in the complex ions (a) $[Fe(bpy)_3]^{2+}$ and (b) $[CrF_6]^{3-}$.

16.7 Draw the geometrical isomers of (a) octahedral $[Mn(H_2O)_2(ox)_2]^{2-}$, (b) octahedral $[Co(H_2O)(NH_3)(en)_2]^{3+}$, (c) square planar $[NiCl_2(PMe_3)_2]$ and (d) octahedral $[CrCl_3(dien)]$.

16.8 Draw the structures of the isomers of the octahedral nickel(II) complex $[Ni(gly)_3]^-$.

16.9 Which of the following octahedral complexes are optically active?

(a) $[Fe(ox)_3]^{3-}$
(b) $[Fe(en)(H_2O)_4]^{2+}$
(c) *cis*-$[Fe(en)_2(H_2O)_2]^{2+}$
(d) *trans*-$[Fe(en)_2(H_2O)_2]^{2+}$
(e) $[Fe(phen)_3]^{3+}$
(f) $[Co(bpy)(CN)_4]^-$

16.10 Suggest possible products (including isomers if appropriate) for the following reactions:

(a) $[Ni(H_2O)_6]^{2+} + 3en \rightleftharpoons$
(b) $[Co(H_2O)_6]^{2+} + 3phen \rightleftharpoons$
(c) $VCl_4 + 2py \rightleftharpoons$
(d) $[Cr(H_2O)_6]^{3+} + 3[acac]^- \rightleftharpoons$
(e) $[Fe(H_2O)_6]^{2+} + 6[CN]^- \rightleftharpoons$
(f) $FeCl_2 + 6NH_3 \rightleftharpoons$

16.11 Figure 16.29 shows the variation in molar conductivity for the series of compounds $[Co(NH_3)_6]Cl_3$, $[Co(NO_2)(NH_3)_5]Cl_2$, $[Co(NO_2)_2(NH_3)_4]Cl$, $[Co(NO_2)_3(NH_3)_3]$ and $K[Co(NO_2)_4(NH_3)_2]$ in aqueous solutions. Suggest reasons for the observed trend. [*Hint*: What is the basis for conductance in aqueous solution?]

16.12 Write down expressions for the equilibrium constants for reactions 16.23–16.27. Use these expressions and that in equation 16.21 to show that the overall stability constant for $[ML_6]^{2+}$, β_6 is given by the equation:

$$\beta_6 = K_1 \times K_2 \times K_3 \times K_4 \times K_5 \times K_6$$

and that

$$\log \beta_6 = \log K_1 + \log K_2 + \log K_3 + \log K_4 + \log K_5 + \log K_6$$

where K_1 is the equilibrium constant for the replacement of the first ligand, and K_2 refers to the second step, etc.

16.29 For problem 16.11

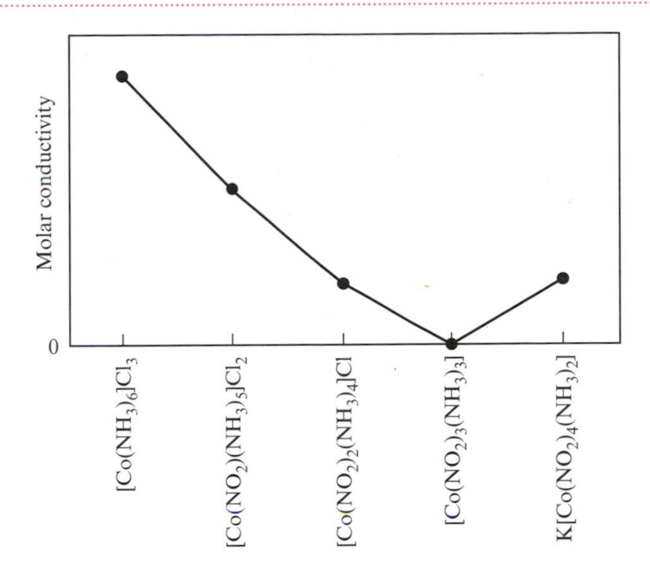

16.13 The overall stability constant for the formation of $[Co(H_2O)_4(NH_3)_2]^{2+}$ from $[Co(H_2O)_6]^{2+}$ is 5.0×10^3, whereas for the formation of $[Co(H_2O)_4(en)]^{2+}$, $K_1 = 1.0 \times 10^6$ (both at 298 K). Determine values of ΔG^o for these reactions. Is it more or less favourable to replace two aqua ligands in $[Co(H_2O)_6]^{2+}$ by two ammine ligands or one 1,2-diaminoethane ligand? Suggest a reason for this difference.

16.14 Classify each of the following coordination complexes in terms of the concept of hard and soft metal ions and ligands, indicating which are the donor atoms in each ligand:

(a) $[AuI_2]^-$
(b) $[Ru(NH_3)_5(MeCO_2)]^{2+}$
(c) $[Y(H_2O)_9]^{3+}$
(d) $[Mn(en)_3]^{2+}$

What shape do you think the $[Y(H_2O)_9]^{3+}$ cation might adopt? [*Hint*: Look back at Figure 11.6.]

16.15 Use the data in Table 16.4 to suggest which linkage isomer of $[Cr(NCS)_6]^{3-}$, might be favoured.

16.16 Explain why there is no high-spin arrangement for an octahedral d^8 or d^9 metal ion.

16.17 Calculate the spin-only magnetic moment for each of the following octahedral complexes:

(a) $[Mn(en)_3]^{3+}$ (high spin)
(b) $[Fe(en)_3]^{2+}$ (high spin)
(c) $[Ni(HOCH_2CH_2NH_2)_2(OCH_2CH_2NH_2)]^+$

16.18 Using the experimentally determined effective magnetic moments of the following octahedral complexes, determine if the metal ion is in a high- or low-spin configuration:
(a) $[Cr(en)_3]^{2+}$ $\mu_{eff} = 4.84$ BM; (b) *cis*-$[MnBr_2L]$ ($L = H_2NCH_2CH_2NHCH_2CH_2NHCH_2CH_2NH_2$) $\mu_{eff} = 5.87$ BM. Draw the structure of the manganese(II) complex.

16.19 What is the oxidation state of the metal centre in each of the following complexes?

(a) $[Co(CO)_4]^-$
(b) $[Ni(CN)_4]^{2-}$
(c) $[Cr(CO)_6]$
(d) $[Mn(CO)_5Br]$
(e) $[Mn(CO)_3(py)_2I]$
(f) $[Mn(CO)_6]^+$

16.20 What do you understand by the synergic effect in metal carbonyl complexes? Suggest reasons for the *trends* in the values shown for the carbon–oxygen bond stretching frequencies listed in Table 16.8.

17 Carbonyl compounds

Topics

- Nomenclature
- Polarity of the C=O bond
- Structural properties
- Infrared and ^{13}C NMR spectroscopies
- Tautomerism
- Syntheses of aldehydes and ketones
- Syntheses of carboxylic acids
- Syntheses of esters
- Syntheses of amides
- Syntheses of acyl chlorides
- Carbonyl compounds as acids
- Carbanions in synthesis
- Nucleophilic attack at the C=O carbon atom

17.1 INTRODUCTION

We have already mentioned several reactions involving aldehydes and ketones – the oxidation of alcohols (Section 14.10) and the reductive amination of ketones (Section 14.11). Carboxylic acids have also featured in our discussions, for example when we introduced weak Brønsted acids and acid dissociation constants (Section 11.9), and as products of the oxidation of primary alcohols (Section 14.10). The structural feature common to aldehydes **17.1**, ketones **17.2**, carboxylic acids **17.3**, amides **17.4**, acyl chlorides **17.5** and esters **17.6** is the C=O double bond and in this chapter we look at the chemistry of compounds containing the *carbonyl* functional group.

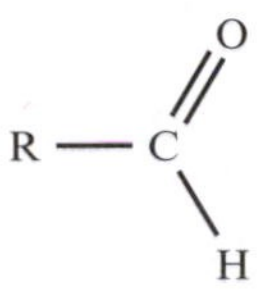

R—C(=O)—H
aldehyde
(17.1)

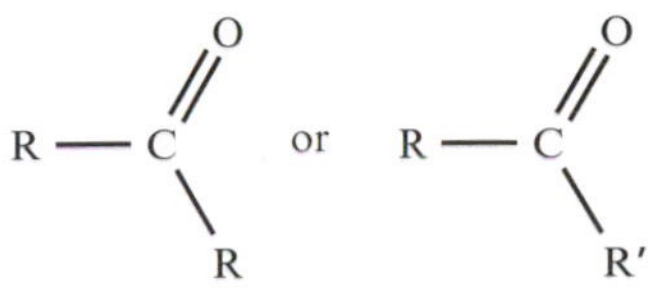

R—C(=O)—R or R—C(=O)—R′
ketone
(17.2)

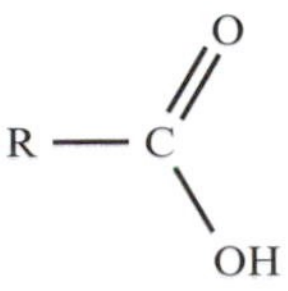

R—C(=O)—OH
carboxylic acid
(17.3)

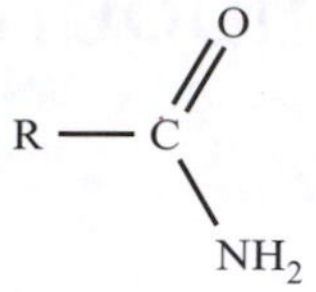

amide
(17.4)

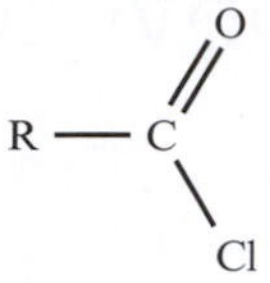

acyl chloride
(17.5)

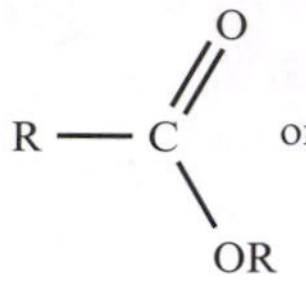

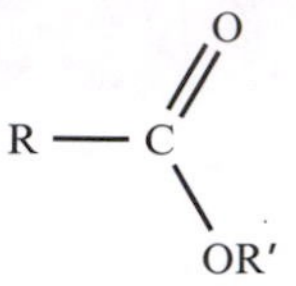

ester
(17.6)

17.2 NOMENCLATURE

Aldehydes

The systematic name for an aldehyde is constructed by taking the root-name for the carbon chain (e.g. methan-, ethan-, propan- for straight chains containing one, two or three carbon atoms respectively) and adding the ending *-al*. Compound **17.7** is methanal and **17.8** is butanal. The trivial names for methanal and ethanal are formaldehyde and acetaldehyde respectively, and both names are still in common use.

HCHO or

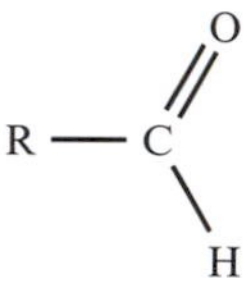

methanal (formaldehyde)
(17.7)

$CH_3CH_2CH_2CHO$ or

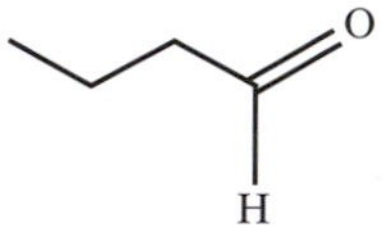

butanal
(17.8)

Notice two particular points:

- the aldehyde functional group is written as CHO and *not* COH in order to avoid confusion with an alcohol group;
- one carbon atom in the chain is part of the functional group; this means that methanal does *not* contain a methyl (CH_3) group, and butanal does *not* contain a butyl (C_4H_9) group.

The benzene derivative **17.9** is called benzaldehyde.

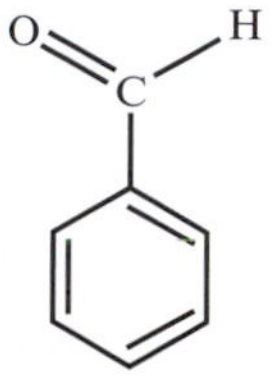

benzaldehyde
(17.9)

Ketones

Ketones are named by using the suffix *-one* (or *-dione* if there are two ketone functional groups). The carbon chain is numbered from the end *nearest* to the functional group – the six carbon atoms in **17.10** provide the root-name hexan-, and to this is added the ending -one accompanied by the position-number to give the name hexan-2-one. In the structural formula for the ketone, brackets are often placed around the O atom to avoid confusion with an ether.

$CH_3C(O)CH_2CH_2CH_2CH_3$ or

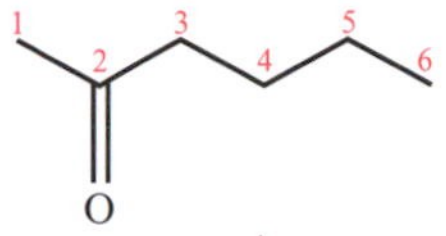

(17.10)

Perhaps the most commonly encountered ketone is acetone, $CH_3C(O)CH_3$. The trivial name is so well recognized, that the systematic name propanone is hardly ever used in the laboratory.

The diketone **17.11** is named pentane-2,4-dione – note the retention of the letter 'e' in the middle of the name; this is the same as in the names of diols, e.g. ethane-1,2-diol. Pentane-2,4-dione is the conjugate acid of the $[acac]^-$ ligand that we encountered in Chapter 16 – the older name for the diketone is acetylacetone, and $[acac]^-$ is the abbreviation for acetylacetonate.

$CH_3C(O)CH_2C(O)CH_3$ or

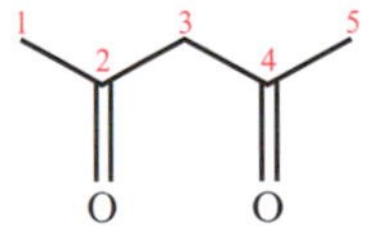

(17.11)

Ketones involving aryl groups are named by recognizing the aryl substituent. For example, diphenyl ketone is Ph_2CO and ethyl phenyl ketone is EtC(O)Ph; the substituent names are placed in alphabetical order – note the spaces between the components of the name. [*Exercise:* Draw the structures of Ph_2CO and EtC(O)Ph.]

Carboxylic acids

Carboxylic acids are named in a similar way to aldehydes. The root-name reveals the number of carbon atoms in the chain (this *includes* the carbon atom of the functional group) and the ending *-oic acid* is added. Compound **17.12** is methanoic acid (although the older name *formic acid* is still used), and compound **17.13** is butanoic acid. The older name for ethanoic acid, CH_3CO_2H, is *acetic acid*, and this name is still used far more widely than the systematic name.

HCO_2H or H—C(=O)—OH

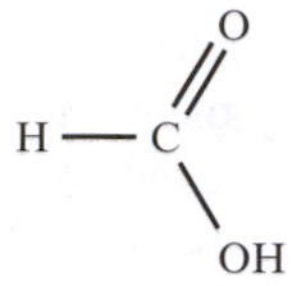

methanoic acid (formic acid)

(17.12)

$CH_3CH_2CH_2CO_2H$ or

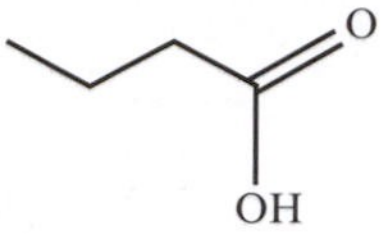

butanoic acid

(17.13)

Box 17.1 Trivial names for carboxylic acids: names with meanings

Whereas systematic nomenclature provides structural information about a molecule, many trivial names have been chosen because they convey some other description of the compound. The common names for some carboxylic acids provide us with interesting examples.

Methanoic acid, HCO_2H, is also called *formic acid* and this name derives from the Latin *formica* meaning ant. Formic acid is the chemical that gives an ant its sting – natural chemical warfare.

Ethanoic acid, CH_3CO_2H, is the component of vinegar responsible for its sour taste. The common name of *acetic acid* comes from the Latin word *acetum* meaning vinegar.

Butanoic acid, $CH_3CH_2CH_2CO_2H$, is also known as *butyric acid* and this name comes from the Latin word *butyrum* which means butter. Rancid butter has a characteristic smell which is caused by butyric acid.

Pentanoic acid, $CH_3CH_2CH_2CH_2CO_2H$, is also called *valeric acid*. *Isovaleric acid* is an isomer of pentanoic acid, $CHMe_2CH_2CO_2H$, and occurs in the roots of the plant valerian.

The trivial names for hexanoic, octanoic and decanoic acids are *caproic*, *caprylic* and *capric acids* which are derived from the Latin words *capra* (meaning female goat) and *caper* (goat). These three acids are together responsible for goaty odours.

Ethanedioic acid, HO_2CCO_2H, is usually called *oxalic acid*; it is toxic and can cause paralysis of the nervous system. The trivial name derives from the Latin name for the plant wood sorrel (*Oxalis acetosella*), the leaves of which contain potassium hydrogen oxalate.

Palmitic acid is the common name for hexadecanoic acid, $CH_3(CH_2)_{14}CO_2H$ and is an important 'fatty acid'. One of the main sources of palmitic acid is palm oil.

The benzene derivative shown in structure **17.14** is called benzoic acid.

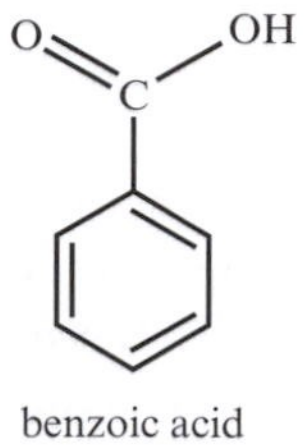

benzoic acid

(17.14)

Salts of carboxylic acids are called *carboxylates*: the methanoate ion is the conjugate base of methanoic acid, and the butanoate ion is the conjugate base of butanoic acid. The salt $Na[HCO_2]$ is sodium methanoate, and $K[CH_3CH_2CH_2CO_2]$ is potassium butanoate.

Amides

We shall only be concerned with *primary amides* of the type shown in structure **17.4**. The name of an amide is constructed in a similar way to that of the corresponding carboxylic acid – the ending *-oic acid* is replaced by *-amide*. Compound **17.15** is butanamide.

$CH_3CH_2CH_2C(O)NH_2$ or

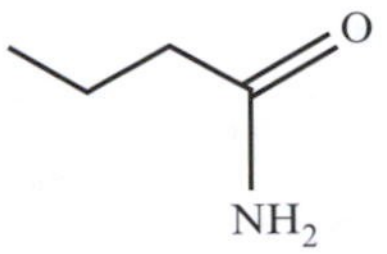

O

NH_2

butanamide

(17.15)

The use of brackets around the oxygen atom in the formula signifies that the carbon atom makes separate bonds to the O atom and the NH_2 group.

Just as ethanoic acid is commonly known as acetic acid, ethanamide **17.16** is often called *acetamide*. The older, and much used, name for methanamide is *formamide*; what is its structure?

$CH_3C(O)NH_2$ or CH_3C O NH_2

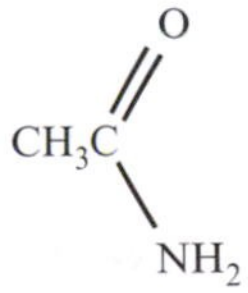

acetamide (ethanamide)

(17.16)

Acyl chlorides

The name of an acyl chloride is derived from that of the corresponding carboxylic acid by replacing the ending *-oic acid* by *-oyl chloride*. Compound **17.17** is butanoyl chloride; compare this structure with that of butanoic acid **17.13**. Ethanoyl chloride **17.18** is more commonly called *acetyl chloride*. The acyl chloride derived from benzoic acid is *benzoyl chloride* **17.19**.

$CH_3CH_2CH_2C(O)Cl$ or O Cl

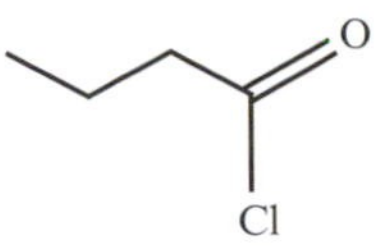

butanoyl chloride

(17.17)

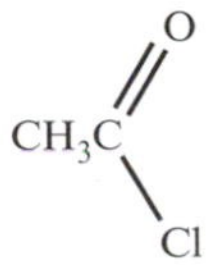

CH_3C O Cl

acetyl chloride (ethanoyl chloride)

(17.18)

O C Cl

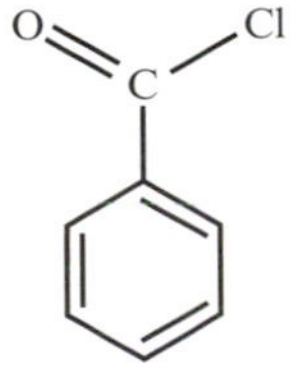

benzoyl chloride

(17.19)

Esters

Esters are derived from carboxylic acids by the replacement of the hydrogen atom of the –OH group by an alkyl or aryl substituent, and the name of the ester contains components of each. Compound **17.20** is derived from butanoic acid, with the –OH group replaced by an –OMe group. The ester is called methyl butanoate – compare this way of naming the compound with that of naming the salts of the corresponding acids. Compound **17.21** is phenyl propanoate, whilst **17.22** is ethyl benzoate. [*Question:* What are the parent acids for **17.21** and **17.22**?]

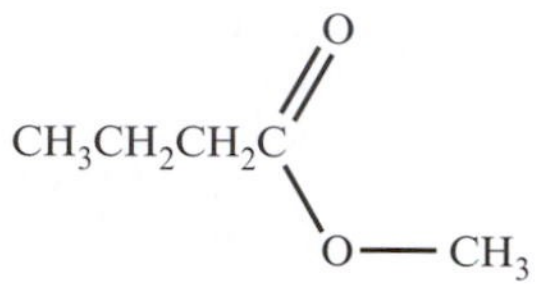

$CH_3CH_2CH_2C(=O)–O–CH_3$

methyl butaoate

(17.20)

$CH_2CH_3C(=O)–O–C_6H_5$

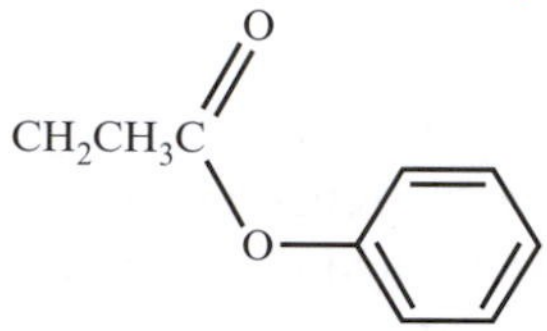

phenyl propanoate

(17.21)

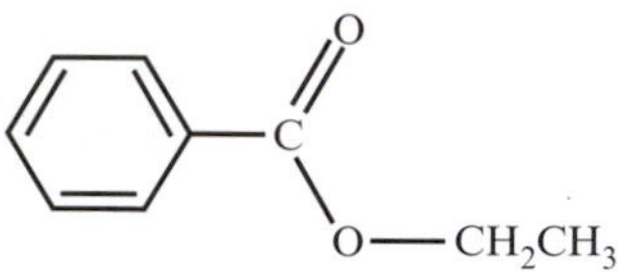

$C_6H_5–C(=O)–O–CH_2CH_3$

ethyl benzoate

(17.22)

17.3 THE POLAR C=O BOND

Polarity

$\chi^P(C) = 2.6$, $\chi^P(O) = 3.4$

The difference between the electronegativities of carbon and oxygen means that the C=O bond is polarized as in structure **17.23**.

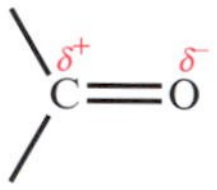

$$\overset{\delta+}{C}=\overset{\delta-}{O}$$

(17.23)

In terms of a molecular orbital bonding description (Figure 17.1), the difference in the effective nuclear charges of the carbon and oxygen atoms means that both the C–O σ-bonding molecular orbital and the C–O π-bonding orbital contain *greater* contributions from the *oxygen* than the carbon atom, and the corresponding σ^* and π^* antibonding MOs possess *more carbon* than oxygen character. This is similar to the situation we described for carbon monoxide in Section 4.14.

The polarization of the C=O bond has important consequences with regard to the reactivity of carbonyl compounds. Electrophiles (most

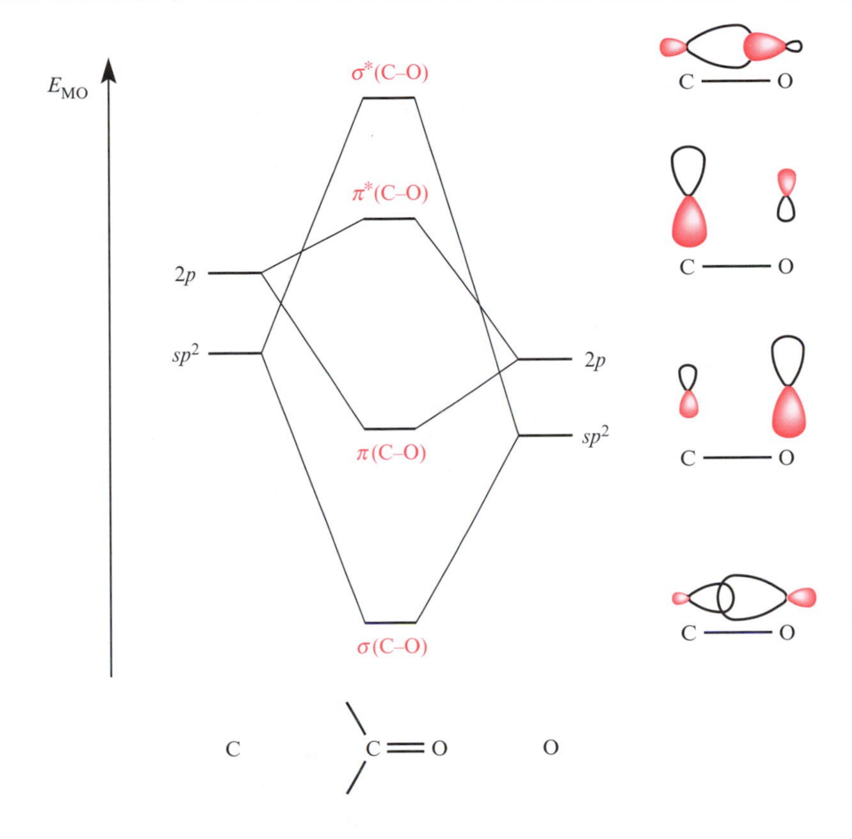

17.1 Part of a molecular orbital diagram for a carbonyl compound; the diagram only shows the orbital interactions which are involved in the carbon–oxygen bond. The C–O σ-bonding molecular orbital and the C–O π-bonding orbital contain *greater* contributions from the oxygen than the carbon atom, while the carbon–oxygen σ^* and π^* antibonding MOs possess *more carbon* than oxygen character. The diagrams at the right-hand side of the figure represent the molecular orbitals.

commonly H^+) attack the oxygen atom, whilst nucleophiles attack the carbon atom (Figure 17.2). But, a word of caution! While H^+ (an electrophile) does indeed attack the oxygen atom, do not become confused when we discuss electrophilic *substitution* reactions (Section 17.12) which lead to bond formation between the *carbon* atom and the electrophile.

In an organic carbonyl compound, the C=O bond is polarized in the sense:

$>C^{\delta+}=O^{\delta-}$

Electrophiles (usually H^+) attack the oxygen atom.

Nucleophiles attack the carbon atom.

Nucleophile → $>C^{\delta+}=O^{\delta-}$ ← Electrophile

17.2 The polarization of the C=O bond means that electrophiles (e.g. H^+) attack the oxygen atom whilst nucleophiles attack the carbon atom.

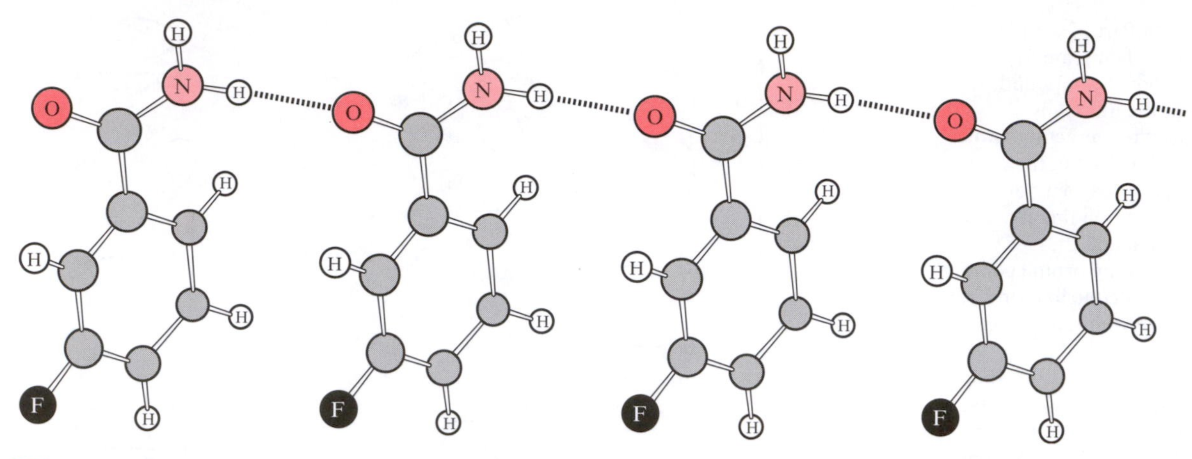

17.3 Intermolecular hydrogen bonding occurs in amides between the oxygen atoms of C=O groups and the hydrogen atoms of the NH_2 groups. The solid state structure of 3-fluorobenzamide illustrates the intermolecular association that results from hydrogen bonding.

Physical properties of carbonyl compounds

The presence of the polar C=O bond in aldehydes, ketones, esters and acyl chlorides results in significant intermolecular interactions and these compounds possess relatively high boiling points. Additionally, carboxylic acids (see below) and amides (Figure 17.3) are able to form hydrogen bonds resulting in molecular association.

Hydrogen bonding: see Section 11.8

Table 17.1 lists boiling points for related carbonyl compounds and compares the data with those for alcohols and an ether of similar molecular masses to the carbonyl compounds.

Many carbonyl compounds possess characteristic odours. The lower molecular mass amides have 'mousy' smells. Esters often smell fruity and some

Table 17.1 Melting and boiling points for selected organic carbonyl compounds. The values for other polar, oxygen-containing compounds of similar molecular masses are given for comparison. Which compounds can form intermolecular hydrogen bonds?

Compound	*Formula*	*Dipole moment / D*	*Relative molecular mass*	*Melting point / K*	*Boiling point / K*
Ethanal (acetaldehyde)	CH_3CHO	2.75	44	152	294
Propanal	CH_3CH_2CHO	2.52	58	192	322
Acetone	$CH_3C(O)CH_3$	2.88	58	178	329
Acetic acid	CH_3CO_2H	1.70	60	290	391
Acetyl chloride	$CH_3C(O)Cl$	2.72	78.5	161	324
Acetamide	$CH_3C(O)NH_2$	3.76	59	355	494
Methyl methanoate	HCO_2CH_3	1.77	60	174	305
Ethanol	CH_3CH_2OH	1.69	46	156	351
Propan-1-ol	$CH_3CH_2CH_2OH$	1.55	60	146	370
Dimethyl ether	CH_3OCH_3	1.30	46	135	248

are used in artificial fruit essences – pentyl ethanoate (with the trivial name of amyl acetate) is called *pear oil*. Many carboxylic acids also possess well-recognized smells – for example, butanoic acid gives rise to the unpleasant smell of rancid butter.

17.4 STRUCTURAL AND SPECTROSCOPIC PROPERTIES

Structure

In each of the compounds **17.1** to **17.6**, the carbon atom of the functional group is in a trigonal planar environment. An sp^2 hybridization scheme is appropriate to describe the formation of three σ-bonds (including one C–O σ-bond); a carbon–oxygen π-bond is additionally formed by the overlap of carbon and oxygen $2p$ orbitals. Figure 17.4 shows the structure of ethanal and the carbon–oxygen bond length of 121 pm is consistent with a carbon–oxygen double bond. The three bond angles around the central carbon atom are close to 120° both in ethanal and in other compounds of this general type.

Dimerization in carboxylic acids

The relative positions of the –OH and C=O groups in a molecule of a carboxylic acid provide it with the potential to form intermolecular hydrogen bonds. As a result, dimerization occurs as shown in structure **17.24**.

(17.24)

Figure 17.5 shows the solid state structure of a dicarboxylic acid – terephthalic acid. The presence of *two* CO_2H groups per molecule allows the formation of an extended structure in the solid state through intermolecular hydrogen bonding.

In the liquid state, evidence for dimerization comes from molecular mass determinations and from the fact that the boiling points of carboxylic acids

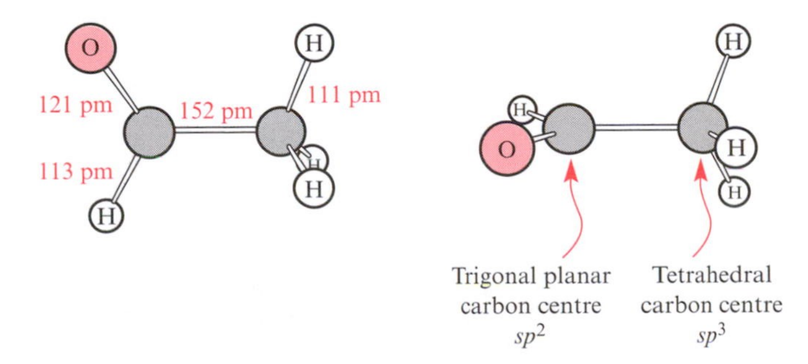

17.4 The structure of ethanal (acetaldehyde) CH_3CHO. The bond lengths have been measured in the gas phase by electron diffraction methods.

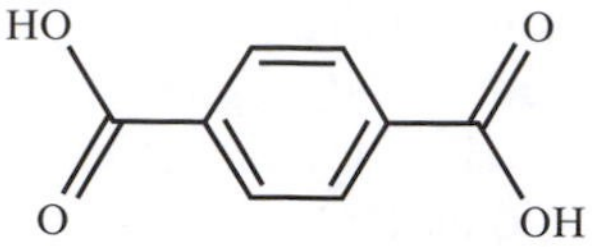

HO O OH O
Terephthalic acid

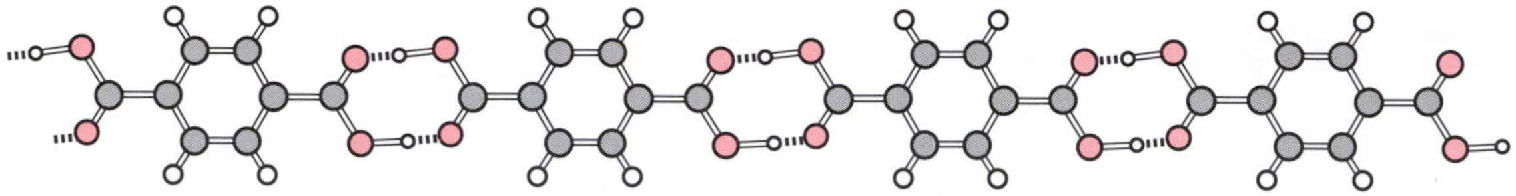

17.5 Terephthalic acid (see Box 17.2) is a dicarboxylic acid. Intermolecular hydrogen bonding in the solid state leads to the formation of a 'ribbon-like' structure.

Hydrogen bonding and Trouton's rule: see Sections 11.8 and 12.9

are relatively high. The boiling points of methanoic and acetic acids (Table 17.2) are significantly higher than those of aldehydes of corresponding molecular masses. Both methanoic and acetic acids show anomalously low values of the molar entropy of vaporization – they do *not* obey Trouton's rule (equation 17.1) – and this can be rationalized in terms of dimer formation. The ratio of the molar enthalpy of vaporization to the boiling point (in kelvin) is the molar *entropy* of vaporization, and the presence of hydrogen-bonded dimers instead of monomers reduces the entropy of the system.

$$\frac{\Delta_{vap}H}{bp} \approx 0.088 \text{ kJ K}^{-1} \text{ mol}^{-1} \qquad \textit{Trouton's rule} \qquad (17.1)$$

Infrared spectroscopy

Typical ranges for C=O stretching frequencies: see Table 9.3

In Section 9.8, we used the carbonyl group as an example of a functionality that could be detected by IR spectroscopy. The stretching of the C=O double bond typically gives rise to a *strong* absorption around 1700 cm^{-1} (Figure 17.6). The IR spectrum of nonanoic acid in Figure 17.6 also con-

Table 17.2 Values of $\Delta_{vap}H$ and boiling points for methanoic and acetic acids compared to the corresponding aldehydes.

Compound	*Formula*	*Relative molecular mass*	*Boiling point / K*	$\Delta_{vap}H$ / kJ mol^{-1}	$\frac{\Delta_{vap}H}{bp}$ / J mol^{-1} K^{-1}
Methanoic acid	HCO_2H	46	374	23	61.5
Acetic acid (ethanoic acid)	CH_3CO_2H	60	391	24	61.4
Acetaldehyde (ethanal)	CH_3CHO	44	293	26	88.7
Propanal	CH_3CH_2CHO	58	321	28	87.2

tains a *broad* absorption around 3000 cm^{-1} due to the O–H stretch – the very broad nature of the band is due to intermolecular hydrogen bonding. Additional sharper absorptions near to 3000 cm^{-1} are assigned to the alkyl C–H stretching modes. Similar absorptions arising from C–H stretches are also present in the IR spectra of acetone and ethyl butanoate. In problem 17.3, additional IR spectra are provided for you to interpret.

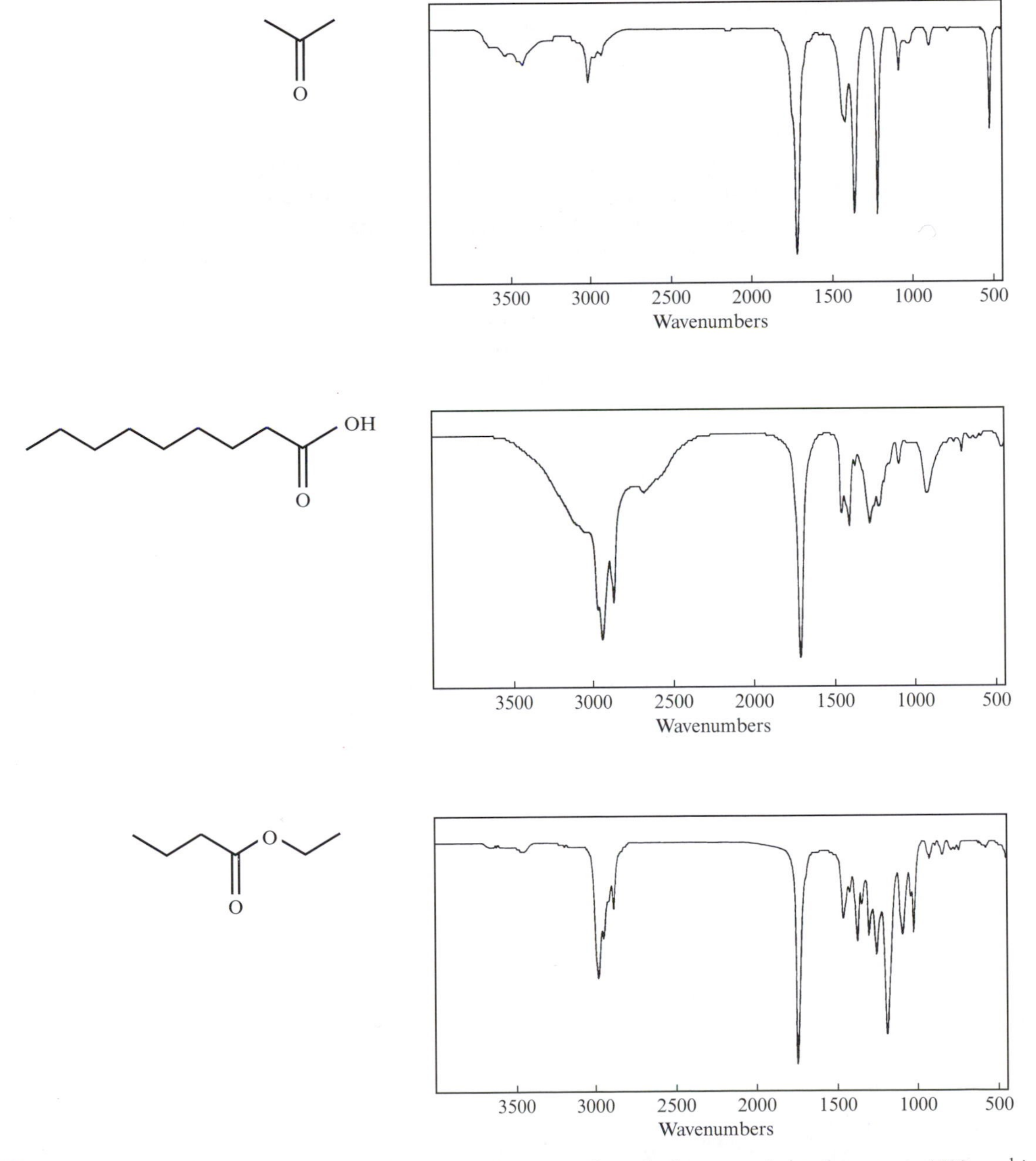

17.6 Infrared spectra of acetone, nonanoic acid and ethyl butanoate. In each, the *strong* absorption near to 1700 cm^{-1} is characteristic of the C=O stretch and is a dominant feature of the spectrum. What other features of the spectra may be assigned and which absorptions lie in the fingerprint region?

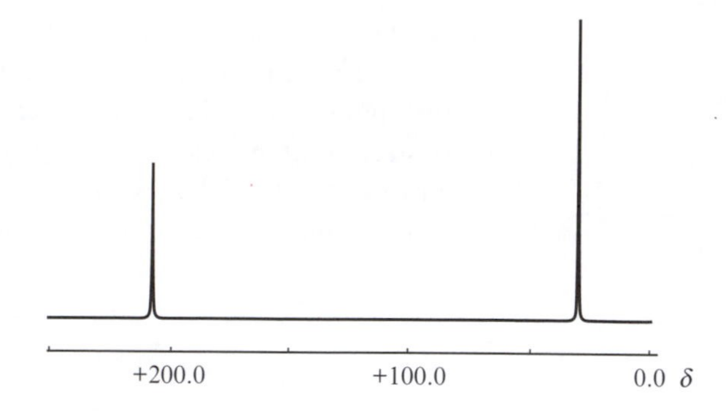

17.7 The ^{13}C NMR spectrum of acetone; the effects of ^{13}C–^{1}H spin–spin coupling have been removed from the spectrum.

^{13}C NMR spectroscopy

The sp^2 hybridized carbon atom of the carbonyl group in a ketone, aldehyde, carboxylic acid, ester, amide or acyl chloride gives rise to a characteristic signal in the ^{13}C NMR spectrum in the region δ +230 to +155. The values listed in Table 9.11 showed that the ^{13}C NMR signal for the C=O group in a ketone or aldehyde tends to be at lower field (δ +230 to +185) than the C=O group in a carboxylic acid, ester, acyl chloride or amide (δ +185 to +155). The ^{13}C NMR spectrum of acetone (Figure 17.7) shows two carbon environments (trigonal planar and tetrahedral) and the signals have markedly different chemical shifts. Problem 17.4 provides you with further practice in the interpretation of ^{13}C NMR spectra.

We shall not discuss ^{1}H NMR spectra here; for typical ^{1}H NMR spectroscopic chemical shifts for aldehydes, amides and carboxylic acids, refer back to Table 9.12.

17.5 KETO–ENOL TAUTOMERS

In the previous section, we issued a warning about the site of electrophilic attack in carbonyl compounds, and now we enlarge upon this discussion.

When an –OH group is attached to a C=C double bond, the species is unstable with respect to rearrangement to a carbonyl compound. This type of rearrangement is called a *keto–enol tautomerism*, and the two tautomers are in equilibrium (equation 17.2) with the *keto*-form usually being favoured. Tautomerism involves the transfer of a proton.

The name 'enol' is a contraction of alk*ene*-alcoh*ol*

$$\mathrm{X(HO)C{=}CR_2} \rightleftharpoons \mathrm{X(O{=})C{-}CHR_2} \qquad (17.2)$$

enol-tautomer   *keto*-tautomer

Tautomerism **involves the transfer of a proton between sites within a molecule with an associated rearrangement of multiple bonds.**

The keto–enol pair is acidic and loss of a proton leads to the conjugate base, known as an *enolate* ion. The negative charge may be drawn on the oxygen

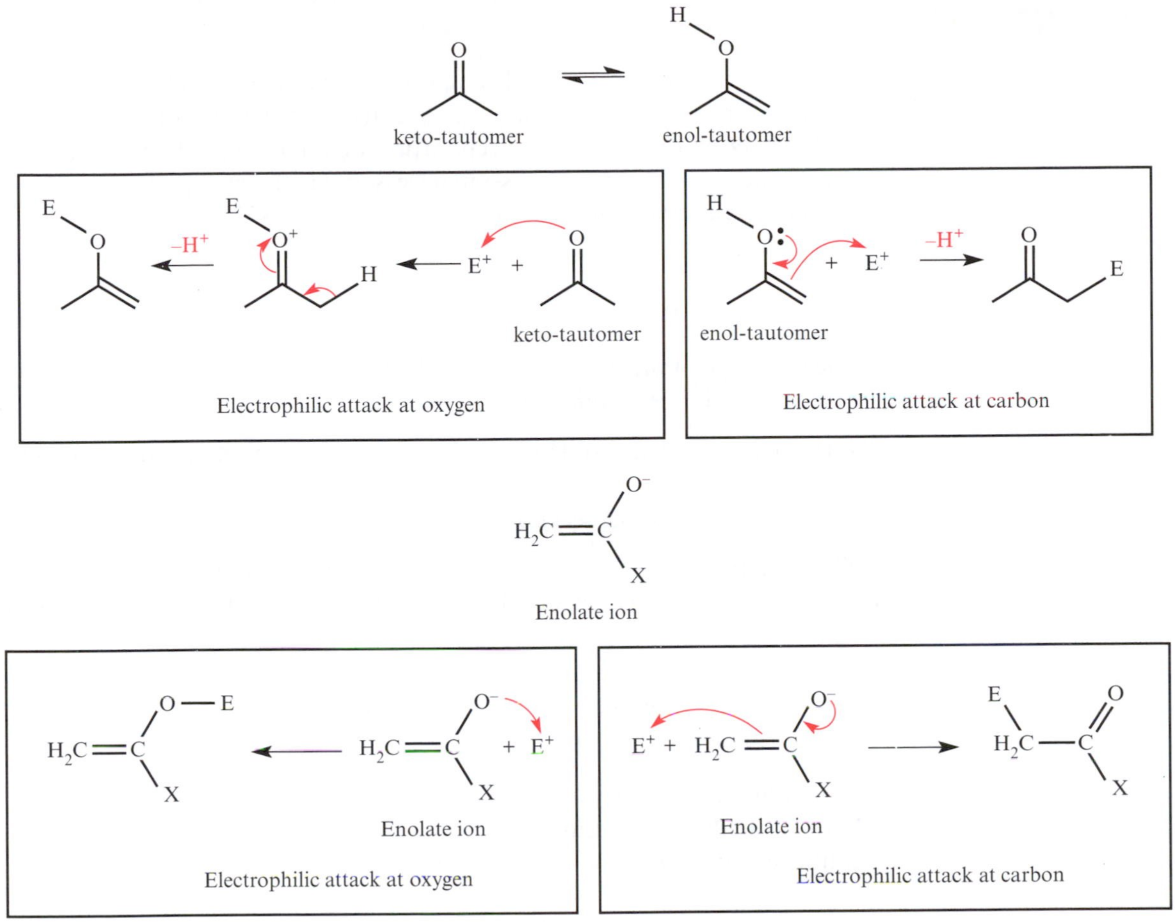

17.8 In enols or enolate ions, electrophilic attack can occur at either carbon or oxygen centres.

atom (structure **17.25**) or on the carbon atom (structure **17.26**). Structures **17.25** and **17.26** are *resonance forms of a single chemical species*, and the resonance pair can be represented by the delocalized structure **17.27**. This is in contrast to the keto and enol forms of the conjugate acid *which are discrete and different chemical compounds*.

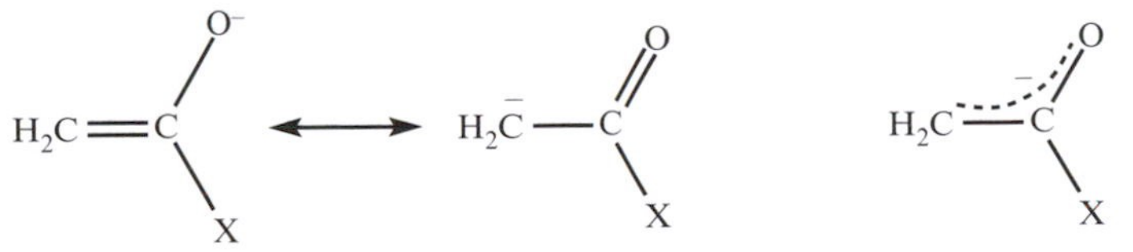

(17.25) **(17.26)** **(17.27)**

Figure 17.8 illustrates that the enol or enolate ion may react with electrophiles at either the carbon or oxygen centres.

17.6 ALDEHYDES AND KETONES: SYNTHESIS

Methods of preparing carbonyl compounds are described in this and the following four sections. Some of the reactions interconvert different types of carbonyl compound (e.g. the preparation of esters or acyl chlorides from carboxylic acids) and the discussion necessarily includes aspects of the reactivity of these compounds.

Oxidation of alcohols

Oxidation of alcohols: see Section 14.10

The oxidation of alcohols can lead to the formation of aldehydes, ketones or carboxylic acids depending upon:

- the position of the –OH group (whether the alcohol is primary or secondary), and
- the conditions of the reaction.

The oxidation of a secondary alcohol using acidified potassium dichromate at room temperature (equation 17.3) leads to a *ketone* which cannot easily be oxidized further.

$$\underset{\text{propan-2-ol}}{CH_3CH(OH)CH_3} \xrightarrow{K_2Cr_2O_7,\ H_2SO_4} \underset{\text{propanone (acetone)}}{CH_3C(O)CH_3} \qquad (17.3)$$

Oxidation of a primary alcohol initially gives an aldehyde (equation 17.4), but with acidified potassium dichromate as the oxidizing agent, the reaction continues to the corresponding carboxylic acid.

$$\underset{\text{propan-1-ol}}{CH_3CH_2CH_2OH} \xrightarrow{K_2Cr_2O_7,\ H_2SO_4} \underset{\text{propanal}}{CH_3CH_2CHO} \xrightarrow{K_2Cr_2O_7,\ H_2SO_4} \underset{\text{propanoic acid}}{CH_3CH_2CO_2H} \qquad (17.4)$$

If the aldehyde is the desired end-product, oxidizing reagents such as Corey's reagent must be used (equation 17.5).

Corey's reagent: see Section 14.10

$$\underset{\text{pentan-1-ol}}{CH_3CH_2CH_2CH_2CH_2OH} \xrightarrow{\text{Corey's reagent}} \underset{\text{pentanal}}{CH_3CH_2CH_2CH_2CHO} \qquad (17.5)$$

Reduction of acyl chlorides

The functional group in an acyl chloride is in a *terminal position* and reduction of the compound yields the corresponding aldehyde. A suitable reducing agent is the lithium aluminium hydride derivative $Li[AlH(O^tBu)_3]$ **17.28**.

tert-Butyl (^tBu) = CMe_3: see Figure 8.7

$$\mathrm{Li^+}\left[\mathrm{AlH(O^tBu)_3}\right]^-$$

(17.28)

This method can be used to prepare both alkyl or aryl aldehydes as in reactions 17.6 and 17.7.

$$\underset{\text{acetyl chloride}}{\mathrm{CH_3C(O)Cl}} \xrightarrow[-78^\circ\mathrm{C}]{\mathrm{Li[AlH(O^tBu)_3]}} \underset{\text{acetaldehyde}}{\mathrm{CH_3CHO}} \qquad (17.6)$$

$$\underset{\text{benzoyl chloride}}{\mathrm{PhC(O)Cl}} \xrightarrow[-78^\circ\mathrm{C}]{\mathrm{Li[AlH(O^tBu)_3]}} \underset{\text{benzaldehyde}}{\mathrm{PhCHO}} \qquad (17.7)$$

Further reduction (to the corresponding alcohol) does not occur with this reducing agent because of the significant steric crowding in the intermediate due to the three *tert*-butyl groups in $\mathrm{Li[AlH(O^tBu)_3]}$.

Friedel–Crafts acylation

Friedel–Crafts acylation of benzene: see Sections 15.6 and 15.7

In Chapter 15, we described in detail the Friedel–Crafts acylation of benzene to give a ketone (equation 17.8).

$$\text{benzene} + \underset{\text{butanoyl chloride}}{\mathrm{CH_3CH_2CH_2C(O)Cl}} \xrightarrow{\mathrm{AlCl_3}\ \text{catalyst}} \underset{\text{phenyl propyl ketone}}{\mathrm{PhC(O)CH_2CH_2CH_3}} + \mathrm{HCl} \qquad (17.8)$$

$$\text{benzene} + \underset{\text{benzoyl chloride}}{\mathrm{PhC(O)Cl}} \xrightarrow{\mathrm{AlCl_3}\ \text{catalyst}} \underset{\text{diphenyl ketone}}{\mathrm{PhC(O)Ph}} + \mathrm{HCl} \qquad (17.9)$$

Diphenyl ketone can be prepared from benzene and benzoyl chloride (equation 17.9) and the synthetic method can be extended to other aromatic compounds. However, the choice of starting compounds may be critical

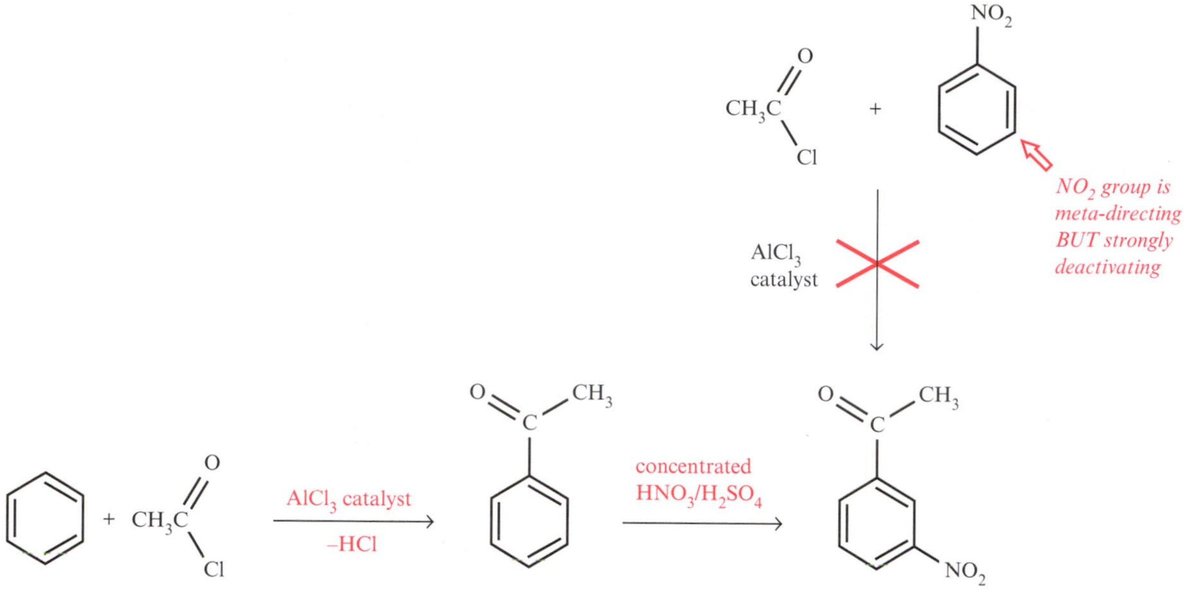

17.9 The presence of a nitro-substituent deactivates the benzene ring towards electrophilic substitution. In order to prepare the target compound $3\text{-}NO_2C_6H_4C(O)Cl$, the nitration step must be carried out *after* the acylation.

Ring activation/deactivation: see Section 15.8

The older name for methyl phenyl ketone is acetophenone

because the substituents in some derivatives of benzene *deactivate* the ring towards electrophilic substitution. Whilst the nitro group in nitrobenzene is *meta*-directing, it is also *strongly deactivating*, and nitrobenzene does *not* react with acetyl chloride (Figure 17.9). If the target compound is **17.29**, then as Figure 17.9 shows, a suitable method of preparation begins with the acylation of benzene followed by the nitration of methyl phenyl ketone. Like the nitro group, the MeC(O)-substituent is *meta*-directing but is less deactivating than NO_2.

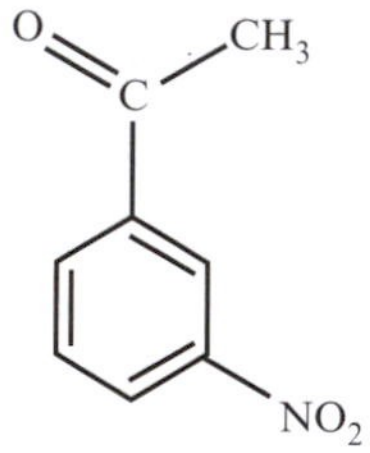

(17.29)

17.7 CARBOXYLIC ACIDS: SYNTHESIS

Oxidation of primary alcohols and aldehydes

We have already seen that the oxidation of primary alcohols by acidified potassium dichromate leads first to an aldehyde and then to the corresponding carboxylic acid (equation 17.4). This method of oxidation, although not often useful for the preparation of aldehydes, *is* convenient for the synthesis of carboxylic acids from primary alcohols. An alternative oxidizing agent is acidified potassium permanganate (equation 17.10).

Benzyl alcohol is a *primary alcohol*, not a phenol

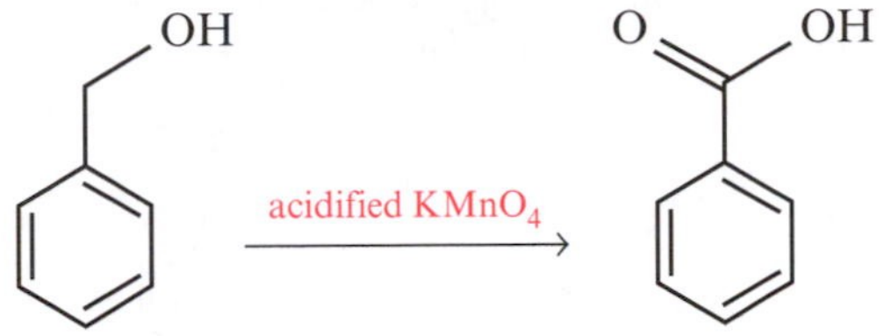

OH

acidified $KMnO_4$

O OH

benzyl alcohol benzoic acid (17.10)

Oxidation of an alkylbenzene

The oxidation of the methyl group requires a stronger oxidant (i.e. *alkaline* $KMnO_4$) than does that of an alcohol (i.e. *acidified* $KMnO_4$), and hot alkaline $KMnO_4$ oxidizes toluene and ethylbenzene to benzoic acid (equation 17.11).

Oxidation of toluene: see Section 15.9

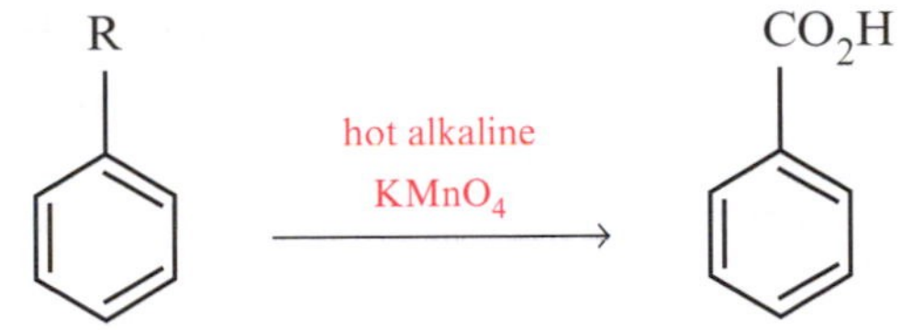

R

hot alkaline $KMnO_4$

CO_2H

R = Me or Et benzoic acid (17.11)

The reaction can be extended to convert more than one methyl group *attached directly to the benzene ring* to carboxylic acid functionalities and reactions 17.12 to 17.14 show the formation of the three isomers of phthalic acid.

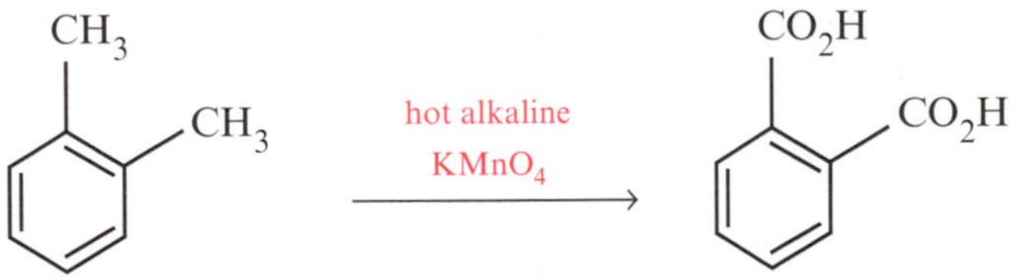

CH_3 CH_3

hot alkaline $KMnO_4$

CO_2H CO_2H

phthalic acid
(benzene-1,2-dicarboxylic acid) (17.12)

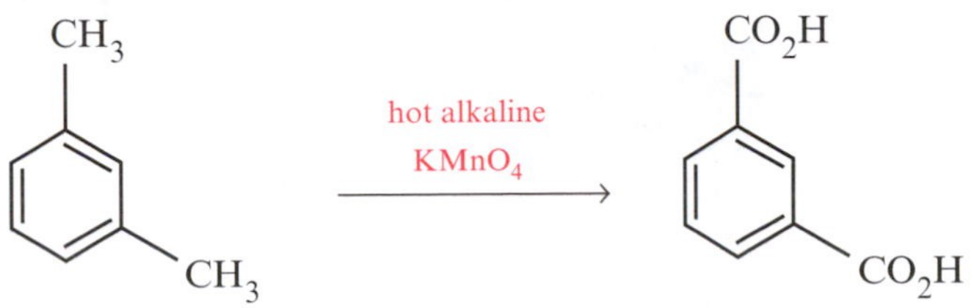

CH_3 CH_3

hot alkaline $KMnO_4$

CO_2H CO_2H

isophthalic acid
(benzene-1,3-dicarboxylic acid) (17.13)

Box 17.2 Benzenedicarboxylic acids: phthalic acid and terephthalic acids

Two isomers of benzenedicarboxylic acid are benzene-1,2-dicarboxylic acid (known as phthalic acid) and benzene-1,4-dicarboxylic acid (known as terephthalic acid). Both compounds are of industrial importance.

phthalic acid terephthalic acid phthalic anhydride

Phthalic anhydride is a derivative of phthalic acid and has applications in the dye-stuffs industry, and is used in the manufacture of certain plasticizers.

Terephthalic acid is twenty-fourth on the list of top-ranking chemicals manufactured in 1994 in the USA – it is used in the manufacture of *polyester* synthetic fibres. Polyesters are formed by the *condensation of a dicarboxylic acid with a diol*, and the role of terephthalic acid in the formation of the polyester Dacron is shown below:

$2n$ HO–CH₂CH₂–OH + $2n$ HO₂C–C₆H₄–CO₂H $\xrightarrow{-H_2O}$

Dacron

The reaction scheme shows two repeat units of the polymer.

In 1993, man-made fibres made up about 46% of the total textile market, and polyester was the major manufactured fibre. The man-made fibre output continues to increase, and in 1994, 11.4 megatonnes of polyester were produced worldwide.

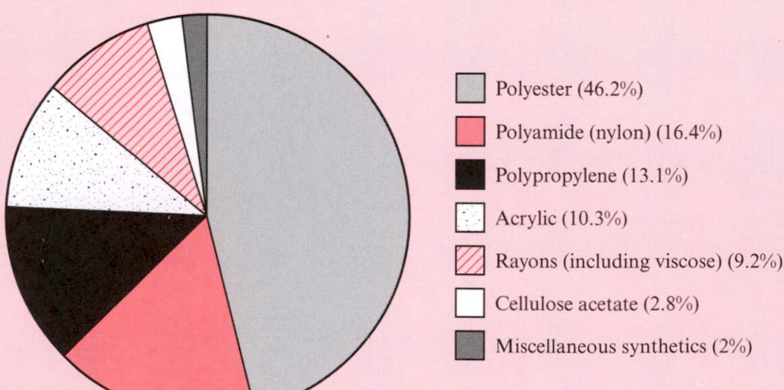

(Data from *Chemistry & Industry*, (1995) p. 870.)

$$p\text{-}CH_3C_6H_4CH_3 \xrightarrow{\text{hot alkaline } KMnO_4} p\text{-}HO_2CC_6H_4CO_2H$$

terephthalic acid
(benzene-1,4-dicarboxylic acid) (17.14)

Use of Grignard reagents

Grignard reagents: see Section 14.4 and Box 14.1

Halogenoalkanes may be converted to carboxylic acids by first transforming the halogenoalkane RX into the corresponding Grignard reagent RMgX (equation 17.15) and then treating it with solid carbon dioxide ('dry ice'). The reaction involves *an increase in the length of the carbon chain* – the precursor is RX and the product is RCO_2H. The Mg–C bond in the organomagnesium compound is polar in the sense $Mg^{\delta+}–C^{\delta-}$ and the $C^{\delta-}$ centre attacks the $C^{\delta+}$ centre of CO_2 (equation 17.16). The product of this step is a carboxylate salt, protonation of which gives a carboxylic acid (equation 17.17).

$$\underset{\text{1-chloropropane}}{CH_3CH_2CH_2Cl} + Mg \xrightarrow{\text{anhydrous } Et_2O \text{ or THF}} CH_3CH_2CH_2MgCl \tag{17.15}$$

$$CH_3CH_2CH_2^{\delta-}—MgCl^{\delta+} + O^{\delta-}{=}C^{\delta+}{=}O^{\delta-} \longrightarrow CH_3CH_2CH_2C(=O)O^{-}\ \overset{+}{M}gCl \tag{17.16}$$

$$\underset{\text{butanoate ion}}{CH_3CH_2CH_2C(=O)O^-} + [H_3O]^+ \rightleftharpoons \underset{\text{butanoic acid}}{CH_3CH_2CH_2C(=O)OH} + H_2O \tag{17.17}$$

The sequence of reactions 17.15 to 17.17 converts 1-chloropropane (a three-carbon chain) to butanoic acid (a four-carbon chain). The method can be used to prepare a wide range of carboxylic acids, including both alkyl and aryl derivatives.

Acid- or base-catalysed hydrolysis of nitriles

Nitriles of the general structure R–C≡N can be produced by the nucleophilic substitution of a halide group in a halogenoalkane by cyanide ion (equation 17.18). The nitrile is readily hydrolysed in the presence of acid or base catalysts and the usual product is the corresponding carboxylic acid (equation 17.19). The sequence of reactions 17.18 and 17.19 results in *an increase in the length of the carbon chain* – 1-chlorobutane (a four-carbon chain) is converted to pentanoic acid (a five-carbon chain).

$$\underset{\text{1-chlorobutane}}{CH_3CH_2CH_2CH_2Cl} + NaCN \rightarrow CH_3CH_2CH_2CH_2CN + NaCl \qquad (17.18)$$

$$CH_3CH_2CH_2CH_2CN \xrightarrow[H^+ \text{ or } [OH]^-]{H_2O} \underset{\text{pentanoic acid}}{CH_3CH_2CH_2CH_2CO_2H} \qquad (17.19)$$

This method can also be used to synthesize amides since these are intermediate products in the hydrolysis step. However, amides are themselves hydrolysed to carboxylic acids and the reaction conditions must be carefully controlled if the amide is to be isolated.

Further routes to amides: see Section 17.9

17.8 ESTERS: SYNTHESIS AND HYDROLYSIS

Synthesis from a carboxylic acid and an alcohol

Esterification: see Section 14.10

An ester is produced by a *condensation reaction* (in which water is eliminated) between a carboxylic acid and an alcohol and this is summarized in reaction 17.20. The reaction is catalysed by acid.

$$\underset{\text{alcohol}}{R{-}O{-}H} + \underset{\text{carboxylic acid}}{H{-}O{-}C({=}O){-}R'} \xrightleftharpoons{H^+ \text{ catalyst}} \underset{\text{ester}}{R{-}O{-}C({=}O){-}R'} + \underset{\text{water}}{H_2O} \qquad (17.20)$$

In a condensation reaction, two reactants combine with the elimination of a small molecule such as H_2O, NH_3, HCl or an alcohol.

Reactions of the type shown in equation 17.20 are quite general and can be used to prepare esters with alkyl or aryl groups, although usually the *alcohol is an alkyl derivative* (equations 17.21–17.23). We showed in Figure 14.12 how to separate ethyl benzoate from the reaction mixture after it had been prepared from benzoic acid and ethanol.

$$CH_3CO_2H + CH_3CH_2OH \xrightleftharpoons{H^+ \text{ catalyst}} CH_3C(=O)OCH_2CH_3 + H_2O \quad (17.21)$$

acetic acid ethanol ethyl acetate

$$CH_3CH_2CO_2H + PhCH_2OH \xrightleftharpoons{H^+ \text{ catalyst}} CH_2CH_3C(=O)OCH_2Ph + H_2O \quad (17.22)$$

propanoic acid benzyl alcohol benzyl propanoate

$$PhCO_2H + CH_3CH_2CH_2OH \xrightleftharpoons{H^+ \text{ catalyst}} PhC(=O)OCH_2CH_2CH_3 + H_2O \quad (17.23)$$

benzoic acid propan-1-ol propyl benzoate

Each of the three equations above is an *equilibrium* – esters react with water to regenerate a carboxylic acid and an alcohol. In the preparation of the ester, the equilibrium must be driven to the right-hand side and one way to achieve this is continually to remove the ester or water by distillation. Alternatively, an excess of the alcohol or carboxylic acid can be added.

Box 17.3 Ester hydrolysis for protection from the sun

Safety in the sun (protection against skin cancer) is a topical issue and the use of sunscreen creams is recommended. High-factor sunscreens should absorb harmful UV radiation and protect the skin, but UV radiation that does reach the skin produces highly reactive free radicals which damage cells irrevocably. An ingredient that has been added to sunscreen lotions is an ester, α-tocopheryl acetate. On the skin, the ester is hydrolysed to the corresponding phenol which functions as a radical quencher.

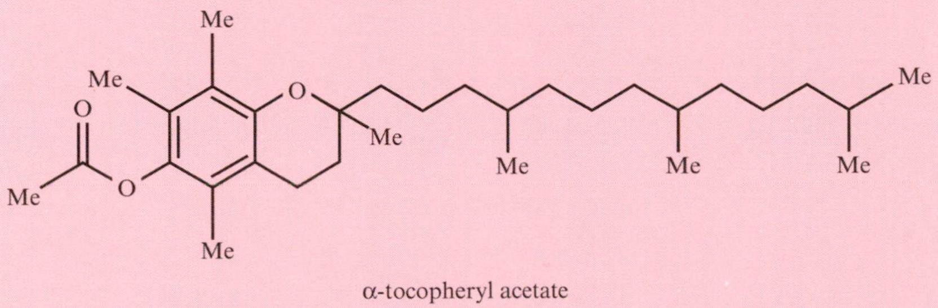

α-tocopheryl acetate

Hydrolysis of esters

The hydrolysis of an ester by aqueous sodium hydroxide is a saponification reaction.

Esters are hydrolysed by aqueous sodium hydroxide. This process is called *saponification* and is the key step in the manufacture of soap from fats which are esters of the triol glycerol. Figure 17.10 shows the esterification reaction between glycerol and stearic acid; this is a *fatty acid* containing a long aliphatic chain. The saponification reaction that follows results in the formation of sodium stearate which is a soap.

Box 17.4 Saturated and unsaturated fats

Fats are esters of carboxylic acids (fatty acids) and play a vital role in nutrition. Those involved in the food industry fall into several categories:

- saturated (C–C bonds in the organic chain);
- monounsaturated (one C=C in the chain);
- polyunsaturated (more than one C=C in the chain).

The relationship between the intake of different fatty acids and blood cholesterol levels is the subject of research and the results are often controversial, providing the consumer with confusing and often contradictory advice.

Saturated fatty acids have varying effects on blood cholesterol levels – lauric and myristic acids appear to be higher-risk components of high-fat diets than palmitic acid. Monounsaturated fatty acids (usually oleic acid) and polyunsaturates (mainly linoleic acid) represent more healthy substitutes to saturated fatty acids. It is not just the *presence* of the unsaturated C=C groups that is important — C=C positional and geometrical isomers possess different nutritional characteristics.

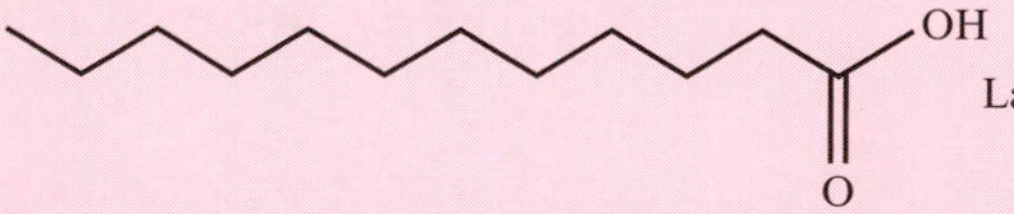

OH
O
Lauric acid

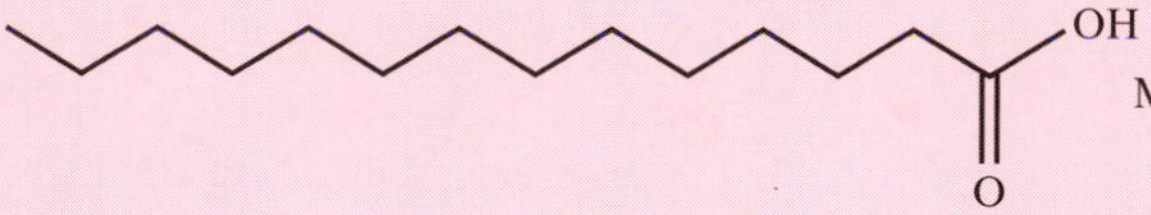

OH
O
Myristic acid

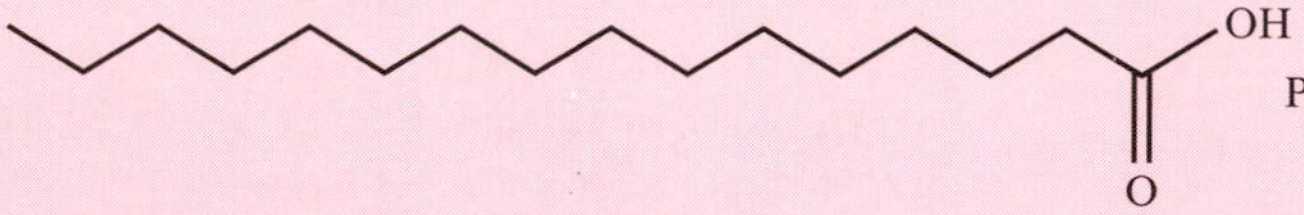

OH
O
Palmitic acid

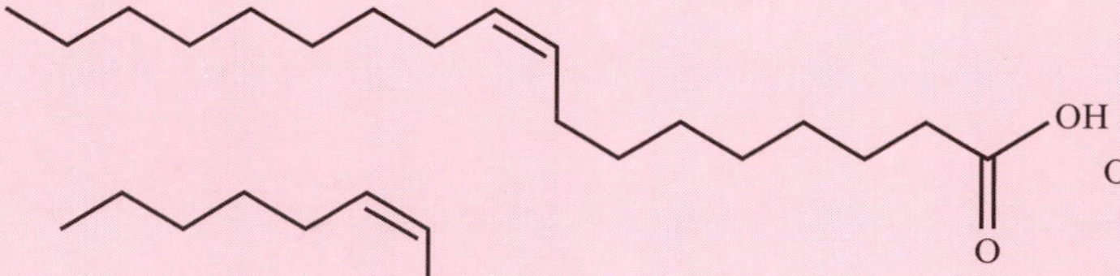

OH
O
Oleic acid

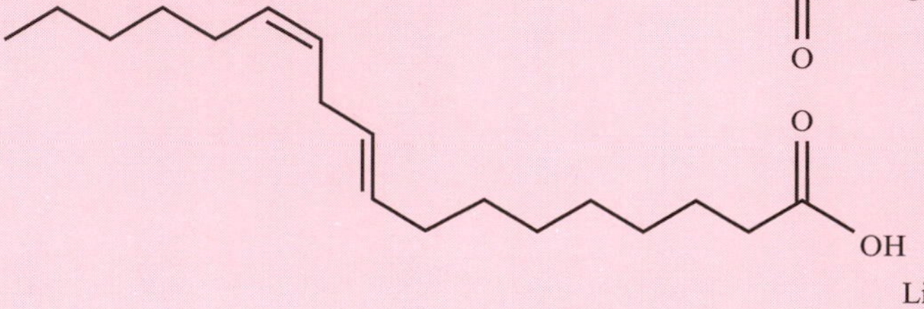

O
OH
Linoleic acid

17.10 The sodium salt of the long chain aliphatic acid stearic acid is a soap and is prepared by the saponication of the appropriate ester of glycerol.

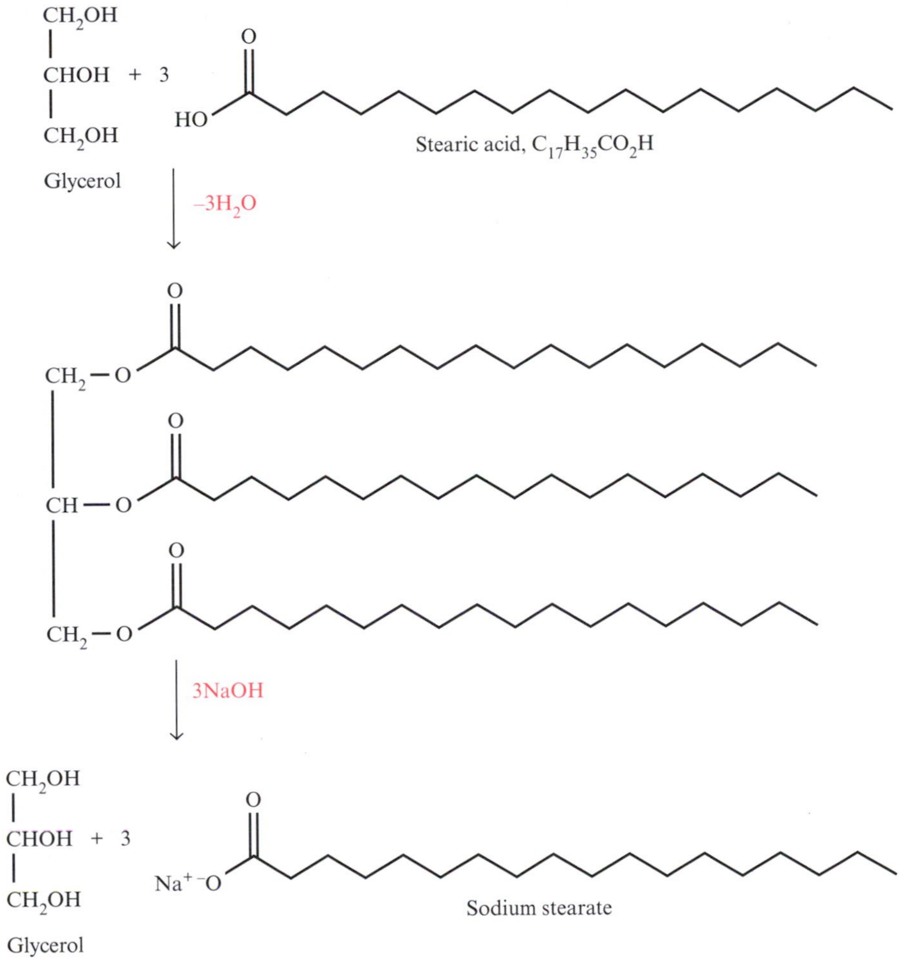

CH_2OH
CHOH + 3
CH_2OH
Glycerol

HO
O

Stearic acid, $C_{17}H_{35}CO_2H$

$-3H_2O$

CH_2-O
O
CH—O
O
CH_2-O
O

3NaOH

CH_2OH
CHOH + 3
CH_2OH
Glycerol

$Na^{+}\ ^{-}O$
O

Sodium stearate

Synthesis from an acyl chloride and an alcohol

An alternative method of preparing esters is the condensation reaction between an acyl chloride and an alcohol in which hydrogen chloride is eliminated. The general reaction 17.24 can be applied to alkyl *and* aryl derivatives including phenols – compare this with the reaction of a carboxylic acid and alcohol.

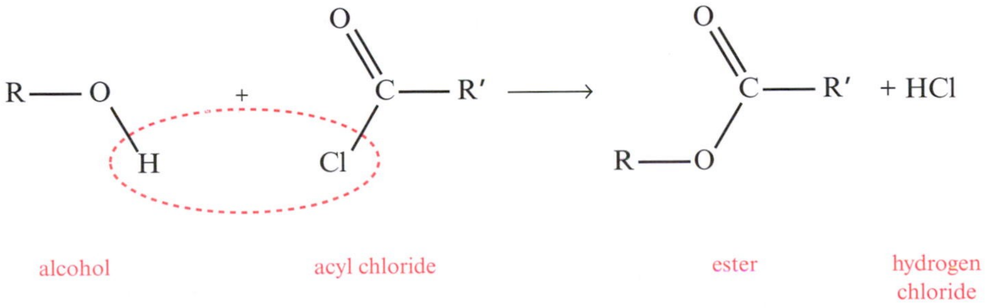

$$R{-}O{-}H + Cl{-}C(=O){-}R' \longrightarrow R{-}O{-}C(=O){-}R' + HCl \quad (17.24)$$

alcohol acyl chloride ester hydrogen chloride

Reactions 17.25 and 17.26 give examples of ester syntheses from acyl chlorides.

$$CH_3C(O)Cl + CH_3OH \longrightarrow CH_3C(=O)OCH_3 + HCl \qquad (17.25)$$

acetyl chloride methanol methyl acetate

$$CH_3C(O)Cl + PhOH \longrightarrow CH_3C(=O)OPh + HCl \qquad (17.26)$$

acetyl chloride phenol phenyl acetate

Box 17.5 Esters as fuels: biodiesel

Biodiesel is an environmentally friendly substitute for diesel fuel. The raw material for biodiesel is rapeseed, instantly recognized as the bright yellow crop in fields throughout Europe. Biodiesel is a mixture of esters and is obtained by the base-catalysed reaction of methanol with rapeseed oil:

$$\text{Rapeseed oil} + \text{MeOH} \xrightarrow{\text{NaOH catalyst}} \text{Biodiesel (rapeseed oil esters)} + \text{Propane-1,2,3-triol (glycerol)}$$

After purification, biodiesel performs well with respect to diesel fuel, and vehicles using it require little or no engine modification. The environmental advantages of biodiesel are significant. It is non-toxic and is almost entirely biodegradable. Compared to diesel fuel, the emissions after combustion contain smaller amounts of SO_2, CO and hydrocarbons.

Biodiesel is a new fuel, but one for which a market is developing rapidly. It is marketed under several names throughout Europe – for example, in France, it is sold as *Diester* (a contraction of diesel and ester).

17.9 AMIDES: SYNTHESIS

Esters undergo nucleophilic substitution reactions with ammonia to produce amides (equation 17.27).

$$R{-}C(=O)OR' + NH_3 \longrightarrow R{-}C(=O)NH_2 + R'OH \qquad (17.27)$$

ester ammonia amide alcohol

Similarly, ammonia displaces chloride ion from acyl chlorides, and examples of amide preparations are shown in equations 17.28 and 17.29. We consider the mechanisms of these reactions in Section 17.12.

$$CH_3C(O)Cl + NH_3 \rightarrow CH_3C(O)NH_2 + HCl \quad (17.28)$$

acetyl chloride → acetamide (ethanamide)

$$PhCO_2Et + NH_3 \rightarrow PhC(O)NH_2 + EtOH \quad (17.29)$$

ethyl benzoate → benzamide

Box 17.6 Feline slumbers

Research has shown that sleep-deprived cats accumulate (*Z*)-octadec-9-enamide in their spinal fluids. It appears that this unsaturated amide is able to function as a signalling molecule in the nervous systems, inducing sleep in our feline companions. The effect is sensitive to molecular structure – the (*E*)-isomer is less effective than the (*Z*)-isomer, and saturation of the C=C double bond or lengthening of the carbon chain also alters the sleep-inducing properties of the compound. These early studies suggest that amides in this family may find potential applications in drugs to treat insomnia.

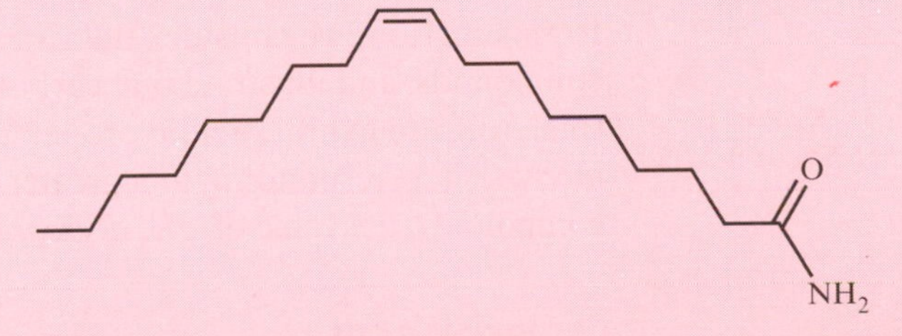

(*Z*)-octadec-9-enamide

17.10 ACYL CHLORIDES: SYNTHESIS FROM CARBOXYLIC ACIDS

Carboxylic acids react with PCl_3, PCl_5 or $SOCl_2$ to give acyl chlorides (equations 17.30–17.32). The advantage of using thionyl chloride ($SOCl_2$) instead of the phosphorus chlorides is that the by-products are gases and the acyl chloride is readily isolated.

$$\underset{\text{acetic acid}}{3CH_3CO_2H} + PCl_3 \xrightarrow{\Delta} \underset{\text{acetyl chloride}}{3CH_3C(O)Cl} + H_3PO_3 \tag{17.30}$$

$$\underset{\text{butanoic acid}}{CH_3CH_2CH_2CO_2H} + PCl_5 \xrightarrow{\Delta} \underset{\text{butanoyl chloride}}{CH_3CH_2CH_2C(O)Cl} + POCl_3 \tag{17.31}$$

$$\underset{\text{benzoic acid}}{PhCO_2H} + SOCl_2 \xrightarrow{\Delta} \underset{\text{benzoyl chloride}}{PhC(O)Cl} + SO_2(g) + HCl(g) \tag{17.32}$$

17.11 CARBONYL COMPOUNDS AS ACIDS

K_a and pK_a: see Section 11.9

In this section, we consider the variation in pK_a values for some carbonyl compounds, and illustrate that carboxylic acids are not the only members of this group to exhibit acidic properties. Remember that as the value of K_a *increases* (corresponding to a greater degree of dissociation of the carbonyl compound), the value of pK_a *decreases*. [Why is this?]

Carboxylic acids

The deprotonation of a carboxylic acid RCO_2H is shown in equation 17.33, and values of pK_a for selected acids are listed in Table 17.3.

$$RCO_2H(aq) + H_2O(l) \rightleftharpoons [H_3O]^+(aq) + [RCO_2]^-(aq) \tag{17.33}$$

Carboxylic acids are weaker acids than inorganic compounds such as hydrochloric, nitric and sulfuric acids but are stronger than water and

Table 17.3 pK_a values for selected carboxylic acids

Compound	*Formula*	pK_a	*Conjugate base*
Methanoic acid (formic acid)	HCO_2H	3.75	$[HCO_2]^-$
Acetic acid	CH_3CO_2H	4.75	$[CH_3CO_2]^-$
Propanoic acid	$CH_3CH_2CO_2H$	4.81	$[CH_3CH_2CO_2]^-$
Butanoic acid	$CH_3CH_2CH_2CO_2H$	4.87	$[CH_3CH_2CH_2CO_2]^-$
Chloroacetic acid	CH_2ClCO_2H	2.85	$[CCl_3CO_2]^-$
Dichloroacetic acid	$CHCl_2CO_2H$	1.48	$[CHCl_2CO_2]^-$
Trichloroacetic acid	CCl_3CO_2H	0.70	$[CCl_3CO_2]^-$
Benzoic acid	$C_6H_5CO_2H$ ($PhCO_2H$)	4.19	$[C_6H_5CO_2]^-$ ($[PhCO_2]^-$)
2-Chlorobenzoic acid	$2\text{-}ClC_6H_4CO_2H$	2.92	$[2\text{-}ClC_6H_4CO_2]^-$
3-Chlorobenzoic acid	$3\text{-}ClC_6H_4CO_2H$	3.82	$[3\text{-}ClC_6H_4CO_2]^-$
4-Chlorobenzoic acid	$4\text{-}ClC_6H_4CO_2H$	3.98	$[4\text{-}ClC_6H_4CO_2]^-$

Box 17.7 Enzymic browning of fruit and salads: Lemon juice comes to the rescue

If you cut an apple or prepare a chicory salad, what originally appeared fresh soon turns brown and, whilst still edible, may not look so appetizing. The process of browning results from the enzymic oxidation of a natural substrate. The enzyme is phenolase (also called catecholase) and both it and the substrate (which varies depending upon the foodstuff) are both present in the fruit but do not react until cell membranes are destroyed when the fruit is cut. Browning can be inhibited by the presence of reducing agents, and sulfites and SO_2 have been used although their toxic properties make them far from ideal.

$$\text{Substrate} \underset{\text{reducing agent}}{\overset{\text{phenolase}}{\rightleftarrows}} \text{Oxidized products}$$

A common method of preventing browning and retaining the fresh look of salad or fruit is to add lemon juice – citric acid alters the pH of the system and, since the enzyme only works in a limited pH range, enzymic browning can be controlled.

CH_2CO_2H

HO — C ····· CO_2H

CH_2CO_2H

citric acid

alcohols. A carboxylic acid RCO_2H is more acidic than an aliphatic alcohol ROH because of the greater stability of the carboxylate anion $[RCO_2]^-$ with respect to the alkoxide ion $[RO]^-$. The structure of the $[RCO_2]^-$ can be represented by the resonance pair **17.30** – the anion is *resonance-stabilized*. [*Question*: Can the ethoxide ion be resonance-stabilized?] An alternative structure for $[RCO_2]^-$ is **17.31** in which there is delocalized bonding. The equivalence of the two carbon–oxygen bonds is confirmed in the solid state structures of many carboxylate salts such as $Na[CH_3CO_2]$ in which both C–O bond distances are 126 pm. This bond length lies in between typical values for C–O single (135 pm) and C=O double (122 pm) bonds involving sp^2 hybridized carbon.

(17.30) (17.31)

Unsubstituted aliphatic acids possess similar pK_a values, but the presence of electronegative substituents such as chlorine may significantly influence the position of equilibrium 17.33, shifting it to the right-hand side – the pK_a

values of CH_3CO_2H and CCl_3CO_2H are 4.75 and 0.70 respectively. Similar effects are observed when electron-withdrawing substituents are introduced into the benzene ring of benzoic acid.

Inductive effect: see Section 14.2

Carboxylate salts are formed when the carboxylic acids react with bases (equations 17.34 and 17.35).

$$\underset{\text{acetic acid}}{CH_3CO_2H} + NaOH \rightarrow \underset{\text{sodium acetate}}{Na[CH_3CO_2]} + H_2O \qquad (17.34)$$

$$\underset{\text{benzoic acid}}{2PhCO_2H} + K_2CO_3 \rightarrow \underset{\text{potassium benzoate}}{2K[PhCO_2]} + CO_2 + H_2O \qquad (17.35)$$

Oxalate anion: see Table 16.2

The oxalate dianion is used commonly as a didentate ligand in coordination chemistry and is derived from the *dicarboxylic acid* $H_2C_2O_4$ **17.32**.

oxalic acid (ethanedioic acid)

(17.32)

Equilibria 17.36 and 17.37 show the first and second dissociation steps of oxalic acid, and the values of pK_a show the usual trend with the first dissociation constant being larger than the second [i.e. $K_a(1) > K_a(2)$, p$K_a(1)$ < p$K_a(2)$].

Polybasic acids: see Section 11.9

$$H_2C_2O_4(aq) + H_2O(l) \rightleftharpoons [H_3O]^+(aq) + [HC_2O_4]^-(aq) \quad pK_a(1) = 1.23 \qquad (17.36)$$

$$[HC_2O_4]^-(aq) + H_2O(l) \rightleftharpoons [H_3O]^+(aq) + [C_2O_4]^{2-}(aq) \quad pK_a(2) = 4.19 \qquad (17.37)$$

Amides

Just as carboxylate anions are resonance-stabilized, so too are the conjugate bases of amides and this results in amides being stronger acids (p$K_a \approx 17$) than amines and ammonia. (Of course, we usually talk in terms of the *basic* properties of NH_3 and amines). Structure **17.33** shows the resonance pair for the $[RC(O)NH]^-$ ion. The partial positive charge on the carbon atom of the carbonyl group stabilizes the anion, but this is only sufficient to render amides *very weak acids*.

(17.33)

Amides themselves are almost non-basic as the lone pair is delocalized as the resonance pair **17.34** illustrates; the lone pair on nitrogen must be available if the molecule is to function as a base. One important consequence of this delocalization is that proteins (which are polyamides of type **17.35**) have well-defined three-dimensional structures as was illustrated in Box 4.4 with the structure of the mammalian O_2-binding protein haemoglobin.

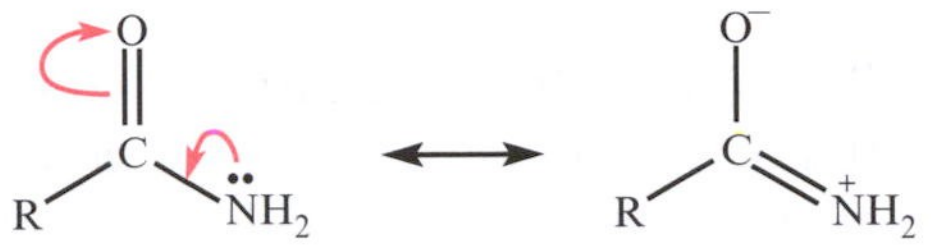

(17.34)

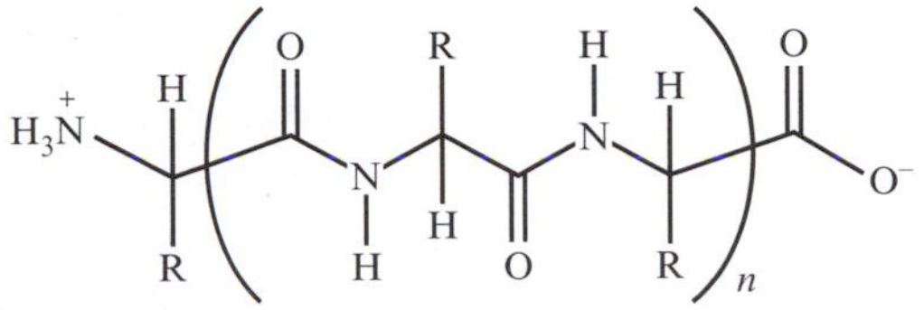

(17.35)

Box 17.8 Bullet-proof vests and spider silk

Kevlar is a man-made fibre which is used in the manufacture of bullet-proof vests and other shock-resistant materials.

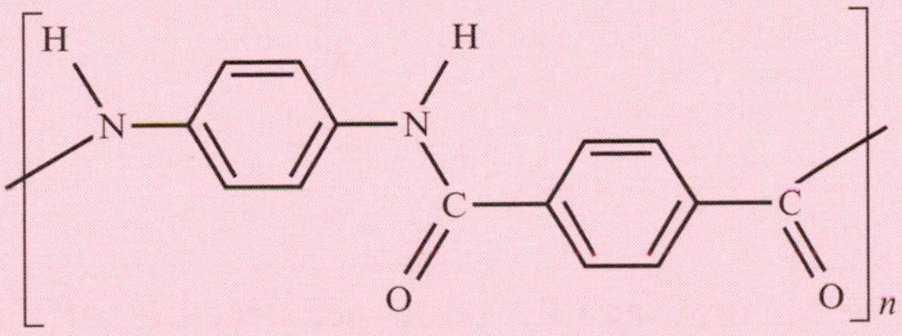

Kevlar

Nature is able to offer a fibre of high tensile strength and great elasticity in the form of spider silk – *Nepila clavipes* is an orb-weaving spider and produces several different forms of silk, among which is dragline silk. This is composed mainly of proteins. The relationship to Kevlar is that a protein consists of a sequence of amino acids which are condensed to give a polymer. This can be represented in a simple manner as follows:

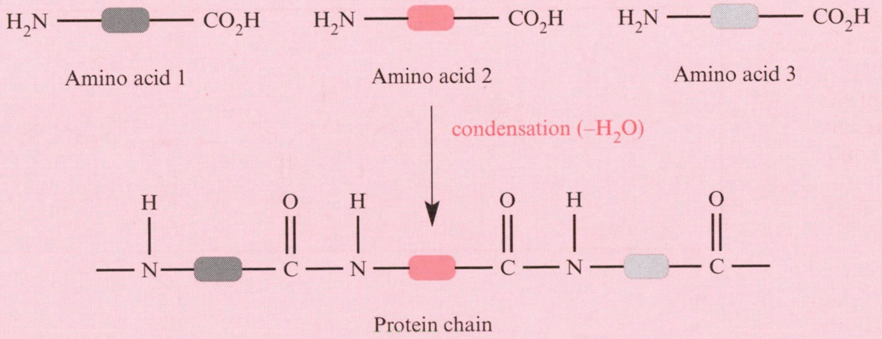

H_2N — — CO_2H H_2N — — CO_2H H_2N — — CO_2H

Amino acid 1 Amino acid 2 Amino acid 3

condensation ($-H_2O$)

Protein chain

The tensile strength of spider silk is only a quarter that of Kevlar, although it takes more energy to break the spider silk (≈ 10 J m^{-3}) than Kevlar (≈ 3 J m^{-3}).

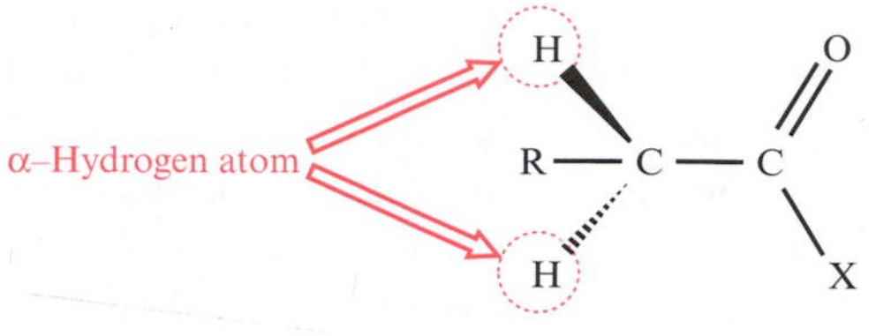

17.11 In a carbonyl compound, an α-hydrogen atom is attached to the carbon atom *adjacent* to the carbon atom of the C=O group.

α-Hydrogen atom

Aldehydes, ketones and esters

We have seen how resonance-stabilization involving the carbonyl group is responsible for the acidic properties of carboxylic acids and, to a lesser extent, of amides. Similar effects are responsible for the acidity of certain ketones and aldehydes – the proton that is lost is initially attached to the carbon atom *adjacent* to the carbon atom of the carbonyl group, the so-called α-hydrogen atom (Figure 17.11). Deprotonation at this site generates a carbanion **17.36** (equation 17.38).

Carbanion: see Section 8.17

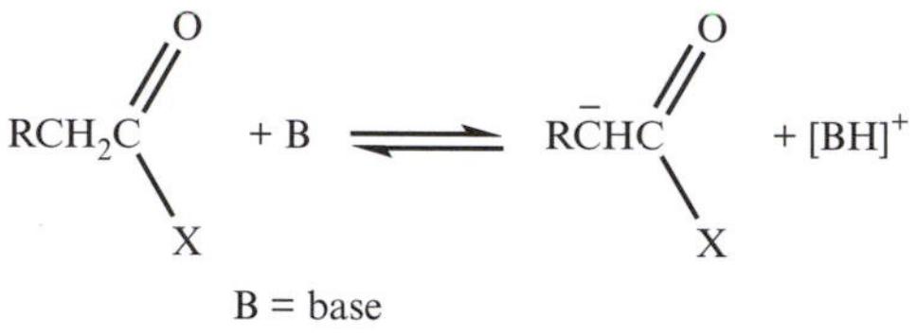

$$RCH_2C(=O)X + B \rightleftharpoons R\bar{C}HC(=O)X + [BH]^+ \qquad (17.38)$$

B = base

(17.36)

We can draw another resonance structure for **17.36** and this is the *enolate ion* **17.37** that we saw in Section 17.5. Alternatively, we can consider the bonding in the carbanion in terms of the delocalization of the negative charge – structure **17.38** is equivalent to the resonance pair **17.36** and **17.37**. The associated resonance stabilization of the conjugate base in equation 17.38 means that ketones and aldehydes which possess α-hydrogen atoms can behave as acids although as the pK_a values in Table 17.4 illustrate, *they are only weakly acidic.*

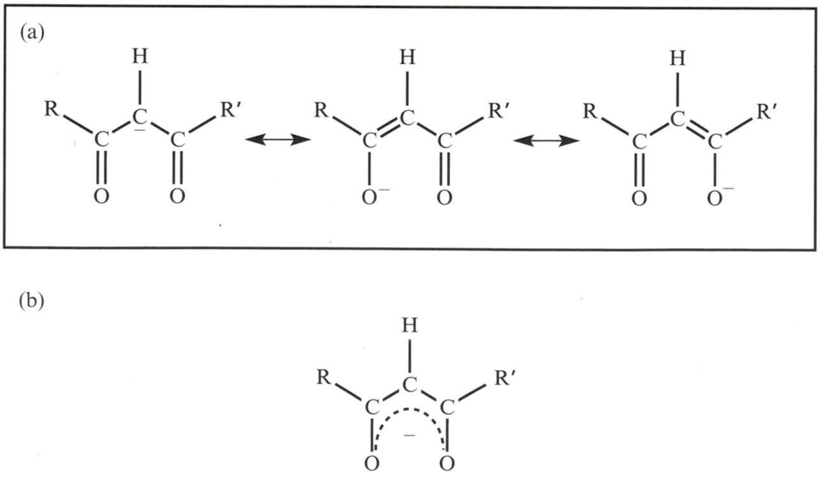

17.12 (a) Resonance structures of (a) β-diketonate $[RC(O)CHC(O)R']^-$, and (b) the structure of the β-diketonate represented with delocalized bonding.

(a)

(b)

Table 17.4 Approximate values of pK_a for carbonyl compounds with α-hydrogen atoms. The R and R′ groups are alkyl substituents. What are the resonance forms that contribute to the stabilization of conjugate base?

Compound type	*Approximate* pK_a	*Conjugate base*
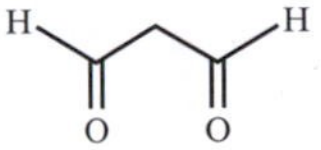	5	$[HC(O)\bar{C}HCHO]$
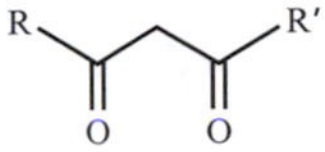	9	$[RC(O)\bar{C}HC(O)R']$
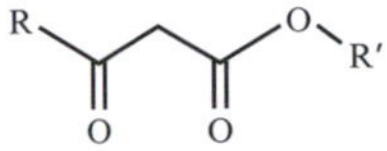	11	$[RC(O)\bar{C}HC(O)OR']$
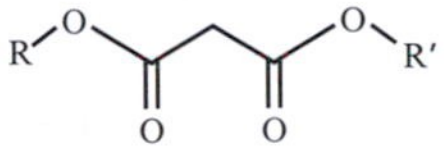	13	$[ROC(O)\bar{C}HC(O)OR']$
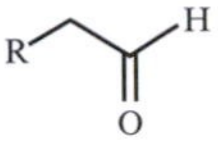	16	$[R\bar{C}HCHO]$
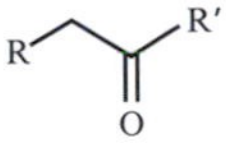	19	$[R\bar{C}HC(O)R']$
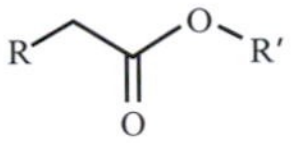	24	$[R\bar{C}HC(O)OR]$

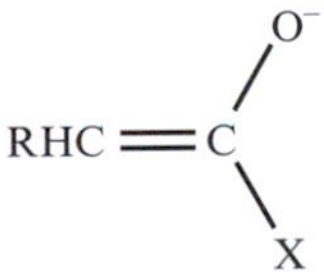

(17.37) 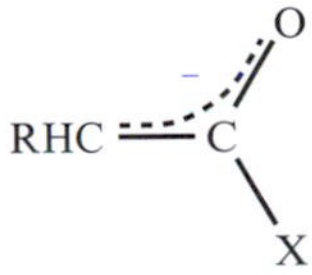 **(17.38)**

A diketone of the type $RC(O)CH_2C(O)R$ or $RC(O)CH_2C(O)R'$ possesses α-hydrogen atoms adjacent to *two* carbonyl groups. Deprotonation (equation 17.39) gives an anion which is resonance-stabilized (Figure 17.12). The conjugate base of such a diketone is called a *β-diketonate* and in Chapter 16, we discussed the ligand behaviour of the β-diketonate $[acac]^-$. Pentane-2,4-dione (Hacac) **17.39** has a pK_a value of 9, and is a weaker acid than a carboxylic acid RCO_2H but a stronger acid than a ketone of type $RCH_2C(O)R'$ (Table 17.4).

The ligand $[acac]^-$: see Table 16.2 and Section 16.3

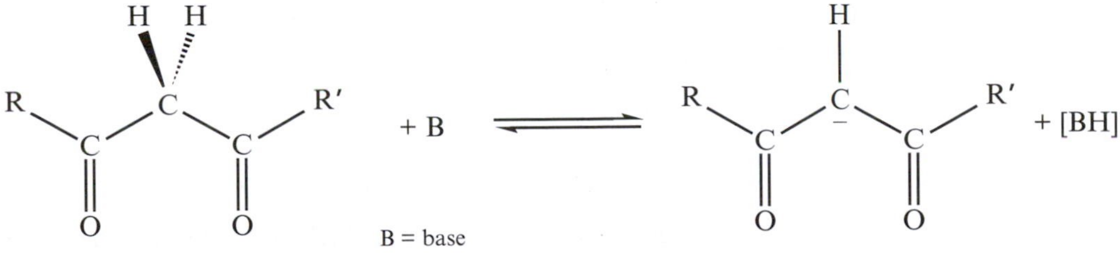

$$RC(O)CH_2C(O)R' + B \rightleftharpoons [RC(O)\bar{C}HC(O)R'] + [BH]^+ \qquad (17.39)$$

B = base

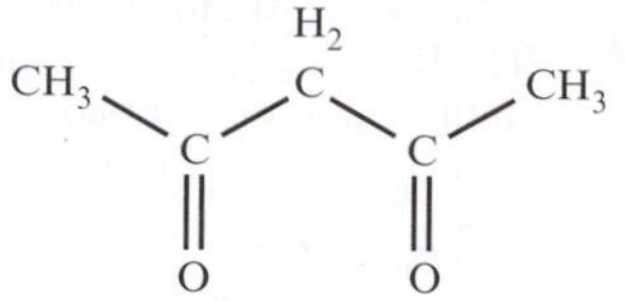

(17.39)

An α-hydrogen atom in an ester such as **17.40** is similarly *weakly* acidic, and so too are esters of the general type $RO_2CCH_2CO_2R'$. Typical values of pK_a for these esters are given in Table 17.4.

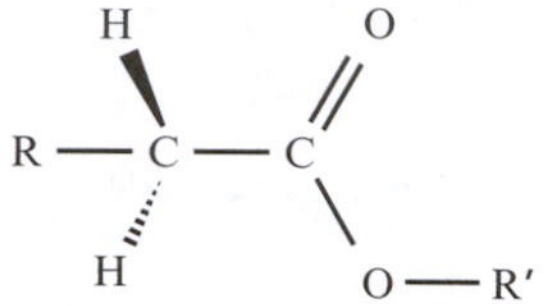

(17.40)

Esters are weaker acids than ketones because the δ^+ charge on the carbonyl C atom is delocalized and this is understood by considering the resonance pair **17.41**.

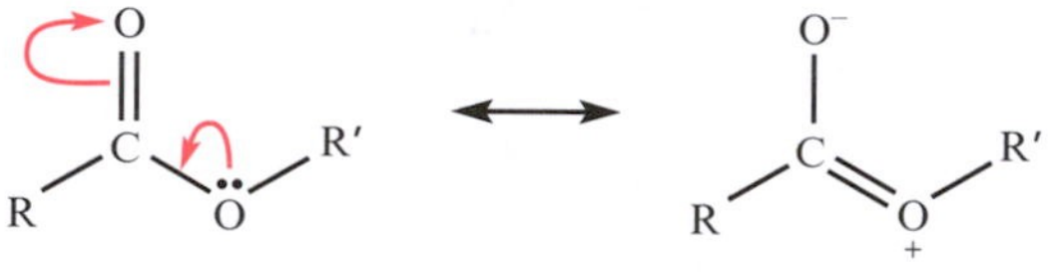

(17.41)

As we see in the next section, the acidic nature of α-hydrogen atoms in some carbonyl compounds produces *carbon centres* which are susceptible to attack by electrophiles.

17.12 CARBANIONS IN SYNTHESIS: ELECTROPHILIC SUBSTITUTION REACTIONS

In Section 17.3 we showed that electrophiles (e.g. H^+) attack the oxygen atom of a C=O group, and nucleophiles attack the carbon atom. In this section we describe some electrophilic substitution reactions which result in bond formation between the ***carbon*** atom and the electrophile. As you work through the section, consider what makes these reactions *different* from the *direct attack* by an electrophile on a carbonyl compound.

Base-catalysed halogenation of aldehydes and ketones

When an α-hydrogen atom is present in an aldehyde or ketone, the reaction of Cl_2, Br_2 or I_2 with the carbonyl compound in the presence of hydroxide ion results in halogenation at the α-position. The bromination of pentan-3-one is shown in equation 17.40 – here the α-hydrogen atoms are all equivalent, but in reaction 17.41, the bromination of hexan-3-one may proceed by substitution at one of two different carbon atoms.

$$CH_3CH_2C(O)CH_2CH_3 + Br_2 \xrightarrow{[OH]^-} CH_3CH_2C(O)CHBrCH_3 + HBr \quad (17.40)$$

$$CH_3CH_2C(O)CH_2CH_2CH_3 \xrightarrow{Br_2,\ [OH]^-} CH_3CH_2C(O)CHBrCH_2CH_3 + CH_3CHBrC(O)CH_2CH_2CH_3 \quad (17.41)$$

Such halogenation reactions proceed by the formation of an enolate (equation 17.42) followed by reaction with the dihalogen X_2 (equation 17.43).

$$[OH]^- + RCH_2C(O)R' \rightleftharpoons \left[R(H)\bar{C}{-}C(=O)R' \longleftrightarrow R(H)C{=}C(O^-)R' \right] + H_2O \quad (17.42)$$

resonance pair

$$X{-}X + R(H)C{=}C(O^-)R' \longrightarrow X^- + X{-}CR(H){-}C(=O)R' \quad (17.43)$$

The overall reaction is the substitution of H^+ by X^+ and the *base-catalysed* halogenation of an aldehyde or ketone is an *electrophilic substitution* reaction.

Although mixtures of products are often observed, halogenation generally occurs at a CH group in preference to a CH_2 group, and at a CH_2 group in preference to a CH_3 group. Additionally, the presence of α-hydrogen atoms in the halo derivative often results in further halogenation. This is generally the case in *methyl* ketones which react to give trihalo derivatives. The electron-withdrawing halides are capable of stabilizing a carbanion, and nucleophilic attack of $[OH]^-$ at the CO group leads to the loss of $[CX_3]^-$ and (after protonation) the formation of HCX_3. This is a general reaction of methyl ketones and is known as the *haloform reaction*. Reaction 17.44 shows the halogenation of butanone.

See problem 17.11

$$CH_3CH_2C(O)CH_3 \xrightarrow{X_2,\ [OH]^-} CH_3CH_2C(O)CX_3 \xrightarrow{[OH]^-} [CH_3CH_2CO_2]^- + [CX_3]^- \quad (X = Cl, Br, I) \quad (17.44)$$

Acid-catalysed halogenation of aldehydes and ketones

In worked example 10.3 we considered the kinetics of the *acid-catalysed* bromination of acetone (reaction 17.45) and found that the rate equation is as given in expression 17.46 – the rate is zero order in bromine, first order in acetone and first order in hydrogen ion. Similar rate expressions are found for other acid-catalysed halogenations of aldehydes and ketones.

$$CH_3C(O)CH_3 + Br_2 \xrightarrow{H^+} CH_3C(O)CH_2Br + HBr \qquad (17.45)$$

$$\text{Rate of reaction} = -\frac{d[Br_2]}{dt} = k[CH_3C(O)CH_3][H^+] \qquad (17.46)$$

A mechanism consistent with these kinetic data involves the generation of an enol. The acid-catalysed rearrangement of the ketone to the enol (equation 17.47) is rate-determining. The enol then reacts with dibromine in fast steps 17.48. The overall reaction is the electrophilic substitution of H^+ for Br^+.

$$\underset{\text{ketone}}{CH_3C(=O)CH_3} \xrightarrow{H^+} CH_3C(=\overset{+}{O}H)CH_3 \xrightarrow{-H^+} \underset{\text{enol}}{CH_3C(OH){=}CH_2} \qquad (17.47)$$

$$CH_3C(OH){=}CH_2 + Br{-}Br \longrightarrow CH_3C(=\overset{+}{O}H)CH_2Br + Br^- \xrightarrow{-H^+} CH_3C(=O)CH_2Br \qquad (17.48)$$

Alkylation of ketones and esters

If a ketone or ester, diketone or diester possesses an α-hydrogen atom, the carbonyl compound can be alkylated at the α-position. The first step in the reaction is the deprotonation of the ketone or ester to form a carbanion and equation 17.49 shows the deprotonation of acetone using an amide ion as the base.

$$CH_3C(O)CH_3 + [{}^iPr_2N]^- \rightleftharpoons H_2\overline{C}C(O)CH_3 + {}^iPr_2NH \qquad (17.49)$$

diisopropylamide ion

Nucleophilic substitution of halogenoalkanes: see Sections 14.4 and 14.5

The carbanion formed in reaction 17.49 can now react with a halogenoalkane RX; the carbanion is a nucleophile and nucleophilic substitution occurs with X^- as the leaving group. Reaction 17.50 illustrates the formation of pentan-2-one from the carbanion generated in reaction 17.49.

$$CH_3CH_2{-}Cl + \overline{C}H_2C(O)CH_3 \longrightarrow CH_3CH_2CH_2C(O)CH_3 + Cl^- \qquad (17.50)$$

The overall reaction converts acetone into pentan-2-one and involves:

- carbon–carbon bond formation, and
- electrophilic substitution of $CH_3CH_2^+$ for H^+.

Such alkylation reactions are successful if the halogenoalkane is primary or secondary but reactions involving tertiary halogenoalkanes are complicated by β-elimination steps as we discussed in Section 14.6.

Similar alkylation reactions occur with esters, diketones and diesters which contain an α-hydrogen atom, but as Table 17.4 indicates, the α-hydrogens positioned *between two carbonyl groups* are more acidic than those adjacent to only one C=O group. This means that weaker bases can be used to generate the carbanion than in equation 17.49, and in reaction 17.51, ethoxide ion is used for the deprotonation step.

$$CH_3C(O)CH_2C(O)OEt + [EtO]^- \rightleftharpoons CH_3C(O)\overline{C}HC(O)OEt + EtOH \qquad (17.51)$$

$pK_a \approx 13$ $pK_a \approx 16$

Equilibrium 17.51 lies to the right-hand side because the dicarbonyl compound on the left-hand side of the equation is a stronger acid (pK_a 13) than ethanol (pK_a 16). The reaction of the carbanion formed in reaction 17.51

with primary or secondary halogenoalkanes leads to alkyl-substituted products (reaction 17.52).

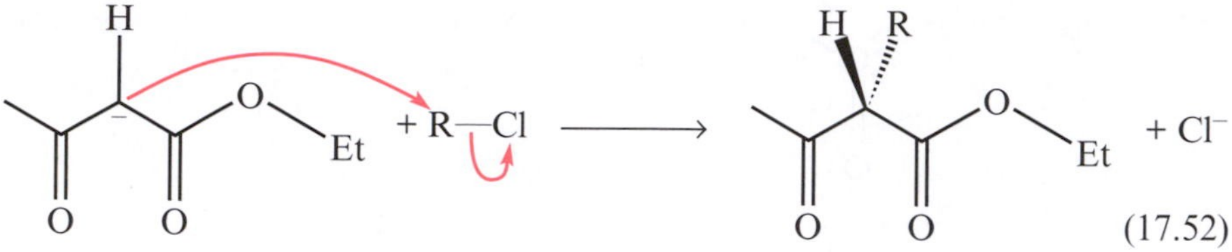

(17.52)

The product still possesses an acidic α-hydrogen atom and the cycle of steps can be repeated, beginning with carbanion formation (equation 17.53). The second alkyl substituent may be the same as or different from that introduced in the first alkylation and equation 17.54 illustrates the possible reactions that can occur in the final stage of the alkylation process.

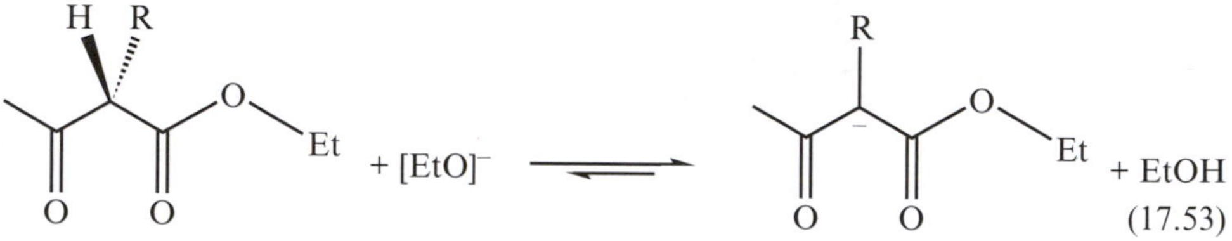

(17.53)

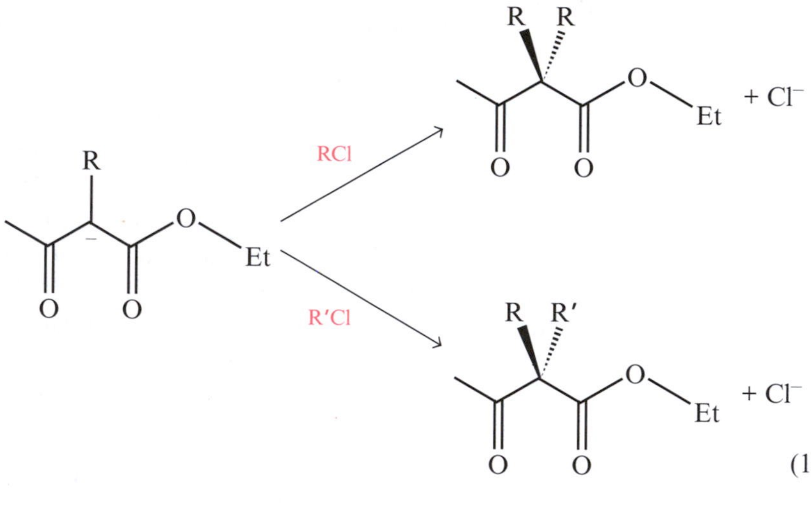

(17.54)

Aldol and Claisen reactions

An *aldol reaction* occurs between two carbonyl compounds and is catalysed by base. The general reaction between two aldehydes – one of which contains an α-hydrogen atom – is shown in Figure 17.13, and involves the formation of a carbanion followed by nucleophilic attack at the C=O carbon atom of the second carbonyl compound. The first carbonyl compound may be an aldehyde, ketone or ester, but the second is usually an aldehyde or a ketone. The aldol reaction is mentioned here because of the role of the carbanion, but is described more fully towards the end of Section 17.13 (where Figure 17.13 is presented).

A Claisen condensation involves the reaction between two esters in the presence of a strong base such as the ethoxide ion. The first step of the reaction involves the formation of a carbanion and condensation proceeds by nucleophilic attack at the carbonyl carbon atom in a similar manner to that shown for the aldehydes in Figure 17.13. We discuss this reaction further at the end of Section 17.13.

17.13 NUCLEOPHILIC ATTACK AT THE C=O CARBON ATOM

Many reactions of carbonyl compounds involve attack by nucleophiles at the $C^{\delta+}$ centre of the carbonyl group and in this section we give selected examples of such reactions, many of which are of synthetic importance. We have already seen examples of such reactions in the haloform reaction (equation 17.44) and the aldol reaction. Equation 17.55 shows a general step which you will see repeatedly in this section; in the reaction, X^- represents a nucleophile and the carbon centre goes from being trigonal planar to tetrahedral.

$$X^- + R(Y)C{=}O \longrightarrow X{-}C(R)(Y){-}O^- \qquad (17.55)$$

Hydrolysis of acyl chlorides

Acyl chlorides hydrolyse readily (equation 17.56) to the corresponding carboxylic acid and many must be stored under water-free conditions.

$$\underset{\text{acetyl chloride}}{CH_3C(O)Cl} + H_2O \rightarrow \underset{\text{acetic acid}}{CH_3CO_2H} + HCl \qquad (17.56)$$

The net result of this reaction is the nucleophilic substitution of Cl^- by $[OH]^-$. The nucleophile is usually water, although some acyl chlorides require aqueous base.

The reaction mechanism is shown in equation 17.57. Initial attack by the water nucleophile leads to O–C bond formation and opening of the C=O double bond. This *is in preference to the direct loss of the leaving group* which takes place in the next step of the mechanism to regenerate the C=O bond. The formation of the O–C bond in the first step places a positive charge on the oxygen atom, and in the last step of the mechanism, neutrality of the carbonyl compound is restored by the loss of a proton.

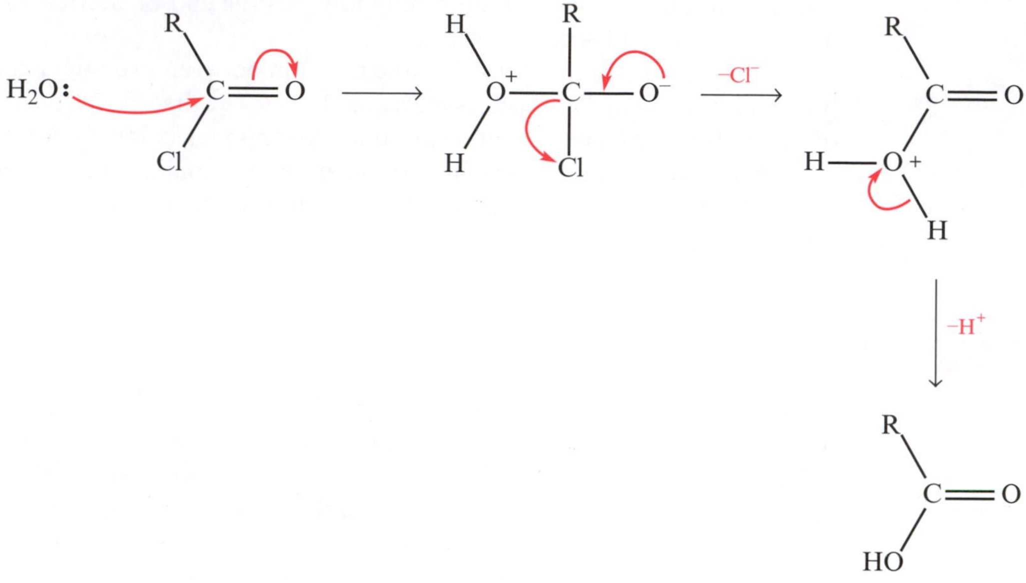

(17.57)

Hydrolysis of amides

Amides are also easily hydrolysed and reaction 17.58 shows the reaction between propanamide and water in the presence of acid. Of course, in acidic conditions, protonation of ammonia occurs and an ammonium salt is formed.

$$\underset{\text{propanamide}}{CH_3CH_2C(O)NH_2} + H_2O \xrightarrow{H^+} \underset{\text{propanoic acid}}{CH_3CH_2CO_2H} + [NH_4]^+ \quad (17.58)$$

In the hydrolysis of the acyl chloride above, the leaving group was Cl^-, but in the case of the amide, $[NH_2]^-$ is a poor leaving group although, if protonated, it can depart as NH_3, a good leaving group. The mechanism of the acid-catalysed amide hydrolysis is given in equation 17.59.

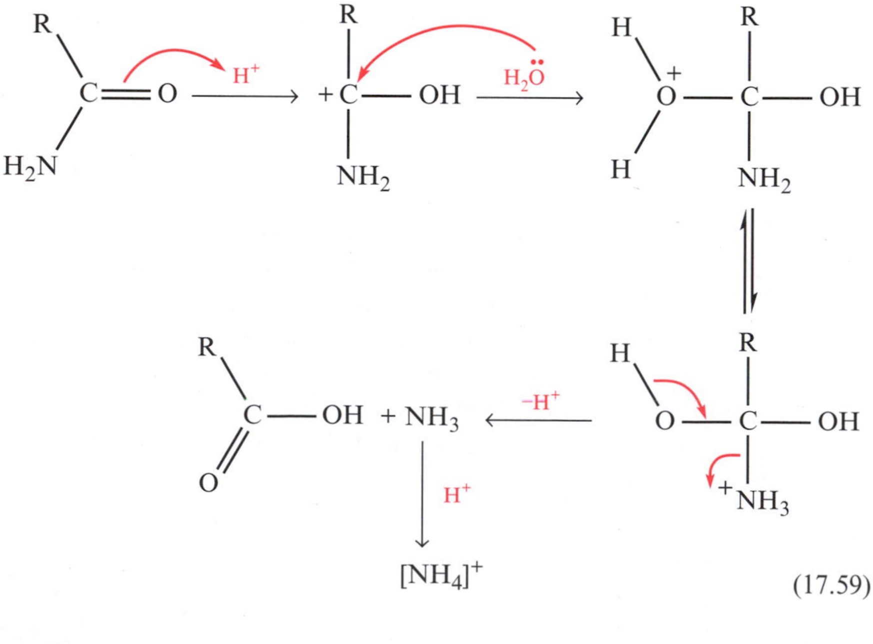

(17.59)

As a *rule of thumb* (but it is *not* infallible), the direction of reactions of the type:

$$RC(O)X + Y^- \rightleftharpoons RC(O)Y + X^-$$

may be determined from the pK_a values of HX and HY. If HX is a stronger acid than HY [i.e. pK_a(HX) < pK_a(HY)], then X^- is more favoured than Y^- and the equilibrium will lie to the right-hand side.

The reaction of aldehydes and ketones with hydrogen cyanide

Hydrogen cyanide is a source of the nucleophile $[CN]^-$ and reacts with aldehydes and ketones (equations 17.60 and 17.61).

$$\underset{\text{propanal}}{CH_3CH_2CHO} + HCN \rightarrow \underset{\text{1-cyanopropan-1-ol}}{CH_3CH_2CH(OH)(CN)} \quad (17.60)$$

$$\underset{\text{acetone}}{CH_3C(O)CH_3} + HCN \rightarrow \underset{\text{2-cyanopropan-2-ol}}{CH_3C(OH)(CN)CH_3} \quad (17.61)$$

The product is a *cyanohydrin* and is characterized by the carbon that was originally part of the C=O group carrying both –OH and –CN groups.

The mechanism of cyanohydrin formation (equation 17.62) involves nucleophilic attack by $[CN]^-$ followed by protonation. This reaction involves carbon–carbon bond formation – a feature that lengthens the carbon chain and can be of importance in synthetic pathways.

$$[CN]^- + RR'C{=}O \longrightarrow NC{-}CRR'{-}O^- \xrightarrow{H^+} NC{-}CRR'{-}OH \quad (17.62)$$

The reactions of Grignard reagents with aldehydes and ketones

Grignard reagents: see Section 14.9 and Box 14.1

The reactions of Grignard reagents RMgX (R = alkyl, X = halide) with aldehydes and ketones are important preparative routes to alcohols. Methanal reacts with RMgX to yield *primary alcohols* in which the terminal $-CH_2OH$ group is derived from the aldehyde. Reaction 17.63 shows the preparation of butan-1-ol.

$$HCHO + CH_3CH_2CH_2MgCl \xrightarrow{\text{anhydrous THF}} CH_3CH_2CH_2CH_2OMgCl$$

$$\xrightarrow{H_2O} CH_3CH_2CH_2CH_2OH + Mg^{2+} + Cl^- + [OH]^- \quad (17.63)$$

As we saw in Section 17.7, the Mg–C bond is polarized in the sense $Mg^{\delta+}–C^{\delta-}$ and the Grignard reagent behaves as a nucleophile, attacking the carbon atom of the carbonyl group in methanal (equation 17.64). The product of this step is an alkoxide, and protonation gives the corresponding alcohol.

$$CH_3CH_2\overset{\delta-}{CH_2}–\overset{\delta+}{Mg}Cl + H_2\overset{\delta+}{C}=\overset{\delta-}{O} \longrightarrow CH_3CH_2CH_2CH_2–O^-\overset{+}{Mg}Cl \xrightarrow{H_2O} CH_3CH_2CH_2CH_2OH + Mg^{2+} + Cl^- + [OH]^- \quad (17.64)$$

If the starting material is an aldehyde *other than methanal* (i.e. of the formula RCHO where R is *not* H), then the reaction with a Grignard reagent yields a *secondary alcohol*. An example is the conversion of propanal to pentan-3-ol (equation 17.65).

$$CH_3CH_2CHO + CH_3CH_2MgCl \xrightarrow{\text{anhydrous THF}} CH_3CH_2CH(OMgCl)CH_2CH_3 \xrightarrow{H_2O} CH_3CH_2CH(OH)CH_2CH_3 + Mg^{2+} + Cl^- + [OH]^- \quad (17.65)$$

The reaction of a *ketone* with a Grignard reagent produces a *tertiary* alcohol and reaction 17.66 shows the preparation of 2-methylpentan-2-ol from acetone.

$$CH_3C(O)CH_3 + CH_3CH_2CH_2MgCl \xrightarrow{\text{anhydrous THF}} CH_3CH_2CH_2C(CH_3)_2–OMgCl \xrightarrow{H_2O} CH_3CH_2CH_2C(OH)(CH_3)CH_3 + Mg^{2+} + Cl^- + [OH]^- \quad (17.66)$$

These reactions have widespread application and can be used to prepare aryl as well as alkyl derivatives; they represent one of the important uses of Grignard reagents in organic synthesis.

Reductive amination of aldehydes and ketones

In Section 14.11 we described the *reductive amination of aldehydes* to give primary amines in which the $-NH_2$ group is attached to the terminal carbon atom of an aliphatic chain and the *reductive amination of ketones* to prepare *primary or secondary amines* in which the amino group is attached to a secondary alkyl group, i.e. the families of compounds shown in structures **17.42** and **17.43** in which R groups may be the same or different.

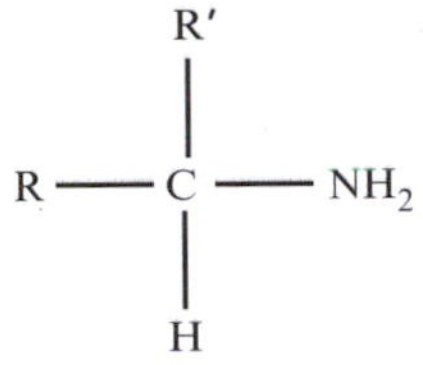

$$RR'CH-NH_2 \quad \textbf{(17.42)} \qquad RR'CH-NHR'' \quad \textbf{(17.43)}$$

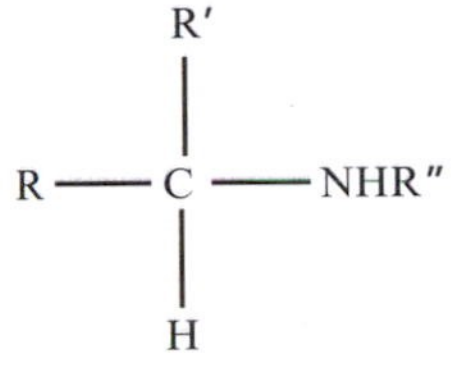

Reductive amination involves the conversion of a carbonyl into an amine group under reducing conditions and uses either NH_3 or a primary amine (RNH_2) as the aminating agent – both are nucleophiles. Equation 17.67 shows the formation of 3-pentylamine from an appropriate ketone.

$$\underset{\text{pentan-3-one}}{CH_3CH_2C(O)CH_2CH_3} \xrightarrow{H_2 \text{ with Ni catalyst, } NH_3} \underset{\text{3-pentylamine}}{CH_3CH_2CH(NH_2)CH_2CH_3} \tag{17.67}$$

The initial step in the reaction is nucleophilic attack by NH_3 or RNH_2 at the carbon atom of the C=O group of the aldehyde or ketone, followed by proton transfer (equation 17.68).

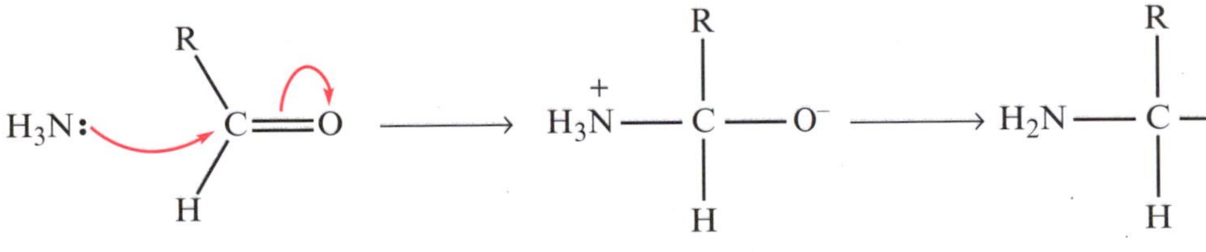

$$H_3N{:} + R(H)C{=}O \longrightarrow H_3\overset{+}{N}-CRH-O^- \longrightarrow H_2N-CRH-OH \tag{17.68}$$

Under appropriate reducing conditions, the alcohol functionality is converted to a C–H bond, and an amine (in this case a primary amine) is formed (equation 17.69). The reaction proceeds by an intermediate unsaturated imine.

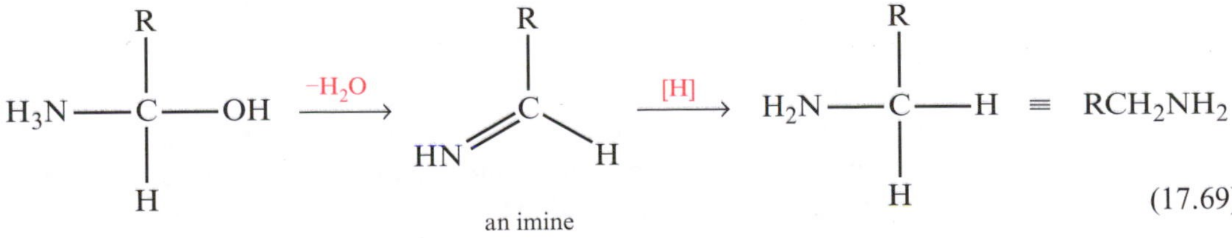

$$H_2N-CRH-OH \xrightarrow{-H_2O} \underset{\text{an imine}}{HN{=}CRH} \xrightarrow{[H]} H_2N-CRH-H \equiv RCH_2NH_2 \tag{17.69}$$

Reactions of acyl chlorides and esters with ammonia

In Section 17.8 we saw that esters can be prepared by nucleophilic substitution reactions between ammonia and esters or acyl chlorides. In the first reaction, an alcohol is eliminated (equation 17.70) and in the second, the elimination product is HCl which reacts with NH_3 to give ammonium chloride (equation 17.71).

$$RCOOR' + NH_3 \rightarrow RC(O)NH_2 + R'OH \quad (17.70)$$

$$RC(O)Cl + 2NH_3 \rightarrow RC(O)NH_2 + NH_4Cl \quad (17.71)$$

The reactions of aldehydes with alcohols: acetal formation

Aldehydes react with alcohols in the presence of an acid catalyst to give *acetals*. An acetal possesses a $C(OR)_2$ functional group – **17.44** shows the generic structure of an acetal.

$$R'\text{—}C(OR)_2\text{—}H$$

(17.44)

An example of acetal formation is the reaction of benzaldehyde with propan-1-ol – water is eliminated during the reaction. As equation 17.72 shows, the reaction is in equilibrium but can be shifted to the right-hand side by removing the water as it is produced.

$$PhCHO + 2CH_3CH_2CH_2OH \overset{H^+}{\rightleftharpoons} PhCH(OCH_2CH_2CH_3)_2 + H_2O \quad (17.72)$$

The reaction proceeds by initial protonation of the oxygen atom of the carbonyl group and this is shown in equation 17.73.

$$R'(H)C{=}O + H^+ \rightleftharpoons \left[R'(H)C{=}O^+{-}H \longleftrightarrow R'(H)C^+{-}O{-}H \right] \quad (17.73)$$

The positive charge now resides on the carbon centre of the original C=O group, and makes this atom even more susceptible to attack by nucleophiles, and in this case the nucleophile is an alcohol (equation 17.74). Initially a *hemiacetal* is formed and the H^+ catalyst is regenerated.

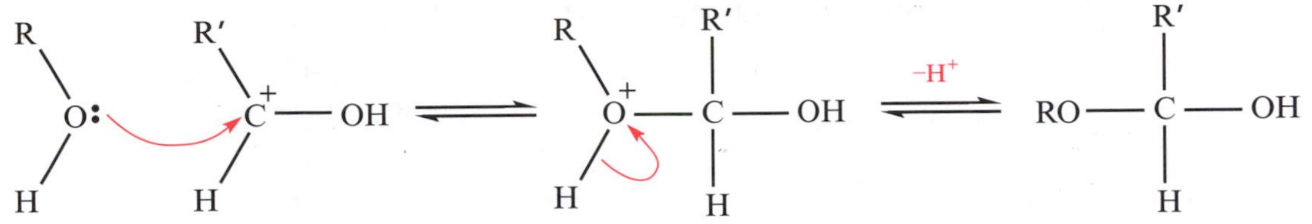

$$\mathrm{R(H)\ddot{O}} + \mathrm{R'(H)\overset{+}{C}{-}OH} \rightleftharpoons \mathrm{R(H)\overset{+}{O}{-}C(R')(H){-}OH} \xrightleftharpoons{-H^+} \mathrm{RO{-}C(R')(H){-}OH}$$

a hemiacetal (17.74)

The reaction continues by the protonation of the OH group and subsequent elimination of water to yield a carbenium ion (equation 17.75) – H_2O *is a good leaving group.*

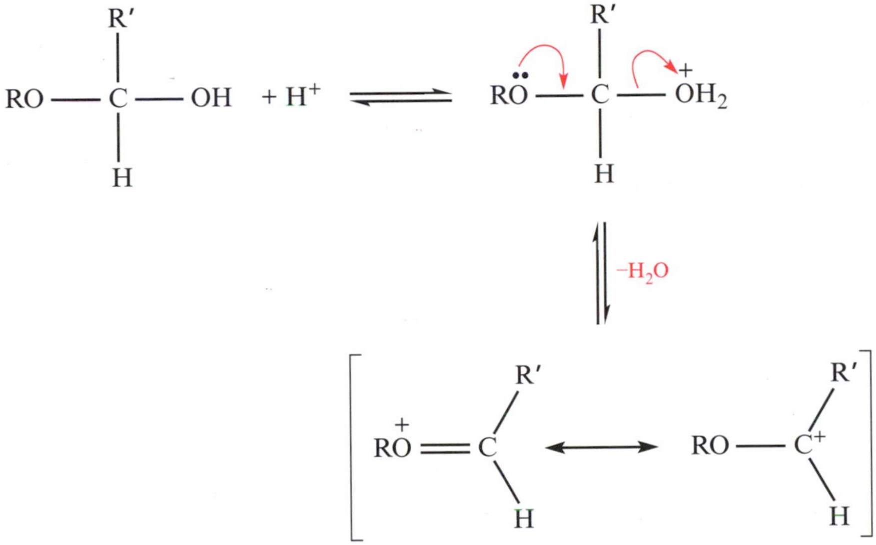

$$\mathrm{RO{-}C(R')(H){-}OH} + \mathrm{H^+} \rightleftharpoons \mathrm{R\ddot{O}{-}C(R')(H){-}\overset{+}{O}H_2} \xrightleftharpoons{-H_2O} \left[\mathrm{R\overset{+}{O}{=}C(R')H} \longleftrightarrow \mathrm{RO{-}\overset{+}{C}(R')H}\right]$$

(17.75)

In the final step of the reaction (equation 17.76), a second alcohol molecule attacks the carbenium ion, and after elimination of H^+ (regeneration of the catalyst used in reaction 17.75), the acetal is produced.

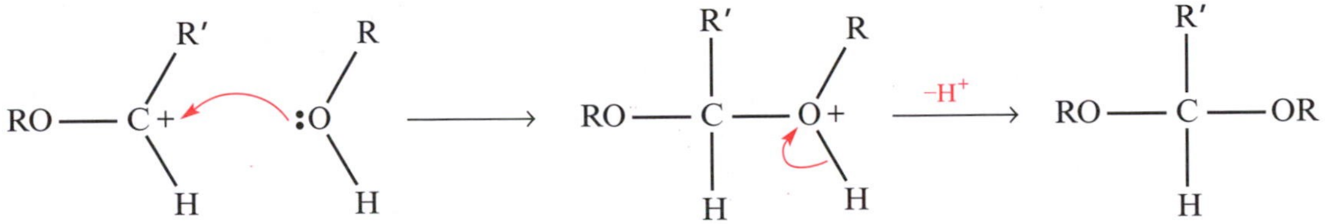

$$\mathrm{RO{-}\overset{+}{C}(R')H} + \mathrm{:O(R)H} \longrightarrow \mathrm{RO{-}C(R')(H){-}\overset{+}{O}(R)H} \xrightarrow{-H^+} \mathrm{RO{-}C(R')(H){-}OR}$$

an acetal (17.76)

Similar reactions do occur between some ketones and alcohols but are not generally as successful as those involving aldehydes.

Acetals are very often used as 'protected' carbonyl compounds in syntheses in which the carbonyl compound itself would react in unwanted ways.

An acetal has the general structure:

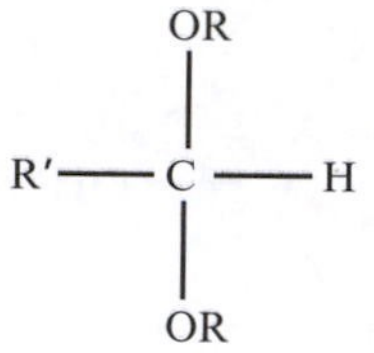

A hemiacetal has the general structure:

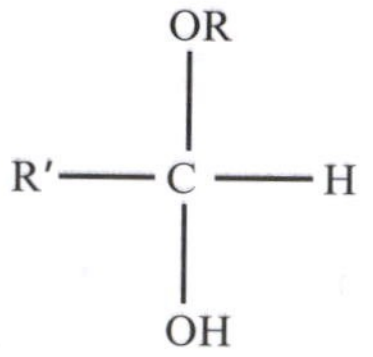

Aldol reaction

We have already mentioned the aldol reaction that occurs between two carbonyl compounds, one of which possesses an α-hydrogen atom. Figure 17.13 summarizes the mechanism for the condensation between two aldehydes, the first of which can be deprotonated to generate a carbanion. Similar reactions can occur between two ketones, aldehydes and ketones, esters and ketones, and esters and aldehydes – the first member of each pair must contain an α-hydrogen atom.

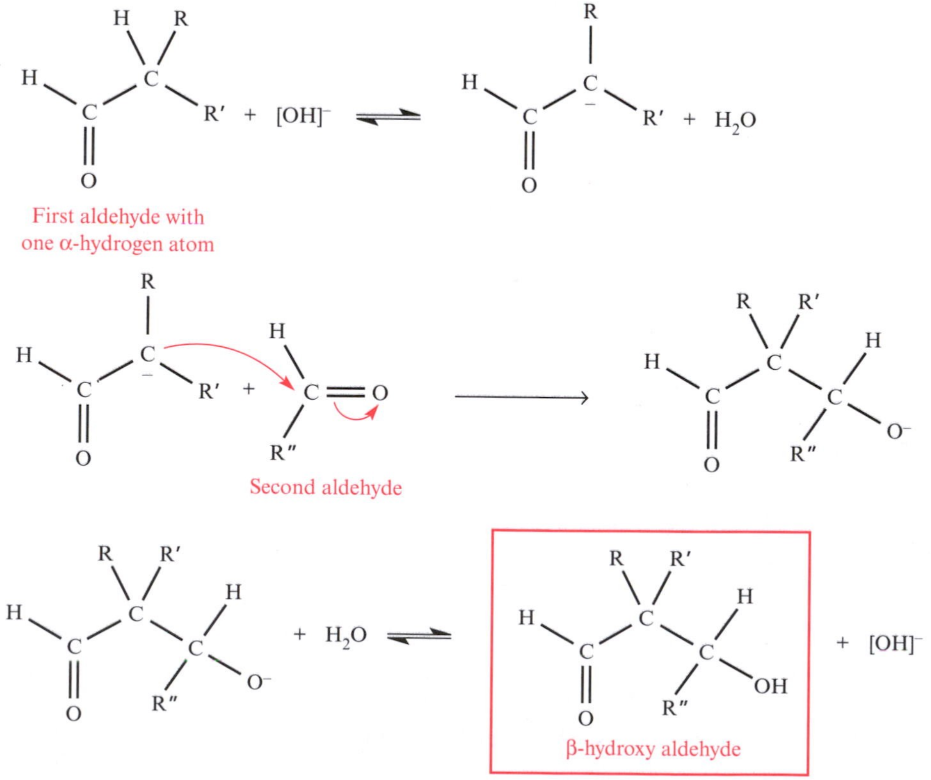

17.13 *An aldol reaction* takes place between two carbonyl compounds, one of which must possess an α-hydrogen atom. It is represented here in a general sense by the reaction between RR′CHCHO and R″CHO. The product is a β-hydroxy aldehyde (an aldol).

First aldehyde with one α-hydrogen atom

Second aldehyde

β-hydroxy aldehyde

17.14 In many aldol reactions, the β-hydroxy aldehyde may undergo dehydration to give an α,β-unsaturated aldehyde. The prerequisite for this is the presence of two α-hydrogen atoms in one of the starting compounds.

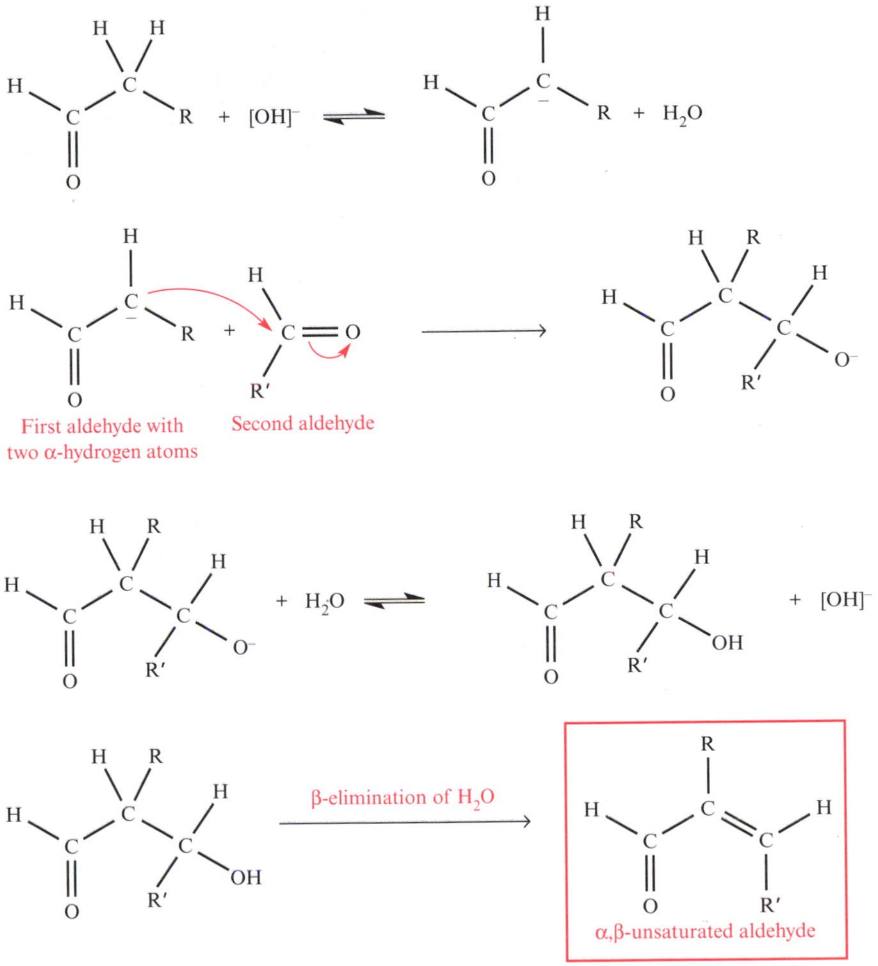

First aldehyde with two α-hydrogen atoms

Second aldehyde

β-elimination of H_2O

α,β-unsaturated aldehyde

In the first step of the reaction shown in Figure 17.13, loss of the acidic hydrogen atom from the first aldehyde produces a carbanion which then acts as a nucleophile attacking the carbon atom of the C=O group of the second aldehyde. In the general example shown in Figure 17.13, the product of the second step does *not* possess an α-hydrogen atom and the reaction proceeds with protonation to give the final reaction product which is a *β-hydroxy aldehyde* – also called an *aldol*. The structural unit of a β-hydroxy aldehyde is shown in structure **17.45** where the α and β carbon atoms are marked.

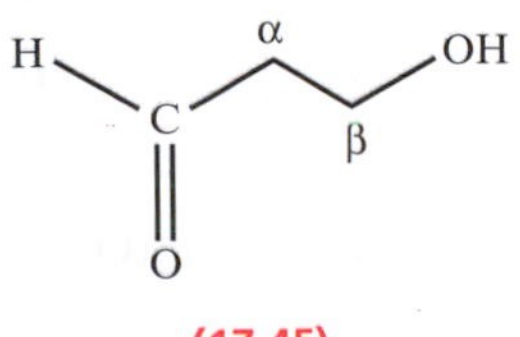

(17.45)

Of course, a problem with such reactions is the possibility that the carbanion formed in step one in Figure 17.13 may attack a molecule of the *first* aldehyde rather than the *second* aldehyde. If the aldehydes happen to be identical, then this does not pose a problem, but if they are different (as in Figure 17.13), then two aldol reaction products are possible. What other β-hydroxy aldehyde could form in Figure 17.13? We return to these complications later.

If instead of starting with the first aldehyde shown in Figure 17.13, we begin with the related ketone **17.46** (which still contains the essential α-hydrogen atom), the final product is a *β-hydroxy ketone*. Follow through the reaction scheme shown in Figure 17.13 but start with ketone **17.46** – what are the characteristic structural features of a β-*hydroxy ketone*?

(17.46)

So far we have considered aldol reaction reactions in which one of the starting carbonyl compounds contains *one* α-hydrogen atom. Now we turn to the more common case of compounds which possess *two* α-hydrogen atoms. The mechanism of such a reaction is shown in Figure 17.14. The first three steps follow the same pattern as we saw in Figure 17.13, *but* the β-hydroxy aldehyde that forms is able to undergo a β-elimination reaction, losing H_2O and forming an α,β-unsaturated aldehyde. Similarly, α,β-unsaturated ketones can be produced if the starting ketone possesses two α-hydrogen atoms. In Section 9.10, we mentioned the π-conjugation that α,β-unsaturated ketones and aldehydes exhibit, and discussed the effects this has on their electronic spectra.

β-Elimination: see Section 14.6

The complications that can arise when, for example, two *different* aldehydes (which possess α-hydrogen atoms) undergo an aldol reaction reaction are significant. In equation 17.77, we show the formation of two products from the reaction of the aldehydes RCH_2CHO and R′CHO.

$$RCH_2CHO + R'CHO \xrightarrow{[OH]^-} R'CH(OH)CH(R)CHO + RCH_2CH(OH)CH(R)CHO \quad (17.77)$$

Both products are able to undergo β-eliminations of H_2O and so the overall reaction is further complicated by alkene formation (equations 17.78 and 17.79).

$$R'CH(OH)CH(R)CHO \xrightarrow{-H_2O} R'CH{=}C(R)CHO \quad (17.78)$$

$$RCH_2CH(OH)CH(R)CHO \xrightarrow{-H_2O} RCH_2CH{=}C(R)CHO \quad (17.79)$$

Some specific examples of aldol reactions are shown in equations 17.80–17.83. For each reaction, identify the α-hydrogen atom(s) in the starting compounds and show how each product arises. The non-specificity means that, in general, such reactions are of no synthetic utility.

$$2CH_3CHO \xrightarrow{[OH]^-} CH_3CH(OH)CH_2CHO \xrightarrow{-H_2O} CH_3CH{=}CHCHO \quad (17.80)$$

$$2CH_3C(O)CH_3 \xrightarrow{[OH]^-} CH_3C(Me)(OH)CH_2C(O)CH_3 \xrightarrow{-H_2O} CH_3C(Me){=}CHC(O)CH_3 \quad (17.81)$$

$$PhCHO + CH_3CH_2CHO \xrightarrow{[OH]^-} PhCH(OH)CH(Me)CHO + CH_3CH_2CH(OH)CH(Me)CHO$$

$$PhCH(OH)CH(Me)CHO \xrightarrow{-H_2O} PhCH{=}C(Me)CHO \qquad CH_3CH_2CH(OH)CH(Me)CHO \xrightarrow{-H_2O} CH_3CH_2CH{=}C(Me)CHO \quad (17.82)$$

$$CH_3CH_2CHO + CH_3CHO \longrightarrow$$

$$\left.\begin{array}{c} CH_3CH_2CH(OH)CH(Me)CHO \xrightarrow{-H_2O} CH_3CH_2CH{=}C(Me)CHO \\ + \\ CH_3CH(OH)CH_2CHO \xrightarrow{-H_2O} CH_3CH{=}CHCHO \\ + \\ CH_3CH(OH)CH(Me)CHO \xrightarrow{-H_2O} CH_3CH{=}C(Me)CHO \\ + \\ CH_3CH_2CH(OH)CH_2CHO \xrightarrow{-H_2O} CH_3CH_2CH{=}CHCHO \end{array}\right\} \quad (17.83)$$

Claisen condensation

The condensation of two esters in the presence of a strong base such as ethoxide ion generates a β-*keto ester* and is called the Claisen condensation. At least one of the esters must possess an α-hydrogen atom and the first step is deprotonation to generate a carbanion. Ethoxide is a suitable base to facilitate the formation of the desired carbanion (equation 17.84).

$$RCH_2C(O)OR' + [EtO]^- \rightleftharpoons R\overset{-}{C}HC(O)OR' + EtOH \quad (17.84)$$

Consider the condensation of two equivalents of the *same* ester. The carbanion from equilibrium 17.84 is a nucleophile and attacks the carbon atom of the C=O group in a molecule of the ester. At this point, the condensation reaction becomes distinct from an aldol condensation – the Claisen condensation proceeds by reforming a C=O group with concomitant loss of an alkoxide (equation 17.85). The product is a β-*keto ester*, named because there is a ketone functionality attached to the carbon atom in the β-position with respect to the ester group.

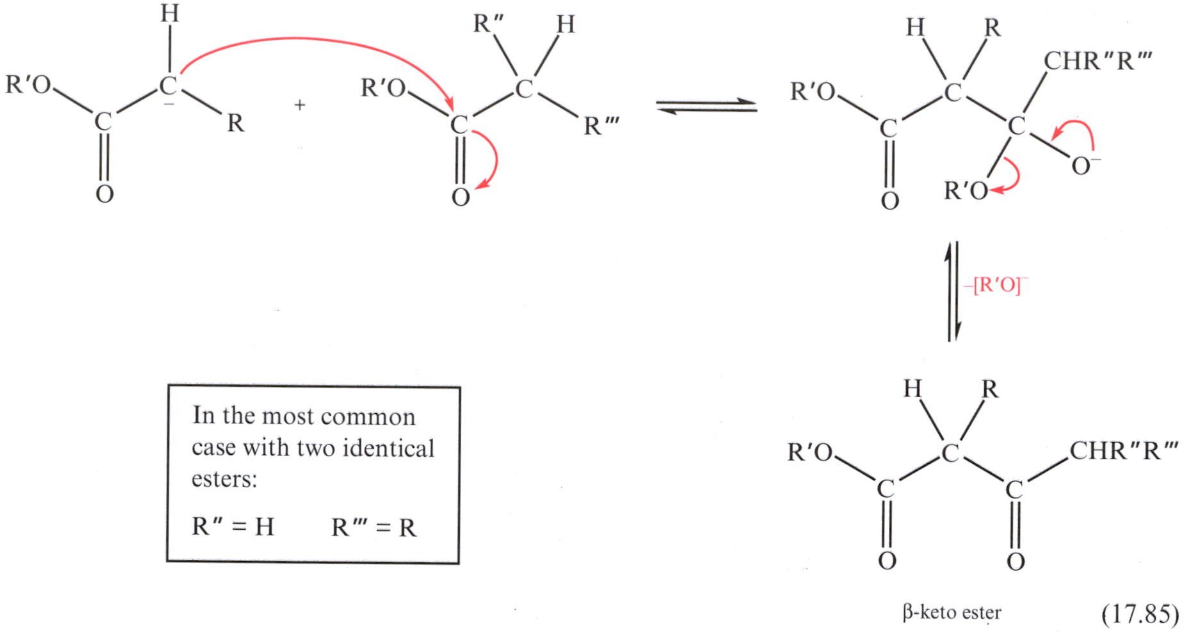

$-[R'O]^-$

In the most common case with two identical esters:

R″ = H   R‴ = R

β-keto ester (17.85)

The β-keto ester in equation 17.85 possesses a weakly acidic α-hydrogen atom and in the presence of alkoxide ion, deprotonation occurs to give the corresponding salt (equation 17.86).

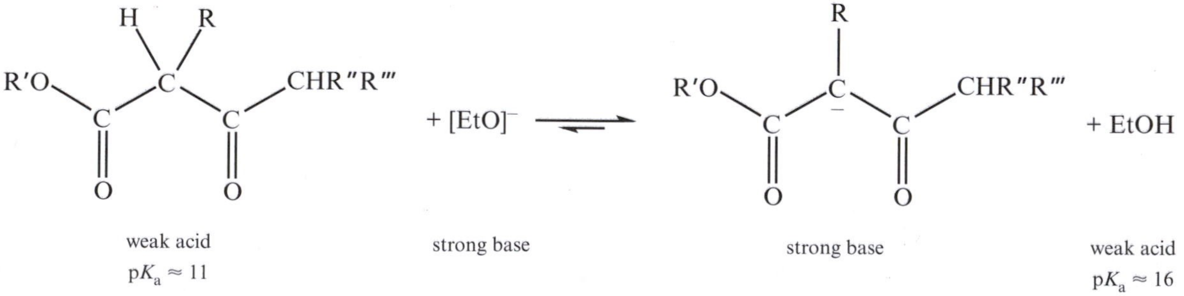

$+ [EtO]^-$ ⇌ $+ EtOH$

weak acid $pK_a \approx 11$ | strong base | strong base | weak acid $pK_a \approx 16$

(17.86)

Equilibria 17.84 and 17.85 lie to the *left-hand side* (not favourable for product formation), but equilibrium 17.86 lies to the *right-hand side* and it is this final step that drives the reaction to a successful end. Why should the products in equilibrium 17.86 be favoured? Although the β-keto ester and alcohol are both weak acids, the presence of the two C=O groups in the β-keto ester stabilizes its conjugate base (Table 17.4). Consequently, (as the pK_a values in Table 17.4 show), the β-keto ester is a stronger acid than the alcohol and equilibrium 17.86 lies to the right-hand side. The presence of two α-hydrogen atoms in the original ester RCH_2CO_2R' is critical to the success of the Claisen condensation. The neutral β-keto ester can be obtained by isolating a salt of the conjugate base in equation 17.86 and adding acid.

Equations 17.87–17.89 provide examples of Claisen condensations. For each, identify the α-hydrogen atoms and write down the mechanisms of the reactions.

$$2CH_3CH_2CO_2CH_3 \xrightarrow{[OEt]^-} [CH_3CH_2C(O)C(Me)CO_2Me]^- \xrightarrow{H^+} CH_3CH_2C(O)CH(Me)CO_2Me \quad (17.87)$$

$$2CH_3CO_2CH_2CH_2CH_3 \xrightarrow{[OEt]^-} [CH_3C(O)CHCO_2CH_2CH_2CH_3]^- \xrightarrow{H^+} CH_3C(O)CH_2CO_2CH_2CH_2CH_3 \quad (17.88)$$

$$2PhCH_2CO_2Me \xrightarrow{[OEt]^-} [PhCH_2C(O)C(Ph)CO_2Me]^- \xrightarrow{H^+} PhCH_2C(O)CH(Ph)CO_2Me \quad (17.89)$$

These examples all involve the condensation of two identical esters. What complications arise if the condensation is between two *different* esters? [*Hint*: look back at the discussion of aldol reactions.]

SUMMARY

This chapter has been concerned with the chemistry of the carbonyl bond in six types of organic compounds: aldehydes, ketones, carboxylic acids, esters, acyl chlorides and primary amides. The susceptibility of the carbon atom of the C=O functional group to undergo nucleophilic attack means that carbon–carbon bond-forming reactions can take place if a carbon nucleophile is present – this is made possible by the acidity of the α-hydrogen atoms in some carbonyl compounds.

Do you know what the following terms mean?

- aldehyde
- ketone
- carboxylic acid
- ester
- acyl chloride
- amide
- keto–enol tautomerism
- enolate ion
- condensation reaction
- fatty acid
- α-hydrogen atom
- haloform reaction
- reductive amination
- cyanohydrin
- acetal
- hemiacetal
- β-hydroxy aldehyde
- β-hydroxy ketone
- α,β-unsaturated aldehyde
- α,β-unsaturated ketone
- β-keto ester
- aldol reaction
- Claisen condensation

You should now be able:

- to draw the functional groups of an aldehyde, ketone, carboxylic acid, ester, acyl chloride and primary amide
- to name simple aldehydes, ketones, diones, carboxylic acids, esters, acyl chlorides and primary amides
- to recognize the trivial names of and draw structures for formic acid, formaldehyde, formamide, acetic acid, acetone, acetaldehyde, acetyl chloride, acetylacetone, acetylacetonate, acetamide, oxalic acid, phthalic acid
- to describe the bonding in the C=O functional group both in terms of valence bond and molecular orbital theories
- to discuss the role of hydrogen bonding in the association of carboxylic acids and amides
- to recognize IR spectroscopic absorptions that are characteristic of the O–H stretch in carboxylic acids and the C=O stretch in carbonyl compounds
- to know the typical values for the ^{13}C NMR spectroscopic chemical shifts of the sp^2 carbon atoms in carbonyl compounds (review Table 9.11)
- to know the characteristic ^{1}H NMR spectroscopic chemical shifts for protons in the functional groups of aldehydes, amides and carboxylic acids (review Table 9.12)
- to know the sense of the dipole moment in a C=O bond
- to understand what effect the polarity of the C=O bond has on the reactivity of carbonyl compounds
- to give methods of preparing aliphatic ketones and aldehydes starting from alcohols, *gem*-dihalides, acyl chlorides (aldehyde synthesis)
- to give methods of preparing aromatic ketones and aldehydes
- to give methods of preparing carboxylic acids from alcohols and alkylbenzene compounds
- to give methods of preparing carboxylic acids by the use of nitriles or Grignard reagents
- to give methods of preparing esters from carboxylic acids or acyl chlorides
- to describe saponification reactions and their industrial significance
- to give methods of preparing amides and acyl chlorides
- to discuss the acidic properties of different types of carbonyl compounds, and compare the relative strengths of these acids
- to give examples of the preparation of carbanions
- to discuss the synthetic utility of carbanions
- to give examples of the base- and acid-catalysed halogenations of aldehydes and ketones, and describe the mechanisms of these reactions
- to give examples of the alkylation of ketones and esters, and describe the mechanisms of these reactions
- to discuss in general terms the susceptibility of the carbon atom of the C=O functional group towards nucleophilic attack

- to give examples of ester and amide hydrolysis, and describe the mechanisms of these reactions
- to show how aldehydes and ketones may be converted to cyanohydrins
- to discuss the synthetic utility of Grignard reagents in the conversion of aldehydes and ketones to alcohols
- to describe how aldehydes and ketones may be converted to amines, emphasizing the particular classes of amines that are formed
- to give examples of the reactions of acyl chlorides and esters with NH_3
- to give examples of aldol reactions, including the mechanisms of these reactions, and discuss the synthetic limitations of this type of reaction
- to give examples of the Claisen condensation and outline the mechanism of this reaction

Using your overall knowledge of organic chemistry:

- you should be able to design a range of multi-step syntheses, and additional practice is provided in the accompanying workbook

PROBLEMS

17.1 Draw the structures of the following compounds:

(a) propanoic acid
(b) pentanamide
(c) hexan-3-one
(d) pentanal
(e) benzaldehyde
(f) butane-2,3-dione
(g) methyl octanoate
(h) phenyl benzoate
(i) diphenyl ketone
(j) dibenzyl ketone

17.2 Name the following compounds:

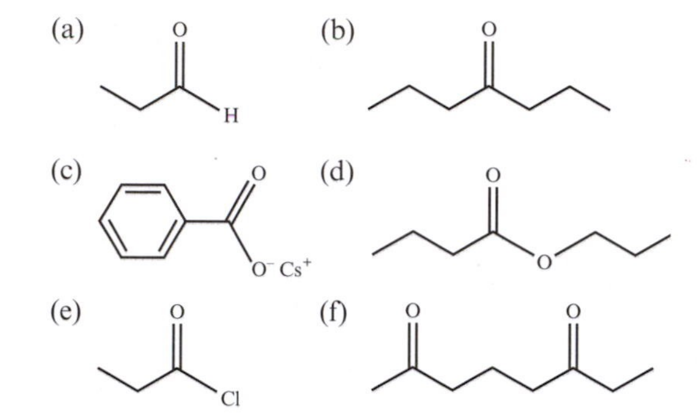

17.3 Figure 17.15 shows three IR spectra labelled **A**, **B** and **C**. The spectra are those of the compounds ethyl acetate, benzyl acetate and hexanoic acid – which spectrum is which? Assign absorptions which are not in the fingerprint region.

17.4 Figure 17.16 shows three ^{13}C NMR spectra (with ^{13}C–^{1}H spin–spin coupling *absent*) labelled **A**, **B** and **C**. The spectra are those of the following compounds:

I II III

Identify which spectrum belongs to which compound and assign the resonances in each spectrum as far as you are able.

17.5 Discuss the trends in the boiling points listed in Table 17.1 in terms of the relative molecular masses and dipole moments of the compounds, and the types of intermolecular interactions that are possible.

17.6 Using your knowledge of electrophilic substitution reactions, suggest a mechanism for the reaction shown in equation 17.9.

17.7 Arrange the following compounds in the likely order of their pK_a values:

CH_3CO_2H $CH_3C(O)CH_2CH_3$
$CH_3C(O)CH_2C(O)CH_3$ CF_3CO_2H
$CF_3C(O)CH_2C(O)CF_3$ CH_3CH_2OH

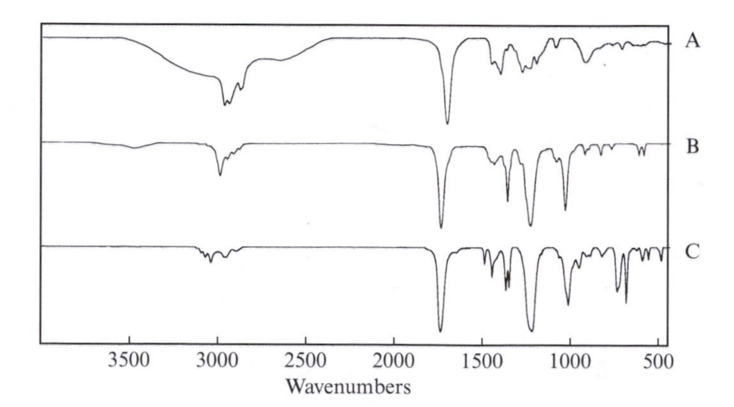

17.15 For problem 17.3.

17.16 For problem 17.4.

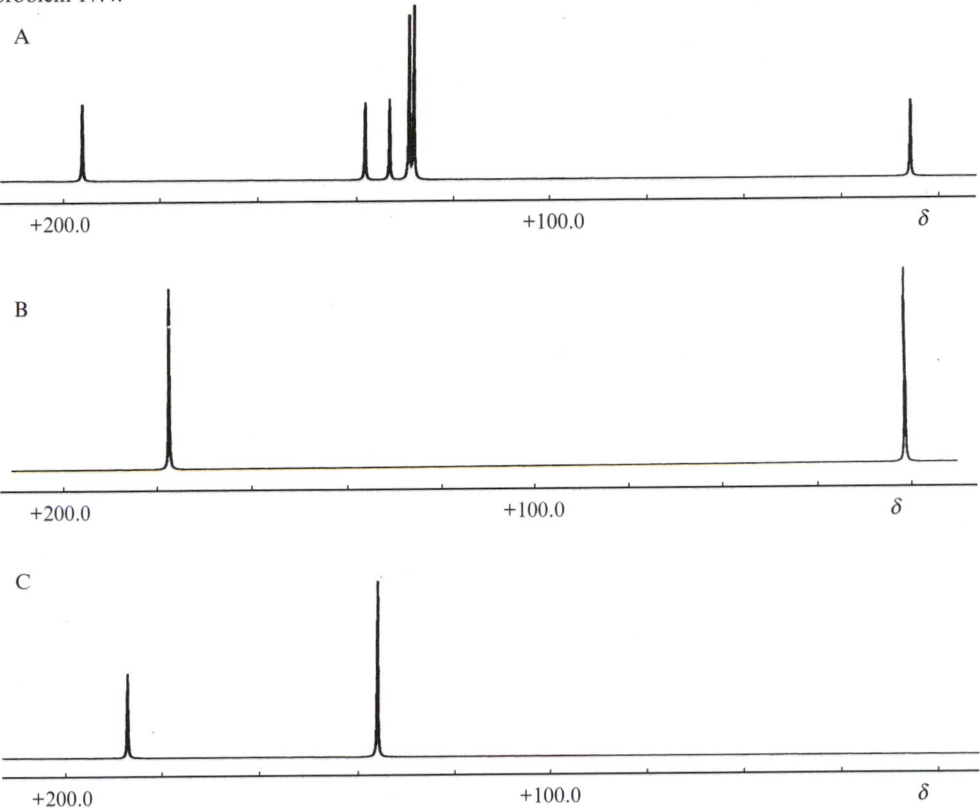

17.8 Suggest likely products for the following reactions:

(a) $PhMe \xrightarrow{\text{hot alkaline } KMnO_4}$

(b) $CH_3CH_2CO_2H + CH_3CH_2OH \xrightarrow{H+}$

(c) $CH_3CH_2CH_2C(O)Cl + CH_3CH_2OH \xrightarrow{H+}$

(d) $PhCH_2CO_2H + CsOH \rightarrow$

(e) $PhC(O)CH_2C(O)Ph + K[Me_3CO] \rightarrow$

(f) $CH_3CH_2CH_2C(O)Cl + NH_3 \rightarrow$

(g) $PhC(O)Cl + H_2O \rightarrow$

(h) $CH_3C(O)CH_3 + I_2 \xrightarrow{H+}$

(i) $CH_3CH_2OH \xrightarrow{\text{acidified } K_2CrO_7}$

(j) $Me_3COH \xrightarrow{\text{acidified } K_2Cr_2O_7}$

(k) $CH_3CH_2Cl \xrightarrow[\text{2. } CH_3CH_2C(O)CH_3,\ H_2O]{\text{1. Mg, anhydrous THF}}$

17.9 How might you achieve the following conversions? More than one step may be required for each.

(a) $CH_3CH_2CH_2Br \rightarrow CH_3CH_2CH_2CO_2H$

(b) $CH_3CO_2CH_3 \rightarrow CH_3CO_2H$

(c) $CH_3CH_2OH \rightarrow CH_3C(O)Cl$

(d) $CH_3CHCl_2 \rightarrow CH_3CO_2H$

(e) $C_6H_6 \rightarrow PhC(O)CH{=}CMePh$

(f) $CH_3CH_2CH_2CH_2OH \rightarrow CH_3CH_2CH_2CO_2Ph$

17.10 Give two synthetic methods by which carbon chains can be lengthened, illustrating your answer with reference to the conversion of 1-chloropentane to hexanoic acid.

17.11 Suggest a mechanism by which the haloform reaction (equation 17.44) may occur. Why are the products of this reaction shown as anions rather than their conjugate acids?

17.12 Write an account of the role of carbonyl compounds in carbon–carbon bond-forming reactions.

Appendix 1 Mathematical symbols

Symbol	*Meaning*
$>$	is greater than
$>>$	is much greater than
$<$	is less than
$<<$	is much less than
$\geq$	is greater than or equal to
$\leq$	is less than or equal to
$\approx$	is approximately equal to
$=$	is equal to
$\neq$	is not equal to
$\rightleftharpoons$	equilibrium
$\propto$	is proportional to
$\times$	multiplied by
∞	infinity
$\pm$	plus or minus
$\sqrt{\ }$	square root of
$\sqrt[3]{\ }$	cube root of
$\lvert x \rvert$	modulus of x
Σ	summation of
Δ	change in (for example, ΔH is 'change in enthalpy'); see also Appendix 3
$\angle$	angle
log	logarithm to base 10 ($\log_{10}$)
ln	natural logarithm, i.e. logarithm to base e ($\log_e$)
$\int$	integral of
$\frac{\mathrm{d}}{\mathrm{d}x}$	differential with respect to x
$\frac{\partial}{\partial x}$	partial differential with respect to x

Appendix 2 Greek letters with pronunciations

Upper case letter	*Lower case*	*Pronunciation*
Α	α	alpha
Β	β	beta
Γ	γ	gamma
Δ	δ	delta
Ε	ε	epsilon
Ζ	ζ	zeta
Η	η	eta
Θ	θ	theta
Ι	ι	iota
Κ	κ	kappa
Λ	λ	lambda
Μ	μ	mu
Ν	ν	nu
Ξ	ξ	xi
Ο	ο	omicron
Π	π	pi
Ρ	ρ	rho
Σ	σ	sigma
Τ	τ	tau
Υ	υ	upsilon
Φ	φ	phi
Χ	χ	chi
Ψ	ψ	psi
Ω	ω	omega

Appendix 3 Abbreviations and symbols for quantities and units

For ligand abbreviations, see Table 16.2. Where a symbol is used to mean more than one thing, the context of its use should make the meaning clear

Symbol	Meaning
A	absorbance
A	frequency factor (in Arrhenius equation)
A	Madelung constant
A	mass number (of an atom)
A_r	relative atomic mass
$A(\theta,\phi)$	angular wavefunction
Å	ångstrom (non-SI unit of length, used for bond distances)
AO	atomic orbital
aq	aqueous
Ar	general aryl group
atm	atmosphere (non-SI unit of pressure)
ax	axial
$\mathbf{B}$	magnetic field strength
bar	bar (unit of pressure)
BM	bohr magneton (unit of magnetic moment)
bp	boiling point
Bu	butyl
c	coefficient (in wavefunctions)
c	concentration (of solution)
c	speed of light
C	coulomb (unit of charge)
cm	centimetre (unit of length)
cm^3	cubic centimetre (unit of volume)
cm^{-1}	reciprocal centimetre (wavenumber)
conc	concentrated
cr	crystal
C_P	heat capacity (at constant pressure)
C_V	heat capacity (at constant volume)
D	bond dissociation enthalpy
$\bar{D}$	average bond dissociation enthalpy
D	debye (non-SI unit of electric dipole moment)
d	distance

Symbol	Meaning
dm^3	cubic decimetre (unit of volume)
E	energy
e	charge on the electron
e^-	electron
E1	unimolecular elimination
E2	bimolecular elimination
EA	electron affinity
$E_{activation}$	activation energy
E_{cell}	electrochemical cell potential
E^o	standard reduction potential (standard electrode potential)
EPR	electron paramagnetic resonance
eq	equatorial
ESR	electron spin resonance
Et	ethyl
eV	electron volt
F	Faraday constant
FID	free induction decay
FT	Fourier transform
G	free energy
g	gas
g	gram (unit of mass)
H	enthalpy
$\mathcal{H}$	Hamiltonian operator
h	Planck constant
$h\nu$	high-frequency radiation (for a photolysis reaction)
HOMO	highest occupied molecular orbital
Hz	hertz (unit of frequency)
I	current
I	nuclear spin quantum number
I	intensity of transmitted radiation
I_0	intensity of incident radiation
IE	ionization energy
IR	infrared
^{i}Pr	isopropyl
J	joule (unit of energy)
J	spin–spin coupling constant

k	force constant
k	rate constant
K	kelvin (unit of temperature)
K	equilibrium constant
K_a	acid dissociation constant
K_b	base dissociation constant
K_c	equilibrium constant for a reaction in solution
K_p	equilibrium constant for a gaseous reaction
K_{sp}	solubility product constant
K_w	self-ionization constant of water
kg	kilogram (unit of mass)
kJ	kilojoule (unit of energy)
kPa	kilopascal (unit of pressure)
L	Avogadro constant or number
l	liquid
l	length
l	orbital quantum number
ℓ	path length
LCAO	linear combination of atomic orbitals
LUMO	lowest unoccupied molecular orbital
M	molarity
m	mass
m	metre (unit of length)
m^3	cubic metre (unit of volume)
m_l	magnetic quantum number
m_s	magnetic spin quantum number
M_r	relative molecular mass
Me	methyl
MO	molecular orbital
mol	mole (unit of quantity)
mp	melting point
N	newton (unit of force)
N	normalization factor
n	Born exponent
n	number of (e.g. moles)
n	principal quantum number
nm	nanometre (unit of length)
NMR	nuclear magnetic resonance
P	pressure
Pa	pascal (unit of pressure)
PES	photoelectron spectroscopy
Ph	phenyl
pm	picometre (unit of length)
Pr	propyl
P_X	partial pressure of component X
q	charge
q	heat
R	general alkyl group
R	molar gas constant
R	resistance
r	internuclear distance
r	radial distance
r	radius
$R(r)$	radial wavefunction
r_{cov}	covalent radius
r_{ion}	ionic radius
r_v	van der Waals radius
r_{metal}	metallic radius
RDS	rate-determining step
RF	radiofrequency
S	entropy
S	overlap integral
S	total spin quantum number
s	second (unit of time)
s	solid
s	spin quantum number
S_N1	substitution nucleophilic unimolecular
S_N2	substitution nucleophilic bimolecular
soln	solution
solv	solvated
T	temperature
T	transmittance
t	time
^{t}Bu	*tert*-butyl
THF	tetrahydrofuran
TMS	tetramethylsilane
TS	transition state
U	internal energy
u	atomic mass unit
UV	ultraviolet
UV–VIS	ultraviolet–visible
V	potential difference
V	volume
V	volt (unit of potential difference)
v	vapour
v	velocity

VB	valence bond
VIS	visible
VSEPR	valence-shell electron-pair repulsion
w	work done
[X]	concentration of X
z	number of moles of electrons transferred in an electrochemical cell
Z	atomic number
Z_{eff}	effective nuclear charge
$\|z_-\|$	modulus of the negative charge
$\|z_+\|$	modulus of the positive charge
α-carbon	C atom adjacent to the one in question
β	stability constant
β-carbon	C atom two along from the one in question
δ	chemical shift
δ^-	partial negative charge
δ^+	partial positive charge
Δ	change in
Δ	heat (in a pyrolysis reaction)
Δ_{oct}	octahedral crystal field splitting energy
$\Delta_a H$	enthalpy change of atomization
$\Delta_c H$	enthalpy change of combustion
$\Delta_f H$	enthalpy change of formation
$\Delta_{fus} H$	enthalpy change of fusion
$\Delta_{hyd} H$	enthalpy change of hydration
$\Delta_{lattice} H$	enthalpy change for the formation of an ionic lattice
$\Delta_r H$	enthalpy change of reaction
$\Delta_{sol} H$	enthalpy change of solution
$\Delta_{solv} H$	enthalpy change of solvation
$\Delta_{vap} H$	enthalpy change of vaporization
$\Delta_f G$	free energy change of formation
$\Delta_r G$	free energy change of reaction
ε	molar extinction (or absorption) coefficient
ε_{max}	molar extinction coefficient corresponding to an absorption maximum (in an electronic spectrum)
ε_0	permittivity of a vacuum
θ	angle (α is also used)
λ	wavelength
λ_{max}	wavelength corresponding to an absorption maximum (in an electronic spectrum)
μ	electric dipole moment
μ	reduced mass
μ(spin-only)	spin-only magnetic moment
μ_{eff}	effective magnetic moment
μ-	bridging ligand
ρ	density
ν	frequency
$\bar{\nu}$	wavenumber
χ	magnetic susceptibility
χ	electronegativity
χ^{AR}	Allred-Rochow electronegativity
χ^{M}	Mulliken electronegativity
χ^{P}	Pauling electronegativity
ψ	wavefunction
Ω	ohm (unit of resistance)
2c–2e	2-centre 2-electron
3c–2e	3-centre 2-electron
^{o}	standard state
‡	activated complex; transition state
°	degree

Appendix 4 The electromagnetic spectrum

The frequency of electromagnetic radiation is related to its wavelength by the equation:

$$\text{Wavelength } (\lambda) = \frac{\text{Speed of light } (c)}{\text{Frequency } (\nu)}$$

where $c = 3.0 \times 10^8$ m s^{-1}.

$$\text{Wavenumber } (\bar{\nu}) = \frac{1}{\text{Wavelength}}$$

with units in cm^{-1} (pronounced 'reciprocal centimetre').

$$\text{Energy } (E) = \text{Planck constant } (h) \times \text{Frequency } (\nu)$$

where $h = 6.626 \times 10^{-34}$ J s.

The energy given in the last column is measured per mole of photons.

Frequency ν / *Hz*	*Wavelength* λ / *m*	*Wavenumber* $\bar{\nu}$ / *cm*$^{-1}$	*Type of radiation*		*Energy E* / *kJ mol*$^{-1}$
	10^{-13}	10^{11}			10^{9}
10^{21}					
	10^{-12}	10^{10}	γ-ray		10^{8}
10^{20}					
	10^{-11}	10^{9}			10^{7}
10^{19}					
	10^{-10}	10^{8}	X-ray		10^{6}
10^{18}					
	10^{-9}	10^{7}			10^{5}
10^{17}					
	10^{-8}	10^{6}	Vacuum ultraviolet		10^{4}
10^{16}				Violet ≈ 400 nm	
	10^{-7}	10^{5}	Ultraviolet	Blue	10^{3}
10^{15}			Visible	Green	
	10^{-6}	10^{4}		Yellow	10^{2}
10^{14}			Near infrared	Orange	
	10^{-5}	10^{3}		Red ≈ 750 nm	10^{1}
10^{13}					
	10^{-4}	10^{2}	Far infrared		$10^{0} = 1$
10^{12}					
	10^{-3}	10^{1}			10^{-1}
10^{11}					
	10^{-2}	$10^{0} = 1$	Microwave		10^{-2}
10^{10}					
	10^{-1}	10^{-1}			10^{-3}
10^{9}					
	$10^{0} = 1$	10^{-2}			10^{-4}
10^{8}					
	10^{1}	10^{-3}			10^{-5}
10^{7}					
	10^{2}	10^{-4}			10^{-6}
10^{6}					
	10^{3}	10^{-5}	Radiowave		10^{-7}
10^{5}					
	10^{4}	10^{-6}			10^{-8}
10^{4}					
	10^{5}	10^{-7}			10^{-9}
10^{3}					

Appendix 5 Naturally occurring isotopes and their abundances

Data from *WebElements* version 2.0b5 (1996) by Mark Winter. Further information on radioactive nuclides can be found on the World Wide Web at:
http://www.shef.ac.uk/~chem/web-elements/nucl/Pm.html

Element	*Symbol*	*Atomic number, Z*	*Mass number of isotope (% abundance)*
Actinium	Ac	89	Artificial isotopes only; mass number range 224–229
Aluminium	Al	13	27(100)
Americium	Am	95	Artificial isotopes only; mass number range 237–245
Antimony	Sb	51	121(57.3), 123(42.7)
Argon	Ar	18	36(0.34), 38(0.06), 40(99.6)
Arsenic	As	33	75(100)
Astatine	At	85	Artificial isotopes only; mass number range 205–211
Barium	Ba	56	130(0.11), 132(0.10), 134(2.42), 135(6.59), 136(7.85), 137(11.23), 138(71.70)
Berkelium	Bk	97	Artificial isotopes only; mass number range 243–250
Beryllium	Be	4	9(100)
Bismuth	Bi	83	209(100)
Boron	B	5	10(19.9), 11(80.1)
Bromine	Br	35	79(50.69), 81(49.31)
Cadmium	Cd	48	106(1.25), 108(0.89), 110(12.49), 111(12.80), 112(24.13), 113(12.22), 114(28.73), 116(7.49)
Caesium	Cs	55	133(100)
Calcium	Ca	20	40(96.94), 42(0.65), 43(0.13), 44(2.09), 48(0.19)
Californium	Cf	98	Artificial isotopes only; mass number range 246–255
Carbon	C	6	12(98.9), 13(1.1)
Cerium	Ce	58	136(0.19), 138(0.25), 140(88.48), 142(11.08)
Chlorine	Cl	17	35(75.77), 37(24.23)
Chromium	Cr	24	50(4.345), 52(83.79), 53(9.50), 54(2.365)
Cobalt	Co	27	59(100)
Copper	Cu	29	63(69.2), 65(30.8)
Curium	Cm	96	Artificial isotopes only; mass number range 240–250
Dysprosium	Dy	66	156(0.06), 158(0.10), 160(2.34), 161(18.9), 162(25.5), 163(24.9), 164(28.2)
Einsteinium	Es	99	Artificial isotopes only; mass number range 249–256
Erbium	Er	68	162(0.14), 164(1.61),166(33.6), 167(22.95), 168(26.8), 170(14.9)
Europium	Eu	63	151(47.8), 153(52.2)
Fermium	Fm	100	Artificial isotopes only; mass number range 251–257
Fluorine	F	9	19(100)

Element	*Symbol*	*Atomic number, Z*	*Mass number of isotope (% abundance)*
Francium	Fr	87	Artificial isotopes only; mass number range 210–227
Gadolinium	Gd	64	152(0.20), 154(2.18), 155(14.80), 156(20.47), 157(15.65), 158(24.84), 160(21.86)
Gallium	Ga	31	69(60.1), 71(39.9)
Germanium	Ge	32	70(20.5), 72(27.4), 73(7.8), 74(36.5), 76(7.8)
Gold	Au	79	197(100)
Hafnium	Hf	72	174(0.16), 176(5.20), 177(18.61), 178(27.30), 179(13.63), 180(35.10)
Helium	He	2	3(<0.001), 4(>99.999)
Holmium	Ho	67	165(100)
Hydrogen	H	1	1(99.985), 2(0.015)
Indium	In	49	113(4.3), 115(95.7)
Iodine	I	53	127(100)
Iridium	Ir	77	191(37.3), 193(62.7)
Iron	Fe	26	54(5.8), 56(91.7), 57(2.2), 58(0.3)
Krypton	Kr	36	78(0.35), 80(2.25), 82(11.6), 83(11.5), 84(57.0), 86(17.3)
Lanthanum	La	57	138(0.09), 139(99.91)
Lawrencium	Lr	103	Artificial isotopes only; mass number range 253–262
Lead	Pb	82	204(1.4), 206(24.1), 207(22.1), 208(52.4)
Lithium	Li	3	6(7.5), 7(92.5)
Lutetium	Lu	71	175(97.41), 176(2.59)
Magnesium	Mg	12	24(78.99), 25(10.00), 26(11.01)
Manganese	Mn	25	55(100)
Mendelevium	Md	101	Artificial isotopes only; mass number range 247–260
Mercury	Hg	80	196(0.14), 198(10.02), 199(16.84), 200(23.13), 201(13.22), 202(29.80), 204(6.85)
Molybdenum	Mo	42	92(14.84), 94(9.25), 95(15.92), 96(16.68), 97(9.55), 98(24.13), 100(9.63)
Neodymium	Nd	60	142(27.13), 143(12.18), 144(23.80), 145(8.30), 146(17.19), 148(5.76), 150(5.64)
Neon	Ne	10	20(90.48), 21(0.27), 22(9.25)
Neptunium	Np	93	Artificial isotopes only; mass number range 234–240
Nickel	Ni	28	58(68.27), 60(26.10), 61(1.13), 62(3.59), 64(0.91)
Niobium	Nb	41	93(100)
Nitrogen	N	7	14(99.63), 15(0.37)
Nobelium	No	102	Artificial isotopes only; mass number range 250–262
Osmium	Os	76	184(0.02), 186(1.58), 187(1.6), 188(13.3), 189(16.1), 190(26.4), 192(41.0)
Oxygen	O	8	16(99.76), 17(0.04), 18(0.20)
Palladium	Pd	46	102(1.02), 104(11.14), 105(22.33), 106(27.33), 108(26.46), 110(11.72)
Phosphorus	P	15	31(100)
Platinum	Pt	78	190(0.01), 192(0.79), 194(32.9), 195(33.8), 196(25.3), 198(7.2)

Element	Symbol	Atomic number, Z	Mass number of isotope (% abundance)
Plutonium	Pu	94	Artificial isotopes only; mass number range 234–246
Polonium	Po	84	Artificial isotopes only; mass number range 204–210
Potassium	K	19	39(93.26), 40(0.01), 41(6.73)
Praseodymium	Pr	59	141(100)
Promethium	Pm	61	Artificial isotopes only; mass number range 141–151
Protactinium	Pa	91	Artificial isotopes only; mass number range 228–234
Radium	Ra	88	Artificial isotopes only; mass number range 223–230
Radon	Rn	86	Artificial isotopes only; mass number range 208–224
Rhenium	Re	75	185(37.40), 187(62.60)
Rhodium	Rh	45	103(100)
Rubidium	Rb	37	85(72.16), 87(27.84)
Ruthenium	Ru	44	96(5.52), 98(1.88), 99(12.7), 100(12.6), 101(17.0), 102(31.6), 104(18.7)
Samarium	Sm	62	144(3.1), 147(15.0), 148(11.3), 149(13.8), 150(7.4), 152(26.7), 154(22.7)
Scandium	Sc	21	45(100)
Selenium	Se	34	74(0.9), 76(9.0), 77(7.6), 78(23.6), 80(49.7), 82(9.2)
Silicon	Si	14	28(92.23), 29(4.67), 30(3.10)
Silver	Ag	47	107(51.84), 109(48.16)
Sodium	Na	11	23(100)
Strontium	Sr	38	84(0.56), 86(9.86), 87(7.00), 88(82.58)
Sulfur	S	16	32(95.02), 33(0.75), 34(4.21), 36(0.02)
Tantalum	Ta	73	180(0.01), 181(99.99)
Technetium	Tc	43	Artificial isotopes only; mass number range 95–99
Tellurium	Te	52	120(0.09), 122(2.60), 123(0.91), 124(4.82), 125(7.14), 126(18.95), 128(31.69), 130(33.80)
Terbium	Tb	65	159(100)
Thallium	Tl	81	203(29.52), 205(70.48)
Thorium	Th	90	232(100)
Thulium	Tm	69	169(100)
Tin	Sn	50	112(0.97), 114(0.65), 115(0.36), 116(14.53), 117(7.68), 118(24.22), 119(8.58), 120(32.59), 122(4.63), 124(5.79)
Titanium	Ti	22	46(8.0), 47(7.3), 48(73.8), 49(5.5), 50(5.4)
Tungsten	W	74	180(0.13), 182(26.3), 183(14.3), 184(30.67), 186(28.6)
Uranium	U	92	234(0.005), 235(0.72), 236(99.275)
Vanadium	V	23	50(0.25), 51(99.75)
Xenon	Xe	54	124(0.10), 126(0.09), 128(1.91), 129(26.4), 130(4.1), 131(21.2), 132(26.9), 134(10.4), 136(8.9)
Ytterbium	Yb	70	168(0.13), 170(3.05), 171(14.3), 172(21.9), 173(16.12), 174(31.8), 176(12.7)
Yttrium	Y	39	89(100)
Zinc	Zn	30	64(48.6), 66(27.9), 67(4.1), 68(18.8), 70(0.6)
Zirconium	Zr	40	90(51.45), 91(11.22), 92(17.15), 94(17.38), 96(2.8)

Appendix 6 Van der Waals, metallic, covalent and ionic radii

The ionic radius varies with the charge and coordination number of the ion.

	Element	Van der Waals radius, r_v/pm	Metallic radius, r_{metal}/pm	Covalent radius, r_{cov}/pm	Ionic radius: Ionic radius, r_{ion}/pm	Ionic radius: Charge on ion	Ionic radius: Coordination number of the ion
Hydrogen	H	120		37[a]			
Group 1	Li		152		76	1+	6
	Na		186		102	1+	6
	K		227		138	1+	6
	Rb		248		149	1+	6
	Cs		263		170	1+	6
Group 2	Be		112		27	2+	4
	Mg		160		72	2+	6
	Ca		197		100	2+	6
	Sr		215		126	2+	8
	Ba		221		142	2+	8
Group 13	B	208		88			
	Al		143	130	54	3+	6
	Ga		122	122	62	3+	6
	In		163	150	80	3+	6
	Tl		170	155	89	3+	6
Group 14	C	185		77			
	Si	210		118			
	Ge		123	122	53	4+	6
	Sn		141	140	74	4+	6
	Pb		175	154	119	2+	6
					65	4+	4
					78	4+	6
Group 15	N	154		75	171	3–	6
	P	190		110			
	As	200		122	58	3+	6
					46	5+	6
	Sb	220		143	76	3+	6
					60	5+	6
	Bi	240		152	103	3+	6
					76	5+	6

[a]Sometimes it is more appropriate to use a value of 30 pm in organic compounds.

	Element	Van der Waals radius, r_v/pm	Metallic radius, r_{metal}/pm	Covalent radius, r_{cov}/pm	Ionic radius		
					Ionic radius, r_{ion}/pm	Charge on ion	Coordination number of the ion
Group 16	O	140		73	140	2–	6
	S	185		103	184	2–	6
	Se	200		117	198	2–	6
	Te	220		135	211	2–	6
Group 17	F	135		71	133	1–	6
	Cl	180		99	181	1–	6
	Br	195		114	196	1–	6
	I	215		133	220	1–	6
Group 18	He	99					
	Ne	160					
	Ar	191					
	Kr	197					
	Xe	214					
First row *d*-block elements	Sc		161		75	3+	6
	Ti		145		67	3+	6
					61	4+	6
	V		132		79	2+	6
					64	3+	6
					58	4+	6
					53	4+	5
					54	5+	6
					46	5+	5
	Cr		137		73	2+	6
					62	3+	6
	Mn		137		67	2+	6
					58	3+	6
					39	4+	4
					53	4+	6
	Fe		124		61	2+	6
					55	3+	6
	Co		125		65	2+	6
					55	3+	6
	Ni		125		44	2+	4 (square planar)
					69	2+	6
	Cu		128		46	1+	2
					60	1+	4
					57	2+	4 (square planar)
					73	2+	6
	Zn		133		60	2+	4
					74	2+	6

Appendix 7 Pauling electronegativity values (χ^P) for selected elements of the periodic table

Values are dependent on oxidation state.

Group 1	Group 2		Group 13	Group 14	Group 15	Group 16	Group 17
H 2.2							
Li 1.0	Be 1.6		B 2.0	C 2.6	N 3.0	O 3.4	F 4.0
Na 0.9	Mg 1.3		Al(III) 1.6	Si 1.9	P 2.2	S 2.6	Cl 3.2
K 0.8	Ca 1.0		Ga(III) 1.8	Ge(IV) 2.0	As(III) 2.2	Se 2.6	Br 3.0
Rb 0.8	Sr 0.9	(*d*-block elements)	In(III) 1.8	Sn(II) 1.8 Sn(IV) 2.0	Sb 2.1	Te 2.1	I 2.7
Cs 0.8	Ba 0.9		Tl(I) 1.6 Tl(III) 2.0	Pb(II) 1.9 Pb(IV) 2.3	Bi 2.0	Po 2.0	At 2.2

Appendix 8 Ground state electronic configurations of the elements (excluding the lanthanoids and actinoids) and ionization energies for the first five ionizations

IE(*n*) in kJ mol^{-1} for the processes:

IE(1) $M(g) \rightarrow M^+(g)$
IE(2) $M^+(g) \rightarrow M^{2+}(g)$
IE(3) $M^{2+}(g) \rightarrow M^{3+}(g)$
IE(4) $M^{3+}(g) \rightarrow M^{4+}(g)$
IE(5) $M^{4+}(g) \rightarrow M^{5+}(g)$

Element	*Ground state electronic configuration*	*IE*(1)	*IE*(2)	*IE*(3)	*IE*(4)	*IE*(5)
H	$1s^1$	1312				
He	$1s^2$ = [He]	2373	5249			
Li	[He]$2s^1$	521	7295	11820		
Be	[He]$2s^2$	897	1756	14850	21006	
B	[He]$2s^22p^1$	801	2432	3657	25030	32826
C	[He]$2s^22p^2$	1090	2354	4622	6224	37834
N	[He]$2s^22p^3$	1399	2856	4574	7478	9446
O	[He]$2s^22p^4$	1312	3387	5297	7468	10990
F	[He]$2s^22p^5$	1679	3377	6050	8404	11019
Ne	[He]$2s^22p^6$ = [Ne]	2084	3956	6117	9369	12177
Na	[Ne]$3s^1$	492	4564	6909	9543	13354
Mg	[Ne]$3s^2$	733	1447	7729	10546	13634
Al	[Ne]$3s^23p^1$	579	1814	2740	11579	14840
Si	[Ne]$3s^23p^2$	791	1573	3232	4352	16095
P	[Ne]$3s^23p^3$	1013	1911	2914	4960	6272
S	[Ne]$3s^23p^4$	1003	2248	3358	4554	7005
Cl	[Ne]$3s^23p^5$	1254	2296	3821	5162	6542
Ar	[Ne]$3s^23p^6$ = [Ar]	1520	2663	3927	5770	7237
K	[Ar]$4s^1$	415	3049	4419	5876	7980
Ca	[Ar]$4s^2$	589	1148	4911	6494	8153
Sc	[Ar]$4s^23d^1$	637	1235	2393	7092	8843
Ti	[Ar]$4s^23d^2$	656	1312	2653	4178	9581
V	[Ar]$4s^23d^3$	646	1418	2827	4506	6301
Cr	[Ar]$4s^13d^5$	656	1592	2992	4747	6706
Mn	[Ar]$4s^23d^5$	714	1505	3252	4940	6986
Fe	[Ar]$4s^23d^6$	762	1563	2962	5288	7237

Element	*Ground state electronic configuration*	*IE(1)*	*IE(2)*	*IE(3)*	*IE(4)*	*IE(5)*
Co	$[Ar]4s^23d^7$	762	1650	3232	4950	7671
Ni	$[Ar]4s^23d^8$	733	1756	3396	5297	7343
Cu	$[Ar]4s^13d^{10}$	743	1959	3551	5539	7700
Zn	$[Ar]4s^23d^{10}$	907	1737	3831	5732	7970
Ga	$[Ar]4s^23d^{10}4p^1$	579	1978	2962	6175	
Ge	$[Ar]4s^23d^{10}4p^2$	762	1534	3300	4410	9022
As	$[Ar]4s^23d^{10}4p^3$	946	1795	2740	4834	6040
Se	$[Ar]4s^23d^{10}4p^4$	946	2046	2972	4139	6590
Br	$[Ar]4s^23d^{10}4p^5$	1139	2103	3474	4564	5760
Kr	$[Ar]4s^23d^{10}4p^6 = [Kr]$	1351	2354	3571	5066	6243
Rb	$[Kr]5s^1$	405	2634	3860	5075	6851
Sr	$[Kr]5s^2$	550	1061	4139	5500	6909
Y	$[Kr]5s^24d^1$	598	1177	1978	5799	7430
Zr	$[Kr]5s^24d^2$	637	1264	2219	3310	7748
Nb	$[Kr]5s^14d^4$	656	1380	2412	3696	4882
Mo	$[Kr]5s^14d^5$	685	1563	2625	4477	5259
Tc	$[Kr]5s^24d^5$	704	1476	2846		
Ru	$[Kr]5s^14d^7$	714	1621	2750		
Rh	$[Kr]5s^14d^8$	724	1746	3001		
Pd	$[Kr]5s^04d^{10}$	801	1872	3175		
Ag	$[Kr]5s^14d^{10}$	733	2075	3358		
Cd	$[Kr]5s^24d^{10}$	868	1631	3618		
In	$[Kr]5s^24d^{10}5p^1$	560	1824	2702	5210	
Sn	$[Kr]5s^24d^{10}5p^2$	704	1409	2943	3927	6976
Sb	$[Kr]5s^24d^{10}5p^3$	830	1592	2441	4265	5403
Te	$[Kr]5s^24d^{10}5p^4$	868	1795	2702	3609	5669
I	$[Kr]5s^24d^{10}5p^5$	1008	1843	3184		
Xe	$[Kr]5s^24d^{10}5p^6 = [Xe]$	1168	2046	3097		
Cs	$[Xe]6s^1$	376	2234			
Ba	$[Xe]6s^2$	502	965			
La	$[Xe]6s^25d^1$	540	1071	1853	4820	5944
Hf	$[Xe]4f^{14}6s^25d^2$	656	1438	2248	3213	
Ta	$[Xe]4f^{14}6s^25d^3$	762	1600			
W	$[Xe]4f^{14}6s^25d^4$	772	1700			
Re	$[Xe]4f^{14}6s^25d^5$	762	1600			
Os	$[Xe]4f^{14}6s^25d^6$	839	1600			
Ir	$[Xe]4f^{14}6s^25d^7$	878				
Pt	$[Xe]4f^{14}6s^15d^9$	868	1795			
Au	$[Xe]4f^{14}6s^15d^{10}$	888	1980			
Hg	$[Xe]4f^{14}6s^25d^{10}$	1003	1814	3300		
Tl	$[Xe]4f^{14}6s^25d^{10}6p^1$	589	1968	2875	4900	
Pb	$[Xe]4f^{14}6s^25d^{10}6p^2$	714	1447	3078	4082	6639
Bi	$[Xe]4f^{14}6s^25d^{10}6p^3$	704	1611	2470	4371	5403
Po	$[Xe]4f^{14}6s^25d^{10}6p^4$	811				
At	$[Xe]4f^{14}6s^25d^{10}6p^5$					
Rn	$[Xe]4f^{14}6s^25d^{10}6p^6 = [Rn]$	1032				

Appendix 9 Electron affinities

Approximate enthalpy changes, $\Delta H(298\ \text{K})$, associated with the gain of one electron to a gaseous atom or anion. The values listed are actually internal *energy* changes $\Delta U(0\ \text{K})$ as explained in Section 6.5. A negative enthalpy (ΔH), but a positive electron affinity (EA), corresponds to an exothermic process.

$$\Delta H(298\ \text{K}) \approx \Delta U(0\ \text{K}) = -EA$$

	Process	$\approx \Delta H / \text{kJ mol}^{-1}$
Hydrogen	$H(g) + e^- \rightarrow H^-(g)$	–73
Group 1	$Li(g) + e^- \rightarrow Li^-(g)$	–60
	$Na(g) + e^- \rightarrow Na^-(g)$	–53
	$K(g) + e^- \rightarrow K^-(g)$	–48
	$Rb(g) + e^- \rightarrow Rb^-(g)$	–47
	$Cs(g) + e^- \rightarrow Cs^-(g)$	–45.5
Group 15	$N(g) + e^- \rightarrow N^-(g)$	≈ 0
	$P(g) + e^- \rightarrow P^-(g)$	–72
	$As(g) + e^- \rightarrow As^-(g)$	–78
	$Sb(g) + e^- \rightarrow Sb^-(g)$	–103
	$Bi(g) + e^- \rightarrow Bi^-(g)$	–91
Group 16	$O(g) + e^- \rightarrow O^-(g)$	–141
	$O^-(g) + e^- \rightarrow O^{2-}(g)$	+798
	$S(g) + e^- \rightarrow S^-(g)$	–201
	$S^-(g) + e^- \rightarrow S^{2-}(g)$	+640
	$Se(g) + e^- \rightarrow Se^-(g)$	–195
	$Te(g) + e^- \rightarrow Te^-(g)$	–190
Group 17	$F(g) + e^- \rightarrow F^-(g)$	–328
	$Cl(g) + e^- \rightarrow Cl^-(g)$	–349
	$Br(g) + e^- \rightarrow Br^-(g)$	–325
	$I(g) + e^- \rightarrow I^-(g)$	–295

Appendix 10 Standard enthalpies of atomization ($\Delta_a H^o$) of the elements at 298 K

Enthalpies are given in kJ mol^{-1} for the process:

$\frac{1}{n}E_n$(standard state) $\rightarrow$ E(g)

Elements (E) are arranged according to their position in the periodic table. The lanthanoids and actinoids are excluded. The noble gases are omitted because they are monatomic at 298 K.

1	2	3	4	5	6	7	8	9	10	11	12	13	14	15	16	17
H 218																
Li 161	Be 324											B 582	C 717	N 473	O 249	F 79
Na 108	Mg 146											Al 330	Si 456	P 315	S 277	Cl 121
K 90	Ca 178	Sc 378	Ti 470	V 514	Cr 397	Mn 283	Fe 418	Co 428	Ni 430	Cu 338	Zn 130	Ga 277	Ge 375	As 302	Se 227	Br 112
Rb 82	Sr 164	Y 423	Zr 609	Nb 721	Mo 658	Tc 677	Ru 651	Rh 556	Pd 377	Ag 285	Cd 112	In 243	Sn 302	Sb 264	Te 197	I 107
Cs 78	Ba 178	La 423	Hf 619	Ta 782	W 850	Re 774	Os 787	Ir 669	Pt 566	Au 368	Hg 61	Tl 182	Pb 195	Bi 210	Po ≈146	At 92

Appendix 11 Thermodynamic properties of selected compounds

The values of $\Delta_f H^o(298\ K)$, $\Delta_f G^o(298\ K)$ and S^o refer to the state specified. The abbreviations for states are: s = solid, l = liquid, g = gas. The standard state pressure is 1 bar (10^5 Pa). In the melting and boiling point columns: sub = sublimes, dec = decomposes.

Compounds containing carbon (with the exception of metal carbonates) are arranged alphabetically *by name*. Compounds *not* containing carbon are arranged according to the non-carbon element under which the compounds are described in the text; for example, metal hydrides were described in Chapter 11 and are listed under hydrogen. The list is necessarily selective, and not all elements are represented.

Compound	*Molecular formula*	*State*	*Melting point / K*	*Boiling point / K*	$\Delta_f H^o(298\ K)$ / kJ mol^{-1}	$\Delta_f G^o(298\ K)$ / kJ mol^{-1}	S^o / J mol^{-1} K^{-1}
Compounds containing carbon (excluding metal carbonates)							
Acetaldehyde (ethanal)	CH_3CHO	g	152	294	–166	–133	264
		l	152	294	–192	–128	160
Acetamide (ethanamide)	$CH_3C(O)NH_2$	s	355	494	–317		115
Acetic acid (ethanoic acid)	CH_3CO_2H	l	290	391	–484.5	–390	160
Acetone (propanone)	$CH_3C(O)CH_3$	l	178	329	–248	200	126
Acetyl chloride (ethanoyl chloride)	$CH_3C(O)Cl$	l	161	324	–274	–208	201
Aniline (phenylamine)	$C_6H_5NH_2$	l	267	457	31	149	191
Benzaldehyde	C_6H_5CHO	l	247	451	–87		221
Benzene	C_6H_6	l	278.5	353	49	124	173
Benzoic acid	$C_6H_5CO_2H$	s	395	522	–385		168
Bromobenzene	C_6H_5Br	l	242	429	61		219
Bromoethane	CH_3CH_2Br	l	154	311	–90	–26	199
(*E*)-But-2-ene	(*E*)-$CH_3CH{=}CHCH_3$	g	167.5	274	–11		
(*Z*)-But-2-ene	(*Z*)-$CH_3CH{=}CHCH_3$	g	134	277	–7		
Buta-1,2-diene	$CH_2{=}C{=}CHCH_3$	g	137	284	162		
Buta-1,3-diene	$CH_2{=}CHCH{=}CH_2$	g	164	269	110		
Butan-1-ol	$CH_3CH_2CH_2CH_2OH$	l	183.5	390	–327		226
Butanal	$CH_3CH_2CH_2CHO$	l	174	349	–239		247
Butane	$CH_3CH_2CH_2CH_3$	g	135	272.5	–126		

Compound	*Molecular formula*	*State*	*Melting point / K*	*Boiling point / K*	$\Delta_f H^o (298\ K)$ / kJ mol^{-1}	$\Delta_f G^o (298\ K)$ / kJ mol^{-1}	S^o / J mol^{-1} K^{-1}
Butanoic acid	$CH_3CH_2CH_2CO_2H$	l	268.5	438.5	–534		222
Butanone	$CH_3C(O)CH_2CH_3$	l	187	353	–273		239
1-Butylamine	$CH_3CH_2CH_2CH_2NH_2$	l	224	351	–128		
Carbon dioxide	CO_2	g	195 sub		–393.5	–394	214
Carbon disulfide	CS_2	l	161.5	319	89	65	151
Carbon monoxide	CO	g	68	82	–110.5	–137	198
Chlorobenzene	C_6H_5Cl	l	227	405	11		
Chloroethane	CH_3CH_2Cl	g	137	285	–112	–60	276
Cyclohexane		l	279.5	354	–156		
Cyclohexene		l	169.5	356	–38.5		215
Cyclopentene		l	179	322	–105		204.5
Cyclopentene		l	138	317	4		201
1,2-Diaminoethane	$H_2NCH_2CH_2NH_2$	l	281.5	389.5	–63		
1,2-Dibromoethane	CH_2BrCH_2Br	l	210	381	–79		223
1,2-Dichloroethane	CH_2ClCH_2Cl	l	238	357	–167		
(*E*)-1,2-Dichloroethene	(*E*)-$CClH{=}CHCl$	l	223	320.5	–23	27	196
(*Z*)-1,2-Dichloroethene	(*Z*)-$CClH{=}CHCl$	l	192.5	333	–26		198
1,1-Dichloroethene	$CH_2{=}CCl_2$	l	151	310	–24	24	201.5
Dichloromethane	CH_2Cl_2	l	178	313	–124		178
Diethyl ether	$CH_3CH_2OCH_2CH_3$	l	157	307	–279		172
Diethylamine	$(CH_3CH_2)_2NH$	l	225	329	–104		

Compound	*Molecular formula*	*State*	*Melting point / K*	*Boiling point / K*	*$\Delta_f H^o$ (298 K) / kJ mol^{-1}*	*$\Delta_f G^o$ (298 K) / kJ mol^{-1}*	*S^o / J mol^{-1} K^{-1}*
Dimethylamine	$(CH_3)_2NH$	g	180	280	–18.5	68.5	273
Ethanal (see acetaldehyde)							
Ethanamide (see acetamide)							
Ethane	C_2H_6	g	90	184	–84	–32	230
Ethane-1,2-diol	$HOCH_2CH_2OH$	l	261.5	471	–455		163
Ethanoic acid (see acetic acid)							
Ethanol	CH_3CH_2OH	l	156	351.5	–278	–175	161
Ethanoyl chloride (see acetyl chloride)							
Ethene	$CH_2=CH_2$	g	104	169	52.5	68	220
Ethyl acetate (ethyl ethanoate)	$CH_3CO_2CH_2CH_3$	l	189	350	–479		258
Ethyl benzene	$C_6H_5CH_2CH_3$	l	178	409	–12		
Ethyl ethanoate (see ethyl acetate)							
Ethylamine	$CH_3CH_2NH_2$	g	192	290	–47.5	36	284
Ethyne (acetylene)	$HC\equiv CH$	g	189	192	228	211	201
Heptane	$CH_3CH_2CH_2CH_2CH_2CH_2CH_3$	l	182	371	–224		
Hexane	$CH_3CH_2CH_2CH_2CH_2CH_3$	l	178	342	–199		
Hexanoic acid	$CH_3CH_2CH_2CH_2CH_2CO_2H$	l	271	478	–584		
Hex-1-ene	$CH_3CH_2CH_2CH_2CH=CH_2$	l	133	336	–74		295
Hydrogen cyanide	HCN	l	260	299	109	125	113
		g			135	125	202
Iodobenzene	C_6H_5I	l	242	461	117		205
Methane	CH_4	g	91	109	–74	–50	186
Methanol	CH_3OH	l	179	338	–239	–167	127
2-Methylpropan-1-ol	$CH_3CH(CH_3)CH_2OH$	l	165	381	–335		215
2-Methylpropan-2-ol	$CH_3C(OH)(CH_3)CH_3$	l	298	355	–359		193
Nitrobenzene	$C_6H_5NO_2$	l	279	484	12.5		
Octane	$CH_3CH_2CH_2CH_2CH_2CH_2CH_2CH_3$	l	216	399	–250		

Compound	*Molecular formula*	*State*	*Melting point / K*	*Boiling point / K*	$\Delta_f H^o(298\ K)$ / $kJ\ mol^{-1}$	$\Delta_f G^o(298\ K)$ / $kJ\ mol^{-1}$	S^o / $J\ mol^{-1}\ K^{-1}$
Oxalic acid	HO_2CCO_2H	s	430 sub		–822		110
Pentane	$CH_3CH_2CH_2CH_2CH_3$	l	143	309	–173.5		
Pentane-2,4-dione (Hacac)	$CH_3C(O)CH_2C(O)CH_3$	l	250	412	–424		
Phenol	C_6H_5OH	s	316	455	–165		144
Phenylamine (see aniline)							
Propan-1-ol	$CH_3CH_2CH_2OH$	l	146.5	370	–303		194
Propan-2-ol	$CH_3CH(OH)CH_3$	l	183.5	355	–318		181
Propane	$CH_3CH_2CH_3$	g	83	231	–105	–23	270
Propanoic acid	$CH_3CH_2CO_2H$	l	252	414	–511	191	153
Propene	$CH_3CH=CH_2$	g	88	226	20	63	267
Propyne	$CH_3C\equiv CH$	g	171.5	250	185		
Pyridine	N	l	231	388.5	100		
Sodium acetate	$Na[CH_3CO_2]$	s	597		–709	–607	123
Tetrabromomethane	CBr_4	s	363	462	19	48	212.5
Tetrachloromethane	CCl_4	l	250	349.5	–128		
Tetrahydrofuran	O	l	165	340	–216		204
Toluene	$C_6H_5CH_3$	l	178	384	12	114	221
1,1,1-Trichloroethane	CH_3CCl_3	l	243	347	–177		227
Trichloromethane	$CHCl_3$	l	209.5	335	–134.5	–74	202
Triethylamine	$(CH_3CH_2)_3N$	l	158	362	–128		

Compound	*Molecular formula*	*State*	*Melting point / K*	*Boiling point / K*	$\Delta_f H^o$ *(298 K) / kJ mol*$^{-1}$	$\Delta_f G^o$ *(298 K) / kJ mol*$^{-1}$	S^o */ J mol*$^{-1}$ *K*$^{-1}$
Compounds *not* containing carbon but including metal carbonates							
Aluminium	**Al**	s	933	2792	0		28
Aluminium oxide	Al_2O_3	s	2345	3253	–1676	–1582	51
Aluminium(III) chloride	$AlCl_3$ or Al_2Cl_6	s	463 (at 2.5 bar)	535 dec	–704	–629	111
Aluminium(III) sulfate	$Al_2(SO_4)_3$	s	1043 dec		–3441		
Arsenic	As (grey)	s	887 sub		0		35
	As (yellow)	s			15		
Arsenic(III) chloride	$AsCl_3$	l	264	403	–305	–259	216
Arsenic(III) fluoride	AsF_3	l	267	330	–821	–774	181
Arsenic(V) fluoride	AsF_5	s	193	220			
Arsenic(V) oxide	As_2O_5	s	588 dec		–925	–782	105
Barium	Ba	s	1000	2170	0		63
Barium chloride	$BaCl_2$	s	1236	1833	–859	–810	124
Barium hydroxide	$Ba(OH)_2$	s	681 dec		–945		
Barium nitrate	$Ba(NO_3)_2$	s	865	dec	–992	–797	214
Barium oxide	BaO	s	2191	≈2273	–553.5	–525	70
Barium sulfate	$BaSO_4$	s	1853		–1473	–1362	132
Beryllium	Be	s	1560	2744	0		9.5
Beryllium chloride	$BeCl_2$	s	678	793	–490	–446	83
Beryllium nitrate trihydrate	$Be(NO_3)_2{\cdot}3H_2O$	s	333	415	–788		
Beryllium oxide	BeO	s	2803	≈4173	–609	–580	14
Bismuth	Bi	s	815	1837	0		57
Bismuth(III) chloride	$BiCl_3$	s	503	720	–379	–315	177
Bismuth(III) hydroxide	$Bi(OH)_3$	s	373 dec – H_2O		–711		
Bismuth(III) oxide	Bi_2O_3	s	1098	≈2163	–574	–494	151

Compound	*Molecular formula*	*State*	*Melting point / K*	*Boiling point / K*	$\Delta_f H^o$ *(298 K) / kJ mol*$^{-1}$	$\Delta_f G^o$ *(298 K) / kJ mol*$^{-1}$	S^o */ J mol*$^{-1}$ *K*$^{-1}$
Boron	B(β–rhombohedral)	s	2453	4273	0		6
Borazine	$[HBNH]_3$	l	215	328	–541	–393	200
Boric acid (boracic acid, orthoboric acid)	H_3BO_3 or $B(OH)_3$	s	442 $-H_2O$		–1094	–969	89
Boron tribromide	BBr_3	l	227	364	–240	–238.5	230
Boron trichloride	BCl_3	l	166	285	–427	–387	206
Boron trifluoride	BF_3	g	144	172	–1136	–1119	254
Bromine	Br_2	l	266	332	0		152
		g			31	3	245.5
Caesium	Cs	s	301.5	978	0		85
Caesium bromide	CsBr	s	909	1573	–406	–391	113
Caesium chloride	CsCl	s	918	1563	–442	–414.5	101
Caesium fluoride	CsF	s	955	1524	–553.5	–525.5	93
Caesium iodide	CsI	s	899	1553	–347	–341	123
Caesium oxide	Cs_2O	s	673 dec		–346	–308	147
Calcium	Ca	s	1115	1757	0		42
Calcium bromide	$CaBr_2$	s	1015	2088	–683	–664	130
Calcium carbonate (aragonite)	$CaCO_3$	s	793 → calcite	1098 dec	–1208	–1128	88
Calcium chloride	$CaCl_2$	s	1055	>1873	–795	–749	108
Calcium fluoride	CaF_2	s	1696	≈2773	–1228	–1176	68.5
Calcium hydroxide	$Ca(OH)_2$	s	853 $-H_2O$	dec	–985	–898	83
Calcium nitrate	$Ca(NO_3)_2$	s	834		–938	–743	193
Calcium oxide	CaO	s	2887	3123	–635	–603	38
Calcium phosphate	$Ca_3(PO_4)_2$	s	1943		–4121	–3885	236
Calcium sulfate	$CaSO_4$	s	1723 dec		–1435	–1332	106.5
Chlorine	Cl_2	g	172	239	0		223
Chlorine dioxide	ClO_2	g	213	explodes	102.5	120.5	257
Oxygen dichloride	Cl_2O	g	253	explodes	80	98	266

Compound	*Molecular formula*	*State*	*Melting point / K*	*Boiling point / K*	$\Delta_f H^o$*(298 K) / kJ mol*$^{-1}$	$\Delta_f G^o$*(298 K) / kJ mol*$^{-1}$	S^o*/ J mol*$^{-1}$ *K*$^{-1}$
Chromium	Cr	s	2180	2944	0		24
Chromium(II) chloride	$CrCl_2$	s	1097		–395	–356	115
Chromium(III) chloride	$CrCl_3$	s	1423	1573	–556.5	–486	123
Chromium(III) oxide	Cr_2O_3	s	2539	4273	–1140	–1058	81
Cobalt	Co	s	1768	3200	0		30
Cobalt(II) chloride	$CoCl_2$	s	997 (under HCl gas)	1322	–312.5	–270	109
Cobalt(II) nitrate hexahydrate	$Co(NO_3)_2{\cdot}6H_2O$	s	328 $-H_2O$		–2216		
Cobalt(II) oxide	CoO	s	2068		–238	–214	53
Cobalt(II) sulfate	$CoSO_4$	s	1008 dec		–888	–782	118
Copper	Cu	s	1358	2835	0		33
Copper(I) chloride	CuCl	s	703	1763	–137	–120	86
Copper(II) chloride	$CuCl_2$	s	893	1266 dec	–220	–176	108
Copper(I) oxide	Cu_2O	s	1508	2073 dec	–169	–146	93
Copper(II) oxide	CuO	s	1599		–157	–130	43
Copper(II) sulfate (anhydrous)	$CuSO_4$	s	473 dec		–771	–662	109
Copper(II) sulfate	$CuSO_4{\cdot}5H_2O$	s	383 $-4H_2O$, 423 $-H_2O$				
Fluorine	F_2	g	53	85	0		203
Dioxygen difluoride	O_2F_2	g	109	216	18		
Oxygen difluoride	OF_2	g	49	128	25	42	247
Gallium	Ga	s	303	2477	0		41
Gallium(III) chloride	$GaCl_3$	s	351	474	–525	–455	142
Gallium(III) fluoride	GaF_3	s	1073 sub		–1163	–1085	84
Gallium(III) hydroxide	$Ga(OH)_3$	s	713 dec		–964	–831	100
Germanium	Ge	s	1211	3106	0		31
Germanium(IV) chloride	$GeCl_4$	l	223	357	–532	–463	246

Compound	Molecular formula	State	Melting point / K	Boiling point / K	$\Delta_f H^o(298\ K)$ / kJ mol^{-1}	$\Delta_f G^o(298\ K)$ / kJ mol^{-1}	S^o / J mol^{-1} K^{-1}
Germanium(II) oxide	GeO	s	983 sub		–262	–237	50
Germanium(IV) oxide	GeO_2	s	1359		–580	–521	40
Hydrogen	H_2	g	13.7	20.1	0		131
Ammonia	NH_3	g	195	239	–46	–16	193
Arsane (arsine)	AsH_3	g	160	218	66	69	223
Calcium hydride	CaH_2	s	1089 dec		–181.5	–142.5	41
Diborane	B_2H_6	g	108	180.5	35.5	87	232
Diphosphane	P_2H_4	g	174	340	21		
Germane	GeH_4	g	108	184	91	113	217
Hydrazine	N_2H_4	g	275	386	95	159	238.5
Hydrogen bromide	HBr	g	185	206	–36	–53	199
Hydrogen chloride	HCl	g	158	188	–92	–95	187
Hydrogen fluoride	HF	g	190	293	–273	–275	174
Hydrogen iodide	HI	g	222	238	26.5	2	207
Hydrogen peroxide	H_2O_2	l	272	425	–187	–120	110
Hydrogen sulfide	H_2S	g	187.5	214	–21	–33	206
Lithium hydride	LiH	s	953		–90.5	–68	20
Lithium tetrahydroborate (Lithium borohydride)	$Li[BH_4]$	s	548 dec		–191	–125	76
Phosphane (phosphine)	PH_3	g	139.5	185	5	13	210
Potassium hydride	KH	s	dec		–58		
Silane	SiH_4	g	88	161	34	57	205
Sodium hydride	NaH	s	1073 dec		–56	–33.5	40
Sodium tetrahydroborate (Sodium borohydride)	$Na[BH_4]$	s	673 dec		–189	–124	101
Stibane (stibine)	SbH_3	g	185	256	145	148	233
Water	H_2O	l	273	373	–286	–237	70
		g			–242	–229	189
Iodine	I_2	s	387	458	0		116
		g			62	19	261

Compound	*Molecular formula*	*State*	*Melting point / K*	*Boiling point / K*	$\Delta_f H^o(298\ K)$ / kJ mol^{-1}	$\Delta_f G^o(298\ K)$ / kJ mol^{-1}	S^o / J mol^{-1} K^{-1}
Iron	Fe	s	1811	3134	0		27
Iron(II) chloride	$FeCl_2$	s	943		–342	–302	118
Iron(III) chloride	$FeCl_3$	s	579	588 dec	–399.5	–334	142
Iron(II) oxide	FeO	s	1642		–272		
Iron(III) oxide (haematite)	Fe_2O_3	s	1838		–824	–742	87
Iron sulfide (pyrite)	FeS_2	s	1444		–178	–167	53
Lead	Pb	s	600	2022	0		65
Lead(II) chloride	$PbCl_2$	s	774	1223	–359	–314	136
Lead(II) iodide	PbI_2	s	675	1227	–175.5	–174	175
Lead(II) oxide (litharge)	PbO	s	1159		–219	–189	67
Lead nitrate	$Pb(NO_3)_2$	s	743 dec		–452		
Lead(IV) oxide	PbO_2	s	563 dec		–277	–217	69
Lead sulfate	$PbSO_4$	s	1443		–920	–813	148.5
Lead(II) sulfide (galena)	PbS	s	1387		–100	–99	91
Lithium	Li	s	453.5	1615	0		29
Lithium bromide	LiBr	s	823	1538	–351	–342	74
Lithium carbonate	Li_2CO_3	s	996	1583 dec	–1216	–1132	90
Lithium chloride	LiCl	s	878	1598	–409	–384	59
Lithium fluoride	LiF	s	1118	1949	–616	–588	36
Lithium hydroxide	LiOH	s	723	1197 dec	–485	–439	43
Lithium iodide	LiI	s	722	1453	–270	–270	87
Lithium nitrate	$LiNO_3$	s	537	873 dec	–483	–381	90
Lithium oxide	Li_2O	s	>1973		–598	–561	38
Magnesium	Mg	s	923	1363	0		33
Magnesium bromide	$MgBr_2$	s	973		–524	–504	117
Magnesium carbonate	$MgCO_3$	s	623 dec		–1096	–1012	66
Magnesium chloride	$MgCl_2$	s	987	1685	–641	–592	90
Magnesium fluoride	MgF_2	s	1534	2512	–1124	–1071	57
Magnesium hydroxide	$Mg(OH)_2$	s	623 –H_2O		–924.5	–833.5	63

Compound	Molecular formula	State	Melting point / K	Boiling point / K	$\Delta_f H^o(298\ K)$ / kJ mol^{-1}	$\Delta_f G^o(298\ K)$ / kJ mol^{-1}	S^o / J mol^{-1} K^{-1}
Magnesium oxide	MgO	s	3125	3873	–602	–569	27
Magnesium sulfate	$MgSO_4$	s	1397 dec		–1285	–1171	92
Manganese	Mn	s	1519	2334	0		32
Manganese(II) chloride	$MnCl_2$	s	923	1463	–481	–440.5	118
Manganese(II) nitrate tetrahydrate	$Mn(NO_3)_2{\cdot}4H_2O$	s	299	402	–576 (anhydrous salt)		
Manganese(IV) oxide	MnO_2	s	808 dec		–520	–465	53
Manganese(II)sulfate	$MnSO_4$	s	973	1123 dec	–1064		
Potassium permanganate	$KMnO_4$	s	513 dec		–837	–738	172
Mercury	Hg	l	234	630	0		76
Mercury(I) chloride	Hg_2Cl_2	s	673 sub		–265	–211	192
Mercury(II) chloride	$HgCl_2$	s	549	575	–224	–179	146
Mercury(II) oxide	HgO	s	773 dec		–91	–58.5	70
Mercury(II) sulfide	HgS	s	856 sub		–58	–51	82
Nickel	Ni	s	1728	3186	0		30
Nickel(II) bromide	$NiBr_2$	s	1236		–212		
Nickel(II) chloride	$NiCl_2$	s	1274 sub		–305	–259	98
Nickel(II) nitrate hexahydrate	$Ni(NO_3)_2{\cdot}6H_2O$	s	330	410	–2223		
Nickel(II) oxide	NiO	s	2263		–244	–216	39
Nickel(II) sulfate	$NiSO_4$	s	1121 dec		–873	–760	92
Nitrogen	N_2	g	63	77	0		192
Ammonium chloride	NH_4Cl	s	613 sub		–314	–203	95
Ammonium nitrate	NH_4NO_3	s	443	483	–366	–184	151
Ammonium nitrite	NH_4NO_2	s	333 explodes		–256.5		
Ammonium sulfate	$[NH_4]_2SO_4$	s	508 dec		–1181	–902	220
Dinitrogen oxide	N_2O	g	182	185	82	104	220

Compound	*Molecular formula*	*State*	*Melting point / K*	*Boiling point / K*	*$\Delta_f H^o$(298 K) / kJ mol^{-1}*	*$\Delta_f G^o$(298 K) / kJ mol^{-1}*	*S^o / J mol^{-1} K^{-1}*
Dinitrogen tetraoxide	N_2O_4	l	262	294	−19.5	97.5	209
Nitric acid	HNO_3	l	231	356	−174	−81	156
Nitrogen dioxide	NO_2	g	262	294	33	51	240
Nitrogen monoxide	NO	g	109	121	90	87	210
Nitrogen trifluoride	NF_3	g	66	144	−132	−91	261
Oxygen	O_2	g	54	90	0		205
Ozone	O_3	g	81	163	143	163	239
Phosphorus	P_4 (white phosphorus)	s	317	550	0		41
	Red phosphorus	s			−18		23
	Black phosphorus	s			−39		
Phosphonic acid (phosphorous acid)	H_3PO_3	s	343	473 dec	−964		
Phosphoric acid	H_3PO_4	s	315	486 −H_2O	−1284	−1124	110.5
Phosphorus(III) chloride	PCl_3	l	161	348	−320	−272	217
Phosphorus(V) chloride	PCl_5	s	440 dec		−443.5		
Phosphorus(III) fluoride	PF_3	g	122	171.5	−958	−937	273
Phosphorus(V) fluoride	PF_5	g	190	198	−1594	−1521	301
Phosphorus(III) oxide	P_4O_6	s	297	447 (under N_2)	−1640		
Phosphorus(V) oxide	P_4O_{10}	s	573 sub		−2984		
Potassium	K	s	336	1032	0		65
Potassium bromide	KBr	s	1007	1708	−394	−381	96
Potassium carbonate	K_2CO_3	s	1164	dec	−1151	−1063.5	155.5
Potassium chlorate	$KClO_3$	s	629	673 dec	−398	−296	143
Potassium chloride	KCl	s	1043	1773 sub	−436.5	−408.5	83
Potassium cyanide	KCN	s	907		−113	−102	128.5
Potassium fluoride	KF	s	1131	1778	−567	−538	67
Potassium hydroxide	KOH	s	633	1593	−425	−379	79
Potassium iodate	KIO_3	s	833	dec	−501	−418	151.5

Compound	Molecular formula	State	Melting point / K	Boiling point / K	$\Delta_f H^o$(298 K) / kJ mol^{-1}	$\Delta_f G^o$(298 K) / kJ mol^{-1}	S^o / J mol^{-1} K^{-1}
Potassium iodide	KI	s	954	1603	–328	–325	53
Potassium nitrate	KNO_3	s	607	673 dec	–495	–395	133
Potassium nitrite	KNO_2	s	713 dec		–370	–307	152
Potassium oxide	K_2O	s	623 dec		–361.5		
Potassium perchlorate	$KClO_4$	s	673 dec		–433	–303	151
Potassium permanganate	$KMnO_4$	s	513 dec		–837	–738	172
Potassium sulfate	K_2SO_4	s	1342	1962	–1438	1321	176
Silicon	Si	s	1687	3538	0		19
Silicon dioxide (silica, α-quartz)	SiO_2	s	1883	2503	–911	–856	41.5
Silicon tetrachloride	$SiCl_4$	l	203	331	–687	–620	240
Silicon tetrafluoride	SiF_4	g	183	187	–1615	–1573	283
Sodium silicate	Na_4SiO_4	s	1291				
Silver	Ag	s	1235	2435	0		43
Silver bromide	$AgBr$	s	705	dec > 1573	–100	–97	107
Silver chloride	$AgCl$	s	728	1823	–127	–110	96
Silver chromate	Ag_2CrO_4	s			–732	–642	218
Silver fluoride	AgF	s	708	1432	–205	–	–
Silver iodide	AgI	s	831	1779	–62	–66	115.5
Silver nitrate	$AgNO_3$	s	485	717 dec	–124	–33	141
Sodium	Na	s	371	1154	0		51
Sodium bromide	$NaBr$	s	1020	1663	–361	–349	87
Sodium carbonate	Na_2CO_3	s	1124		–1131	–1044	135
Sodium chloride	$NaCl$	s	1074	1686	–411	–384	72
Sodium cyanide	$NaCN$	s	837	1769	–87.5	–76	116
Sodium fluoride	NaF	s	1266	1968	–577	–546	51
Sodium hydroxide	$NaOH$	s	591	1663	–426	–379.5	64.5
Sodium iodide	NaI	s	934	1577	–288	–286	98.5
Sodium nitrate	$NaNO_3$	s	580	653 dec	–468	–367	116.5
Sodium nitrite	$NaNO_2$	s	544	593 dec	–359	–285	104

Compound	*Molecular formula*	*State*	*Melting point / K*	*Boiling point / K*	$\Delta_f H^o$ *(298 K) / kJ mol*$^{-1}$	$\Delta_f G^o$ *(298 K) / kJ mol*$^{-1}$	S^o */ J mol*$^{-1}$ *K*$^{-1}$
Sodium oxide	Na_2O	s	1548 sub		–414	–375.5	75
Sodium sulfate	Na_2SO_4	s	1157		–1387	–1270	150
Sulfur	S_8 (orthorhombic)	s	388	718	0		32
	S_8 (monoclinic)	s			0.3		
Disulfur dichloride	S_2Cl_2	l	193	409	–59		
Sulfur dioxide	SO_2	g	200	263	–297	–300	248
Sulfur(IV) fluoride	SF_4	g	149	233	–763	–722	300
Sulfur(VI) fluoride	SF_6	g	222.5	337	–1221	–1116.5	291.5
Sulfur trioxide	SO_3	s	290	318	–454	–374	71
		g			–396	–371	257
Sulfuric acid	H_2SO_4	l	283	603	–814	–690	157
Thionyl dichloride (thionyl chloride)	$SOCl_2$	l	168	413 dec	–246		
Tin	Sn (white)	s	505	2875	0		52
	Sn (grey)	s			–2	0.1	44
Tin(II) chloride	$SnCl_2$	s	519	925	–325		
Tin(IV) chloride	$SnCl_4$	l	240	387	–511	–440	259
Tin(IV) oxide	SnO_2	s	1903		–578	–516	49
Zinc	Zn	s	693	1180	0		42
Zinc bromide	$ZnBr_2$	s	667	923	–329	–312	138.5
Zinc carbonate	$ZnCO_3$	s	573 $-CO_2$	–813	–731.5	82	
Zinc chloride	$ZnCl_2$	s	556	1005	–415	–369	111.5
Zinc oxide	ZnO	s	2248		–350.5	–320.5	44
Zinc sulfide (wurtzite)	ZnS	s	1458 sub		–193		
Zinc sulfide (zinc blende)	ZnS	s	1293 → wurtzite		–206	–205	58
Zinc sulfate	$ZnSO_4$	s			–983	–871.5	111

Appendix 12 Selected standard reduction potentials (298 K)

The concentration of each aqueous solution is 1 mol dm^{-3} and the pressure of a gaseous component is 1 bar (10^5 Pa). (Changing the standard pressure to 1 atm (101 300 Pa) makes no difference to the values of E^o at this level of accuracy.)

Reduction half-equation	E^o / *V*
$Li^+(aq) + e^- \rightleftharpoons Li(s)$	–3.04
$K^+(aq) + e^- \rightleftharpoons K(s)$	–2.93
$Ca^{2+}(aq) + 2e^- \rightleftharpoons Ca(s)$	–2.87
$Na^+(aq) + e^- \rightleftharpoons Na(s)$	–2.71
$Mg^{2+}(aq) + 2e^- \rightleftharpoons Mg(s)$	–2.37
$Al^{3+}(aq) + 3e^- \rightleftharpoons Al(s)$	–1.66
$[HPO_3]^{2-} + 2H_2O + 2e^- \rightleftharpoons [H_2PO_2]^- + 3[OH]^-$	–1.65
$Ti^{2+}(aq) + 2e^- \rightleftharpoons Ti(s)$	–1.63
$Mn^{2+}(aq) + 2e^- \rightleftharpoons Mn(s)$	–1.19
$Te(s) + 2e^- \rightleftharpoons Te^{2-}(aq)$	–1.14
$[SO_4]^{2-}(aq) + H_2O(l) + 2e^- \rightleftharpoons [SO_3]^{2-}(aq) + 2[OH]^-(aq)$	–0.93
$Se(s) + 2e^- \rightleftharpoons Se^{2-}(aq)$	–0.92
$2[NO_3]^-(aq) + 2H_2O(l) + 2e^- \rightleftharpoons N_2O_4(g) + 4[OH]^-(aq)$	–0.85
$Zn^{2+}(aq) + 2e^- \rightleftharpoons Zn(s)$	–0.76
$S(s) + 2e^- \rightleftharpoons S^{2-}(aq)$	–0.48
$[NO_2]^-(aq) + H_2O(l) + 3e^- \rightleftharpoons NO(g) + 2[OH]^-(aq)$	–0.46
$Fe^{2+}(aq) + 2e^- \rightleftharpoons Fe(s)$	–0.44
$Cr^{3+}(aq) + e^- \rightleftharpoons Cr^{2+}(aq)$	–0.41
$Ti^{3+}(aq) + e^- \rightleftharpoons Ti^{2+}(aq)$	–0.37
$PbSO_4(s) + 2e^- \rightleftharpoons Pb(s) + [SO_4]^{2-}(aq)$	–0.36
$Tl^+(aq) + e^- \rightleftharpoons Tl(s)$	–0.34
$Co^{2+}(aq) + 2e^- \rightleftharpoons Co(s)$	–0.28
$H_3PO_4(aq) + 2H^+(aq) + 2e^- \rightleftharpoons H_3PO_3(aq) + H_2O(l)$	–0.28
$V^{3+}(aq) + e^- \rightleftharpoons V^{2+}(aq)$	–0.26
$Ni^{2+}(aq) + 2e^- \rightleftharpoons Ni(s)$	–0.25
$Sn^{2+}(aq) + 2e^- \rightleftharpoons Sn(s)$	–0.14
$Pb^{2+}(aq) + 2e^- \rightleftharpoons Pb(s)$	–0.13
$Fe^{3+}(aq) + 3e^- \rightleftharpoons Fe(s)$	–0.04
$2H^+(aq) + 2e^- \rightleftharpoons H_2(g, 1 bar)$	0
$[NO_3]^-(aq) + H_2O(l) + 2e^- \rightleftharpoons [NO_2]^-(aq) + 2[OH]^-(aq)$	+0.01
$[S_4O_6]^{2-}(aq) + 2e^- \rightleftharpoons 2[S_2O_3]^{2-}(aq)$	+0.08
$S(s) + 2H^+(aq) + 2e^- \rightleftharpoons H_2S(aq)$	+0.14
$2[NO_2]^-(aq) + 3H_2O(l) + 4e^- \rightleftharpoons N_2O(g) + 6[OH]^-(aq)$	+0.15

Reduction half-equation	*E°/ V*
$Cu^{2+}(aq) + e^- \rightleftharpoons Cu^+(aq)$	+0.15
$[SO_4]^{2-}(aq) + 4H^+(aq) + 2e^- \rightleftharpoons H_2SO_3(aq) + H_2O(l)$	+0.17
$Cu^{2+}(aq) + 2e^- \rightleftharpoons Cu(s)$	+0.34
$[ClO_4]^-(aq) + H_2O(l) + 2e^- \rightleftharpoons [ClO_3]^-(aq) + 2[OH]^-(aq)$	+0.36
$O_2(g) + 2H_2O(l) + 4e^- \rightleftharpoons 4[OH]^-(aq)$	+0.40
$Sn^{4+}(aq) + 2e^- \rightleftharpoons Sn^{2+}(aq)$	+0.51
$Cu^+(aq) + e^- \rightleftharpoons Cu(s)$	+0.52
$I_2(aq) + 2e^- \rightleftharpoons 2I^-(aq)$	+0.54
$H_3AsO_4(aq) + 2H^+(aq) + 2e^- \rightleftharpoons HAsO_2(aq) + 2H_2O(l)$	+0.56
$[MnO_4]^-(aq) + 2H_2O(aq) + 3e^- \rightleftharpoons MnO_2(s) + 4[OH]^-(aq)$	+ 0.59
$[BrO_3]^-(aq) + 3H_2O(l) + 6e^- \rightleftharpoons Br^-(aq) + 6[OH]^-(aq)$	+0.61
$O_2(g) + 2H^+(aq) + 2e^- \rightleftharpoons H_2O_2(aq)$	+0.70
$[BrO]^-(aq) + H_2O(l) + 2e^- \rightleftharpoons Br^-(aq) + 2[OH]^-(aq)$	+0.76
$Fe^{3+}(aq) + e^- \rightleftharpoons Fe^{2+}(aq)$	+0.77
$Ag^+(aq) + e^- \rightleftharpoons Ag(s)$	+0.80
$[ClO]^-(aq) + H_2O(l) + 2e^- \rightleftharpoons Cl^-(aq) + 2[OH]^-(aq)$	+0.84
$2HNO_2(aq) + 4H^+(aq) + 4e^- \rightleftharpoons H_2N_2O_2(aq) + 2H_2O(l)$	+0.86
$[NO_3]^-(aq) + 3H^+(aq) + 2e^- \rightleftharpoons HNO_2(aq) + H_2O(l)$	+0.93
$[NO_3]^-(aq) + 4H^+(aq) + 3e^- \rightleftharpoons NO(g) + 2H_2O(l)$	+0.96
$HNO_2(aq) + H^+(aq) + e^- \rightleftharpoons NO(g) + H_2O(l)$	+0.98
$[IO_3]^-(aq) + 6H^+(aq) + 6e^- \rightleftharpoons I^-(aq) + 3H_2O(l)$	+1.09
$Br_2(aq) + 2e^- \rightleftharpoons 2Br^-(aq)$	+1.09
$2[IO_3]^-(aq) + 12H^+(aq) + 10e^- \rightleftharpoons I_2(aq) + 6H_2O(l)$	+1.20
$O_2(g) + 4H^+(aq) + 4e^- \rightleftharpoons 2H_2O(l)$	+1.23
$Tl^{3+}(aq) + 2e^- \rightleftharpoons Tl^+(aq)$	+1.25
$2HNO_2(aq) + 4H^+(aq) + 4e^- \rightleftharpoons N_2O(g) + 3H_2O(l)$	+1.30
$[Cr_2O_7]^{2-}(aq) + 14H^+(aq) + 6e^- \rightleftharpoons 2Cr^{3+}(aq) + 7H_2O(l)$	+1.33
$Cl_2(aq) + 2e^- \rightleftharpoons 2Cl^-(aq)$	+1.36
$2[ClO_4]^-(aq) + 16H^+(aq) + 14e^- \rightleftharpoons Cl_2(aq) + 8H_2O(l)$	+1.39
$[ClO_4]^-(aq) + 8H^+(aq) + 8e^- \rightleftharpoons Cl^-(aq) + 4H_2O(l)$	+1.39
$[BrO_3]^-(aq) + 6H^+(aq) + 6e^- \rightleftharpoons Br^-(aq) + 3H_2O(l)$	+1.42
$[ClO_3]^-(aq) + 6H^+(aq) + 6e^- \rightleftharpoons Cl^-(aq) + 3H_2O(l)$	+1.45
$2[ClO_3]^-(aq) + 12H^+(aq) + 10e^- \rightleftharpoons Cl_2(aq) + 6H_2O(l)$	+1.47
$2[BrO_3]^-(aq) + 12H^+(aq) + 10e^- \rightleftharpoons Br_2(aq) + 6H_2O(l)$	+1.48
$HOCl(aq) + H^+(aq) + 2e^- \rightleftharpoons Cl^-(aq) + H_2O(l)$	+1.48
$[MnO_4]^-(aq) + 8H^+(aq) + 5e^- \rightleftharpoons Mn^{2+}(aq) + 4H_2O(l)$	+1.51
$2HOCl(aq) + 2H^+(aq) + 2e^- \rightleftharpoons Cl_2(aq) + 2H_2O(l)$	+1.61
$PbO_2(s) + 4H^+(aq) + [SO_4]^{2-}(aq) + 2e^- \rightleftharpoons PbSO_4(s) + 2H_2O(l)$	+1.69
$Ce^{4+}(aq) + e^- \rightleftharpoons Ce^{3+}(aq)$	+1.72
$H_2O_2(aq) + 2H^+(aq) + 2e^- \rightleftharpoons 2H_2O(l)$	+1.78
$Co^{3+}(aq) + e^- \rightleftharpoons Co^{2+}(aq)$	+1.92
$XeO_3(aq) + 6H^+(aq) + 6e^- \rightleftharpoons Xe(g) + 3H_2O(l)$	+2.10
$H_4XeO_6(aq) + 2H^+(aq) + 2e^- \rightleftharpoons XeO_3(aq) + 3H_2O(l)$	+2.42
$F_2(aq) + 2e^- \rightleftharpoons 2F^-(aq)$	+2.87

Appendix 13 Trivial names still in common use for selected inorganic and organic compounds, inorganic ions and organic substituents

Many of the trivial names are accepted by the IUPAC. The list is selective but includes many chemicals used in routine laboratory work.

Trivial name	*IUPAC name*[a]	*Formula*
Acetamide	Ethanamide	$CH_3C(O)NH_2$
Acetic acid	Ethanoic acid	CH_3CO_2H
Acetone	Propanone	$CH_3C(O)CH_3$
Acetonitrile	Ethanenitrile	CH_3CN
Acetylacetone (Hacac)	Pentane-2,4-dione	$CH_3C(O)CH_2C(O)CH_3$
Acetyl chloride	Ethanoyl chloride	$CH_3C(O)Cl$
Acetylene	Ethyne	$HC{\equiv}CH$
Allene	Propadiene	$CH_2{=}C{=}CH_2$
Aniline	Phenylamine	$C_6H_5NH_2$
Anisole	Methoxybenzene	C_6H_5OMe
Arsine	Arsane	AsH_3
Bicarbonate	Hydrogentrioxocarbonate(IV)	$[HCO_3]^-$
Boric acid (orthoboric acid, boracic acid)	Trioxoboric acid	$B(OH)_3$ or H_3BO_3
Borohydride	Tetrahydroborate(1–)	$[BH_4]^-$
^{t}Butyl (*tert*-butyl) (substituent)	1,1-Dimethylethyl	Me_3C
Carbon tetrachloride	Tetrachloromethane	CCl_4
Carbonate	Trioxocarbonate(IV)	$[CO_3]^{2-}$
Chloroform	Trichloromethane	$CHCl_3$
18-Crown-6	1,4,7,10,13,16-Hexaoxacyclooctadecane	see structure **16.12**
Cumene	Isopropylbenzene	$C_6H_5CHMe_2$
Dimethylacetylene	But-2-yne	$CH_3C{\equiv}CCH_3$
Ethylene	Ethene	$H_2C{=}CH_2$
Ethylenediamine (en)	1,2-Diaminoethane	$H_2NCH_2CH_2NH_2$
Ethylene glycol	Ethane-1,2-diol	$HOCH_2CH_2OH$
Formaldehyde	Methanal	$HCHO$
Formamide	Methanamide	$HC(O)NH_2$
Formic acid	Methanoic acid	HCO_2H

Trivial name	*IUPAC name*[a]	*Formula*
Hydrogen sulfide	Sulfane	H_2S
Isopropyl (substituent)	2-Methylethyl	Me_2CH
Mesitylene	1,3,5-Trimethylbenzene	1,3,5-$Me_3C_6H_3$
Methylene chloride	Dichloromethane	CH_2Cl_2
Nitrate	Trioxonitrate(V)	$[NO_3]^-$
Nitrite	Dioxonitrate(III)	$[NO_2]^-$
Nitronium	Nitryl	$[NO_2]^+$
Perchlorate	Tetraoxochlorate(VII)	$[ClO_4]^-$
Permanganate	Tetraoxomanganate(VII)	$[MnO_4]^-$
Phosgene	Carbonyl dichloride	$COCl_2$
Phosphate	Tetraoxophosphate(V)	$[PO_4]^{3-}$
Phosphine	Phosphane	PH_3
Phosphite	Phosphonate	$[HPO_3]^{2-}$
Phthalic acid	Benzene-1,2-dicarboxylic acid	1,2-$(HO_2C)_2C_6H_4$
Picric acid	2,4,6-Trinitrophenol	2,4,6-$(O_2N)_3C_6H_2OH$
Quicklime or lime	Calcium oxide	CaO
Silica	Silicon dioxide	SiO_2
Slaked lime	Calcium hydroxide	$Ca(OH)_2$
Stibine	Stibane	SbH_3
Sulfate	Tetraoxosulfate(VI)	$[SO_4]^{2-}$
Sulfite	Trioxosulfate(IV)	$[SO_3]^{2-}$
Thionyl chloride	Thionyl dichloride	$SOCl_2$
Toluene	Methylbenzene	C_6H_5Me
Water glass	Sodium silicate	$Na_4[SiO_4]$
meta-Xylene	1,3-Dimethylbenzene	1,3-$Me_2C_6H_4$
ortho-Xylene	1,2-Dimethylbenzene	1,2-$Me_2C_6H_4$
para-Xylene	1,4-Dimethylbenzene	1,4-$Me_2C_6H_4$

[a] This may not necessarily be the *only* accepted IUPAC name. For fuller details see: G.J. Leigh (ed.), *IUPAC Nomenclature of Inorganic Chemistry: Recommendations 1990*, Blackwell Scientific Publications, Oxford; *A Guide to IUPAC Nomenclature of Organic Compounds*, Blackwell Scientific Publications, Oxford (1993).

Appendix 14 1994 ranking of the top 25 chemicals manufactured in the USA

Ranking in 1994	*Chemical*	*Formula*
1	Sulfuric acid	H_2SO_4
2	Dinitrogen	N_2
3	Dioxygen	O_2
4	Ethene	$H_2C{=}CH_2$
5	Calcium oxide (lime)	CaO
6	Ammonia	NH_3
7	Propene	$CH_3CH{=}CH_2$
8	Sodium hydroxide	$NaOH$
9	Phosphoric acid	H_3PO_4
10	Dichlorine	Cl_2
11	Sodium carbonate	Na_2CO_3
12	1,2-Dichloroethene	$HClC{=}CHCl$
13	Nitric acid	HNO_3
14	Ammonium nitrate	NH_4NO_3
15	Urea	$O{=}C(NH_2)_2$
16	Chloroethene (vinyl chloride)	$H_2C{=}CHCl$
17	Benzene	
18	Methyl *tert*-butyl ether	CH_3OCMe_3
19	Ethylbenzene	CH_2CH_3
20	Phenylethene (styrene)	$CH{=}CH_2$
21	Carbon dioxide	CO_2
22	Methanol	CH_3OH
23	1,2-Xylene (and 1,3- and 1,4- isomers)	CH_3 CH_3
24	Terephthalic acid	CO_2H CO_2H
25	Methanal (formaldehyde)	$HCHO$

Data from *Chemical & Engineering News*, (1995) p. 5.

Appendix 15 Some regular polyhedra

In a tetrahedral molecule such as CH_4, the H atoms define a tetrahedron; in an octahedral complex such as $[Co(H_2O)_6]^{2+}$, the O donor atoms of the water ligands define an octahedron, and so on.

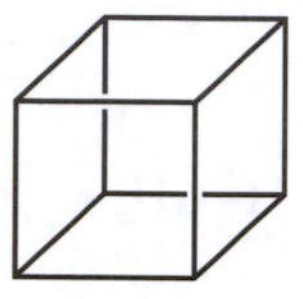

Cube

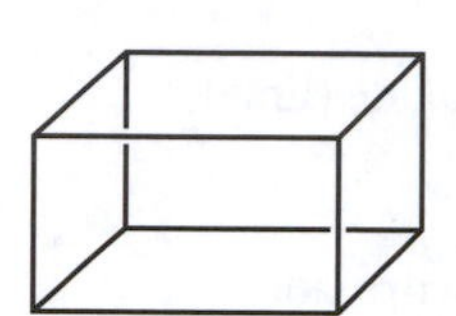

Cuboid

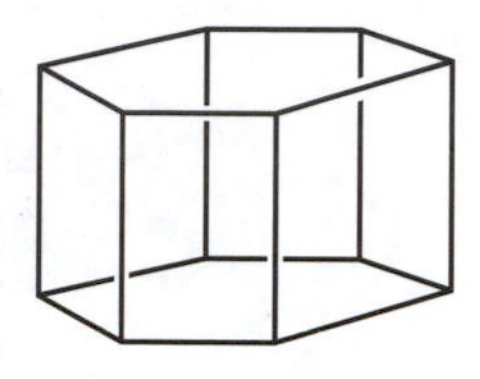

Hexagonal prism

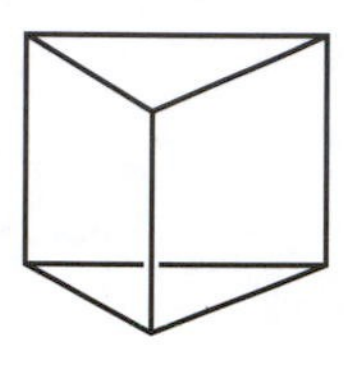

Trigonal prism

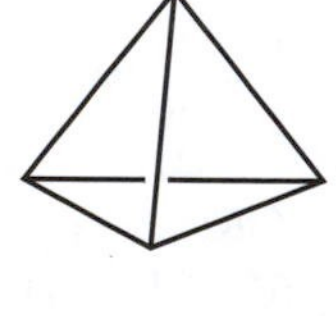

Tetrahedron

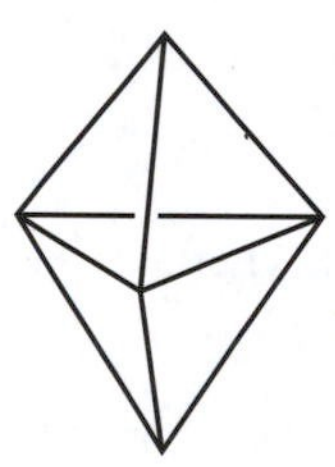

Trigonal bipyramid

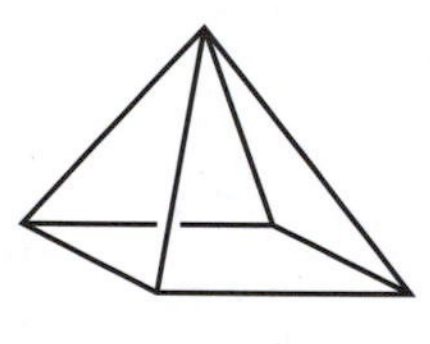

Square-based pyramid

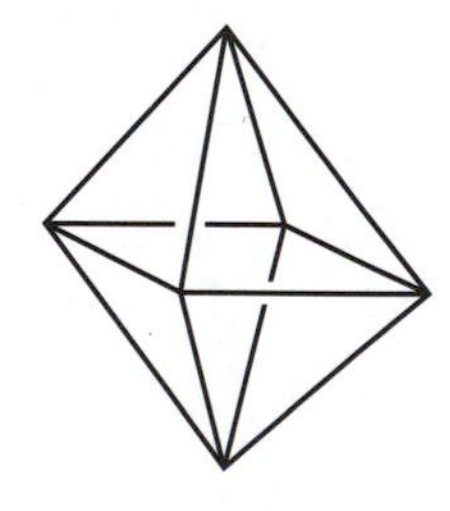

Octahedron

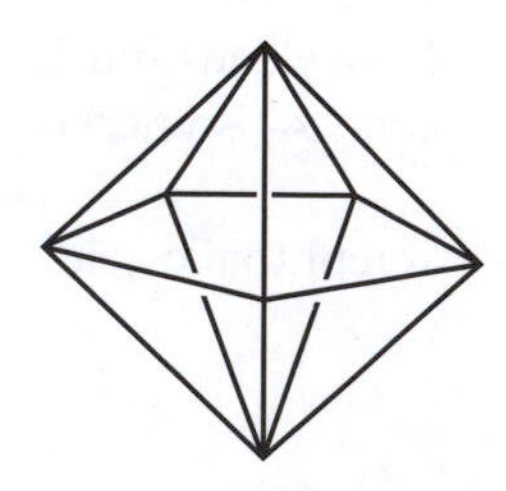

Pentagonal bipyramid

Answers to non-descriptive problems

CHAPTER 1

1.1 0.0006 m = (a) 0.6 mm,
(b) 6×10^8 pm,
(c) 0.06 cm,
(d) 6×10^5 nm.

1.2 0.122 nm.

1.3 2.704×10^{-5} m^3.

1.4 J s.

1.5 LHS: energy = J = kg m^2 s^{-2};
RHS: mass = kg; velocity = distance per unit time = m s^{-1}.

1.6 10.8.

1.7 158 (relative intensity = 1),
160 (relative intensity = 2),
162 (relative intensity = 1).

1.8 (a) 2.86×10^{-5} m^3;
(b) 4.85×10^{-3} m^3;
(c) 0.318 m^3;
(d) 407 m^3.

1.9 0.84 moles.

1.10 N_2 0.57 bar, Ar 0.80 bar. This leaves 0.63 bar pressure due to other gases.

1.11 (a) 5×10^{-3} moles,
(b) 0.082 moles,
(c) 0.04 moles,
(d) 5×10^{-4} moles.

1.12 (a) 0.166 g,
(b) 1.17 g,
(c) 0.71 g.

1.13 (a) NaI; (b) $MgCl_2$;
(c) MgO; (d) CaF_2;
(e) Li_3N; (f) Ca_3P_2;
(g) Na_2S; (h) H_2S.

1.14 Al_2O_3, $AlCl_3$, AlF_3, AlH_3.

1.18 (a) +1; (b) +2;
(c) +4; (d) +3;
(e) +4; (f) +5.

1.20 Zero in O_2 and O_3; –1 in H_2O_2 (see Box 8.8); –1 in $[O_2]^{2-}$; $+\frac{1}{2}$ in $[O_2]^+$.

1.22 (a) –1 411 kJ per mole of ethene;
(b) +44 kJ per mole of ethanol;
(c) –133 kJ per mole of ethene.

1.23 0.317 bar^{-2}.

1.24 0.97 moles.

1.25 (a) Sodium carbonate or sodium trioxocarbonate(IV);
(b) iron(III) bromide or iron tribromide;
(c) cobalt(II) sulfate or cobalt(II) tetraoxosulfate(VI);
(d) barium chloride;
(e) iron(III) oxide;
(f) iron(II) hydroxide;
(g) lithium iodide;
(h) potassium cyanide;
(i) potassium thiocyanate;
(j) calcium phosphide.

1.26 (a) NiI_2; (b) NH_4NO_3;
(c) $Ba(OH)_2$; (d) $Fe_2(SO_4)_3$;
(e) $FeSO_3$; (f) AlH_3;
(g) PbO_2, (h) SnS.

1.27 (a) 8; (b) 6;
(c) 3; (d) 10;
(e) 4.

CHAPTER 2

2.2 (a) 1.5×10^{-5} m,
(b) 6.7×10^{-8} m,
(c) 1.4×10^{-9} m.

2.3 For $n = 4$: $l = 0, 1, 2, 3$; for $l = 0$, $m_l = 0$; for $l = 1$, $m_l = -1, 0, +1$; for $l = 2$, $m_l = -2, -1, 0, +1, +2$; for $l = 3$, $m_l = -3, -2, -1, 0, +1, +2, +3$; the 4*f* atomic orbitals correspond to $n = 4$, $l = 3$ and since $m_l = -3, -2, -1, 0, +1, +2, +3$, there are seven 4*f* orbitals.

2.4 (a) The two electrons occupy the 1*s* orbital: for one electron $n = 1$, $l = 0$, $m_l = 0$, $m_s = +\frac{1}{2}$ and for the other electron $n = 1$, $l = 0$, $m_l = 0$, $m_s = -\frac{1}{2}$.

(b) The ground state electronic configuration is $1s^2 2s^2 2p^1$.
For the $1s$ electrons:
$n = 1, l = 0, m_l = 0, m_s = +\frac{1}{2}$
$n = 1, l = 0, m_l = 0, m_s = -\frac{1}{2}$
For the $2s$ electrons:
$n = 2, l = 0, m_l = 0, m_s = +\frac{1}{2}$
$n = 2, l = 0, m_l = 0, m_s = -\frac{1}{2}$
Let the $2p$ electron be in one of the three possible $2p$ orbitals, and the set of quantum numbers will be:
$n = 2, l = 1, m_l = -1, m_s = +\frac{1}{2}$ (or $-\frac{1}{2}$)
or $n = 2, l = 1, m_l = 0, m_s = +\frac{1}{2}$ (or $-\frac{1}{2}$)
or $n = 2, l = 1, m_l = +1, m_s = +\frac{1}{2}$ (or $-\frac{1}{2}$)

2.5 –1312 kJ, –328 kJ, –146 kJ.

2.6 1310 kJ mol^{-1}.

2.8 Be $1s^2 2s^2$; F $1s^2 2s^2 2p^5$; P $1s^2 2s^2 2p^6 3s^2 3p^3$; K $1s^2 2s^2 2p^6 3s^2 3p^6 4s^1$.

2.10 PCl_3 is predicted but PCl_5 is not.

CHAPTER 3

3.4 (a) 1200,
(b) 2128,
(c) 1062,
(d) 151 kJ mol^{-1}.

3.5 +60 kJ mol^{-1}.

3.6 $\bar{D}$(C–H) = 416 kJ mol^{-1} and D(C–C) = 331 kJ mol^{-1}.

3.9 Two structures:

:C═C: C≡C

3.12 $[Li_2]^+$.

3.13 $[N_2]^-$ paramagnetic; $[N_2]^+$ paramagnetic.

CHAPTER 4

4.1 $Z_{eff}(A) > Z_{eff}(B)$.

4.4 262 kJ mol^{-1}.

4.5 (a) 430,
(b) 329,
(c) 401 kJ mol^{-1}.

4.6 (a) 860,
(b) 1316,
(c) 1604 kJ mol^{-1}.

4.7 χ^M values: N 7.25, O 7.53, F 10.40, Cl 8.31, Br 7.59, I 6.75.

4.11 $q = 0.02$

CHAPTER 5

5.2 All linear; central atoms: $[ICl_2]^-$ (I), $[IBr_2]^-$ (I), $[ClF_2]^-$ (Cl), $[BrF_2]^-$ (Br).

5.3 (a) linear, (b) bent,
(c) bent, (d) linear,
(e) linear, (f) linear.

5.4 (a) trigonal planar,
(b) trigonal pyramidal,
(c) bent,
(d) bent,
(e) T-shaped,
(f) trigonal bipyramidal,
(g) trigonal bipyramidal (Cl in the equatorial plane),
(h) tetrahedral,
(i) planar,
(j) linear.

5.5 (a) linear,
(b) octahedral,
(c) linear,
(d) octahedral,
(e) tetrahedral,
(f) tetrahedral,
(g) octahedral,
(h) tetrahedral,
(i) trigonal bipyramidal (Cl atoms in the equatorial plane).

5.7 $[NiBr_3(PMe_2Ph)_2]$: look back at structure **5.40**.

5.8 (a) trigonal planar, $\mu = 0$,
(b) trigonal pyramidal, $\mu > 0$,
(c) trigonal bipyramidal, $\mu = 0$,
(d) trigonal planar but asymmetrical, $\mu > 0$,
(e) square-based pyramidal, $\mu > 0$,
(f) T-shaped, $\mu > 0$,
(g) bent, $\mu > 0$,
(h) trigonal planar, $\mu = 0$.

5.12 (a) sp^3, (b) sp^3,
(c) sp^3d, (d) sp^2,
(e) sp^3d^2, (f) sp^3d,
(g) sp^3.

CHAPTER 6

6.2 Group 2.
6.4 +575 kJ mol^{-1}.
6.5 –5 kJ mol^{-1}.
6.6 Reaction (i) ΔH = +984 kJ per mole of reaction; reaction (ii) ΔH = +1606 kJ per mole of reaction.
6.7 +439 kJ mol^{-1}.
6.14 (a) 8,
(b) 7,
(c) 8.5,
(d) 7.
6.15 ΔU(0 K)(Born–Landé) = –631 kJ mol^{-1}.
6.16 ΔU(0 K)(Born–Haber) ≈ ΔH(298 K) = –3834 kJ mol^{-1};
ΔU(0 K)(Born–Landé) = –3925 kJ mol^{-1}.

CHAPTER 7

7.4 (a) 12,
(b) 12,
(c) 6,
(d) 8.
7.8 $x = y = 1$.

CHAPTER 8

8.1 (a) Propane,
(b) Heptane,
(c) Dodecane,
(d) Icosane.
8.2 (a) C_6H_{14},
(b) C_9H_{20},
(c) $C_{10}H_{22}$,
(d) $C_{14}H_{30}$.
8.3 (a) 5, (b) 8,
(c) 4, (d) 7.
8.4 2-Methylbutane.
8.5 2,6,7-Trimethylnonane.
8.6 (a) methyl,
(b) ethyl,
(c) butyl,
(d) *tert*-butyl,
(e) isopropyl.
8.8 (a) C(1) sp^3, C(2) sp^2, C(3) sp^2, C(4) sp^3, C(5) sp^3;
(b) C(1) sp^3, C(2) sp, C(3) sp, C(4) sp^3;
(c) C(1) sp^3, C(2) sp^2, C(3) sp, C(4) sp^2, C(5) sp^3, C(6) sp^3, C(7) sp^2, C(8) sp^2;
(d) C(1) sp^3, C(2) sp^2, C(3) sp^2, C(4) sp^3, C(5) sp^3, C(6) sp^3, C(7) sp^3.
8.10 (a) Na, oxidation; C, reduction;
(b) C oxidation; Cl, reduction;
(c) N, oxidation; O reduction;
(d) C, reduction; H, oxidation;
(e) K, oxidation; H, reduction.
8.16 From bond enthalpies: $\Delta_r H$ = –136 kJ per mole of ethene. From values of $\Delta_f H^o$: $\Delta_r H$ = –136.3 kJ per mole of ethene.

CHAPTER 9

9.1 Energy of transitions in NMR < vibrational < electronic.
9.2 Red light (λ lies in the range 750–620 nm) is absorbed.
9.3 (a) 0.0009 mol dm^{-3};
(b) 0.014 mol dm^{-3}.
9.4 1903 cm^{-1}.
9.5 $\mu(H_2) = 8.3 \times 10^{-28}$ kg; μ(HD) = 1.11×10^{-27} kg; $\mu(H^{35}Cl) = 1.61 \times 10^{-27}$ kg; $\mu(H^{37}Cl) = 1.62 \times 10^{-27}$ kg.
9.7 (a) all IR active,
(b) symmetric is IR inactive,
(c) all IR active,
(d) symmetric is IR inactive. (All the molecules are linear but the differences lie in whether the molecule possesses a dipole moment or not.)
9.10 (a) 2100–2260 cm^{-1},
(b) 3600–3200 cm^{-1} (broad),
(c) 3500–3300 cm^{-1},
(d) 1740–1720 cm^{-1},
(e) 1680–1620 cm^{-1}.
9.12 (a) toluene,
(b) phenol,
(c) cyclohexane,
(d) benzene.
9.14 $\psi_1 = [c_1 \times \psi(2p)C_1] + [c_2 \times \psi(2p)C_2] + [c_3 \times \psi(2p)C_3]$

$\psi_2 = [c_4 \times \psi(2p)C_1] - [c_5 \times \psi(2p)C_3]$
$\psi_3 = [c_6 \times \psi(2p)C_1] - [c_7 \times \psi(2p)C_2] + [c_8 \times \psi(2p)C_3]$

9.16 1.1 : 1.0; no dependence on path length.

9.19 ^{31}P 162 Mz, ^{13}C 62.5 MHz.

9.20 δ +178.1 (sp^2 C, aldehyde), +20.6 (sp^3 C).

9.21 $(CH_3)_3COH$: $R_3\underline{C}OH$ carbon is downfield of the methyl carbons. Relative intensities of the signals should assist assignment.

9.22 δ +170.0 (sp^2 C), +80.8 (sp C), +52.0 (methylene), +20.5 (methyl).

9.23 $CH_3CH(CH_3)CH_3$ All methyl groups are equivalent; therefore two ^1H environments: doublet (rel. intensity = 9) and decet (1:9:36:84:126:126:84:36:9:1) (rel intensity = 1).

9.24

Me *a*
b H—C—C≡N
Me *a*

δ +1.3 (*a*), +2.7 (*b*)

O
a *b* //
CH_3CH_2C
\ *c*
NH_2

δ +1.1 (*a*), +2.2 (*b*), +6.4 (*c*)

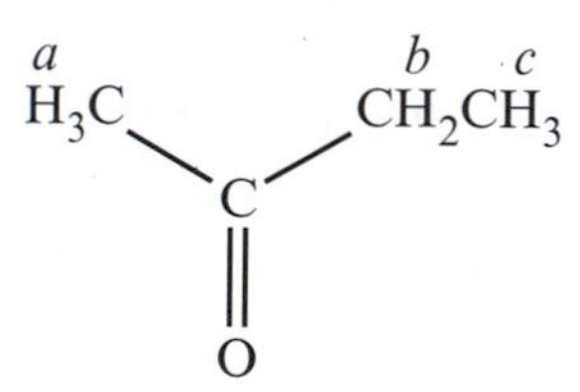

δ +1.1 (*c*), +2.1 (*a*), +2.5 (*b*)

a *b* *c* *d*
$CH_3CH_2OCH_2CO_2H$

δ +1.3 (*a*), +3.7 (*b*), +4.1 (*c*), +10.9 (*d*)

a *b*
CH_3CHBr_2

δ +2.5 (*a*), +5.9 (*b*)

a *b* *a*
$CH_2ClCH_2CH_2Cl$

δ +2.2 (b), +3.7 (a)

9.25 Coupling between ^1H and ^{19}F nuclei (both $I = \frac{1}{2}$). CF_3CH_2OH: broad signal for OH proton, and quartet (J_{HF}) for the CH_2 group. Probably no J_{HH} observed.

CHAPTER 10

10.1 (a) Rate = gradient = 0.001 mol dm^{-3} s^{-1}.
(b) Rate of reaction at t_1 = gradient at $X(t_1) \approx 9 \times 10^{-4}$ mol dm^{-3} s^{-1}; rate of reaction at t_2 = gradient at $X(t_2) \approx 6 \times 10^{-4}$ mol dm^{-3} s^{-1}.

10.5 Rate of reaction = $k[Fe^{3+}][I^-]^2$.

10.6 (a) First order;
(b) $k = 0.087$ s^{-1};
(c) no, water may be involved and k may in fact be the pseudo-first order rate constant.

10.7 Reaction I: first order with respect to I_2; $k' = 2.77 \times 10^{-4}$ s^{-1}; reaction II : second order with respect to I_2; $k' = 1.14 \times 10^{-2}$ dm^3 mol^{-1} s^{-1}.

10.8 First order with respect to $C_6H_5N_2^+Cl^-$.

10.9 (a) Second order with respect to **I**;
(b) $-\frac{d[\mathbf{I}]}{dt} = k[\mathbf{I}]^2$
(b) 4.3 mol dm^{-3} min^{-1}.

10.10 (b) First order;
(c) first order;
(d) ε_{max} for $[P]^{2-}$.

10.11 (a) Varies, (b) varies,
(c) constant, (d) varies.

10.12 $E_{activation}$ = 79 kJ mol^{-1}.

CHAPTER 11

11.1 X = C, 2275 cm^{-1}; X = N, 2264 cm^{-1}; X = O, 2256 cm^{-1}.

11.3 Part 1: (a) Cu, yes, (b) Zn, no,
(c) Fe, no, (d) Ni, no,
(e) Mn, no.

Part 2: (a) Mg, yes, (b) Ag, no, (c) Ca, yes, (d) Zn, yes, (e) Cr, yes.

11.4 Methanol: -726.5 kJ mol^{-1}; octane: -5472 kJ mol^{-1}.

11.7 (a) -3, (b) -3, (c) -2 (the N–N bond does not affect the oxidation state), (d) -3.

11.11 pH = 3.38.

11.12 pH = 10.98.

11.13 pH = 1.93.

CHAPTER 12

12.2 (a) -248 J (per 0.1 mole of O_2 produced from 0.2 moles H_2O_2); (b) -99.4 kJ (per mole of H_2O_2).

12.5 -408.0 kJ mol^{-1}.

12.7 -603 kJ mol^{-1} (per mole of reaction written).

12.8 (a) Zn, Al, Sr and Ca; (b) $\Delta G^o \approx -30$ kJ mol^{-1} for the reaction $C + FeO \rightarrow Fe + CO$.

12.10 1.2×10^{-12}.

12.11 $\Delta_{fus}S = 9.4$ J mol^{-1} K^{-1}; $\Delta_{vap}S = 77.6$ J mol^{-1} K^{-1}.

12.12 SO_3: ΔS^o(298 K) = -83.1 J mol^{-1} K^{-1}; $\Delta_f G^o$(298 K) = -370.9 kJ mol^{-1}.

12.13 (a) $E^o_{cell} = 0.13$ V; $\Delta G^o = -25$ kJ (per mole of reaction).
(b) $E^o_{cell} = 0.13$ V; $\Delta G^o = -12.5$ kJ (per mole of reaction).
(c) $E^o_{cell} = 0.56$ V; $\Delta G^o = -324$ kJ (per mole of reaction).

12.15 $E = +1.01$ V.

12.16 1.05×10^{-5} mol dm^{-3}.

12.17 (a) $K_{sp} = 1.1 \times 10^{-12}$ mol^3 dm^{-9}.

CHAPTER 13

13.1 (a) +2; (b) +5; (c) +4; (d) +5; (e) +5; (f) +3; (g) +4; (h) +6; (i) +7; (j) +7; (k) -1
(remember that the S–S bond does not contribute, see Box 8.8).

13.2 -513 kJ (per mole of B_2O_3).

13.10 (a) Paramagnetic;
(b) diamagnetic;
(c) paramagnetic;
(d) diamagnetic;
(e) diamagnetic.

13.11 $K \approx 5.1 \times 10^{-12}$.

13.20 (a) N_2O_3, (b) N_2O_5, (c) P_4O_{10}, (d) P_4O_6, (e) Cl_2O, (f) I_2O_5.

CHAPTER 14

14.1 (a) $C^{\delta+}–F^{\delta-}$, (b) $N^{\delta+}–O^{\delta-}$, (c) $C^{\delta+}–O^{\delta-}$, (d) $O^{\delta-}–H^{\delta+}$, (e) $N^{\delta-}–H^{\delta+}$, (f) C–S non-polar.

14.5 (a) 2,3-diiodobutane;
(b) 1,1,1-trichloropropane;
(c) 2-bromo-5,5-dimethylhexane;
(d) 1,3,4-trifluoropentane;
(e) 1,3-dibromo-2-methylbutane.

14.12 (a) Methyl propyl ether;
(b) 3-methoxypentane;
(c) ethyl isopropyl ether.

14.16 **X** has the molecular formula $C_5H_{12}O_2$ and contains an –OH group ($\approx$3500 cm^{-1}) and an ether group ($\approx$1100 cm^{-1}). The actual spectrum is that of 1-methoxypropan-2-ol:

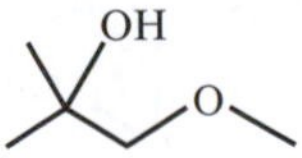

but other isomeric structures are possible, including:

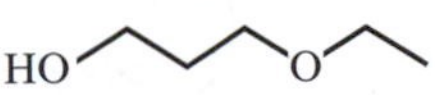

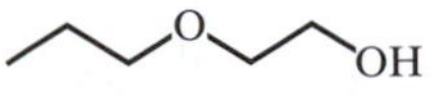

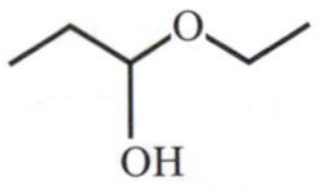

14.18 (a) But-2-ene,
(b) pent-2-ene,
(c) pent-2-ene,
(d) 2,3-dimethylbut-2-ene.

14.19 Major product is expected to be 2-chloro-2-methylbutane (minor, 2-chloro-3-methylbutane).

14.20 (a) Acetone;
(b) propanal (CH_3CH_2CHO), further oxidized to propanoic acid ($CH_3CH_2CO_2H$).

14.22 (c) pH = 12.0.

14.23 $[H_3NCH_2CH_2CH_2NH_3]Br_2$.

CHAPTER 15

15.4 Cyclopentane, cyclononane.

15.5 (a) 1,2-Dipropylcyclobutane,
(b) 1,2,4-trimethylcyclopentane,
(c) 1-ethyl-2-methylcyclohexane.

15.6 Spectrum (a) = methylcyclopentane,
(b) = 1,2-dimethylcyclopentane,
(c) = ethylcyclopentane.

15.7 (a) 1-Ethyl-2-methylcyclobutene,
(b) cyclopentadiene,
(c) 2,3,5-trichlorocyclopentadiene,
(d) cycloocta-1,5-diene.

15.9 (a) Non-aromatic: appears to obey the Hückel rule ($n = 1$) but is not delocalized;
(b) aromatic: planar, delocalized and $n = 2$;
(c) three separate aromatic C_6 rings: three planar rings, $n = 1$;
(d) non-aromatic: is not planar, does not obey the Hückel rule and is not delocalized.

15.11

(a) Br

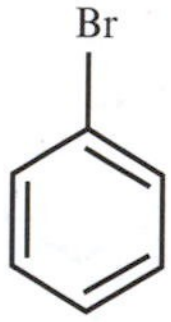

(b) Me, Me, Me, C

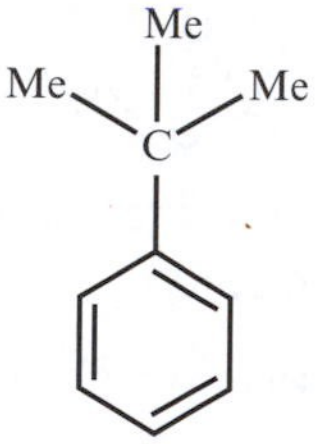

(c) H, Me, Me, C

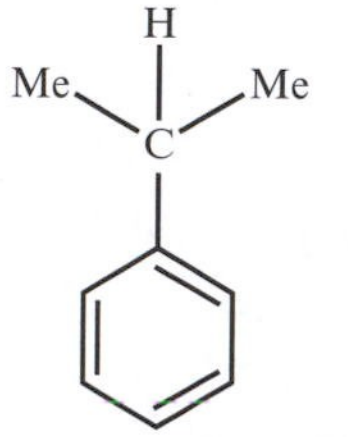

(after rearrangement)

(d) H_2, O, C, C, CH_3, C, H_2

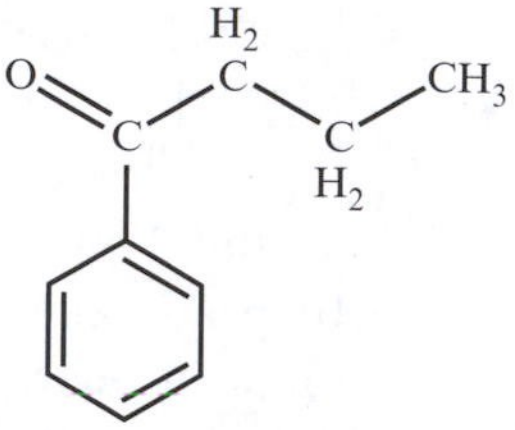

(no rearrangement)

15.12 Suitable catalysts (because they are Lewis acids) are BF_3, $GaCl_3$, $SbCl_5$.

15.13 Ph_2CH_2.

15.14 (a) CH_3, NO_2 and CH_3, NO_2

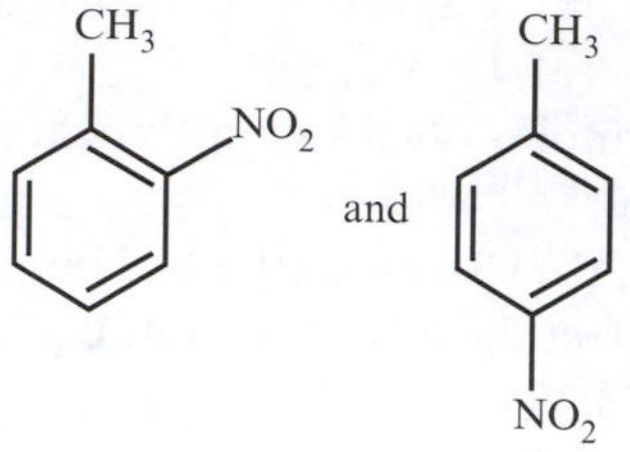

(b) CO_2H, CO_2H

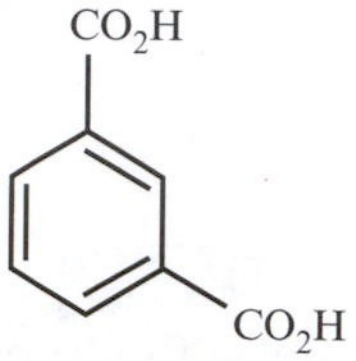

(c) [PhO]K

(d) CH_2Ph, O, Me, Me

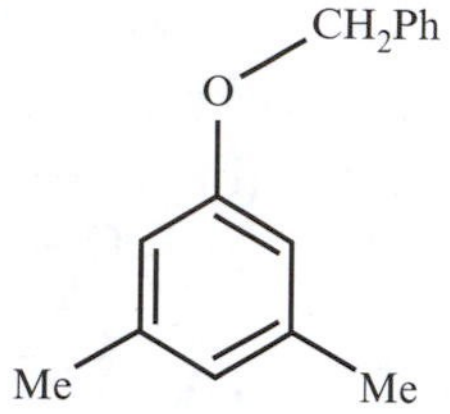

15.15 Phenol pH 5.4.

15.16 (a) Aromatic;
(b) aliphatic;
(c) aromatic;
(d) aliphatic.

15.17 (a) NH_2

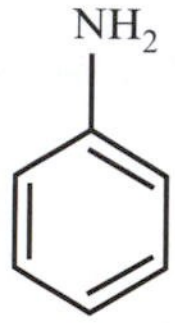

(b) OH, , N, N, OH

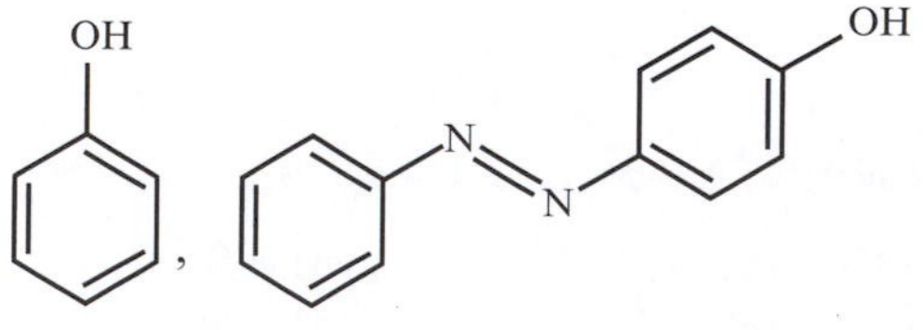

(c)

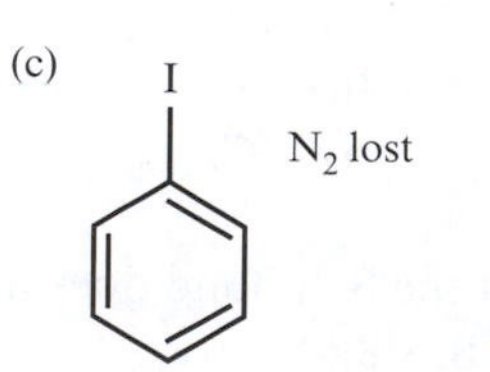

15.18 (a) pH = 8.97,
(b) pH =9.28.

15.19 (a) [pyH][ClO_4]

(b)

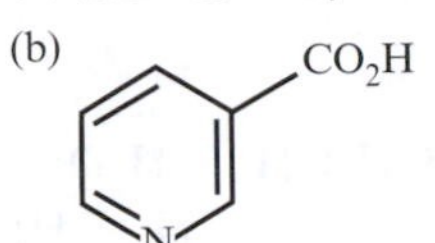

(c)

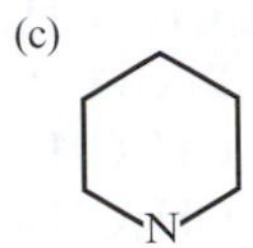

(d)

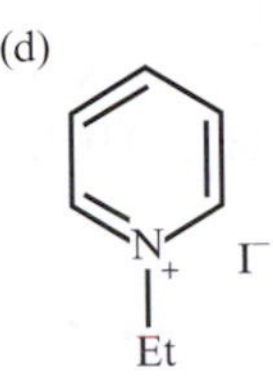

15.21 (a) Pyridine,
(b) cycloheptatriene,
(c) THF,
(d) cyclohexa-1,4-diene,
(e) cyclohexa-1,3-diene.

CHAPTER 16

16.2 (a) $[Ar]4s^23d^6$; (b) $[Ar]3d^5$;
(c) $[Ar]4s^13d^{10}$; (d) $[Ar]3d^{10}$;
(e) $[Ar]3d^2$; (f) $[Ar]4s^23d^8$;
(g) $[Ar]3d^3$; (h) $[Ar]3d^{10}$.

16.3 In each, *indicates a potential donor atom.

(a) (b)

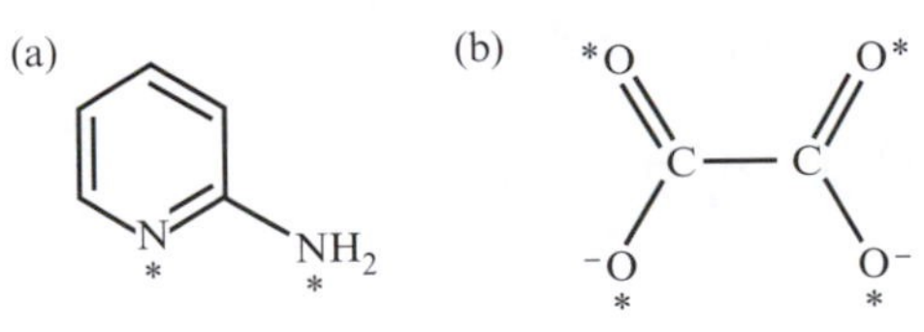

(c) 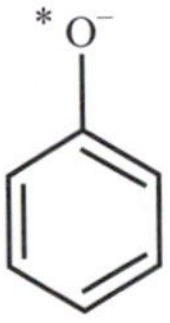 (d)

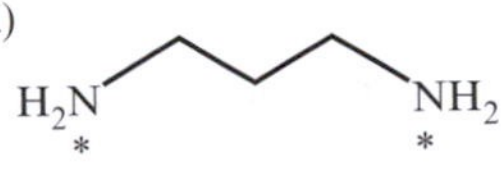

(e)

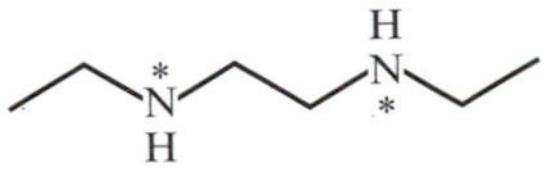

(g) 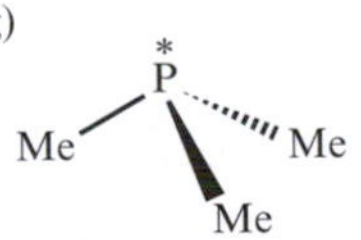 (g)

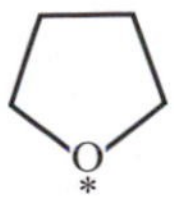

(h)

Possible chelating ligands are (a), (b), (d), (e) and (h).

16.4 (a) +2, (b) +3,
(c) +2, (d) +3,
(e) +3, (f) +2,
(g) +2, (h) +2,
(i) +2, (j) +3.

16.5 (a) Tetrahedral (no isomers);
(b) octahedral (no isomers);
(c) tetrahedral (no isomers);
(d) octahedral (*cis*- and *trans*-isomers);
(e) octahedral (*mer*- and *fac*-isomers);
(f) trigonal planar;
(g) trigonal bipyramidal (three isomers, steric factors suggest amido groups favour axial positions).

16.6 (a) Fe, 0; each N $+\frac{1}{3}$;
(b) Cr, 0; each F $-\frac{1}{2}$.

16.9 Optically active:
(a) yes, (b) no,
(c) yes, (d) no,
(e) yes, (f) no.

16.10 Products:
(a) $[Ni(en)_3]^{2+} + 6H_2O$,
(b) $[Co(phen)_3]^{2+} + 6H_2O$,
(c) $[VCl_4(py)_2]$,
(d) $[Cr(acac)_3] + 6H_2O$,
(e) $[Fe(CN)_6]^{4-} + 6H_2O$,
(f) $[Fe(NH_3)_6]Cl_2$.

16.13 For $[Co(H_2O)_4(NH_3)_2]^{2+}$ $\Delta G^o = -21.1$ kJ mol^{-1}; for $[Co(H_2O)_4(en)]^{2+}$ $\Delta G^o = -34.2$ kJ mol^{-1}.

16.15 *N*-bonded (hard–hard).

16.17 (a) 4.90, (b) 4.90
(c) 2.83 BM.

16.18 (a) High spin, (b) high spin.

16.19 (a) –1; (b) +2;
(c) 0; (d) +1;
(e) +1; (f) +1.

CHAPTER 17

17.2 (a) Propanal;
(b) heptan-4-one;
(c) caesium benzoate;

(d) propyl butanoate;
(e) propanoyl chloride;
(f) octane-2,6-dione.

17.3 Spectrum **A** (hexanoic acid), **B** (ethyl acetate) and **C** (benzyl acetate). Assignments: **A**: broad absorption centred ≈3000 cm^{-1} ν(OH) superimposed with $\nu(CH_{alkyl})$, ≈1700 cm^{-1} ν(C=O); **B**: 3000 cm^{-1} $\nu(CH_{alkyl})$, ≈1700 cm^{-1} ν(C=O); **C**: >3000 cm^{-1} $\nu(CH_{benzyl})$, ≈3000 cm^{-1} $\nu(CH_{alkyl})$, ≈1700 cm^{-1} ν(C=O).

17.4 Spectrum **A** is compound **II**, **B** is **III**, **C** is **I**. Assignments of resonances is possible except for unambiguous assignments of the ring carbon atoms in compound **II**.

17.7 pK_a values: $CF_3CO_2H < CH_3CO_2H < CF_3C(O)CH_2C(O)CF_3 < CH_3C(O)CH_2C(O)CH_3 < CH_3CH_2OH < CH_3C(O)CH_2CH_3$.

17.8 (a) $PhCO_2H$;
(b) $CH_3CH_2CO_2CH_2CH_3 + H_2O$;
(c) $CH_3CH_2CH_2CO_2CH_2CH_3 + HCl$;
(d) $Cs^+[PhCH_2CO_2]^- + H_2O$;
(e) $K^+[PhC(O)CHC(O)Ph]^- + Me_3COH$;
(f) $CH_3CH_2CH_2C(O)NH_2 + HCl$;
(g) $PhCO_2H + HCl$;
(h) $CH_3C(O)CH_2I + HI$;
(i) CH_3CO_2H;
(j) no reaction;
(k) $CH_3CH_2C(Me)(OH)CH_2CH_3$.

17.9 (a)
$$CH_3CH_2CH_2Br \xrightarrow{[CN]^-} CH_3CH_2CH_2CN \xrightarrow{H_2O,\ H^+} CH_3CH_2CH_2CO_2H$$
or
$$CH_3CH_2CH_2Br \xrightarrow{\text{Mg in anhydrous THF/diethyl ether}} CH_3CH_2CH_2MgBr \xrightarrow[2.\ H^+]{1.\ CO_2} CH_3CH_2CH_2CO_2H$$

(b) $CH_3CO_2CH_3 \xrightarrow[2.\ H^+]{1.\ \text{base hydrolysis}} CH_3CO_2H$ (hydrolysis of an ester)

(c) $CH_3CH_2OH \xrightarrow{\text{acidified } K_2Cr_2O_7} CH_3CO_2H \xrightarrow{PCl_3} CH_3COCl$
(Alternative reagents for the last step are PCl_5 and $SOCl_2$).

(d) $CH_3CHCl_2 \xrightarrow{H_2O,\ H^+} CH_3CHO \xrightarrow{\text{acidified } K_2Cr_2O_7} CH_3CO_2H$

(e) $C_6H_6 \xrightarrow{MeCOCl,\ AlCl_3} PhC(O)Me \xrightarrow[2.\ H^+]{1.\ Na[OEt]} PhC(O)CH{=}CMePh$

(f) $CH_3CH_2CH_2CH_2OH \xrightarrow{\text{acidified } K_2Cr_2O_7} CH_3CH_2CH_2CO_2H$

$\xrightarrow{PCl_3} CH_3CH_2CH_2C(O)Cl \xrightarrow{PhOH} CH_3CH_2CH_2CO_2Ph$

Index

Further information about a particular element and its compounds may also be found under the relevant group entry.